Estimating Numbers of Terrestrial Birds

Reprinted by

THE BLACKBURN PRESS

This book is a complete reprint of the volume originally published by
the Cooper Ornithological Society, Copyright 1981
Studies in Avian Biology No. 6.
The cover is a new design.

Estimating Numbers of Terrestrial Birds

ISBN: 1-930665-77-6

Library of Congress Card Number: 2003103131

Cover photographs: Eastern Kingbird (*Tyrannus tyrannus*) taken on the Snake River near Lewiston, Idaho, by R.D. Dixon; observer in forest in Idaho Panhandle National Forest, Idaho, by R.D. Dixon.

THE BLACKBURN PRESS
P. O. Box 287
Caldwell, New Jersey 07006
U.S.A.
973-228-7077
www.BlackburnPress.com

Estimating Numbers of Terrestrial Birds

Editors

C. JOHN RALPH and J. MICHAEL SCOTT

U.S. Forest Service
Institute of Pacific Islands Forestry
1151 Punchbowl Street, Rm. 323
Honolulu, Hawaii 96813

U.S. Fish and Wildlife Service
Mauna Loa Field Station
P.O. Box 44
Hawaii National Park
Hawaii 96718

Proceedings of an International Symposium
held at Asilomar, California
October 26–31, 1980

Sponsors:

California Natural Resources Federation
Cooper Ornithological Society
International Association of Fish and Wildlife Agencies
International Union of Forestry Research Organizations
National Audubon Society
National Wildlife Federation
UNESCO, U.S. Man and Biosphere Program
U.S. Bureau of Land Management
U.S. National Park Service
U.S. Fish and Wildlife Service
U.S. Forest Service
U.S. Water and Power Resources Service
Wildlife Management Institute
The Wildlife Society
World Wildlife Fund (U.S.)

Studies in Avian Biology No. 6

A PUBLICATION OF THE COOPER ORNITHOLOGICAL SOCIETY

STUDIES IN AVIAN BIOLOGY

Studies in Avian Biology, as successor to *Pacific Coast Avifauna*, is a series of works too long for *The Condor*, published at irregular intervals by the Cooper Ornithological Society. Manuscripts for consideration should be submitted to the Editor at the above address. Style and format should follow those of previous issues.

CONTENTS

READER'S GUIDE

The proceedings of this Symposium offer a plethora of material on the subject of estimating the abundance of birds. We offer the following as a subject guide to be used in conjunction with the Table of Contents as a quick means of locating sections of the proceedings dealing with specific topics.

The Editors

(Page numbers in **bold face** type refer to primary or comprehensive references)

METHODS, COMPARISON

Method	Relative	Capture-recapture	Fixed width line transects	Fixed diameter circles	Mapping	Variable distance circle plots	Variable distance strip plots
Absolute Measures	76, 393–394, **554–558**	—	—	—	—	—	—
Capture-Recapture	—	—	—	—	—	—	—
Fixed-width Line Transects	—	230–231	—	—	—	—	—
Fixed-diameter Circles	—	—	555–556	—	—	—	—
Mapping	209–214	**119–120**	**152–159**, 193–197, 199–201, 377–381, 554–555, 559–565	554–558	—	—	—
Variable-distance Circle Plots	—	—	377–381	—	170–176, 177–185, **189–190**, 377–381	—	—
Variable-distance strip plots	—	—	139–145, 377–381	—	165–169, 189–190, 193–198, 199–201, 202–208, 377–381	**170–176**, **186–192**, 377–381	—

Studies in Avian Biology No. 6:1–2, 1981.

INTRODUCTION

J. MICHAEL SCOTT[1] AND C. JOHN RALPH[2]

Counting birds has a long tradition. Since early in human history, man has noted and recorded the presence, absence, and abundance of birds.

This long, and presumably honorable, pursuit that we all engage in, to a greater or lesser extent, is the common currency of many ornithological studies. These studies range from multiple regression analyses of habitat variables to life history studies.

Counts have been the raw material of many discoveries in ornithology and its lineal descendent, ecology. Bird counts have been seminal in our knowledge of bird migration, competition, community ecology and structure, population dynamics, environmental adaptation, impact of human alterations, and island biogeography.

Interest in estimating bird numbers is extremely high today, but many questions relating to sampling methods remain unasked or unanswered. Increasingly, workers are finding that the sophisticated questions asked of the data and the precision required in the data analysis are at odds with the methods of data collection and the behavior of the birds being sampled. It was felt that a symposium bringing together all those interested in the problem would facilitate the communication of solutions to some problems, perhaps lead to some novel solutions, and at least clearly define other problems.

Thus we gathered together biologists and statisticians to assess critically the methods and assumptions we use in data gathering and analysis. We hope that this effort will lead to a better understanding of what can and cannot be done with data sets, and of ways to increase the sophistication and accuracy of our analytical and sampling methods.

There are many methods now used to estimate bird numbers. The methods vary considerably in their preciseness and accuracy. However, there are three things common to all: observers to count, birds to be counted, and habitats to be surveyed.

How variations in methods, observers, training of observers (or lack of training), species, habitats, and other environmental variables all affect the accuracy and precision of bird counts must be known and assessed if the field is to progress.

Bird numbers are used in many discussions of biology today. If these discusssions are to have validity, we need to consider seriously the reliability of their data base. Far too often, biologists have given little attention to their counting methods, yet have made sweeping statements based on rather fragile numbers. If these numbers are to have any validity, then we as researchers and fields workers have to:

(1) be more precise in defining the questions we want answered;

(2) determine whether we need numbers per unit area or only an index of relative abundance to answer those questions;

(3) determine which method best meets our needs and recognize that each has its own advantages, disadvantages, and assumptions;

(4) pay careful attention to sampling design; and

(5) recognize that observers are important and spend more time and money improving their quality and performance through selection, training, and modification of data gathering methods.

In reading the papers in these proceedings, you will find a good many differences of opinion. It was our intent to bring together diverse points of view. You will also find a large gap between the state of the art and actual field practices. It is our hope that the papers in this book will serve to highlight these differences and motivate the field biologists to close the gap.

ACKNOWLEDGMENTS

Two years of planning and work went into this symposium. During that period we received much help and encouragement from our agencies and many individuals. Our ten session chairmen edited manuscripts, conferred with authors, and generally made our work much easier. We are deeply indebted to all these talented individuals who gave their valuable time to seeing the symposium through to its conclusion. Dr. Robert Z. Callaham and Paul Guilkey of the U.S. Forest Service, and Dr. David L. Trauger of the U.S. Fish and Wildlife Service provided impetus and support at every stage. Ralph Raitt, Frank Pitelka, and Joe Jehl of the Cooper Ornithological Society provided active help, encouragement, and support. John T. Emlen provided helpful suggestions during the developmental stages of the symposium.

In the planning and organization, Richard Hubbard (USFS) and Susan DeTreville did much of the work with the meeting arrangements. Betty Lusk, Ann Tuscany, and Jonette Kaapana provided much clerical help, organizational talents, and insight at all stages. Clearly, our work would have been impossible and

[1] U.S. Fish and Wildlife Service, Patuxent Wildlife Research Center, Mauna Loa Field Station, P.O. Box 44, Hawaii National Park, Hawaii 96718.

[2] USDA Forest Service, Institute of Pacific Islands Forestry, 1151 Punchbowl St., Honolulu, Hawaii 96813.

these proceedings much delayed without their constant effort and good humor in face of impossible demands from us. During the symposium itself, Anthony Gomez (USFS) ran a tight ship, coordinating poster presentations, social events and projecting sharp and upright slides.

Finally, we are grateful to our wives, Carol Pearson Ralph and Sharon Scott who provided help and encouragement. We trust that the thousands of hours on the part of the authors, summarizers, session chairmen, and ourselves will, in small part, be repaid through these proceedings.

Volcano, Hawaii
June 1981

Studies in Avian Biology No. 6:3–4, 1981.

GENESIS OF THE SYMPOSIUM

ROBERT Z. CALLAHAM[1]

We in the Forest Service of the U.S. Department of Agriculture are proud of the leading roles our employees have played in creating and arranging this symposium. That it meets a significant need is attested to by the registration of over four hundred specialists from many countries. My role is to explain the origins, objectives, and challenges of this conference.

FOREST SERVICE AS A SPONSOR

Some may think Forest Service sponsorship of a symposium dealing with birds to be unique. A Treaty of the Potomac (River) was signed in 1953 by the Forest Service (U.S. Department of Agriculture), and the Fish and Wildlife Service (U.S. Department of the Interior). This agreement assigned to the Forest Service the primary responsibility for research and management programs on avian habitats in forests and rangelands. The Fish and Wildlife Service was assigned primary responsibility for research on avian biology, ecology, and management. It was agreed that either agency could fill the other's role, if the other was unable to do so, after giving due notice and providing coordination. This is a partnership of long standing to improve the management and knowledge of inter-relationships of wildlife with forest and range habitat.

During the last decade, Federal expenditures for wildlife management and particularly for bird-related programs have increased dramatically. The National Forest System and Bureau of Land Management have greatly expanded their cadres of wildlife biologists. The Endangered Species Act of 1973 caused a dramatic increase in concerns and expenditures particularly for birds listed as threatened and endangered. Yet management of avian habitats and populations in forests and rangelands is still based on woefully inadequate inventories and technical information.

The Forest Service, having a big stake in generating and applying such technology for management, eagerly and fully supported this conference.

OBJECTIVES

The first purpose of this conference was to bring together, for stimulating and fruitful interaction, an international mix of specialists from divergent disciplines. We wanted them to mingle, not just with their friends and peers, but with specialists in other disciplines as well. I urged participants to be willing to expose for constructive criticism their past studies and their plans for the future.

The second purpose of this conference was to evaluate both the state of knowledge and the state of practice. By state of knowledge, I mean the scientific and technical basis of our understanding. Gaps in this base must be filled by additional research and investigation. By state of practice, I mean how things are being done in actual programs. Obviously, there can be quite a disparity between what we know and what we do. From this conference emerged an appreciation of the status of knowledge and of practice, as well as suggestions for moving available technology into practice.

To fulfill our third purpose for this conference, we hoped that participants would identify what needs to be done to improve the utilization of available knowledge. From this conference came suggestions for training programs relevant to on-the-job needs of practitioners or of scientists. These suggestions cover training courses that would be useful to Federal and State agencies, as well as educators. Suggestions for guidelines, handbooks, manuals, or audio-visual aids are also a part of this effort.

The fourth purpose of this conference was to describe what should be done to fill the gaps in knowledge and methodology. From this conference will come ideas about gaps in the frontiers of knowledge that need to be filled by research. Participants helped us by identifying researchable problems and questions, and suggesting investigative approaches.

BEYOND THIS CONFERENCE

Our purposes could not be achieved by this conference alone. We plan to make this the starting point for a series of ongoing activities. We tried to learn what should follow in the way of technical seminars, training sessions, or whatever else might be needed.

CONCLUSION

I trust that you sense my personal pleasure at having this chance to discuss the Forest Service and its interest in this subject, and about our aspirations for it. Particularly do I appreciate the support that our many co-sponsors have provided. They have been very generous. I especially

[1] Director, Pacific Southwest Forest and Range Experiment Station, Forest Service, U.S. Department of Agriculture, Berkeley, California 94701.

want to thank the authors and those who organized these sessions. All worked very hard before the conference, and they worked even harder before this publication was finished. I challenge you the reader not just to enjoy this as another proceedings of a conference, but to use the proceedings for work, for planning, and to help to start the activities that will follow beyond this symposium.

Studies in Avian Biology No. 6:5–6, 1981.

THE NEED FOR CENSUSES IN POLICY MAKING

DAVID L. TRAUGER[1]

We of the Fish and Wildlife Service of the U.S. Department of the Interior are proud to be one of the primary sponsors of this Symposium. This Symposium was supported by funds from both the Migratory Birds and the Endangered Species Programs of the Service. We're especially proud to be sponsoring and participating in this Symposium. We feel that these proceedings have the potential of being an event of singular importance in the conservation of terrestrial birds, with important implications for future research on avian biology and ecology. We're delighted in the interest and participation in the Symposium.

Although Dr. Scott requested that I address the need for bird censuses in policy making, from what I know about the subject, we'd better withhold judgment. As you know, policy-making has evolved into a rather exact science, particularly in Washington.

Rather than discuss policy, we should talk about counting birds, because that's what the Symposium was all about. One of the key uses of population data is to make predictions about various biological responses to environmental phenomena. Without making too many unwarranted assumptions, I predict that if you all spend five days focused on this topic, there will be a lot of unstandard deviations by mid-week and a lot of multiple regression by the end of the week.

In my opening remarks, I was asked to give you a clear and concise statement of what policy makers need from those who are estimating bird numbers. At first blush, this represented a formidable challenge to me because, I'm not a policy maker. If I were a policy maker, however, I would want the most relevant and accurate information available for consideration, along with an objective assessment of what were the assumptions, limitations, alternatives, and implications of the data collected and actions recommended. This would be the ideal. Unfortunately, few policy makers have this luxury, primarily because the data and the expertise are generally not availble for this type of review and analysis.

I come to this subject and to the Symposium from a slightly different background and perspective than Dr. Callaham, yet I appreciate and share some of his skepticisms and concerns. If we all don't know it by now, we will by the end of the week. Bird censuses, that is, total counts of birds in a predescribed area of natural habitat, are probably impossible at the present state of the art.

The complexities and intricacies of avian biology and ecology in combination with the vagaries and dynamics of environmental factors in their spatial and temporal dimensions pose formidable obstacles for students of bird populations, to say nothing of the influences of the human factors (ability, experience, perception, and persistence) in relation to the design and conduct of experiments or surveys which may or may not have had review and analysis by a competent statistician.

In the last decade of the 20th Century, wild life and wild places are under massive, escalating and accelerating assaults from a multiplicity of human agents and activities. Exploitation and extirpation of many wildlife populations as a result of, and in combination with, the deterioration and destruction of habitats throughout the world are matters of grave concern with all biologists and conservationists. Several bird species are already in serious jeopardy; many more are becoming threatened or endangered at alarming rates. Extinction overshadows the future of many species. Correspondingly, administrators of natural resource agencies are under increasing pressure to defend their actions in the protection and preservation of species and habitats in the highly political milieu of competing and conflicting societal values and goals. At all professional levels, there is need for accurate and reliable census methodology for management and conservation of avian populations. Biologists need accepted and standardized methodology for conducting studies and analyzing data, contributing to the information on which management decisions can be based. Managers need to have the confidence that the information they have for prescribing practices and evaluating actions to influence environmental factors regulating or enhancing populations is scientifically sound. Administrators need reliable information for developing policies and implementing programs to direct the management of populations and protection of habitats. In our relationships with the private sector, natural resource researchers, managers, and administrators must strive to establish and maintain the highest levels of professionalism,

[1] Chief, Division of Wildlife Ecology Research, U.S. Fish & Wildlife Service, Washington, D.C. 20240.

confidence, and credibility to generate and perpetuate the level of public support and involvement essential to achieving our conservation goals.

We are here to participate in an international symposium on a subject of critical importance for providing the basis to develop, implement, enforce, and evaluate a myriad of public policies affecting the management of avian populations and conservation of natural habitats. I want to commend all of those, especially Drs. J. Michael Scott and C. J. Ralph, as well as all the sponsors, the Forest Service, and all of you who have come great distances and who have played a role in making this Symposium happen. I challenge all of you to build on the knowledge presented in the proceedings.

Photographs of Participants
by Anthony Gomez

The Asilomar Conference Grounds on Monterey Bay, California provided an invigorating and beautiful setting for the Symposium. Shown here is Merrill Hall, site of the talks.

The Symposium's organizing committee and sessions chairmen: (L to R) front row: Robert Ohmart, Ralph Raitt, Harry Recher; 2nd row: Fred Ramsey, John Weins, Michael Scott; 3rd row: C. J. Ralph, Joseph Hickey, John Emlen, David Anderson; Back row: Chandler Robbins, Jared Verner and Cameron Kepler.

Excellent meals and attractive grounds made the full five days pass pleasantly.

Poster Papers were presented each evening and provided a social focus and a site for intensive dialogue between participants.

The shores of Monterey Bay, only a short walk away through the sand dunes, provided a ready source of sea and shore birds as well as a place for more solitary walks.

Although late in the fall, some migrant birds, as well as residents, provided good birding opportunities on the conference grounds. (on left—Soren E. Svensson).

As participants walked through the stands of native Monterey Pine (*P. radiata*), there were many opportunities for informal discussions (L to R, Jan Ekman and A. R. Sen).

(L to R) Frances James, David DeSante, Wayne Arendt and Mrs. Arendt.

Frequent refreshment breaks provided the opportunity for much discussion.

(L to R) A. R. Sen and Barbara Diehl.

(L to R) David Dawson and Raymond J. O'Connor.

(L to R) Martin Erdelen, Ann Maben and Fred L. Ramsey.

Participants ranged in age from six months to 90 years.

Joseph J. Hickey and two
other participants.

(L to R) John Wiens and
J. Michael Scott.

(L to R) David Anderson, J. Michael Scott,
Larry Pank, and Kenneth Burnham.

(L to R) C. J. Ralph, A. S. Adegoke,
and Fillibus Bature

Pyong-oh Wong

(L to R) Frank A. Pitelka
and Olli Jarvinen.

(L to R) C. J. Ralph, Jan Ekman,
and J. P. Meyers.

(L to R) Camille Ferry, Soren Svensson, and Hans Oelke.

General scene during refreshment break.

(L to R) Ken Burnham, Pat Gould, Dave Anderson, Frank Pitelka, Hans Oelke, Duncan MacLulich, Martin Edelen, and another participant.

(L to R) J. Michael Scott and C. J. Ralph on last day of Symposium.

Studies in Avian Biology No. 6:11, 1981.

INTRODUCTORY REMARKS: ESTIMATING RELATIVE ABUNDANCE (PART I)

JOSEPH J. HICKEY,[1] CHAIRMAN

At the start of this historic symposium, it is appropriate to recall the great pioneers who broke away from the traditional shotgun approach to field ornithology and started us on quantitative studies of bird distribution and avian ecology. (Pertinent references for this brief review are in Kendeigh 1944.)

The first bird census taken in this country was carried out by Alexander Wilson on 8 acres of a botanic garden in Philadelphia apparently in 1811. During that summer, Wilson felt he had not less than 51 pairs. Excluding three species that probably foraged off this tract, the density would be not less than 3.4 pairs/acre (7.4 pr/ha). This density is the only one we have on record before the arrival of the House Sparrow (*Passer domesticus*) and Starling (*Sturnus vulgaris*) on this continent.

Nine decades later F. L. Burns censused the breeding birds on 1 square mile (2.6 km^2) of mixed habitats at Berwyn, Pennsylvania. In spite of the large size of his area, Burns obtained a density of 1.1 pairs/acre (2.7 pr/ha). Wilson depended on nests found, but Burns also relied on some sort of mapping. There was around 1900 in the United States some counting for daily lists, an activity giving rise to today's National Audubon's Christmas Bird Count; but the great breakthough occurred on 29 August 1906 when A. O. Gross and H. A. Ray began a series of transects which they carried out across the state of Illinois until September 1909 under the direction of S. A. Forbes. Gross and Ray always walked 30 yd (27 m) apart in open country, counting birds out to a distance of 10 yd (9 m) on each side and up to 100 yd (90 m) in front. In dense habitat their distance apart was 20 yd (18 m) and the census strip 30 yd (27 m). Graber and Graber (1963), who repeated this remarkable census 50 years later, found that the results of this method compared extremely well with those obtained by territorial mapping of passerines, but they noted that both methods underestimated the numbers of nesting pheasants (*Phasianus colchicus*). Such a narrow fixed-width transect does not, of course, lend itself to censusing owls and hawks.

In 1914, the U.S. Bureau of Biological Survey, led by W. W. Cooke, began "a census of the birds of the United States." In 1916–20 this project involved 256 areas censused 1 or 2 years and 32 censused 3 or more years. The technique used was fairly crude: the "census" was to be taken at the height of the breeding season, beginning at daylight, zigzagging back and forth across tracts of 40–80 acres, counting singing males, the count to be repeated at least once or checked out by subsequent observations. This cooperative project lasted only about 10 years. Although published, it never matured in technique or ecological insight, and it never gave sufficient credit and identification to its cooperative amateurs.

The binocular had now replaced the shotgun in field ornithology. Eliot Howard had convinced the scientific community of the existence of territory in birdlife. In Australia, J. B. Cleland was reporting counts based on transects of a known length but uncertain width. In Germany Gottfried Schiermann, an experienced egg collector who knew how to find nests, worked out the density of breeding birds on 28 km^2 (10.8 sq mi.) by means of 16 study areas. In Finland, Pontus Palmgren resorted to the mapping method. In Greenland, members of an Oxford University expedition censused the nesting birds on 21.5 km^2 (8.3 $mi.^2$). Finally in 1932, E. M. Nicholson published *The Art of Bird-Watching* with 40 pages devoted to bird-census work.

We were on our way!

[1] Department of Wildlife Ecology, University of Wisconsin, Madison, Wisconsin 53706.

Studies in Avian Biology No. 6:12–16, 1981.

COUNTING BIRDS FOR A RELATIVE MEASURE (INDEX) OF DENSITY

DAVID G. DAWSON[1]

ABSTRACT.—Counts of birds from points or transect lines give an index of population density, even when distances have not been accurately estimated. Factors which influence the counts include the species, age, sex or reproductive group of each bird, the season, habitat, time of day, weather, environmental noise, the observer, the number of other birds being recorded and details of the counting technique. If valid deductions about bird densities are to be made, such influences must be standardized, or their effects removed.

In this paper I review the characteristics of techniques which use counts of birds from points or transect lines, and which do not map territories or estimate accurately the distance of every bird from the point or line. The population density of a species (individuals per hectare, d) may be inferred from the total counted, c, in the area under consideration by the relationship $d = kc$, where k is a coefficient of conversion. This coefficient can be considered as the reciprocal of the effective area sampled, and is large for inconspicuous birds and small for conspicuous ones. The use and applicability of these techniques are determined by the influences which affect the value of k.

INFLUENCES ON THE COUNT

SPECIES

Species differ in how easy they are to see or hear, so the effective area sampled from a point or a line will differ between species. Each species is therefore measured on a separate scale, and the counts of different species may not be added together or used in species-diversity calculations (Sammalisto 1974, Dawson et al. 1978)—a common error.

Within a species, each age, sex or reproductive group may also have its own value of k. For example, Slagsvold (1973b) found that male Song Thrushes (*Turdus philomelos*) in spring could be divided into two groups: mated birds with a very low singing rate, and unmated birds that sang much more. Thus the counts may be weighted heavily towards one section of the population and affected by movement of birds into and out of that section. Some counting techniques register singing males twice (i.e., as representing a pair) and interpret other observations in terms of 'pairs' (e.g., Ferry 1974, Purroy 1974, Järvinen and Väisänen 1976c). I have seen no study to justify the implicit assumption that such weighting corrects for differences in conspicuousness. Only detailed studies such as that of Slagsvold (1973b) can yield useful weighting.

SEASON

Many authors (Kimball 1949, Howell 1951, Amman and Baldwin 1960, Davis 1965, Gates 1966, Robbins and Van Velzen 1967, Blondel et al. 1970, J. T. Emlen 1971, 1977a; Sammalisto 1974, Weber and Theberge 1977, Bibby 1978, Dawson et al. 1978, Lancaster and Rees 1979, Shields 1979, Gill 1980) have demonstrated that for many species the numbers counted, c, varies with the time of the year. We may expect bird densities, d, to vary seasonally with breeding, mortality, immigration and emigration, but we can also expect seasonal changes in conspicuousness, and hence in k. These will be due to changes in behavior, such as singing, and also to changes in the habitat, such as the length of grass or the loss of foliage from deciduous trees. Thus, the effects of seasonal variation in d and in k are impossible to distinguish without information beyond the count alone. At times the numbers counted change much more than density could be expected to (Robbins and Van Velzen 1967, and Fig. 1), so we can deduce a change in k, but not its magnitude. Järvinen et al. (1976, 1977b) found little variation in their measure of k or in total counts over five weeks in the breeding season, but Slagsvold (1973b) demonstrated that at least some species have a very short singing period which may also differ in timing between years, so that k may vary most awkwardly.

The most common response to the problem of seasonal variation in k (and in d when an annual index of density is sought) is to confine field work to a few months of the year or less (Kimball 1949, Purroy 1974, Robbins and Bystrak 1974, Robbins and Van Velzen 1974, Sammalisto 1974, Raynor 1975, Järvinen and Väisänen 1976c, Crook et al. 1977, Källander et al. 1977, Ratowsky and Ratowsky 1977, 1978; Svensson 1977a, Shields 1979). The French Indice Ponctuel d'Abondance method (I.P.A.) uses at each point the highest of two counts several weeks apart in the breeding season (Ferry 1974, Blondel 1975, see also Purroy 1974); and Shields (1979) and Connor and Dickson (1980) recommend counting in experimental and control areas both before and after some treatment is applied

[1] Ecology Division, Department of Scientific and Industrial Research, Private Bag, Lower Hutt, New Zealand.

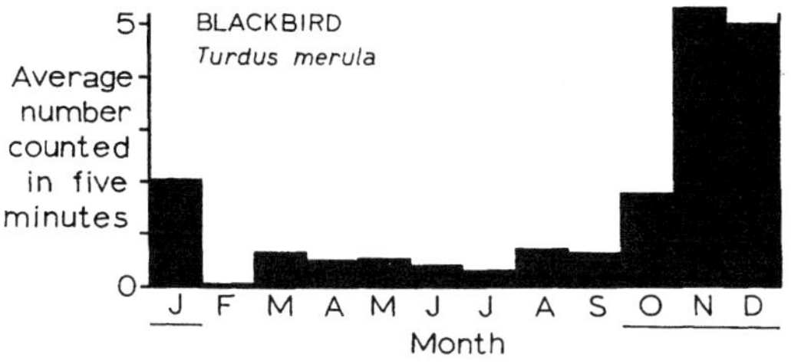

FIGURE 1. Monthly averages of 5-min counts of Blackbirds in New Zealand forest. Based on a minimum of 120 counts a month in each of four study areas in Fiordland, by K. Morrison (pers. comm.), counted as described by Dawson and Bull (1975). The months when young are leaving the nest are underlined. Falla et al. (1970) give the period of full song as August to November in New Zealand. The counts in January and February are much lower than would be expected if the birds were as conspicuous as they were in November and December.

to the experimental one, to overcome seasonal problems.

Some workers have treated seasonal changes as if they might be due to changes in d alone, and have only acknowledged that k may change when their results made no sense (e.g., Sammalisto 1974). This approach could lead to serious error and should be avoided.

Habitat

Surprisingly few authors have noted that the effective area sampled will vary with such things as the density of the surrounding vegetation, or topography (Howell 1951, Blondel et al. 1970, J. T. Emlen 1971, Svensson 1977a, Weber and Theberge 1977, Shields 1979), perhaps because many studies were confined to a limited range of habitat types and the (often tacit) assumption was made that the limited variation in the habitats did not materially affect the value of k (Ferry 1974, Crook et al. 1977, Dawson et al. 1978, Shields 1979, Conner and Dickson 1980, Gill 1980). Others have tried to sample habitats representative of the geographical area, and stratified their analysis by broad geographical habitat types (Robbins and Van Velzen 1970, Järvinen and Väisänen 1976c). There seems no way of studying the effect of habitat on k without having good independent estimates of d, but something may be made of the ratio of counts between two species in different habitats, or between the counts of a single species in two seasons and different habitats, if one can assume that there is a characteristic (although unknown) value of k for each habitat in each season (Fig. 2).

Another effect of habitat is on the onset of singing. Slagsvold (1973b) found a correlation between the time that Song Thrushes began singing and the leafing of birch trees (*Betula* spp.); there was a delay in song of 2–3 days for every 100 m rise in altitude. This interaction between habitat and the seasonal course of k could be studied and corrected for.

Time of Day

As with season and habitat, in most published work the time of day has been kept to a limited range, often round and a few hours after dawn, to maximize the numbers counted (Kimball 1949; Blondel et al. 1970; Robbins and Van Vel-

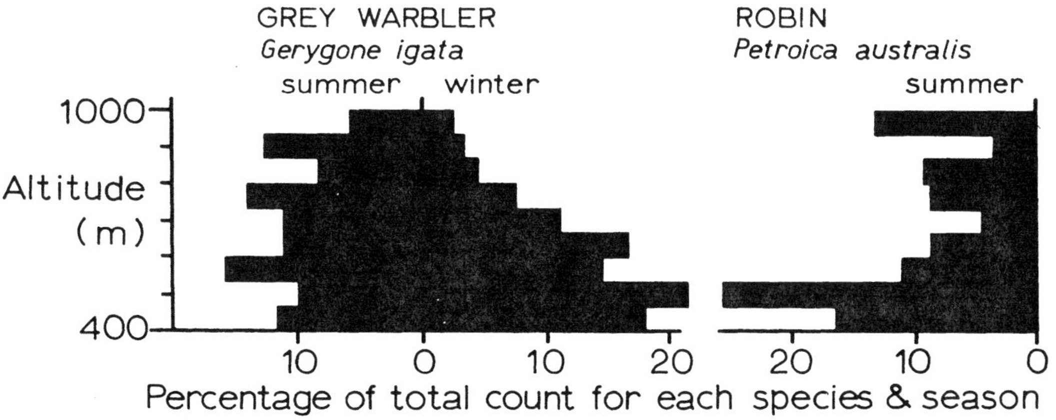

FIGURE 2. Altitudinal distribution of Grey Warblers and Robins in summer (December to February, 33 counts) and of Grey Warblers in winter (June to August, 42 counts) in a mountainside forest in New Zealand. (H. A. Best, pers. comm.). Nine counting stations in forests on the Victoria Range were counted each time, using the technique of Dawson and Bull (1975). In summer the Robin clearly favors lower altitudes than does the warbler. The warbler favors lower altitudes in the winter than it does in the summer.

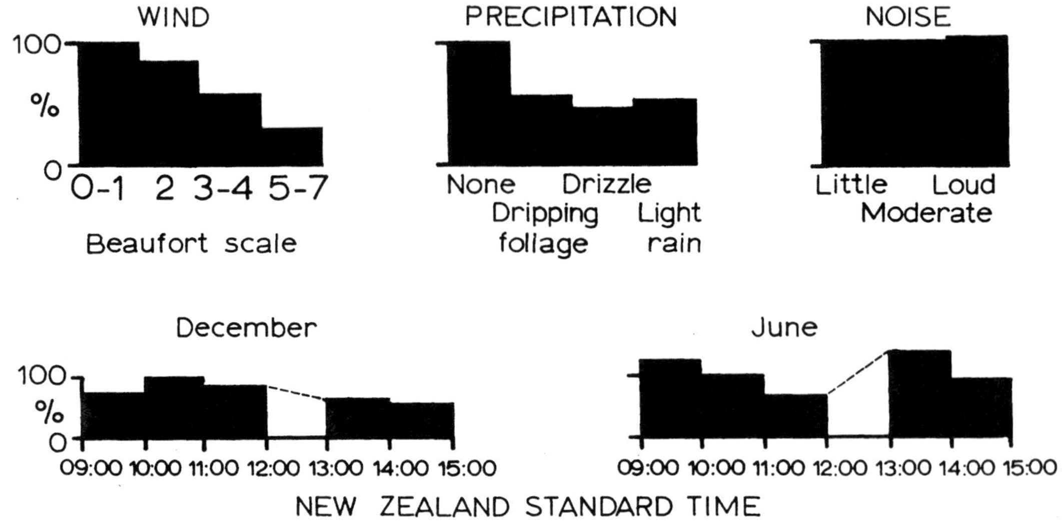

FIGURE 3. Effect of wind, precipitation, noise, and time of day on the number of Grey Warblers counted in New Zealand forests. Data from the study described by Dawson et al. (1978). For each of wind, precipitation, and noise the analysis of each factor was confined to counts with minimal values of the others. Only wind and precipitation affected the counts significantly ($P < 0.01$ in analyses of variance). The results are expressed on a relative vertical scale, with 100% being arbitrarily set at minimum values of wind, precipitation, and noise and at 10:00–11:00 for the time.

zen 1970, 1974; Slagsvold 1973b; Nilsson 1974b; Järvinen and Väisänen 1976c; Ratowsky and Ratowsky 1977, 1978; Connor and Dickson 1980), but some species are more vocal later in the day (Shields 1977). Some studies do not mention time of day (e.g., Amman and Baldwin 1960, Lancaster and Rees 1979). Dawson and Bull (1975) counted between 09:30 and 15:30 to avoid "the rapid change in birds' conspicuousness near dusk and dawn," and Yapp (1956) gave similar advice. Figure 3 is typical of the numbers counted of five common New Zealand species, and it shows that the numbers counted vary little in the middle hours of the day. The effect of time of day may be expected to vary with season (Shields 1979), and in very hot regions such as deserts.

There are few good studies of the effect of time of day, where the counts of individual species have been examined over a range of hours (Kimball 1949; Davis 1965; Robbins and Van Velzen 1967; Slagsvold 1973b; Järvinen et al. 1976, 1977b; Shields 1977, 1979). Analyses of the total number of individuals regardless of species (e.g., Figure 2 *in* Hogstad 1967, Figure 2 *in* Robbins and Van Velzen 1970, and Figure 18 *in* Svensson 1977a) are misleading because species may peak at differing times and cancel each other. The better studies suggest that time of day near dawn will make such a substantial contribution to the variance of the counts of many species that it should be studied, so that it may be allowed for (e.g., Shields 1977, 1979); most work is deficient in this regard. The alternative of counting over a very short (e.g., half hour) standard period is usually impractical, as a large number of such samples would be needed to yield enough data. The best solution may be to count over a longer period near the middle of the day (Dawson and Bull 1975).

Weather

The effect of weather on k, and hence on the numbers counted, is mentioned as an anecdote in several studies and the prescription often mentions that high wind, rain or cold are to be avoided (Yapp 1956, Robbins and Van Velzen 1967, Blondel et al. 1970, J. T. Emlen 1971, Dawson and Bull 1975, Järvinen and Väisänen 1976c, Svensson 1977a, Shields 1979, Conner and Dickson 1980). Analysis of the effects of weather is particularly difficult because weather variables are often intercorrelated and also correlate with season, habitat and time of day. Ratowsky and Ratowsky (1979) found that wind reduced the number of species recorded in transect counts in Tasmanian forest, but that rain had no detectable effect; Hogstad (1967) suggested that both high wind and rain depress counts. Figure 3 summarizes some typical results from my unpublished study of the effects of weather on bird counts and confirms that the usual advice to avoid strong wind and rain is well founded.

Bad weather is likely to affect the performance of the observer as well as that of the birds, but the two effects cannot be distinguished in the counts.

Environmental Noise

Like weather, this factor is covered only anecdotally in published work. Figure 3 illustrates a typical example of its effects, which appear not to be important. It, too, could act through effects both on the observer and on the birds.

Observer

Differences between the counts of individual observers could reflect judgement (e.g., of whether a new sound came from a bird already counted or not), ability to follow a prescribed plan, acuity of hearing and vision, knowledge and experience. For these reasons several studies have been careful to keep the same observers, where possible (Robbins and Van Velzen 1967, Sammalisto 1974, Lincoln 1975, Källander et al. 1977, Dawson et al. 1978, Shields 1979, Conner and Dickson 1980). If observers are rotated between study areas they can be included as a factor in the analysis and their differences estimated and allowed for (e.g., Dawson et al. 1978). Svensson (1977a) suggests 'calibrating' observers by having them count simultaneously on the one area.

Number of Registrations

Walankiewicz (1977) reported that, in a study of five forest areas in Poland, an estimate of k for I.P.A. counts of several individual species increased with increasing total number of birds counted. This suggests a saturation effect: more cues are missed when an observer is busy recording a large number of birds. Frochot et al. (1977) reached a similar conclusion. Järvinen and Väisänen (1976b) suggested that the proportion of records from an inner transect belt to the total counted was greater the more birds there were, although they later commented that their correction factor to allow for this was not necessarily applicable elsewhere and could sometimes lead to error (Järvinen et al. 1978b). Table 1 illustrates an attempt to examine this question, and shows that a high total number recorded did not seem to reduce the number of distant birds detected.

Another problem is that there may be an upper limit to the number of any one species that can be distinguished when many of that species are singing. This is certainly the subjective impression of some New Zealand observers, but

TABLE 1
Effect of the Total Number of Birds Counted on the Percentage of All Records That Were "Far" Records[a]

Species	Percentage 'far' records when the total number of birds counted was: Few[b]	Many
Myna (*Acridotheres tristis*)	67	78
Silvereye (*Zosterops lateralis*)	30	16
Goldfinch (*Carduelis carduelis*)	14	14

[a] 'Far' records were those estimated to be 50–200 m from the observer. Data from 112 transects (of 200 m) made in pastoral and orchard land in Hawke's Bay, New Zealand, March 1980. If a large number of records impedes an observer's ability to discriminate, this would result in a lower percentage of far records, but the two percentages did not differ significantly for any of the three species.
[b] "Few" was less than the median total number.

it will probably have to await calibration studies for confirmation.

The Technique

The number of birds counted will vary with details of the technique used. I leave a comparison of transect and point counts, and an examination of the effect of the observer's speed to another paper in this symposium.

Dawson and Bull (1975) compared point counts of 5 and 10 min duration, and Dawson and Robertson (unpubl.) compared 2 min with 5 min; in both comparisons the alternatives gave measures of closely comparable efficiency. The first few species were detected quickly with short counts, and many sites could be sampled per unit time—but these advantages were offset by the greater time spent moving between counts. The duration of a point count therefore seems an arbitrary choice, at least over the range 2–10 min. However, a long duration does not permit many points to be sampled and so impedes replication of study areas. For this reason most techniques use a short duration (Robbins and Van Velzen 1967, Dawson and Bull 1975, Svensson 1977a). I consider the 20 min count used in the Indices Ponctuels d'Abondance (Ferry 1974) and in the Échantillonnages Fréquentiels Progressifs (Blondel 1975) to be too long.

The spacing of point counts is usually regular, along transect lines or on a rectangular grid, at intervals of between 0.2 km and 0.8 km (Robbins and Van Velzen 1967, Ferry 1974, Crook et al. 1977, Svensson 1977a, Dawson et al. 1978). If the counts are too close to each other, the chances of a bird being counted twice are increased, but if they are too far apart, much time is spent moving between them. For this latter reason, and if the observer is to walk between the points, I prefer a short spacing (0.2 km).

Both point and transect counts may use a cut-off distance beyond which no birds are recorded. Burnham et al. (1980) argue against such truncation, except to remove a small percentage of outliers, because the distant birds aid in estimating numbers. Dawson and Bull (1975) used a long (0.2 km) cut-off for this reason. Smaller, or variable, cut-off distances may be needed in the study of habitat selection, so that the birds may be associated with the correct habitat.

Most techniques do not specify what the observer should do when uncertain whether or not a new cue comes from a bird already counted. Järvinen and Väisänen (1976c) counted only in front of the observer, to minimize the risk of double records. Dawson and Bull (1975) proposed an arbitrary rule that unless the observer is reasonably sure the cue comes from the same individual it is taken as new. Some such rule should always be used, so as to lessen the variation between observers that would otherwise occur.

CONCLUSION

This paper has reviewed a large number of influences on the number of birds counted. Some of them (technique, season, and time of day) may be avoided by standardizing the observations, others (observer, weather, and noise) may be documented as the counts are done, so that their effects may be studied and corrected or avoided. The effects of species, age, reproductive group, habitat and cues from other birds require special study.

ACKNOWLEDGMENTS

A. E. Aldridge, P. C. Bull, M. Harrison, E. B. Spurr, and R. H. Taylor made constructive comments on the manuscript.

Studies in Avian Biology No. 6:17–23, 1981.

THE CHRISTMAS BIRD COUNT AND AVIAN ECOLOGY

Carl E. Bock[1] and Terry L. Root[1]

Abstract.—The Christmas Bird Count (CBC) is an enormous but weakly standardized avian count. Observers spend thousands of hours annually, counting as many species and individuals as possible inside hundreds of 15 mile (24 km) diameter circles in North America. CBC data are an inappropriate substitute for more controlled census work associated with local projects. Scientists probably would ignore CBC data altogether, were it not for their potential application to large-scale studies. CBC results have proven to be good descriptors of continent-wide patterns of avian geographical ecology—patterns which otherwise often would remain undetected. Nevertheless, CBC data should be used with caution. Large sample sizes are very important. The data are better indicators of real patterns among common and well-dispersed species than for rare and/or highly social species. For species spread relatively evenly across count circles, CBC results should be standardized by dividing raw numbers by total party-hours of count effort. Species which are social and restricted to unusual habitats (e.g., waterfowl) are likely to be counted or estimated totally regardless of the overall count effort. The best standardization for these birds probably is to compute raw numbers per count. Most CBC studies have been concerned with population trends, but the data are equally valuable indicators of spatial abundance patterns. Application of clustering techniques to CBC data results in a numerical biogeography, the power of which lies in its being quantitative, objective, and based upon abundance patterns.

Certainly the largest and oldest bird census in the New World is the annual Christmas Bird Count. Each CBC is a day-long tally of birds seen inside a circle 15 miles (24 km) in diameter just prior to or following Christmas Day. Results, published in *American Birds,* include latitude-longitude coordinates of each census, lists of species seen and their numbers, numbers of observers, party-hours and party-miles as indices of effort, and information about weather and habitats surveyed. Each year thousands of observers participate in hundreds of counts across North America. We estimate that just in the decade 1962–71 Christmas counters spent well over one million hours afield and recorded about 635 million birds.

If the CBC ranks as the world's largest bird-population count, it probably is the least structured. Since they are involved in a type of birding contest, participants will do what they can to see as many species and individuals as possible. This can involve staking out rare birds ahead of time and sowing bird seed in likely places. Counting birds at backyard feeders is common. Organizers try to assure uniform coverage, but observers naturally spend most of their time in the best spots.

Given these conditions of data collection, scientists might be expected to avoid CBC's altogether. Doubtless this would be the case, were it not for the potential that they offer to students of avian ecology. With CBC's we can at least *ask* questions about winter bird-population fluctuations and about bird-abundance patterns on a geographic scale not possible for any other organisms.

[1] Department of Environmental, Population and Organismic Biology, University of Colorado, Boulder, Colorado 80309.

This, of course, presumes that CBC data are realistic indicators of winter bird-population distribution and abundance. There is no way to test the quality of the data, except as they conform to expected patterns or are confirmed by independent sources of information. In this regard, however, results have been quite good. CBC data revealed an unexpected synchrony and cyclic regularity to southward eruptions of normally boreal birds (Bock and Lepthien 1976d), a finding subsequently supported by banding data (Kennard 1976, 1977). Many apparent and generally recognized population changes have been documented with CBC data (e.g., Davis 1937, 1974; Wing 1943; Brown 1973, 1975; DeHaven 1973; Bock and Lepthien 1976c; Larson 1980; Pruett-Jones et al. 1980). CBC patterns of species richness and diversity are strongly correlated with certain climatic variables (Tramer 1974; Bock and Lepthien 1975b). Recent comparisons of CBC data with results of migration studies at the Long Point Observatory led Hussell and Risley (1978:98) to conclude that "Christmas Bird Count indices can be used to monitor populations of many species with greater precision than had previously been suspected."

Despite these encouraging results, it is obvious that CBC data must be used with great care and caution (Arbib 1967). They can be a powerful analytic tool; but there are circumstances under which they ought not to be used; and there are some methods of analysis that are much more effective than others. The purposes of this paper are to discuss: (1) means of standardizing CBC data, (2) effects of weather on count results, (3) the problem of adequate sample sizes, (4) the problems presented by rare and by highly social species, and (5) the application of cluster analyses to CBC data.

STANDARDIZING CHRISTMAS COUNT DATA

Measures of Count Effort

CBC results must be normalized to be meaningful indicators of winter bird population sizes (Kenaga 1965, Raynor 1975). Davis (1974 and earlier papers cited therein) divided total birds seen by the number of counts per year in his studies of Northern Shrike (*Lanius excubitor*) eruptions. Birds per census would be a reliable index of population trends, as long as average effort per census did not change over the years or areas being considered. Most authors have been unwilling to make this assumption, although it probably holds if enough counts are considered at once. Some workers have divided bird numbers by party-miles (Brown 1971, 1973) or by numbers of observers (Raynor 1975), but most have chosen party-hours as the best measure of count effort (e.g., Schreiber and Schreiber 1973, Bystrak 1974, Stahldecker 1975, Plaza 1978).

DeHaven (1973) found that dividing by party-hours and party-miles gave essentially the same picture of Starling (*Sturnus vulgaris*) population growth in California. Raynor (1975:628) concluded that "it probably makes little difference which measure of effort is used to normalize count data," since all are strongly correlated. Falk (1979) analyzed 15 count sites in the north-central United States and found that, overall, total party-hours was the best predictor of both the number of species and the number of individuals recorded on those counts. Party-hours seems to be the best and most widely accepted factor for CBC standardization.

The Problem of Unusual Habitats

One of the greatest difficulties with CBC data concerns their application to aquatic species or other birds restricted to special habitats inside count circles. Observers will be aware of these areas and will cover them each year regardless of overall count effort. If there is one pond in an otherwise terrestrial count circle, it will be covered for ducks each year. If total party-hours doubled over a 10-year period, then "ducks per party-hour" would fall to one half its original value over the same period, despite a stable duck population on the pond. Raynor (1975) recommends calculation of "effective party-hours" for such circumstances, which in the above example would be the party-hours actually spent at the pond counting ducks. This solution has several problems. First, it is not possible to determine just where a particular species was seen. Published count results include lists of habitats visited (by percent of total count time spent in each), but the birds are not listed by those habitats. Even if one could logically connect the species of interest with a particular habitat type, and then calculate effective party-hours spent in it, such calculations would soon become prohibitively tedious in any study involving large numbers of counts. Most importantly, we have found it very difficult to compare lists of habitats from one count to another, since there is no standard terminology in use.

Morrison and Slack (1977) found no correlation between raw numbers of Olivaceous (*Phalacrocorax olivaceus*) and Double-crested Cormorants (*P. auritus*) and any measures of effort among Gulf Coast CBC's. They used actual numbers per count as the best measure of population trends. This probably is the simplest and most meaningful way to standardize CBC data for birds which are restricted to aquatic habitats, which are social, and which therefore are easily censused.

Standardization by Comparison of Similar Species

A less-used but potentially powerful approach to CBC analysis is to use the number of birds themselves as a means for data standardization. Suppose, for example, we are interested in comparing abundance patterns of Eastern (*Sturnella magna*) vs. Western Meadowlarks (*S. neglecta*). We simply divide the raw numbers of the eastern species by the total number of meadowlarks of both species. No additional standardization is necessary. The problem of unusual habitats is eliminated by this method, since dividing by total meadowlarks presumably compensates for the fact that different counts involved different amounts of time spent in meadowlark habitat.

This method of calculating relative abundance tells us nothing about absolute changes in numbers, but it does tell us a great deal about patterns of geographic complementarity among related or similar species, and it should be as applicable to readily counted species (e.g., waterfowl) as to species spread more evenly across count circles. Bock et al. (1977) and Root et al. (in press) used this approach in studies of the geography of flickers (*Colaptes*), juncos (*Junco*), meadowlarks (*Sturnella*), bluebirds (*Sialia*), phoebes (*Sayornis*), and thrashers (*Toxostoma*) across the central U.S. in winter. The technique dates to the work of Wing (1943), who examined ratios of Mallards (*Anas platyrhynchos*) and Black Ducks (*A. rubripes*) across the eastern U.S. between 1900 and 1939. We have computed similar ratios for the 1962–71 counts, grouped in five degree blocks. Figure 1 shows blocks in which Mallards and Black Ducks were most abundant. Wing's line of equal ratios is superimposed, showing how the Mallard has expanded its numerical dominance eastward in the past 30 years (see also Johnsgard and Di-Silvestro 1976).

WEATHER

It is common for users of CBC data to be concerned about weather, not as it affects the numbers of birds present in a count circle, but as it might influence their detectability and/or observers' enthusiasm (Arbib 1967). Compensation for the weather may be necessary in some cases. For example, W. H. Brown (1971) found that more Red-shouldered Hawks (*Buteo lineatus*) were counted on clear than on cloudy days. Morrison and Slack (1977) found that numbers of cormorants counted varied inversely with cloud cover and wind. However, Falk (1979:689) analyzed 15 counts in the north-central U.S. and found that (1) "weather conditions do not consistently affect measures of count effort," and (2) "there are relatively few significant values when selected weather conditions are correlated with number of bird species and individuals."

THE PROBLEM OF SAMPLE SIZES

The best safeguard against spurious effects of weather or any other stochastic event, such as observer skill and numbers, or movements of flocking birds, is to analyze a sufficient number of counts over a sufficient number of years so that these variables will cancel out or equilibrate. What is a sufficient sample size? We can give no universal answer except to state the obvious: the more counts that are analyzed, the greater the confidence that can be placed in any emerging trend or pattern. Bock and Smith (1971) analyzed population trends for 20 selected species in Colorado for the years 1940 to 1970. Most species curves fluctuated wildly until about 1950, when the sample size per year jumped rapidly from four or fewer counts to eight or more counts. Schreiber and Schreiber (1973) found a similar pattern in Florida counts.

The importance of large samples is illustrated by the following simple but real examples. Root et al. (in press) analyzed 124 count sites in Texas, Oklahoma, and eastern New Mexico for abundances of bluebirds. They found that the Eastern Bluebird (*Sialia sialis*) was significantly more abundant east of the 100th meridian than it was to the west. Yet, between 1962 and 1971 three east Texas CBC's (Ft. Worth, Dallas, Travis County South) together counted fewer Eastern Bluebirds per party-hour than did three west Texas counts (Davis Mts., Lubbock, and San Angelo). This does not mean that the six counts in question were "wrong," but neither does it

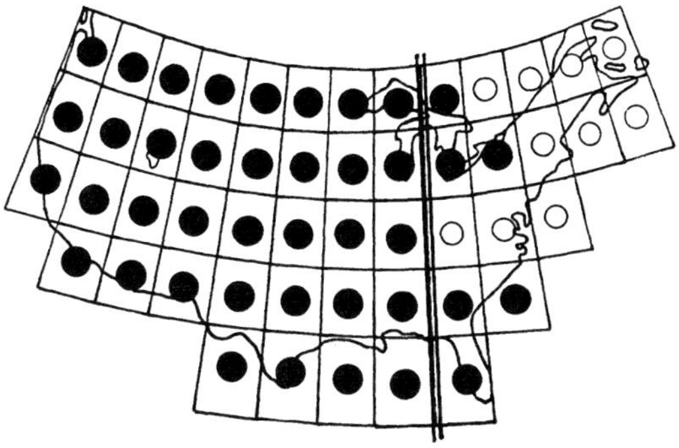

FIGURE 1. Geographic blocks in which Mallards (dots) or Black Ducks (circles) were most common on 1962–1971 CBCs. Between 1900 and 1939 Black Ducks outnumbered Mallards east of the heavy double line (Wing 1943).

invalidate the general conclusion about bluebird geography in Texas.

Bock and Lepthien (1976d) found that, between 1962 and 1971, synchronous and continent-wide southward eruptions of boreal seed-eating birds occurred in the winters of 1963–64, '65–66, '68–69, '69–70, and '71–72. Sample size for this study exceeded 7000 counts. CBC's in a 5 degree block including parts of Minnesota and Wisconsin (Block 7 in Fig. 6A) showed this pattern clearly for the strongly eruptive Common Redpoll (*Carduelis flammea*); yet, in that block, the Duluth count conformed to the continental pattern only in some of those years (Fig. 2).

Some workers have attempted to improve the "quality" of CBC data by accepting only those counts with certain ranges of party or observer numbers (e.g., Graber and Golden 1960, Stahldecker 1975). Others have included only those counts conducted continuously for the span of years being considered, or only counts occurring in some minimum number of years (e.g., DeHaven 1973, Brown 1975). Such selectivity may be necessary for studies involving a small number of years, a restricted geographic area, or a rare and/or flocking species. *A priori* exclusion of counts is a legitimate step, but it can lead to a powerful and dangerous urge to exclude certain counts *a posteriori,* when those counts are obscuring an otherwise clear trend.

RARE AND FLOCKING SPECIES

Problems associated with very rare species may be so severe that no meaningful CBC analysis is possible. Observers will work hard to find at least one individual of any rare species which might occur in a count circle. The result is that some rare species appear to occupy the country

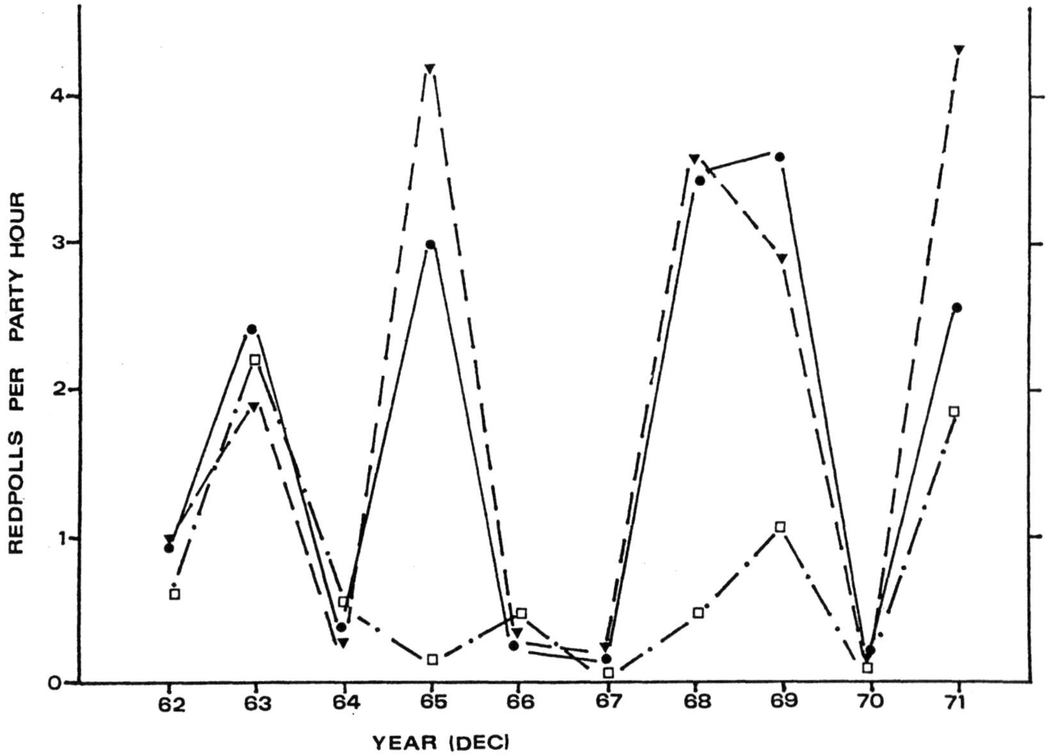

FIGURE 2. Number of Common Redpolls counted per party-hour on all counts (solid line), Block 7 counts (dashed line), and on the Duluth, Minnesota, count (broken line). See Figure 6A for location of Block 7. Multiply abundance scale by five for Block 7 and Duluth count data.

in the ornithological equivalent of a monomolecular layer. Chance encounters of rare and narrowly distributed species can significantly affect results. We found no correlations between carefully considered population estimates of the California Condor, *Gymnogyps californianus* (Sidney et al. 1968) or Whooping Crane, *Grus americana* (Olsen 1980) and our computerized 1962–71 CBC data for these species.

Flocking and communally roosting species can be equally difficult. One of the first projects we attempted with our CBC data bank was an examination of nationwide population changes in the Starling. We discovered that three or four individual counts in the Southeast, if they occurred in a particular year and if observers found roosts, could double the average number of Starlings counted per party-hour for the entire country. Our data for blackbirds and grackles are equally variable and appear beyond any sort of meaningful analysis.

BIOGEOGRAPHY

Species' Abundance Patterns

Most CBC ornithology has been concerned with temporal changes in species populations. The data are equally valuable for studies of the spatial abundance patterns of birds.

Biogeography traditionally has considered only the presence and absence of species on continents, since these are the only data available on such a scale. Yet presence versus absence is recognized to be a grossly simplistic

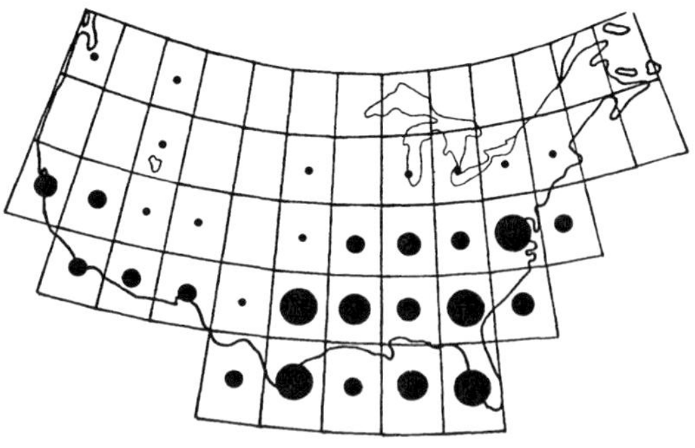

FIGURE 3. Numbers of vultures counted per 100 party-hours, 1962–1971 CBC data. Five sizes of dots represent <1, 1–33, 34–65, 66–100, and >100 birds/100 party hours. Open blocks = no vultures counted.

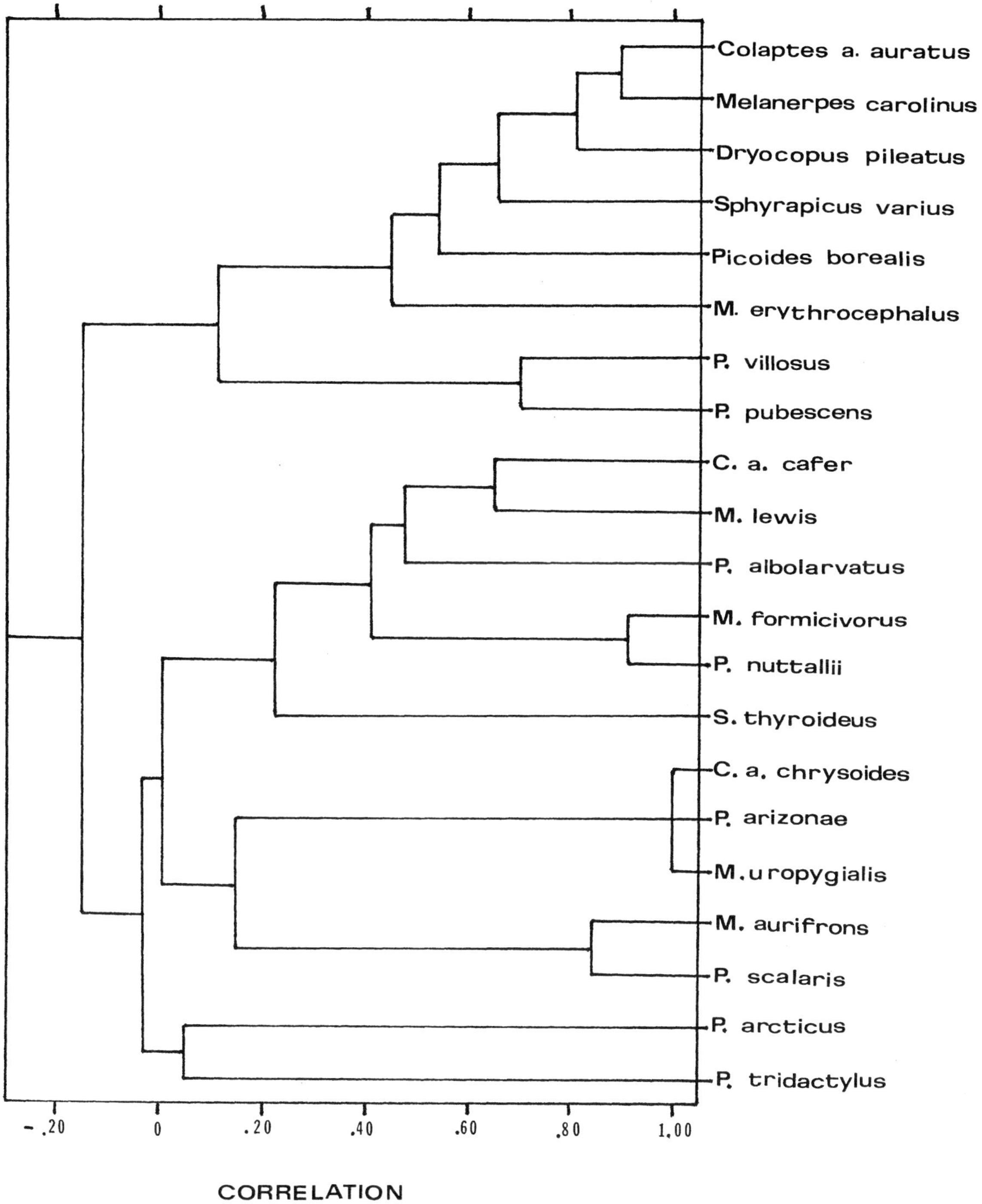

FIGURE 4. Dendrogram showing relationships among 21 taxa of woodpeckers, based upon their CBC abundance patterns in 51 latitude-longitude blocks (Fig. 6A). Matrix correlation = 0.891.

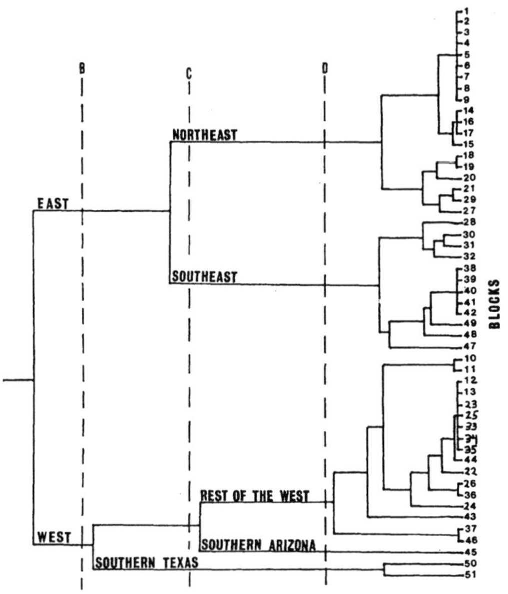

FIGURE 5. Dendrogram of 51 latitude-longitude blocks, based upon CBC abundance patterns of 21 woodpecker taxa. Matrix correlation = 0.917.

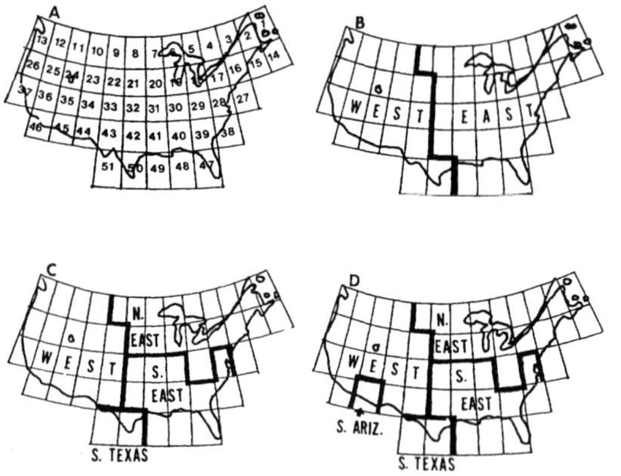

FIGURE 6. A. Locations of 51 5-degree latitude-longitude blocks. Maps B–D show geographic positions of clusters of blocks corresponding to branches of dendrogram transected by dashed vertical lines B–D in Figure 5.

view of the reality of species distributions (e.g., Udvardy 1969, Rotramel 1973). Simply stated, species are common in some places and rare in others. There is good reason to suspect that the last outpost of a species' range tells us less about what is environmentally important to it than would data showing us where, inside its range, that species switches from being abundant to being scarce. With CBC's we anticipate that we can plot a species position in space in terms of contours of declining density away from centers of abundance (Grinnell 1922). We could then compare these with known patterns of climate, topography, and vegetation and learn what factors of the environment are important to the bird populations of interest.

One common criticism of CBC data is that they are collected in early winter, before some species have fully settled on their wintering grounds (Arbib 1967). If we used CBC's as indicators of the limits to winter ranges of species, this would indeed be a problem. But the unique aspect of CBC information is that it shows us where species concentrate in winter, so that the occurrence of stragglers becomes largely unimportant.

The most extensive geographic analysis to date using CBC data has been the winter-range-mapping project (Bystrak 1971, Bystrak et al. 1974), which resulted in publication of abundance maps for 140 species, each prepared by a volunteer. The maps are of good quality for the most part. If they have a weakness, it is that they usually were based on only one year's data. Recently, Plaza (1978) has shown the potential of applying computer mapping programs to CBC information.

We have built a computerized data bank containing the results of 7891 counts occurring between the winters of 1962–63 and 1971–72. We combined the counts into 5-degree blocks of latitude and longitude, then computed the mean number of birds seen per party-hour for each block, for each of the ten winters, for 627 species. Figure 6A shows locations of the blocks, by number; sample sizes were very uneven across blocks, ranging from 11 in Block 48 to 864 in Block 16.

CBC abundance maps are interesting and valuable indicators of species' centers of abundance. Through correlation analysis we have compared species' abundance patterns with climatic variables (e.g., Lepthien and Bock 1976), and with vegetation (e.g., Bock and Bock 1974). Figure 3 shows the combined 10-year mean winter abundance pattern of the North American vultures (*Cathartes aura* and *Coragyps atratus*). Applying environmental data to the same grid, we find that this pattern is positively correlated with estimates of primary productivity ($r = .601$), number of frost-free days ($r = .746$), and annual precipitation ($r = .458$).

CBC's and Numerical Biogeography

By applying techniques of cluster analysis to CBC data (Sneath and Sokal 1973), it is possible to describe geographic patterns for whole groups of species in a quantitative manner. Beginning with a raw-data matrix of *n* species by *n* geographic areas, we can group species which share centers of abundance (e.g., Bock and Lepthien 1976a), or recognize groups of blocks which are highly correlated in terms of their winter avifaunas (Bock et al. 1978). We present here a single-linkage cluster analysis (Sneath and Sokal 1973) of the family Picidae to illustrate the value of this approach. Figure 4 is a cluster of 21 taxa of woodpeckers, based upon correlations of their abundances in 51 latitude-longitude blocks (Fig. 6A). We can cluster the blocks by inverting the matrix, so that the characters become the objects to be classified. The resulting dendrogram (Fig. 5) shows groups of blocks with high faunal similarities, but which differ from one another. The dashed vertical lines labelled B, C, and D in Figure 5 transect stems of the dendrogram which include groups of blocks shown geographically in Figures 6B, C, and D, respectively. The maps then show areas of high internal avifaunal homogeneity, and the boundaries between them. The strengths of this approach to avian biogeography are that (1) it is quantitative, repeatable and objective, and (2) it is based upon the abundances of species and not simply their presence or absence.

CONCLUSIONS

Eugene Odum (1950:227) said of the CBC: "One has the feeling that there is more gold buried in the mass of data than has yet been uncovered." There has been considerable mining activity since then, and some rich veins have been explored. Clearly the strength of Christmas count data lies in their quantity more than in their quality. There may be ways to improve the nature of CBC information (Stewart 1954, Arbib 1967, Arbib 1981), but we agree with Hickey (1955) that one ought not to tamper with the event very much, lest the thousands of volunteers who make it happen, and who do it largely for fun, stop *having* fun and quit. Also, any change in CBC rules which would make it impossible to compare past and future results would defeat the very purpose of the census.

Christmas count data do not appear to work well until they include a critical mass of years and count circles, but it is only in pursuit of such long-term and continentally scaled patterns that we need them. Christmas count data should not be substituted for careful local censuses in environmental impact studies, or in any other sort of field ornithology within the strategic capabilities of an individual investigator. For large-scale studies, CBC analysis is very cost-effective. The petroleum has already been combusted, and the hours have been expended in the field. No group of paid professionals could ever mobilize the time or dollars to gather such data. CBC's appear to be surprisingly good indicators of pattern in avian geographical ecology, if they are used carefully and conservatively, and especially if they are used in large numbers.

ACKNOWLEDGMENTS

Research for this paper was supported by NSF grants GB 36860 and DEB 79-10784. We thank Michael Grant and Larry Lepthien for assistance with statistics and computer analyses, and Joseph J. Hickey for reviewing the manuscript.

Studies in Avian Biology No. 6:24–29, 1981.

THE CHRISTMAS BIRD COUNT: AN OVERLOOKED AND UNDERUSED SAMPLE

SUSAN RONEY DRENNAN[1]

ABSTRACT.—The Audubon Christmas Bird Count represents the most extensive, longest-term, continuous, and most geographically comprehensive data set in American ornithology. It provides the empirical basis for an increasing number of research studies, particularly longitudinal analyses of relative abundance of species and their dynamics, and definition and spatial abundance of the early winter ranges of the avian species of North America. This paper assesses the numerous assumptions involved in making statistical inferences from Christmas Bird Count data. Both biological and statistical constraints inherent in those data are addressed. The results suggest that the Christmas Bird Count has properties comparable to a valid probability sample, when latitudinally stratified. Further, that robust estimation methods should accommodate deviations intrinsic in the data.

The Constitution of the United States established the decennial census of human population numbers in 1790. In 1900 the Audubon Societies, parent organization of the National Audubon Society, inaugurated the Christmas Bird Count. Neither was originally conceived as a data base for statistical inference to increase knowledge of human or avian populations, but the world has changed; and they have been so used. Owing to their higher degree of reliability, the United States Census Bureau demographers prefer utilizing their sample estimates, based upon age-cohort survival tables and fertility rates, rather than the census "official" population count. Similarly, researchers should be wary of inferring too much from Christmas Bird Count (hereafter, CBC) raw numerical data, as it is not a census by strict definition of that term.

The Audubon Christmas Bird Count is the single, most popular, voluntary, early winter bird continental inventory in the world. The 1979–80 CBC had 33,022 participants who censused 1320 count units in North, Middle and South America, and Hawaii. A count unit is defined as that area contained within a circle of 24.1 km diameter. With few exceptions, each count unit is discrete, having no common parts. Counts must be made during a single calendar day within the official CBC period, two weeks centered around Christmas Day. Searching the count unit is accomplished by parties of observers of varying numbers. Every individual bird encountered is included in the inventory. Results of each count unit are reported on standardized forms, which also solicit details on weather, habitat coverage, methods of canvassing the area, natural food resources, and hours afield and miles covered. Count statistics have been published annually for the past 80 years. The CBCs represent the most extensive, longest-term, continuous, and most geographically comprehensive data set in American ornithology.

With increasing frequency, ornithologists interested in winter population trends, winter range extensions, and winter bird distribution have been turning to those data accumulated in the CBCs. Preston (1980) has used CBC data to analyze distributions of commonness and rarity. Major patterns of avian species diversity and abundance have been analyzed by Bock and Lepthien (1974, 1975b, 1976d), Bock, Mitton and Lepthien (1978), Tramer (1974), Smith (1979), and Falk (1979), to name but a few. Several recent longitudinal studies of single species density have been done by Brown (1973, 1975, 1976b), Bock and Lepthien (1975a, 1976b), Stahldecker (1975), Anderson and Anderson (1976), Johnsgard and DiSilvestro (1976), Kennard (1977), Morrison and Slack (1977), Neidermyer and Hickey (1977), Aldrich and Weske (1978), Bonney (1979), Raynor (1980), Rosahn (1980), and Stewart (1980).

Relative abundance maps seemingly provide viable results when the density of a fairly ubiquitous species is plotted by using the number of individuals recorded per 1, 10, or 100 party-hours as the basic standard (Bystrak and Drennan 1975, Plaza 1978). In 1974, the United States Air Force funded a project to map the winter distribution and relative abundance of 143 selected species considered the greatest potential threat to low-level-aircraft operation in the United States (Bystrak et al. 1974). These maps were plotted by using CBC data.

The CBC provides the empirical basis for a growing number of studies, particularly longitudinal analyses of definition and spatial abundance of the early winter ranges of the avian species of North America. Critics (e.g., Stewart 1954, Kenaga 1965, Robbins and Bystrak 1974) question the extent of scientific usefulness of CBC data, citing the lack of rigid controls as an inherent limitation. Others (Hickey 1955) argue that the CBC should be disregarded as a scien-

[1] National Audubon Society, 950 Third Ave., New York, New York 10022.

tific tool and be considered ". . . essentially a recreational activity in which distinct elements of competition, surprise, rarities and the big list are bright and personally thrilling."

This paper evaluates the present CBC and finds that it may provide useful and valid statistical estimates. This is suggested by its size, design (unplanned), and consistency of results. This paper does not address the issue of unstandardized count methods, at which most of the criticism has been directed. However, the results here suggest that that lack of standardization and imperfections in some areas of data collection may not be critical for use in certain types of analysis.

THE CBC AS A SAMPLE

Inasmuch as a census is a total count of all the members in an unambiguously defined universe, or a measure of a characteristic of all the members in a population, the CBC is not a census of North American birds. There have been no bird censuses of any continent for any widespread or numerous species, a fact which needs emphasis at this symposium. Practicality limits the use of a census to situations where the population being counted is very narrowly defined. Otherwise, drawing inferences about the population under study is accomplished through use of a sample, or subset of the individual items in an unambiguously defined population. By this means, unknown population parameters (e.g., its size: N, the mean: μ, and standard deviation: σ, of some population characteristic), can be estimated and the likely range of error of such estimates computed, provided that certain conditions are satisfied.

For example, let K equal the total number of subareas or count units, each of fixed area A, into which North America can be partitioned, such that each point on its surface lies within one and only one uniquely identified count unit. Let k equal a subset or sample from the K count units, such that $0 < k < K$. Let n_{ij} equal the number of birds of species j observed in the ith count unit ($i = 1, \ldots k$), on a given day in December. The mean absolute density of species j per area A (recalling that all count units are of equal size, A), for the sample of k count units is computed as

$$\bar{n}_j = \frac{\sum_{i=1}^{k} n_{ij}}{k} \qquad (1)$$

Then the estimate of N_j, the total number of birds of species j in North America, is: $\hat{N}_j = K(\bar{n}_j)$. Because $\bar{n}_j$ is computed from a sample, it is subject to error, and consequently, the estimate of N_j is subject to error. The standard error (SE), for n_j is given by:

$$SE = \sqrt{\frac{\sum_{i=1}^{k} (n_{ij} - \bar{n}_j)^2}{k(k - 1)}} \qquad (2)$$

So the 95% confidence interval for N_j is: $\hat{N}_j \pm 1.96K$ (SE). That is, there is a 95% probability that the true total number of wintering birds of species j in North America lies within the stated numerical limits.

At the outset of this example it was noted that such sample estimates of unknown population values are valid provided that certain conditions are met. The main conditions are: (1) the k count units selected for the sample represent a random sample from the total K count units (i.e., every count unit in K has an equal likelihood of being selected for the sample), and (2) the coverage and method of data collection in each of the k sample count units is the same, and every sample count unit is, in fact, covered.

The issue at hand is whether or not the CBC meets these conditions.

Regarding condition (1), the randomness of the sample, we have already established that the CBC is not a continental census. Additionally, it is not a simple random sample, because not every possible count circle in North America has an equal chance of being selected for the sample. The use of circles instead of hexagons or rectangles as the actual CBC count unit does not bias the results, as all of the circles are of equal size and are virtually mutually exclusive. Condition (2), method of coverage, is discussed elsewhere in this volume by Arbib (1981). Although the set of CBC circles covered each year is an accident of history, highly correlated with human population centers, it has properties comparable to a valid probability sample. A probability sample (including simple random samples) is a sample design in which the sampling units are selected according to laws of chance, such that the probability of any unit being included is known and not zero for every unit in the population.

As stated, condition (1) is not met by the CBC because the count areas included annually do not constitute a random sample from the set of all possible North American count areas. However, because geographical distribution of bird species is so uneven (nonrandom), a simple random sample would be inappropriate. The more appropriate sample design called for is a stratified sample, in which the probability that any count unit is selected is higher for high species density areas and lower for low species density areas. The most efficient distribution of sam-

TABLE 1
1979–80 CBC DISTRIBUTION: NUMBER OF SPECIES BY LATITUDINAL BELTS

	Number of counts by latitudinal belts							
Number of species	25–29°	30–34°	35–39°	40–44°	45–49°	50–59°	60+°	Total counts
0–9	0	0	0	0	2	2	1	5
10–19	0	0	2	5	13	4	6	30
20–29	0	2	7	30	46	11	2	98
30–79	6	45	207	376	87	20	2	743
80–89	2	25	31	29	5	0	0	92
90–99	2	36	16	12	4	1	0	71
100–149	35	59	49	38	17	0	0	198
150+	22	14	17	2	0	0	0	55
Total counts	67	181	329	492	174	38	11	1292

pling effort is contingent on concentration of effort increasing as avian density increases. As Caughley (1977) and others have shown, it borders on the absurd to intensively sample an area containing few animals.

It can be shown that, although the present CBCs are not the result of a rigid sample design, sampling effort is clearly more intensive in high species density areas than in low density areas. Table 1 presents the distribution of CBCs by reported species density (i.e., number of species) and latitude belt for 1979–80. Of 1292 CBCs (excluding the handful of counts south of latitude 25°N), 324 or 25.1% had 90 or more species. By contrast, 133 CBCs or 10.3% had 29 or fewer species. Thus, the sampling effort is more concentrated in high density areas.

If the goal of the CBC was to estimate density and absolute number of a single avian species, the ideal stratification of count areas would be based upon density of that species. However, if the goal was to estimate density and absolute number for as many different species as possible, then the ideal stratification of count areas would be based upon density of all species. The implication here is that sampling effort would not be randomly distributed over North America but would tend to be concentrated on coastal areas and in lower rather than higher latitudes. As we cannot know at the outset that which we would like to measure, density of species for every area of North America, the stratification should be based upon one or more characteristics which are measurable and are known to be correlated with number of species, e.g., latitude. Table 1 certainly supports the hypothesis of an inverse relationship between species density and latitude (see also Bock and Lepthien, 1974). Table 2 compares the distribution of the 1979–80 CBC with stratification based upon latitude belts. With the population domain defined as North America (excluding Greenland), the total area of 20.5 million km^2 is partitioned into latitude belts, and areas within each are approximated in column 2. Every CBC circle encompasses an area of 457.9 km^2. If A_1 equals the area of any latitude belt (in thousands km^2), then column 3 shows the population of count units (K) within each latitude belt, i.e., $K_1 = A_1/0.4579$. Column 4 shows the actual number of CBCs conducted in each latitude belt (k) in 1979–80. The total of 1292 sample count units conducted represents 2.9% of the land area of the continent. In other words, one out of every 34 possible count units was included in the sample.

With regard to flocking birds, an observation by Caughley (1977) is entirely applicable: that the higher the sum of sampled units, the more accurate the estimate; and the more clustered the animals, the greater the number of sampling units required to render a density estimate that is reasonably accurate. Note that percent of land area included in the sample tends to increase with decreasing latitude. The historical reasons for this are, of course, that: (1) the geographic distribution of birders is similar to the geographic distribution of people, and (2) birders tend to pick CBC areas relatively near their homes. Those two unchallenged principles are fundamental to past criticisms of the CBC as not representing a random selection of count areas. But the historical reasons are not germane. The de facto outcome is a sample in which the sampling effort (i.e., the percent of area covered) tends to increase with decreasing latitude. That is a highly desirable feature as avian winter populations are not randomly distributed, and their distribution is inversely correlated with latitude. Therefore, although ca. 54% of the area of North America lies above 50°N latitude, it would be inefficient and wasteful to devote ca. 54% of the

TABLE 2
STRATIFIED SAMPLE DESIGN: 1979–80 CBC

North latitude	Area, million km² (±)	Potential CBC units	Actual CBC units	% Potential CBC units sampled	Number of species		Coefficient of variation ($s/\bar{X}$)
					Mean ($\bar{X}$)	Std. Dev. (s)	
60°+	5.4	11,790	11	0.1%	19.6	13.2	0.673
50–59°	5.7	12,450	38	0.3	32.9	16.1	0.489
45–49°	2.6	5680	174	3.1	46.2	30.1	0.652
40–44°	2.3	5020	492	9.8	58.8	24.3	0.413
35–39°	2.1	4590	329	7.2	77.8	34.5	0.443
30–34°	1.6	3490	181	5.2	99.8	32.5	0.326
25–29°	0.8	1750	67	3.8	129.9	33.8	0.260
Total average	20.5	44,770	1292	2.9%	70.3[a]	37.1	0.529

[a] Weighted average.

sampling effort, namely ca. 697 CBC circles, to that area.

Thus the problem of latitudinal nonrandomness of sampling is not as critical as has been previously supposed. Nonetheless, it is important to note other elements of bias, e.g., the concentration of effort expended sampling urban or semi-urban habitats and coastal habitats. Through statistical weighting of disproportionate habitats (biased samples), perhaps errors in estimation can be minimized.

Means ($\bar{X}$) and standard deviations (s) of species density for each belt indicate that: (1) species density increases with decreasing latitude, and (2) variability is decidedly higher below latitude 50°N than above. Because variability is not homogeneous, a stratified sample is more appropriate, assuming that our aim is to measure species density or absolute density, for each of many species. In the last column of Table 2, the gain from stratification is clearly seen in the coefficients of variation ($s/\bar{X}$). The coefficient of variation for the total area is .529, which is larger than the coefficient in five of the seven latitude belts, indicating that variability is reduced by stratification. When the absolute density of every species in an area to be sampled is heterogeneous, some stratification is preferable to none. It is certainly appropriate to a sample design that estimates species density by direct counting. An advantage that stratified sampling has over simple random sampling is that it divides broad heterogeneously dense areas into discrete sampling units, within which species density is roughly homogeneous. Species densities of each unit are inventoried. The mean density of inventoried units is used as an estimate of mean density for inventoried and uninventoried units combined. This consequently increases the precision of the estimate, as it is now a function of density within stratified zones, as opposed to density over a broad area. Precision of density estimates for each species is thus inversely related to the density variability within the entire area under survey.

The CBC sample size and sample design (planned or accidental) may then be useful. What about the sampling results? Tables 3 and 4 compare broad measures of results from the previous two CBCs (1978–79 and 1979–80). They show that the allocation of the sample is closely consistent for these two different periods. In Table 3, the mean number of species observed by latitude belts are almost identical in all but the lowest latitude belt. In Table 4, the percent distributions of counts by reported species-density classes are practically identical.

Thus the evidence supports the hypothesis that CBC data may be useful when treated as a stratified sample. There are three reasons supporting that conclusion. First, the de facto sample design closely conforms to a stratified sample, in which sampling effort is more intensive in high-density areas than in low-density areas. Second, the sample is impressively large (one out of 34 potential count units are included). Third, the allocation of the sample for two different periods is highly consistent.

One might argue that the validity of the sample design is limited because the sample count units, within each latitude belt, are not selected by random process. The response to that objection is pragmatic. A random selection of count units would of course, require assigning birders to those sample areas selected. Theoretically, under such conditions, it is entirely possible that the number of willing participants would plummet dramatically. Perhaps 200± count units would be covered. This would result in a significant rise in the standard error (SE) of estimates. If ca. 200 CBCs were covered, then SE would tend to increase by a factor of ±2.6, as shown.

TABLE 3
CBC RESULTS: MEAN NUMBER OF SPECIES BY LATITUDE BELTS

North latitude	Mean number of species	
	1978–79	1979–80
60°+	20.3	19.6
50–59°	33.7	32.9
45–49°	45.4	46.2
40–44°	57.7	58.8
35–39°	77.6	77.8
30–34°	97.5	99.8
25–29°	120.0	129.9

TABLE 4
CBC RESULTS: COUNT DISTRIBUTION BY NUMBER OF SPECIES

Number of species	Percent distribution CBCs	
	1978–79	1979–80
0–9	0.6	0.4
10–19	2.4	2.3
20–29	7.4	7.6
30–79	57.6	57.5
80–89	7.9	7.1
90–99	5.4	5.5
100–149	14.1	15.3
150+	4.6	4.3
	100.0	100.0

The basic formula for SE is

$$SE = \frac{s}{\sqrt{k}}\sqrt{1 - \frac{k}{\Sigma K}} \tag{3}$$

With the present (1979–80) CBC, the number of sample count units, k, is 1292; ΣK, the total number of potential count units in North America, is 44,770. The standard deviation, s, for some variable measured with the sample, such as a particular species density, need not be numerically specified to illustrate the point. So with the present count,

$$SE = \frac{s}{\sqrt{1292}}\sqrt{1 - \frac{1292}{44{,}770}} \tag{4}$$

$$SE = 0.0274s$$

That is, SE is ±2.7% of s. With a random selection of count units within each latitude belt and the assignment of participants to sample areas resulting in a drop of k to ca. ±200 (which may be optimistically high), the new standard error (SE′) would be

$$SE' = \frac{s}{\sqrt{200}}\sqrt{1 - \frac{200}{44{,}770}} \tag{5}$$

$$SE' = 0.0706s$$

That is, SE′ is ±7% of s. Thus, by substitution, SE′ = 2.575 SE. The SE′ is >2.6 times larger than SE, which is a marked reduction in precision of sample estimates. It is difficult to imagine that any gains from instituting a random selection process would be worth that loss of precision.

Another objection is that the tendency for birders to select count areas of high habitat variation and, consequently, avian density, may introduce an upward bias in sample estimates within any latitude belt. However the size of that bias could be estimated with some controlled field work and, in the interest of standardization, could be used to derive more satisfactory sample estimates.

An additional point is worth making here: although the CBC data appear to be useful as an estimation scheme when treated as a stratified sample, stratification by latitude belts is neither the only nor necessarily the optimal stratification. As already noted, bias which may be introduced by disproportionate habitat coverage (e.g., urban vs rural, coastal vs interior) is not eliminated by latitudinal stratification. To determine the best stratification requires further work. Researchers might try stratification schemes based on distinct physiographic regions, riverine systems or coastal zones. Stratification based on fixed or variable distances from urban centers may help in removing the well known urban sampling bias. Stratification based on botanical or ecological homogeneity may provide stimulating insights into avian densities. These few designs suggest the potential for CBC analysis according to stratification.

CONCLUSIONS

On balance then, the CBC is an enormously rich data source which may be useful for estimating population parameters, provided that researchers are aware of its limitations. Biostatisticians sometimes caution that only random sampling ensures true representation of the population under study. That is, only randomness ensures consistency, efficiency, sufficiency, and unbiasedness. The data of the CBC do not perfectly conform with the underlying assumptions (e.g., random selection of count areas, standard procedures of coverage, equal effort in all count areas, etc.) assumed in applying statistical procedures such as calculating mean number of species or absolute density for a species. But those statistical procedures are known to be so robust that departures from the assumptions within the data can be encompassed, especially when the sample is large. The methods of statistics are exact; the real world data to which they are applied, here CBCs, are inexact. So although we know that within, say the 40°–45°N latitude belt, some counts had two or three parties, others six or more parties, and not all had the same number of parties as last year, much less the same weather conditions, the same routes, the same participants, party composition, or collective skill, nevertheless, the mean number of species for the belt was 58.8 in 1979–80, almost identical to the mean of 57.7 in the prior year (1978–79). In other words, the lack of standardization may be so much statistical "noise" that tends to disappear when results are aggregated for broad areas. Curiously, the limitations of the CBC, this "noise," have received so much casual attention that the baby has too often been discarded with the bath water. There is no other branch of field zoology which has any sample comparable in size, scope, and regularity to the CBC, and yet ornithologists have hardly begun to exploit it.

Studies in Avian Biology No. 6:30–33, 1981.

THE CHRISTMAS BIRD COUNT: CONSTRUCTING AN "IDEAL MODEL"

ROBERT S. ARBIB, JR.[1]

ABSTRACT.—In increasing numbers, research studies in early-winter bird abundance and distribution in North America derived from Christmas Bird Count data are being published. It has long been argued that weaknesses and inconsistencies in the method of data collection and reporting cast doubt on the reliability of these data. This paper addresses itself to a 12-point appraisal of existing practices and procedures for assembling and reporting the annual data, and compares their potential for research with a suggested "Ideal Model." The assessment serves both to highlight present weaknesses and to suggest their remedies. The practical problems of implementing steps towards the "Ideal Model" are discussed.

It is now apparent that increasing use in scientific studies is being made of the long-neglected mountains of distributional data provided by the annual Audubon Christmas Bird Count (CBC). This field-work phenomenon is now in its 81st year, and in 1978–79 involved 33,020 named participants in 1320 published counts, of which 1289 were located in continental North America north of Mexico. These studies have proved the general validity of the CBC data in spite of, or in ignorance, of certain weaknesses and flaws in CBC practices both in the field and in the reporting procedure. This paper will explore ways in which both can be improved to provide the researcher not only with more accurate and more reliable raw data, but even afford new areas of analysis.

The refinements suggested will be incorporated in a limited "Ideal Model" for CBC procedures, limited because it recognizes that this proposal must accommodate the real world—with strong traditions of competition, recreation, amateur involvement, and social interaction.

THE IDEAL MODEL

The author is in a unique position to propose changes in CBC practices and procedures. For 10 years he has written the CBC regulations and instructions, and has been editor-in-chief and final court of appeals for 11,200 counts in that time. The proposed "Ideal Model" is therefore not merely theoretical; it can be effectuated, as further refined, perhaps as early as 1981–82.

In this paper I will consider those current problems that may affect the presentation and understanding of raw data. I am not concerned here with censusing *techniques* or their relative effectiveness, or the effectiveness of the human being as a receptor of bird registrations. The problems I consider are intrinsic to the CBC process. Some are presently of minor statistical significance, others of *potential* import, but some are of major importance. Twelve specific areas will be considered.

1. *Count circle adherence.*—I believe from personal knowledge that boundary stretching is widespread; it biases all totals, but until now is ignored. Ideal Model compilers will be required to verify that the count boundary was not violated nor the count circle opportunistically shifted.

2. *Overlapping count circles.*—A 1979–80 study shows that of 576 eastern United States counts 57, or 10%, have overlaps from 5% to 80% of their areas. Researchers have ignored or are unaware of this source of error. Compilers would be required to verify that no overlap occurred, or to segregate overlap and non-overlap totals.

3. *Habitat analysis.*—The data reporting the percentages of various habitat types are not widely used, but if refined have great potential for habitat/population studies. The present error involves the reporting of the actual habitat percentages in an area rather than the actual percentages covered in that area (Table 1). If practical a national or universal habitat classification would be provided each count, to foster uniformity of definitions.

4. *Elevation.*—Only high and low map elevations for the count circle are now required. Compilers will be asked to give altitudinal high, low, and means of the CBC area actually *covered.* Of potential research value only and not presently of frequent use.

5. *Weather.*—The effects of weather on bird presence, detectability, and on observer effort are real and important variables, but difficult to quantify. Long-term, they may average out. However, better information can be furnished than is now published. It would be informative if counts rated Count Day weather on a subjective scale of 1 (worst) to 10 (best) as to its effect on bird finding and observer effort. The "Ideal Model" count would also report, using the same subjective scale, on weather factors for 1) the

[1] National Audubon Society, 950 Third Ave., New York, New York 10022.

TABLE 1
FICTITIOUS TYPICAL EXAMPLE OF POTENTIAL ERROR IN HABITAT ANALYSIS REPORTING

Habitat type	Reported	Covered	Actual
Woodland	40%	50%	27.8%
Fields	30	80	33.3
Roadsides	10	100	13.9
Residential	20	90	25.0
Totals	100%		100%

TABLE 3
SPECIES TOTALS RELATED TO OBSERVER NUMBERS, CALIFORNIA, 1979–80.

Observers	Counts	Species/Count
1–19	27	76.0
20–49	30	126.7
50–99	21	158.8
100+	5	190.8

week preceding Count Day, and 2) the 3-week period prior to that week.

6. *Party-hours and party-miles.*—Since most studies today are based on the factors birds per 1, 10, or 100 party-hours or party-miles, improved calculating and reporting of these data would be required. The "Ideal Model" accounting would not only assess miles and hours logged by basic parties, but calculate miles and hours added by split parties. Basic parties which divide part-time shall be considered multiple parties for that time, if they are counting different birds. Correcting this badly flawed factor may be the most significant contribution of the Ideal Model. I suggest that many of the papers based on the assumption of accuracy of party-hour and party-mile statistics heretofore published may be in substantial error, on the low side for party-hour totals, high for birds/party-hour.

7. *Observer numbers and effort.*—Total participants listed and total parties afield relate strongly to the adequacy of count-area coverage and the discovery of birds. A wide spectrum exists in CBC participation (Table 2). In 1979–80, participants per count varied from one (10 counts) to 213 (one count).

An analysis of 83 counts taken in California (1979–80) shows an increase in species totals with increases in observers and parties (Tables 3 and 4). However, there is no optimum to the number of parties afield if maximum species (and individual totals) are sought.

While optimum numbers of participants and parties for meaningful data will vary with the type, access and topography of the terrain, except for special situations (ferry transects, pelagic counts, desert areas with oases, etc.), the great majority of present CBCs do not adequately cover their 176.6 mi.^{2} (457.4 km^2) circles.

To qualify, Ideal Model Counts would be required to meet individually specified minima for observer and party coverage. In 1979–80, less than 9.2% of all counts fielded 30 or more observers and 12 or more parties.

A further, unrecognized bias is introduced by compilers who incorrectly report participant totals, adjusting party-miles and party-hours accordingly. Circumstantial evidence strongly suggests that this practice is widespread, and for some counts participants and parties may be understated by 50% or more. This bias can be eliminated if, for Ideal Model Counts, the participant fee is waived.

8. *Observer credibility.*—To improve the credibility of observers' reports, heretofore the single most questioned of CBC data reliability factors, a method for assessing observer reliability must be developed. At present we rely on three review stages and documentary evidence for questionable reports. The Ideal Model would propose to strengthen the process by ad-

TABLE 2
OBSERVERS PER COUNT, 1979–80 CBCs

Observers	Counts	Per Cent
1	10	0.8%
2–4	83	6.3
5–9	222	16.8
10–24	552	41.8
25–49	332	25.2
50–99	102	7.7
100+	19	1.4

TABLE 4
SPECIES TOTALS RELATED TO PARTY NUMBERS, CALIFORNIA, 1979–80

Parties	Counts	Species/Count	Species/Party
1–5	19	69.4	19.7
6–11	24	100.4	13.1
12–19	17	134.4	9.5
20–29	14	144.0	5.8
30–39	6	172.0	4.9
59–74	2	180.5	2.7

vancing an observer "Reliability Index" as a rough gauge for measuring observer experience, to help balance party composition.

The following formula, while imperfect, has the advantage of being easy to calculate, applicable to all, and of giving scores which seem to test out surprisingly well. The formula is calculated as follows:

Each observer estimates the number of hours afield birding per year during the last 5 years (a measure of experience) and multiplies this total by the percentage of his state's (or province's) currently accepted living bird list (a measure of expertise) and (for convenience only) divides by 100. Results in the normal range will be between 0.5 and 40. Three examples:

Calculating the Reliability Index

A. *Infrequent birder*
12 days per year. 6 hours afield each.
45% of state list.
12 × 6 × 5 × .45 × .01 = R.I. of 1.62

B. *Average birder*
26 days per year. 7 hours afield each.
70% of state list.
26 × 7 × 5 × .70 × .01 = 6.37 R.I.

C. *Keen, dedicated birder*
50 days per year. 9 hours afield each.
86% of state list.
50 × 9 × 5 × .86 × .01 = 19.35.

Obviously, any arbitrary index of credibility must be evaluated by the compiler, who may waive the rule on low R.I.'s owing to unusual factors. But the Ideal Model proposes *no parties afield* without at least one participant with a 5.0 or better rating. A possibility now under study is the preparation of identification test kits consisting of slides and tapes which CBC groups may use either as educational or evaluational tools.

9. *Numerical estimation.*—The Ideal Model will propose more stringent accuracy in counting and estimating based on expanded training and testing of participants. Parties will be instructed to keep running counts of scatter-type species, instead of end-of-day estimates. For flock counting, training workshops and tests would be programmed. With adequate training and practice, errors in large-numbers estimation can be reduced to 5% or less. Further, observers can discover their own habitual counting bias, and compensate for them. Compilers will be urged to carefully evaluate party routes and bird lists to avoid possible duplication.

10. *Feeding station counts.*—Species totals at feeding stations would be segregated from those afield, removing an ignored biasing factor. Where several feeders are so closely spaced that their patrons circulate between them, averages—not totals—will be presented.

11. *Linear or other transects.*—Carefully *censused* tracts based either on straight transects, point counts, or following topographic features such as streambeds, trails or roads, might be incorporated into the CBC format in certain Ideal Model Counts. These might serve as more accurate sampling yardsticks for comparing year-to-year numbers. But it would be wholly unrealistic to expect any present CBC group to base its statistics solely on transects. The Ideal Model may suggest but will not require counts to experiment with various forms of census sampling techniques.

12. *Summary statistics.*—Ideal Model Counts will expand the information presently given in the summary, to give added dimension to the data presented. The present accounting is: "Total: 135 species, 101,450 individuals." The Ideal Model accounting would be: "Totals: 135 species; 10-yr ave., 128 species, cum. total (26 years) 201 species; 1.61/party-hr, 10-yr ave., 1.52 party-hr; 101,450 individuals, 10-yr ave., 123,456; 1207/party-hr, 10-yr ave., 1469/party-hr."

We propose to test an Ideal Model, refined from this outline, perhaps as early as the 1981–82 season. It would be limited at first to counts fielding statistically significant numbers of observers and basic parties, meeting all other stated conditions, with observers of high R.I.'s and dedicated compilers. We would encourage a sampling from various latitude belts both coastal and inland. We would be pleased to have 25 CBCs run on Ideal Model lines the first year. A long-term goal might be 100 per year. But every count of any size could comply with most of the Ideal Model reporting procedures.

Two problems suggest themselves. An Ideal Model count obviously demands greater effort and care by every participant, especially by compilers. And the editing and publishing of Ideal Model Counts would require more time, effort, and funds.

Two possible solutions to these problems offer themselves. Various direct incentives might be offered Ideal Model Counts: forgiveness of all participant fees, the honorary designation of "Elite Counts," special publicity and awards, such as scrolls or insignias, free reprints, and perhaps even financial assistance. As for the publishing problem, experience will determine whether special funding is required. The incentives of pride in leadership, of pioneering into new frontiers, of acquiring reputations of superiority are powerful motivating forces.

Going one step beyond the limited Ideal Model proposed herewith, study will be given to the possibility of designing new CBC report forms which can be more easily edited or converted to direct entry into a computer system. This prospect, however, suggested many times in recent years, may run counter to our caveat concerning the CBC and real world possibilities.

ACKNOWLEDGMENTS

Among those to whom the author is indebted for helpful suggestions are Joseph J. Hickey, Chandler S. Robbins, Susan R. Drennan and Lois H. Heilbrun.

Studies in Avian Biology No. 6:34–41, 1981.

THE NORTH AMERICAN BREEDING BIRD SURVEY

DANNY BYSTRAK[1]

ABSTRACT.—A brief history of the North American Breeding Bird Survey (BBS) and a discussion of the technique are presented. The approximately 2000 random roadside routes conducted yearly during the breeding season throughout North America produce an enormous bank of data on distribution and abundance of breeding birds with great potential use. Data on about one million total birds of 500 species per year are on computer tape to facilitate accessibility and are available to any serious investigator.

The BBS includes the advantages of wide geographic coverage, sampling of most habitat types, standardization of data collection, and a relatively simple format. The Survey is limited by placement of roads (e.g., marshes and rugged mountainous areas are not well sampled), traffic noise interference in some cases and preference of some bird species for roadside habitats. These and other problems and biases of the BBS are discussed.

The uniformity of the technique allows for detecting changes in populations and for creation of maps of relative abundance. Examples of each are presented.

In response to the need for a reliable index of North America bird populations, the U.S. Fish and Wildlife Service and the Canadian Wildlife Service initiated the North American Breeding Bird Survey. In 1965 the roadside technique was tested along 60 routes in Maryland and Delaware to determine its feasibility. Based on this pilot effort, the decision was made to sample the United States and Canada east of the Mississippi River in 1966, and about 600 routes were conducted that year. Coverage was expanded to include the Great Plains states and provinces in 1967 and the entire continent in 1968. The number of routes has slowly increased to approximately 2400 by 1980, with between 1800 and 1900 covered each year, entirely by volunteer observers.

Every effort was made to minimize biases on the BBS so the data could be used without modification and would be as widely useful as possible. In the course of the 15 years of the BBS, problems and questions have emerged, some of them anticipated, some not. The purpose of this paper is to discuss the uses, problems and criticisms of the BBS and the technique.

METHODS

A detailed description of the methods appears in Robbins and Van Velzen (1967), so this section will be brief. The basic unit of the BBS is the route. Each route is conducted on secondary roads and consists of 50 3-minute counting locations 0.8 km (½ mile) apart. In order to apply standard statistical methods to the results, it was necessary to insure random selection of routes; thus the starting point and direction of each route were selected from a table of random numbers.

A sampling scheme based on 1-degree blocks of latitude and longitude (latilongs) was devised for the distribution of routes. Throughout North America the number of routes per latilong varies according to availability of qualified observers, but is uniform across a state or province to prevent clustering of routes. In most of the West, the level of coverage is one route per latilong. In the East there are four to eight per latilong and two in most central states and provinces. The number of routes in a state or province is increased when coverage is complete and increased cooperation assured.

Each route is assigned a stratum number based on a stratification system I devised (Fig. 1), largely from the U.S. Forest Service's "Natural Land Use Area of the United States" (Barnes and Marschner 1933), also from Fenneman (1931, 1938) and Kuchler (1964). The Canadian portion was based on Aldrich's (1963) life zones, Munro and Cowan (1947) and the Atlas of Saskatchewan (Richards and Fung 1969). Fine adjustment of boundaries was done by examination of Canadian and U.S. topographic maps, using relief, elevation, tree cover and land use as a guide. In all analyses, these strata are used as the basic unit, on the assumption that the populations within each stratum are similar and that they differ from adjacent strata (see Peterson 1975). Table 1 briefly describes the 62 strata currently used.

In each state and province there is a volunteer coordinator who is in contact with a large portion of the local amateur ornithologists. The coordinators receive copies of each year's results for their respective areas and often prepare summaries for publication. Many of these highly dedicated individuals also run several routes.

Observers are supplied with rules and all necessary forms and maps, and are instructed to choose a day in June with good weather conditions on a date that is as close as possible to previous runs. Each observer starts at exactly ½ hour before local sunrise, counting and recording all birds detected in three minutes at the starting point. The counting is repeated at the remaining 49 stops. Only birds counted during the 50 3-minute stops are included in the totals. A route should take from 4 to 4½ hours to complete. It is important to finish in this time-frame because on most mornings total bird song decreases rapidly after the first four hours, and for many species, the first three hours.

Each observer summarizes the results and returns all forms to the Migratory Bird and Habitat Research

[1] Migratory Bird and Habitat Research Laboratory, U.S. Fish and Wildlife Service, Laurel, Maryland 20811.

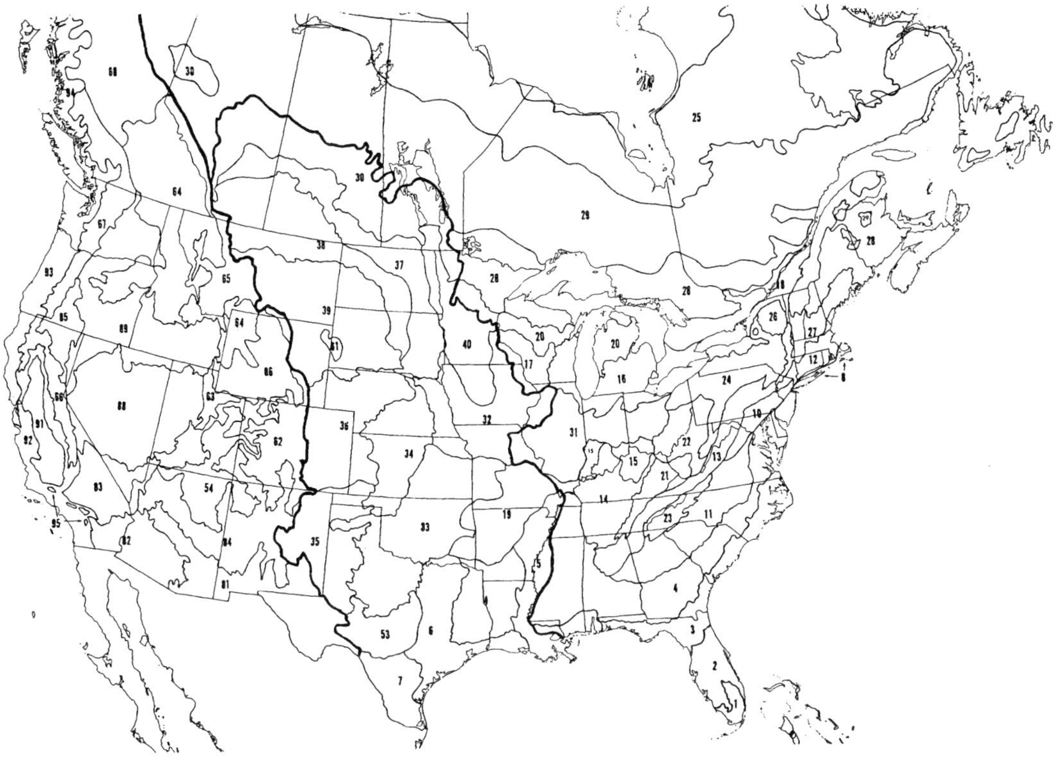

FIGURE 1. Physiographic stratification of North America as used in BBS analyses. Heavy lines delimit Eastern, Central and Western analysis regions. See Table 1 for legend.

Laboratory where biologists and clerks carefully edit them, comparing field to summary sheets and questioning observers on any discrepancies or unverified reports of rare species. All data are transferred to magnetic tape and subjected to various computer edits. Printouts of the results are sent to the observers to be compared to copies of the forms retained by them. After final corrections are made, the data for that year are ready for the preparation of various listings and analyses.

RESULTS

Monitoring Populations

Currently, the main analytical program for BBS data is an analysis for year to year change in population of 140 species and species groups. In most years only a few species exhibit significant annual population changes, and many of these are not meaningful. The Yellow-billed Cuckoo (*Coccyzus americanus*), for example, can fluctuate wildly in number between years from aberrant migration alone. Occasionally, a weather event drastic enough to cause a serious decline in the population of certain species occurs. After the extended cold of the winter of 1976–77, it was no surprise that the Carolina Wren (*Thryothorus ludovicianus*) population showed a significant decline in the summer of 1977. These wrens require open ground on which to feed by scratching for insects. Thus, the snow and ice prevented access to their food source. Conversely, warm winters mean an increased population. Figure 2 demonstrates graphically the increase in population during five warm winters and the drastic drop after one severe winter. Based on the 1978 through 1980 survey results, the population seems to be very slow to recover. The winter of 1977–78 was also harsh in the East, which kept the population low another year. Only in 1980 (based on preliminary data) does there seem to be a slight increase.

Another species that responded to these harsh winters is the Eastern Bluebird (*Sialia sialis*) (Bystrak 1979). The decrease was not, however, as drastic as that of the Carolina Wren. Despite chronic decreases in the population (Zeleny 1976), Eastern Bluebirds have recovered well from these latest winter disasters. In 27 states composing the bulk of the bluebird's range, the mean birds per route has increased from a low

TABLE 1
EXPLANATION OF PHYSIOGRAPHIC STRATIFICATION USED IN BREEDING BIRD SURVEY ANALYSES. (HIERARCHICAL DESIGNATIONS ARE TENTATIVE.)

I. Northern Boreal Forest
- 25 Open Boreal Forest
- 28 Northern Spruce-Hardwoods
- 29 Closed Boreal Forest

II. Eastern Deciduous Forest
- A. Appalachians
  - 8 Glaciated Coastal Plain
  - 10 Northern Piedmont
  - 12 Southern New England
  - 13 Ridge and Valley
  - 21 Cumberland Plateau
  - 22 Ohio Hills
  - 23 Blue Ridge Mountains
  - 24 Allegheny Plateau
  - 26 Adirondack Mountains
  - 27 Northern New England
- B. Interior Plains
  - 14 Highland Rim
  - 15 Lexington Plain
  - 16 Great Lakes Plain
  - 17 Driftless Area
  - 18 St. Lawrence River Plain
  - 19 Ozark-Ouachita Plateau
  - 20 Great Lakes Transition
  - 31 Till Plains

III. Southeastern Forest
- A. Coastal Plain
  - 1 Subtropical
  - 2 Floridian
  - 3 Coastal Flatwoods
  - 4 Upper Coastal Plain
  - 5 Mississippi Alluvial Plain
  - 6 East Texas Prairies
  - 7 South Texas Brushlands
- B. Foothills
  - 11 Southern Piedmont

IV. Great Plains
- A. Northern Plains
  - 30 Aspen Parklands
  - 32 Dissected Till Plains
  - 37 Drift Prairie
  - 38 Missouri Coteau
  - 39 Great Plains Roughlands
  - 40 Black Prairie
- B. Southern Plains
  - 33 Osage Plain—Cross Timbers
  - 34 High Plains Border
  - 35 Staked and Pecos Plains
  - 36 High Plains
  - 53 Edward's Plateau

V. Rocky Mountains
- A. Basins and Deserts
  - 54 Colorado, Uinta Basins
  - 84 Pinyon-Juniper Woodlands
  - 85 Pitt-Klamath Plateau
  - 86 Wyoming Basin
  - 88 Great Basin
  - 89 Columbia Plateau
- B. Forested Mountains
  - 61 Black Hills
  - 62 Colorado Rockies
  - 63 High Plateaus of Utah
  - 64 Northern Rockies
  - 65 Dissected Rockies
  - 68 Canadian Rockies

VI. Pacific Mountains
- A. Cascade-Sierra Axis
  - 66 Sierra Nevada
  - 67 Cascade Mountains
- B. Pacific Ranges
  - 91 Central Valley
  - 92 California Foothills
  - 93 S. Pacific Rainforests
  - 94 N. Pacific Rainforests
  - 95 Los Angeles Ranges

VII. Southwestern Arid
- 81 Mexican Highlands
- 82 Sonoran Desert
- 83 Mojave Desert

of 1.88 in 1978 to 2.89 in 1980. This almost compares to the pre-disaster mean of 3.48 in 1976.

Long-term population increases and decreases can also be plotted. Figure 3 gives some examples of declines and increases of selected species. These four species all exhibited statistically significant long-term trends at the continental level over the 12-year period shown. The data are also analyzed by three major regions (Fig. 1) and by strata and state and province. The species in Figure 3 each exhibited significant trends in the three regions as well as on the continental level.

MAPPING DISTRIBUTION AND ABUNDANCE

For some species, maps can be prepared showing changes in distribution. By examination of yearly range maps prepared from BBS data the fluctuations in the Dickcissel (*Spiza americana*) breeding distribution can be followed (Robbins and Van Velzen 1969 and 1974). This is perhaps the only North American species that regularly shows significant annual fluctuations in breeding range. The southward spread of the breeding range of the Barn Swallow (*Hirundo rustica*) has been well documented during the 14 years of the BBS (Bystrak 1979). Figure 4 shows a similar expansion in breeding range of the House Finch (*Carpodacus mexicanus*) in the eastern United States, where it was introduced in the 1940's.

Because the data are all gathered uniformly, range maps showing relative abundance can be prepared. Yearly maps (Robbins and Van Velzen 1967 and 1969) are difficult to interpret and rather incomplete because peripheral portions of a species' range show largely as zeros, and adjacent routes in the center of the range may have very different counts in any one year just from chance. When long-term route means are used, high and low counts are averaged out, yielding smoother isolines and filling in peripheral portions. Thus, maps based on several years are more easily interpreted. Figure 5 shows the breeding range of Scissor-tailed Flycatcher (*Muscivora forficata*) prepared from 13-year route means. Such maps are potentially useful to biogeographers, taxonomists, population biologists and birders (see also Bystrak 1979).

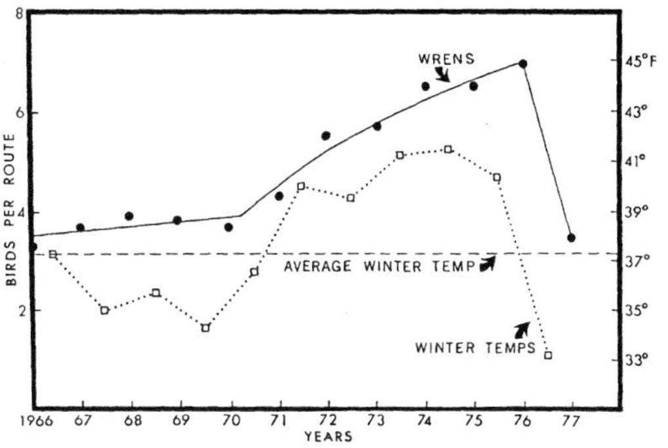

FIGURE 2. Population index of Carolina Wren for states and provinces east of the Mississippi River from BBS data, plotted against winter temperatures for Washington, D.C. (from Bystrak 1979).

DISCUSSION

USES OF BBS DATA

The BBS has demonstrated its usefulness as an effective index of bird population levels, both temporally and spatially. In addition to these intended uses, the BBS has been used for many other purposes. Data requests are received weekly from other Department of Interior agencies, the U.S. Forest Service, State Fish and Game departments, Environmental Impact consulting firms, university personnel and other researchers.

The BBS technique has been used in other more intensive studies such as Thompson's (1980) Circle West where it was shown to be an effective method of baseline data gathering, and in the form of mini-routes to aid with data gathering on Atlas projects (Klimkiewicz and Solem 1978). Rotenberry and Wiens (1976) used an adaptation of the technique to estimate species dispersion within various grassland habitats. Rotenberry and Wiens (1978) also used the data from 60 BBS routes to define avifaunal regions of the Pacific Northwest. These regions agreed rather closely with the stratification system used for the BBS. This implies that birds are a useful measure of ecoregions, considering that the BBS stratification system is based on physical characteristics and land use.

Nearly every state ornithological journal has included articles using its state's BBS data (e.g., Van Velzen 1966, Whitney 1967, Zimmerman 1968, Davis 1969, Monroe 1970, Robbins 1971, Evans and Dawson 1976, Cortelyou 1978). BBS data have also been incorporated into state bird books (Imhof 1976, Hall in prep., Robbins, in prep.). Numerous other papers, either summarizing or using BBS data, have been published.

The BBS has even made interesting contributions to state ornithological records, by encouraging bird observations in remote areas during June, when most such activity normally slackens. Several pioneer records have been discovered on BBS routes. Examples include the first Great-tailed Grackle (*Quiscalus mexicanus*) in Arkansas, the first Cassin's Sparrow (*Aimophila carpalis*) in South Dakota, the first Gray Kingbird (*Tyrannus dominicensis*) and White-winged Dove (*Zenaida asiatica*) in Maryland, and the first Willow Flycatchers (*Empidonax traillii*) in Nova Scotia.

Results of the BBS routes and the technique have been used often in Master's and Ph.D. thesis work. Examples include Wallace's (1970) evaluation of the technique with particular references to seasonal changes, weather, time of day, and conspicuousness and Baker's (1977) analysis of California routes relative to environmental parameters.

The BBS is the only program of its kind in North America. The only similar program in the world known to the author is that of Winterbottom (1972) in South Africa. Because there is no effort as comprehensive and uniform, the BBS has often been used as a standard against which to compare other studies. Hussell (1981), for example, compared migration data with BBS data for determining population trends.

PROBLEMS

Differences in detectability.—Although the BBS appears to be an effective, versatile technique, it does suffer from certain problems. Per-

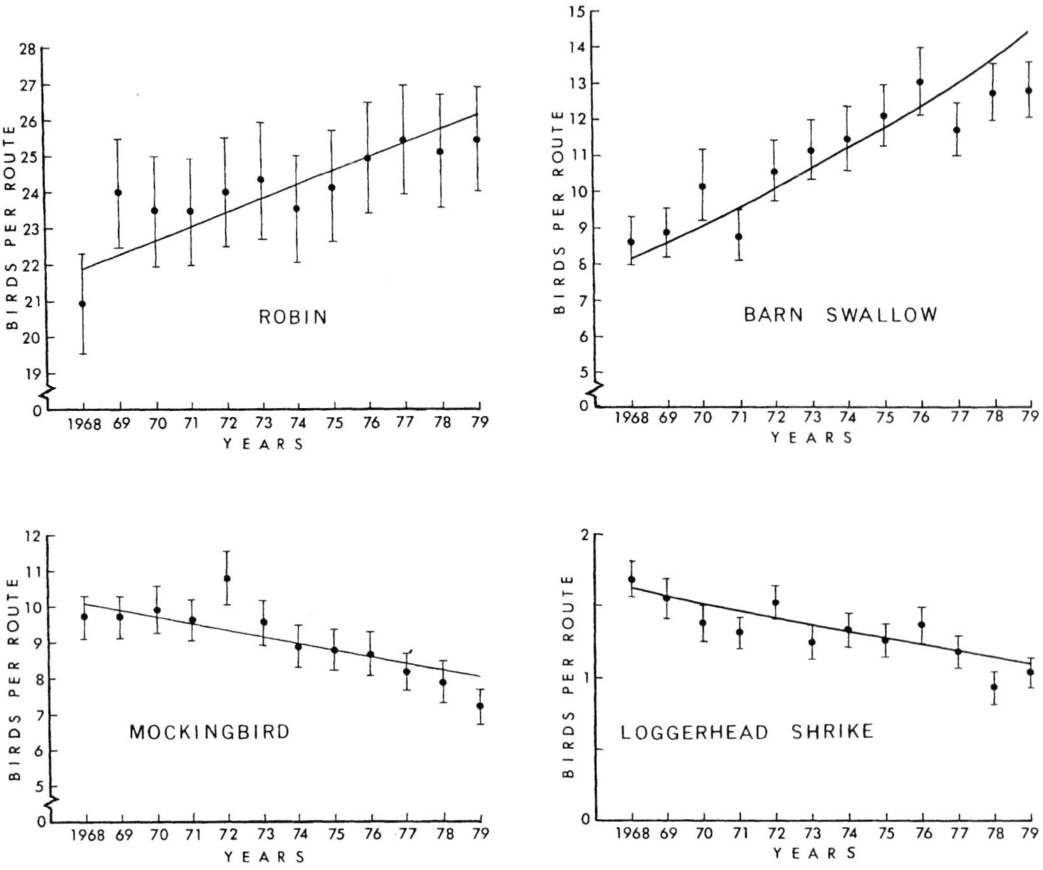

FIGURE 3. Continental population trends for Barn Swallow, Mockingbird (*Mimus polyglottos*), American Robin (*Turdus migratorius*) and Loggerhead Shrike (*Lanius ludovicianus*). Annual means are depicted with 95% confidence limits.

haps the most potentially serious relate to the credibility of the technique. One major criticism offered is that only species characteristic of roadsides are accurately counted. Related to this is the aural detectability difference among species. Wallace (1970) calculated conspicuousness indices for several species, comparing among species and across time. Because each species sings at a different volume and rate, some are easier to detect than others. Because the main purpose of the BBS is to detect population trends within species from year to year, these are not serious problems. The BBS is an index derived from randomly selected routes, all conducted uniformly; therefore, conclusions can be made when comparing one species to itself from year to year or geographically. No pretense is made that one species can be compared to another, although even this may be possible if a conspicuousness index is employed.

Habitat.—Because BBS routes are along roadsides, an unavoidable built-in sampling bias is that certain habitats such as marshes and steep mountainsides tend to be avoided in road construction. As a result, the BBS seriously undersamples these habitats. Related to this is the criticism that habitat parameters along BBS routes are not measured. Weber and Theberge (1977) suggested that the value of the BBS would be enhanced by collection of land use and habitat data along routes. They believed that land use will be the most important single factor responsible for long-term changes in bird numbers, and without recording these data, changes attributable to subtle causes like pesticides may be masked by land use changes. This is a legitimate suggestion, but one not easily incorporated into the Survey. A simple classification of North American habitats is not available, and we are reluctant to ask too much of an all-volunteer crew. An experiment with a raptor survey along BBS routes in 1980 (Patterson, in prep.) indicated the reluctance of most observers to make a return trip over their routes, usually citing time, gasoline prices and exhaustion as the main reasons. Because habitat evaluation would extend the coverage time excessively, it would have to be done on an additional trip, and thus present the same problems. In Maryland, Robbins (unpubl. data) has experimented with a simple habitat form, which may used on a voluntary trial basis on the 1981 BBS. Satellite imagery and/or aerial photography, as proposed by Skaley (pers. comm.), may be useful tools, if these techniques will provide the necessary resolution and not be too expensive. Skaley suggested testing such a classification scheme for avian habitats with BBS data.

Unmated birds.—Berthold (1976) catalogues several papers that approach the problem of unpaired singing males in bird populations. Opinions vary on the extent to which unpaired males sing and the effect they would have on population sampling if they did represent a large percentage of the population. If the percentage is consistent throughout a species' range, however, it should have no adverse effect on the BBS.

Overloading.—Another problem is that of high densities masking change. A BBS stop at which eight Red-eyed Vireos (*Vireo olivaceus*) are recorded may actually have 25 singing males present, but the human ear cannot resolve that many. If the population decreased to 12 the next year, eight might still be the number recorded. Fortunately, for most species, few stops have more than three singing males of any species, so those at which this problem occurs are probably an insignificant proportion of the total. The "dawn chorus" presents a more serious problem on most routes. The flurry of activity by most species makes sorting and counting of birds difficult. These facts suggest that the first several stops produce the least reliable and repeatable data on BBS routes, especially considering that this is when the observer is least prepared.

Operational problems.—Operationally, the BBS also has its share of problems, traffic being perhaps the most serious. In addition to distractions and potential dangers, the traffic noise on some routes can seriously hamper hearing. Other noise factors include streams, dogs, livestock, and curious residents. Unfamiliarity with routes can also present problems. Much time may be lost making wrong turns or missing stops, especially on first runs. This situation is often amplified by poor maps, which frequently are the only maps available. Time is also lost to unexpected detours caused by situations such as bridge washouts, stripmines and locked gates. Many observers complain that 50 stops are too many and that fatigue affects their results on the last 10 or 20.

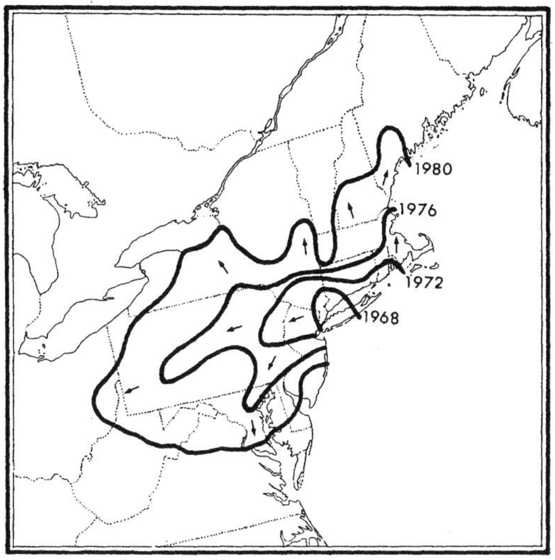

FIGURE 4. Range expansion of the House Finch in the eastern United States as recorded by the BBS. Isolines encompass most extreme records during each 4-year period indicated.

Comparable results.—Maintaining comparable results on routes from year to year also can be a problem. Because of traffic noise, missing bridges and road closings, it is occasionally necessary to make changes on some routes. Minor changes affecting a few stops are not considered serious, but because comparability is important, a route that would require major changes affecting many stops is, instead, replaced by a new route on adjacent roads in similar habitat. Because many routes were drawn hurriedly from poor maps, these substitutions have been all too common. Replacement routes are, of course, given new names and numbers so results are not compared directly. Exact stop locations are not recorded on many routes, so when observers change, stop locations often do too. This also happens when an observer changes cars or even tires, because odometers are notoriously inaccurate. Fortunately, changes in stop locations seem to have little effect on total counts, although no experimentation has been done with this. Nontheless, observers are encouraged to mark exact stop locations and to always use the marked stops. Changing observers presents another problem in maintaining route comparability because we have no way to know the extent to which observers are competent or of sim-

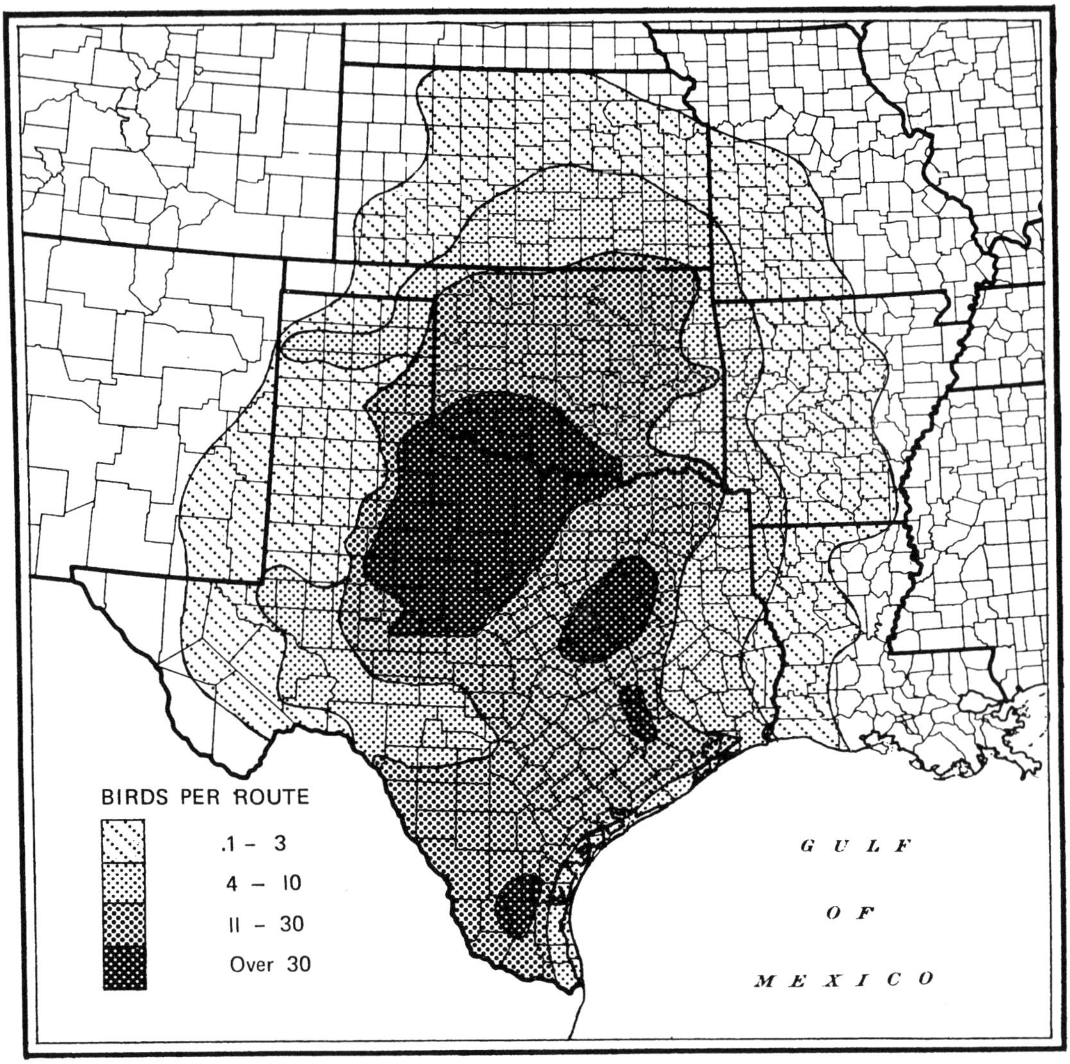

FIGURE 5. Breeding range of Scissor-tailed Flycatcher from BBS data showing relative density.

ilar ability. Two highly qualified observers produce only slight differences in results, but underqualified observers submit results that are less comparable even to themselves (Faanes and Bystrak 1981). Because of the volunteer nature of the BBS, rejection of underqualified observers has been a difficult approach. Careful selection of observers and even more careful use of data are the best solutions to this problem.

Weather.—Variable weather also contributes to a lack of comparability, so observers are urged to choose weather conditions as similar as possible to previous years. Often there is only one opportunity to run a particular route, so it may be run under less than ideal conditions. Again, data obviously affected by weather must be rejected from most analyses. They are useful, however, in analyzing effects of weather on counts (Robbins 1981).

Logistics.—The volume of the data presents some logistical problems. The number of records on magnetic tape is staggering, with approximately 100,000 new records added each year. Such a huge data bank limits the extent to which the data can be manipulated. Massive sorts are necessary for many otherwise simple programs, and storage of data in a more accessible format is prohibitively expensive. Locating and correcting errors is also difficult because the data bank is so large and analysis procedures so complex.

Analysis problems.—Analysis for year-to-year changes is straightforward, but the results rarely show significant changes. Response to extreme conditions is usually the cause of any meaningful yearly aberrations. However, analysis for long-term trends is far more complicated. Bird population trends rarely behave in any predictable fashion. Some populations may be cyclic over time periods longer than that of the history of the BBS and an erroneous conclusion of a significant decrease or increase could be made during a downward or upward portion of the cycle. Geissler and Noon (1981) have experimented extensively with parametric and non-parametric approaches to long-term analysis of BBS data. In addition to the unpredictability inherent in bird populations, there is the problem of uneven coverage. Because routes are covered voluntarily, efforts to reduce clumping are effective on the state level, but many areas of the continent remain severely wanting in coverage. Likewise, individual routes are rarely covered for extended time periods without a break of one or more years. Weighting of data is necessary to compensate for both of these deficiencies. Geissler and Noon (1981) have also experimented with proper weighting techniques. It cannot be overstressed how desperately help is needed in many remote parts of the continent. As the BBS continues to prove itself, I feel it will become more obvious that a system of compensating observers for conducting routes in remote areas is necessary to make the BBS more meaningful and reliable, especially in the increasingly crucial intermontane west.

Stratum crossers.—Another minor problem is that of routes crossing stratum boundaries. Most of the routes were first plotted before the present stratification system was developed. As a result, approximately four percent of the routes cross the present stratum boundaries. In each instance, a route is classified according to its predominant stratum. Some of these crossings are serious because the adjacent stratum is very different in bird composition. Because most analyses involving BBS data use these strata as a basic unit, these routes can contribute misleading data.

The Future

The future of the BBS is, unfortunately, closely tied to the continued availability of affordable transportation. A few observers have already dropped out, using gasoline prices as the reason. This is certain to have an adverse affect on the more remote routes, but, until outdoor recreation in general is seriously hampered by gasoline prices, a serious decrease in BBS coverage is not likely. Short-term crises like the "gas shortage of 1979," which occurred during the height of the breeding season, will probably continue to plague us. However, even in 1979, few routes were sacrificed because of gasoline unavailability.

If we can assume that gasoline will be unavailable or prohibitively expensive some day, experimentation with an alternative such as routes covered on foot should begin soon. One problem such routes might present is a lack of random route distribution, with clustering near population centers. I personally feel that as long as there is a basic need for Americans to be mobile, the BBS will go on, although it may suffer a depression as gasoline prices continue to increase. "Mobile American Thinking" will surely produce an alternative to petroleum-dependent transportation, and the BBS should continue to prosper.

Studies in Avian Biology No. 6:42–51, 1981.

ESTIMATES OF AVIAN POPULATION TRENDS FROM THE NORTH AMERICAN BREEDING BIRD SURVEY

PAUL H. GEISSLER[1] AND BARRY R. NOON[1]

ABSTRACT.—One of the major purposes of bird population studies is to document changes in population size over a period of years. The traditional method used in Europe and North America to detect population change is to calculate annual ratios. However, this method can produce spurious results when ratios are accumulated over many years. Consequently, new methods of computing trends are needed. Several new methods of estimating population trends are developed and illustrated with data from the North American Breeding Bird Survey (BBS). Each method is compared in terms of its assumptions, biases, and limitations. On the basis of these comparisons we recommend one method that we feel most accurately detects true population trends. Both the biological and statistical justifications for the model selection are presented. Trends estimated with this model are then presented for two species.

The estimation of changes in the sizes of migratory bird populations provides important information for the management of these species. For many species, estimates of the absolute population size are not available, and the estimation of changes must be based on an index to the population size. In the North American Breeding Bird Survey (BBS) (Robbins and Van Velzen 1969), the numbers of birds heard or seen on randomly selected routes are counted under standardized conditions each year. These counts are used as an index to the population size in the vicinity of that route. The routes were selected as a stratified random sample and have been used each year since 1965 without drawing a new sample of routes. This survey is stratified by latitude degree blocks but does not have defined primary sampling units. Instead, the coordinates of the start and the direction of each route were selected at random. There are 50 stops on a route spaced at 0.8 km intervals. Routes are post-stratified into physiographic strata and into State and Province strata. Physiographic strata estimates are combined to obtain estimates for the three regions and for the continent. Bystrak (1981) gives a brief description of the physiographic strata and continental regions.

Much has been written on the use of population indices (e.g., Overton and Davis 1969, and Seber 1973). Many papers have focused on the relationship between the true population size (density) and the index (discussed in Caughley 1977). Although this is an important subject, it must be emphasized that this is not the subject of the present study. Here the index will be assumed to be proportional to the population size along a route although it is subject to some random measurement error.

It is generally assumed that the proportion of birds detected (the index) is independent of population size. However, the efficiency of the index likely changes with an increase in population size leading to a biased estimate of population change over time (Bart MS). We have not corrected for this bias in our analyses. As a result, our estimates of the number of significant population changes (increases or declines) are probably conservative.

In the development that follows, all of the models are restricted to considering species one at a time. The species index is implicit in the formulation.

AN OVERVIEW OF PREVIOUS METHODOLOGIES

A common method of estimating population change employs a proportional base year adjustment to allow for missing values resulting from the failure to run all routes every year (e.g., Erskine 1978). Starting with the number of birds recorded in some base year or from an arbitrary index value (e.g., 100), the adjusted number for succeeding and preceding years is calculated from the proportional change in comparable routes for each pair of years, working forward and backward from the base year. Routes are considered comparable if they are run in consecutive years or in some cases only if run by the same observer in consecutive years.

Let C_{iy} be the number of birds detected on the ith route in the yth year. Next the mean number of birds detected in year y on routes comparable to the year after is defined as

$$A_y = \sum_i [C_{iy} I_{iy(y+1)}] / \sum_i I_{iy(y+1)} \qquad (1)$$

and the mean number on routes comparable to the year before as

$$B_y = \sum_i [C_{iy} I_{i(y-1)y}] / \sum_i I_{i(y-1)y} \qquad (2)$$

[1] U.S. Fish and Wildlife Service, Migratory Bird and Habitat Research Laboratory, Laurel, Maryland 20811.

where

$I_{iy(y+1)}$ = 1 if route i was run in years y and $y + 1$.
= 0 otherwise.
$y = m, \ldots, n$ years

The adjusted mean number of birds per route (base year index) for the base year (b) is

$$Z_b = \sum_{y=p}^{q} (A_y + B_y)/2(q - p + 1) \quad (3)$$

and the index values for other years are

$$Z_{y+1} = Z_y B_{y+1}/A_y \text{ for } y \geqslant b \text{ and} \quad (4)$$

$$z_{y-1} = Z_y A_{y-1}/B_y \text{ for } y \leqslant b \quad (5)$$

where

b = base year and
p, q = first and last years (respectively) used to calculate the base year adjusted call counts, Z_b, $m \leqslant p \leqslant q \leqslant n$.

A problem with the proportional adjustment is that the adjustment makes annual indices dependent upon the adjacent year's index. As a result, the sampling errors accumulate and the annual indices tend to behave in the fashion of a random walk. Ten separate random series of artificial count means were generated with the same mean (9.2) and standard deviation (0.51) as the continental Mockingbird (*Mimus polyglottos*) mean counts per route. Four of these projected trends are presented in Figure 1. In the simulation, a new set of routes was used for each pair of years, demonstrating the maximum effect of the base year adjustment method. If the routes are run every year, the proportional adjustment multiplies each annual count by a constant, leaving the trend unchanged. In situations where the proportional adjustment does not change the counts, no distortions are introduced.

Although 95% of the generated annual means were between 8.2 and 10.2 birds per route, 95% of the adjusted index values were between 0.7 to 20.5 birds per route. The magnitude of the fluctuations seems to be greatly exaggerated by the base year adjustment. Although the artificial annual means were generated without any trends, the base year adjusted indices calculated from them seem to show realistic looking "trends." This phenomenon is similar to the infamous moving average, which generates apparent population "cycles" (Cole 1954).

Kozicky et al. (1954) estimated the trend of woodcock singing ground counts using a balanced analysis of variance. This model viewed the year effects as being predominant, with route effects nested within years, assuming that

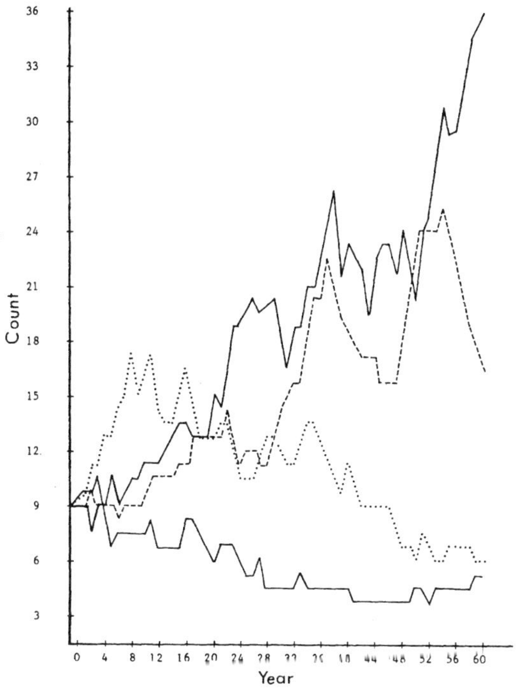

FIGURE 1. Simulated population fluctuations in the base year proportionally adjusted indices for the Mockingbird. See text for additional details.

the counts on a route in successive years are uncorrelated. We take the opposite view that the route effects are important and that the year effects are nested within routes. In our view, estimating population trends is similar to estimating animal growth curves. Here each route has a "population trend," determined at least in part by the habitat (and its changes), drawn from some population of "population trends." In this view the error term for testing trends should be calculated among the route trends.

Schultz and Muncy (1957), studying the counts of deer along transect lines, used an analysis of variance to compare months prior to the hunting season with months after the hunting season; a similar comparison could be made between groups of years. Duncan's multiple range test was used to compare the monthly counts.

Zimmerman (1979) provided a ten year summary of trends from the Kansas BBS. The mean number of birds per route for routes on which the species had been recorded were analyzed by the theory of runs (Dixon and Massey 1957). This method treated the entire State of Kansas as the sampling unit and therefore was insensitive to more local changes in populations. In addition, the runs test is only a qualitative index of population change and does not reflect the magnitude of population trends.

Dolbeer and Stehn (1979) estimated population trends of blackbirds and starlings from the North American BBS. In addition to using the base year adjustment, they used a paired t-test to compare each consecutive pair of years and to compare the first three years with the last three years. They also used an analysis of variance to fit a common slope to all routes within a physiographic stratum while allowing each route to have a separate intercept. Fitting a common slope with the model

$$c_{iy} = a_i + by + \epsilon_{iy} \quad (6)$$

where

c_{iy} = count on ith route in yth year
a_i, b = parameters to be estimated, and
ϵ_{iy} = error terms

assumes that successive error terms ($\epsilon_{iy}, \epsilon_{i(y+1)}$) are uncorrelated. These correlations will be unimportant only if counts are controlled primarily by stratum-wide effects (e.g., weather) that are independent of individual routes. However, if counts are controlled by route specific effects such as habitat, successive error terms for a route will be correlated and the common slope model would be inappropriate. If the primary controlling effect on local population size is habitat change, the variance should be calculated among routes.

DEVELOPMENT OF ALTERNATIVE METHODOLOGIES

Several alternative approaches are developed below. These methods are currently being used to analyze BBS data collected since 1965. The analysis is being conducted by staff at the Migratory Bird and Habitat Research Laboratory, U.S. Fish and Wildlife Service, Laurel, MD. An overview of their findings will appear as a 15 year summary of the BBS (Chandler Robbins, pers. commun.). Implicit reference is made in what follows to the results output from the computer program developed by Geissler.

ANNUAL MEANS

Annual mean counts are estimated to depict the dispersion about the fitted trends and to show possible systematic departures from these trends. Stratification in the analyses is based on physiographic regions (Bystrak 1981), but the routes themselves are allocated on a State (U.S.) or Province (Canada) basis, resulting in unequal probabilities of selection of routes within a stratum. However, probabilities of route selection are equal for individual States and Provinces within strata, but these areas are too variable in size and are often too small to constitute satisfactory strata.

An estimate of the stratum mean obtained by considering States and Provinces within strata as substrata is

$$\begin{aligned}\bar{c}_{iy} &= \sum_j \left[(N_{ij}/N_i) \sum_k^{n_{ijy}} c_{ijky}/n_{ijy}\right] \\ &= \sum_j \sum_k \left[c_{ijky}N_{ij}/(N_i n_{ijy})\right] \quad (7)\end{aligned}$$

where N_{ij} = area of the jth State or Province within the ith stratum

$N_i = \sum_j N_{ij}$ = total area of the ith stratum

n_{ijy} = number of routes in the sample from jth State or Province within the ith stratum with counts in the yth year

c_{ijky} = count from kth route in the jth State or Province within the ith stratum in the yth year.

(Note that throughout this paper, capital letters will used for population values and small letters for sample values.) The starting points of routes for the BBS were selected at random within a stratum without reference to well defined sampling units. Consequently, N_{ij} is taken to be the area in the jth State or Province within the ith stratum instead of the number of primary sampling units.

The variance of $\bar{c}_{iy}$ can be estimated by

$$\begin{aligned}\nu(\bar{c}_{iy}) &= \sum_j \sum_k \left[(t_{ijky}N_{ij}/N_i n_{ijy})^2 \nu(c_{ijky})\right] \\ &= \left[\sum_j \sum_k (t_{ijky}N_{ij}/N_i n_{ijy})^2\right]\nu(c_{ijky}) \quad (8)\end{aligned}$$

where $\nu(c_{ijky}) = \sum_j \sum_k (c_{ijky} - \bar{c}_{i..y})^2/(n_{i.y} - 1)$

$\bar{c}_{i..y} = \sum_j \sum_k c_{ijky}/n_{i.y}$

$n_{i.y} = \sum_j n_{ijy}$

t_{ijky} = 1 if count c_{ijky} was made
= 0 otherwise.

The means for regions or the continent over strata are

$$\bar{c}_y = \sum_i (N_i/N)\bar{c}_{iy} \quad (9)$$

with variance

$$\nu(\bar{c}_y) = \sum_i (N_i/N)^2 \nu(\bar{c}_{iy}) \quad (10)$$

where $N = \sum_i N_i$.

The annual means of the routes that were run each year can be influenced by which routes happen to be surveyed in a particular year. For example, if new routes were added in areas with few birds, the mean counts would be reduced although the population may not have changed. The other estimates that follow are calculated within each route to avoid this effect and to take advantage of blocking on routes.

To eliminate the effect on the annual mean counts of routes not run in certain areas, adjusted annual means can be calculated using predicted values whenever a route is not run. The predicted values are estimated from an analysis of variance for each stratum using the model

$$c_{ry} = a + b_r + d_y + \epsilon_{ry} \quad (11)$$

where c_{ry} = count on rth route in yth year
a = intercept
b_r = effect of rth route
d_y = effect of yth year
ϵ_{ry} = error term.

When a predicted value cannot be calculated, the mean count on that route is substituted. Covariables such as weather can easily be included in this model, but care should be taken to avoid including a covariable that shows a time trend as this would remove desired time trends from the counts.

It is important to determine the relative stability of the bird populations as well as their relative sizes. The coefficient of variation calculated among the yearly point estimates for the strata, regions and continent provides an indication of their relative stability.

Quenouille Estimator and Jackknife Variance

In the development that follows, Quenouille's estimator (jackknife) is used to estimate parameters in order to reduce the bias from order n^{-1} to order n^{-2} (Cochran 1977:175–177). The basic form of Quenouille's estimator of a parameter p is

$$\hat{p} = mp' - (m - 1)\bar{p}' \quad (12)$$

where $\bar{p}' = \sum_g^m p'_{(g)}/m$
p' = estimate based on all the data
$p'_{(g)}$ = estimate based on the remaining data after leaving out data from the gth group.

The jackknife variance is used to obtain variance estimates of ratios and of parameters which require weighting by random variables. An estimate of this variance of $\hat{p}$ is

$$v(\hat{p}) = [(m - 1)/m] \sum_g^m (p'_{(g)} - \bar{p}')^2. \quad (13)$$

To calculate these estimates, the routes in each stratum are randomly grouped as evenly as possible into m groups. These groups are formed by sorting the routes within each stratum into random order and then assigning the routes to groups in rotation.

Ratios of Mean Annual Counts

The ratio of the mean annual counts in one span of years y to the mean annual counts in another span of years y' indicates the relative change in the populations between the two time periods. The estimator of this ratio for stratum $i(r_{iyy'})$ is obtained by substituting

$$r'_{iyy'} = \left(\sum_j \sum_k c_{ijky} t_{ijkyy'} N_{ij}/N_i n_{ij}\right) \Big/ \left(\sum_j \sum_k c_{ijky'} t_{ijkyy'} N_{ij}/N_i n_{ij}\right) \quad (14)$$

for p' in equations 12 and 13 where

$t_{ijkyy'}$ = 1 if the counts c_{ijky} and $c_{ijky'}$ are available and not both equal to zero
= 0 otherwise
n_{ij} = number of routes in sample from jth State or Province within the ith stratum which have counts in years y and y'.

Here $r'_{iyy'}$ is the ratio of $\bar{c}_{iy}$ to $\bar{c}_{iy'}$ with $t_{ijyy'}$ selecting only those routes which have counts in years y and y'. The combined ratio estimator is used to obtain the estimate for regions and the continent over strata. This is equivalent to a separate ratio estimator using the estimated bird counts $\sum_j \sum_k c_{ijky'} t_{ijkyy'} N_{ij}/N_i n_{ij}$, as the stratum weight. Quenouille's estimator and jackknife variance are also used to obtain estimates for regions and the continent $(r_{yy'})$ over the strata by substituting

$$r'_{yy'} = \left[\sum_i (N_i/N)\left(\sum_j \sum_k c_{ijky} t_{ijkyy'} N_{ij}/N_i n_{ij}\right)\right] \Big/ \left[\sum_i (N_i/N)\left(\sum_j \sum_k c_{ijky'} t_{ijkyy'} N_{ij}/N_i n_{ij}\right)\right] \quad (15)$$

for p' in equations 12 and 13.

Geometric Mean of Ratios of Counts in Successive Years

This quantity estimates the average annual rate of change in the size of the bird population. The ratio of the counts in successive years is estimated and their geometric mean calculated. The geometric mean $[(c_2/c_1)(c_3/c_2) \ldots (c_p/c_{p-1})]^{1/(p-1)}$

reduces to $(c_p/c_1)^{1/(p-1)}$ if all the routes were run each year. In this situation the geometric mean depends only on the first and last count and would be highly variable. Because $r_{iy(y+1)}$ and $r_{iy'(y'+1)}$ are not independent, the variance cannot be calculated as the variance of a linear combination without including their covariances. However the variance can be estimated by expressing the geometric mean as a function of the c_{ijky}'s and jackknifing it. Here

$$a'_i = \left[\prod_{y}^{p-1} r'_{iy(y+1)}\right]^{1/(p-1)} \quad (16)$$

is substituted for p' in equations 12 and 13 to obtain the strata estimates and

$$a' = \left[\prod_{y}^{p-1} r'_{y(y+1)}\right]^{1/(p-1)} \quad (17)$$

is substituted to obtain the estimates for the regions and the continent; where $r_{iy(y+1)}$ and $r_{y(y+1)}$ are defined in equations 14 and 15, respectively.

Slope on Logarithmic Scale

Another estimator of the rate of change in the size of the bird population is the slope on the logarithmic scale, obtained by fitting the model

$$c_{ijky} = b_{ijko} b^{y}_{ijk} \epsilon_{ijky} \quad (18)$$

where c_{ijky} = the count in year y on route k in State or Province j within stratum i

b_{ijko} = the intercept on route k in State or Province j within stratum i

b_{ijk} = the population trend on route k in State or Province j within stratum i

ϵ_{ijk} = random error term associated with the predicted count. The error terms are assumed to be lognormally distributed with mean = 0 and variance = 1.

A multiplicative rather than an additive model is used because: (1) it is likely that population changes affect a proportion of the population (multiplicative model) rather than a specific number of individuals (additive model); (2) observers probably see or hear a proportion of the birds present; and (3) a multiplicative model (logarithmic transformation) has the advantage of stabilizing the variance for those data sets examined. Taking logarithms, the model becomes

$$c^*_{ijky} = b^*_{ijko} + b^*_{ijk} y + \epsilon^*_{ijky} \quad (19)$$

where the asterisk indicates the natural logarithms of the quantities and $c^*_{ijky} = \ln(c_{ijky} + 0.5)$. Because the logarithm cannot be taken of a zero count an arbitrary positive constant is added to c_{ijky}. The value 0.5 is used because it is half way between the smallest observable count and zero.

We wish to estimate the rate of change of the total bird population. Note that this is a different parameter than a "per area" rate of change which would give equal weight to each route. For example if half the routes doubled their counts from 50 to 100 birds and the others halved their counts from 10 to 5 in a year with equal probability sampling, the geometric mean rate of change would be 1 (no change). But the rate of change in the bird population would be 105/60 = 1.75.

To develop a justification for the estimate of the rate of change of the bird population, consider

$$b'_i = \left(\sum_j \sum_k b_{ijk} \bar{\bar{c}}_{ijk} N_{ij}/N_i n_{ij}\right) \Big/ \left(\sum_j \sum_k \bar{\bar{c}}_{ijk} N_{ij}/N_i n_{ij}\right) \quad (20)$$

where

$b_{ijk} = \exp[b^*_{ijk}]$ (estimated trend on route k)

$$b^*_{ijk} = \sum_{y=1}^{n_{ijk}} (c^*_{ijky} - \bar{c}^*_{ijk})(y - \bar{y}) / \sum_{y=1}^{n_{ijk}} (y - \bar{y})^2$$

$$\bar{c}^*_{ijk} = \frac{1}{n_{ijk}} \sum_{y=1}^{n_{ijk}} \ln(c_{ijky} + 0.5)$$

$\bar{\bar{c}}_{ijk} = (c_{ijk1} c_{ijk2} \ldots c_{ijkp})^{1/p}$ is the geometric mean of the counts on the kth route in the jth State or Province within the ith stratum.

Representing $\bar{\bar{c}}_{ijk}$ and b_{ijk} by functions of the predicted values from (18),

$$\bar{\bar{c}}_{ijk} \doteq \hat{b}_{ijko}(\hat{b}^1_{ijk} \hat{b}^2_{ijk} \ldots \hat{b}^p_{ijk})^{1/p} = \hat{b}_{ijko} \hat{b}^{(p+1)/2}_{ijk} = \hat{c}_{ijkq}$$

and

$$\hat{b}_{ijk} = \hat{c}_{ijk(q+1)} / \hat{c}_{ijkq}$$

where $\hat{c}_{ijkq}$ is the predicted count in year $q = (p + 1)/2$ where there are p annual counts. Substituting these values in (20)

$$b'_i \doteq \left(\sum_j \sum_k \hat{c}_{ijk(q+1)} N_{ij}/N_i n_{ij}\right) \Big/ \left(\sum_j \sum_k \hat{c}_{ijkq} N_{ij}/N_i n_{ij}\right). \quad (21)$$

This is the estimated ratio of counts in successive years, the quantity we wish to estimate. If the counts are proportional to the bird population along the route, this estimate is also the ratio of the bird populations in successive years.

The jackknife estimate is used to reduce bias, substituting

$$b'_i = \left(\sum_j \sum_k b_{ijk}\bar{\bar{c}}_{ijk} w_{ijk}/N_{ij}/N_i n_{ij}\right) \Big/ \left(\sum_j \sum_k \bar{\bar{c}}_{ijk} w_{ijk} N_{ij}/N_i n_{ij}\right) \quad (22)$$

for p' in equations 12 and 13 where

$\bar{\bar{c}}_{ijk} = \left[\prod_y^p (c_{ijky} + 0.5)\right]^{1/p}$ (0.5 added to avoid multiplying by a zero count)

n_{ij} = number of routes in jth State or Province within the ith stratum with two or more counts

w_{ijk} = 1 (will be redefined later).

The estimates for regions or the continent over strata are obtained by substituting

$$b' = \left[\sum_i (N_i/N)\left(\sum_j \sum_k b_{ijk}\bar{\bar{c}}_{ijk} w_{ijk} N_{ij}/N_i n_{ij}\right)\right] \Big/ \left[\sum_i (N_i/N)\left(\sum_j \sum_k \bar{\bar{c}}_{ijk} w_{ijk} N_{ij}/N_i n_{ij}\right)\right] \quad (23)$$

for p' in equations 12 and 13.

Slope on Logarithmic Scale, Weighted to Reduce Variance

Weighting the counts on individual routes by the number and dispersion of sampled years will reduce the variance of the trend estimate. However, if the decision to run or not run a route is related to the population trend on that route, a bias will be introduced. BBS routes are run by unpaid volunteers, and whether or not a route is run depends on the availability of volunteers. All routes are scheduled to be run each year but many routes are not run for reasons that appear unrelated to the bird populations or their trends. Routes are often added evenly across a State or Province as more observers volunteer. Therefore we feel that weighting BBS routes based on the years they were run will substantially reduce the variance of area estimates, but will introduce little bias into these estimates.

To develop the weighting factor consider the estimate of variance of the trend, which is

$$v(b^*_{ijk}) = \sum_y (c^*_{ijky} - \hat{c}^*_{ijky})^2 \Big/ \left[(n_{ijk} - 2)\sum_{y=1}^{n_{ijk}} (y - \bar{y})^2\right].$$

Because this variance is proportional to

$$\left[(n_{ijk} - 2)\sum_y^{n_{ijk}} (y - \bar{y})^2\right]^{-1},$$

the route estimate was weighted by

$$w_{ijk} = \left[(n_{ijk} - 1)\sum_y^{n_{ijk}} (y - \bar{y})^2\right]^{0.5} \quad (24)$$

to stabilize the variance. This gives less weight to routes sampled only a few years which would be expected to have a much greater variance than routes sampled many years. Here $(n_{ijk}-1)$ is used instead of $(n_{ijk}-2)$ to avoid giving zero weight to a route with two annual observations. Estimates of the weighted parameters are obtained by substituting their weight (w_{ijk}) into equations 22 and 23 prior to their substitution into equations 12 and 13.

Theil's Nonparametric Slope Statistics

This estimate of the rate of change of the bird population is less affected by extreme counts than are the parametric slope estimates. Individual slope estimates for each pair of years for a route are formed according to

$$d_{ijkyy'} = (c^*_{ijky'} - c^*_{ijky})/(y' - y),\ y < y'$$

and let

$$d_{ijk} = \exp[\text{median}\ (d_{ijkyy'})] \quad (25)$$

(see Hollander and Wolfe 1973:205–206). Estimates are obtained by replacing b_{ijk} with d_{ijk} in equations 22 and 23 before substituting them into equations 12 and 13.

Theil's Nonparametric Slope Statistic, Weighted to Reduce Variance

The variance of a median is approximately proportional to $p/(p + 1)^2$ (Gibbons 1971:36 eq. 6.6) where p is the number of data points. Because there are $p_{ijk} = n_{ijk}(n_{ijk} - 1)/2$ pairs of points on a route which are used to calculate the $d_{ijkyy'}$'s, the variance of d_{ijk} is proportional to $p_{ijk}/(p_{ijk} + 1)^2$. Thus the route estimates are weighted by

$$w_{ijk} = (p_{ijk} + 1)/(p_{ijk})^{0.5} \quad (26)$$

to stabilize the variance. Estimates are obtained by replacing b_{ijk} with d_{ijk} and substituting the above weight w_{ijk} in equations 22 and 23 prior to substitution into equations 12 and 13.

RESULTS AND DISCUSSION

It is extremely difficult to select among available models the one that best reflects the true dynamics of the population of interest. To make an unambiguous selection requires valid, up-to-date information on a species' population status from one or more independent sources. In the

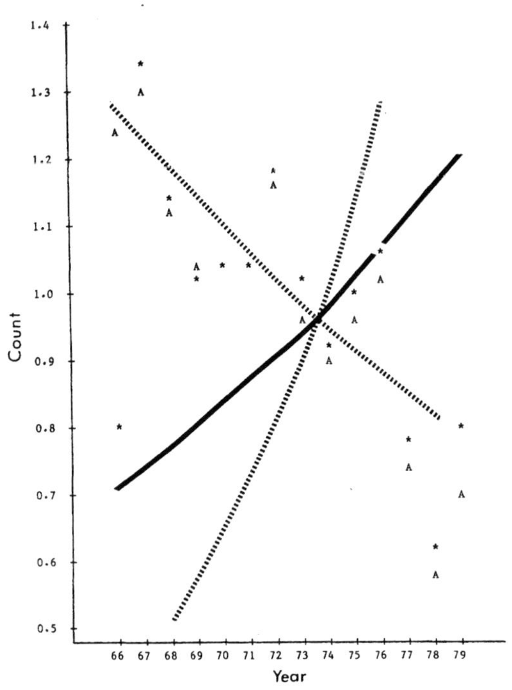

FIGURE 2. Continental population trend (solid line) and its 95% confidence limits (interrupted lines) of the Eastern Bluebird as estimated by a parametric slope (equation 23). The mean count on the routes that were run each year (*) and the adjusted mean counts (A) (when they differ from the mean counts) are also indicated. Data are from 1227 Breeding Bird Survey routes.

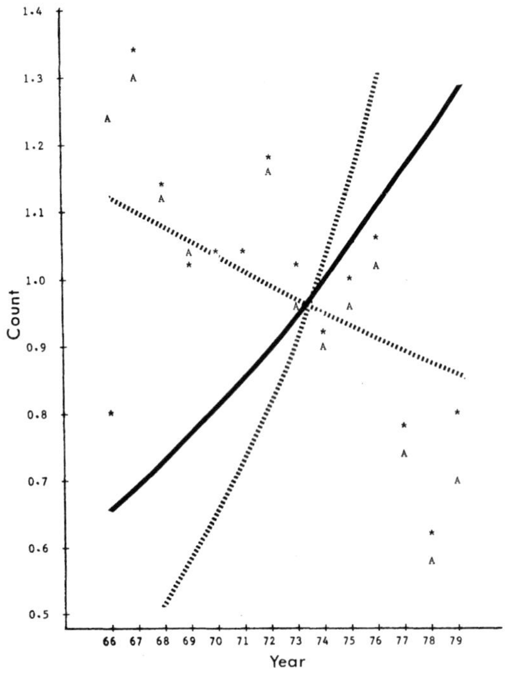

FIGURE 3. Continental population trend (and its 95% confidence limits) of the Eastern Bluebird as estimated by a nonparametric slope (equation 25). Mean (*) and adjusted mean counts (A) as in Fig. 2.

absence of such information other criteria are used. For the analysis of population trends considered here we have used both statistical and biological criteria. Estimates of the slope of the population growth curve on the logarithmic scale were weighted by the estimated species' population on that route. This weighting results in an estimate of the ratio of the total bird population in one year to the population in the previous year, the ratio desired. In addition, we expect routes with higher counts to be more centrally located with respect to the species' distribution pattern. As a result, these routes should have more weight than routes on the periphery of the species' range which are more prone to random fluctuations and which represent a smaller fraction of the species' population. Estimates that were weighted to reduce the variance consistently appear to have narrower confidence intervals on their slope estimates and importantly, for those cases examined, also showed very little bias as judged by the annual means (Geissler, unpublished analysis).

Biologically we looked for independent sources of corroboration for our trend estimates. Our search, restricted to the ornithological literature, particularly state field journals, supplied only anecdotal information. The lack of independent sources of avian trend estimates is not surprising. The BBS data are unique in that they represent the only data set that indexes the status of many species' populations over a large geographical area for a long period of time.

Proportional base year adjustment methods have been used to estimate population change. However, when we examined the performance of this method for random series of artificial mean counts we noted that realistic looking trends appear even when no trend exists in the original data (Fig. 1). The magnitude of these trends is even greater for species we examined with larger coefficients of variation in their annual mean counts.

Another approach to model selection is to compare the performance of the models with known artificial populations using computer simulation. We have not done any simulations, although we hope to in the future.

To make a tentative selection among the models developed we have examined in detail

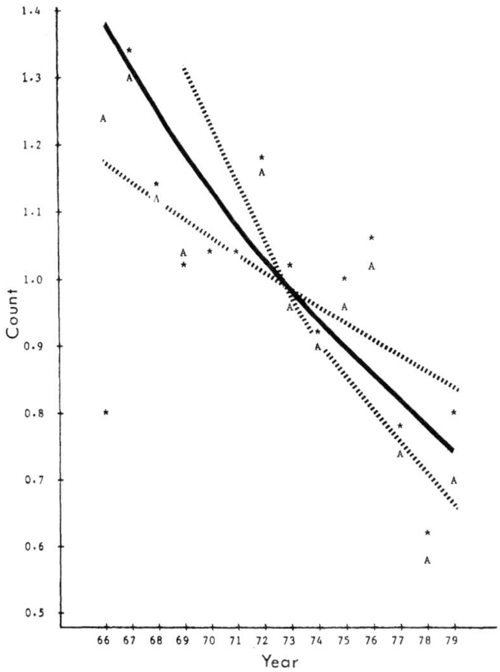

FIGURE 4. Continental population trend (and its 95% confidence limits) of the Eastern Bluebird as estimated by a parametric slope scale weighted to reduce the variance (equation 24). Mean (*) and adjusted mean counts (A) as in Fig. 2.

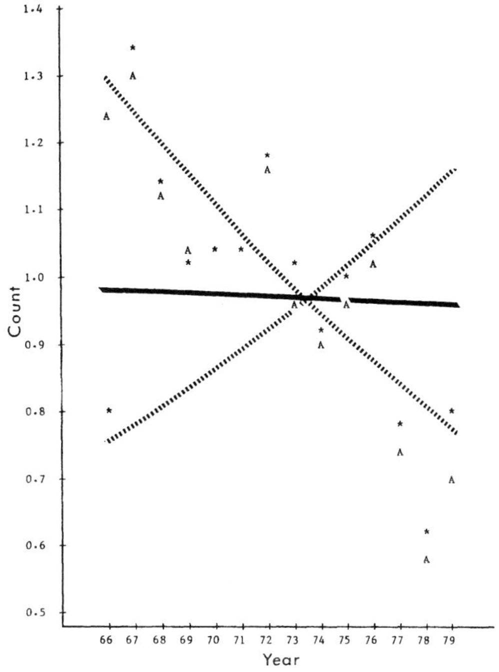

FIGURE 5. Continental population trend (and its 95% confidence limits) of the Eastern Bluebird as estimated by a nonparametric slope on the logarithmic scale weighted to reduce the variance (equations 25 and 26). Mean (*) and adjusted mean counts (A) as in Fig. 2.

the population trends for two species, the Eastern Bluebird (*Sialia sialis*) and the Loggerhead Shrike (*Lanius ludovicianus*). These species were selected for several reasons: (1) both species are well sampled by the BBS methods; (2) both species are well represented in the data set; (3) they have extensive geographic distributions; and (4) considerable anecdotal information indicates that these species have undergone substantial declines in population over the past 10 years. Anecdotal information on regional population declines is perhaps best summarized by the Blue List published since 1971 in *American Birds*. During this nine year interval the Loggerhead Shrike has been listed every year and the Eastern Bluebird, four of the nine years, most recently in the 1980 Blue List (also see Monroe 1978, Zimmerman 1979).

We present a comparison of the continental (Canada and U.S.A.) trend estimates for the Eastern Bluebird (Figs. 2–6). Figure 2 illustrates the unweighted, parametric slope estimate on the logarithmic scale (from equation 23); Figure 3, the unweighted, non-parametric slope on the logarithmic scale (equation 25 substituted into equation 23); Figure 4, the weighted, parametric slope on the logarithmic scale (equation 23 with weighting factor from equation 24); Figure 5, the weighted, non-parametric slope on the logarithmic scale (equations 25 and 26 substituted into equation 23); and Figure 6, the geometric mean model (equation 17).

In all figures both actual (equation 7) and adjusted mean counts (equation 11) are presented. Adjusted mean counts have the advantage of not being directly influenced by which routes were run in a particular year. If routes on the periphery of a species' range are added in later years of the survey it is possible for mean counts to show a decrease even if counts are increasing on most routes. However, adjusted mean counts will show this increase because predicted values are substituted whenever a count is missing. We make note of the fact that the difference between adjusted and actual mean counts for 1966–67 (e.g., Fig. 2) are a result of the algorithm used to calculate regional mean counts coupled with the poor coverage for the BBS in these years. These differences do not affect the trend estimates.

The difference in the trend estimates among the models is extensive. Both the parametric and

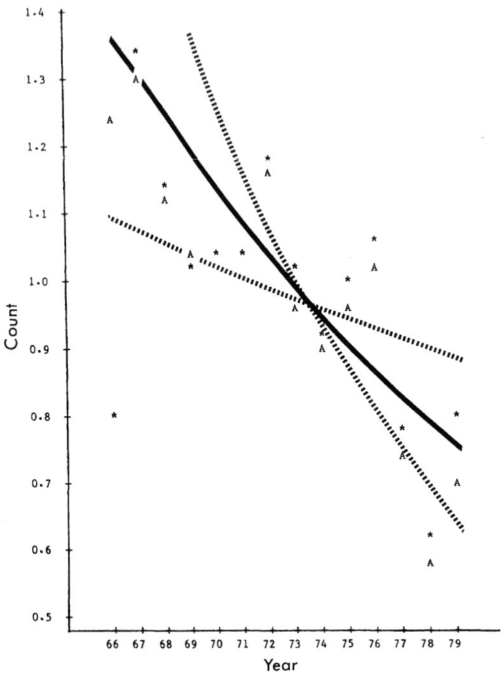

FIGURE 6. Continental population trend (and its 95% confidence limits) of the Eastern Bluebird as estimated by the geometric mean of the annual ratios of the counts in successive years (equation 17). Mean (*) and adjusted mean counts (A) as in Fig. 2.

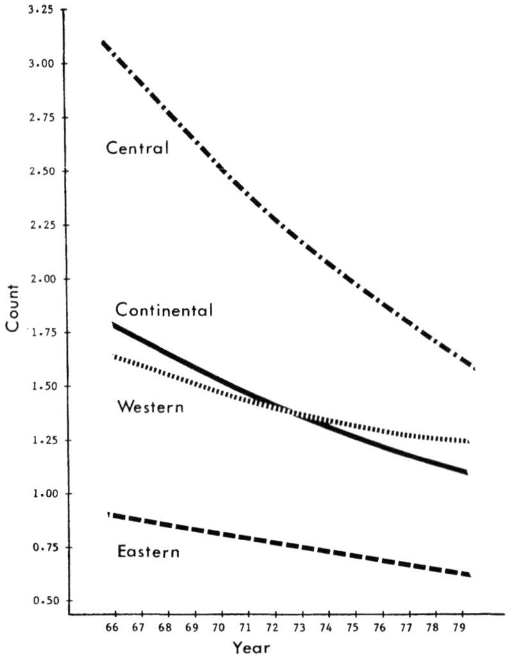

FIGURE 7. Continental (solid line) and regional population trends for the Loggerhead Shrike as estimated by a parametric slope weighted to reduce the variance (equation 24). Data are from 1021 Breeding Bird Survey routes.

non-parametric, unweighted models (Figs. 2 and 3) differ from the others in that they: (1) show an increasing population trend; (2) have large variances associated with the slope estimate; (3) fail to follow the adjusted and unadjusted mean counts; and (4) fail to corroborate the anecdotal evidence which indicates that the Eastern Bluebird is declining. The weighted nonparametric and geometric mean trend estimates are quite similar and closely follow the annual means. However, note that the confidence interval around the slope of the weighted parametric model is substantially smaller than for the other models (cf. Fig. 4 with Figs. 2, 3, 5, and 6). An identical pattern, with the weighted parametric model performing best, was noted on close examination of trends for the Loggerhead Shrike.

Judging by the performance of the weighted parametric model, we feel that it may accurately reflect the population trend of the Eastern Bluebird. It also seems intuitively reasonable to weight each route by the number and spread of years sampled. Routes sampled many years would be expected to give a much better index of the local population stability of a species. In addition, routes sampled many years would be expected to have a lower variance than routes sampled only a few years.

Weighting the route by the number of years sampled has the potential of introducing a bias if the decision to survey a route is based on some a priori concept of a species' trend along that route. To investigate this potential bias we calculated the correlation between the weighted parametric and weighted non-parametric route slope estimates and their weights. For the Eastern Bluebirds the correlations were significant ($P < .001$) but so low ($r = -0.14$) that we are confident that the weighting introduced little bias.

Continental and regional population trends for the Loggerhead Shrike, as estimated by the parametric slope on the logarithmic scale weighted to reduce the variance, are presented in Figure 7. Note that the trends for the three regions are quite different. All slope estimates, except for the Western Region, are different from zero ($P < .01$).

Partitioning of the continental estimate into regional estimates represents only a first-level breakdown. In addition, our analyses breakdown each regional estimate into strata estimates (see Bystrak 1981) and each stratum estimate into individual route estimates. As a result, we are able to look at very specific areas within a species' range to investigate any local effects that may be contributing to the species' changing population status. The power of the analysis as an investigative tool rests on the use of individual routes as the basic unit of analysis.

CONCLUSIONS

Three different approaches are used to estimate the annual rate of change in the bird populations c_{y+1}/c_y. The geometric mean has an intuitive appeal because of its simplicity, being the average of the annual ratios. It does not require weighting because the routes automatically contribute their proportional share to the numerator and denominator of the ratio. However it does not use all of the available data because ratios of successive years are restricted to routes that were run both years. In the extreme case where all routes were run every year, only the first and last years of data are used. In spite of this, the geometric mean performs surprisingly well.

Another approach to estimating the annual rate of change is to estimate the slope of the population growth curve on the logarithmic scale. Both parametric and nonparametric slope estimators are used. In either case, the route slope estimate must be weighted by the estimated bird population on that route to obtain an estimate of the change in the total bird population. Weighting according to the years a route was run was effective in increasing the precision of the estimates, but has the potential of introducing bias. To investigate the effect of the weighting on the slope estimate, we calculated the correlation between the weighted parametric and weighted non-parametric route slope estimates and their weights. For the Eastern Bluebird the slope estimate was significantly correlated with the weight but was so small that there was little opportunity to introduce a bias. In addition, the fitted trends do not show any obvious bias as judged by the adjusted annual means.

Nonparametric slope estimates have both the advantage and disadvantage of being less influenced by extreme points than the parametric slope estimates. This is an advantage if the extreme points are mistakes resulting from recording error, mistaken identification, etc. However, the lack of sensitivity to extreme points means that the nonparametric slope estimate will not be as sensitive as the parametric slope estimate to sharp changes in the bird population.

Some may argue that it is not logical to fit an average rate of change to a population for which the rate of change is fluctuating. This is certainly true in some situations. These situations can be identified by observing the fluctuations of the annual mean counts and the adjusted annual mean counts, and their failure to conform.

Based on our limited experience, the parametric slope, weighted to increase the precision, appears to be the best estimator. It has the smallest confidence intervals and is more sensitive to sharp population changes than the nonparametric estimates. However, more work is needed to investigate the properties of these estimators. Both an examination of their performance on several sets of data and a Monte Carlo study are planned. Early results of these studies indicate that the parametric slope estimates are positively biased but that the bias can be easily corrected.

ACKNOWLEDGMENTS

We appreciate the helpful suggestions and comments made by D. H. Johnson, J. D. Nichols, C. S. Robbins, G. E. J. Smith, S. L. Stokes, and B. K. Williams. The technical assistance of L. M. Moyer and R. C. Perry is also appreciated.

Studies in Avian Biology No. 6:52–57, 1981.

REAPPRAISAL OF THE WINTER BIRD-POPULATION STUDY TECHNIQUE

CHANDLER S. ROBBINS[1]

ABSTRACT.—Standardized morning and afternoon counts were conducted in two deciduous forest plots during seven winters. Morning counts yielded more species than afternoon counts, as well as higher counts of most species. A minimum of eight visits is recommended. Six visits yielded 2% to 5% fewer species, depending on the size and bird population of the study plot. Cumulative means for individual species varied erratically for most species when the number of visits was less than seven.

This study was conducted to determine the relative effectiveness of morning and afternoon counts in winter, and to determine the optimum number of visits. Unlike the Audubon Breeding Bird Census, the totals published are the seasonal means for each species, which are typically somewhat less than the actual population. Thus the number of visits may greatly affect the total number of species recorded, and the timing of these visits influences the number of individuals and their reported density.

METHODS

Four morning and four afternoon visits were made to each of two deciduous forest plots along the Middle Patuxent River in Howard County, Maryland, in seven winters from 1972 to 1980. The Audubon Winter Bird-Population Study instructions (Anon. 1950) were followed. These specify making at least six visits during the period 20 December to 10 February, mapping all birds observed on each visit, and determining as closely as possible the number of each species present on each trip. The mean number per trip is then computed and rounded to the nearest half bird. Finally, the density per 100 ha is computed and the means for each species are summed to get the total mean population.

The 56 morning visits started at about sunrise; the 56 afternoon trips ended shortly before sunset. Visits to the floodplain plot averaged about three hours each, those to the smaller upland plot, about two hours. The same route was followed on each visit. All birds observed on each trip were plotted on census maps, from which the trip totals for each species were determined. Results of each winter's study were published in the respective issues of *American Birds* under the titles Hickory-Oak—Ash Floodplain Forest, and Upland Tulip-tree—Maple-Oak Forest, as was also a preliminary appraisal of the method based on the 1972 results (Robbins 1972). Because the two plots were of different size (18.4 ha for the floodplain plot, 12.0 ha for the upland plot) and different bird densities, and because of year-to-year changes in these populations, the counts for each of 19 common species were analyzed by analysis of variance in order to detect any differences attributable to time of day.

Cumulative species totals for each year in each plot were fitted to the equation $y = a - be^{-cx}$ to learn what percentage of the estimated number of species using each of the plots was detected in any given number of visits.

To determine the influence of number of visits on the counts of individual species, the 8-trip mean for each year was used as a standard with which the means of lesser numbers of visits were compared. Percentage departures from the 8-visit mean were computed and the absolute values averaged for each species over the 7-year period (excluding any years when a species was not observed on the plot).

RESULTS

TIME OF DAY

Results of the analysis of variance for 19 of the more common species are summarized in Table 1. For all except the junco, the 56 morning trips yielded a higher mean than did the 56 afternoon trips, but only eight species showed a significant ($P < .05$) difference between the morning and afternoon. The mean number of species recorded was also higher ($P < .01$) in the morning. The greatest differences noted were for the Evening Grosbeak (95% fewer in the afternoon), Blue Jay (59% fewer), and Purple Finch (42% fewer).

NUMBER OF VISITS

Estimates of both the species composition and the mean numbers of individuals vary with the number of visits. As visits increase, the number of species detected increases toward an asymptote that varies from year to year. Also, as visits increase the cumulative mean number of birds of each species tends to stabilize. The number of visits that will be made is, of necessity, a compromise between the number needed to obtain stable estimates and the manpower, cost, and number of days with favorable census weather available prior to beginning of spring migration.

Species richness

Cumulative species estimates were generated separately for each year in each plot. In six out of seven years in the 18.4 ha floodplain plot, 84–

[1] Migratory Bird and Habitat Research Laboratory, U.S. Fish and Wildlife Service, Laurel, Maryland 20811.

TABLE 1
COMPARISON OF MORNING AND AFTERNOON VISITS

Principal species	Mean of 56 visits: Morning	Afternoon	% difference
Common Flicker (*Colaptes auratus*)	3.70	2.80	−24
Red-bellied Woodpecker (*Melanerpes carolinus*)	4.11	3.66	−11*
Hairy Woodpecker (*Picoides villosus*)	1.63	1.57	−4
Downy Woodpecker (*Picoides pubescens*)	7.27	6.18	−15
Blue Jay (*Cyanocitta cristata*)	1.95	0.79	−59**
Carolina Chickadee (*Parus carolinensis*)	7.43	6.39	−14
Tufted Titmouse (*Parus bicolor*)	4.79	3.46	−28*
White-breasted Nuthatch (*Sitta carolinensis*)	2.89	2.18	−25*
Brown Creeper (*Certhia familiaris*)	.73	.45	−38*
Winter Wren (*Troglodytes troglodytes*)	.39	.21	−46
Carolina Wren (*Thryothorus ludovicianus*)	2.18	1.70	−22
Golden-crowned Kinglet (*Regulus satrapa*)	.64	.55	−14
Yellow-rumped Warbler (*Dendroica coronata*)	18.79	15.55	−17
Northern Cardinal (*Cardinalis cardinalis*)	7.34	5.91	−19
Evening Grosbeak (*Hesperiphona vespertina*)	5.18	.25	−95**
Purple Finch (*Carpodacus purpureus*)	2.88	1.68	−42**
American Goldfinch (*Carduelis tristis*)	1.23	1.02	−17
Northern Junco (*Junco hyemalis*)	.96	1.30	+35**
White-throated Sparrow (*Zonotrichia albicollis*)	7.70	7.46	−3
Number of species	16.68	14.88	−11**

* = $P < .05$.
** = $P < .01$.

TABLE 2
PERCENT OF SPECIES DETECTED EACH YEAR AS NUMBER OF VISITS INCREASED

Plot and year	Species recorded	Number of visits: 4	6	8	10	12
Floodplain						
1972	33	93	99	99	100	100
1973	36	54	61	67	72	76
1975	39	96	96	100	100	100
1976	41	89	99	99	100	100
1977	28	84	91	98	98	99
1978	38	87	92	98	99	100
1979	31	94	97	97	100	100
6-yr. mean	35.4	90.5	95.7	98.5	99.5	99.8
Upland						
1972	30	95	95	99	100	100
1973	27	86	90	97	99	99
1975	29	77	86	89	95	97
1976	27	47	55	65	71	77
1977	24	82	91	99	98	99
1979	19	79	96	96	99	100
1980	19	81	81	91	96	98
6-yr. mean	24.7	83.3	89.8	95.2	97.8	98.8
All years combined						
Floodplain		89	96	98	99	100
Upland		77	88	94	97	98

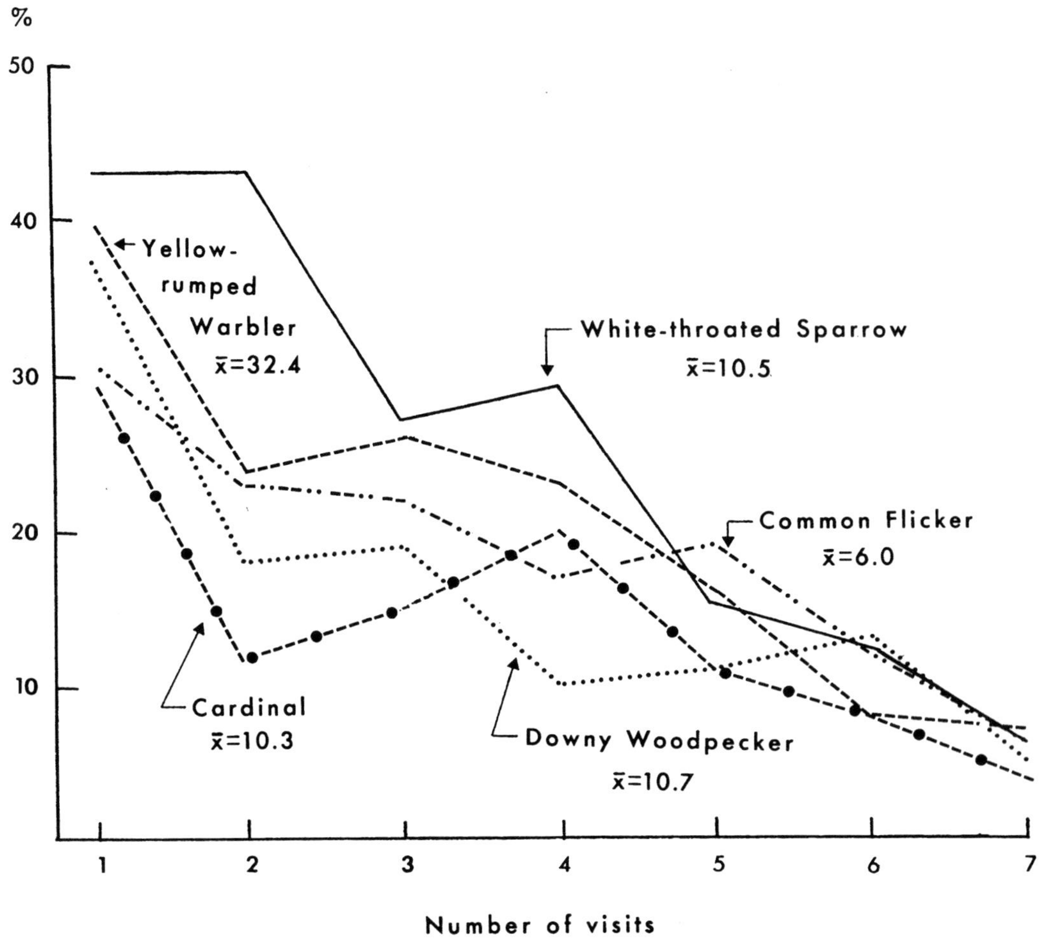

FIGURE 1. Cumulative effect of number of visits on the means for five common species in the 18.4 ha floodplain plot. Percentage departures from the eight-visit mean become less as the number of visits increases.

96% of the projected number of species were detected in the first four visits, 91–99% were detected in six visits, and 97–100% in eight visits (Table 2). In six out of seven years in the 12.0 ha upland plot, 77–95% of the species were detected in four visits, 81–96% in six trips, and 89–99% in eight visits. In one year in each plot, the cumulative species totals were so erratic that eight trips were insufficient to generate valid predictions. This resulted in the estimates being too low. The results for these two years, 1973 in the floodplain plot and 1976 in the upland plot, are not included in the means in Table 2.

Counts of individual species

Figures 1, 2, and 3 show, for selected species, how the results of one through seven visits compare with the 8-visit mean. For five common species (Fig. 1), with mean winter populations of 6 to 72 individuals on the floodplain study plot, the first four visits resulted in a mean that averaged 10% to 29% away from the 8-visit mean. With five visits the departures narrowed to 11% to 19%; with 6 it decreased to 8% to 13%, and with seven visits departures ranged from 4% to 7% away from the 8-visit mean.

For flocking fringillids, however, the estimates from small numbers of trips were extremely variable (Fig. 2). Departures from the 8-visit mean computed from the mean of 4, 5, 6, and 7 visits were 20–34%, 24–28%, 18–27%, and 6–30%, respectively.

In the upland plot, where bird populations were much smaller (Fig. 3), variability was much greater than in the floodplain. For example, departures for the Yellow-rumped Warbler were about four times as great in the upland plot, where the mean was 1.72 (Fig. 3), as in the

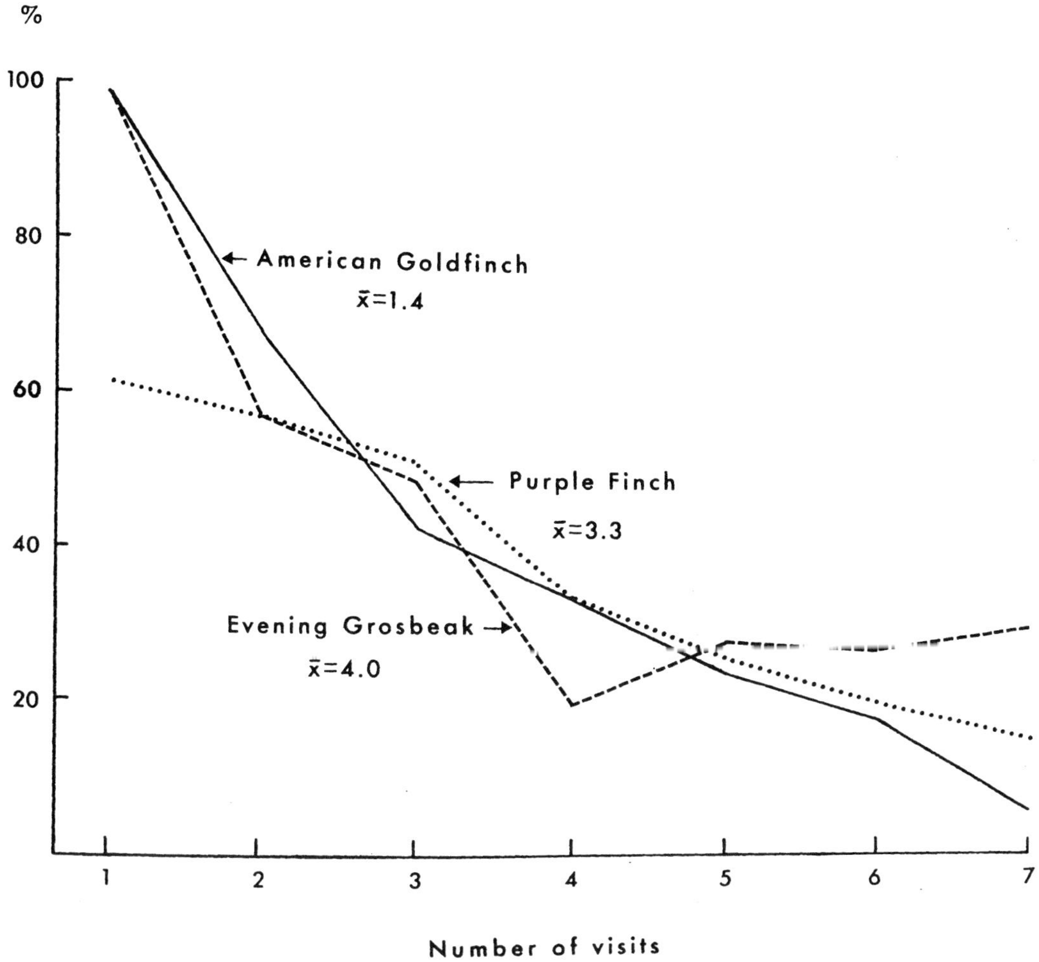

FIGURE 2. Cumulative effect of number of visits on means for flocking fringillids in the floodplain plot.

floodplain, where it was 32.4 (Fig. 1). For the Cardinal they averaged about three times as great in the upland, where the mean was 2.59, as in the floodplain, where the mean was 10.3. For the White-throated Sparrow they averaged about twice as great in the upland, where the mean was 4.68, as in the floodplain, where it was about twice as high (10.5). For a non-flocking species such as the Downy Woodpecker, however, the small number of visits gave fairly dependable results even with a small population.

DISCUSSION

In the comparison of morning and afternoon counts based on only a single year (Robbins 1972), significantly higher numbers of total individuals were recorded in the morning ($P < .01$): 20% higher in the 18.4 ha floodplain, 47% higher in the 12.0 ha upland. Although only two species, Blue Jay and Evening Grosbeak, revealed significant ($P < .05$) differences with this small sample of data, the mean values of 11 of the 13 species tested were higher in the morning than in the afternoon in both plots (if present in both). The other two species, Common Crow and Carolina Wren, had higher morning counts in one plot, while morning and afternoon tallies were identical in the other plot.

With the addition of six more years of data from the same two plots, the differences between morning and afternoon counts become more obvious. The Blue Jay, Evening Grosbeak, and Purple Finch now show highly significant differences ($P < .01$), and four other species have joined the significant list. Of the 19 species examined, only the junco had higher afternoon counts, as a result of birds entering the plots from nearby fields late in the day. With the pos-

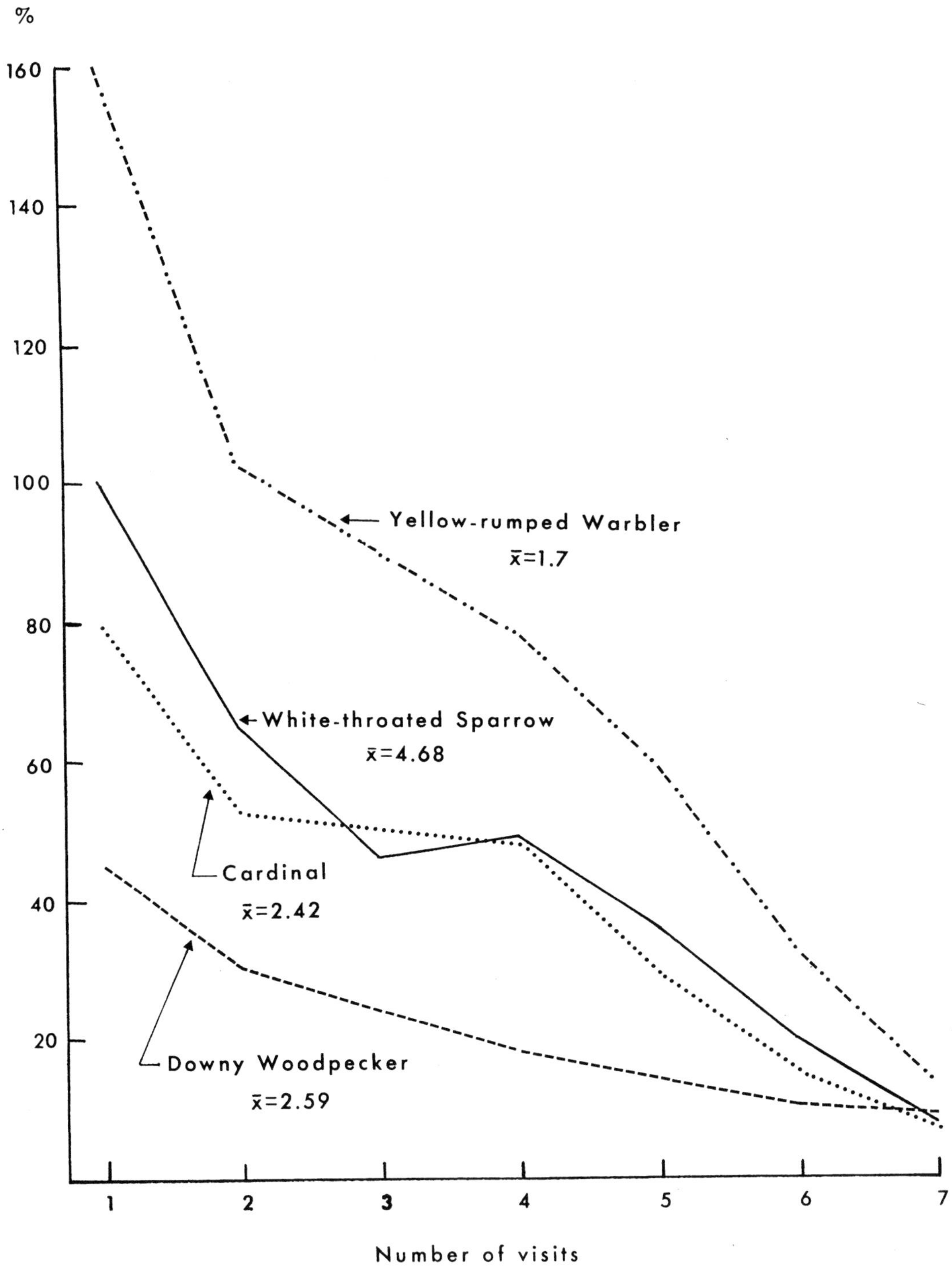

FIGURE 3. Cumulative effect of number of visits on means for species with low populations in the 12.0 ha upland plot.

sible exception of the Evening Grosbeaks, I believe the differences noted resulted from lower detectability in the afternoon rather than any important difference in the numbers of birds present in the study plots. I conclude, therefore, that the morning trips are more productive than the afternoon trips, both for total number of species and for the numbers of individuals of each. As no counts were made in mid-day, the comparative efficiency of mid-day counts is not known.

The Winter Bird-Population Study instructions call for a minimum of 6 visits, preferably more. In practice, eight is the number most frequently used. Of the 64 studies published in *American Birds* in 1980, 2 had 5 visits, 12 had 6, 5 had 7, 16 had 8, 10 had 9 or 10, 5 had 11 or 12, 13 had 13, 14, or 15, and 1 had 22.

Results of the present study indicate that in forest habitats in the 12–18 ha range, which is about the average size of Winter Bird-Population Study plots, six visits should be the very minimum, and eight or more would be preferable. Whereas an estimated average of 5% of the species in the 12.0 ha plot were missed with eight visits (Table 2), 10% were missed with six visits, and 17% with only four visits. In the 18.4 ha plot, an average of 2% of the species were missed with eight visits, 4% with six visits, and 10% with four visits. If all visits had been made in the early morning, fewer species would have been missed.

On an individual species basis, we must recognize that only the commoner species, those averaging at least one and preferably five or more birds per visit, can be satisfactorily estimated with eight visits. And it must be re-emphasized here that a satisfactory estimate is not a valid estimate of the actual population, but rather a stable index that can be used for comparison with other years or possibly other study plots.

Day-to-day variability in winter was so high that eight visits per year for seven years could not adequately define the number of visits required for various species. This variability was caused in part by birds moving in and out of the plots, but also in part by weather factors and sampling error. Figures 1, 2, and 3 do suggest, however, that non-flocking species and common species are more reliably estimated than others, and that studies based on fewer than seven trips can give misleading results for many species.

CONCLUSIONS

The Winter Bird-Population Study method is adequate for describing average use of a plot by wintering birds if published instructions are followed, and if the following recommendations are considered:

(1) The study plot should be large enough or rich enough so the more important bird species are represented by an average of five or more individuals each.

(2) Plan to make at least eight visits unless results show that counts of the major species have stabilized and no additional species are being found.

(3) If recommendation 1 cannot be met, extra visits (more than eight) are recommended to obtain better population estimates.

(4) Plotting of a cumulative species "discovery curve" or calculating the estimated total species present using the formula presented above will show whether enough visits have been made to detect most of the species present.

(5) Expect morning visits to be more productive than afternoon visits.

(6) Consider making extra visits for nocturnal or crepuscular species. Mid-day trips may reveal presence of vultures, buteos, or other birds not found in early morning.

ACKNOWLEDGMENTS

I thank D. Bystrak, D. K. Dawson, and B. K. Williams for their constructive comments on the manuscript.

Studies in Avian Biology No. 6:58–59, 1981.

SUMMARIZING REMARKS: ESTIMATING RELATIVE ABUNDANCE (PART I)

DOUGLAS H. JOHNSON[1]

Ralph Schreiber will be making some general comments on the papers from a biological standpoint; I will simply note a few statistical items I think worthy of consideration.

Bock and Root (1981) present a useful discussion of techniques for analyzing results of Christmas Bird Counts (CBC's). Of statistical interest is their proposal for standardizing CBC data. For common and widespread species, they suggest dividing the number of birds seen by the number of party-hours involved in the count, a measure of effort. For species restricted to special habitats, which are likely to yield the same total regardless of effort, they recommend considering the actual number seen per count. I suggest a more general approach. If E is the number of party-hours effort in a particular CBC, then the total number of birds seen could be standardized by division by $1 - \alpha + \alpha E$, where α is a constant between 0 and 1. Values near 0 would give total birds seen, whereas values near 1 would give the number seen per unit effort.

The merit of this approach lies in the possibility of developing useful values of α for various groups of birds within a CBC area. For example, if a CBC area contained about 90% deciduous forest, 10% open field, and a single pond, we would anticipate counts of forest birds to increase almost linearly with effort, and α for those species might be nearly 1. For birds of open fields, α might be about ½, and for waterbirds, which are likely to show the same total whether there are five observers or 50, α would be near 0.

These values could be estimated from an analysis of a number of years of CBC's in an area. This approach may appear too difficult for routine application, but I suspect it could be worthwhile for detailed analyses of a few CBC areas.

Arbib (1981) offers a good critique of current CBC practices and recommends several improvements. Among other analyses, he shows (his Table 2) that CBC's with more observers tend to identify more species. The implication, no doubt correct, is that more species are likely to be found if more observers are involved. It is true that the 22 observers in the Jamestown (North Dakota) count saw 40 species in 1979, and that 51 observers on the Monterey Peninsula (California) count tallied 194, but I doubt that much of the difference in species totals was due to the number of observers. Although Arbib's table is limited to California, I think a better comparison would involve an examination of the number of species versus number of observers across years for a particular CBC area, rather than across areas in a particular year.

Arbib also suggests that training sessions and examinations be used to develop proficiency in estimating numbers of birds in flocks. We have found that even professional observers tend to underestimate the number of animals in large groups, and the bias increases with the size of the groups. A. R. E. Sinclair (1973) found the same relationship, and provided evidence that training can in fact work. A 20-minute training exercise, which involved showing the observers color slides of various groups of animals, asking them to estimate the number in the group, and providing them with the correct answer, caused a rather pronounced underestimation bias virtually to vanish.

Bystrak (1981) gives an overview of another popular and productive bird survey performed in large part by amateurs, the Breeding Bird Survey (BBS). He points out some of the difficulties in interpretation and analysis; see the paper by Geissler and Noon (1981) for statistical details.

A complaint voiced by participants in the BBS is the length of the survey routes: 50 stops at 0.5 mile intervals, each requiring three minutes of observation. Participants suffer from fatigue, which affects their performance on the last 10 or 20 stops. In addition, the long time span required for 50 stops covers intervals when birds are extremely vocal and conspicuous and intervals when they are relatively retiring and inconspicuous.

From a statistical viewpoint, I suspect that shorter surveys, perhaps 30 stops, would not result in serious loss of information. It is a generally held belief that sample size is more important than the size of the sample unit. I feel confident that three surveys of 30 stops each would be more valuable than two surveys of 50 stops each. Even two of the shorter routes would probably be nearly as good as two longer ones and could actually be better if the longer one happened to straddle a stratum boundary. The potential loss of information could be assessed rather simply by examination of the current data base. The data are tallied by 10-stop

[1] Northern Prairie Wildlife Research Center, U.S. Fish and Wildlife Service, Jamestown, North Dakota 58401.

summaries, so the first 30 stops could be analyzed instead of all 50. If a stratum boundary is crossed, the 30 stops most clearly contained in a stratum should be used.

Järvinen and Väisänen (1981) give a worthwhile review of the methodology used in an extensive survey in Finland and the thinking that went into the selection of the method. I noted that they sampled all relevant habitats in approximately correct proportions. This is called proportional allocation, which results in a self-weighting sample design. Although that choice of allocation may be far different from one that is optimal for a particular purpose, it is easy to use and discuss, and can usually be recommended.

A valuable control in their design is to insist that each region is covered by more than one observer, to minimize the effect that differences in observers may have on results for wide areas.

Dawson (1981) touches on a great many key points when he discusses the factors affecting the detectability of birds. He mentions two ways to account for the effects of those variables, such as season, time of day, and weather, that influence the counts. His first method is to standardize the counts by holding those variables as constant as possible; this can be viewed as controlling them. The second method allows those variables to vary, but their effects are estimated and accounted for; this is more in line with modern methods of experimental design and permits analysis of variance or analysis of covariance to be employed. I would define a third method, which in fact is probably the most commonly used: Ignore those variables, take large samples, and hope their effects "average out." This might appropriately be termed the "Pollyanna approach." Throughout this symposium we will see all three methods used, but with little discussion about their relative merit.

Studies in Avian Biology No. 6:60, 1981.

SUMMARIZING REMARKS: ESTIMATING RELATIVE ABUNDANCE (PART I)

RALPH W. SCHREIBER[1]

Perhaps C. J. Ralph and Mike Scott invited me to summarize this session to obtain a "pelican's eye view" of studies of little brown birds, or perhaps, since I am a curator in a major natural history museum so I could provide a place to put all the specimens that you are censusing, and thus provide a true count of the population. In any case, the following is the ornithologist's view to complement the statistical summary view of Doug Johnson.

I believe that the analyses of the data presented by the speakers in this session are in good hands. These methods will be continually worked over and refined by researchers, and especially those who are interested in fiddling with computers, correlations, and fudge factors. If I can find a common thread in these papers, it is that there is a need to influence the field observers in the Christmas Bird Counts, who are frequently "amateurs," to somehow make the data collection more "scientific." I wish to make three observations from listening to the paper presentations and then provide a suggestion for future work.

(1) The CBCs and Breeding Bird Surveys (BBS) provide an invaluable index to population changes but probably only within an order of magnitude and only if large geographic areas are analyzed together. I believe that species composition and relative abundance may prove to be the most useful, derivable index from these data. Further, it is obvious that only through analysis of 10 or more years of data will accurate determination of avian population trends be possible. Probably a decade is a minimum and 20 to 30 years are required for anything approaching reality. This fact needs to be emphasized to funding agencies and government bodies attempting to use the data for management recommendations.

(2) I believe that obtaining more data relevant to the non-breeding season and/or non-breeding biology of birds is absolutely critical. We know a great deal about breeding in birds but "nothing" about the remainder of the year. The few studies that are available and the inferences that can be made about total biology of species clearly indicate that any and all data we can collect about the time birds spend away from the nest will be extremely valuable.

(3) It has been my observation that most observers participating in CBCs are primarily emphasizing "ticking" another species on the list. They tend to ignore numbers, especially for common-abundant birds, and totally ignore age and sex data when they are available. The need to make the CBCs more scientifically-biologically useful is obvious and one step would be for more instructions to be given to the organizers and more emphasis to be placed on counting, sexing, and aging the birds observed.

Dr. Callaham has asked us to provide suggestions for activities that need to be done. Thus, I make the following suggestions:

We need to apply some of the techniques and methodology of the BBS and those presented by our Finnish and New Zealand colleagues to the CBCs. One obvious factor would be to provide a detailed habitat map of the count circle so that actual habitats can be accurately determined. This would greatly assist in analysis of bird populations relevant to the habitats in which they are found.

I believe that we should not give up on the observers in the CBCs. They are intelligent people interested in birds and a proper training program would be extremely useful and effective. In contrast to the beliefs of Drs. Hickey (1981) and Bock and Root (1981), I believe that we can and must do something to improve thd data collected. We need an effort to balance the fun of the CBCs with the great need to make the data more relevant for bird population studies. Ornithologists may need to spend more time with the "amateurs" to educate and cajole more reliable data from them. The conservation policy making possibilities may be a valuable tool in this regard.

We need to view the bird watchers, listers, and participants in CBCs and BBSs as an untapped resource. It may seem like an enormous job but if one researcher-scientist can influence one or two CBC compilers or one field worker in a count circle, or one birder who consistently visits a region, we could have a valuable effect on the census data. The better the field data the more accurately the analysis will reflect the actual avian populations.

[1] Natural History Museum, Los Angeles County, 900 Exposition Blvd., Los Angeles, California 90007.

Studies in Avian Biology No. 6:61, 1981.

INTRODUCTORY REMARKS: ESTIMATING RELATIVE ABUNDANCE (PART II)

ROBERT D. OHMART,[1] CHAIRMAN

The papers contained in this session are heterogeneous with respect to avian surveys. Some deal with the use of taped avian calls to quantify avian numbers; one paper presents the technique of estimating relative densities with the use of mist nets; and still another paper presents the method of recording species presence or absence data through the use of an atlas.

The only unifying component in these papers is that we are dealing with avian studies. This does not reduce the importance of these papers, but simply illustrates the breadth and complexity of the topics that are related to surveying avian populations.

[1] Department of Zoology, Arizona State University, Tempe, Arizona 85281.

Studies in Avian Biology No. 6:62–67, 1981.

SURVEYING BIRDS WITH MIST NETS

JAMES R. KARR[1]

ABSTRACT.—The role of mist nets in the arsenal of field equipment of ornithologists has increased in recent years, especially in studies of banded birds and migration. Advantages of mist nets include reduced variability in data when compared to procedures which depend on extensive experience with sight and sound identification, or judgment in compilation and analysis of field data. Finally, mist nets can provide a wealth of data in a relatively short period.

Use of mist nets has increased in studies of between-season and among-year patterns in avian populations during the past decade. In addition, mist nets can be used to detect differential use of subtly different habitat types in small geographic areas (e.g., forest on dry exposed ridge vs. in moist, sheltered valley). Other uses of nets include studies of avian use of treefall gaps vs. nearby undisturbed forest. My research group recently initiated a study of reproductive success in forest islands in central Illinois. Mist nets operated after the nesting season, but before migration, yield data on the relative abundances of young and adult birds over a range of island sizes.

These and other recent uses of mist nets in avian studies illustrate the kinds of quantitative data amenable to statistical analysis which can be obtained through judicious use of mist nets. However, use of nets, like any counting procedure, must be approached with caution. Potential difficulties of interpretation of mist net results are discussed.

Mist nets were introduced into the United States after World War II. Without doubt, they have revolutionized the study of birds in their natural habitats, especially efforts requiring banding and monitoring of individuals and in collection of specimens for museums. However, their use as a counting procedure has not been great.

Although nets are not a panacea to solve all counting problems, their judicious use can provide considerable insights into dynamics of avian populations and communities. In the present paper I summarize advantages and disadvantages of mist nets in bird count work, and demonstrate inferences that can be developed from use of mist nets. In addition, I illustrate the potential for misuse of results from mist net studies.

ADVANTAGES AND DISADVANTAGES OF NETS

Like any count procedure, nets have both advantages and disadvantages. Important points to keep in mind in selection of a count procedure are objective of the study and type of data required to meet that objective. No single procedure is suitable for all habitats and research objectives.

The primary advantage of nets is that they do not require familiarity with songs and field marks of birds of an area. In addition, nets can be used to standardize sample size (or sample effort). Capture rates for a community or for individual species can be expressed in number of captures per net hour or net day. However, it is important to standardize sample times because capture rates vary with time of day (see below). Another standard sample involves use of a specific number of captures. Some researchers prefer to exclude all recaptures while others include both original and recaptures in a standard sample. I prefer the latter as a measure of bird activity, independent of individuals involved.

In addition, use of a standard time or number of individuals sampled avoids the problem of extrapolating abundance information to standard areas such as 100 acres (40 ha). This advantage is especially attractive in studies of patches of habitat which are small or vary significantly in size.

Like any survey procedure that involves handling of organisms, those organisms tend to avoid the nets after a few days (when most individuals have been captured). Capture rates become vanishingly small after the third day of net operation (when only permanent residents are present). During periods with considerable day-to-day turnover in individuals (e.g., migration periods), capture rates are less likely to decline so obviously; indeed, they may markedly increase as waves of migrants pass through an area.

In my experience a 100-capture sample is the best compromise between the number of captures and time, especially in view of the fact that capture rates decline throughout the period of netting. For a more detailed discussion of these and other related subjects, see Karr (1979, 1980).

Finally, with mist nets it is possible to accumulate quantitatively reliable information in a relatively short period. Repeatability of results of mist netting for several major population and

[1] Department of Ecology, Ethology, and Evolution, 102A Vivarium, 606 E. Healey, University of Illinois, Champaign, Illinois 61820.

community metrics is also an advantage (Karr 1980).

Reduction of observer-related biases and rapid accumulation of standard samples are major advantages. The latter is especially important in areas with high species richness or where many of the assumptions of more classical procedures (territoriality, monogamy, etc.) are not met.

However, like all procedures, netting is not without problems. The value of mist nets is minimized in inclement weather, especially during periods of rain and/or, in more open habitats, high winds. Other general problems are placement of nets in the field, and variability in tension and spacing of shelf strings when nets are erected. Finally, it is important to avoid inferences which are not supported by the data base provided by mist-netting; e.g., avoid reference to densities.

In some cases it seems clear that mist nets are useful, perhaps only to provide a mechanism to band many of the birds in a local population for studies of population dynamics and behavior. In other cases, mist nets are useful for provision of population and community data. However, great care should be used in the application of mist nets to counting problems.

FIELD METHODS

A wide variety of factors should be considered in establishment of field protocols involving use of mist nets in collection of ecological data. Perhaps the most significant net attribute to determine capture rates is mesh size. Generally, larger meshes capture larger birds (Heimerdinger and Leberman 1966). I find 36 mm mesh nets to be the most effective for the widest range of birds found in most terrestrial habitats. Birds less than 5 to 8 g are not efficiently captured, nor are birds above about 100 g (Karr 1979). In addition to standardizing mesh size, it is essential that net length and height (thus net area) be standardized for production of the most valuable comparative data.

Habitat type also is an important factor affecting efficiency of mist nets and thus capture rates. In grassland, for example, nets tend to be more visible than in forest. In second-growth areas a greater proportion of the fauna (species and individuals) is likely to be captured in ground level nets. Thus, it is unwise to compare capture rates between forest and second growth as if they are equally good indexes of the relative densities in the two habitats (Karr 1979).

Deployment of nets also can play a major role in the success of a data collection effort. Generally, I place nets in a loop covering about 2 ha. Others use a line of nets placed end to end (Terborgh and Faaborg 1973, Wright 1979). The major deployment problem to be avoided is concentration of nets in too small an area. Capture rates are depressed under such conditions.

One major disadvantage of mist nets is the difficulty and expense of using them in canopy and subcanopy levels. A number of systems have been developed to operate nets in the canopy and subcanopy (Greenlaw and Swinebroad 1967, Humprey et al. 1968, Whitaker 1972, Sappington and Jackson 1973, Karr 1979). They vary with respect to net mobility, cost, and ease of use.

Time of day also is an important determinant of net capture rates. Bird activity varies throughout the day and nets change in their visibility with shifts in sun angle. Generally bird activity peaks in morning and in late afternoon or early evening. Precise time of the peaks varies with day length, length of twilight and, in some areas, evening and morning temperatures. Habitat type and season may also be important variables (see below).

Frequency with which nets are checked must also vary with several factors. I find that nets left for up to 90 minutes are not a problem in forests or other habitats where birds are not exposed to direct sunlight. In more open areas, nets must be checked as frequently as every 15 to 30 minutes to avoid major mortality. Another disadvantage of long intervals between net runs is that birds get increasingly tangled and difficult to extract with increase in time between runs. For most of my research, I check nets every 60 minutes. Careful, regular checking of nets typically keeps mortality below one percent.

Like any other procedure, an important factor in use of mist nets is regular and reliable measurement and recording of data. Recording of time of day that nets are opened and closed to the nearest 5 minutes is essential to allow precise determination of the duration (in hours or days) of sampling. Additionally, records of age, sex, moult, and other natural history information can be valuable in interpreting many aspects of population and community characteristics. Since vegetation affects avian use of an area, careful measurement of vegetation attributes must be made.

DATA ANALYSIS AND INTERPRETATION

In this section I select a few examples of the kinds of biological insights that can be derived from studies employing mist nets as a census technique. These are meant to be illustrative examples only.

No counting procedures can be expected to provide absolute density information without total disturbance of the organisms under study. Capture rates from mist net studies are simply relative population estimates; their use should be tempered with that realization in mind. Relative density data provided by mist-net sampling can be used to evaluate changes in populations in both space and time.

TEMPORAL PATTERN

During the past two years I have collected 100-bird mist net samples from each of four forest study plots in central Panama. Each site has been sampled four times—twice in wet and twice in dry seasons. The four study plots are located in an area of about 2 km^2 on the Pipeline Road, Parque Nacional La Soberania. One area, Limbo Hunt Club, has been described in more

TABLE 1

MINIMUM AND MEAN HUMIDITIES FOR DRY AND WET SEASONS IN THE UNDERGROWTH OF THREE FOREST STUDY PLOTS IN CENTRAL PANAMA. MEANS DETERMINED FROM HOURLY OBSERVATIONS ON ONE DAY AT SEVEN SITES IN EACH STUDY PLOT

	Minimum		Mean (n)	
Study plot	Dry	Wet	Dry	Wet
Ridgetop	50	94	67 (72)	98 (80)
Limbo	62	95	74 (28)[a]	97 (80)
Valley	79	98	81 (54)	100 (88)

[a] Observations every three hours.

detail elsewhere (Karr 1971, 1976b). Local rainfall averages about 2600 mm per year with a rainy season that begins in April and extends, in most years, into December. The late December to early April period is typically dry, with less than 100 mm of rain per month. Temperatures average 27°C throughout the year. Macroclimate does not vary among areas. However, because of topographic differences among the areas, microclimate of the undergrowth does vary (Table 1). Only three areas will be described here. At one extreme is a flat elevated area with no stream within the study plot. Temperatures are higher and relative humidities lower (as low as 50%) at midday during the dry season at this site (Ridgetop). An intermediate area (Limbo Hunt Club) is a mixture of an upland and stream-edge environment with a maximum topographic relief of about 8 meters. Dry-season temperature and humidity profiles in the undergrowth are intermediate between those at Ridgetop and the next study plot (Valley). The Valley plot is adjacent to the Limbo Hunt Club plot but is a deep stream valley with a narrow basin and high, steep slopes. Dry-season temperature and humidity profiles at midday in the undergrowth are different from the two earlier sites. Humidities rarely go below 60% at any location. Daily rains during the wet season modify the microclimate

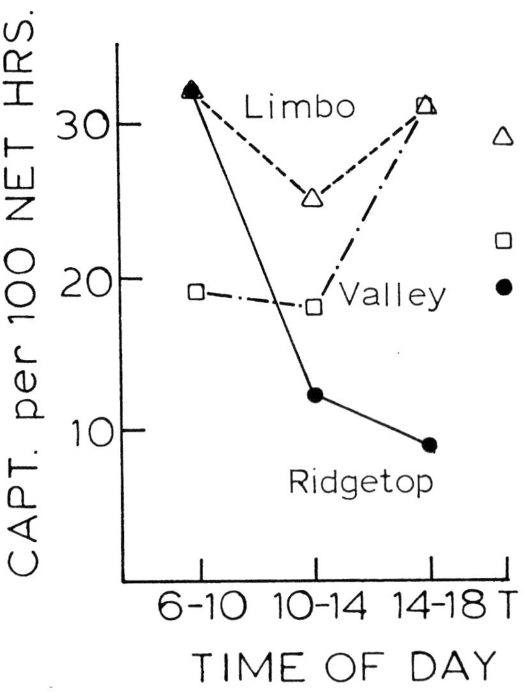

FIGURE 1. Capture rates in birds per 100 net hours for first three days of netting on three study areas in Central Panama, dry season 1979. T = average per day.

of the sites to the extent that temperature and humidity profiles are nearly identical. A fourth site not described here, intermediate between Ridgetop and Limbo, is somewhat more similar to Limbo. When bird capture data for each season are combined from all four study sites, capture rates are strikingly consistent within seasons and between years (Table 2). However, they are lower in wet season than in dry season.

TABLE 2

SUMMARY OF MIST NET RESULTS (COMPOSITE OF 4 SAMPLE PLOTS) IN LOWLAND WET FOREST IN CENTRAL PANAMA

	1979		1980	
	Dry season	Wet season	Dry season	Wet season
Number of				
Captures	415	364	416	370
Net hours	1809	2102	1851	2096
Captures per 100 net hours	22.9	17.3	22.5	16.7

TABLE 3

DRY-SEASON CAPTURE RATES AND SPECIES RICHNESS FOR SEVERAL LOWLAND WET FOREST SITES IN CENTRAL PANAMA. NOTE THAT 1979 WAS A RELATIVELY DRY DRY SEASON WHILE RAINFALL DURING 1980 DRY SEASON WAS ABOVE NORMAL

	Study site		
Year	Ridgetop	Limbo	Valley
Capture rates (capt./100 net hours)			
1979	19.1	29.3	21.9
1980	26.7	30.5	15.5
Species richness (spp./100 captures)			
1979	32	38	33
1980	36	30	33

TABLE 4
DRY SEASON CAPTURE RATES FOR BIRDS IN UNDERGROWTH MIST NETS IN "UPLAND" AND "LOWLAND" LOCATIONS AT LIMBO HUNT CLUB, PARQUE NACIONAL LA SOBERANIA, PANAMA

Year	Season	Region	Number of Net hours	Number of Captures	χ^2	Captures per 100 net hours
1979	Dry	Lowland	149	59	9.18[a]	39.6
		Upland	209	46		22.0
	Wet	Lowland	247	50	0.04	20.2
		Upland	256	50		19.5
1980	Dry	Lowland	161	59	3.79	36.6
		Upland	187	47		25.1
	Wet	Lowland	244	49	1.47	20.1
		Upland	208	53		25.5

[a] Sign at $P < 0.05$ at $\chi^2 = 3.84$.

TABLE 5
NUMBER OF WET AND DRY SEASON CAPTURES OF *PIPROMORPHA OLEAGINEA* AT 5 STUDY AREAS IN CENTRAL PANAMA—1979–80

	Number of captures: Dry season	Number of captures: Wet season	Probability level[a]
Study plot			
Limbo Hunt Club	20	7	0.0006
Ridgetop	4	16	0.0018
Roadside	17	9	0.0019
Valley	28	4	0.0001
Barro Colorado Island (1980 only)	11	0	0.0017

[a] Based on binomial probability test with expected values (p and q) based on number of net hours at each site in wet and dry seasons.

Diurnal variation in capture rates is also obvious (Fig. 1). Thus, mist nets can be used to examine year-to-year, seasonal, and diurnal variation in avian activity.

SPATIAL PATTERN

Diurnal variation in capture rates depicted in Figure 1 clearly shows that bird use varies among microhabitats. Bird activity is highest in early morning at the driest site (Ridgetop). The wettest site (Valley) shows essentially the reverse pattern. The area is avoided in early morning due to steep slopes and increased energy demands of considerable vertical movement by foraging birds. However, as the day progresses, birds may be attracted to this area because insects remain active in the sheltered, more humid environment. There is also some evidence that permanent water supply attracts birds to the valley during late afternoon. Several species that are rarely seen below canopy levels have been captured during dry season at the Valley site as they descended to drink and bathe from the stream. The intermediate site (Limbo) illustrates an intermediate pattern.

Although regional activity of birds is consistent for each season between years (Table 2), capture rates vary independently among the four study plots. Patterns of variation are clearly explicable in light of microhabitat patterns. For example, note that bird activity was higher at the Ridgetop site in the wetter 1980 dry season than in the drier 1979 dry season (Table 3). The reverse was true in the wetter Valley site. The intermediate Limbo site had essentially the same capture rates in the two years.

But even this consistent intermediate pattern masks habitat selection dynamics within the study plot. The Limbo Hunt Club plot consists of a flat area along a stream, a small rise of about 8 m, and a flat upland. Capture rates vary between upland and lowland areas between years. Capture rates were significantly higher in lowland than in upland during the especially dry dry season of 1979 (Table 4). Only three months later, during wet season, capture rates were identical (Table 4). During the following relatively wet dry season of 1980, there were no differences in the capture rates between the two regions of the 2 ha study plot. Again, recall that these capture rates are for areas only 8 m apart in elevation. Overall, birds tend to shift their spatial use of habitat in response to temporal variation in microclimatic conditions.

Species richness of the samples also varies among the three sites (Table 3). At Limbo Hunt Club, number of species in 100-bird samples (range 30–38) is inversely correlated with March (sample month) rainfall (Karr 1980). At Ridgetop, species richness (like capture rates) was higher in the wetter year while the reverse was true at Limbo. The Valley samples contained the same number of species in both years.

In addition to general community metrics, it is possible to discern variation in abundance of species with mist-net sampling. For example, Karr et al. (in press) found that 10 species were captured at significantly ($P < 0.05$) different rates in samples collected nearly a decade apart. In some cases, shifts were due to habitat changes (especially increased area of treefall gaps) while in others seasonal movements of temperate zone migrants were important.

Variation between seasons in capture rates of birds also has been documented (Karr et al. in press). Insectivore-nectarivores were more common during the dry season, a period of peak

TABLE 6

COMPARISON OF OBSERVED AND PREDICTED NUMBERS OF RESIDENT SPECIES IN FOREST ISLANDS IN EAST-CENTRAL ILLINOIS (SUMMER 1979; BLAKE AND KARR, UNPUB. DATA).

		Point census		Mist net samples	
Study area	Size (ha)	Observed	Predicted	Observed	Predicted
Rittenhouse Woods	1.8	8	9.6	7	7.9
Brownfield Woods	24.3	20	20.2	10	18.4
Hart Woods	28.0	31	31.3	21	28.2
Funks Grove B	65.0	26	28.2	17	21.4
Allerton Park	600.0	34	34.1	29	38.2

flowering activity, while species that follow army ants were more common in the wet season.

The frugivorous Ochre-bellied Flycatcher (*Pipramorpha oleaginea*) showed distinct seasonal changes in capture rates in the past two years. On four of five study plots the species is captured at higher rates in dry than in wet season (Table 5). The only exception is Ridgetop, which apparently is not suitable for high densities of many species during dry season (see above).

In another study in central Panama, Schemske and Brokaw (in press) tested the hypothesis that bird communities of treefall gaps in tropical forest differ from those of adjacent intact forest. They found that species richness was greater in gaps but capture rates were nearly identical in gaps and intact forest. Of 31 species with sample sizes adequate for analysis, 5 were caught more regularly in gaps (3 species) or in intact forest (2 species).

An ongoing study of forest islands as habitat for birds is providing additional examples of the use of mist nets in studying birds (Blake and Karr, unpubl. data). When the number of species observed in a forest tract is plotted against number of 15 minute observation periods (or number of mist net captures), an asymptote is approached. To achieve some degree of statistical confidence in the estimated number of species in a particular forest tract, these values can be fitted to an equation for a hyberbola:

$$\frac{1}{S} = K\left(\frac{1}{T}\right) + \frac{1}{S_{\max}}$$

where S = number of species observed (captured)

T = number of 20-minute observation periods (or number of species captured)

and $S_{\max}$ = the predicted maximum number of species present.

This equation is equivalent to the Lineweaver-Burk equation, a transformation of the Michaelis-Menten equation, the rate equation for one-substrate, enzyme-catalyzed reactions (Lehninger 1975). The predicted maximum should be equal to or greater than the number actually observed. Deviation from the predicted number could be due to sampling error (not all species were seen or captured) or to an ecological deficiency of the forest tract, such that the predicted maximum will not actually be achieved.

Number of species predicted from netting data was significantly correlated with both number of species predicted from census data and number of species actually observed (Table 6) during census periods ($r = 0.940$ and 0.956, respectively; $P < 0.01$ in both cases). Predicted number of species derived from netting data is based on lower observed (captured) totals than is the predicted number derived from census periods. The advantage of this model is that it may allow predictions of species richness with small samples, or conversely, an indication of the minimum number of census periods needed to achieve a given level of precision.

An additional component of the island study is use of mist nets after the fledging period to try to determine adult-juvenile ratios before dispersal. It is our hope that these can be used as a measure of reproductive success. If such success varies within and between species over a range of island sizes, it may be possible to more clearly understand effects of island size on small population survival.

In this section I have tried to provide examples of the uses of mist nets in the study of avian ecology. Other population and community metrics have been studied by other researchers. These include guild signatures (Karr 1980), turnover dynamics (Terborgh and Faaborg 1973, Karr in press), migrant abundances (Karr 1976a, 1979, Terborgh and Faaborg 1980a), community saturation (Terborgh and Faaborg 1980b), and habitat selection (Willson and Moriarty 1976). Clearly, use of nets as a procedure for counting birds is limited only by the originality of future generations of ornithologists.

FINAL COMMENTS

Lest the reader conclude that mist nets provide a foolproof way to count birds and study population and community dynamics, I hasten to add a few words of caution. It is essential that researchers recognize deficiencies and biases of mist nets as a sampling tool.

It is not, in my opinion, appropriate to use mist nets (or any other procedure) for "fishing expeditions." Researchers should have precise study objectives (hypotheses) in mind and select sampling protocols to yield highest quality data for those objectives.

The most important caution is that mist net capture rates do not constitute a measure of absolute density. Thus, scientists should avoid discussing them as if they were. As an example, I earlier concluded that the Ochre-bellied Flycatcher is less regularly captured in wet than dry season on four of my study plots. Several possible explanations of that pattern could be advanced. Perhaps the abundance of the species does indeed shift. If so, where do the birds go? Perhaps the birds are less active in wet season (due to nesting, more uniformly dispersed food supplies, or water) and thus are captured less frequently. It is not possible at this time to clearly distinguish among these and other alternatives. The fact that the dry site is out of phase with the others suggests that it is indeed a shift in spatial use of habitat. This example reinforces the principle that there is no substitute for knowledge of the organisms under study. Mist nets can be used to provide reliable quantitative data but interpretation of results requires caution. They can be even more valuable if backed up with other quantitative and qualitative observations about the birds under study.

Readers should note that rate of capture of birds in ground level mist nets will be in proportion to the percent of activity by species in the sample space (within 3 m of the ground). Comparisons of abundances of species with different activity levels in the ground layer obviously should be avoided. Similarly, two species should not be directly compared if their activities (flight distance, flight frequencies, etc.) are not similar.

Mist nets combined with color banding of birds also allow more precise determination of movement patterns and the extent of overlap of territories. Red-capped Manakins (*Pipra mentalis*), for example, are typically the most regularly captured species in the undergrowth of Central American lowland forest. Capture rates of the species vary significantly from month to month (Karr et al. in press) with extensive almost day-to-day turnover in individuals (Karr 1971, unpubl. data). Careful examination of capture-recapture rates provides considerable insights into the populations dynamics and movements of this species relative to others. But they must be interpreted carefully.

In summary, mist nets are valuable tools for bird counting. They should be used more extensively, but with precision, if the greatest possible yield of scientific conclusions is to be forthcoming. They can be especially useful when familiarity with birds in the field is minimal, when many shy and/or secretive species are present, and in areas (or seasons) where birds rarely sing. Further, they are valuable where assumptions of other procedures (territorial systems, monogamy) are not met.

ACKNOWLEDGMENTS

I have used nets for sampling tropical and temperate avifaunas for over a decade. Far too many individuals to mention individually here have aided me in those studies; to all of them I say thanks. The Smithsonian Tropical Research Institute, National Geographic Society, EARTHWATCH, U.S. Fish and Wildlife Service (Grant #INT-14-16-0009-78-092), and the University of Illinois Research Board have provided funds for these efforts.

Studies in Avian Biology No. 6:68–75, 1981.

PLAYBACK RECORDINGS AS A SPECIAL AVIAN CENSUSING TECHNIQUE

R. ROY JOHNSON,[1] BRYAN T. BROWN,[2] LOIS T. HAIGHT,[3] AND JAMES M. SIMPSON[4]

ABSTRACT.—The literature on the efficiency of tape recorders in detecting relative or absolute population densities of birds is reviewed. This review and field investigations by the authors showed at least 51 species that are responsive to the use of playback recordings as a census technique. This represents only a small percentage of the species which could be censused by this method. Examples of detailed field techniques and their results are outlined for several species that present particular censusing problems, such as nocturnal species and others, e.g., Lucy's Warbler, found in the rich riparian avifauna of the southwestern United States. The highest breeding density of Screech Owls in North America, reported herein, was discovered by this technique.

Tape-recorded bird calls have been used as an avian censusing technique for more than two decades. Although this technique has been most commonly used to census species which present special problems, our findings demonstrate that it can and should be in much wider usage. This applies to a large percentage of the species that are standardly censused by conventional methods. In fact, both the literature and our research demonstrate that thorough, accurate breeding censuses are rarely conducted without the aid of this research tool.

Counts of spontaneous auditory signs, or call counts, of birds have been used since Stoddard (1931) reported the use of counts of male vocalizations as a technique for measuring the relative abundance of quail. Call-count censusing differs from tape-recorded censusing in that a call count does not use a playback recording to elicit responses but instead counts the number of spontaneous calls. This technique was originally used as a population index of game birds along established survey routes. Call-count sampling has been used in censusing the wild Turkey (*Meleagris gallopavo*) (Overton and Davis 1969), Ruffed Grouse (*Bonasa umbellus*) (Hungerford 1953), quail (Phasianidae) (Smith and Gallizioli 1965), Chukar (Williams 1961), Ring-necked Pheasant (*Phasianus colchicus*) (Kimball 1949), American Woodcock (*Philohela minor*), and doves (Columbidae) (McClure 1939, Foote et al. 1958).

Although it has been noted that vocally imitated calls could be used to detect the presence of nocturnal species such as owls (Bent 1938, Miller and Miller 1951, Foster 1965), the use of tape-recorded calls as a count technique was not experimented with until the late 1950's. Earlier recordings depended on reel-to-reel tapes and heavy, expensive recorders and equipment. It was not until the 1960's that compact, lightweight cassette recorders with tape counters were developed to provide reference points for specific recordings. Thus, only during the last decade has the technology developed to economically allow investigators to routinely carry playback recording equipment to the field for use in a count.

Bohl (1956) originally carried game farm Chukars to the field in portable pens, anticipating that their calling would stimulate answering calls from Chukars in the wild. However, he discovered it was more practical to locate and census the wild Chukars by the use of tape recordings of their calls. This technique soon found acceptance in censusing several game species that had previously been censused by the call-count method (Levy et al. 1966, Stirling and Bendell 1966). Tape-recording census techniques are now widely used for both game and nongame species. The special applications of playback recordings in avian censusing include the following: (1) nocturnal species (e.g., owls); (2) species found in inaccessible habitats or habitats with limited visibility, such as marshes, tropical forests, or dense brushland (e.g., rails or Plain Chachalacas); (3) species which may occur in low densities (e.g., Yellow-billed Cuckoos in southwestern riparian habitats); (4) species occurring in high densities (e.g., Lucy's Warblers, Screech Owls, and Elf Owls, in southern Arizona velvet mesquite, *Prosopis velutina*, bosques [woodland] and saguaro, *Cereus giganteus*, forests); (5) species with large territories (e.g., most raptors); and (6) species with soft or barely audible calls (e.g., Least Bittern, *Ixobrychus exilis*, and Black-tailed Gnatcatchers, *Polioptila melanura*).

[1] Cooperative National Park Resources Studies Unit, School of Renewable Natural Resources, and the Department of Ecology and Evolutionary Biology, University of Arizona, Tucson, Arizona 85721.

[2] Cooperative National Park Resources Studies Unit and School of Renewable Natural Resources, University of Arizona, Tucson, Arizona 85721.

[3] Cooperative National Park Resources Studies Unit, University of Arizona, Tucson, Arizona 85721.

[4] Department of Biology, Museum of Northern Arizona, Flagstaff, Arizona 86001.

Although we speak here primarily in terms of the use and application of playback recordings, it is recognized that mimicry and noisemakers, both generalized and specific, are applicable for avian censusing under varying circumstances. This includes vocal mimicry and manufactured squeakers and calls, with duck (Anatidae), crow (*Corvus* spp.), and Turkey calls being the most commonly used.

EXAMPLES OF SPECIAL FIELD TECHNIQUES

The playback recording techniques used to census birds may differ, depending on the species and its behavior patterns, response traits, and territory size. Techniques share a high degree of similarity within families, for example within the families Rallidae and Strigidae. Table 1 illustrates the wide application of these techniques, with examples from the literature and the authors' field investigations. As there is not space here to give detailed techniques for each species listed under authors' experience, some of the more cogent points regarding the specific techniques for two sample species are outlined below. The following remarks refer to territorial breeding birds unless noted otherwise.

As playback recording census techniques will vary for different species under different conditions, we shall generalize regarding techniques we have found to be the most accurate from our experience during the past 11 years. For most species, one person with a portable tape recorder can conduct an accurate census. Standard censusing rules should be followed. Comparable times should be kept for various plots. However, when conducting breeding bird censuses, times are academic if one is not recording all of the birds present. That is, keeping a predetermined schedule is secondary to an accurate census. Stops should generally be made every 25 to 100 m, depending on the thickness of cover and avian densities. Taped refrains of bird calls should be separated by intervals of time comparable to those for the species. Intervals should be even longer, if necessary, to allow the observer to listen for answering songs between the recorded songs. At regular intervals it is often desirable to set the tape recorder on the ground or in a tree or shrub and allow it to play while walking around it at a distance of 20 to 30 m and listen for answering calls. The volume for playing the tape recorder can be determined by trial and error. Ideally, an optimum volume would be used where birds answer from the greatest possible distance while still allowing the observer to hear responses above the noise of the tape recorder. With species which exhibit a supernormal response (Tinbergen 1960) where "the louder the tape, the better," the tape recorder can be turned to nearly full volume and set down at frequent intervals while the observer walks away from the recorder to listen for responses.

NOCTURNAL BIRDS

Nocturnal birds are commonly not included in figures of population densities. This is because of both the mechanical difficulty of censusing the birds and the lack of the necessary technical knowledge of most investigators. Caprimulgids, for example, seem to be vocal enough so that playback recordings are rarely needed. This is especially true for species which seem to call regularly on successive nights, e.g., Poor-wills (*Phalaenoptilus nuttallii*), Whip-poor-wills (*Caprimulgus vociferus*), and Ridgeway's Whip-poor-wills (*C. ridgwayi*). Common and Lesser Nighthawks (*Chordeiles minor, C. acutipennis*) are often seen in flight during crepuscular hours and are very vocal during the breeding season.

The most often overlooked avian species in censusing are the owls. This is due in part to the fact that owls often are vocal only if censused with the use of playback recordings or vocal imitations (Table 2). In addition, few researchers enjoy tramping around in the dark in rattlesnake country. Table 2 shows a Screech Owl breeding density of 9 pairs/4 ha (90 pairs/40 ha). This high density is from extremely productive riparian woodland habitat in southern Arizona. This is the habitat type in which Screech Owls have been reported to occur "100 yards" apart (Phillips et al. 1964) or less (Miller and Miller 1951). On the Salt River in central Arizona, a cottonwood, *Populus fremontii*—mesquite grove measuring 125 × 50 m was censused with playback recordings on March 24, 1972 (J. M. Simpson and I. Simpson), and again on March 30, 1972 (J. M. Simpson and R. R. Johnson). Nine Screech Owls were found on five territories. A nearby grove, 100 × 60 m, contained eight owls on four territories. Adjacent mesquite bosques had paired Screech Owls spaced as closely as 50 m apart. This is the highest reported breeding density for Screech Owls in North America.

These extreme population densities in mesquite bosques of central and southern Arizona necessitate special efforts to obtain accurate counts. The dense riparian woodland presents a visibility problem, even near full moon in the spring when vocal activity is at its peak. We commonly use two persons, one near the tape recorder, which is played at high volume, and another to walk in a circle around the recorder and count owls. The second observer usually stays approximately 50 to 100 m from the recorder. The reason for this unorthodox method

TABLE 1
SPECIES RESPONSIVE TO PLAYBACK RECORDINGS AS A COUNTING TECHNIQUE

Family Species	Type of census[a]	Source
Podicipedidae		
Pied-billed Grebe (*Podilymbus podiceps*)	b	authors' experience[b]
Ardeidae		
Least Bittern (*Ixobrychus exilis*)	b	authors' experience[b]
Accipitridae		
Goshawk (*Accipiter gentilis*)	b	James Mosher, pers. comm.[c]
Cooper's Hawk (*A. cooperii*)	b	James Mosher, pers. comm.[c]
Red-tailed Hawk (*Buteo jamaicensis*)	b	James Mosher, pers. comm.[c]
Red-shouldered Hawk (*B. lineatus*)	b	James Mosher, pers. comm.[c]
Broad-winged Hawk (*B. platypterus*)	b	James Mosher, pers. comm.[c]
Cracidae		
Plain Chachalaca (*Ortalis vetula*)	b	Marion 1974b
Tetraonidae		
Blue Grouse (*Dendragapus obscurus*)	b	Stirling & Bendell 1966
Spruce Grouse (*Canachites canadensis*)	b	MacDonald 1968
White-tailed Ptarmigan (*Lagopus leucurus*)	b	Braun et al. 1973
Phasianidae		
Masked Bobwhite (*Colinus virginianus ridgwayi*)	b	Gallizioli 1964, Tomlinson 1972, Brown & Ellis 1977
Scaled Quail (*Callipepla squamata*)	b	Levy et al. 1966
Gambel's Quail (*Lophortyx gambelii*)	b	Levy et al. 1966
Montezuma Quail (*Cyrtonyx montezumae*)	b	Levy et al. 1966
Chukar (*Alectoris chukar*)	b	Bohl 1956, Oelklaus 1976, Mudd et al. 1979 & 1980
Aramidae		
Limpkin (*Aramus guarauna*)	b	Marion et al. 1981
Rallidae		
King Rail (*Rallus elegans*)	b	Maehr 1980
California Clapper Rail (*R. longirostris obsoletus*)	b, w	Gill 1979
Sonora Clapper Rail (*R. l. rhizophorae*)	b, w	Tomlinson & Todd 1973
Yuma Clapper Rail (*R. l. yumanensis*)	b, w	Tomlinson & Todd 1973, Smith 1974, Jurek 1975, Gould 1975, Todd 1976
Virginia Rail (*Rallus limicola*)	b, w	Glahn 1974, Todd 1976, Griese et al. 1980, authors' experience
Sora (*Porzana carolina*)	b, w	Glahn 1974, Todd 1976, Griese et al. 1980, authors' experience
Black Rail (*Laterallus jamaicensis*)	b, w	Jurek 1975, Todd 1976, Manolis 1977 & 1978, Repking & Ohmart 1977
Cuculidae		
Yellow-billed Cuckoo (*Coccyzus americanus*)	b	Gaines 1974a & 1974b, Gaines 1977 (unpubl. rpt., Calif. Game and Fish Dept., Sacramento)
Strigidae		
Screech Owl (*Otus asio*)	b	Heintzelman 1979, authors' experience
Great Horned Owl (*Bubo virginianus*)	b	Springer 1978; Fuller & Mosher, this volume
Pygmy Owl (*Glaucidium gnoma*)	b	authors' experience
Ferruginous Owl (*G. brasilianum*)	b	authors' experience

TABLE 1. (CONTINUED)

Family / Species	Type of census[a]	Source
Elf Owl (*Micrathene whitneyi*)	b	Cardiff 1978, Gould 1979, authors' experience
Barred Owl (*Strix varia*)	b	Fuller & Mosher, this volume
Spotted Owl (*Strix occidentalis*)	b	Gould 1974, 1977, 1979; Whisler and Horn 1977, Forsman et al. 1977, Marcot 1978 (unpubl. rpt., Six Rivers Natl. Forest, Calif.), Cardiff 1978, Delamore 1979, Garcia 1979
Boreal Owl (*Aegolius funereus*)	b	authors' experience
Saw-whet Owl (*A. acadicus*)	b, w	authors' experience
Trogonidae		
Elegant (Coppery-tailed) Trogon (*Trogon elegans*)	b	Taylor 1978 & 1979 (unpubl. rpt., Coronado Natl. Forest, Ariz.)
Tyrannidae		
Least Flycatcher (*Empidonax minimus*)	b	Oech & Oech 1960
Eastern Wood Pewee (*Contopus virens*)	b	Oech & Oech 1960
Troglodytidae		
Long-billed Marsh Wren (*Cistothorus palustris*)	b	authors' experience
Mimidae		
LeConte's Thrasher (*Toxostoma lecontei*)	b	Rea 1977
Turdidae		
Veery (*Catharus fuscescens*)	b	Oech & Oech 1960
Vireonidae		
Bell's Vireo (*Vireo bellii*)	b	authors' experience
Gray Vireo (*V. vicinior*)	b	Barlow & Johnson 1969
Red-eyed Vireo (*V. olivaceus*)	b	Oech & Oech 1960
Parulidae		
Lucy's Warbler (*Vermivora luciae*)	b	authors' experience
Yellow Warbler (*Dendroica petechia*)	b	authors' experience
Ovenbird (*Seiurus aurocapillus*)	b	Oech & Oech 1960
Mourning Warbler (*Oporornis philadelphia*)	b	Oech & Oech 1960
Common (Salt Marsh) Yellowthroat (*Geothlypis trichas*)	b	Foster 1977a, authors' experience
Yellow-breasted Chat (*Icteria virens*)	b	authors' experience
Icteridae		
Hooded Oriole (*Icterus cucullatus*)	b	authors' experience
Thraupidae		
Summer Tanager (*Piranga rubra*)	b	authors' experience
Fringillidae		
Cardinal (*Cardinalis cardinalis*)	b	Dow 1970
Blue Grosbeak (*Guiraca caerulea*)	b	authors' experience

[a] b = breeding, w = wintering.
[b] Authors' personal experience. The length of this paper prohibits a detailed explanation of each species and the technique used in the authors' investigations. Researchers should use this as a guide to species which are responsive, while devising their own methods based on available literature.
[c] Appalachian Environmental Laboratory, University of Maryland, Frostburg.

is that so many owls may answer at once that one stationary person cannot ascertain how many individuals are responding. In optimum habitat Screech Owls are commonly spaced at intervals of approximately 50 m. Thus, a single person may be listening to a dozen or more pairs (with both birds calling) from any given spot. It is impossible for a single observer to accurately census Screech Owls in this situation. On occasions we have used two tape recorders and three or four observers to help determine densities. In addition, territorial boundaries may be determined by persons with playing recorders moving toward each other. Excited territorial

TABLE 2
SAMPLE OWL BREEDING CENSUSES FOR COMPARING VARIOUS TECHNIQUES

	Transect[a]	Transect size (m)	Species	Censusing technique	No. of territories identified
Transect 1[b]	a.	400 × 100	Screech Owl	Playback recording	9
	b.	400 × 100	Screech Owl	Call count	0
Transect 2[b]	a.	200 × 100	Screech Owl	Playback recording	5
	b.	200 × 100	Screech Owl	Call count	0
	c.	200 × 100	Screech Owl	Vocal mimicry	4
Transect 3[c]	a.	400 × 50	Screech Owl	Call count	0
	a.	400 × 50	Screech Owl	Playback recording	4
Transect 4[d]	a.	400 × 100	Screech Owl	Playback recording	4
	a.	400 × 100	Screech Owl	Vocal mimicry	3[e]
	a.	400 × 100	Screech Owl	Call count	0
	a.	400 × 100	Screech Owl	Playback recording	3[e]
Transect 5[d]	a.	200 × 100	Elf Owl	Call count	0
	b.	200 × 100	Elf Owl	Playback recording	5

[a] When transect "a" appears more than once under a given number, the same transect was censused repeatedly by the different methods.
[b] Cottonwood-mesquite (*Populus fremontii—Prosopis velutina*) habitat, Blue Point Cottonwoods on the Salt River near Phoenix, Arizona, April 27 (Transect 1) and May 25 (Transect 2), 1980. One census was conducted on this and all the following transects.
[c] Riparian mesquite woodland habitat, Saguaro National Monument (East) near Tucson, Arizona, June 28, 1980.
[d] Palo verde-saguaro-mesquite (*Cercidium* spp.—*Cereus giganteus—Prosopis velutina*) habitat, Saguaro National Monument (East) near Tucson, Arizona, July 1, 1980. The Screech Owl transects were censused at 20:15, 20:30, 20:40 and 21:00 respectively.
[e] Adult calling on two territories, "stationary" young on one of those territories and scattered young on a third territory.

owls commonly follow the recorder through their own territories until they meet, often engaging in fierce disputes.

SPECIES OCCURRING IN HIGH DENSITIES

The use of tape recorders in censusing riparian and other wetland habitats is particularly important as bird densities in these habitats may exceed 1000 nesting pairs/40 ha in the Southwest (Carothers and Johnson 1971). Lucy's Warbler is an example of a species which may occur in high breeding densities in optimum habitat. Densities of approximately 12.5 pairs/ha (500 pairs/40 ha) reported by Russell and Johnson (1973) were verified by the use of playback recordings on the lower Verde and Salt rivers near Phoenix, Arizona, in mature mesquite bosque. In this situation, territorial disputes can be triggered by playback recordings between two or more males approximately every 20 to 35 m, with territories averaging approximately 30 m in diameter. These populations seem difficult to explain but similar densities for Yellow Warblers were found on plots of mature cottonwood forest by Carothers and Johnson (1971) in the Verde Valley of central Arizona. In willow-alder (*Salix-Alnus* spp.) brush near Old Crow, Alaska, Irving (1960) found a territorial pair of Yellow Warblers every "50 feet" for a distance of 100 m along the Yukon River system.

Lucy's Warbler exhibits a supernormal response to playback recordings and under optimum conditions in dense mesquite bosques occurs in even greater densities than Screech Owls. Since censusing is done during the day for the former, it is not so difficult to follow the movements of individual birds. We found that by walking 50 m between stops under these conditions one might move through segments of two to four territories between stops. In addition, birds on adjacent territories were drawn into territorial squabbles by the tape recorder, thereby adding to the confusion. Our eventual technique used two persons. One would play the tape recorder at high volume, stopping at 30 m intervals, while the second would range out from the recorder at distances of approximately 20 to 40 m. Both observers counted responding birds in conjunction with one another.

DISCUSSION OF CENSUS TECHNIQUES

REASON FOR AND VALIDITY OF METHODS

A playback recording census can increase the total number of species counted or increase the total numbers of birds seen or heard for a given species in comparison to a conventional census. This is especially true for species with low song activity (Robbins 1978a). Dow (1970) reported an increase of 37 to 160 percent in the numbers of Cardinals responding to playback recordings over the use of call count sampling without recordings. However, he noted that spontaneous singing during this period decreased as the season progressed from April to July, while the responses to recordings remained about the same.

Oech and Oech (1960) documented an increase of 40 to 370 percent in the response of six common passerines in Minnesota with playback recordings compared to the use of call-count sampling. Stirling and Bendell (1966) obtained population densities for male Blue Grouse on Vancouver Island that were essentially the same for both a conventional call count and a playback recording census, but the taped census was four times as fast as the conventional search. Glahn (1974) found that a playback recording census of Sora and Virginia Rail populations increased the number of territories located by 71 percent in comparison with a standard nest survey.

Problems with the Playback Technique

In rare instances, the use of playback recordings has not been advantageous. Robbins (1978a) stated that indiscriminate use of tape recordings on repeated visits during the breeding season can bias the results as birds may alter their habits or their territorial boundaries if they believe a competing member of the same species is holding territory nearby. Although this may be true in some instances, we would like to see better evidence for this hypothesis. Preliminary information from some of our Screech Owl studies suggests that if censused too often some individuals and/or species may become less responsive. In an Oregon study area, four Spotted Owl pairs were located through extensive ground searches and the pairs subsequently located through radio telemetry (Forsman et al. 1977). A simultaneous census with playback recordings located only three of the four pairs. Springer (1978) reported similar findings in Ohio populations of Great Horned Owls. With the Ohio owls, the relative effectiveness of a foot survey (95 and 95.8%) was higher than with a playback recording survey (72 and 87.5%) in locating 66 pairs of owls. Marion (1974) found that between 44 and 59% of Plain Chachalacas being censused by playback recordings were not responding to the recordings. A correction factor of 2.0 was then necessary to adequately estimate Plain Chachalaca numbers. It is not known if experimental manipulation in one of these three cases (Spotted Owl) affected the playback censusing. Conventional census techniques, however, in these rare instances prove no better. The only remaining technique is a series of methodical, time-consuming visual searches.

Conversely, the use of playback recordings can result in supernormal responses in some species resulting in exaggerated territorial activity. The use of playback recordings can also attract some individuals away from their territories, as in Elegant (Coppery-tailed) Trogons (Taylor 1978 and 1979—unpubl. rpt., Coronado Natl. Forest, Ariz.), and result in inflated population density estimates. Further work is needed to identify the reliability of the technique with many species. For example, little is known concerning differences in density estimates that may occur when censusing with alarm calls versus territorial song.

Factors Influencing Response Rates

Factors which may influence the rates of response to tape-recorded songs include wind, rain, time of day, temperature, seasonality, species response traits, lunar cycles, and disturbance by man or predators. These factors are often interrelated in various combinations. Climatic factors that act as probable influences of vocal response are wind and rain (Dow 1970, Oelklaus 1976, Whisler and Horn 1977). Observed responses decrease as winds increase due to the facts that both bird activity and the observer's hearing ability are decreased. Likewise, rain seems to inhibit singing and can make listening impossible (Dow 1970). Dow also found that very dense fog appears to have had no influence on male Cardinal responses, although spontaneous singing may have been suppressed. Stirling and Bendell (1966) noted the positive response of male Blue Grouse to recordings of a female grouse whinny call even during a snowstorm.

Oech and Oech (1960) and Robbins (1978a) suggested that birds will respond to tape recordings at times when they would otherwise remain silent. We have found this to be especially true in the fall with Screech Owls at Saguaro National Monument in Arizona and to a lesser extent with Elf Owls. Importantly some species may be censused by this method late in the breeding season or later when spontaneous vocalizations normally decline. Spring and summer response for the Cardinal reaches a peak in the early morning, drops to a low level in the mid-afternoon and rises slightly in the evening (Dow 1970). Optimum censusing time for most species seems to be at sunrise. This response pattern of a morning peak, a mid-afternoon low, followed by a rise in evening activity which is not quite as high as the morning peak is similar to the pattern of spontaneous singing noted in many passerine birds by Van Tyne and Berger (1959). However, the daily response patterns of some passerines and nonpasserines will differ slightly. A recording of the female Blue Grouse whinny call is effective at all times of day in eliciting a response from territorial males (Stirling and Bendell 1966). The optimum response of Gambel's and Scaled Quail to tape recordings was in the morning and evening, while Montezuma Quail answer tape recordings equally well throughout the day (Levy et al. 1966). However, male Gambel's Quail with their mates present will not respond to tape recordings of a female call.

The use of tape recordings is normally confined to a census of males in the breeding season, although Tomlinson and Todd (1973) reported the usefulness of recordings in censusing breeding and wintering populations of both male and female Yuma Clapper Rails, for a minimum population index. Both male and female Elf Owls (Cardiff 1978) and Screech Owls are known to respond to taped recordings. Owls are apparently affected by lunar cycles with the optimum response to taped recordings occurring on nights with a bright, waxing moon (Johnson et al. 1979). The daily peaks of response by owls to tape recordings generally coincide with their crepuscular activity patterns, in that just after dark and just before sunrise are the best times to elicit responses. In censusing Chukars with tape recordings, Oelklaus (1976) found that disturbance by avian predators, coyotes, or man was followed by a short period in which the Chukars' response rate to tape recordings was not consistent, necessitating a lapse in the census.

Research Considerations

Several comments should be made regarding the use of recorders, results and special techniques. The use of playback recordings, as with any other tool, is only as accurate as the person carrying out the census. Secondly, the source and quality of the recorded vocalizations used are important. Due to racial and dialectual variations, the use of local recordings is best. In the absence of local recordings, the use of good commercial recordings such as Cornell's *Field Guide to Bird Songs* and *A Field Guide to Western Bird Songs* are adequate. Even then the observer will have varying degrees of success, depending on the species used. One of the drawbacks of most commercial or mass produced recordings is the limited repertoire for a given species. For example, the above recordings have two basic sounds for the Saw-whet Owl, where we know of at least five clearly distinguishable sounds (Johnson et al. 1979). The most successful tape we have used in several dozen attempts with Screech Owls during the past 11 years is of a caged female. Background noise in this taped recording, including parakeets and street noise, is ignored by Screech Owls in the mesquite bosques as they come to investigate the recording itself which varies from soft, coy, and coaxing to loud, strident, and aggressive. On at least one occasion using this tape, a Screech Owl landed on the ground a few feet from the recorder and stomped demandingly up to this "territorial invader," strutting like a miniature turkey gobbler. Although background noise in the above Screech Owl recording apparently had no ill effect, excessive background noise may have a negative influence on the response of some species, as Mudd et al. (1979) suggested occurs in Chukars.

Our findings suggest that only a fraction of the responsive species have been censused by playback recordings. Many non-colonial territorial birds which rely on song as a territorial proclamation should be censused or have supplemental data gathered about them using this method. The literature on bird song, while not directly related to the application of tape-recorded censuses, can provide important background information regarding the response of certain species to this method. There is enough auditory response information available, for example, regarding crows and gulls (*Larus* spp.) (Frings et al. 1958), to suggest that they could be censused using recordings of the appropriate attractant call. Sonograms and observations on the structure and function of many bird songs exist and are too numerous to present.

For some species which are more easily and accurately censused by tape recorded methods, the taped call technique could be useful in obtaining an annual or periodic index to a species' abundance. This possibility was mentioned by Tomlinson and Todd (1973) for Yuma Clapper Rails along the lower Colorado River, although other methods such as habitat inventories were noted as being less expensive. Taped censusing may also be useful to supplement the information provided through a conventional breeding bird census. Owls, commonly not included in even some of the better population studies, could be standardly censused with these techniques.

Management and Conservation Implications

Playback recordings are widely used by both professional and amateur ornithologists as they are so effective in calling out secretive birds. However, a certain amount of controversy exists over the use of tape recordings. The Coronado National Forest in southeast Arizona has banned the use of playback recordings in locating the Elegant (Coppery-tailed) Trogon, as it is thought their use causes nest failures (R. Taylor 1978 and 1979, unpubl. rpt., Coronado Natl. Forest, Ariz.) and other problems (Glinski 1976). The male trogons are highly responsive to taped recordings and may be lured long distances from the nest. As males share in the incubation responsibilities, any male that leaves the eggs to fight a tape recording of another male trogon risks loosing the year's clutch.

Several rare species mentioned previously are threatened or endangered (sensu U.S. Fish and Wildlife Service). Often these species are found in inaccessible habitats or in low densities. The use of playback recordings for responsive endangered species would be an important management tool. For example, the Washington State Game Department is presently experimenting with the use of playback recordings to census breeding Peregrine Falcons (*Falco peregrinus*) (Frederick Dobler, pers. comm.). If successful, this would provide information for the management of the species.

CONCLUSIONS

The use of playback recordings is probably the most overlooked major technique for avian censusing. It has been used primarily for augmenting conventional censuses in searching for problematic species (e.g., nocturnal birds) or birds in dense vegetation (e.g., marsh and woodland). Demonstrated advantages of the playback technique include: (1) increased numbers of individuals detected, both per single census and per census area over time; (2) time efficient sampling; and (3) detection of the aforementioned problematic species.

The lack of use of this technique apparently is attributable to the need for additional equipment such as recorders and tapes, and the necessity of mastering the mechanics of the playback technique. Neither of these is an overly difficult problem. The necessary equipment can be purchased for less than a pair of medium-priced binoculars and is roughly comparable in weight to carrying a gun for collecting. The real problem, as with any other technique, is in becoming proficient through practice. The J. T. Emlen (1971) technique, for example, is excellent when used by John Emlen. However, the results of many of the "modified Emlen techniques" are questionable, to say the least. Thus, as with other techniques, the results are only as good as the user.

The idea that birds come from territories some distance away to the recorder, thereby biasing the results, is generally false. Where are the birds that should be defending their territories as these distant intruding males "flock to be counted" by the census taker? It is true that the use of the recorder can increase the number of birds counted. We maintain that this increase is the result of heightened responses from otherwise silent or hidden birds, those that would have been overlooked by a more conventional census.

If standardization is a goal for a particular censusing program, it is argued that unless everyone uses it, no one should use it. Granted, that philosophy does help to "standardize" the system, a system already fraught with variables, e.g., weather, time of day and year, moon cycles and behavioral peculiarities of specific species or individual birds. In our judgment, however, standardization is rarely an acceptable substitute for using every available tool to increase censusing accuracy and efficiency. The best all-around tool that we have used thus far is the playback technique.

ACKNOWLEDGMENTS

We would like to thank Frederick Dobler, Richard Greer, Ronald Jurak, Wayne Marion, James Mosher, and Richard Todd for assistance in locating references. Wayne Marion and Robert Ohmart reviewed an early draft of this paper.

Studies in Avian Biology No. 6:76–80, 1981.

INDIRECT ESTIMATES OF ABUNDANCE OF BIRDS

EVELYN L. BULL[1]

ABSTRACT.—Relative density can answer many questions regarding bird populations, precluding the necessity of taking the additional time and expense to determine absolute density. Indirect indices of relative density include auditory signals, feeding and dusting sites, and track, roost, fecal, and nest counts. Their use assumes these indicators are related to the population size.

Population censusing is a methodological problem in ecology, particularly for inconspicuous, mobile animals which are distributed over a large area (Marten 1972). Knowledge of the number in a population is a prerequisite for effective wildlife resource management (Andrewartha 1971). All management techniques require information about the total or relative number so the effects of the management effort can be assessed. This information is essential to establish the relationship of a population to its habitat and to determine the changes in the population level over time (Talbot 1970).

Two abundance estimates include numbers per unit area (absolute density) and population densities relative to one another (relative density). Some studies such as sustained-yield harvesting, and those relating density to behavior, reproduction, survival, emigration, and immigration require estimates of absolute density, while studies concerning habitat use, rate of increase, dispersal, and population reaction to manipulation can be considered using relative density (Caughley 1977). The estimate selected depends on study purpose, species, season, and habitat.

ABSOLUTE DENSITY

Counting birds yields an accurate measure of absolute density if area size is known. Disadvantages of using absolute density include (1) high cost, (2) disturbance to the population, (3) difficulty in counting secretive or nocturnal species, and (4) high time requirements in counting birds with large home ranges (Scattergood 1954). In populations too large to count, a sample of the entire population is taken. Sampling is less costly and disturbing, but representative samples are sometimes difficult to obtain.

RELATIVE DENSITY

Relative density is an index to population size and is used when the actual size of a population is not needed. Indices are derived based on the assumption that the sample represents a constant but unknown proportion of the population. When appropriate conversion factors are used, such indices can be converted to absolute density. This relationship can be variable, however, and estimates of that variability difficult to determine.

There are two types of indices: direct and indirect. Direct indices are derived from counts of birds in a sampling scheme. A direct population estimate is not obtained. Examples include migrating birds seen flying between observer and the moon per hour and birds seen per kilometer of transect walked. Accuracy depends on standard census conditions (e.g., weather, time) and the observer's skill.

Counts of variables associated with animal presence produce indirect indices. Examples include tracks, calls, and fecal counts. Advantages are: (1) less skilled observers are required, (2) it is easier to develop standard techniques, (3) results are affected less by viewing conditions, (4) less disturbance is created, and (5) effectiveness in studying secretive species is increased (Caughley 1977). There is a time lag between creation of the sign and its observation. Signs, then, provide indices to density over time and are not indices of current density (Caughley 1977). As a result, there may be a less direct relationship to density than in the case of direct counts. I present a variety of indirect indices in the remainder of this paper.

AUDITORY SIGNALS

The use of auditory signals (e.g., singing or calling males) to estimate bird numbers is suited to territorial, noncolonial species. This technique assumes each singing male is mated and that the count reflects the number of breeding pairs in the area (Davis 1965). Because these assumptions are not always valid, this index is best suited to make comparisons in bird use between areas or for the same area between years. Correction factors used in deriving density estimates may be gained by simultaneously counting birds for comparison to the call counts.

Variables to be considered include: weather, effects of terrain and vegetation on sound, time of day, season, territoriality, breeding condition, duplication of counts, and variation between ob-

[1] USDA Forest Service, Pacific Northwest Forest and Range Experiment Station, La Grande, Oregon 97850.

servers (Davis 1965). Variability should be minimized and identified, and counts adjusted to increase reliability.

Using auditory signals to obtain population estimates has several advantages. Birds are disturbed less. Relatively few observers can cover a large area and obtain a large number of observations rather inexpensively even when the density of a species is low (Gates and Smith 1972). Some species like the Greater Prairie Chicken (*Tympanuchus cupido*) (Silvy and Robel 1967) and White-tailed Ptarmigan (*Lagopus leucurus*) (Braun et al. 1976) respond readily to recorded calls making it easier to locate the bird.

Auditory signals that are hard to hear yield poor population estimates, and distances from the observer to the bird are difficult to calculate. This hinders density calculations. Counts must be made under standard conditions, because wind, temperature, season, time of day, and precipitation influence singing and some affect audibility. Territorial diurnal birds sing most profusely shortly before and after sunrise throughout the breeding season. The intensity depends on the stage of the breeding period. Kendeigh (1944) and Lack (1937) reported unmated males singing more prior to acquiring a mate.

Auditory signals include: crowing, hoots, songs, calls, and drumming. I present examples of surveys using these signals below.

At least since 1939 (McClure 1939), Mourning Dove (*Zenaida macroura*) call counts have been used to provide an index to spring population levels (Sayre et al. 1978). Calling doves are counted for 4 minutes at 20 plots at 1.6 km (1 mile) intervals along predetermined routes (Cohen et al. 1960). The radius of the audible plot varies with terrain and vegetation. Foote et al. (1958) reported a plot radius of 0.6 km (⅜ mile). Nonrandom selection of route was biased toward higher populations compared to those selected through stratified random sampling (Foote et al. 1958).

This survey is used primarily as an index of population density showing shifts in the density rather than estimating absolute density (Gates and Smith 1972). However, Petraborg et al. (1953) presented a quantitative approach to estimating density, and Lowe (1956) found 1.74 breeding pairs for each calling bird heard.

This species is a good example of a large scale survey with a large sample size and standard techniques and analytic procedures. Recent findings, however, suggest that factors influencing the calling activity may affect validity. Sayre et al. (1978) identified pairing as the primary influence on cooing rates. Unmated males called more than mated males. Laperriere and Haugen (1972) reported higher levels of cooing when more than one bird called and different levels of calling activity associated with weather conditions.

Calls have been used to survey male Ring-necked Pheasants (*Phasianus colchicus*). Kozicky (1952) recorded crowing along a 10 mile circular route and found temperature, cloud cover, and presence of dew to have little effect on crowing behavior, although wind greater than 8 mph and time past sunrise decreased the counts.

Brown et al. (1978) counted calling Scaled Quail (*Callipepla squamata*) along 28-km routes stopping at 1.6 km intervals for 3 minutes. He recorded the number of single calls, the number of birds calling, and calculated a call-count index from the mean of the highest count.

Robel et al. (1969) investigated factors affecting the number of Bobwhite (*Colinus virginianus*) whistles heard. They found time of year, time of day, wind velocity, temperature, and relative humidity influenced calling rate.

Woodcocks (*Philohela minor*) occupy singing grounds in the spring where their vocalizations can be reported by stopping at points along routes for a predetermined amount of time (Stroll 1980).

Bergerud and Mercer (1966) found a becking census of Willow Ptarmigan (*Lagopus lagopus alleni*) was the only technique other than aerial surveillance suitable for extensive surveys. They assumed a 0.8 km (½ mile) audibility radius and calculated cocks per square mile. They cautioned that phenological, meteorological, and density factors affect the calling behavior.

Drumming counts have been used to determine population trends and relative abundance of Ruffed Grouse (*Bonasa umbellus*) (Petraborg et al. 1953, Dorney et al. 1958, Stroll 1980). Observers record number of drummings heard at plots 1.6 km (1 mile) apart along a 16.1–24.1 km (10 to 15 mile) route. These counts start before sunrise and last several hours. Routes are duplicated at least three times, and the highest count is used. Gullion (1966) concluded that the amount of drumming heard on roadside counts may have little relation to the actual size of the breeding population. The frequency and persistence of the drumming activity varies from bird to bird and between years and is influenced by date of snowmelt, temperature, and precipitation. He thinks the total census of drumming activity centers provides the best population estimate but cautions that any estimate based on drumming behavior has the problems of an unknown sex ratio and an unknown number of nondrumming males. Woodpeckers and sapsuckers (Picidae) drum on trees as part of their territorial display. These auditory signals can be used in the same manner as grouse drumming, dove call, and pheasant crowing counts to calculate relative abundance. Sapsucker drummings are distinct from those of woodpeckers, but distinguishing among the woodpecker species is difficult (Rushmore 1973, Jackman 1974). Rushmore (1973) surveyed forests for Yellow-bellied Sapsuckers (*Sphyrapicus varius*) by imitating their drumming and feeding sounds along transects.

Owls, being nocturnal and secretive, are difficult to locate. Yet, many species can be surveyed aurally because they respond readily to taped calls (Forsman 1976).

Nest Count

Relative densities are often calculated from nest tallies (Robbins 1978a). This technique works best with species that have conspicuous nests, colonial nesters, and species nesting in open country (Kendeigh 1944, Oetting and Dixon 1975). Problems include: finding enough nests, individuals within species do not nest at the same time (Lowe 1956), some pairs have more than one brood (Kendeigh 1944), and nests are often abandoned or unsuccessful. So, all individuals that actually breed during the season may not be counted.

Eagle populations are frequently assessed by locating nest sites. Grier (1974, 1977) surveyed nesting Bald Eagles (*Haliaeetus leucocephalus*) with aerial searches as nest trees were conspicuous from the air. McGahan (1968) and Boeker (1971) located nests of Golden Eagles (*Aquila chrysaetos.*). Supernumerary nests could complicate calculating a nesting density as McGahan (1968) found an average of 1.8 supernumerary nests per pair ranging from a few meters to 6.1 km (3.8 miles) apart.

Nest counts of colonial nesting birds serve as an index showing changes in the population over time or between areas. Great Blue Herons (*Ardea herodias*) (Williams 1957), Rooks (*Corvus frugilegus*) (Birkhead 1974), Cliff Swallows (*Petrochelidon pyrrhonota*) (Emlen 1941), and albatrosses (*Diomedea immutabilis, D. nigripes*) (Rice and Kenyon 1962) have been surveyed in this manner. Nettleship (1976) presented methods of surveying seabirds of Arctic and eastern Canada. Most techniques included counting the number of nests for species like gulls (*Larus* spp.), cormorants (*Phalacrocorax* spp.), and terns (*Sterna* spp.); although counting burrows in the ground or rock scree worked for the Leach's Storm-Petrel (*Oceanodroma leucorhoa*) and Common Puffin (*Fratercula arctica*).

Nest counts are well suited to cavity nesters, because the number of cavities is correlated with the number of cavity nesters (Haartman 1957, Beebe 1974, Jackman 1974, Balda 1975b, Thomas et al. 1979a). During the breeding season, nest sites of cavity dwellers are readily located by checking available cavities. Activity around a cavity can be used to verify it as a nest. During the postbreeding season and up to one year later, active nest sites of excavators can be identified by the presence, abundance, and coloration of the chips on the ground and the coloration of the wood at the cavity entrance. After one year, aging cavities is difficult.

Evaluating old cavities can give an index to cavity nester populations if several factors are considered. Cavities of most woodpecker species can be distinguished on the basis of size. Because it is difficult to distinguish among the sapsuckers (*Sphyrapicus* spp.) and species in the genus *Picoides,* they should be combined. Only a certain percentage of apparent cavities actually are completed so a correction factor can be developed by climbing some of the trees to verify completed cavities. Between 40 and 60 percent of nests excavated by Pileated Woodpeckers (*Dryocopus pileatus*) in northeastern Oregon are not completed the same year (Bull, unpubl. data). Jackson (1977) reported inflated estimates of Red-cockaded Woodpecker (*Picoides borealis*) abundance based on the presence of cavity trees.

Roost Count

Roost sites can be used as an abundance index assuming that number of roosts correlates with abundance. This technique applies particularly to species with conspicuous roost sites or communal roosts.

Conspicuous roosts are left by a variety of species. White-tailed Ptarmigan roost in burrows below the surface of the snow (Braun et al. 1976). Some grouse burrow in the snow to roost (Glover 1948). Barwick et al. (1970) reported roosts of young wild Turkey (*Meleagris gallopavo*) broods to consist of a depression in the grass with numerous poult droppings in the vicinity. Generally adult wild Turkeys roost in trees and are identified by the droppings underneath (Boeker and Scott 1969). Many cavity nesters roost in holes (Jackman 1974), so number of roost cavities indicates abundance.

Communal roosting species lend themselves well to censusing. Sometimes thousands of birds congregate at roosts which are used repeatedly. Emlen (1938, 1940) estimated the mid-winter distribution of Common Crows (*Corvus brachyrhynchos*) in New York and California by locating all the roosts. Stewart (1973) calculated numbers of Starlings (*Sturnus vulgaris*), Red-winged Blackbirds (*Agelaius phoeniceus*), Common Grackles (*Quiscalus quiscula*), and Brown-headed Cowbirds (*Molothrus ater*) using a roost based on the amount of fecal material present.

Owl (Strigiformes) roosts are often scattered with pellets of undigested food which has been regurgitated (Welty 1975). These pellets help locate birds, indicate number of birds using a particular roost, and suggest species distribution in an area. I will discuss this topic in greater detail in the section on feeding sites.

Track Count

Track counts of birds, particularly gallinaceous birds (Overton 1971), serve as indices assuming the number of tracks correlates with the number of birds. Tracks, however, are remote from the animal in time; and it is difficult to determine how many birds made the tracks (Stearns 1970). Soil and weather conditions affect the visibility of tracks. During the winter, some types of snow conditions make track counts readily visible. In areas of high concentrations, track counts on kymograph paper may be feasible (Seber 1973).

Flocks of wild Turkeys (Glover 1948, Eaton et al. 1970) and White-tailed Ptarmigan (Braun et al. 1976) have been located by following tracks. The tracks indicate approximate numbers in the flocks.

Buller (1967) and Guthery (1975) demonstrated that the larger race of the Sandhill Crane (*Grus canadensis*) can be distinguished from the lesser race by footprint measurements. The track measurements taken in the central flyway revealed the race composition during different periods of the fall migration.

Fecal Count

The fecal-count method of estimating relative numbers of animals is used most extensively with ungulates (Neff 1968) but has been used occasionally in the study of gallinaceous birds (McClure 1945). Although presence or absence of feces is commonly recorded, the number of droppings can be counted on plots along transects. The number of droppings correlates with the number of birds present if the durability of scat and resistance to weather, diet, behavior, and time are considered. McClure (1945) thought this method was best for determining relative pheasant populations, particularly during the winter when the fecal pellets freeze and remain intact longer than in the summer when they are readily attacked by insects or dissolved by rain. Diet also affects longevity of pellets. For these reasons, it is difficult to calculate absolute density from pellet counts even though the defecation rate is known.

The presence of fecal material is used to identify roost sites of species including the Turkey (Hoffman 1968, Boeker and Scott 1969, Barwick et al. 1970, Eaton et al. 1970) and Pileated Woodpecker (Bull 1978). A Pileated Woodpecker had been using a roost cavity for at least four months based on the accumulation of fecal material at the base of the tree (Bull 1978). Gullion (1966) determined the active status of Ruffed Grouse drumming logs by the fresh accumulation of droppings at the drumming stages. Stewart (1973) calculated that 2,294,713 blackbirds and Starlings used a roost by determing the amount of fecal material deposited overnight by individual birds and by all birds in the congregation.

Czekala and Lasley (1977) developed a technique to determine the sex of birds by comparing the amount of sex steroid excreted in the fecal material. They found females had higher values of estrogen/testosterone than males. Some of the species they investigated included American Kestrel (*Falco sparverius*), Bobwhite Quail, and Rock Dove (*Columba livia*).

Feeding Sites

Indices of abundance can be derived for species that leave conspicuous evidence of feeding activities. Examples include scratch marks, plucking posts, excavations, and pellets.

Brown (1976a) used the scratch marks of Montezuma Quail (*Cyrtonyx montezumae*) to obtain population estimates. He searched along transects for scratching. It was difficult to distinguish other soil disturbances from scratch marks particularly in areas covered by litter and where there was rodent activity.

Wild Turkeys leave evidence of their feeding activities particularly during the winter by scratching through the snow to obtain food (e.g., beechnuts, old acorns, dried fruit) (Glover 1948, Eaton et al. 1970).

Goshawks (*Accipiter gentilis*) perch on logs or stumps to pluck prey (Reynolds 1978). These "plucking posts" can be used as an index to relative abundance over large areas.

Hook-billed Kites (*Chondrohierax uncinatus*) extract and feed on snails at an extracting perch and leave characteristically damaged snail shells. The presence and density of these sites are an index to occurrence and relative abundance of this kite in the area within the previous year or two (S. A. Temple, pers. commun.).

Some woodpeckers, particularly Pileated and sapsuckers, leave characteristic excavations at foraging sites. Pileated Woodpeckers make large rectangular holes into the interior of dead and down woody material (Jackman 1974). Sapsuckers drill rows of small holes in the bark of living trees (Rushmore 1973). The relative abundance of each species can be determined by observing feeding sites along transects in different areas. It is difficult, however, to accurately age the excavations other than distinguishing between a current year's activity and older feedings.

At least eight families of birds form pellets (Rea 1973). Pellets have been analyzed to identify prey remains for food habitat studies of some birds, particularly owls (Forsman 1976). Hawks generally eat less roughage than owls and digest bones more thoroughly making their pellets less useful in determining food habits (Welty 1975). Birds regurgitate distinct pellets, so pellet numbers indicate presence or absence of a species and numbers using a particular area.

Dusting Sites

Dustbathing behavior is characteristic of several taxa of birds. Originally this behavior was thought to aid in removing parasites, but recent work indicates that dustbathing reduces excess lipid substances on the bird's plumage and prevents the feathers from becoming matted (Borchelt and Duncan 1974, Borchelt 1975).

Bailey and Rinell (1968) reported wild Turkeys dusting frequently in the summer in the dry residue of rotten logs, anthills, and newly tilled soil. Bobwhite Quail dust regularly (Borchelt 1975). Dusting leaves telltale soil disturbances which can be used as an abundance index. Because dustbathing regulates the amount of lipid substance on the feathers, however, the amount of dusting may be in response to environmental factors (e.g., diet), so caution should be used in comparison between populations.

Studies in Avian Biology No. 6:81–85, 1981.

USE OF PLAYBACK RECORDINGS IN SAMPLING ELUSIVE OR SECRETIVE BIRDS

WAYNE R. MARION, TIMOTHY E. O'MEARA, AND DAVID S. MAEHR[1]

ABSTRACT.—The playback technique has been used successfully to detect the presence of many bird species and to study the behavior of others. Few reports, mostly involving the family Rallidae, were reviewed in which the playback of avian vocalizations was used in population estimation. Our application of this technique to field studies of three species of rails, Limpkins, and Plain Chachalacas was successful. The detection of rails and Plain Chachalacas without stimulation from playback recordings would be nearly impossible, as these birds are otherwise difficult to find and observe. Playback techniques augment studies of Limpkins, making them more readily observed in the field. Since Limpkins seem to be attracted by recorded calls and typically approach the observer, it is necessary to record a bird's location at initial response to avoid a bias in density estimation.

Rails and Plain Chachalacas are best sampled in the early morning hours and in the case of the latter species, during the breeding season. Limpkins could be sampled successfully with recordings either morning or evening. Repeatability of the recordings in eliciting calling responses from Plain Chachalacas was found to be excellent over short time intervals.

Researchers have not taken full advantage of the characteristics of birds which could assist in detecting or censusing inconspicuous species. The playback technique has great potential for use with highly vocal avian species that are otherwise difficult to detect in the field.

Tape recordings of avian vocalizations have been employed in a variety of studies over the last two decades to elicit behavioral or vocal responses from birds. Playback techniques have proven advantageous in eliciting responses from otherwise detectable birds, thereby increasing the number of observations possible per unit of time and increasing sampling efficiency in the field. Playback recordings of bird sounds have been used in several ways, including (1) detection of secretive, elusive or nocturnal birds by scientists and birdwatchers (e.g., Christmas Bird Counts), (2) investigation of avian social behavior and territoriality, and (3) estimation of population size. The increase in use of natural recordings and a thorough review of available equipment were presented by Bradley (1977).

Tape recorded sounds have been used to aid in the detection of birds by Bohl (1956) for Chukar (*Alectoris chukar*), Stirling and Bendell (1966) for Blue Grouse (*Dendragapus obscurus*), MacDonald (1968) for Spruce Grouse (*Canachites canadensis*), Braun et al. (1973) for White-tailed Ptarmigan (*Lagopus leucurus*), Marion (1974a, b) for Plain Chachalacas (*Ortalis vetula mccalli*), and Glahn (1974) for Virginia Rails (*Rallus limicola*). Levy et al. (1966) used recorded female calls to detect male Gambel's Quail (*Lophortyx gambelii*), Harlequin Quail (*Cyrtonyx montezumae*), and Scaled Quail (*Callipepla squamata*). In addition, recorded sounds have been used successfully in trapping Greater Prairie Chickens (*Tympanuchus cupido*) on their booming grounds (Silvy and Robel 1967) and female Sharp-tailed Grouse (*Pedioecetes phasianellus*) during the brood-rearing period (Artmann 1971). With a variety of songbirds, recordings have been used to investigate intraspecific recognition of territorial boundaries (Weeden and Falls 1959, Falls 1969, S. T. Emlen 1971, Krebs 1971, Verner and Milligan 1971, Goldman 1973, Kroodsma 1976a, Patterson and Petrinovich 1978, and Robbins 1978a) and to stimulate reproductive development in females (Kroodsma 1976b).

Glinski (1976) cited a number of potential problems associated with the repeated use of tape-recorded territorial calls employed by birdwatchers and recommended that these uses be minimized when they involve certain rare species. He was concerned about unnecessary disturbance of birds at their nest sites and the possible consequences of extra energy drain on birds responding to taped vocalizations.

Playback recordings apparently have not been employed extensively to estimate populations of elusive or secretive birds. The technique has been used at sunrise and sunset during the breeding season to study the presence, distribution, and density of rails in Colorado (Glahn 1974, Griese et al. 1980) and Kansas (Baird 1974). We could not find previous evidence of the use of playback recordings with Limpkins (*Aramus guarauna*) and only two references (Marion 1974a, b) to its use in studying Plain Chachalacas. Despite a paucity of published information on their use, playback recordings show good potential for use with highly vocal species of birds. In this paper, we use King Rails (*Rallus elegans*), Virginia Rails, Soras (*Porzana carolina*), Limpkins, and Plain Chachalacas to assess the value of this technique.

[1] School of Forest Resources and Conservation, University of Florida, Gainesville, Florida 32611.

TABLE 1
DENSITY ESTIMATES FOR THREE SPECIES OF RAILS DETECTED USING THE PLAYBACK RECORDING TECHNIQUE ON THE TWO WETLANDS IN NORTHERN FLORIDA, 1979–80

	Density (birds/ha)			
Wetlands	Spring	Summer	Fall	Winter
King Rails				
A	0.6	0.6	0.6	0.0
B	0.0	1.1	0.6	0.0
Virginia Rails				
A	2.1	0.0	0.0	0.6
B	0.6	0.6	0.0	0.3
Soras				
A	5.3	0.9	1.2	0.6
B	3.5	0.3	0.9	0.0

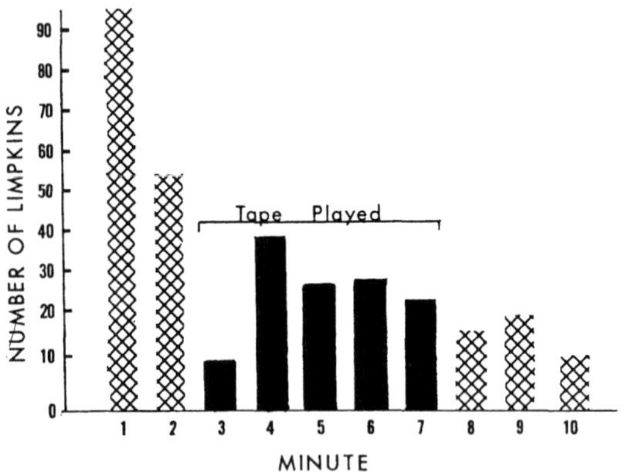

FIGURE 1. Number of Limpkins detected by 1-minute intervals during 10-minute sampling periods, Lake Ocklawaha, Florida, June 1980. Recorded vocalizations were played during minutes 3–7.

METHODS

RAILS

Rails were studied as part of a larger investigation of bird communities in habitats created by phosphate mining in Hamilton County, Florida. Since the three species of rails present on these areas are secretive inhabitants of freshwater marshes or other densely vegetated wetlands, they are often difficult to census. To facilitate detection of these species, recordings of their calls were played at six sample points along the periphery of densely vegetated, diked impoundments. Recordings were played for approximately one minute every morning for 10 days each season from spring 1979 through winter 1980. Distances to rails that responded were estimated and recorded. Ramsey and Scott's (1979) variable circular plot method was used to estimate rail densities in these impoundments.

LIMPKINS

Playback recordings of vocalizations were used to determine their effectiveness in detecting and estimating populations of Limpkins on Lake Ocklawaha, Marion and Putnam counties, Florida. Lake Ocklawaha is a shallow, man-made reservoir flooded in the early 1970's in preparation for its inclusion in the unfinished Cross-Florida Barge Canal. Limpkins were sampled on two areas of the lake, both of which were located along the former channel of the Ocklawaha River. The "downriver" area was characterized by an open water channel (approximately 30 m wide) bordered by a 200 m wide band of flooded dead timber with extensive mats of floating water hyacinth (*Eichhornia crassipes*) and scattered emergent vegetation. The "upriver" area was similar to the downriver area with the exceptions that standing dead timber was virtually absent and a higher proportion of emergent vegetation (e.g., *Cicuta maculata*) was present.

Permanent sample points were marked along the former river channel (13 upriver and 12 downriver). These were sampled mornings and evenings by boat for five consecutive days in early June 1980. Each sample included 10 minutes of observation while we were anchored at a point. Samples consisted of two minutes prior to playing recorded Limpkin vocalizations on a portable cassette recorder, five minutes during which the tape was played, and three minutes subsequent to playing the tape. Distances (up to 100 m) to all detected Limpkins were estimated and recorded. The order in which points were sampled was rotated daily to avoid confounding effects of time of day and sample location with the number of birds detected.

The distance from sampling points at which the number of birds detected began to decline (inflection point) was determined by plotting the density of birds observed in 10 m annuli around each sample point. Densities were estimated as the number of birds observed within the basal radius divided by the area of a circle with radius equal to the distance to the inflection point. Densities between the upriver and downriver areas and number of detections between morning and evening counts were compared using *t*-tests.

PLAIN CHACHALACAS

The most comprehensive data available to us on the use of playback recordings was included in a survey of the distribution and abundance of Plain Chachalacas in southern Texas (Marion 1974a). A total of 880 census points was established at 0.4 and 0.8 km intervals adjacent to tracts of suitable habitat throughout the Lower Rio Grande Valley of Texas. These points were censused at least once during 1971; the majority of these censuses occurred within the breeding season (late March–June). A tape recording of Plain Chachalaca vocalizations was played at each census point and the distance to all responding chachalacas was recorded. The maximum distance at which Plain Chachalacas consistently responded was estimated from these data. The percentage of birds present within a circle of this radius that did not respond was estimated from censuses at two relatively distinct, isolated tracts of habitat where population sizes were known from counts at local feeders. Counts of birds at other points were adjusted to account for nonresponding birds using this correction factor.

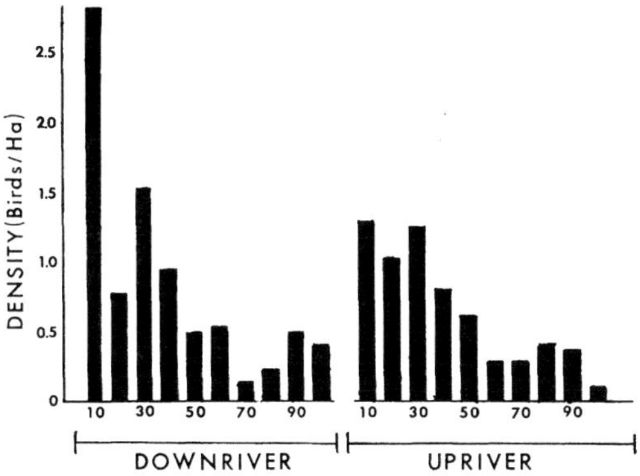

FIGURE 2. Density of Limpkin detections in 10-m annuli around sample points, Ocklawaha River, Florida, June 1980.

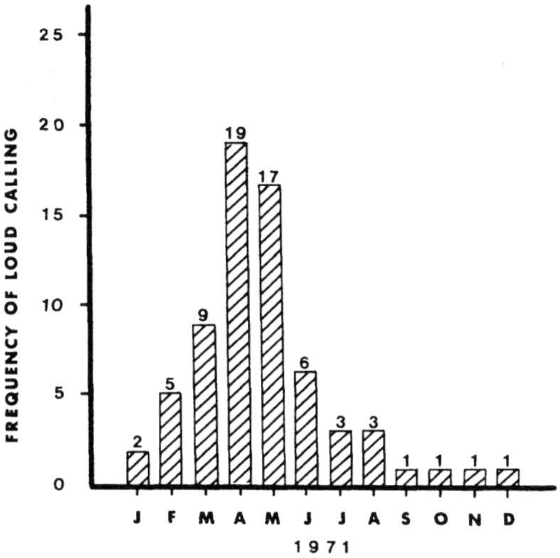

FIGURE 3. Seasonal variation in calling frequency of Plain Chachalacas at Santa Ana National Wildlife Refuge, Hidalgo County, Texas. Data shown are the total number of days each month on which loud calling was heard in 1971.

A total population estimate for Plain Chachalacas in Texas was calculated using two correction factors; one for areas known to contain Plain Chachalacas but not included in the survey and the other for nonresponding birds in the population. All suitable Plain Chachalaca habitat was not sampled during the survey due to limitations on time and access to private property. Area correction factors were calculated for each county involved using the ratio of known occupied habitat to the area sampled at census points.

RESULTS AND DISCUSSION

Rails

Playback recordings were effective in detecting each of the three species of rails on our study areas. Density estimates resulting from the variable plot estimator of Ramsey and Scott (1979) are shown in Table 1. Overall, Soras appeared to be year-round residents and had the greatest density of the three species, with a peak in density occurring in the spring. Densities of Virginia Rails also were recorded as being greatest on our study areas during the spring, but these rails were not detected during the fall. King Rails occurred in relatively lower densities in the spring, summer, and fall, but they were not detected during the winter. Detection of rails appeared to be strongly dependent upon the use of playback recordings and these recordings were successfully used to augment variable plot census techniques.

Limpkins

Limpkin counts were enhanced appreciably by the use of playback recordings of their calls. Frequency of observations declined rapidly during the first three minutes of observation, but increased markedly within two minutes of initiation of the recorded calls (Fig. 1). Rate of new detections then diminished at a slower rate during and after the time the tape was played. Most Limpkins observed responded vocally to the tape; however, some birds were observed which apparently heard the tape but did not respond vocally. Also, a typical response to the tape was for Limpkins to fly toward the tape recorder and circle or perch near the sample point. Care must be taken to record a bird's location at its initial response to the tape to avoid a possible bias in density estimation.

Importance of using a technique for determining radii of similar detectability (i.e., inflection points) was exemplified by a comparison of absolute and relative densities between areas. Inspection of detection curves indicated inflection points at 30 m for both areas (Fig. 2). Mean densities per point (within 30 m) by this method did not differ ($P > 0.05$) between the upriver and downriver areas with estimates of 1.4 and 1.1 birds per ha, respectively. However, when mean number of birds per point (all detections) were compared between areas, a higher density ($P < 0.01$) was indicated for the downriver area than for the upriver area with means of 2.9 and 1.5 birds per point, respectively. The apparently erroneous conclusion of greater densities on the downriver area resulting from the relative index was probably a result of vegetation differences on the two areas. Numerous standing dead trees on the downriver area apparently increased Limpkin detectability at greater distances by providing perches which elevated birds above the obscuring vegetation. This demonstrates the

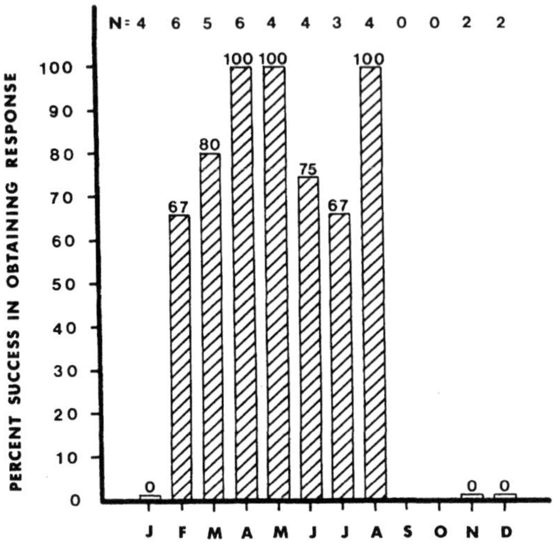

FIGURE 4. Seasonal variation in effectiveness of recorded calls in obtaining responses from Plain Chachalacas on four areas in southern Texas. Number of attempts (N) during each month are also shown. Data were lacking for September and October due to the extensive flooding and inaccessability of study areas.

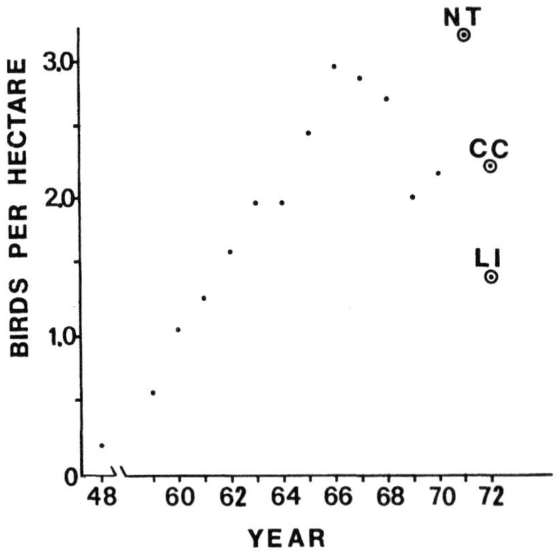

FIGURE 5. Plain Chachalaca population estimates at Santa Ana National Wildlife Refuge, Texas. Estimates for years prior to 1970 were made by refuge managers. Estimates from this study in 1971 came from nest transects (NT), and in 1972 came from call counts (CC) and Lincoln Index (LI).

importance of obtaining comparable estimates of density when censuses from two areas with differing vegetation characteristics are to be compared.

No difference ($P > 0.05$) was found between the number of birds detected during morning and evening counts. Limpkins calling naturally without stimulation from the tapes vocalize extensively during June; they call at various times throughout the day, and occasionally at night. It seems therefore, that "time of day" is not as important in sampling this species as it is with other avian species.

Our overall assessment of these results was that playback techniques are useful for increasing both detections and observations of Limpkins but are not absolutely necessary to obtain observations of this species.

TABLE 2
POPULATION ESTIMATES FOR PLAIN CHACHALACAS AT THE 648-HA SANTA ANA NATIONAL WILDLIFE REFUGE, HIDALGO COUNTY, TEXAS, 1971–72

	Population estimate	
Method	Total birds	Density (birds/ha)
Lincoln	998	1.5
Nest transects	1993	3.1
Call counts	1593	2.2

PLAIN CHACHALACAS

Plain Chachalacas responded readily to tape-recorded vocalizations, particularly just prior to and during the spring breeding season when the frequency of natural calling was highest (Fig. 3). Recorded calls were from 75–100% effective in eliciting calling responses from wild birds during the breeding season (Fig. 4) and these were used in detecting the birds in the field.

Data were available from nine separate areas with a total of 21 census points where samples were repeated on alternate mornings. Comparisons of responses revealed no significant difference ($P > 0.05$) between days in number of birds responding to calls at the nine locations.

TABLE 3
ESTIMATES OF PLAIN CHACHALACA POPULATIONS IN THE LOWER RIO GRANDE VALLEY OF TEXAS, 1972

County	Chachalacas counted	Population[a] size	Area sampled (hectares)	Density (birds per hectare)	Total population[b]
Cameron	1701	3402	1195	2.9	8845
Hidalgo	971	1942	777	2.5	9322
Starr	71	142	121	1.2	880
Willacy	30	60	66	0.9	204
Total	2773	5546	2159	2.6	19,251

[a] Number counted times the correction factor for nonresponding birds, i.e., 2.0.
[b] Population size times area correction factor.

These data indicated that, at least over short time intervals, the repeatability of responses obtained using recorded calls was excellent. Observations recorded at these nine locations also indicated that Plain Chachalacas respond more readily to recorded calls during early morning hours (06:00–09:30) than during late morning hours (09:30–12:00).

The maximum distance at which Plain Chachalacas consistently responded to recorded calls was estimated as 180 m. Pairs of Plain Chachalacas generally responded together within this distance, but apparently not all birds responded to recorded calls. The proportion of nonresponding birds within this distance was estimated on two isolated tracts of known density. On one tract, 22 out of 50 (44%) Plain Chachalacas responded to recorded calls. On the other tract, 10 out of 17 (59%) responded. A correction factor for nonresponding birds was calculated as the ratio of the total number of Plain Chachalacas present to the number responding to recorded calls. In the two observations, approximately half of the Plain Chachalacas responded to recorded calls; therefore an average correction factor of 2.0 was used to account for nonresponding birds within 180 m of the census point.

Specific examples to illustrate the practical use of playback recordings in population estimation follow. On April 18, 1972, 111 Plain Chachalacas responded to recorded calls at 10 census points at Santa Ana National Wildlife Refuge in southern Texas. Use of the correction factor for nonresponding birds (2.0) gave a corrected total of 222 birds on the approximately 101.8 hectares sampled, for a density of 2.2 birds/ha. This density extrapolated to a total population estimate for the Refuge of 1,426 birds per 648 hectares.

This estimate of 2.2 birds per hectare was compared with two other population estimates for the same area, the Lincoln Index and estimates based upon nest density (Table 2) as measured along transects. For the Lincoln Index, which is based upon proportions of marked birds in the population, the density of birds was estimated as 1.5 birds per hectare. Using nest density data obtained from nest transects which were extrapolated to the entire area, a value of 3.1 birds per hectare was obtained. The population estimate based upon the call counts was between the above two estimates and was probably the most accurate of the three methods of population estimation. Also, our estimates based upon call counts closely approximated annual estimates made by managers at Santa Ana National Wildlife Refuge; these estimates, apparently based upon observations at photo blinds and other feeding locations, were obtained from unpublished annual reports. The estimates were 2.5, 2.8, 2.8, 2.8, 2.0, and 2.2 birds per hectare for 1965–70, respectively. These estimates yielded a 6-year average of 2.5 birds per hectare, comparable to the call count estimate obtained for the same area during 1972 (Fig. 5).

Another example on a broader scale includes the following estimation of Plain Chachalaca populations throughout the Lower Rio Grande Valley of Texas based entirely upon responses obtained using the playback technique. A total of 2773 Plain Chachalacas responded to recorded calls at 447 out of 880 census points. After correction for nonresponding birds in the population and for habitat area not sampled, the total Plain Chachalaca population in Texas was calculated to be approximately 19,000 birds (Table 3).

Generally, our evaluation of the playback technique for use with this species was very favorable. It seems to be the only practical way of estimating populations of these elusive birds in their dense habitats.

ACKNOWLEDGMENTS

This paper is contribution No. 2807 of the Journal Series, Florida Agricultural Experiment Station, Gainesville.

Studies in Avian Biology No. 6:86–91, 1981.

MAPPING TERRITORIES WITH PLAYBACK: AN ACCURATE CENSUS METHOD FOR SONGBIRDS

J. BRUCE FALLS[1]

ABSTRACT.—A playback method for mapping territories of songbirds is described, including suggestions for overcoming commonly encountered difficulties. Because birds are stimulated to sing and only boundary points are determined, this technique is more efficient for a single species than methods based on observation. Territory determinations are accurate in dense populations but may be unrealistically large where birds have no close neighbors and may follow a speaker into unoccupied areas. As a census method, about 5–15 birds in a 10–15 ha plot can be mapped with a single pass through the area in one morning.

Mapping methods of censusing breeding birds involve the enumeration of territories and the assignment of space to each singing male. Behavioral studies of territoriality often require more detailed maps. Since 1955, my students and I have mapped territories of several passerines using playback of recorded song. Similar methods have been used by Dhont (1966) and Krebs (1971). Here I describe our method, list our main findings concerning territorial behavior, and evaluate the usefulness of this technique for censusing songbirds.

METHODS

Traditional observational methods require repeated passes through surveyed plots or following individual birds for extended periods. Although territories are defined as defended areas, these methods only occasionally record instances of active defense. Most of the "points" gathered represent singing or foraging and, since these activities typically show marked central tendencies (Zach and Falls 1978, 1979) few boundary locations are obtained. Mapping the extent of defended areas in this way is a slow business.

We use playback of conspecific song to provoke territory holders into active defense and attempt to draw them to the boundaries of their territories. The method is relatively efficient because little time is wasted on interior locations.

We start near a singing bird which typically responds by calling, approaching rapidly, flying about the loudspeaker, and singing. In short, the territory holder responds as it would to an intruder. As soon as a bird approaches we move the speaker. Continuing in one direction, a point is soon reached where the bird will no longer approach, although a neighbor may do so. This establishes a point on the boundary of the original bird, either where it approached most closely or (more conservatively) at the last location where it sang. Alternatively, a boundary may be drawn midway between points of closest approach or song of two neighbors. Continuing with the original bird, the speaker may be taken back into its territory and moved as before but in a different direction. Thus, we follow a zig-zag path with locations inside and outside the territory until we have mapped the defended area of one bird. Some information will have been gathered on neighboring birds and we can then explore their other boundaries. The rapidity of mapping depends on the responsiveness of the birds and the accuracy desired. With territories in the order of 1 ha, about 1 h per bird usually suffices (Fig. 1).

PROBLEMS AND SUGGESTIONS

Some birds do not respond readily and others lose interest rapidly. A few simple precautions help to minimize these difficulties. Songs of strange individuals (recorded some distance away) evoke the strongest responses (Weeden and Falls 1959). They should be played at normal volume which will require a recording at high level if played directly from a Uher recorder. Intervals between songs used for playback should simulate a bird singing rapidly (say about 10 sec). Birds that do not respond initially often do so when a neighbor is attracted to a common boundary; thus, they can be mapped along with a more responsive neighbor. However, failing this, a visit on a later day may be necessary to fill in a "hole." Some birds (including recently banded ones) may not readily approach an observer. In dealing with such individuals (and for detailed mapping generally) it may be useful to place the speaker away from the observer on a long cable (say 20 m). Once a bird has responded, it will usually continue to do so if the speaker is moved quickly, playing only as many songs as are needed at each location. If a bird's response wanes it may be revived by using a different song. Unless a neighbor appears, the speaker should be taken some criterion distance (say 20–30 m) beyond the last point where a bird approached. To be sure that a bird which stops responding has not simply habituated to the playback, the speaker should be moved back to a point where it responded before. This can be done before moving off in another direction.

Confusion may result if several birds approach the speaker at once. On such occasions it helps to have more than one observer. First, any pairs should be identified. Beyond that, it is often possible to separate the voices of different individuals by ear. The different singers can be recorded using a small cassette recorder and identified later from "voice prints" (sonagrams). Problems of this kind arise along boundaries and can be resolved as the mapping proceeds. Sometimes neighbors invade each other's territories in pursuit of the playback. Usually, they are chased out again and a point of equilibrium is soon reached which can be

[1] Department of Zoology, University of Toronto, Toronto Ontario, M5S 1A1 Canada.

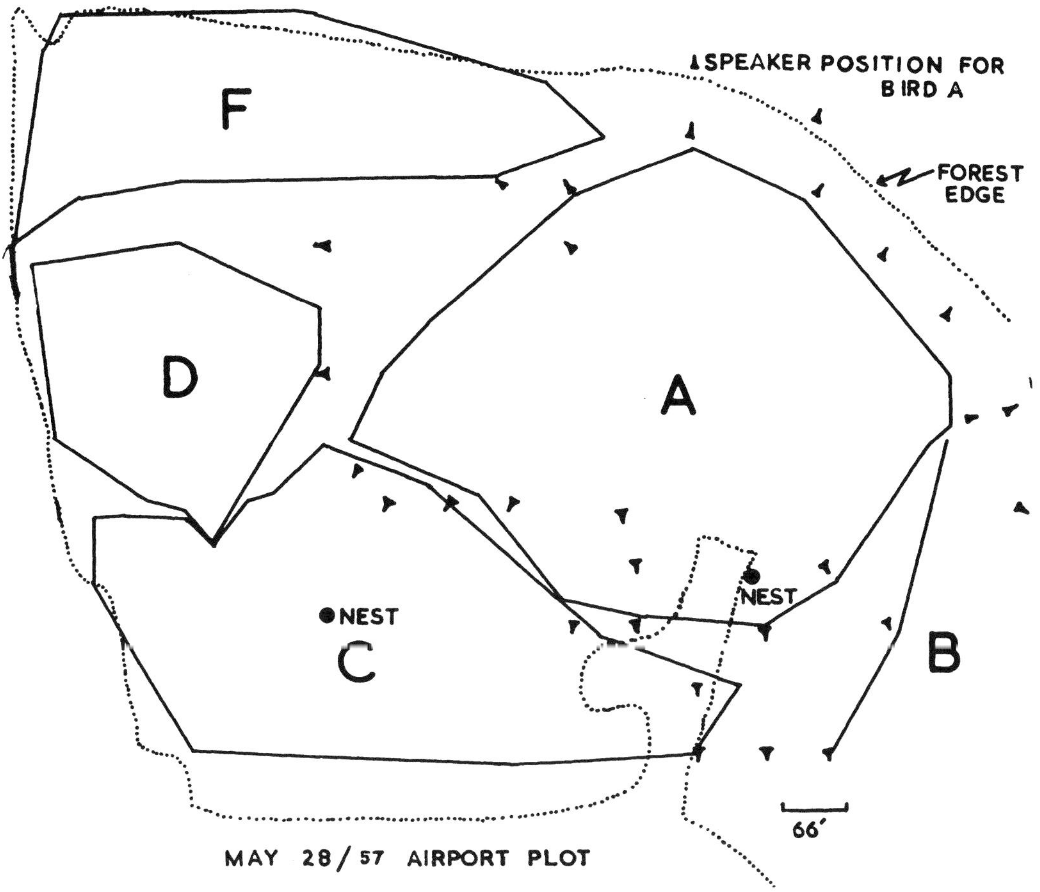

FIGURE 1. Territories of Ovenbirds determined in one morning using playback. (Speaker positions used for bird A are shown. Boundaries are based on closest approach.)

mapped as a common territory boundary. If these problems are severe, they may be alleviated by mapping each bird using its own song. Most birds respond fairly strongly to their own songs, while neighbors tend to ignore them (Weeden and Falls 1959).

Since responses wane over long periods, it is not advisable to return to the same area for several days. Indeed, the best map may be obtained on the first occasion so it pays to choose a day early in the breeding season, after territories have been established and when there is frequent song. Although mapping by playback can be done at any time of day, the morning is usually best. Rainy or windy weather should be avoided. As with other mapping methods, a grid of marked points is helpful and locations should be entered on a corresponding chart as the work proceeds. We have also used overlays on aerial photographs but even then some points of reference are needed.

Equipment

We use Uher (Report/Monitor) portable tape recorders, either directly or with a portable 10W amplifier and speaker (either Nagra DH Amplifier-Speaker or an amplifier with a separate mid-range horn). For easy changing, tapes are mounted in Cousino Audio-vendor cartridges, held in place by electrical tape.

Species and Habitats

Most of our mapping with playback has been done with three species: the Savannah Sparrow (*Passerculus sandwichensis*) which lives in rough grassland, the White-throated Sparrow (*Zonotrichia albicollis*) a bird of coniferous woodland and forest edge, and the Ovenbird (*Seiurus aurocapillus*) which only occurs in closed canopy forest. Our main study areas are in Algonquin Park, Ontario.

RESULTS AND DISCUSSION

Results Concerning Territorial Behavior

(1) Birds respond strongly to playback, apparently attempting to evict the intruder. This is consistent with the notion that song is partly a threat display (Falls 1978) and with the concept of territories as defended areas.

(2) At any one time, the birds we have studied defend non-overlapping "song territories." Like

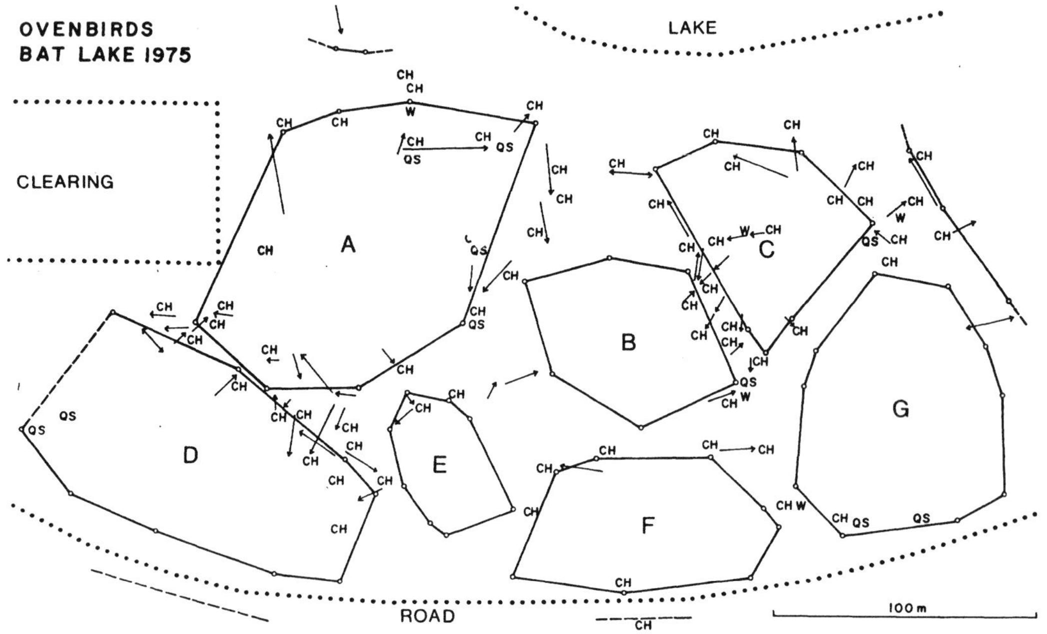

FIGURE 2. Ovenbird Territories determined over three mornings using playback. (Boundaries are based on closest song. Locations of chipping (CH), quiet song (QS) and warbling song (W) are also shown.)

other methods based on song, ours does not detect movements of silent birds beyond their defended boundaries. Thus, territory maps may not reflect the distribution of activities unrelated to defense (Zach and Falls 1979, Jones and Falls, MS).

(3) Neighbors may be separated by a narrow "buffer strip" in which they will not sing but may approach a speaker (Fig. 2). This suggests that they refrain from provoking each other, although they will still attack a strange singer in this area.

(4) Playback territories are usually larger than those determined by observation in comparable periods of time (day or season) (Figs. 3 and 4). This reflects both the greater efficiency of the playback method and a tendency for birds to attempt to defend larger areas than those in which they normally sing.

(5) Where birds lack close neighbors (in sparse populations, patchy habitats), they may follow a speaker into unoccupied areas giving extremely large estimates of territory size. For example, Savannah Sparrows occupying a narrow strip of rough grass beside an airport runway, followed speakers across the runway or into shrubbery where there were no other birds (Fig. 4, Table 1). We have obtained similar results with Ovenbirds. At Churchill, Manitoba, where White-throated Sparrows are scarce, a bird followed a speaker for a mile. Clearly, in such cases territory measurements are unrealistic. However, these and similar observations show that an intruder will be challenged even where song territories appear to be widely spaced. Territory boundaries do not exist in a vacuum but are defined by points of equilibrium between intrusion and defense (Melemis and Falls, MS).

(6) Observation territories are usually contained within playback territories but boundaries obtained by the two methods may overlap (Fig. 4). This may result from shifting over time (different days), from boundary changes in response to playback, or from wandering beyond defended boundaries.

(7) When playback results are accumulated over a long period, adjacent territories often appear to overlap (Fig. 3). This reflects shifting of boundaries over time. Nevertheless, some territories remain very stable (Fig. 5). Playback territories of Ovenbirds do not seem to fluctuate in size with the breeding cycle as do areas in which birds sing spontaneously (Stenger and Falls 1959).

PLAYBACK AS A CENSUS METHOD

For censusing, the playback method can be speeded up by relaxing the accuracy of mapping. Thus, the recorder can be hand held and used directly and fewer boundary points (four to six)

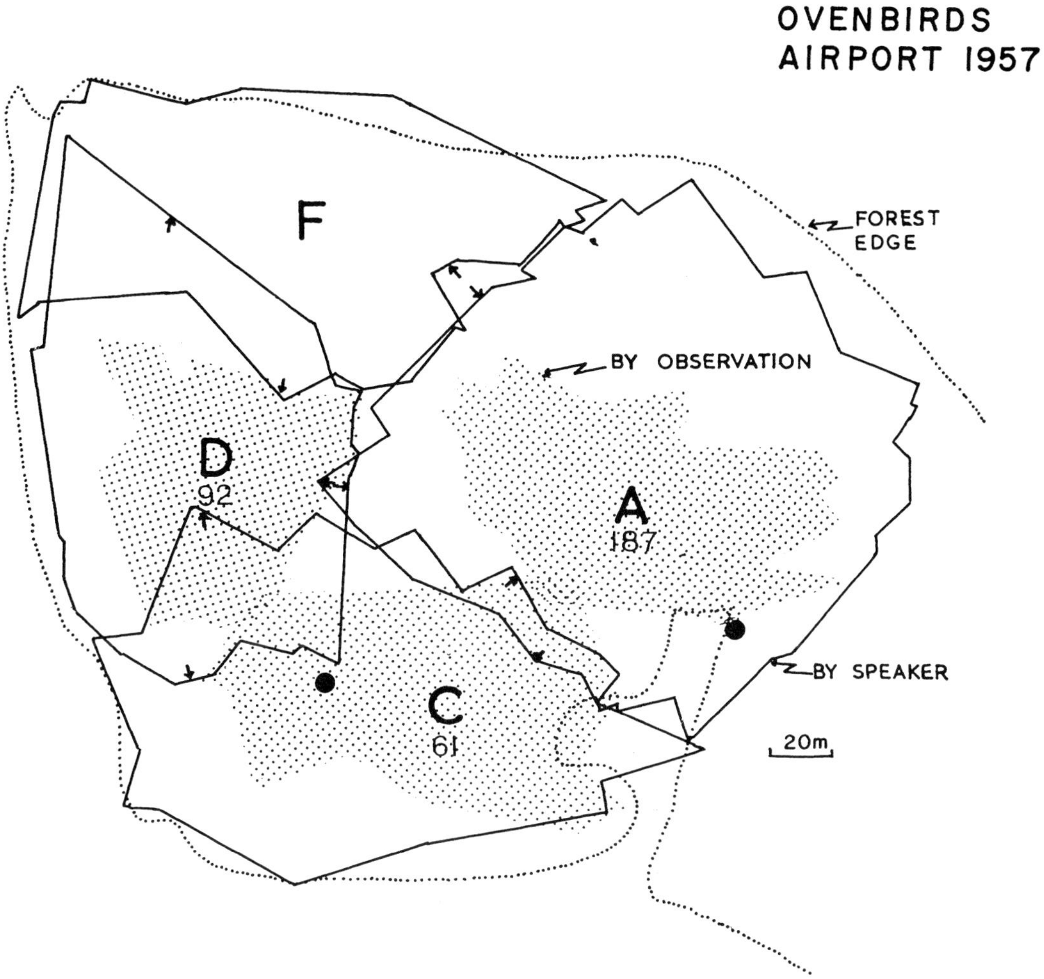

FIGURE 3. Total Territories mapped by playback over the breeding season for Ovenbirds in Fig. 1. (Note smaller stippled areas obtained by observation for three birds and overlaps of playback territories.)

can be determined for each territory. We have been able to map (census) 5–15 males in 10–15 ha plots in one morning (Fig. 6).

The essence of this method is to account for the birds in all the occupied ground. The tendency (referred to above) of birds to defend unoccupied areas may pose problems for density estimation depending on two things: the amount of unoccupied ground and the size of the plot. In dense populations (little unoccupied ground) this should not be a problem. However, if the population is sparse (rare species, patchy habitat) there may be considerable movement of birds and the population could be overestimated if sample plots are small in relation to this movement. Since this is an edge effect it can be minimized by using larger plots. Thus, while playback helps to locate rare species, discretion must be used in interpreting territorial maps and estimating density in such cases.

The usefulness of playback for censusing is not confined to songbirds. The method described here should be applicable to any species using sound signals to advertise exclusive territories that are accessible to the investigator. Where these conditions are not met (less territorial species, inaccessible habitats) variants of this method may still be useful alone or in conjunction with other census techniques.

In cases where it works well, the following points summarize the features of this technique as compared with other mapping methods:

(1) Only one species can be censused at a time.

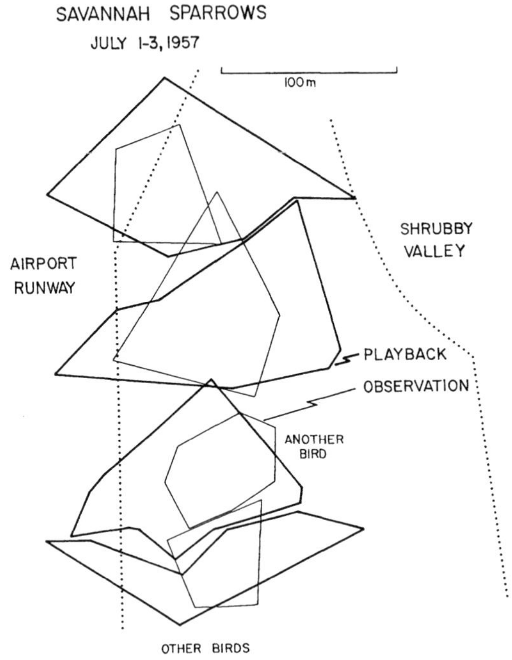

FIGURE 4. Territories of Savannah Sparrows determined over three mornings by playback (speaker) and by observation (overlaps on different days). (Birds followed the speaker into unoccupied areas at right angles to runway.)

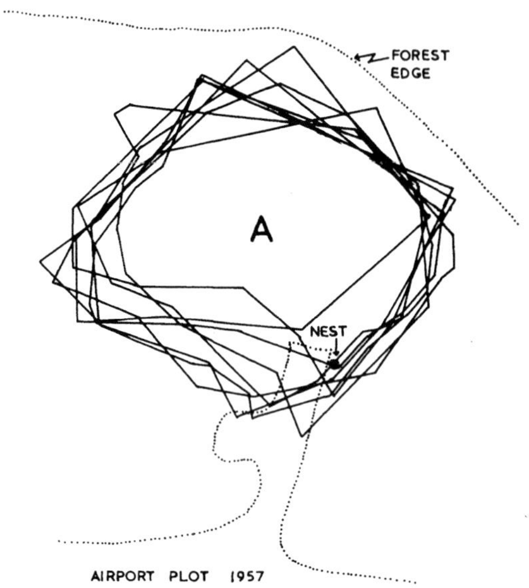

FIGURE 5. Territory Maps determined by playback for Ovenbird A in Fig. 1 over 8 weeks. (Note enlargement, to include the nest in later weeks.)

(2) Usually a single pass through the area is required. Taking (1) and (2) together, the time required to census a few common species is comparable to that needed for other mapping methods.

(3) Birds that might not sing spontaneously during a census can be stimulated to do so. For example, we have been able to census two morphs of the White-throated Sparrow which have very different rates of singing (Table 2). Members of the tan-striped morph would almost certainly be overlooked by observational mapping methods.

(4) Since birds are attracted to the vicinity of the observer, they can be heard and, if desired, recorded easily. Taking (3) and (4) together, problems of cue production and detection are minimized.

(5) Boundaries are obtained rather than clusters of points. This eliminates some problems of interpretation encountered in mapping methods based on observation.

(6) Because birds are stimulated to sing, this method is less sensitive than observational

TABLE 1

DIMENSIONS OF SAVANNAH SPARROW TERRITORIES[a] AT RIGHT ANGLES AND PARALLEL TO AN AIRPORT RUNWAY, DETERMINED BY TWO METHODS

	Observation method			Playback method		
(1) Bird	(2) Right angle to runway	(3) Parallel to runway	(4) Ratio (2)/(3)	(5) Right angle to runway	(6) Parallel to runway	(7) Ratio (5)/(6)
A	78	79	0.98	152	118	1.29
B	71	70	1.01	149	73	2.04
C	69	54	1.29	141	52	2.71
E	76	72	1.06	214	99	2.15
All	(3) > (4) in 11/20			(5) > (6) in 12/13		

[a] Average of three to five trials per bird in meters. Figure 4 shows a typical trial.

TABLE 2

PAIRS OF WHITE-THROATED SPARROWS[a] DURING AN INSECT OUTBREAK, CLASSIFIED BY THE MORPH OF THE MALE

Year	WS[b]	TS[b]	Total	% Change[c]
1969	12	6	18	0
1970	10	8	18	0
1971	12	12	24	33.3
1972	15	14	29	61.1
1973	17	10	27	50.0

[a] Camp Road plot, Algonquin Park, Ontario.
[b] Male morph: WS = white-striped, TS = tan-striped.
[c] 1969 as base (= 18 pairs).

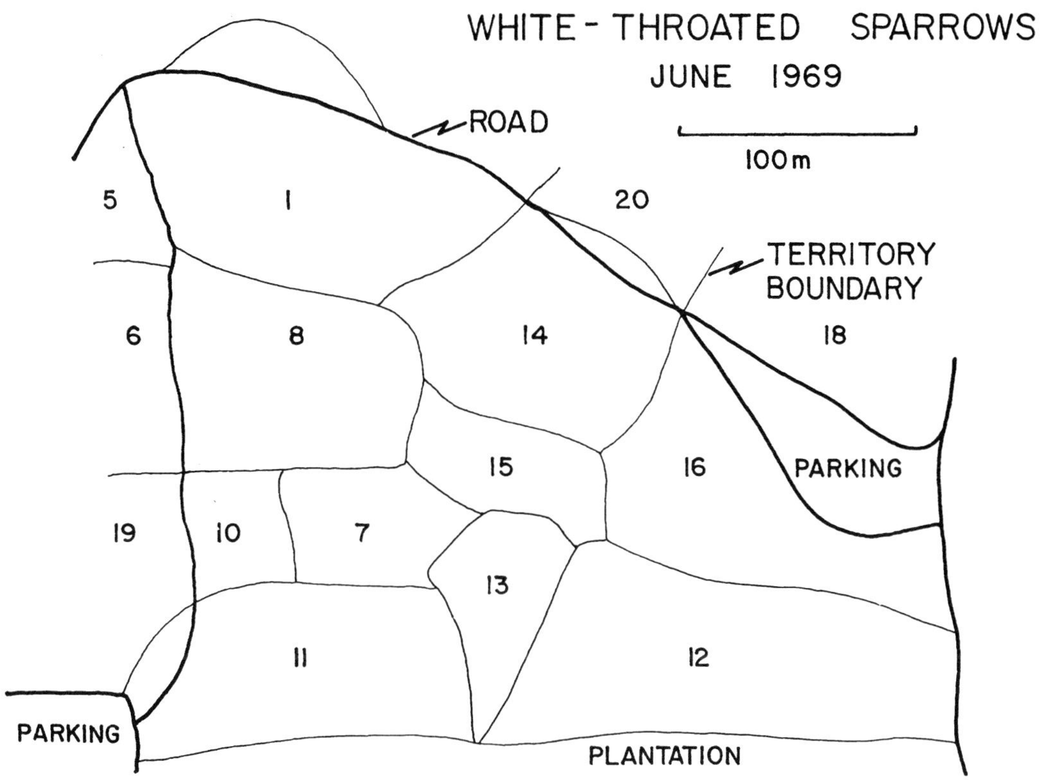

FIGURE 6. Census of White-throated Sparrows by playback in one morning. (Boundaries are drawn between points of closest approach by neighbors.)

methods to daily or seasonal effects. Thus, censusing need not be confined to the early morning or to the early part of the breeding cycle.

(7) In dense populations, we can check that all territorial birds have been counted by accounting for occupancy of all the ground. In such cases reasonable estimates of territory sizes are also obtained. This method does not, of course, enumerate non-territorial members of the population.

(8) Because playback can provide a total count of territorial birds in a known area, it is useful as a background for other studies. Thus, it could be used to calibrate more rapid census techniques. We have used it to measure numerical responses of several species to an outbreak of spruce budworm (*Christoneura fumiferana*) (Table 2) and as a background for removal and behavioral experiments (Falls and Loncke, MS; Falls and Brooks 1975).

(9) Combined with individual marking (or recording), playback censuses can be used to measure population turnover. This may be important, for example, in studies of environmental impacts, where rapid replacement of birds could mask effects if observational census methods were used before and after treatments.

(10) Like other census methods, mapping by playback must be carefully done by an experienced investigator to obtain accurate results. In particular, knowledge of the social structure of populations and the response behavior of each species is important.

Clearly, the methods described here should be tested on more species.

ACKNOWLEDGMENTS

Robert Walker, John Madore and other students assisted with field work. Peter Marler made helpful suggestions about the manuscript. The Ontario Ministry of Natural Resources provided facilities in Algonquin Park. Financial support was provided in part by the National Research Council of Canada.

Studies in Avian Biology No. 6:92–102, 1981.

THE USE OF MIGRATION COUNTS FOR MONITORING BIRD POPULATION LEVELS

DAVID J. T. HUSSELL[1]

ABSTRACT.—Previous use of migration counts for monitoring bird population levels has been based largely on indices derived from a summation of counts over part or all of one or more migration seasons. A new method is described in which multivariate regression techniques are used to assign variability in counts at one or more sites to year, date, weather factors and other variables. Variability attributable to year provides a relatively reliable index of annual migration volume and allows statistical tests of differences between years. The method is illustrated by examples of indices calculated from spring counts of migrants at the Long Point Bird Observatory, Ontario, in the years 1962–79 and is validated as a population indicator by correlating migration indices with breeding bird survey indices for 1968–79. Pros and cons of migration indices as population indicators are discussed as well as the applicability to other groups of birds of the methods described here.

Although migration counts integrate information on bird populations over wide areas and often sample relatively large numbers of individuals at a single observation site, they have been little used to monitor changes in population levels. There are two main reasons for this: (a) it is often difficult to associate particular migrant populations with corresponding breeding and wintering populations, and (b) many factors other than population change contribute to variability in migration counts. The purposes of this paper are to present a method which attempts to overcome the second of these difficulties, to suggest how it can be applied elsewhere, and to discuss the pros and cons of migration indices.

Previous attempts to measure annual or longer-term changes in numbers of migrants include a variety of situations and objectives (e.g., Mueller and Berger 1967b; Hackman and Henny 1971; Busse 1973; Williamson 1975; Berthold and Schlenker 1975; Mueller et al. 1977; Nagy 1977; Langslow 1977, 1978; Hjort and Lindholm 1978; Berthold and Querner 1979). Svensson's (1978c) study is notable because he showed that migration indices for several species at Swedish bird observatories were correlated with independently-derived results from the Swedish Breeding Bird Census. He concluded, however, that the Breeding Bird Census was a more efficient method for detecting population changes because of high variability in the migration indices, which he attributed to the effects of weather factors.

Although the studies cited above differ in the level of standardization of field procedures and in the details of their methods, in essence all base their indices of migration volume on summation of counts over a period of days or weeks in one or more migration seasons. Because of the well-documented effects of weather on migration, such indices are often regarded as more or less unsatisfactory, except for demonstrating gross long-term changes in population level. Apart from the early attempt by Ulfstrand (1958) to compensate for the effects of wind on counts of migrating hawks, no methods have been described to correct migration indices for the effects of weather, nor have the statistical attributes of the data been examined carefully with a view to developing appropriate indexing procedures.

The relationships between weather and migration volume have been studied for decades and multivariate regression techniques have been used extensively to examine the effects of weather factors on migrating birds, particularly in radar studies (Richardson 1978). Here I extend these procedures to provide a method for detecting annual population change at one or more observation sites, while simultaneously compensating for the effects of date, weather factors and other variables. In its present form the method should be regarded as a preliminary attempt to correct migration indices for the effects of date of observation and weather factors; further study may lead to improvements and refinements. The method is presented first in the form of a general model which may be applicable to a variety of situations. As an example of its application, it is then used to determine migration indices from counts of nocturnal migrants at Long Point Bird Observatory, Canada, and the indices are validated as population indicators by comparison with independently-derived indices of breeding population size. In the Discussion section I examine the potential usefulness of migration indices derived from this procedure.

[1] Long Point Bird Observatory, P.O. Box 160, Port Rowan, Ontario, Canada N0E 1M0.

THE MODEL

DESCRIPTION

The general model relating bird numbers to date, year, site, and environmental factors is

$$\ln(N_{ijk} + 1) = A_j + \sum_{m=0}^{M} c_{im}k^m + \sum_{v=0}^{V} b_{vi}X_{vijk} + e_{ijk} \quad (1)$$

where N_{ijk} is the number of birds at site i, in year j, on date k; A_j is a year factor specific to year j; $\sum_{m=0}^{M} c_{im}k^m$ is an Mth power polynomial in k (date), that is specific to site i and in which c_{im} are constants; X_{vijk} is the value of environmental variable v at site i, in year j, on date k, and b_{vi} are constants specific to environmental variable and site; and e_{ijk} is an error factor representing unexplained variation. Multiple regression techniques are used to estimate A_j, c_{im} and b_{vi} with c_{i0} for one of the sites arbitrarily set to zero. Certain characteristics of the model and the rationale for its use are discussed in the following paragraphs.

(1) The regression model assumes homoscedasticity (equal variances), normal distribution of residuals, and additive effects of variables. When (a) standard deviation of the residuals varies directly as the means, (b) the distribution of residuals is skewed (to high values), and (c) the effects on the original scale are multiplicative, a logarithmic transformation is appropriate in order to meet the assumptions of the regression (Snedecor and Cochran 1967:141–144, 329–330; for a good discussion of the assumptions of multiple regression in relation to analysis of migration data see Richardson 1974). Extensive examination of bird count data from Long Point shows skewed distributions and variances that increase with the means. This may prove to be a general rule with bird count data (cf. Alerstam 1978, Blokpoel and Richardson 1978, Prater 1979). Moreover, it is logical to assume that effects on the original scale are multiplicative. For example, if the population doubles between year j and year $(j + 1)$, we would expect the number of birds counted on day k in year $(j + 1)$ at site i to be twice that on day k in year j, if all other conditions remain constant. If the number of birds on day $(k + 5)$ in year j at site i is twice that on day k for the same year and site, however, we would expect four times as many birds at that site on day $(k + 5)$ in year $(j + 1)$ as on day k in year j, if all other factors remain the same. Logarithmic transformation converts these multiplicative effects to additive ones which can be analysed by multiple regression techniques. Note that one is added to N_{ijk} prior to taking logarithms because it is impossible to take a logarithm of zero. This introduces some distortion into the multiplicative-additive conversion, especially when there are many observations of zero or small numbers (less than 10) of birds.

(2) A_j is a year factor common to all sites, an assumption that is appropriate only for sites in the same local area or which for other reasons can be assumed to be sampling the same migrant bird populations. A_j is a measure of annual migration volume, which can be used to derive an annual migration index (see below).

(3) $\sum_{m=0}^{M} c_{im}k^m$, $i = 1, 2, \ldots, I$, is a series of I polynomials in k, each of which represents the seasonal pattern of migration at site i (cf. Alerstam 1978). No assumptions are made concerning the similarity or otherwise of the patterns at different sites.

(4) $\sum_{v=0}^{V} b_{vi}X_{vijk}$, $i = 1, 2, \ldots, I$, is a series of I sets of terms for different environmental variables v, whose coefficients b_{vi} are specific to each site i. Thus, no assumptions are made concerning the similarity or otherwise of the effects of environmental variables at different sites. In principle, the X variables need not be confined to environmental factors but can include any factor that is related to bird numbers. Thus, measures of sampling effort can be included here, provided that they meet the assumptions of the regression procedure (see Discussion section).

MIGRATION COUNT INDICES

The A_j values represent the effects of year on $\ln(N_{ijk} + 1)$. If $\ln(N_{ijk} + 1) = Y_{ijk}$, then the adjusted means for year, $\hat{\bar{Y}}_{.j.}$, provide a measure of migration volume in year j. The adjusted means are calculated as

$$\hat{\bar{Y}}_{.j.} = A_j + \sum_{i=1}^{I} \sum_{m=0}^{M} \left[\frac{n_i}{n} c_{im}(\bar{k^m})_{i..}\right] + \sum_{i=1}^{I} \sum_{v=1}^{V} \left[\frac{n_i}{n} b_{vi}\bar{X}_{vi..}\right] \quad (2)$$

Where n_i is the number of observations at site i (over all dates in all years), $n = \sum_{i=1}^{I} n_i$, and

$(\overline{k^m})_{i..}$ and $\bar{X}_{vi..}$ are the means of all values (over all dates in all years) at site i of k^m and X_{vijk}, respectively. An index of annual migration volume, expressed in terms of untransformed bird numbers is: $A_j' = e^{\hat{Y}_{.j.}} - 1$.

MATERIALS AND METHODS

THE DATA

Bird Migration Counts

The migration counts were taken from the records of the Long Point Bird Observatory for 16 March–15 June, 1962–1979. No counts were available for 1965 and data for 1973 and 1974 were sparse. Migration counts were estimates of the number of each species occurring each day in specified areas at two sites on Long Point, a 32 km peninsula on the north shore of Lake Erie. Site 1 is at the eastern tip of the peninsula and consists mainly of dunes sparsely vegetated with cottonwoods (*Populus deltoides*). Site 2 is at the southwestern end of a wooded dune ridge 19 km west of site 1 (see Figure 1 in Hussell and Stamp 1965). Each morning that the Observatory stations were manned, a census of about 1 h duration was conducted over an approximately 2.0 km circuit covering a representative sample of the habitat at each site. On most days, Heligoland traps and/or mist nets were used to capture birds for banding (Hussell and Woodford 1961). At the end of the day, all observers present conferred and agreed on estimates of the totals of each species occurring within the specified area at the site. Estimates were based on the census, birds captured, and any other observations during the day. These estimates for six nocturnal migrant land birds were used as the migration counts in this analysis.

For most species, I used all available data spanning the period from the first observation to the last spring observation of that species in any of the years. For species with small summer resident populations in the sample areas, however, the data were inspected and an arbitrary cut-off date was selected for the end of the spring migration period. Sample sizes for each species are in Table 2.

Weather

Weather data were from weather stations at the Long Point lighthouse (within the site 1 area) and at Simcoe, Ontario, about 35 km N of site 2. I used the following weather factors measured at 07:00 Eastern Standard Time: (1) wind direction at Simcoe, recorded on a 16-point scale, N, NNE, NE, etc., and reduced to an eight-point scale by combining N and NNE to become 'N,' NE and ENE to become 'NE,' etc.; (2) wind speed at Simcoe in miles per hour; (3) dry bulb air temperature at Simcoe in °F; (4) cloud cover (total cover) at Long Point recorded as eighths of sky covered; (5) visibility at Long Point recorded on a nine-point scale and converted to km.

Weather data from Long Point were missing for 1976 (cloud cover only) and 1979 (cloud cover and visibility) and for scattered dates in other years. I preferred to use Long Point cloud and visibility data, however, because I suspect that these factors may directly influence the numbers of migrants terminating their flights on Long Point. Therefore, I estimated missing values of factors (4) and (5) from multiple regression equations obtained from regressing known values of each of these variables on 12 other weather variables and date. For cloud cover (4), $R^2 = 0.71$ ($n = 1169$, $P << 0.001$) and for visibility (5), $R^2 = 0.48$ ($n = 1266$, $P < 0.001$). As expected, the most important predictor of cloud cover at Long Point was opacity (opaque cloud cover) at Simcoe, while the most important predictor of visibility was the square of visibility at Simcoe.

REGRESSION PROCEDURE

A_j, c_{im} and b_{vi} in equation (1) were estimated using a backward stepwise regression procedure with Biomedical Computer Program P-series BMDP2R (Dixon and Brown 1979). The dependent variable was the natural logarithm of (migration count + 1), named LN(N + 1) in the computer program. Sixty-one independent variables were used in the regression analysis: these were made up of 1 dummy variable for site, 16 dummy variables for year, 14 site-date interaction variables, and 30 site-weather interaction variables (Table 1).

One year (1970) was designated the reference year and its variable was excluded from the regression. The other 16 year variables were forced into the regression at the start and retained throughout, since determination of all values of A_j is the objective of the analysis. Likewise, site 1 was made the reference site and the dummy variable for site 2 was forced into and retained in the regression to provide a unique intercept for site 2.

The only environmental variables used in this study were weather variables and the data were the same for both sites, since only one suitable set of data was available. In some situations it might be preferable to use weather data specific to site, for at least some of the weather variables. Second and third order terms were used in the temperature and wind variables because experience showed that bird count numbers were often nonlinearly related to these variables. Date variables and weather variables used in the regressions were always in the form of interactions with the dummy variable for site and they were made available for entry and removal by the stepwise procedure.

A backward stepping procedure was used in order to detect the effects of interactions between polynomial terms. I used the stepwise procedure 'F' in BMDP2R (Dixon and Brown 1979:405–406). By setting F-to-remove and F-to-enter at very low values (0.10 and 0.11, respectively) all or nearly all available variables were entered. The F-to-remove and F-to-enter values were then reset to higher values and backward stepping began. By setting F-to-enter at 2.71 and 2.72, respectively, only those variables with $P < 0.10$ in a standard F-test were retained in the regression. Significance of variables selected in a stepwise procedure should be treated with caution, however, as their true probability levels may be substantially higher by an unknown amount (Freund and Minton 1979:129, 149; Hall 1979:7–8).

Plots of residuals (observed-predicted) showed that their dispersion was not uniform over the range of values of the dependent variable predicted by the regression. For example, in a residual plot for the Ruby-crowned Kinglet (*Regulus calendula*) (Fig. 1a)

TABLE 1
INDEPENDENT VARIABLES USED IN REGRESSION ANALYSIS

Factor or variable	Variable names[a]	Explanation[b]
Year	Y62, Y63, Y64, Y66, Y67, Y68, Y69, *Y70,* Y71, Y72, Y73, Y74, Y75, Y76, Y77, Y78, Y79.	Dummy variables. Example: Y62 = 1 if j = 1962, otherwise Y62 = 0.
Site	*A1*, A2.	Dummy variables. Example: *A1* = 1 if i = 1, otherwise *A1* = 0.
Date[c]	*DRT*, *D1*, *D2*, *D3*, *D6*, *D7*, *D10*, *D11*.	$DRT = \sqrt{\lceil k/50 \rceil}$, $D1 = k/50$, $D2 = (k/50)^2$, etc.
Site-date interaction[d]	A1DRT, A1D1, A1D2, A1D3, A1D6, A1D7, A1D10, A1D11, A2DRT, A2D1, A2D2, A2D3, A2D6, A2D7.	For each case A1DRT = *A1* × *DRT*, A1D1 = *A1* × *D1*, etc.
Temperature[e]	*TP*, *TP2*, *TP3*	*TP* = (dry bulb air temperature − 45), $TP2 = (TP)^2$, $TP3 = (TP)^3$.
Cloud	*CL*	*CL* = cloud cover.
Visibility[f]	*VSRT*	*VSRT* = square root of visibility.
Wind[g]	*E*, *SE*, *S*, *SW*	*E* = 1 if wind direction is E, *E* = −1 if wind direction is W, otherwise *E* = 0; etc.
	EV, *SEV*, *SV*, *SWV*	*EV* = *E* × (wind speed/10), *SEV* = *SE* × (wind speed/10), etc.
	EV2, *EV3*, *SEV2*, *SEV3*, *SV2*, *SV3*, *SWV2*, *SWV3*	$EV2 = (EV)^2$, $EV3 = (EV)^3$, etc.
Site-temperature interaction	A1TP, A1TP2, A1TP3, A2TP, A2TP2, A2TP3.	A1TP = *A1* × *TP*, A1TP2 = *A1* × *TP2*, etc.
Site-cloud interaction	A1CL, A2CL	A1CL = *A1* × *CL*, etc.
Site-visibility interaction	A1VSRT, A2VSRT	A1VSRT = *A1* × *VSRT*, etc.
Site-wind interaction[d]	A1EV, A1EV2, A1EV3, A1SEV, A1SEV2, A1SEV3, A1SV, A1SV2, A1SV3, A1SWV, A1SWV2, A1SWV3, A2EV, A2EV2, A2SEV, A2SEV2, A2SV, A2SV2, A2SWV, A2SWV2.	A1EV = *A1* × *EV*, A1EV2 = *A1* × *EV2*, etc.

[a] Names of variables used in the computer program. Variables not used as independent variables in the stepwise regression analysis are italicised.
[b] i = site, j = year, k = date. For a discussion of the use of dummy variables and interaction variables, see Nie et al. (1975:373–383). See text for further explanation.
[c] Because of the tolerance limitations of BMDP2R, it was necessary to reduce correlations among date variables by setting $k = 0$ to a date near the midpoint of the season for each species and by omitting some terms from the polynomial series. $k/50$ was used to avoid large values and small coefficients in the polynomial terms.
[d] Tenth and eleventh order site-date interaction variables and third order site-wind interaction variables for site 2 were omitted to reduce the possibility of overfitting of the site 2 data, which make up only about one third of the observations. Overfitting tends to occur if the number of cases does not greatly exceed the number of variables.
[e] Temperature difference from normal is preferable (Richardson 1974, 1978) but was not used in this study.
[f] Square root of visibility was used, following Richardson (1974).
[g] Wind speed/10 was used to avoid large values and small coefficients in the polynomial terms.

the distribution of residuals becomes increasingly distorted at predicted values below 1.5 because observations of zero place a lower limit on the value of the residual (cf. Blokpoel and Richardson 1978:357). The lowest diagonal band of points in Figure 1a represents observations of zero birds. When the predicted value is less than zero, the residuals and their means are necessarily positive, a condition which is a serious violation of the assumptions of the regression. This problem is most pronounced in species that occur in small numbers and have many observations of zero.

To mitigate this situation, I removed cases with predicted values of zero or lower and recalculated the regressions from the reduced data set. Indices and other results quoted in this paper are always from this second calculation. A plot of residuals for the second

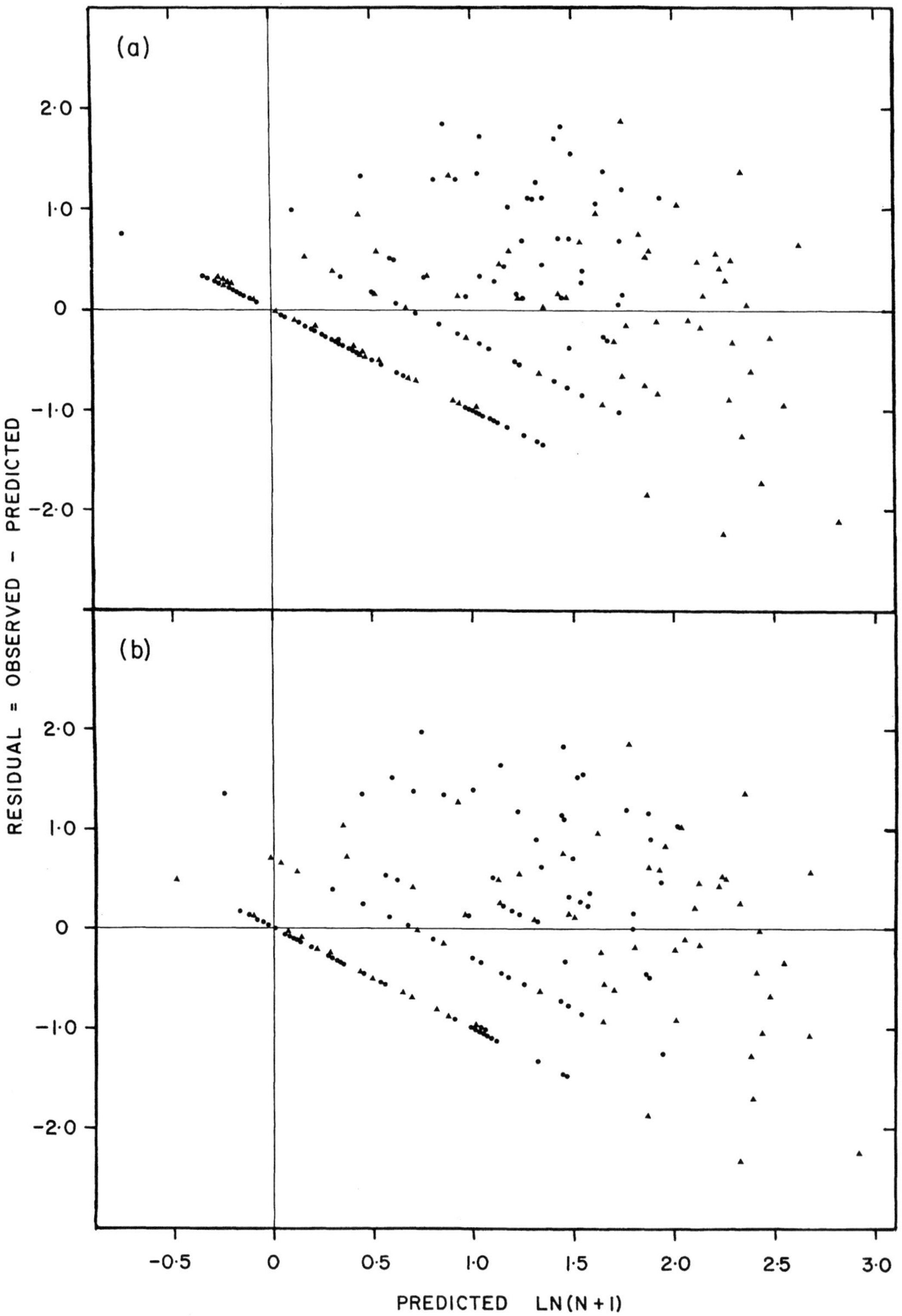

FIGURE 1. Plot of residuals against predicted values of LN(N + 1) for the Ruby-crowned Kinglet. (a) First calculation using all data. (b) Second calculation with reduced data set (see text). Only data for 1976 and 1977 are shown (computer plots of all data show a similar dispersion). Circles = site 1; triangles = site 2.

calculation in the Ruby-crowned Kinglet is shown in Figure 1b. The case removal procedure does not entirely eliminate predicted values of less than zero in the second calculation, because it often results in lower predicted values for the remaining cases; but it does provide an objective method for removing many observations which contribute to distortion in the distribution of the residuals. In most instances the excluded observations are from extremely early or late in the migration season or represent conditions that are otherwise relatively unfavorable for occurrence of the species in question.

Migration Count Indices (MCI)

The adjusted mean for each year was calculated by replacing each variable (except the year variables) in the regression equation by its mean value for all cases, and adding the coefficient of the dummy variable for that year. This gives the same result as equation (2) since the means of site interaction variables are weighted means of observations at each site with weights equal to n_i/n. The indices A_j' were calculated from the adjusted means, as described previously, then rescaled so that the MCIs have an average value of 100 for the years 1975–79.

The significance of differences between indices for different years is determined by testing the differences between adjusted means for year. The significance of the difference between the adjusted mean for any year and the adjusted mean for the reference year in the regression can be determined from the F-to-remove value for the dummy variable for that year with 1 and $(n - v - 1)$ degrees of freedom, where n is the number of cases and v is the number of independent variables in the regression. To determine the significance of differences between all successive years 1962–79, I used Program BMDPIR (Dixon and Brown 1979) to recalculate the regression with different reference years (i.e., omitting another year variable instead of Y70).

Validation

To determine whether migration indices reflect population trends, the MCIs were correlated with indices from an independent method for monitoring population change, the Breeding Bird Survey (BBS). The BBS indices were for two regions generally to the north of Long Point, designated as "central Ontario and Quebec" and "southern Ontario and Quebec" (Erskine 1978, Freemark et al. 1979, Finney et al. 1980). These regions presumably contain the summer ranges of many of the spring migrants that pass through Long Point. The six species considered here are common migrants at Long Point that are well represented in the central region; they are Common Flicker (*Colaptes auratus*), Winter Wren (*Troglodytes troglodytes*), Hermit Thrush (*Catharus guttatus*), Ruby-crowned Kinglet, Yellow-rumped Warbler (*Dendroica coronata*), and White-throated Sparrow (*Zonotrichia albicollis*). BBS indices are available for these species for the central region for the years 1969–79 and for the Common Flicker and White-throated Sparrow for the southern region for 1968–79. (Populations of the other four species in the southern region are too small to give useful BBS indices.) Unfortunately, the BBS indices are believed to be subject to inaccuracies except for relatively short-term comparisons (G. H. Finney, pers. commun.), but these indices are the best indicators of breeding bird population levels that are currently available to me.

If either the BBS index or the MCI for any species does not reflect population change or if the two indices are measuring uncorrelated changes in different populations, the expected correlation coefficient between the two indices is zero. If the two indices measure changes in the same or overlapping populations or correlated changes in different populations of a species, then r should be positive. Considering several species, average r should be positive or zero, respectively, if the indices do or do not track the same or correlated population changes. Demonstration of positive r provides evidence that MCIs (and BBS indices) measure population change; lack of such a demonstration provides contrary evidence only if there are independent reasons to indicate that the two indices are sampling the same, overlapping or correlated populations.

RESULTS

Regression Results

Using 1970 as the reference year, the regression equation for the White-throated Sparrow is given below (see Table 1 for definitions of variables). One and two asterisks indicate variables whose coefficients have F-to-remove values greater than 3.85 and 6.67, respectively ($P < 0.05$ and $P < 0.01$, respectively, in a standard F-test). Regression coefficients of all other date and weather variables have F-to-remove greater than 2.71 (P between 0.10 and 0.05).

LN(N + 1) = 3.55 − 0.13 A2 + 0.82 Y62** + 0.86 Y63** + 0.05 Y64 + 0.78 Y66** + 0.78 Y67 + 0.03 Y68 + 0.24 Y69 + 0.62 Y71** + 0.40 Y72** + 0.12 Y73 + 0.16 Y74 + 1.02 Y75** + 0.54 Y76** + 0.38 Y77* + 0.26 Y78 + 0.04 Y79 − 1.82 A1D1* − 24.0 A1D2** − 30.5 A1D3* + 476 A1D6** + 1340 A1D7* − 5550 A1D10* − 15300 A1D11* − 2.81 A2D1** − 1.86 A2DRT − 11.7 A2D2** + 0.071 A1TP** − 0.0038 A1TP2** − 0.0008 A1TP3 + 0.029 A1CL − 0.089 A1VSRT** + 0.340 A1EV* − 0.157 A1EV3* + 0.093 A2TP** − 0.0018 A2TP3** − 0.442 A2EV2** + 0.488 A2SEV** − 0.308 A2SWV*

R^2 for the regression is 0.537, which is highly significant ($P << 0.01$). For site 1, seven date variables and seven weather variables had large enough effects for inclusion in the regression whereas for site 2 three date variables and five weather variables were included. The date variables alone accounted for 37.3% of the variation, weather variables alone for 8.9% and year variables alone for 3.2%. When entered in sequence after the date variables, however, the weather and year variables explained an additional 13.1% and 3.3% of the variation, respectively. In a similar analysis for the Ruby-crowned Kinglet, date variables alone accounted for 25.8% of the vari-

TABLE 2
REGRESSION RESULTS FOR SIX SPECIES

Species	Sample size[a] (n)	Mean birds/day[b]	R^2	No. of variables in regression[c] Site 1 Date	Site 1 Weather	Site 2 Date	Site 2 Weather
Common Flicker	971 (1016)	5.25	0.630	6	4	6	6 (1)
Winter Wren	828 (972)	0.66	0.266	2	7	3	1
Hermit Thrush	892 (1080)	0.83	0.410	6	8	3	9 (1)
Ruby-crowned Kinglet	957 (1090)	2.06	0.469	3	5	4	8 (2)
Yellow-rumped Warbler	967 (1177)	0.96	0.391	7	3	5	6
White-throated Sparrow	955 (1002)	8.36	0.537	7	7 (2)	3 (1)	5

[a] Figures in parentheses are original sample sizes used in the initial regression calculation, prior to exclusion of cases with predicted values less than or equal to zero (see text).
[b] Geometric mean of $(N + 1)$, minus 1.
[c] Figures in parentheses are number of variables of marginal significance (with $0.10 > P > 0.05$ in a standard F-test) included in the total. In addition to variables shown, site variable A2 was included in all regressions but was not significant ($P > 0.10$) in Yellow-rumped Warbler or White-throated Sparrow. A2 was significant with $P < 0.05$ in the other 4 species.

ation, weather variables alone for 10.4% and year variables alone for 5.1%, but year variables explained more variation when entered second than did weather variables. Year variables and weather variables explained an additional 9.8% and 11.3% of the variation when entered in that sequence to give a total of 46.9% of the variation explained by all variables in the regression. The greater percentage of variability explained by year variables in the Ruby-crowned Kinglet compared with the White-throated Sparrow is presumably a reflection of greater year-to-year variability in populations of the former species. In both species year variables explained a highly significant amount of the variation ($P << 0.01$) when entered last. This provides evidence that in these migratory populations there are measurable annual fluctuations that are unrelated to the other variables in the regressions.

Significance levels of year variables in the regression equation for the White-throated Sparrow indicated that in 1962, 1963, 1966, 1967, 1971, 1972, 1975, 1976, and 1977 the level of migration was significantly greater than in the reference year (1970). All of the coefficients of year variables are positive because the reference year had the lowest migration level of any year; negative coefficients would indicate years with lower migration levels than the reference year.

Regression results for 6 species during spring migration are summarized in Table 2. R^2 varied from 0.268 in the Winter Wren to 0.630 in the Common Flicker. Mean birds/day gives a rough indication of the relative abundance of each species and in general the more abundant species had higher R^2 values. Two to seven date variables (mean 4.6) and one to nine weather variables (mean 5.8) for each site were included in the regressions. Every date and weather variable except A1SEV2 and A1SEV3 was included in a regression for at least one of the six species. The most frequently included variables for date were A1D2 (6 species), A2D1 (6), A1D1 (5), and A2D6 (5); and for weather were A1TP (6), A2TP (6), A1TP2 (5), and A2SWV (5). Interpretation of the significance of individual date and weather variables in relation to migratory behavior is often difficult because of correlations between variables, and is outside the scope of this paper.

MIGRATION COUNT INDICES

The Migration Count Indices shown in Figure 2 indicate that migration levels at Long Point fluctuate substantially: in the period 1962–79, 28 significant differences were detected among 77 possible comparisons between successive years in six species. The Winter Wren and Ruby-crowned Kinglet, two species believed to be subject to high mortality in cold winters, showed wide fluctuations in numbers with coefficients of variation (CV) of 63.7% and 66.8%, respectively ($n = 14$ and 16, respectively, using only indices based on 20 or more cases). Both species had low numbers in 1963–64 and 1977–79, and relatively low numbers in 1970. The Hermit Thrush and White-throated Sparrow also occurred in low numbers in 1970, but over the long-term the indices for these species have been relatively stable with CVs of 42.1% and 38.6%, respectively. The Yellow-rumped Warbler indices also fluctuate in a relatively narrow range (CV = 36.0%), but in this species there are indications of a decline in numbers, especially in the last five years. In the Common Flicker the overall variation is greater (CV = 61.4%) and indices have averaged substantially lower in the last five years than in the period 1962–74.

Indices for 1962–70 were compared with those

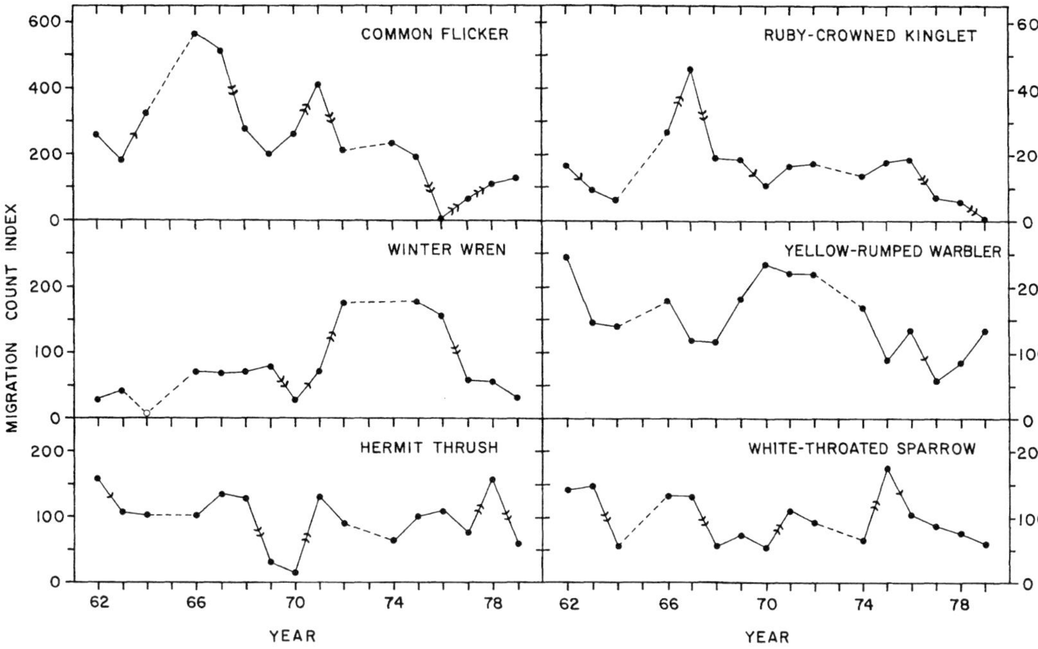

FIGURE 2. Migration Count Indices (MCIs) for six species during spring migration at Long Point, 1962–79. Solid circles = MCIs based on 20 or more cases; open circle = MCI based on 10–19 cases. Indices based on fewer than 10 cases were excluded, and those years are spanned by broken lines. Single and double arrows indicate significant differences between successive years at the 5% and 1% level, respectively.

for 1971–79 for each of the six species in Figure 2 (Wilcoxon 2-sample Rank Test). The only significant difference was in the Common Flicker in which indices for 1971–79 averaged lower than for 1962–70 ($P < 0.05$).

Validation of MCIs as Population Indicators

Correlation between MCI and BBS indices is shown in Table 3. Seven of the eight simple correlation coefficients (r) are positive and two are significant. The mean r is 0.429 and is significantly greater than zero ($P < 0.01$, one tailed t-test), providing evidence that the two indices vary in parallel. It might be argued that correlation coefficients for the central and southern regions in the same species are not independent and both should not be included in the tests. If the two southern region coefficients are excluded, the mean r becomes 0.396 and remains significantly greater than zero with $P < 0.05$.

In two species it was possible to calculate multiple correlation coefficients (R) between the MCI and BBS indices for two regions. In the Common Flicker R differed little from r for the central region for the same years, but in the White-throated Sparrow two BBS indices together explained about 24% more of the variation than either one alone.

DISCUSSION

This paper describes a method for measuring year-to-year changes in numbers of migrants at one or a series of sites on the migration route of a species. The index of migration level (MCI) is corrected for effects of date, site, and weather factors and allows tests of significance of differences between indices in different years. The method is illustrated here for counts of small nocturnal landbird migrants, but the general model is probably applicable to a wide range of situations that involve counts or other samples of migrants.

Although the computations are quite complex, the indexing method is designed to use data that are derived from field procedures that are simple and straightforward and that are already available at many migration stations. Several years of data will be necessary from any site to properly assign variability to various factors. Although this is a disadvantage for new migration stations, it is an advantage for established ones such as many European bird observatories and North American hawk migration lookouts. Once

TABLE 3
CORRELATION BETWEEN MIGRATION COUNT INDICES AND BREEDING BIRD SURVEY INDICES

Species	BBS region[a]	Sample size[b]	Correlation coefficient[c]
Common Flicker	C	9	0.641
	S	11	0.415
	CS	9	0.642
Winter Wren	C	8	0.168
Hermit Thrush	C	9	0.464
Ruby-crowned Kinglet	C	9	0.864**
Yellow-rumped Warbler	C	9	−0.331
White-throated Sparrow	C	9	0.572
	S	11	0.636*
	CS	9	0.786*

[a] Breeding Bird Survey region: C = central Ontario and Quebec; S = southern Ontario and Quebec (Erskine 1978).
[b] Years in which MCIs were based on fewer than 20 cases or BBS indices were based on fewer than 20 routes were excluded.
[c] When one region is given (C or S) the coefficient is the simple correlation coefficient between MCI or BBS index. When two regions are given (CS) the coefficient is the multiple correlation coefficient between MCI and the two BBS indices. One and two asterisks indicate coefficients that are significant at the 5% and 1% levels, respectively.

the necessary computer programs and data handling procedures have been set up, entering new data and calculating indices each year should prove to be a relatively simple process.

The MCI is a measure of the migration level at the observation site(s) in a particular year, corrected for some of the confounding effects of environmental factors, but it is not necessarily an index of population level. The Long Point MCIs reflect trends in another presumed population index, the Breeding Bird Survey Index, thus providing evidence that these MCIs do track population changes at least to some degree. Nevertheless it must be borne in mind that factors other than population change may influence MCIs, and that it may be difficult or impossible to assign variation to them. Such factors may include year-to-year changes in the site (including changes in characteristics, vegetation or food supply), changes in the migratory behavior of the species sampled (e.g., change in speed or route of migration), changes in sampling procedures and other consistent errors between years, and effects of environmental factors not used in the regression analysis. Sampling the same population at several sites will help to reduce the effects of random year-to-year changes at individual sites and strict standardization of counting or other sampling procedures over long periods of time is clearly desirable. Because consistent errors in sampling between years will introduce biases into the indices, field procedures should avoid practices that might lead to such errors or should include methods for correcting them. An example of a possible source of consistent errors is the use of counts made by different observers in different years. Use of data from many observers at several sites is likely to reduce such effects, even if some individual sites are subject to such errors.

In the examples given here, 27–63% of variability in counts was explained by year, date, site, and weather variables (Table 2). Better standardization of field procedures or skillful choice of additional or alternative weather variables for inclusion in the regressions might lead to higher values of R^2 (explained variation) and lower standard errors. In turn this would give improved resolution of differences between years. In lieu of standardization of field methods it may be possible to explain additional variation by including one or more variables for sampling effort, e.g., number of observers, hours of observation, net-hours. To meet the assumptions of the regression, however, the values of sampling variables must be independent of bird numbers, i.e., sampling effort must not be influenced by bird numbers. Except for sampling effort variables that are clearly independent of bird numbers, a safer approach is to measure correction factors in some way and apply them to the data before starting the regression analysis. Whenever possible, however, it is preferable to standardize procedures so that such correlations are unnecessary.

Because factors other than population change may influence MCIs, it is prudent to be cautious in drawing conclusions about apparently significant year-to-year and short-term changes in migration levels, at least until we have had more experience with MCIs. Examination of the behavior of MCIs in relation to other population indices and to short-term changes in avian environments will allow us to develop a better understanding of the relationships between population dynamics and migration levels. There is less reason to believe that nonpopulation factors would consistently influence MCIs over longer periods of time, however, especially if the indices are based on data from more than one site. In this paper, I was able to show that indices for 1962–70 were significantly higher than for 1971–79 in the Common Flicker, but not in five other species. Based on Breeding Bird Survey results, Finney et al. (1980) also noted the recent decline of the Common Flicker in central Canada. Ability to corroborate such long-term trends is one of the objectives of migration indexing.

In the data used in the central region correlations in Table 3 there were 17 statistically significant changes in the MCI between successive years compared with seven in the BBS index, from a possible total of 41. For the two species with southern region populations the corresponding figures were eight and three, respectively, from a possible total of 18. That the MCI shows more significant differences than the BBS index is probably due to relatively large fluctuations in migration levels at Long Point and does not necessarily indicate greater sensitivity to population change. The coefficients of variation of the MCIs averages 53.0% (range 39.7–67.0%) whereas for the central region BBS indices in the same years they averaged 26.4% (range 9.3–42.9%). Svensson (1978) argued that such differences between coefficients of variation indicate that migration indices are less effective at detecting changes in population levels than are breeding bird indices. In the present state of the art, I regard this as an untenable argument since we do not know enough about the characteristics of either type of index in relation to true population changes to state what the coefficients of variation mean. A plausible hypothesis, for example, is that breeding bird indices vary less than do bird populations because of the inability of observers to detect very high or very low breeding densities and/or because populations of territorial birds vary less than the total population. More analysis of data from as many sources as possible is needed to elucidate the behavior of populations.

This brings me to a discussion of the potential value of migration indices as measures of population change. As far as I can determine wide-scale population censusing or indexing is usually undertaken for one or both of two purposes: (a) to monitor the condition of the birds' environment and (b) to study population dynamics.

To monitor environmental quality there must be a clear association between the bird species and the habitat or geographic area we wish to monitor. At first sight it would appear that breeding bird indices have a distinct advantage over migration indices in this respect, but we must remember that most species only spend a small proportion of the year on their breeding grounds, especially at high latitudes. An extreme example is the Least Flycatcher (*Empidonax minimus*) which is estimated to spend an average of no more than 64 days or 17.5% of the year on the breeding range (Hussell 1981b). If we are to use such species to monitor breeding habitat quality we must also have knowledge of conditions in the wintering areas and along the migration route. Thus, breeding bird indices alone are inadequate for monitoring habitat quality except for purely sedentary species.

Although migration indices can be associated with a particular migratory population (e.g., the population that migrates through Long Point), we are often uncertain which breeding and wintering areas are represented in those populations. From band encounters, we know that some of the White-throated Sparrows that migrate through Long Point winter in Alabama (unpubl. data), but we have almost no precise information about where they spend the summer. More detailed analyses of band encounters, perhaps supplemented by other approaches (e.g., Kelsall and Calaprice 1972, Taylor 1980), are needed to determine breeding areas, migration routes and wintering areas of subpopulations if information from migration indices, and indeed from breeding and wintering censuses, is to be fully utilized. Once distinct wintering and breeding ranges are known for migrant populations, migration indices will provide a way to examine the structure of and fluctuations in populations at a time during the annual cycle for which such information was not previously available.

Migration indices may prove to be most useful for those species whose populations are not easily monitored in other ways, because of low density, inaccessibility or difficulty in detecting them in the breeding or wintering areas. In Canada many song birds such as the Gray-cheeked Thrush (*Catharus minimus*), Cape May Warbler (*Dendroica tigrina*), Blackpoll Warbler (*Dendroica striata*), and White-crowned Sparrow (*Zonotrichia leucophrys*) are undersampled by the Breeding Bird Survey because their ranges are relatively inaccessible or their songs difficult to detect or both. Moreover many of the same species winter south of the United States where their populations are difficult to monitor. Most raptors are also undersampled because of low densities and/or low detectability. The method described here should be suitable not only for small nocturnal migrants, however, but also for diurnal migrants such as hawks, that concentrate along leading lines in response to weather conditions. It may also be applicable to coastal waterbird migrants such as loons, eiders and scoters.

Finally, it should be noted that all of our methods for monitoring bird populations are relatively primitive and uncertain when compared, for example, to those for determining human population changes. When preservation of bird species, populations or habitats is involved, we need to marshall all the evidence we can from as many independent sources as possible to make a strong case. None of the present methods has been validated against an absolute measure of population. Under these circumstances two independent methods that show the same trends will always be more than twice as convincing and useful as one, even if one of the methods alone seems more "efficient" than the other.

ACKNOWLEDGMENTS

This paper is a publication of the Long Point Bird Observatory.

The research developed out of earlier work conducted for the Canadian Wildlife Service (C.W.S.) under contract 05SZ.KL210-6-5849, Department of Supply and Services, Canada. Recent work was assisted by grants from the Ontario Ministry of Culture and Recreation and the Canadian National Sportsmen's Fund. I thank H. Boyd, F. G. Cooch, G. E. J. Smith and S. G. Wendt of C.W.S. for their help and advice. G. E. J. Smith provided an earlier version of equation (1) applicable to a single site. C. J. Risley and A. B. Lambert contributed importantly to the development of the research particularly through discussion of the methods and assistance with computing and compilation of data. E. Nol and D. Fieseler also assisted with data compilation. My thanks are extended to H. Boyd, E. H. Dunn, P. H. Geissler, R. D. Ohmart, C. J. Ralph, W. J. Richardson and G. E. J. Smith for their comments on an earlier draft of this paper. I wish to express my appreciation also to A. Salvadori, Department of Information and Computing Science, University of Guelph, to Dr. and Mrs. A. D. Brewer, and to Dr. and Mrs. A. L. A. Middleton for their assistance in various ways during this study.

Unfortunately it is not possible to acknowledge individually all of the observers, banders and assistants who gathered the field data at the Long Point Bird Observatory and thereby indirectly stimulated this research. The field work was supported at various times by grants from the Canadian National Sportsmen's Show and the Federation of Ontario Naturalists, as well as by many individual donors. I thank them all.

Studies in Avian Biology No. 6:103–109, 1981.

AN OVERVIEW OF GRID-BASED ATLAS WORKS IN ORNITHOLOGY

MIKLOS D. F. UDVARDY[1]

ABSTRACT.—Distributional atlas work began in the British Isles in 1950 and in Scandinavia about the same time. These efforts by botanists were very successful, indicating on an area basis not only the occurrence but also frequency of the flowering plants. The British Trust of Ornithology soon followed suit; by 1978 the British, French and Danish Atlases of bird distribution have been published and several others are under way in Europe. The International organisation of European atlas work has its centers in Leeds, England and Gambleaux, Belgium. A survey will be presented telling about the status of these and of the Australian, New Zealand, and USSR atlas works. All above undertakings have in common that they use the 10 km^2 square, viz. 50 km square, grids. The second half of the presentation will be devoted to discussing the advantages of the grid system for gaining meaningful population estimates on an area basis.

GRID-BASED ATLASES AND BIRD ATLASES

What is a grid-based atlas?—A grid is a "network composed of two sets of uniformly spaced straight lines intersecting in right angles." An atlas is a "collection of maps bound together." The first atlas was published in 1595 by Gerhard Kremer, Flemish geographer. Kremer also introduced the projection of the globe which was named after his Latin pseudonym: Mercator.

The most widespread grid used on maps is the intersection of latitude and longitude (meridian) lines, or smaller, rectangular squares formed by divisions of the latitude/longitude (for short, "latilong") grid. From this grid the geographic position, the "coordinates" of each locality (town, collecting place, study plot, etc.) can easily and accurately be expressed, though the calculation is cumbersome since degrees, and their divisions in minutes and seconds, are used. Latilong-based grids were used lately in the gross mapping of biological localities on large scale maps, e.g., those of the Atlas of Speciation in African Passerine Birds (Hall and Moreau 1970). The Breeding Bird Survey of North America uses 1° and ½° squares (Robbins 1977); the Zambian Bird Atlas project uses 30′ squares (Dowsett 1979).

Long after the introduction of the metric system of measurements by the great majority of the world's political entities, the world atlas has been enriched by a metric grid with its basic unit the 10 km square (or block, but geographers prefer the word *square*). Mercator used his projection with the Equator as base; this way areas up to 15° north and south of the Equator (within which many of the important trade routes of his time were situated) showed true areas without distortion. The transverse Mercator projection uses segments of the globe along the meridians 6° of latitude apart as basic units. The position of a locality in this system is given by its distance, in kilometers, from the Equator, and, latitudinally, from the nearest base meridian. Thus two numerals describe each position e.g., for each 10 km square: the distance, of its southwestern corner, from the sixth meridian east of it (in the Western Hemisphere), and the distance from the Equator.

The Universal Transverse Mercator (UTM) grid is international. Besides uniformity, its main advantage is that it uses the decimal system and its units may be a division or multiplier of the basic unit, i.e., 1, 2, 2½, 5, 10, 50, 100 km etc. all easily converted. The position of any locality on the globe is thus precisely stated. Asilomar, California, for instance, is 595:4053 in Zone 10 of the UTM system: it is situated 595 km east of the zone boundary, the 126° W meridian, and 4053 km north of the Equator. A rare collecting site can be described to the nearest meter on the two axes of the numbering system.

In Canada as in Europe and most other countries the topographic maps are based on the UTM grid. In the USA this grid developed in the years following World War II by the Geological Survey; it is the basic grid of many states, and is used exclusively for military mapping. Many new topographic maps or those recently corrected show the UTM grid either in black, in blue, or at least the metric coordinates are marked on the margins of the map sheets by blue ticks. On the latter the completion of any desired metric grid can be accomplished either by a "master" overlay or by the use of a T-shaped ruler and India ink or pencil.

The first biological use of a metric grid on a wide geographical base known to me (Udvardy 1969) was the Atlas of the Distribution of vascular plants in NW Europe (Hultén 1950). The UTM grid (10 km square grid) was first used by the Atlas of the British Flora (Perring and Walters 1962). One example of its research advantages is demonstrated by Järvinen and Väisänen

[1] Department of Biological Sciences, California State University, Sacramento, California 95819.

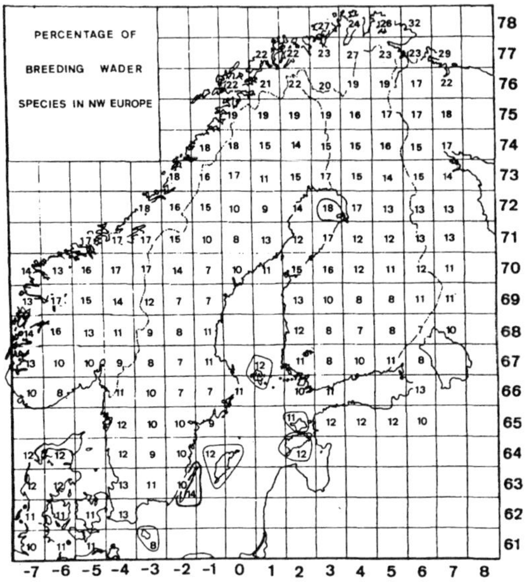

FIGURE 1. Percentage of wader species (of all land birds) in NW Europe. The S to N coordinates of the squares indicate distance, in 100 km, from the Equator. The W to E coordinates indicate, also in 100 km units, the distance from the base meridian of the Finnish National Grid, i.e., 20°E Latitude. Negative 100 km values indicate merely that they are measured westward from the base meridian, an ad hoc procedure for the sake of the wader study only. From Järvinen and Väisänen (1978), (reproduced by permission from Oikos).

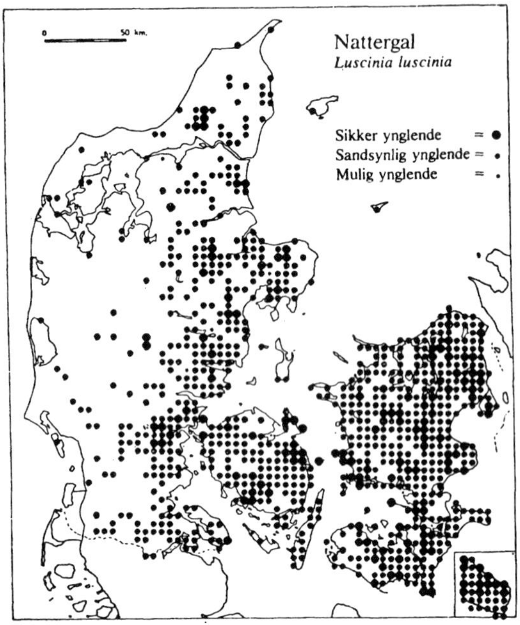

FIGURE 2. Breeding distribution of the Thrush Nightingale in Denmark during 1971–1974. Diminishing dot sizes indicate confirmed, probable, and possible breeding in the 5 km square blocks of the Danish atlas, from Dybbro (1976).

(1978) in their work on the zoogeography of wading birds in Northern Europe (Fig. 1). Several countries are following the British Isles in recording biological data by the UTM grid coordinates rather than latitude and longitude. Coordinates in degrees, minutes and seconds are difficult to compute and compare, and place names change frequently and are restricted locally, as everyone knows who has tried to decipher a specimen label from a foreign collecting locality. Enormous amounts of time, labor and uncertainty are saved by the use of this system and its adoption is a "must" for international science.

The first avian monograph using faunistic maps throughout was the Birds of the Soviet Union (Dementiew and Gladkow 1951–1954). The first avian atlas was that of the European Birds by Voous (1960). These firsts were based on the often scarce and incomplete data of distribution by outlining or shading (Voous) the general area involved. Therefore they often exaggerate distribution (cf. Dybbro 1976: 26). The Palearctic bird distribution atlas initiated by Stresemann and Portenko (1960) connects the distal points of documented breeding localities but rarely shows disjunctions or gaps of the area. Yet there are large expanses of unsuitable habitat where the species cannot and does not live or has died out. Most such atlases use data from various past periods and do not show fluctuations of distribution. Yet these fluctuations may be substantial. In Denmark, for example, 42% of the breeding avifauna of a 110-year period had fluctuating borders; in Hungary, 19% in a 100 year period had fluctuating borders (Udvardy 1970).

The above show clearly that reliable, clear distributional knowledge can only be gained by concentrated effort over relatively large areas during a limited observation period (a few consecutive years) to be repeated at greater time intervals, e.g., several decades. The exploration of any country is haphazard and this applies to breeding bird distribution as well. Willing data collectors have to be organized—enthusiastic amateurs have to be led by scientists to spend their efforts wisely; this can best be achieved by using an uniformized geographic scheme, i.e., a grid.

The five years' project dealing with the breeding birds of Britain and Ireland (1968–1972) was the first such accurate project involving over 1500 regular and about 15,000 ad hoc observers covering 3862 grid units of 10 km square (Sharrock 1976). Subsequently France (Yeatman

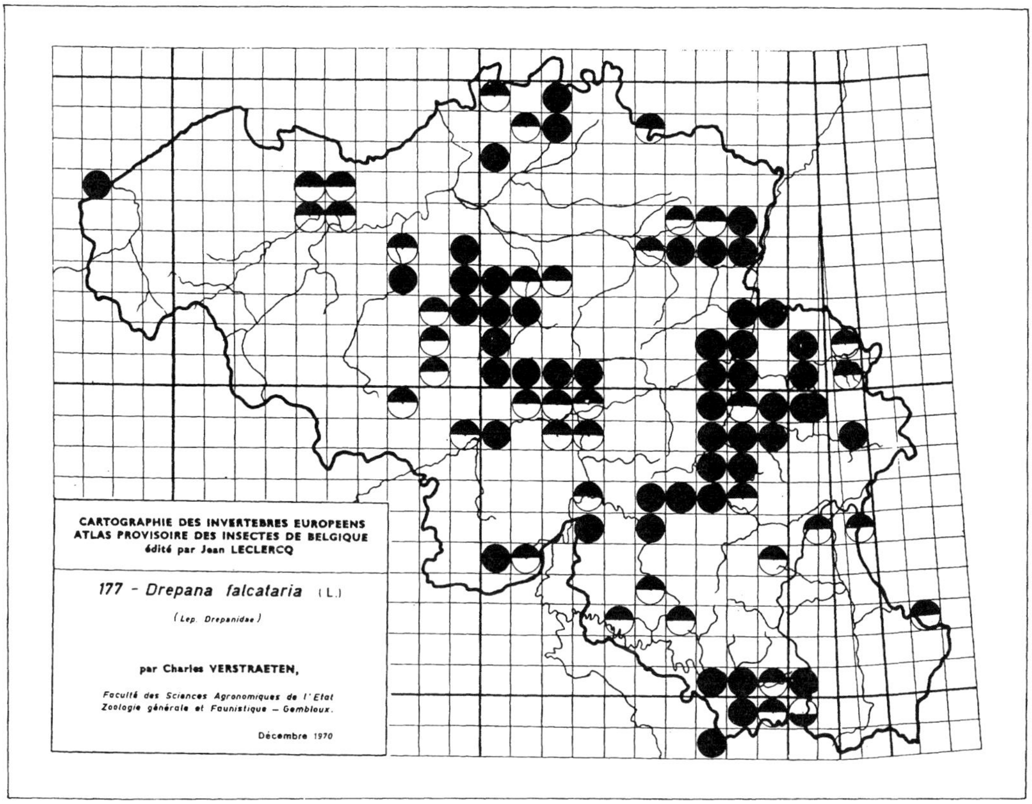

FIGURE 3. Distribution of the butterfly *Drepana falcataria* (L.) (Drepanidae, Lep.) in Belgium as shown by Ch. Verstraeten on the UTM grid map of Europe. Full dots are observations later than 1950, half-dots above are those before 1950, and half dots below are literature data or data from before 1940 where the author has not verified the material. Note that near the eastern border of the country the European grid changes zone to compensate for the earth's curvature. From Leclercq (1970).

1976), Denmark (Dybbro 1976; see Fig. 2), and the Netherlands (Texeira 1980) published, Belgium, Czechoslovakia, Estonia, Lettonia and Lithuania (Kumari in litt. 1980), Finland, Poland, Sweden, Switzerland, West Germany and others started their national breeding bird atlases (Pinowski and Williamson 1974, Pinowski et al. 1977). In the southern hemisphere, New Zealand has published a grid-based provisional atlas (Bull et al. 1978) but breeding occurrence records are still incomplete. This is an important example of the fact, emphasized already by Leclercq (1967), that no matter how spotty the data, they have to be brought together, for documentation of our state of knowledge at the time but chiefly because it spurs interest in further field collecting. The Australians are working in several parts of their large continent on atlas projects, four of which have been completed (van Tets in litt.). There are several atlas projects under way in Africa, notably in Ethiopia, Kenya, Malawi, Rwanda and Burundi, Somaliland, South African Republic, Tanzania, Uganda, and Zambia (Dowsett and Dowsett-Lemaire 1979, Dowsett 1979).

The botanists of Europe are working on their united plant distribution atlas. The European Invertebrate Survey (e.g., Heath 1970) notes all data previously collected or published as well, though for these different symbols are used on their maps. A typical example is provided by the atlas of Belgian insects (Leclercq 1970) using the UTM grid for that country (Fig. 3). The European Ornithological Atlas (Sharrock et al. 1977) is an undertaking planned and prepared for 1985–88. The UTM grid is to be used with 50 by 50 km squares, since some of the countries are large and have few observers.

The North American bird atlas projects were adequately summarized by Robbins (1977). He mentions nine states of the USA where atlases or mapped quantitative studies are in progress. Since that date several more have started there, and one in Canada. Owing to the use of the "imperial" system of land measurement, the metric grid has generally not been utilized. Consequently few of these are easily comparable with one another or with distribution atlases elsewhere. Two adjacent, smaller areas (counties), however, in the state of Maryland happen to be situated at the latitude of 39°N where the standard size of the U.S. detailed topographic map (scale 1:24,000) i.e., 8′45″ × 7′30″ almost is divisible into six 5 km squares. Here in these counties immediately adjacent to Washington, D.C. a 3-year "pilot project" was carried out and coordinated by Klimkiewicz and Solem (1978), in 60 plots of 10 km square and 136 plots of 5 km square each. So far this project is the only one somewhat comparable with the atlas works elsewhere on the km square basis. Unfortunately, due to the spheroid surface of the earth this method of obtaining 'atlas blocks' is not satisfactory at other latitudes. The 1:24,000 map sheet of Monterey, California, for example (at 36°30′N and 122°00′W) is divisible into six squares, each of the size: 5.61 × 5.3933 km, yielding blocks of 30.2566 km² rather than 25.0000 km² of the 5 km square blocks in the UTM system or the 5 km square ±0.3% as in the case of Montgomery and Howard counties of Maryland in the above mentioned study. To illustrate the discrepancies in using the latitude/longitude based grid system, consider that a 1°00′ distance of two meridians is:

100.93 km at 25°N—about the Florida Keys
85.37 km at 40°N—about Philadelphia, PA
78.82 km at 45°N—about Seattle, WA
47.16 km at 65°N—about Fairbanks, AK

—thus a grid based on one degree distance is more than twice as large at Fairbanks as at the Florida Keys.

BIRD ATLAS PROJECTS USED FOR POPULATION ESTIMATES

An atlas project generally is executed during three, five or more breeding seasons. Observers are assigned to squares, by the planning and supervisory personnel and cover their grid square several times in the field until sufficient coverage is reached. Difficult squares are covered during subsequent years, or new observers cover them again. In Europe criteria for breeding evidence have been agreed upon previously, following international rules. Upon completion of the field seasons, editing personnel scrutinize the data gathered on breeding occurrence.

Population estimates (or counts) can theoretically be obtained in several ways (Table 1). In the following, the categories of Table 1 will be illustrated by examples.

Categories 1a and 1b.—The Polish workers (1976–80, Bogucki 1977) estimated the numbers of breeding pairs of each species in 10 km squares in 5 categories: 1–10, 11–100, 101–1000, 1001–10,000, and over 10,000. Where the actual number of breeding pairs could be estimated, these were also to be recorded. These categories are very wide and inaccurate, but their decimal nature enabled Bogucki and other editing personnel to arrive at average frequencies for larger (50 or 100 km square) units by a simple mathematical formula. So far verification of the validity of these frequency values by comparison with actual censuses is wanting.

The Dutch atlas project (1973–77, Texeira 1980) started a purely qualitative effort, but for the last three years the field workers of the 5 km squares were asked to census the breeding pairs of territories of 50 selected species. This change in goal occurred upon the insistence of the Dutch nature conservation organizations which recognized the value of the atlas project in recognizing and estimating populations of species either themselves potential objects of conservation measures or indicators of threatened habitats. One finds on this list the eight big raptors, 17–18 wetland breeders, 15 songbirds, and some others. Estimated and actually counted pairs were noted by different symbols. The Dutch atlas/census years would also fit into category 2b of Table 1.

Category 1b.—The Danish atlas work (1971–74, Dybbro 1976) in 5 km squares lists the 20 commonest species arrived at in three different ways: species registered in most squares; those registered as sure breeders in most squares; and those 20 species whose registration as sure breeders is highest (percent) of all the squares where they occurred. A total of 30 species comprise the three lists, and, out of 189 breeding species the commonest were the Barn Swallow (*Hirundo rustica*), the House Sparrow (*Passer domesticus*) and the Starling (*Sturnus vulgaris*), followed by the Blackbird (*Turdus merula*) and the Great Tit (*Parus major*).

The Swedish atlas data (1974–83, Svensson 1977, 1979c) are planned to be, and partially already are, evaluated for frequency of occurence by the percentual value of the 5 km squares covering a larger area unit. Svensson's grid frequency map of the Icterine Warbler (*Hippolais icterina*) demonstrates well the usefulness of atlas maps in indicating density of dispersion and distributional limits (Fig. 4).

Bezzel and Utschick (1979), evaluating the German atlas projects, recommend that census would follow upon the atlas field work. As the Swedes, these authors also use grid frequency,

TABLE 1
POPULATION ESTIMATES BASED ON ATLAS PROJECTS

(1) Based on the actual data supplied by the atlas project	(1a)	By atlas contributors themselves: noting onto their field record cards their estimates of numbers breeding in their square
	(1b)	By the editing personnel: Summarized for larger area units which contain several squares
(2) Atlas field work supplementd by simultaneous census work done by the field workers	(2a)	Concerning very common species
	(2b)	Concerning species, usually uncommon, possessing characteristics of their breeding habits which make their censusing easy
	(2c)	Concerning very rare species
(3) Atlas project data are used by the editing personnel in combination with results of previous censuses		
(4) Atlas project followed by census work which is based on atlas results		

e.g., the percentage of squares occupied by one species compared with the total number of surveyed squares. Grid frequency is a value corresponding to the probability of finding a species on an area equal to the grid size: it is determined by the frequency of occurrence and the evenness of dispersion of the breeding pairs, which again is partly dependent on the nature of the habitat. They analyze the relation of grid size and grid frequency with some simple but elegant diagrams, and conclude that the more even the dispersion of the species and its habitat, the smaller is the optimum plot with 100% grid frequency value. Wink (1980) further analyzes the 1974–78 grid-mapping project (very small, 2.2 km squares on an area of 2.400 km^2) for population size estimates and fluctuational trends using the grid frequency concept. A pilot project of this size and intensity proved good for calculations of optimal grid size. Wink concludes that grid frequency values serve population size estimates especially in finding trends in subsequent years and other time periods. It is not usable for too common species (when grid frequency reaches 100, i.e., all squares contain the species, the map shows merely distribution). Likewise when grid frequency is too low, in rare species, such calculations are not feasible.

Category 2a.—The British Breeding Bird Survey, a long-range census project, preceded, paralleled and followed the 1968–72 atlas project years, and its results augmented those derived directly from the atlas data in yielding general estimates about the magnitude of each species' population on the British Isles (Sharrock 1976).

Category 2c.—Counting of single, rare or endangered species in the grid squares was carried out also by the Danes and Swedes. Both these projects, as can be seen from the Danish atlas and the preliminary Swedish work, seem to have estimates of adequate accuracy of populations of larger raptors, i.e., *Accipiter gentilis* and *Buteo buteo,* over areas as large as Denmark (little less than 44,000 km^2) and Sweden (about 450,000 km^2).

Category 3.—All three major atlases published in the 1970s (the British, French and Danish) gave a population estimate of most breeding species and these were based on a combination of grid frequencies, estimates or accurate counts performed by atlas field workers, and censuses executed previously by various agencies or groups regarding certain single species. All atlases emphasize that such estimates are rough at best, and serve as indicators of the magnitude of population size, to be followed by more accurate censuses.

Category 4.—Grid frequencies, complete atlases or incompletely censused grids all act in stimulating follow-up field work in the form of censuses etc. Because even the oldest published atlases are fairly recent they do not yet provide data for evaluation.

DISCUSSION

How does atlas work serve quantitative population ecological studies? Answers fit into the following objective framework:

Autecology

Population Data Study (statics):

(1) Total population of a species over its whole distribution area (range).

(2) Total population of a species within a certain geographic area (continent, island, mountain chain, country, state, province, etc.).

Study of Population Changes (dynamics)

(3) Trends of population increase/decrease/stability of a species, throughout its distribution area.

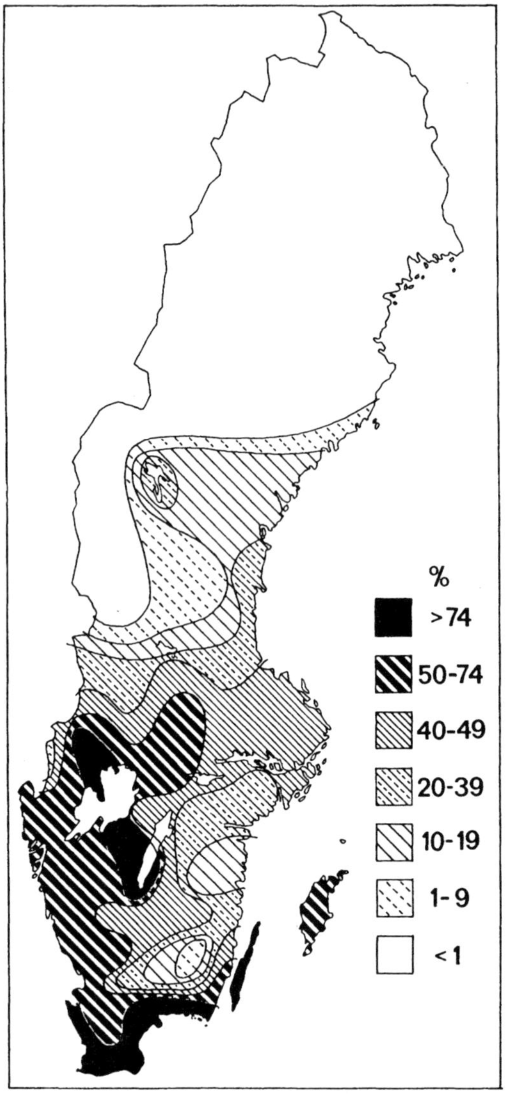

FIGURE 4. Grid Frequency classes of the Icterine Warbler based on the 1974–78 data of the Swedish breeding bird atlas project. The percent values indicate percentual occurrence in the 5 km square blocks of larger area units. From Svensson (1979).

(4) Local trends of the dynamics of species populations.

Synecology (Comparative population dynamics from the point of view of)

(5) The total breeding bird population of a geographic area.

(6) The numerical relations of "bird communities," "guild," etc. within certain habitat types and ecosystems.

The most accurate estimates come from census works based on area units. Censusing any larger area (conforming to points (1) and (5) above) necessitates the use of smaller samples and project their results, proportionately, to larger areas. In a summarizing paper, Merikallio (1961) applied this method first to bird population estimates of large areas, in his case, the total bird population of species nesting in Finland.

The only efforts made to count or estimate the total (world) population of species, objectives (1) and (2), concern dwindling, rare populations of endangered species on the brink of extinction; these usually are populations ranging from less than 100 to some 10,000's of breeding pairs. Game birds (galliforms and anseriforms) are censused mainly for managemental purposes, by indirect methods, and not necessarily over their total area but usually over the area of managemental responsibility.

For population ecological as well as conservational reasons (widespread agreement is found in the atlas literature here reviewed about the role of birds as biological indicators) the early clarification of objective (1) is a "must." Whereas the primary task is achieved by censuses, the subject of this symposium, the geographic aspect can only be achieved by accurate distribution mapping of *species* and habitat. I would suggest that the combined goal of *all* census and atlas projects should be to clarify by the year 2000: What is a common, uncommon, scarce, or rare passerine, or nonpasserine bird?

In objective (3) the atlas initiatives and their theoretical foundations, as we have narrated above, would enable the ornithologist of the area to read population dynamical trends from the grid frequency data of subsequent time periods: direct estimates by the field workers serve the same end and since these mainly include conspicuous and large, therefore often endangered species, the goals of the ornithologist-scientist and of the ornithologist-conservationist often coincide and enhance one another. The same applies to objective (4), local trends.

Objectives (5) and (6) suggest the same problems as the previously treated ones, but in a synthetic way. Area-based censuses preferably preceded by area mapping and habitat mapping are the answer. In this respect again, the goals of the avian scientist and of the "environmentalist" coincide. A good example might be the California subspecies of the Yellow-billed Cuckoo (*Coccyzus americanus*). By 1972 it was apparent that this bird was rare (Leach et al. 1974). It has been studied, together with other rare species (Gaines 1977) and these studies show that the total habitat needs to be studied (R. E. Warner unpubl. MS).

In North America the U.S. Geological Survey recently issued a series of large-scale (1:250,000) special maps which show "land use and land cover." Thirty-seven types of land use/cover are superimposed on a 10 km square UTM grid, and on the network of water, roads, boundaries and villages or towns. These maps could provide the common and uniform basis of coast-to-coast census and atlas work of the breeding birds of North America. Avian biology, biogeography and ecology as well as local and global conservational goals would benefit from such cooperation.

ACKNOWLEDGMENTS

This paper has been greatly improved by the generous editorial help of Albert L. Delisle, Jane Durham, and Robert D. Ohmart. Financial help was provided by the Foundation of California State University, Sacramento, in form of a Faculty Research Grant.

Studies in Avian Biology No. 6:110–111, 1981.

SUMMARIZING REMARKS: ESTIMATING RELATIVE ABUNDANCE (PART II)

P. V. RAO[1]

The papers presented in this session highlighted a variety of statistical problems needing attention from statisticians and biologists alike. Because there is neither time nor space to go into these problems in great detail, I shall confine my summary to some general comments on statistical contents of these papers.

In his interesting paper, Karr (1981) presents an excellent exposition of the use of mist nets as a tool for counting birds. In addition to noting advantages and disadvantages of using mist nets, Karr discusses several examples of data analysis and interpretation. The main thrust of his examples is that mist net count data can be used for studying spatial and temporal patterns in capture rates and species richness.

The methods of analysis found in Karr's paper provide good examples of techniques for handling data of the type considered. Use of binomial and regression models is certainly reasonable, but the following suggestions, if implemented, might improve the analysis further.

My first suggestion concerns the choice of binomial distribution to model the number of captures shown in his Tables 3 and 4. I have some reservations about this choice because: (1) these numbers represent counts over time intervals of different lengths; and (2) the total count is itself a random quantity. A more appropriate model in my view is a model which regards the number of captures as Poisson random variables. The expected number of captures would then represent the capture rate.

My second suggestion is to consider the model:

$$S = \frac{1}{S_{\max}} \exp(-K/T) \qquad (1)$$

for predicting species richness instead of the hyperbola used by Karr. Fitting this model does not cause any new problems because a logarithmic transformation will reduce the model into a simple linear regression model. Very often, exponential models, such as the one given by (1) are found to be effective in transforming count data to data suitable for regression analysis.

[1] Department of Statistics, University of Florida, Gainesville, Florida 32611.

The papers by Johnson et al. (1981) and Marion et al. (1981) deal with play back recording as a technique for censusing avian population. The main point in both papers is that play back recording often increases sample size by eliciting responses from birds which are elusive and not easy to detect visually. Johnson et al. (1981) provide a general review of field techniques suitable for surveying with play back recording, while Marion et al. discuss the results of using play back recording to survey five species of birds in Texas and Florida.

Several sources of bias in estimates obtained from play back recording census are noted in the two papers. Most important among these is the failure of a certain proportion of the population to respond to auditory signals. Marion et al. (1981) provide an example of a survey of Plain Chachalacas in which they develop a correction factor to correct for bias due to non-response. Because a survey of two tracts of known density yielded 44% and 59% response rates, they assume an average response rate of 50% and adjust all density estimates by multiplying by 2.0.

Certainly, the idea of adjusting for bias using appropriate correction factors is a good one. Indeed the idea of Marion et al. (1981) can be extended a step further to establish an interval of plausible values of the correction factor. Such an interval may be preferable over the subjective method of selecting an average value to represent the observed values of 44% and 59% response rates.

An interval of appropriate values of the correction factor for the chachalaca data may be calculated as follows. First, use the fact that 22 + 10 = 32 out of 50 + 17 = 67 chachalacas responded to play back recording to calculate the estimated response rate as 32/67 = .48 with a standard error of $\sqrt{(.48)(.52)/67} = .06$. Thus the true response rate may be estimated to (at approximate 95% confidence level) lie between .36 (.48 − 2(.06)) and .60 (.48 + 2(.06)), yielding a range of 1.7 to 2.8 for the correction factor. This range of the correction factor, when applied to the observed response of 1.1 birds/ha in the April 18, 1972 survey, results in the approximate 95% confidence interval of 1.9 birds/ha to 3.1 birds/ha for the population density.

It is not clear why Marion et al. (1981) did not incorporate a correction factor in their calculation of the density estimate for rails, but I am glad to see standard errors next to the estimates in their Table 1. Note the relatively large magnitudes of the standard errors, pointing out the need for more refinements in techniques of data collection and/or data analysis. Of course this last comment is not meant as a criticism of the excellent work accomplished thus far. Rather it is an expression of the fact that there are problems yet to be solved.

In his paper, Hussell (1981) describes how the multiple regression technique can be used to develop migration count indices to measure yearly bird population changes. His method is to regress the variable $Y = \text{Ln}(N + 1)$, where N is the number of birds observed, on independent variables such as year, site, date, etc. If $\hat{Y}$ denotes the adjusted estimate of Y for a given year, then the migration count index for that year is defined as $I = \exp(\hat{Y}) - 1$.

Hussell's paper provides yet another example of the power of multiple regression technique to solve practical data analysis problems. However, because of its popularity and availability of numerous computer packages for its implementation, there are many instances of improper use of regression analysis. Great care must be taken to insure the validity of assumptions underlying the model as well as to correctly interpret the model parameters. Hussell's model is more complicated than warranted by the situation, but I am pleased by his careful analysis of the underlying assumptions. In this connection, it is likely that the error structure in his data is more suitable for a time-series analysis.

The paper by Udvardy (1981) contains an overview of grid-based atlas as a tool for assessing spatial and temporal variation of avian population density.

Finally, because papers presented in this session contained several statistical tests of significance (e.g., t-tests, χ^2-tests), I would like to conclude this summary by making some comments on possible misinterpretation of such tests.

If a t-test shows statistically significant difference between two means, then all that one can conclude is that the population difference is not zero. It is quite possible that the real difference may be quite small to make it practically insignificant. If only the significance of a test is reported, there is the danger of interpreting it as indicating a practically significant difference.

A method to evaluate the practical significance of an observed difference is to construct a confidence interval. For example, the intervals (.01, .02) and (4.0, 6.0) both indicate statistically significant difference because both intervals exclude zero difference. Yet, the difference implied by the first could be considered practically unimportant in some situations.

To enable the user to examine confidence intervals, it is best to report the estimated difference and its standard error whenever possible. Indeed, as a general rule, it is important to report the standard error of every estimate. Without the standard error it is not possible to evaluate the reliability of the estimate.

Studies in Avian Biology No. 6:112, 1981.

SUMMARIZING REMARKS: ESTIMATING RELATIVE ABUNDANCE (PART II)

STANLEY A. TEMPLE[1]

We can begin this summary by reviewing the basic premises of studying relative abundances of birds. First, there should be a clear understanding that such estimates of abundance differ from absolute abundance because of biases that are often difficult to assess. It is usually hoped, however, that the estimate is at least nearly proportional to absolute abundance. Secondly, it should be recognized that estimates of relative abundance are useful primarily, if not exclusively, for making certain types of comparisons. They are most useful for comparisons within the same species between different time periods and localities, particularly when methodology has been consistent during the study. Although not as straight forward, comparisons between different methods are possible; comparisons between species are more difficult.

I can think of at least three situations in which a researcher would use measurements of relative abundance in preference to the more difficult measurements of absolute abundance. First, there are instances where data have been collected, or are fortuitously available, in such a form that it is impossible to determine absolute abundance. In this session we have heard about Christmas Bird Counts, Breeding Bird Surveys, migration counts, and atlas work.

Secondly, there are instances where a specific experimental design or objective can be most easily achieved by comparisons of relative abundance. We have heard of studies using mist-net captures to detect seasonal changes and using audio play backs to detect secretive birds.

Finally, there are situations when comparisons between time periods, localities, methods, observers (or interactions between these) cannot be justifiably made in a direct fashion, and only relative abundance can usefully be compared.

There is usually a challenge of making relative abundance data as comparable as possible before the actual comparisons are made. Even when we have not had strict control over how the data were collected, there are some simple ways to make data comparable.

One simple method is to use correction factors, not necessarily to make the estimate closer to absolute abundance, but to make comparisons more easily interpretable. We have heard about attempts to use correction factors to make migration counts more comparable between years when we know that weather, in addition to actual population change, has influenced observed numbers.

Another approach is to use a statistical method such as an analysis of covariance to adjust for sources of bias. I am surprised that we have not seen this approach being used more often. Data from migration counts and Christmas Bird Counts are certainly amenable to such analyses.

Another approach that has not been mentioned directly, but one that I feel has broad utility, is reducing data to simple terms before making comparisons (e.g., in terms of presence or absence, frequency of occurrence, rank abundance). This is a rather conservative basis for comparisons that can facilitate comparisons between time periods, localities and methods. For example, I used changes in frequency of occurrence on weekly checklists over a 37-year period to detect population changes in various species (Temple and Temple 1976). I have also compared mist-netting capture rates with results of nearby line-transect estimates. Although frequencies of capture and frequencies of detection were only weakly correlated, the rank correlation between the two measures of relative abundance was very strong.

I propose that Christmas Bird Count data could be analyzed in a similar way. If each party's checklist was treated separately instead of being combined into the count-circle total, it would be possible to calculate frequency of occurrence on party lists within the count-circle. In this way it would be possible to calculate the variance within the count-circle each year, and comparisons with other localities and years would be facilitated. An analytical change like this would avoid changes in field methodology that might discourage volunteer observers.

It seems clear that many, if not most, of our objectives in surveying bird populations can be adequately achieved by using relative-abundance data. If measures of relative abundance will allow detection of population changes that are interesting to study, perhaps we should not think of relative abundance as less appropriate a measurement than absolute abundance. I suspect that as we discuss other methods, we will find that all of them produce results that are really nothing more than relative abundances.

[1] Department of Wildlife Ecology, University of Wisconsin, Madison, Wisconsin 53706.

Studies in Avian Biology No. 6:113, 1981.

INTRODUCTORY REMARKS: ESTIMATING BIRDS PER UNIT AREA

DAVID R. ANDERSON,[1] CHAIRMAN

Many ornithological investigations require estimates of density in order to make inferences about hypotheses being tested. The most common method used to estimate density is line transect sampling and I will make a few comments on this general technique because of its importance in this symposium.

At least 26 papers have appeared on line transect sampling since 1970. I suggest the term *line transect* be reserved for the data gathering, sampling, and field measurement activities. These are becoming somewhat standardized, e.g., grouped or ungrouped perpendicular distance data, with or without a fixed width or boundary. Alternatively, sighting distances and angles can be recorded and then the perpendicular distances can be calculated.

The analysis of these distance data is far from standardized and about three dozen analysis methods can now be found in the literature. It is no longer sufficient to remark "density was calculated by the line transect method." The majority of the methods are ad hoc; they lack a firm foundation, little is known about their small sample properties, sampling variance estimators are not available, and so on. Unfortunately, several of these methods are still seeing heavy use by ornithologists.

Well based methods did not appear until 1968 (see Eberhardt 1968 and Gates et al. 1968) and approximately a dozen have appeared since then. A few of these dozen form a class that represent very good analysis methods, but are just beginning to see widespread use (see Burnham et al. 1980). These methods have estimators of sampling variance, the small sample properties are known, they are very efficient, they allow broad assumptions about the unknown detection function, and goodness of fit tests are available. Finally, it is important to note that strip transects and circular variable plot surveys are special cases of line transect theory and methodology.

[1] Utah Cooperative Wildlife Research Unit, Logan, Utah 84322.

Studies in Avian Biology No. 6:114–118, 1981.

LIMITATIONS OF THE MAPPING METHOD

HANS OELKE[1]

ABSTRACT.—The evaluation of bird territories by means of the mapping method has a number of serious limitations: the excessive effort; the rather sophisticated, nearly "bureaucratic" recommendations (= international and national rules of mapping); and the variety of personal and bird- or bird-community specific mapping errors. My discussion of errors is focused on the difficulties of mapping small study plots in tropical woodlands, on the inappropriateness of maps per se (through problems of scale, symbols and saturation) and on the limits set by time and memory for the interpretation of the mapping.

The traditional method of area-specific, quantitative bird surveying in central and northwest Europe (Germany, Czechoslovakia, Poland, part of Austria, Switzerland, Denmark, Sweden, Norway, and the United Kingdom) is the so-called mapping method (reviewed in Palmgren 1930, Enemar 1959, Oelke 1966, Williamson 1972, Berthold 1976, Zenker 1980). The numerous and mosaic-like, man-made habitats in these parts of Europe are normally suitable for the mapping method.

COMMON LIMITS OF THE METHOD

There are a great number of limitations of the mapping method which recently became apparent because of experience with ever increasing and lasting environmental impacts, and because of the results of studies on the population biology of several species. There have been a large number of mapping recommendations published after the 1969 Ammarnäs-Symposium of the International Bird Census Committee (IBCC) and nationally modified in many ways (for the Federal Republic of Germany see Oelke 1970, 1974; for the United Kingdom see Williamson et al. 1976). This proliferation is becoming an increasing obstacle. The detailed recommendations will not only distract participation or cooperation of new bird watchers, they are limiting the number of study areas. The atlas and grid net programs, although mostly qualitative, must be regarded as a solution to the difficulties of the mapping method because of their simplicity and efficiency. More data from more observers may be summarized for wider geographical areas on clear maps within a relatively short time. For examples in central and northwest Europe see Yeatman (1976), Dybbro (1976), Rheinwald (1977), Bezzel et al. (1980), Schifferli et al. (1980).

METHODOLOGICAL DIFFICULTIES

The methodological difficulties of mapping are summarized by Berthold (1976):

1) Song registrations cannot be used as basic mapping units because song activities are highly influenced by (a) inter- and intraspecific differences; (b) seasonal, yearly, and daily changes; (c) by weather conditions; (d) by differences in abundance of birds; and (e) by simultaneous singing of neighbouring territory holders. This results in marked mistakes that are masked in the registrations.

2) The plotting of bird territories using the registrations of singing males or calling birds is hindered by: (a) the extreme territory areas occupied by some species; (b) the simultaneous occupancy of several territories; (c) the displacement of territorial border lines; (d) the territorial behaviour of migrants and of unmated, non-territorial birds, and abnormal and variable pair bonds.

3) Quite normally the efficiency of detecting species differs species-specifically.

4) The size of the study area is not precisely laid down either internationally or nationally with the result that study areas differ in size and thus they differ in the accuracy of the bird population estimates.

5) Day by day activity differences are not only characteristic of song but of the behavior, especially the movement activities as a whole.

6) Maxima of certain species due only to phases of migration can lead to wrong conclusions.

7) As with any set of observations, there are large observer deviations, and even errors.

8) The evaluation of data is too susceptible to subjective interpretation.

9) Usable or reliable correction coefficents more or less do not exist.

Unfortunately, these difficulties and inconsistencies are even surpassed by the disadvantages of the nest search method favoured by Berthold (1976) (see Oelke 1977). On the basis of long experience in the practical use of the mapping method, Tomialojć (1980) has added additional weak points. For example, he discusses the limitations with the differences between territorial and non-territorial songbirds and other species; the standardization number of visits for each observer; the undescribed or overlooked duration of visits within the recommendations; and especially to the thesis that birds are singing in the center, not along the pe-

[1] Zoologisches Institut, Universität Göttingen, D-34 Göttingen Federal Republic of Germany.

TABLE 1
DIFFERENCES BETWEEN A PRAGMATIC AND A THEORETICALLY NECESSARY TERRITORY MAPPING OF A TROPICAL WOOD (KAKAMEGA FOREST[a], KENYA, EAST AFRICA, 1978, 1979)

	Achieved	Theoretically necessary
Area (ha)	2.25	10
Edge length (m)	600	1265
Quotient 2/1	267	127
Number of visits	9	10
Duration (min/ha)	100	100–150
Daily distribution between:		
Sunrise and sunset (in h)	07:30–19:00	03:00–22:00 (Arctic/Temperate Zone)
Sunrise–sunset (h)	06:30–19:00	24 (Arctic)
Maximum observation time (h)	appr. 12	up to 24
Form of control routes	fixed on trails	unfixed

[a] Characteristics of the tropical study plot: Kakamega Nature Wood Reserve, neighbouring north side of Kakamega Forest Station (sheet 102/4 Kaimosi, East Africa, Kenya, coordinates YR 707.5/026.5); semitropical rainforest, elevation 1580 m with luxuriant tree, shrub, ground cover, numerous epiphytes and lianes, approximately 25 km N of the equator; 23 tree species with a height of 38.5 m (diameter up to 101 cm), a cover of the tree layer of 51%, of the shrub layer of 50–75%, of the ground layer of 84%, on the average; approximately 2740 shrubs per ha.

riphery of the territories. Tomialojć (1980) suggests: a greater flexibility in the international rules; the inclusion of all species of the bird community; the more accurate determination of territories by contemporary contacts; more tests on the reliability in extreme mapping situations; increase of length of each visit; a revival of the discussion of edge territories; and a revision of the international recommendations of the IBCC to improve the effectivity of the mapping method. Some additional mapping experiences of mine favor this revision.

SPECIAL PROBLEMS

PLOT SIZE AND MORE COMPLEX HABITATS

The IBCC recommends that the minimum size of a study plot should be 40–100 ha in an open habitat, and 10–30 ha in a more complex habitat. These recommendations are based on experiences in holarctic study plots such as the tundras, steppes/prairies, woods/forests and man-made habitats in North America and Europe. However, these recommendations are met by nearly unsurpassable methodological difficulties

TABLE 2
COMPARISON OF BIRD CENSUSES CARRIED OUT BY ZIMMERMAN (1972) AND OELKE (1978 AND 1979) IN THE KAKAMEGA WOOD RESERVE, KENYA, EAST AFRICA

	Zimmerman	Oelke
Size of study plot (ha)	8.1	2.25
Year(s) of study	1963, 65, 66	1978, 79
Time spent (in h)	431	17
Time factor (h/ha/a)	17.7	4.1
Number of control (days/a)	appr. 16	9
Methods	spot-mapping, mist-netting, tape replay	mapping, mist-netting
Mist nets, number, length	6 × 14 m	2 × 6 m
Operation time	continuously	2–3 h/day
Abundance (pairs resp. males/10 ha)	165	196
Territorial species (*n*)	64[a]	32
Total of identified species (*n*)	125[b]	73[c]
Proportion (%) of species occurring (206)	60.7	35.4
Woodland bird species (*n*)	92	64
Percent of occurring species (154)	60.0	41.6

[a] Minor human interference as lacking road and trail nets, more distant settlements, reduced cattle grazing, minor poaching, no bird collecting incl. bird studies should be kept in mind to get higher bird species numbers in 1963–66.
[b] Zimmerman includes even species of open woodland and outside the woods.
[c] Edge species, feeding or visiting species are included. The species numbers of the Kakamega area are from the lists of Cunningham-van Someren (1979).

TABLE 3
DIFFERENTIATION OF METHODS FOR STUDYING BIRD POPULATIONS OF A TROPICAL STUDY PLOT (KAKAMEGA NATURE WOOD RESERVE, KENYA, EAST AFRICA, SEE TABLE 1, 2)[a]

	1978		1979		"Normal" mapping results[b]	
Dominant method	No. species (%)	No. territories (%)	No. species (%)	No. territories (%)	% species	% territories
Registration of singing males	2 (8.3)	6 (12.4)	2 (8.3)	7 (17.5)	50	50(–70)
Song/sight observations	12 (50.0)	31 (64.0)	11 (45.8)	20 (50.0)	30	(10–)30
Sight observations only	1 (4.2)	1 (2.1)	5 (20.8)	5–6 (12.5)	10	10(–20)
Nest records	—	—	—	—	—	10(–30)
Mist netted	9 (37.5)	11 (22.7)	6 (25.0)	7–8 (17.5)	—	—
Other identifications[c]	—	—	—	—	5	5
Sum	24	48–49	24	39–41		
Additional unproved records	5	5	2–3	3–5		

[a] Only the stationary birds of the population are included, not feeding species, visitors, and migrants.
[b] Based on bird censuses in more complex woodland study plots in North American and Germany (Oelke 1963, 1967, 1977b).
[c] e.g., feather samples, pellets, fecal rests, tracks left from running, swimming, or feeding birds.

in monitoring populations of tropical woodland bird populations (see Tables 1–3).

The examples discussed below from Kakamega Forest area are derived from study plots with previous ornithological inventories. Thus, the results of Zimmerman's (1972) population studies, the bird skin collections of the National Museums of Kenya, Nairobi, and the experiences of British banding groups have contributed supplementary and necessary information.

Without tape recorders and mist nets for catching the birds of the ground and shrub layer, an exact identification of bird species is nearly impossible. The upper canopy of the wood reaching 60 m in height in the Kakamega Forest area restricts, too, the identification of smaller birds, mostly passerines. Small study plots also present difficulties in censusing tropical woodland areas. These difficulties are (a) acoustic and visual species identification problems due to extreme species diversity and complex differentiation of habitats, (b) temporally non-fixed or nearly unknown territorial periods, and (c) unknown interspecific relations between resident bird populations and palaearctic migrants. Methodological tests on the effectiveness of the mapping method as carried out recently by Svensson (1978) or in a number of important German dissertations (Cyr 1977, Blana 1978, M. Erdelen 1978, B. Erdelen 1979) are still lacking in the tropics, at least of Africa. These tests are increasingly important along with the need to monitor population size and dynamics of birds in western industrial nations. Tropical areas are more or less affected by pollution affecting birds both during migration and on their migratorial rest grounds. The task of monitoring has not yet really started.

MAPS FOR MAPPING

The IBCC recommends maps scale 1:1250–1:2500 for mapping in woodland. In open country maps scale 1:2500–1:5000 may be used. (The German recommendations are based on maps scale 1:5000–1:10,000; see Oelke 1974.) Even when using an optimal scale and by use of special markings for orientation, putting down any location of a bird on a map is subject to error (Table 4). These difficulties increase in dense habitats or densely populated areas such as woodlands with more than 100 breeding pairs or territorial males per 100 ha with a greater number of mappings even during one visit. Each mapping (daily visit) is limited by the capacity of mapping symbols within one map. De facto, this capacity is even lower, probably 25–50%, as calculated in Table 4. This is because there are always concentrations of birds in some parts of the plot and therefore on the map, or simply because intervals must be left between the symbols/numbers for interpreting purposes.

The optimum of mapping with maps of a scale of 1:1250 can only be reached in a few instances because these maps are unwieldy. The normal map format generally coincides with the commercial typewriter formats (in Germany it is

TABLE 4

MAP SIZE (IN CM) OF A 10 HA STUDY PLOT QUADRATE IN REGARD TO DIFFERENT MAP SCALES[a]

Scale	Paper size of study plot (in cm)	Area cover of an abbreviated symbol (in m)	Differences (in m) when moving the symbol for			Capacity for mapping symbols[a]	
			1 mm	5 mm	10 mm	In theory *n*	In reality *n*
1:1250	25.4 × 25.4	2.5 × 3.75	1.25	6.25	12.5	1344	1000
1:2500	12.7 × 12.7	5 × 7.5	2.5	12.5	25	336	250
1:5000	6.34 × 6.34	10 × 15	5	25	50	83	60
1:10,000	3.17 × 3.17	20 × 30	10	50	100	21	15
1:25,000	1.3 × 1.3	50 × 75	25	125	250	4	3

[a] Area cover (in m × m) of a normal written, abbreviated species symbol (e.g., B = Buchfink—*Fringilla coelebs*, appr. 2 × 3 mm) and area differences (in m) when moving the abbreviated species symbol 1/5/10 mm on the map.

[b] Area of a symbol inclusive details on bird activities: appr. 4 × 6 mm; the same area is calculated for the interval between the symbols necessary for reading (interpretation).

29.5 × 21 cm). Normally an observer does not carry with him any further enlargements of the maps but tries to complete the registrations on *one* single map of the study plot. Rarely 1000 registrations (scale 1:1500) can be put down on such a map the size of a typewriter paper. The scale 1:1500 is sufficient in most cases for one visit (100–200 registrations on the average, in my own experience); difficulties arise during prolonged visits, and because of clustering of bird symbols on some parts of the map. In these cases even the normal daily visit map is marked by unclear and hardly interpretable parts.

Not only do the daily visit maps offer some problems, but more often the species maps, summary and combination of all daily registrations of one species per observation period, result in problems. The main purpose of the species map is to determine territories. This cannot, or can only arbitrarily, be solved in the case of high densities or clusters of registrations. For example, it is unclear in which way Erdelen (1978) and Blana (1978) established territories with a maximum of 23 visits to a 10 and 5 ha thicket and 33 visits to a 25 test study plot in mature oak forest. Both authors, however, made use of their registrations to evaluate possible additional errors as correlations between territory numbers and number of visits as well as intraspecific differences in daily and seasonal registrations.

TABLE 5

DIFFERENCES BETWEEN FIRST AND REPEATED TERRITORY EVALUATIONS (1980) FROM SPECIES MAPS OF THE CHAFFINCH DERIVED FROM MAPPING A 10 HA MATURE MIXED DECIDUOUS WOODLAND PLOT (STAATSFORST HÄMELERWALD, COMPARTMENT 128/129, PEINE COUNTY, LOWER SAXONY, FEDERAL REPUBLIC OF GERMANY, 1968–1977)

Year	Number of territories		% differences	New territory boundaries (*n*)
	Study year	Reanalysis		
1968	7	5	−29	2
1974	5–6	8	(+)25–(+)60	5
1975	4–5	4	0–(+)20	1
1976	4	3–4	0–(−)25	2

LENGTH OF INTERPRETATION OF REGISTRATIONS

The territorial mappings are normally summarized in a few parameters such as number of species, territories or territorial birds, abundance, dominance, frequency, and diversity indices. The process of interpretation has to start with these data for the normal reader. Normally, access to the primary registrations of the field data is not possible. Many raw data are lost forever because the census takers are no longer living. But it is often not sufficient even if all raw data are made available, and the necessary personal knowledge of a habitat is achieved. The census taker himself is always in the best position to analyze data because of a bulk of unmentioned, unwritten "intimate" details of a mapping process. This is quite normal and is due to: a lack of space and time; difficulties of transforming the complex of behavioural activities into graphical symbols; and very often because of the ever changing locations of birds. These unwritten details are available during the period of active research and shortly afterwards when analyzing the territories. But they will be forgotten in the course of time. After 2 or more years, perhaps even earlier, the observer hardly remembers the intricate reasons for his decisions to determine the territories. This is especially true in study plots with many registrations, or with high bird densities. I have found major differences between my species maps from 1968 and 1974–1976 of the Chaffinch

(*Fringilla coelebs*), a species with rather good detectability. This species was mapped on a 10 ha plot in mixed deciduous forest (Table 5). The discrepancies are even higher when determining the size and border lines of the Chaffinch territories.

CONCLUSION

Regardless of all attempts to register "absolute" density figures of birds, territorial mappings have to be regarded as good or sometimes better approximations. A wide spectrum of different errors will always be inherent in this ecological field method.

ACKNOWLEDGMENTS

The African (Kenya) field data were enabled by the support of Drs. G. R. Cunningham-van Someren, A. W. Diamond, H. Hagen; Mrs. J. F. Horne, Forester J. R. Oruko, Drs. H.-U. Reyer, J. Steinbacher; Mr. D. A. Turner, but especially, too, by my field assistants J. Gutwa and J. R. Ocholla. I am grateful to these persons and the Government of Kenya which issued the necessary research permit.

Studies in Avian Biology No. 6:119–120, 1981.

TERRITORY AND HOME RANGE OF THE BLACKCAP (*SYLVIA ATRICAPILLA*) AND SOME OTHER PASSERINES, ASSESSED AND COMPARED BY MAPPING AND CAPTURE-RECAPTURE

C. FERRY,[1] B. FROCHOT[1] AND Y. LERUTH[2]

ABSTRACT.—Capture-recapture by netting of banded breeding birds permits an estimate of individuals' home ranges. For five territorial species, these ranges were much larger (from two to twelve times) than their territories, estimated simultaneously by a mapping plot.

The question of the relationship between territory (the defended area) and home range (total occupied area) of a breeding bird has not been adequately addressed by ornithologists. In this paper we report on a first approach to studying this question.

MATERIAL AND METHODS

In the 1976 breeding season, the bird community was censused simultaneously by capture-recapture and by a mapping plot in an area of oak (*Quercus pedunculata*) forest (Fig. 1). The capture-recapture study used 127 nets equally spaced over 51 ha of forest (Fig. 2). Eight 4-day sessions of netting were carried out between May 17 and July 31. One-quarter of the nets were in use on each of the four days, so that each net was in use eight times. During netting days, an average of 90 min elapsed between the successive checks of each net. All the netted birds were banded on their first capture, and a record was kept of subsequent recaptures.

The mapping plot covered 14 ha inside the netting area. This plot was censused ten times in April and May. Applying the I.B.C.C. technique (International Bird Census Committee 1969) we obtained density estimates for 17 species of birds. The average area of each species territory was assessed by dividing 14 ha by the number of identified territories; this is a maximum value. We assumed that the territories of each species covered the whole 14-ha area.

To estimate the home range of birds, we used the intercatch distances of recaptured, banded individu-

[1] Faculte des Sciences, Universite, 21 000 Dijon, France.
[2] Ecologie Theorique, U.C.L., 1348 Louvain La Neuve, Belguim.

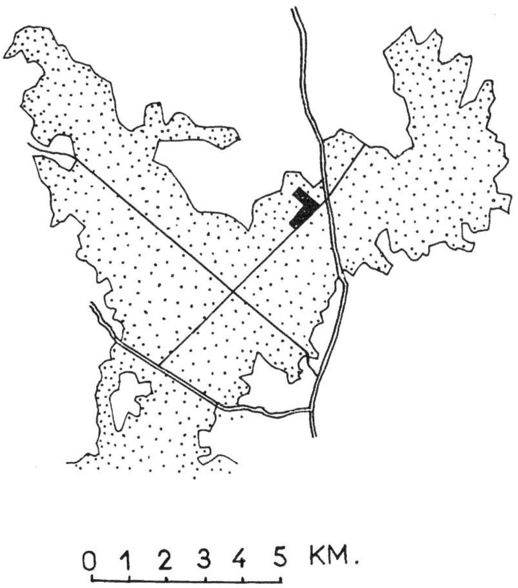

FIGURE 1. Map of the "Forêt de Citeaux," 20 km south of Dijon (Burgundy), with the netting plot (shaded area).

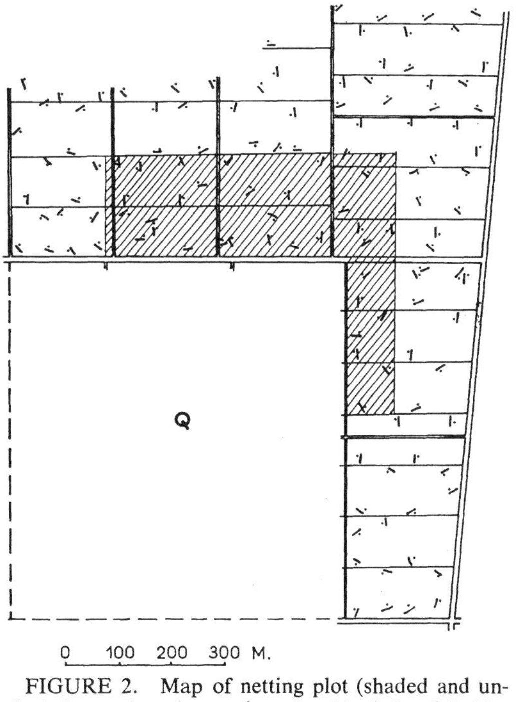

FIGURE 2. Map of netting plot (shaded and unshaded areas) and mapping plot (shaded only). Net locations are designated by the symbol (./). These areas are surrounded by vast expenses of forest of the same type; the part Q was excluded from the present study because the forest there is older.

TABLE 1
DISTRIBUTION OF 121 RECAPTURE DISTANCES OF BANDED MALE *Sylvia atricapilla*

Intercatch distance (m)	Number of recaptures
0–50	16
50–100	28
100–150	34
150–200	20
200–250	7
250–300	4
300–350	1
350–400	5
400–450	2
450–500	3
500–550	1

als. This calculation was possible only for those species which yielded enough recaptures. Two theoretical approaches permit this estimation (Luu–mau Thanh 1962, Taylor 1966). Both conclude that the quadratic mean of intercatch distances (x) is the best estimation of the home range radius (R) by the formula:

$$R = \sqrt{\frac{\Sigma x^2}{N}}$$

where N equals the number of recaptures.

RESULTS

We had the most recaptures for the Blackcap (*Sylvia atricapilla*). Table 1 shows the distribution of the 121 recaptures of 34 males. From this we can estimate a mean home range radius of 154 m, thus a home range area of 7.4 ha. In this species we determined 12.5 territories on 14 ha, for an average territory size of 1.12 ha. Thus the estimated home range was more than six times larger than the estimated average territory.

Four other passerine species gave enough recaptures to make the same calculations for the males home range compared to their estimated territories (Table 2). We see that the size ratios of home ranges to territories vary from two to twelve.

The above estimates of home ranges are derived from recaptures from May to July. We had enough recaptures for three species to calculate the home range size for May alone, the month of maximum territoriality. We obtain the following ratios of home range to territory size: Blackcap 4.6, robin 2.2, and Chiff-chaff 14.6. For all species, home ranges of conspecifics overlapped substantially, and the home ranges also overlapped into adjacent territories.

We were also able to estimate the home ranges of the females of the same three species for the period May to July. In each case it was much larger than the home range of the males: varying between 3 to 4 times the male's home range area.

CONCLUSIONS

This first approach shows that for five territorial species the estimated home ranges of the males were much larger than their estimated territories in all cases. The same conclusion holds for the females of three studied species. Furthermore, territory sizes we undoubtedly smaller than our estimates, because it is unlikely that all available space within our mapping plot would be defended by males of any given species. Thus the actual ratios of home range to territory size are likely greater than suggested by our analysis.

TABLE 2
RATIO OF HOME RANGE SIZE TO TERRITORY SIZE

Species	Territory size (ha)	Home range of adult males: Size (ha)	No. of indiv.	Recaptures per indiv.	Ratio
Blackcap (*Sylvia atricapilla*)	1.1	7.5	34	3.26	6.8
Chiff-chaff (*Phylloscopus collybita*)	0.8	9.9	18	2.56	12.4
European Robin (*Erithacus rubecula*)	2.1	7.6	16	3.50	3.6
European Blackbird (*Turdus merula*)	4.7	10.0	11	2.45	2.1
Wren (*Troglodytes troglodytes*)	5.1	12.6	9	4.22	2.4

Studies in Avian Biology No. 6:121–136, 1981.

REMARKS ON THE USE OF MARK-RECAPTURE METHODOLOGY IN ESTIMATING AVIAN POPULATION SIZE

JAMES D. NICHOLS,[1] BARRY R. NOON,[1] S. LYNNE STOKES,[1] AND JAMES E. HINES[1]

ABSTRACT.—Mark-recapture models are classified according to requisite assumptions about population closure. The resulting classes of models are briefly discussed and the experimental situations to which they apply are described. Model assumptions are presented, with emphasis being placed on those which are most likely to be violated in avian population studies. Comments are provided on experimental design, and previous ornithological studies in which the various models have been used are briefly reviewed. Special attention is devoted to the Jolly-Seber model which was developed for open populations and which perhaps has the greatest potential applicability to detailed, long-term population studies. A number of examples of the use of the Jolly-Seber model with avian mark-recapture data are presented in summary fashion.

The literature of field ornithology is much more extensive than that dealing with field studies of other terrestrial vertebrates. However, mark-recapture methods of estimating population size have seen only limited use in the ornithological literature, but are frequently employed in published studies of mammals, reptiles and amphibians. For example, reviews of small mammal population estimation are dominated by mark-recapture methodologies (e.g., see Smith et al. 1975), while such methods are barely mentioned in methodological reviews of avian population estimation (Kendeigh 1944, Berthold 1976, Shields 1979). We believe that the neglect of mark-recapture methods in avian studies results from the generally high visibility and audibility of birds and the relative ease with which they can be directly enumerated. This ease of enumeration has naturally and justifiably resulted in an emphasis by ornithologists on estimation methods that involve actual counts of birds (or nests), which are then expanded in various ways to estimate total population size or density. However, no single population estimation method is universally appropriate for avian studies, and there is still much debate about census methodology (see J. T. Emlen 1971, Berthold 1976, Shields 1979, this symposium).

We do not suggest that mark-recapture methods will generally provide a preferable alternative to direct observational methods of estimating avian population size. However, we do believe that mark-recapture studies can provide estimates that are useful in assessing the appropriateness of other estimation methods, and that they may provide the most reasonable means of estimating population size in some situations. In this paper we attempt to introduce the subject of mark-recapture methods and to briefly review their use in previous avian studies. We will be concerned only with the estimation of population size, and will thus omit discussion of models for estimating survival rate from band recoveries of dead birds (e.g., see North 1978 and reviews in Taylor 1966, Seber 1973, Brownie et al. 1978). Our discussion will be introductory and brief, and we urge the interested reader to consult the excellent general reviews of mark-recapture models and their associated literature provided by Cormack (1968, 1979) and Seber (1973).

Mark-recapture models can be usefully classified according to their requisite assumptions about population closure. We define a closed population as one which remains unchanged over the period of investigation and which is thus not influenced by mortality, recruitment, or migration (both emigration and immigration) during this time. An open population is one that does change over the period of investigation as a result of either mortality and emigration, recruitment and immigration, or both sets of factors.

CLOSED POPULATIONS

TWO-SAMPLE EXPERIMENT

The first attempt to estimate the size of a bird population using mark-recapture methods is generally attributed to Lincoln (1930; also see Winkler 1930), and the "Lincoln index" or "Petersen estimate" has been the most widely-used mark-recapture method for estimating animal population size. Although the assumption of population closure can be relaxed, the Lincoln index or estimate is generally modeled and discussed assuming a closed population. The Lincoln estimate is obtained from a two-sample experiment. A sample of n_1 birds is taken from a population of size N. Individuals in the sample are marked, returned to the population, and allowed time to mix freely with the unmarked birds. A second sample of size n_2 is then taken, and the proportion marked in this sample (m_2/n_2, where m_2 denotes the number of marked

[1] U.S. Fish and Wildlife Service, Migratory Bird and Habitat Research Laboratory, Laurel, Maryland 20811.

birds in the second sample) is assumed to estimate the proportion of marked birds in the population. The total population, N, is then estimated based on this assumption:

$$(m_2/n_2 = (n_1/N), \text{ so } \hat{N} = (n_1n_2/m_2).$$

The above estimator is generally said to require the following assumptions (e.g., see Seber 1973:59): (1) the population is closed, (2) all animals have equal capture probabilities in the first sample, (3) marking does not affect subsequent catchability, (4) the second sample is a simple random sample, (5) animals do not lose their marks, and (6) all marked animals occurring in the second sample are reported. If n_1 and n_2 are fixed and if these six "ball-and-urn model" sampling assumptions hold, then the number of recaptures, m_2 (given n_1, n_2, and N), can be modeled with the hypergeometric distribution or (with either replacement sampling or large N) the binomial distribution. In both cases the maximum likelihood estimator of N is the same as the intuitive estimator (however this estimate is biased and can be modified as suggested by Chapman 1951 or Bailey 1951). If sample sizes are not fixed, then it seems reasonable to regard capture probabilities as fixed (either because of characteristics of the sampled animals or expenditure of fixed effort by the experimenter) and to model the numbers of individuals with each of the four possible capture histories (caught in period 1, caught in period 2, caught in both periods 1 and 2, not caught) as a multinomial random vector. The maximum likelihood estimator of N based on the multinomial model is again the same as the intuitive estimator, and its asymptotic variance is essentially the same as that obtained for the hypergeometric model (Cormack 1979). These two general types of models (fixed sample size hypergeometric and random sample size multinomial) also represent the most common approaches to modeling other mark-recapture experiments. Cormack (1979:220) briefly compares these two approaches and notes that "strict adherence to either sampling rule is seldom possible." Although the biologist should be aware of these two common approaches (other approaches are also possible, such as the loglinear models of Fienberg 1972, Cormack 1979), the important practical result is that they have yielded essentially the same estimates for experiments to which they have both been applied.

The practical value of any estimate based on mark-recapture models will depend on how closely the field experiment and resulting data correspond to the assumptions of the chosen model and how precise (and accurate) the estimate is (e.g., as indicated by its estimated sampling variance), given that model assumptions are met. Both of these considerations, precision and assumption validity, are important when designing mark-recapture experiments and when interpreting their results. The problem of designing a two-sample experiment to estimate population size with specified levels of precision and accuracy is addressed thoroughly by Robson and Regier (1964; also see Seber 1973:64–70), and we recommend these references to ornithologists interested in planning such a study.

Assumptions generally stated for the Lincoln estimate were listed above, and a complete discussion of these assumptions and tests of their validity is presented by Seber (1973:70–104). Certain specific deviations from the assumption of population closure are permissible when using the Lincoln estimate. Of particular interest to ornithologists is the fact that members of a population may be subjected to mortality during the intersample period and, as long as the average mortality rates of marked and unmarked birds are the same, the Lincoln estimate, $\hat{N}$, is still a reasonable estimate of initial (at the time of the first sample) population size (see Seber 1970b). The presence of both recruitment and mortality results in overestimation of both initial and final population size. Probably the best practical means of insuring that violations of closure are negligible involve using a short time period between samples and choosing an appropriate time of the year for sampling. Naturally, short intervals between samples provide less opportunity for mortality, recruitment, and migration than longer intervals. Similarly, experiments should be conducted at a time of the year when migration and recruitment of young birds to the population are not occurring.

Assumptions 2, 3 and 4 involve catchability, which is an important consideration in any mark-recapture study. Assumptions 2 (all animals having equal capture probabilities in the first sample) and 4 (the second sample is a simple random sample) provide good examples of the difficulties involved in applying ball-and-urn statistical models to biological populations. Among avian studies, there are numerous examples in which all individuals in a population are not equally catchable. Differences in catchability or the probability of being sampled are sometimes associated with classes or subgroups of birds. For example, evidence of sex-specific differences in trappability has been provided for numerous waterfowl species by Petrides (1944) and Bellrose et al. (1961). Sex-specific differences in the probabilities of both capturing and resighting Northern Fulmars (*Fulmarus glacialis*) have been noted by Dunnett and Ollason (1978). Band recoveries by hunters are sometimes used in two-sample mark-recapture experiments (e.g., Lincoln 1930, Bergerud and Mercer 1966, Geis 1972) and recovery rates of many species of waterfowl are known to differ sex-specifically (e.g., see Bellrose et al. 1961). Many bird species exhibit age-specific differences in the probability of being taken in both trap and hunting samples (e.g., see Farner 1949, Bellrose et al. 1961, Kautz 1977, Sulzbach and Cooke 1978). Catchability can also be associated with breeding status in some species (e.g., see Orians 1958, Robel 1969, Sulzbach and Cooke 1978). In any case, when capture probabilities differ among identifiable subgroups of animals, then it is appropriate to treat each subgroup separately when estimating population size.

Catchability assumptions can also be violated by individual variation in capture probability that is not associated with identifiable subgroups (see assumptions 2 and 4). It is very difficult, if not impossible, to draw inferences about individual variation in capture probability using data from two-sample experiments, but recapture data on individuals from experiments involving a number of sample periods (K-sample experiments) have been used to examine catchability in avian studies. Evidence of individual variation in capture and/or resighting probabilities has been provided for several bird species (e.g., Borror 1948; Orians 1958; Young 1958, 1961; Taylor 1966; Carothers 1979). When inherent variation in capture probability exists, then the first sample will consist of a high proportion of the more trappable animals. Therefore, the average catchability of the marked animals will be higher than that of the unmarked animals, and marked animals will tend to be overrepresented in the second sample, resulting in biased estimates, $\hat{N}$. However, it is conceivable that an inherent high probability of being sampled by one method would not necessarily insure a high sampling probability for another method. As Seber (1973) has pointed out, two selective samples can provide an unbiased estimate, $\hat{N}$, if the methods of selection are independent. In practice, this argument suggests that the use of completely different capture techniques for the first and second samples can provide some protection against bias resulting from variation in catchability. With birds, it is sometimes possible to use traps to obtain the first sample and to apply highly visible markings (e.g., colored leg or neck bands, patagial tags, body dyes, backtags; see Cottam 1956, Taber and Cowan 1969, Marion and Shamis 1977, for reviews of avian marking techniques). The second sample is then taken by observing birds and tallying resightings of marked individuals as well as sightings of unmarked birds. This approach has been used successfully with Mallards (*Anas platyrhynchos*), Blue-winged Teal (*Anas discors*) and Wood Ducks (*Aix sponsa*) (Cowardin and Higgins 1967) and with Willow Ptarmigan (*Lagopus lagopus*) (Bergerud and Mercer 1966). When dealing with hunted species it is possible to obtain the first sample via trapping and to then use band recoveries returned by hunters as the second sample (e.g., see Lincoln 1930, Bergerud and Mercer 1966, Moisan et al. 1967, Goudy et al. 1970, Geis et al. 1971, Robel et al. 1972, Whitcomb 1974).

Effects of trapping and handling on future capture probability can also result in model assumption violations (assumptions 3 and 4), and such effects have been noted in birds. Sometimes such effects result from a behavioral response to trapping. Evidence of "trap-happy" birds that tend to return frequently to baited traps (presumably to take advantage of easily obtained food) has been provided by Borror (1948), Taylor (1966), and Murton et al. (1972). Trap and net aversion or "shyness" have been noted in a number of species (see Borror 1948, Young 1958, Stamm et al. 1960, Taylor 1966). Presumably, this involves an adverse reaction to being trapped and handled that results in unusual wariness of the bird or that causes the bird to avoid the immediate vicinity of the traps (e.g., see Owen and Morgan 1975). In addition it is possible that birds could be injured during the capture and handling processes, and that this might result in lower survival probabilities for marked birds. Aversive behavioral responses and injuries can be minimized through the use of proper capture and handling techniques (see reviews in Taber and Cowan 1969, U.S. Fish and Wildlife Service and Canadian Wildlife Service 1977). Apparent cases of trap-happy and trap-shy birds can also be produced by certain methods of trap placement. For example, Swinebroad (1964) mist-netted a Wood Thrush (*Hylocichla mustelina*) population and obtained evidence of unequal capture probabilities which he attributed to net placement. Apparently, some birds in the study area were exposed to nets within their "maximum activity areas" while other birds were only exposed to nets on the fringes of their ranges.

The marks or tags that are applied to birds can also affect survival and the probability of future recapture. For example, the accumulation of ice on neckbands and nasal saddles of geese during severe winter weather is a potential problem, but is not believed to affect survival to a great extent (e.g., see Greenwood and Bair 1974, Craven 1979). Neckbands have been reported to contribute to starvation in Snow Geese (*Anser caerulescens*) (Ankney 1975). Nasal discs and saddles have been suspected of increasing mortality of diving ducks (see discussion in Bartonek and Dane 1964), and patagial tags have been suspected of increasing susceptibility of birds to predation and altering behavior patterns of tagged birds (Anderson 1963). In addition to causing mortality differences between marked and unmarked birds, some marking methods can produce unequal catchabilities of the two groups at the time of the second sample. For example, higher resighting probabilities have been suspected for patagial-tagged (Cowardin and Higgins 1967) and backtagged and dyed birds (Bergerud and Mercer 1966), than for unmarked birds.

Assumption 5 is that animals do not lose their marks. In most two-sample experiments the period between samples will be sufficiently short that loss of marks commonly applied to birds should not be a problem. Retention times for several types of short-term avian markers (e.g., dyes, backtags) are presented by Taber and Cowan (1969) and Marion and Shamis (1977). Avian leg bands are generally considered to be good long-term markers but can be lost because of excessive wear or removal by the banded bird. The problem of band loss is thought to be important in long-lived birds and has been reviewed by Farner (1955), Ludwig (1967), and Marion and Shamis (1977). Leg band "survivorship curves" were developed by Ludwig (1967) for gulls and terns based on wear and weight loss data on known age bands from recaptured birds. A method for estimating the probability of tag loss from double-tagged animals (i.e., animals to which two tags are applied) is provided by Seber (1973:94–96). A practical means of reducing band loss problems is to replace worn bands on captured birds.

Assumption 6 is that all marked animals occurring in the second sample are reported, and this assumption is generally relevant only to experiments in which the second sample is based on band recoveries made by the general public (e.g., hunting recoveries). In cases where all recovered bands are not reported, some workers have estimated the total number of recovered bands for use in Lincoln index estimates (e.g., see Moisan et al. 1967, Goudy et al. 1970, Geis et al. 1971). This procedure requires an estimate of the "reporting rate" or the proportion of recovered bands that is reported. Reporting rate has been estimated using either additional information on the number of recovered bands obtained from hunter questionnaire surveys (Geis and Atwood 1961, Martinson 1966, Martinson and McCann 1966), or "reward band" studies in which some bands are marked with a message that a reward is offered for their return and are assumed to have a reporting rate of 1.0 (Bellrose 1955, Tomlinson 1968, Henny and Burnham 1976).

There are a number of reports in the avian literature of the use of the Lincoln index to estimate population size, and here we will only list some representative examples. Borror (1948), Boyd (1956), Stamm et al. (1960), Nunneley (1964), Taylor (1966) and Robel et al. (1972) have computed Lincoln index estimates from trapping and netting samples. All of these authors exhibited appropriate concern for at least some of the assumptions of the method, and none of them rejected the method as being completely inappropriate for their experimental situations. In an interesting comparison of methodologies Stamm et al. (1960) obtained general agreement between Lincoln index estimates and spot-mapping (Williams 1936) estimates for a number of eastern deciduous forest bird species. As noted earlier, Lincoln index estimates have also been computed from resightings of dyed or tagged birds (e.g. Bergerud and Mercer 1966, Cowardin and Higgins 1967, Readshaw 1968). Bergerud and Mercer (1966) compared estimates from several methods with "direct counts" believed to be accurate within 5% of the true population. The Lincoln index based on resightings provided estimates that agreed well with the direct counts for three of four years, while other population estimation methods proved less reliable. Cowardin and Higgins (1967) thoroughly examined Lincoln index assumptions and concluded that emigration of marked birds and increased visibility of marked birds may have resulted in biased estimates. Nevertheless, they concluded that Lincoln index estimates of population size were more realistic than total counts of birds. Hewitt (1963, 1967) suggested an interesting technique for obtaining Lincoln index estimates for Red-winged Blackbirds (*Agelaius phoeniceus*) which does not require the actual capture of birds. Territorial males seen along roads are "marked" using tape-recorded descriptions of their exact sighting locations. The road is traveled again, and the presence of a territorial male in a location which was occupied at the time of the first sample is considered as a resighting. Consideration of requisite assumptions and comparison of estimates with estimates obtained using other methods led Francis (1973) and Albers (1976) to conclude that Hewitt's methodology will generally underestimate population size. Lincoln index estimates based on hunting recoveries of banded birds have been used frequently (e.g., Lincoln 1930, Crissey 1963, Bergerud and Mercer 1966, Moisan et al. 1967, Goudy et al. 1970, Geis et al. 1971, Robel et al. 1972, Whitcomb 1974). Crissey (1963) and Bergerud and Mercer (1966) have reported that such estimates for Mallards and Willow Ptarmigan, respectively, agreed well with other estimates of population size. However, additional comparisons of Lincoln index estimates with aerial survey estimates for the Mallard population in North America have resulted in large discrepancies in recent years (Munro and Kimball, pers. comm.).

K-Sample Experiment

Here we discuss experiments involving two or more trapping or sampling periods. The two-sample Lincoln index experiment is simply a special case of this type of experiment, and we treated it separately only because of its frequent use relative to other mark-recapture experiments. Otis et al. (1978) have presented a unified treatment of population estimation from K-sample experiments with closed populations. We highly recommend this monograph to readers interested in such experiments, and our brief review of models here will follow their approach and terminology (see also Seber 1973:130–195, Pollock 1974).

The sampling scheme is very similar to that of the two-sample experiment. Animals are captured during an initial sampling period, marked, and returned to the population. A second sample is then taken (e.g., on the following day) and recaptures of marked animals are noted. New captures are also given marks and all animals are returned to the population. The procedure is repeated for K sampling periods. Perhaps the greatest operational difference between the K-sample and the two-sample experiments is that animals generally must be given individual marks (e.g., serially numbered leg bands) in the K-sample experiment. The models used to describe recapture data from K-sample experiments generally require complete capture histories of individual animals for estimating population size. The probability distribution for the set of possible capture histories is then expressed using a multinomial model (hypergeometric approaches have also been used in some cases) treating population size and capture probabilities as parameters.

Assumptions required by the models reviewed here are that (1) the population is closed, (2) animals do not lose their marks during the experiment, and (3) all marks are correctly noted and recorded at each trapping occasion. In addition to these three assumptions, each of the models discussed by Otis et al. (1978) embodies a different set of assumptions about sources of variation in catchability or probability of capture. As indicated for the two-sample experiment, proper choice of the time (season of the year) and duration of the experiment are two of the most important practical means of approaching closure (assumption 1). Regardless of the precautions taken in designing the experiment, however, it is desirable to test this assumption after the experiment is completed in order to insure that closed population models are appropriate. Closure tests based on specific closed population models are provided by Pollock et al. (1974) and Otis et al. (1978:66–67). Care should be used in interpretation of the results of the Otis et al. (1978) closure test, however, because it is sensitive to various sorts of variability (e.g., over time) in capture probabilities as well as to non-closure. It is thus prone to false rejection of the closure assumption. The assumption (2) that animals retain marks should be easily met in most closed population experiments because of their short duration relative to retention times of most types of marks. The correct recording of marks (assumption 3) will be a natural consequence of careful field work and can be insured by the use of well-designed field data sheets and appropriate editing procedures for keypunched or summarized data.

The simplest and least realistic model described by Otis et al. (1978) is M_0. In addition to the three assumptions listed previously, M_0 assumes that all individuals in the population have identical capture probabilities for each of the K sampling periods. M_0 thus includes only two parameters; population size, N, and capture probability, p. The general maximum likelihood estimator for N under M_0 cannot be written in closed form, but must be computed numerically. As noted earlier, there is a great deal of evidence of variation in capture probability among subgroups and individuals of various bird species. In addition, it is reasonable to expect capture probabilities to vary from one sampling period to another as a result of such factors as weather conditions, changes in sampling effort, etc.

Model M_t of Otis et al. (1978; also see Schnabel 1938, Darroch 19) has historically been the most commonly used model for K-sample closed population mark-recapture experiments (see reviews in Cormack 1968, Seber 1973). M_t permits capture probabilities to vary from one sampling period to another, but assumes that within a sampling period, all individuals have the same capture probability. The general maximum likelihood estimate of N is again not available in closed form. While M_t is somewhat more realistic than M_0, its assumption of equal capture probability within a sampling period will still be inappropriate for the many bird species that exhibit individual variation in catchability and variation associated with nonidentifiable subgroups.

Model M_b (Otis et al. 1978) corresponds to the situation in which the initial capture of an individual affects its probability of capture on subsequent sampling periods. As noted earlier, trap and net responses have been well-documented in birds, and this model may thus be realistic for some bird species. The model contains three parameters: population size, capture probability for initial captures (i.e., for unmarked animals), and capture probability for subsequent captures (i.e., for marked animals). The capture probabilities for marked and unmarked animals are assumed constant for all time periods. A numerical procedure is again needed to compute the estimate, $\hat{N}$.

Model M_h (Burnham and Overton 1978, Otis et al. 1978) corresponds to the situation in which each member of a population is characterized by a distinct capture probability. These individual capture probabilities are assumed to remain the same over all sampling periods. In the development of the estimation procedure, Burnham and Overton (1978) recommend treating the set of individual capture probabilities as a random sample of size N from some probability distribution. They then develop an estimator, $\hat{N}$, using an extension of the jackknife method of bias reduction.

Models M_t, M_b, and M_h of Otis et al. (1978) represent efforts to model specific sources of variation in capture probability. However, we might reasonably expect more than one of these sources of variation to be important in a given experiment. For this reason, Otis et al. (1978) developed a set of models corresponding to the various combinations of these sources of variation in capture probability. An estimator for N can be obtained for one of these models, M_{bh}, which assumes that each member of a population has a specific probability of capture prior to its initial capture and another specific probability of capture after it has been marked. No estimators for N are currently available for the other combination models (M_{tb}, M_{th}, M_{tbh}).

Assumptions 1–3 for all of the models have been briefly discussed. In addition, the assumptions distinguishing between the models concern the type of variability present in the capture probabilities. Otis et al. (1978) suggest seven tests for evaluating these assumptions. Some tests compare two competing models or hypotheses (one of which is more general than the other), while the others assess the goodness-of-fit of a particular model to the data. Otis et al. (1978) have also developed a discriminant classification function based on data simulated from each of the eight models. This classification function can be used to provide an objective selection of the appropriate model (and thus the appropriate catchability assumptions) for a given data set. We note that other tests dealing with variation in catchability are available (e.g., see Young 1958, 1961, as well as later discussion of open population models). Suggestions about the practical aspects of designing K-sample experiments for closed populations are provided by Otis et al. (1978:74–80).

We are aware of only one published mark-recapture experiment with birds in which one of the K-sample closed population models described above has been used (see Hewitt 1967). Most of these models are relatively new and would not be expected to have received much use, but others (e.g., M_t) have been available for some time. Otis et al. (1978) have developed a comprehensive computer program (see White et al. 1978) which computes estimates for their models, computes test statistics, and uses the classification function to select the appropriate model. We suspect that the availability of this program will result in use of the Otis et al. (1978) models in future mark-recapture studies of birds. T. J. Dwyer (pers. comm.) is currently using the program in conjunction with his mark-recapture study of Woodcock (*Philohela minor*) at the Moosehorn National Wildlife Refuge, Maine. Preliminary results suggest that model M_t may be appropriate for some of Dwyer's within-season data sets.

Under certain assumptions, models for K-recapture experiments on closed populations can be based on the observed frequencies of capture (i.e., on the numbers of animals captured 1, 2, . . . K times). For example, if probability of capture remains constant over time (the assumption of M_o), then the probability that an animal is captured x times is given by the binomial distribution (e.g., see Eberhardt 1969, Seber 1973:169–170). Since the number of animals in the population that are never caught ($x = 0$) is unknown, the observed frequency of recaptures follows the zero-truncated binomial distribution. Under this assumption the number of unobserved animals can be estimated (Seber 1973:169–170) and the fit of the data to the model can be evaluated using a χ^2 goodness-of-fit test. If capture probability is constant and small, then the zero-truncated Poisson distribution provides a reasonable approximation to the binomial model for capture frequencies. Using the Poisson model, the zero class can again be estimated (see Eberhardt 1969, Seber 1973:170) and the fit of the model tested. If probability of capture is not the same for each individual, then certain assumptions can lead to different truncated models of recapture frequency. For example, different hypotheses about the distribution of catchability can lead to geometric and negative binomial models (see reviews in Eberhardt 1969, Seber 1973).

Although, as Seber (1980) has suggested, model M_h of Burnham and Overton (1978) will probably replace other capture frequency approaches, these methods have seen limited use in avian studies. Bergerud and Mercer (1966) used a truncated Poisson model to estimate ptarmigan population size. Eberhardt (1969) examined the fit of the truncated Poisson and geometric distributions to published trapping and sighting data of several bird species. Condor (*Gymnogyps californianus*) sighting data and Bobwhite Quail (*Colinus virginianus*) capture data fit both Poisson and geometric models (Eberhardt 1969). North (1978) examined the fit of Manx Shearwater (*Puffinus puffinus*) recapture data within years to Poisson, geometric and negative binomial models. North (1978) also used a "double Poisson" model to deal with trap response and a "heterogeneous model" to handle two classes of birds that differ in their response to initial capture. The geometric and heterogeneous models both produced reasonable estimates. Both Eberhardt (1969) and North (1978) suggested that in cases where two models appear to fit the data well, selection of the model to use must be based on a detailed consideration of the biological assumptions involved. However, there is not always a biological basis for choosing among competing models. Cormack (1979:228) points out that several possible models may fit capture frequency data well but yield vastly different estimates of population size. For example, Dwyer (pers. comm.) found that both geometric and Poisson models fit his Woodcock capture frequencies well in many cases, but that the two models produced very different estimates of population size. Because of this problem and because the variety of models in Otis et al. (1978) correspond to most of the different biological situations that have been used to derive other capture frequency models, we suspect that the Otis et al. models will generally be preferred in future K-sample closed population experiments.

OPEN POPULATIONS

Completely Open Populations

In this section we examine K-sample experiments that are conducted in the same general manner as those already discussed, but where population gains and/or losses are allowed to occur between sampling periods through birth/immigration and death/emigration. Early models for these open populations incorporated the view that population changes were deterministic, and that randomness was associated only with the sampling process (see review of these models in Cormack 1968). However, it was recognized that the change in population size and composition from one time period to another could be more realistically viewed as a stochastic process itself (e.g., see Robson 1969). This problem was partially solved by Darroch (1959), when he developed models for partially open populations; i.e., ones in which either gains or losses in population size, but not both, are allowed to occur. The solution for the completely open population stochastic model was presented simultaneously by Jolly (1965) and Seber (1965) (denoted the Jolly-Seber model), and their model has now virtually replaced the deterministic models in experiments on open populations.

Assumptions that are generally listed for the completely open model are: (1) every animal in the population has the same probability, p_i, of being caught in sample i, given that it is alive and in the population during sampling period i, (2) every marked animal in the population has the same probability, ϕ_i, of surviving from sampling periods i to $i + 1$, given that it is alive and in the population immediately after the time of release in sample i, (3) every animal caught in sample i has the same probability, v_i, of being returned to the population, (4) marked animals do not lose their marks, (5) all marks are reported and correctly recorded on recovery, (6) all samples are instantaneous (i.e., sampling time is negligible), and (7) losses to the population from emigration and death are permanent. Assumptions 1–6 are listed in Seber (1973:196) and assumption 7 is noted by Robson (1969:126) and Seber (1973:199).

Both Jolly (1965) and Seber (1965) used multinomial approaches to modeling. In Jolly's (1965) general formulation, p_i, ϕ_i, and v_i are considered as unknown parameters, and N_i, M_i (the number of marked individuals in the population at sampling period i), B_i (the number of new individuals entering the population between sampling periods i and $i + 1$) and $\rho_i(M_i/N_i$, or the proportion of individuals in the population at time i which is marked) are treated as unknown random variables.

The estimate for population size under the Jolly-Seber model is:

$$\hat{N}_i = M_i n_i / m_i \qquad (i = 2, 3, \ldots, K - 1)$$

where n_i is the number of animals caught at time i and m_i is the number of marked animals caught at time i. This is an intuitive Lincoln-type estimator in which the proportion of marked animals in the entire population (M_i/N_i) is simply equated with the proportion of marked animals in the sample (m_i/n_i). However, while n_i and m_i are observable random variables; M_i must be estimated. As pointed out by Jolly (1965; also see Cormack 1968, 1972) the estimator for M_i is also intuitively appealing:

$$\hat{M}_i = \frac{R_i z_i}{r_i} + m_i \quad (i = 2, 3, \ldots, K - 1)$$

where R_i is the number of marked animals released into the population after sampling period i, r_i is the number of marked animals released after sampling period i and caught subsequently, and z_i is the number of animals caught before sample i, but not in sample i, and caught again after sample i. The marked individuals in the population just after sampling period i consist of two groups: the animals captured during period i (R_i), and the number not captured during period i ($M_i - m_i$). Of the former group, r_i are subsequently recaptured, and z_i from the latter group are subsequently recaptured. The two ratios, r_i/R_i and $z_i/(M_i - m_i)$, should be approximately equal and the solution of the equality for M_i yields $\hat{M}_i$.

We are aware of no published guidelines for planning a mark-recapture experiment to be analyzed with the Jolly-Seber model. However, the form of the asymptotic variances, as well as the simulation studies of Manly (1970, 1971a), Gilbert (1973), Bishop and Sheppard (1973) and Kreger (1973), demonstrate the importance of high capture probabilities, p_i, to obtaining accurate estimates (having low or negligible bias) of population size with low estimated variances. Gilbert's (1973) simulations also demonstrate that substantial reductions in bias of $\hat{N}_i$ can result from increasing the length of the experiment (i.e., the number of sampling periods). The asymptotic variance estimators of Jolly (1965) are known to produce confidence intervals with poor coverage in some cases (Manly 1971a). Specifically, in small sample size situations there is a high positive correlation between N and its estimated variance, causing underestimates of population size to appear to be more precise than they really are (Manly 1971a). This problem has led to a search for alternate methods of estimating variances for this model (Manly 1977a). Nevertheless, estimated coefficients of variation (standard error of estimate/estimate) of population size do provide some indication of how precise and reliable the estimates are (Kreger 1973). Carothers (1973) points out that an estimate with large "small-sample" bias can normally be recognized as unreliable by its large estimated variance.

We prepared Figures 1 and 2 in an effort to provide some indication of the sampling effort required to achieve various levels of precision of $\hat{N}_i$. Figure 1 assumes a constant intersample survival rate of $\phi_i = 0.50$ while Figure 2 assumes $\phi_i = 0.75$. It was assumed that there were no accidental deaths during banding and handling. In both Figures, the N_i were assumed constant at values of 100, 200, 500 and 1000, and $B_i = (1 - \phi_i)N_i$ in order to insure a stable population. All experiments were assumed to include six sampling periods. The solid lines in the Figures represent approximations to the expected value of the estimated coefficient of variation of $\hat{N}_3$, $E[\widehat{CV}(\hat{N}_3)]$. These were obtained by approximating the expected values of $\hat{N}_3$ and the standard error of $\hat{N}_3$, $\widehat{SE}(\hat{N}_3)$. This was done by using the expected values, $E(R_i)$, $E(n_i)$, $E(m_i)$, $E(z_i)$, and $E(r_i)$, given the N_i, ϕ_i, B_i, and p_i, in conjunction with the equation for $\hat{N}_i$ presented earlier and the equation for $\widehat{var}(\hat{N}_i)$ presented by Jolly (1965). The approximation for $E[\widehat{CV}(\hat{N}_3)]$ was then obtained as $E[\widehat{SE}(\hat{N}_3)]/E[\hat{N}_3]$. The plotted points in Figures 1 and 2 each represent the mean value of $\widehat{CV}(\hat{N}_3)$ computed from 500 simulated data sets. Simulated data sets were constructed by treating capture and survival of each individual for each sampling period as independent Bernoulli trials using pseudorandom numbers. If the selected values of ϕ and N are thought to be reasonable guesses for certain populations, then Figures 1 and 2 may be of use in planning experiments.

Figures 1 and 2 illustrate the importance of high capture probabilities (and corresponding large sample sizes) to obtaining precise estimates of population size. Cormack (1979) has noted that the generality of the Jolly-Seber model, with its separate parameters for each survival and capture probability, may sometimes limit its practical utility. However, Jolly (1979, in prep.) has recently developed models (and appropriate model testing procedures) in which survival probability or capture probability or both are constant over time (i.e., $\phi_i = \phi$, $p_i = p$). We suspect that these reduced-parameter models will be of considerable practical value.

Assumption 1 of equal catchability has been discussed with respect to closed population models, and we listed a number of examples of unequal catchability in bird populations. When identifiable subgroups of animals exhibit different probabilities of capture, then the groups can be treated separately. However, when variation in capture probability is present within a subgroup, problems can arise. Carothers (1973) examined the relative bias of the Jolly-Seber $\hat{N}_i$ resulting from variation in catchability among individuals. When capture probability varied

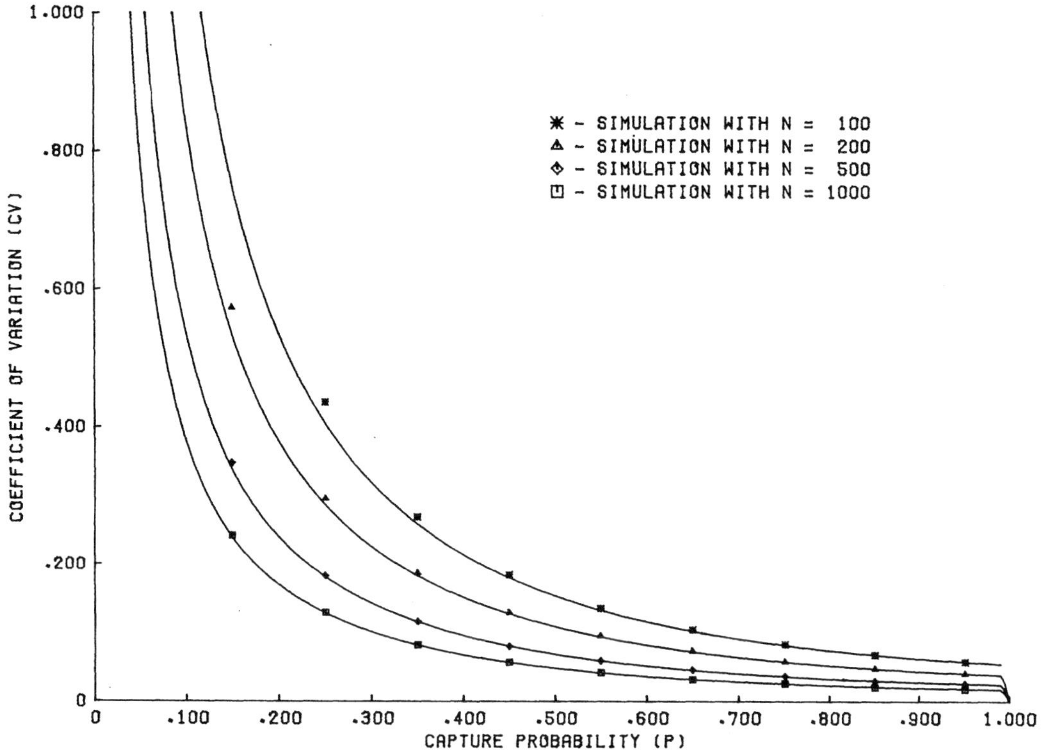

FIGURE 1. Relationship between capture probability, P, and the estimated coefficient of variation of $\hat{N}_3$, $\widehat{CV}(\hat{N}_3)$, assuming a six-period experiment with $\phi_i = 0.50$.

moderately among individuals, the resulting relative bias of $\hat{N}_i$ was fairly small for the situations examined, but when capture probability varied widely, large negative biases of $\hat{N}_i$ resulted (Carothers 1973). Gilbert (1973) also used analytic approximations and computer simulation to examine the bias of $\hat{N}_i$ associated with unequal catchability. Gilbert's work emphasized the importance of the average capture probability. If this average is above 0.50, then bias resulting from variation among individuals will be small (Gilbert 1973). Gilbert (1973:524) concluded that "an experimenter need not attempt to design an experiment so that all animals have the same probability of capture, but only that the distribution be shifted nearer one so that nearly all animals have probabilities of capture say $\geq$ 0.50." A test of the equal catchability assumption in a K-sample experiment was provided by Leslie (1958) and extended by Carothers (1971) (also see the test of Cormack 1966). Practical recommendations for trying to avoid unequal catchability have been discussed relative to the two-sample experiment, although some of these suggestions (e.g., different capture techniques) may not be as applicable to the K-sample situation.

Assumptions 2 and 3 require a homogeneous population with regard to probability of surviving the intersample and sampling periods, respectively. Homogeneous sampling period survival probability can be assured by use of proper capture and handling techniques. However, the population can always be stratified if accidental deaths are thought to be higher among some groups of animals than others. Heterogeneity in intersample survival, ϕ_i, may be more difficult to account for or control. Age- and sex-specific variation in survival rates is thought to be common among birds (see Ricklefs 1973 and references therein). High predation rates on nesting females (e.g., see Johnson and Sargeant 1977) may also result in different survival probabilities of breeding and non-breeding adults. In any case, when survival probabilities differ among identifiable subgroups within a population, stratification is again appropriate. Certain handling and marking techniques are thought to influence survival probabilities of various bird species (see examples provided in the discussion of two-sample model assumptions). Tests of the hypothesis that survival is independent of capture probability and mark status were developed by Robson (1969), Manly (1971b), and Seber (1973:230–

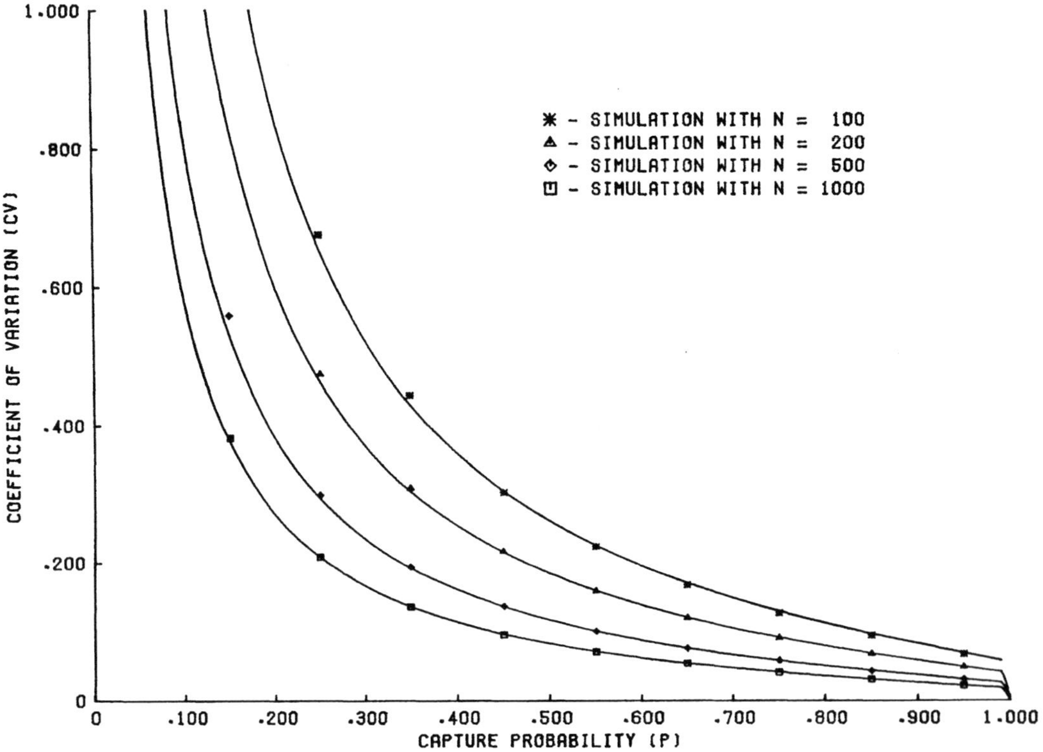

FIGURE 2. Relationship between capture probability, *P*, and the estimated coefficient of variation of $\hat{N}_3$, $\widehat{CV}(\hat{N}_3)$, assuming a six-period experiment with $\phi_i = 0.75$.

232). Robson (1969) discussed models that permit survival probabilities to be different for one or two sampling intervals after initial capture. These models were designed for field experiments in which marking or handling somehow stresses the captured animals. Pollock (1975a) generalized this model to permit not only different survival probabilities but also different capture probabilities for various numbers of intervals after initial capture. This very general model includes the Jolly-Seber model as a special case.

We have already noted that when subgroups (e.g., based on age, sex, breeding status) differ in capture or survival probability, they can be analyzed separately. However, when variation is age-specific young animals often have to be omitted from analysis until they become adults. Manly and Parr (1968) suggested one approach to estimating size of age-stratified populations. More recently, Pollock (1981b) and Stokes (1980) have developed general models for open, age-stratified populations. A model permitting geographic stratification, with animals of different areas having different capture and survival probabilities, has also been developed (Arnason 1973).

Assumptions 4 and 5 regarding retention of marks and reporting of recaptures have been discussed in the section on closed models. The assumption 6 of instantaneous sampling will of course never be strictly met, but efforts should be made to keep the sampling period fairly short relative to the intersample period. Mortality during the sampling period should be negligible. Assumption 7 regarding non-permanent emigration has not been well-studied but may be extremely important. Cormack (1981, pers. comm.) has had some success in recognizing nonpermanent emigration with his loglinear models approach. In addition, capture and/or resighting efforts in areas peripheral to the main study area could be used to gain insight to the magnitude of the problem. Detailed studies of movements of small samples of birds via radio telemetry could also provide an indication of the extent to which non-permanent emigration might be occurring.

In addition to individual tests of specific underlying assumptions, tests designed to assess the reasonableness of the Jolly-Seber model for a given data set are also available. Seber (1973:223–224) suggested a goodness-of-fit test for the Jolly-Seber model based on the expected numbers of individuals having each possible capture history. We have found this test to be somewhat impractical because considerable cell pooling is generally required for the data sets we have examined and because large experiments require computation of a large number of expected cell values. Jolly (in prep.) has recently suggested a goodness-of-fit test based on the expected number of individuals released at time i and next caught at times $i + 1$, $i + 2$, . . . K. This test also requires some cell pooling, but we have found it useful with actual data sets (see later examples). Another approach to assessing the reasonableness of the underlying model was suggested by Leslie et al. (1953; also see Seber 1973:224–226). The method basically involves using data from second and subsequent captures of individuals to estimate the number of "first captures". The variance of this estimate can also be computed, and the actual number of first captures can be compared with the 95% confidence interval around the estimated number. In addition, survival rates, ϕ_i, and population size, N_i, can be estimated using only second and subsequent captures, and these estimates can then be compared with ϕ_i and $\hat{M}_i$ (the estimate of the marked population size) based on the full data set.

There has been very little use by ornithologists of open population mark-recapture models to estimate population size. Orians (1958) and Readshaw (1968) used the deterministic model of Leslie (1952) to estimate population sizes of Manx Shearwaters and Pied Currawongs (*Streperu graculina*), respectively. Orians (1958) estimated numbers of adult Shearwaters using Leslie's (1952) approach and obtained estimates that agreed with independent estimates obtained by expanding counts of Shearwater burrows. However, Orians (1958) used tests developed by Leslie et al. (1953) and Leslie (1958) to infer that Shearwaters were not being randomly sampled. Orians (1958) concluded that his estimates were not "trustworthy" and cautioned against using mark-recapture models without carefully examining assumptions. Readshaw (1968) estimated size of a wintering Currawong population but found the standard errors of the estimates too large to permit detection of variation with time. Hammersley (1953) used a stochastic mark-recapture model to estimate population size of two Alpine Swift (*Apus melba*) colonies. Hammersley concluded that the birds had not been sampled randomly because of the concentration of the bander on different groups of birds in different years. Darroch (1959) criticized Hammersley's (1953) model.

The Jolly-Seber model was used by Anderson and Sterling (1974) to estimate the number of drake Pintails (*Anas acuta*) on two molting marshes in Saskatchewan. Despite large banded sample sizes, estimated capture probabilities were very small, ranging from 0.009 to 0.024, and the resulting population estimates were not very precise (Anderson and Sterling 1974). Sulzbach and Cooke (1979) used the Jolly-Seber model to estimate the number of adult Snow Geese in a nesting colony in Manitoba and obtained fairly precise estimates of population size for some years. They carefully considered underlying assumptions and used the method of Leslie et al. (1953) to estimate numbers of first captures and thus indirectly assess model fit. Estimated numbers of first captures agreed fairly well with actual values in most years, and it was concluded that the model provided a reasonable description of the data set. Agreement of Jolly-Seber estimates with "visual estimates" was poor, and the authors cautioned against the sole use of either method (Sulzbach and Cooke 1979). Kautz (1977) pointed out problems associated with the use of the Jolly-Seber model on Band-tailed Pigeon (*Columba fasciata*) data combined from a number of different trap sites. He recommended that parameter estimation should be restricted to data from a specific trapping site (or from sites in close proximity to each other). Kautz (1977) then used data from a Colorado banding site to estimate population size, and suggested that variation in these estimates might be indicative of variation in the area from which birds were drawn to the baited trapping station.

We are interested in the potential applicability of the Jolly-Seber model to avian mark-recapture studies, but this has been difficult to evaluate because of the paucity of published avian studies in which it has been used. Here we will attempt to provide some information on the potential utility of the method by presenting selected parameter estimates from Jolly-Seber analyses of a number of mark-recapture data sets for birds. The data sets were obtained from both published and unpublished sources and represent a variety of species, capture methods, sampling intervals, etc. Information on study locations and methodologies is presented in Table 1. Jolly's (1965) estimates were computed for parameters of interest with all data sets. χ^2 goodness-of-fit test statistics as suggested by Jolly (in prep.) were computed for all data sets in which adequate degrees of freedom remained after cell pooling. Some data sets were available only in Leslie Method B Table summary form (Leslie and Chitty 1951), and these sets were only subjected to the χ^2 goodness-of-fit test. Data sets for which individual capture history data were available permitted estimation of the number of first captures occurring in each sampling period as suggested by Leslie et al. (1953). For these data sets we recorded the proportion of the sampling periods (for which estimates of first captures could be computed) in which the estimated 95% confidence interval for first captures covered the actual known number.

Arithmetic mean estimates of capture probability, survival probability and population size, and the range of estimated coefficients of variation of population size are presented in Table 2. These summary statistics provide some indication of the relationship between these parameters and the precision of the resulting population size estimates. The two Canada Goose (*Branta canadensis*) data sets (sets 2 and 3) had the highest capture probabilities, and the resulting $\widehat{CV}(\hat{N}_i)$ were low. The Pintail data set (4) exhibited the lowest estimated capture probability, but because of the large population size and resultant banded sample sizes, the $\widehat{CV}(\hat{N}_i)$ were not the highest encountered. Similarly, the Manx Shearwater data set (1) had relatively low capture probabilities, but because of the large sample sizes the $\widehat{CV}(\hat{N}_i)$ were also low. The data sets containing smaller sample sizes (e.g., fewer than 500 banded individuals) tended to yield a wide range of values for $\widehat{CV}(\hat{N}_i)$, with low coefficients of variation associated with sampling periods having high capture probabilities, and vice versa, as expected.

The χ^2 goodness-of-fit test statistics indicated significant ($P < 0.05$) rejection of the model in 3 of the 7 data sets for which they could be computed. The most obvious rejection was obtained for the Manx Shearwater data set (1), for which sampling was thought to be nonrandom (Orians 1958). The proportion of estimated confidence intervals covering the actual number of new captures was less than 0.95 for six of seven data sets, indicating possible violations of model assumptions.

A useful method of evaluating the utility or reasonableness of a parameter estimate is to compare it with an independent estimate of the same, or perhaps a similar, parameter. For example, the suggestion from the χ^2 test that the male Canada Goose data set (2) from Old Hickory Lake, Tennessee, did not fit the Jolly-Seber model was disturbing because of the high capture probability and general quality of this data set. The female data set (3) from the same location provided a means of checking both sets of estimates. Male and female Canada Geese are generally thought to exhibit equivalent survival probabilities and the estimated mean Jolly-Seber survival rates for the 2 sexes did not differ significantly. If hatching sex ratio is approximately 50% males and if first year survival rates are approximately equal for the sexes (e.g., see summary data in Bellrose 1976), then male and female population sizes should be roughly equivalent. Annual population size estimates and associated 95% confidence intervals for adult males and females are plotted in Figure 3. There is good correspondence between the male and female estimates for all years. We have also plotted mid-winter aerial estimates of total population size for the Old Hickory Lake Canada Goose flock (from Cromer 1978:54). These mid-winter estimates apply to a different time of the year than the Jolly-Seber estimates, and they contain both sexes and all age classes, so correspondence between these and the sex-specific adult estimates is not expected to be close. Nevertheless, the similar patterns of population increase are reflected in the plots, and we believe that the comparisons presented in Figure 3 increase the credibility of the Jolly-Seber estimates for this population.

Partially Open Populations

Mark-recapture experiments on populations that are open to both gains and losses perhaps have the greatest potential applicability to studies on the dynamics of avian populations. However, if the population is partially rather than completely open, it is advantageous in the interest of parsimony to use one of the earlier models of Darroch (1959), which are shown by Jolly (1965) to be special cases of the Jolly-Seber model. A population that is closed to death and emigration but open to recruitment and immigration probably represents an extremely rare experimental situation. Thus, we have chosen to consider populations that experience only losses and no gains during the experimental period. A banding study of birds during a nonbreeding period, or, if young recruits to the population can be distinguished from older residents, even the breeding season could be reasonably modeled with this partial closure assumption.

Assumptions generally listed for the "death but no recruitment" model are the same as those of the Jolly-Seber model but with the addition of (8) the population is closed to recruitment. Both Darroch (1959) and Jolly (1965) considered models based on the multinomial distribution and treated ϕ_i, p_i, and (in Jolly's case) v_i as unknown parameters to be estimated. Population size, N_i, at time i is treated as an unknown random variable. The population size estimate under this model is:

$$\hat{N}_i = \left(\frac{R_i Z_i}{r_i}\right) + n_i \quad (i = 1, 2, \ldots, K - 1)$$

where R_i is the number of marked animals released after sample i, Z_i is the number of ani-

TABLE 1
INFORMATION REGARDING THE DATA SETS ANALYZED FOR TABLE 2

Data set	Species	Study location	Study period	Sampling time	Sampling period duration	Number sampling periods	Inter-sample period	Data source
1	Manx Shearwater	Skokholm Island, Wales	1946–53	April–Sept.	6 mo.	9	1 yr.	Orians (1958: Table 3)
2	Canada Goose	Old Hickory Lake, TN	1967–77	20 June–14 July	3 days	11	1 yr.	Cromer (1978: Table 6)
3	Canada Goose	Old Hickory Lake, TN	1967–77	20 June–14 July	3 days	11	1 yr.	Cromer (1958: Table 7)
4	Pintail	Pel & Kutawagan Marshes, Saskatchewan	1955–58	9 July–24 July	2 weeks	4	1 yr.	Anderson & Sterling (1974: Table 3)
5	American Woodcock	Canaan Valley, WV	1965–69	July–Sept.	3 mo.	5	1 yr.	Clark (1978: Table 13)
6	American Woodcock	Canaan Valley, WV	1965–69	July–Sept.	3 mo.	5	1 yr.	Clark (1978: Table 13)
7	Band-tailed Pigeon	Longmont, CO	1970–75		2–3 mo.	6	1 yr.	Kautz (1977: Table 13)
8	Acadian Flycatcher (*Empidonax virescens*)	Patuxent Wildl. Res. Center, Laurel, MD	1959–72	Late May–early July	2 mo.	14	1 yr.	C. S. Robbins (unpubl. data)
9	Least Flycatcher (*Empidonax minimus*)	Hubbard Brook Experimental Forest, West Thornton, NH	1969–79	Late May–early July	2 mo.	11	1 yr.	R. T. Holmes (unpubl. data)
10	Barn Swallow (*Hirundo rustica*)	Patuxent Wildl. Res. Center Laurel, MD	1948–54	Late May–early July	2 mo.	7	1 yr.	C. S. Robbins (unpubl. data)
11	Tufted Titmouse (*Parus bicolor*)	Patuxent Wildl. Res. Center Laurel, MD	1959–72	Late May–early July	2 mo.	14	1 yr.	C. S. Robbins (unpubl. data)
12	Swainson's Thrush (*Catharus ustulatus*)	Hubbard Brook Experimental Forest, West Thornton, NH	1969–79	Late May–early July	2 mo.	11	1 yr.	R. T. Holmes (unpubl. data)
13	Red-eyed Vireo (*Vireo olivaceus*)	Hubbard Brook Experimental Forest, West Thornton, NH	1969–78	Late May–early July	2 mo.	10	1 yr.	R. T. Holmes (unpubl. data)
14	Red-eyed Vireo	Patuxent Wildl. Res. Center Laurel, MD	1959–70	Late May–early July	2 mo.	12	1 yr.	C. S. Robbins (unpubl. data)
15	Yellow Warbler (*Dendroica petechia*)	Point Pelee, Ontario	1958–64	May–June	6–23 days	7	1 yr.	Roberts (1971: Tables 1, 2)
16	American Redstart (*Setophaga ruticilla*)	Hubbard Brook Experimental Forest, West Thornton, NH	1969–79	Late May–early July	2 mo.	11	1 yr.	R. T. Holmes (unpubl. data)
17	Seaside Sparrow (*Ammospiza maritima*)	Stone Harbor, NJ	1969–78	Late May–July	2–3 mo.	10	1 yr.	C. B. Worth (unpubl. data)
18	White-throated Sparrow (*Zonotrichia albicollis*)	Columbus, OH	1946	14 Oct–4 Nov	1 day	20	1 day	Borror (1948: Fig. 1)

TABLE 2

SUMMARY STATISTICS OF JOLLY-SEBER ANALYSES OF DATA SETS DESCRIBED IN TABLE 1[a]

Data set	Species	Age[b]	Sex[c]	Number individuals banded	Number recaptures	$\hat{\bar{p}}$	$\hat{\bar{S}}$	$\hat{\bar{N}}$	Range of $\widehat{CV}(\hat{N}_i)$	Goodness-of-fit test[d] χ^2	df	p	Proportional CI coverage[e]
1	Manx Shearwater	A	M&F	20,017	5175	0.18	0.68	16,046	0.05–0.08	105.87	21	0.00	—
2	Canada Goose	A	M	622	712	0.80	0.81	142	0.05–0.09	4.96	1	0.03	—
3	Canada Goose	A	F	611	764	0.76	0.84	147	0.04–0.11	5.95	3	0.11	—
4	Pintail	A	M	18,820	238	0.02	0.59	272,105	0.22–0.25	0.02	1	0.88	—
5	American Woodcock	A	M	127	17	0.23	0.73	190	0.67–1.45	—	—	—	—
6	American Woodcock	A	F	201	23	0.43	0.34	197	0.03–0.89	—	—	—	—
7	Band-tailed Pigeon	A	M&F	1186	217	0.19	0.65	1312	0.18–0.33	4.23	5	0.52	—
8	Acadian Flycatcher	A	M&F	345	121	0.62	0.42	61	0.13–0.47	—	—	—	0.38
9	Least Flycatcher	A&Y	M&F	237	56	0.73	0.42	51	0.04–1.23	—	—	—	—
10	Barn Swallow	A	M&F	134	74	0.63	0.75	61	0.08–0.73	—	—	—	0.67
11	Tufted Titmouse	A	M&F	330	119	0.66	0.53	62	0.08–1.02	—	—	—	0.67
12	Swainson's Thrush	A&Y	M&F	149	68	0.54	0.57	46	0.22–1.25	—	—	—	—
13	Red-eyed Vireo	A&Y	M&F	201	95	0.42	0.57	77	0.14–0.67	—	—	—	—
14	Red-eyed Vireo	A	M&F	896	401	0.46	0.59	241	0.13–0.28	17.32	7	0.02	0.44
15	Yellow Warbler	A&Y	M&F	208	35	0.27	0.62	171	0.62–0.80	—	—	—	0.33
16	American Redstart	A&Y	M&F	232	55	0.37	0.67	66	0.08–1.13	—	—	—	—
17	Seaside Sparrow	Y&A	M&F	230	112	0.60	0.55	60	0.14–0.42	—	—	—	0.83
18	White-throated Sparrow	A&Y	M&F	114	150	0.23	0.92	67	0.20–0.70	18.74	12	0.10	1.00

[a] $\hat{\bar{p}}$ denotes arithmetic mean capture probability estimate, $\hat{\bar{S}}$ denotes arithmetic mean survival probability estimate, $\hat{\bar{N}}$ denotes arithmetic mean population size estimate, and $\widehat{CV}(\hat{N}_i)$ denotes the estimated coefficient of variation for population size in sampling period i.

[b] Y denotes young (first year) and A denotes adult.

[c] M denotes male and F denotes female.

[d] Test suggested by Jolly (in prep.)

[e] Method of Leslie et al. (1953) where ν_i denotes the number of new captures in period i, and $\hat{\nu}_i$ and $\widehat{SE}(\hat{\nu}_i)$ denote the estimated number of new captures and its estimated standard error. $[\hat{\nu}_i - 1.96\widehat{SE}(\hat{\nu}_i) \leq \nu_i \leq \hat{\nu}_i + 1.96\widehat{SE}(\hat{\nu}_i)]$.

mals not caught in sample i but caught subsequently, r_i denotes the number of marked animals released after sample i and subsequently recaptured, and n_i is the number of animals caught in sample i. This $\hat{N}_i$ can be shown to be an intuitively reasonable estimator using logic analogous to that used previously for the Jolly-Seber estimator, $\hat{M}_i$.

This model has not been widely used by ecologists. Robson (1979) has discussed the planning of experiments designed to estimate survival rates using this model, but we are aware of no similar work emphasizing estimation of N_i. We do note that examination of the estimator for the variance of $\hat{N}_i$ (see Jolly 1965:242) shows the importance of a large number of recaptures, r_i, to the precision of $\hat{N}_i$. For this reason, the suggestion of Jolly (1965:239) for his completely open model that release and recapture operations can be operated independently is relevant to this model also. While estimates of N_i can only be obtained for sampling periods in which animals are released, recaptures could be obtained continuously (e.g., via resightings) by biologists or even volunteers in order to improve the precision of $\hat{N}_i$.

The discussion of assumptions 1–7 for completely open models is relevant to the "death but no immigration" model also. Pollock et al. (1974) provide a series of tests which can be used to examine assumption 8 of partial closure. Proper choice of time of the year and duration of the experiment can serve to improve the chance that this assumption is met. Finally, we note that a χ^2 goodness-of-fit test for the model can be computed in a manner analogous to that suggested by Jolly (in prep.) for the general Jolly-Seber model.

To our knowledge, the only use of this model in ornithological work has been that of Dwyer (pers. comm.). He has used the model to estimate population size of Woodcock based on summer mark-recapture experiments. The model was thought to produce reasonable estimates, and goodness-of-fit test statistics indicated that most of the data sets did fit the model.

DISCUSSION

As indicated in this review, ornithologists have not made much use of mark-recapture methodologies for estimating population size. A modest number of reviewed studies employed the two-sample Lincoln index experiment which unfortunately does not permit much of the testing of assumptions which is possible with K-sample experiments. Ornithologists have made virtually no use of the models developed for K-sample experiments on closed populations. A small number of examples of the use of K-sample open population models with bird populations was found. However, only three papers were seen in which the stochastic Jolly-Seber model had been used to estimate size of a bird population. We found no published study in which the stochastic "death but no immigration" model had been used with an avian population.

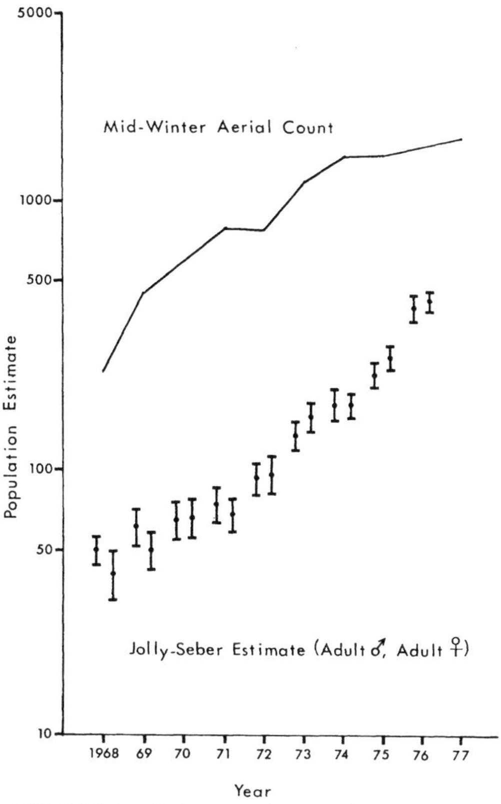

FIGURE 3. Estimates of population size of Canada Geese on Old Hickory Lake, Tennessee (data from Cromer 1978).

As previously noted, we suspect that the high visibility and audibility of birds and the resulting potential for use of other estimation methods has been at last partly responsible for the neglect of mark-recapture methods by ornithologists. However, we also suspect that the level of statistical training required to understand these models and the complexity of the numerical computations required to obtain some estimates may contribute to the neglect of mark-recapture models by biologists. This suspicion is reinforced by the dramatic increase in the use of the band recovery model developed by Seber (1970a) and Robson and Youngs (1971) following the publication of Brownie et al. (1978), which was written for biologists and which contains instructions, output explanation, and examples of the use of two user-oriented FORTRAN programs for carrying out necessary computations.

A number of recent efforts have been directed at making mark-recapture methods both understandable and accessible to biologists. Seber's (1973) book (a new edition is due in 1980) provides an excellent review of methodology. Methods are illustrated with numerous field examples and much of the book should be understandable to biologists. More recently, Begon (1979) has written a monograph on capture-recapture methods aimed specifically for biologists. Two papers by Cormack (1972, 1973) are directed at the logic of mark-recapture methods and the intuitive nature of the associated estimators. The monograph of Otis et al. (1978) on closed population models is also written with the biologist in mind. A more elementary treatment of the material presented in Otis et al. (1978) is provided by White et al. (1981).

With respect to computational algorithms, computer programs providing Jolly-Seber estimates have been available for nearly a decade (Davies 1971, White 1971, Arnason and Kreeger 1973). The new version (see Arnason and Baniuk 1980) of one of these programs has extensive data management capabilities and also computes estimates based on the two partially open population models of Jolly (1965) and the closed population model of Darroch (1958). A comprehensive computer program was developed by Otis et al. (1978) to provide estimates and conduct tests leading to selection of appropriate closed population models (see White et al. 1978). Crosbie (1979) has developed a computer program which computes estimates based on a number of open population models (e.g., the standard Jolly-Seber model and similar reduced-parameter models) and computes test statistics to aid in the selection of an appropriate model. We have developed a program for the age-stratified open population model of Stokes (1980) which is available to interested researchers now and which should be sufficiently user-oriented for general distribution in the near future (Hines, Stokes and Nichols, unpubl.).

It is difficult to make general statements about the potential applicability of mark-recapture methods to avian population estimation because of the small number of relevant studies which have been conducted. In particular, there has been a complete lack of K-sample closed population experiments. We hope that the models and program of Otis et al. (1978) will be used by ornithologists for short-term experiments (or in conjunction with long-term investigations comprised of a number of relatively short sampling periods), and we will be interested to learn how well the various models seem to fit bird recapture data. With respect to open populations, the analyses presented in Table 2 provide some indication of the applicability of the Jolly-Seber model to avian studies. Results from some of the data sets (e.g., the Canada Geese) were very encouraging. However, assessments of model fit provided indications of problems with underlying assumptions in a number of the data sets. In addition, precision of population size estimates varied widely within a number of the data sets. We suspect that the age-stratified open population models of Pollock (1981b) and Stokes (1980) will be useful in avian studies, and it will be interesting to learn whether the general models of Robson (1969) and Pollock (1975a) are useful with bird recapture data. The large variance estimates associated with many of our Jolly-Seber analyses emphasize the need to consider the parsimonious, reduced-parameter models of Jolly (1979, in prep.) and Crosbie (1979). We hope that ornithologists interested in long-term population studies will examine these various open population models and make use of them as they become available.

ACKNOWLEDGMENTS

We thank T. J. Dwyer, R. T. Holmes, C. S. Robbins, C. B. Worth, J. R. Cromer and the Tennessee Wildlife Resources Agency for use of unpublished data. R. M. Cormack, G. M. Jolly and K. H. Pollock kindly provided copies of unpublished manuscripts and manuscripts in press. D. R. Anderson, T. J. Dwyer, H. F. Percival, and B. K. Williams reviewed the manuscript and provided useful suggestions. We thank C. Samuel, R. Wilcox and P. Bowley for typing the manuscript.

Studies in Avian Biology No. 6:137–138, 1981.

MARK-RECAPTURE—WHAT NEXT?

G. M. JOLLY[1]

ABSTRACT.—The direction in which mark-recapture methods are likely to develop in the immediate future is discussed briefly. More sophisticated models are envisaged which will approximate more closely to real-life situations, and the biologist has an important role to play in helping the statistician to define these. A hypothetical example is used to illustrate.

A reading of Seber (1973) shows mark-recapture as having developed originally around insect populations. For example, it is commonly assumed that ages of individuals cannot be determined, that survival is independent of age, and that birth or immigration has no known distribution over time. With birds or mammals these assumptions may all be false, and the additional information on age and birth processes requires more complex models than have been available in the past. Movement patterns may also be of interest. Brownie and Robson (1976) is one of many examples of the need to create specific models for specific situations.

An excellent review of recent developments in mark-recapture with reference to avian populations is given by Nichols et al. (1981).

A HYPOTHETICAL BANDING STUDY

DESCRIPTION

Let us consider the imaginary goose (*Anser mysticus*) which, though rare, is plentiful on certain remote Arctic islands on one of which, called *S*, banding has taken place as convenient towards the end of the breeding season for the last three years. The population on *S* is heterogeneous, consisting of at least two intermingling segments, *S*1 and *S*2. Both young and adult are banded in an intensive effort over a two-month period.

Segments *S*1 and *S*2 migrate to their winter residences at the same time, *S*1 to an island, *W*1, where banded birds are observed and some further banding occurs, and *S*2 to a rocky island, *W*2, where banding or close observation are considered impracticable on account of landing difficulties. Some data are also available from injured or resting birds on a small island situated on the migratory route of *S*1. Invariably the *S*1 segment return to their summer residence before *S*2 and, among these earlier arrivals, some birds have been consistently observed that have never been found on *W*1. This suggests the possibility of yet another, unknown wintering area.

The aim of the study is to assess population numbers and movements as well as to estimate survival rates. It is therefore desirable to extend the survey to include all segments of the population of *Anser mysticus* and so obtain an integrated picture of the species' behaviour. In the current year, *S*1 and *S*2 have been augmented by an influx from a neighbouring island. Virtually the complete, unknown population of this island is thought to have left in response to industrial developments. The number of these arriving to join *S*1 and *S*2 will be estimated. Since the species is strictly protected most of the data are from observations on live birds (banded or otherwise), but information on dead birds discovered will also be utilized. An attempt may also be made to initiate a small sampling scheme on *W*2.

ASSUMPTIONS AND PROBLEMS

The following include the main initial assumptions, but these will be continually monitored as data accumulate.

Randomness.—At present data on *S* are taken only from the few large concentrations of birds. A quarter to a third of the population however occur in smaller pockets. In the future sampling of these pockets will be undertaken but at a low intensity since banding there will be less cost-effective. Policy thereafter will depend on how representative is the main sample and how much information is lost if birds banded on *W*1 and spending the summer in these pockets are not recorded.

Segments.—Different yearly survival rates will be assumed for each segment but the same survival rate will be regarded as applicable during the summer period when the birds are together on *S*. Equal probability of capture will be assumed for all segments.

Sex.—Yearly survival could be different for male and female except possibly during the summer period. Probability of capture is unlikely to be associated with sex.

Age.—Up to four age classes can be determined approximately. Exact age will be used for birds banded in their first year. An age-survival curve will be estimated. Although young and old are likely to have the same chance of being netted while the birds are immobile, there may be circumstances when a greater proportion of immature birds are caught.

[1] Agricultural Research Council Unit of Statistics, University of Edinburgh, Mayfield Road, Edinburgh EH9 3JZ, United Kingdom.

DISCUSSION

Over a thousand birds have been banded during the last three years at moderate cost. To avoid over-interference with the population on *S* it is thought desirable to limit the effort there while extending the study to other localities as already mentioned.

Fully efficient methods of analysis will be used, and these will be modified as necessary to take account of changes in assumptions. Whenever possible the number of parameters must be reduced to give maximum precision, as, for example, when survival and probability of capture are assumed constant over the two-month banding period (Jolly, 1981a). No general model is yet available to allow for varying probability of capture among individuals in open populations, although Burnham and Overton (1978) give a method for closed populations.

CONCLUSIONS

The above example is intended to illustrate some of the theoretical and practical problems that can arise in a mark-recapture study. Wherever possible the aim of the investigator should be to simplify the assumptions over which he can exercise some control, in particular equal catchability over time or among classes of individuals. When this is not possible a model should be defined to take account of the facts.

At present the many recent developments have not yet had time to be sufficiently co-ordinated to enable data from the above hypothetical example to be analysed exactly as envisaged, and some further theory is still required. However, within a few years it should, in the author's opinion, be possible to have comprehensive computer programs incorporating an ever increasing range of theoretical flexibility and coming gradually closer to biological and technical reality.

Studies in Avian Biology No. 6:139–145, 1981.

THE DETERMINATION OF AVIAN DENSITIES USING THE VARIABLE-STRIP AND FIXED-WIDTH TRANSECT SURVEYING METHODS

KATHLEEN E. FRANZREB[1]

ABSTRACT.—This study assesses the extent to which the variable-strip and fixed-strip transect methods satisfy the assumptions upon which they are based. Mathematical as well as verbal descriptions of both sampling methods are provided.

The variable-strip transect method involves the observer traversing a transect of predetermined length and recording the lateral distance from the transect of each bird observed. Avian densities are calculated by counting the number of individuals found in strips on both sides of the transect from the base to the point of inflection on the distribution curve of the results. This transect method can be used at any time of the year and enables an observer to quickly census relatively large areas for all birds including breeding birds, non-breeding birds, and fledglings.

Several modifications in the variable-strip transect method are suggested including using the additional category of "all observations" in the density calculation. It is also suggested that measurements be recorded as precisely as possible and pooled at a later time into smaller increments than those recommended by J. T. Emlen (1971).

In the fixed-strip or belt transect method, a transect of known length and width are sampled. Species may be assigned belts of different widths depending upon each species' detectability characteristics.

This study computed avian densities in a mixed-coniferous forest in the White Mountains, Arizona, using the variable-strip transect method. Data were also segregated on the basis of strips of fixed-widths (15 m, 30 m, 60 m, and 125 m wide belts located on both sides of the transect).

The variable-strip transect data indicated a total avian community density of 835.4 birds per 40 ha. Of the four fixed-widths, the highest density was calculated for the 60 m strip on either side of the trail (519.3 birds per 40 ha).

Until relatively recently probably the most widely used sampling technique to estimate population size of breeding birds was the spot-map (or territorial mapping) method (Williams 1936). This approach has several limitations, namely, it is time-consuming, and is only applicable during the breeding season, since it is only then that most avian species maintain territories. A number of innovative approaches have been proposed which offer alternatives to the spot-map technique.

One such alternative is the variable-strip transect method developed by J. T. Emlen (1971) which is now a widely used censusing technique and has provided a rapid, relatively easy way to sample large areas during any season of the year. The objectives of this investigation were to: (1) explore the theory of line sampling more fully by examining the assumptions upon which the variable-strip and fixed-width transect methods are based, (2) discuss the shortcomings and advantages of both these techniques, (3) compare the results of the variable-strip transect method to those of transects of fixed-widths, and (4) suggest modifications of the variable-strip transect method to enhance its reliability.

[1] U.S. Fish and Wildlife Service, Endangered Species Office, 1230 N Street, 14th Floor, Sacramento, California 95814.

MATERIALS AND METHODS

STUDY AREA

The study was conducted in the Willow Creek watershed, a mixed-coniferous forest located in the Apache-Sitgreaves National Forest, White Mountains, Arizona, during the summer of 1974. Elevation ranged from 2682 m to 2805 m. The vegetation is dominated by Douglas fir (*Pseudotsuga menziesii*), ponderosa pine (*Pinus ponderosa*), and southwestern white pine (*Pinus strobiformis*). A total of eight tree species were present, of which only quaking aspen (*Populus tremuloides*) was deciduous. A detailed description of the study area derived from the plotless point-quarter sampling method (Cottam and Curtis 1956) is provided in Franzreb and Ohmart (1978).

AVIAN SPECIES DENSITIES

Assumptions

Line transect methods, in general, employ a variety of assumptions which include the following (the accuracy and validity of the variable-strip and fixed-width sampling methods depend on the degree to which the assumptions are satisfied and will be addressed later in the Discussion): (1) birds are uniformly and randomly distributed; (2) the probability of observing a bird decreases with distance from the transect, or remains constant to a given distance and then declines rapidly; (3) the behavior of birds in one portion of the band width does not influence those in another; (4) the probability that a bird is observed if it is at right angles from the transect at a distance (w) is given by the simple function $g(x)$ such that $g(0) = 1$ (Burnham et al. 1980, Seber 1973) (This simply means that birds directly on the line will never be overlooked); (5) the bird does not move in response to the observer's presence prior to being detected; (6) no bird is counted more than once; (7) there are no measurement errors; (8) the response behavior of the avian community does not change appreciably throughout the sampling period; and (9) the response behavior of individuals of a species is similar regardless of sex or age.

In this study a transect line 1.6 km long was established using plastic flagging, a steel tape, and compass. Sampling began ½ hr after sunrise and was completed within two hours. This line consisted of four individual parallel transects each extending 400 m in length. The starting point of the first transect was randomly located. Six surveys concentrated at the beginning of each month were conducted beginning 1 June and ending 9 August 1974. The weather during each survey was either clear or with less than 30% overcast and with little, or no wind. Results are represented in terms of the mean of the monthly values. The results of each month's six surveys were pooled and the density values then calculated to provide a monthly figure. The overall density value was computed by averaging the June, July, and August values.

Data analysis

The accumulated data for each species are plotted on a graph with distance on the abscissa and number of observations on the ordinate. The density value for each species is calculated by counting the number of individuals encountered in strips located on both sides of the transect from the base to the point of inflection on the distribution curve of the results. Given that detectability declines with distance from the transect, if the area in these particular strips is multiplied by the appropriate value, the resulting figure will be the number estimated to occur within the 125.6 m (412 ft) or another appropriate value, on both sides of the transect line. The procedure and computation are more fully described in J. T. Emlen (1971).

According to the established technique, data are generally tabulated separately for singing males and for all other observations (J. T. Emlen 1971). Results from the singing male data are multiplied by two (assuming each male is paired) and compared to those from all other observations, with the adoption of the larger value. Instead, I utilized the highest number of observations encountered in either twice the singing male data, or all other observations, or all total observations (male data plus all other observations). Data are expressed as density per 40 ha as this is a standard size in avian studies.

J. T. Emlen (1977) proposed refinements in his variable-strip transect method as described in 1971. During the breeding season, he suggests that locality specific cue frequency values based on song frequency be determined for each species which are then used to calculate breeding density. A further explanation regarding the data collection and analysis process involved in the derivation of cue frequency values and the computation of avian densities appears in J. T. Emlen (1977). Results of this study were analyzed following the procedures as outlined in J. T. Emlen's 1971 paper.

Mathematical representation

The variable-strip transect method as described by J. T. Emlen (1971, 1977a) was not mathematical in its development but is similar to the method developed by Anderson and Pospahala (1970). Mathematically the model is represented as follows (Burnham and Anderson 1976): if W is the fixed strip width, a characteristic proportion of birds of a given species will be detected within $2LW$ where L is the length of the transect. It is assumed that $g(0) = 1$ which indicates that all birds on the actual transect line will be observed (probability of 1); then the coefficient of detectability $(\widehat{CD}_w) = n/(W\hat{h}(0)) = 1/W\hat{f}(0)$ where $\hat{h}(0)$ is the estimator of $n\hat{f}(0)$ which was determined for a smoothed frequency histogram; and finally the density estimator $(\hat{D})$ is: $\hat{D} = n/2LW\widehat{CD}_w) = (n\hat{f}(0))/(2L)$.

A similar method to the variable-strip technique as developed by Emlen was described by Kelker (1945). Both methods rely upon density estimates for bands within which it is assumed there is 100% coverage. Kelker discards observations falling beyond the distance from the transect at which observations begin to decline, whereas Emlen uses all the data (though the actual density calculation is similar). Anderson and Pospahala (1970) developed an elaboration of Kelker's method which involves fitting a regression curve for the frequency distribution data to allow calculation of an estimate of the objects observed in the belt of attempted coverage. Robinette et al. (1974) compared 10 census methods including those of Kelker and Anderson and Pospahala and found the results were within 15% of the correct density.

Data analysis fixed-width method

The fixed-strip survey method (Kendeigh 1944, Emlen 1974, and others), whereby belts of given widths are sampled on either side of the established transect, was applied to the data collected using the variable-strip transect data for the month of July. Data were segregated according to 15.2 m (50 ft), 30.5 m (100 ft), 61.0 m (200 ft), and 125.6 m (412 ft) wide belts of strips extending on either side of the transect.

Mathematical representation fixed-width method

Mathematically the fixed-width technique's density estimate $(\hat{D})$ is indicated by $\hat{D} = n/(2LW)$ where n is the number of observations within the strip of width W and transect of length L.

Other models

Numerous models representing the distribution curve of the sampling results have been proposed which graphically and mathematically portray the relationship of the number of observations versus distance from the transect line. For a detailed discussion of such models the reader is referred to the monograph by Burnham et al. (1980).

RESULTS

Analysis of data derived from the variable-strip transect sampling method indicated an avian community density of 835.4 birds per 40 ha (Table 1). Among the most abundant species

TABLE 1
AVIAN SPECIES DENSITIES (NUMBER PER 40 HA) DERIVED USING THE VARIABLE-STRIP AND FIXED-STRIP CENSUSING METHODS

Species	Density				
	Variable-strip transect	Fixed-strip width			
		15 m	30 m	60 m	125 m
Goshawk (*Accipiter gentilis*)	1.8	0.0	1.4	0.7	0.3
Broad-tailed Hummingbird (*Selasphorus platycercus*)	20.8	2.7	9.6	4.8	2.4
Common Flicker (*Colaptes auratus*)	18.3	8.2	11.0	13.8	7.5
Yellow-bellied Sapsucker (*Sphyrapicus varius*)	5.2	0.0	2.7	2.1	1.0
Williamson's Sapsucker (*S. thyroideus*)	5.5	2.7	2.7	2.1	1.0
Hairy Woodpecker (*Picoides villosus*)	6.7	0.0	4.1	3.5	1.7
Downy Woodpecker (*P. pubescens*)	4.2	2.7	2.7	2.1	1.4
Northern Three-toed Woodpecker (*P. tridactylus*)	12.8	5.5	5.5	4.8	2.4
Western Flycatcher (*Empidonax difficilis*)	71.6	21.9	32.9	21.4	10.5
Olive-sided Flycatcher (*Nuttallornis borealis*)	0.6	0.0	0.0	0.7	0.3
Violet-green Swallow (*Tachycineta thalassina*)	8.6	0.0	0.0	1.4	0.7
Steller's Jay (*Cyanocitta stelleri*)	16.5	5.5	9.6	7.6	5.1
Common Raven (*Corvus corax*)	3.4	2.7	1.4	0.7	0.3
Clark's Nutcracker (*Nucifraga columbiana*)	2.4	0.0	2.7	1.4	0.7
Mountain Chickadee (*Parus gambeli*)	64.8	43.8	37.0	26.2	13.9
White-breasted Nuthatch (*Sitta carolinensis*)	5.2	2.7	2.7	1.4	0.7
Red-breasted Nuthatch (*S. canadensis*)	23.8	2.7	6.9	6.2	3.4
Pigmy Nuthatch (*S. pygmaea*)	27.2	0.0	20.6	16.6	8.2
Brown Creeper (*Certhia familiaris*)	46.4	33.6	31.3	23.5	11.6
House Wren (*Troglodytes aedon*)	5.7	0.0	2.7	2.1	1.4
American Robin (*Turdus migratorius*)	3.0	0.0	0.0	1.4	1.0
Hermit Thrush (*Catharus guttatus*)	42.8	30.1	48.0	43.5	34.0
Golden-crowned Kinglet (*Regulus satrapa*)	51.4	30.1	45.2	23.5	11.6
Ruby-crowned Kinglet (*R. calendula*)	88.6	38.4	49.3	33.8	18.4
Warbling Vireo (*Vireo gilvus*)	17.7	8.2	8.2	6.2	3.4
Olive Warbler (*Peucedramus taeniatus*)	3.7	8.2	5.5	2.8	1.4
Yellow-rumped Warbler (*Dendroica coronata*)	136.5	82.2	87.7	57.3	28.2
Grace's Warbler (*D. graciae*)	4.9	8.2	5.5	2.8	1.4
Red-faced Warbler (*Cardellina rubrifrons*)	40.3	13.7	15.1	8.3	4.1
Western Tanager (*Piranga ludoviciana*)	3.6	2.7	1.4	1.4	0.7
Black-headed Grosbeak (*Pheucticus melanocephalus*)	3.7	8.2	4.1	2.1	1.0
Pine Siskin (*Carduelis pinus*)	14.0	2.7	9.6	5.5	2.7
Green-tailed Towhee (*Pipilo chlorurus*)	1.8	0.0	2.4	1.4	0.7
Gray-headed Junco (*Junco caniceps*)	66.6	35.6	43.8	29.7	15.0
Chipping Sparrow (*Spizella passerina*)	5.5	0.0	2.7	2.1	1.4
TOTAL	835.6	405.0	519.3	364.9	199.5
Species richness	35	24	32	35	35

were the Yellow-rumped Warbler (136.5 birds/40 ha), Ruby-crowned Kinglet (88.6 birds/40 ha), and Gray-headed Junco (66.6 birds/40 ha).

Of the four fixed-width surveys the 30.5 m strip on each side of the transect yielded the highest density (519.3 birds/40 ha) (Table 1). By widening the effective width of the strip to 61.0 m the maximum number of species (35) was included. Fixed-strip width data indicated the Yellow-rumped Warbler, Mountain Chickadee, Ruby-crowned Kinglet, and Gray-headed Junco were the most abundant species.

The majority of species were most numerous within the 15.2 m and 30.5 m wide strips on either side of the transect. Observations declined rapidly within the 61.0 m and 125.6 m strips.

Relative results were similar between the fixed-width and variable-strip sampling methods. However, none of the various belt widths approached either the overall density of the majority of individual species' densities derived from the variable-strip transect method. In assessing the similarity in results from these two techniques, it was assumed that a species' density value, determined for any of the four fixed-strip widths was similar to that of the variable-strip transect results if the two values were within 10 percent (an arbitrarily selected value). Using this criterion, in this study eight species had approximately similar dentities, whereas 27 did not. Of those which did not, the density values computed from the variable-strip transect data were higher for 24 of the 27 species.

DISCUSSION

Various techniques have been devised to enable investigators to compute avian species' densities. An acceptable method must provide relatively reliable results, be reasonably efficient to use in the field, and rely upon as few assumptions as possible. The following discussion focuses on the various assumptions of the variable-strip and fixed-strip sampling methods (as previously stated in the Assumptions section) and the extent to which each satisfies the assumptions. Transect methods of variable as well as fixed-width strips are advantageous in that they embrace all individuals, not just breeding birds, and can be utilized during any season of the year. Yet do these methods provide reasonably good predictions of the actual absolute densities of members of the avian community?

The design of the transect route should consider the size and shape of the area to be sampled, the terrain, the type of habitat, and most importantly, biological features of the avian community (e.g., a species with a large territory may require a longer transect(s) to obtain a sufficient sample size than for a numerous species and/or one with a small territory).

A single long transect or a series of parallel transects may be established. If the latter approach is utilized then care should be taken to assure that lines are sufficiently far apart as to preclude counting the same individual from more than one line. Several lines can be combined into one sampling unit (Eberhardt 1978). Transects may be of various configurations as necessitated by terrain and do not have to be parallel.

Gates et al. (1968), Eberhardt (1978), and Seber (1973) noted that animals should be distributed uniformly and independently (assumption 1) but that this was rarely the case under natural circumstances. In view of this, Eberhardt (1978) believes that establishment of randomly placed transect lines is needed. A systematic design will satisfy this requirement in some cases as long as the beginning of the first transect is randomly located (Anderson et al. 1979). The study should also be designed so that the transect is sufficiently long and wide to provide (if possible) at least 40 observations for each species (Burnham et al. 1980).

Assumption 2 (pertaining to the decrease in probability of observing a bird as the distance from the transect line increases) is generally no problem nor is assumption 3 (relating to the behavior of birds in one transect band not influencing the behavior of birds in another), at least not in this study. However, situations could arise where, for example, an alarm call issued close to the transect could silence the other singing birds in the vicinity, and thereby affect the results.

In addressing assumption 4 (a bird on the transect has a probability of 1 of being observed) and assumption 5 (the bird does not move in response to the observer prior to being detected), J. T. Emlen (1971) noted on the distribution curve of the results that the number of observations increases with distance from the transect to a maximum point and then declines. Even though the observer's ability to detect a species should be maximal in the strips immediately adjacent to the transect, the birds' response to the observer may effectively preclude this. If a bird is attracted to the observer, shies away, or "drives" in front of the observer, assumptions 4 and 5 will not be satisfied. In other work it has been noted that only approximately 20 percent of the total observations occurred within 25 m of the transect (Järvinen and Väisänen 1975). It is therefore not surprising that in the fixed-width survey results the highest densities for most avian species did not occur in the 15 m wide belt on either side of the transect.

In other animal surveys, short lateral movements caused by the observer's approach have been observed or suspected (Eberhardt 1978, Hirst 1969, and Dassmann and Mossman 1962). Eberhardt (1978) suggested a modification of the variable-strip method by using a width sufficiently wide as to include at least two-thirds of the total observations. If so, then shifts in movement of this sort may not influence the results. However, this modification may make the width measurement used for density computation purposes unnecessarily wide and thereby reduce the density value. Emlen (1977a) suggested that birds be counted if they are first detected within a distance of 61.0 m (200 ft) before and behind the advancing observer which may minimize this problem (although double-counting may then be a consideration). In this study some lateral movement was observed in response to the observer, but it was limited to short distance changes in position to nearby trees. Movement itself is not critical if it is independent of the observer and slow with respect to the observer's speed (Anderson et al. 1979).

Another problem is the potential for double-counting the more mobile individuals (assumption 6) and hence, overestimating such species' densities. Some individuals may be attracted to the observer while others may move ahead ("drive") of the observer along the transect. Either situation may result in counting the individual more than once. If this situation is not detected and recognized, an erroneously high density estimate will be obtained. This difficulty is alleviated, to a certain extent, by the observer moving along the transect at a moderately-fast, constant pace. Recording only those birds at approximately right angles to the observer traversing the transect line also aids in reducing the probability of double-counting. In most cases transboundary movement should even out if enough tranverses of the transect are conducted.

With either the variable-strip or fixed-width transect methods, the quality of the results depends, in part, upon the degree to which the observer accurately determines the distance measurements (assumption 7). Some investigators have argued that strip surveys do not permit density calculations because an observer can not estimate distance measurement by eye with enough accuracy (Enemar and Sjöstrand 1967). However, with experience and the use of a properly calibrated range-finder or steel tape, distance measurement in most habitats should not pose an insurmountable obstacle. Pacing, since it tends to be quite variable, especially in rough terrain, should be avoided. In densely forested situations when one is attempting to gauge the distance of singing, non-visual males, estimation of distance becomes more difficult and more susceptible to error.

Results may be biased in that distance measurements may be rounded off to convenient numbers (e.g., 0, 5, 10 m). This phenomenon has been noted by Gates et al. (1968), Anderson and Pospahala (1970), and Robinette et al. (1974). Judicious selection of distance class intervals and more thorough instructions to field personnel may alleviate this problem.

The extent to which assumption 8 (similar avian behavior throughout the course of the study) is satisfied is difficult to assess; however, with the exception of perhaps a limited degree of habituation to the observer's presence, it probably holds for both of these sampling methods. Further, the degree of detectability will vary between individuals, sexes, and season. This is directly contradictory to assumption 9. In some species males become less conspicuous and ardent in territorial advertisement as the breeding season progresses. Since females are generally considerably less obvious than the males, owing to their lack of song, usually drab coloration, and larger proportion of the maternal duties such as incubation, the likelihood of detecting them is substantially less than for the males. Such problems are inherent in any transect method and in part, are ameliorated by concentrating on sampling an area during a short, carefully specified time frame.

In analyzing the nine basic line transect assumptions to satisfy objective 1, there appears to be basically little difference in the degree to which these assumptions are met by the variable-strip and fixed-width sampling methods. However, it should be noted that although both rely upon estimating distances, a distance error is considerably more critical (and more likely) with the variable-strip method. This is because each observation needs an accurate distance measurement, whereas with the fixed-width, all one must do is accurately decide if the bird is within the belt (a significantly easier undertaking). Also short lateral movements (assumptions 4 and 5) are less meaningful to the precise calculation of the fixed-width results than is the case with the variable-strip method. Other than these two differences the methods are similar in their assumptions and the degree to which they meet them (objective 1).

At least four major factors influence the success of transect censusing methods and include the competence of the observer, weather conditions, habitat type, and inherent nature of the avian species being sampled. If we assume the observer is experienced and the weather is propitious, the type of habitat greatly impacts the level of censusing accuracy in that a dense heavily-vegetated forest situation will present more detectability problems than will, say, an open, sparsely-vegetated habitat. The last, and probably most important, factor is the species' inherent behavior, which in concert with the sparseness or denseness of the vegetation determines the detectability of the species. If a species is conspicuous either because of its foraging behavior (e.g., flycatching from a clearly visible branch), frequent and or readily audible songs or calls, striking plumage coloration, limited fear of the observer, or other behavioral characteristics (such as drumming, wing-flashing, tail-bobbing, or aerial courtship displays), then the probability of encountering the individual is enhanced and the computed density will more closely approach reality. In this study species with high probabilities of detection included the Mountain Chickadee (frequent call, little apparent shyness toward the observer), Yellow-rumped Warbler (frequent call and song, prefers relatively open tree foliage in which to forage), Ruby-crowned Kinglet (frequent and strident song), and Gray-headed Junco (ground forager using more open areas).

There are various advantages and adverse aspects of each of these two sampling methods (objective 2). With the variable-strip transect method it is assumed that all individuals are detected within the strips on either side of the transect line bounded by the point of inflection on each species' distribution curve of the results. In this study's dense mixed-coniferous forest, this assumption was probably not valid. Therefore, the actual density was undoubtedly higher, at least for some species. As degree of conspicuousness of a species decreases, an even larger disparity between the results and the real density will be realized. Recognizing this dilemma, J. T. Emlen (1971) suggested that a basal detectability adjustment factor be applied to the results to take into consideration the incompleteness of the surveys. However, the adjustment value must be obtained by using another sampling method which itself is subject to limitations and increases the amount of time necessary to sample the plot.

Coefficient of detectability (CD) values as described by J. T. Emlen (1971) are designed to enable similar habitats to be sampled quickly. However, they are not necessary in order to calculate avian density. CD values may vary as the season progresses and degree of conspicuousness declines, and also on a yearly basis as densities change.

A highly conspicuous species will be observed more frequently than a less readily observable species. Thus the density computation for the latter species once the transect results are averaged will be far below its actual value. The same argument applies to sparsely distributed species in that the probability of encountering them is reduced. Hence, results for the less conspicuous and/or uncommon species are probably not as reliable as are those of more conspicuous and/or densely distributed species.

Fixed-width strip transects present similar problems to those encountered with the variable-strip transect method. The results are susceptible to detectability difficulties and errors in distance estimations.

As evidenced in this study, since some species will have highest densities in the narrowest belt width, whereas others may only be observed in the furthest belts, it is best to choose the strip width wisely, keeping in mind each species' detectability characteristics.

In comparing the variable-strip transect results to those of the various fixed-strip surveys (objective 3), it becomes apparent that they will be in full accord only in those instances whereby the peak (point of inflection) on the distribution curve corresponds to the exact width of the fixed-width strip survey. Otherwise the transect method should provide higher densities than those of the fixed-width transect survey because the data increments are so much smaller (i.e., 3 m belts vs 15 m or larger belts). In the variable-strip transect technique the density calculation depends on the curve's point of inflection. In constrast, with the fixed-strip method, even though the number of observations may reach a peak and then decline to the far edge of the belt, the entire area encompassed by the belt is used in the final density calculation, thus reducing the density value from what it would have been had only the area from the transect to the peak of observations been considered.

The main asset of a fixed-width survey is its simplicity in recording observations and analyzing data which allows for a rapid density computation. However, because the variable-strip transect method provides a more reliable estimate of species and population densities, at least on theoretical grounds, it is preferable except in cases of narrow habitat strips which lend themselves more readily to a fixed-width analysis.

I recommend several modifications in the data collection and analysis process for the variable-strip transect method (objective 4). Instead of comparing male data times two to all other observations and selecting the higher figure as proposed by J. T. Emlen (1971), I suggest that a third category encompassing "total observations" be included in the comparison. This is because the total observations may equal a higher value than either the male data times two, or all the other non-male data. If so, then the total observations category represents a more accurate reflection of the population density than would either of the other two groups.

Another possible modification of the variable-strip method involves the determination of a more precise distance measurement for observations beyond 30.5 m (100 ft) of the transect instead of lumping all observations lying within the 30.5 m–61.0 m (100–200 ft) strip on either side of the transect and all those from 61.0 m–125.6 m (200–412 ft) as described by J. T. Emlen (1971). I recommend recording each bird's distance as accurately as possible and then grouping results according to 6 m substrips from 30–60 m from the transect, and 12 m intervals for substrips out to 126 m of the transect. For observations within 30 m of the transect, the use of 3 m substrip intervals is suggested in a similar fashion to that indicated by J. T. Emlen (1971). The consideration of additional substrips for distances beyond 30.5 m of the transect allows for the demarcation of the point of inflection of the curve for those species whose peak lies within the 30–60 m range, and the albeit few species displaying an inflection point beyong 60 m. Without such a modification in the prescribed procedures, it is difficult to obtain a reliable estimate of densities for those species peaking at the relatively greater distances from the transect. Furthermore, at a later date the actual measurements can be segregated into a prescribed number of intervals with particular interval widths. The number of intervals and their widths will depend upon the width of the transect, the number of observations obtained, and the accuracy of the measurements (Anderson and Paspahala 1970).

It has been recommended that when analyzing data from intervals, it is best to use the average of all the exact measurements falling within the interval rather than using the interval midpoint for the estimation, as it provides a more precise value (Pollock 1978).

ACKNOWLEDGMENTS

I wish to thank R. D. Ohmart, J. T. Emlen, J. Verner, W. F. Laudenslayer, and others for providing insight and ideas for analyzing avian species densities and communities. I am especially grateful to D. R. Anderson for his valuable suggestions on improving the manuscript. This study was supported through cooperative aid agreements from the U.S. Forest Service, Rocky Mountain Forest and Range Experiment Station, Tempe, Arizona (No. 16-402-CA and 16-382-CA).

Studies in Avian Biology No. 6:146–151, 1981.

METHODOLOGY FOR CENSUSING LAND BIRD FAUNAS IN LARGE REGIONS

OLLI JÄRVINEN[1,2] AND RISTO A. VÄISÄNEN[1]

ABSTRACT.—As the biological context determines the accuracy needed in bird censusing, no ideal all-purpose methods exist. In particular, many problems in the borderland between ecology and biogeography can, for economical reasons, only be solved with the aid of rapid, one-visit census methods. Using biologically meaningful examples, we here review methodological problems encountered by us in an extensive line-transect project in Finland and adjacent countries in 1973–77. As line transects do not aim at giving absolutely accurate estimates of density, the methodological program is simply to minimize error variance, and minimize bias. The following points are discussed:

(1) The methods adopted in field-work must be well standardized, for example, with respect to dates, census hours, and the time used per unit area. Further, standards for weather conditions must be sufficiently strict. (2) Field tests for determining the accuracy of the census compared with other standardized methods are necessary. (3) All areas should be sampled on phenologically comparable dates in all years. (4) The censuses should be maximally dispersed over the region. (5) The censuses should sample all relevant habitats in approximately correct proportions. (6) Owing to interobserver variation, the consistency among censuses made by different observers should be carefully checked. (7) All major observers should cover as wide areas as possible, and all regions should be covered by more than one observer. (8) In interpreting the results, distrust deviating points. (9) In long-term comparisons, where interobserver variation cannot be checked in field tests, devise tests examining the null hypothesis that the patterns observed can be accounted for by changes in the ability to census.

Finally, we list several problems connected with analysis of transect data.

Population ecologists study the distribution and abundance of organisms in relation to different factors, while biogeographers usually focus on broad patterns of geographical distribution. However, population ecology and biogeography do not seem to be as close to each other as would be desirable, but the patterns studied in the two disciplines appear to be separated by a substantial gap. We can illustrate this best by means of examples.

Example 1.—It is presumably generally agreed that densities tend to decrease towards the geographic periphery of the species range. But the data available are scattered, often inconclusive, and in many cases simply nonexistent. *Densities may decrease towards the range boundary, but how much? What are the typical patterns in different species? Are smooth declines or abrupt drops the dominant pattern?*

Consider the most abundant passerine breeding in southern Finland, the Chaffinch (*Fringilla coelebs*). Its densities (Fig. 1) show a consistent decrease towards the northern range boundary in Finland. Maximum regional densities, as determined from transect data, exceed 50 pairs/km^2, while the species becomes very scarce near the Arctic Circle, about 500 km north of the southern peak densities in Finland. As the density classes used in the map are logarithmic, the decrease is actually very steep; the range of the Chaffinch thus seems to come to a fairly abrupt end in the north. In southern Finland our censuses (Haila et al. 1980a) have often revealed densities of over 100 pairs/km^2 in favorable habitats for the Chaffinch (maxima near 200 pairs/km^2). Reports from southern Scandinavia or Central Europe (e.g., Enemar 1966, Grempe 1973, Williamson and Williamson 1973, Witt 1976) indicate similar densities, implying that the densities of the Chaffinch are fairly high in a large region extending from Central Europe to southern Finland, but then the densities suddenly decrease. This decrease coincides with the increase of the ecologically similar congener, the Brambling (*F. montifringilla*); the density ratio of the two species changes about 10,000-fold within 600 km in Finland (Fig. 2; for additional data and discussion, see Järvinen and Väisänen 1979a).

We conclude that data on quantitative distribution patterns should be available for elucidating details of geographical distribution and for finding out possible ecological causes for range limitation. Let us take another example.

Example 2.—Data on the quantitative aspects of faunal dynamics are meager, but they are often essential in understanding ecological or zoogeographical patterns. An instructive example is provided by the remarkable range expansion of the Scarlet Rosefinch (*Carpodacus erythrinus*) in Finland, studied in detail by Stjernberg (1979). The species breeds both in closed forest habitats, particularly edges of luxuriant forests, and in various bushy habitats created by man. The proportion of birds breeding

[1] Department of Zoology, University of Helsinki, P. Rautatiekatu 13, SF-00100 Helsinki 10, Finland. Permanent address of both authors.

[2] Department of Biological Science, Florida State University, Tallahassee, Florida 32306.

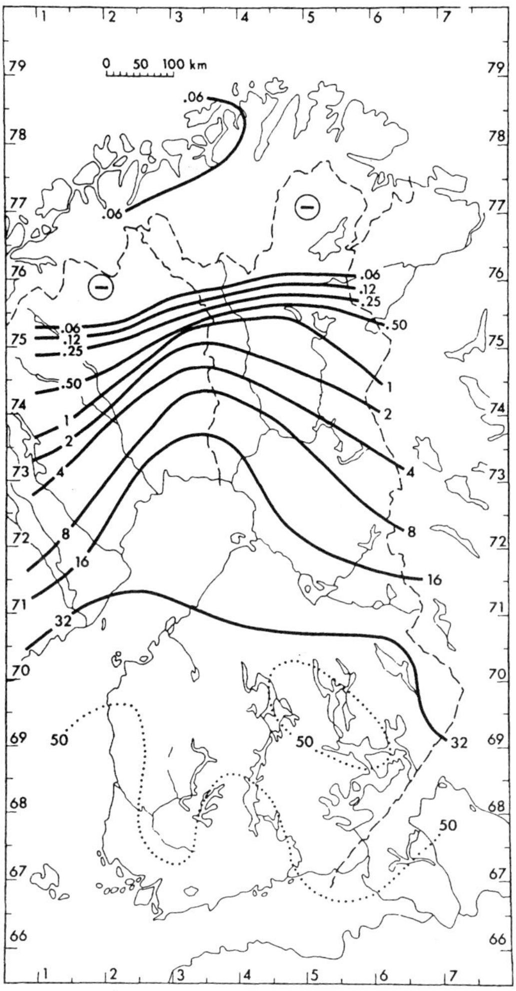

FIGURE 1. Transect density (pairs/km²) of the Chaffinch in Finland and adjacent areas (total 495,000 km²) in 1973–77. The lowest density given is 0.06 pairs/km², and the highest 32 pairs/km², in geometric progression (ratio of adjacent curves 1:2). Densities of 50 pairs/km² and greater are also shown. The encircled minus signs in the north indicate that Chaffinches were not observed in the censuses, though they may breed in the area in low densities. The coordinates refer to the 100-km squares of the Finnish uniform grid.

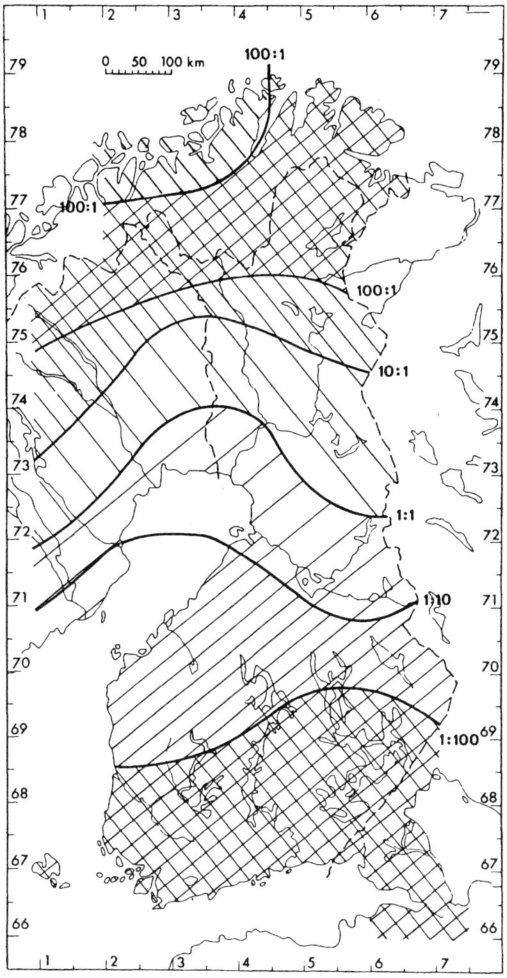

FIGURE 2. The ratio of the density of the Brambling to that of the Chaffinch according to line transects censused in 1973–77. From Järvinen and Väisänen (1979a).

in open habitats has increased considerably in recent decades; the breeding success was twice as high in the new open habitat as in closed forest habitats. *But is this a sufficient explanation for the range expansion observed in Finland?* Stjernberg could, on the basis of quantitative estimates on changes in the breeding numbers in Finland during the three past decades (Järvinen and Väisänen 1976c, 1979b), show that the changes he had observed in breeding success were a sufficient explanation for the range expansion and population increase of the species in Finland.

Similarly, it has been possible to show that several species associated with spruce forests have increased roughly in proportion with the increased coverage of spruce in Finnish forests, and that species associated with old forests have decreased more dramatically than the area of old forests (e.g., Järvinen and Väisänen 1979b and references there). As an example, consider the guild of foliage-gleaning resident insectivorous passerines of coniferous forests in Finland (Fig. 3). Their densities have crashed in northern Finland where old forests have been extensively cut, but, as the forestry statistics show, consid-

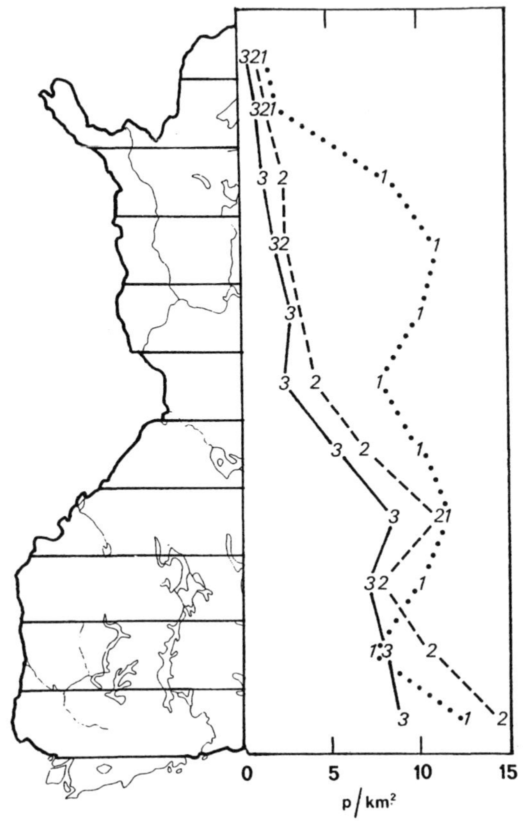

FIGURE 3. Density (pairs/km²) of the guild of foliage-gleaning resident insectivorous passerines of coniferous forests (*Parus cristatus, P. ater, P. montanus, P. cinctus, Certhia familiaris,* and *Perisoreus infaustus*) in 100-km zones in Finland in the 1940s (1), 1950s (2), and 1970s (3). The population crash in the north is due to similar decreases in those species of the guild which are abundant in northern Finland. From Järvinen and Väisänen (1979b). See also Järvinen and Väisänen (1979a).

erable areas of old forests are still standing in northern Finland (Järvinen et al. 1977a, Järvinen and Väisänen 1979b).

It is obvious that censusing land-bird faunas in large regions cannot be so accurate as censusing birds in a 10 ha woodlot. The problem is to develop a methodology for eliminating errors or at least estimating quantitatively the probable magnitude of the errors involved. The program is certainly simple—*minimize error variance and minimize bias*—but it is abundantly clear that the problems are complicated.

In the following, we shall sketch the major methodological problems we encountered in an extensive project in Finland in 1973–77. We used the line-transect method (Järvinen and Väisänen 1976c), but we attempt to discuss the problems on a more general level, paying special attention to problems that are still, in our opinion, poorly understood.

CHOICE OF THE METHOD

The first problem is clearly to choose the census method. This tends to be an economical problem, as sampling large regions implies high costs. In our transect project, more than 120,000 pairs of land birds were censused in 1973–77, and only the time used for censusing in the field required about 5000 hours. Therefore, the only economically feasible alternative seems to be to accept rapid one-visit census methods, even if more accurate choices, such as mapping or multiple-visit censusing, exist. This is especially so because the study area must be covered fairly evenly.

STANDARDIZATION OF THE FIELD-WORK

It is an essential requirement that the census method be standardized as well as possible. Before discussing specific problems, we first describe the field procedure (for details, see Järvinen and Väisänen 1976c).

In Finnish line transects the observer records all birds (pairs) observed, those within 25 m of the transect separately. The transects are previously planned on a map, and an attempt is made to include all major terrestrial habitats of the region in approximately correct proportions; the observer also reports the coverage of different habitats on the transect, as determined in the field. The censuses are made in early morning from 04:00 to 09:00, with little flexibility; censusing is not permitted if wind or rain impair detectability. The recommended census dates are June 1–20 in southern Finland and June 10–30 in northern Finland, but somewhat earlier and later censuses are accepted, according to exact rules depending on latitude, that is, on phenological differences in different parts of the country (see Järvinen and Väisänen 1977c). Two features of Fenno-Scandia make the transect method especially favorable: the breeding season of land birds is much more compact than in more southern regions, and there are, thanks to a long Fenno-Scandian tradition, only few legal restrictions to conducting bird censuses—or other undamaging and nondisturbing activities—in areas of one's free choice.

ACCURACY OF THE METHOD

Line transects, as one-visit censuses in general, do not give absolute densities, but underestimate the true numbers. Three experiments have compared the efficiency of the line transects with mapping, which is certainly a more accurate method (see Enemar et al. 1976, 1979) than any of the one-visit methods, though not faultless (Svensson 1974b, Nilsson 1977b, and references there). One experiment was made in a Polish forest, another in a south Finnish forest area, and a third one in north Swedish mountain birch forest. Two of the experiments (Järvinen et al. 1978a, 1978b) were made in optimal conditions and suggested an average efficiency of 80% or more of the mapping result, while an experiment in a boreal forest area in southern Finland made under more typical conditions (Tiainen et al., in press) suggested an average efficiency of 60–65%, which agrees with estimates derived from mapping studies (Järvinen 1978b and references there). However, too few experiments have still been made. There are no definite data on whether census efficiency varies latitudinally or according to habitat, and data on interspecific differences in detectability are scanty (see Järvinen 1978b). Well-conducted comparisons between standardized methods are thus badly needed, and we urge that primary data be published as extensively as possible.

DISTRIBUTION OF CENSUSES IN SPACE AND TIME

As phenological differences are a major source of error in bird censusing (Järvinen et al. 1977b and especially Slagsvold 1977), these should be eliminated as completely as possible. For example, we compared older censuses, especially those of Merikallio (1958), with new censuses made in the 1970s. Because he continued his censusing to about mid-July, parts of Merikallio's data were obviously not comparable. After imposing identical constraints on census dates, however, the average census dates became very comparable: the average date for the censuses made in 1936–49 was June 22, while it was June 17 for 1952–63 and June 19 for 1973–77; incidentally, the average for 1936–63 was also June 19. In these calculations we eliminated the bias that different latitudinal zones were studied with variable intensity. At the same time, the censuses should be phenologically comparable within the region; for example, in our censuses the average dates for southernmost Finland are June 15, but for northernmost Finland June 25. Finally, alleged annual population fluctuations involve a considerable source of error if different areas are sampled in different years. The methodological rule is thus clear: *Sample all areas on phenologically comparable dates in all years.*

In studying population changes, it is usually not realized that the spatial distribution of the censuses has a prominent role. This comment applies both to the regional and to the habitat scale, as population trends in different habitats and different parts of the range may be different. An example illustrates the importance of this methodological remark.

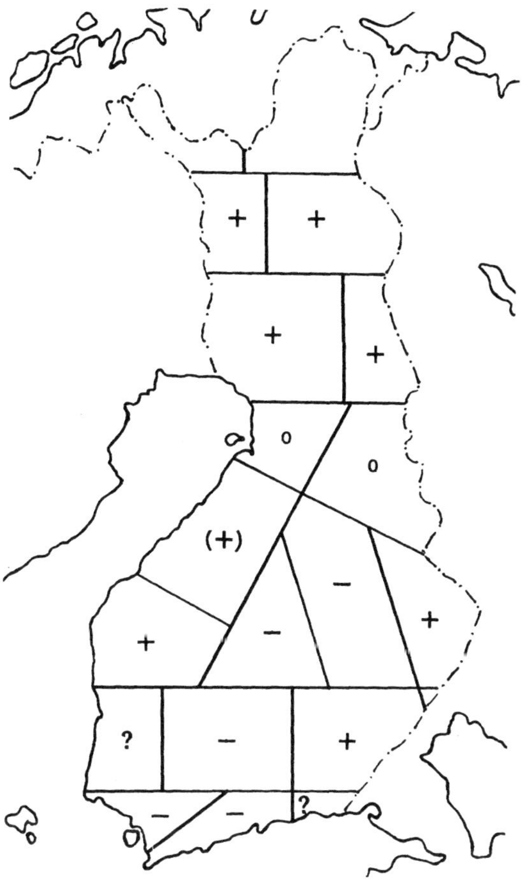

FIGURE 4. The direction of changes in the numbers of the Curlew in different provinces of Finland from the 1940s to the 1970s. Even in total numbers in Finland have changed little, decreasing trends (minuses) dominate in the south and increasing trends (pluses) in the north.

The range dynamics of the Siberian Tit (*Parus cinctus*) and the Crested Tit (*P. cristatus*) have been cited as an example of the climatic amelioration influencing the balance of two presumed competitors (Lack 1954). Quantitative censuses from recent decades reveal, however, that the actual pattern is much more complicated (Järvinen and Väisänen 1979a). Both species seem to have decreased owing to forestry, and—quite to the contrary as expected from competition theory—the region of *maximum* decrease has been in the overlap zone of the two species.

It is probably agreed by the majority of ornithologists that peripheral populations fluctuate more than central populations, but it is a fact that quantitative studies of population changes seldom consider the regional patterns. It is certainly true that country-wide population indexes can be useful, but it should be realized that they mask regional differences in population trends. Fig. 4 shows population trends of the Curlew (*Numenius arquata*) in Finland. The species is classified as stable on the basis of country-wide data, but what has happened is that the decrease in the south has been approximately compensated by increasing numbers in the north.

Similarly, it is probably generally agreed that avian numbers in suboptimal habitats fluctuate more than in optimal habitats (Fretwell and Lucas 1969, von Haartman 1971), but it is a fact that quantitative studies of population changes seldom make an attempt to cover all habitats in approximately true proportions. We thus insist on the following methodological rule: *Censuses should be maximally dispersed over the region, and they should sample all relevant habitats in approximately correct proportions.*

INTEROBSERVER VARIATION

A major problem is that different ornithologists must be used in censusing large areas, so interobserver variation is introduced into the results. Part of the differences can be eliminated easily. For example, the census reports sometimes clearly show that the standard rules have not been followed; such censuses should naturally be discarded. In our own work we have analyzed the results of each transect census (total number about 1000) separately and checked whether or not the results conform to the general pattern: are the densities reported similar to those reported by other observers censusing similar habitats in the same region? It has been our experience that very few censuses deviate on the basis of this criterium; less than 1 per cent of all censuses were discarded on this basis. So we suggest the following rule: *Search for consistency among censuses made by different observers.*

The following rule guarantees that no larger area merely reflects the effect of an exceptional observer: *Cover all regions by more than one observer.* Another important rule is clearly: *See that all major observers cover very different areas.* For example, most of the major observers in our transect project made censuses both in southernmost and northernmost parts of Finland, and all major observers traveled hundreds of kilometers owing to the census work. This methodological rule gives a solid basis for evaluating geographical trends in the results, as it can be checked that different observers report the same trends.

A peculiarity of the Finnish line transects is that each census report has two parts: the observer must report all birds observed, but give a separate list for the so-called main belt, that is, the birds observed within 25 m of the transect. In analyzing the data for each transect we have thus different possibilities for evaluating interobserver differences: we calculate, for each transect separately, the results for the main belt and the results based on all observations, using two different methods (see Järvinen and Väisänen 1980:68). We see two main types of discrepancy. Firstly, certain observers, none among the most experienced, reported average densities based on all observations, but their main-belt data indicated substantially higher densities. We attributed this type of discrepancy to errors in estimating the width of the main belt and ignored the main belt data. Secondly, we have developed certain correction methods for analyzing transect data (Järvinen and Väisänen 1976b), but the applicability of the correction method has been checked for each transect separately (for details, see Järvinen and Väisänen 1980:68).

A final point in eliminating errors due to the effect of single observers was our interpretation of the final results: we ignored all patterns based on deviating values in single 100-km squares, which were the basis used in analyzing the data. As adjacent squares were generally censused by a considerably different set of observers, interobserver differences were certainly decreased on the basis of the following rule: *In interpreting the results, distrust deviating points.*

The problems become more intricate when population changes over long periods of time are studied. We have studied long-term trends in Finnish land bird populations on the basis of transect data, but, of course, some of the trends may be artifacts caused by interobserver differences. It may be argued that we should expect to see a bias towards increased numbers in the results of the census, because field ornithological skills have undoubtedly hugely improved in recent decades.

If this hypothesis is a scientific one, it is testable. It is clear that direct tests in the field cannot be made. But, for example, the hypothesis would predict that population trends do not show geographical patterns, but they do (Figs. 3–4 and our unpubl. data on many other species); and it would be predicted that most population increases are observed from the 1950s to the 1970s, but, in actual fact, many population changes occurred from the 1940s to the 1950s, according to our analyses (e.g., Järvinen et al. 1977a, Järvinen and Väisänen 1978, 1979b). This is significant, because most censuses in the 1940s and the 1950s were made by a single man, Einari Merikallio. Tests based on methodological ideas can also be devised.

In line transects, birds within 25 m of the transect can certainly be censused more easily than those outside the 25 m belt. For example, two European pipits, *Anthus pratensis* and *A. cervinus,* breed in similar open habitats, and their song is similar. However, *pratensis* is much more common than *cervinus*. In view of their similar behavior and habitats we would expect that the proportion of the close (within 25 m) observations is similar in both species, but this is not true (Järvinen and Väisänen, unpubl. data): if the bird is singing far from the transect, many observations on *cervinus* are overlooked or the species is confused with *pratensis*. In other words, the proportion of close observations is an efficient index of observer ability. Essentially this effect has been experimentally demonstrated by Hutto and Mosconi (1981).

Therefore, if the ability of census-makers has improved in recent decades, we expect that the proportion of close observations is higher in the old censuses of Merikallio than in present censuses; but the contrary is true (Järvinen and Väisänen 1975). Another version of this test omits all species observed more often than 10 times in Merikallio's censuses—we may expect that the species observed rarely by Merikallio were especially difficult for him. There were 46 land bird species observed at most 10 times in Merikallio's censuses. Data for single species are, of course, not testable because the sample size is at most 10 by definition. However, the percentage of close observations was more often *lower* than higher in Merikallio's censuses, as compared with the corresponding percentage in modern censuses (lower in 30 species, identical in 4, and higher in 12; 30 is significantly different from 12, $\chi^2 = 7.71$, $P < 0.01$). We also calculated the expected numbers of close observations for Merikallio's censuses, assuming that the percentage of close observations is identical with that observed for the same species in present censuses. We would have expected 48.8 close observations and 156.2 far observations on the 46 rare species, but Merikallio had 29 close observations and as many as 176 far observations. The difference is significant ($\chi^2 = 10.54$, $P < 0.01$), but in the *opposite* direction as predicted from the "observer ability" hypothesis.

The above tests are actually tests of the null hypothesis that no population changes have really occurred but all changes observed are merely artifacts due to interobserver variation. As indicated, these tests, as well as direct comparisons of our data with von Haartman's censuses in SW Finland (see Haila et al. 1980b), allow us to reject the "observer ability" hypothesis for this data set, although we agree that interobserver variation is an important potential source of error in transect studies.

DATA ANALYSIS

Analyzing data is certainly also a problem, but it is not specific to studying large areas. Some of the open problems in analyzing transect data should, however, be mentioned.

(1) The proportion of close observations changes during the census hours and during the season (Järvinen et al. 1976 and our unpubl. data). We use averages in our analysis; but what are the sources of error involved?

(2) Our analytical method (Järvinen and Väisänen 1975) involves the assumption that detectability decreases linearly from the transect, but other functions might better represent reality. Carefully devised experiments are certainly needed here.

(3) Are there realistic possibilities for devising reliable species-specific methods in order to correct for the incompleteness of the census?

Of course, these or other methodological problems should not be studied in isolation, without consideration of the specific needs of the research problem. The science of bird censuses may be regarded as an art of developing the perfect method, and studies in this direction are helpful in illuminating potential sources of error in census work. But bird censuses are also a tool, and we should sometimes give serious consideration to the fact that the ultimate problem is not perfecting the tool, but using it for meaningful purposes.

ACKNOWLEDGMENTS

We thank J. J. Hickey for many useful comments on the manuscript, K. Mikkola for valuable discussion on the effect of interobserver variation on transect data, and the Academy of Finland for financial support of our work.

Studies in Avian Biology No. 6:152–159, 1981.

SOURCES OF ERROR INVOLVED IN THE FINNISH LINE-TRANSECT METHOD

OLAVI HILDÉN[1]

ABSTRACT.—The line-transect method has been used extensively in Finland for estimating numbers of land birds, but only a few attempts have been made to test its efficiency and reliability. The present paper examines the applicability of this census method to (1) estimating absolute densities, and (2) studying changes in bird populations.

Single line-transect censuses were compared with careful mapping of pairs and searching for nests on two study plots in Finland. In both tests only 46–49% of the breeding pairs were recorded in the single censuses. Similar tests restricted to seven selected species in three study areas gave similar results: the census efficiency ranged from 33 to 67%, averaging 48%. The reliability of the line-transect method was tested further by letting one person census the same transect several times throughout the breeding season: the numbers of pairs of most species varied greatly from one census to another, the maximum being often 3–5 times higher than the minimum. Consequently, the published density values, biomass and energy flow calculations, and estimates of the total numbers of pairs in Finland, based on line-transects, must be considered unreliable; most of them are serious underestimates.

The unpredictable outcome of single line-transect censuses also reduces the reliability of apparent annual population fluctuations detected by this method. This is especially true if there are between-year differences in (1) census takers, (2) dates of censuses, (3) weather conditions, or (4) proportions of different habitats. A good example is provided by the recent population trend of Finnish Starlings: the species is known to have decreased catastrophically in the 1970s, yet the extensive line-transect material did not reveal any such change. Least safe are comparisons between old and current censuses because of several additional factors, like (1) advances in field ornithology, (2) different working methods, (3) scanty data, and (4) influence of exceptional census years, all of which could bias the results.

The line-transect method has been used extensively in Finland for estimating numbers of land birds. A pioneer in this field of ornithology was Prof. E. Merikallio, who censused more than 1000 km of transects in the 1940s and 1950s (Merikallio 1946, 1951, 1958). In the 1970s, Drs. O. Järvinen and R. A. Väisänen revived the line-transect censuses in Finland and collected material covering over 3000 km of transects. In many stimulating papers they have presented and discussed their data (e.g., Järvinen and Väisänen 1980, and the literature cited there).

In contrast to the impressive amount of work providing the Finnish line-transect material, our knowledge of the efficiency and reliability of the method is poor. Although its weaknesses have been generally recognized, only a few attempts have been made to test quantitatively the influence of the various sources of error upon the results. This lack of information naturally reduces the confidence that can be placed in them. How valid are, for instance, the bird density values for different habitats, estimates of the total numbers of pairs in large regions, or long-term trends in the avifauna, based on line-transect censuses?

In his extensive review of bird census methods, Berthold (1976) has emphasized correctly that only reliable methods should be used and that the sources of error involved in the methods should be critically tested. The same views had already been clearly expressed by Palmgren (1930). The aim of the present paper is to examine the applicability of the line-transect method to (1) estimating absolute densities, and (2) studying annual and long-term changes in bird populations.

METHODS

The methods of the Finnish line-transect censuses have been described in detail by Järvinen and Väisänen (1976c), so only a few facts need to be stressed here. Each transect is counted only once, during the month of June, between 04:00 and 09:00. The birds observed within 25 m on both sides of the transect are included in the *main belt,* those registered farther away belong to the *supplementary belt*; together, the two belts form the *survey belt.* Bird densities are estimated in general from the survey belt data, using a correction method based on a linear model (Järvinen and Väisänen 1975, 1976c).

The efficiency of the line-transect method, or any other census method based upon a single visit to a study area, can be tested most reliably by comparing the census results with the true numbers of stationary birds. The true composition of the bird community within a certain area, in its turn, can be figured out best by careful mapping of pairs and searching for nests throughout the breeding season, preferably combined with color-ringing.

In this paper, five such tests are reported. In two of them the entire community of a study plot was censused, while three tests were confined to two or three dominant species of the habitat. In all these investigations, the independent single transect counts were

[1] Department of Zoology, University of Helsinki, P. Rautatiekatu 13, SF-00100, Helsinki, Finland.

TABLE 1
CENSUS EFFICIENCY OF THE LINE-TRANSECT METHOD ON A STUDY PLOT OF 5 HA OF MIXED WOODLAND IN KIRKKONUMMI, S FINLAND (O. HILDÉN & L. J. LAINE). THE RESULT OF A SINGLE CENSUS ON 17 JUNE IS COMPARED WITH THE TRUE NUMBERS OF PAIRS IN 1980

Species	True numbers	Single census
Fringilla coelebs	6	4
Ficedula hypoleuca	6	4
Phylloscopus trochilus	5	3
Erithacus rubecula	5	2
Apus apus	4	—
Parus major	3	3
Turdus philomelos	3	1
Prunella modularis	3	2
Carduelis spinus	3	1
Other species (15)	19	8
Total no. of pairs	57	28 (49%)
Total no. of species	24	14 (58%)

TABLE 2
CENSUS EFFICIENCY OF THE LINE-TRANSECT METHOD ON A STUDY PLOT OF 60 HA LUXURIANT MARSH IN KARIGASNIEMI, FINNISH LAPLAND (O. HILDÉN). THE COMBINED RESULTS OF ANNUAL SINGLE CENSUSES IN LATE JUNE ARE COMPARED WITH THE TRUE PAIR TOTALS IN 1969–71

Species	True numbers	Single census	Efficiency (%)
Anthus pratensis	90	25	28
Phylloscopus trochilus	59	33	56
Calcarius lapponicus	57	31	54
Motacilla flava	35	27	77
Luscinia svecica	35	6	17
Carduelis flammea	32	14	44
Emberiza schoeniclus	20	11	55
Limicola falcinellus	19	6	32
Phalaropus lobatus	19	6	32
Tringa glareola	16	9	56
Other species (11)	52	32	62
Total no. of pairs	434	200	46
Total no. of species	57	50	80

made by experienced census takers, in favorable weather conditions and using the standard rules of the Finnish line-transect method. Information on slight modifications of the normal field procedure used in some of the tests, as well as relevant details of the locality, habitat, census, etc., are given in the text.

The efficiency of the line-transect method can also be tested in another way, by counting the same transects several times in the course of the season and comparing the results of the successive censuses with each other and with the maximum numbers recorded along the routes. Such experiments were organized in 1979 by L. J. Laine at three localities in southern Finland. Each transect was surveyed by the same person (T. Ahlström, P. Koskimies, L. J. Laine), using standardized methods and in optimal weather at about 10-day intervals from mid-May to early July, six times in all. Survey belt data were used in this case, partly to avoid the biases caused by small samples, partly because the main objective of these tests was to examine the constancy of successive transects in the course of the season.

CENSUS EFFICIENCY

One of the two community censuses was made on my own property, consisting of 5 ha of mixed woodland in southern Finland, about 30 km west of Helsinki. In 1980, I determined the numbers of its breeding birds very carefully by daily observation throughout the breeding season; about two-thirds of all nests or broods were found. On 17 June, between 04:55 and 05:40, an independent transect count was conducted by L. J. Laine, who walked along a zigzag route so that the total area was covered as well as possible by the main belt. As shown in Table 1, only 49% of the pairs and 58% of the species were recorded in this single census.

The other study plot in which a similar test was made is of completely different habitat—a luxuriant marsh of 60 ha in Finnish Lapland. In 1966–72, it was the main research area during my study on subarctic bird communities and was surveyed almost daily by several students from early June to mid-July. The estimate of the numbers of pairs was based on nests found in almost half the instances, otherwise on careful observation of the birds. In three summers, in late June, an independent single census was conducted jointly by three students, who crossed the marsh walking side by side along parallel transects, first through one half and then back through the other so that the whole area was surveyed. The distances between the counters (30–80, average 60 m) were slightly greater than the main belt width (50 m) in normal line-transects, but this was compensated by the open habitat, which made it easy to observe the birds. The results, summarized in Table 2, show that, on average, 46% of the pairs were recorded in the single censuses; for some species the efficiency was as low as 17% in the Bluethroat (*Luscinia svecica*) and 28% in the Meadow Pipit (*Anthus pratensis*). Far fewer species were missed than in the wooded habitat.

The three other tests were confined to selected species whose numbers of pairs in the study areas could be estimated accurately by means of nests found, color-ringing, and careful observation of the birds. The results are summarized in Table 3, which also gives some additional details of the censuses. It should be noted that

TABLE 3

CENSUS EFFICIENCY OF THE LINE-TRANSECT METHOD FOR SELECTED BIRD SPECIES IN THREE STUDY AREAS IN FINLAND; THE MEANS OF SINGLE TRANSECTS (NO. IN BRACKETS NEXT TO AREA) IN JUNE ARE COMPARED WITH THE TRUE NUMBERS OF PAIRS

Area and species	True no. of pairs	Mean of transects	Efficiency (%)
Pori 1968 (6)[a]			
Emberiza schoeniclus	16	7.5	47
Acrocephalus schoenobaenus	8–9	2.8	33
Kirkkonummi 1979 (2)[b]			
Parus major	19	9.5	50
P. caeruleus	6	2.0	33
Ficedula hypoleuca	10	5.5	55
Valassaaret 1980 (3)[c]			
Anthus pratensis	>18	9.3	<52
Oenanthe oenanthe	9	6.0	67

[a] Census taker: Haukioja (1968). Size of the area: 5 ha. Habitat: willow thickets, meadows and reeds. Method: normal line-transect, only main belt data used.

[b] Census takers: O. Hildén & L. J. Laine. Length of the transect: 3.5 km, with 39 nest-boxes placed within the main belt and 3 slightly outside. Habitat: mixed woodland. Method: normal line-transect, survey belt data used.

[c] Census taker: T. Pahtamaa. Size of the area: c. 50 ha. Habitat: scrub heathland. Method: line-transect along a zigzag route, average width of the census strip 65 m.

survey belt data (i.e., all observations) were used in the census of box-nesting species in Kirkkonummi, although almost all their nests were situated within the main belt. If only main belt registrations were used, the efficiency percentage for these three species would drop to 32, 17 and 45, respectively. In the study plot censuses at Valassaaret, the average width of the strip (65 m) slightly exceeded the main belt (50 m) in ordinary line-transects but, on the other hand, the openness of the habitat, low bird density and concentration on only two species considerably facilitated the counts. Although the three tests concerned different species, living in different habitats, the results are fairly consistent, showing that between 33 and 67% (average 48%) of the stationary pairs were recorded in the single line-transects in June.

The results of successive counts of the same transects are shown in Table 4. For almost all species, the numbers of pairs counted varied greatly from one census to another, the maximum being often 3–5 times higher than the minimum. The numbers of Chaffinches (*Fringilla coelebs*) proved least variable between censuses, while those of Robins (*Erithacus rubecula*) were most variable (maximum: minimum = 1.4 and 8.4, respectively). This finding reveals the degree of unpredictability of a single census for each species, but does not tell us much about its accuracy in relation to a true population estimate. A rough calculation of this can be made in the following way:

The length of all the line-transects was 4 km, and observations were recorded separately for each kilometer. By summing the highest numbers recorded within these quarters in any census, an estimate of the maximum numbers of pairs along the whole transect was obtained; this value was 19% higher, on average, than the highest value for a single count. As even the best censuses are likely to be underestimates, some birds being always overlooked, these "maximum numbers" may serve as rough estimates of the true populations. The census efficiency was then estimated by comparing the results obtained in June (the recommended period for line-transects in Finland) with the maximum numbers of pairs. This gave a mean efficiency of 47%.

This result is in good agreement with those from the study plot censuses reviewed earlier. The conclusion is that even in favorable conditions an experienced observer will record, on average, about half of the stationary birds present in a study area, if using a single line-transect census. Thus, the efficiency of single transects is comparable to that of single counts in the mapping method, which has been estimated at about 50% (Enemar et al. 1978). The low efficiency of the line-transect method should not surprise ornithologists familiar with the poor detectability of most bird species during certain phases of their breeding cycle. Indeed, an even lower efficiency has been reported by Lehtonen (1979). For about 30 years, he has made extensive tests to compare the accuracy of the different methods used to census land birds, and concluded that in forest habitats in southern Finland at best 25–40% of the stationary birds within the main belt are recorded in line-transects.

All the tests reviewed here thus prove convincingly that the conclusion of Järvinen et al. (1978a) that between ⅔ and ⁵/₆ of the birds within the main belt are recorded in single line-transect censuses is far too optimistic. Such a high efficiency can be reached only in exceptionally suitable habitats under optimal conditions, as in the test made by Järvinen et al. (1978a) in mountain birch forest. In addition, the line-transects in this test were compared only with the mapping method, which tends to underestimate population densities (e.g., Nilsson 1977b, and the literature cited there), not with the true numbers of pairs; this probably also contributed to the high apparent census efficiency obtained.

In itself, the low efficiency of the method would not be a serious argument against the use of single line-transects, if the efficiency remained more or less constant (1) from species to species, (2) from one observer to another, and (3) during the census period. If this was the case, the results could be transformed easily to real densities by using a correction coefficient. But this cannot be done, because there are such striking differences in detectability among species (Table 4) and in the capacity to observe birds between census takers; moreover, the detectability of the species fluctuates in different ways during the course of the season. The latter fact was shown convincingly in successive censuses of the same transects, and some examples are depicted in Figure 1. Pronounced seasonal patterns in census efficiency of certain species were also found by O'Connor (1980c) in an experimental investigation of the effects of census date on the results of Common Birds Census surveys.

What is particularly striking, when one analyzes the results of the repeated counts of the same transects, is that the best census period for most species in southern Finland is the latter half of May, i.e., before the time recommended for line-transects. With few exceptions (*Turdus merula, Fringilla coelebs*), this was true of all sedentary species and all migrants arriving by mid-May. For some resident birds, such as titmice, a reliable census presupposes still earlier counts, started in March-April (Nilsson 1977b). The only species that are censused better in June are the few late migrants, such as *Phylloscopus sibilatrix, Sylvia borin* and *Muscicapa striata,* which arrive in Finland from mid-May onward. The superiority of May counts for most species is also indicated by the fact that the results of the three simultaneous transects were in general more consistent in May than in June, obviously because higher detectability reduces the effect of mere chance. This finding also strongly suggests that transient birds were not included in the May counts, as then, on the contrary, the numbers should have varied more from sample to sample than later in the season.

A logical consequence of the line-transect tests reported here is that all the published density values, biomass and energy flow calcultions, and estimates of the total pair numbers in Finland, based on this census method, are unreliable; most of them are serious underestimates.

ANNUAL CHANGES IN BIRD POPULATIONS

In recent years, line-transect data have been used in several papers by Järvinen and Väisänen (e.g., 1977b, 1977d, 1978b, 1978c, 1978d, 1978e, 1979a) also for studying annual changes in the avifauna. This may appear a useful approach, even if the weaknesses of the method are recognized, as the sources of bias and error can be expected to remain more or less constant from year to year. However, a detailed consideration reveals that results obtained by comparing annual line-transect data may give a seriously misleading picture of population trends. In the following paragraphs, I will comment briefly on the most important sources of error involved in this approach.

1. Unpredictability of single line-transects.—As shown by successive censuses of the same transects, the numbers of pairs counted vary greatly, not only between the different phases of breeding cycle but also between two consecutive counts conducted a few days apart. Particularly when small amounts of data are compared, considerable apparent differences between the annual density values may be attributable to this factor alone.

2. Differences between census takers.—There are considerable differences among ornithologists in their capacity to detect and identify birds, as revealed by several tests concerning both censuses of breeding birds (e.g., Enemar 1962, Snow 1965, Hogstad 1967, Berthold 1976, Enemar et al. 1978) and counts of migrants (e.g., Enemar 1964, Källander et al. 1972, Källander and Rydén 1974). Consequently, apparent annual differences in the numbers of birds recorded by different persons on line-transects may in fact reflect differences between the census takers rather than real changes in bird populations.

3. Different dates of censuses.—As shown in this paper and by several earlier students (e.g., Slagsvold 1973c, 1977; Berthold 1976; and the literature cited in these), the song activity and thus the census efficiency for a particular species depends greatly on the phase of its breeding cycle. Even small annual differences in the timing of the censuses relative to the breeding cycle may thus affect considerably the results obtained, and longer time differences can be expected to mask completely the true population changes of most species.

4. Different weather conditions during the censuses.—The detectability of birds is greatly influenced by weather conditions (e.g., O'Connor and Hicks 1980). Although the standard rules for line-transects presuppose good census weather, complete accordance in this respect is never reached. The effect of this factor cannot be measured reliably.

5. Different proportions of various habitats in the samples.—Unless permanent line-transects are used, the different habitats are seldom represented by the same proportions in successive annual samples. This is especially true of more scarce habitats showing a patchy distribution. The result is that the occurrence of a number of the more locally distributed species on the transects is affected.

TABLE 4
CONSTANCY AND EFFICIENCY OF SINGLE LINE-TRANSECT CENSUSES IN THE COURSE OF THE BREEDING SEASON; COMBINED RESULTS FROM THREE TRANSECTS IN S FINLAND, 4 KM EACH, COUNTED SIX TIMES FROM MID-MAY TO EARLY JULY AT ABOUT 10-DAY INTERVALS; ONLY THE 20 MOST ABUNDANT SPECIES ARE CONSIDERED

Species	Maximum numbers of pairs[a]	Recorded numbers of pairs		Census efficiency in June[b]		
		Lowest	Highest	Lowest	Mean	Highest
Fringilla coelebs	213	140	202	71.8	79.8	86.9
Phylloscopus trochilus	133	61	121	45.9	66.7	82.0
Erithacus rubecula	72	7	59	9.7	31.0	56.9
Turdus iliacus	59	17	52	47.5	56.5	67.8
Turdus philomelos	56	11	42	19.6	42.3	58.9
Carduelis spinus	52	11	43	38.5	49.4	63.5
Anthus trivialis	51	17	37	35.3	42.5	51.0
Phylloscopus sibilatrix[c]	45	15	39	40.0	60.0	86.7
Turdus merula	37	7	31	29.7	49.5	64.9
Columba palumbus	35	6	25	17.1	34.3	51.4
Regulus regulus	30	5	22	20.0	35.6	53.3
Muscicapa striata[c]	29	5	20	17.2	42.5	69.0
Sylvia borin[d]	26	11	22	42.3	56.4	76.9
Prunella modularis	24	4	16	16.7	38.9	58.3
Loxia sp.	24	6	24	25.0	47.2	70.8
Cuculus canorus	23	5	18	26.1	46.4	78.3
Parus major	23	3	18	21.7	34.8	43.5
Phylloscopus collybita	23	5	16	21.7	40.6	60.9
Corvus corone cornix	20	4	12	30.0	41.7	55.0
Emberiza citrinella	20	4	16	20.0	41.7	65.0
Mean	49.8	17.2	41.8	29.8	46.7	65.1

[a] Estimated by summing the highest numbers recorded within each kilometer of the transects.
[b] Estimated by comparing the results of the nine censuses in June with the maximum numbers of pairs.
[c] The first count on 14–15 May not included, as only a small part of the population had arrived.
[d] Both May counts excluded due to the late arrival of the species.

All the sources of error listed above will be accentuated in small samples, and reduced as more data are gathered. With extensive data, covering hundreds of line-transect kilometers each year, their effect might be expected to approach zero. But we have a good example which shows convincingly that, unfortunately, this is not always so. The example concerns the recent population trend of the Starling (*Sturnus vulgaris*) in Finland.

In the 1970s, Finnish field ornithologists and even farmers interested in nature noted a rapid decrease in the numbers of Starlings, which culminated in a crash during the last years of the decade. This catastrophic decline was documented by several long-term censuses of populations nesting in boxes and was reported from different parts of the country (von Haartman 1978a, 1978b; Ojanen et al. 1978; von Knorring 1978; Korpimäki 1978; Tiainen and Solonen 1979; Orell and Ojanen 1980). At the same time, the annual numbers of nestlings ringed in Finland showed a continuous steep decrease (Saurola 1978). However, when the extensive Finnish line-transect material from the years 1973 to 1977 was analyzed by Järvinen and Väisänen (1978d), no trend of decline in the Starling population was found. On the contrary, the authors concluded that the Finnish Starling population was fairly stable in the period in question, and even increased in 1977. The striking discrepancy between the two sets of data is shown by Fig. 2. If line-transect data collected from more than 2000 km and concerning one of the commonest species fail to reveal even such dramatic and well documented changes in numbers, how can this method be considered reliable when smaller quantities of data, scarcer species or lesser changes in populations are concerned?

When we remember that even the mapping method, in spite of its high effort (10 visits to the study plot), may fail to reveal marked population changes (Berthold 1976; Nilsson 1977b, 1977c), it is not surprising that the Finnish line-transect method, based on single counts, is considerably less successful.

LONG-TERM CHANGES IN AVIFAUNA

The biases involved in the method of monitoring bird population changes by means of line-transect data grow even more serious when long-term trends are concerned. In this instance

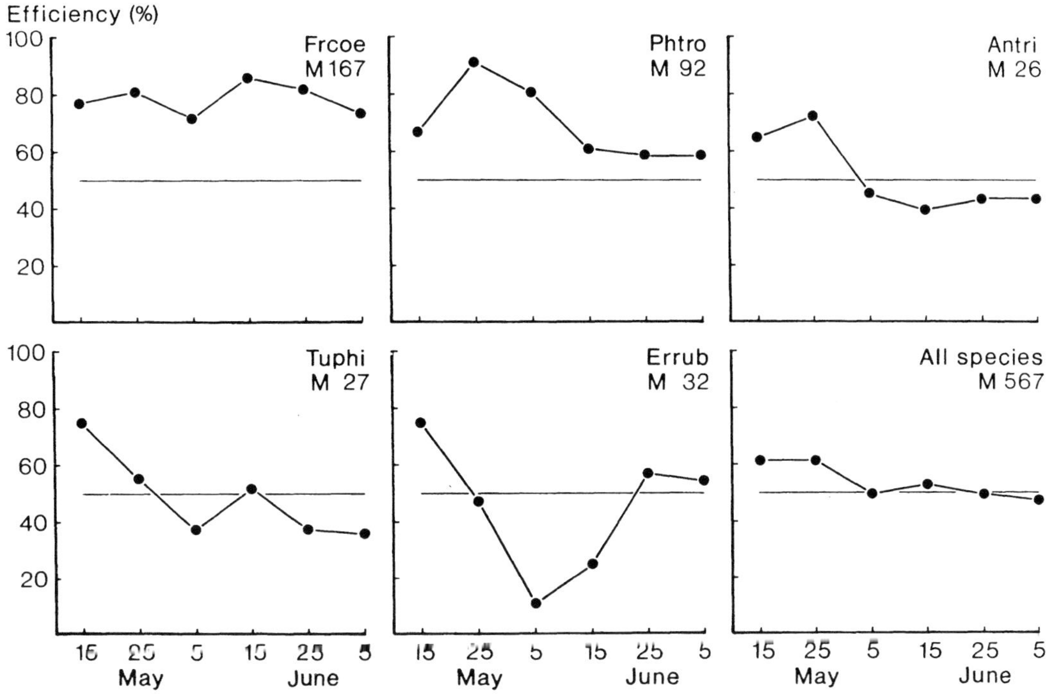

FIGURE 1. Seasonal fluctuations in the census efficiency for five selected species (*Fringilla coelebs, Phylloscopus trochilus, Anthus trivialis, Turdus philomelos, Erithacus rubecula*) and the entire bird community (excluding 12 species arriving late) in S Finland. The data are based on combined numbers of pairs of three line-transects, censused at ca. 10-day intervals from mid-May to early July. M = mean of all six censuses.

one has to compare old and current censuses, which invariably brings forth a number of new sources of error. In addition to the factors treated above, at least the following weaken still further the reliability of the results obtained.

(1) Present-day ornithologists are much more skillful and better equipped when identifying birds in the field than were their predecessors.—The tremendous advances in field ornithology are self-evident, but how much this factor has affected the line-transect censuses in different periods can only be guessed (for details, see Hildén 1979).

(2) Distant visual records and acoustic records are utilized today to a much greater extent than formerly.—Probably most ornithologists are ready to accept this statement, but only a few seem to have realized its powerful impact on field ornithology. I have considered this problem in many ways and concluded that the whole attitude to field observations has changed during recent decades (Hildén 1979). Formerly one had to see or hear a bird well and at close range before its identification was accepted, but nowadays even distant birds are assigned to a species from a hasty glimpse or faint call-note, and just as easily with the help of ears as eyes. This change in ease of identification of birds must have influenced the results of censuses, but by how much is, of course, impossible to evaluate.

(3) Individual differences between ornithologists are accentuated when the data of one early student are compared with the average of dozens of present-day census takers.—The bulk of earlier line-transect data in Finland was gathered by one single man, E. Merikallio. He was born in 1888 and thus a genuine representative of the old ornithologist generation; the line-transect material he collected at the advanced age of 53 to 68 years. It is hard to believe that Merikallio's census results from the 1940s and 1950s could be directly comparable with the current ones, compiled by mainly young, modern ornithologists in the 1970s.

(4) The working methods and the timing of censuses have somewhat changed.—The standard rules for line-transect censuses have remained roughly the same from Merikallio's time to the present, but there are some slight differences between the practices followed formerly and now; e.g., in the dates and time of day of the censuses, the speed of walking on the transects, the use of supplementary belt observa-

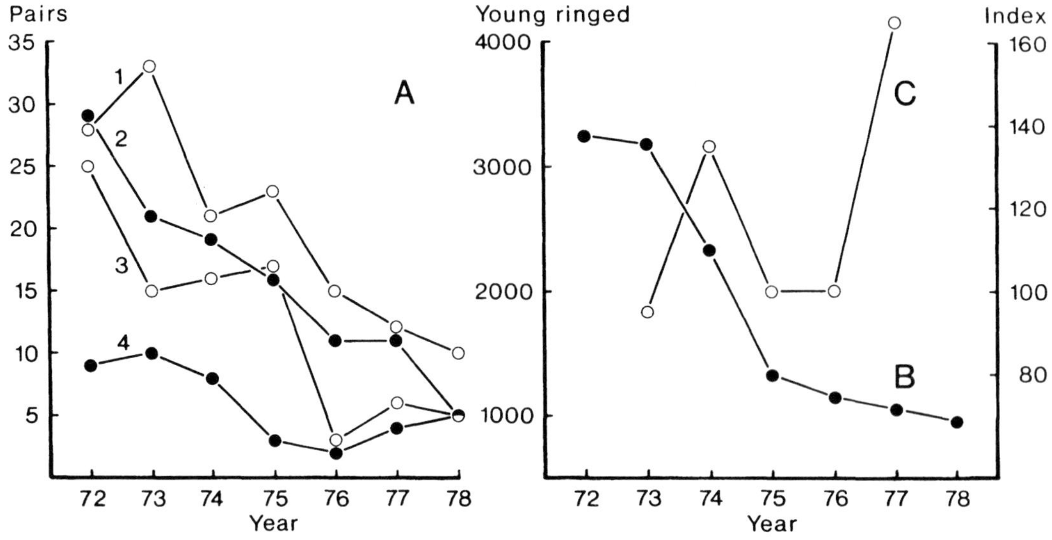

FIGURE 2. Trends in the numbers of Starlings (*Sturnus vulgaris*) in Finland in 1972–1978. A. Annual counts of pairs nesting in boxes in four areas in different parts of the country: 1 = Lammi, S Finland (Tiainen and Solonen 1979, pers. comm.), 2 = Oulu area, N Finland (Ojanen et al. 1978, Orell and Ojanen 1980), 3 = Lemsjöholm, SW Finland (von Haartman 1978a, 1978b), 4 = Salo, SW Finland (von Knorring 1978, pers. comm.). B. Annual numbers of nestlings ringed in Finland (Saurola 1978, pers. comm.). C. Annual index values of densities according to Finnish line-transect data collected from 2163 km (Järvinen and Väisänen 1978d).

tions, etc. These differences are likely to have affected the results to some extent, at least for certain species, but again their significance cannot be measured.

(5) The data for many species are sparse and thus liable to wide limits of error.—Although impressive when considered as a whole, the Finnish line-transect material is relatively small with respect to scarce species. In view of all the sources of error involved in the method, particular care is needed when conclusions are drawn from sparse data. Can a species be said to have decreased if it is represented in two samples from different years by, say, 6 and 3 observations? In my opinion, such a conclusion is not justified. But some of the long-term trends reported by Järvinen and Väisänen (1978c) for an area in southern Finland are, in fact, based on such small species.

(6) Long-term changes may be masked by annual fluctuations, as most data for the periods to be compared often are confined to 1–2 years only.—In their study of long-term changes in the Finnish avifauna, Järvinen and Väisänen had divided the line-transect material into three periods, 1936–49, 1952–63 and 1973–77. Within these periods, however, the data are not evenly distributed between the years. Thus, in the first period most data were collected during four summers between 1942 and 1947 (Merikallio 1946, 1951); i.e., in the years following the extremely severe winters at the beginning of the 1940s. Similarly, in the second period no less than 39% of all line-transects were censused in 1955 (Järvinen and Väisänen 1979a:265), which happened to be a year with an exceptionally cold spring. In the third period also, half the material is from one year, 1977 (Järvinen and Väisänen 1978d), and the whole period either overlaps or immediately follows the warmest five-year period ever recorded in Finland, 1971–75.

The populations of most small passerine birds are known to fluctuate considerably from year to year, peak densities being often 2–3 times higher than the troughs. As the Finnish line-transect censuses are so clearly concentrated in certain, often climatically exceptional years, the population indices obtained for the three periods hardly represent reliable averages of the entire periods, but rather the situations that prevailed in the main (atypical) census years. Consequently, the indices, even if real, are unlikely to show the true long-term trends in the populations of different species.

Järvinen and Väisänen have come to the general conclusion that most of the common landbirds in Finland have increased in number during the last 30 years. According to them, no less than 72.5% of the 40 most abundant south Finnish forest bird species have shown a steady increase from 1936 to 1977, 17.5% have fluctuated irregularly, and only 10% have decreased (Järvinen and Väisänen 1978b). These results have been met with a certain skepticism in Finland, as such a strong increase of the entire bird fauna, including species from a variety of different habitats, appears puzzling. To me, the key to the riddle seems clear: the "general increase" is to a considerable extent only apparent and explicable on methodological grounds. First, the average census efficiency (points 1–4 above) is likely to have improved in parallel with the general advances in field ornithology, resulting in more birds being observed on the transects now than formerly. Second (point 6), the censuses of the first two periods were confined to years following exceptionally severe winters or cold springs when many species had low population densities, whereas the opposite was true of the third period.

To conclude, I wish to make two proposals. First, single line-transect counts should be abandoned in bird census work because of their unreliability. Instead, each transect should be censused three times in different phases of the breeding season, and only the highest numbers recorded for each species should be taken into account. Second, more absolute methods of censusing based on mapping of territories, searching for nests and observing adult birds should be used whenever possible (cf. Berthold 1976).

ACKNOWLEDGMENTS

My warmest thanks are due to Peter Berthold, Peter Evans, Hans Källander, Pontus Palmgren, Chandler S. Robbins, and Sören Svensson for their helpful comments on the draft; Dr. Evans also improved my English. I am also grateful to Lasse J. Laine and Tuukka Pahtamaa for letting me use their unpublished census data, and to Kauri Mikkola for many fruitful discussions.

Studies in Avian Biology No. 6:160–161, 1981.

SUMMARIZING REMARKS: ESTIMATING BIRDS PER UNIT AREA

CHARLES E. GATES[1]

Evelyn Bull (1981) presented a very interesting discussion on the estimation of indirect measures of abundance for birds. I have no quibble with her rather thorough coverage of the many indices that have been suggested in the literature. However, the paper does not recognize that methods have been suggested in the literature for obtaining not only relative, but absolute measures of abundance (indeed, one method has been suggested specifically for birds). Three instances of attempts made to estimate absolute population densities are: the use of aural information requiring estimates of both the number of calls and the calling rate per time period. This development is exemplified by Gates and Smith (1972) for Mourning Doves (*Zenaida macroura*). A second example is the use of pellet group counts to estimate the size of deer and elk populations. The third example is the estimation of absolute hare populations from tracks in the snow (Hayashi et al. 1966). Both the latter two methods could be adapted to birds, e.g., tracks in the dust for some species. It almost goes without saying that the assumptions for absolute densities from indirect measures are even more stringent than for direct measures (see, for example, Gates and Smith 1972).

Capture-mark-recapture methodology has been widely used in small mammals and fisheries. Here Nichols et al. (1981) examine its uses in estimating avian populations. The paper is lengthy, comprehensive and comprehensible. I highly recommend reading it in its entirety. I do note that the paper does not deal specifically with density even though this is a density session. A minor quibble with the paper is that the relative advantages of mark-recapture methodology vis-a-vis other sampling methods are not discussed. This would permit ornithologists and other potential users of the methodology to make more rational decisions. For example, compared to the line transect method, the mark-recapture method will be much more time consuming, (i.e. expensive), birds must be handled (except for specialized situations such as Hewitt's (1963) Red-winged Blackbird procedure), but the final results may be achieved with better precision.

As I understand the paper by Oelke (1981), it is concerned with the controversy of whether bird density should be determined by territorial mapping, obtained by observing bird behavior, or by discovering the nests of nesting birds. Without a clearer understanding of what the controversy is all about, I will not enter the fray.

The manuscript by Franzreb (1981) has to do with an empirical evaluation of the strip vs. the line transect methods of sampling. I observe there is a great deal of variation in the use of sampling terminology; not only here at the conference, but in the literature as well. I should like to make a very strong appeal to this Symposium that standardized terminology be adopted in transect sampling as well as other areas suggested by the Symposium organizers. I make a strong plea for adopting the standardized terminology set forth by Eberhardt (1978). In his terminology "variable-strip transect" and "fixed-width transect" simply become the *line* and *strip transects,* respectively.

Franzreb's (1981) Figure 1 and Table 1 appear to demonstrate direct violations of the assumptions underlying line or strip transect methodology given by the author. The theoretical curve of right angle flushing distances must be non-increasing. In Figure 1 there is clear-cut evidence by any standard of an increase in the number of birds sighted at a right angle distance away from the line. I do not believe the author addresses the question of whether the "excess" of birds at 30 m is due to birds moving out from the line in areas in which they are subsequently seen or is due to birds making themselves more inconspicuous on the line. (The behavior is undoubtedly species dependent.) For those species that move away from the observer, and are subsequently seen along with all other birds at that distance, a method that appears to have merit is the *spline* method. The spline method is specifically designed for situations in which all birds are observed at some unknown distance from the transect at which point the sightings begin to fall-off. The spline procedure fits by least squares a horizontal straight line intersecting the ordinate ($Y = \hat{f}(0)$) with a quadratic curve approximating the downward trend of the observations in the right tail. Because the intersection point of the horizontal line and the quadratic curves are assumed unknown, either non-linear least squares or some special spline technique must be used to solve the equations and hence to estimate density. The method is outlined in Gates (1980). The advantages to the spline pro-

[1] Texas A&M University, College Station, Texas 77483.

cedure, where it is appropriate, appear to be considerable. Least squares, rather than a subjective estimation of the "point of inflection," determines $\hat{f}(0)$ and the right angle distance to the left of which it is assumed all birds are seen. The spline method would appear to be an attractive method of analysis for data gathered by the J. T. Emlen (1971) procedure.

If, however, the paucity of observations near the transect is due to birds becoming less conspicuous, rather than moving, the situation appears to be very difficult.

In conclusion, I heartily agree with Franzreb's (1981) recommendation of recording each bird's distance as accurately as possible, but possibly for different reasons. If distances are recorded rather than intervals, analysis of the resulting data sets is vastly more flexible. One can then subsequently group the data and use spline or Emlen procedures or use one of the robust estimation procedures mentioned by Burnham et al. 1980:125–127.

Studies in Avian Biology No. 6:162, 1981.

SUMMARIZING REMARKS: ESTIMATING BIRDS PER UNIT AREA

LARRY F. PANK[1]

Techniques to enumerate populations fall into two categories: (1) relative estimators that provide some measure of abundance or activity to assess change or make comparisons, and (2) absolute estimators that provide direct measures of densities or population numbers. Density, the subject of this session, requires both an accurate count of objects and a defined area associated with the count. Two of the presentations partially addressed both parameters. The remaining presentations were indirectly related to the subject or addressed only one of the parameters. Rather than depart from the subject to review the presentations, I'd prefer to motivate your thought processes on the "how many?" and the "why?" questions associated with estimating densities.

The dynamics of avian communities (i.e., demographic variables), the difficulty of detecting all individuals (i.e., physical and behavioral barriers) and the problems of defining the spatial and temporal sampling frame (i.e., relating observations to objectives and geographical boundaries) all confound the estimation of bird densities. All of the density estimators attempt to remove these variables by somehow freezing birds or objects for a count within a defined space-time frame. Techniques to achieve this range from quantitatively removing or accounting for the variables via sampling and experimental design to developing estimators that are robust to or are unaffected by the variables. When this fails, we impose restrictions on the conclusions by constricting objectives and defining assumptions. Even after all of this a gnawing concern persists because we have never validated the estimator on known populations that match ours; an effort that is grossly deficient in the field.

Ideally, the instantaneous location of all objects of concern in three dimensional space would: (1) provide a precise estimate of density within any selected plane or strata, and (2) permit a quantitative description of the spatial pattern of the objects. The former answers the obvious question of "how many per unit area?" and the latter provides insight into the "why?" that inevitably follows. Spatial patterns are mentioned to encourage all of you involved with defining objectives and developing estimators to go beyond the tunnel visioned approach of answering only "how many?" The possibility of understanding the underlying distributions and regulatory mechanisms behind the objects with the same data set used to estimate density should be sufficient incentive.

Objectives, assumptions and effort associated with the applied density estimators can also be compared in terms of the spatial and temporal resolution of objects within the sampling frame. For example, the quadrat type methods (strip and circular plot) utilize only the number of objects observed within a fixed search area. The variable search area methods (line transect, circular variable plot) require spatial resolution of the objects in one dimension; perpendicular distance from transect line to object or distance from observer to object respectively. Mapping methods, an extension of the quadrat methods, increase spatial resolution to two dimensions by plotting objects on a horizontal plane. The increase in spatial resolution is generally interrelated with: (1) a greater knowledge about the population (interspecific relationships, packing, etc.), (2) an extension of the temporal sampling frame, (3) a change in objective from the density of all individuals to territorial individuals, (4) an increase in effort, and (5) altered assumptions. It is apparent that many budgets are wasted because these interrelationships are poorly understood. A prime example is the failure to recognize and utilize the differences in the detection related assumptions between the variable search area and quadrat-type methods.

Similarly, modern technology has more to offer than binoculars, tape measures, pencils, paper and adding machines to collect and analyze the data. Excellent "state of the art" publications and computer programs already exist and with a little creativity, technological breakthroughs in data collection are just around the corner.

To summarize, the ultimate method for estimating density is yet to be developed. In the meantime, you users have the responsibility of selecting the method that answers well-defined objectives within acceptable limits of accuracy, precision, effort and funding. If no method fits, stay in bed.

[1] U.S. Fish and Wildlife Service, P.O. Box 244, Hilo, Hawaii 96720.

Studies in Avian Biology No. 6:163, 1981.

INTRODUCTORY REMARKS: COMPARISON OF METHODS

CHANDLER S. ROBBINS,[1] CHAIRMAN

Up to now we have been hearing interesting papers on estimating relative abundance and estimating actual densities. Now we are going to shift emphasis and discuss comparison of methods. As you listen to the papers for the remainder of the morning and the first half of the afternoon keep in mind that we are not just testing one method against another, but rather, trying to determine which method or combination of methods may be best for a particular purpose. So be thinking what your objectives are, what you are trying to estimate or compare. What are the constraints of time, funding, weather, and other variables?

Do not neglect such basic decisions as selection of proper study plots, points, or transects. Do you want your results to apply only to your specific study site or will you use a random or systematic sampling plan so your results will be applicable to a larger ecological or geographical area? We no longer have such simple choices as between spot mapping, transects, or point counts, but must decide which variation of a method will be used. I hope you noticed Tilghman and Rusch's (1981) excellent paper on twelve variations of the transect method.

How much can we standardize our procedures? What size grid are we going to use? What rate of travel, what time of day, how long a listening period, and how will distances be estimated? How do we document what we did so the study will be repeatable in a subsequent year?

Give some thought as to exactly what you are trying to estimate. Do you want to know what the population is on one particular day in the nesting season? Or do you want the average for the entire breeding season? Do you want the maximum at any one time, or the total number of species that utilize the area during an entire breeding season? Will a method that is satisfactory near the Arctic circle, where all species nest at essentially the same time, be applicable to an area in the mid latitudes or in the tropics? And finally, should you routinely apply some additional methodology to improve your estimates or reveal your deficiencies? These are some of the things it will pay us to keep in mind as we listen to the following papers.

[1] Migratory Bird and Habitat Research Laboratory, U.S. Fish and Wildlife Service, Laurel, Maryland 20811.

Studies in Avian Biology No. 6:164–169, 1981.

A COMPARATIVE ANALYSIS OF TERRITORIAL MAPPING AND VARIABLE-STRIP TRANSECT CENSUSING METHODS

KATHLEEN E. FRANZREB[1]

ABSTRACT.—Both the territorial mapping (spot-map) and variable-strip transect methods are widely used censusing techniques. This study compared the results obtained using these two censusing methods to determine avian densities during the breeding season in a mixed-coniferous forest in the White Mountains, Arizona. Overall density was higher with the spot-map method. Although densities were similar for the majority of species, the foraging guild results indicated that the tree-foliage searcher guild had a higher value with the variable-strip transect method, whereas the ground or slash (downed tree debris) guild had a higher value with the spot-map technique. Nesting guild analysis revealed a higher density for the hole-nesting guild and a lower density for the ground nesting guild using the spot-map method.

The success of each sampling method is dependent upon the skill of the observer, observation conditions such as weather, and time of year. If an accurate count of territorial males is made, then a reasonable estimate of the number of breeding individuals can be achieved with the mapping method. Unlike the territorial mapping method, the variable-strip transect method requires fewer replications than does the mapping technique. However there are a number of potential problems in calculating the density values. In moderately or densely vegetated habitats, or with species which are not particularly conspicuous, it is not unusual to underestimate species' densities.

Quantitative investigations of avian population ecology require the use of efficient and reasonably reliable censusing techniques. Much valuable information on avian community structure can be obtained by utilizing methods to determine species densities. In addition, such densities form an integral part of the procedures to monitor the status of avian communities and their responses to various land management practices.

The purpose of this paper is to compare two prominent density sampling methods—the spot-map (territorial mapping) and variable-strip transect censusing technique—and to illustrate advantages and disadvantages of each method. An understanding and appreciation of the weaknesses and assets of these methods should enable the investigator to determine which technique is preferable under given circumstances.

METHODS, MATERIALS, AND DESCRIPTION OF STUDY PLOT

STUDY PLOT

The study plot was located on the Willow Creek watershed in the Apache-Sitgreaves National Forest, White Mountains, Arizona (elevation 2667–2805 m). This mixed-coniferous forest was dominated by Douglas fir (*Pseudotsuga menziesii*), ponderosa pine (*Pinus ponderosa*), and southwestern white pine (*Pinus strobiformis*). Other tree species present included: alpine fir (*Abies lasiocarpa*), white fir (*Abies concolor*), blue spruce (*Picea pungens*), Engelmann spruce (*Picea engelmanni*), and a deciduous species, quaking aspen (*Populus tremuloides*).

VEGETATION ANALYSIS

I used the plotless point-quarter sampling method (Cottam and Curtis 1956) to analyze the vegetation in a 15.5 ha study plot. Further details of the vegetation sampling procedure may be found in Franzreb and Ohmart (1978).

DETERMINATION OF AVIAN DENSITIES

In the spot-map method (Williams 1936) density values are obtained by determining for each species the number of territories encompassed within the study plot. The breeding bird density for a species is derived by multiplying the number of territories of that species by two and relating this figure to the size of the study plot. Specific details of spot-map field techniques are provided by the International Bird Census Committee (1970).

I recorded each observation on a map of the grid pattern and coded each to signify singing male, nonsinging male, female, fledgling, or nest. Data for each day's observations were recorded using a different color ink to facilitate estimating number of singing males and to delineate territorial boundaries. Symbols reflecting particularly critical behavioral information were used and included symbols for: nonsinging male; singing male; female; individual of undetermined sex; nest location; two observations of the same individual during the same day (an arrow may be used to indicate the bird's direction of movement); concurrent observations of two different birds of the same species (i.e., two males singing simultaneously); and an intraspecific aggressive encounter. Data on simultaneous contacts are particularly significant in both defining boundaries between adjacent territories and in assessing the number of territories.

The distance between flags and between transect lines is dependent upon the density and type of vegetation. In dense forest situations flags should generally not be more than 25 m apart and transect lines not more than 50 m apart. As the openness of the habitat increases, the distance between flags and transect lines may be expanded.

Grid size will also be determined by habitat type and vegetation density. In dense habitats a plot of 10–30 ha represents a grid size that can be comfortably censused in a reasonable amount of time (i.e., less than 3 hr).

[1] U.S. Fish and Wildlife Service, 1230 N Street, 14th Floor, Sacramento, California 95814.

In this study censuses were begun ½ hr after sunrise and completed within three hours. Censusing was conducted 1 June–9 August 1974. The plot was censused six times each month for a total of 18 censuses. The number of territories multiplied by two gave the number of breeding birds per 15.5 ha which was converted to number per 40 ha, a standard reference area for avian studies. Means of the monthly values for June and July were used because for most species breeding activity had been completed by August.

Partial territories were addressed by estimating the proportion of the total territory that lay within the study plot. This figure was then incorporated into the results for the rest of the plot.

With each observation obtained using the variable-strip transect method, I recorded the species' right angle distance from the transect line, sex of bird (if possible), behavior (i.e., singing), and age (adult vs. fledgling). Censusing of the 1.6 km long transect line began ½ hr after sunrise and was completed within 2 hrs. Censusing began 2 June and ended 9 August 1974. The transect line was sampled six times per month for a total of 18 surveys. Results are in terms of a mean of the monthly values.

For each species a coefficient of detectability (CD) value was derived, which represents the proportion of the population that an observer can detect during the course of sampling the transect. I calculated the density and CD values for each species by counting the number of observations on strips on both sides of the transect from the base to the point where the distribution curve began to decline rapidly. Because detectability varies with distance from the transect line, the area in the strips when multiplied by an appropriate value will yield the expected number to be found within the prescribed width (commonly 125.6 m) on both sides of the transect line. An observer walking a 1.6 km (1 mi) transect and surveying a 125.6 m width on both sides of the transect line will cover 40 ha (100 ac). A more detailed explanation of this method may be found in Emlen (1971). Bird species diversity (BSD) values were calculated using Shannon's formula (Shannon 1948).

The *t*-test was used to determine if there was a statistically significant ($P \leq 0.05$) difference in species' densities using the spot-map and variable-strip transect method.

RESULTS

Vegetation

The plotless point-quarter vegetation analysis indicated a total tree density of 626.2 trees per ha. Douglas-fir and ponderosa pine had high densities and importance values. A complete vegetational profile is described in Franzreb and Ohmart (1978).

Avian Species Densities

The total density for the avifauna derived from the spot-map procedure was 865.9 breeding birds per 40 ha in contrast to 835.6 birds per 40 ha for the variable-strip transect method (Table 1). For the majority of species, density values using the two methods were similar (Table 1). Greatest variation was noted in the Hermit Thrush, Golden-crowned Kinglet, and Yellow-rumped Warbler.

The most prominent difference in foraging guild results was found in the tree-foliage searchers in which the variable-strip transect method yielded higher overall densities (Table 2). In contrast, higher densities were derived for the ground or slash (downed tree debris) foragers using the spot-map method.

Higher densities for members of the hole-nesting guild were obtained using the spot-map technique (Table 3). However, for ground nesters high densities were calculated using the variable-strip transect method. Percent distribution values were similar.

T-tests indicated that ten species had significantly different densities using the two censusing techniques. Of the ten, six had higher densities with the spot-map method.

DISCUSSION

No censusing method is applicable in every instance or without its drawbacks. The particular technique selected should be dictated by the set of circumstances encountered. Many workers regard the spot-map method as providing a reliable absolute density estimate which, in fact, has been found to be above 90% in accuracy for the majority of species examined in one study (Stewart et al. 1952). It has also been used as a basis for assessing the accuracy of other censusing methods (Enemar 1959). If the observer can obtain an accurate count of the number of territorial males, then a reasonably good estimate of breeding bird density can be calculated. However, several factors impinge upon the investigator's ability to achieve a reliable measurement.

The limitations of the spot-map method result, in part, because an appropriate delineation of territories can be a difficult undertaking. For most avian species territory sizes are poorly known and tend to vary with a variety of factors such as breeding bird density, habitat, and degree of individual aggressiveness (Schoener 1968). Outlining of territorial boundaries is relatively easy for low-density species that have widely spaced territories. However, as densities increase and territories become more compressed, such boundary delimitations can be laborious and occasionally arbitrary. In this study the problem for the high density Yellow-rumped Warbler and Ruby-crowned Kinglet was partly alleviated by using a map four-times larger than for the other species, employing the symbolism already discussed, and using different pen colors for each day of censusing. It is of utmost importance to be as accurate as possible because

TABLE 1
COMPARISON OF AVIAN DENSITIES (NUMBER PER 40 HA) DERIVED USING THE SPOT-MAP AND VARIABLE-STRIP TRANSECT CENSUSING METHODS

Species	Foraging method[a]	Nest type[b]	Density	
			Spot-map[c]	Variable-strip transect[d]
Goshawk (*Accipiter gentilis*)	P	C		1.8
Flammulated Owl (*Otus flammeolus*)	P	H	10.2	
Great Horned Owl (*Bubo virginianus*)	P	C	5.1	
Broad-tailed Hummingbird (*Selasphorus platycercus*)	N	C	30.8	20.8
Common Flicker (*Colaptes auratus*)	GS	H	25.6	18.3
Yellow-bellied Sapsucker (*Sphyrapicus varius*)	TD	H	10.2	5.2
Williamson's Sapsucker (*S. thyroideus*)	TD	H	5.1	5.5
Hairy Woodpecker (*Picoides villosus*)	TD	H	10.2	6.7
Downy Woodpecker (*P. pubescens*)	TD	H	10.2	4.2
Northern Three-toed Woodpecker (*P. tridactylus*)	TD	H	15.4	12.8
Western Flycatcher (*Empidonax difficilis*)	F	C	48.7	71.6*
Olive-sided Flycatcher (*Nuttallornis borealis*)	F	C	5.1	0.6*
Violet-green Swallow (*Tachycineta thalassina*)	A	H	10.2	8.6
Steller's Jay (*Cyanocitta stelleri*)	TFS	C	25.6	16.5*
Common Raven (*Corvus corax*)	P	C	5.1	3.4
Clark's Nutcracker (*Nucifraga columbiana*)	C	C	5.1	2.4
Mountain Chickadee (*Parus gambeli*)	TFS	H	58.9	64.8
White-breasted Nuthatch (*Sitta carolinensis*)	TG	H	15.4	5.2*
Red-breasted Nuthatch (*S. canadensis*)	TG	H	25.6	23.8
Pigmy Nuthatch (*S. pygmaea*)	TG	H	25.6	27.2
Brown Creeper (*Certhia familiaris*)	TG	H	51.3	46.4
House Wren (*Troglodytes aedon*)	GS	H	7.7	5.7
American Robin (*Turdus migratorius*)	GS	C	5.1	3.0
Hermit Thrush (*Catharus guttatus*)	GS	C	76.9	42.8*
Townsend's Solitaire (*Myadestes townsendi*)	GS	G	5.1	
Golden-crowned Kinglet (*Regulus satrapa*)	TFS	C	30.8	51.4*
Ruby-crowned Kinglet (*R. calendula*)	TFS	C	74.4	88.6*
Warbling Vireo (*Vireo gilvus*)	TFS	C	25.6	17.7
Olive Warbler (*Peucedramus taeniatus*)	TFS	C		3.7
Yellow-rumped Warbler (*Dendroica coronata*)	TFS	C	89.8	136.5
Grace's Warbler (*D. graciae*)	TFS	C	25.6	4.9*
Red-faced Warbler (*Cardellina rubrifrons*)	TFS	G	25.6	40.3
Western Tanager (*Piranga ludoviciana*)	TFS	C	12.8	3.6
Black-headed Grosbeak (*Pheucticus melanocephalus*)	TFS	C	5.1	3.7
Pine Siskin (*Carduelis pinus*)	TFS	C	25.6	14.0*
Green-tailed Towhee (*Pipilo chlorurus*)	GS	C		1.8
Gray-headed Junco (*Junco caniceps*)	GS	G	51.3	66.6
Chipping Sparrow (*Spizella passerina*)	GS	C	5.1	5.5
Total			865.9	835.6
Bird species diversity			3.19	2.92
Species richness			35	35

[a] Foraging method (guild): A = aerial forager, F = flycatcher, GS = ground or slash forager, N = nectar feeder, P = predator on vertebrates or carrion feeder, C = cone forager, TD = timber driller, TFS = timber-foliage searcher, TG = timber gleaner.
[b] Nest type: H = hole nester, C = cup nester (non-ground), G = ground.
[c] Spot-map data were previously published in Franzreb and Ohmart (1978).
[d] An asterisk (*) indicates that t-test results showed a significantly different ($P \leq 0.05$) difference in species' density between the spot-map and variable-strip methods.

an underestimate of territory size can result in a population density that is too high, and vice-versa.

Species with very large territories such as raptors represent another problem as it is virtually impossible to predict what portion of their territory is encompassed by the relatively small study plot. Hence, raptor densities may have been over-estimated. There are at least two alternatives to remedy this problem. One is to represent the density of such species in the table of results only by a "P" for present, density undetermined. Second and less practical, a very large study plot could be estiblished for the pur-

TABLE 2
COMPARISON OF FORAGING GUILDS WITH RESPECT TO NUMBER OF SPECIES AND DENSITIES (NUMBER PER 40 HA) OBTAINED USING THE SPOT-MAP AND VARIABLE-STRIP CENSUSING METHODS

	Number of species		Density		Percent distribution	
Foraging guild	Map	Transect	Map	Transect	Map	Transect
Aerial forager	1	1	10.2	8.6	1.2	1.0
Flycatcher	2	2	53.8	72.2	6.2	8.6
Ground or slash	7	7	176.8	143.7	20.4	17.2
Nectar feeder	1	1	30.8	20.8	3.5	2.5
Predator/carrion feeder	3	2	20.4	5.2	2.4	0.6
Cone forager	1	1	5.1	2.4	0.6	0.3
Timber-driller	5	5	51.1	34.4	5.9	4.1
Tree-foliage searcher	11	12	399.8	445.7	46.2	53.3
Timber gleaner	4	4	117.9	102.6	13.6	12.3
Total	35	35	865.9	835.6	100.0	99.9
Bird species diversity (BSD)			3.19	2.92		

pose of more accurately censusing the large-territory species. In most instances the latter approach would entail an unreasonable amount of time and effort. Radio-telemetry, although expensive and time-consuming, in many cases will provide the most reliable results for raptor surveying work.

Inter-observer variability in interpreting spot-maps has been examined by Best (1975). Five workers were asked to independently interpret a spot-map of Field Sparrow data. Best concluded that the spot-map method provided highly variable estimates of breeding density for this example of a species with a desne population and small territories. Among the factors which undoubtedly made the interpretation of the results more difficult and hence may have influenced the results were: some males may have totally escaped detection; various observations were inappropriately assigned to adjacent territories; Field Sparrows sing more intently and earnestly when unmated; the Field Sparrow is not sexually dimorphic; and territories were small and contiguous (Best 1975). Many of these difficulties may have been overcome had data on simultaneous registrations been provided to the five interpreters (Robbins, pers. comm.).

Other workers have noted that counting singing males may not give a reliable estimate of true breeding males. Kendeigh (1944) noted that 9% of the singing House Wrens were unmated. Degree and strength of vocalizations varies among species, and tends to decline as the breeding season progresses. Hence, using singing males an an indicator of number of breeding pairs may be unwarranted (at least for some species) under certain circumstances and may tend to overestimate the true breeding population.

The accuracy of spot-mapping may be reduced for sparse species that have few stationary males, and hence, are more prone to be influenced by factors of chance (Enemar and Sjöstrand 1969). In such instances the population would tend to be underestimated.

Several drawbacks are apparent with the spot-map in that a considerable expenditure of time and effort is necessary to map territories, requiring at least three readings through the study area per month. Kendeigh (1944) recommended five per breeding season, whereas the International Bird Census Committee (1970) suggests 10 visits in a closed canopy habitat and eight in an open habitat. Spot-mapping is applicable only during the breeding season because for most

TABLE 3
RELATIONSHIP OF NESTING GUILD DENSITIES (NUMBER PER 40 HA) DERIVED FROM THE SPOT-MAP AND VARIABLE-STRIP TRANSECT CENSUSING METHODS

	Density		Number of species		Percent distribution	
Nesting guild	Map	Transect	Map	Transect	Map	Transect
Hole	281.6	234.4	14	13	32.5	28.1
Cup (non-ground)	502.3	494.3	18	20	58.0	59.1
Ground	82.0	106.9	3	2	9.5	12.8
Total	865.9	835.6	35	35	100.0	100.0

species it is only then that territories are maintained.

With the variable-strip transect method, on the other hand, one individual can sample large areas easily. As few as three replications are needed (J. T. Emlen 1971). Unlike the spot-map procedure, the transect method incorporates data for breeding and non-breeding birds as well as fledglings.

Theoretically, at least upon initial inspection, the variable-strip transect method results should be equal to or higher than those obtained from spot-mapping because non-breeders and fledglings are incorporated into the results. However, as evidenced in this study and elsewhere (Franzreb 1976), this often is not the case. What factors influence whether or not the transect method results approach reality? Given that the observer is competent and experienced, and the weather conditions are suitable, the primary determining factor is the inherent behavior of the birds. This is chiefly the result of the degree of conspicuousness of a species, which relates directly to the probability of an observer detecting an individual during any particular reading of the transect. Species with high variable-strip transect values (or surpassing those obtained from spot-mapping) have high detectability probabilities. These individuals may perch on a clearly visible branch, sing intently and/or frequently, be less shy of the observer, be strikingly colored, may forage in an open situation, or may have other behavioral traits such as wing-flashing that draws our attention. In this study individuals particularly exemplifying this included the Western Flycatcher (conspicuous perch, strident call), and Yellow-rumped Warbler (frequent song and call, forages in relatively open vegetation).

However, males of some species may become less conspicuous as the breeding season progresses once females start incubating or young hatch. Likewise, females may become less detectable once incubation begins. Shyness of certain species may reduce the chance of observing them. These factors affect what J. T. Emlen (1971) refers to as the coefficient of detectability (CD) value. In this study, I assumed that all birds were detected within the strips on either side of the transect line bounded by the point on the distribution curve where the results began to decline rapidly; however, in this densely vegetated habitat, this assumption was probably invalid. Hence, the total estimated density will be less than the actual value, and more so for the less detectable species. In part, this accounts for the lower densities observed for 24 of the 35 species using the variable-strip transect method in this study. To adjust for incomplete surveys a basal detectability adjustment factor should be applied to the results (J. T. Emlen 1971), but this value must be obtained from another sampling method, which itself is subject to error.

It is also noteworthy that if three readings of a transect are conducted and the members of a particular species have been observed, on the average, on only two out of the three runs, the density value for that species will be estimated at 66% of its true value. Hence, values for less detectable species, in general, are not as reliable as are those for more conspicuous ones. It is also difficult to differentiate whether an individual bird is missed on any given survey or if the bird was simply not present on the plot at that particular time.

On the other hand, with the spot-map method if one-third of the individuals of a species are missed during any one examination of the study plot, the results are not necessarily detrimentally affected, provided that sufficient replications are done to delimit territorial boundaries.

In this study, it appears that members of the flycatcher and timber foliage-searching foraging guilds were more conspicuous in relation to the other guilds. This factor and the inclusion of fledglings accounted for the higher density values for these guilds using the variable-strip transect method than from spot-mapping.

CD values are purported to have value in permitting density determinations to be calculated quickly for species in habitats similar to the ones in which they were derived. Yet they are not necessary in order to arrive at density values and presumably will vary yearly and seasonally as the species' densities change.

It is important to recognize that the similarity, or lack thereof, in results derived from these two censusing methods may vary depending on habitat and year. A similar comparison of the two methods in this same study plot during 1973 revealed that spot-mapping resulted in higher densities for 25 of 31 species and differences of larger magnitude than reported in this study (Franzreb 1976). I suggest this is related to the overall community density which was considerably higher in 1974 than 1973, thus necessitating enhanced vocalizations and other territorial defense and advertisement by males. A given replication of the transect would have a higher probability of detecting any particular male.

In summary, the method selected will be dependent upon the individual set of circumstances. Interest in a particular foraging or nesting guild may also influence one's choice of technique. Dense, heavily vegetated habitats present detectability problems at least for certain species and may more comfortably be sampled using the spot-map approach. This decision, however, must consider the commitment of time and personnel involved in spot-mapping and the smaller area that may be conveniently censused relative to the variable-strip transect method. The variable-strip transect technique is applicable at all times of the year, and is not therefore, limited by the breeding season. It incorporates fledglings, migrants, and other nonbreeders as well as breeding birds.

ACKNOWLEDGMENTS

I am grateful for the valuable advice and ideas which were provided during discussions with J. T. Emlen, R. D. Ohmart, J. Verner, and W. F. Laudenslayer. I thank C. S. Robbins, C. van Riper III, and G. Golly for their beneficial suggestions on improving the ms. This research was funded through cooperative aid agreements with the U.S. Forest Service, Rocky Mountain Forest and Range Experiment Station, Tempe, Arizona (No. 16-402-CA and 16-382-CA).

Studies in Avian Biology No. 6:170–176, 1981.

A COMPARISON OF THREE AVIAN CENSUS METHODS

DANIEL K. EDWARDS, GEOFFREY L. DORSEY AND JOHN A. CRAWFORD[1]

ABSTRACT.—Sample plot, line transect, and variable circular-plot census methods were used during a 14 month study of avian communities on an island in the Columbia River, Oregon. Fifteen areas in four types of habitat, both aquatic and terrestrial, were examined. Certain avian community parameters (number of species, density, diversity (H') and evenness (J')) were derived from each census method. These parameters were analyzed on four season, six season, monthly, and habitat bases. More species were detected with the variable circular-plot technique; few differences were found in regard to other community parameters among the three methods.

Various methods for estimating absolute density of avian populations have been described and considerable discussion has occurred regarding the most effective methods. The territorial or spot-mapping method (Williams 1936, Kendeigh 1944) has been the principal technique for estimating density of small, non-flocking, terrestrial birds (J. T. Emlen 1971). This method, however, has several major deficiencies: (1) it is applicable only during the breeding season; (2) it requires considerable time and effort; and (3) it is applicable only to relatively small tracts of habitat.

The sample plot method was also used to estimate bird density (Bond 1957, Anderson 1970b, Anderson 1972, Anderson and Shugart 1974). Such censuses were conducted either with the observer traversing an area or remaining at a point and recording birds within a fixed distance. Density estimates derived by these methods from structurally different habitats may not be comparable because species detectability in each habitat was not considered. Thus, these data are frequently reported as relative rather than absolute densities.

Line transect procedures have been used to census birds in many different habitats under a variety of conditions. Shields (1979) indicated that the methods described by J. T. Emlen (1971, 1977a) and by Järvinen and Väisänen (1975, 1976b) were the most promising of the line transect procedures; both methods incorporate detectability of species under varying habitat conditions.

Reynolds et al. (1980) described a variable circular-plot technique where the observer remains stationary and estimates the horizontal distance to each bird observed. Thus, this method also accounts for detectability of species. The authors felt that by remaining stationary they could concentrate more on censusing birds and less on trail navigation.

[1] Department of Fisheries and Wildlife, Oregon State University, Corvallis, Oregon 97331. Present address of first author: 950 Ave. F, Seaside, Oregon 97138.

Several studies have compared various censusing methods both in practice and in theory. Emlen (1977a) compared the line transect and spot-map census methods. He concluded that the transect method was more efficient, could be applied to a larger area, and was as accurate as the spot-map method. Franzreb (1976) compared line transect and spot-map methods in coniferous forests and concluded that line transects were more efficient and just as accurate. Most authors who compared various methods agreed that tests where data were collected simultaneously in one area were necessary for adequate comparison of various methods and the ultimate determination of the best method.

The objective of this study was to census the bird communities of 15 areas in four habitats with three different avian censusing methods and to compare these methods with respect to number of species, density, diversity (H'), and evenness (J').

STUDY AREA AND METHODS

The study was conducted on Miller Sands Island in the Columbia River, approximately 20 km east of Astoria, Oregon. Miller Sands, about 225 ha in size [island (150 ha), sand spit (50 ha), and marsh (25 ha)], was formed from deposition of dredged material which began in the 1930's. Four habitats were identified. Beach areas consisted of sand with little or no vegetation. Marsh habitats were dominated by tufted hairgrass (*Deschampsia cespitosa*), Lyngby's sedge (*Carex lyngbyei*), and spike rush (*Eleocharis palustris*). Upland sites were characterized by horsetail (*Equisetum* spp.), stream lupine (*Lupinus rivularis*), and a variety of grasses and forbs. Willow (*Salix* spp.), black cottonwood (*Populus trichocarpa*), and red alder (*Alnus rubra*) were common species in the tree-shrub habitat. Marsh and beach habitats were characterized by ducks (*Anas* spp.), gulls (*Larus* spp.), sandpipers (*Calidris* spp.) and swallows (Hirundinidae). Upland areas were dominated by Savannah Sparrows (*Passerculus sandwichensis*) and a few other passerine species. Tree-shrub areas were characterized by Common Crows (*Corvus brachyrhynchos*), Black-capped Chickadees (*Parus atricapillus*), Bewick's Wrens (*Thryomanes bewickii*), and Song Sparrows (*Melospiza melodia*).

Three methods were used to collect information on number of species and density in each area. Data were collected between sunrise and 2.5 hr after sunrise. One observer conducted all censuses to avoid multi-observer biases. The sample plot method (Bond 1957, Anderson 1970) involved an observer identifying the species and recording all birds observed within a circular plot with a radius of 56.4 m and an area of 1 ha. Three such plots were located within each study area. The average number of individuals of each species for the three plots yielded the density (birds/ha) for each species. A line transect method that involved an observer walking along three 100-m transects within each study area also was used. The observer recorded the lateral distance from the line to each bird. Lateral boundaries beyond which no birds were recorded were established at 50 m. Observations were recorded in 10 m bands parallel to the transect line, similar to the method described by J. T. Emlen (1971). The number of individuals of each species was plotted for each 10 m band within each habitat and for each season. The distance at which detections began to decrease was determined, a basal plateau established, and a coefficient of detectability was determined for each species during each season (J. T. Emlen 1971). Density was calculated for each species from the coefficient of detectability and the number of individuals of each species recorded.

The variable circular-plot method (Reynolds et al. 1980) was the third method employed. With this method the edge of the habitat or the distance at which a bird was identifiable constituted the outer boundary of the study site. An area with a certain detectability, $P = 1.0$, for each species within each habitat during each season was estimated by plotting the density by species within each 10 m band radiating around the observation point out to 150 m and within each 25 m band beyond 150 m. Density for each species within study areas during each month was then calculated (Reynolds et al. 1980). No birds observed beyond the distance where $P = 1.0$ were used for density determination. Ramsey and Scott (1978) described an alternative method for species density determination where all observations were utilized to calculate density. Data from sample plot and variable circular-plot censuses were collected simultaneously for 10 min at points mid-way along each transect line. Approximately 4 min were spent collecting data along each 100 m segment of transect.

Fifteen study areas were established in four habitats; beach (5), marsh (5), upland (3), and tree-shrub (2). Data collection began in July 1976 and was completed in August 1977. Each month the results from six replications (two censuses/month) of each censusing method in each of 15 study areas were used to calculate number of species and bird density. Bird species diversity (BSD) was calculated by the Shannon formula $H' = \sum_{i=1}^{s} p_i \log_e p_i$ where s = number of species and p = the proportion of the total number of individuals that belong to the ith species. Evenness (J') was obtained by dividing the diversity value for an area by the maximum possible for the area (H'/H_{max}), where $H_{max} = \log_e s$ (s = the number of species observed).

Analysis of Data

We tested the data to determine if significant differences in number of species, density, diversity and evenness existed among the three methods on a monthly basis (data from the 15 study areas were used for replications). Data were arranged in a randomized block design and tested by analysis of variance techniques (Snedecor and Cochran 1967:299–338). A Bartlett's Test for Homogeneity of Variance revealed that variances of the means of the three methods were not equal for number of species or density. Thus, these data were transformed by $\sqrt{x + 1}$ (Snedecor and Cochran 1967). Diversity data were not transformed, as Hutcheson (1970) demonstrated that such values were asymptopically normal. Evenness values were not transformed.

Two approaches were used to test for seasonal effects: (1) data were arranged into four seasonal categories: spring (March–May), summer (June–August), fall (September–November), and winter (December–February); and (2) data were analyzed with respect to a modified six-season arrangement (Anderson 1970b) which involved early spring (March–April), late spring (May), early summer (June), late summer (July–August), fall (September–October), and winter (November–February). The six seasons were based on change in bird activity. Additionally, data were arranged according to habitat (data from the 14 months of study were used as replications) to determine if differences existed among the three methods. The Newman-Keuls mean separation test was used whenever significant F-values were obtained from analysis of variance tests. Fiducial limits were set at the 95% confidence level for Newman-Keuls tests and at the 99% level for significant F-values.

Simple correlation analysis (Snedecor and Cochran 1967) was used to determine if, despite any numerical differences, the results from the census methods provided similar trends and, consequently, similar conclusions from the data. Each method was compared to the others in a series of tests which involved number of species, density, diversity, and evenness. In this analysis, data were arranged according to four-seasons (data from each of the 15 study areas served as replications) and according to habitat (time, 14 months were used as replications). Fiducial limits for t-tests associated with the correlation analysis were established at the 99% level of confidence.

RESULTS

Analysis of Variance

Number of Species

Analysis of variance indicated that significantly more species were detected with the variable circular-plot method than with the other procedures when data were analyzed on monthly, four-season, six-season, and habitat bases (Table 1). Results of the analysis of variance test also indicated that significantly more species were detected with the sample plot method than with the line transect method in: 4 of 14 months, 2 of 4 seasons, 2 of 6 seasons, and 2 of 4 habitats (Table 1).

Density

Few instances (5) of significant differences among density estimates derived from the three census methods were detected (Table 1). Where differences were detected the density estimates derived from the variable circular-plot method were significantly higher than line transect estimates (Table 1). In four of these five instances densities derived from the variable circular-plot method did not differ significantly from those obtained from sample plot (Table 1).

Diversity

Significantly higher diversities were detected with the variable circular-plot method than with the other techniques in 6 of 14 months, 4 of four seasons, and five of six seasons (Table 1). Diversity values derived from the variable circular-plot method were significantly higher than those determined from line transect and sample plot methods in four and three instances respectively when compared on a habitat basis (Table 1). The sample plot method yielded significantly higher diversities than the line transect in two of four habitats (Table 1).

Evenness

No significant differences were detected among evenness values yielded by the three methods in monthly and habitat analyses; significant differences were found for one of four seasons and two of six seasons.

Correlation Analysis

Number of species

Line transect and sample plot methods were significantly correlated in all seasons (Table 2). All permutations of census methods were significantly correlated during summer (Table 2). The number of species detected in each habitat by each method was significantly correlated in all comparisons except line transect and variable circular-plot in marsh habitats (Table 3).

Density

All correlations of variable circular-plot with sample plot and line transect on seasonal bases were significant (Table 2). Densities derived from line transect and sample plot methods were significantly correlated during spring and fall only (Table 2). Comparisons of census methods on a habitat basis were significant in 11 of 12 instances; the exception occurred for the comparison of line transect and sample plot methods in marsh habitats (Table 3).

Diversity

Seasonal comparisons of diversity, calculated from the results of each census method, revealed significant correlations in spring and summer for all permutations of the census methods (Table 2). The census methods were not significantly correlated in fall and winter except sample plot and variable circular-plot diversity values were significantly correlated during winter (Table 2). Among comparisons of census methods on a habitat basis only line transect and variable circular-plot methods for aquatic habitats (beach and marsh) were not significantly correlated (Table 3).

Evenness

Evenness values derived for each census method during spring were significantly correlated for all comparisons (Table 2). Sample plot and variable circular-plot were significantly correlated during all seasons except winter (Table 2). No significant correlations were detected among comparisons of census methods for other seasons (Table 2). Sample plot and variable circular-plot exhibited the only significant correlations when evenness values were arranged on a habitat basis (Table 3).

DISCUSSION

The greater number of species detected with the variable circular-plot method was attributable to the greater distances at which birds were recorded; greater observation distances allowed more uncommon species to be recorded. Similarly, Järvinen and Väisänen (1975) and Shields (1979) noted that rare species were detected when greater observation distances were used. If birds observed at more than 50 m from the transect line had been recorded, more species may have been detected with the line transect method.

Time was another factor that may have added to a greater number of species being detected with the variable circular-plot technique than with the line transect method. Approximately 4–5 min were spent collecting line transect data, whereas 10 min were spent at each of the variable circular-plot and sample plot stations. Significantly more species were found with the sample plot method than with the line transect method in 33% of the tests (Table 1). Thus, time appeared to have an influence on the number of species detected. J. T. Emlen (1971, 1977a) and Shields (1979) noted that standarization of censusing time was essential. Tabor (pers. comm. 1980) noted that number of species recorded on a site increased significantly when the time period was increased from 5 to 10 min when using the variable circular-plot method.

Density values determined from the three techniques generally did not differ significantly. Thus, even though more species were observed

TABLE 1
COMPARISONS OF MEAN VALUES FOR EACH AVIAN COMMUNITY PARAMETER DERIVED FROM TRANSECT, SAMPLE PLOT, AND VARIABLE CIRCULAR-PLOT CENSUS METHODS ON MONTHLY, 4- AND 6-SEASON, AND HABITAT BASES; DATA CONNECTED BY UNDERLINES ARE THOSE THAT ARE *not* SIGNIFICANTLY DIFFERENT

	Number of species			Density			Diversity			Evenness		
	Transect	Sample plot	Variable circular-plot	Transect	Sample plot	Variable circular-plot	Transect	Sample plot	Variable circular-plot	Transect	Sample plot	Variable circular-plot
Monthly												
July	4.40	5.07	8.27	13.32	8.21	10.56	1.04	1.08	1.19	0.76	0.68	0.57
August	4.07	4.40	8.73	5.88	6.91	6.93	0.93	0.94	1.27	0.63	0.61	0.58
September	3.13	3.33	9.00	3.56	3.78	4.40	0.72	0.90	1.26	0.51	0.75	0.62
October	2.93	3.00	9.20	4.97	4.51	5.23	0.54	0.67	1.35	0.40	0.62	0.65
November	2.26	2.73	8.86	3.48	6.85	6.51	0.64	0.69	1.09	0.56	0.63	0.54
December	2.53	2.46	7.53	5.55	6.11	6.94	0.66	0.58	0.80	0.55	0.51	0.40
January	2.07	2.40	8.13	2.39	4.98	4.65	0.70	0.59	1.12	0.50	0.47	0.52
February	2.73	2.80	8.40	2.55	3.96	4.74	0.87	0.87	1.46	0.62	0.65	0.66
March	3.26	3.73	9.33	3.92	6.10	5.15	0.86	0.85	1.51	0.63	0.55	0.66
April	4.66	5.60	11.53	19.05	16.92	13.18	0.94	1.06	1.30	0.67	0.66	0.54
May	4.73	6.40	11.26	9.35	13.90	10.24	0.86	1.03	1.16	0.52	0.55	0.49
June	4.73	5.40	9.27	3.85	4.44	5.37	1.11	1.28	1.57	0.72	0.79	0.71
July	4.53	5.93	9.30	9.63	12.27	15.25	1.10	1.24	1.34	0.73	0.73	0.62
August	5.06	8.40	11.13	8.85	42.51	20.50	0.94	1.24	1.23	0.61	0.61	0.53
Four seasons												
Spring	4.22	5.24	10.71	10.77	12.31	9.53	0.89	0.99	1.31	0.60	0.59	0.56
Summer	4.56	5.84	9.35	8.31	14.87	11.72	1.03	1.16	1.32	0.69	0.68	0.60
Fall	2.78	3.02	9.02	4.01	5.05	5.38	0.63	0.76	1.25	0.48	0.67	0.61
Winter	2.44	3.56	8.02	3.50	5.02	5.44	0.74	0.68	1.12	0.59	0.54	0.52
Six seasons												
Early spring	3.97	4.67	10.43	11.48	11.51	9.17	0.91	0.97	1.39	0.65	0.61	0.59
Late spring	4.73	6.40	11.26	9.35	13.90	10.24	0.86	1.03	1.16	0.52	0.55	0.49
Early summer	4.73	5.40	9.27	3.85	4.44	5.37	1.11	1.28	1.57	0.72	0.79	0.71
Late summer	4.52	5.95	9.37	9.42	17.47	13.31	1.01	1.13	1.26	0.69	0.66	0.58

TABLE 1
CONTINUED

	Number of species			Density			Diversity			Evenness		
	Transect	Sample plot	Variable circular-plot	Transect	Sample plot	Variable circular-plot	Transect	Sample plot	Variable circular-plot	Transect	Sample plot	Variable circular-plot
Fall	3.03	3.17	9.10	4.27	4.15	4.82	0.63	0.78	1.31	0.45	0.68	0.63
Winter	2.40	2.60	8.23	3.49	4.71	5.18	0.72	0.68	1.11	0.58	0.57	0.53
Habitat												
Beach	3.29	3.96	9.86	6.70	9.34	8.65	0.74	0.81	1.14	0.56	0.62	0.51
Marsh	4.00	4.54	10.84	9.08	15.79	9.85	0.86	0.86	1.25	0.60	0.57	0.53
Upland	2.48	3.28	6.95	3.17	3.78	3.86	0.72	0.89	1.30	0.58	0.62	0.64
Tree-shrub	5.46	6.86	7.46	7.43	7.30	12.09	1.32	1.54	1.56	0.79	0.81	0.79

when the variable circular-plot was used, the observation distances and numbers of individuals recorded were such that very little was added to density values (i.e., common species, which were recorded by all three methods, accounted for the bulk of density values). Density values obtained from the three methods were significantly correlated with each other in most instances.

Diversity values derived from the variable circular-plot method generally were significantly higher than those detected with the other methods. This resulted from the greater number of species detected by the variable circular-plot method.

The paucity of differences detected among evenness values derived from the three methods further substantiated the idea that increased diversity was a result of the increase in the number of species detected (Table 1). Additionally, in the instances where differences occurred among evenness values the variable circular-plot method resulted in the lowest values. More rare species were observed with the variable circular-plot technique than with the others, and the addition of several species with only a few individuals increased diversity and decreased evenness.

A high incidence of significant correlations among numbers of species and diversity values yielded by the three methods and the low degree of correlations among evenness values further illustrated the contention that the number of species detected was the major difference among the methods.

Analyses of data on a habitat basis also indicated that the number of species detected was the major difference among the three methods. The difference in the amount of time spent censusing probably accounted for differences in density values determined for beach habitats as the sample plot and variable circular-plot did not differ significantly (Table 1). Diversity and evenness values within habitats also indicated that the greater number of species detected by the variable circular-plot method (i.e., greater values of diversity would result from more species and evenness would possibly go down or remain the same depending on the number of individuals detected for each species) was the most important difference among the methods.

One of the most interesting results of this study was the lack of differences among density values yielded by the three methods, especially in view of the time and distance variables involved in data collection. Our results differed from those found in a study in western Oregon where eight different habitats were censused during both winter and the breeding season; densities of 63% of the species observed during the breeding season and 66% during winter increased significantly when the time spent at each station was increased from 5 to 10 minutes (J. Tabor, pers. comm. 1980).

One other interesting result was the relative lack of differences among evenness values yielded by the methods. We expected lower evenness values to result from the variable circular-plot method in light of the greater number of rare species observed when it was used. Addition-

TABLE 2
SEASONAL CORRELATION COEFFICIENTS FOR COMPARISON OF CENSUS TECHNIQUES WITH RESPECT TO NUMBER OF SPECIES, DENSITY, DIVERSITY, AND EVENNESS

Community parameter	Correlation coefficient: Line transect and sample plot	Line transect and variable circular-plot	Sample plot and variable circular-plot
Number of species			
Spring	0.66**	0.42**	0.37
Summer	0.53**	0.43**	0.64**
Fall	0.64**	0.03	0.01
Winter	0.63**	0.36	0.35
Density			
Spring	0.68**	0.71**	0.88**
Summer	0.03	0.37**	0.61**
Fall	0.43**	0.46**	0.59**
Winter	0.34	0.68**	0.57**
Diversity			
Spring	0.62**	0.59**	0.51**
Summer	0.45**	0.40**	0.73**
Fall	0.42	−0.14	0.35
Winter	0.42	0.38	0.43**
Evenness			
Spring	0.54**	0.49**	0.46**
Summer	0.10	0.24	0.73**
Fall	0.08	−0.08	0.53**
Winter	0.15	0.17	0.26

** $P < 0.01$.

TABLE 3
HABITAT CORRELATION COEFFICIENTS FOR COMPARISON OF CENSUS TECHNIQUES WITH RESPECT TO NUMBER OF SPECIES, DENSITY, DIVERSITY, AND EVENNESS

Community parameter	Correlation coefficient: Line transect and sample plot	Line transect and variable circular-plot	Sample plot and variable circular-plot
Number of species			
Upland	0.82**	0.73**	0.74**
Tree-shrub	0.55**	0.55**	0.93**
Marsh	0.53**	0.14	0.44**
Beach	0.69**	0.57**	0.46**
Density			
Upland	0.63**	0.64**	0.95**
Tree-shrub	0.82**	0.64**	0.70**
Marsh	0.04	0.37**	0.70**
Beach	0.69**	0.72**	0.88**
Diversity			
Upland	0.59**	0.46**	0.55**
Tree-shrub	0.63**	0.58**	0.90**
Marsh	0.36**	0.13	0.49**
Beach	0.35**	0.23	0.41**
Evenness			
Upland	0.29	0.01	0.11
Tree-shrub	0.43	0.06	0.61**
Marsh	0.03	0.01	0.43**
Beach	0.16	0.23	0.62**

** $P < 0.01$.

ally, very few significant correlations were detected between evenness values revealed with the three methods, which possibly indicated that evenness was not a good parameter for evaluating these methods.

CONCLUSIONS

(1) Significantly more species were detected with the variable circular-plot than with the other procedures within all habitats and during all seasons, probably because of the theoretically unlimited boundaries of this method.

(2) Density values calculated from the three methods did not differ frequently and were not greatly affected by season or habitat.

(3) Diversity measurements, calculated from the variable circular-plot method, were greater, which likely reflected the greater number of species detected by the variable circular-plot method.

(4) Evenness values did not reveal a significant pattern.

(5) Generally, the parameters calculated from any one method tended to show correlation to other methods for density, number of species, and diversity. Evenness values were infrequently correlated between any two methods.

(6) Plot size, seemingly, was of greater importance in regard to the number of species detected and diversity than was time spent sampling. Nevertheless, time probably resulted in some significant differences.

RECOMMENDATIONS

Based on the application of the three census methods as used in this study, use of the variable circular-plot will result in the detection of a greater number of species and, as a consequence, higher diversity values. The three methods will yield similar estimates for density and evenness. The sample plot method is probably more efficient with respect to time required for training of the observer. With the sample plot method only one distance must be learned; whereas, with the variable circular-plot, the observer must develop proficiency at distance estimation. Also, Reynolds et al. (1980) have noted that larger areas can be covered by the line transect method.

ACKNOWLEDGMENTS

The data for this paper were collected in conjunction with a study of avian communities, which was funded under contract DACW57-76-C-0180 of the Waterways Experiment Station of the U.S. Army Corps of Engineers, Vicksburg, Mississippi.

Studies in Avian Biology No. 6:177–185, 1981.

A FIELD TEST OF THE VARIABLE CIRCULAR-PLOT CENSUSING TECHNIQUE IN A CALIFORNIA COASTAL SCRUB BREEDING BIRD COMMUNITY

DAVID F. DESANTE[1]

ABSTRACT.—The density and distribution of eight species of breeding birds were determined in a California coastal scrub habitat. These breeding bird determinations were made by an intensive program of color banding, spot-mapping, and nest monitoring. Variable circular-plot censusing methods were employed in this same location by four trained observers, naive as to the actual density and distribution of birds. The variable circular-plot method underestimated the total density of these species by about 18%; individual species were variously underestimated by from less than 2% to about 70%. The mean absolute error for these eight species was 25.3% ± 22.8 (SD). For the four individual observers, the mean absolute percent error in the density determinations of these eight species ranged from 30.6 to 38.8. The variable circular-plot method was also found to be capable of determining the breeding distributions of most of these species. The method tended somewhat to overestimate the density of a species where it was sparse, and to underestimate the density of the same species where it was dense.

Most ecological studies of birds use abundance or number of birds as their basic information. A measure of the relative abundance of most birds is generally easy to obtain because of their relatively large size, conspicuous behavior, and diurnal habits. An accurate estimate of the absolute abundance of birds, however, is much more difficult to obtain because they are highly mobile and have the ability to interact with (approach and investigate or hide from) an observer. A number of census techniques have been proposed that endeavor to measure absolute density. These vary considerably in the amount of time and effort they require, in their applicability to various habitats, and in their comparability among various habitats. They include standard territorial spot-mapping methods (Williams 1936, Kendeigh 1944), methods using small plots of fixed size (Fowler and McGinnes 1973, Anderson and Shugart 1974), and methods using variable area transect counts (J. T. Emlen 1971). More recently, Reynolds et al. (1980) have proposed a variable circular-plot method that makes use of timed counts at a number of stations located along a transect line through the habitat. This method is reported to offer distinct advantages for surveying large geographical areas, for comparing different habitats, and for working in rugged and remote terrain (Scott et al. 1981a).

A basic problem with all of these methods is that their accuracy and reproducibility remain unknown. While several of these methods have been compared with each other (Robinette et al. 1974, Franzreb 1976), all of the methods produce only estimates of density. The actual numbers and distributions of the birds are generally not known. Without such information, the absolute accuracy of the methods cannot be determined; only their relative differences can be described.

We have, however, a 36 ha study plot in the coastal scrub habitat of central California where the population dynamics and life history strategies of eight permanent resident species have been continuously and intensively studied year-round since January 1979 (less intensive studies of certain of these species have been conducted at this location since 1966). These eight species are Scrub Jay (*Aphelocoma coerulescens*), Bushtit (*Psaltriparus minimus*), Wrentit (*Chamaea fasciata*), Bewick's Wren (*Thryomanes bewickii*), Rufous-sided Towhee (*Pipilo erythrophthalmus*), Brown Towhee (*Pipilo fuscus*), White-crowned Sparrow (*Zonotrichia leucophrys*), and Song Sparrow (*Melospiza melodia*). The absolute numbers and breeding distributions of all individuals of these eight species were determined, as accurately as possible in a dynamic community, in both 1979 and 1980. These determinations were made by a combination of extensive color banding, detailed spot-mapping, and daily nest monitoring. The variable circular-plot method was tested on these eight species during the breeding season of 1980. That work forms the basis of this paper.

METHODS

THE STUDY PLOT

The 36 ha irregularly shaped study plot was located in coastal scrub habitat at the Palomarin Field Station of the Point Reyes Bird Observatory (PRBO), just inside the southern end of the Point Reyes National Seashore. A map of the study plot is presented in Figure 1.

The study plot consisted of a relatively level 60 m elevation marine terrace of about 18 ha, a moderate to steep southwest-facing slope of about 14 ha, and another level marine terrace of about 4 ha at the 180

[1] Point Reyes Bird Observatory, 4990 Shoreline Highway, Stinson Beach, California 94970.

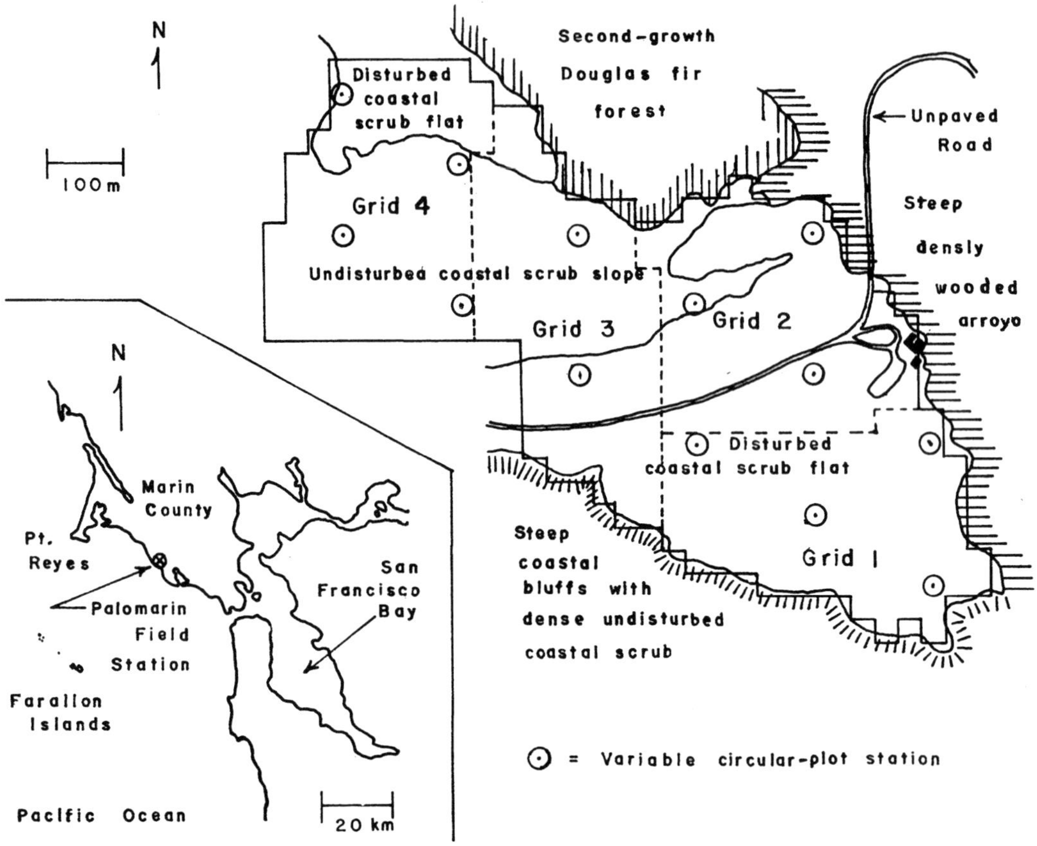

FIGURE 1. Map of the 36 ha coastal scrub study plot located at the Palomarin Field Station of the Point Reyes Bird Observatory and its location in central coastal California (inset).

m elevation. The vegetation of the slope consisted of dense 1 m tall undisturbed mature coastal scrub, composed primarily of California sage (*Artemisia californica*), coyote bush (*Baccharis pilularis*), bush monkey flower (*Mimulus aurantiacus*), poison oak (*Rhus diversiloba*), and coffee berry (*Rhamnus californica*). A less dense successional stage coastal scrub, composed primarily of 1.5 m tall coyote bush and interspersed with patches of introduced annual grasses (*Avena, Bromus, Stipa*), thistles (*Cirsium*), and radish (*Raphanus sativa*), occupied the flat areas, which were cultivated until 1965. A few small (1.5–6 m tall) widely scattered Douglas firs (*Pseudotsuga menziesii*) occurred throughout the study plot. The plot was bordered on the south by steep ocean bluffs that extend down to the beach, on the east by a steep heavily wooded arroyo, on the north by a second-growth Douglas fir forest, and on the west by habitat roughly similar to the study plot.

Breeding Bird Densities and Distributions

The 36 ha study plot was divided into four 9 ha subplots (grids) which were rectilinearly gridded at 30.5 m (100 ft) intervals by means of 2 m tall metal stakes color coded by row and column. One observer was responsible for all data collected and compiled on each grid. These data were gathered by three techniques.

Color banding.—A concentrated effort was made to capture and color band all individuals of the eight study species. This was accomplished by: (1) a trapping program in which twelve 4-cell "potter" traps, baited with cracked corn ("chick scratch"), were opportunistically placed and operated about one day/week for six hours/day on each grid; (2) a netting program in which one or two 6 or 12 m mist nets were opportunistically run at favorable locations on each grid, often in conjunction with the playback of the songs of territorial males or the distress calls of nestling birds; (3) the color banding of all nestlings; and (4) the standardized daily operation of a battery of twenty 12 m mist nets along the edge of the arroyo that bounds the east edge of the study plot.

Spot-mapping.—Detailed spot-mapping censuses, in which were recorded the individual identities, exact locations, and coded behaviors of all birds encountered in the grids, were run on the average of four times/week by each observer. These data were then transcribed onto individual color banded bird encounter sheets (which provided a temporal history of each individual bird), onto individual color banded bird grid maps (which provided a spatial history of each individual bird), and onto species grid maps (which provided a record of territorial boundary disputes, countersinging, pair bond associations, and flock associations). A total of 512 man-hours was spent by four observers censusing the study plot between mid-March and mid-May 1980.

Nest monitoring.—A concentrated effort was made to locate and monitor daily all nests of all eight study species in the study plot. This combination of techniques generated breeding distributions for all individuals of the eight study species that were felt to be accurate and complete.

THE VARIABLE CIRCULAR-PLOT TECHNIQUE

The details of this method have been fully described by Reynolds et al. (1980). Basically, stations (points) are established within a habitat at intervals along a transect or are scattered in such a manner as to minimize the probability of encountering the same bird at several stations. Each bird heard or seen during a fixed time period from each station is counted and the horizontal distance to its location when first encountered is estimated. The basal radius $\hat{r}$ for each species is then determined as the distance from the stations where the density of birds first begins to decline, presumably because they are being overlooked or not heard by the observer. Finally, the density of each species is determined from the total number of birds encountered within the circles of radius $\hat{r}$. This method therefore rejects, in the final analysis, all encounters outside the circle of basal radius $\hat{r}$, although they are used to establish the value of $\hat{r}$. Statistical methods employed for determining the basal radius are discussed below under Results.

Three parameters must be established at the onset of a variable circular-plot censusing program: the length of time at each station, the distance between stations, and the number of stations to be censused. The length of time at each station should correspond to the minimum time required to count all birds within the basal radius $\hat{r}$. Ideally it is a time long enough for every bird within $\hat{r}$ to display its presence by sight or sound, but short enough that the probability of counting the same bird twice (by recording it in two separate locations or at two different times) is minimized. Since various species differ greatly in their rates of movement or vocalization, this parameter must necessarily be derived by compromise. A period of eight minutes following a one minute "rest" period for "equilibration of bird activity after arrival at each station" was found to be sufficient for closed canopy rain forest in Hawaii as well as for more open habitats elsewhere (Reynolds et al. 1980). This time period was employed throughout this study.

The distance between stations must likewise be chosen by compromise. Ideally one would want the stations to be far enough apart to keep them statistically independent. However, the farther apart are the stations, the fewer can be placed within any one habitat and the fewer can be censused in any given day or within any given time period (because of increased travel time between stations). The distance between stations must therefore, to some extent, be a function of the number of stations required to establish statistical significance of the results and the size of the area being sampled. Because this method was to be tested within a circumscribed area wherein the actual number of birds was known, and because the actual distribution of birds within the study plot was considerably heterogeneous, it was decided to establish as many stations as possible in the study plot and to distribute the stations as uniformly as possible. With this in mind, a minimum distance between stations was established to be approximately twice the largest basal radius, $\hat{r}$, expected for any of the species under study. Basal radii for species in forest habitats have been found to range generally from 25 to 75 m (J. M. Scott, pers. comm.). The distance between stations was therefore chosen to be approximately 180 m (actually 177.7 m [583.1 ft], thereby allowing the stations to fall on established grid points).

The actual distribution of stations in the study plot is shown in Figure 1. Thirteen stations were established. It should be noted that three of the stations fell near the edges of the study plot. For these stations, only the birds encountered in the half of the circle that lay within the study plot were used in the final calculations of density. Thus only 11.5 effective stations were used to calculate density. Similarly, bird encounters from any station that were clearly outside the study plot were excluded from the final density calculations. All bird encounters, however, were used to determine the basal radius $\hat{r}$ for each species.

Each station was censused four times by each of four different observers. Two observers censused the 13 stations on any given day so that a total of eight days was spent on the study plot and 208 variable circular-plot counts were obtained. Reynolds et al. (1980) have indicated that 10–15% variations in density could be significantly detected in five Hawaiian forest species by roughly this number of counts. The census dates were 19 and 22–28 April 1980. Strong winds and light rain precluded censuses on 20–21 April. The weather on the eight census days ranged from clear to overcast with WNW or NW winds ranging from 2–28 kmph (1–15 knots). Temperatures were slightly below normal on most of the census days.

On alternate days the 13 stations were visited counterclockwise with observers beginning at stations 1 and 8, and clockwise with observers beginning at stations 7 and 13. This procedure insured that all stations were censused, on the average, at the same time of day, thereby facilitating comparisons among stations. As the travel time between stations averaged about six minutes, adjacent stations were censused by a single observer about 15 minutes apart. However, about 90–105 minutes elapsed between visits of the first and second observers to any given station. Other activities in the study plot were carefully scheduled on census days so that no one was within 100 m of any station for at least 60 minutes preceding the time that station was censused. The starting time at the first station on all census days was 05:45 PST, about 15 minutes after local sunrise, and all censuses were completed by about 09:00.

Individuals of all species encountered (except swallows), not just study species, were recorded. The

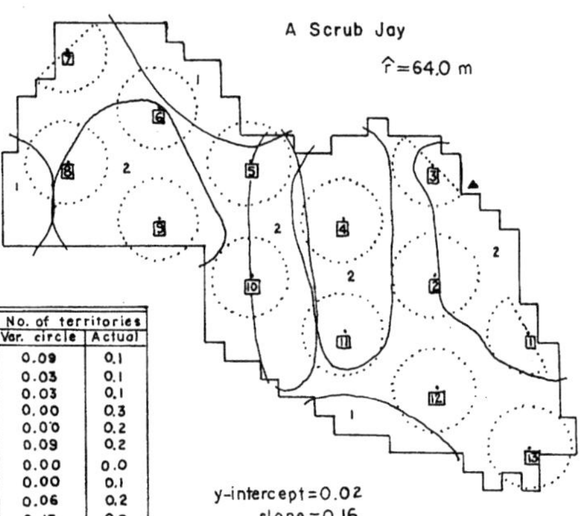

Station number	No. of territories Var. circle	Actual
1	0.09	0.1
2	0.03	0.1
3	0.03	0.1
4	0.00	0.3
5	0.00	0.2
6	0.09	0.2
7	0.00	0.0
8	0.00	0.1
9	0.06	0.2
10	0.13	0.2
11	0.03	0.2
12	0.00	0.0
13	0.00	0.0
Total	0.47	1.7

y-intercept=0.02
slope=0.16
r=0.34
p>0.1

FIGURE 2. Distribution of the 1980 first brood breeding territories in the coastal scrub study plot by species (A–H). Numbers within territories indicate the number of color banded birds in each territory. Known nests are indicated by triangles. Stations for the variable circular-plot method are indicated by squares. Circles of basal radius $\hat{r}$ are indicated by dotted lines. The experimental and actual number of territories contained within each circle of basal radius $\hat{r}$ are presented in the accompanying tables. The results of linear regressions between these numbers are also shown.

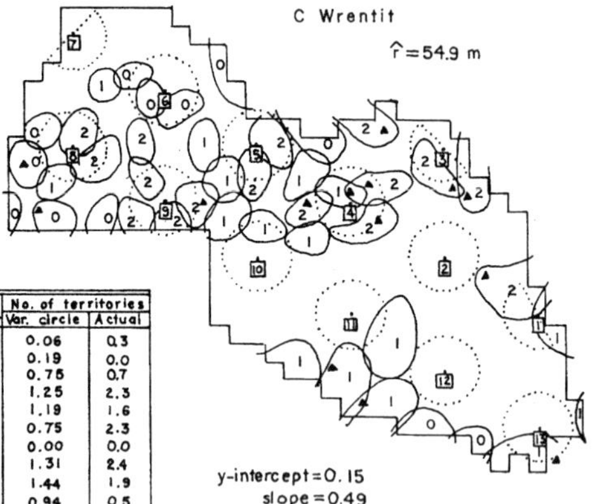

Station number	No. of territories Var. circle	Actual
1	0.06	0.3
2	0.19	0.0
3	0.75	0.7
4	1.25	2.3
5	1.19	1.6
6	0.75	2.3
7	0.00	0.0
8	1.31	2.4
9	1.44	1.9
10	0.94	0.5
11	0.06	0.2
12	0.06	0.0
13	0.13	0.3
Total	8.13	12.5

y-intercept=0.15
slope=0.49
r=0.86
p<0.001

FIGURE 2, continued.

numbers of singing males and "all other observations" were recorded separately. The density of territorial males was determined using only the number of singing males unless twice that number was less than the number of singing males plus the number of "all other observations." In the latter case, the density of territorial males was determined from half the sum of the number of singing males plus the number of "all other observations." This procedure follows Franzreb (1971) and Reynolds et al. (1980). Distances were estimated in feet.

Since the variable circular-plot method necessitates the accurate identification of all birds seen or heard as well as accurate estimation of the horizontal distance from the observer to the birds, persons well-experienced in the censusing of birds in general and with the sounds of all the birds in the study plot were chosen. All the observers, however, were naive as to the actual distribution and abundance of birds in the study plot. An intensive four-day training period involving all four observers was completed in the study plot just prior to testing. The first day was spent insuring that all observers were equally and accurately identifying all the calls and songs of each species. The second day was spent estimating and verifying the distances to visual and aural bird encounters. This was greatly aided by the existing grid system in the study plot. The

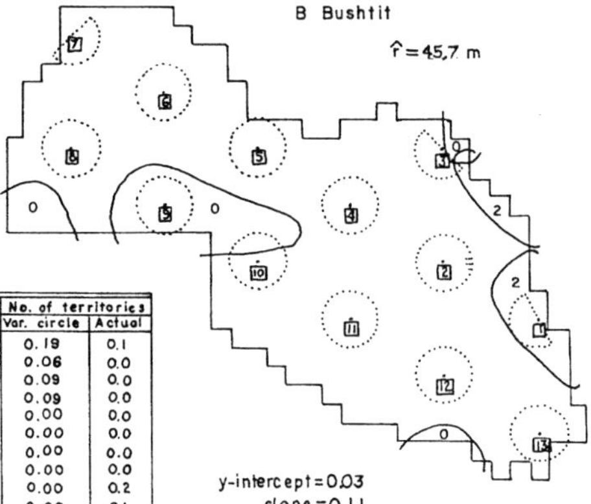

Station number	No. of territories Var. circle	Actual
1	0.19	0.1
2	0.06	0.0
3	0.09	0.0
4	0.09	0.0
5	0.00	0.0
6	0.00	0.0
7	0.00	0.0
8	0.00	0.0
9	0.00	0.2
10	0.00	0.1
11	0.00	0.0
12	0.00	0.0
13	0.00	0.0
Total	0.44	0.4

y-intercept=0.03
slope=0.11
r=0.12
p≫0.1

FIGURE 2, continued.

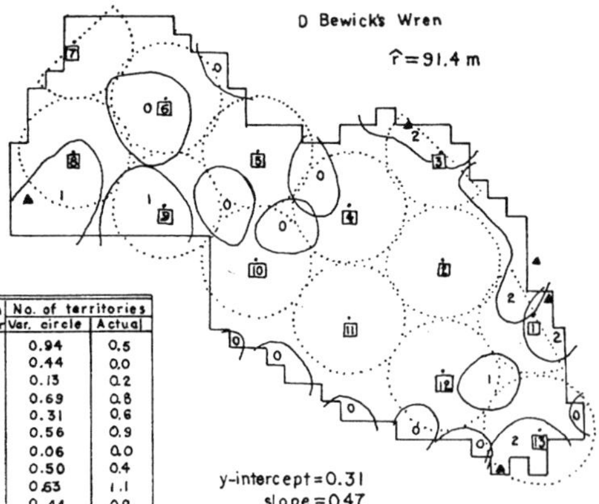

Station number	No. of territories Var. circle	Actual
1	0.94	0.5
2	0.44	0.0
3	0.13	0.2
4	0.69	0.8
5	0.31	0.6
6	0.56	0.9
7	0.06	0.0
8	0.50	0.4
9	0.63	1.1
10	0.44	0.9
11	0.50	0.0
12	1.25	0.9
13	0.94	0.9
Total	7.37	7.2

y-intercept=0.31
slope=0.47
r=0.57
p<0.05

FIGURE 2, continued.

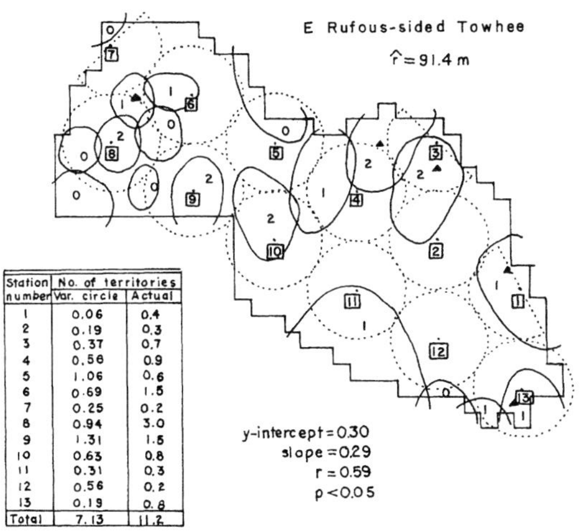

Station number	No. of territories Var. circle	Actual
1	0.06	0.4
2	0.19	0.3
3	0.37	0.7
4	0.56	0.9
5	1.06	0.6
6	0.69	1.5
7	0.25	0.2
8	0.94	3.0
9	1.31	1.5
10	0.63	0.8
11	0.31	0.3
12	0.56	0.2
13	0.19	0.8
Total	7.13	11.2

FIGURE 2, continued.

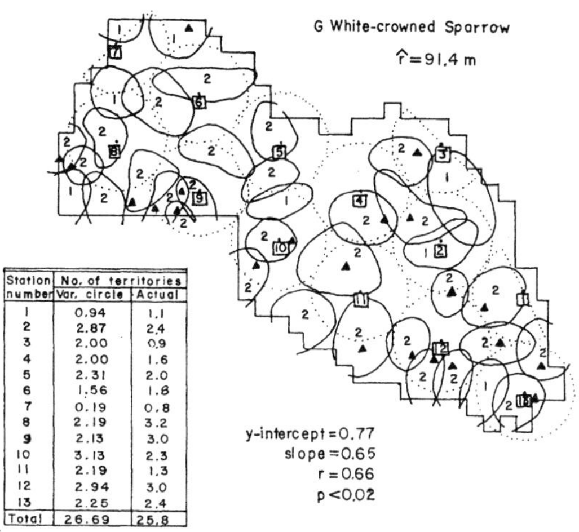

Station number	No. of territories Var. circle	Actual
1	0.94	1.1
2	2.87	2.4
3	2.00	0.9
4	2.00	1.6
5	2.31	2.0
6	1.56	1.8
7	0.19	0.8
8	2.19	3.2
9	2.13	3.0
10	3.13	2.3
11	2.19	1.3
12	2.94	3.0
13	2.25	2.4
Total	26.69	25.8

FIGURE 2, continued.

third and fourth days were spent doing simultaneous station counts and checking and correcting errors or discrepancies between the observers on-the-spot. By the end of the fourth day of training it was felt that all observers were proficient in detecting and identifying birds and in estimating their distances.

RESULTS

The actual distribution of the 1980 first-brood breeding territories for each of the eight study species is shown in Figures 2A–H. The territorial boundaries were determined primarily from the spot-mapping of color banded singing males between mid-March and mid-May and represent the territories as they existed in late April. The total number of territories, any part of which touched the study plot, are presented in Table 1. The thoroughness of the effort to determine these breeding distributions can be gauged by the percent of the total number of individuals in these territories that were color banded and by the percent of these territories for which nests were found (Table 1). For all species combined, 64.2% of the individuals were color banded and 43.0% of the nests were found. These data provide confidence that Figures 2A–H accurately represent the actual distribution of first-brood territories.

The proportion of each of these breeding territories that was contained within the 36 ha study plot was estimated to the nearest tenth. These estimates were based both upon the area of the territory contained within the study plot and upon the time spent by the pair within the study plot. These proportions were summed for

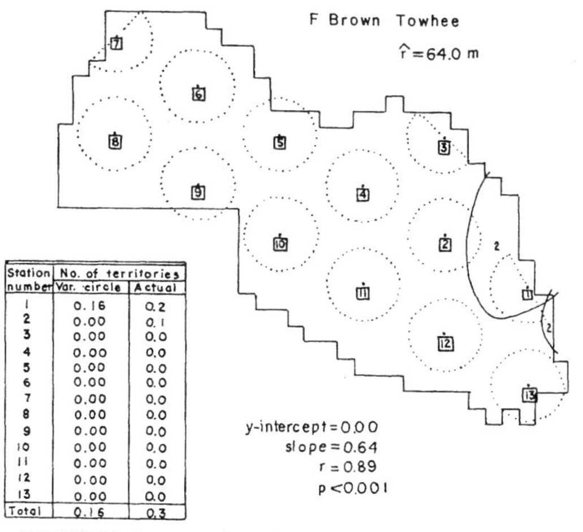

Station number	No. of territories Var. circle	Actual
1	0.16	0.2
2	0.00	0.1
3	0.00	0.0
4	0.00	0.0
5	0.00	0.0
6	0.00	0.0
7	0.00	0.0
8	0.00	0.0
9	0.00	0.0
10	0.00	0.0
11	0.00	0.0
12	0.00	0.0
13	0.00	0.0
Total	0.16	0.3

FIGURE 2, continued.

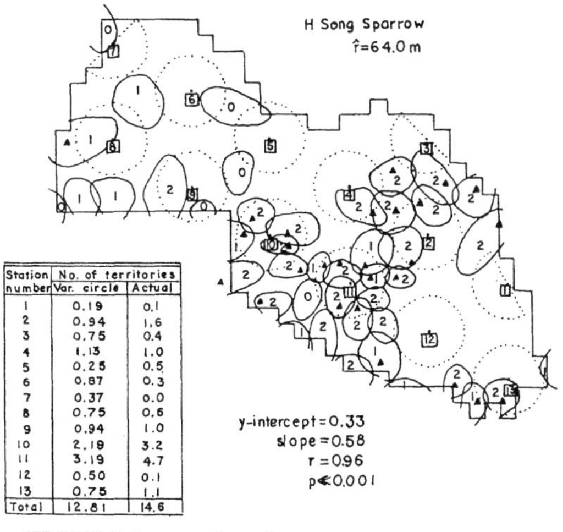

Station number	No. of territories Var. circle	Actual
1	0.19	0.1
2	0.94	1.6
3	0.75	0.4
4	1.13	1.0
5	0.25	0.5
6	0.87	0.3
7	0.37	0.0
8	0.75	0.6
9	0.94	1.0
10	2.19	3.2
11	3.19	4.7
12	0.50	0.1
13	0.75	1.1
Total	12.81	14.6

FIGURE 2, continued.

TABLE 1
ACTUAL DETERMINATION OF BREEDING TERRITORIES IN A CALIFORNIA COASTAL SCRUB HABITAT

Species	No. of terr.[a]	% of ind. color banded[b]	% of nests found[c]	Density (/36 ha)
Scrub Jay	7	78.6	14.3	3.8
Bushtit	6	33.3	0.0	2.2
Wrentit	44	55.2	32.6	36.3
Bewick's Wren	18	30.6	27.8	9.4
Rufous-sided Towhee	19	44.7	26.3	14.0
Brown Towhee	2	100.0	0.0	0.6
White-crowned Sparrow	37	93.0	64.7	32.4
Song Sparrow	44	72.4	62.8	35.5
Total	177	64.2	43.0	134.2

[a] Total number of breeding territories that touched any part of the 36 ha study plot.
[b] Based on the total number of individuals in the study plot, including all females of bigamous males and all unmated territory-holding males.
[c] Based on one nest/female in the study plot.

TABLE 2
FIELD TEST OF THE VARIABLE CIRCULAR-PLOT CENSUSING METHOD

	Actual	Variable circular-plot		
Species	Density (/36 ha)	Basal radius (m)	Density (/36 ha)	% error[a]
Scrub Jay	3.8	64.0	1.1	−70.0
Bushtit	2.2	45.7	2.1	−5.0
Wrentit	36.3	54.9	26.9	−25.9
Bewick's Wren	9.4	91.4	8.3	−11.4
Rufous-sided Towhee	14.0	91.4	8.5	−39.4
Brown Towhee	0.6	64.0	0.4	−36.7
White-crowned Sparrow	32.4	91.4	31.8	−1.9
Song Sparrow	35.5	64.0	31.2	−12.2
Total	134.2		110.3	−17.8

[a] Negative values indicate underestimates by the variable circular-plot method.

all breeding territories contained in the 36 ha plot, and are the values against which the results of the variable circular-plot method are to be compared (Tables 1 and 2).

The results of the variable circular-plot method are presented in Table 2. Except for Scrub Jay, Bushtit, and Brown Towhee, which rarely or never sang, densities were determined from the number of singing males. The first step was the determination of the basal radius, $\hat{r}$, for each species. For this calculation the entire data set, including birds seen or heard out of the study plot, was summed for all observers, census days, and stations. A distribution of densities based on 9.14 m (30 ft) bands from 0 to 182.9 m (0–600 ft), and 18.29 m (60 ft) bands from 182.9 to 365.8 m (600–1200 ft) was produced. The basal radius $\hat{r}$ was determined as the inside radius of the first band that had a density significantly less than the density of the previous bands. Significance was determined by likelihood ratio testing with a critical value of 4 (Ramsey and Scott 1979).

The density within the study plot was determined from the total number of birds encountered within the circles of radius *$\hat{r}$ that were inside the study plot* divided by 184, the number of effective stations censused within the study plot. This density, when converted to birds/36 ha, can be directly compared to the actual density. These data (Table 2) show that the variable circular-plot method *underestimated* the densities of all eight study species. In half of the species, Bushtit, Bewick's Wren, White-crowned Sparrow, and Song Sparrow, the error was less than about 12%. This must be considered excellent agreement. In fact, the density of White-crowned Sparrows was underestimated by less than 2%. In three of the remaining four species, errors varied from about 25% to 40%. Only the Scrub Jay density was badly underestimated, by 70%. The mean underestimation for the eight species was 25.3% ± 22.8 (SD). The total density of the entire community (of the eight study species) was underestimated by less than 18%. A linear regression between the variable circular-plot method densities and the actual densities for all eight species produced a highly significant correlation ($r = 0.982$; d.f. = 6; $P < .001$). The y-intercept (−0.88) and slope (0.87) of the regression line indicate the bias toward underestimation.

The variable circular-plot method, therefore, appears capable of producing reasonably good estimates, biased toward the low side, of the *densities* of most species in the coastal scrub habitat. It is informative to inquire how well the method describes the *distribution* of these species among the 13 stations. Circles of basal radius, $\hat{r}$, for each of the study species were drawn around each station. These are shown in Figures 2A–H. The mean number of territories detected by the variable circular-plot method at each station and the actual number of territories (measured to the nearest tenth) contained in these circles are presented on each of the figures 2A–H. Linear regressions were then performed between these two sets of values.

Six of the eight study species showed significant ($P < 0.05$) positive correlations. Two species, Scrub Jay and Bushtit, showed no significant correlation between these values. Thus the variable circular-plot method was capable of describing the density of Bushtits in the study plot, but not their distribution. The method could do neither for the Scrub Jay. It is notable

TABLE 3
RESULTS OF THE VARIABLE CIRCULAR-PLOT METHOD FOR FOUR INDIVIDUAL OBSERVERS

Species	Range of basal radii[a] (mean)	Range of density[b] (mean)	Range of % error (mean)	No. of signif. correl.
Scrub Jay	36.6 to 155.4 (91.4)	1.0 to 2.8 (1.5)	−73.9 to −26.6 (−59.5)	0
Bushtit	27.4 to 73.2 (48.0)	1.7 to 3.6 (2.4)	−24.5 to +62.7 (+10.5)	1
Wrentit	27.4 to 64.0 (52.6)	23.1 to 43.0 (30.6)	−36.3 to +18.5 (−15.7)	3
Bewick's Wren	45.7 to 100.6 (70.9)	6.5 to 19.1 (11.8)	−30.7 to +102.9 (+25.9)	2
Rufous-sided Towhee	64.0 to 91.4 (82.3)	5.4 to 14.3 (9.5)	−61.7 to −2.4 (−32.1)	1
Brown Towhee	54.9 to 155.4 (91.4)	0.1 to 0.8 (0.5)	−83.3 to +38.3 (−23.3)	3
White-crowned Sparrow	45.7 to 109.7 (70.9)	23.8 to 37.7 (29.3)	−26.5 to +16.4 (−9.4)	2
Song Sparrow	45.7 to 91.4 (64.0)	23.2 to 58.8 (34.9)	−34.5 to +65.5 (−1.5)	4

[a] Meters.
[b] Territories/36 ha.

that Scrub Jays and Bushtits are not strictly territorial in the fullest sense. The males of neither species have true territorial songs and individuals occasionally wander far outside their normal "territorial boundaries." Thus the actual distributions shown for these species in Figures 2A–H may be somewhat suspect. As a result, the poor estimate of Scrub Jay density and the excellent estimate of Bushtit density shown in Table 2 should be viewed with caution.

In the six remaining species, where significant correlation was obtained, it is notable that all y-intercepts were equal to or greater than zero and all slopes were less than one. Since perfect correlation between the methods should produce a y-intercept of zero and a slope of one, the indication is that the variable circular-plot method tends to overestimate density where a species is sparse and to underestimate density where a species is dense. The effect of overestimation where sparse was greatest in the White-crowned Sparrow (y-intercept = 0.77), a species with a notably far-carrying song. The result in this species was that the overestimate (where sparse) balanced the underestimate (where dense) to produce a very accurate density estimate for the total study plot. Song Sparrows and Wrentits, with songs that carry for shorter distances, were more consistently underestimated (y-intercept = 0.33 and 0.15 respectively) and thus showed greater errors in their total density estimates.

Thus far we have considered the variable circular-plot method utilizing the combined data from all four observers. It is also of interest to

TABLE 4
RESULTS OF THE VARIABLE CIRCULAR-PLOT METHOD ON THE TOTAL DENSITY OF THE EIGHT STUDY SPECIES

	Total density		Regression analysis			
Observer	terr/36 ha	% error	y-int.	slope	r	P
1	135.3	+0.8	−2.32	1.15	0.882	<.01
2	119.3	−11.1	−0.80	0.94	0.921	<.01
3	98.4	−26.7	−0.51	0.76	0.998	<.001
4	129.8	−3.3	+2.71	0.81	0.897	<.01
Mean	120.7	−10.1	−0.23	0.91	0.987	<.001
Comb. data	110.3	−17.8	−0.88	0.87	0.982	<.001
Actual	134.2					

examine the reproducibility of the method by comparing the results obtained from each of the four individual observers. As perhaps expected, the basal radii obtained for the various species differed among the observers. The range (and mean) of basal radii found by the four observers are presented for each species in Table 3. These individual basal radii were then used to calculate the densities of each species for each of the four observers. The range (and mean) of densities thus obtained are also presented, along with the range (and mean) of the resulting percent errors, in Table 3. As perhaps expected, some of the errors thus obtained were considerably greater than those obtained by utilizing combined data, presumably because of the reduced sample size. In contrast to the combined data, individuals occasionally overestimated the densities of certain species. Bushtit, Bewick's Wren, and Brown Towhee densities were each overestimated by two observers, and Wrentit, White-crowned Sparrow, and Song Sparrow densities were each overestimated by one observer. The mean absolute percent error in the density determinations of the eight species, however, were found to be quite consistent among the four observers. They ranged from 30.6% to 38.8% and compared favorably to the mean absolute percent error obtained by utilizing combined data from all four observers, 25.3%.

The mean (of the four observers) densities for most species agreed quite well (Table 3) with the actual densities (Table 1). Five species were underestimated by less than about 32%, two species (Bushtit and Bewick's Wren) were overestimated by less than about 25%, and only one species (Scrub Jay) was badly underestimated by about 60%. The mean absolute percent error of these mean densities was 22.2% ± 18.0 SD, slightly less than that obtained from combined data.

The total density of the entire community (of the eight study species), as determined by each observer, is shown in Table 4. Surprisingly, three of the four observers determined this total density more accurately than did the combined data for all observers. Errors among the four varied from only −26.7% to +0.8% and averaged −10.1%. Linear regressions (Table 4) between the variable circular-plot densities and the actual densities for the eight species produced significant correlations ($P < .01$) for all four individual observers as well as for the mean (of the four observers) densities. The preponderance of negative y-intercepts and fractional slopes in these data again indicate the bias toward underestimation. In general, the mean of the four individuals' results produced more accurate estimates than did the combined data.

Finally, we can inquire as to how well each observer could describe the distribution of each study species among the 13 stations. Linear regressions for each species for each observer were performed between the number of territories detected by the variable circular-plot method at each station (within circles of the basal radius for that species for that observer) and the actual number of territories (contained in those circles and measured to the nearest tenth) at those stations. The number of observers (of the four) that produced significant ($P < .05$) positive correlations are shown for each of the eight species in Table 3. Individual observers were able to significantly describe the breeding distributions of the eight species in half (16 of 32) of the cases.

DISCUSSION

The variable circular-plot censusing method, based on four 8-min visits to each of 13 circular plots, was found to be capable of determining the breeding densities of most species in a California coastal scrub habitat to an accuracy of better than about 25–30%. In general, the densities of most species were underestimated. The apparent failure of all individuals of a species within its basal radius to display their presence by sight or sound was presumably responsible for this underestimate. It should be pointed out, however, that for five of the eight species the density determinations were based on counts of singing males. The number of "all other observations" of these species was minimal. This suggests that the variable circular-plot method, while reasonably accurate during the height of the breeding season, may be quite inaccurate at other times of the year.

The variable circular-plot method was also capable of describing the breeding distributions of most species in the coastal scrub habitat. The method generally tended to overestimate density where a species was sparse and to underestimate density where it was dense. This is presumably because birds tend to hold larger territories where their density is sparser. (This is empirically evident from Figures 2A–H, especially for Wrentit and Song Sparrow.) Birds with larger territories tend to move between more greatly separated singing posts than do birds with smaller territories. Thus there is a greater probability that they will be counted twice at the same station. On the other hand, birds in dense territorial situations with smaller territories tend to move over smaller distances between singing posts and often countersing with their nearest neighbor on the edge of their territory. Such pairs of countersinging birds can easily be counted as one, thus producing an underestimate of their density.

Finally, the variable circular plot method is quite sensitive to observer error in distance es timation, particularly so to underestimates. A 25% underestimate of distance, for example, will produce a 78% overestimate of density, but a 25% overestimate of distance will only produce a 36% underestimate of density. These errors can presumably be minimized by extensive ob server training and experience. The open, nearly two dimensional nature of the coastal scrub hab itat and the presence of an easily observed grid of known distance markers greatly aided dis tance estimates in this field test. One should therefore expect this method to be less accurate than indicated here in more closed, complex, three dimensional habitats without such easily observed distance markers.

ACKNOWLEDGMENTS

This paper owes much to the efforts of the full time volunteers working at the Palomarin Field Station. The actual determinations of breeding territories were made by Janet Kjelmyr (Grid 1), Geoff Geupel (Grid 2), Kathy Keane and Tad Theimer (Grid 3), and Mike Bonoff and David Sherman (Grid 4). The variable cir cular plot censusing was performed by Brett Engs trom, Peter Pyle, Ken Warheit, and the author. Tad Theimer aided greatly with the preparation of the fig ures. Computer analyses were kindly provided by C. J. Ralph. The manuscript was much improved by com ments from James R. King, L. Richard Mewaldt, C. J. Ralph, and Chandler S. Robbins. I thank you all for your assistance. This is PRBO Contribution No. 225.

Studies in Avian Biology No. 6:186–192, 1981.

COMPARISONS OF AVIAN CENSUS RESULTS USING VARIABLE DISTANCE TRANSECT AND VARIABLE CIRCULAR PLOT TECHNIQUES

BERTIN W. ANDERSON[1] AND ROBERT D. OHMART[1]

ABSTRACT.—We compared avian census results obtained with the variable circular plot and variable distance transect techniques. Data are from riparian vegetation along the lower Colorado River, Arizona-California border. Censusing was done in spring and early summer.

The two techniques provided estimates of densities, BSD, evenness, and species richness that were similar when we used a radius of 30 m for the first band with the circular plot technique. Using 10 m as the width of the first band yielded significantly dissimilar results for these variables.

We conclude that the transect technique is the more feasible method, if stands of vegetation are large enough to establish transects 700 m to 800 m in length and the topography allows ambulation. These transects must be adequately cleared and marked so that the censuser does not waste time trying to find the transect. The circular plot method is desirable in areas where the vegetation occurs in small stands, or where clearing transects is not feasible.

Reynolds et al. (1980) have recommended censusing of birds by a variable circular plot technique. They found that a stationary observer spent more time searching for birds and less time watching the path of travel than one using the transect technique. They reasoned that stationary observers should have less effect on bird activity and provide better estimates of density. The use of stations should also allow better correlation between habitat variables and abundance and occurrence of bird species.

In this report we compare avian census results obtained by the variable circular plot method and the J. T. Emlen (1971) version of the variable strip transect count. We also compare detection rates per unit of time with the two methods. We considered transportation costs in our evaluation of these techniques. As birds close to the observer are more likely to be detected, we compare the total area censused at close distances with each technique.

Data were collected monthly from March through June 1980 in riparian vegetation along the Colorado River north of Blythe, Riverside County, California, and north of Ehrenberg, Yuma County, Arizona.

METHODS

Censusing was done the same way each month. On the first day, each of three observers was assigned to one of three census routes. Two used the circular plot technique, and one used the Emlen version of the transect technique. On the second day, each observer censused the same area but changed technique. On the third day the observers rotated. At the end of six days, each observer had censused each route using each technique for a total of three censuses per technique. This procedure eliminated the bias of observer variability.

March and April census results were from a 283 ha stand of honey mesquite (*Prosopis glandulosa*). In this plot we established 11 transects in 1974, and we have censused each transect two to four times monthly to date. Circular plots were laid out along regular census routes that totaled 8878 m. In March all transects were used, and along the same census routes we were able to establish 41 circular plots, the perimeters of which were 100 m apart. The 11 transects were divided into four census routes; thus four censusers and eight days were required to census all transects. Three routes included three transects each, and one route had two transects.

In April another test was conducted in honey mesquite in basically the same manner as in March, except that three rather than four routes were included. Each route had three transects, and three observers were used. A total of 33 circular plots were used in this test. An additional test was conducted in stands of mixed salt cedar (*Tamarix chinensis*), willow (*Salix gooddingii*), and cottonwood (*Populus fremontii*). This test involved six census routes, each with one to three transects, for a total of nine transects and 31 circular plots. All plots and transects were censused three times.

In March and April all circular plots were censused for eight minutes, with time beginning as soon as the center of the plot was reached. Although it is recommended that some pause time be allowed before censusing begins, several birds present upon arrival were not redetected. We therefore elected to have no pause time. In May the census time at each station was reduced to six minutes. The May test was conducted on the same nine honey mesquite transects as in April.

In June censusing was done on four census areas, each with one to three transects, for a total of ten transects. These transects included about 219 ha of dense riparian vegetation, 8 or 9 m in height. These areas had the highest bird densities found along the lower Colorado River and included high densities of the Mourning Dove and the semi-colonial White-winged Dove. Each transect and circular plot was censued once by each of four observers over an eight-day period.

It is important to stress that all censusing for both methods began and ended in the same place; the total

[1] Department of Zoology and The Center for Environmental Studies, Arizona State University, Tempe, Arizona 85281.

distance that had to be walked was exactly the same for each method. The observers were all highly qualified, averaging five years of experience in judging distances to points of detection. All observers had average or above-average sight and hearing and each had about eight years of field experience.

Birds were detected laterally on transects out to 120 m. The radius of circular plots was 120 m. Detections of greater distances were not possible without danger of including birds from vastly different habitat types, such as agricultural areas or creosote bush desert. The radius was divided visually into bands; the first three were 5 m wide, the next three were 15 m wide, and the last two were 30 m wide. All observers agreed that recording detections in 5 m bands beyond 15 m was guesswork. In our study area, more than 90% of the detections of passerines beyond 15 m are auditory. As many of the birds are constantly moving (Verdin) or have very weak calls (Black-tailed Gnatcatcher), precision in distance estimates beyond 15 m is very difficult. Although detections were recorded in 5 m bands in the first 15 m, density estimates calculated using the bands separately or combined into a single 15 m wide band were similar. When differences did occur, use of the narrower bands resulted in inflated values, often absurdly so (based on spot map data in the same area). Densities were also calculated using a 30 m width for the first band; this wider band appeared to help in eliminating distorted estimates caused by extreme outliers. See Ramsey and Scott (1979) for further discussion of outliers. The method for calculating densities was basically that of Reynolds et al. (1980) and corresponds to method M1 of Ramsey and Scott (1979).

Reynolds et al. (1980) noted that their circular plots were censused twice per day. In our area this would be unacceptable because as diurnal temperatures increase, bird activity decreases dramatically (Anderson and Ohmart 1977).

Spot mapping was conducted 9 to 12 times monthly on three 40 ha plots in honey mesquite. Bird species diversity (BSD) was calculated as $H' = -\Sigma p_i \log n\ p_i$ where p_i is the proportion of the ith species in the total (MacArthur and MacArthur 1961). Evenness (J) was calculated as H'/H'max where H'max equals log n of the total species present. Chi-square tests were used in determining statistically significant differences in bird densities obtained with different techniques or observers. Significant differences in bird species diversity were determined using a t-test (Zar 1974). The accepted level of statistical significance in this report was $P \leq 0.05$.

RESULTS AND DISCUSSION

Vegetation

Foliage profiles revealed that the tests in honey mesquite (March and April) were through vegetation that averaged moderately dense up to a height of about 5 m. The considerable horizontal diversity varied from open patches with little vegetation to patches with very dense vegetation reaching a height of 7 to 8 m. In the April test, the vegetation in areas of mixed cottonwood, willow, and salt cedar was both denser and taller than in honey mesquite, with visibility often severely restricted in the former habitats. The test in June was also in very dense vegetation, with visibility severely restricted over much of the censused area.

TABLE 1
LOGISTIC COMPARISONS OF CIRCULAR PLOT AND VARIABLE DISTANCE TRANSECT CENSUSES

	Total area censused (ha)	Minutes spent: Walking census route	Minutes spent: Actually censusing	Total detections	Detections per hour[a]	Visual detections
March						
CP	567.6[b]	498[b]	468[b]	1185[b]	38	249[b]
T	884.4	396	396	1368	52	315
April—Honey Mesquite						
CP	456.8[b]	462[b]	432[b]	1594[b]	55[b]	255[b]
T	702.4	348	348	2145	123	579
April—Cottonwood—Willow						
CP	380.0[b]	420[b]	390	1168[b]	45[b]	164[b]
T	613.6	342	342	1759	77	405
May						
CP	456.8[b]	414	372	1857[b]	75[b]	409[b]
T	702.4	360	360	2335	97	701
June						
CP	664.4[b]	428	396	3180[b]	120[b]	571[b]
T	1015.2	444	444	4849	164	1600

[a] Actual census time, excluding time walking from plot to plot.
[b] $P \leq 0.05$ that observed difference due to chance with 1 df and $\chi^2 \geq 4.0$.

Comparison Of Techniques

Census time

The number of minutes necessary to walk the census routes was always significantly longer using the circular plot method (Table 1) when each plot was censused for eight minutes. There was less difference when each plot was censused for six or seven minutes (May and June). Even though the length of time spent in the field was usually longer with the circular plot technique, the total area censused was significantly less than with the transect method because of the 100 m interval between plots that was not censused (Table 1).

Total detections

Total detections were always significantly greater with the transect technique (Table 1). This was to be expected because an average of 36 percent more area was censused with the transect technique. If detections were proportionate to the area censused, there should, by chance alone, have been 36% more detections made with the transect technique. However, even when the difference in area censused was

TABLE 2
DENSITY ESTIMATES AND NUMBERS OF SPECIES AFTER EACH CENSUS AND AFTER EACH DAY IS ADDED TO PREVIOUS CENSUS RESULTS

A. Circular plot method	Width of first band							
	15 m				30 m			
Day	Mar	Apr	May	Jun	Mar	Apr	May	Jun
Densities (n/40 ha)								
1	242	271	388	882	163	271	362	556
2	436	345	225	768	292	281	277	478
3	160	343	352	981	144	264	337	356
4	—	—	—	602	—	—	—	328
1–2	337	519	205	813	206	346	199	423
1–3	211	285	226	982	152	230	240	368
1–4	—	—	—	817	—	—	—	502
Species[a]								
1	28-21	22-22	19-18	23-20	28-21	22-22	19-18	23-22
2	27-24	28-21	19-18	23-23	27-24	28-21	19-18	23-23
3	26-24	25-23	20-20	24-23	26-24	26-24	20-20	24-23
4	—	—	—	22-21	—	—	—	22-21
1–2	32-26	31-24	22-17	24-24	32-26	30-23	22-17	24-24
1–3	32-21	31-25	22-19	26-25	32-23	31-24	24-17	26-24
1–4	—	—	—	28-22	—	—	—	28-22

B. Transect method Day	Mar	Apr	May	Jun
Densities (n/40 ha)				
1	86	296	273	518
2	192	274	286	373
3	—	253	268	672
4	—	—	—	531
1–2	139	289	287	462
1–3	—	276	276	523
1–4	—	—	—	525
Species[a]				
1	26-24	33-25	20-20	23-20
2	28-26	33-28	24-20	23-21
3	—	25-23	24-22	29-24
4	—	—	—	27-26
1–2	31-25	33-30	25-22	23-23
1–3	—	34-26	28-21	29-21
1–4	—	—	—	32-24

[a] First number is total species detected; second is number of species with density ≥1.

taken into account, the transect technique always had as many detections as expected and in one case more than expected.

Detections per hour were usually significantly greater with the transect technique (Table 1). Even when an adjustment was made for the difference in area censused per hour, using the transect technique, an average of 20% more birds were detected per hour spent on the census routes. The difference in detection rates was less in March than at other times, presumably because birds were less vocal. Had the transect censusing rate (1.83 kmph) been decreased, the number of detections presumably would have increased. If birds in the first lateral strip were missed, it would have been reflected in density estimates. That this did not happen indicates that although the birds were less detectable, it was in lateral strips at greater distances from the transect that they were missed.

More detections were made visually with the transect technique (Table 1). Distance estimates of detections made visually at close distances are likely to be more accurate than estimates from aural detections made from great distances. Given census areas of the same size, only six percent of the area of circular plots was within 30 m of the observer, whereas with the transect technique, more than 25% of the area was within 30 m. Since visual detections are more likely to be made at close distances, it seems logical that the transect method would have somewhat fewer mistakes in identification.

Day-to-day differences

The day-to-day fluctuation in population estimates was more erratic with the circular plot technique when the width of the first band was 15 m or 30 m (Table 2). Logistically this is very important. The density estimates at the end of the second census were not significantly different ($P > 0.05$, 1 df, $\chi^2 \leq 0.7$) from estimates at the end of the third transect census, suggesting that fewer than three censuses are necessary with the transect technique. But with 33 circular plots, even three visits to each plot may not be adequate; density estimates with two visits to each plot were significantly different ($P < 0.01$, 1 df, $\chi^2 \geq 15.9$) from estimates at the end of three censuses in three of four cases when the initial band was 15 m wide, and in two cases ($P < 0.01$, 1 df, $\chi^2 \leq 8.1$) when the initial band was 30 m wide.

In general, the addition of a third census did very little for either method in terms of adding new species (Table 2). There was a slight but consistent tendency for more species to be detected by two or three censuses than by a single census. The number of species in both categories tended to increase after three censuses.

Density estimates

Estimates of total density using all the visits were significantly different in four of the five tests, being higher with the circular plot technique (Table 3) when the first band was 15 m. When the initial band was 30 m for the circular plot technique, results from the two techniques were not significantly different.

BSD, calculated from the census results of both techniques, was significantly different for each test when the first band for the circular plot method was 15 m wide (Table 3). The pattern for BSD and evenness (Table 3) was only roughly similar; that is, the highest BSD found with the circular plot method did not correspond to the highest BSD with the transect method. Evenness was greatest in March with the transect technique and in May with the circular plot technique. When the width of the first band was 30 m for circular plots, the patterns of BSD and J were similar, and BSD was not significantly different in three of the compared censuses. Species richness was similar with both techniques.

TABLE 3
AVIAN POPULATION VARIABLES DERIVED FROM THE VARIABLE CIRCULAR PLOT AND VARIABLE DISTANCE TRANSECT METHODS

		Species detected			
	Density	Total	With density ≥1	BSD	J
March					
CP 15	211[a]	32	21	2.16[b]	.695
CP 30	152	32	23	2.51[b]	.801
T	139	31	25	2.75	.853
April—Honey Mesquite					
CP 15	285	31	25	2.60	.808
CP 30	283	31	24	2.57	.810
T	276	34	26	2.63	.808
April—Cottonwood—Willow					
CP 15	334[a]	33	29	2.51[b]	.745
CP 30	253	33	27	2.68	.812
T	263	31	27	2.66	.806
May					
CP 15	226[a]	22	19	2.55[b]	.866
CP 30	230	24	17	2.28	.804
T	276	28	21	2.33	.766
June					
CP 15	817[a]	28	22	2.08[b]	.672
CP 30	502	28	22	2.26[b]	.732
T	525	32	24	1.94	.610

[a] $P \leq 0.05$ that observed difference between circular plot and transect estimates due to chance with 1 df, $\chi^2 \geq 8.4$.
[b] $P \leq 0.05$ that observed difference between circular plot and transect BSDs due to chance with ≥290 df and $t \geq 2.0$.

Comparisons of three techniques

Average population estimates obtained with three techniques (spot map also) for 10 bird species in honey mesquite (Table 4) showed that only the spot map and transect totals were significantly ($P < 0.05$, 1 df, $\chi^2 = 5.7$) different. Differences between the transect and circular plot techniques were not significant ($P > 0.05$, 1 df, $\chi^2 \leq 1.0$), but mean densities with the circular plot technique were higher. For the combined average, the spot map technique yielded consistently higher densities than either of the other techniques.

Comparison of biological conclusions from the two data sets

Only species richness was similar with the two techniques when the first band was 15 m wide. This leads inexorably to the conclusion that the two census methods do not yield similar conclusions about population densities and diversities, and that when using 15 m wide bands, one, or even both, techniques are insensitive to the real structure of the bird community. However, with

TABLE 4

DENSITY ESTIMATES OBTAINED WITH THREE DIFFERENT CENSUSING METHODS IN HONEY MESQUITE HABITAT ALONG THE LOWER COLORADO RIVER. DENSITIES ARE N/40 HA AVERAGED FOR MARCH, APRIL, AND MAY 1980[a]

Species	Density estimates			
	T	CP-1	CP-2	SM
Gila Woodpecker (*Centurus uropygialis*)	2	2	2	2
Ladder-backed Woodpecker (*Picoides scalaris*)	4	3	6	8
Ash-throated Flycatcher (*Myiarchus cinerascens*)	11	12	10	12
Black-tailed Gnatcatcher (*Polioptila melanura*)	14	28	26	24
Verdin (*Auriparus flaviceps*)	7	8	10	10
Cactus Wren (*Campylorhynchus brunneicapillus*)	4	4	4	8
Lucy's Warbler (*Vermivora luciae*)	37	32	28	41
Northern Oriole (*Icterus galbula*)	12	15	14	13
Crissal Thrasher (*Toxostoma dorsale*)	2	1	2	14
Abert Towhee (*Pipilo aberta*)	21	25	23	21
Totals	114	130	125	153

[a] Methods: T = Transect; CP-1 = Circular plot when the first band was 15 m; CP-2 = Circular plot when the first band was 30 m; SM = Spot map. Numbers represent densities per 40 ha.

the circular plot technique, if one considers the first band as being 30 m wide, the calculated densities and diversities compare favorably with the strip census technique.

Other possible inferences can be drawn from each of the data sets collected for this evaluation. A summary of these inferences indicates concurrence in nine of twelve (75%) of the instances, when the first circular plot band was 15 m or 30 m wide (Table 5). Disagreement is almost exclusively restricted to conclusions involving BSD and evenness. Perhaps these differences represent more of a vindication of criticism in the recent literature of BSD and evenness calculations than any serious indictment of either census method. Using species richness as a measure of diversity, there is close agreement between results gathered with the two census techniques.

Although density estimates differed somewhat, the patterns were similar; that is, the greatest densities were found in June and the lowest in March with both techniques. Furthermore, a Spearman-rank correlation between densities obtained with the two techniques was $R_s \geq 0.8$ ($P < 0.01$) for each test. This means that species abundant using one technique were also abundant using the other technique. Simi-

TABLE 5

CONCLUSIONS WHICH MIGHT BE DRAWN FROM THE CENSUS DATA AND THE EXTENT OF CONCURRENCE (+) WITH THE DATA SETS OBTAINED WITH THE TRANSECT AND CIRCULAR PLOT CENSUSING TECHNIQUES

Possible conclusions	Technique		
	Transect	Circular Plot: Width of first band	
		15 m	30 m
1. June—greatest densities.	+	+	+
2. March—lowest densities.	+	+	+
3. Density—greater in cottonwood-willow than honey mesquite.	−	+	+
4. Honey mesquite—April densities greater than May.	+	+	+
5. March—lowest BSD.	−	+	+
6. Honey mesquite and cottonwood-willow—BSD about equal.	+	+	−
7. June—evenness lowest.	+	+	+
8. March—evenness highest.	+	−	+
9. Species detected—lowest in May.	+	+	+
10. Species with densities ≥1—lowest in May.	+	+	+
11. Species with densities ≥1—highest in cottonwood-willow.	+	+	+
12. Number of species—greater in April than in March or June.	+	+	+

TABLE 6
MINUTE-BY-MINUTE DISTRIBUTION OF DETECTIONS FOR 240 SEPARATE CENSUSES OF CIRCULAR PLOTS FOR EIGHT MINUTES EACH

	Minute								
	1	2	3	4	5	6	7	8	Totals
Total detections	679	246	165	113	121	105	75	94	1598
Percent of total	42.5	15.4	10.3	7.1	7.6	6.6	4.7	5.9	100.1

larities in the densities of each species were tested further with the Kolmogorov-Smirnov test (Sokal and Rohlf 1969). If the species with the greatest density was the same species with both census methods, but for one technique the density estimate was 50 and for the other technique it was 100, the distributions would still have a high rank correlation. The K-S test would indicate if one or more species accounted for a significantly greater proportion of the total density using one technique than would result from using the other technique. As it turned out, the results differed in four of the five data sets when the initial circular plot band was 15 m wide. This is a reflection of the great daily variation in density estimates with the circular plot technique (Table 2) and indicates that it is pronounced for certain species rather than resulting from a similar change for all species. When the first band was 30 m wide, nonsignificant K-S tests resulted in four of five tests.

COMMENTS ON THE CIRCULAR PLOT TECHNIQUE

Detections per minute

Censusing for eight minutes in each circular plot gave densities about the same as, or greater than, those obtained with the transect technique. In May, when census time per plot was cut to six minutes, the estimated densities were slightly lower in circular plots than in transects. A reduction of two minutes results in a decrease in detections of 11 percent at the center of the plot (Table 6). Apparently a census time of 8 minutes yields different results due to a more thorough census. However, if some of the detections are actually recounts, less accurate density estimates may result.

Species with fewer than 50 detections

Nearly 70% of all species had fewer than 50 detections, even with nearly 100 circular plot censuses (Table 7). When a species had fewer than 50 detections, the procedure was to calculate the density by using the effective radius of a species judged to have about the same detectability as the scarcer species. About 80% of all the species censused in riparian habitats using 100 circular plot censuses had mean density values that were too low to reduce the confidence limits to 25% (Table 6). If the number of plots were doubled in order to include most of these low detection species, it would mean doubling the work of the staff or reducing the overall objectives of the study.

CONCLUSIONS

With respect to total densities, the two techniques did not consistently yield similar results, except when circular plot calculations were made using a width of 30 m for the first band. With the transect technique, the day-to-day variations were usually small, and addition of a third visit added little. In contrast, with 30–35 circular plots visited daily, day-to-day results often varied by 50% or more. At least three censuses of each plot were necessary. When the first circular plot band was 15 m wide, species richness values were generally similar, but BSD and evenness frequently varied widely from transect results. Among ten of the common passerine species, the two techniques gave similar results, but overall, the two techniques were only marginally comparable. Greater similarity

TABLE 7
NUMBER OF CIRCULAR PLOT CENSUSES, TOTAL NUMBER OF SPECIES DETECTED, NUMBER OF SPECIES WITH LESS THAN 50 DETECTIONS, AND NUMBER OF SPECIES WITH INSUFFICIENT NUMBER OF DETECTIONS TO OBTAIN 25% VARIATION AROUND THE MEAN

		Number of species (%)		
Month	Number of plots	Total detected	With <50 detections	With >25% variation around the mean
March	123	32	24 (75)	29 (91)
April Honey mesquite	99	33	26 (79)	28 (85)
April Cottonwood-willow	96	34	21 (62)	30 (88)
May	99	28	13 (46)	20 (71)
June	144	32	14 (44)	21 (66)
Average	131	33	21 (64)	27 (80)

of results was obtained by using a width of 30 m in the first band. Extrapolation from such a small area when narrow inner bands are used often leads to the ballooning effect, i.e., high densities for those species having individuals detected very close to the observer. If calibration of the transect and circular plot techniques is desired, we recommend that the first band have a width of 30 m, although it remains to be demonstrated whether these densities are correct.

Even when adjusted for the larger area censused, detection rates were as great or greater with the transect technique. The idea that censusers spend less time actually looking for birds because they have to watch the transect path is not supported by our data. However, our transects are in level terrain and are well marked.

Six minutes was found to be insufficient for censusing circular plots; seven and eight minutes apparently were better. This violates the assumption that the count period is short enough for birds to occupy fixed locations during count periods.

The transect technique is logistically more feasible in flat land than is the circular plot technique. An equivalent level of accuracy with the circular plot method would require a larger work force and more vehicles. A major advantage of the transect method is that along a given census route, a greater proportion of the area can be censused more quickly and distance estimated more accurately than with the circular plot method. The major advantage of the circular plot method in our study area is that it can be used effectively where patches of different types of vegetation are too small for a transect.

ACKNOWLEDGMENTS

We wish to thank Bea Anderson, J. Anderson, J. Durham, and Drs. J. Meents and C. Robbins for reviewing and editing various drafts of the manuscript. We are grateful to Bea Anderson for typing early drafts and to C. Zisner for typing the final manuscript. We thank Dr. J. Meents, K. Conine, D. Laush, L. Pyc, and especially J. Drake for assistance in collecting and analyzing the data for the report. The work was supported by Water and Power Resources Service Contract No. 7-07-30-V0009.

Studies in Avian Biology No. 6:193–196, 1981.

A FIELD TEST OF TWO DENSITY ESTIMATORS FOR TRANSECT DATA

TIMOTHY E. O'MEARA[1]

ABSTRACT.—Bird populations were studied during the breeding season on two areas of planted slash pine (*Pinus elliottii*) in north-central Florida. Density estimates were calculated for all bird species detected from transects using the spot-mapping method, Emlen's technique, and the three estimators of Järvinen and Väisänen. Chi-square goodness of fit tests indicated that, of the three detectability models tested (linear, negative-exponential, normal), the normal model consistently provided the best fit to the distribution of observed distances. Maximum differences in density estimates averaged only 18% for the three estimators of Järvinen and Väisänen. Density totals were greater for the spot-mapping method than for the transect techniques, yet the transect techniques provided estimates for both breeding and non-breeding species. Emlen's technique and the normal estimator of Järvinen and Väisänen provided similar estimates of species densities. Järvinen and Väisänen's method was the more efficient of the two because it required only determination of bird presence within a specified distance of the transect.

Methods widely used for surveying birds during the breeding season have included spot-mapping and transect techniques. The spot-mapping method (Williams 1936) entails mapping bird observations during several visits to a marked plot. Territorial boundaries are subsequently delineated to estimate the number of territorial males per plot. This is the technique most frequently used by census workers and is generally accepted as the most accurate of the methods available (Robbins 1978a). It has been used in comparative studies of other techniques (Enemar and Sjostrand 1967, Haukioja 1968, J. T. Emlen 1971, Franzreb 1976, Brewer 1978, Dickson 1978, Järvinen et al. 1978).

Counts of birds along transects have been used as indices of relative abundance and for estimating absolute densities (Robbins 1978a, Shields 1979). Numerous estimators for determining density have been proposed. J. T. Emlen's (1971, 1977a) techniques for converting transect data to density estimates have been used widely. With these methods, bird detections are grouped into successive parallel strips on either side of the transect. The distance from the transect to the strip in which the number of detections makes a marked decline is determined; detectability is assumed to be constant out to this point. Density is calculated as a function of the number of birds counted within the area of constant detectability.

Järvinen and Väisänen (1975) developed a method of estimating density from transect data that has apparently received little use in the United States. It assumes detectability is perfect at the transect and declines with lateral distance according to some function. The decline is estimated from the proportion of all detections that occur within a specified distance of the transect. Average detectability and density are determined using this proportion and one of three declining functions (linear, negative-exponential and normal).

In his review of avian census techniques, Shields (1979) indicated the need for simultaneous counts comparing Emlen's and Järvinen and Väisänen's techniques with an independent, more accurate control to facilitate development of a standard technique. To make the comparison, breeding-bird censuses were conducted on two areas of planted slash pine (*Pinus elliottii*) in north-central Florida. Resulting data were analyzed using the two transect estimators and the spot-mapping technique.

STUDY AREA AND METHODS

Two tracts of planted slash pine were chosen for study in Alachua County (Lochloosa area) and Bradford County (Bradford area), Florida. Areas selected had similar overstory characteristics but differed in the amount of understory present. Average DBH of overstory trees on the Bradford area was 20 cm with a basal area of 26 m^2/ha. Woody plant cover within 5 m of the ground was approximately 46% as determined by the line-intercept method (L. F. Conde, pers. commun. 1980). Dominant shrub species included bitter gallberry (*Ilex glabra*), ground blueberry (*Vaccinium myrsinites*), saw palmetto (*Serenoa repens*), and wax myrtle (*Myrica cerifera*).

Average DBH of trees on the Lochloosa area was also 20 cm with a basal area of 22 m^2/ha as determined by distance and diameter measurements of randomly selected trees. This area had been prescribe-burned four months prior to initiation of the study. Woody understory was virtually absent except for scattered oaks (*Quercus* spp.), wax myrtle bushes, and remnants of saw palmetto.

Census plots of 13.25 ha and 12.25 ha were delineated on the Bradford and Lochloosa areas, respectively. Transects were spaced at 50 m intervals crossing the plots and were marked with numbered flags at 25 m intervals along transect lengths. Total transect

[1] School of Forest Resources and Conservation, University of Florida, Gainesville, Florida 32611.

TABLE 1
RESULTS OF χ^2 GOODNESS OF FIT TESTS COMPARING OBSERVED DISTRIBUTIONS OF DISTANCES FROM TRANSECTS ON THE LOCHLOOSA AREA WITH THREE HYPOTHETICAL DISTRIBUTIONS

Species[a]	*N*	Linear[c]	Exponential[c]	Normal[c]
Pine Warbler	114	<0.05	<0.01	NS[b]
Rufous-sided Towhee	75	<0.10	NS	NS
Carolina Wren	33	<0.01	<0.01	<0.10
Cardinal, *Cardinalis cardinalis*	23		NS	NS
Great Crested Flycatcher	22	<0.01	<0.05	<0.10
Red-bellied Woodpecker	19		<0.01	<0.10
Carolina Chickadee, *Parus carolinensis*	9		NS	NS

[a] Names from American Ornithologists' Union (1957, 1973, 1976).
[b] $P > 0.10$.
[c] Probability of encountering a larger χ^2.

lengths were 2.5 km on the Bradford area and 2.3 km on the Lochloosa area.

Transects were sampled eight times on each area in 1979; transects on the Bradford area were also sampled eight times in 1980. Counts were completed within three hours after sunrise in late May and early June. The order in which transects were sampled was rotated daily to avoid confounding time-of-day effects with transect differences.

Locations of all detected birds were recorded on field maps of the study plots. Densities of all breeding species were estimated from these data following the guidelines of the International Bird Census Committee (1970). The perpendicular distance of each bird "pair" from the transect was estimated. Järvinen and Väisänen (1976c) defined a "pair" as a male heard singing or otherwise observed, or if a male is not observed, as a female, group of fledglings or inhabited nest.

Chi-square goodness of fit tests (Snedecor and Cochran 1967:236, Burnham et al. 1980:49) were used to determine which of three models (linear, negative-exponential, normal) best fit the observed distribution of perpendicular distances. Preliminary results indicated that detection frequencies for some species increased slightly with distance from the transect before declining. This phenomenon also was observed by J. T. Emlen (1971) and apparently resulted either from birds near the transect "freezing" as the observer approached, decreasing their detectability, or from a tendency for birds to flush and move away from the transect (Shields 1979). Because the resulting peak in detectability curves occurs within 30 m of the transect for most species (J. T. Emlen 1971), and Järvinen and Väisänen's estimators lump all detections within 25 m of the transect, data were grouped by distance into classes of 0–25 m, 25–35 m, 35–45 m, etc., for the Chi-square tests.

The proportion of all detections that occurred within 25 m of the transect was calculated using data from all transects on each area. Species efficiencies (ratio of birds observed to birds present within 25 m of the transect) were calculated from these proportions (Järvinen and Väisänen 1975). Density estimates were calculated with these efficiencies, using data from alternate transects on each area. Data were restricted to alternate transects to minimize multiple observations of the same bird from more than one transect (Anderson et al. 1979).

Detections of "pairs" were grouped by distance from transect into 5 m intervals for density estimation using Emlen's technique. Basal distances (distance of "inflection" points) were determined for each species using data from all transects (for each area) and the "M5" estimator of Ramsey and Scott (1979). This estimator uses likelihood ratio testing procedures to determine the shortest basal distance such that the likelihood of differing densities within and without the basal distance exceeds four times the likelihood of

TABLE 2
RESULTS OF χ^2 GOODNESS OF FIT TESTS COMPARING OBSERVED DISTRIBUTIONS OF DISTANCES FROM TRANSECTS ON THE BRADFORD AREA WITH THREE HYPOTHETICAL DISTRIBUTIONS

Species[a]	*N*	Linear[b]	Exponential[b]	Normal[b]
Rufous-sided Towhee	261	<0.01	<0.01	<0.05
Pine Warbler	135	<0.01	<0.01	<0.10
Common Yellowthroat	120	<0.01	<0.01	NS[a]
Carolina Wren	118	<0.01	<0.01	<0.10
Great Crested Flycatcher	110	<0.01	<0.01	NS
White-eyed Vireo	46	<0.01	NS	NS
Blue Jay	26	<0.01	<0.10	NS
Red-bellied Woodpecker	24	<0.01	<0.05	NS

[a] $P > 0.10$.
[b] Probability of encountering a larger χ^2.

TABLE 3

DENSITIES (PAIRS/KM^2 ± SE) OF BIRD SPECIES ON THE LOCHLOOSA AREA IN 1979 AS ESTIMATED BY THE SPOT-MAPPING METHOD, JÄRVINEN AND VÄISÄNEN'S THREE ESTIMATORS, AND EMLEN'S TECHNIQUE

Species	Spot-mapping	Järvinen and Väisänen Linear	Järvinen and Väisänen Exponential	Järvinen and Väisänen Normal	Emlen
Pine Warbler	49.0	59.5 ± 8.6	68.8 ± 11.9	55.2 ± 7.7	53.6 ± 15.5
Rufous-sided Towhee	32.7	28.7 ± 7.1	32.3 ± 8.6	26.8 ± 6.5	15.9 ± 15.7
Carolina Wren	12.2	2.4 ± 3.6	2.5 ± 3.9	2.5 ± 3.4	10.2 ± 8.4
Cardinal	8.2	5.5 ± 6.7	5.8 ± 1.0	5.2 ± 14.3	3.6 ± 2.6
Downy Woodpecker, *Picoides pubescens*	8.2	4.0 ± 6.5	4.5	3.7 ± 14.0	7.9
Red-bellied Woodpecker	8.2	2.4 ± 1.2	2.5 ± 1.5	2.3 ± 1.1	2.4 ± 3.0
Great Crested Flycatcher	8.2	1.7 ± 2.6	1.8 ± 3.5	1.6 ± 2.4	4.0 ± 2.3
Other species (n = 7)		18.5	20.9	17.3	23.1
Total	126.7	122.7	139.1	114.6	120.7

equal densities. On the Bradford area, where data from two breeding seasons were available, coefficients of detectability were calculated for birds within 100 m of the transect. Density estimates were subsequently obtained with these coefficients using data from alternate transects only. On the Lochloosa area, where fewer data were available for calculating coefficients of detectability, the "M5" estimator was used for determining basal distance. Densities were estimated as the number of birds divided by the area within the basal distance using data from alternate transects. On both areas, where insufficient data were available to estimate basal distance for a species, the method of Balph et al. (1977) was used to estimate density.

RESULTS AND DISCUSSION

The normal model was the only one of the three tested which did not differ ($P > 0.05$) from the distribution of observed distances for most species (Tables 1 and 2). The only exception was the Rufous-sided Towhee (*Pipilo erythrophthalmus*) on the Bradford area. Decline in detectability most closely followed the normal model despite the fact that understory densities were markedly different on the two areas, visibility was impaired less by vegetation on the Lochloosa area, and total bird densities on the Bradford area were almost double those on the Lochloosa area.

Although goodness of fit differed significantly for the three models, maximum differences in density estimates based on the three models averaged only 18%. Estimates were greatest with the negative-exponential estimator and smallest with the normal estimator with the exception of species for which all detections occurred within the main belt (≤25 m) (Tables 3, 4, and 5). These results are similar to those of Järvinen and Väisänen (1975) and suggest that consistent results can be obtained regardless of the model used. Järvinen and Väisänen (1975) observed differences when using the negative-exponential model with "short-distance" species and advised caution when using this model with this group.

Density totals were generally greater for the spot-mapping technique than for the transect

TABLE 4

DENSITIES (PAIRS/KM^2 ± SE) OF BIRD SPECIES ON THE BRADFORD AREA IN 1979 AS ESTIMATED BY THE SPOT-MAPPING METHOD, JÄRVINEN AND VÄISÄNEN'S THREE ESTIMATORS, AND EMLEN'S TECHNIQUE

Species	Spot-mapping	Järvinen and Väisänen Linear	Järvinen and Väisänen Exponential	Järvinen and Väisänen Normal	Emlen
Rufous-sided Towhee	71.7	54.9 ± 3.8	63.1 ± 5.0	50.7 ± 3.4	50.1 ± 8.8
Pine Warbler	41.5	29.9 ± 3.6	35.0 ± 4.4	27.5 ± 3.3	22.4 ± 6.6
Great Crested Flycatcher	34.0	36.6 ± 6.5	42.0 ± 8.7	33.8 ± 6.0	32.9 ± 13.3
Carolina Wren	34.0	16.7 ± 3.9	17.8 ± 4.3	16.0 ± 3.6	17.0 ± 8.7
Common Yellowthroat	15.1	17.7 ± 2.7	19.6 ± 3.1	16.6 ± 2.5	17.2 ± 5.7
White-eyed Vireo	11.3	5.8 ± 1.1	6.4 ± 1.2	5.4 ± 1.0	5.7 ± 2.2
Blue Jay	11.3	4.5 ± 2.1	5.0 ± 3.7	4.2 ± 2.0	7.0 ± 7.3
Brown Thrasher	11.3	2.9 ± 1.3	3.9 ± 1.3	2.7 ± 2.6	2.4 ± 1.8
Summer Tanager, *Piranga rubra*	7.6	3.7 ± 0.9	4.1 ± 1.1	3.4 ± 0.8	1.4 ± 1.5
Other species (n = 9)		31.6	29.8	36.2	34.7
Total	237.8	204.3	226.7	196.5	190.5

TABLE 5

DENSITIES (PAIRS/KM^2 ± SE) OF BIRD SPECIES ON THE BRADFORD AREA IN 1980 AS ESTIMATED BY THE SPOT-MAPPING METHOD, JÄRVINEN AND VÄISÄANEN'S THREE ESTIMATORS, AND EMLEN'S TECHNIQUE

Species	Spot-mapping	Järvinen and Väisänen			Emlen
		Linear	Exponential	Normal	
Rufous-sided Towhee, *Pipilo erythrophthalmus*	98.1	69.9 ± 4.2	80.4 ± 5.7	64.5 ± 3.7	67.8 ± 14.2
Pine Warbler, *Dendroica pinus*	67.9	43.8 ± 3.2	51.4 ± 4.2	40.3 ± 2.8	32.9 ± 12.3
Carolina Wren, *Thyrothorus ludovicianus*	37.7	12.1 ± 2.4	12.9 ± 2.7	11.5 ± 2.2	14.2 ± 7.8
Common Yellowthroat, *Geothlypis trichas*	34.0	22.6 ± 1.6	25.0 ± 2.0	21.2 ± 1.5	25.5 ± 8.0
White-eyed Vireo, *Vireo griseus*	30.2	10.3 ± 0.6	11.4 ± 0.7	9.6 ± 0.6	10.1 ± 2.8
Great Crested Flycatcher, *Myiarchus crinitus*	22.6	18.3 ± 2.4	21.0 ± 3.6	16.9 ± 2.1	16.8 ± 9.1
Red-bellied Woodpecker, *Melanerpes carolinus*	22.6	8.4 ± 5.1	9.4 ± 1.6	7.8 ± 5.3	9.7 ± 3.9
Brown Thrasher, *Toxostoma rufum*	7.5	7.3 ± 1.3	9.7 ± 1.3	6.7 ± 11.7	6.1 ± 2.0
Bobwhite, *Colinus virginianus*	7.5	3.3 ± 1.4	3.9 ± 0.6	3.0 ± 3.3	11.1 ± 14.1
Blue Jay, *Cyanocitta cristata*	7.5	3.2 ± 1.4	3.5 ± 2.6	3.0 ± 1.4	5.0 ± 5.5
Brown-headed Nuthatch, *Sitta pusilla*	7.5	1.2 ± 0.5	1.4 ± 0.6	1.1 ± 0.4	2.8 ± 2.5
Other species (n = 10)		23.9	8.2	39.6	11.1
Total	343.1	224.3	238.2	225.2	213.1

methods, yet the spot-mapping technique provided density estimates for only 50% of the species for which densities were available from the transect data. Density estimates for breeding species were greater from the spot-mapping technique than from the transect techniques in almost all cases, accounting for differences in totals. The transect techniques apparently provided estimates for both non-breeding and breeding species, while underestimating species densities relative to the spot-mapping technique.

J. T. Emlen's (1971) technique and the normal estimator provided similar estimates of density on both areas. Basal detectability adjustments (the factor required to convert transect estimates to spot-mapping estimates) averaged over both areas were 2.5 for Järvinen and Väisänen's technique and 1.9 for Emlen's technique. Järvinen and Väisänen (1975) predicted basal detectability adjustments should average 1.3 to 1.6 and Järvinen et al. (1978a) reported a value of 1.4 for censuses in mountain birch forests. Their values are based on the linear model which gives greater estimates of density than the normal estimator. Both values reported here are within the range of 1.1 to 2.5 predicted by J. T. Emlen (1971). Franzreb's (1976) comparison of Emlen's technique and the spot-mapping technique in a mixed coniferous forest indicated an average basal detectability adjustment of 2.0. The tendency for the transect techniques to give smaller estimates than the spot-mapping technique, resulting in basal detectability adjustments greater than one, is apparently a result of incomplete detection of birds at or near the transect (J. T. Emlen 1971, Järvinen and Väisänen 1975).

Although density estimates were similar for the normal estimator and Emlen's technique, the normal estimator gave more precise estimates based on the same data. Standard errors for the normal estimator were generally less than half those for Emlen's technique.

These data indicate that similar results can be obtained between Emlen's and Järvinen and Väisänen's techniques. Järvinen and Väisänen's technique is less laborious to use in the field, however, as it requires only determination of whether detected birds are within a specified distance from the transect. Emlen's technique requires estimating distance to all birds detected so they can be grouped into successive "belts" by distance. Järvinen and Väisänen's technique is more efficient in that less time is required for distance estimation and resulting density estimates are more precise with the same sampling effort. Whichever transect technique is used, the census taker should bear in mind that both underestimate actual densities and both may require species specific correction factors to make estimates reflect true densities.

ACKNOWLEDGMENTS

I am indebted to L. A. Rowse for her unrewarded assistance in conducting censuses in 1979. R. C. Littell and P. V. Rao provided advice on statistical analyses. R. J. Grable assisted with computer programming. M. W. Collopy, W. R. Marion, and T. E. Rogers reviewed the manuscript. Contribution 4001, Journal Series, Florida Agricultural Experiment Station, Gainesville.

Studies in Avian Biology No. 6:197–201, 1981.

AN EVALUATION OF BREEDING SEASON CENSUS TECHNIQUES FOR LONG-BILLED CURLEWS (*NUMENIUS AMERICANUS*)

ROLAND L. REDMOND, THOMAS K. BICAK, AND DONALD A. JENNI[1]

ABSTRACT.—We compare results of spot-mapping, variable distance strip transect, and Finnish line transect census techniques for Long-billed Curlews on a 1600 ha study plot during the 1978 and 1979 breeding seasons. Spot-mapping of curlews seen during the arrival through incubation period (late March through April) provided estimates of 6.53 and 6.38 territorial males/100 ha in 1978 and 1979, respectively. Calculated densities based on one km wide strip transects (covering about 51% of the study plot) were 7.02 and 6.60 males/100 ha for the same respective periods. Densities calculated according to the Finnish line transect method (using the same data base as the strip transects) were 6.42 and 5.94 males/100 ha for the same periods.

Prior to hatching of their clutches, female curlews are less detectable than males and cannot be accurately censused by either transect method until the brood rearing period (mid May to late June). During this later period, however, both transect methods overestimate male density because of male mobbing behavior.

Assuming that our spot-map results best estimated male curlew densities at the start of these breeding seasons, then density of males attempting to breed in an area can be most reliably and efficiently estimated by the Finnish line transect method. Proper counting of females during the brood rearing period can provide an estimate of successfully nesting individuals, and hence an index of annual nesting productivity.

Long-billed Curlews are large and conspicuous shorebirds which breed on patches of short grass prairie across western North America. Management of this important non-game species could be facilitated and improved if breeding densities and distribution were better known. However, censusing of Long-billed Curlews is complicated by aspects of their breeding behavior and social organization.

The objectives of this paper are: (1) to compare and evaluate results from spot-mapping, variable distance strip transect, and Finnish line transect census methods for Long-billed Curlews during two complete breeding seasons; and (2) to outline an efficient and accurate technique for censusing these birds during their breeding season.

METHODS

The study area was a short grass rangeland located between the Boise and Payette River valleys about 11 km NNW of Middleton, Idaho (SW ¼ of T. 6 N., R. 3 W., Boise Meridian). The area was characterized by choppy to rolling topography, a mean elevation of 810 m, and a semi-desert type of vegetation. Dominant plant species were cheatgrass brome (*Bromus tectorum*) and medusahead wildrye (*Taeniatherum asperum*) as well as several perennial buchgrass species.

Soon after the birds' arrival in late March of 1978 and 1979, we began driving a standardized route (21.7 km) along pre-existing roads through the study area. We drove the route daily (avoiding periods of rain and high wind) until egg laying began in early April, and then twice weekly for the remainder of the season. One person drove the route, and we varied starting time equally among early morning (07:00–09:00), midday (11:00–13:00), and afternoon (15:00–17:00) periods. Locations of all curlews sighted along the route were recorded on maps according to sex and by date.

From this standardized route, we selected six relatively straight segments or transects totalling 8.23 km such that: (1) there was no overlap among 0.5 km strips on either side of these transects (Fig. 1), and (2) all habitats within the study area were sampled in approximate proportion to their occurrence. In spite of the non-random nature of this sampling procedure, we justify the approach for practical reasons given the size and homogeneity of the study area, as well as the uniform distribution of Long-billed Curlews within it.

We measured the lateral distance of each sighting perpendicular to a transect. From these lateral distances we calculated coefficients of detectability at 500 m (CD_{500}) (J. T. Emlen 1971) as well as ratios (MB) of sightings within a 25 m "main belt" to all sightings regardless of lateral distance (Järvinen and Väisänen 1975).

The Finnish line transect method (Järvinen 1976, Järvinen and Väisänen 1973) specified just one sampling of a transect. In departure from this method, however, we pooled all sightings from our transect surveys within each time period to determine single MB ratios. We then calculated density estimates according to Järvinen and Väisänen (1975). For the arrival through incubation period we based the calculations on maximum numbers on males and females seen on any one day. During the brood rearing period, however, the territorial dispersion pattern changed dramatically with brood movements, and daily tallies for both males and females tended to be more variable. Thus, for this period our calculations were based on an average of the three highest tallies obtained.

Investigation of curlew breeding biology and behavior was concentrated within this 1600 ha plot during 1978 and 1979. During the arrival through incubation periods (~32–40 days), we surveyed different portions of the area on 20 and 25 days in 1978 and 1979, respectively. Such a survey was carried out coincidently with other research objectives and entailed registering

[1] Department of Zoology, University of Montana, Missoula, Montana 59812.

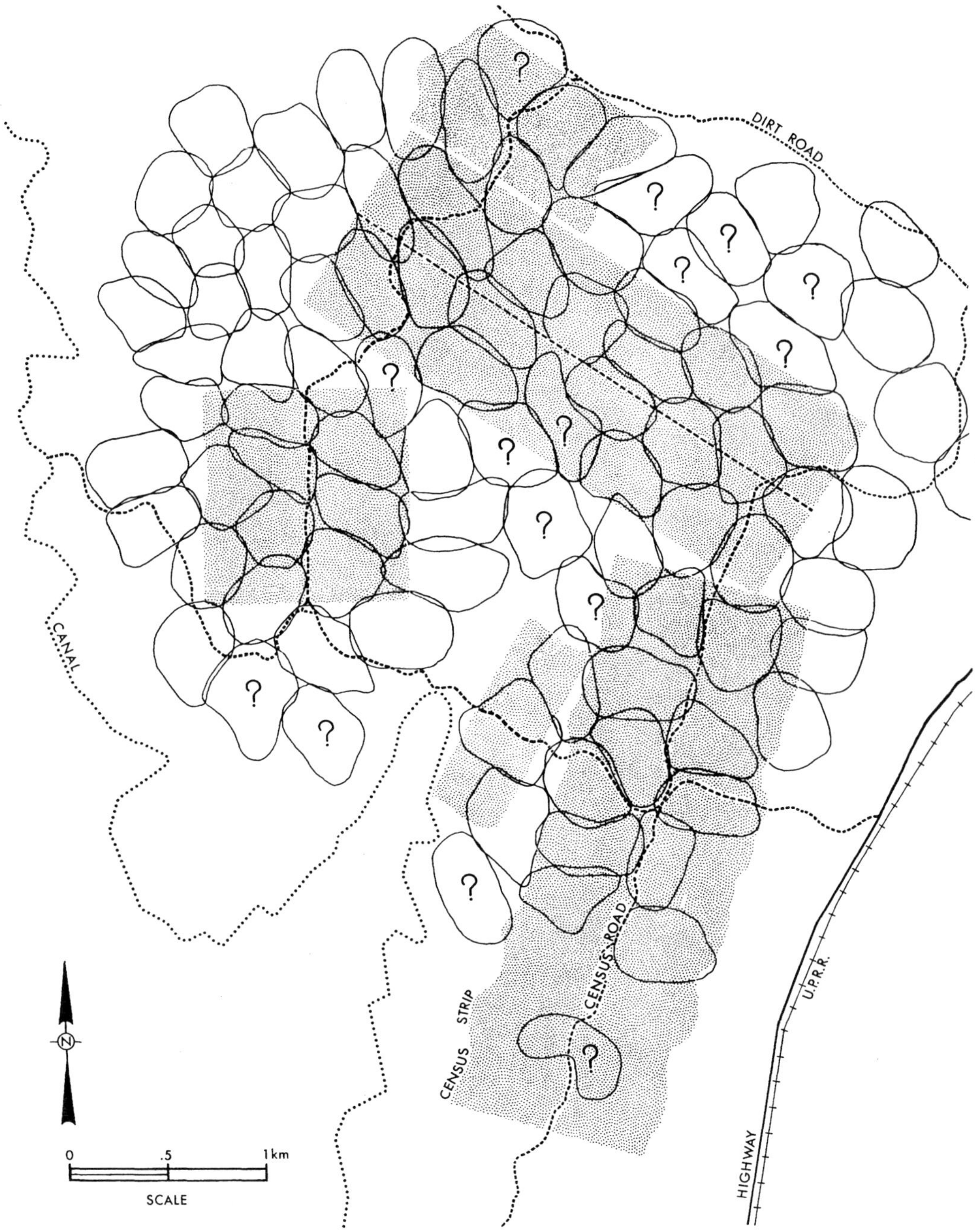

FIGURE 1. Major portion of Long-billed Curlew study plot showing census route, transects, and spot-mapped territories (for 1979) (? indicates that no female was ever observed on a territory).

all daily curlew sightings by sex and date onto visit maps. We tried to walk or drive through all portions of the study area on a regular basis (every 7–10 days), but this was not always possible because of other responsibilities and a lack of manpower. While some regions were surveyed more frequently and intensively than others, we were careful to sample all regions at least twice during the arrival through incubation periods each year. Again we avoided rain and high winds when conducting these surveys.

From these daily visit maps we compiled weekly composite maps of the curlew observations. By in-

TABLE 1
DENSITY ESTIMATES OF LONG-BILLED CURLEWS—SPOT-MAPPING

Year	Ha sampled	Territorial males	% males paired[a]	Males per 100 ha	Females per 100 ha
1978	1592	104	79	6.53	5.16
1979	1584	101	85	6.38	5.43

[a] If a female was observed at least once in an area delineated as a territory, then the male occupying that area was classified as paired.

corporating data on known pairs (color-marked as well as radio-marked), nest sites (33 and 25 in 1978 and 1979, respectively), topography, vegetation, and our intimate knowledge of the area, we interpreted these composite maps and delineated "territories" (sensu Robbins 1970) for at least those males displaying over the study plot. The range of registrations used in delimiting a territory was 3–18.

RESULTS

Density estimates from the different methods are given in Tables 1–4. There is little difference among the calculated densities for male curlews during the arrival through incubation period. Female densities during this period, however, were severely underestimated by all methods.

During the brood rearing periods of 1978 and 1979, both the male and female density estimates from both transect methods increased dramatically above those of the earlier time period. The magnitude of both increases was greater with the Finnish line transect method. Absolute density estimates from the Finnish method were greater in 1978; whereas in 1979, estimates from the variable distance strip transect method were greater.

For comparative purposes we selected the linear model as best representing the decline in Long-billed Curlew detectability with increasing lateral distance from the transect line. We found no significant differences among the density estimates (within a time period and by sex) calculated from the linear, negative exponential, or normal models of Järvinen and Väisänen (1975). It may be, however, that the normal model is a more appropriate choice given the openness of the habitat and the large, conspicuous nature of the birds.

DISCUSSION

Assuming that the spot-map results provide the best annual density estimate of territorial male curlews, then both transect methods permit a comparable yet far more efficient estimation of this parameter. The Finnish line transect technique, as described by Järvinen and Väisänen (1973 and 1976c), is easier to perform because one need only tally observations according to whether they occur inside or outside a 25 m belt. The variable distance strip transect method requires that all lateral distances be estimated in order to determine an inflection point for the calculation of CD ratios (J. T. Emlen 1971). It is also more reasonable to assume complete detection of curlews within a strip 25 m wide than 125 m (determined as our inflection point).

Density of female curlews attempting to breed is an elusive parameter. Female detectability tends to be low prior to the brood rearing period because females do not perform display flights upon arrival, instead they spend most of their time on the ground feeding. Furthermore, after completion of their clutches, females incubate by day and are very difficult to flush from their nests (Redmond and Jenni, unpubl. data).

We were unable to quantify a non-breeding component to this population, but we did confirm that some males were unsuccessful at attracting a mate. It may be that males return to the breeding ground at an earlier age than females, and that they are present as non-breeders for one or two years. If so, then there may be a slight excess of males at least until mid May.

We did observe females at least once on 79% and 85% of the spot-mapped territories in 1978 and 1979, respectively. These data are not reliable, however, because of incomplete sampling

TABLE 2
PARAMETERS AND DENSITY ESTIMATES OF LONG-BILLED CURLEWS—VARIABLE DISTANCE STRIP TRANSECT METHOD

Year	Ha sampled	Dates	No. seen ♂♂	No. seen ♀♀	CD_{500}[a] ♂♂	CD_{500}[a] ♀♀	Nos. per 100 ha ♂♂	Nos. per 100 ha ♀♀
1978	819.4	24 Mar–25 Apr	32	13	0.557	0.592	7.02	2.68
		13 May–18 Jun	40.7	24.0	0.435	0.436	11.4	6.71
1979	811.7	25 Mar–4 May	30	15	0.560	0.604	6.60	3.06
		16 May–26 Jun	24.3	15.0	0.432	0.409	6.93	4.52

[a] Coefficient of detectability at 500 m: number seen within 500 m/4 × number seen within 125 m.

TABLE 3
PARAMETERS AND DENSITY ESTIMATES OF MALE LONG-BILLED CURLEWS—FINNISH LINE TRANSECT METHOD (LINEAR MODEL)

Year	Dates	N[a] (#♂♂)	MB[b] (s_{MB})	$1/k$[c] (E_1)	♂♂/100 ha	SD[d]
1978	24 Mar–25 Apr	198 (32)	0.0808 (0.0194)	606 (0.979)	6.42	1.58
	13 May–18 Jun	322 (40.7)	0.121 (0.0182)	400 (0.969)	12.4	1.90
1979	25 Mar–4 May	251 (30)	0.0797 (0.0171)	614 (0.980)	5.94	1.30
	16 May–26 Jun	204 (24.3)	0.0882 (0.0199)	554 (0.977)	5.33	1.24

[a] N = total no. of males seen along transects during time period; #♂♂ = no. of males on which estimate is based (see Methods text).
[b] MB = no. sighted within 25 m main belt/N; s_{MB} = SD of MB ratio.
[c] k is a constant and E_1 the efficiency of the normal model.
[d] SD of estimates based on s_{MB} values (see Järvinen and Väisänen 1975).

of the study plot each year. The 85% figure for 1979 probably reflects increased manpower and sampling effort more than any real difference in breeding pair density between the years. We suspect that breeding pair density lies somewhere between 85% and 100% of the figure for male density obtained by censusing during the arrival through incubation period.

Accurate censusing of curlews by either transect method during the brood rearing period is complicated by several features of the birds' behavior. Male densities were consistently overestimated by both transect methods during brood rearing because of cooperative mobbing behavior. Any intruder resembling a potential predator is mobbed aggressively by adults attending chicks, and males in particular may be attracted from 500 m or more to participate in these efforts. Detectability of females changes abruptly after the hatching of their clutch, but females whose nests are destroyed may or may not depart promptly from the breeding ground (Redmond and Jenni, unpubl. data). These non-parental females do not appear to participate in mobbing efforts, and therefore, any that do remain past late May should be distinguishable from those females attending broods. In 1978 more females remained on our study plot during the brood rearing period than did so in 1979. Nest success rate was slightly greater in 1978 (Redmond and Jenni, unpubl. data), but not enough to account for the difference in female density estimates between years. We suggest that the 1979 female density estimates for the brood rearing period reflect number of broods produced rather than number of breeding females. The transect data for females from the 1978 brood rearing period must be reanalyzed, omitting the non-parental females before any conclusions can be drawn. We do expect, however, that careful counting of females during brood rearing with the Finnish line transect

TABLE 4
PARAMETERS AND DENSITY ESTIMATES OF FEMALE LONG-BILLED CURLEWS—FINNISH LINE TRANSECT METHOD (LINEAR MODEL)[a]

Year	Dates	N (♀♀)	MB (s_{MB})	$1/k$ (E_1)	♀♀/100 ha	SD
1978	24 Mar–25 Apr	90 (13)	0.0667 (0.0263)	737 (0.983)	2.14	0.87
	13 May–18 Jun	150 (26)	0.113 (0.0259)	428 (0.971)	7.38	1.74
1979	25 Mar–4 May	116 (15)	0.0517 (0.0206)	954 (0.987)	1.91	0.77
	16 May–26 Jun	131 (17)	0.0687 (0.0221)	715 (0.983)	2.89	0.94

[a] See Table 3 for explanation of column headings.

method can provide at least an annual productivity index, and perhaps a meaningful estimate of broods per 100 ha.

We conclude that male Long-billed Curlews can be counted accurately and efficiently during the arrival through incubation period by a modified Finnish line transect method. We suggest that a minimum of five counts be conducted on separate days and in fair weather to insure an adequate sample size. The sequence should begin at the peak of curlew arrival (late March or early April) and end within two weeks. To index nesting productivity, the same transects should be sampled for parental females beginning in late May or early June. Some experience with female parental behavior may be necessary before one can easily distinguish between parental and nonparental females. Finally, care should be exercised in extrapolating density estimates over large areas, particularly if the sampling effort is relatively small.

ACKNOWLEDGMENTS

This research was supported by the U.S.D.I., Bureau of Land Management, Contract No. YA-512-CT7-54, administered through the Boise District Office. We are particularly grateful to A. Sands, H. D. Sherrets, M. Roth, and A. Tripp for their invaluable encouragement and cooperation with all aspects of this research.

We thank M. W. Shantz and J. W. Parker for assistance with some of the fieldwork and D. Williams for preparation of the figure. R. L. Hutto and C. S. Robbins provided helpful comments on earlier drafts of this paper.

Studies in Avian Biology No. 6:202–208, 1981.

COMPARISON OF LINE-TRANSECT METHODS FOR ESTIMATING BREEDING BIRD DENSITIES IN DECIDUOUS WOODLOTS

NANCY G. TILGHMAN[1] AND DONALD H. RUSCH[2]

ABSTRACT.—We compared the relative bias, sample variance, and cost of 12 line-transect methods used to estimate the density of 10 breeding bird species in woodlots of south-central Wisconsin. The relative bias and variance of bird density estimates varied with the method and species involved; no method gave estimates within an order of magnitude of our best estimate of density of breeding Eastern Kingbirds as determined by the mapping of territories. Five methods provided density estimates for all 10 species with average relative bias of less than 20%. Estimates provided by all transect methods were fairly precise; estimates from only three methods gave coefficients of variation (CV) of more than 40%. CVs varied more among species than among methods. Line-transects provided estimates of breeding bird densities at considerable savings in effort over that required by mapping of bird territories. Our data suggest that several line-transect methods could consistently provide estimates of the breeding density of selected species which were within an order of magnitude of actual densities, but no method provided relatively unbiased, precise estimates of breeding densities of each of the 10 species.

Data on the abundance of wildlife populations are critically important in the preparation of environmental assessments including impact statements. With the passing of the National Environmental Policy Act in 1969 and new impacts of accelerated energy development, the demand for reliable techniques for detection of areal and temporal differences in wildlife population levels has greatly increased. Present methods of wildlife impact assessment generally are either qualitative (based only on species lists) or subjective (based only on the expertise and knowledge of the field worker; Leopold et al. 1971, Daniel and Lamaire 1974). If practical methods were available, evaluation of impacts associated with proposed projects would be improved by quantitative and objective measurements of the wildlife resource.

Although techniques for censusing game species have received the most emphasis in the past century, several techniques have also been developed to measure the absolute and relative abundances of songbird populations. These methods vary widely in both degree of bias and sample variance. Generally, the mapping of territories of singing males (Williams 1936, Enemar 1959, Robbins 1970) and locating all nests (Schiermann 1930, 1934; Lack 1935) are considered the least biased methods. These methods are presumed to be accurate because they require many hours of field work and familiarity with an area and its bird populations, but true population densities, and hence degree of bias, are in fact unknown. The time required to employ these methods is extensive. For example, the time needed to census a 9.7 ha plot by the territorial mapping method is about 30 hours (Emlen 1977a). Line-transect methods are generally acknowledged to be less time-consuming, but the bias and variance associated with line-transect estimates is often deemed unacceptable (Enemar 1959). In general, however, the acceptability of an estimate depends upon how it will be used; bias, variance and cost must be evaluated in relation to the data needs.

Some authors have examined the relative merits and inadequacies of line-transect techniques, but few of these evaluations were based on actual field work with animals and even fewer dealt with songbirds, which are subject to problems of detectability and mobility. In an early study, Amman and Baldwin (1960) compared the accuracy and variation of five line-transect methods when used to estimate woodpecker density. Gates (1969) used computer simulation to evaluate the bias and variance of six line-transect methods. J. T. Emlen (1971) reviewed the general characteristics and applicability of seven transect methods to estimate songbird numbers, but conducted a limited field evaluation on only three methods. Robinette et al. (1974) evaluated the bias of 10 different line-transect methods, but most of their field studies involved censuses of wooden blocks. Seber (1973), Eberhardt (1978), and Gates (1979) have provided detailed reviews of the development and theory of several line-transect methods. Of these only the latter included an evaluation of the methods by examination of original field data; he did not, however, examine any data from extremely mobile populations such as songbirds. Burnham et al. (1980) discussed the properties and attributes of several line-transect methods and have described some of the more robust methods in detail. Both artificial and field data were used to illustrate the bias and sampling variance of these methods; but, again, the illustrations did not in-

[1] Present address: U.S. Forest Service, Northeastern Forest Experiment Station, Amherst, Massachusetts 01003.

[2] Present address: Wisconsin Cooperative Wildlife Research Unit, University of Wisconsin, Madison, Wisconsin 53706.

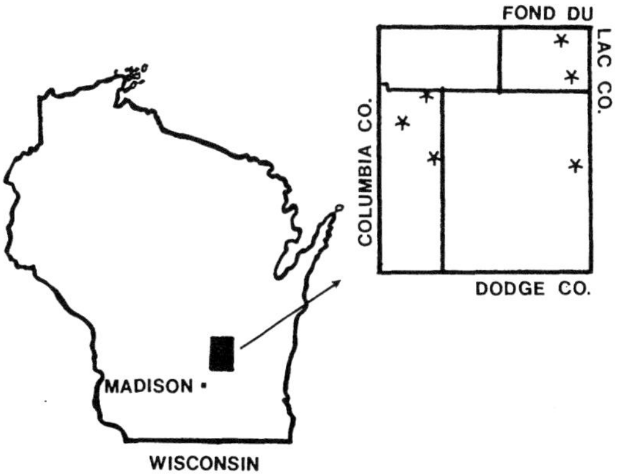

FIGURE 1. The study area in south-central Wisconsin showing the location of six deciduous woodlots (*).

clude any data demonstrating the applicability of these methods to songbirds.

In recent years, use of line-transect techniques to estimate breeding bird densities has increased, but the bias and variance of many of the methods employed have yet to be adequately evaluated in different habitats. The bias and/or variance of some line-transect estimates of breeding bird density have been measured and evaluated in a deciduous forest (J. T. Emlen 1971), a mixed-coniferous forest (Franzreb 1976) and sagebrush and grassland areas (Mikol et al. 1979). The present study was designed to evaluate bias, variance and cost of several transect techniques used to estimate the densities of several breeding bird species in deciduous woodlots of south-central Wisconsin.

STUDY AREAS AND METHODS

The field work for this study was conducted on six deciduous woodlots in Columbia, Dodge and Fond du Lac counties in south-central Wisconsin (Fig. 1). The ramdomly selected woodlots were all privately owned and small (8.9 to 16.2 ha). Less than 12% of the land in these counties is forested (Spencer and Thorne 1972). These wooded areas can be characterized as small scattered islands of trees surrounded by agricultural lands; these woodlots are either remnants of the presettlement forests or abandoned land reverting to forest (Curtis 1956). The dominant tree species of these woodlots are shagbark hickory (*Carya ovata*), white oak (*Quercus alba*), bur oak (*Q. macrocarpa*), black oak (*Q. velutina*), American elm (*Ulmus americana*), black cherry (*Prunus serotina*), sugar maple (*Acer saccharum*), red maple (*A. rubrum*), and silver maple (*A. saccharinum*).

We compared the abilities of 12 line-transect techniques (Table 1) to estimate breeding bird densities. The relationships among the variables given in the formulas of these methods are shown in Fig. 2. The first 10 methods listed were also evaluated by Robinette et al. (1974) and include methods originally developed to

TABLE 1
LINE-TRANSECT METHODS USED TO ESTIMATE DENSITY OF BIRDS IN WOODLOTS OF SOUTH-CENTRAL WISCONSIN

Method	Formula[a]	Source
	Based on sighting distances and angles	
King	$\hat{D} = n/2L\bar{R}$	Leopold (1933)
Hayne	$\hat{D} = n/2L\bar{H}$	Hayne (1949)
Gates II	$\hat{D} = (2n - 1)/2L\bar{R}$	Gates (1969)
Gates III	$\hat{D} = n/2L\bar{G}$	Gates (1969)
Webb	$\hat{D} = n/2L\bar{R} \sin \bar{T}$	Webb (1942)
	Based on perpendicular distances	
Leopold	$\hat{D} = n/2L\bar{Y}$	Leopold et al. (1951)
Gates I	$\hat{D} = (n - 1)/2L\bar{Y}$	Gates et al. (1968)
Frye	$\hat{D} = n_a/2L\bar{Y}$	Overton (1971)
Kelker	$\hat{D} = n_b/2L\Delta$	Kelker (1945)
Anderson & Pospahala	$\hat{D} = n_c/2LW$	Anderson & Pospahala (1970)
Emlen I	$\hat{D} = (n_d/2LW)/C$	Emlen (1971a)
Emlen II	$\hat{D} = (n_e/2LS)/F$ or $\hat{D} = n_f/2LS$	Emlen (1977a)

[a] Definitions: C = coefficient of detectability, a correction factor representing the proportion of animals detected; $\hat{D}$ = estimated density of the animal population; F = frequency of cue production for the species under study (ratio of number of visits on which a particular bird was detected to the total number of visits to that bird's territory); $\bar{G}$ = geometric mean of sighting distances; $\bar{H}$ = harmonic mean of sighting distances; L = total length of transect lines; n = number of animals detected; n_a = number of animals detected within $\bar{Y}$ units of the transect; n_b = number of animals detected within Δ units of the transect; n_c = (y intercept of the negative regression of the midpoint of perpendicular distance classes on the number of animals detected per class) multiplied by the number of classes; n_d = either the number of animals detected within W units of the transect, or [the number of singing males (birds) detected within W units of the transect] multiplied by 2, whichever is larger; n_e = (number of singing males within S units of the transect) multiplied by 2; n_f = number of animals (birds) detected within S units of the transect; $\bar{R}$ = mean sighting distance; S = species specific perpendicular distance within which most individuals were detected; $\bar{T}$ = mean sighting angle; W = width of strip surveyed (on either side of the transect); $\bar{Y}$ = mean perpendicular distance; Δ = estimated perpendicular distance within which all animals were detected.

census deer, hares, grouse, quail, deer carcasses and nests. We evaluated two additional transect methods designed specifically for birds—Emlen I, the "coefficient of detectability" method (J. T. Emlen 1971) and Emlen II, the specific strip method (Emlen 1977a).

To evaluate the bias of these estimates of breeding bird densities, we needed baseline data on the actual density of birds in these woodlots. The mobility of birds and the lack of practicable methods render this goal unattainable, thus we used two times the number of territorial males (as estimated by the spot-mapping method, Robbins 1970) as our best estimate of the actual density of breeding birds in the woodlots.

The 10 breeding bird species that we studied (Table 2) were selected for their territorial habits and high frequency of occurrence. Bird surveys in each woodlot were completed during 10-day periods beginning with the first woodlot in mid-May and continuing until

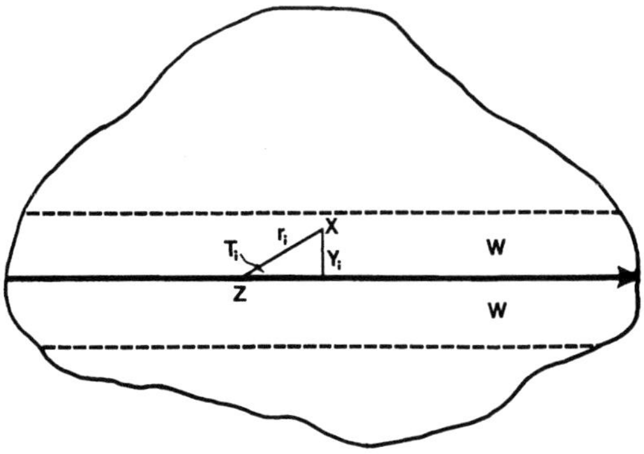

FIGURE 2. Terminology and notation used in line-transect methods. (Solid line = transect of length L; W = maximum width within which birds were recorded; Z = position of observer; X = location of bird detected; r_i = sighting distance; T_i = sighting angle; Y_i = perpendicular distance.)

the last woodlot was censused in mid-July 1976. All surveys were run on clear calm days, during the period from one-half hour before to five hours after sunrise.

Transect bird counts were conducted during the first five days of the survey period for each woodlot. Data for all line-transect techniques were gathered simultaneously. Individual observers walked transect lines originating and ending at sites identified on aerial photos; compass bearings were followed and all birds heard or seen within an estimated lateral distance of 50 m from the transect were recorded. Sighting distance, sighting angle and number of birds were recorded. The other values needed to calculate estimates from the various transect methods were calculated from these data collected in the field. Small size of the woodlots and consequent low numbers of breeding birds forced us to pool the woodlots into one sample area in order to calculate the transect estimates. Data gathered on the first visit to each of the six woodlots were combined and used to derive a density estimate for the 12 transect methods. Data from each of the other four days' visits were treated similarly. The total length of transects in all six woodlots averaged 8.6 km per visit. Thus for every mapping estimate of the density of a given species on all six woodlots, there are up to five corresponding density estimates (daily replicates) from each transect method. Bird density could not always be estimated from transect data collected for all five days' visits to the areas. If fewer than five individuals of a given species were recorded on the *n*th visit to all six areas, the data were considered insufficient and no estimates were calculated for that species and day.

Intensive mapping surveys were conducted on all 10 mornings of the survey period for each woodlot. The location and behavior of breeding birds were recorded and, later, interpreted according to the guidelines set forth by Robbins (1970). The mapping method cannot be expected to give completely accurate bird density estimates (see Bell et al. 1973, Best 1975 for discussion of the limitations of this method), but we assume that these mapping estimates closely approximated the actual breeding bird densities. Observational and interpretational biases were minimized by having the same observer conduct all mapping surveys and interpret all species maps.

The 12 line-transect methods were evaluated by comparison of relative biases, coefficients of variation and costs associated with each method. Relative bias (RB) of the transect estimates versus the mapping estimates of bird densities were calculated by RB = $100 \times (\hat{D} - D)/D$, where $\hat{D}$ is the transect estimate of bird density and D is the density determined from the mapping of territories. We also examined the degree to which the transect density estimates were sensitive to differences in the behavior (detectability) of the 10 breeding bird species by testing for a significant interaction of methods and species. Because of unequal subclass sizes, the two-way analysis of variance was completed by the method of fitting constants (Steel and Torrie 1960:257–265).

The sample variance for density estimates derived from each transect method was calculated from the daily replicates. Coefficients of variation (CVs) were used to compare sample variances within methods and

TABLE 2

SPOT-MAPPING ESTIMATES OF DENSITY[a] FOR 10 SELECTED BREEDING BIRD SPECIES AND MNEMONIC CODES

Species	Bird density	Mnemonic code[b]
Eastern Kingbird (*Tyrannus tyrannus*)	7.6	EAKI
Eastern Wood Pewee (*Contopus virens*)	55.8	EWPE
Black-capped Chickadee (*Parus atricapillus*)	17.8	BCCH
White-breasted Nuthatch (*Sitta carolinensis*)	25.4	WBNU
House Wren (*Troglodytes aedon*)	124.2	HOWR
Gray Catbird (*Dumetella carolinensis*)	81.8	GRCA
Wood Thrush (*Hylocichla mustelina*)	35.5	WOTH
Red-eyed Vireo (*Vireo olivaceus*)	55.8	REVI
Common Yellowthroat (*Geothlypis trichas*)	17.8	COYE
Indigo Bunting (*Passerina cyanea*)	86.2	INBU
All 10 species	507.9	

[a] These density estimates (birds/km^2) were used as our best estimates of actual bird density in the evaluation of relative bias of the various transect methods.
[b] After Klimkiewicz and Robbins (1978).

TABLE 3

RELATIVE BIAS[a] OF TRANSECT DENSITY ESTIMATES FROM 12 LINE-TRANSECT METHODS FOR 10 BREEDING BIRD SPECIES

Methods	Species[b]										Average for 10 species	Aver. of absolute values[c]
	EAKI	EWPE	BCCH	WBNU	HOWR	GRCA	WOTH	REVI	COYE	INBU		
King	244.1	−54.0	31.1	−22.7	−35.6	−38.1	−42.5	−72.2	−25.6	−54.9	−7.0	62.1
Hayne	421.7	−31.2	303.0	4.0	32.3	−9.6	−17.4	−66.9	0.3	−25.5	61.1	91.2
Gates II	546.1	−11.1	147.7	46.0	26.8	20.4	9.5	−47.3	36.3	−12.2	76.2	90.3
Gates III	326.3	−47.5	80.8	−12.4	−18.3	−27.1	−32.7	−70.0	−14.4	−44.6	14.0	67.4
Webb	259.2	−36.6	77.3	−17.4	−15.8	−32.0	−27.8	−64.5	−18.3	−50.5	7.4	59.9
Leopold	276.3	−24.9	82.9	15.9	0.0	−20.9	−18.5	−54.8	−14.4	−40.4	20.1	54.9
Gates I	250.0	−30.0	62.6	3.3	−3.2	−25.2	−26.3	−59.6	−28.5	−43.4	10.0	53.2
Frye	177.0	−65.3	−18.5	−33.6	−46.1	−59.4	−59.6	−75.4	−42.5	−68.6	−29.2	64.6
Kelker	173.7	−63.1	−11.4	−32.7	−48.1	−60.5	−57.2	−74.3	−36.9	−68.9	−27.9	62.7
Anderson & Pospahala	156.6	−56.7	9.9	−30.7	−39.7	−51.6	−57.7	−68.4	−50.4	−69.7	−25.8	59.1
Emlen I	294.7	−27.3	−11.4	−32.7	−13.2	−52.2	−27.9	−50.0	26.8	−39.6	6.7	57.6
Emlen II	294.7	16.3	−11.4	−49.2	10.8	31.6	64.9	12.3	194.6	−9.4	55.5	69.5

[a] Relative bias = $100 \times (\hat{D}_i - D)/D$, where D is our best estimate of actual bird density, and $\hat{D}_i$ is the mean estimate of bird density from the ith line-transect method.
[b] See Table 2 for interpretation of mnemonic code.
[c] Average of relative bias for all 10 species disregarding the direction of the bias.

within species. The number of hours required to obtain a density estimate was used as an index to the cost associated with each of the line-transect methods. These costs were then compared with those associated with the spot-mapping method.

RESULTS AND DISCUSSION

Estimates of bird densities varied considerably among the 12 line-transect methods and among the 10 bird species. Total density estimates for all 10 species combined ranged from 237 birds/km^2 to 619 birds/km^2; mapping of territories indicated total bird density of these species to be about 508 birds/km^2 (Table 2).

RELATIVE BIAS

Bird density estimates for all 10 species from the Gates II, Hayne, and Emlen II methods had the highest average relative bias (76, 61, and 56%, respectively) of all methods investigated (Table 3). Much of this positive relative bias arose from high density estimates for Eastern Kingbirds and Black-capped Chickadees, both very active, conspicuous birds. Density estimates for the other species tended to exhibit relatively small positive and negative biases. One major exception was the high positive bias in the density estimate for the Common Yellowthroat which was derived using the Emlen II method (Table 3). The overestimation of density for this species suggests that the adjustment for cue frequency associated with this method tended to overcompensate for undetectable birds.

The average relative bias of the Leopold, Gates III, Gates I, Webb, Emlen I, and King methods, for all 10 bird species combined, tended to be quite small (from +20% to −7%, Table 3). As with the previous group of methods, densities of highly conspicuous species tended to be overestimated, while densities of most of the other species were underestimated.

Three methods tended to underestimate the combined density of all 10 species by about 25% (Table 3). These methods—Anderson and Pospahala, Kelker, and Frye—use only those observations within a particular perpendicular distance of the transect (Table 1). The negatively biased estimates derived from these methods suggest that the proposed methods for determining the perpendicular distance within which all birds were detected were inaccurate.

The line-transect techniques that gave the least biased estimates for all 10 species combined were the Emlen I, King, Webb, Gates I, Gates III and Leopold methods. None of these methods, however, provided relatively unbiased density estimates for all individual species. Depending on the method and species involved, transect estimates of density of any one species deviated from 0 to 546% from density estimates derived from the more time-consuming spot-mapping method. No significant interaction was found between the transect methods and bird species ($F = 0.5$, df $= 99, 360$; $P > 0.05$). Even though the conspicuousness of the bird species varied (frequency of cue production ranged from 0.30 to 0.66; see footnote on Table 1 for further explanation), the methods responded similarly to changes in detectability of the species. The densities of highly conspicuous species such as

TABLE 4
COEFFICIENTS OF VARIATION OF BIRD DENSITY ESTIMATES FROM 12 LINE-TRANSECT METHODS FOR 9[a] BREEDING BIRD SPECIES

Method	Species[b]									Average for 9 species
	EAKI	EWPE	BCCH	WBNU	HOWR	GRCA	WOTH	REVI	INBU	
King	18.7	22.1	55.7	37.2	29.6	19.8	33.0	51.8	37.8	34.0
Hayne	49.8	46.2	99.7	36.8	80.5	23.4	43.0	68.7	56.6	56.1
Gates II	18.7	22.8	57.0	38.9	30.0	20.1	34.5	54.4	38.5	35.0
Gates III	35.8	27.7	68.6	38.2	41.8	20.7	35.8	59.3	48.7	41.8
Webb	22.8	18.7	57.9	34.0	40.7	19.5	34.0	48.9	38.7	35.0
Leopold	4.9	22.4	45.3	45.9	34.1	22.9	31.1	57.1	33.0	33.0
Gates I	13.3	23.9	48.5	49.5	34.8	23.4	34.3	61.9	34.4	36.0
Frye	25.9	40.5	19.1	77.4	31.5	28.9	28.3	70.0	30.0	39.1
Kelker	15.6	15.7	35.1	61.6	28.9	26.1	24.4	60.3	25.9	32.6
Anderson & Pospahala	28.3	29.6	49.7	56.8	30.3	26.7	28.6	64.8	53.7	40.9
Emlen I	32.5	13.6	35.1	61.6	29.0	34.7	30.7	57.4	23.1	35.3
Emlen II	32.5	18.0	35.1	42.7	25.8	38.5	50.9	57.4	27.9	36.5
No. density estimates	2	5	3	5	5	5	5	4	5	—

[a] Coefficient of variation could not be calculated for the Common Yellowthroat. We had only one daily estimate because of lack of sufficient encounters on the other four days.
[b] See Table 2 for interpretation of mnemonic code.

the Eastern Kingbird or Black-capped Chickadee were generally overestimated by these methods, whereas the density of a species such as the Wood Thrush which generally has a lower frequency of cue production was usually underestimated (Table 3). The densities of species such as the Red-eyed Vireo and Indigo Bunting tended to be underestimated because observers had difficulty in distinguishing individuals on adjacent territories.

Another way of looking at the average relative bias of estimates by a particular method for all 10 species is to disregard the direction of the bias and take the average of the absolute values of relative bias. These values (Table 3) are considerably higher than a simple average of relative bias because an overestimate of the density of one species is not offset by the underestimate of another. The Hayne and Gates II methods gave bird density estimates which had the highest percent deviations from the mapping estimate of bird density (91 and 90%, respectively; Table 3).

SAMPLE VARIANCE

Sample variances as measured by coefficients of variation were similar for 11 of the 12 line-transect methods; the coefficients of variation for daily density estimates averaged over species ranged from 33 to 42% (Table 4). Density estimates calculated according to Hayne's method were more variable (CV = 56%); the higher variability among estimates obtained with this method may be a result of greater sensitivity to small changes in estimates of the detection distances. These changes may be real changes in the detectability of the birds or may result from differences in observer ability to estimate these distances.

COSTS

The amount of time involved in estimating the average breeding bird density of a specific woodlot was considerably lower for each of the line-transect techniques than for the spot-mapping method (Table 5). Breeding bird density estimates from the transect methods were derived from the average of five surveys of the study area, whereas the mapping method involved 10 surveys of the study area. All transect methods except the Emlen II method required about 25% of the time necessary to complete the spot-mapping method. The Emlen II method required 32% of the time involved in the spot-mapping method because of the increased amount of time required to record and evaluate the territories of singing males so that the frequency of cue production by each species could be determined. Vegetation sampling which usually is a part of any intensive bird study will add to the time spent in the field, but these hours will be a greater proportion of total field time when the transect methods are used than when the spot-mapping method is used. Estimates of the cost of the 12 transect methods and the spot-mapping method illustrate the great advantage of the transect methods over the mapping method. Use of these methods should make it easier and less time-consuming to include quantitative estimates of bird populations in environmental impact work. These transect methods, however, may miss an unknown proportion of the birds, especially the rarer or quieter species, due to the lesser amount of time spent in the field.

CONCLUSIONS

An examination of the bias and variance of density estimates from the 12 line-transect methods suggests that these methods should be useful as indices to the density of individual species in deciduous woodlots. Although some methods tended to overestimate or underestimate densties, they did so to a fairly consistent degree. Our results suggest that no one method will provide the least biased estimates for all species. Thus inferences about the structure of bird communities from line-transect density estimates can be dangerous. For example, our line-transect estimates suggest that densities of Eastern Kingbirds and Eastern Wood Pewees were about the same, whereas mapping estimates indicated that pewees were seven times more abundant than kingbirds.

Estimates of the cost of the 12 transect methods illustrate the great advantage of the transect methods over the spot-mapping method. Use of transect methods should make it easier and less time-consuming to include quantitative estimates of bird populations (indices) in studies such as environmental impact assessments which are often subject to time constraints. An added advantage of replicated estimates from line-transects is that density variances can be calculated.

Several of the basic assumptions of line-transect methods (see Burnham et al. 1980) were not met in our studies, and will generally not be met when these techniques are used to estimate densities of songbirds in wooded habitats. Birds located on or near the line of travel may have either moved away as the observer approached or may have stopped singing and escaped detection. In forested habitats, birds in the canopy may not have been detected even though they were in the vertical plane described by the transect line. Whatever the reason, the number of birds detected in a narrow strip along the transect line was often less than the number detected at greater distances from the line. In the case of curious birds such as the Black-capped Chickadee, some individuals may have been counted twice. And finally, the estimation of distances in closed woodland habitats was subject to error. Most birds were detected by song and the observer had to estimate the location of the bird from these cues. The magnitude of error in the estimate of bird location may have varied depending on the density of habitat between the observer and the bird, and on the orientation of the bird as it sang. In addition, distances were estimated; use of measuring tapes or range finders was impractical in the closed habitat of these woodlots. Despite these failures to meet the basic assumptions of line-transect theory, line-transect methods provided reasonable indices to density of most bird species. We feel that these methods are useful in situations in which time and/or money constraints limit the choice of census techniques. We stress that this study was conducted on ten common passerine bird species in deciduous woodlots, and our conclusions may not apply to other groups of birds and different habitat types.

Our results do not clearly show that one line-transect method or group of methods would be significantly better than another in providing bird density estimates. Since most of the transect methods tested in this study exhibited relatively similar amounts of bias, sample variance and cost, the selection of a transect estimator may better be made on the basis of the theoretical background and assumptions of the different transect methods. Eberhardt (1978), Gates (1979), and Burnham et al. (1980) have examined the theoretical basis for several transect methods and suggest that some methods are more

TABLE 5

COSTS[a] OF THE SPOT-MAPPING METHOD AND THE 12 LINE-TRANSECT METHODS USED TO ESTIMATE THE DENSITY OF BREEDING BIRD POPULATIONS IN SOUTH-CENTRAL WISCONSIN

Method	Bird sampling: Hours spent in field	Bird sampling: Hours spent on calculations	Bird sampling: Total number hours required	Hours spent on vegetation sampling
Spot-mapping	56[b]	6[c]	62	8[d]
All transects exc. Emlen II	15[e]	1	16	8
Emlen II	18[f]	2[g]	20	8

[a] Number of hours necessary to survey all birds on a 14 ha woodlot by the various methods. The length of transects in this study area is 1688 m.

[b] Hours include those spent on flagging the area and preparation of a cover map (2 people × 10.5 hours = 21 hours) and visits in which singing males were mapped (10 visits × 3.5 hours/visit = 35 hours).

[c] Hours include the transfer of observations from daily field maps to individual species maps (5 hours) and the delineation of territories on the completed species maps (1 hour).

[d] Hours include time spent collecting data on the composition, density and structure of the vegetation at 20 randomly selected points in the woodlot.

[e] Time spent walking and recording observations along transects (5 visits × 3 hours/visit = 15 hours).

[f] Field time for Emlen II requires 5 visits to the area, mapping territories, and at the same time recording observation data along transects (5 visits × 3.5 hours/visit = 17.5 hours).

[g] Calculations involve the evaluation of territory maps to determine the frequency of cue production for each species sampled. These values are then used in the calculation of density estimates on the specific strip.

robust than others. These robust methods may well provide density estimates which are less subject to the problem of meeting certain assumptions as to the shape of the distribution curve of detection distances, even though all transect methods are subject to the observational errors described above.

ACKNOWLEDGMENTS

We thank K. Dubois, K. Hale, E. Judge, S. Lauzon, J. Nishio, and K. Peoples for their assistance with the fieldwork; J. R. Cary for his advice and assistance in computer analyses; G. Bartelt, D. Katsma, M. Mossman, and S. A. Temple for their suggestions concerning this manuscript; and the landowners of the woodlots for their hospitality. This is a contribution of the Wisconsin Cooperative Wildlife Research Unit (University of Wisconsin, Wisconsin Department of Natural Resources, U.S. Fish and Wildlife Service, and Wildlife Management Institute, cooperating) and the UW Department of Wildlife Ecology. Funds were provided by the U.S. Fish and Wildlife Service, contract numbers 14-16-008-1156 and 14-16-008-2012; and the UW College of Agriculture and Life Sciences.

Studies in Avian Biology No. 6:209–214, 1981.

DO TRANSECT COUNTS MONITOR ABUNDANCE TRENDS IN THE SAME WAY AS TERRITORY MAPPING IN STUDY PLOTS?

SÖREN E. SVENSSON[1]

ABSTRACT.—In 1975–1979 two methods were used in the Swedish Breeding Bird Census for monitoring changes in the number of common birds: territory mapping in study plots and point counts along transects. The results from a comparison of 27 species showed that the agreement between the methods was good for those species with changing population numbers between the years. In species with only small changes in numbers, the correlation between the two methods was not significantly different from zero. In a few species the two methods did not agree: different explanations are suggested. In conclusion, transect methods, even those based on only one count every year, seem to be powerful tools for population monitoring, provided some caution is applied when interpreting the results for certain species (e.g., flocking species, species with a considerable non-breeding population, species with a variable and brief peak song period).

Large scale monitoring programs for land birds usually rely on either territory mapping in study plots (as in the Common Birds Census of Great Britain; Williamson and Homes 1964) or transect counts (e.g., point counts, as in the North American Breeding Bird Survey; Robbins and Van Velzen 1974). In Sweden we have used both methods: territory mapping in study plots since the start of the Swedish Breeding Bird Census in 1969, and counts along point transects since 1975 (Svensson 1975). These counts sample the bird fauna of the southern 40% of Sweden, an area of 160,000 km². In this paper I will compare these two methods.

The Breeding Bird Census of Sweden is an activity now permanently included in the Program of Environmental Quality Monitoring operated by the Swedish Environmental Protection Board (Bernes 1980). The bird census program will expand considerably over the next few years, especially in the north of Sweden, where until now very little work has been done. The intention is to use both ambitious, detailed counts in small study plots and less labor-demanding transect counts over much wider areas. The aims of the two methods will be somewhat different.

The study plots will be located in stable habitats, habitats that are artificially maintained as stable or are naturally stable (mature). Many will be located in protected areas.

The transect counts will cover wide areas with all the major habitats represented, including those under more or less heavy human influence. Thus the transects will sample the bird fauna of what we could call the "normal," dynamic landscape.

In spite of these two different aims it is essential to know to what extent the two methods monitor the same aspects of bird species fluctuations. Otherwise it will not be possible to use the censuses in the study plots as a reference for the transect counts. Additionally, a great deal of the future data for the program will be collected by amateur ornithologists, who are allowed to choose their study plots as well as their transects arbitrarily, making it necessary to find ways by which the two kinds of information can be combined.

Until now almost all the data have been collected by amateur ornithologists. This is the case with the census data used in this study, with the exception of the censuses in Lapland.

It is obvious that when the aim of a census is to determine the true number of resident birds, a multi-visit schedule to a precisely delimited sample plot, including efficient search at each visit with detailed mapping of all observations, will produce a more nearly accurate result than point or line transects in the same area. This is partly because the latter must involve the use of experimentally derived efficiency coefficients to convert time or length based indices to area-based density estimates. On the other hand, the two methods should measure population changes between years equally well, provided that proper methodological standardization is applied. However, the methods are different in principle, because in territory mapping we count resident birds almost exclusively, whereas in transect counts we usually count all birds observed. Another difference is that in territory mapping censuses the several visits will cover most of or the whole breeding season, whereas in point or line transect counts normally only one count is performed each year.

In the present study I have used mainly unpublished data from the Swedish Breeding Bird Census for the years 1975–1979. I have also included data from a smaller study in Lapland with both territory mapping in study plots and line transects in 1972–1979.

[1] Department of Animal Ecology, University of Lund, Sweden.

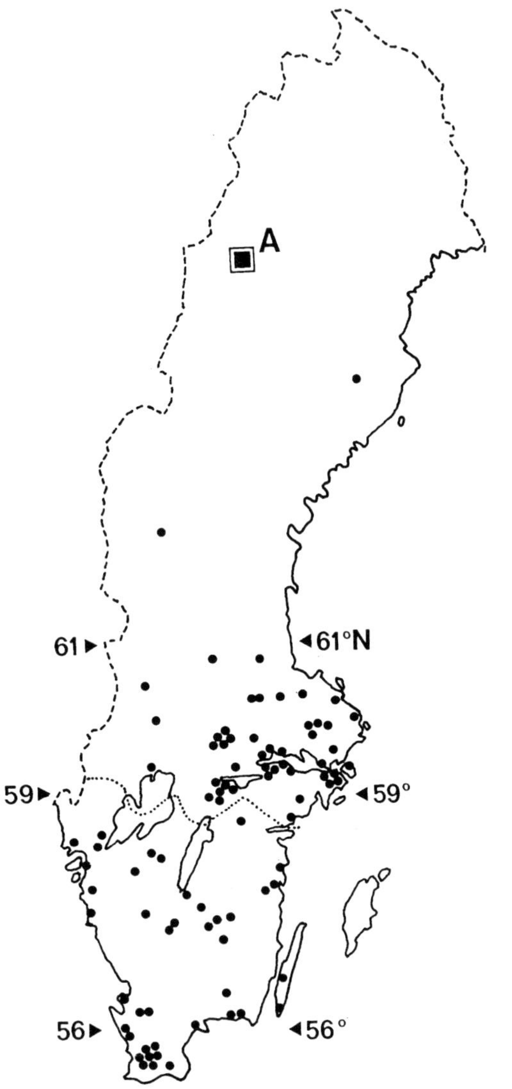

FIGURE 1. Location of Swedish study plots. (Plots censused at least two adjacent years in the period 1975 through 1979. The dotted line is the border between the southern and central provinces. A = location of the special study in Lapland.)

METHODS AND STUDY AREAS

Territory Mapping in Study Plots

The method we use in the Swedish Breeding Bird Census, as well as in the special study in Lapland, follows the recommendations given by the International Bird Census Committee (1969). The size of the study plots varies between 10 and 100 ha, being most often 10–30 ha in woodland and 30–100 ha in open habitats. The distribution of the study plots is almost completely confined to south Sweden (Fig. 1). The number of study plots is given in Table 1. The special study in Lapland involved two plots of 100 ha each.

TABLE 1
Number of Study Plots (Territory Mapping) and Point Transects Censused in Sweden in 1975–1979

	Study plots		Point transects	
Year	In a single year	In each of two years	In a single year	In each of two years
1975	99		89	
1975/76		69		67
1976	85		139	
1976/77		61		99
1977	73		154	
1977/78		56		111
1978	64		142	
1978/79		51		99
1979	56		119	

About 10 visits are paid to a study plot over the breeding season. Each visit lasts for about 2 hrs/10 ha in a woodland plot and for about 1 hr/10 ha in an open plot. All birds observed are entered at each visit on a visit map with special marks for different activities and movements. Special care is given the registration of simultaneous contacts with birds from neighboring territories. These entries are then transferred to species maps, which are evaluated by the census taker, sent to the central office and reevaluated there.

Point Counts

The point count method we are using is standardized in the following way. The census taker selects a route. Along the route 20 stops are chosen, sufficiently distant from each other to prevent double counts of the same birds from adjacent stops. At each stop the observer counts all birds seen or heard during exactly five minutes. The date of the count and the starting hour are standardized from year to year. The date may vary plus or minus five days from the date of the count in the first year. The hour of start may vary plus or minus 30 minutes from the hour of the first year's count. Rainy and windy days are avoided. Counts are accepted only if made by the same person in all the years.

The number of point counts is given in Table 1. They are distributed over Sweden in the same way as the study plots in Fig. 1.

Line Transects

The line transects in Lapland ("A" in Fig. 1) were carried out along permanent lines of a total length of about 100 km. No lateral limit was used for the collection of observations, but measurements showed that a very small proportion of observations was more distant than about 150 m. The counts were always made in fair weather, and the date and hour of start were chosen with the same limitations as above for the point counts.

Comparison Between Years

Because, with the exception of the counts in Lapland, the number of mapping study plots and point counts varied from year to year, only those plots and

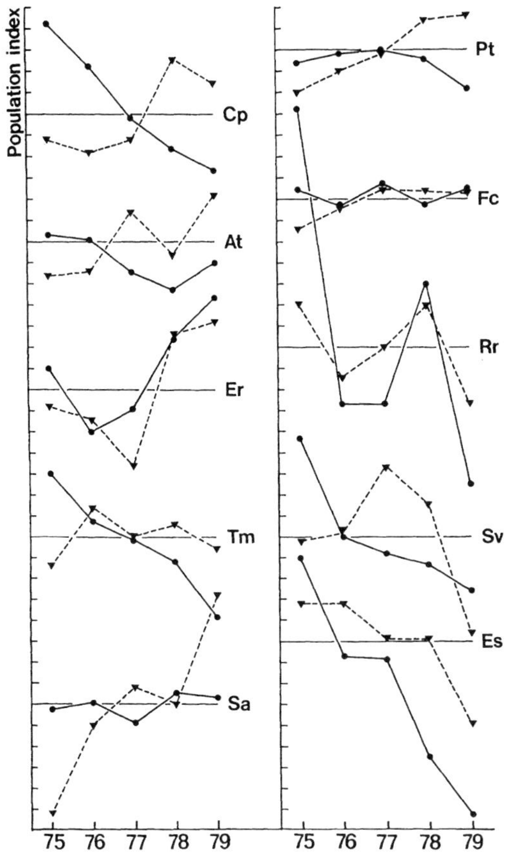

FIGURE 2. Population index for ten species in 1975–1979, separately for study plots and point transects. (Solid lines with dots = mapping study plots. Broken lines with triangles = point transects. The species are: Cp = *Columba palumbus,* At = *Anthus trivialis*, Er = *Erithacus rubecula*, Tm = *Turdus merula*, Sa = *Sylvia atricapilla*, Pt = *Phylloscopus trochilus*, Fc = *Fringilla coelebs*, Rr = *Regulus regulus*, Sv = *Sturnus vulgaris*, Es = *Emberiza schoeniclus*. Index marks are ten percentage units apart.)

TABLE 2
SAMPLE SIZE (*N*) AND COEFFICIENT OF VARIATION (CV) IN THE STUDY PLOTS AND THE CORRELATION (*r*) BETWEEN THE INDICES DERIVED FROM MAPPING STUDY PLOTS AND POINT TRANSECTS

Species	*N*	CV %	*r*
Columba palumbus	79	13.8	−0.77
Dendrocopus major	18	34.3	+0.75
Anthus trivialis	157	5.6	−0.48
Prunella modularis	63	15.4	+0.46
Troglodytes troglodytes	20	53.4	+0.80
Erithacus rubecula	190	12.2	+0.88
Luscinia luscinia	57	8.1	−0.01
Saxicola rubetra	67	8.1	+0.80
Turdus merula	171	12.8	−0.15
Turdus philomelos	87	9.5	+0.16
Sylvia communis	66	14.2	+0.06
Sylvia borin	210	6.1	+0.36
Sylvia atricapilla	109	2.8	+0.28
Phylloscopus trochilus	575	3.7	−0.49
Regulus regulus	75	33.8	+0.87
Muscicapa striata	65	8.3	−0.09
Parus palustris	45	13.6	+0.42
Parus ater	19	12.4	+0.34
Parus caeruleus	117	9.4	+0.33
Parus major	238	7.4	+0.34
Sitta europaea	41	19.2	−0.14
Certhia familiaris	40	6.3	−0.30
Sturnus vulgaris	238	14.0	+0.19
Fringilla coelebs	662	2.3	+0.16
Carduelis chloris	86	11.6	−0.03
Emberiza citrinella	114	7.9	−0.25
Emberiza schoeniclus	88	26.5	+0.83

transects common to two adjacent years were used. In these the percentage change in the number of birds was calculated, and then these figures were used to develop an index for the whole period (cf. Fig. 2).

RESULTS

Fig. 2 shows how the population index changed for ten species in southern Sweden. For some species it is easy to see that there is agreement between the results from mapping study plots and point counts, e.g., in the European Robin (*Erithacus rubecula*) and the Goldcrest (*Regulus regulus*). For the Wood Pigeon (*Columba palumbus*) and the Starling (*Sturnus vulgaris*) it is apparent that the results of the two methods are quite different.

For many species, such as the Willow Warbler (*Phylloscopus trochilus*) and the Chaffinch (*Fringilla coelebs*), the changes from year to year have been so small that one would expect sampling effects to dominate the true population changes.

I first calculated the correlation coefficient between the two methods for each of the 27 species included in the study (Table 2). There were 10 negative and 17 positive correlations. This does not, however, give much information since we do not know in how many, if any, species there have been actual population changes of measurable magnitude.

It seems reasonable to assume that we would find high positive correlation in the most variable species. I have therefore plotted (Fig. 3) the coefficient of variation (standard deviation divided by the mean) against sample size (average for the five year period). It is clear that seven species are relatively more variable than the rest.

If we then examine how the correlation coefficient is related to the deviation of the observed coefficient of variation from the expected one (Fig. 4), we see that these same seven species have very high correlations: in five of the seven

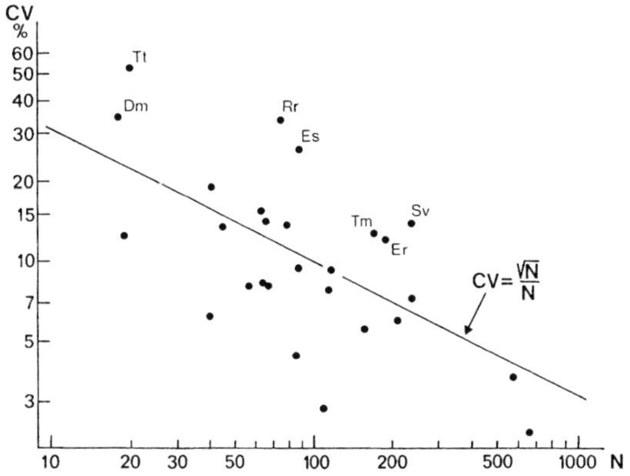

FIGURE 3. Coefficient of variation (CV) in relation to the mean number of pairs (*N*) in the study plots in 1975–1979. (Seven species with comparatively high CVs are indicated: Dm = *Dendrocopus major*, Er = *Erithacus rubecula*, Es = *Emberiza schoeniclus*, Rr = *Regulus regulus*, Sv = *Sturnus vulgaris*, Tm = *Turdus merula*, Tt = *Troglodytes troglodytes*. The straight line is the expected coefficient of variation under the assumption of random fluctuations, so that the variance equals the mean.)

species *r* is greater than +0.50, whereas in the other 20 species there is only one species with *r* greater than +0.50.

In the Lapland study the comparison between data from the study plots and line transects gave the following results: For two species, the Bluethroat (*Luscinia suecica*) and Willow Warbler (Fig. 5) the agreement was excellent. For the Lapland Bunting (*Calcarius lapponicus*) the correlation coefficient was high, but only because both methods gave high counts in one

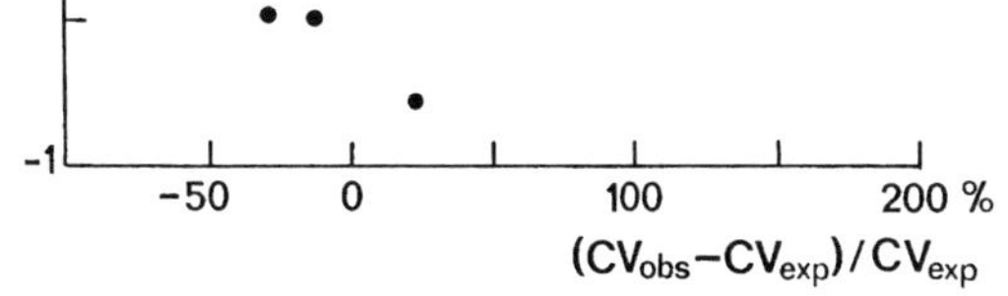

FIGURE 4. Correlation (*r*) between point and study plot counts in relation to the deviation of the observed coefficient of variation from the expected one in the study plot counts. (Key to the seven indicated species in Fig. 3.)

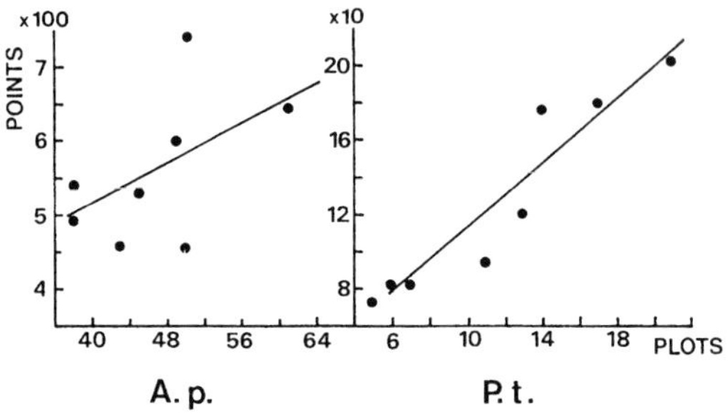

FIGURE 5. Comparison between the number of birds in two study plots (*x*-axes) and the number along line transects (*y*-axes) in an alpine region in Lapland. (Ap = *Anthus pratensis*, Pt = *Phylloscopus trochilus*.)

year. The correlation was not significant if this year was excluded. Rather weak and non-significant correlation coefficients were obtained for the Meadow Pipit (*Anthus pratensis*) (Fig. 5), the most abundant of the species, and the Wheatear (*Oenanthe oenanthe*). Diagrams for all species and the whole community, and other details of the study are given in Svensson (1980).

DISCUSSION

Swedish Breeding Bird Census

The view that territory mapping in study plots and transect counts monitors population changes equally well is supported by two lines of evidence: (1) the assumption that high variability of the population index indicates a high frequency of "true" population changes of some magnitude; and (2) the observation that high variability is connected with high correlation between the results of the two methods. That the majority of species shows no significant correlation can be explained by the assumption that any changes in these species were so small that they were masked by sampling errors. If that is the case, we would expect what we did find, i.e., the same number of positive as negative correlation coefficients.

TABLE 3
Distribution of Expected and Observed Number of Species in Relation to *r* from Comparison of Results of Mapping and Transect Counts

	Value of *r*			
	<−.8	−.8–0	0–+.8	>+.8
Expected %	5	45	45	5
Exp. no. species	1.35	12.15	12.15	1.35
Obs. no. species	0	10	14	3

TABLE 4
CORRELATION BETWEEN POPULATION INDICES FROM STUDY PLOTS AND POINT COUNTS IN SOUTHERN AND CENTRAL PROVINCES SEPARATELY

Species	South. prov.	Centr. prov.	Mean of S & C prov.	Both regions
Anthus trivialis	−0.46	−0.47	−0.46	−0.48
Erithacus rubecula	+0.98	+0.52	+0.75	+0.88
Turdus merula	+0.28	−0.66	−0.19	−0.15
Sylvia borin	+0.31	+0.82	+0.56	+0.36
Phylloscopus trochilus	−0.20	−0.71	−0.46	−0.49
Parus caeruleus	−0.48	+0.30	−0.09	+0.33
Parus major	+0.45	+0.46	+0.45	+0.44
Sturnus vulgaris	+0.38	+0.05	+0.21	+0.19
Fringilla coelebs	+0.81	−0.17	+0.32	+0.16

The distribution of r is known. If the expected value of r is zero and the number of degrees of freedom is three we obtain the distribution for the 27 species shown in Table 3. If we exclude the seven deviating species there are 11 positive and 9 negative coefficients. Therefore I think the two methods do not agree in the majority of the species because these species have not fluctuated very much in the study period.

Particular attention must be paid to those species in which we have found important changes between certain years or clear trends over a series of years. There are three such species among those shown in Fig. 2. The number of Wood Pigeons has decreased uninterruptedly in the mapping study plots over the five-year period. The number of Blackcaps (*Sylvia atricapillus*) has been very stable in the plots, but shows a considerable increase in the transect counts. The number of Starlings has decreased every year in the mapping study plots, whereas there is a peak in the middle of the period in the transect counts.

In the first two instances there are no independent counts that could be used for determining which of the methods gave the correct information. But in the Starling such information exists: It has been reported from a number of nest-box studies in south Sweden that the number of Starlings has declined since the early seventies. This decline is parallelled by a similar decline in the number of Starlings in the mapping study plots since 1973. Olavi Hildén has found exactly the same decline in nest-box studies in Finland. I therefore believe that there has been a long term decline in the number of breeding Starlings in Scandinavia over the past ten years.

Why, then, do the transect counts not show this decline, but instead even an increase in 1976–77? Interestingly, line transect data from Finland show the same pattern as the Swedish point counts in 1975–77. Because of this, the easiest explanation, that there is some error involved in the transect counts, cannot be accepted without qualifications. Perhaps the two methods simply do not count the same birds. We know that in the Starling the counts in the mapping study plots refer to breeding birds. The species is usually not censused by actual territory mapping, but by counts of active nests while adults are feeding young. The close agreement with data from nest-box studies is therefore no surprise. In the transect counts nonbreeding birds are also included, e.g., flocks that spend their first summer in open habitats without breeding. A similar explanation can be put forward for the discrepancy in the Wood Pigeon. It is, like the Starling, a species with a nonbreeding fraction of the population that often gathers in flocks rather early in the summer. If there are shifts in the proportion of breeding and non-breeding birds in a population, detailed counts of breeding birds in mapping study plots could show population declines while transect counts showed population increases.

The Blackcap picture is more difficult to explain. Phenological effect is one possibility. If the period of peak song activity shifts from year to year and the date of the count is stable, as it was in this instance, an increase in the number of birds observed may occur even if the true number did not change. Another explanation for an increase could be the training of observers. Such training or experience, however, should not be restricted to a very few species but should show up in a general upward trend. This is not the case.

There might still be a training effect: the number of agreements between the two methods has increased over the five-year period. In 1975/76 the direction of change was the same in the two methods in only 40% of the 27 species. This percentage increased to 74% in 1978/79. But such a trend can have other sources too, e.g., a greater number of actual changes in 1978/79, which in fact seems to be the case. The winter of 1978/79 was rather severe and a number of wintering species and short distance migrants decreased between 1978 to 1979.

I have assumed that the changes were the same in the whole of southern Sweden. If this was not the case, one would perhaps obtain higher correlation coefficients if the material was divided into geographical subareas. This has been done for some of the most abundant species. The country was divided into two parts (the dotted line in Fig. 1) and correlation coefficients were calculated separately for each part

(Table 4). But there is no overall trend for higher correlation coefficients for the separate parts of the country; some are higher and some are lower than the pooled data.

The Counts in Lapland

In the Lapland study the mapping study plots and the line transects were in close proximity to each other, and parts of the line transects were carried out within the study plots. It is therefore surprising that the data from the two methods were not in better agreement. The only explanation I can offer for the poor agreement in, for example, the most common species, the Meadow Pipit, is that its conspicuousness varied because of weather conditions or slight shifts in its phenology.

CONCLUSIONS

Territory mapping in study plots and transect counts show the same population changes in many species. Point counts and line transects can, consequently, be considered powerful tools in monitoring bird population trends. However, in some species there are considerable disagreements. Great care must therefore by applied in interpreting results obtained by the simple and rapid methods involving only one point or line transect a year. It is particularly disturbing that the line transects and study plot results of the Lapland study did not agree in all species because they were all carried out in the same small area of homogeneous habitat.

Studies in Avian Biology No. 6:215–216, 1981.

SUMMARIZING REMARKS: COMPARISON OF METHODS

G. M. JOLLY[1]

METHODS

The methods compared are summarized in Table 1. Although Anderson and Ohmart (1981) and Edwards et al. (1981) restricted the width of their transects, there is no essential difference between their procedure and Franzreb's (1981); all three used J. T. Emlen's (1971) method. Svensson's (1981) Swedish data are from point counts at points along a transect, and his Lapland data are from continuous line transects; he did not attempt to adjust his counts to estimate absolute density, regarding them only as indices. Hildén (1981) conducted repeated censuses on the same transects with the aim of detecting as many birds as possible and compared the results with each other and with the maximum numbers recorded in sections along the routes; he also compared single fixed-width transects with careful, long term censuses of the same study plots; he did not use an Emlen or other type of adjustment.

CONCLUSIONS OF THE PAPERS

Of the four comparisons of spot mapping with transects, both Franzreb (1981) and Anderson and Ohmart (1981) obtained lower density estimates on average from transects, although some species gave reasonably good agreement and a few gave higher densities. Hildén (1981) concluded that a single count of a line transect was inadequate, but that even after several repeated counts, most species were seriously underestimated. Both Hildén (1981) and Svensson (1981), looking respectively at trends over time and correlations between simultaneous counts, found considerable disagreement between methods for some species. There was general recognition that spot mapping itself tends to underestimate density, also that the methods may be observing different populations, either because of being conducted at different times or because, for example, non-breeders may be excluded in a spot mapping census; interesting anomalies were quoted for Starlings.

Anderson and Ohmart's (1981) comparison of variable circular plots with transects gave some inconsistencies in density estimates. In part, this appeared to result from the lower precision of the circular plot data which came from a smaller total sample area. When the census time for circular plots was 8 minutes, higher densities were obtained than from transects, whereas reducing the time to 6 minutes tended to produce the opposite effect. On the average, circular plot densities were slightly below those for spot mapping. The authors recommend transects in preference to circular plots on grounds of efficiency, except where differing patches of vegetation occur.

For two out of the four habitats, Edwards et al. (1981) found that the variable circular plot produced substantially higher densities than did transects, as also happened with the fixed circular sample plot for one habitat. These effects are not brought out in their summary remarks where lack of consistent (although significant) differences seems to be interpreted as absence of any difference. The authors have not shown results for separate species.

By an intensive program of banding, spot mapping and nest monitoring, DeSante (1981) claims to have achieved virtually a completely accurate enumeration of eight species. A comparison with density estimates from variable circular plots shows the latter to be an underestimate by amounts ranging from 2% to 70% for different species (doubtless the smaller discrepancies would be neither significant nor important). Similarly, Hildén (1981), from his study plots, estimates that about 50% of the stationary

[1] ARC Unit of Statistics, University of Edinburgh, United Kingdom.

TABLE 1
METHODS USED IN THIS SESSION

Author(s)	Spot mapping	Line Transect: Fixed	Line Transect: Variable	Circular Plot: Fixed	Circular Plot: Variable
Franzreb	×		×		
Svensson	×		× + point counts		
Hildén	×	×			
De Sante	× + banding				×
Anderson and Ohmart	×	×			×
Edwards et al.		×		×	×

birds are missed in a line transect census when only a single visit is made.

It is difficult to see any justification in Hildén's (1981) use of maximum numbers recorded along the route as an estimate of mean density—for example, by dividing a 4 km transect into four 1 km sections and using the largest of the four observations as the general estimate. This maximum would depend on the length of the sections into which the transect was divided, on the actual distribution of the birds, as well as on the distribution of the proportion missed.

Number of species and associated parameters are necessarily related to total observing effort and are therefore less effective than density for comparing methods. Nevertheless, on all their four habitats, Edwards et al. (1981) detected considerably more species with variable circular plots than with the other methods tested; on marsh the average difference was 11 species compared with 4 to 5.

GENERAL REMARKS ON PAPERS

Each of the six papers has provided a valuable set of results, conclusions and comments. They complement each other in various ways and exhibit no alarming differences of opinion. Comparisons among the three types of method are made difficult by the fact that the different shapes of sampling units and the varying intensity of observation within them make it impossible for the methods to be tested on identical sample material. Thus, precision of estimates is low. More liberal use of standard errors would have aided interpretation. Graphical presentation of results was helpful. Correlation coefficients, however, are not a good measure of agreement between methods. For example, two methods that give almost identical results in each of several samples will not produce a significant correlation if the variation in true density among samples is small. It is preferable to work with straight differences.

There is little general discussion of methodology. As Svensson (1981) comments, "it may seem surprising . . . that the two methods should not measure population changes between years equally well, provided that proper methodological standardization is applied." The many discrepancies encountered show that assumptions are not met in most instances. It would be easy to try fitting alternative models to the line transect data presented, since much work has been done recently in this field, a comprehensive guide being that of Burnham et al. (1980).

The major cause of underestimation, however, is likely to be that birds are being missed "close" to the observer. If so, all the methods listed in Table 1 will break down unless supplemented by correction factors derived from a comparison with an accurate census such as De Santes' intensive program. For methods in which detectability is adjusted for distance from observer, as in line transects and variable circular plots, the correction factor need relate only to data "close" to the observer since the purpose is merely to compensate for the fact that probability of sighting "close" to the observer is not unity. Various possibilities exist. For example, if such correction factors were available for specific species or groups of species under stated conditions, the necessity emphasised by Hildén (1981), of repeated censuses for line transects might be relaxed, allowing effort to be spread over more independent transects. It is undoubtedly neither feasible nor desirable to calibrate in this way, for example, by applying an accurate method to a sub-sample of units for *every* individual survey. Rather, information for a particular habitat could be built up gradually for at least the principal species and correction factors established whose accuracy would improve with time and increasing experience.

Studies in Avian Biology No. 6:217–218, 1981.

SUMMARIZING REMARKS: COMPARISON OF METHODS

CHARLES VAN RIPER III[1]

The evolution of censusing methodology involves continual refinement of techniques. The comparison of different census methods outlined in this session, to my mind makes this the most valuable session of the symposium. When different techniques were compared here, the results showed that density estimates varied depending upon which method was tested. However, deeper probing into each paper reveals similar themes which wind through the session. These may allow us to tie together some of the fundamental censusing problems which are being encountered today. In this summary I will present thoughts and suggestions on the biological ramifications of the methods and ideas presented in this session: first will be an analysis of the logistical comparisons; and second, some biological implications that have been made apparent by these comparisons.

The best available density estimate for a population is obtained when all the birds in an area are banded. In decreasing accuracy this method is followed by the spot mapping technique, the circular plot method, line transect counts, and least effective—the guess. But all of the sampling methods that were compared in this session invariably underestimated total population numbers. It was also pointed out that bird density estimates were greatly modified by the type of habitat in which the census was made. As Anderson and Ohmart (1981a) showed, open vegetation types lend themselves better to the line transect technique. If Edwards et al. (1981) had expanded the area of their line transects (as suggested by one of the questioners), they also would have found this to be true. For a general rule of thumb, line transects are best used in open areas such as savannah or scrub, whereas the circular plot method seems to be more applicable to closed canopy forests. This is particularly true for tropical areas where there is either a very high canopy or a dense understory.

There was unanimous agreement among all the participants that it is logistically more expedient to use the line transect technique. The circular plot and mapping methods take longer; and this problem is magnified when there is behavioral modification of cues, such as a decrease in vocalization rates over a short time period. A way around this problem might be to replicate circular plots with multiple teams. But a cautionary note—there is a disturbance factor when one person follows closely behind another on a transect (Scott and Ramsey 1981a, Scott et al. 1981b).

It was fairly well agreed upon that, regardless of technique, the more intensive the effort, the greater the number of species counted. The participants also agreed that replications are needed, particularly when using the line transect method. As Jolly suggests, in most situations three replications of a transect should be an absolute minimum. Perhaps the use of an index, as pointed out in Franzreb (1981b), where preliminary censuses are run to determine that point after which new species are no longer encountered, might be useful. This information could then be used to determine the number of replications needed for future census work in that particular habitat. All of the techniques used were adequate to determine densities of common and vociferous species, but silent and/or rare birds were always badly underestimated. The technique that proved best able to deal with the rare species problem was the circular plot.

The final major problem brought out in this session was that of swamping. In an avifauna in which a few species are very abundant, the common birds will mask the presence of silent and/or rare species. This proves especially true for areas such as tropical forest habitat, where there may easily be 250 species to be included in the census (Karr 1981). One suggestion made in regard to the Christmas Bird Count might be applied to this problem—that the census be stepped-down to a simpler level by using only presence-absence data. Or, in an effort to increase the reliability of density estimates, censuses might be broken down into separate species groups based on abundance or possibly guilds (Franzreb 1981b, Scott and Ramsey 1981a). Possible divisions of a population to be sampled include: vociferous versus non-vociferous species, foraging guilds, horizontal vegetation strata, or distinct vegetation types. In summary of this first section, know what objectives you want to accomplish before going out, and be aware of what logistics need to be taken into account to accomplish those objectives.

The second section of this summary will deal with the biological implications made during the comparison of papers presented in this session. These studies have all shown that in order to decrease sampling problems one must intimately

[1] Department of Zoology & Cooperative National Parks Study Unit, University of California, Davis, Davis, California 95616.

know the ecosystem to be sampled. This includes awareness of the vegetation structure and the responses of each species to be counted to the biotic and abiotic parameters of the habitat. In short, know the animals you will be censusing. Some of the biological problems brought up during this session follow:

Song cycles.—Eighty to 90% of the detections during a count period are from aural cues (Ramsey and Scott 1981a, Cyr 1981), yet very little background work has been done to incorporate variances of vocalizations into censusing methodology. One needs to be aware of daily and annual song cycles of the birds that will be counted. Much of this information is available in the literature and can be put to good use (e.g., Thorpe 1961, Hinde 1969, Armstrong 1963). Dawson (1981a) showed that singing rates of birds in New Zealand are fairly uniform throughout the day, whereas Anderson and Ohmart (1981a) found a decided early morning vocalization peak. A predawn chorus may pose a censusing problem in certain areas, and in colder climates birds may not vocalize until later in the day. Differences in vocalization patterns must be taken into account if the census results are to be meaningful.

Breeding patterns.—Hildén (1981) and Svensson (1981) pointed out that the initiation of the breeding season is quite variable between years, particularly at higher latitudes. The number of cues recorded during a count period will be greatly modified by the stage of the breeding cycle during the census. Therefore, if the objective is to compare censuses between years, there must be some flexibility in the initiation date of the census. Some point in the breeding cycle should be selected that will enable censuses to be repeated at the same "biological time" year after year, such as an interval after the arrival of a certain species or after the first nest is found.

Banded birds.—The elegant comparison of census techniques done by DeSante (1981) was made possible only because he had a banded population of birds to work with. In conducting comparative censuses, advantage should be taken of situations, such as the one at Point Reyes Bird Observatory, where banded populations of birds are present. This is one of the most expedient ways in which to refine censusing techniques.

Cavity-nesting birds.—There are some real problems with the censusing techniques which are currently available to us when attempting to determine accurate numbers of cavity-nesting birds. Hildén (1981), Svennsson (1981) and Järvinen and Väisänen (1981) all showed discrepancies between the numbers of European Starlings (*Sturnus vulgaris*) counted in nest boxes versus those recorded on line-transect censuses over a period of years. Population estimates derived from nest box data revealed a dramatic decrease in Starling numbers, whereas no declining trend was found from the line-transect census results. It is possible that the Starlings are utilizing different nesting locations, or perhaps there is presently a high proportion of nonbreeding birds in the population. In any event, this problem might be inherent to cavity-nesting species, and censuses dealing with them should take into account this potential problem.

Nonbreeding birds.—Large groups of nonbreeding birds can contribute a bias to census results. Males without mates (which have not yet obtained a mate or have lost a mate) will vocalize more than breeding males (Nolan 1978). The number of "floaters" in a population is usually ignored, but in certain areas these birds can make up a significant portion of the population being counted (Smith 1978). Karr (1981) has suggested that large numbers of floating birds exist in tropical forests, Recher (1977) has documented numerous floating birds in Australia, and my own experience in Hawaii has shown that many individuals sighted within an area are floaters (van Riper 1978, 1980). However, as DeSante (1981) found at Point Reyes, some areas do not have large percentages of floating birds in the population. Before censusing an area, it must be determined if a floating population is present, how large it is, and to what degree it might bias the census results.

In summary, each censusing method has its weaknesses and limitations. Researchers should be aware of the limitations before applying a technique to a censusing problem. The area to be censused should be carefully surveyed *prior* to embarking on a study. This preliminary information should then be analyzed so that the censusing method best suited logistically to the area can be chosen. In addition, the species which are to be censused should be researched to determine if any biological or behavioral variances will modify the cues to be recorded during the count periods. The study should be planned to minimize logistical and biological variances. The papers presented at this symposium and the discussion arising from their comparison should better enable workers to use the censusing tools now available to scientists in the field.

Studies in Avian Biology No. 6:219, 1981.

INTRODUCTORY REMARKS: SPECIES VARIABILITY

CAMERON B. KEPLER,[1] CHAIRMAN

This session on species variability marks an important shift in direction for the symposium, carrying it from a discussion of the methods used to assess avian populations, to a consideration of some of the important, often thorny, variables that stand between the biologist and an understanding of those populations. These variables affect the results of all the methods so far discussed, be they designed to generate indices of abundance or estimates of absolute density. They include the conspicuousness of the birds themselves, the physical and biological characteristics of their habitats, and variation inherent in the observers.

Birds vary enormously in their detectability. Not only is there an inherent diversity in conspicuousness between species (Mayfield 1981), there also exists considerable intraspecific variation that depends upon such factors as time of day, season, age, sex, stage in the breeding cycle, foraging strategy, dominance relations, and many others. Biologists working with some groups, such as nocturnal birds or raptors, have had to devise unique sampling strategies that consider the peculiarities of their study animals (Fuller and Mosher 1981). Others, especially those sampling forest ecosystems, often simply exclude these birds from consideration. Raptors and nocturnal birds, however, only serve to illustrate in a dramatic way these universal problems in detectability.

As we seek to refine our methods in what is clearly an inexact science, we will increasingly confront these factors in avian conspicuousness. As species such as Ekman's (1981) tits, or Diehl's (1981) shrikes become better understood, we will be forced to include the known behavioral and ecological idiosyncracies of each of them in our experimental design before we attempt to sample them. And after our data is in hand, we will have to apply this knowledge, in the form of correction factors, to it. At that time we may finally procure a foundation of reliable information sufficient to understand the dynamics of avian populations and community structure.

[1] U.S. Fish & Wildlife Service, Patuxent Wildlife Research Center, Maui Field Station, 248 Kaweo Pl., Kula, Hawaii 96790.

Studies in Avian Biology No. 6:220–224, 1981.

PROBLEMS IN ESTIMATING POPULATION SIZE THROUGH COUNTS OF SINGING MALES

HAROLD F. MAYFIELD[1]

ABSTRACT.—Counts of singing males in breeding season appear to be an easy way of estimating bird populations, but detailed studies of such typical songbirds as Kirtland's Warbler and Prairie Warbler show enormous variability in song frequency for different species, different indiviuduals, different stages of reproduction, different hours of the day, and different weather conditions; and these uncertainties are compounded by the fact that sex ratios are seldom known accurately. Consequently, censuses often contain errors much larger than customarily encountered in scientific measurements. Under ideal conditions the probability of hearing a male Kirtland's Warbler within a 5-minute period is about .85, and the probability of hearing a male Prairie Warbler is about .55. To test my census efficiency on an assortment of familiar species, I conducted transect counts on ten consecutive days in early June at the 2½ hectare tract surrounding my own home. This test showed census efficiencies for different species ranging from zero to 90%, and the average efficiency for all species of 40–50%. Counts of males are commonly translated into population totals by assuming one female for each male. Yet the difficulty of determining exact sex ratios is illustrated by the Brown-headed Cowbird, for which published estimates vary by 30% or more, and my own 5-year sample of 18,000 birds taken on the breeding grounds showed a high predominance of females early in the season and a high predominance of males at later dates, suggesting large differences in mobility but leaving the true ratio in question. From my experience I have concluded (1) some species cannot be counted effectively by brief listening periods, (2) difficulties with each species can be appraised only through lengthy study of each, and (3) people studying the behavior and reproduction of birds should direct attention to problems of censusing to be expected by others.

J. T. Emlen (1971, 1977a) has written comprehensively on methods of censusing bird populations and their shortcomings. Of all the methods available, one of the most attractive for use with small land birds is the count of singing males. It looks easy because pairs during the nesting season are anchored to exclusive territories, and the males advertise their presence loudly. I have used it for many years with the Kirtland's Warbler (*Dendroica kirtlandii*), and through experience have become increasingly conscious of sources of error with the best of subjects and appalled by the potential error with more difficult subjects. The errors are much larger than customarily expected in scientific measurements, and usually they are not quantified or even acknowledged.

Here I will focus attention on two sources of uncertainty that I have examined: (1) the probability that a male will not be detected on its territory in a brief census period, and (2) the ratio of males to females needed to calculate the total population size. This information is not available for most species with precision. My examples are drawn, first, from closely-related warblers familiar to me and studied in depth, particularly the Kirtland's Warbler studied by me for 18 years and the Prairie Warbler (*Dendroica discolor*) studied for 21 years by Val Nolan; second, from ten consecutive daily censuses on my own property where the resident birds were already known; and third, from very large samples of Brown-headed Cowbirds (*Molothrus ater*) collected over five breeding seasons in a unique effort at total removal of this parasite from the nesting grounds of the Kirtland's Warbler.

VARIATIONS IN WARBLER SONG

In my first census of the entire population of Kirtland's Warblers in 1951 (Mayfield 1953:18–20), I reported 432 singing males, and since that time I have cringed at calculations based on this exact number. At the time I expressed the reservation that the count might have been understated by as much as 25% for various reasons, and in later accounts I usually rounded the count to 500. Nevertheless, the exact figure persists, and the reservations are usually forgotten.

With thought to census needs, I attempted to assess the probability a Kirtland's Warbler would be heard by a person walking slowly through its territory during nesting season (Mayfield 1960:130–135). Under ideal conditions the song can be heard at a distance of 400 m, the full width of a male's territory. Like most songbirds, it gives its song in courses. Songs lasting 1–1½ seconds are uttered 6–9 times a minute, but these courses of song may be interspersed with periods of silence lasting many minutes. When I analyzed detailed records of song gathered at various nesting stages, dividing the periods of time into segments of 5 minutes each, I found that 85% of 480 time segments contained at least one song, and thus the probability a male would be heard in one passage through his territory was about .85. The Kirtland's Warbler proves to be an almost ideal subject for censusing, and through repeated counts on familiar

[1] 9235 River Road, Waterville, Ohio 43566.

areas in succeeding years, we believe our accuracy has improved. Still we cannot place complete confidence in these stated probabilities because of possible selectivity in our song samples. Poor singers are less likely to get into the records.

Kirtland's Warbler males behaving in standard fashion can be counted fairly accuragely by visiting each "colony" a few times or walking back and forth through it several times at the height of the nesting season, provided the weather is good and the hour of the day is early. This belief has been confirmed by intensive field work on censused areas. But even under ideal conditions, we may wonder how many males are *not* behaving in standard fashion. Although the species sings persistently through all stages of the nesting process, males are sometimes unaccountably silent for hours at a time. More than once I have searched in vain for an hour where a male was believed to be present, and then later have found it readily. I have concluded that males leave their territories more often than generally realized. For example, I once found a banded male (believed unmated) singing as though on territory 2 km from his home site. Only by accident was he discovered. How many such did I miss? Also during my work in a small isolated "colony" where all the resident males were banded and recognized by idiosyncracies of song, I occasionally detected unknown males slipping through the area silently. Previous authors have named such males "floaters," implying that they are nonbreeding males without territories drifting through occupied regions, perhaps ready to fill any vacancies that occur. However, Norman Ford (unpubl. data) has expressed doubt that such floaters exist among the Yellow Warblers (*Dendroica petechia*) he has studied on one tract of land for several years. Instead, he believes these are mated, territorial males reconnoitering away from home. Nolan (1978:362) reached similar conclusions about Prairie Warblers. Whatever the mechanisms, it is clear there is flexibility in movements and bonds, allowing lost mates to be replaced quickly in many cases. We also have instances where males have occupied two territories simultaneously, with or without polygyny. Ordinarily it is assumed such anomalies are so rare they can be ignored in calculations. Yet Nolan's (1978:364–365) study of the Prairie Warbler is not reassuring on this point. He found polygyny occurring in about 15% of male Prairie Warblers, and sometimes it involved separate or elongated territories that could have been disentangled only by prolonged study.

In his transect counts J. T. Emlen (1971) employed a "coefficient of detectability" to deal with differences in the distances at which species reveal themselves, as it affects the strip-width being censused in a transect. The distances species may be heard vary enormously, and so do the singing habits of species and individuals within a species. The uncertainties caused by periods of silence, immobility, and absence are very large in some species. In addition to the characteristic elusiveness of some birds, we need to consider variations for different stages of the nesting season, for different hours of the day, for different weather conditions, and for individual differences within each species. Consequently, to proceed from a count or a series of them to an accurate determination of population size may require knowledge based on a prodigious amount of field work.

The Prairie Warbler is a persistent singer, but Nolan (1978:64, 71) found some of them silent more than twice as much of the time as others on the same date, and at different reproductive stages he found song frequencies for the same individual varying in the ratio of four to one. He has supplied me with details on seven extended periods of song by males with active nests in various stages. He divided the time into 5-minute segments, noting whether or not song took place in each of these. Considering only the first five hours of daylight (before 10:00), I have analyzed a sample of 405 segments. Of these, 224 (55%) contained at least one song. Hence, the probability of detecting one of these birds in a 5-minute listening period was about .55. The variability among individual birds, however, was startling. One bird sang for 15 minutes very early in one morning and then was silent for three hours, while another was almost silent during the first two hours and then sang frequently later in the morning. One bird sang in 90% of one morning's 5-minute segments; yet another male sang in only 25% of them.

In his study of Yellow Warblers, Ford (unpubl. data) found males to be relatively quiet during days of nest building. On the other hand, males of this species and many others are most vocal when unmated, but their conspicuousness at one location may be offset by a greater tendency to stray and sing elsewhere.

CENSUSING AN ASSEMBLAGE OF SPECIES

Up to this point I have considered census problems with species whose behavior has been studied thoroughly. Usually the person conducting a census is faced with the more difficult problem of dealing with a variety of birds whose habits are not known to him in detail. One way to appraise the accuracy of such a count is to conduct it on a tract that is already under such

TABLE 1
COUNTS OF SINGING MALES ON TRACT NEAR WATERVILLE, OHIO, 1980

Song freq.[a]	Species	Date in June										Total times counted	Pairs actually present	Maximum possible count	% eff.
		6	7	8	9	10	11	12	13	14	15				
.40	Mourning Dove		1		1		1	2	1	2	2	10	3	30	33
	Yellow-billed Cuckoo									1		1	1	10	10
	Ruby-thr. Hummingbird											0	1	10	0
	Downy Woodpecker			1	1			1				3	1	10	30
	E. Wood Pewee	1			1	1	1	1	1	1	1	8	1	10	80
	Blue Jay		1									1	1	10	10
	Tufted Titmouse								1		1	2	1	10	20
	White-br. Nuthatch				1				1			2	1	10	20
.66	House Wren	4	3	4	3	4	3	4	4	4	3	36	4	40	90
.44	Catbird		1		1	1	1	1	1	1	1	8	1	10	80
.19	Robin	1	1	3	1	2	3	3	2	2	2	20	4	40	50
	Cedar Waxwing									1	1	2	1	10	20
	Starling					1			1		1	3	1	10	30
.10	Northern Oriole	1	1		1	1			1		1	6	1	10	60
	Common Grackle		1	1	2	1	1	1	1	2	2	12	4	40	30
.51	Brown-h. Cowbird								1			1	1	10	10
.59	No. Cardinal	1	1	1			2		1	3	1	10	3	30	33
.53	Indigo Bunting	1	1	1			1	1		1	1	7	1	10	70
	Chipping Sparrow									1		1	1	10	10
.75	Song Sparrow		1							1	1	3	1	10	30
Hours after sunrise		3	2	2	2	1	2	1	2	1	1				

[a] Song frequency from Emlen (1977b:461) approximately equivalent to percent efficiency here.

close scrutiny, preferably by different observers, that the residents are thoroughly known (see DeSante 1981, Hildén 1981). Excellent opportunities for such tests are presented at research stations where several people are already engaged in separate projects.

To explore this problem in a preliminary way, I conducted ten censuses of the birds at my own home. I was familiar with the birds present, having observed them before and after they set up territories. On 10 consecutive mornings between one and three hours after sunrise, June 6–15, 1980, I walked slowly down the middle of my property for its entire length. The strip was 250 m in length, and I took 10–14 minutes for the route. Thus, my walking speed was about 1 km per hour, roughly comparable to Emlen's transect walking speed and Robbins' (1979b) 3-minute listening stops, since I progressed less than 100 m in each 3 minutes. I believe no song escaped me within 50 m on either side of my path and some birds were heard at greater distances, and I thus considered the area covered to be roughly 2½ hectares.

This tract was mostly covered with mature trees, under which lay mowed lawn, many shrubs, and, at the end near a river, undisturbed underbrush. It had sharp ecological boundaries at each each end—a highway and cultivated field at one end and a 100 m wide river at the other. The sides, however, had no natural boundaries. One of the long sides overlapped a brushy abandoned orchard, and the other adjoined lawns and woodland like my own. An ecological island would have been better. At this time of year the vegetation was in full leaf, and visibility was severely limited in the canopy and in the brushy understory.

Since my ultimate interest was to determine the total population, I noted all birds seen as well as heard, and recorded the numbers in terms of pairs rather than individuals (Table 1); that is, a family out of the nest, a pair seen together, or a singing male were each recorded as one pair. I judged 33 pairs of 20 species to be resident, but I had some troublesome decisions in arriving at these arbitary numbers. Although not every nest was found, and some known nests were being lost and replaced at new locations, all of these residents were believed to be nesting on or immediately adjacent to the census area, with territories overlapping it. However, even in this familiar situation, I was troubled with uncertainties about which birds were properly to be considered residents or merely visitors. I excluded from the set of birds "actually present" several species known to be in the vicinity but not believed to be occupying it regularly at the date of the censuses. One such was the Wood Thrush (*Hylocichla mustelina*), which nested on the tract regularly in previous years, and sang here before and after the test period but was not detected at any time during the 10-day interval. Others also excluded after internal debate were water and shorebirds, swallows and swifts, Common Flicker (*Colaptes auratus*), Hairy Woodpecker (*Dendrocopus villosus*), Rose-breasted Grosbeak (*Pheucticus ludovicianus*), and American Goldfinch (*Spinus tristis*). Still other species were seen here from time to time, and including them or not seriously affects the calculated census efficiency.

My census efficiency ranged from zero to 90% for different species. Lumping all the species I admitted into my sample, my efficiency for the entire set was 40%. When I omitted certain species judged unsuitable or doubtfully suitable for a census of this kind, my efficiency approached 50%; that is, I detected about half the pairs actually present and reasonably expected to be found in such a count. The birds of my set presenting difficulties because of wide-ranging habits or inconsistent song were as follows: Yellow-billed Cuckoo (*Coccyzus americanus*), Ruby-throated Hummingbird (*Archilochus colubris*), Downy Woodpecker (*Dendrocopus pubescens*), Blue Jay (*Cyanocitta cristata*), Tufted Titmouse (*Parus bicolor*), White-breasted Nuthatch (*Sitta carolinensis*), Cedar Waxwing (*Bombycilla cedrorum*), Starling (*Sturnus vulgaris*), and Brown-headed Cowbird. The most reliable subjects, with every male detected on more than 70% of the counts, were House Wren (*Troglodytes aedon*), Catbird (*Dumetella carolinensis*), Eastern Wood Pewee (*Contopus virens*), and Indigo Bunting (*Passerina cyanea*). Others more likely to be seen than missed in a quick count, with every male found on more than half the censuses, were Northern Oriole (*Icterus galbula*) and Robin (*Turdus migratorius*). However, since the numbers of birds in this sample are very small, these figures may reflect individual as much as species characteristics.

Surprises lay at the other end of the scale. It seemed remarkable that such a conspicuous and noisy species as the Blue Jay, which was seen many times each day, appeared only once in the test counts. It was also surprising to hear the song or call of the Chipping Sparrow (*Spizella passerina*) only once, although it was feeding fledglings on the area, and I could find it by deliberate search every time. In many species it was apparent a count earlier in the season would have yielded different results. Starlings and Common Grackles (*Quiscalus quiscula*) were already feeding large flying young and were ranging far beyond the census strip. Some Mourning Doves (*Zenaida macroura*) were between nestings, and Cedar Waxwings had not yet begun. Any date chosen will be wrong for some pairs.

No bird sings continuously, and a bird in heavy foliage is not likely to be detected if silent during a brief period of observation. For example, the Northern Oriole and Catbird, which were among the most reliable subjects in this test, were never seen on any of the counts. Had they been less vocal, like the Yellow-billed Cuckoo or Ruby-throated Hummingbird, they would not have been detected at all. These last two species I judged unsuitable for this kind of census at any season. Even birds we regard as conspicuous, like the Northern Cardinal (*Cardinalis cardinalis*) and Tufted Titmouse, were seldom detected when not singing. The female Brown-headed Cowbird was seen only once and the male not at all, although she laid several eggs on the tract during this period.

My results showed very loose correspondence for some of the species of Emlen's (1977b:461–462) study of a 48 acre tract of woodlands in Wisconsin, which also identified the House Wren, Catbird, Cardinal, and Indigo Bunting among the more dependable singers, but some differences in our findings were notable. For example, he heard the Northern Oriole much less often than I did; I detected the Brown-headed Cowbird rarely (female only), and he found it half the time; my pair of Song Sparrows (*Melospiza melodia*), which had nests on this tract before and after the study period, revealed themselves to me in only one-third of my counts, while his Song Sparrows sang on three-fourths of his transects. This enormous variability reminds us to expect large errors in brief surveys and small samples.

DIFFICULTIES WITH SEX RATIOS

Thus far I have been concerned with uncertainties in the counts themselves and have not mentioned a further step required to go from the number of singing males to the total number of birds in a population. This step also is fraught with problems not likely to be appreciated except by those engaged in intensive study of a species.

Since we usually find songbirds in pairs, we conveniently assume there are about as many females as males. Observers, however, commonly record more males then females but doubt this indicates an excess of males since they are more conspicuous. For example, in the Kirtland's Warbler unmated males are found occasionally, but unmated females are never seen. Are there no unmated females or are they just impossible to find?

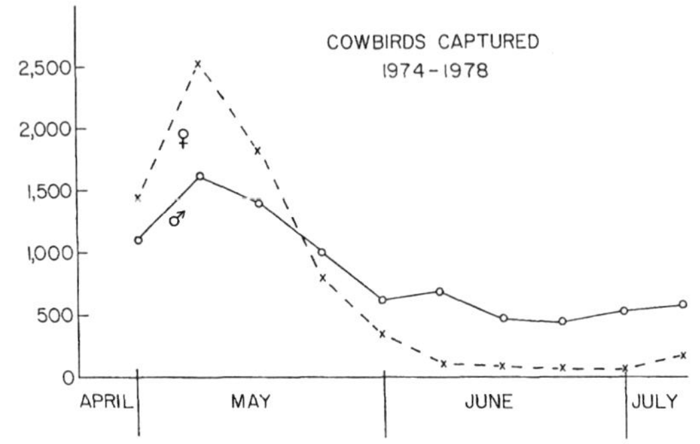

FIGURE 1. Points on chart show five-year totals for number of Brown-headed Cowbirds captured during each week of the breeding season in northern Lower Michigan.

Intuitively we expect females to be fewer than males because they are judged to be at greater risk sitting on nests than males flying freely. Indeed, in long-term studies of nesting populations, we often record higher survival rates for males, but we know also that males of some species are more likely to return to the same nesting site year after year, and thus more females survive than are counted.

The fact that nearly all males and females we see are mated and nesting is reassuring at first glance, but it leaves open the possibility some behave in a different way and do not conform to our assumptions. For example, how many yearlings breed and hold territories? We know that many yearling songbirds nest, but we find so few of them we cannot be sure all of them do so.

Among Prairie Warblers, Nolan (1978:359) did not find males outnumbering females. He found the apparent ratio changing as the season progressed, with females slightly outnumbering males during the height of the nesting season when all of the females were in breeding condition.

Sex ratios in Brown-headed Cowbirds vary widely as the season progresses, and observers have reached different opinions about the true sex ratio. I have analyzed a sample of 18,000 taken in five breeding seasons through trapping and removal of cowbirds in the Kirtland's Warbler nesting range (Shake and Mattsson 1975). This trapping came close to achieving the goal of the 100% sample, since it removed cowbirds so completely from the locality that we rarely saw this species outside the traps, and parasitism of Kirtland's Warbler nests dropped nearly to zero in most years.

The sex ratio changed by weeks in the five years, 1974–1978 (Fig. 1). In the total sample males outnumbered females by 22%. The ratio, however, varied greatly through the season and yet consistently from year to year. Clearly, males late in the season tended to move more than females into the vacuum created by trapping, and the variations may reflect differences in mobility rather than the true ratio of the sexes. In the first three weeks of each season, April 25 to May 14, females outnumbered males 5:3, but the ratio reversed in the last five weeks of the nesting season, June 5 to July 12, when males outnumbered females 4:1. These shifting ratios are difficult to interpret, even though the totals are comparable to those reported by other investigators using different methods in different regions, notably Darley (1971:563), who concluded males outnumbered females by 30–50%. These findings, with discrepancies of 30% or more, illustrate the uncertainties about sex ratios even among common and well-studied birds.

CONCLUSIONS

1. Censuses of singing males yield efficiencies below 50% for many species, and some common nesting birds cannot be censused effectively by the usual transect and spot-listening methods. Cuckoos and hummingbirds might head such a list, and woodpeckers also are candidates.

2. Singing behavior differs so much and sex ratios are so poorly known that generalizations from one species to another are untrustworthy, and the special problems with each species can be appraised adequately only after prolonged and detailed study of each.

3. People studying song and reproductive behavior should address themselves to the censusing problem and give informed judgments about the sources and magnitude of errors to be expected in standard methods of population counts.

Studies in Avian Biology No. 6:225–229, 1981.

BIRD POPULATIONS CONSIST OF INDIVIDUALS DIFFERING IN MANY RESPECTS

BARBARA DIEHL[1]

ABSTRACT.—Differences in the detectability of individual breeding pairs were analyzed, with special emphasis on least detectable birds. The mapping method was used to estimate numbers. The analysis is restricted to that part of the population whose nests had been located. It was found that the detectability of particular pairs ranged from less than three to almost ten records per ten visits. The proportion of least detectable birds depends on many interrelated factors, including habitat type, population density and breeding success. It also varies with time.

There is growing evidence that large differences exist among individual birds of the same population, and that these differences can play an important part in population processes responsible for stability (Dröscher 1974, Best 1977, Kuroda 1977, Payne and Payne 1977).

If so, it would be useful to recognize the nature of these differences, and to determine their range under various habitat conditions. I suggest that differences in the detectability (conspicuousness) of individual birds might be considered as one of the indices resulting from underlying behavioral or ecological factors varying between conspecifics within a given population. In this regard, it would be important to know whether clusters of few records (less than three per 8–12 visits when using the standard mapping method to estimate the density of breeding pairs) really represent surplus registrations, as suggested in the IBCC (1969) recommendations. Perhaps they represent special breeding bird categories, one of the characteristics of which is inconspicuousness to the observer. In this paper I attempt to shed some light on this problem.

That individual differences among birds are very large, and can have major effects on their detectability, was demonstrated when I observed the responses of Red-backed Shrikes (*Lanius collurio*) to nest inspection (B. Diehl, unpubl. data). At some nests there were no signs of the parents. They were, however, hidden not far from the nest. This became evident in a few cases when a nestling started to vocalize loudly when handled or touched; the parents revealed their presence at once. At the other extreme were those birds, both males and females, which attacked me vigorously as soon as I approached their nests. I could see and hear them at every nest inspection.

METHODS AND MATERIALS

The study was carried out on a meadow in Kampinos National Park near Warsaw, Poland. This meadow includes a mosaic of sites ranging from marshes with standing water more than 0.5 m deep in spring to dry sites submerged only after heavy rains. Clumps of trees, remnants of the forest cleared early in this century, are scattered throughout the area. The meadow had been grazed and mown up to the early 1960s, when a nature reserve was established. From that time the meadow has been in the first stages of forest succession with such trees as alders, willows, and birches the dominant plant species in the undergrowth.

Two adjacent plots, of 20.5 ha and 23 ha, were censused from 1964 to 1980 (the years 1966–1968 are excluded from the analysis because of an insufficient number of visits). On one plot clumps of trees and shrubs were more densely and more uniformly distributed than on the other, and this was reflected in the density of particular bird species.

The mapping method was used to estimate the number of breeding pairs. The number of visits per plot ranged from 19 in 1977 to 52 in 1964 (Table 1). They were generally made from late April to late July, although in some years they continued through August. Visits ranged from about three to four hours, sometimes five hours, early in the morning.

Adult birds were not individually marked, thus in the analysis of differences between individual pairs only those whose nests had been located were used. Due to this, it was possible to reduce the error that may occur when delimiting boundaries among clusters of registrations on the species maps, particularly with birds with few registrations that otherwise might be considered surplus birds. The number of nests located is shown in Table 1: only first broods are considered.

Since species detectability varies largely with the phase in the breeding cycle, four phenological periods have been distinguished, and most results are calculated for each of them separately. These are (1) the prelaying period, (2) the laying and incubation period, (3) the brooding period, and (4) the first two weeks after fledging. These periods were individually timed for each pair, and the number of censuses falling within each of these periods was then determined for each pair, the number of registrations on these censuses was then counted from the species map. The date of the first observation of any given bird was accepted as the date of its arrival from the wintering grounds. This could bias the results toward increased detectability indices in the prelaying period since some time could elapse between the actual date of arrival and the date of the first observation. Some birds were not registered at all over the prelaying period. In such cases an assumption was made that they were present in their territories at least one week preceding the onset

[1] 03388 Warszawa, Wysockiego 22 m 82, Warsaw, Poland.

TABLE 1
THE NUMBER OF CENSUSES AND BREEDING PAIRS WITH LOCATED NESTS IN PARTICULAR STUDY YEARS

Year	Total number of visits per plot	Red-backed Shrike: Successful breeders	Red-backed Shrike: Unsuccessful breeders	Red-backed Shrike: Total	Barred warbler: Successful breeders	Icterine warbler: Successful breeders
1964	52	12	9	21	2	0
1965	38	10	9	19	2	0
1969	31	13	5	18	12	0
1970	23	13	11	24	8	0
1971	28	13	17	30	5	2
1972	27	8	28	36	7	1
1973	28	16	10	26	5	0
1974	32	7	25	32	2	0
1975	23	10	16	26	7	0
1976	32	5	18	23	2	3
1977	19	10	14	24	6	4
1978	34	8	6	14	4	1
1979	32	1	13	14	5	1
1980	36	2	5	7	2	4
Total	435	128	186	314	69	16

of laying, and the number of censuses made during this period was used in further calculations for the prelaying period.

My visits were not evenly distributed over the breeding season, ranging from one to 14 for individual pairs in different phenological periods over all the study years. Thus the data for all years have been pooled to increase the reliability of the results, and an average number of registrations per breeding pair per 10 visits was calculated for each of the phenological periods. Also, the frequency distributions of birds differing in detectability were computed as mean values for all the study years.

This analysis was done only for successful breeders, as it is frequently impossible to precisely time phenological periods for unsuccessful breeders.

Three species are analyzed here: Red-backed Shrike, Barred Warbler (*Sylvia nisoris*), and Icterine Warbler (*Hippolais icterina*). All of them are late breeders, returning from migration either in the first half of May (Red-backed Shrike and Barred Warbler) or in the second half of May (Icterine Warbler). The four phenological periods total about two months.

The three species differ in their behavior. Red-backed Shrikes display very little vocal activity. They utter special vocalizations when establishing territorial boundaries, and thereafter are almost completely silent until the young fledge. Nevertheless, they are very conspicuous because, when hunting, they perch motionless on peripheral parts of shrubs at a height of 1–3 m and look for invertebrates moving in the grass around them. At that time they are well exposed to an observer, even over large distances, since the white-brown coloration of the male contrasts sharply with

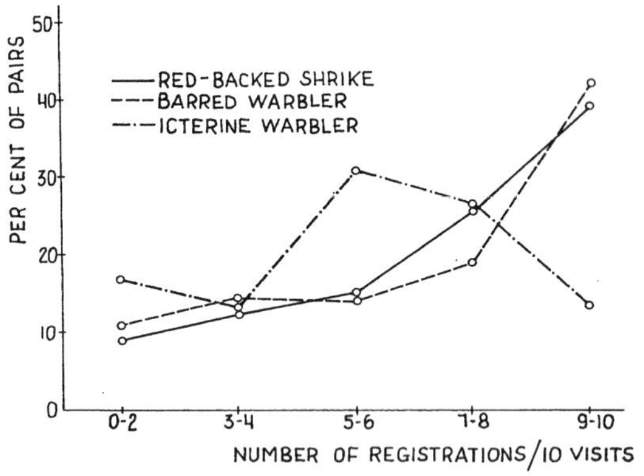

FIGURE 1. Frequency distribution of the detectability for successful breeders in the Red-backed Shrike, Barred Warbler and Icterine Warbler in the prelaying period (all the years pooled).

the background of green foliage. Consequently, this species was mostly recorded by sight. The other two species are typical song birds, and thus I primarily recorded singing males.

RESULTS

RANGE OF INDIVIDUAL DIFFERENCES IN DETECTABILITY

The frequency distribution of registrations for successful breeders in the prelaying period ranged from 0–10 records per 10 visits (Fig. 1). Between 10 and 20% of pairs had less than three registrations.

The proportion of birds with less than three registrations largely increased in the incubation, brooding and fledgling periods (Fig. 2), except for the Red-backed Shrike in the fledgling period.

These results suggest that at least some clusters of less than three registrations per 10 visits should not be considered as surplus ones. The number of such clusters tends to increase with time, thus it is advantageous to census these species in the prelaying period. The decrease observed for the Red-backed Shrike concerns only successful breeders, not unsuccessful ones (see below).

THE EFFECT OF BREEDING SUCCESS ON DETECTABILITY

Figure 3 shows the frequency distribution of registrations for successful and unsuccessful Red-backed Shrike breeders, calculated as an average for the whole breeding season. The proportion of birds with less than three registrations was about 40% for unsuccessful pairs, compared to 12% for successful ones. This proportion increased as the season advanced, reaching a peak of about 50% in the brooding and fledgling pe-

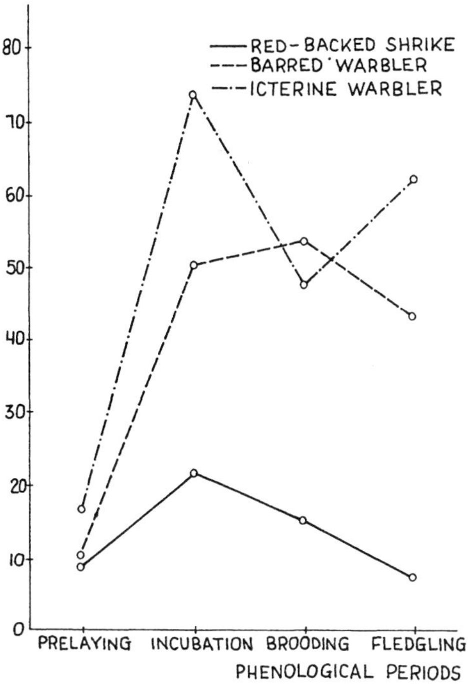

FIGURE 2. Frequency distribution of successful breeders with less than three records per 10 visits in four phenological periods (all the years pooled).

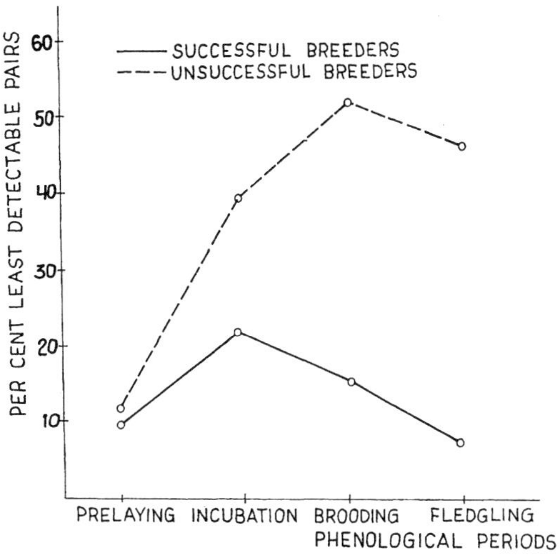

FIGURE 4. Frequency distribution of successful and unsuccessful Red-backed Shrike breeders with less than three records per 10 visits in four phenological periods (all the years pooled).

riods of successful breeders (Fig. 4). Thus individual differences in the conspicuousness of birds were additionally modified by their success at breeding.

The rate of brood mortality was density-dependent when a threshold population density was exceeded (Diehl 1976). For the Red-backed Shrike, this occurred on one of the study plots, which I call the high-density plot. On the low-density plot, average shrike density was lower by half. Figure 5 shows the relationship between brood mortality and the proportion of low-detectable birds in the Red-backed Shrike population. Thus the danger of underestimating the population increases with growing population density as a result of increasing predation on broods, followed by changes in adult activity affecting their conspicuousness.

Weather conditions including air temperature also affect breeding success (Diehl 1976), with the same effects on the detectability of adults.

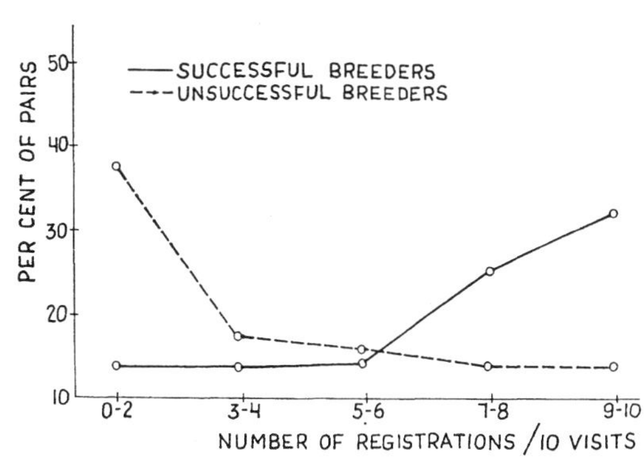

FIGURE 3. Frequency distribution of the detectability of successful and unsuccessful Red-backed Shrike breeders as an average for the whole breeding season (all the years pooled).

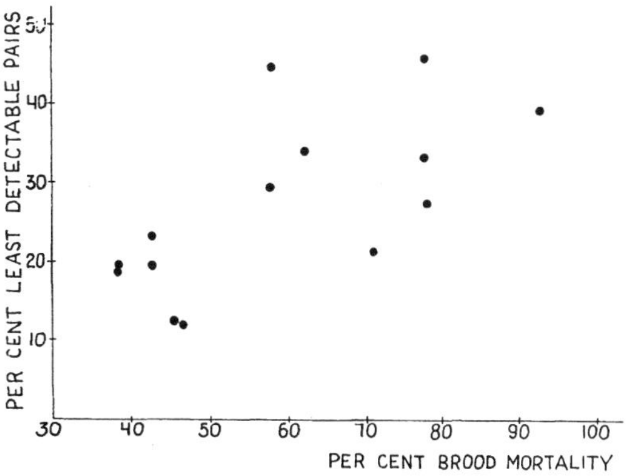

FIGURE 5. Scatter diagram of the proportion of breeding pairs with less than three registrations per 10 visits against brood mortality in the Red-backed Shrike in different years.

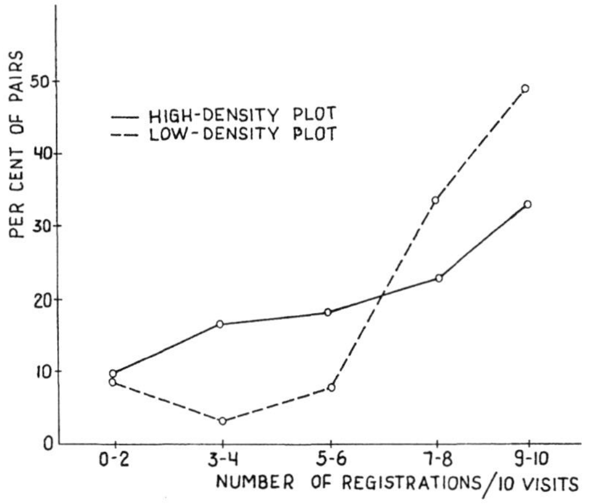

FIGURE 6. Frequency distribution of the detectability of successful breeders in the low- and high-density plots for the Red-backed Shrike in the prelaying period (all the years pooled).

Habitat-related Changes in Detectability

Apart from the differences in detectability related to breeding success, there were also differences within the group of successful breeders between the two plots. In the low-density plot the proportion of birds with few registrations was lower, and the proportion with a large number of registrations was greater, than in the high-density plot, particularly in the prelaying period (Fig. 6). As a result, shrike detectability was

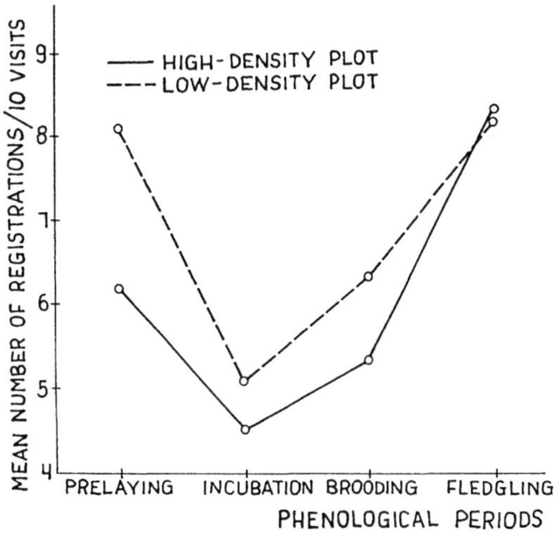

FIGURE 7. Mean detectability of successful breeders in the plots of low and high population density for individual pairs of the Red-backed Shrike in four phenological periods (all the years pooled).

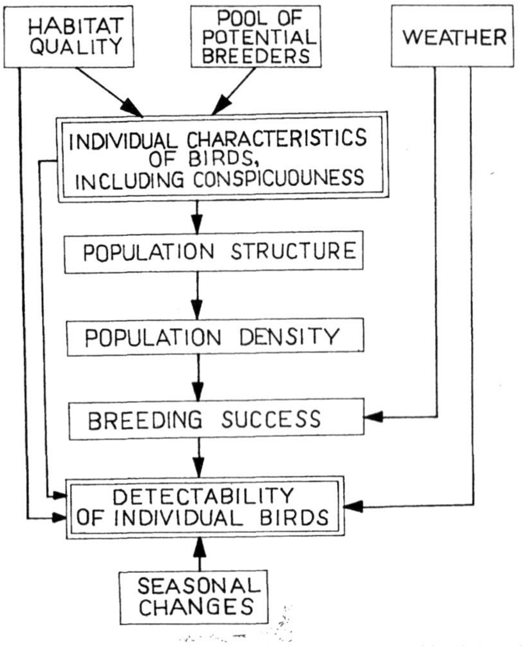

FIGURE 8. A scheme for the relation of individual variability to other factors influencing the detectability of birds in a population.

generally lower in the high-density plot compared with low-density plot (Fig. 7). This difference was most pronounced in the prelaying period (statistically significant at $P = 0.05$), then disappeared near the end of the breeding season.

The disappearance of this difference with time indicates that it is not simply related to habitat structure (more open in the low-density plot), instead, it suggests that there are some specific differences between the birds living in high and low densities. Again, the underestimation of population density as a result of rejecting clusters of less than three records can be greater in habitats with dense populations, at least for the Red-backed Shrike.

DISCUSSION

The present analysis reveals that there are large individual differences in the detectability of birds. Consequently, the effectiveness of the mapping method depends on the proportion of birds of various detectability in the population. The crucial point is the proportion of least detectable birds represented by clusters usually considered as surplus registrations. It can be very high during some periods in the breeding season. It does not seem possible to estimate their number without a greatly increased field effort. Is it worthwhile? The answer may vary depending on the purpose of the study. I would say yes if one wants to get a deeper insight into population processes.

The matter would probably be easier to cope with if we knew what other avian characteristics are associated with conspicuousness. In small mammals, for instance, individuals showing low trappability have low social rank, and are usually young (Gliwicz 1970). To my knowledge nothing of this kind is known for inconspicuous birds. There are, however, indications that great differences can exist between birds living in high and low densities. For instance, Red-backed Shrike males occupying the low-density plot were generally slightly less aggressive in response to the presence of an observer at their nests than those living in the high-density plot. Females were much more aggressive on the low-density plot (B. Diehl, unpubl. data). No correlation was found, however, between the aggressiveness and the detectability of those birds.

To characterize the possible importance of individual variation between birds to census efficiency against the background of other, usually interrelated factors, I propose a scheme in which this variability is an inherent component of the population structure (Fig. 8).

Studies in Avian Biology No. 6:230–234, 1981.

PROBLEMS OF UNEQUAL OBSERVABILITY

JAN EKMAN[1]

ABSTRACT.—Generally the assumption of equal observability (probability of capture) forms the basis for the application of census models: (a) observability must remain stable if results from direct counts are to be compared and give a true picture of changes in number; and (b) capture-recapture models require equal capture probability of individuals (homogeneity) to yield unbiased estimates.

Errors arising when the conditions under (a) are not met could be seasonally changing behavior. Line transect indices of Willow Tits increased in late winter to early spring, although the population was apparently stable and no immigration occurred, as shown by detailed capture-recapture estimates from individually color-banded birds.

Conditions under (b) may not be met when behavior differs between individual categories. In groups of color-banded Willow Tits using the same area, some individuals were consistently less liable to observation and identification than others. These differences in observability were traced to height separation while foraging in trees.

Census models usually require that capture probability (observability) does not vary in time, between habitats (line transect census), or between individuals (capture-recapture models). The application of capture-recapture models may produce serious negative biases when the capture probability differs between individuals (heterogeneity) (Gilbert 1973). As biologists, however, we acknowledge individual variation in morphological traits and behavior as the very basis of natural selection and evolution. On this basis we should hardly expect the "equal catchability" assumption to be met in reality, and our confidence in census data has to be founded upon how robust models are to violations of their underlying assumptions. Gilbert (1973) and Carothers (1973) could, for instance, demonstrate that some capture-recapture models produce accurate estimates under certain conditions even when the individuals differ in their capture probabilities.

Unequal capture probability is not itself the only cause for the lack of accuracy in census estimates. The magnitude of the difference in individual capture probability, as well as the average capture probability for the entire population, are further critical attributes (Gilbert 1973). Successful application of census models will therefore not only require information on the heterogeneity itself, but also information on the distribution of capture probabilities, and preferably its behavioral bases.

Methods to identify and quantify heterogeneity are, however, poorly developed. Tests to reveal heterogeneity (Leslie 1958, Keith and Meslow 1968, Carothers 1971) are insensitive and their statistical justification has been queried (Roff 1973), but only recently have advances been made to quantify the effects of unequal catchability (Carothers 1979).

This work centers on attempts to identify and quantify heterogeneity in the capture probability of Willow Tits (*Parus montanus*) during census work in a population study. The approach developed here was to independently study the behavior of known individuals, rather than to start from the census data themselves. Further, different census techniques were operated simultaneously, and their results analyzed for temporal heterogeneity in capture probability.

METHODS

Willow Tits were studied in mature (about 70 years old) coniferous forest some 40 km east of Göteborg (Gothenburg), SW Sweden. The populations were censused either by line transect counts or by capture-recapture.

LINE TRANSECT

Willow Tits were censused along a 12.3 km trail. Between November 1968 and November 1975 this trail was censused by up to eight observers at the end of each month. From November 1975 to November 1978 censuses were conducted only every third month, but the trail was censused more times on each occasion. During census periods at the end of the months, only two observers censused on the same day. They usually started at the same place and time, moving in opposite directions. The census trail formed a loop, and both observers censused the entire trail. Censuses started within one hour after dawn, and, depending on weather conditions, took between four and six hours to walk (roughly 35 to 50 m/min). At the halfway point there was a lunch and resting break for 15 to 30 minutes. During periods with thick snowlayer the censuses were conducted on skis. As only two observers censused on the same day, each census period consisted of several days. When weather conditions allowed, census days were consecutive. However, no censuses were conducted when wind velocities exceeded 10 m/sec. We had no lower temperature limit where census work ceased. In practice the lowest temperatures encountered were around −15°C. All birds heard or seen were noted, regardless of their distance from the trail.

[1] Department of Zoology, University of Göteborg (Gothenburg), Box 250 59, S - 400 31 Göteborg, Sweden.

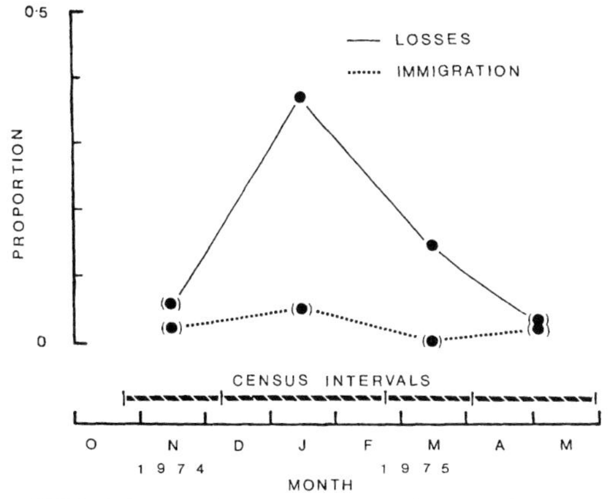

FIGURE 1. Disappearance and immigration rates of Willow Tits in winter (Jolly-Seber estimates. Brackets mark estimates not significantly larger than zero).

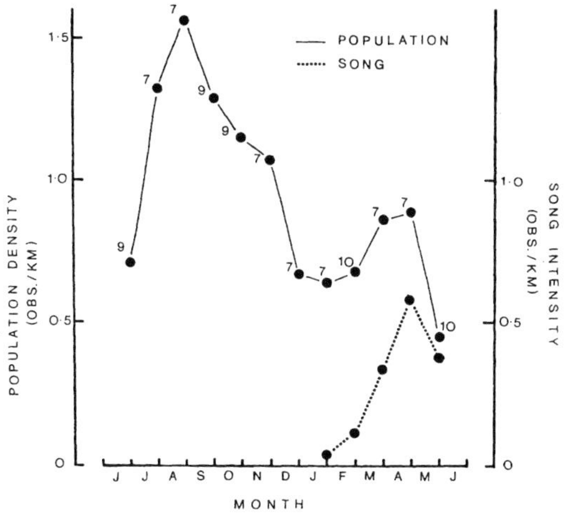

FIGURE 3. Pattern of seasonal change in line transect data for the Willow Tit. (Mean values for ten years census data. The figures show how many years data each point is based on.)

Willow Tits in the census trail area were never captured (or banded), so their observability was unaffected by human handling.

CAPTURE-RECAPTURE

Beginning in 1974, all Willow Tits were continuously banded in an 8 km² area whose southern border was roughly 400 m to the north of the census trail loop. Each Willow Tit received a unique combination of colorbands. This population was visited at bimonthly intervals, and samples of recaptures were collected by remote identification of their color combinations. Individuals could be, and usually were, observed and identified several times during each sampling occasion. However, no account was made for such repeated identifications in the capture-recapture sequence; it was only noted if an individual had or had not been identified. From the re-observation sequences, population size, survival and immigration was estimated, together with their variances, by the Jolly–Seber capture-recapture model (Jolly 1965, Seber 1965) using a computer and a slightly modified FORTRAN program (Davies 1971).

Winter survival in Willow Tits was age-specific (Ekman et al. in press) to an extent that necessitated separate calculations of adult and juvenile estimates (Manly 1970). Therefore no overall estimates of variance are available for population parameters as a whole, and tests for population changes had to be performed separately for adults and juveniles. Significant changes refer to t-tests with $P < 0.05$ within any of these groups.

Adult Willow Tits are highly sedentary (Ekman 1979), and all immigrants were therefore treated as juveniles. At the age of one year Willow Tits were considered recruited to the adult cohort.

The study area was provided with nest boxes; these were checked and natural nests searched for. The total number of nests found each year provided an independent control of our capture-recapture estimates of the breeding population.

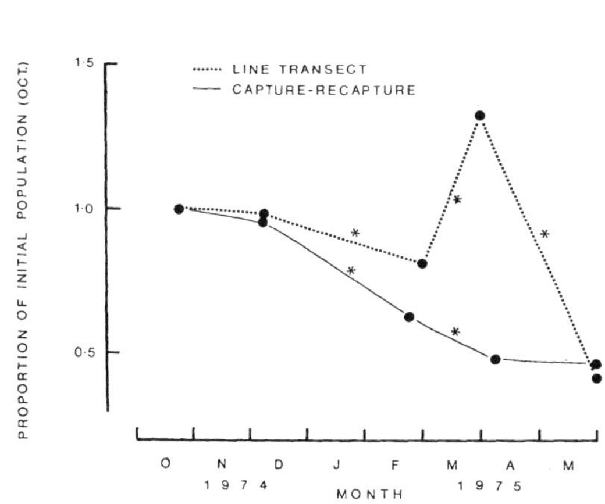

FIGURE 2. Seasonal changes in Willow Tit numbers measured by line transect census and capture-recapture. (Significant change between two samples indicated by asterisk by the line.)

BEHAVIORAL STUDIES OF OBSERVABILITY

Willow Tits organize into small groups in winter. The groups contain a stable set of individuals and use restricted bordering, but non-overlapping, winter ranges (Ekman 1979). These groups, due to their sedentariness and stability in composition, provide excellent opportunities for studies of individual behaviors and mutual relationships. Registrations of observability were collected from such groups with known composition of individuals by keeping them under continuous surveillance for some hours while collecting identifications at random. After the identification of a banded bird, no further registrations were collected until the last identified bird had left the tree where it had originally been spotted. I then deliberately looked away so as not be see where it landed.

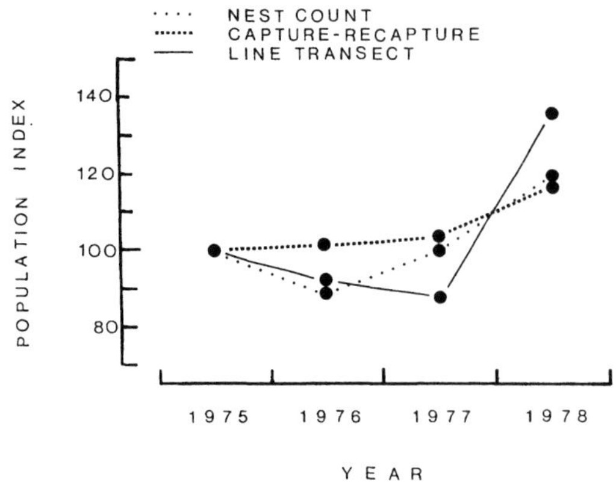

FIGURE 4. Annual fluctuations in Willow Tit numbers measured by three different methods.

TABLE 1

OBSERVABILITY DIFFERENCES BETWEEN WILLOW TIT GROUP MEMBERS (ONE SELECTED EXAMPLE)

Individual category	Relative capture probability	Number of identifications[a]
Female, adult	0.183	37
Male, adult	0.208	42
Female, juvenile	0.277	56
Male, juvenile	0.332	67

[a] Distribution of identifications differs significantly from a random expectation ($\chi^2 = 12.04$, $P < 0.001$).

I then resumed searching and the very first flock member spotted was identified. Identifications were registered on a tape recorder, which also gave the sequence in which group members were identified.

Simultaneously with the identification, the height at which the birds were first spotted was recorded, with height described as one of five height classes of the tree in which a bird was found.

RESULTS

SEASONAL DIFFERENCES IN OBSERVABILITY

Jolly-Seber estimates from the color-banded Willow Tit population could not verify any immigration in winter (Fig. 1). Hence, the population censused by capture-recapture steadily declined as the winter progressed, and the rate of this decline was determined solely by losses in the study population (Figs. 1 and 2). Presumably these losses were caused by natural mortality, as several tits were recovered dead (Ekman et al. in press), but no emigrants were found in neighboring areas.

Line transect censuses in the immediately adjacent forest simultaneously yielded increasing indices for Willow Tits in late winter and early spring (Fig. 2). Conceivably, the increasing line transect indices were caused by enhanced observability among local survivors, as we have no evidence for immigration to the banded population during the same period. Neither had there been any emigration from the banded population to the north into the census trail area, as no banded Willow Tits were ever found along the census trail. Hence, there are good reasons to believe that we have a bias in the line transect count censuses with seasonally changing observability. The conclusion that this is a consistent bias is reinforced by the fact that the spring increase of Willow Tit line transect indices is a regular phenomenon clearly reflected in the mean of ten years of census data (Fig. 3).

The increased observability of Willow Tits in spring could conceivably be connected to the onset of sexual activities and singing. The increase is fairly nicely timed to the increase in singing activities (Fig. 3), but other factors may also be involved, since song intensity reaches its peak in April/May without any corresponding steep increase in the population index in April. As the arthropod populations start to build up after the winter, a more mobile hunting strategy could pay for the tits (Norberg 1977), which would make encounters with a censusing observer more likely. Further, territorial defense may also call for conspicuous movements between different sections of the borders.

Data from line transect censuses are susceptible to changes in observability. This apparently makes them less powerful a tool to follow seasonal changes of Willow Tits. This objection does not, however, detract from the usefulness of line transect data in reflecting annual changes. Data obtained at the same phase of a seasonal cycle should still give an accurate picture of annual fluctuations. This point can be illustrated by comparing annual fluctuations in breeding numbers (early May) of Willow Tits estimated by different methods; (a) line transect census, (b) capture-recapture, and (c) nest counting. These three methods provide concordant patterns of annual fluctuation (Fig. 4).

INDIDIVUAL DIFFERENCES IN OBSERVABILITY

Winter social groups of Willow Tits are fairly uniform in composition, usually consisting of an adult pair and a juvenile pair (Ekman 1979). Group members are together most of the time, but are not equally liable to identification (Table 1). The difference in observability between group members did not vary at random. Old birds, particularly females, were observed and identified less often than young birds. Old females yielded the fewest identifications in all six four-member groups studied ($P = 0.0009$, Randomization test), and adults consistently yielded fewer observations than younger birds ($P = 0.03$, Randomization test).

The lower number of identifications for old females is due to low observability. An alternative hypothesis—that she might stick less to the group—is rejected because the identification sequences of other group members between consecutive identifications of old females do not depart from a random distribution (goodness-of-fit test; variance/mean-ratio = 1.025), indicating that she was present during the entire observation period. Otherwise we would have expected a bimodal distribution of the observation sequences of other group members, with one peak representing observations when old females were around, and one representing observations when they were away.

Low female observability is also suggested by their tree use pattern. While juvenile group members usually forage in the lower parts, old females are more frequently found in the upper sections, where the birds are more difficult to spot (Table 2).

TABLE 2
OBSERVATION HEIGHT OF DIFFERENT WILLOW TIT GROUP MEMBERS

	Tree height pentile	Frequency distribution of identification height[a]		
		Female, ad.	Male, ad.	Juveniles
Top	5	0	0	0
	4	0.15	0.01	0.02
	3	0.64	0.45	0.11
	2	0.16	0.44	0.48
Bottom	1	0.05	0.11	0.38
Total number of observations		61	85	89

[a] Differences in distribution of heights differ significantly from a random expectation ($\chi^2 = 80.99$, df = 6, $P < 0.001$).

DISCUSSION

Census models can produce misleading and seriously biased estimates when the underlying assumptions fail. In this study it is obvious that Willow Tit line transect data cannot be used for studies of population processes occurring within one season. For instance, it is not possible to apply *k*-factor analysis since the slope of the line where survivors are regressed against initial population will be altered from the true relationship if population estimates do not have the same bearing on numbers.

Other methods, like the capture-recapture model, will not break down entirely when the assumptions fail. Unequal catchability will introduce only insignificant bias in Jolly-Seber estimates when the capture probability exceeds 0.5 (Gilbert 1973). Proper adjustment of the sampling design, however, requires that the heterogeneity can be identified and, preferably, also quantified. In this context, the failure of traditional "equal catchability" tests (Roff 1973) is distressing. If heterogeneity can be identified and sampling design is adjusted properly, unbiased estimates can still be obtained for separate population strata. For Willow Tits, direct behavioral observations, instead of tests on census data, proved a covenient and powerful tool to demonstrate unequal observability. The reobservations from a population with strata differing in capture probability, but where capture probability follows a Poisson distribution within strata, will produce a compound Poisson distribution when pooled (Feller 1970). Roff (1973) stressed that "equal catchability tests" will fail to identify strata with differing capture probability for this reason. The tests performed on data collected from Willow Tit flocks, where members of known age, sex and rank are compared on an individual basis, evade this objection since the only variation included is between single individuals from different strata.

On basis of what is known about the biological basis of unequal catchability in Willow Tits it is possible to elaborate the sampling (here reobservation procedures) to reduce the effect of heterogeneity. Sampling Willow Tit observations involves two steps: (a) localizing the flock, and (b) identifying the flock members. As an overwhelming majority of flocks contain the strata treated here (adults, juveniles, males, females), and have largely the same composition, differences in catchability are encountered during the identification of flock members. Hence, it is the procedures used once a flock is found which determine how observability differences will affect the total recapture material. Assume that our sampling schedule allows for repeated identification of the same individual. From the data embodied in Table 1 on the relative difference in observability of flock members, it is possible to calculate the number of observations necessary in a flock of given (or guessed) size to yield observations of the least observable individuals with a given probability. For instance, if we want to observe adult female Willow Tits with a probability of 0.5 in flocks of four individuals, we can extract the relative adult female observability from Table 1 (0.183) and calculate the required number to approximately four by substituting these values into the expression for a Poisson distribution ($0.5 = 1 - e^{-0.183\ n}$, where

$e^{-0.183\ n}$ is the probability of finding no adult female (zero term of Poisson distribution) in n identifications). At capture, probabilities exceeding 0.5 estimates will be only marginally biased by unequal catchability (Gilbert 1973).

With calculations such as these, it is possible to elaborate sampling procedures and data processing, by e.g., increasing sampling intensity or treating subgroups separately, to alleviate the consequences of heterogeneity. The procedural rationale used here is to study the behavior of individuals in situations where environmental "noise" can be eliminated, because unequal catchability is an attribute of individuals, and has a behavioral basis. The power of this approach is demonstrated by the ease by which heterogeneity was demonstrated for the Willow Tit. With the biological basis of heterogeneity known, the logical corollary of quantifying the individual differences would simply entail a sampling of the kind of data presented in Table 1 for a number of groups. Further, identification of the behavioral basis of heterogeneity provides an opportunity to consider the behavioral plasticity of individuals. Therefore, this approach also has the potential of understanding the dynamics of heterogeneity. For instance, how would observability of Willow Tits alter if flocks broke up? In our census models we will have to represent the dynamics of heterogeneity and these dynamics cannot be represented realistically by simple correction factors based on comparisons of different census models applied on the same population. What we need to know is under which conditions behavior, and heterogeneity, change.

The approach of looking at behavior proved powerful for Willow Tits where several population strata (or presumably all, since flocks are usually of identical composition) are present in the same flock, and can be readily compared. The same method should apply to many social species where population strata can be found in company. Differences in behavior between individuals of species living solitarily are less tractable.

ACKNOWLEDGMENTS

My thanks are due to C. Askenmo, G. Cederholm, A. Enemar, and U. Unger at the Department of Zoology, University of Gothenburg, who all had a large share in planning and conducting these studies, and to all those ornithologists who in their spare time indulged in line transect census in blazing sunshine or in the cold of the winter. I owe a special thanks to C. Kepler for his patience in turning my writing into readable English.

Studies in Avian Biology No. 6:235–246, 1981.

METHODS OF DETECTING AND COUNTING RAPTORS: A REVIEW

MARK R. FULLER[1] AND JAMES A. MOSHER[2]

ABSTRACT.—Most raptors are wide-ranging, secretive, and occur at relatively low densities. These factors, in conjunction with the nocturnal activity of owls, cause the counting of raptors by most standard census and survey efforts to be very time consuming and expensive. This paper reviews the most common methods of detecting and counting raptors. It is hoped that it will be of use to the ever-increasing number of biologists, land-use planners, and managers that must determine the occurrence, density, or population dynamics of raptors.

Road counts of fixed station or continuous transect design are often used to sample large areas. Detection of spontaneous or elicited vocalizations, especially those of owls, provides a means of detecting and estimating raptor numbers. Searches for nests are accomplished from foot surveys, observations from automobiles and boats, or from aircraft when nest structures are conspicuous (e.g., Osprey). Knowledge of nest habitat, historic records, and inquiries of local residents are useful for locating nests. Often several of these techniques are combined to help find nest sites. Aerial searches have also been used to locate or count large raptors (e.g., eagles), or those that may be conspicuous in open habitats (e.g., tundra). Counts of birds entering or leaving nest colonies or colonial roosts have been attempted on a limited basis. Results from Christmas Bird Counts have provided an index of the abundance of some species. Trapping and banding generally has proven to be an inefficient method of detecting raptors or estimating their populations. Concentrations of migrants at strategically located points around the world afford the best opportunity to count many raptors in a relatively short period of time, but the influence of many unquantified variables has inhibited extensive interpretation of these counts.

Few data exist to demonstrate the effectiveness of these methods. We believe more research on sampling techniques, rather than complete counts or intensive searches, will provide adequate yet affordable estimates of raptor numbers in addition to providing methods for detecting the presence of raptors on areas of interest to researchers and managers.

The present paper reviews methods that have been used to detect and count raptors in a variety of geographic areas and habitats. The term "raptor" refers collectively to species of the orders Falconiformes and Strigiformes. In the past, most intensive studies of raptors were conducted by specialists who learned, through experience, methods to locate and count these birds. The methods used were often specific for certain species and limited in application to specific objectives or study areas. The techniques employed for studies of other groups of birds are often unsuitable for raptors. Additionally, the study area is often not large enough to obtain a meaningful sample of the raptor population.

In recent years people other than raptor specialists have been required to detect and count raptors. For example, information about raptors is necessary for comprehensive investigations of avian communities and studies of zoogeography (e.g., Cody and Diamond 1975). Governments at the national, state or provincial, and local levels are more intensively managing tracts of land as parks, reserves, or refuges on which birds of prey may be of special concern (e.g., Olendorff and Kochert 1977, Mathisen et al. 1977). Because raptors occur at low densities relative to birds at other trophic levels, the stability of their populations has been susceptible to persecution and man-caused changes in the environment (see Newton 1979). Laws and international agreements for protection of birds now require governments to monitor the status of avian populations (e.g., Conder 1977, Hilton 1977). In some nations, land-use planners, developers, and resource managers are required to assess the potential or actual impact of their practices on birds (e.g., Mathisen 1968, White et al. 1977). These requirements have led to increased study and management of raptors by people with diverse backgrounds and experience (or lack of it). It is, therefore, useful to provide a review of techniques for locating and counting these birds.

Furthermore, there is a need for much more development of methods for more accurately and efficiently estimating numbers of raptors. At the 1976 meeting (Ithaca, New York) of the Raptor Research Foundation, a special workshop was convened to discuss survey and counting techniques. Specialists encouraged publication of existing methods and research of new methods. More recently, the opinions of ecologists and wildlife managers in North America have focused attention on the need for improved survey and estimating techniques in order to address questions concerning population dynamics and management of raptors (Fuller, in press). The present review will emphasize some aspects of development of methodology that we hope researchers will pursue.

[1] U.S. Fish and Wildlife Service, Migratory Bird and Habitat Research Laboratory, Laurel, Maryland 20811.

[2] Appalachian Environmental Laboratory, Frostburg State College, Frostburg, Maryland 21532.

APPLICATIONS AND REQUIREMENTS OF METHODS

Occurrence

A survey to determine presence or absence is often the first step in evaluating potential impact of land-use on the status of a species. Surveys to detect raptors are also used to associate species with general habitats and resources and to delineate the geographic range of species and describe changes in distribution (e.g., Galushin, in press). These uses of information about the presence or absence of a species in an area do not require careful enumeration of the birds.

Numbers

If it is necessary to know the size of a population on a prescribed area, a census (complete count) or sampling (estimate of numbers) technique must be used. Sample counts are more efficient for large areas or for species that are very difficult to detect, but it is important to understand the variables that affect the accuracy of the estimate of the population. One should consider factors affecting the precision of counts in order to separate actual fluctuations in the size of the populations from those variables resulting from limitations in techniques. Papers published in Hickey (1969), Murphy et al. (1975), and Chancellor (1977) demonstrate the variety of data used to evaluate the status of species or trends in populations.

Population Dynamics

To be most useful, studies of population dynamics, including such factors as reproductive success, natality, and mortality, require censuses or estimates of population size for which the variability of the estimates has been calculated (see Brown 1974, Postupalsky 1974, Fraser 1978). Careful counts are also necessary for thorough descriptions of avian communities, evaluation of predator-prey relations, competitive interactions, and studies of ecosystems.

DIFFICULTIES AND GENERAL VARIABLES

Birds of prey are not, for the most part, easy to study. They nest at relatively low densities; they are usually wide-ranging and rapid-moving; many species habitually avoid areas of human activity; and most owls are more active at night than during the day. These characteristics make it difficult to gather quantitative data about raptors. In preparation for locating and counting birds, one usually assembles general information about the area and the habitats in which the species of interest are known to occur. The possibility for biasing search efforts exists if surveys are conducted only in habitats in which one expects to find birds. Some raptors are among the most widely distributed birds. Barn Owls (*Tyto alba*), Ospreys (*Pandion haliaetus*), and Peregrine Falcons (*Falco peregrinus*) are found over most of the world, and even many species that are more restricted in their distribution, such as Common Buzzards (*Buteo buteo*), Black Kites (*Milvus migrans*), Red-tailed Hawks (*Buteo jamaicensis*), and Great-horned Owls (*Bubo virginianus*), occur over vast areas and in a variety of habitats (see Burton 1973, and Brown and Amadon 1968). Also, some more localized species, such as the Ferruginous Hawk (*Buteo regalis*), have adapted to using a vast array of nesting sites (Call 1978).

The strategy of becoming familiar with habitats in which one is likely to find birds can be very useful, but should be applied with caution. We have met individuals who have developed excellent "search images" for the local habitat in which a species will most likely be found. Some of these experts are so specialized, however, that they may overlook other habitats in which the birds occur.

Observer ability, experience in searching for and identifying raptors, and knowledge of behavior are important factors influencing survey results (Call 1978) and a source of potentially great variability. Furthermore, differences among participants can be compounded when more than one species, in more than one habitat or terrain, is involved. There are often similarities in the "ecological role" of many species of raptors and many share a common susceptibility to certain impacts. Additionally, searches for raptors are often costly, labor intensive efforts. Consequently, surveys and counts frequently include more than one observer looking for several species over large areas of diverse habitat and terrain (e.g., Craighead and Craighead 1956, Rowan 1964, Murphy et al. 1969, Nagy 1977).

Other factors that influence surveys and counts include differences in visibility of birds because of seasonal changes in vegetation, in habitat use, and behavior. The behavior of birds also changes during the course of the day. For example, early in the morning, soaring birds such as buzzards and vultures may remain perched until the air warms and thermals or updrafts are created. Inclement weather may reduce the activity and thus the visibility of raptors if they seek shelter. Variables and assumptions will be addressed in relation to specific methods. It is important to realize that a survey or count conducted by different observers, under different conditions, can produce a great deal of variability in the results. Since an adequate yet affordable census of raptor populations is difficult to obtain under the best of conditions (Grier et al., in press), it is necessary to choose an appropriate sample method in order to avoid variability that may render results useless.

METHODS

ROAD COUNTS

To cover the large areas necessary to sample raptor populations, investigators have often used automobiles for transportation and roadways for transect routes (e.g., Hicks 1933, Winterbottom 1933, Nice 1934, Leopold 1942, Nankinov 1977). This method usually involves driving slowly (10–25 mph; 17–40 kmph) and counting the birds that one or two observers detect, usually within a specified distance (0.25–1.0 mi; 4–1.6 km), on each side of the road (e.g., Craighead and Craighead 1956). Most investigators will stop momentarily if a bird cannot be identified from the moving vehicle. Some road counts include regular stops (e.g., Vian and Bliese 1974, Bystrak 1979). Data can be summarized as number of birds seen per distance driven or the reverse (e.g., km/bird).

Several investigators (e.g., Hiatt 1944, Craighead and Craighead 1956, Cade 1969, L. Brown 1971, Craig 1978) have noted many of the variables that may affect the results of road counts. Most investigators acknowledge that the capability of observers is not uniform for factors such as detection and identification of birds or ability to judge the distance to the edge of the transect. The extent to which differences among observers affect counts has not been determined.

The structure of vegetation, terrain, the roadway, and developments nearby (e.g., fences, power poles, open ground, buildings, human activity) are "habitat" variables that influence the use of the area by birds and their visibility there. For example, *Buteo* hawks, which soar and perch in open habitats, are more likely to be detected than a forest-dwelling *Accipiter* (see L. Brown 1971, Marion and Ryder 1975). Owls generally cannot be counted along roadways. Craighead and Craighead (1956) compared the results of road counts with results of other techniques and established correction values to account for differences in visibility of several species of Falconiformes. Inclement weather and seasonal changes in vegetation and bird behavior may affect the results of road counts. The activity, and thus the visibility, of some birds of prey varies on a daily basis and should be accounted for if routes require several hours to drive (e.g., Bildstein 1978). When comparing routes on different areas, or over periods of time, it is also important to consider that communal roosts (e.g., in winter) and temporal abundances of food may result in "inflated" estimates of density. Similarly, one must be aware of the influx of fledglings on routes or the arrival of migrants. In addition, the presence of one species on an area may influence the behavior and visibility of other species (Craighead and Craighead 1956).

Despite the potential impact of these numerous variables, many investigators have been able to establish routes and conduct counts in a manner that makes the results comparable. Road counts can then provide values of relative abundance of birds of prey (e.g., Siegfried 1966, Rowan 1964, L. Brown 1971, Smeenk 1974, Bart 1977, Woffinden and Murphy 1977, Craig 1978, Phelan and Robertson 1978) or estimates of populations on a given area (Craighead and Craighead 1956, Woffinden and Murphy 1977). Road counts have been conducted to obtain data for particular seasons (Rudebeck 1963, Enderson 1964, 1965; Schnell 1967, Nankinov 1977, Bildstein 1978), or to detect changes in species diversity and numbers of birds during different seasons (Allan and Sime 1943, Call 1975). In several instances the results of road counts have been used to compare the occurrence of raptors in different geographic areas (Hiatt 1944, White 1965, Mathisen and Mathisen 1968, Call 1975). Data from road counts have also provided insight into long-term trends in the numbers of raptors in an area (e.g., Cade 1969, L. Brown 1971, Johnson and Enderson 1972).

A road count survey is a useful method when large areas or many species need to be counted. The Breeding Bird Survey (BBS) of North America is a relatively standardized road count, employed over a vast area (Bystrak 1979). Regrettably, not all raptors are observed often enough on BBS routes to allow the use of statistical analysis for demonstrating changes in numbers. For example, counts of only five species of Falconiformes occurring in the eastern United States and nine species from the western United States were sufficient to test for significant changes in number of birds detected from 1967 to 1971 (U.S. Department of the Interior 1971). We encourage research that will develop techniques, and/or correction factors that will increase the usefulness of road count data for a greater variety of raptors. In the meantime, road counts will be conducted as supplements to other methods for counting birds of prey (Craighead and Craighead 1956, Southern 1963, 1964; Murphy et al. 1969, Misztal 1974, McKay 1976, Rogers and Dauber 1976, Petersen 1979, Sykes 1979).

VOCALIZATIONS

The vocalizations of birds allow many species to be detected along road counts or on study plots. Detection of the calls of raptors, especially strigiforms, has been used in many studies. One approach is to simply listen for the calls of the birds and note or plot their approximate location (e.g., Baumgartner 1939, Bell 1964, Hinz 1969, Petersen 1979, Bystrak 1979). Vocalizations of some species may be elicited by imitating the call(s) of the species (e.g., Miller 1930, Marshall 1939, Simpson 1972, Rogers and Dauber 1977) or by broadcasting a recording of a vocalization (Beatty 1977). It is with this latter technique that the most "standardization" has occurred. The locations from which recordings are played have been from 0.4 to 1.6 km (0.25–1.0 mi) apart. The shorter intervals are used for counting the smaller birds which presumably range over less area. Generally, several calls are played, followed by a period of silence (15–60 seconds) after which this sequence is repeated. Investigators usually stay at each broadcast stop for 4–15 minutes. If the vocalization of more than one species of raptor is being played, it is suggested that the call of the smallest birds be played first because response behavior of some birds may be inhibited by the "presence" of larger competitors or predators. Nowicki (1974), Cink (1975), Forsman et al. (1977), Springer (1978), Ortego (1979), and Johnson et al. (1979), among others, have discussed the rationale for protocols of broadcasting vocalizations.

A number of variables must be considered when attempting to detect or count raptors by listening for calls or eliciting responses. Comparatively little work has been conducted with Falconiformes, so the points discussed below generally refer to owls. Peterson (1979) noted day-to-day variability in whether birds on an area gave territorial calls. Several species are more vocal or more responsive to recordings at certain times during the year, and then, within certain hours of the day (Grant 1966, Nowicki 1974, Smith 1978, Siminski 1976, Berggren and Wahlstedt 1977, Forsman et al. 1977, Springer 1978, Postovit 1979). Johnson et al. (1979) found trends of calling associated with lunar cycles.

Many owls have a repertoire of several vocalizations, not all of which can be heard by a human more than a few meters from the bird, and many of which are given in a behavioral context not likely to be elicited by the investigator. Both females and males of some species are known to respond to recordings (e.g., Marshall 1939, Smith 1978, Forsman et al. 1977); however, for some owls, sexual differences have been found in the number of broadcast calls necessary to stimulate a vocal response, the number of calls given in response, and the distance within which a bird will approach the loud-speaker (Siminski 1976, Springer 1978). Other investigators have questioned whether both sexes call and whether only mated birds respond (Nowicki 1974, Cink 1975). In some instances a bird will approach the source of a call, but not respond by vocalizing (Nowicki 1974). Many species respond to a human's imitation of their calls and to the vocalizations of other species (Miller 1930, Foster 1965, Fitzpatrick 1973, Ortego 1979). At the other extreme, Siminski (1976) elicited fewer responses by Great-horned Owls in Ohio to recordings of calls of the species from New York and Oregon than from the call of another Ohio owl. Thus, local dialects may influence the degree to which a bird responds. There is some literature about behavior and vocalizations (e.g., Marshall 1939, Haverschmidt 1946, Ligon 1968, Emlen 1973, Martin 1974, van der Weyden 1975) with which one should be familiar if surveys or counts of calls are to be conducted, but more research is needed about the factors associated with vocalizations, especially elicited calls.

When broadcasting of a recorded call is used to elicit a response, the variables associated with behavior are compounded by factors related to equipment. Field workers have used many kinds of tape recorders, amplifiers, and speakers, all of which may affect the accuracy with which a call is broadcast and the distance the sound will transmit. "Background" noise produced by the equipment, or in the recording, and environmental noise (e.g., traffic noise, wind) may interfere with efficient transmission. Additionally, the structure of vegetation and terrain will influence sound transmission.

If objectives include an estimate of the number of birds present in an area, it is important to know the range over which the sample is being taken. Several people have compared the number of birds responding to broadcasts with other estimates of the number of owls in the same area and found that from 75.0 to 82.6% responded (Siminski 1976, Forsman et al. 1977, Springer 1978). Numerous investigators have based estimates of the number of owls in an area on responses to broadcasts or imitated calls, or counts of hoots (e.g., Nowicki 1974, Cink 1975, Smith 1978, Garcia 1979) and others have used these techniques as supplements to road counts or nest searches (e.g., Hinz 1969, Rusch et al. 1972, Call 1978, Hennessy 1978, Petersen 1979).

We are presently testing the feasibility of using responses to broadcast vocalizations to estimate numbers of breeding raptors in forested habitats. To date, we have played the calls of Red-shouldered Hawks (*Buteo lineatus*), Red-tailed Hawks, Broad-winged Hawks (*Buteo platypterus*), Goshawks (*Accipiter gentilis*), Cooper's Hawks (*Accipiter cooperii*), and Barred Owls (*Strix varia*) on the study areas of the Central Appalachian Raptor Ecology Program in western Maryland, and in northern Wisconsin, northern New Hampshire, and northern Connecticut. The calls are broadcast from roadside routes that run through the center of study areas encompassing about 32 km². From March through June, the study areas are systematically and completely searched on foot for all raptor nests. Additionally, field workers record all contacts made during various searching activities (driving, walking, sit-and-watch, etc.) in order to provide data about the distribution of birds on the study areas, and to allow evaluation of activities that lead to the most contacts for each species.

Based on preliminary results, all species were responsive to calls recorded from commercial bird song records, and these responses enabled us to make more contacts than by only stopping to look and listen for birds on our roadside counts. For example, in 1980 only two contacts were recorded for Cooper's Hawks or Red-shouldered Hawks in western Maryland by stopping to look and listen. However, during and after the sequence of broadcasting calls, 10 Cooper's Hawk and 18 Red-shouldered Hawk contacts were made. The test species appear to be most responsive during the period from arrival on their nest area until egg laying, less responsive during incubation, and moderately responsive during the fledgling period. For example, during the three weeks prior to incubation, eight contacts with Cooper's Hawks (10 with Red-shouldered Hawks) were recorded, compared to two contacts during the first three weeks of incubation (one contact for incubating Red-shouldered Hawks), and one during the two weeks following incubation (four contacts with Red-shouldered Hawks) (Mosher, Fuller, Kopeny, unpublished data). Thus far, the results from responses to the broadcasts are consistent with the distribution of the target species, in that birds are contacted at those stops along the routes near raptor nests. We will be continuing this work during the 1981 field season and establishing additional study areas in northern New York and southern Michigan. We emphasize that these are preliminary results and we expect that standardization and testing of this technique will require several more field seasons, but it may ultimately provide an efficient method of detecting and counting many species of forest-inhabiting hawks and owls.

Searches for Nests

Data from surveys to locate active nests are useful for management and protection of breeding birds (e.g., Mathisen et al. 1977) and have often been used as the basis for estimating the number of birds on an area. In addition, once nests have been located, observation of adults and monitoring of egg-laying, hatching, and fledging, can provide data for studies of population dynamics. Consequently, much of the effort devoted to surveys or counts of raptors has been concentrated on nest searches. Before presenting the nest-search techniques, it is useful to review some of the factors that affect the search process and influence the results.

As always, there are potential differences among the abilities of observers to detect nests. Grier et al. (1981) found that with three experienced observers, the proportion of nests found on a study area varied from 67–87%. Few surveys of birds of prey have evaluated observer bias, but studies have been conducted to determine visibility bias associated with detecting nests on different search forays (Fraser 1978), in different habitats (Henny et al. 1977), in different seasons (Craighead and Craighead 1956, Grier et al. 1981), and in finding nests of different species (Craighead and Craighead 1956, Call 1978, Postovit 1979). Light conditions, altered by time of day or weather, may also alter the visibility of nest structures or of evidence to indicate the presence of a nest (Call 1978, Grier et al. 1981). Logistic limitations of some search methods (e.g., searching from a boat) may inhibit or restrict the observer from seeing areas in which nests may also occur (Craighead and Craighead 1956, Wiemeyer 1977, Call 1978). After a nest has been detected, the structure may blow down (Mathisen 1977), the pair may use an alternate nest, or the nest may have been built by a non-breeding pair (e.g., Ratcliffe 1962, Boeker 1974, Stocek and Pearce 1978). Therefore, many surveys or counts must rely on observations of the behavior of birds at the nest to confirm use of a site, and to verify which species is using it.

The observation of raptors near a nest, especially early in the breeding season, is not always evidence the nest is active (Stocek and Pearce 1978, Hodges et al. 1979). The practices of renesting or multiple clutching by some species further complicate the interpretation of limited observations at a nest (Call 1978). The behavior of raptors near a nest creates visibility biases because birds of some species perch and roost nearby, whereas others are not often seen near the nest. Some species react to intrusion by flying away quietly; others may vocalize or fly about or attack an investigator (e.g., Call 1978, Postovit 1979). Many Falconiformes are territorial at the nest site (Newton 1979), thus, a pair of birds can be counted for each active nest which is found (for species that do not maintain alternate nests). However, large or overlapping hunting ranges (Picozzi 1978, Newton 1979) and the existence of semi-colonial nesting, for example by Ospreys (Mathisen 1977), polygamy, and nest-helpers of some species (Hamerstrom 1969, Mader 1979, Faaborg et al. 1980) complicate the use of nest counts for estimating the number of birds present in an area.

Searching for nests and attempting to obtain observations of raptors at the nest can lead to disturbance and nesting failure. This problem has been addressed by several authors (e.g., Hickey 1969, Fyfe and Olendorff 1976) and is discussed in many of the papers cited below.

The objectives of a study may greatly influence the strategy and time allocated for locating nests. If a study area has been delineated and the objective is to locate *all* nests, of some or all of the birds of prey, there is little choice but to initiate a systematic search of the entire area staggered over the nesting period (e.g., Craighead and Craighead 1956, Murphy et al. 1969, White et al. 1977). A more efficient approach in terms of time and people required is to sample the study area in either a random or stratified manner (Postovit 1979). When only certain species are of interest, and a census of nests is not practical, many investigators preselect certain habitat types in which to conduct searches. Call (1978), Reynolds (1975), Kennedy (1977a), and others suggest learning the habitats of birds and using maps and photos of the study area to identify those habitats in which the birds are most likely to occur. The availability of topographic maps, air photos, satellite imagery, and soil and forest cover-type maps for many areas permits careful reconnaissance before entering the field. Detailed descriptions of nesting habitat exist for many species (e.g., Hickey 1942, Call 1978), but one must be aware of the variety of habitats in which raptors breed (e.g., Hickey and Anderson 1969, Jones 1979) and the existence of local variability by members of some species (Call 1978, Jones 1979). Counts and estimates of population size for areas in which only the "most likely" habitat was searched should acknowledge this bias (e.g., Grubb et al. 1975, Stocek and Pearce 1978, Titus and Mosher, in press).

Historic records

The literature, of course, provides information about general nesting habitat use by birds. For some species or populations which have relatively restricted nest site requirements and use or build structures that last many years (e.g., caves, ledges, large trees), historic data can lead one to specific nest areas or nest sites. Ratcliffe (1972), Lindberg (1977), and others (see Hickey 1969) were able to document the decline of nesting Peregrine Falcons by visiting eyries that had been described in the literature and in records of museums, ornithologists, and falconers. Historic records have played an important role in understanding the status of the California Condor (*Gymnogyps californianus*; Wilbur 1978a) and many other large, conspicuous raptors that are very traditional in their use of nest sites (Newton 1979). The status of nesting populations of Bald Eagles (*Haliaeetus leucocephalus*) and Ospreys has often been monitored with the aid of historic nest site data (e.g., Howell and Heinzman 1967, Newman et al. 1977, Sindelar 1977). In addition, valuable information about other species, such as Barn Owls (Smith and Marti 1976), Common Buzzards (Tubbs 1974) and Harriers (*Circus cyaneus*) (Watson 1977) has also been obtained by examining records (e.g., nest record programs of the British Trust for Ornithology, and the Laboratory of Ornithology at Cornell University) and by contacting people in the area of the study.

Local inquiries and questionnaires

Often, some people who live in an area are familiar with the location of nests of birds of prey. Inquiries of these people (e.g., Brown 1974, Saurola 1976, Roberts and Lind 1977, Sindelar 1977) and questionnaires sent to local wildlife managers and amateur ornithologists have often formed the basis for nest searches or an evaluation of the status of local nesting birds (e.g., Baldwin et al. 1932, Prestt 1965, Oberheu 1977). In the Soviet Union, local conservationists and birdwatchers contribute to a "bounty" fund for those people who locate raptor nests (M. S. Dolbik, A. M. Dorofeev, and V. M. Galushin, pers. comm.). In this system, greater rewards are paid for locations of the rarer species. Foresters, farmers, and other interested people have found previously unrecorded nests of Ospreys, eagles, and other uncommon raptors. Villagers helped Kennedy (1977a) find Philippine Eagle (*Pithecophaga jefferyi*) nests and volunteers participated in locating California Condors (Wilbur 1978b). Utilization of information provided by people in the area of study and historic records may reduce the time and expense spent with other techniques in searching for nests.

Aerial searches

The use of aircraft, though seemingly expensive, is an efficient method of searching for nests over large areas. Planning for use of aircraft, evaluation of types of aircraft, safety precautions, and costs have been discussed in several papers (Hickman 1972, White and Sherrod 1973,

Carrier and Melquist 1976, Grier et al. 1981). The nests of species that nest in relatively open situations (e.g., cliffs or tundra) such as Golden Eagles (*Aquila chrysaetos*), Rough-legged Hawks (*Buteo lagopus*), Peregrine Falcons, and Gyrfalcons (*Falco rusticolus*) are often easily detected from the air (e.g., Boeker and Ray 1971, Swartz et al. 1975, Pennycuick 1976, White et al. 1977). Osprey nests, according to Henny et al. (1978), are ideally suited for air searches because they are conspicuous, the nest habitat is limited, and nest cycles are synchronous. Henny et al. (1974) developed aerial visibility rates to correct for differences in ability to detect nests located on different structures (e.g., trees, channel markers) and have since used this technique for estimating numbers of Osprey in several regions of North America (Henny and Noltemeier 1975, Henny et al. 1977, Henny et al. 1978a, Henny et al. 1978b, Henny and Anderson 1979). Wetmore and Gillespie (1977) and Prevost et al. (1978) have also used air searches to locate the nests of Ospreys.

Leighton et al. (1979) derived correction factors for detecting Bald Eagle nests from the air as they searched randomly selected units containing a uniform amount of "primary" nesting habitat. Grier (1977) estimated the population size of Bald Eagles and number of breeding areas on a study area in Canada with 95% confidence intervals, and by stratifying the samples, was able to reduce the variance of the mean estimates by about 22%. Grier and Hamilton (1978) subsampled clusters on a stratified basis and used optimum allocation of samples in an effort to further reduce variance. No significant reduction in variance was achieved, but the survey did reveal a significantly different number of nests in different habitat strata, and the sampling scheme reduced the flight time needed for the surveys by about 15%. Grier et al. (1981) found that about 76% of the nests and 85% of the breeding areas of Bald Eagles were detected by one air search, and a total of 94% and 98% were found on a second flight over the same area. Fraser (1978) also found that not all nests were seen on all flights.

When the nests on a study area have been found, there is usually an opportunity to gather data about reproductive parameters. Fraser (1978) conducted "experimental" two-flight surveys over a well-studied Bald Eagle population and found that errors occur in classification of nest occupancy and activity, in judging the chronology of nesting, in counting young, and in estimating other parameters. He discussed the importance of timing of flights, between year differences in results, statistics for monitoring reproduction, and other factors one should consider before initiating air searches.

Howell (1973), Whitfield et al. (1974), Swenson (1979), and others have used air surveys in conjunction with locating and monitoring nests from the ground, counting active nests, and checking productivity (e.g., Herman 1971). Several people have used air searches to complement other techniques for finding nests. Before the leaves emerge on deciduous trees, the stick nests (or old nests) of forest-dwelling species can sometimes be located or raptors may be flushed and sighted (Luttich et al. 1971, McGowan 1975, Petersen 1979). Visible signs associated with raptor presence, such as the white excreta deposited on the cliff face at nests or perches and the orange foliose lichen associated with these sites, can be useful for locating nests from the air and from the ground (Call 1978).

Searches from automobiles

As with searches from the air, location of bulky stick nests from vehicles can be accomplished in some habitats before leafout of deciduous trees (e.g., Hinz 1969, Boswell 1974, Fitch and Bare 1978, Kirkley and Springer 1980). In open habitats such as shrublands, deserts, grasslands, or cliffs, nest sites or signs such as excreta may be detected throughout the nesting season (e.g., Platt 1971, Call 1978). Also, while traveling relatively quickly over large areas by automobile, the behavior of raptors can be noted (e.g., courtship displays, food carrying, and food exchanges) and sightings of birds can be mapped. In this way it is possible to delineate the area in which one is most likely to find a nest (e.g., Craighead and Craighead 1956, Hamerstrom 1969, Call 1978).

Searches from boats

The nests of falcons (e.g., Cade 1969, Oliphant and Thompson 1978), eagles (Whitfield et al. 1974, Hansen 1977, Call 1978), and Ospreys (Reese 1975, Kennedy 1977b, Wiemeyer 1977) which occur along cliffs, lakes, and rivers, have often been found during surveys from boats. Nest surveys from a boat were used by Sykes (1979) to supplement Snail Kite (*Rostrhamus sociabilis*) nest searches and could be effectively used to locate the nest sites of many other species that nest near the shorelines, bays, lakes, and rivers, and in swamp and marsh habitats.

Searches on foot

Walking through areas, looking for nests, and pausing at vantage points to watch for birds, is probably the most common method of finding the nests of raptors (e.g., Newton et al. 1977, Picozzi 1978). Though time-consuming, searching on foot provides more opportunities to see well-concealed nests, to tap a stick against trees containing cavities in which an owl or small falcon may be incubating, to use a dog to sniff for a ground nest, or to elicit vocalizations or defensive flights of a bird (Call 1978). Areas in which a concentration of sightings has occurred, or where the nest of a previous year has been plotted, can be searched carefully on foot. When walking, one can look for molted feathers, the "butcher-block" or prey plucking areas, or excreta under roosts and nests (Craighead and Craighead 1956, Call 1978). Knowledge of nest habitat and the location of one or two nests can lead to the likely location of other nests. Based on information about spacing between nest sites, or the size of the area used by a pair of breeding birds, it is possible to estimate density for an area larger than one's study area. Newton (1979) reviewed the major factors that are likely to influence the dispersion of breeding raptors. These factors (e.g., nest structure, habitat, prey base) are likely to vary over space and time, therefore caution is recommended in regard to extrapolating densities from a local study plot to larger areas. Sampling of density over this larger area of concern should provide a relatively accurate estimate of breeding density.

Estimates of the number of nests on large areas can be obtained relatively efficiently on foot if one samples a portion of the area. Postovit (1979) used simple random and stratified random samples to search up to 33% of his total study area (233 km^2), and was able to locate nests with 63% precision. He believed that sampling a larger proportion of the area (more field workers to conduct the searches in the short period of nesting) and a correction factor for visibility biases for some of the 13 species he observed, would allow increased precision.

Multiple techniques

To survey large areas in short periods of time, many investigators have used more than one technique to locate nests. The objectives of studies in which two or more techniques were used have included gathering basic information about nesting habitat, breeding chronology and reproductive parameters (e.g., Lahti 1972, Misztal 1974, Kennedy 1977b, Newton et al. 1978, Bednarz 1979, Sykes 1979, Titus and Mosher, in press), determining the status of, or monitoring a population of a species (Brown 1964, Howard et al. 1976, Fyfe et al. 1976, Brown 1977, French and Koplin 1977, Stocek and Pearce 1978, Mattox et al. 1980), and evaluating the relationships between prey density and raptor density (e.g., Craighead and Craighead 1956, Murphy et al. 1969, Phelan and Robertson 1978, Smith and Murphy 1978, 1979; Newton, 1979, Petersen 1979). Counts of prey species, that in some instances are less time-consuming than detecting raptors, may provide indices that are useful for estimating raptor densities.

Aerial Counts

During one breeding season Bald Eagles were counted from aircraft flown over a random sample of 30 blocks (166 km^2 each) in preselected habitats (King et al. 1972). This survey was repeated 10 years later using the same flight techniques and sample plots (Hodges et al. 1979), so that a statistical comparison could be made between the two surveys. Randomly selected aerial transects constituting about 7% of each study area were used to monitor the yearly winter population of Golden Eagles in the southwestern United States. Similarly, the Golden Eagle population in Wyoming was counted by flying over randomly selected transects (Higby 1975). Boeker and Bolen (1972), and Boeker (1974) described flight techniques and discussed variables such as rough terrain, which may affect the aerial counts. Hancock (1964) described his flight techniques for counting Bald Eagles over the major shorelines of British Columbia. He also discussed visibility differences between adult and immature birds and the affects of seasonal movements, and various assumptions on population estimates.

Other aerial surveys include those by Wrackestraw (1973) to count Golden Eagles, Lish and Lewis (1975) and Southern (1964) to supplement ground tallies of Bald Eagles. Presently the National Wildlife Federation (Washington, DC) is coordinating surveys of wintering Bald Eagles. Air surveys flown in conjunction with this effort have revealed several areas where birds were previously not known to winter (M. Pramstaller, pers. comm.). Enderson et al. (1970) counted eagles and Miller et al. (1975) recorded numbers of Snowy Owls (*Nyctea scandiaca*) while on aerial surveys of large mammals.

Counts at Colonies and Roosts

There are several groups of Falconiformes (e.g., vultures, kites, harriers, and small falcons) in which some species nest colonially or semicolonially, at least in certain parts of their breeding range (L. Brown 1971, Newton 1979). Despite their density, nests in some colonies may not be easy to find (e.g., Eleonora's Falcon (*Falco eleonorae*), harriers), and may be difficult, dangerous and time consuming to reach (e.g., Parker 1975). Because of these factors, and the disturbance to many birds caused by moving about in the colonies, it is often more desirable to estimate the number of breeding birds by counting individuals as they fly to and from the colony. Erwin and Ogden (1979) had an error rate of 13% when estimating the number of four species of nesting wading birds in colonies. Walter (1979) believed that only about 10% of the breeding population may be seen above or near colonies of Eleonora's Falcons, even at the height of reproductive activity.

If we can learn about those variables affecting flight rates (e.g., weather, food requirements, food availability), it may be possible to make good estimates of nesting birds by counting raptors as they fly in and out of breeding colonies. For example, Sykes (1979) made counts of Snail Kites going to night roosts. Many investigators, including Southern (1963, 1964) and McClelland (1973), have counted Bald Eagles at winter roosts. Both Schnell (1967, 1969), counting Rough-legged Hawks, and Bildstein (1979a) counting Hen Harriers, estimated local numbers of those raptors using roost counts and discussed some variables that may affect estimates. Weller et al. (1955) counted Hen Harriers and Short-eared Owls (*Asio flammeus*) at winter roosts. Population estimates by roost count techniques may also be applicable for other species of harriers (L. Brown 1971, Watson 1977), some vultures, Long-eared Owls (*Asio otus*), and several other raptors that roost communally at certain times of the year.

Christmas Bird Counts

Many raptors are found in greater densities on the wintering grounds than in their breeding areas, thus making it a bit easier to count them in winter. In the early 1900's, ornithologist Frank M. Chapman encouraged Christmas-time bird "censuses" as a substitute for traditional annual hunts for raptors, crows, and other "vermin" (Stewart 1954). In the past years, participation in Christmas Bird Counts (CBC) has become very popular, and has resulted in the collection of a great deal of information on local winter bird populations. Results of these winter counts have been published in local birding publications and in *American Birds* (formerly *Audubon Field Notes*). Inherent in the way the counts are conducted are many variables influencing the use of CBC results for estimating avian populations. The implications of these variables for estimating numbers of birds or detecting trends in populations have been discussed (e.g., Stewart 1954, Arbib 1967, Raynor 1975).

The number of parties searching in the prescribed circle (12.1 km radius) and the number of experienced people per party may greatly influence the number of raptors detected on a CBC. In recent years, the number of participants has increased in most areas and frequently some people make special efforts to find birds of prey by searching certain habitats, using tape recordings to elicit responses, etc. Because the extent and type of coverage varies on counts from year to year, it becomes very difficult to interpret CBC data.

Bystrak (1971) and Renaud and Wapple (1977), among others, have drawn winter distribution maps based on CBC data. Brown (1964) noted, however, that a lack of even distribution of counts over the state of Iowa precluded accurate mapping of winter ranges of two buteos. In an effort to reduce variability associated with different numbers of participants over the years, Graber and Golden (1960) included only counts conducted by 10 or fewer people in parties of four or fewer participants. The late W. H. Brown (1971) analyzed count data to detect trends in the number of Red-shouldered Hawks. He "normalized" counts by tallying the number of hawks seen per distance traveled by a party. He also grouped counts according to different geographic regions, but was unable to detect any pattern in the decline in numbers across the country. Brown found that the number of Red-shouldered Hawks and other species of raptors seen on cloudy days was less than observed on clear days, but that the plots of yearly counts were the same shape for data obtained on clear or cloudy days (W. H. Brown 1971, 1973, 1975, 1976b). His analyses of CBC results revealed that a substantial increase in effort and consequently, in numbers of birds counted in a state or province, could greatly affect the shape of curves from national counts. The interpretation of winter counts of vultures (Brown 1976b) was also affected by interruptions in a series of yearly counts. Bildstein (1979b) limited his analysis to data from CBC circles which had been searched each year during the 6-year period with which he was concerned.

Raynor (1975) emphasized the importance of comparing counts conducted in similar habitats, and Stahldecker (1975) chose only circles containing similar proportions of the same habitats. Stahldecker's analyses revealed another limitation of the use of CBC data for some birds of prey; that is, for the plots he used, there were too few sightings of Goshawks, Cooper's Hawks, and Sharp-shinned Hawks (*Accipiter striatus*), so the counts of those species had to be combined before analysis could produce a trend in *Accipiter* numbers. For one area, Bildstein (1978) found that his intensive counts on one study area did not correlate with CBC counts in the same area. A 5-year increase in raptors detected with his estimates was not correlated with weather, number of participants, nor with his mean estimate of the number of birds present in December. He believed day-to-day counts probably reflect previous weather as well as weather on the count day.

These problems with CBC data provide examples of the type of factors one must consider when using counts made by a diverse group of participants, over large areas and long periods of time, and with relatively vague instructions or directions. In addition to these concerns are many variables affecting the distribution of the birds from year to year. In most instances we do not know the specific geographic origins of wintering populations. We do know that the movements of these birds are affected by prey availability, weather, and interspecific interactions not only on the wintering grounds, but also on the breeding range and along the migratory routes (Craighead and Craighead 1956, L. Brown 1971, Newton 1979). Until we learn more about these factors, we must interpret winter count data cautiously and restrict use of trend data derived from these counts to identifying areas for further study and/or to supplement evidence from other estimates of numbers (e.g., U.S. Dept. of Interior 1971).

Trapping

Capture of birds of prey has seldom been used as a counting technique because trapping raptors under most circumstances is very time-consuming. For example, Doerr and Enderson (1965) and Doerr (1968), in their efforts to count Goshawks, captured only one raptor per 40.5 and 31.2 trap-days, respectively. Even when a variety of trapping techniques are used on relatively small study areas (e.g., 1,000–10,000 hectares), capture rates may not be high (1 bird/15.4 days) and recapture of raptors is even more time-consuming (Fuller and Christenson 1975). Because of these difficulties, and the relatively "data hungry" nature of most population estimator models, capture-recapture approaches to raptor population studies have not generally been pursued (see Nichols et al. 1981).

The greatest opportunity to capture large numbers of birds of prey exists during migration, but not all locations are well-suited for capture. Bartelt and Orde (1976) captured only 1 bird/19.4 trap-days in South Dakota. Along migration routes where raptors become concentrated, hundreds of birds may be captured during an autumn season (Evans 1975, Newton 1979). Several people have been conducting trapping and banding projects for a number of years and have provided information about raptor migration and the usefulness of trapping for counts of birds of prey (Gray 1961, Mueller and Berger 1961, Field 1971, Clark 1973, Berry and Ward 1975, Evans 1975). Trapping results combined with observations have been used to address questions about the status of a species or trends in populations (e.g., Berry 1971, Enderson 1965, Ward and Berry 1972, Rogers and Hunt 1975, Mueller et al. 1977).

There are many variables affecting the movements of migratory raptors (Newton 1979) and additional factors affecting trapping results such as effectiveness of types of traps and bait for different species, and different age or sex of birds (e.g., Mueller and Berger 1970). In many instances the trapping data are not used for counts per se but as supplements to sightings and to provide information about the seasonal and daily timing of migration for different species and different ages and sex groups within species (e.g., Mueller and Berger 1967a, 1968, 1973; Catling 1971, Ward and Berry 1972, Mueller et al. 1977, 1979; Rosenfield and Evans 1980). Furthermore, the results of trapping efforts have contributed to our understanding about the effects of weather on the movements of migrating raptors (e.g., Evans 1980). Finally, recoveries and returns from birds of prey banded along migration routes have provided some very valuable data about the origin and destination of migratory raptors (e.g., Enderson 1965, Mueller and Berger 1969, Clark 1976). More data of this nature are needed before counts made of migrating birds can be fully utilized to estimate the size of, and to detect trends in, populations.

Counts of Migrants

Newton (1979) discussed the major factors that influence the movements of Falconiformes. Several of these factors relate to migratory movements and have important implications for interpretation of counts of migrants: (1) populations may remain longer on the breeding grounds in years when food is plentiful there, (2) for some species, birds of different age or sex may not migrate, or they may migrate at different times depending on the availability of food, (3) birds may migrate farther if food is scarce, and (4) separate populations of the same species may migrate along different routes and winter in different areas. Counts of nomadic and cyclic populations (e.g., Shelford 1945, Galushin 1974, Mueller et al. 1977) may vary greatly from one year to the next at any one location on the migratory route. In these instances long-term changes in the status of populations can be evaluated only after accumulating counts for many years. In some areas it may be difficult to count migrants because the distribution of migrants overlaps with that of resident birds (Brown 1971a, Thiollay 1978, Newton 1979).

Movements of raptors are also influenced by many factors along their migration routes. Many species of raptors migrate along certain features of the terrain and may become further concentrated in space and time by weather conditions (e.g., Mueller and Berger 1967c). Both local weather and weather over a regional or continental area may affect movements of the birds (e.g., Bagg 1950, Mueller and Berger 1961, Haugh and Cade 1966, L. Brown 1971, Heintzelman 1975, Haugh 1975, Evans 1980). Not all migratory raptors pass along concentration points, but rather some portions of many populations are spread over larger areas (e.g., Hamerstrom 1969, Hopkins 1975, Dekker 1979). In some areas the autumn flights are concentrated, whereas in other locations the spring migration becomes concentrated (see Newton 1979). The degree to which birds become more or less concentrated in association with various weather or biological phenomena is only generally known. Thus, observers at a specific site cannot know whether a low count is attributable to a lower population or to movements of the birds over a different area.

There are also many unquantified variables involved in the observation and counting of passing raptors; for example, (1) the ability of the observer to detect birds, (2) the ability to accurately count birds, (3) the ability to identify migrants, (4) the effects of multiple observers, (5) the use of optical aids (from binoculars to high-powered telescopes), (6) counting from more than one station in a local area, (7) recounting birds that remain in the area, (8) the effects of weather on visibility, and (9) the extent to which counts are conducted in inclement weather. These factors and others have been discussed by several investigators (e.g., Enderson 1969, Spofford 1969, Heintzelman 1975, Dunne and Clark 1977, Nagy 1977, Harwood and Nagy 1977, Fuller 1979). In an effort to reduce some of the variability of counting techniques and of recording count data and some other relevant information, the use of standardized forms has been encouraged (Harwood 1975, Heintzelman 1975, Robbins 1975, Fuller and Robbins 1979). It is hoped that standardized instructions and forms will facilitate gathering of data which can be pooled for comparisons of different locations, species, years, weather conditions, etc. Certainly not all the pertinent variables can be accommodated on these forms nor will they be applicable in all situations (especially in regions where birds of prey are not concentrated). Rather, the information from the form should be considered as a starting point. The gathering of additional data relevant to factors influencing movements of raptors and initiation of more research on raptor migration is needed.

The methods that can be adopted for studies of raptor migration are diverse. Stearns (1949) viewed hawks from a blimp (dirigible) and Hopkins (1975) used airplanes and subsequently, a motor-glider as suggested by Pennycuick (1975). Radar has been used in several areas (Alerstam and Ulfstrand 1972, Evans and Lathbury 1973, Richardson 1975), and revealed that birds of prey often fly higher and over broader areas than had been detected by observers. Smith (1980) photographed raptors overhead and subsequently made accurate counts of the number of birds in each "sample." These methods reveal basic information on raptor migration that permits more reasonable evaluation of counts.

Numerous authors have emphasized the value of counts made along migration routes for providing relative numbers that can be compared from year to year (e.g., Edelstam 1972, Ulfstrand et al. 1974, Robbins 1975). Indeed, migration counts have been used to help assess the status of certain species (Kruyfhooft 1964, Spofford 1969, Hamerstrom 1969), to help detect population trends (Snyder et al. 1973, Nagy 1977), and as supplements to other estimates of numbers of birds or their status (Hackman and Henny 1971, Robbins 1975, U.S. Dept. of Interior 1971). These counts must be interpreted cautiously (e.g., Harwood and Nagy 1977) until we gather more basic data about raptor migration. Problems associated with enumerating migrating hawks seem worth further study becouse it is relatively easy to count birds of prey during migration and because there are many locations around the world at which concentrations of migrant raptors occur.

Counts of migrating birds of prey have been conducted in Africa (e.g., L. Brown 1971, Moreau 1972, Elgood et al. 1973, Thiollay 1978), in the Mediterranean region (e.g., Simmons 1951, Nisbet and Smout 1957, Evans and Lathbury 1973, Beaman and Galea 1974, Thiollay 1977), in the Middle East (e.g., Safriel 1968, Nielsen and Christensen 1970), in Europe (e.g., Edelstam 1972, Ulfstrand et al. 1974, Roberts 1979), in Central America (e.g., Skutch 1945, Hicks et al. 1966, Smith 1980), in Mexico (Purdue et al. 1972, Thiollay 1980), and in the United States and Canada (see Harwood 1975, Heintzelman 1975). No doubt other concentration points exist along the eastern coast of the Soviet Union and mainland China, and along river valleys, mountain ranges, and at mountain passes on the Asian continent. Thus, there is ample opportunity for making raptor counts during the migration periods, and the potential exists for applying these counts to estimates of raptor population sizes in many parts of the world.

CONCLUSIONS

Birds of prey are generally widely spaced (especially during the breeding season), rapid-moving, and wide-ranging, and are therefore difficult to detect and count. Often, the most reliable estimates of numbers of breeding raptors are the result of intensive searches for nests of breeding pairs. The numbers of some birds of prey, or their nests, have been estimated with more efficient sampling techniques such as counts along transects or searches on random or stratified plots. Generally, however, success with sampling has been limited to those species or nest structures that are conspicuous by virtue of their large size or occur in open habitats or on prominent structures. Even with these species, it is difficult to obtain adequate yet affordable samples (Grier et al. 1981).

The usefulness of partial counts of birds and detection by indirect methods (e.g., vocalizations) remains limited because few studies have been able to relate the "sample" to the actual (or statistically estimated) population size. This is especially true with forest-dwelling Falconiformes and a majority of owls. Most relative abundance indices have also been of limited value because counts were not conducted in a standardized manner or frequently enough to evaluate comparisons of different areas or different times. Birds of prey are often less widely dispersed during the winter and particularly during migration when many species become concentrated along various routes in many parts of the world. Counts are made more easily at these times than during the breeding season; however, the origin of most birds is unknown and therefore count data cannot be related to the population dynamics of particular demes or geographic areas. Interpretation of surveys and counts conducted during the non-reproductive season are hindered by our lack of knowledge about the degree to which variables associated with climatic conditions and biological variability (e.g., food availability) affect the movements of birds. Thus, we cannot differentiate a change in actual population size from our inability to count birds that may have moved elsewhere. Trapping and banding projects, yearly counts conducted at migration concentration points, and studies employing radar, radio-telemetry, air surveys, and other specialized techniques are providing information critical for our understanding of bird movement. However, much more research on the variables affecting counts is needed.

Since birds of prey occur at comparatively low densities, the loss of relatively few individuals may change the status of a population. Unfortunately, raptors are sensitive to the contamination, disturbance, and loss of habitat that often accompany development. Consequently, the conservation of raptors has become the concern of a variety of people involved in land-use planning and resource management. These people, in addition to biologists specializing in studies of birds of prey, need data about the ecology, distribution, and status of raptors.

We encourage more efforts to develop reliable and efficient sampling techniques suitable for use with the diversity of raptor species and their habitats. Because most surveys and counts are limited by time and funds to a relatively small portion of the range of a species, we believe knowledge about raptors, and their management and conservation is best served when the results of many studies can be pooled and compared. As Fraser (1978) has emphasized, the key to successful comparison is often not complete standardization of techniques, but rather standardization of the estimated parameters and inclusion of measures of the variability (e.g., confidence intervals, standard error) associated with the estimates.

ACKNOWLEDGMENTS

We thank those colleagues who suggested references for this review and D. Bystrak, B. Dowell, C. Robbins, and J. Partelow for comments on the manuscript. This is contribution 1117 of the Appalachian Environmental Laboratory and Technical Report 2 of the Central Appalachian Raptor Ecology Program.

Studies in Avian Biology No. 6:247–248, 1981.

SUMMARIZING REMARKS: SPECIES VARIABILITY

DAVID L. OTIS[1]

In this summary, a few statistical aspects of each paper will be briefly discussed, followed by some general comments on the need for statistical methodology in ornithological research.

Diehl (1981) was concerned with the IBCC recommendation that a pair observed less than three times per 10 occasions not be counted as a breeding pair when the standard mapping method is used to estimate density. She presented evidence supporting the contention that such inconspicuous pairs can in fact be viable breeding pairs, but that the probability of this occurrence varies with time, density, habitat type, breeding success, and species. The use of statistics is minimal; in fact, only one test of hypothesis is performed and, because data were pooled over years, no variances for the various responses of interest were presented. This lack of quantitative treatment of the data is unfortunate in many instances, e.g., the apparent relationship between percent brood mortality and the percent of least dectectable pairs is not quantified. The reader could have been presented with a more objective, quantitative treatment that would have allowed the inferences suggested by the author to be more easily evaluated and interpreted.

Mayfield (1981) presented a very sobering demonstration of the potential biases involved in conducting censuses using counts of singing males. The most disturbing point arising from the study is not that there are differences in detectability among species, but that the reliability of a single census count is very low. That is, if a count conducted at 07:00 on one day produces very different results than one conducted at 09:00 on the same day or 07:00 the next day, can there be any confidence in inferences made from such counts?

Such diversity in census results can be expected, based on the author's data that show that only 42% of the total population present on the study area is identified on a single average count, and that the average probability of detecting a single member of a given species on a single count is 0.40 with a range of 0.00–0.90. Because no estimates of reliability (variance) can be made from transects on which only a single count is made, multiple counts on the same transect seem necessary. Mayfield (1981) also presented evidence of the danger of assuming a 1:1 sex ratio for songbirds, although he admits that apparent differences in the sex ratio within a season may be due to the fact that the sexes vary in their trappability within a season.

Ekman (1981) was concerned with the effects of violating the assumption of equal observability among individuals, as well as in time, on Jolly-Seber (JS) capture-recapture estimates as well as census counts. Ekman documented such heterogeneity in capture probabilities (observability) among age and sex classes, and lamented the fact that there are no powerful tests for detecting unequal catchability when using the JS model. In this particular capture-recapture experiment, however, I suspect that estimates will not be seriously biased because capture probabilities of all classes are relatively high and do not vary extremely.

In general, however, stratification of the data, as Ekman suggests, into strata containing members with relatively homogeneous catchability is an excellent idea if sufficient data are available. As Ekman points out, sampling design can sometimes be adjusted so that good estimates of stratum parameters can be obtained. Unfortunately, the paper fails to report sampling variances for parameter estimates generated by the JS model; these variances can be easily computed and should have been included so that the reader can appreciate the amount of precision associated with the estimates. Ironically, I am curious as to how the author computed sampling variances for the census counts tested for differences in Fig. 2, because there is no replication in time or space. Finally, Ekman concludes that unequal observability prevents census counts from being accurate indexes of seasonal change in density, but does not preclude the use of such a method for assessing annual changes. I would argue that such between year comparisons are also dangerous, because habitat and environmental conditions in the same area at the same time of year could significantly change between years, and hence the observability of the species could change and therefore bias census count results. I believe it is a general axiom that use of index methods which do not have some kind of theoretical model supporting them is dangerous, and should not automatically be viewed as a logical alternative when density estimation is not practical or possible.

The review by Fuller and Mosher (1981) of the methodology currently available for counting raptors reveals the tremendous difficulties associated with obtaining reliable raptor density

[1] Section of Bird Damage Control, U.S. Fish and Wildlife Service, Building 16, Denver Federal Center, Denver, Colorado 80225.

estimates from relatively small scale censuses or surveys. They emphasize, and I agree, that research on sampling techniques is badly needed, but I am not sure that improvements in such sampling methods will render adequate estimates "affordable." Because of the wide range and low density of these birds, large scale surveys will be necessary. I disagree with them that pooling many small studies is a preferable strategy, for such pooling can often introduce extraneous sources of error and bias into the estimates. With standardized and efficient statistical techniques the cost of such surveys can be minimized, but, depending somewhat on the objectives of a given study, reliable estimates will be expensive.

The papers in this session typify both the growing awareness by ornithologists of the assumptions involved in various census/survey techniques and of the consequences of making inferences from such techniques when assumptions are violated. I believe this awareness brings with it a sense of frustration—we are becoming aware of the deficiencies and inappropriateness of some of our methods but we often have no alternatives! If these frustrations are to be relieved, it is imperative that statisticians and ornithologists begin to work together to develop and improve methodologies for estimating parameters of bird populations. This partnership should involve a joint learning process.

The statistician must have an appreciation for the constraints (monetary, manpower, logistical) under which the researcher is operating, and the biologist must understand the basic concepts (sampling variance, experimental error, bias, robustness) that provide the framework for good statistical practice. For example, a statistician working with researchers using IBCC mapping methods could perhaps develop models and associated methods that use all the data collected in such a census in an efficient manner, and thereby avoid the problems associated with essentially ad hoc recommendations for editing the data, which was the main concern in Diehl's (1981) paper. The same statistician might also assist Ekman (1981) in following up on his excellent documentation of heterogeneous observability by using the computer to simulate a multitude of survey situations so that some general conclusions concerning the practical application of capture-recapture methods in ornithological research could be stated. The raptor biologist could explain, by dragging the statistician out from behind his desk and into the airplane, why it is not possible to take a 10 % sample of the area contained within the range of the species of concern.

These hypothetical examples are designed to emphasize that communication between the two disciplines is the key element necessary for progress. These proceedings make it very clear that ornithologists can no longer use methodologies subject to known, yet not quantified, biases in poorly designed studies whose objectives are not clearly stated. The statistician's help should be solicited from the initial planning stages to the analysis of the collected data, and the statistician must respond to this challenge by working with the researcher to develop practical yet rigorous approaches that will facilitate the performance of such research.

Studies in Avian Biology No. 6:249–250, 1981.

SUMMARIZING REMARKS: SPECIES VARIABILITY

ROBERT E. RICKLEFS[1]

Biologists have been using census techniques for years without fully understanding the potential errors in estimating relative or absolute abundances of populations. The papers in this session demonstrate some of the inefficiencies and biases of standard census techniques and, more importantly, suggest some of the underlying bases for these difficulties. The papers have restated a theme heard frequently during this symposium concerning the reliability of bird censuses. The detectability of birds on transect or singing male counts is extremely variable with respect to season, species, individual variation, time of nest cycle, and perhaps other variables not explicitly dealt with in this session (e.g., habitat, weather, interobserver variation). The overall efficiency of observing individual birds or pairs of birds on single transects was found to average 40% (range, 0–90%) in deciduous forest in Ohio (Mayfield 1981), and about 50% for the Red-backed Shrike and two warblers in old-field habitat in Poland (Diehl 1981). Moreover, on small census areas, or for species with low population densities, the efficiency of sampling species can be quite poor. During each of his transects, Mayfield recorded only between 6 and 13 of 20 species present. This inability to sample species fully in brief periods has important implications for the comparative study of bird communities, in which large numbers of censuses are required.

Detectability clearly is an important factor in estimating the abundance of populations and the composition of communities. But the factors influencing detectability appear to be so numerous and complexly interrelated that it is not yet possible to judge the detectability of particular species under particular conditions simply by generalizing past experience. Therefore, the error inherent in the census techniques employed in a particular study can be estimated accurately only by direct validation.

The papers presented in this session have revealed some of the factors that influence detectability of birds, and they have implicitly pointed future research in some useful directions. Census techniques are usually validated by comparing their results with thorough, exhaustive censuses, often incorporating nest finding (Diehl 1981, Mayfield 1981), or with unbiased procedures (e.g., capture-recapture method used by Ekman 1981). Comparisons of these techniques are not the main topic of this session and have been treated in detail elsewhere in the symposium.

The speakers in this session have left no doubt that there are major differences between species and individuals in detectability, but they have provided us with few clues to the particular characteristics that make some species and individuals conspicuous and others cryptic. The speakers have provided more details concerning the influence of season and stage of nesting cycle on detectability. For example, Ekman (1981) determined that the detectability of Willow Tits on transects varied seasonally by as much as two- to three-fold relative to population densities estimated by the capture-recapture method. In his presentation, Ralph (1981) demonstrated for the 'Elepaio seasonal fluctuations in the number of individuals observed. In six of 10 Hawaiian species, the effective detection distances exhibited seasonal patterns of variation, presumably related to behavior or habitat cycles. Diehl (1981) showed a strong correlation between detectability and stage of the nesting cycle, it being highest during the prelaying period and, in the Red-backed Shrike, lowest during the incubation and nestling periods.

For me, the most interesting observations in this session related detectability directly to the behavior of individual birds or to seasonal variation in behavior patterns. In winter flocks of Willow Tits of known composition, Ekman (1981) showed that females, especially adults, are less detectable than males and younger birds because they feed higher in the trees and are thus more frequently hidden from the view of observers. Diehl's (1981) observation on the Red-backed Shrike that successful breeders were more detectable than unsuccessful ones was intriguing, and suggests that there are behavioral changes associated with nest failure. Ralph (1981) applied a correlation analysis to determine the relationship of census results to seasonal changes in patterns of vocalizations and movements. This summary is not the proper place to discuss the application of multivariate techniques to the detectability problem except to note their potential for sorting out many interrelated variables and to urge others to follow Ralph's example. In his study, several species, including the 'Elepaio, revealed correlations between behavior and census results.

The papers in this session seem to me a good start toward understanding the biological bases

[1] Department of Biology, University of Pennsylvania, Philadelphia, Pennsylvania 19104.

of biases and errors involved in estimating the numbers of birds. It is reasonable that additional effort be directed to more detailed validation studies of census techniques, and especially to studies of individual activity patterns, in order to devise a general theory of census efficiency and to improve our ability to estimate errors associated with particular censuses. I suspect that detectability can be related to such factors as foraging behavior, mating system, season, stage of nesting cycle, weather, time of day, and habitat. But whether knowledge of these factors and either their "experimental" control or entry into analyses as covariates can substantially improve estimates of numbers will be determined only by extensive comparative study. It may be possible to relate detectability to such indirect measures as plumage brightness and complexity or to certain morphological characteristics that are related to behavior and movement patterns. But suitable correction factors for census data that take into account such considerations will probably accumulate only through the experience gained in systematic attempts to relate detectability coefficients to other easily measured attributes of the species.

Studies in Avian Biology No. 6:251, 1981.

INTRODUCTORY REMARKS: ENVIRONMENTAL INFLUENCES

HARRY F. RECHER,[1] CHAIRMAN

An accurate estimate of the number of birds in a community or population is the goal of all census work. We may argue whether absolute or relative numbers are needed, but we can not escape the need for accuracy and precision. Ultimately the accuracy of a census and our ability to replicate results depends on how well we control for variation in methods and the environment in which we work.

The effects of weather and time of day on avian activity are well known and easily corrected for in the design of census procedures. Other variables, such as the structure of vegetation, topography, environmental acoustics, and changes in the detectability of birds at different stages of the reproductive cycle, are not as well known, nor are they easy to control. Accounting for these variables requires a detailed understanding of the system in which we work and the biology of individual species.

I can illustrate some environmental problems from my own efforts to count birds in Australian heathlands and forests. At first I thought that Australian birds were like those in the temperate regions of North America. I assumed that only males sang, that a reproductive unit consisted only of a mated pair, and that each unit was territorial. In other words, my birds were good Christians.

I now know that I was wrong. The bird communities that I work with in southeastern Australia are very different from those first naive assumptions. Nesting is asynchronous; at any time more than half the individuals in a community may be non-breeding. Not all species sing—a fair number just make noise, and of those that do sing, both sexes may be equally vocal. There is a high level of song mimicry, fewer than half the species are territorial, and about a quarter of the species are group or social nesters. Moreover, the pattern of social or cooperative nesting can change between successive broods. Multiple broods are regular. For some species the first nesting is by a pair, but subsequent broods are reared with the cooperation of their older siblings. Unless these details are known for each species, it is difficult to obtain an accurate census.

The lesson here is that Australia may not be different from the warmer and less seasonal parts of the world. Persons attempting to count birds in these places must keep in mind that census methods have been developed largely in the cold north where my original assumptions are generally correct. We should be careful therefore to ensure that our knowledge of the avifauna and its adaptation to regional environments keeps pace with the level of sophistication of census procedures and means of data analysis. We also need to be aware of the different ways that the physical environment affects our ability to detect birds and understand how the human social environment may temper our freedom to conduct censuses.

In this section, the biological and physical factors which affect the activity of birds and our ability to detect individuals are described and suggestions made on ways to minimize their effects on census results. Methods can be standardized, observers trained and excesses of environmental change avoided, but corrections for the full range of environmental variation requires compensation on a species by species and site by site basis. Such attention to details may appear excessive, but it is necessary for accurate and repeatable results.

[1] Department of Vertebrate Ecology, The Australian Museum, 6-8 College Street, P.O. Box A285 Sydney South, NSW 2000 Australia.

Studies in Avian Biology No. 6:252–261, 1981.

SEASONAL CHANGES IN DETECTION OF INDIVIDUAL BIRD SPECIES

LOUIS B. BEST[1]

ABSTRACT.—Changes in frequency and distance of detection during the breeding season (considering both singing-male and all observations) are documented for selected species occupying deciduous forest habitats. Several general patterns are identified in the seasonal profiles of detection and in mean detection distances. The following factors influencing seasonal changes in detection are explored: frequency of song, stage of the breeding cycle, nesting synchrony, breeding season length, flux in community composition, habitat, and weather. The function of song influences persistence of singing throughout the breeding season. Singing frequency and likelihood of escaping detection during counts vary with stage of the breeding cycle; unmated males are most conspicuous. Nesting synchrony depends upon arrival times of males and females and on nesting success. Clumping patterns of count observations are influenced by season length. Seasonal profiles of detection can be used to determine the optimal period to schedule counts for individual species. General application of a series of such profiles depends upon additional research to develop profiles for other species and habitats and to determine the consistency of profiles for individual species.

Estimating the size of avian populations is an integral part of many field studies. Usually, the results and conclusions drawn from such studies depend heavily upon the accuracy of the census procedure. Important considerations in developing census methodology are the number and distribution of counts during the study period. Seasonal changes in kinds of cues emitted by birds, emission frequency, and attenuation have been alluded to in the literature, but efforts to quantify these changes and to use this knowledge in designing census schedules are rare.

The objectives of my study were to (1) document changes in detection during the breeding season of selected species occupying deciduous forest habitats, (2) develop seasonal profiles of detection and explore factors responsible for seasonal changes, and (3) discuss the general application of such information in planning census schedules. The discussion will focus primarily on passerines and other passerine-like species that breed in North America, although much of the information has broader application.

STUDY AREAS AND METHODS

Count data used in this report were obtained from two separate studies. Most information came from an investigation of avian communities breeding in habitats adjoining streams in central Iowa (Stauffer and Best 1980). Specific habitats included open-canopy (noncontiguous tree crowns) upland woodland, closed-canopy (contiguous crowns) upland forest, and closed-canopy floodplain forest. The vegetation composition of these habitats is given in Stauffer and Best (1980). (The two upland habitats mentioned here are included in the upland woodland category described by Stauffer and Best.) Each study plot consisted of 1–5 transects, marked at 25 m intervals, that paralleled the stream channel; successive transects were 50 m apart. The length and number of transects per plot were determined by the extent of relatively homogeneous habitat. Breeding birds were censused on all plots by using the Spot-mapping Method. Counts were begun 15–30 minutes before sunrise and were not conducted on days with rain or strong wind. During a count, transect lines were followed until the plot had been completely traversed. The location and behavior of birds observed on either side of the transects were recorded on grid maps of the plots. Counts were conducted on a rotational basis from mid-April to mid-July until each plot had been covered 12 times. Perpendicular distances of count observations from transect lines were measured on the grid maps for each study plot and then scaled to actual ground distances.

The second study was of a marked population of Field Sparrows (see Table 1 for scientific names) occupying old field (shrub-grassland) habitat in central Illinois (Best 1975). The study plot was gridded throughout at 25-m intervals; vegetation composition of the plot is described in Best (1977). Counts were conducted from 06:00 to 09:00 several times each month from May through August. During each count, the grid lines were followed in a north-south direction until the plot had been completely covered, alternately walking the even- and odd-numbered lines on successive counts. All Field Sparrows seen during each count were recorded on a grid map. Because this was part of an intensive study of Field Sparrow breeding ecology (Best 1977, 1978), the mating and nesting status was known for all territorial males/pairs during each count.

For determining seasonal patterns of detection, only initial observations were used; sightings made after an initial observation were not considered. I reasoned that the circumstances under which birds are first observed are influenced less by the observer's presence, and consequently, provide the best index to seasonal changes in detection.

Seasonal Profiles of Detection (hereafter abbreviated as SPDs) were constructed for selected species on the basis of the number of singing-male observations and all observations recorded for each species during half-month intervals within a given habitat (Fig. 1). For the Iowa study, each interval included results from two counts, except for late April, which included

[1] Department of Animal Ecology, Iowa State University, Ames 50011.

only one, and late June, which included three. Only species where at least 10 territorial males/pairs were estimated to be in a given habitat (using the Spot-mapping Method) were included in the analysis. For some species, this involved using count data from a single study plot; for others, it required combining results from two or more plots of similar habitat. All SPDs were standardized to the same scale; the number of observations made of each species during each half-month interval was expressed as a proportion of the total for the time interval with the highest count. Thus, curves for all species reach values of 1.0.

RESULTS AND DISCUSSION

General Patterns of Seasonal Profiles of Detection

The SPDs for 16 bird species that commonly nest in deciduous forest are illustrated in Figure 1. Frequencies of detection from late April through early July vary substantially among the species. Greater variation in a SPD represents greater disparity in the number of birds observed per count. The SPDs provide information on both the general seasonal pattern as well as the magnitude of variation in frequency of detection.

Seasonal changes in frequency of detection may be associated with fluctuations in population density, but they also may occur independently of population size and may be caused by factors such as variations in behavior associated with different stages of the breeding cycle, foliation of vegetation, or seasonal weather patterns. The degree to which SPDs reflect actual change in population size cannot be ascertained directly. Usually, this is not critical because the objective of most breeding-bird studies is to determine population density at the peak of the breeding season when the number of territorial males/pairs is largest. SPDs indicate clearly the period when the greatest number of individuals of each species is likely to be observed and, hence, the optimal time to schedule counts. A decline from this optimum, whether caused by flux in population density or by some other factor, reduces efficiency in detecting the total number of breeding birds.

Most SPDs take one of several general forms. All, except those of year-round residents, begin at zero, provided that counts are started early enough in the year. Probably the most common pattern is for frequency of detection to increase gradually to a peak, followed by a decline. The period over which detection remains at or near the peak (duration of optimal detection) varies greatly among species. The Great Crested Flycatcher and Rose-breasted Grosbeak are examples of species with relatively brief peaks in detection. In other species (e.g., EWP, HW, GC, C, IB; abbreviations for common names are giv-

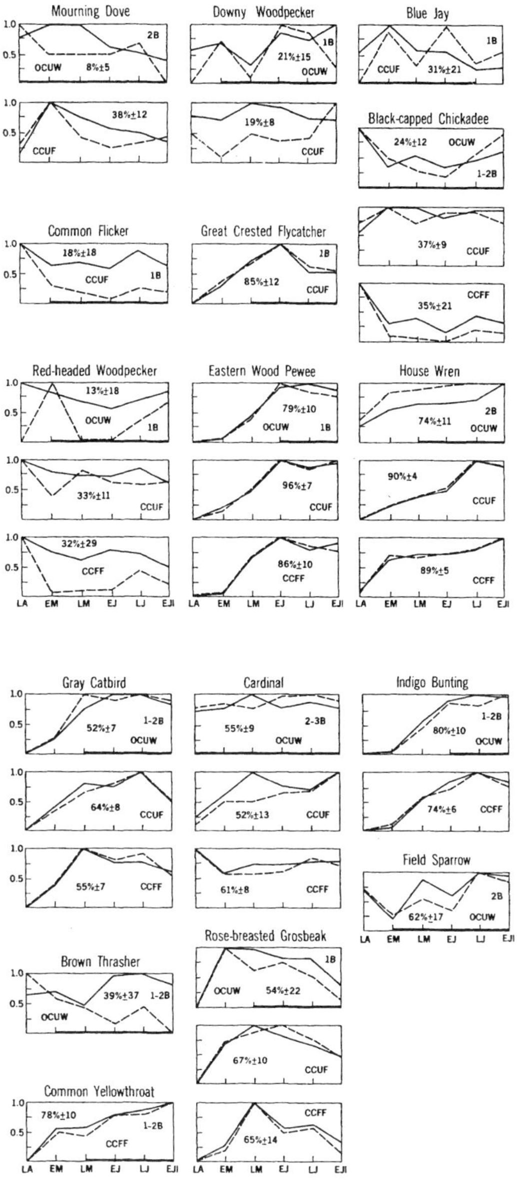

FIGURE 1. Seasonal Profiles of Detection for selected species breeding in open-canopy upland woodland (OCUW), closed-canopy upland forest (CCUF), and closed-canopy floodplain forest (CCFF). The sample units are half-month intervals covering from late April through early July. Values are proportions of the maximum half-month count of singing males (dashed lines) and of all observations combined (solid lines). The percentage that singing-male observations constituted of all observations ($\bar{X} \pm$ SD), the usual number of broods per season (designated by "B"), and the nesting season length (horizontal bar) are also given. Broods per season were obtained from Forbush (1927, 1929).

en in Table 1) frequency of detection increases to a relatively high level and then remains quite constant (at least over the interval that I censused). Cardinals and, to a lesser degree, House Wrens and Gray Catbirds, have relatively protracted periods of optimal detection (OD). In nearly all cases, SPDs ultimately decline, but the more important concern is duration of OD.

Some species show rather consistent patterns of seasonal decline or seasonal increase in detection. The Brown Thrasher is a good example of the former (on the basis of singing observations), and the Common Yellowthroat exemplifies the latter.

The Field Sparrow SPD illustrates the effect of multiple nestings. The somewhat cyclic pattern of detection coincides with stages of the nesting cycle that are repeated throughout the breeding season (see "Factors influencing seasonal changes in detection"). Highs and lows in the SPDs of other species also may be associated with different stages of the nesting cycle.

Some species seemingly lack any consistent seasonal pattern of detection. This is particularly true for the Red-headed and Downy Woodpeckers and the Black-capped Chickadee, and also may be true for the Mourning Dove. It is noteworthy that, while the SPDs of Red-headed and Downy Woodpeckers lack consistency among habitats, they are somewhat similar within the same habitat.

The disparity in detection profiles for singing-male versus all observations vary considerably among species. In some instances, the two curves essentially coincide (e.g., GCF, EWP, HW, GC, CY, IB), whereas in others, they seem to fluctuate independently (e.g., MD, RHW, BT). In general, the greater the percentage that singing males constitute of all observations, the more congruent are the two curves. This is evident even within the same species, comparing among different habitats (e.g., HW, RBG). Singing-male observations generally constitute less of all observations in open-canopy woodlands than in closed-canopy forests. This probably is because birds in more open habitat are more visible, resulting in less reliance on aural observations.

The percentage that singing-male observations constitute of all observations varies throughout the season, although the amount of seasonal variation differs among species. For some species, seasonal variation is considerable (e.g., BT), whereas for others, it varies little (e.g., HW, GC). The general pattern of this variation is indicated by the direction and magnitude of disparity between the curves for singing-male and all observations. The more the singing-male curve is displaced below that for all observations, the smaller is the percentage that the former constitutes of the latter.

The degree of similarity in SPDs for the three habitats differs among species. For some species (e.g., EWP, GC, IB) SPDs are nearly identical among the habitats; for others (e.g., MD, RHW, DW, BCC), they differ greatly. Dissimilar SPDs among habitats for a species could indicate that (1) habitat per se has a significant influence on behavior and observability of that species or (2) the species does not exhibit any predictable pattern of detection regardless of habitat. The woodpeckers and Mourning Dove (all nonpasserines), in particular, suggest the latter. Not only are their SPDs inconsistent among habitats, but the curves for singing-male and all observations differ considerably.

Seasonal Changes in Detection Distances

Not only may the number of birds observed change seasonally, but also the distance at which they are detected. The magnitude and pattern of seasonal change in mean distance of detection varies among species. Measurements for all observations are presented in Table 1. Mean detection distances of singing males generally were slightly longer than those for all observations, but rarely by more than 3 m. Small variations in mean distances probably are attributable to sampling error.

Detection distances for some species remain essentially constant throughout the breeding season (e.g., BJ, GC, CY, RBG). For others, there is an initial decline, followed by relatively stable distances (e.g., BCC, HW, C). The decline probably results from plant foliation and growth. Several species begin breeding after most plant foliation and growth have occurred; thus, plant development does not appreciably influence their detection. The shortest distances of detection for some species (e.g., MD, RHW, GCF, BT) occur midway through the breeding season. In some instances (particularly single-brooded species), shorter distances may coincide with the presence of fledglings.

Factors Influencing Seasonal Changes in Detection

Frequency of song

In many breeding-bird studies, singing-male observations constitute the most important criterion for determining population size (Enemar 1959:89). Song has different functions for various species, and the persistence and frequency of song throughout the breeding season depend upon its function(s). Two functions particularly pertinent to bird detection during counts are (1) song used to advertise and defend a territory and

TABLE 1

MEAN DISTANCES OF DETECTION (M) OF SELECTED SPECIES IN THREE HABITAT TYPES DURING DIFFERENT PERIODS OF THE BREEDING SEASON[a]

	OCUW						CCUF						CCFF					
Species	LA	EM	LM	EJ	LJ	EJl	LA	EM	LM	EJ	LJ	EJl	LA	EM	LM	EJ	LJ	EJl
Mourning Dove—MD[b] (*Zenaida macroura*)	—[c]	18	16	13	18	—	—	23	21	14	17	—						
Common Flicker—CF (*Colaptes auratus*)							23	23	18	24	19	19						
Red-headed Woodpecker—RHW (*Melanerpes erythrocephalus*)	—	24	20	14	18	18	23	17	17	12	16	20	29	18	21	17	17	17
Downy Woodpecker—DW (*Picoides pubescens*)	—	21	—	18	18	16	—	24	23	19	20	18						
Great Crested Flycatcher—GCF (*Myiarchus crinitus*)							—	—	20	24	15	20						
Eastern Wood Pewee—EWP (*Contopus virens*)	—	—	15	18	19	16	—	—	24	21	18	22	—	—	11	19	22	16
Blue Jay—BJ (*Cyanocitta cristata*)							22	22	22	18	21	22						
Black-capped Chickadee—BCC (*Parus atricapillus*)	19	17	15	13	15	15	—	18	15	16	18	16	26	18	14	14	15	14
House Wren—HW (*Troglodytes aedon*)	24	15	15	17	16	17	—	17	16	17	18	19	—	20	16	18	18	18
Gray Catbird—GC (*Dumetella carolinensis*)	—	16	17	14	17	16	—	20	18	20	16	16	—	16	13	14	15	15
Brown Thrasher—BT (*Toxostoma rufum*)	—	18	15	11	13	14												
Common Yellowthroat—CY (*Geothlypis trichas*)													—	19	16	18	15	17
Cardinal—C (*Cardinalis cardinalis*)	22	19	17	18	16	15	27	21	23	21	21	20	20	20	19	17	19	18
Rose-breasted Grosbeak—RBG (*Pheucticus ludovicianus*)	—	17	15	14	16	16	—	19	20	17	20	21	—	17	16	17	18	13
Indigo Bunting—IB (*Passerina cyanea*)	—	—	17	14	20	18							—	—	17	20	17	16
Field Sparrow—FS (*Spizella pusilla*)	26	22	18	17	21	15												

[a] Based upon all observations. See Figure 1 for explanation of habitat and period abbreviations.
[b] Abbreviation for common name.
[c] Fewer than 10 distances sampled.

(2) song used to attract a mate. For a given species, song may have either one or both functions. Although woodpecker vocalizations were recorded as singing-male observations, they were not advertisement songs (characteristic of passerines) but rather location calls (Lawrence 1967:18). Location calls function to locate other conspecifics and thus are given in a different context from that of advertisement songs.

Singing frequency declines dramatically after pair formation in species in which song has evolved primarily for mate attraction (Tinbergen 1939:80; see also "Stage of the breeding cycle"). An exception to this may be polygynous species that mate sequentially (e.g., Armstrong 1963: 118). In Figure 1, the Brown Thrasher is the most graphic example of song attenuation after pairing (see also Kroodsma and Parker 1977). Colquhoun (1940b) commented that cessation of song in woodland species that keep low in the vegetation may render them so inconspicuous that conventional census procedures would be futile, the alternative being a species-by-species study of marked individuals. When song is primarily for territorial advertisement, singing may be much more consistent throughout the breeding season (e.g., Smith 1959), although there often is a gradual seasonal decline. In some species, both sexes sing, and the song frequency may be quite constant throughout the breeding season (e.g., Laskey 1944). This may account for the relatively protracted period of OD for the Cardinal (Fig. 1). Knowing the function of song in individual species would facilitate interpreting their SPDs, although information on this subject is generally lacking.

The time of first song and the duration of singing in early morning change throughout the breeding season. Song initiation is stimulated by increasing light intensity at dawn. Not only does the time of sunrise vary seasonally, but singing for many species becomes progressively earlier relative to civil twilight until the north solstice and then gradually grows later (e.g., Allard 1930, Nice 1943:109, Davis 1958, Leopold and Eynon 1961, Nolan 1978:62). Song initiation in the House Wren occurs at civil twilight throughout the breeding season (Leopold and Eynon 1961). The duration of singing also is influenced by the rise in daytime temperatures (Robbins and van Velzen 1970), which changes seasonally. Planning the time of day to make counts should take these changes into consideration.

Differences among species in their spring arrival times may influence the period and duration of frequent singing. Slagsvold (1977) reported that species arriving early begin singing earlier in the season, have a longer interval between arrival and maximum song activity, and have a longer period of peak song activity than those arriving late.

Stage of the breeding cycle

The breeding behavior and activity level of birds may change as they progress from an unmated to mated status and then from nest building through fledging of the young. Sometimes these behavioral shifts are abrupt, particularly with respect to singing frequency. Such changes may affect census results dramatically, and Järvinen et al. (1977b) have cautioned that census results used in comparative studies should be from similar parts of the breeding season. Reported yearly fluctuations in population estimates may be purely artifactual (see Slagsvold's [1973c] comments relative to Enemar's [1966] findings) if the time when counts are taken each year varies relative to species' periods of OD.

The effects that stage of the breeding cycle can have on bird detection are illustrated by count data collected from the Field Sparrow population in Illinois. Singing observations constituted 92% of all observations made of unmated males, whereas only 13% of the observations made of pairs during other stages of the breeding cycle were singing males (Fig. 2). Dramatic declines in singing frequency of males once they have paired also have been documented in many other species (e.g., Michener and Michener 1935, Quaintance 1938, Tinbergen 1939:77, Lack 1943, Nice 1943:172, von Haartman 1956, Durango 1956, Davis 1958, Frankel and Baskett 1961, Armstrong 1963:152, Bell et al. 1968, Kroodsma and Parker 1977, Nolan 1978:63, Samson 1978). Once the nesting cycle has begun, singing frequency in the Field Sparrow evidently is greater during incubation than the nestling stage (see also Nice 1943:119, Davis 1958, Smith 1959, Armstrong 1963:153, Verner 1965, Falls 1969, Nolan 1978:63). An opposite pattern has been reported by others (Colquhoun 1940b, Clark 1947). No male Field Sparrows were heard singing during the interim after a successful nest and before egg laying in the subsequent nest. In some species, vocalization increases noticeably after the young leave the nest (e.g., Lawrence 1967:23, Falls 1969, Slagsvold 1973b, 1977). Stage in the nesting cycle evidently has little effect on song frequency in the House Wren (Gross 1948), thus accounting for the relatively flat SPDs for this species once breeding is initiated (Fig. 1).

Likelihood of escaping detection during a count also is influenced by stage of the breeding cycle. In the Field Sparrow study, the total number of times that pairs (or unmated males) were at given stages of the breeding cycle during counts was determined and used to calculate the

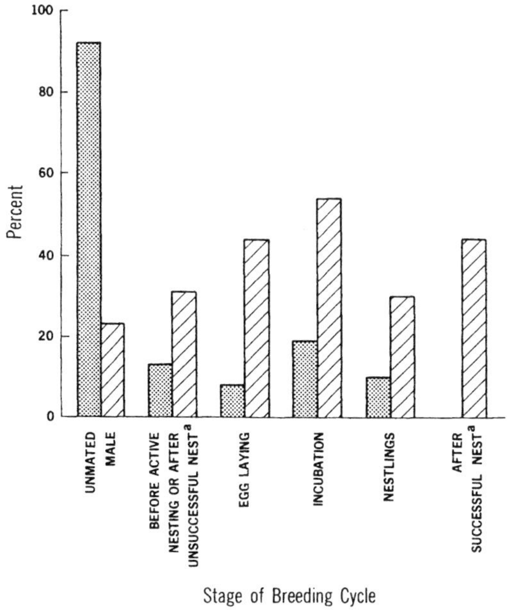

FIGURE 2. The percentage of all Field Sparrow observations during counts that were singing males (stippled bars) and the percentage of the time that pairs (or unmated males) were unsighted during counts (crosshatched bars). Categories with superscript "a" included only the interval until egg laying in a subsequent nest.

percentage of the potential observations during which birds remained unsighted (Fig. 2). Unmated males were least frequently missed during counts, whereas pairs were most likely overlooked during incubation. The undulating SPD for the Field Sparrow (Fig. 1) probably is associated with different stages of the nesting cycle that are repeated throughout the breeding season. Lows in the curves could be associated with a preponderance of pairs in the incubation stage, and the peaks may correspond with more in the nestling stage. The pattern is cyclic because this species is a persistent renester. The midseason low in the song SPD for the Blue Jay reflects its inconspicuousness during nesting (Tyler 1946).

In most species, males arrive before females in the spring (Dorst 1962:252). Thus, there is a period of frequent song during territory establishment, followed by reduced singing associated with pair formation (at least for species where song functions primarily for mate attraction). Males on territories of inferior quality often are unmated longer before acquiring mates, or in some cases, they remain unmated throughout the breeding season (see also Samson 1980). Once the main period of territory es-

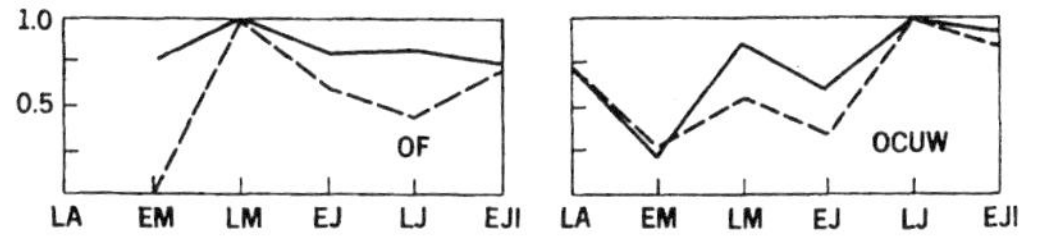

FIGURE 3. Seasonal Profiles of Detection for a Field Sparrow population with extremely low nesting success (old field in Illinois) and one where nesting success presumably was considerably higher (open-canopy upland woodland in Iowa).

tablishment and pair formation has occurred, males still unmated may be represented disproportionately in counts. Also, males may be deserted by their mates during the breeding season, resulting in increased song activity. All these factors influence seasonal detection of individual territorial males, and unfortunately, males most easily observed during a count are those without mates that likely occupy inferior sites.

Although nesting chronologies have been reported for many species, changes in detectability associated with different stages of the breeding cycle are generally unknown, and, if they are known, rarely have they been quantified. Such information can be obtained only when marked populations with known nesting histories are counted regularly. This is a productive area for future research, and the results of such efforts will greatly facilitate interpreting seasonal patterns of detection.

Nesting synchrony

The degree of synchrony in arrival of males and of females on their breeding sites influences the period over which unmated males prevail in the population. The more asynchronous the arrival, the more protracted is the period of pair formation. This, in turn, can influence the seasonal pattern of singing by males (e.g., Bell et al. 1973).

If females arrive synchronously on the breeding grounds, members of a population will be at a similar stage in the nesting cycle (at least until nesting efforts are disrupted). This would maximize the effects that stage of the nesting cycle has on seasonal detection patterns. Where members of a population lack synchrony in their nesting, effects of stage of the nesting cycle are attenuated (see also Slagsvold 1977). Migrant species that arrive late may nest more synchronously than those that arrive early (Slagsvold 1977). The abbreviated nesting seasons of species that breed at higher altitudes or latitudes also may cause greater synchrony, both intra- and interspecifically. And finally, greater nesting synchrony would be expected in colonially nesting species than in solitary nesters because of social stimulation (Orians 1961). Species that exhibit successive polygyny nest asynchronously in as much as pair bonds are formed over a time interval (von Haartman 1951).

Nesting success also may influence nesting synchrony. If predation, desertion, or other factors inducing nest failure occur regularly, then members of a population would soon be at all stages of the nesting cycle. Although the first nesting may be synchronous, renestings would not be. This is particularly true of multibrooded species later in the breeding season.

Effects of nesting success on SPDs are evident in the Field Sparrow data. Figure 3 illustrates SPDs obtained from counts of the Illinois population, where nesting success was known to be extremely low (Best 1978), and from the Iowa population, for which the form of the curves suggests that nesting success was considerably higher. Detection was low for both populations in early May, probably associated with the incubation stage of first nestings. As the breeding season progressed, nesting became asynchronous in the Illinois population. Synchrony persisted to some degree in the Iowa population, as evidenced by the cyclic pattern of detection throughout the breeding season, undoubtedly associated with second and possibly third nestings. Thus, variability in nesting success may affect the predictability of seasonal changes in detection (see also Haukioja 1968, Slagsvold 1973b).

In censusing entire avian communities, the degree of synchrony among species in their breeding seasons becomes more important than intraspecific variations in nesting chronology (see also Haukioja 1968). Although breeding seasons of some species overlap considerably, those of others are nearly mutually exclusive. In my study, the Eastern Wood Pewee and Indigo Bunting began their nesting season later than the other species, whereas the Mourning Dove, Cardinal, and Blue Jay were particularly early. Species differ, not only in the beginning of their breeding seasons, but also in the period when nesting reaches a peak. Also, the factors that influence nesting synchrony generally would not affect all species within a community equally.

Breeding season length

Length of the breeding season and propensity to renest after nest failure differ among bird species. Some species produce only one brood per season, whereas others may rear two or three broods yearly (Fig. 1). At higher altitudes or latitudes, breeding seasons generally are shorter, as well as is the period between spring arrival and start of breeding (Slagsvold 1976b). If stage of the breeding cycle significantly influences frequency of detection, then cyclic patterns would be more pronounced in SPDs of multibrooded species than those of single-brooded species (see also Slagsvold 1977).

Length of the breeding season influences nesting synchrony. Nesting becomes less synchronous as the breeding season progresses and as more nesting attempts are disrupted. The SPD for the Iowa Field Sparrow population (Fig. 1) evidently illustrates this, where the amplitude of oscillation in total observations dampens as the breeding season advances.

The degree of "clumping" of count observations also may depend upon length of the breeding season. In multibrooded species, successive nests may be considerable distances apart. If pairs concentrate their activities near the nest, observations would tend to be clustered around nests active when the counts were taken. Over the season, this could result in several clumps of observations. Thus, individual territorial pairs would be more difficult to identify, and population size might be overestimated. Restricting counts to a single nesting cycle would reduce the likelihood of this bias, although it still could occur if nesting chronology within the population was asynchronous. Nesting success may influence the extent of clumping of observations about nests. If nests are frequently disrupted early in the nesting cycle, count observations will be more dispersed than when nesting attempts are successfully completed. Multiple clumpings for a pair pose the greatest problem for the Spot-mapping Method because it depends on clusters of observations to enumerate territorial males (International Bird Census Committee 1970).

The Field Sparrow data (Fig. 4) illustrate clumping of observations near active nests; a relatively large and a relatively small territory (Best 1977) are included for comparison. Clumping is quite discernible in the large territory, but less evident in the small one. In smaller territories, nests are closer together; thus, observations associated with each nest overlap more, and the likelihood of multiple clusters for a single breeding pair is less. In contrast to my findings, Enemar et al. (1976, 1979) reported no relationship between the distribution of nests and clusters of observations.

Flux in community composition

Counts conducted early in the season more likely include migrants and transients than do those made later (Enemar 1959:20). Also, early in the season, territories still are being established, and site tenacity of males may not be fully developed. As the breeding season progresses, some females will desert their mates, and males will begin abandoning their territories. A high rate of nest failure may increase the frequency and advance the date of mate desertion and territory abandonment. Territory fidelity varies among species. Males of most species usually persist on a single territory for the entire breeding season, although others (particularly males of polygynous species) often shift territories during the season (e.g., Kendeigh 1941:37, von Haartman 1951, Bell et al. 1968, Robins 1971). Species with more protracted breeding seasons, by virtue of the time span involved, probably incur greater change in population composition than those with relatively short seasons. All these factors influence composition of individual populations and, consequently, community composition.

The more extended the period for counts, the more results will be confounded by flux in community composition. Shortening the count period would reduce this confounding effect, and knowledge of species' SPDs would facilitate determining the most appropriate periods for counts.

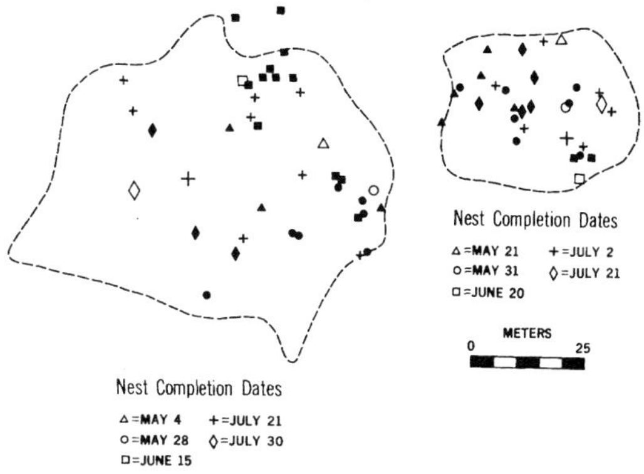

FIGURE 4. Spatial distribution of observations during counts and nests for two Field Sparrow pairs. Each open symbol represents a nest, and closed symbols of similar shape represent observations made when that particular nest was active. Only nests with observations from more than one count are included; designated locations are initial sightings (see "Study area and methods"). Dashed lines represent territory boundaries.

Habitat

Foliation and growth of plants during the breeding season, at least in temperate regions, change the degree that vegetation visually screens potential bird observations. Seasonal changes in the "vegetative screen" vary among habitats and probably are less pronounced in arid habitats than in mesic ones characterized by lush plant growth during the breeding season. My counts were confined to forest habitats exposed to the same regional climate; thus, comparisons among habitats would not be expected to show differences. Within deciduous forests, the most dramatic increase in visual screening occurs in spring when leaves first emerge. (The reverse would be true at leaf fall in autumn.) The early-season reductions in mean detection distance noted for some species in Table 1 likely are attributable to plant foliation and growth.

Habitat may influence nesting chronology. Erskine (1976b) noted that nesting in artificially warmer urban environments may begin earlier than in nearby rural areas. Length of the breeding season also may differ with habitat. Forest interior species tend to produce only one brood per breeding season, whereas species associated with more open habitats may produce two or three broods yearly (Brewer and Swander 1977).

In dense habitats, such as a closed-canopy forest, greater reliance is placed on aural observations during counts than in more open habitats. Thus, species with marked seasonal variation in song frequency are more difficult to accurately census in dense cover than in open areas. Interestingly, communication by sound (versus by sight) is more widely distributed among birds of closed habitats than of open ones (Armstrong 1963:227).

Habitat may influence the function of song and, consequently, persistence of singing throughout the breeding season. Songs of species occupying open areas, where sight advertisement of territories could suffice, may serve primarily to attract mates, whereas songs of birds in dense habitats may function also to proclaim territories (e.g., Brown Towhee, *Pipilo fuscus*, vs. Rufous-sided Towhee, *P. erythrophthalmus*, Quaintance 1938; Sedge Warbler, *Acrocephalus schoenobaenus* vs. Reed Warbler, *A. scirpaceus*, Catchpole 1973). The relationship between function of song and SPDs has been discussed (see "Frequency of song").

Weather

Year-to-year variations in weather influence arrival times of migrants and may cause seasonal shifts in plant phenology and nesting chronology. In many species, particularly early-spring migrants, arrival time (Nice 1937:43, Williams 1950, Dorst 1962:237, Slagsvold 1977) and nesting chronology (Nice 1937:98, von Haartman 1963, James and Shugart 1974, Slagsvold 1976a, 1977) are advanced by warmer temperatures. Weather has relatively little effect on migrants of late spring (Dorst 1962:232). Thus, within a community, nesting chronologies of different species may not be affected equally by yearly variations in weather. Yearly variations in breeding chronology may not necessarily modify the general form of SPDs, but they would cause temporal displacement.

Inclement weather (particularly early in the season) can modify seasonal patterns of detecting species either by checking or retarding breeding (Armstrong 1963:214) or by increasing nest failure rate, consequently disrupting nesting synchrony. Weather also may influence the length of the breeding season. Slagsvold (1977) reports that warm and wet weather may prolong the period of song activity in some species.

Extremes in weather, whether cold (Alexander 1931, Smith 1959, Garson and Hunter 1979) or hot (Robbins and van Velzen 1970), inhibit song; the former would be more likely early in the season, the latter, more likely later. These probably would have minor effects on SPDs if counts are restricted to days with moderate weather and are conducted only during early morning. Both are recommended procedures.

Application of Seasonal Profiles of Detection

In this study, SPDs were developed for 16 representative species that breed in deciduous forests of Iowa. Some SPDs in Figure 1 undoubtedly will require additional refinement and verification, and SPDs need to be developed for habitats and species not addressed in this report. At present, little information is available on seasonal patterns of detection (or song) of avian species during counts; the best documentation comes from European studies (Colquhoun 1940b, Cox 1944, Slagsvold 1973b, 1977; Järvinen et al. 1977b, Nilsson 1974b, 1977; O'Connor 1980c; O'Connor and Hicks 1980). Weber and Theberge (1977) presented weekly counts of common birds breeding in Canada. To my knowledge, no one has attempted to standardize information on seasonal patterns of detection into a series of profiles that could be made available for widespread use in determining the most appropriate times to count birds.

One major factor influencing the general applicability of SPDs is their consistency. That they may differ for a given species among habitats need not be an overriding concern, as long as they are consistent within each habitat. Differences among habitats would require developing a series of profiles covering the habitats occupied by the species and of interest to census takers. Where habitat does not significantly influence seasonal detection of a species, a single SPD would suffice. SPDs would have limited applicability for species whose seasonal patterns of detection are inconsistent even within the same habitat. More research is required to determine whether or not the SPDs in Figure 1 are consistent patterns for the respective species and to make similar assessments for other species. Assuming that patterns of detection for most species prove to be relatively consistent, then it would be practical to develop a series of SPDs for the species of a given region and to use this information in planning census schedules.

Two characteristics of SPDs that should be considered in their interpretation and implementation for censusing are the time and duration of OD. Identifying the exact time period when observations reach their peak probably is not as important as demarking the interval over which observations remain relatively high. Determining the lower limits of OD is arbitrary and could be set at any percentage of the maximum. If a 75% limit were applied to the SPDs in Figure 1, considerable variation among species in both time and duration of OD is evident. Once periods of OD are defined for the species to be studied, then counts can be planned. The smaller a species' range of OD, the more precisely its census schedule must be timed to avoid underestimating population density (see also Bell et al. 1973). This is especially critical when either the period for counts is short or few counts are taken, such as the North American Breeding Bird Survey (Robbins and van Velzen 1970). Range of OD could be used to identify "sensitive species" that require special attention in planning their counts. For example, Cardinals probably could be censused during any period of the breeding season, whereas Brown Thrasher populations probably should be censused only in the brief period before pair formation (see also Haukioja 1968). Single-brooded species with a single, sharp peak in their SPDs (e.g., GCF, RBG) also would require confining counts to relatively brief intervals. The corollary of a more confined census schedule for sensitive species is a reduction in the interval between successive counts such that the period of OD is adequately sampled and not missed between counts. Thus, not only the number of counts but also their timing and spacing should be considered when planning census schedules (Bell et al. 1973).

Planning counts for populations of a single species requires determining the species' period of OD and then adjusting the census schedule accordingly. Censusing efforts are most productive when confined to the period of OD. Censusing entire avian communities presents a much more complicated situation. Both the length and the seasonal timing of the period of OD vary among species. Thus, overlap in OD is highly variable within a community. A general principle in planning community counts would seem to be: A single census schedule, unless it is both *intensive* and *extensive*, cannot do all things for all species. To adequately sample an entire avian community, frequent counts over the entire breeding season may be necessary. Then, subsets of all the counts could be selected to estimate population sizes of individual species on the basis of their periods of OD (see also Shields 1979).

Once a general series of SPDs was developed, implementation still would require adjusting for factors such as year-to-year differences in weather and differences in altitude and latitude. Yearly variations in weather cannot be anticipated in advance, but their relationship to time of arrival and chronology of the breeding cycle is somewhat predictable (see "Weather"). According to Slagsvold (1973b, 1977), temperature, snowmelt, and plant development (specifically birch leaf emergence) are correlated with the start of song and the time of song maxima, particularly for "early" species. By using these or other predictors of yearly variation in weather, SPDs could be temporally adjusted accordingly. Such "predictors" would require field verification for different species and different geographical regions before their widespread use.

Higher altitudes or latitudes delay spring arrival and the onset of nesting (e.g., James and Shugart 1974) and shorten the season length in a predictable way. The former is more pronounced in species that arrive early than in later ones (von Haartman 1963, Slagsvold 1975, 1976b). A general series of SPDs probably could be corrected for altitude and latitude, but developing more than one SPD for species that breed over a broad altitudinal or latitudinal range might be more practical.

Other factors, such as effects of nesting success on nesting synchrony and flux in population or community composition, influence seasonal patterns of detection in more subtle ways, and adjusting for these factors is unrealistic. Their effects can be evaluated only by intensively monitoring the breeding activities of the species involved. Usually, this requires marking the birds for individual identification, and having done so, the need to census is eliminated. At the community level, such intensive study is unwieldy and impractical; thus, such factors affecting seasonal change in detection frequency are largely ignored. It would seem, then, that some sources of error in interpreting seasonal changes in counts can be remedied only by modifying the research design such that censusing is no longer necessary. For species in which these sources of error significantly influence seasonal patterns of detection, generalizable SPDs would be difficult, if not impossible, to develop.

Attempts to "standardize" the number and seasonal distribution of counts may be overly optimistic, particularly if census results are used for interspecific comparisons. Opinions vary widely relative to these two aspects of census planning (for a review, see Berthold 1976). Characteristics of seasonal detection for individual species must be considered when planning a census schedule, especially at the community level; otherwise, species populations will be sampled disproportionately (see also Slagsvold 1973a). Some species may require fewer counts than others to adequately sample their populations. Even for a single species, the number of counts necessary to estimate population size may vary throughout the season; during periods when individuals are more detectable, fewer counts would be required (see also Slagsvold 1973c, Svensson 1978b). And, as already discussed, periods of OD for individual species are highly variable within a community. Thus, neither the period in the season when counts should be conducted nor the minimum number of counts required may be standardizable, at least at the community level (see also Jensen 1974, Nilsson 1977). Standardization may be most practical in developing a series of SPDs that could be used to make enlightened decisions about the temporal spacing and number of counts required and to increase effectiveness of censusing efforts.

ACKNOWLEDGMENTS

Special thanks go to Dean F. Stauffer for his efforts in collecting the Iowa field data. He was assisted by Thomas Rosburg and Robert Deitchler. Diane Boellstorff worked untiringly tabulating census results. James J. Dinsmore, Mark R. Ryan, Dean F. Stauffer, Harry F. Recher, and C. John Ralph provided helpful comments on an earlier draft of the manuscript. The Illinois study was funded by the Department of Zoology, University of Illinois; the Iowa study by the U.S. Fish and Wildlife Service, Office of Biological Services.

Journal Paper No. J-10048 of the Iowa Agriculture and Home Economics Experiment Station, Ames, Iowa. Project No. 2085.

Studies in Avian Biology No. 6:262–264, 1981.

SEASONAL CHANGES IN AVIAN DENSITIES AND DIVERSITIES

BERTIN W. ANDERSON, ROBERT D. OHMART, AND JAKE RICE[1]

ABSTRACT.—We examined changes in avian species richness and densities over each season for four and one-half years. Data are from a variety of riparian areas along the lower Colorado River. We found that species richness and density estimates tended to vary through time and space in a nonrandom fashion, resulting in skewed or abnormally peaked distributions. Investigating this further with an analysis of variance, we found that significant variation is introduced through differences in the kinds of dominant vegetation present, vertical structure being nearly the same, and by seasonal variation. Both of these factors were further affected by annual changes; that is, the extent of variation in species richness and densities differed, not only from one type of vegetation to another and from season to season, but also from year to year. We concluded that one must be cautious in making inferences from comparative data collected in: (1) the same year but in different seasons; (2) vegetation differing in plant species composition even though structurally similar; and (3) the same season and vegetation but in different years. Before meaningful habitat evaluations can be made, data should be collected over several seasons and years.

It is important to biologists and crucial for natural resource managers to determine the factor(s) that account for variation in avian densities and diversities. It is also useful to know if a given habitat is of equal value to birds during all seasons, if density and species richness vary, and whether similarly structured vegetational communities, differing in species composition of the dominant vegetation, also differ with respect to avian numbers and species richness. While addressing these issues, we consider whether conclusions about avian use of the vegetation would differ if drawn from a single season or a number of seasons over a period of years. In this report we examine variation in avian numbers and species richness at specific sites over several seasons and years. We investigated these factors from 55 months of avian censusing (4950 censuses) in riparian ecosystems.

STUDY AREA

Our studies were conducted along the lower reach of the Colorado River from Davis Dam, Arizona-Nevada border, to the Mexican boundary south of Yuma, Arizona. Water flow in this area of the Colorado River is controlled for production of electrical power and irrigation. Natural flooding has been eliminated since the 1930s; consequently litter accumulates in the vegetation along the river, and fires are common. For this reason most of the vegetation consists of pure stands of the exotic fire-adapted salt cedar (*Tamarix chinensis*), or salt cedar mixed with arrowweed (*Tessaria sericea*), honey mesquite (*Prosopis glandulosa*), screwbean mesquite (*Prosopis pubescens*), cottonwood (*Populus fremontii*), or willow (*Salix gooddingii*). All of these stands of vegetation have been burned at some time during the last 10 to 20 years. Because burning has not been uniform, the vegetation has a patchy horizontal profile of dense to moderately open vegetation, with the bulk of the foliage being from 3 to 6 m high, except in recently burned areas.

[1] Department of Zoology, and The Center for Environmental Studies, Arizona State University, Tempe, Arizona 85281.

Most riparian plant species are deciduous, with leaf drop occurring in November and December. Climate in the winter months of December through February varies annually, from years that are frost-free to years that have 45 or more nights of frost, with temperatures dropping to as low as −9°C.

Spring (March and April) temperatures along the lower Colorado River are also variable, with some years having numerous cold days with frost, whereas other years are mild and frost-free. These variable spring temperatures, combined with precipitation, play a major role in the timing of phenological events.

The summer months of May, June, and July are least variable since they are consistently hot and dry with varying amounts of wind. If rainfall occurs, it is generally during August and September, when humidity is also higher. Both day and night temperatures are relatively hot. The fall months of October and November are usually mild and dry with low temperatures occasionally dropping below 0°C by the end of November.

For the purpose of this report we recognized five types of vegetation. The species composition varied from pure salt cedar and pure honey mesquite to stands of approximately half salt cedar-half screwbean mesquite, half salt cedar-half honey mesquite, or salt cedar with scattered cottonwood and/or willow. In all vegetation types considered in this report, the vertical configuration was similar; about 25% of the volume was between 0 and 0.6 m, 50% was between 0.6 and 4.5 m, and about 25% was between 4.5 and 7.5 m. This configuration is typical of about 63% of the riparian vegetation in the lower Colorado River valley.

METHODS

In each of five types of vegetation, we established six avian census lines which totaled 4 to 8 km in length. On each of the 30 transects, three avian censuses were conducted each month, either in mid-month or during each third of the month from December 1974 through July 1979, using the variable distance transect technique developed by J. T. Emlen (1971). The data presented are based on these 4950 censuses. Censusers were rotated to balance possible observer differences across vegetation types.

The year was divided into five seasons: winter (December–February), spring (March–April), summer

TABLE 1
RESULTS OF THREE-WAY ANALYSIS OF VARIANCE FOR AVIAN DENSITIES

	DF	*F*	*P*	R^2
Main effects	11	45.7	<0.001	75.5
Years	4	32.0	<0.001	14.4
Vegetation	4	53.6	<0.001	32.2
Season	4	48.0	<0.001	28.8
Two-way interaction	40	2.9	<0.001	17.2
Year-vegetation	12	0.3	NS	0.5
Year-season	12	2.2	<0.030	3.9
Vegetation-season	16	5.3	<0.001	12.8
Total R^2				92.7

	N	Deviation
Years		
1975	25	−0.14
1976	25	−0.03
1977	25	0.04
1978	25	0.13
Vegetation		
Cottonwood-willow-salt cedar	20	0.11
Honey mesquite	20	0.17
Salt cedar	20	−0.25
Salt cedar-honey mesquite	20	−0.07
Screwbean mesquite-salt cedar	20	0.04
Seasons		
Winter	20	−0.21
Spring	20	−0.09
Summer	20	0.19
Late summer	20	0.10
Fall	20	0.00

TABLE 2
RESULTS OF A THREE-WAY ANALYSIS OF VARIANCE FOR SPECIES RICHNESS

	DF	*F*	*P*	R^2
Main effects	11	48.2	<0.001	76.1
Years	3	47.5	<0.001	20.3
Vegetation	3	78.5	<0.001	44.9
Season	4	18.5	<0.001	10.9
Two-way interactions	40	3.1	<0.001	17.5
Year-vegetation	12	4.4	<0.001	7.6
Year-season	12	2.6	<0.010	4.4
Vegetation-season	16	2.4	<0.010	5.5
Total R^2				93.6

	N	Deviation
Years		
1975	35	−4.65
1976	25	−1.01
1977	25	2.23
1978	25	3.43
Vegetation		
Cottonwood-willow-salt cedar	20	7.73
Honey mesquite	20	1.18
Salt cedar	20	−6.12
Salt cedar-honey mesquite	20	−3.12
Screwbean mesquite-salt cedar	20	0.33
Seasons		
Winter	20	−3.77
Spring	20	2.38
Summer	20	−1.37
Late summer	20	1.28
Fall	20	1.48

(May–July), late summer (August–September), and fall (October–November). The population estimates from the three censuses were averaged for each census line each month. Monthly avian densities in each type of vegetation were calculated as the mean of the monthly estimates of all transects within that type. Seasonal avian densities were derived by taking the monthly means and computing a mean of means to represent the seasonal value for each vegetation-structural type.

Distributions were analyzed by a three-way analysis of variance and were normalized with $\log_{10}$ transformations.

RESULTS

DENSITIES

Main effects.—Initially we determined if seasons, years, and types of vegetation had a significant and systematic effect on the observed variation in avian densities. The combined effect of seasons, years, and types of vegetation was significant ($P < 0.001$) and explained 76% of the variance (Table 1). All three effects were also independently significant ($P < 0.001$). The greatest proportion of the variance was due more to differences between types of vegetation than to variation between years. The deviation from the "overall" or "grand" mean density indicated that 1975 densities were lowest among the years; among the types of vegetation, in salt cedar; among the seasons, in winter (Table 1). Densities were highest among the years, in 1978; among the types of vegetation, in honey mesquite; among the seasons, in summer.

Two-way interactions.—Combined, the two-way interactions explained an additional 17% of the variance (Table 1). Avian densities differed between years, but the amount of difference depended on which season (but not which type of vegetation) was considered. A year of high density in one type of vegetation was a year of high density in others as well. Not surprisingly, from the main effects we determined that the avian densities were different between seasons, but the amount of difference depended on both the year (year-season) and the type of vegetation (vegetation-season). The avian densities in the types of vegetation differed overall, and the amount of difference depended on season, but not on year; i.e., a favored type of vegetation in a good year was still favored in a poor year.

Species Richness

Main effects.—The effects of years, vegetation type, and season collectively explained 76% of the variance in the species richness data (Table 2). All three of these effects independently explained a significant ($P < 0.001$) amount of the variance. Habitat was the overwhelmingly most important factor, followed by annual variation, with seasonal variation of less (but still significant) importance. Species richness was lowest in 1975 among the years; in salt cedar, among the types of vegetation; and in winter, among the seasons (Table 2). Species richness was highest in 1978 among the years; in cottonwood-willow, among the vegetation types; and in spring among the seasons.

Two-way interactions.—Two-way interactions were all significant ($P < 0.001$) and accounted for an additional 18 percent of the variance (Table 2). Species richness varied annually, but the amount of difference depended on the type of vegetation and on the season. Similarly, richness varied seasonally, but the extent of difference depended on the year and type of vegetation. Finally, species richness varied with the type of vegetation, but the extent of the difference depended on the year and season.

DISCUSSION

Data presented demonstrate that, at least in the lower Colorado River area, significant differences in densities and diversities between vegetation types occur and should be looked for even though these communities differ very little in height and foliage volume. Differences can also be expected in the same vegetation type from season to season. Finally, given the same vegetation type and season, differences can be expected to occur between years. This suggests that considerable caution must be exercised when comparing census data. We need to be cautious in making inferences from comparative data with the following characteristics: (1) same year but different seasons; (2) vegetation differing in plant species composition even though structurally similar; and (3) same season and type of vegetation but different years.

It may be reasoned that limited data provide a poor basis for making fundamental comparisons and management decisions. Some examples of misleading comparisons include:

(1) Average avian density in mixed cottonwood-willow-salt cedar communities in spring 1975 was 128 birds per 40 ha. In salt cedar in spring 1979, the density was also 128 birds per 40 ha. One might conclude, erroneously, that salt cedar supported as many birds as the mixed communities, whereas in most years, salt cedar consistently contained fewer birds, and 1975 was simply a year of low avian densities.

(2) In summer 1979 there were 429 birds per 40 ha in salt cedar and 361 birds per 40 ha in cottonwood-willow-salt cedar mixes. One might conclude that pure salt cedar supported a higher density of birds than the mixed communities. However, over five summers the mean density was 276 birds per 40 ha in salt cedar and 339 birds per 40 ha in the mixed communities. For the entire study, salt cedar averaged 135 birds per 40 ha, and the mixed communities averaged 247 birds per 40 ha. In general, salt cedar did not support as large a population as the mixed communities. These data stress the biological importance of the significant interactions we report.

Numerous additional examples involving species richness in addition to densities could be cited. It seems clear that before habitat evaluations can be made, data should be collected over a considerable time frame. Attention should also be given to censusing during more than one season. Comparisons of avian population density estimates, like all other types of comparisons in science, should be based on sufficient replications of seasons and years to give credence to conclusions.

Legitimately one can ask: What kinds of comparisons can be made which will lead to worthwhile conclusions? We have not fully and finally answered this question. It is apparent that careful consideration must be given to avian density or species richness variations which might result from even slight variations in vegetation (species composition, foliage volume, diversity), differences in seasonality, and annual variation.

ACKNOWLEDGMENTS

We wish to thank Bea Anderson, Jeannie Anderson, Jane Durham, Drs. Julie Meents and Harry Recher for reviewing and editing various drafts of the manuscript. We are grateful to Bea Anderson for typing early drafts and to Cindy D. Zisner for typing the final manuscript. We thank Dr. Julie Meents, Kathleen Conine, Diane Laush, Larry Pyc, and especially Jeff Drake for assistance in collecting and analyzing the data for the report. The work was supported by Water and Power Resources Service Contract No. 7-07-30-V0009.

Studies in Avian Biology No. 6:265–270, 1981.

AN INVESTIGATION OF THE EFFECT OF SEASONAL ACTIVITY LEVELS ON AVIAN CENSUSING

C. JOHN RALPH[1]

ABSTRACT.—Intensive variable distance circular-plot censuses and timed activity budget data were used to compare the effects of conspicuousness upon census results. In six of ten species no correlation was found, suggesting that all birds within the "Effective Detection Distance" (EDD) were seen. In four species there were significant correlations. Multiple regression analyses confirmed these results. Generally larger correlations were found with the number of birds per station than with density per ha; the latter is presumed to correct for changes in conspicuousness, possibly through changes in the EDD. However, a completely unexpected result showed that the EDDs themselves, while showing cycles, did not correlate with the activity variables. This indicates that, although they probably have biological meaning and are relevant to censusing, EDDs are not simply related to obvious changes in conspicuousness as measured by activity patterns.

In dense forests censusing of birds is thought to have less reliability than in open habitats due to individuals being overlooked. This problem is probably greatest in a fixed distance transect. Theoretically, a variable distance count (e.g., J. T. Emlen 1971) corrects for the lessened visibility at greater distances. A researcher does this by calculating a distance at which all birds are detected, or a given proportion is detected. A species that is very conspicuous will have a larger detection distance, on the average, than inconspicuous species. However, variable distance methods assume that within some distance from the observer that is calculated from the data, all birds are detected. If this assumption does not hold, if a proportion of the birds close to the observer go undetected, the census will underestimate the density of birds.

The purpose of this paper is to investigate the problem of inconspicuous birds. That is, are the seasonal changes in abundances calculated from variable distance methods due to actual changes in density, or to changes in conspicuousness? Since it is difficult at best to measure the presence of the "invisible birds," I have used indirect measures to determine the effects of conspicuousness on censuses. I will test the efficacy of the count method by two analyses:

(1) I examined monthly trends in count results and compared them to measures of conspicuousness. Positive correlation would make us suspect the count results are responding to behavior. Lack of correlation can indicate a variety of things, among them the possibility that the population density is changing and the count method is measuring these changes.

(2) The factor in the variable distance count methods that responds to changes in conspicuousness is presumably the detection distance. Positive correlations between distance and conspicuousness would indicate that the method is satisfactorily compensating for changes in conspicuousness. That is, the seasonal changes in calculated density are due to actual changes in abundance, rather than to changes in conspicuousness.

MATERIALS AND METHODS

Monthly censuses were conducted between July 1977 and April 1980 in a wet (ca. 1500 mm rain per year) forest at approximately 1600 m elevation on Keauhou Ranch, near Hawaii Volcanoes National Park on the island of Hawaii. On a 16-ha gridded plot, 8-minute variable distance station counts (Reynolds et al. 1980) were taken between dawn and about 10:00 on 25 stations, 100 m apart, usually three times each month, for a total of 50 to 100 station counts per month. The census data were analyzed using a method from Ramsey and Scott (1979) giving a density of birds per ha based on the assumption that all birds are detected within some minimum distance from the observer. For this study, I grouped all observation in 10 m increments, setting the minimum distance for calculation of density at 20 m. The census data also provided a measure of uncorrected abundance, the number of birds per station. This latter set of data should be responsive to conspicuousness, and the density data less so.

The conspicuousness of the species was determined by timed activity budgets taken throughout the day. Each month approximately 35 individuals of each species were followed for a minimum of 10 seconds (average 20 to 25) for each individual. We recorded the number of calls, hops, and flights, and the total length of song bouts, hops and flights. These are considered indices of conspicuousness. Data from three years were combined by month, and the resulting monthly averages of these activities were compared with the census data. Average sample size was between 150 to 250 individual budgets per species per month. The square roots of all quantities were used in the following analyses to correct for their non-normality.

RESULTS

Although 10 species are presented in this study, for the sake of space, detailed analyses

[1] USDA Forest Service, Institute of Pacific Islands Forestry, 1151 Punchbowl Street, Honolulu, Hawaii 96813.

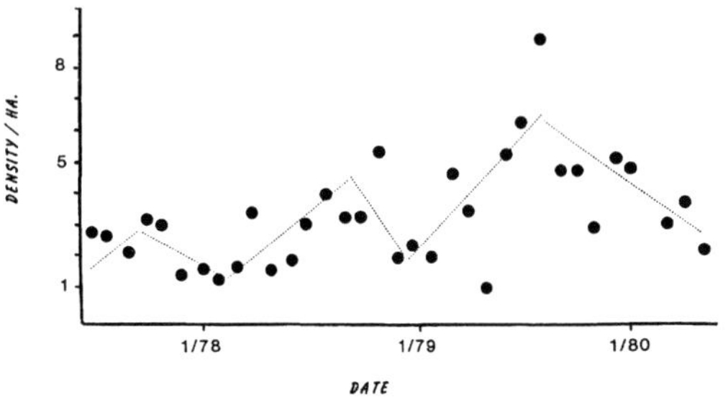

FIGURE 1. Calculated density of 'Elepaio, presumably corrected for effects of conspicuousness.

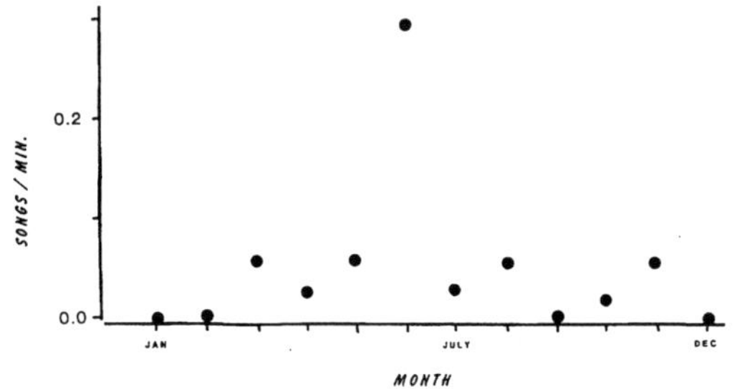

FIGURE 3. Monthly mean length of song bouts per minute of the 'Elepaio.

will be shown for only the 'Elepaio (*Chasiempis sandwichensis*).

Results of Censuses

Seasonal variations in both density (Fig. 1) and number per station (Fig. 2) are apparent for the 'Elepaio, differing in amplitude and somewhat in timing. The number observed per station varied from a low in the early part of the year to a high shortly after the breeding season. The density figures have a similar pattern, showing a pattern of gradually increasing density over the course of the three years.

Changes in Activity Levels

Vocalizations.—This species has few prolonged song bouts. What few there are peaked in June (Fig. 3). Calls, on the other hand, are more evenly distributed throughout the year, being most frequent from April through October (Fig. 4). The average calling rate of between 0.75 and 2.0 calls per minute provides the observer with between four and ten opportunities to identify and record a bird during a single count period.

Movements.—The number of hops and flights showed a regular seasonal pattern (Fig. 5), peaking between December and June, and reaching a low point in August. These movements are possibly related to food availability. In some of the other species studied, where food resources could be quantified, decreased food (e.g., fewer flowers with nectar) resulted in more and faster movements. The average distance flown per minute (Fig. 6) showed two peaks in the 'Elepaio, one in June and the other in early winter. Distance traveled in hops showed little trend, with only a minor peak in October.

Correlations between Abundance and Activity

I calculated Pearson correlation coefficients between the six activity variables and the 10 species' census data. Significant correlation might suggest that apparent changes in abundance merely reflect changes in activity levels. A low correlation could indicate, among other possibilities, that there is no linear relationship, and that variable distance counts are immune to changes in conspicuousness resulting from the birds' behavior. The analyses resulted in 34 significant correlations in the 120 comparisons (Table 1—"Dens" and "#/Sta" columns). In six species, the 'Amakihi (*Loxops virens*), Hawaiian Thrush (*Phaeornis obscurus*), 'Akiapola'au (*Hemignathus wilsoni*), Hawaiian Creeper (*Loxops maculatus*), Japanese White-eye (*Zosterops japonicus*), and Hawaiian 'Akepa (*Lox-*

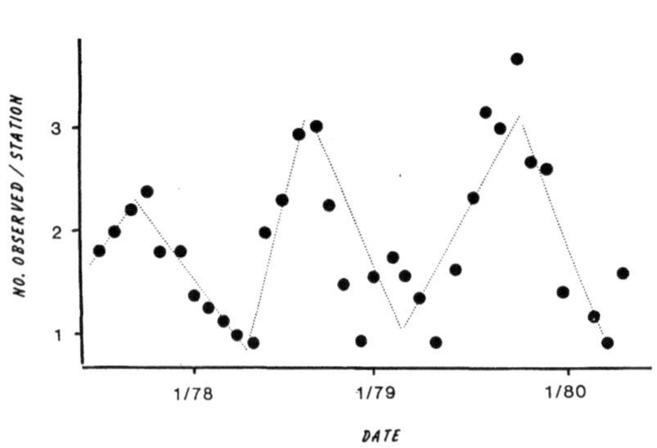

FIGURE 2. Number of individual 'Elepaio over three years per station censused.

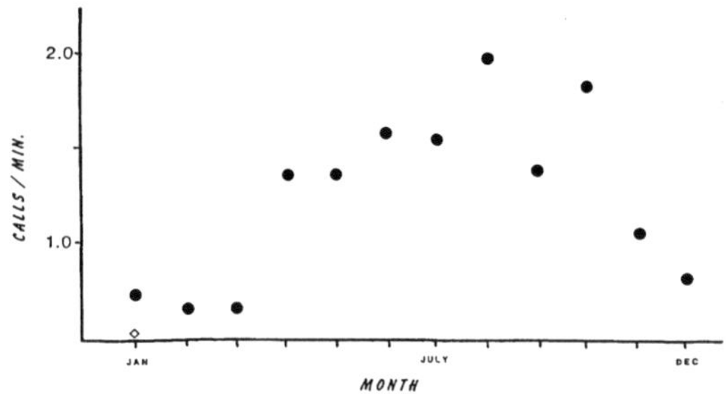

FIGURE 4. Monthly mean number of calls per minute of the 'Elepaio.

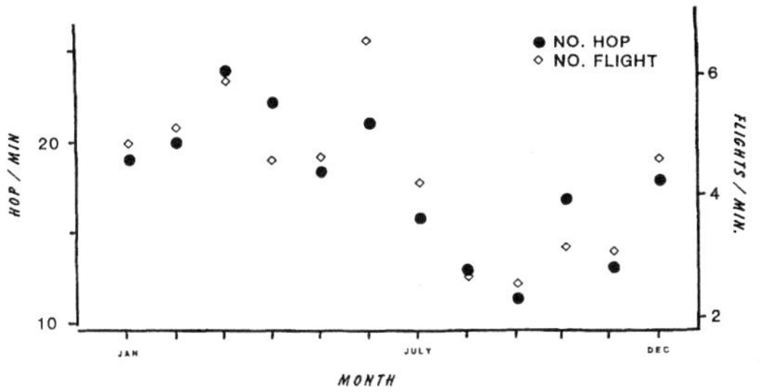

FIGURE 5. Monthly mean number of hops and flights per minute of the 'Elepaio.

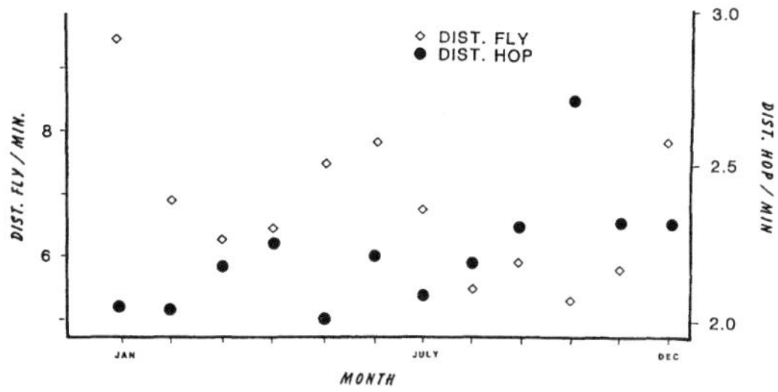

FIGURE 6. Monthly mean total distance of hops and flights per minute of the 'Elepaio.

ops coccineus), there were only two or fewer significant comparisons between activity levels and the two density measures, and these only at the 0.05 level.

However, in four species, 'Elepaio, 'Apapane (*Himatione sanguinea*), Red-billed Leiothrix (*Leiothrix lutea*), and 'I'iwi (*Vestiaria coccinea*), there were several significant correlations. It can therefore be concluded that activity variables possibly influence population counts in these species. Since about half of the correlations are negative (19 of 34), it is apparent that no simple relationship exists. For instance, an increased number of hops could presumably make a bird more conspicuous, but it could also move a bird out of view more rapidly.

We can determine if the density figures are less correlated with the activity variables than are numbers per station, as we would expect if the former do indeed correct for conspicuousness. This was not clearly the case. Of the 60 comparisons between conspicuousness and one or both abundance measures, 20 had at least one significant correlation. Only in one of these (distance hopped in the thrush) was there actually a significant ($P \leq 0.05$) difference between the sample correlations. A trend was, however, apparent. Of the 20 significant correlations of activity variables (Table 1): (1) in 12 cases there was a greater correlation with number per station than with density; (2) in six cases the level of significance was the same; and (3) in only two cases was there a more significant correlation with density than with number per station. One interpretation is that the density calculations (derived from the variable distance method) at least partially correct for changes in activity variables.

Multiple Regression Analysis

Simple correlations may be insensitive to the underlying influences, so I did a multiple regression analysis. For this I first used a principal components analysis to define between one and three (e.g., Fig. 7) new variables from the original six. The new variables (called factors 1–3), unlike the original, are independent and orthogonal, and therefore their order of entry into the statistical model does not alter their significance to the model. These were then used to predict the numbers of birds per station and the density. Factors 1–3 are linear combinations of the original six and have no immediate and simple biological meaning. They are best thought of as abstractions of the birds' behaviors. But, of course, the original variables are also abstractions—those that are easy to quantify. Moreover, the relationship between any specific behavior and detectability is not at issue; only whether behavior, in some abstract sense, contributes to differences in the numbers of birds observed.

This analysis showed that in five species relatively little of the variation in either the density or the number per station is explained by the activity patterns (Table 1—"Model" row, R^2 value). The percent of the variation explained in these species ranges from only 0 to 8%. However, in five other species, the model explains significant amounts of the variation. As was expected, conspicuousness usually explains more of the variation in the number per station than in density.

Variation in Effective Detection Distances

As a bird becomes more conspicuous, the average distance at which detections are made should increase, and vice versa. The "Effective Detection Distance" (EDD) is essentially this distance (calculated by the method in Ramsey and Scott 1979).

Of the 10 species studied, six showed definite annual cycles of EDDs (e.g., Fig. 8), while two were questionable, and two did not (Table 1). Because of their cycling, it appears that EDD is measuring (or is sensitive to) some aspect of the birds' behavior in many cases, perhaps some aspect of conspicuousness.

The next obvious analysis was to determine

TABLE 1
SIGNIFICANCE OF CORRELATION[a]

Species x̄ density/ha	'Apapane 14.4			White-Eye 5.3			Leiothrix 3.4			'Elepaio 3.3			Thrush 3.0			'I'iwi 2.8		
	Dens	#/Sta	EDD	Dens	#/Sta	EDD	Dens	#/Sta	EDD	Dens	#/Sta	EDD	Dens	#/Sta	EDD	Dens	#/Sta	EDD
Activity variables (x̄ per min)																		
Number calls	—	*	—	—	—	—	—	—	—	*	***	—	—	—	—	—	—	*
Length songs	—	****	—	—	—	—	—	—	—	—	—	—	—	—	—	—	—	*
Number hops	**	**	—	—	—	—	xx	xx	—	x	xxx	—	—	—	—	*	**	—
Number flights	**	**	—	—	—	—	xx	x	—	xx	xxx	—	—	—	—	**	**	—
Dist. flown	—	—	—	—	—	—	x	x	—	x	x	—	—	—	—	—	—	—
Dist. hopped	x	xxx	—	—	—	—	—	x	—	*	**	—	x	*	—	—	x	—
Model {P	*	***		—	—		**	*		**	***	—	—	—		—	**	
{R^2	.20	.40		.00	.00		.26	.24		.27	.44		.04	.08		.12	.22	
EDD cycle?			Yes			Yes			Yes			?			Yes			Yes

Species x̄ density/ha	'Amakihi 2.2			'Akepa 0.6			Creeper 0.3			'Akiapola'au 0.1		
	Dens	#/Sta	EDD	Dens	#/Sta	EDD	Dens	#/Sta	EDD	Dens	#/Sta	EDD
Activity variables (x̄ per min)												
Number calls	—	—	—	—	—	—	—	—	—	—	—	—
Length songs	—	—	—	—	—	—	—	—	—	—	—	—
Number hops	—	—	—	—	—	—	—	—	—	—	x	—
Number flights	—	—	—	—	—	—	—	—	—	—	—	—
Dist. flown	x	—	—	—	—	—	—	—	—	—	—	—
Dist. hopped	—	—	—	—	—	—	—	—	—	—	—	—
Model {P	—	—		—	—		—	—		—	*	
{R^2	.03	.07		.04	.07		.04	.03		.05	.13	
EDD cycle?			No			No			?			Yes

[a] * = positive correlation; x = negative correlation; * and x = <0.5; ** and xx = <.01; *** and xxx = <.001; **** and xxxx = <.0001.

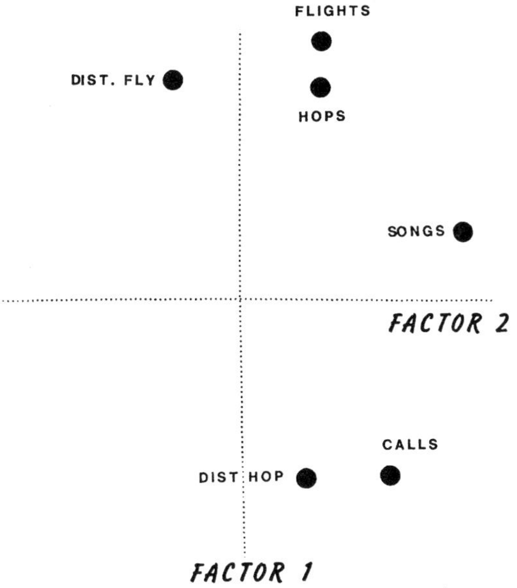

FIGURE 7. Result of two factor Principal Components Analysis on activity variables of the 'Elepaio.

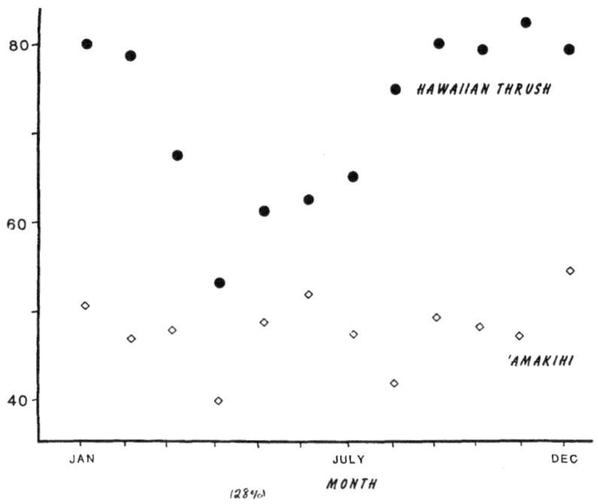

FIGURE 8. Monthly mean of Effective Detection Distances (EDD) for a cycling species (Hawaiian Thrush) and a non-cyling species ('Amakihi).

if these changes in EDDs were related in any simple way to the activity variables. I predicted that many species would be highly correlated. Pearson correlation coefficients (Table 1—"EDD" columns) yielded, however, only two significant ($P \leq 0.05$) relationships. By chance alone, one would have expected three. Multiple regression analyses yielded no significant models based upon the activity variables themselves predicting EDDs. This completely unexpected result throws some doubt upon a very basic tenant of variable distance methods: i.e., differences between species' conspicuousness are accounted for by differences in their detection distances.

DISCUSSION

In six of the 10 species, conspicuousness changing over the season had little relation either to the number of birds seen at each station or to the density calculated from the data. This indicates that these species are possibly active or vocal enough that rather few birds are missed during the 8-minute count period. Therefore, the census method probably accurately measures the population changes. However, in four species, conspicuousness possibly plays a role, since changes in abundance were correlated with activity variables. In these four species: (1) significant correlations exist between the birds' activity and their apparent abundance; and (2) a multiple regression model, based on their activity, significantly predicted their abundance. However, on the bright side, even in these species the multiple regression analysis still only explained less than half of the variation. The rest of the variation was presumably due to other factors, paramount among them being, one would hope, actual changes in the birds' density.

It is suggestive that, in most cases, the level of significance of correlation (and the percent of the variation explained by the activity level changes) are greater when one is dealing with the number observed per station as opposed to density. That is, although one is likely to see and hear more birds when they are conspicuous, either empirical (J. T. Emlen 1971) or calculated (e.g., Ramsey and Scott 1979) densities do tend to remove the effects of conspicuousness.

However, an examination of the relation of the EDDs and the activity variables does not make it clear why these calculations do succeed. The fact that between six and eight of the species' EDDs cycle indicates that some aspect of the birds' behavior is being measured by the EDDs. However, the lack of relationship between the changes in EDDs and the activity measures indicates that EDDs reflect: (1) some abstraction of behavior; (2) an interplay of factors not easily measured; and/or (3) something irrelevant to detection of birds and, therefore, to censusing.

Although the distance measure used in the variable distance census methods provides an aesthetic and intuitively pleasing quantity, my data do not clearly show that changing EDDs reflect changes in conspicuousness. This is a very basic assumption in any variable distance method and quite obviously merits further investigation.

ACKNOWLEDGMENTS

I am very grateful to many field researchers in the Native Hawaiian Forest Ecosystem Project of the U.S. Forest Service, who gathered most of the data for this paper. The manuscript benefited greatly by comments of J. Michael Scott, Carol P. Ralph, and James A. Baldwin. Baldwin and Stuart Pimm generously provided astute statistical counseling.

Studies in Avian Biology No. 6:271–274, 1981.

EFFECT OF TIME OF DAY AND TIME OF SEASON ON THE NUMBER OF OBSERVATIONS AND DENSITY ESTIMATES OF BREEDING BIRDS

AARON A. SKIRVIN[1]

ABSTRACT.—In 1978 and 1979, a study was conducted to assess hourly and biweekly changes in number of detections and density estimates of birds during the breeding period. Bird detections, obtained from variable-circular-plot censuses, tended to decline from the first hour after sunrise to the fourth hour. Changes in bird detections among census hours are presented for selected species. Changes in detection distance estimates in relation to changes in the number detected are discussed.

Basic to all studies requiring estimates of bird abundance are census techniques that produce accurate, or at least, relative results. For most bird census methods, the timing of the counts is one factor that influences accuracy and comparability (J. T. Emlen 1971, Shields 1979). Bird detectability (the proportion of the population that is detected) changes during the day (Shields 1977) and varies among species (Robbins and Van Velzen 1967). Additionally, the number and kinds of birds observed on an area change within seasons (Holmes and Sturges 1975).

During the the breeding season, the number of birds detected and density estimates may be highest during the "middle" of the season (Järvinen et al. 1977b); detectability of many species peaks in the early morning (Robbins and Van Velzen 1967). Many observers restrict breeding bird censuses to morning hours during the "peak" of the breeding period when detections are highest and populations are presumably most stable. However, the duration of the morning census in relation to changes in the number detected requires further clarification. Also, information on changes in the number of birds detected throughout the breeding period would be useful in identifying periods when the detectable population is relatively stable. In 1978 and 1979, a study was conducted to determine if the number of detections and resulting density estimates of birds differed among four consecutive 1-hr periods after sunrise and among biweekly periods in the breeding season.

STUDY AREA

Censuses were conducted in a 65-ha ponderosa pine (*Pinus ponderosa*) stand located on a southwest slope between 1450 and 1600 m elevation in the Blue Mountains about 15 km north of John Day, Grant County, Oregon. Ponderosa pine seedlings and trees up to 10 m tall were abundant; pine trees up to 1.2 m dbh and 30 to 40 m tall dominated the site. Western juniper (*Juniperus occidentalis*) was common and widespread. Edges of some forest openings supported stands of mountain mahogany (*Cercocarpus ledifolius*) shrubs 2 to 7 m tall. Douglas fir (*Pseudotsuga menziesii*), grand fir (*Abies grandis*), and western larch (*Larix occidentalis*) were present but confined to the southwest portion of the site. Conspicuous herbs and low shrubs included elk sedge (*Carex geyeri*), heart-leaf arnica (*Arnica cordifolia*), shiny-leaf spiraea (*Spiraea betulifolia*), and snowberry (*Symphoricarpos* sp.).

METHODS

Ten fixed stations were established approximately 100 m apart along a transect through the study area. Each year, in each of seven consecutive biweekly periods between late April and the end of July, six variable-circular-plot censuses (Reynolds et al. 1980) were conducted at the stations.

Censuses began at sunrise at an endmost station on the transect. The starting station for the first census in a 2-week period was determined by coin toss; starting stations were alternated for the subsequent censuses in the period. Each morning the transect was traversed twice, the second traverse in the reverse direction. Thus, censusing was conducted at 20 stations each morning (120 stations per biweekly period). Five consecutive stations (one station group) were visited within each census hour; hence, the censuses took about 4 hours to complete.

Wind velocity and air temperature were measured, and percent cloud cover was estimated at the beginning, middle, and end of the census. Censusing during cloudy weather was avoided as much as possible; however, to achieve six censuses per 2-week period, some censuses were conducted under cloudy skies. Most censuses were performed when wind velocity was less than 4.8 km/hr. Censusing was terminated if wind exceeded about 16 km/hr, or if rain or snow fell. Results of incomplete censuses were excluded from the analysis.

I counted birds for 10 min at each station but did not census during the 2- to 3-min walk between stations. Birds were counted if they were detected within or below the tree canopy. Birds that typically foraged in flight, such as swallows and swifts, were counted where observed flying above the trees. Species, estimated horizontal distance, behavior (singing or nonsinging), and mode of detection (visual, auditory, or both) were recorded for each detection. I was unable to distinguish songs from other vocalizations of Steller's Jays, nuthatches, Red Crossbills, and Pine Sis-

[1] Department of Fisheries and Wildlife, Oregon State University, Corvallis, Oregon 97331.

TABLE 1
MEAN NUMBERS OF BIRD DETECTIONS AND DENSITY ESTIMATES PER CENSUS HOUR

Detection and density categories by year	Census hour 1	2	3	4
Total detections, 1978	73.2a[a]	71.1ab	67.4bc	62.8c
Total detections, 1979	88.2a	89.1a	84.6ab	79.5b
Total number of singing birds, 1978	55.1a	51.9ab	49.3bc	45.2c
Total number of singing birds, 1979	63.8a	62.2ab	57.0bc	55.4c
Total density of singing birds, 1978	59.4a	56.9a	57.1a	52.8a
Total density of singing birds, 1979	72.5a	70.4a	71.7a	69.6a

[a] Within rows, means not sharing the same letter differ significantly ($P < 0.05$).

kins. All detections of these species and drumming woodpeckers and Ruffed Grouse were recorded as singing birds.

Within each year, the singing-bird data for each species were categorized according to time of detection (first two or second two census hours). For each category an effective radius of detection, required for estimating densities, was derived following the procedure developed by Ramsey and Scott (1979). The number of birds per 40.5 ha (D) was calculated with the formula, $D = N(405{,}000)/5\pi(\rho)^2$, where N is the total number detected in each census hour (five stations) and ρ is the calculated effective radius in meters.

Factorial analysis of variance for a completely randomized design (Sokal and Rohlf 1969:343–356) was used to determine if mean numbers detected and mean density estimates differed significantly among census hours and biweekly periods. The two station groups were included as a factor in the analysis. Student-Newman-Keuls multiple comparison test (Sokal and Rohlf 1969:240–241) was employed to separate means in significant ANOVAs. Mean separation tests were performed at 5% level of significance.

RESULTS AND DISCUSSION

TIME OF DAY

For 1978, 11,526 bird detections were recorded; 74.4% (8515) were singing birds. In 1979, 14,336 detections were tallied; 70.5% (10,114) were singing birds.

Both years, total detections and total singing-bird detections decreased significantly ($P < 0.05$) from the first hour to the fourth hour (Table 1). Of the species with at least 100 detections, several showed relatively large (greater than 20%) changes in number detected between hourly periods (Table 2). Although detections usually decreased from hour 1 to hour 4, not all species showed statistically significant differences between mean hourly detections. All significant differences represented decreases in the number detected from the earlier hour. Numbers of some species remained remarkably constant among census hours (e.g., Chipping Sparrow in 1979). In 1979, detections of Steller's Jays increased from earlier to later census hours. Analysis of variance of the jay data indicated significantly different ($P < 0.05$) mean hourly detections, but the mean separation test failed to identify different hourly means.

Total density of singing birds declined during the morning; however, hourly means did not differ significantly (Table 1). Effective radii for most species decreased from the first two to the second two hours. Although fewer birds were detected in the last two hours, the estimated size of the area (area = $\pi\rho^2$) in which birds were counted also decreased. Some species did not exhibit the trend of shorter effective radius with fewer detections, or conversely, longer distances with more detections. For example, in 1979, detections of Clark's Nutcrackers declined by 41% in the last two hours, but the estimated census area increased by 34%. In 1978, Hermit Thrush detections increased by 47% while the estimated census area decreased by 27%. Relatively large hourly changes in numbers detected for both species were not significantly different ($P > 0.05$); mean density estimates, however, differed significantly among census hours.

Large changes in detectability during the morning census would confound results of studies designed to monitor absolute or relative changes in observed bird abundance caused by factors other than time of census. However, changes in the number of detections during the morning may not dictate corresponding changes in density estimates unless effective radius is directly and positively correlated to number of detections. Results for some species indicate that there might be a compensatory relationship between number of detections and effective radius.

Changes in frequency and loudness of songs during the census could account for changes in effective radius, assuming that techniques for estimating distances to birds and calculating radii are sensitive to subtle changes in bird behavior and are accurate. Ramsey and Scott's (1979) model for estimating effective radii assumes that birds are "independently" and "uniformly" distributed over the census area "according to a

TABLE 2
PERCENT CHANGE IN THE NUMBER OF SINGING BIRDS DETECTED BETWEEN CENSUS HOURS[a]

Species	*n*	Hours after sunrise 1–2	2–3	3–4	1–3	2–4	1–4
Ruffed Grouse	107	−19	−7	−44	−25	−48	−58[b]
(*Bonasa umbellus*)	148	−33	+11	−44	−25	−37	−58[b]
Common Flicker	164	−4	−6	−61[b]	−10	−63[b]	−65[b]
(*Colaptes auratus*)	89	—	—	—	—	—	—
Empidonax Flycatcher	1189	+1	−11	−11	−10	−21[b]	−20[b]
(*Empidonax* sp.)	1202	−11	−12	−4	−22	−16	−25[b]
Steller's Jay	68	—	—	—	—	—	—
(*Cyanocitta stelleri*)	186	+109	+19	+2	+148	+21	+152
Clark's Nutcracker	11	—	—	—	—	—	—
(*Nucifraga columbiana*)	154	+26	−50	+11	−37	−44	−30
Mountain Chickadee	356	−44[b]	−6	−16	−48[b]	−22	−56[b]
(*Parus gambeli*)	289	−26[b]	−29	+4	−48[b]	−26	−56[b]
White-breasted Nuthatch	306	−10	−19	−10	−27	−27	−34
(*Sitta carolinensis*)	318	−18	−2	−19	−20	−21	−35
Red-breasted Nuthatch	115	+21	−20	−18	−3	−34	−21
(*Sitta canadensis*)	1247	+6	+3	−6	+9	−2	+3
Brown Creeper	193	+7	−28	−30	−23	−49[b]	−46[b]
(*Certhia familiaris*)	205	−18	+4	−19	−15	−16	−31
American Robin	271	+1	−3	−28	−1	−30	−29
(*Turdus migratorius*)	170	−21	−7	+3	−27	−29	−25
Hermit Thrush	222	+57	+31	−17	+106	+9	+71
(*Catharus guttatus*)	238	+19	+11	−26	+32	−17	−2
Solitary Vireo	354	+8	0	−12	+8	−12	−5
(*Vireo solitarius*)	274	+16	−26	+12	−14	−18	−4
Yellow-rumped Warbler	740	+1	−6	−6	−5	−11	−10
(*Dendroica coronata*)	802	+17	−16	−7	−2	−22	−9
Western Tanager	615	−15	−1	−14	−16	−15	−28
(*Piranga ludoviciana*)	490	−25	−5	−9	−29	−14	−35
Cassin's Finch	1228	−12	−6	+1	−17	−4	−16
(*Carpodacus cassinii*)	1291	−7	−6	+8	−12	+2	−5
Pine Siskin	145	−3	+15	+5	+12	+21	+18
(*Carduelis pinus*)	66	—	—	—	—	—	—
Red Crossbill	328	−11	+18	−24	+5	−10	−26
(*Loxia curvirostra*)	715	−13	−12	−5	−23	−16	−27
Green-tailed Towhee	210	−10	−18	+9	−26	−11	−20
(*Pipilo chlorurus*)	228	+29	−31	+8	−11	−25	−4
Dark-eyed Junco	845	−13	+1	+10	−14	+11	−3
(*Junco hyemalis*)	731	−9	−8	+26	−16	+16	+6
Chipping Sparrow	665	+23	−2	−3	+20	−6	+16
(*Spizella passerina*)	922	+3	−2	−3	0	−5	−2

[a] Percentages depict changes from the earlier hour; within species, values in upper row are for 1978, lower row for 1979.
[b] Significantly different ($P < 0.05$) mean numbers detected.

random process." Further, they assume that distances to birds are estimated "without error." Certainly these assumptions are not strictly met; moreover, not only is distance estimation fallible, the magnitude and direction of the errors are probably inconsistent among observations within species. The influence of unfulfilled assumptions on estimates of effective

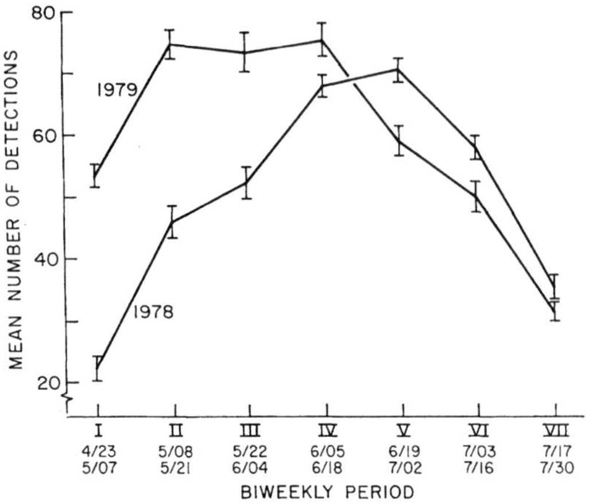

FIGURE 1. Mean numbers and standard errors of singing birds detected in seven biweekly periods. Dates shown are for 1978; 1979 dates were two days later.

radius may mask any likely compensatory effects due to changes in bird behavior. Results for species that deviated from the compensatory relationship may represent varying degrees of the effects of violations of the assumptions. Further investigation of interactions between number and effective radius seems warranted because several species exhibited a compensatory relationship.

TIME OF SEASON

For 1978, peak numbers of singing birds occurred in June; peak numbers in 1979 were recorded between mid-May and mid-June (Fig. 1). Yearly patterns of biweekly changes in total density of singing birds and total detections were similar to the patterns shown in Figure 1. Within years, mean detections of singing birds were similar ($P > 0.05$) among peak periods (IV and V in 1978 and II, III, and IV in 1979; Fig. 1), but were significantly greater ($P < 0.05$) than means for other periods.

Species listed in Table 2 exhibited statistically significant ($P < 0.05$) changes in number of detections among biweekly periods. However, patterns of observed changes in detections through time varied considerably among species. Generally, detections of resident species were highest in May; numbers of migrants usually peaked in June. Ruffed Grouse (a resident species), for example, were undetectable after mid-June in both years. Detections of singing Cassin's Finches, another resident species, decreased dramatically after the second or third biweekly period. Yet Western Tanager, a migrant, did not arrive on the area until mid-May.

To obtain data representative of the structure of a breeding bird community, censuses should be conducted throughout most of the breeding season. Abundance estimates for some species would be severely underestimated (and other species missed entirely) if censusing was restricted to periods of peak detections. Additionally, because of rather large changes in detections among biweekly periods (and perhaps among weeks), an estimate of average seasonal abundance for most species would be misleading. Results of this study indicate that if abundance estimates are to be relative among species in the same community or within species among communities, the researcher must account for changes in detectability during the morning census and changes in abundance estimates during the breeding period.

ACKNOWLEDGMENTS

Reviews by D. S. deCalesta, M. S. Eltzroth, and K. W. Skirvin improved the manuscript and are sincerely appreciated. L. Mauer typed the manuscript. The project was funded by the U.S. Forest Service (FS-PNW-Grant No. 40). The OSU Computer Center provided funds for data analysis.

Studies in Avian Biology No. 6:275–286, 1981.

EFFECT OF TIME OF DAY ON BIRD ACTIVITY

CHANDLER S. ROBBINS[1]

ABSTRACT.—Breeding season activity, based on detections recorded on more than a million 3-minute Breeding Bird Survey stops, reaches a peak for most species during the hour centered at sunrise or in the following hour. Activity of most species then declines gradually as the morning progresses. When large samples are considered, activity patterns for a given species are quite constant from year to year; but each species has its own characteristic pattern and there is much similarity among members of the same genus.

Activity reaches a low point in midday, and may almost cease in some habitats (e.g. deserts); but in deciduous forests, activity of many species continues at a reduced rate. By reducing walking rate or lengthening listening periods, productive censusing of many species could be extended into midday. Winter activity is even more strongly oriented toward the early morning.

Bird activity through the day is predictable to a degree, and varies greatly from species to species. Knowledge of peak times of activity or conspicuousness can be helpful in planning the timing of field work and even in the selection of the most productive method. This paper summarizes data from literally millions of sight observations and tens of thousands of net-hours of breeding season banding activity.

METHODS

BREEDING BIRD SURVEY

All North American Breeding Bird Survey (BBS) data for 1965–1979 were summed annually by species for each 10-stop interval to determine what percentage of the birds were recorded in stops 1–10, 11–20, . . . 41–50. These periods are roughly equivalent to five hourly intervals with the first one centered at sunrise. If a species were equally conspicuous in all five intervals, 20% of the total recorded would be observed in each interval.

ALL DAY "IPA" POINT COUNTS

Twenty-minute point counts, recorded as four consecutive 5-minute counts, were taken periodically throughout several days at a single location on the wooded Patuxent River bluff at my home near Laurel, Maryland to plot singing activity and general conspicuousness through the day. Of interest were changes in activity as reflected in number of species and individuals detected and number of singing birds.

HOURLY BREEDING SEASON BANDING TOTALS

These totals are based on dawn-to-dusk operation of a grid of 44 mist nets, operated generally on 12 days per season for 12 summers. The nets, which sampled about 40 ha of lowland deciduous forest on the Patuxent Wildlife Research Center near Laurel, Maryland, were operated on alternate days and visited on a 2-hour schedule. Time of capture was considered to be one hour prior to removal from the net.

WINTER BIRD SURVEY

This experimental sampling technique consisted of a grid of 46 8-km transects, one located at the center of each 7½′ quadrangle (USGS topographic map) in central Maryland. The area sampled included all of Anne Arundel, Baltimore, and Howard Counties and Baltimore City as well as parts of adjacent Carroll, Frederick, Montgomery, and Prince Georges Counties. Transects were 8 km in length, with 2 km covered on foot in each hour. Coverage began at sunrise and lasted exactly 4 hours. Data from all five years of the Survey, 1970–74, were used in this study.

RESULTS

BREEDING SEASON

Early morning activity

Morning breeding season activity patterns for woodland species, which are detected primarily by voice, are shown in Figure 1. Breeding Bird Survey coverage begins at one-half hour before local sunrise and ends between 3½ and 4½ hours after sunrise, thereby bracketing the period of greatest activity. As shown by Figure 1, the number of individual birds recorded in a series of 3-minute stops decreases more rapidly as the morning progresses than does the number of species detected.

Each species has its own distinctive diurnal activity pattern, as illustrated for representative species in Figure 2 and Table 1, based on the percentage of individuals recorded during each ten-stop interval on the BBS. The percentages are remarkably constant from year to year as shown by the 95% confidence limits in Figure 2.

Herons.—Although there tends to be consistency in activity patterns within a genus, there are some striking differences within families. The large and conspicuous Great Blue Heron (*Ardea herodias*) shows almost no change in conspicuousness as the morning progresses (Fig. 2). Several herons, including Green (*Butorides striatus*), Louisiana (*Hydranassa tricolor*), and Snowy Egret (*Leucophoyx thula*), reach a peak in the first hour after sunrise. Other herons are most conspicuous in the sunrise hour as they fly to or from their nests; these include the Little Blue Heron (*Florida caerulea*), Cattle Egret (*Bubulcus ibis*), Common Egret (*Casmerodius albus*), Black-crowned Night Heron (*Nyc-

[1] Migratory Bird and Habitat Research Laboratory, U.S. Fish and Wildlife Service, Laurel, Maryland 20811.

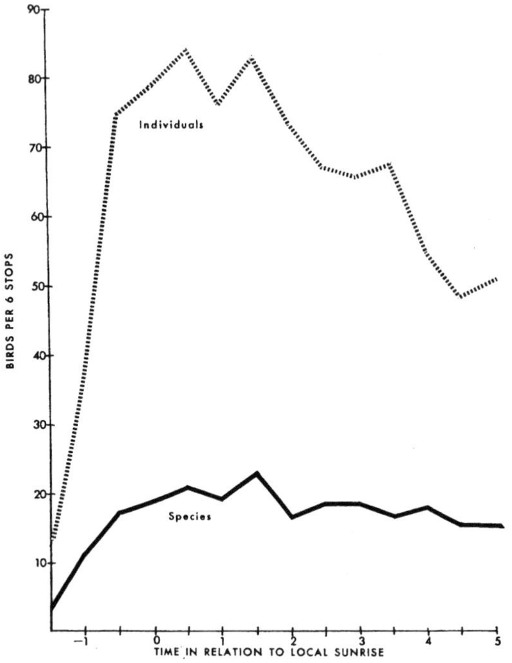

FIGURE 1. Numbers of species and individuals recorded by the author during 3-minute intervals at six points in floodplain forest habitat (mean of four days). Counts were repeated every half hour.

ticorax nycticorax), and especially the three ibises. The American Bittern (*Botaurus lentiginosus*) has its greatest vocal activity period during the sunrise hour, as shown in Figure 2.

Waterfowl, vultures, hawks.—Waterfowl activity patterns are varied and unpredictable, except that observations tend to decrease in the fifth hour. Turkey (*Cathartes aura*) and Black (*Coragyps atratus*) Vulture observations increase to a strong peak in the final hour (Fig. 3, Table 1). Hawks also increase in conspicuousness as the morning progresses, though not so dramatically.

Gallinaceous birds, limpkins, rails.—Some gallinaceous birds, including Turkey (*Meleagris gallopavo*), Ring-necked Pheasant (*Phasianus colchicus*, Fig. 2), and especially Greater Prairie Chicken (*Tympanuchus cupido*) decline sharply in activity after sunrise, while others, such as Ruffed Grouse (*Bonasa umbellus*), Gambel's Quail (*Lophortyx gambelii*), and Gray Partridge (*Perdix perdix*), follow the pattern of the Bobwhite (*Colinus virginianus*, Fig. 2). The Scaled (*Callipepla squamata*), California (*Lophortyx californica*), and Mountain (*Oreortyx picta*) Quail have a pattern between those of the Ring-necked Pheasant and the Bobwhite (Fig. 2). Limpkin (*Aramus guarauna*) and rail observations decrease gradually after passing a peak in the sunrise hour.

Shorebirds.—Killdeer (*Charadrius vociferus*, Table 1), Marbled Godwit (*Limosa fedoa*), Upland Sandpiper (*Bartramia longicauda*), Willet (*Catoptrophorus semipalmatus*), Spotted Sandpiper (*Actitis macularia*), and Common Snipe (*Capella gallinago*) exhibit low activity in the sunrise hour, and peak in the first three hours after sunrise. American Woodcock (*Philohela minor*), however, declined very sharply after a sunrise peak of 72%, and Wilson's Phalarope (*Steganopus tricolor*) showed an increase from 10% at sunrise to 24% in the final period.

Doves, cuckoos.—Pigeons and doves reached a peak in the first hour after sunrise, except for the White-crowned Pigeon (*Columba leucocephala*) and Mourning Dove (*Zenaida macroura*, Table 1), which were equally conspicuous in the sunrise period. The Yellow-billed (*Coccyzus americanus*, Table 1) and Black-billed (*C. erythropthalmus*) Cuckoos also reached a peak in the hour after sunrise, then declined 37% and 48% respectively, by the final period.

Owls, goatsuckers.—Screech (*Otus asio*), Great Horned (*Bubo virginianus*), and Barred (*Strix varia*) Owls were equally clustered in the sunrise hour, with 77 to 79% of observations recorded then. Not so restricted to the dawn period were the following owls, listed by decreasing percentage detected in the sunrise hour: Pygmy (*Glaucidium gnoma*, 43%), Short-eared (*Asio flammeus*, 30%), and Burrowing (*Athene cunicularia*, 22%). Except for the nighthawks, the goatsuckers were almost entirely restricted to the sunrise hour; see Chuck-will's-widow (*Caprimulgus carolinensis*), Table 1. The Lesser Nighthawk (*Chordeiles acutipennis*) was more restricted to the sunrise period (71% of observations) than was the Common Nighthawk (*C. minor*, 58%, Table 1).

Swifts, hummingbirds.—Counts of all species of swifts were depressed in the sunrise hour. The pattern shown for the Chimney Swift (*Chaetura pelagica*, Table 1) is typical. Hummingbirds, except for the Broad-tailed (*Selas-*

→

FIGURE 2. Activity patterns from BBS, shown as the percentage of the total birds (with 95% confidence limits) detected that were recorded in each of the five 10-stop intervals, corresponding approximately to hourly periods beginning one-half hour before sunrise.

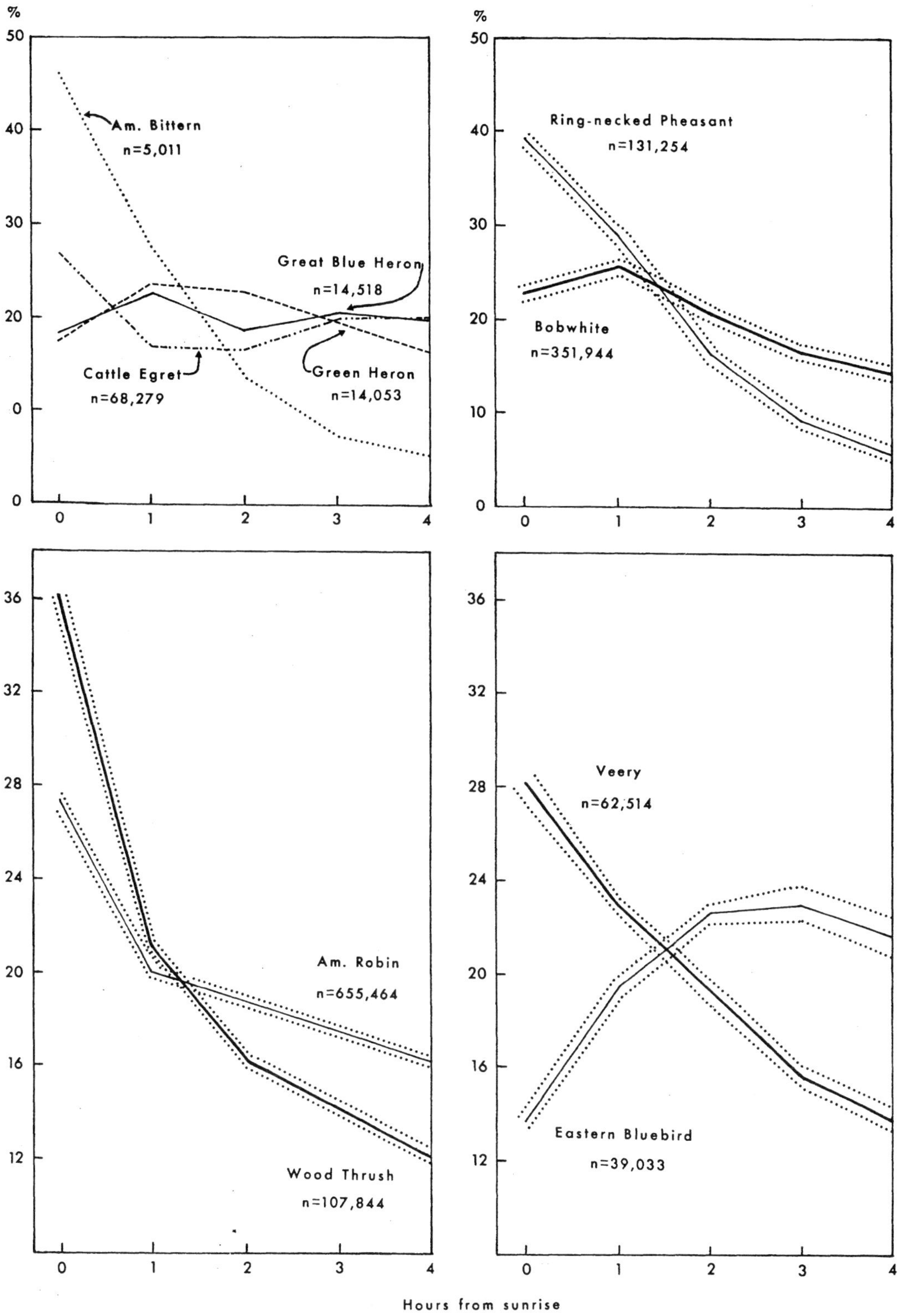
%
50
40
30
20
0
0
Am. Bittern
n=5,011
Great Blue Heron
n=14,518
Cattle Egret
n=68,279
Green Heron
n=14,053
0 1 2 3 4
%
50
40
30
20
10
0
Ring-necked Pheasant
n=131,254
Bobwhite
n=351,944
0 1 2 3 4
36
32
28
24
20
16
12
Am. Robin
n=655,464
Wood Thrush
n=107,844
0 1 2 3 4
36
32
28
24
20
16
12
Veery
n=62,514
Eastern Bluebird
n=39,033
0 1 2 3 4
Hours from sunrise

TABLE 1
PERCENT OF TOTAL OBSERVATIONS THAT FELL IN EACH 10-STOP BBS INTERVAL, 1965–1979

Species	Total recorded	% each hour after sunrise				
		0	1	2	3	4
Mallard	58,055	20.7	24.1	19.9	19.2	16.1
Black Vulture	6350	4.1	10.7	11.9	21.4	52.0
Red-tailed Hawk	12,806	10.3	16.1	19.3	24.0	30.3
American Kestrel	17,374	14.7	19.9	20.8	22.3	22.3
Killdeer	112,701	16.1	22.9	22.1	20.3	18.6
Rock Dove	161,004	7.4	30.2	26.3	21.1	15.0
Mourning Dove	600,763	25.0	24.6	19.5	16.3	14.6
Yellow-billed Cuckoo	71,506	21.5	24.6	20.4	18.0	15.5
Chuck-will's-widow	9577	98.3	1.1	0.3	0.2	0.1
Common Nighthawk	32,039	58.2	13.9	9.0	9.3	9.5
Chimney Swift	138,701	13.5	21.0	21.8	20.9	22.8
Ruby-throated Hummingbird	1195	11.1	19.4	21.6	24.1	23.9
Eastern Kingbird	90,303	19.5	21.9	20.6	19.0	18.9
Eastern Phoebe	36,478	29.6	18.2	18.2	17.5	16.2
Barn Swallow	367,013	11.9	19.4	21.8	23.7	23.1
Bank Swallow	68,515	7.0	17.4	23.3	23.5	28.7
Black-capped Chickadee	43,603	22.1	19.9	21.8	19.5	16.8
Gray Catbird	102,092	25.5	21.4	20.1	17.3	15.7
Swainson's Thrush	52,379	29.6	21.8	17.6	15.4	15.8
Blue-gray Gnatcatcher	18,426	25.3	21.3	20.0	18.0	15.5
Cedar Waxwing	45,939	10.3	20.0	22.4	23.5	23.8
Red-eyed Vireo	180,795	18.3	22.4	20.5	19.9	18.8
Yellow Warbler	80,874	19.0	21.6	21.1	19.8	18.4
Ovenbird	76,168	22.2	23.0	19.3	18.0	17.5
Common Yellowthroat	212,523	22.0	22.4	19.8	18.2	17.6
Western Meadowlark	438,513	22.4	23.6	20.7	17.7	15.6
Northern Oriole	80,333	14.1	20.6	22.3	21.6	21.5
Scarlet Tanager	28,341	19.5	22.2	20.8	19.4	18.1
Indigo Bunting	223,071	26.7	19.8	18.6	17.6	17.3
Chipping Sparrow	164,677	25.3	19.6	19.2	18.4	17.5

phorus platycercus), are late risers, with counts for the sunrise hour only about half as great as in the next hour; see Ruby-throated Hummingbird (*Archilochus colubris*) in Table 1.

Woodpeckers.—The desert-inhabiting Gila Woodpecker (*Melanerpes uropygialis*) was the only member of this family with significantly more birds ($P < 0.05$) recorded in the sunrise hour than later in the morning. The Lewis' Woodpecker (*M. lewis*) was unique in this family in becoming more conspicuous as the morning progressed. The other woodpeckers had activity patterns similar to that shown for the Red-bellied Woodpecker (*M. carolinus*) in Figure 3.

Flycatchers.—The Eastern (*Tyrannus tyrannus*, Table 1) and Western (*T. verticalis*) Kingbirds, which are conspicuous by both voice and sight, were among the most constant species throughout the morning. The Scissor-tailed Flycatcher (*Muscivora forficata*), however, was more conspicuous (25%) in the sunrise hour and then declined by about 30% by the last two hours. The genus *Myiarchus* showed a peak in the first hour after sunrise, then declined sharply as the morning progressed. The three species of phoebes, however, peaked in the sunrise hour and continued to decline thereafter; see Eastern Phoebe (*Sayornis phoebe*) in Table 1. Most members of the genus *Empidonax* peaked in the first period and decreased thereafter, but the Yellow-bellied (*E. flaviventris*) and Acadian (*E. virescens*) Flycatchers had significantly higher ($P < 0.05$) totals in the hour after sunrise than

→

FIGURE 3. Comparison of summer (solid lines) and winter (dotted lines) patterns from BBS and Winter Bird Survey. See text for explanation. Scientific names not in the text are: Blue Jay (*Cyanocitta cristata*), Common Crow (*Corvus brachyrhynchos*), Carolina Chickadee (*Parus carolinensis*), Tufted Titmouse (*P. bicolor*), and Carolina Wren (*Thryothorus ludovicianus*).

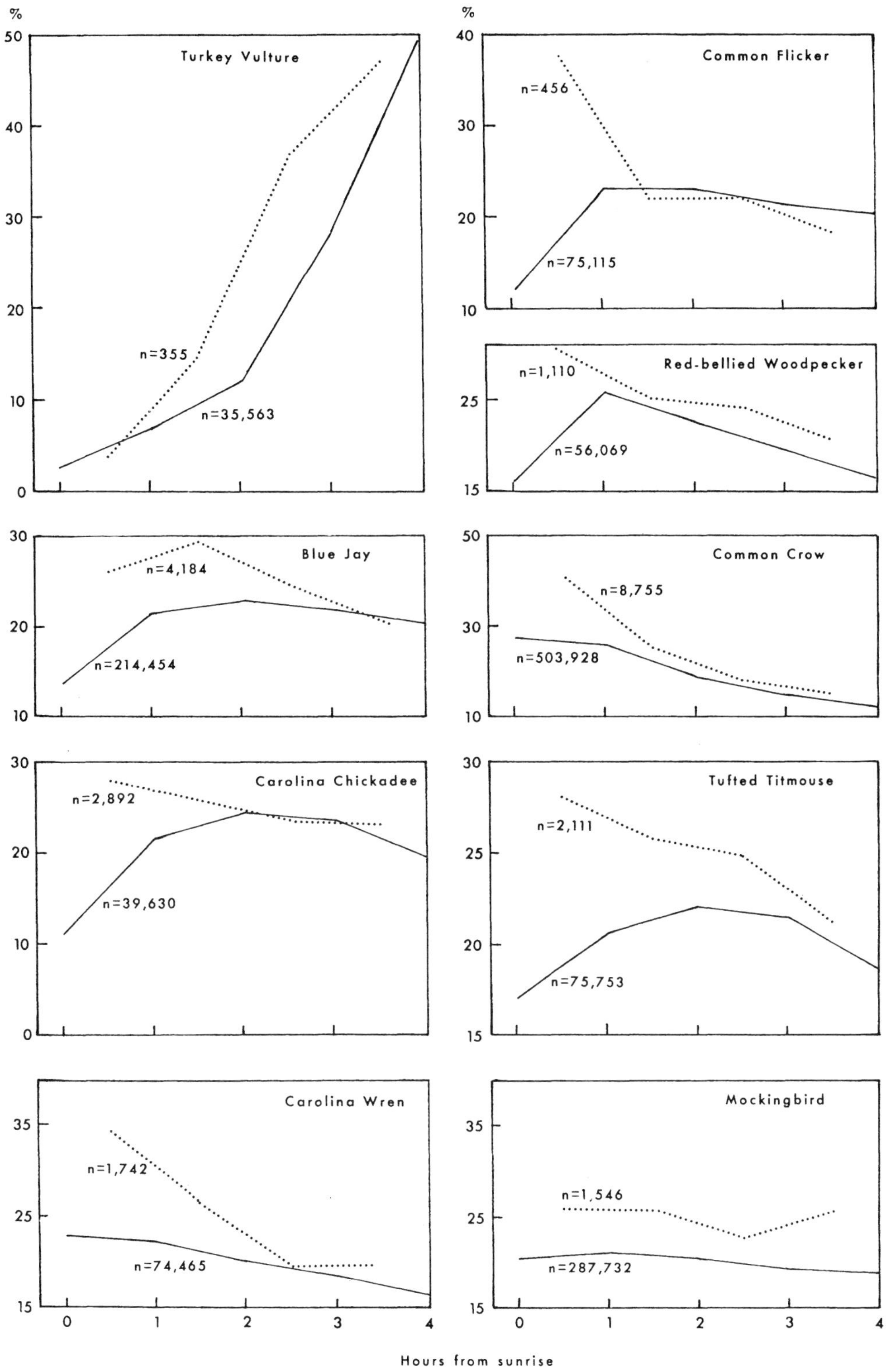
%
Turkey Vulture
n=355
n=35,563
Common Flicker
n=456
n=75,115
Red-bellied Woodpecker
n=1,110
n=56,069
Blue Jay
n=4,184
n=214,454
Common Crow
n=8,755
n=503,928
Carolina Chickadee
n=2,892
n=39,630
Tufted Titmouse
n=2,111
n=75,753
Carolina Wren
n=1,742
n=74,465
Mockingbird
n=1,546
n=287,732
Hours from sunrise

in any other period. The Eastern Wood Pewee (*Contopus virens*) varied little with time, whereas the Western Wood Pewee (*C. sordidulus*) decreased slightly from a sunrise peak, as did the Olive-sided Flycatcher (*Nuttallornis borealis*).

Larks, swallows.—Horned Larks (*Eremophila alpestris*), with a sample of a quarter of a million observations, decreased smoothly from 22.2% in the first period to 17.6% in the last. Except for the Purple Martin (*Progne subis*), which maintained a constant detectability, the swallows were consistently low in the sunrise hour, then gradually increased to peaks in the third or fourth hour after sunrise; see Barn Swallow (*Hirundo rustica*) and Bank Swallow (*Riparia riparia*) in Table 1.

Jays, crows, titmice.—Jays were low in the sunrise hour, but steady thereafter, while crows steadily declined in observations after a sunrise peak that was twice as high as their final hour (Fig. 3). Chickadees and titmice tended to peak in one of the middle periods (Fig. 3).

Nuthatches, creepers, wrens.—Nuthatch observations were especially low in the sunrise period, then nearly doubled in the next hour and maintained their high totals through the last hour. One quarter of the entire Brown Creeper (*Certhia familiaris*) count, however, was tallied in the sunrise period. Wrens declined gradually from a dawn peak, except for the Long-billed Marsh (*Cistothorus palustris*), Short-billed Marsh (*C. platensis*), and Canyon (*Catherpes mexicanus*) Wrens, which peaked strongly at sunrise (34%, 30%, and 29% of total observations), then declined to between 11% and 14% by the final period.

Mockers, thrushes.—The Mockingbird (*Mimus polyglottos*), which is a loud and persistent singer as well as a conspicuous roadside bird easily identified in flight, had one of the most consistent records throughout the morning (Fig. 3). The Gray Catbird (*Dumetella carolinensis*), on the other hand, frequently sings from within dense cover and is less often seen; catbird observations dropped steadily (Table 1), as did those of thrashers. Graphs for representatives of four thrush genera are depicted in Figure 2. The American Robin (*Turdus migratorius*), after a dawn song peak, maintained a high detectability because it feeds in the open and is easily detected visually. The more secretive Varied Thrush (*Ixoreus naevius*) has a pattern much like that of the Wood Thrush (*Hylocichla mustelina*, Fig. 2). The Hermit (*Catharus guttatus*) and Swainson's (*C. ustulatus*) Thrush patterns are similar to that shown for the Veery (*C. fuscescens*, Fig. 2). The Mountain Bluebird (*Sialis currucoides*) pattern is similar to that depicted for the Eastern Bluebird (*S. sialis*, Fig. 2), but flatter, while the Western Bluebird (*S. mexicana*) has a sunrise peak followed by a lull the next hour.

Gnatcatchers, waxwings, shrikes, starlings.—The Blue-gray Gnatcatcher (*Polioptila caerulea*, Table 1), which is detected at close range, primarily by its calls, continued to decline in activity after its sunrise peak. Cedar Waxwings (*Bombycilla cedrorum*, Table 1), on the other hand, were low in the sunrise hour, doubled in the next hour, then gradually rose to peak activity in the last two hours. Loggerhead Shrikes (*Lanius ludovicianus*) and Starlings (*Sturnus vulgaris*), which are visually conspicuous, followed a pattern similar to that of the Yellow-shafted Flicker (*Colaptes auratus*) in Figure 3.

Vireos.—Vireos are detected almost entirely by voice, but they are loud and persistent songsters and vary little in detectability during early and mid-morning. The Red-eyed Vireo (*Vireo olivaceus*) pattern shown in Table 1 is also typical of the White-eyed (*V. griseus*), Yellow-throated (*V. flavifrons*), and Solitary (*V. solitarius*) Vireos. The Warbling Vireo (*V. gilvus*), however, differed by reaching its highest total in the sunrise hour, from which it declined only 10.7% by the last period; this was one of the smallest changes noted in any species.

Wood warblers.—The great majority of the warblers followed the pattern shown for the Yellow Warbler (*Dendroica petechia*) in Table 1, with the peak in the first hour after sunrise, followed by a gradual decline in detections. Notable exceptions were those species that were most conspicuous in the sunrise hour and steadily decreased in vocal activity thereafter; these were the Tennessee (*Vermivora peregrina*), Orange-crowned (*V. celata*), Nashville (*V. ruficapilla*), Louisiana Waterthrush (*Seiurus motacilla*), Mourning (*Oporornis philadelphia*), MacGillivray's (*O. tolmiei*), Yellow-breasted Chat (*Icteria virens*), and Canada Warbler (*Wilsonia canadensis*).

House Sparrow, icterids, tanagers.—The House Sparrow (*Passer domesticus*), Eastern Meadowlark (*Sturnella magna*), and the various blackbird species had activity patterns similar to that shown in Table 1 for the Western Meadowlark (*S. neglecta*), with the highest count in the first hour after sunrise. The orioles, however, started low and built up to an activity peak in the middle period that was maintained through the final hour, as shown for the Northern Oriole (*Icterus galbula*) in Table 1. Except for the Scarlet Tanager (*Piranga olivacea*) shown in Table 1, the native tanagers showed a strong activity peak in the sunrise hour, followed by a gradual decline.

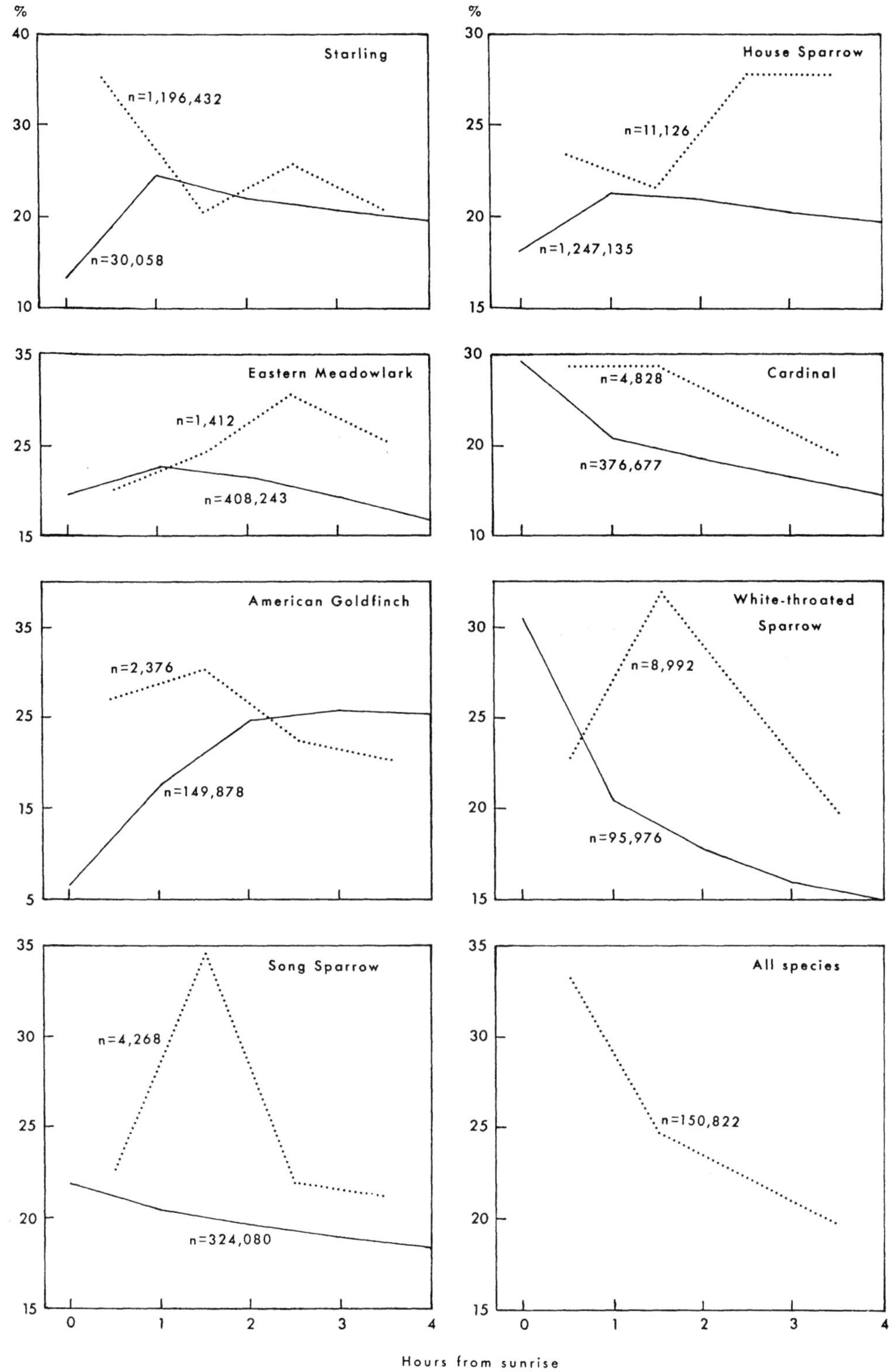

FIGURE 4. Comparison of summer (solid lines) and winter (dotted lines) patterns from BBS and Winter Bird Survey. See text for explanation.

Grosbeaks, finches.—The Cardinal (*Cardinalis cardinalis,* Fig. 4) was the only grosbeak with a sharp activity peak in the sunrise hour; twice as many birds were recorded in this first hour as in the final hour. The pattern for the Pyrrhuloxia (*C. sinuatus*) was similar to that for the Painted Bunting (*Passerina ciris*) and for the Indigo Bunting (*P. cyanea*) shown in Table 1. Blue (*Guiraca caerulea*) and Black-headed (*Pheucticus melanocephalus*) Grosbeaks gradually declined from sunrise peaks, while the Rose-breasted Grosbeak (*P. ludovicianus*) and Lazuli Bunting (*Passerina amoena*) maintained equal activity throughout the five hours. The Lazuli Bunting varied less than any other species, with a difference of only 8.5% between the highest and lowest hourly counts. The Dickcissel (*Spiza americana*) reached a slight but significant ($P < 0.05$) peak in the hour after sunrise, then declined gradually. The cardueline finches, represented in Figure 4 by the American Goldfinch (*Carduelis tristis*), typically had very low counts in the sunrise hour, reached a peak in the second or third hour after sunrise and maintained an above-average count into the final hour.

Towhees, sparrows.—A strong sunrise peak involving more than half of the observations was characteristic of the Olive Sparrow (*Arremonops rufivirgata*). Similar but weaker early peaks were recorded for the towhees. Sparrows were typically most conspicuous in the sunrise hour. Species with a peak of 30% or more in the sunrise hour, as illustrated by the breeding season graph for the White-throated Sparrow (*Zonotrichia albicollis*) in Figure 4, were Henslow's (*Ammodramus henslowii*), Le Conte's (*Ammospiza lecontii*), Sharp-tailed (*A. caudacuta*), Seaside (*A. maritima*), and Bachman's (*Aimophila aestivalis*) Sparrows, and members of the genus *Junco*. Lower sunrise peaks followed by more gradual decreases were recorded for the Savannah (*Passerculus sandwichensis*), Grasshopper (*Ammodramus savannarum*), Baird's (*A. bairdii*), Vesper (*Pooecetes gramineus*), Rufous-crowned (*Aimophila ruficeps*), Cassin's (*A. cassinii*), Black-throated (*Amphispiza bilineata*), Clay-colored (*Spizella pallida*), Brewer's (*S. breweri*), Black-chinned (*S. atrogularis*), White-crowned (*Zonotrichia leucophrys*), Lincoln's (*Melospiza lincolnii*), and Swamp (*M. georgiana*) Sparrows and are represented by the Chipping Sparrow (*S. passerina*) in Table 1. A few species were lowest in the sunrise hour: Lark Bunting (*Calamospiza melanocorys*), Lark Sparrow (*Chondestes grammacus*), Sage Sparrow (*Amphispiza belli*), Fox Sparrow (*Passerella iliaca*), and Chestnut-collared Longspur (*Calcarius ornatus*). Only the Field Sparrow (*Spizella pusilla*) and Song Sparrow (*Melospiza melodia,* Fig. 4) varied little as the morning progressed.

All-day activity

Activity measured by point counts.—The results of five all-day series of 20-minute point counts are summarized in Figure 5. The counts were conducted on 12, 13, 15, 19, and 20 July 1980 from a single point on the Patuxent River bluff near Laurel, Maryland. Local sunrise during this period ranged from 04:53 to 04:59, E.S.T., sunset from 19:30 to 19:34. Minimum temperatures at the observation point in the woods ranged from 16° to 21°C on these five days and maxima ranged from 26° to 33° (3.5°C above normal). Figure 5A shows the mean number of singing males (and 95% confidence limits) recorded per 20-minute period. Figure 5B shows (with 95% confidence limits) the number of species recorded per 20-minute period (light line, above) and per 5-minute period (heavy line, below). Figure 5C indicates the mean number of individuals observed per 5-minute period.

Singing activity was at a morning peak from 05:00 to 07:00, declined 37 percent by noon, then rose gradually to a brief evening peak. The number of species recorded per 20 minutes peaked at 06:00, declined to a low five hours later and maintained about the same level for the rest of the day; there was no evening peak in number of species observed. Each 20-minute observation period was broken into four 5-minute segments. The 5-minute counts, although lower than the 20-minute counts, were more consistent throughout the morning, but dropped more rapidly in the afternoon. The number of species recorded was lowest at 13:00, E.S.T. The mean number of individuals per 5-minute period had a broad morning peak, 06:00–09:00, then decreased gradually to a low at 17:00 before recovering for the evening peak. Broadness of the morning peak results from the greater likelihood of detecting more individuals in 5 minutes versus the 3-minute interval used in the BBS.

Because the number of individuals detected at a single point is too small to show all-day activity levels for most species, only a few representative species are considered. Singing patterns (5-day means of 20-minute totals) are shown in Figure 6 for the Wood Thrush, Scarlet Tanager, and Cardinal. Note the prominent evening peak for Wood Thrush, the early morning peak for the Cardinal (see also Fig. 4), the brief pre-dawn peak for the Scarlet Tanager followed by the sunrise low (see Table 1), and the evening peak for this same species. Note also that the lowest counts in mid-day did not coincide.

Activity detected by banding.—Systematic use of Japanese mist nets (within 2 m of the ground) reveals a very different diurnal activity

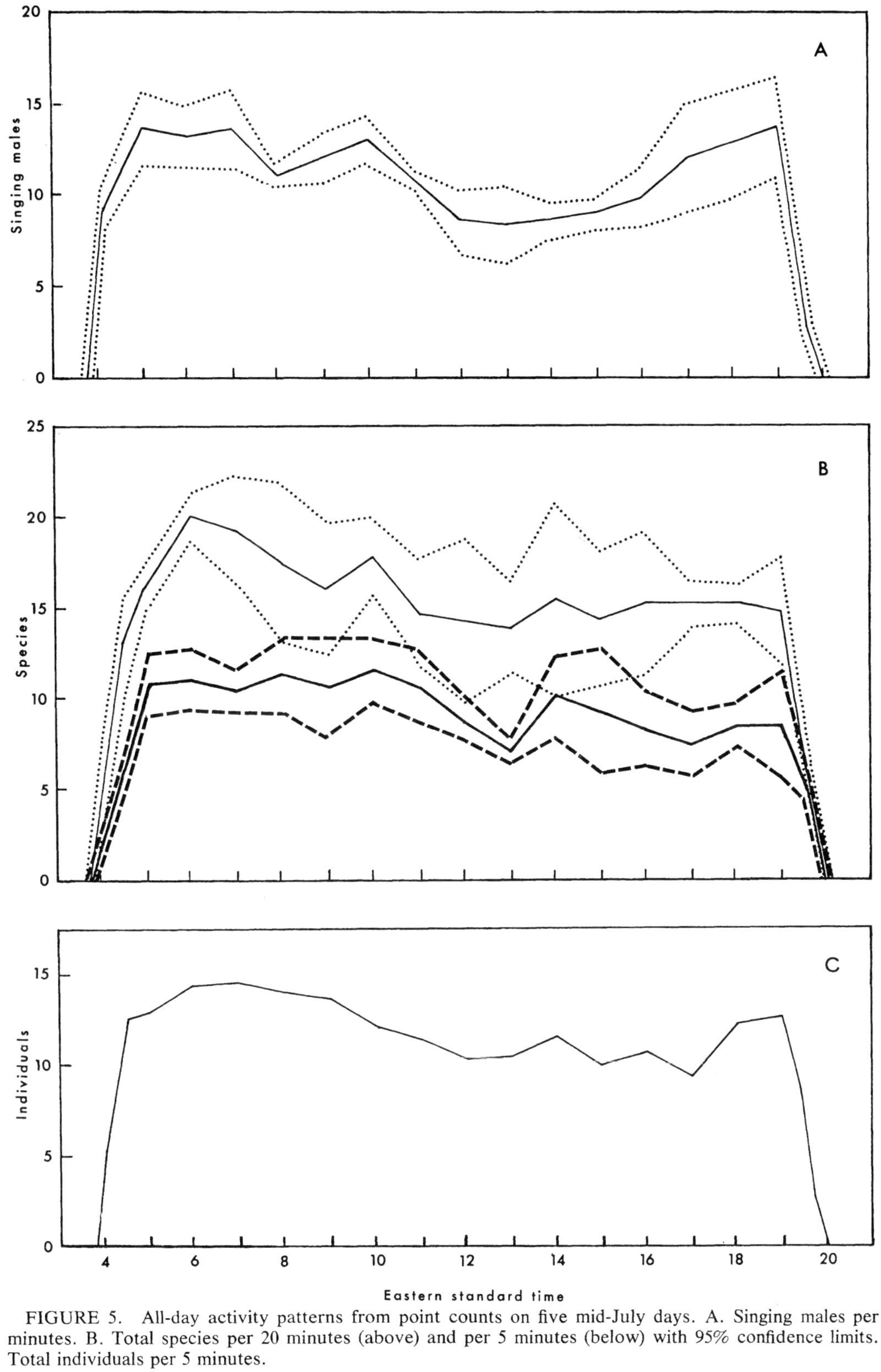

FIGURE 5. All-day activity patterns from point counts on five mid-July days. A. Singing males per 20 minutes. B. Total species per 20 minutes (above) and per 5 minutes (below) with 95% confidence limits. C. Total individuals per 5 minutes.

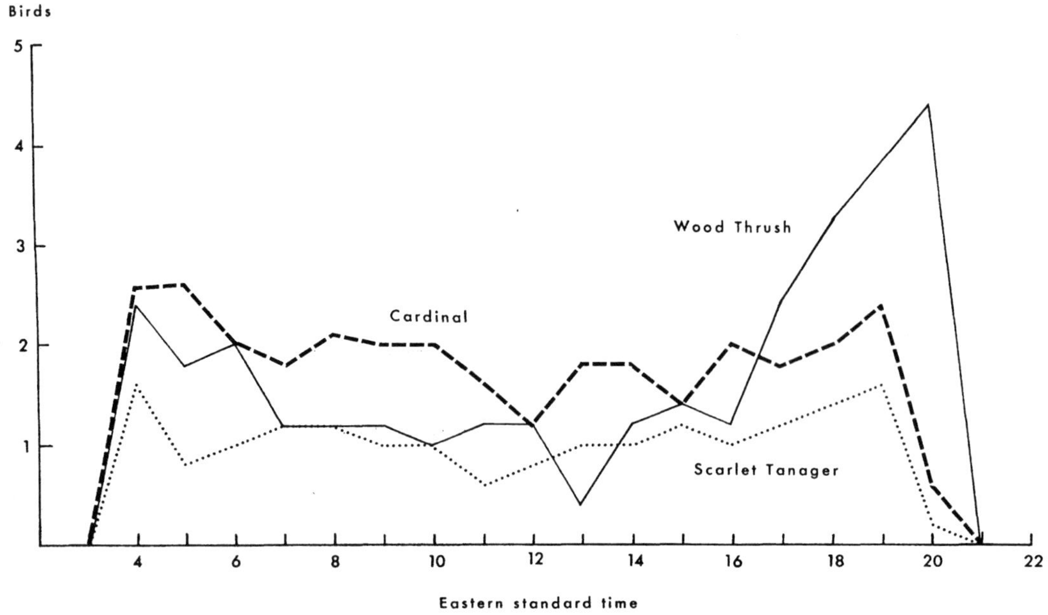

FIGURE 6. Singing patterns of three species from 20-minute point counts on hot days in mid-July (5-day means).

period than that observed by the census taker. Capture totals for 14 species are shown by one-hour intervals in Figure 7. Note the high early peaks for the thrushes, waterthrushes, and most of the other warblers, and the rapid decline that follows. Note also that even for the Wood Thrush there is only a very minor recurrence of activity in the evening, and the pattern for the day is almost the reverse of the singing activity shown in Figure 4. Woodpeckers, titmice, and flycatching birds, on the other hand, remained active through most of the day. Patterns for summer resident species were very similar to those for close relatives that occurred only as transients (see thrushes and waterthrushes, Fig. 7).

Winter Season

A comparison of breeding season and winter activity is presented in Figures 3 and 4. The breeding season figures are the percentages of total birds recorded in each of the five 10-stop BBS intervals. The winter figures are percentages recorded in each of the four one-hour Winter Bird Survey intervals starting at sunrise. Because of the difference in number of intervals, the breeding season data (solid lines) are centered about a mean of 20%, while the winter data are centered about a mean of 25%. Thus it is the shape, not the position of the line that is important. As in the breeding season, the Mockingbird demonstrates fairly constant activity through the morning. The Turkey Vulture in both winter and summer becomes progressively more active as the morning progresses. Most other species in Figure 3 are more frequently recorded in the first hour in winter than in summer, indicating the greater importance of early censusing in winter. The high early morning count of Starlings in winter is probably related to dispersal from their roosts. Fringillids, unlike most other families, show a low count in the first hour in winter; but when all species are considered (Fig. 4), the first hour after sunrise is by far the most productive.

DISCUSSION

There are remarkably few literature references to effect of time of day on conspicuousness or on singing activity of North American birds other than the beginning of dawn song.

→

FIGURE 7. Activity patterns from all-day mist netting during the breeding season. Scientific names not in the text are: Hairy Woodpecker (*Picoides villosus*), Downy Woodpecker (*P. pubescens*), Ovenbird (*Seiurus aurocapillus*), Northern Waterthrush (*S. noveboracensis*), Kentucky Warbler (*Oporornis formosus*), Hooded Warbler (*Wilsonia citrina*), and American Redstart (*Setophaga ruticilla*).

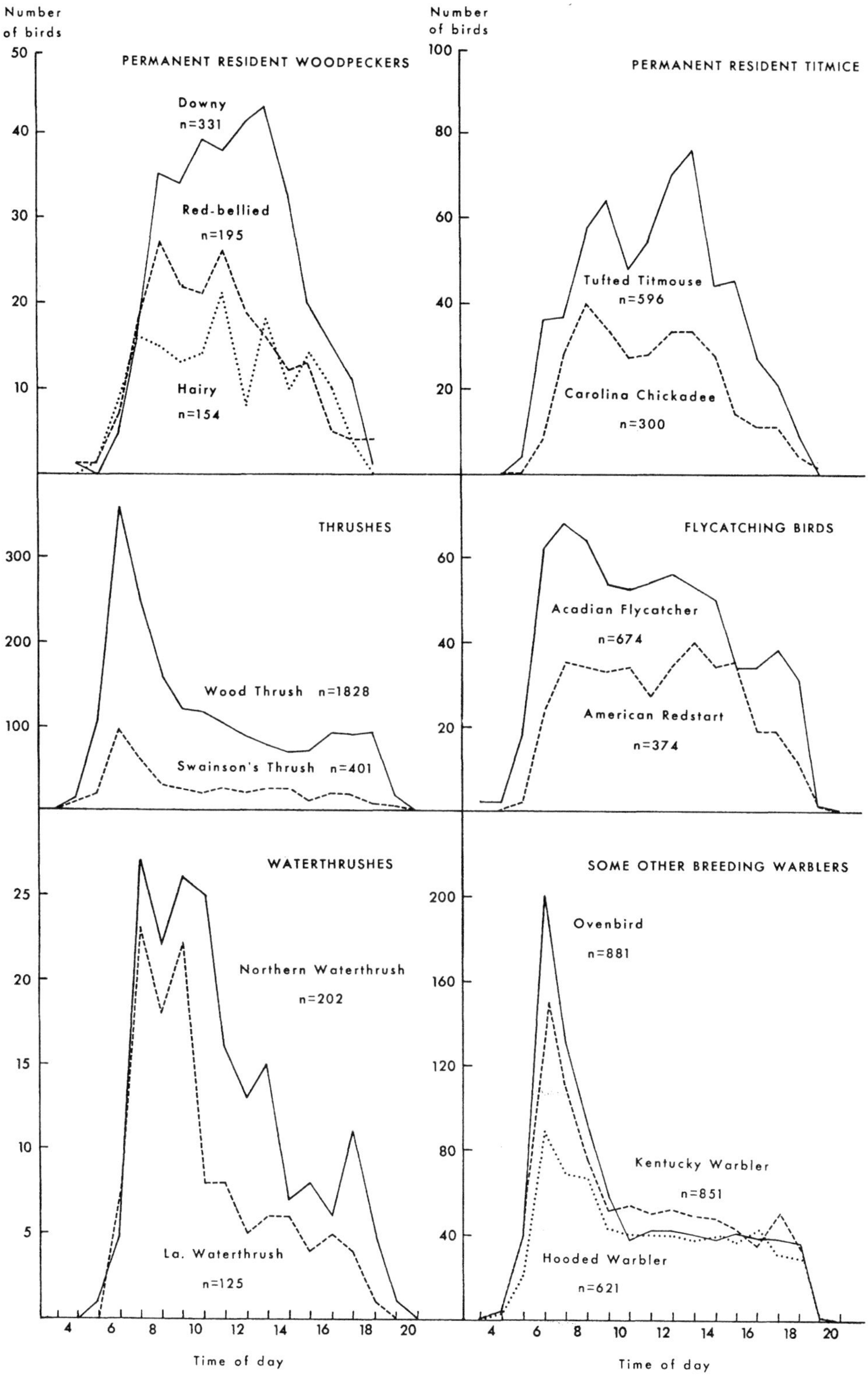
Number of birds
PERMANENT RESIDENT WOODPECKERS
Downy n=331
Red-bellied n=195
Hairy n=154
PERMANENT RESIDENT TITMICE
Tufted Titmouse n=596
Carolina Chickadee n=300
THRUSHES
Wood Thrush n=1828
Swainson's Thrush n=401
FLYCATCHING BIRDS
Acadian Flycatcher n=674
American Redstart n=374
WATERTHRUSHES
Northern Waterthrush n=202
La. Waterthrush n=125
SOME OTHER BREEDING WARBLERS
Ovenbird n=881
Kentucky Warbler n=851
Hooded Warbler n=621
Time of day

Nice (1964) made an all-day count on 11 May 1935 of 2305 songs from one Song Sparrow that had lost its mate four days before; these showed a peak of 278 and 277 songs in the first two hours beginning at 04:45, the time of the first song. After seven hours with totals of 200 or more songs each, the subsequent hourly totals were 150, 182, 121, 60, 52, 16, 12, and 20. This early morning breeding season activity is similar to that shown in Figure 4. Fortunately, point counts, especially those that last as much as 20 minutes, do not register nearly as sharp a decline as do the number of songs given.

Mayfield (1960) gave 5-minute song counts for a Kirtland's Warbler (*Dendroica kirtlandii*) on 21 June 1956, one day before its first egg hatched. When converted to 05:00 sunrise for comparison with the BBS data, there were 80 (7%) songs in the hour centered at sunrise, and 156 (14%), 291 (27%), 271 (25%), and 287 (26%) in the next four hours. Unfortunately, there are no BBS data for this species; but other *Dendroica* warblers do not have such low detectability in the first one or two hours. Again, the point counts are less sensitive to changes in singing intensity than are counts with shorter exposure to each individual bird. Note also that total individuals decline much more slowly on 20-minute point counts (Fig. 5C) than on 3-minute BBS stops (Fig. 1) as the morning progresses. Saunders' (1929:67) statement that "song in the middle of a hot day in June or July is a rarity" should not be taken literally in view of the results of 20-minute point counts made on five abnormally hot days in mid-July (Fig. 5A). It must be kept in mind, however, that even though some birds, such as vireos, orioles, tanagers and finches, continue to sing at least occasionally through the middle of the day, others such as pheasants, doves, thrashers, gnatcatchers, and many of the warblers may sing very little after mid-morning, especially in the latter part of the breeding season (late June and July).

CONCLUSIONS

When large amounts of data are examined, diurnal activity patterns are consistent from year to year. These patterns vary with species, but birds in the same genus tend to have similar patterns.

Most of the species that are detected primarily by voice are recorded in largest numbers in the hour centered at sunrise or the next hour following; after that they are less frequently detected as the morning progresses. Birds frequently detected by sight show less decline later in the morning than do species detected by ear.

Activity patterns can be used in planning field work so as to concentrate censusing during periods of maximum activity or of most consistent activity. When, because of weather delays or insufficient field personnel, it is necessary to conduct some censusing during other than the most productive early morning hours, slower walking or longer listening periods can compensate in part for decreasing bird activity. In some instances, knowledge of activity patterns may be helpful in adjusting census data to account for time of day, especially when very large samples are involved.

ACKNOWLEDGMENTS

It is a pleasure to acknowledge the contribution of the BBS cooperators without whose data this paper could not have been written. I thank D. Bystrak, C. Kepler, B. Noon, and B. K. Williams for their critical comments on the manuscript.

Studies in Avian Biology No. 6:287–291, 1981.

DIURNAL ACTIVITY PATTERNS AND POPULATION ESTIMATES OF BREEDING BIRDS WITHIN A DISTURBED AND UNDISTURBED DESERT-SCRUB COMMUNITY

CHRISTIAN E. GRUE, RUSSELL P. BALDA, AND CLARENCE D. JOHNSON[1]

ABSTRACT.—We censused breeding birds along two 1.6 km transects, one within a transmission-line right-of-way and the other within an adjacent undisturbed desert-scrub community. Our objectives were to compare the diurnal activity patterns of breeding birds along the two transects, and to determine the effects of changes in activity on population estimates. Censuses were conducted between 20 April and 17 May 1974 using a modified strip-transect method. Birds within 63 m of each transect were censused 10 times between 06:00 and 08:00, and 12:00 and 14:00, and 5 times between 17:00 and 19:00. We used the average number of detections (birds/20 ha) as an index to avian activity. Activity patterns and projected densities (birds/40 ha) of permanent and summer residents were compared within and between transects using chi-square or paired *t*-tests. Similar comparisons were made for projected densities within foraging guilds.

The number of birds detected within the disturbed and undisturbed habitats between 12:00 and 14:00 was 32 to 49% of the number detected in the morning. Activity on the two transects increased in the evening, but with the exception of Mourning Doves, remained lower than that observed between 06:00 and 08:00. Reductions in activity were greatest within the transmission-line right-of-way. Similar differences were observed in projected densities between time intervals on both transects; reductions in population estimates were greatest within the right-of-way. Neither resident status nor foraging guild appeared to alter observed trends. Composition and structure of the two communities varied with time of day. Species richness was greatest on both transects in the morning whereas evenness was greatest between 12:00 and 14:00. Diversity was lowest in the evening. Data indicate census results may be significantly affected by diurnal changes in avian activity. Censuses conducted early in the morning appear to provide the most accurate estimates of habitat utilization by breeding birds within disturbed and undisturbed communities.

Ornithologists have noted changes in avian activity with time of day for decades. However, only recently have effects of diurnal activity patterns on avian census results been quantified (for review, see Shields 1979). During the breeding season, activity and detectability of most bird species are greatest at dawn, decrease to a diurnal minimum at mid-day (Robbins and Van Velzen 1967, 1970; Weber and Theberge 1977; Shields 1977), and increase in the late afternoon (Järvinen et al. 1977b). These changes in avian activity appear to be correlated with weather conditions, particularly air temperature (Shields 1979).

Census techniques are often used to assess the impact of human activities on bird populations (e.g., Grue 1977, Bock and Bock 1978, Cantle 1978, Franzeb and Ohmart 1978). These activities often involve the removal of ground and canopy vegetation. Environmental conditions (e.g., air temperature and solar radiation) may, therefore, be more severe within disturbed habitats. Knowledge of the diurnal activity patterns of birds within disturbed and undisturbed habitats is essential to adequately assess the impact of habitat modification on bird populations using census techniques. However, a comparison of the diurnal activity patterns of breeding birds within disturbed and undisturbed habitats and their effect on population estimates is lacking. We provide here such a comparison for breeding birds within a transmission-line right-of-way and an adjacent undisturbed desert-scrub community in which daytime temperatures normally exceed 38°C.

STUDY AREA

Our study area was located 25.6 km north of Phoenix, Arizona, in Maricopa County (latitude 33°48′N, longitude 112°15′W) at 515 m in elevation. The terrain was flat. Several dry washes traversed the 80 ha study area. The habitat on the study area was relatively open. Ground cover of grasses and forbs was ca. 35%; trees, shrubs, and cacti covered ca. 11% of the study area. Dominant plant species which occurred within the desert-scrub community were saguaro (*Carnegiea gigantea*), cholla and prickly-pear cactus (*Opuntia* spp.), ocotillo (*Fouquiera splendens*), palo verde (*Cercidium microphyllum*), creosote bush (*Larrea tridentata*), burr sage (*Franseria deltoidea*), and catclaw acacia (*Acacia greggii*). Annuals included plantain (*Plantago purshii*), globe mallow (*Sphaeralcea fendleri*), heron bill (*Erodium circutarium*), and milk vetch (*Astragalus* spp.). (Plant names correspond to those given by Kearney and Peebles, 1960.) The study area consisted of two parallel transects, 1.6 km long. One transect was located within the right-of-way of the newly constructed double 500 kV Navajo Project Southern Transmission Line (disturbed transect), the other 250 m to one side of the right-of-way (control).

[1] Department of Biological Sciences, Northern Arizona University, P.O. Box 5640, Flagstaff, Arizona 86011; present address of CEG: U.S. Fish and Wildlife Service, Patuxent Wildlife Research Center, Laurel, Maryland 20811.

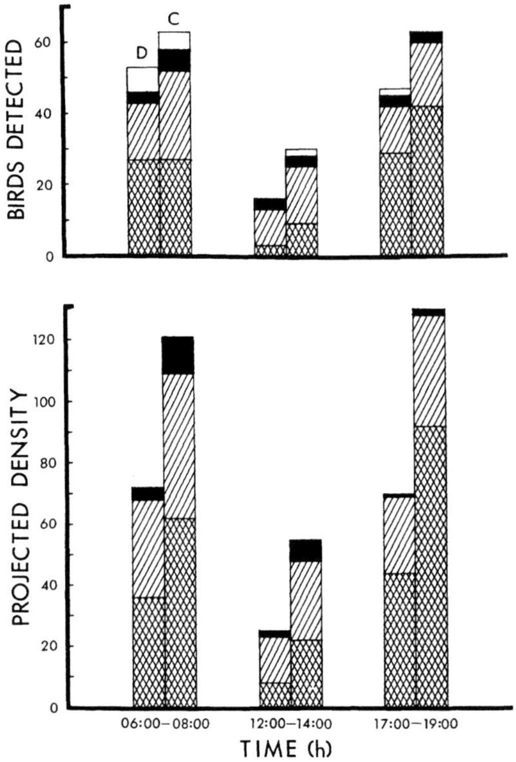

FIGURE 1. Diurnal changes in the average number of detections (birds/20 ha, including birds flying over) and projected densities (birds/40 ha) within a disturbed (D) and undisturbed (C) desert-scrub community during the breeding season. Diagonal lines represent permanent residents (cross-hatching = Mourning Doves), solid bars represent summer residents, and open bars represent transients.

Location of the study area was based on homogeneity of the habitat, accessibility, and lack of disturbances other than those associated with powerline construction. Approximately 5.7 ha of desert-scrub vegetation was removed from the disturbed transect for construction of the transmission line.

METHODS

Censuses of breeding birds were conducted by Grue along both transects between 20 April and 17 May 1974. A modified strip-transect method (J. T. Emlen 1971) was employed. Ten censuses were conducted between 06:00 and 08:00 and 12:00 and 14:00, and five censuses were conducted between 17:00 and 19:00. The two transects were alternated as to which was censused first. Censuses were conducted by slowly walking the transects in succession and recording (with a tape recorder) all birds detected by visual or auditory cues within a lateral distance of 63 m (maximum width of disturbance within the right-of-way). Progression along transects was continuous with stops only to observe and listen. We used the average number of detections (birds/20 ha) on each transect within each time interval as an index to avian activity. The activity index was equal to the total number of birds detected within 63 m divided by the number of censuses conducted; birds flying over were included.

Population estimates (projected densities) were derived using methods described by J. T. Emlen (1971). The mode lateral distance interval was found for each species during each time interval by summing the number of detections within seven lateral distance intervals. The first five intervals were 3.0 m wide, the sixth was 15.2 m wide, and the seventh was 32.6 m wide. Both visual and auditory cues were used to determine the total number of detections per interval; birds flying over were not included except for species which forage from the air. The number of birds detected up to and within the mode lateral distance interval (usually $\geq$15.2 m due to the open habitat and short ground cover) was then extrapolated to a lateral distance of 126 m. This value was then divided by the number of censuses conducted to give the projected number of individuals of a species within either 40 ha of transmission-line right-of-way or undisturbed desert scrub. All densities were rounded up to whole birds.

The activity index and projected densities for each time interval were compared within transects using chi-square tests (Snedecor and Cochran 1967:231) and between transects using paired *t*-tests (Snedecor and Cochran 1967:91). Similar comparisons were made for the activity and projected densities of permanent and summer residents, and for projected densities within different foraging guilds (ground; picker or gleaner; hawker, hoverer or aerial; and pecker, hammerer, or tearer; after Root 1967). The composition and structure of the disturbed and control avian communities were also compared using the community comparison index (*C* value, Kulcynski 1937, *in* Oosting 1956) and diversity (Shannon 1948) and evenness (Pielou 1975) indicies.

We placed three constantly recording Honeywell hygrothermographs, each within a standard white weather box, on the study area. One weather box was placed on the access road within the right-of-way. The remaining two weather boxes were placed along the control transect; one in the open on existing ground cover, and the other beneath a palo verde. Operation of the hygrothermographs coincided with morning, noon, and evening bird censuses. Mean temperatures were calculated at 2 h intervals (06:00–18:00) and compared using paired *t*-tests.

RESULTS

The number of birds detected on the disturbed and control transects between 12:00 and 14:00 was 32 to 49 percent of the number detected in the morning (Fig. 1). Activity on the two transects increased in the evening, but, with the exception of Mourning Doves (*Zenaida macroura*) remained lower than that observed between 06:00 and 08:00. These differences were statistically significant ($P < 0.05$). Differences between time intervals in the number of permanent and summer residents detected were also statistically significant. The majority of transients were also detected between 06:00 and 08:00;

TABLE 1

PROJECTED DENSITIES (RESIDENT BIRDS/40 HA) WITHIN A DISTURBED AND UNDISTURBED DESERT-SCRUB COMMUNITY AT DIFFERENT TIMES OF THE DAY DURING THE BREEDING SEASON

	Disturbed			Undisturbed		
Species	06:00–08:00	12:00–14:00	17:00–19:00	06:00–08:00	12:00–14:00	17:00–19:00
Turkey Vulture (*Cathartes aura*[a]; PR, HHA[b])					1 (1)	
Red-tailed Hawk (*Buteo jamaicensis*; PR, HHA)	1 (2)[c]					
Gambel's Quail (*Lophortyx gambelii*; PR, G)	2 (4)	1 (1)		2 (4)	1 (2)	4 (4)
White-winged Dove (*Zenaida asiatica*; SR, G)	1 (1)			2 (4)		
Mourning Dove (*Zenaida macroura*; PR, G)	36[d] (176)	8 (16)	44 (57)	62[c] (191)	22 (54)	92 (106)
Black-chinned Hummingbird (*Archilochus alexandri*; SR, HHA)			1 (1)	2 (2)	1 (1)	2 (1)
Common Flicker (*Colaptes auratus*; PR, PHT)	1 (4)	2 (6)	2 (3)	3 (6)	1 (1)	1 (1)
Gila Woodpecker (*Centurus uropygialis*; PR, PHT)				2 (6)		1 (2)
Ladder-backed Woodpecker (*Picoides scalaris*; PR, PHT)				1 (1)	1 (1)	
Cassin's Kingbird (*Tyrannus vociferans*; SR, HHA)				1 (1)		
Ash-throated Flycatcher (*Myiarchus cinerascens*; SR, HHA)	2 (5)	2 (3)		6 (16)	3 (9)	
Rough-winged Swallow (*Stelgidopteryx ruficollis*; SR, HHA)	1 (4)				3 (4)	
Verdin (*Auriparus flaviceps*; PR, PG)	5 (16)	3 (9)	2 (1)	3 (14)	2 (5)	7 (5)
Cactus Wren (*Campylorhynchus brunneicapillus*; PR, G)	6 (28)	3 (11)	7 (8)	17 (64)	9 (32)	12 (23)
Curved-billed Thrasher (*Toxostoma curvirostre*; PR, G)	8 (28)	4 (17)	6 (15)	11 (48)	9 (25)	12 (15)
Loggerhead Shrike (*Lanius ludovicianus*; PR, HHA)	4 (15)	2 (5)	4 (8)	1 (2)	1 (1)	1 (1)
Starling (*Sturnus vulgaris*; SR, G)				1 (1)		
House Finch (*Carpodacus mexicanus*; PR, G)	1 (5)		2 (4)	3 (11)	1 (2)	3 (5)
Black-throated Sparrow (*Amphispiza bilineata*; PR, G)	4 (8)		2 (2)	4 (11)	3 (6)	1 (1)
Total Density[e]	36[c]	17	26	59[c]	33	38
Number of Species	13	8	9	16	14	11
Diversity[f]	1.80	1.90	1.36	1.80	2.01	1.17
Evenness[g]	0.70	0.91	0.62	0.65	0.76	0.49

[a] Scientific names, American Ornithologists' Union (1957, 1973, 1976).

[b] Resident status: PR = permanent resident, SR = summer resident, ST = summer transient; after Phillips et al. (1964); and foraging guild: G = ground; HHA = hawker, hoverer, or aerial; PG = picker or gleaner; PHT = pecker, hammerer, or tearer; after Root (1967).

[c] Total number of observations excluding flyovers except for hawkers, hoverers, or aerial feeders.

[d] Differences between times of day significant, chi-square test, ($P < 0.05$).

[e] Excluding Mourning Doves.

[f] Shannon (1948).

[g] Pielou (1975:15).

TABLE 2
PROJECTED DENSITIES (RESIDENT BIRDS/40 HA) WITHIN FORAGING GUILDS AT DIFFERENT TIMES OF DAY WITHIN A DISTURBED AND UNDISTURBED DESERT-SCRUB COMMUNITY DURING THE BREEDING SEASON

	Disturbed			Undisturbed		
Foraging guild	06:00–08:00	12:00–14:00	17:00–19:00	06:00–08:00	12:00–14:00	17:00–19:00
Ground[a]	22[b] (6)[c]	8 (3)	17 (4)	40[b] (7)	20 (5)	26 (5)
Hawker, hoverer, or aerial	8 (4)	4 (2)	5 (2)	11 (4)	8 (5)	2 (2)
Picker or gleaner	5 (1)	3 (1)	2 (1)	3 (1)	2 (1)	7 (1)
Pecker, hammerer, or tearer	1 (1)	2 (1)	2 (1)	6 (3)	2 (2)	2 (2)

[a] Excluding Mourning Dove.
[b] Differences between times of day significant; chi-square test, $P < 0.05$.
[c] Number of species.

very few were detected during noon and evening censuses. Reductions in activity were greatest within the transmission-line right-of-way. The total number of birds detected on the disturbed transect was 16% lower than the number detected on the control between 06:00 and 08:00, 47% lower between 12:00 and 14:00, and 25% lower between 17:00 and 19:00. These differences were statistically significant.

Similar diurnal changes were observed in projected densities (Fig. 1, Table 1). Total projected densities on the disturbed and control transects between 12:00 and 14:00 were 35–46% of densities projected from morning censuses. Projected densities on both transects increased in the evening, but, with the exception of Mourning Doves, remained lower than results of censuses conducted between 06:00 and 08:00. Diurnal changes in projected densities were greatest on the disturbed transect. Total projected densities within the right-of-way were 40% lower than the control in the morning, 54% lower between 12:00 and 14:00, and 46% lower in the evening. All differences were statistically significant. Neither resident status (Fig. 1) nor foraging guild (Table 2) appeared to alter observed trends.

The structure and composition of the avian communities, as indicated by the census results, also changed with time of day. The number of species observed was greatest on both transects between 06:00 and 08:00 (Table 1). Diversity indicies were, however, greatest on both transects between 12:00 and 14:00 due to greater equitability among densities of the species observed. Both species richness and evenness were lowest in the evening on the two transects. Community comparison indices indicate the avian communities observed between 12:00 and 14:00 on the disturbed ($C = 0.28$) and control ($C = 0.33$) transects were the least similar to those depicted by the morning censuses. The disturbed and control communities were also the least similar at this time ($C = 0.40$).

DISCUSSION

Diurnal changes in activity and population estimates similar to those we observed have been reported for breeding birds within cooler and more mesic habitats. With the exception of Järvinen et al. (1977a), these studies have dealt only with changes during the morning. Davis (1965) observed a decrease in singing by male Rufous-sided Towhees (*Pipilo erythrophthalmus*) of 36% within ca. 30 min in the morning. Robbins and Van Velzen (1970) reported reductions in the number of species and individuals detected of ca. 37 and 25%, respectively, within 5 h after sunrise. Shields (1977) observed a 21% decrease in detectability between 06:00 and 07:30. Järvinen et al. (1977a) present data for breeding birds within fields and forests of southern Finland similar to those we collected. These authors reported reductions in density estimates of forest birds of 13% between early and late morning, 49% between early morning and afternoon, and 38% between early morning and evening. The number of species detected decreased from 64 in the early morning to 55 in the afternoon, and increased to 57 by evening. The decrease in population estimates between early and late morning (−38%) was greater for birds within fields. Järvinen et al. (1977a), however, found little difference in diversity indices for forest birds with time of day (diversity = 3.30–3.51, evenness = 0.79–0.83). The changes we observed in the diurnal activity of breeding birds within disturbed and undisturbed desert scrub appear to have been more pronounced than those reported for these cooler and more mesic habitats. Though the decrease in the number of species we observed on both transects between morning and evening censuses may have been due to a reduction in sample size, results are in agreement with those of others (Robbins and Van Velzen 1970, Järvinen et al. 1977a).

The diurnal changes in the activity of breeding birds we observed, as well as those observed by others, are probably due to changes in weather conditions (e.g., air temperature and solar radiation; for review, see Shields 1979). Evidence that activity patterns are in response to changes in air temperature and solar radiation and not some other correlate of time of day is circumstantial. The apparent increase in avian activity in the late afternoon and evening (Järvinen et al. 1977a, and this study) is associated with a decrease in air temperature and solar radiation. In addition, several authors (e.g., Robbins and Van Velzen 1970, Austin 1976) report that declines in activity are more rapid and of greater magnitude on very warm days. That the decreases we observed in the activity of breeding birds within undisturbed desert scrub were greater than those reported for cooler, more mesic habitats supports the hypothesis (Shields 1979) that increasingly stressful weather conditions negatively affect detectability. This hypothesis is further supported by the reported effects of extreme weather conditions on bird song (Welty 1962:199–200, Dorst 1974:162–163) and locomotor behavior (Grubb 1975, 1977, 1978; Austin 1976).

The "stressful weather hypothesis" may also explain the greater decrease in activity of breeding birds within the transmission line right-of-way compared to undisturbed desert scrub. The mean temperature on the access road between 06:00 and 18:00 (24.7 ± 6.4°C) was, however, lower than that recorded on ground cover on the undisturbed transect (25.5 ± 7.1°C). Therefore, increased exposure of birds to solar radiation following removal of 5.7 ha of desert-scrub vegetation for the access road and tower sites appears to have been a major factor in the greater decrease in activity on the disturbed transect compared to the control. Shade appears to be critical to the survival of breeding birds within this habitat, this is true even for permanent residents such as the Verdin (*Auriparus flaviceps*) and Cactus Wren (*Campylorhynchus brunneicapillus*). Foraging activity of Verdins is greatly reduced at temperatures above 35°C and long periods of inactivity are predominant (Austin 1976). The diversity of microhabitats used by Cactus Wrens also decreases as temperature increases (Ricklefs and Hainsworth 1969).

Our data, as well as those of others, indicate results of censuses conducted early in the morning most accurately describe habitat utilization by breeding birds within the habitats studied to date. However, as Shields (1979) notes "it is not enough to haphazardly limit censusing to a random portion of an extended (e.g., 4 h) census period." Guidelines for designing censuses to eliminate biases due to environmental conditions have been presented by Shields (1979) and Conner and Dickson (1980). Replicate or comparative censuses should be conducted at the same time of day. If more than one transect is to be sampled per day, starting times should be alternated, and transects should be started at opposite ends on alternate days. Time of day can also be included as a factor in subsequent statistical analyses removing the variation in densities of breeding birds due to regular changes in census starting times (e.g., Shields 1977).

ACKNOWLEDGMENTS

We gratefully acknowledge the financial support of the Arizona Public Service Company, The Salt River Project, and the Tucson Gas and Electric Company which made this study possible. Thanks are also due C. M. Bunck and G. L. Hensler for help with the statistical analyses, T. A. Grunwald and L. M. Palmer for preparing the illustration, and L. M. Thomas for secretarial assistance. C. J. Ralph and H. F. Recher kindly reviewed the manuscript.

Studies in Avian Biology No. 6:292–296, 1981.

INTERPRETING POPULATION ESTIMATES OF BIRDS FOLLOWING PESTICIDE APPLICATIONS—BEHAVIOR OF MALE STARLINGS EXPOSED TO AN ORGANOPHOSPHATE PESTICIDE

CHRISTIAN E. GRUE[1] AND BRYON K. SHIPLEY[2]

ABSTRACT.—We determined activity budgets for 10 pairs of captive male Starlings between 7 May and 18 July 1980. Our objective was to quantify changes in behavior after exposure to an organophosphate (OP) pesticide and to assess the impact of changes in behavior on the interpretation of population estimates of birds following pesticide applications. We observed each pair of males for an hour at 07:30 and 09:30 for four days and classified their behavior into one of four categories: flying, perching, foraging, or singing and displaying. At 06:30 on day 2, one male received a single oral dose of 2.5 mg dicrotophos (3-hydroxy-*N*, *N*-dimethyl-*cis*-crotonamide dimethyl phosphate) per kg of body weight; the other male received an equivalent exposure of corn oil. Changes in the activity budgets of OP-dosed and control males were compared using *t*-tests.

Activity of OP-dosed males was significantly ($P \leqslant 0.05$) reduced within the 2–4 h following exposure. OP-dosed males spent more time perching (46.1%) than controls and less time flying (−96.6%), foraging (−28.5%), and singing and displaying (−49.5%). The frequency of perching (−75.3%), flying (−83.8%), foraging (−54.1%), and singing and displaying (−59.2%) was significantly reduced. Activity in OP-dosed males returned to normal by 26–28 h posttreatment. Results suggest that movement and vocalization may be significantly reduced in birds exposed to organophosphate and carbamate pesticides. Conventional censusing techniques and population estimating procedures may, therefore, be inadequate to assess changes in bird populations after pesticide applications because of the difficulty in separating decreases in density due to mortality or emigration from reductions in activity.

Organophosphates (OP's) and carbamates are becoming increasingly important as insecticides because of their low potential for accumulation in the environment (Andrilenas and Eichers 1977, Lamoreaux and Newland 1977, Fowler and Mahan 1978). Applications of these pesticides often coincide with peaks in avian abundance and reproductive activity. Census techniques dependent on visual and auditory cues (e.g., Williams 1936, J. T. Emlen 1971) are often used to monitor impacts of these pesticides on bird populations (Finley 1965, McLeod 1967, Doane and Schaefer 1971, Pillmore et al. 1971, Fowle 1972, Moulding 1976, Pearce et al. 1976, Bart 1979, DeWeese et al. 1979, Richmond et al. 1979). Exposure to organophosphates has, however, been shown to reduce activity in captive (Hill 1971, Pope and Ward 1972) and wild (Edwards and Graber 1968) birds. Whether reported reductions in bird populations following OP or carbamate applications (Finley 1965, McLeod 1967, Doane and Schaefer 1971, Fowle 1972, Moulding 1976, Pearce et al. 1976, Bart 1979) were due to emigration, lethargy, or mortality is not clear. Published studies quantifying the effects of sublethal OP or carbamate exposure on bird behavior, particularly song production, are lacking. The objective of this study was to quantify changes in the behavior of captive male Starlings (*Sturnus vulgaris*) after sublethal exposure to an organophosphate pesticide and to assess the impact of changes in behavior on the interpretation of population estimates of birds following pesticide applications.

METHODS

We determined activity budgets for 10 pairs of male Starlings, 1 pair per week, between 7 May and 18 July 1980. Males were housed individually within 2.4 × 3 × 12 m open-wire pens containing a wood nest box (Kessel 1957), two perches, a water pot, and a hanging feeder supplied with commercial turkey starter. Burlap on the sides of the pens provided visual isolation. Birds were acclimated to the pens for 12 days. Grass in the pens was mowed 3 days prior to each trial. Following the acclimation period, we observed each pair of males for an hour at 07:30 and 09:30 for 4 days from blinds and classified their behavior into one of four categories: (1) flying, (2) perching, (3) foraging, or (4) singing and displaying. All periods of inactivity (i.e., time not spent in one of the three other behaviors) were classified as perching; preening was included in this category. The frequency and duration of each behavior were recorded using an Esterline-Angus event recorder. Hours when observations were made corresponded to extremes within which censuses of wild birds have been conducted following pesticide applications (Moulding 1976, DeWeese et al. 1979, Richmond et al. 1979). Food consumption was monitored daily at 06:30. At 06:30 on day 2, one male was given a single oral dose of 2.5 mg dicrotophos (3-hydroxy-*N*, *N*-dimethyl-*cis*-crotonamide dimethyl phosphate) dissolved in corn oil per kg of body weight, a sublethal exposure which has been shown to cause a 50% reduction in brain cholinesterase (ChE) activity in female Starlings (Grue, Powell, and McChesney MS). The remaining male received an equivalent exposure of only the corn-oil carrier. Males were sacrificed at 10:30 on day 4 and frozen (−20°C) prior to brain ChE assays. Differences between changes ($\Delta = \text{day}_1 - \text{day}_{2,3,\text{ or } 4}$)

[1] U.S. Fish and Wildlife Service, Patuxent Wildlife Research Center, Laurel, Maryland 20811.
[2] 13649 Highland Road, Clarksville, Maryland 21029.

TABLE 1
BRAIN CHOLINESTERASE ACTIVITY IN OP-DOSED AND CONTROL MALE STARLINGS

	Time postexposure (h)	
	28	52
Control males		
n		11
Mean		21.82[a]
SD		1.77
OP-dosed males[b]		
n	11	10
Mean	11.03	10.78
SD	2.12	1.91
% inhibition	49.5	50.6

[a] μmoles acetylthiochloline iodide hydrolyzed per min per g tissue.
[b] 2.5 mg/kg dicrotophos.

in the activity patterns of OP-dosed and control males were compared using t-tests ($H_0{:}\mu_D = 0$, Snedecor and Cochran 1967:98). Average differences in frequency and total duration of each behavior across the two 1-h observation periods were used in calculating the test statistics for 2–4, 26–28, and 50–52 h postexposure. Differences between changes in the behavior of OP-dosed and control males were considered significant if the probability associated with the test statistic was less than or equal to 0.05.

Immediately following the 10 trials, we dosed an additional 10 male Starlings with 2.5 mg dicrotophos per kg of body weight to determine the level of brain ChE inhibition at 28 h postexposure. Males were housed together in one of the four adjacent pens used for the behavioral trials. These birds were sacrificed 28 h after exposure and frozen. Brain ChE activity was determined colorimetrically using methods described by Ellman et al. (1961), as modified by Deiter and Ludke (1975) and Hill (1979).

RESULTS

Brain ChE activity of OP-dosed males was inhibited an average of 49.5% 28 h following exposure and an average of 50.6% 52 h postexposure (Table 1). Compared to controls, the activity of OP-dosed males was significantly reduced within the 2–4 h after exposure (Table 2). Frequency of perching (−75.3%), flying (−83.8%), foraging (−54.1%), and singing and displaying (−59.2%) was significantly lower in OP-dosed males than in controls (Fig. 1). OP-

TABLE 2
ACTIVITY BUDGETS OF OP-DOSED AND CONTROL MALE STARLINGS

	Pretreatment average		Average change posttreatment					
			2–4 h		26–28 h		50–52 h	
Activity	OP-dosed[a]	Control	OP-dosed	Control	OP-dosed	Control	OP-dosed	Control
Perching								
Frequency[b]	173.4 ±90.0	190.4 ±109.5	−123.1[d] ±96.0	8.3 ±99.0	−32.3 ±83.6	−9.4 ±128.2	−1.9 ±79.1	−3.4 ±98.3
Total duration[c]	1721.3 ±579.0	1983.9 ±603.7	607.0[d] ±1135.9	−214.6 ±566.6	−140.5 ±635.0	−167.3 ±562.5	−85.2 ±674.5	−268.7 ±567.0
Foraging								
Frequency	44.3 ±17.5	42.3 ±20.5	−24.0[d] ±19.3	−0.1 ±13.9	−2.2 ±19.7	2.5 ±17.1	−7.0 ±13.1	−0.4 ±15.1
Total duration	1261.2 ±669.4	1134.7 ±564.0	−76.6 ±1111.0	255.0 ±589.6	242.6 ±630.4	170.9 ±557.0	45.1 ±730.6	236.2 ±565.6
Flying								
Frequency	135.2 ±67.6	156.4 ±112.2	−104.2[d] ±73.1	10.5 ±112.0	−24.5 ±61.6	−3.2 ±126.5	−8.4 ±53.9	4.8 ±104.6
Total duration	207.4 ±121.5	243.3 ±195.0	−159.9[d] ±126.9	48.2 ±216.3	−47.3 ±96.4	−5.4 ±230.2	−25.1 ±87.3	7.7 ±211.6
Singing and displaying								
Frequency	42.1 ±51.7	22.8 ±31.6	−36.8[d] ±47.8	−6.4 ±20.2	−9.2 ±23.9	−1.6 ±12.3	4.2 ±26.3	−0.3 ±12.0
Total duration	433.7 ±544.0	262.2 ±397.8	−395.1[d] ±523.7	−109.0 ±250.6	−74.5 ±238.1	−25.5 ±224.8	45.4 ±195.3	8.3 ±191.5

[a] 2.5 mg dicrotophos per kg of body weight.
[b] Frequency per hour ± standard deviation.
[c] Total duration per hour in seconds ± standard deviation.
[d] Change significantly different from controls, t-test, $P \leq 0.05$.

TABLE 3
DAILY CONSUMPTION OF FOOD (G) BY OP-DOSED AND CONTROL MALE STARLINGS

	Pre-treatment	Time postexposure (h) 24	48
Control males			
n	10	10	10
Mean	25.5	25.0	28.0
SD	5.8	5.8	5.9
OP-dosed males[a]			
n	10	10	10
Mean	23.1	13.7[b]	26.1
SD	5.9	10.5	6.0

[a] 2.5 mg dicrotophos per kg of body weight.
[b] Difference between OP-dosed and control males significant, $P \leq 0.05$, paired *t*-test.

dosed males also spent significantly more time perched (46.1%) than controls, and significantly less time flying (−83.8%) and singing and displaying (−49.5%, Fig. 1). Though males exposed to the organophosphate spent less time foraging on the ground (−28.5%) compared to controls, the difference was not statistically significant. However, consumption of food within the hanging feeders by OP-dosed males was significantly lower (−38.7%) than that of controls within the 24 h after dosing (Table 3). Differences between changes in the behavior of OP-dosed and control males 26–28 and 50–52 h postexposure were not statistically significant.

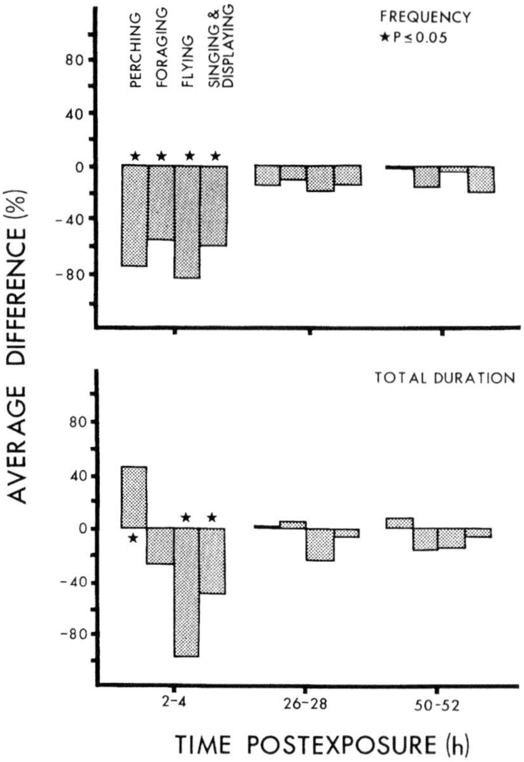

FIGURE 1. Average difference (%) in changes in total duration (seconds per hour) and frequency (per hour) of perching, foraging, flying, and singing and displaying of OP-dosed (2.5 mg dicrotophos per kg of body weight) male Starlings relative to controls.

DISCUSSION

Organophosphates and carbamates act by inhibiting ChE with subsequent disruption of nerve activity caused by accumulation of acetylcholine at nerve endings (O'Brien 1967:55, 88). Since behavior is dependent on nerve function, alterations in neural transmission may be expected to result in changes in behavior. Though no comparable data exist for birds, results of most studies with laboratory rats (for review, see Banks and Russell 1967) indicate a close correspondence between ChE activity and behavioral change with a critical level of ca. 40–60% inhibition (for exception, see Kurtz 1977). The level of brain ChE inhibition (ca. 50%) in the male Starlings we dosed with dicrotophos was within this critical range and was comparable to that observed in wild birds following applications of organophosphates and carbamates (Elder and Henderson 1969; Seabloom et al. 1973; Richmond et al. 1979; Zinkl et al. 1979, 1980). Though a reduction in brain ChE activity of 50% may be considered severe (diagnostic of poisoning by ChE inhibitors in dead birds, Ludke et al. 1975), OP-dosed males suffered no apparent muscular incoordination or impairment of flight.

Physiological and behavioral effects of ChE inhibitors reported in captive birds and other vertebrates may account for the changes in activity we observed in OP-dosed male Starlings. Intoxication of passerines following sublethal OP exposure under laboratory conditions is usually characterized by a reduction in activity and anorexia followed by a state of lethargy (Hill 1971, Pope and Ward 1972). Studies with laboratory rats (Adams 1977) suggest that inhibition of brain ChE is associated with a reduction in food seeking behavior. Sublethal OP exposure has also been associated with reduced visual acuity (Oba and Oto 1976), information processing, and psychomotor speed (Levin and Rodnitzky 1976) in man, and auditory detection in squirrel monkeys (*Saimiri sciureus*) (Reischl et al. 1975).

Why we did not observe a statistically significant reduction in the amount of time OP-dosed males spent foraging within the 2–4 h after treatment is not clear. OP exposure in birds has been associated with decreased food intake (Keith and Mulla 1966; Mehrotra et al. 1967; Hill 1971; Pope and Ward 1972; and this study). The majority ($\bar{X}$ = 88.2%) of all foraging activity by our male Starlings during pretreatment observations occurred on the ground. Difficulty in differentiating active searching for food from movement on the ground not associated with foraging may account for this discrepancy.

Reductions in activity, particularly song production, similar to that we observed in OP-dosed male Starlings have been reported in wild birds during censuses conducted after applications of organophosphates or carbamates (Finley 1965, McLeod 1967, Edwards and Graber 1968, Giles 1970, Doane and Schaefer 1971, Fowle 1972, Pearce et al. 1976, Bart 1979). With the exception of Edwards and Graber, these investigators and others have considered differences between pre- and postspray census results to be indicative of pesticide-induced changes in population density due to emigration or mortality. However, our data suggest that changes in behavior related to pesticide exposure may reduce detectability and make interpretation of census results difficult. Though changes in behavior associated with OP poisoning appear to be short-lived after exposure ceases (Keith and Mulla 1966; Mendelssohn and Paz 1977; and this study), effects may be present weeks after field applications. Zinkl et al. (1979) reported brain ChE inhibition of greater than 40% in birds 33 days after an aerial application of the OP, acephate. Probably these birds were still being exposed to the OP several weeks after treatment as brain ChE activity in birds exposed to ChE inhibitors may be expected to reach 20% of normal within ca. 26 days after exposure ceases (Fleming and Grue, MS).

The interpretation of results of censuses conducted after pesticide applications may be further complicated by the movement of birds in and out of treated areas. Several authors (McEwen et al. 1965, Giles 1970, Doane and Schaefer 1971, Moulding 1976, Bart 1979) have suggested out-of-area feeding as an avian response to insecticide-induced food shortages. As in the case of pesticide-induced reductions in activity, long sorties for food by adult birds with young would decrease the probability of detecting individuals still utilizing treated areas. Conversely, the immigration of birds into areas following pesticide applications may be rapid and mask treatment effects on population density. Within breeding bird populations, there appear to be "floaters," silent nonterritorial birds, and vacant territories are quickly reoccupied (Stewart and Aldrich 1951, Robbins 1964). These replacements may be more active and vocal than their predecessors (Stewart and Aldrich 1951). Conventional censusing techniques and population estimating procedures (e.g., Williams 1936, J. T. Emlen 1971), therefore, appear inadequate to assess changes in bird populations after pesticide applications. Pesticide-induced changes in behavior (e.g., reduced detectability) may result in overestimation of decreases in density due to mortality or emigration, whereas immigration of birds into treated areas may mask pesticide effects. The "disappearance" of birds after pesticide applications should not be considered synonymous with death (Heinz et al. 1979).

Use of mist nets to capture, mark, and recapture individual birds may be an effective way to more accurately determine the effects of pesticide applications on resident bird populations. Potential difficulties outlined by Heinz et al. (1979) and Richmond et al. (1979) should be considered. Handling of birds during the nesting season may adversely affect reproductive success. If pesticide effects are species specific, examination of only those species most trappable may lead to erroneous results. In addition, the number of man hours required to capture, mark, and recapture large numbers of birds may be prohibitive.

Several other techniques have been used in conjunction with census methods to assess the impact of pesticide applications on bird populations. These techniques also have their drawbacks. Brain ChE determinations appear to be an excellent means of monitoring exposure of birds to ChE inhibitors and diagnosing related mortality (Ludke et al. 1975). However, brain ChE assays have only recently been included in field investigations of the effects of applications of organophosphates and carbamates (Zinkl et al. 1977, 1979, 1980; DeWeese et al. 1979; Richmond et al. 1979) and relationships between sublethal ChE inhibition (≥20%) and changes in bird behavior are poorly known.

Carcass searches may be a necessary tool (Heinz et al. 1979). If dead birds are observed after a pesticide application, it is essential to obtain samples so that the cause of the mortality can be confirmed and the magnitude of the kill estimated. Searching for carcasses is seldom easy and requires much time, skill, and motivation (Heinz et al. 1979). Considering the difficulty in locating carcasses and the rapidity with which they may disappear (Davis 1970), location of a small number of dead birds may be reason to suspect some unusual cause of mortality (Rosene and Lay 1963, Heinz et al. 1979).

Nesting studies are probably the most effective technique used in evaluating the impact of pesticide applications. None of the methods discussed previously is sufficiently sensitive to assess the potential subtle effects of pesticide exposure on bird behavior and reproduction. Though we are aware of only two studies which have examined the effects of ChE inhibitors on avian reproductive behavior in detail (Grue et al., MS; and this study), both have reported significant effects. In the former study, care of nestlings by OP-dosed wild female Starlings was significantly reduced. Under most field situations, however, it is difficult to obtain adequate reproductive data on enough nests of one or more species to permit statistical analysis. Time and manpower may be limited and nests may be scarce, hard to locate, or inaccessible. We believe the use of nest boxes may enhance the capability of investigators to collect reproductive data before and after pesticide applications. Though nest boxes have been used effectively in studies of avian ecology (e.g., Dahlsten and Copper 1979) and the effects of DDT on passerine reproduction (Mitchell et al. 1953, McCluskey et al. 1977), few investigators (Black and Zorb 1965, Bednarek and Davidson 1967, Powell and Gray 1980) have utilized them to study the effects of ChE inhibitors. In addition, reproductive data may be collected automatically from nest boxes using a variety of electronic devices (Royama 1959, Dahlsten and Copper 1979).

Considering the potential difficulties in interpreting the results of conventional censusing techniques following pesticide applications, we recommend, as have others (DeWeese et al. 1979, Richmond et al. 1979), that future studies concentrate on quantifying pesticide exposure and its effects on avian behavior and reproductive success. Carcass searches may provide valuable additional information if mortality of adult birds is suspected.

ACKNOWLEDGMENTS

We thank G. H. Heinz and G. L. Hensler for their help in designing the study. C. Vance constructed the blinds, control panels, and outdoor storage cabinets for the event recorders. C. L. Christian, C. H. Gorsuch, and L. M. Palmer assisted with data collection, and along with P. J. Decker, K. C. Groves, and T. B. Wright helped with data analysis. K. L. Spaulding assisted with animal care. Thanks are also due L. M. Palmer for preparing the illustration, and L. M. Thomas for providing the necessary secretarial assistance. Critical reviews by E. H. Dustman and C. J. Henny improved the manuscript.

Studies in Avian Biology No. 6:297–300, 1981.

ENVIRONMENTAL ACOUSTICS AND CENSUSES OF SINGING BIRDS

DOUGLAS G. RICHARDS[1]

ABSTRACT.—The tasks in a census of singing birds include: (1) identification of the species of a singer, and (2) estimation of the location of the singer. The acoustical properties of the environment interfere with both these tasks; effects depend on the type of habitat. In open fields the primary sources of degradation are attenuation and amplitude fluctuations. Attenuation limits the maximum range of detection, and alters the frequency composition of the sound as heard by the observer. Fluctuations are random, and interfere with amplitude modulation. They may also cause error in estimation of distance. In forests attenuation has a similar effect, though the exact values depend on the type of forest. Reverberation is of major importance as a source of sound degradation, and virtually obliterates the fine details of songs at moderate distances, interfering with recognition. Sound scattered by trunks and foliage also makes estimation of the location of the singer more difficult. Adaptations by birds to combat the effects of song degradation by the environment may both help and hinder species identification and location estimation.

A census of singing birds has three tasks: (a) detection of the presence of a singer, (b) identification of the species of the singer, and (c) estimation of the location of the singer so that a distribution map may be compiled. Many factors, including hearing ability, extraneous noise and the acoustical properties of the environment, affect the performance of human observers on these tasks. The acoustical properties of the environment, which may be unknown to persons who census birds, will degrade bird song in a variety of predictable and unpredictable ways.

The song emitted by a bird is degraded by attenuation, fluctuation, and reverberation, all of which are both frequency and distance dependent, and vary among different habitats. The problem is not simply an increase in the difficulty of detecting a song with distance; a sound may be detectable, but accurate species identification may be difficult owing to degradation of the structure of the song. In addition, with increasing distance, error in estimation of the position of a singer may also increase. This paper will discuss the sources of environmental degradation of bird song, help define favorable acoustic conditions, and suggest empirical studies aimed at minimizing and controlling these effects in censuses.

ATTENUATION

Attenuation, the decrease in the intensity of sound with distance, is influenced by a number of environmental factors. In an ideal free field (with no sources of absorption or scattering), sound will spherically diverge from a point source, and the intensity will decrease by 6 dB with each doubling of distance. In the environment of a bird there are sources of excess attenuation which further reduce the intensity of a sound at a given distance. These include molecular absorption, which varies with humidity and air temperature, absorption and scattering by heterogeneities in the air (eddies), absorption and scattering by soil, and absorption and scattering by foliage. All become more troublesome with higher frequencies. The presence of wind and thermal stratification in the air may also have an effect, sometimes to the extent of creating a sound shadow which results in a sharp attenuation of sound beyond a certain distance.

Attenuation creates several problems, both for birds trying to communicate, and for ornithologists attempting censuses. First, it limits the maximum distance at which a song can be heard. This maximum range depends on the characteristics of the sound at the source and the acoustics of the habitat. Morton (1975), Marten and Marler (1977), Marten et al. (1977), Bowman (1979), and Linskens et al. (1976) have measured attenuation in various habitats. Their measurements do not agree closely, and are probably of little use as an indicator of maximum detection distance of bird song. For that, it is necessary to perform measurements in the habitat in which the census is being carried out. The previous studies do agree that attenuation is frequency dependent, with maximum transmission in the midrange frequencies from 1 to 4 kHz, those frequencies found in most passerine bird songs.

Two important factors affecting attenuation vary even within a given habitat: the time of day and the presence or absence of foliage. Time of day affects wind and thermal stratification, as these are largely dependent on solar heating. In north temperate forests, sudden growth in foliage occurs in the spring (usually in the middle of the census season) and can increase the absorption of higher frequencies. These factors differ in importance in forest and field habitats. Under the canopy of a forest, where little thermal

[1] Kewalo Basin Marine Mammal Laboratory, 1129 Ala Moana Boulevard, Honolulu, Hawaii 96814.

stratification develops, change in foliage is likely to have a much greater effect than time of day. In an open field there is a much greater time of day effect, and little effect from growth of foliage.

A more subtle factor, affecting species recognition rather than detection, is the frequency-dependent character of the attentuation. Attenuation greatly increases above the 1–4 kHz mid-range frequencies. Birds with a high-frequency component in the song useful for identification will be more difficult to recognize even if the low-frequency song component has been detected.

FLUCTUATIONS

Random fluctuations in the received level of sound are present owing to microclimatic heterogeneities in the transmission path (air eddies). These are undoubtedly responsible for some of the scatter in measurements of attenuation by various researchers. They are important to censusers for two reasons: they cause variability in detection distances and distance estimates, and they interfere with species identification.

The important parameters of fluctuations are the amplitude and the periodicity. Measurements by Richards and Wiley (1980) indicate that the amplitude of the fluctuations increases with frequency and distance, and is influenced by wind and thermal conditions in the air. The periodicity of the fluctuations determines the type of interference with communication. Long period fluctuations, changes in attenuation with time of day, for example, will affect the distance at which a song can be detected, but will not affect recognition of that song. Fluctuations with periodicities similar to those of amplitude modulations in the song can cause difficulty in species recognition, while having little effect on detection distance. The results of Richards and Wiley (1980) show that fluctuations are most severe at frequencies below 20 Hz, and interfere with amplitude modulation in that range. For this reason amplitude modulation is probably not used for species recognition in most open field species, but is nevertheless characteristic of many species songs. Fluctuations could mislead ornithologists using this characteristic for identification.

Birds have evolved a strategy for minimizing the effects of both attenuation and fluctuations: sing in the early morning hours before the effects of solar heating become important. Ornithologists generally census at this time as well. Though this strategy will maximize detection distance, and minimize the effects of fluctuations on species recognition, it cannot decrease the differential attenuation of the higher frequencies with distance. Observers should be trained not only in recognition of high-quality songs recorded in close proximity to a bird, but also in recognition of songs recorded from a considerable distance, at which the intensity relationships of the various frequency components may be changed.

REVERBERATION

Potentially the most severe problem for song recognition and localization in a forest, one which has been largely neglected in favor of studies of simple attenuation, is reverberation. Reverberation is also the most complex form of acoustic degradation in natural habitats, and is difficult to measure and study.

Scattering and reverberation of sound can result from the presence of trunks, foliage, and ground. The results of Richards and Wiley (1980) show it to be strongly frequency dependent. The effects of distance have not yet been studied. The frequency dependence is due to the relationship between the size of the scatterers and the wavelength of the sound; wavelengths greater than the dimensions of the scatterer are not scattered. Wavelengths approaching the dimensions of the scatterer and smaller are deflected to varying degrees. Since the higher frequencies in bird song have wavelengths comparable to the size of leaves in deciduous forests, they are degraded more severely by reverberation than the lower frequencies.

Scattering of sound in the path of the beam will have the initial effect of reducing the intensity of the beam. This energy is not lost, and some may be scattered again to re-enter the beam. At great distances from the source, the energy reaching the receiver is almost all scattered sound, out of phase with the direct sound, and arriving from a variety of directions. Consider a brief pulse of sound; for example, a single note in the trill of a bird song. At a distance from the source the direct wave from the original song will be weak. Most of the energy received will be scattered, arriving after the direct wave, and decaying with time. It is possible for the scattered sound to be *more* intense than the direct sound. Thus information useful for species identification encoded in note duration or internote interval is bound to be degraded. Wiley and Richards (1978), Richards and Wiley (1980), and Richards (1981a, 1981c) give examples of forest reverberation both in artificially produced sounds and in actual bird songs.

We localize sound by comparing time differences of onset and termination of a pulse at different ears, phase differences between ears, and intensity differences between ears. Reverberation interferes with determination of the location of a singer in two ways. First, as most of the received sound is scattered, the apparent locus of emission is broadened. Second, phase and intensity differences are confounded by the variety of complex phase and intensity fluctuations produced by multiple scattering. Since a pulse with sharp onset and termination arrives at the receiver with progressive onset and long reverberations, this cue for localization also becomes difficult to use. Eyring (1946) studied the ability of humans to localize sound in a forest. He used random firing of gunshots in selected locations, and found an error of about 20 degrees in judging the bearing of the sound. Contrary to theoretical expectation he found that error *decreases* with distance from the source in the range 300–600 feet over which his tests were conducted. Gunshot is an inherently locatable sound, with its sharp onset, and frequency transients. Bird song is probably more difficult to localize, but no studies of human ability to locate bird song have been performed. Of course, in a census, estimation of not only bearing, but distance, is important.

The effects of reverberation are not likely to be linear with distance from the source. Unfortunately an observer cannot choose the habitat to minimize the effects of reverberation (unless he only conducts censuses in open fields). Empirical measurement of the ability of observers to judge distance and bearing to playback of recorded song is necessary. Such playbacks should be done at heights above ground similar to those used by singing birds, since height can also affect sound reception. Only with measurements of this type can reliable limits be placed on transect size and number of listening points. Table 1 summarizes some of the acoustical properties of the environment. More detailed discussions of environmental acoustics and avian song communication may be found in Wiley and Richards (1978, 1981a) and Richards and Wiley (1980).

TABLE 1
EFFECTS OF ATTENUATION, FLUCTUATIONS, AND REVERBERATIONS ON BIRD SONG IN OPEN AND FORESTED HABITATS

Open

1. *Attenuation:* increases with distance, increases with sun-produced temperature stratification, decreases in midrange frequencies (1–4 kHz); limits maximum detection distance for song.
2. *Fluctuations:* increase with frequency and distance, dramatic increase with presence of air eddies caused by solar heating; cause variability in detection distance and interfere with recognition.
3. *Reverberation:* of little importance in open fields owing to absence of scattering surfaces.

Forest

1. *Attenuation:* no consistent differences from open conditions, but much less pronounced increase with solar heating.
2. *Fluctuations:* no consistent differences from open conditions, but much less pronounced increase with solar heating.
3. *Reverberation:* causes interference with both localization and recognition of songs, becomes more severe with seasonal regrowth of foliage in spring.

SONG RECOGNITION

The effects discussed thus far are often thought to interfere primarily with song *detection*. More subtle, but potentially more serious for censuses, are problems with song *recognition* caused by environmental acoustics. As a communication signal, bird song is adapted to the acoustics of the environment. It may not be the case, however, that the adaptations are for maximum locatability or discriminability. Many calls, not only alarm calls, are convergent across species, and are specialized for short distance transmission and difficulty in localization (Wiley and Richards 1981). The consequences of a mistaken identification of a few songs are less severe for a bird than for an ornithologist.

Territorial males are commonly able to individually recognize their conspecific neighbors. They sing for hours each morning, and only need listen for unusual *changes* in the singing patterns of neighbors (see Schleidt 1973). A human observer, has a more difficult recognition task. He must, in a short period, identify *all* species of singing birds from a given location. Recognition of multiple signals arriving at unknown times is one of the most difficult signal detection tasks (Green and Swets 1974, Richards 1981a). The decisions of the observer are biased by his expectations; common birds are easily identified, rare birds with songs degraded by distance may be misidentified or ignored. The task of recognition of multiple species and the time frame in which recognition takes place are so different from the communication problems faced by birds that adaptations for combating degradation may not be of great use to human observers. Other papers at this symposium address the problems of accuracy of species identification—my intent here is to draw attention to some problems of recognition occurring as a result of the acoustics of the environment.

One adaptation of bird song, facilitating detection by birds, but potentially interfering with recognition by human observers, is the separation of songs into alerting and message components (Richards 1981a). Numerous species of birds have initial song notes which converge in structure to a relatively pure-tone form. In the same species the remainder of the song is frequently extremely variable. Such species include the Song Sparrow (*Melospiza melodia*), White-crowned Sparrow (*Zonotrichia leucophrys*), Rufous-sided Towhee (*Pipilo erythrophthalmus*), and at least 15 other North American passerines. For birds the easily detectable initial note facilitates detection at a distance and may elicit an approach for recognition of the more easily degraded rapid, complex trills. In a census the observer does not have the option of a close approach. Reverberation may virtually obliterate the component of the song containing most of the species-specific information. Many of the species with this adaptation have such diverse songs that such detail is crucial for a positive species identification. To reduce the problem particular attention must be paid to training of observers in identification of the songs of multi-themed birds with alerting-note song structures, using tapes of both normal and distance-degraded songs.

Even more troublesome may be birds with songs that are easily distinguished at close range, but whose species-specific characteristics converge when the songs are degraded by distance. Again, the problem is most severe in birds with multiple and diverse themes. A Carolina Wren (*Thryothorus ludovicianus*), for example, often has at least 20 distinguishable song themes, and may have many more. Some of these are rather similar to those of Tufted Titmice (*Parus bicolor*), Northern Cardinals (*Cardinalis cardinalis*), and Kentucky Warblers (*Oporornis formosus*). Species identification by observers is often made on the basis of song "quality," rather than the specific sequence of notes. The acoustic variables in judgements of song quality are primarily the amplitude envelopes of individual notes, tremolo (rapid frequency or amplitude modulation), and high-frequency harmonic content. These are also the acoustic parameters most susceptible to degradation over a distance. Observer training on degraded as well as high quality song will help reduce misidentifications of distant song.

The acoustic problems discussed, though unavoidable, are often sufficiently predictable that they need not interfere with a census conducted by trained observers. Empirical measurements of detection distance and locatability must be made for each census area. Observers must be tested for reliability in estimation of distance and direction, and for ability to recognize degraded song.

ACKNOWLEDGMENTS

I thank R. H. Wiley, J. T. Emlen, and M. DeJong for their helpful comments on this research.

Studies in Avian Biology No. 6:301–310, 1981.

BIRD ACTIVITY LEVELS RELATED TO WEATHER

CHANDLER S. ROBBINS[1]

ABSTRACT.—The Breeding Bird Survey data bank serves as a primary source for studying effects of sky cover, wind speed, and temperature on bird census results. Other standardized methods, such as spot-mapping (Breeding Bird Census), point counts, banding, and the Winter Bird Survey, provide additional, but limited, means of assessing effects of weather.

Numbers of songbirds detected are generally inversely correlated with wind speed, but hawks often are seen in larger numbers on windy days. Rain greatly reduces the numbers of birds detected. Cloud cover has relatively little influence on early morning bird counts during the peak of the breeding season. Fog selectively favors auditory detections of some species. Counts of many species are correlated with temperature, but effects are minor unless temperatures are extreme.

Under marginal weather conditions, total species observed may be nearly normal, whereas number of individuals observed is reduced, as is the opportunity to record simultaneous registrations.

This paper is concerned with effects of weather on counting bird populations during the breeding and winter seasons. The effects of weather on activity during migration periods are a different subject and beyond the scope of this paper. I discuss primarily the effects of weather on bird activity, especially on singing, but reference is made to the effects on the observer and detection ability.

Bird censusers as a rule are conscientious and devoted to achieving the best estimates of the birds on their study plot(s). Close estimates of the actual populations or of an index to these populations are essential if the investigator is making comparisons between habitats or over time. Accuracy is especially important if the results are to enter the literature or data banks used for subsequent investigations.

Census workers, therefore, tend to conduct their fieldwork under the best possible environmental conditions. In particular, strong winds, rain, and excessive heat are avoided. Most instructions for the spot-mapping method (Hall 1964, International Bird Census Committee 1970, Berthold 1976) do not stipulate acceptable weather conditions, and instructions for the Common Birds Census of the British Trust for Ornithology specify only that visits should be made during favorable weather when song is not reduced by wind or heavy rain. Instructions for the Breeding Bird Survey (available from U.S. Fish and Wildlife Service) say to avoid fog, steady drizzle, prolonged rain, and winds greater than Beaufort 3 (13–19 kmph), and further state that winds of Beaufort 2 (6–12 kmph) or less are preferable.

There are few published references to measured effects of weather on bird censusing, so the inexperienced observer has no guidelines as to which conditions should be avoided. Similarly, the experienced observer has no guide on ways to compensate for counts that are unusually low (or high) because of wind, fog, rain, or of snow cover.

[1] Migratory Bird and Habitat Research Laboratory, U.S. Fish and Wildlife Service, Laurel, Maryland 20811.

METHODS

The following data sources were examined for purposes of this paper.

BREEDING BIRD SURVEY

Breeding Bird Survey (BBS) data are suitable for analysis of the effects of weather on census results. The large sample size, broad geographic distribution, the large number of species sampled, and the fact that temperature, wind speed, and sky condition are recorded at both the beginning and the end of each survey make these data especially useful. For this paper I selected BBS counts from four geographic regions (Figure 1). These were selected so that within each region there would be uniformity of normal weather conditions—especially a narrow range of average minimum temperatures (13–16°C in regions 1, 2, and 4, and 7–13°C in region 3 for the week 4–10 June; Visher 1954:68). I had to be sure that differences in birds detected were a result of local temperatures during the counts rather than reflecting different parts of the range where abundance of the species was different. Altogether, BBS totals for 46 species that occur commonly in one or more of the four regions were selected for analysis.

Routes were sorted into groups according to the sky condition at the start of each route (clear, partly cloudy, overcast, fog, drizzle, and showers). Starting rather than ending sky condition was used because (1) more birds are recorded in the first half of BBS routes than in the second half, and (2) starting rather than ending conditions influence the observer's decision whether to conduct a count. Wind speed recorded in the Beaufort scale at the end of each coverage was selected for analysis because wind has a greater effect in mid-morning than near sunrise. Starting temperatures were grouped by 2.78°C categories (5°F). Mean counts (and standard errors) for the 46 species in each of the four regions were computed and plotted for each weather interval, and the data for some of these species were examined by analysis of variance to remove any geographical or route effect.

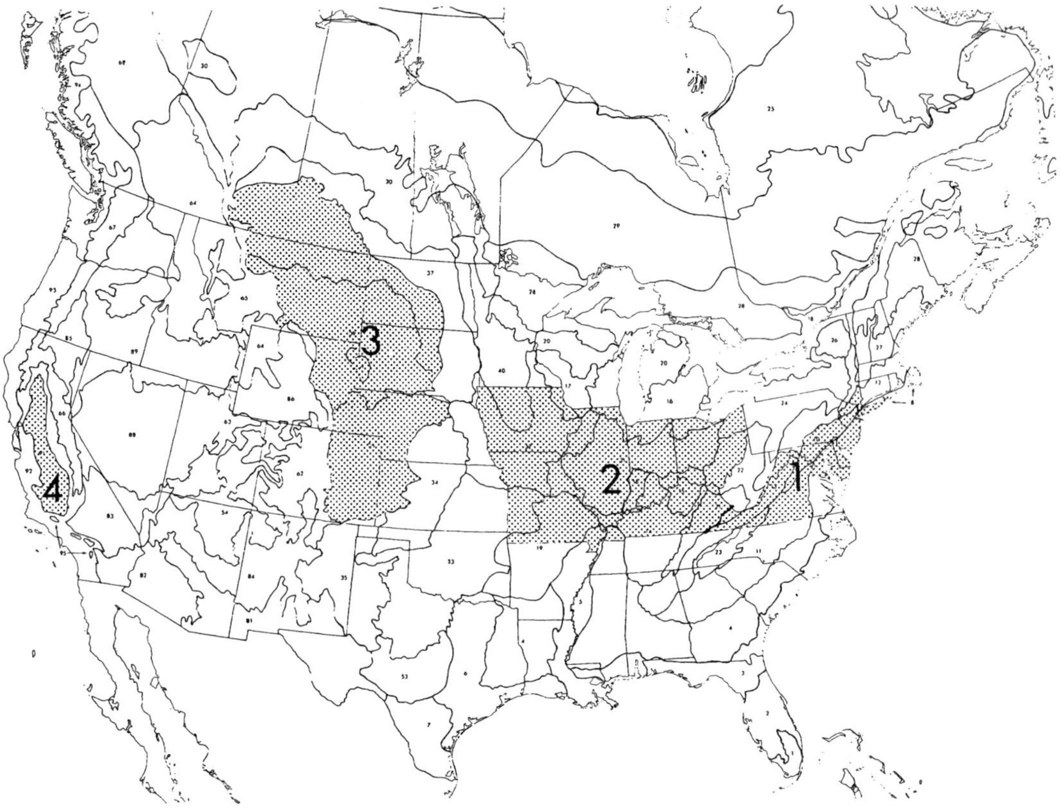

FIGURE 1. The four regions used for analysis of effect of weather conditions on BBS results.

Breeding Bird Survey Check Routes

There are several BBS routes in Maryland that have been covered numerous times in the same year by the same or different observers in the course of various types of observer tests. Two of these 50-stop routes provided an opportunity to study weather effects. Because these routes had been used to study observer bias, it was easy to eliminate observations of participants who had hearing problems. Accordingly, 58 comparable coverages of the Beltsville BBS route (46-022) and 26 of the Harmony route (46-038) were examined for effects of temperature, wind, and sky cover on 30 common bird species.

Other Breeding Season Counts

Spot-mapping.—On 8 July 1980 I took a census trip in steady light rain through an upland hardwood plot in Howard County, Maryland, that I have been censusing for eight years. Normally I would have canceled my plans, but this was an opportunity to measure the effect of light rain in comparison with a similar count I had made in good weather on the previous day.

Breeding season banding.—Results of intensive all-day banding in a floodplain study plot on the Patuxent Wildlife Research Center in Laurel, Maryland, over a period of 10 years were examined.

Point counts.—Because very little point count data were available for poor weather conditions, I conducted 20-min. point counts under marginal conditions when I had satisfactory counts available from an adjacent day for comparison.

Winter Counts

Transect counts.—Transect counts from all seasons of the year made through a variety of habitats on the Patuxent Wildlife Research Center were examined to see whether species totals were correlated with wind, temperature, or cloud cover.

Winter Bird-Population Study.—Personal records from the Audubon Winter Bird-Population Study were examined to see whether there were consistent effects of low temperatures on the birds of two woodland study plots in Howard County, Maryland.

Christmas Bird Counts.—Twenty-five years of data from three Christmas Bird Count areas on the Eastern Shore of Maryland were examined as were data from 20 years from a fourth area. Two of these areas were selected because I have been the compiler for the entire period and have personal knowledge of how the weather conditions have been recorded and how the coverage (party-miles) has been calculated. I have also participated in the other two counts, although they had been compiled by others. Five species were selected

TABLE 1
Species Affected by Cloud Cover

Species	Mean with clear sky (n = 1911)	% change when overcast (n = 917)
Bobwhite	32.4	−15.7**
Eastern Kingbird	16.8	−23.8**
Tufted Titmouse	8.0	−14.9**
Gray Catbird	7.0	+14.3**
Brown Thrasher	5.9	+12.0**
American Robin	39.4	+10.9**
Yellow Warbler	1.5	+22.0*
Eastern Meadowlark	32.2	−10.3*
Rufous-sided Towhee	8.1	+14.4*
Song Sparrow	19.4	+9.7*

* $P < 0.05$.
** $P < 0.01$.

for study and in each year the number of individuals of each of these species was divided by the number of party-hours of coverage in that year. The resulting birds per party-hour were then grouped by weather categories and examined by stepwise regression analysis to sort out year-to-year changes from changes influenced by weather conditions.

RESULTS

Breeding Bird Survey

Data from each of the four regions shown in Figure 1 were analyzed separately for each species. It soon became apparent that hundreds if not thousands of records for each species would be needed to demonstrate effects of weather conditions. One reason large samples are required is because the great majority of counts were taken in a narrow range of satisfactory conditions. Thus fog, drizzle, showers, high winds, and extreme temperatures are poorly represented in the sample. For most species in region 4, samples were too small (n = 186 route-years) to show significant effects of weather, and even in region 3 (n = 788) the results were meager. Most of the following comments, therefore, relate to the more common species that nest in both region 1 (n = 1417) and region 2 (n = 2435).

Sky condition

Cloud cover.—After finding only five significant differences ($P < 0.05$) between number of birds found on clear vs. overcast mornings in any single region, I combined regions 1 and 2, which had the most species in common. This increased the number of significant differences to 10 (Table 1), but note that it required nearly 3000 route-years of data to demonstrate these differences (without counting the days that were partly cloudy, foggy, or rainy).

Fog.—Fog has a greater effect on bird counts than overcast vs. clear skies (Table 2). With much smaller samples (173 foggy days), 18 species showed significant differences ($P <$ 0.05). The pattern that emerges from Table 2 is that birds detected primarily by sight (hawks, swifts, swallows, Starlings, blackbirds) are found in smaller numbers on foggy mornings

TABLE 2
Species Affected by Fog

Species	Mean with overcast sky (n = 917)	% change when foggy (n = 173)
Red-tailed Hawk (*Buteo jamaicensis*)	0.33	−42.8**
Mourning Dove (*Zenaida macroura*)	27.4	−18.5**
Chimney Swift (*Chaetura pelagica*)	13.2	−26.7**
Eastern Kingbird (*Tyrannus tyrannus*)	4.5	−34.9**
Eastern Wood Pewee (*Contopus virens*)	5.0	+17.1*
Barn Swallow (*Hirundo rustica*)	23.7	−25.5**
Tufted Titmouse (*Parus bicolor*)	7.1	+33.6**
American Robin (*Turdus migratorius*)	43.7	−13.2*
Wood Thrush (*Hylocichla mustelina*)	9.2	+64.7**
Starling (*Sturnus vulgaris*)	110.9	−19.4**
Red-eyed Vireo (*Vireo olivaceus*)	5.9	+82.9**
Ovenbird (*Seiurus aurocapillus*)	1.7	+39.8*
Red-winged Blackbird (*Agelaius phoeniceus*)	124.1	−28.0**
Brown-headed Cowbird (*Molothrus ater*)	11.2	−21.0**
Cardinal (*Cardinalis cardinalis*)	23.0	+18.3*
Indigo Bunting (*Passerina cyanea*)	22.8	+27.4**
Chipping Sparrow (*Spizella passerina*)	8.8	+31.2**
Field Sparrow (*Spizella pusilla*)	14.5	+25.7*

* $P < 0.05$.
** $P < 0.01$.

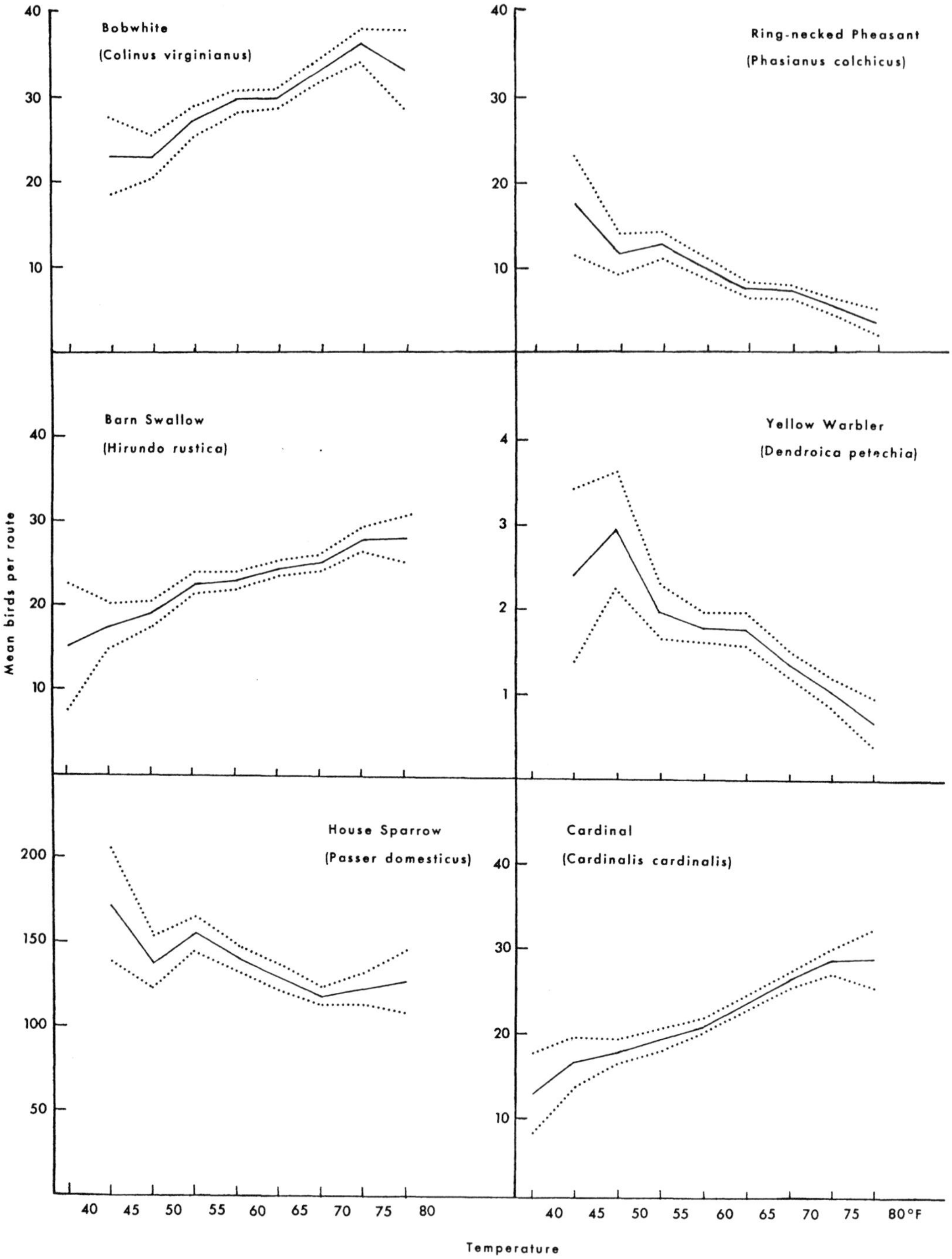

FIGURE 2. Effect of temperature on six species of breeding birds in Regions 1 and 2 based on 3852 route-years of BBS data. Temperatures (°F), taken one-half hour before sunrise, were grouped by 5° intervals for the analysis. Mean birds per 50-stop route and 95% confidence limits are shown. Note that small samples at the temperature extremes (n = 63 for the 41–45° temperature group and n = 82 for 75–80°) broaden the confidence limits.

TABLE 3
SIGNIFICANT CORRELATIONS WITH TEMPERATURE
($P < 0.05$)

Species	Areas	Correlation
Red-tailed Hawk	2, 3, 4	–
Mourning Dove	2, 3	+
Whip-poor-will	1, 2	+
Chimney Swift	1, 2	+
Eastern Kingbird	1, 2	+
Horned Lark	2, 3	–
Mockingbird	1, 2, 4	+
American Robin	1, 2, 4	–
Red-eyed Vireo	1, 2	–
Eastern Meadowlark	2, 3	+
Indigo Bunting	1, 2	+
Dickcissel[a]	3	+
Grasshopper Sparrow	1, 3	+
Vesper Sparrow	2, 3	–
Song Sparrow	1, 2	–

[a] *Spiza americana.*

TABLE 4
NUMBER OF SPECIES (FROM A SAMPLE OF 30 SPECIES) SIGNIFICANTLY CORRELATED WITH WEATHER CONDITIONS ON THE BELTSVILLE AND HARMONY BBS ROUTES

Weather variable	Number of significant correlations, $P < 0.05$	
	Positive	Negative
Temperature at start	9[a]	2[b]
Temperature at end[c]	3	0
Wind at start (Beaufort)[c]	0	1
Wind at end	3[d]	3[e]
Wind, sum of start and end	1[f]	4[g]
Wind at end vs. count at last 20 stops	2[h]	6[i]
Sky at start[c]	3	0
Sky at end	2[j]	2[k]
Sky, sum of start and end[c]	0	2
Sky at end vs. count at last 20 stops[c]	2	1

[a] Eastern Kingbird, Barn Swallow, American Robin (both routes), Eastern Meadowlark (*Sturnella magna*), Cardinal, Rufous-sided Towhee, and Grasshopper Sparrow (*Ammodramus savannarum*; both routes).
[b] Starling, Red-winged Blackbird.
[c] From a sample of 30 species on each of 2 BBS routes, 3 significant correlations with this variable would be expected from chance alone.
[d] Starling, Red-winged Blackbird, Field Sparrow.
[e] Mourning Dove, Eastern Wood Pewee, Song Sparrow.
[f] Wood Thrush (Harmony route).
[g] Barn Swallow, Wood Thrush (Beltsville route), Rufous-sided Towhee, Field Sparrow.
[h] Chimney Swift, Grasshopper Sparrow.
[i] Gray Catbird (*Dumetella carolinensis*; both routes), Brown Thrasher (*Toxostoma rufum*), House Sparrow, Cardinal, Chipping Sparrow.
[j] Red-eyed Vireo, Common Grackle (*Quiscalus quiscula*).
[k] Chimney Swift, Mockingbird.

whereas fringillids and some of the other passerines with loud songs are found in larger numbers. Fog improves ability to hear distant birds, and in addition, when an observer cannot see distant birds he naturally devotes more attention to detecting auditory cues.

Wind

Of 18 species analyzed from region 4 in California (corrected for route effect), one-third were significantly correlated with wind speed, the Red-tailed Hawk (scientific names not in the text appear in Table 2 or in footnotes of Table 4) positively, the others negatively: Horned Lark (*Eremophila alpestris*), Mockingbird (*Mimus polyglottos*), House Sparrow (*Passer domesticus*), and Red-winged Blackbird (*Agelaius phoeniceus*).

Of more practical interest than whether the numbers recorded are correlated with wind is the wind speed at which there is a significant change in the number recorded. In the following list the numbers in parentheses are the wind speeds (kmph) that correspond to the highest wind speed (originally recorded in the Beaufort scale) at which there was not a significant ($P <$ 0.05) decline in registrations as compared with calm days: Bobwhite, *Colinus virginianus* (11); Mourning Dove (19); Whip-poor-will, *Caprimulgus vociferus* (11); Eastern Wood Pewee (11); Mockingbird (5); Gray Catbird, *Dumetella carolinensis* (11); Red-eyed Vireo (5); Yellow Warbler, *Dendroica petechia* (11); Ovenbird, *Seiurus aurocapillus* (11); Common Yellowthroat, *Geothlypis trichas* (11); Eastern Meadowlark, *Sturnella magna* (11); Scarlet Tanager, *Piranga olivacea* (2); Cardinal (5); Indigo Bunting (11); Rufous-sided Towhee, *Pipilo erythrophthalmus* (11); Chipping Sparrow (11); and Field Sparrow (11). These data are from region 2, which had the largest sample (n = 2435). Observations of almost all species declined more sharply at Beaufort 4 and 5, indicating that wind speeds above Beaufort 3 (13–19 kmph) should be avoided when possible.

Temperature

Effects of temperature were discernible only with large samples. In region 4, the Central Valley of California (n = 186), significant temperature trends ($P < 0.05$) were detected only after route effect was removed by analysis of variance.

Temperature trends for six species are shown in Figure 2, based on the combined results of regions 1 and 2. Some other significant correlations are listed in Table 3. Analyses for other species had to be abandoned because the large samples submitted for analysis of variance exceeded the capacity of the computer.

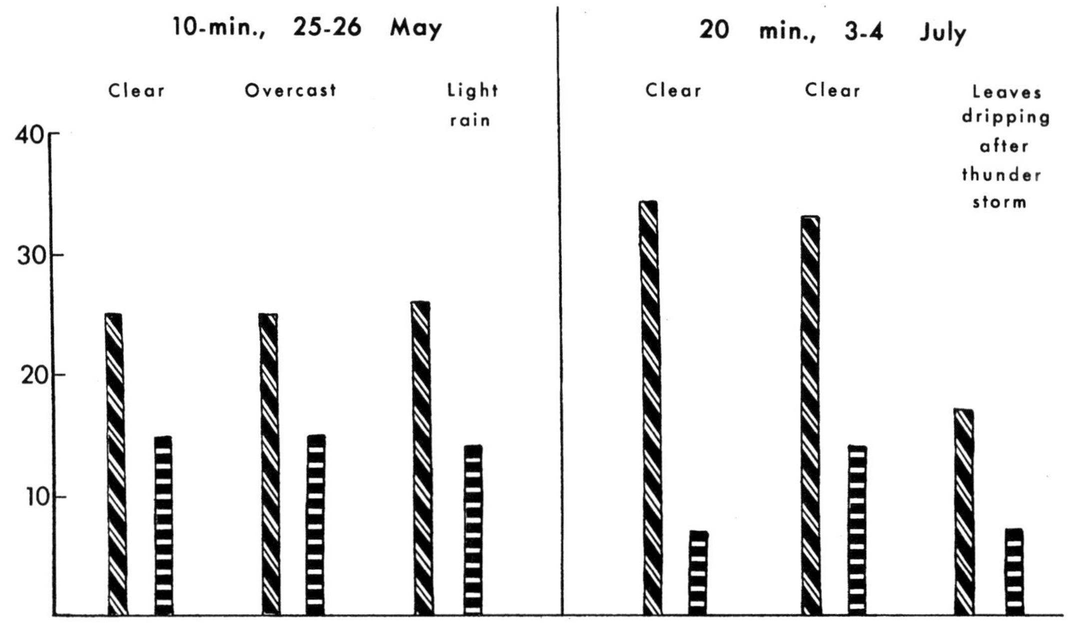

FIGURE 3. Effect of light rain and of dripping leaves on total individuals (diagonal barring) and total species (horizontal barring) recorded on IPA point counts. Although light rain did not depress singing, dripping leaves after a nocturnal thunderstorm did.

Breeding Bird Survey Check Routes

I used linear regression analyses to explain the variation in the counts of 30 species on the Beltsville and Harmony, Maryland, BBS routes. The results and independent variables used in these analyses are given in Table 4. Note that with a sample of 30 species on each of two BBS routes, three significant correlations with each weather condition variable would be expected from chance alone. It was surprising, therefore, to find that with a large sample of records from experienced observers on two BBS routes, the number of correlations with weather conditions was so low.

As an example of effects of rain that occurred on just part of a BBS route (number 46-047), the right half of Table 5 shows the percentage decline, by family, caused by rain. This route was covered on two consecutive days by the same observer. Numbers given are the 50-stop totals. On the first day, moderate rain occurred at 9 of the first 18 stops, followed by overcast skies; skies were clear when the count was repeated the next day. Although the total number of species observed was only 4 (6%) fewer on the rainy day, total individuals dropped 19% and 8 of the major families registered declines of more than 35%. On another route (46-038), where coverage was cancelled after 20 stops with rain, comparison with the first 20 stops of the rerun (under partly cloudy skies) showed a reduction of 36% in species and 61% in individuals on the rainy day.

Other Breeding Season Counts

Spot-mapping.—Although comparative data are few, they demonstrate that even light rain can have a strong effect on the numbers of birds recorded (Table 5, left half). These observations are similar to those of Hogstad (1967), who reported that the number of registrations in rainy weather drops to about half that in dry weather. The effect becomes greater as precipitation becomes heavier, and especially if precipitation is accompanied by high winds. Any sharp decrease in singing activity also reduces the likelihood of simultaneous registrations, which in turn decreases the effectiveness of a spot-mapping visit.

Breeding season banding.—Because rainy days were avoided, and because the study plot was heavily wooded and well sheltered from the wind, extreme weather conditions were not encountered. The closed canopy of the mature forest shaded the nets from the sun at all times. Lack of correlations between weather and numbers of birds netted serves as a recommendation for using banding in conjunction with other tech-

TABLE 5
EFFECT OF LIGHT RAIN ON TWO COUNT METHODS

	Spot mapping			BBS		
Family	Clear	Rain	Difference	Clear	Rain (9 stops)	Difference
Columbidae	4	4	0	13	7	−46%
Cuculidae	2	2	0	21	10	−52%
Picidae	7	3	−57%	28	15	−46%
Tyrannidae	10	6	−40%	15	17	+13%
Hirundinidae				39	31	−21%
Corvidae	0	1	+	57	44	−23%
Paridae	4	1	−75%	49	26	−47%
Troglodytidae				21	10	−52%
Mimidae	4	1	−75%	12	15	+25%
Turdidae	13	12	−8%	54	34	−37%
Sturnidae				114	96	−16%
Vireonidae	10	3	−70%	43	39	−9%
Parulidae	13	10	−23%	44	40	−9%
Ploceidae				60	33	−45%
Icteridae	4	0	−100%	176	179	+2%
Thraupidae	5	1	−80%	8	5	−38%
Fringillidae	11	6	−45%	64	63	−2%
All species	87	51	−41%			

niques in closed canopy habitats. In an open situation either sun or wind could severely reduce the catch.

Point counts.—From a sample of over 500 20-minute point counts I have selected three examples to illustrate effects of weather conditions. The first two, shown in Figure 3, illustrate different effects of rain on total adults recorded and total species recorded. The first count, taken on 25 May 1980 in deciduous woods at Laurel, Maryland, had 10 minutes of very light rain followed by 10 minutes of solid overcast. The numbers of species and individuals recorded were almost identical during and following the light rain. The count was repeated the next day under a clear sky and broken into two 10-minute segments to be comparable with the two segments of 25 May. The mean of the two segments is plotted at the left in Figure 3 and shows that in this instance the light rain had no effect on the bird count.

On the right half of Figure 3 are the results of 20-minute counts taken in another upland deciduous forest area in Maryland on 3–4 July 1980. The count shown at the right was made while water was still dripping from the leaves after a thunderstorm that occurred during the night. In comparison with this are two consecutive 20-minute counts taken at the same point on the following day. Time, temperature, and wind conditions were essentially the same both days. Although singing activity was reduced by about 50%, judged by the number of adult birds recorded, the number of species detected during the 20-minute period was not reduced.

In the third example, the effect of wind on bird activity was recorded in an upland deciduous forest in Garrett County, Maryland, on two clear days, 6–7 July 1980. The wind was estimated at Beaufort 3 (13–19 kmph) when the first 20-minute count was taken at 0915 EDT, and it was calm on the following day at the same time. The wind caused a 40% drop in total adults recorded, a 41% drop in the number of species detected, and a 54% drop in the number of singing males.

These examples, based on selected pairs of observation periods, are purely illustrative, but they indicate that counts may be seriously affected by such common environmental factors as dripping leaves and winds of about 15 kmph.

In a 13-day series of early morning 20-min. point counts made 12–24 July 1980 in deciduous forest in Laurel, Maryland, the number of species recorded was inversely correlated both with local temperatures ($r = -0.872$, $P < 0.001$) and with relative humidity recorded 30 km away at National Airport ($r = -0.781$, $P < 0.005$).

WINTER COUNTS

Transect counts.—The most consistent source of transect data was from repeated coverage of an 8-km Winter Bird Survey route on the Patuxent Wildlife Research Center in Maryland. This transect was covered at least three times each winter by each of two observers in late December or early January for 5 consecutive years. Data from counts taken on the coldest winter mornings were compared with matching data from seven warm mornings the same winter. Starting temperatures on the cold mornings ranged from −16° to −7°C; starting temperatures on the warm mornings ranged from −4° to +3°C. Because of the small sample size, birds were grouped in families for comparison. Sample sizes ranged from 66 wrens to 370 woodpeckers. No significant differences were found for any of the groups or individual species tested: woodpeckers (Picidae), titmice (Paridae), wrens (Troglodytidae), kinglets (Sylviidae), and the Cardinal. No effects of weather conditions could be detected on a series of breeding season transects either.

Winter Bird-Population Study.—Data from eight years of Audubon Winter Bird-Population Studies on two forest plots in Maryland were examined for differences between results on warm mornings compared with cold mornings in the same winters. More birds were recorded on warm mornings (starting temperatures −8° to 0°C) than on cold mornings (−27° to −14°C), but the differences were not significant ($P > 0.05$).

Christmas Bird Counts.—I selected five species (Red-bellied Woodpecker, *Melanerpes carolinus*; Carolina Chickadee, *Parus carolinensis*; Brown Creeper, *Certhia familiaris*; Hermit Thrush, *Catharus guttatus*; and Cardinal) for weather analysis using Christmas Bird Count data from the Ocean City and Southern Dorchester County, Maryland, counts. Year-to-year changes in totals of these five species on these two counts, and on two other counts from the Eastern Shore of Maryland, showed relatively little annual variation when corrected for party-hours of coverage. Of the five species examined, the Red-bellied Woodpecker was negatively correlated ($P < 0.05$) with the starting wind speed at Ocean City, the Brown Creeper was negatively correlated with the maximum wind speed at Ocean City, and the Brown Creeper was negatively correlated with the minimum temperature at Southern Dorchester County. Examination of Christmas count data for other species and for other areas, especially when there are high contrasts in weather conditions and when a large comparable sample is available, would reveal the limiting conditions beyond which counts become too variable to have comparative value.

DISCUSSION

TEMPERATURE

There is close, but not perfect, agreement among the various data sets examined. Comparing the effects of starting temperature, five of the seven species in footnote (a) of Table 4 also appear as positively correlated with temperature in Table 3 or Figure 2. The two exceptions are the Rufous-sided Towhee, whose positive correlation was not statistically significant ($P > 0.05$), and the American Robin, which was negatively correlated with temperature in regions 1 and 2. O'Connor and Hicks (1980), in a study of breeding birds on a nature reserve in southern England, found one negative and seven positive correlations with temperature based on 40 visits in the spring and summer of 1978. They had no species in common with the present study, nor any consistent temperature effects within families that can be compared with this study.

One pattern that emerges is the tendency for birds near the southern limit of their breeding range (American Robin; Vesper Sparrow, *Pooecetes gramineus*; Song Sparrow, *Melospiza melodia*) to be negatively correlated with temperature, whereas those near their northern limit (Mockingbird, Cardinal) are positively correlated with temperature (i.e., each sings more regularly at temperatures more like those at the center of its breeding range). Birds that feed on flying insects (Whip-poor-will, Chimney Swift, Barn Swallow) were positively correlated with temperature, as their prey also would be. Other species that showed a consistent trend tended to be positively correlated with temperature.

In regions 1 and 2, activity of the Mourning Dove and many other species declined when starting temperatures were above 23°C (73°F). McClure (1939:325) found that Mourning Doves did not coo much at temperatures above 25°C, and Wimmer (1961:38) said that calling was reduced above 29°C. Three of four authors cited by Baskett et al. (1978), however, stated that weather factors within the normal range encountered on call-count surveys did not significantly affect cooing of Mourning Doves.

When Mourning Dove data from 3852 route-years in areas 1 and 2 were divided into 10-stop (ca. 1-hr) intervals, temperature had no effect on activity during the hour centered at sunrise. In the second hour, however, activity increased slightly with increasing temperatures, reaching a peak near 23°C. In the third hour the increase with temperature was more pronounced, peaking again near 23°. The fourth hour was similar to the third, but with a peak at 20°. The fifth hour showed an even sharper increase to a still cooler peak at 17°. The count in each hour was lower than that of the preceding hour. Thus, counts may be variously affected by temperature, depending on when they are started and how long they continue.

I have found (unpublished data) that the number of phrases sung per minute by the Red-eyed Vireo is directly proportional to air temperature, and Curio (1959) found the same thing for the Pied Flycatcher (*Ficedula hypoleuca*). Higgins (1979) showed that there was a positive correlation between temperature and the duration of the morning song period of the Song Thrush

(*Turdus ericetorum*) in central England in December and January.

Anderson and Ohmart (1977) concluded, as we did for the Breeding Bird Survey instructions, that winds of less than 20 kmph (12½ mph) interfere very little with accurate censusing. They censused 22 transects on days when winds were from 24 to 50 kmph and compared these with the same transects on adjacent calm days. In about half of these tests they found a significant diminution of counts on windy days. Higgins (1979), summarizing work of Thorpe (1961), Armstrong (1973), and Astrom (1976), reported that onset of first song in the morning is delayed by both wind strength and cloud cover.

Sky Condition

Heavy overcast delays the dawn chorus and causes early cessation of evening activity. Whether activity continues later in the morning on an overcast day depends largely on other conditions, such as temperature and wind. Solid overcast prevents thermals and it delays or reduces soaring flight by vultures, hawks, and other large soaring birds. Dark sky and haze increase identification problems because of poor visibility. Bright sun can be a problem if the observer is looking toward it.

Although fog reduces visibility and will seriously affect visual counts of distant birds, fog tends to improve transmission of sound, and unless accompanied by wind it frequently increases the distance over which birds can be heard. Light falling snow affects visibility but may not seriously interfere with detecting calls. Because falling snow is dry and nearly silent, it has less effect than rain on the observer and his equipment.

My findings for Bobwhite disagree with those reported by Bennitt and Elder. Bennitt (1951:19–21) found no significant effect of cloudiness, and a negative correlation with temperature; and Elder (1956:650) reported that neither wind nor temperature had a significant effect. In areas 1 and 2, BBS data showed a strong positive correlation with temperature (Fig. 2), a highly significant decrease with cloud cover (Table 1), and a sharp drop with wind speeds above Beaufort 4.

Wind

O'Connor and Hicks (1980) reported that out of 27 British species tested, registrations of two, Great Tit (*Parus major*) and Blue Tit (*P. caeruleus*) were positively correlated with wind speed ($P < 0.05$), whereas two others, the Wren (*Troglodytes troglodytes*) and Chiffchaff (*Phylloscopus collybita*), were negatively correlated. In the BBS data I did not find any consistent indication that any small passerine species were found in larger numbers on a windy day, but counts for Starlings, blackbirds, and swallows, most of which were recorded by sight, were independent of wind speed.

Winter Conditions

Results from the winter analyses are far from conclusive. They have demonstrated that large samples of data collected under controlled conditions are required. On Christmas Bird Counts, where the most extreme weather conditions are apt to occur because dates are set in advance, the weather data tend to be imprecise, to apply to only one point in the 458 km^2 circle, and cannot be related to any particular time in the day; for example, the maximum wind might occur during the early morning and seriously interfere with songbird counts or, on the contrary, it might appear briefly during a thunderstorm gust or frontal passage and have little overall effect.

General Conditions

Even when counts are carefully timed, accurate measurements of local temperature, wind, and precipitation generally are not taken. Few observers are equipped to measure rainfall or wind speeds at ground level, let alone treetop level. Conditions recorded at the beginning and the end of a mapping or transect census do not necessarily apply throughout the course of the activity. Similarly, wind and temperature conditions recorded at the 50th BBS stop may not be similar to those prevailing at the 49th stop under entirely different habitat conditions.

CONCLUSIONS

A steady hard rain probably has a greater effect on bird counts than any other weather condition. This affects not only singing and other behavior of birds but also affects the observer and his equipment, including binoculars, eyeglasses, and notepaper. Part of the disturbing effect of rain is the noise created, and even the noise of dripping leaves after the rain has stopped can affect census results. Effects of light or intermittent rain at times are negligible. Drizzle, unless prolonged, will not necessarily interfere with census results; this conclusion is based on examination of 173 BBS routes started during drizzle. Effects of cloud cover are minimal.

Strong winds certainly affect bird counting. The generally accepted limit for satisfactory count results is 20 kmph (12 mph). During higher wind, singing decreases, birds tend to take shelter (keeping out of sight or concentrating in sheltered areas), and side effects, such as rustling leaves or rough water, affect detection by sound or sight. Blowing dust or drifting snow adds to the problems of high winds.

Effects of temperature are minimal within the range normally experienced by census takers, but unusually low temperatures tend to inhibit activity and unusually high temperatures in summer shorten the activity period.

Taken in combination, effects of unusual weather conditions are compounded. On the one extreme, hot calm air causes sounds to be diverted upward away from the observer so that singing birds are not detected as far away as under normal conditions, and distances are likely to be overestimated. On the other hand, a combination of low temperatures and strong wind produces a serious chill factor that will have a profound effect on the observer if not on the birds themselves.

ACKNOWLEDGMENTS

I thank the several thousand BBS cooperators whose work has contributed so substantially to this study, and especially Willet T. Van Velzen, Danny Bystrak, and Daniel Boone who participated in various phases of the WBS and BBS fieldwork. B. Kenneth Williams wrote computer programs for analysis of the BBS data. Kathy Marois prepared the illustrations and Romell Decker typed the manuscript. I thank Deanna Dawson, Mark Fuller, Barry Noon, and B. Kenneth Williams for their critical review of this paper.

Studies in Avian Biology No. 6:311–315, 1981.

SAMPLING IN RUGGED TERRAIN

DEANNA K. DAWSON[1]

ABSTRACT.—Work in rugged terrain poses some unique problems that should be considered before research is initiated. Besides the obvious physical difficulties of crossing uneven terrain, topography can influence the bird species' composition of a forest and the observer's ability to detect birds and estimate distances. Census results can also be affected by the slower rate of travel on rugged terrain. Density figures may be higher than results obtained from censuses in similar habitat on level terrain because of the greater likelihood of double-recording of individuals and of recording species that sing infrequently.

In selecting a census technique, the researcher should weigh the efficiency and applicability of a technique for the objectives of his study in light of the added difficulties posed by rugged terrain. The variable circular-plot method is probably the most effective technique for estimating bird numbers. Bird counts and distance estimates are facilitated because the observer is stationary, and calculations of species' densities take into account differences in effective area covered amongst stations due to variability in terrain or vegetation structure. Institution of precautions that minimize the risk of injury to field personnel can often enhance the observer's ability to detect birds.

Relatively few avian studies have been conducted in areas with rugged terrain. A number of reports based on observational data are available on the altitudinal limits and distribution of birds, particularly in the Appalachian Mountains, where the breeding ranges of certain species extend considerably farther south at high elevations (Oberholser 1905, Wetmore 1939, Murray 1946, Stevenson and Stupka 1948, Tanner 1955). More recent studies have focused on avian community structure along elevational gradients (Alexander 1973, Able and Noon 1976, Terborgh 1977, Noon and Able 1978, Sabo 1980). Other studies and a number of Breeding Bird Censuses have been conducted to investigate the organization of avian communities in montane, bog, or swamp habitats where terrain is uneven or unstable (e.g., Snyder 1950, Salt 1957, Brewer 1967, Breeding Bird Censuses in American Birds 1971–1980).

In the great majority of instances, however, knowledge of avian distribution and habitat preference is limited to habitats and regions easily accessible to the observer. With additional research in areas with rugged terrain, we may find even the more common, well-studied bird species occupying a broader range of habitats than was previously known. For example, in a study in progress in Upper Michigan (unpublished data, U.S. Fish and Wildlife Service), Ovenbirds (*Seiurus aurocapillus*) exhibiting territorial behavior have been located in moist lowlands with dense shrub cover as well as in mature deciduous and coniferous forests. At this time, it is unclear whether these individuals are actually breeders or are unmated males occupying suboptimal habitats. In addition, the Black-and-white Warbler (*Mniotilta varia*), a summer resident of mature, dry deciduous forests throughout much of its breeding range, was commonly found in wooded swamps with high shrub densities, particularly those in which northern white-cedar (*Thuja occidentalis*) was dominant (Noon et al. 1980).

The study of certain species of birds may make work in rugged terrain unavoidable. Palm Warblers (*Dendroica palmarum*) and Northern Waterthrushes (*Seiurus noveboracensis*) nest in bogs or shrubby swamps where travel is difficult; rosy finches (*Leucosticte* spp.) in the Rocky Mountains nest almost exclusively above timberline. A number of the endangered Hawaiian species (e.g., ʻOʻu (*Psittirostra psittacea*), Kauai ʻOʻo (*Moho braccatus*), ʻAkiapolaʻau (*Hemignathus wilsoni*)) inhabit high elevation rain forests far from roads or trails. Heavy rainfall, dense vegetation, and steep slopes can make working conditions extremely strenuous.

In addition, the increased pressure to develop fossil fuel resources has necessitated more research in rugged terrain to determine the impact mining and the resulting conversion to a different vegetation type will have on bird populations. Strip mining of peat and coal may center in extensive boglands or on steep slopes where, because of site conditions, terrain, and climate, disturbance has been minimal in the past.

This paper deals with sampling in rugged terrain. Potential problems are identified, and considerations are discussed concerning the choice of a sampling technique, location of study sites, and safety measures. Although emphasis is placed on areas with steep slopes, some of the same problems are encountered wherever terrain is uneven or unstable, or where slash, vegetation, rocks, or other obstacles make foot travel difficult or obscure vision and hearing.

[1] U.S. Fish and Wildlife Service, Migratory Bird and Habitat Research Laboratory, Laurel, Maryland 20811.

SAMPLING BIASES INDUCED BY TERRAIN

Work in rugged terrain poses some unique problems that should be considered before research is initiated. Attempts should be made to minimize the effects of these problems because, in conjunction with other variables such as weather conditions, daily or hourly changes in activity and behavior, and differences among observers, considerable bias may be introduced into census results.

RATE OF TRAVEL

When the rate of travel along a census transect is slowed because of terrain, the length of time the observer is exposed to each bird increases. Care must be taken to avoid double-recording of individuals, particularly when no opposition singing is detected. In addition, the likelihood of recording species that sing infrequently is greater. Colquhoun (1940) and Shields (1979) found that the walking speed of the observer influences census results. A fast walker records more species and individuals per unit of time. However, if figures are converted to densities, the numbers will generally be lower than those of an observer traveling over the same area at a slower speed. These factors should be taken into account if comparisons are to be made with census results obtained on level terrain. If a transect is to be covered several times, the time bias can be reduced to some extent by taking extra care in clearing and flagging the routes of travel.

DETECTABILITY OF BIRDS

Besides the physical difficulties associated with work in uneven or unstable terrain, topography can influence the observer's ability to detect birds. When paying attention to the terrain underfoot, it may be difficult to concentrate on observing birds. Additionally, large boulders, ridges, or other physiographic features may obscure bird song and conceal activity. The frequent streams in hilly or mountainous terrain present additional detection problems; census results may show an inaccurate preponderance of relatively loud-voiced species. This factor may be an especially serious bias in roadside counts, as many mountain roads follow streams. Myberget and Strömme (1974) concluded that it is unlikely that obstructive terrain features will cause large errors in population estimates from randomly located line-transects when data collected are based on visual observations. Emlen (1977a) addresses the problem of varying detectability of species and individuals by basing breeding season estimates of density exclusively on song. In areas where terrain obscures sound, the only effective means of alleviating detection problems may be to require that transects cross rather than parallel obstructions, or to deviate from the prescribed route of travel.

EFFECT OF PHYSICAL FACTORS ON CENSUS RESULTS

Physical factors (e.g., slope, aspect, elevation, presence of streams) may also influence the bird species composition of a forest. Species such as the Louisiana Waterthrush (*Seiurus motacilla*), Northern Parula (*Parula americana*), Cerulean Warbler (*Dendroica cerulea*), and Acadian Flycatcher (*Empidonax virescens*) are often found near water, though vegetation characteristics may not be measurably distinct. If the effect of a treatment on bird populations is being studied, study areas should be similar with respect to these physical factors as well as vegetational composition and structure. In studies in which associations are to be made between bird species' presence or abundance and habitat characteristics, physical factors should be included as variables unless no differences exist among areas.

ESTIMATES AND MEASUREMENTS OF DISTANCE

Uneven terrain or dense vegetation can introduce error into estimates of distance, particularly when estimates are based on aural detections of birds. Birds singing across a valley may sound considerably closer, while birds over a ridge or downslope from the observer may sound more distant. Estimating distances and plotting locations of singing birds can be facilitated by plotting landmarks and vegetation changes to scale on field maps of census plots or transects. Setting up a finer grid (e.g., 25 m, as opposed to the more usual 50 or 100 m) in tracts censused by the territory mapping method and prominently marking plot or transect boundaries can also increase the accuracy of plotted locations and distance estimates. Kepler and Scott (1981) also emphasize the value of training observers in minimizing errors in estimates of distance.

If the avian data gathered are to be related to areas defined on maps or aerial photos, measurements made on slopes should be corrected to accurately reflect the horizontal projection. Since the slope can easily be determined with the use of a clinometer, Abney level, or Haga altimeter, the required correction is straightforward; the desired horizontal distance is divided by the cosine of the angle of slope, a figure readily available from a hand calculator or a prepared table. For example, to be equivalent to a map distance of 50 m, the field measurement on a 15°

slope should be corrected to 51.8 m ($x = 50$ m/cos 15° = 50/0.966 = 51.76 m). The significance of such corrections becomes apparent as distance or slope increases. For example, if a 1-km transect is being laid out on a 15° slope, an additional 35 m (a correction of 3.5%) should be added so that the measurement corresponds to the horizontal distance. On a 30° slope, a correction of approximately 15.5% is required. However, if mapping quantitative data is not the objective of the study, slope corrections should not be made, since an overestimation of the area censused, and hence an underestimation of the density of birds, will result.

OTHER CONSIDERATIONS

Time

Since the period during which avian breeding studies can be conducted is limited, the time involved in collecting census data is a major concern of the researcher. The extra time and physical exertion required for travel in rugged terrain generally mean that daily and seasonal goals of how much can be accomplished must be lowered from those normally established. This should be a prime consideration in the selection of study areas and a census technique, especially when little or no field assistance is available or when field work cannot be extended beyond one breeding season.

Safety

The risk of personal injury is inherent in any field situation; however, the chances of injury increase as terrain becomes more rugged. If the project leader cannot meet the necessary provisions for safety of field personnel, a study area should be chosen where hazards are minimal. Institution of safety precautions can often enhance the observer's ability to detect birds.

(1) Field personnel should be in good physical condition.

(2) All field personnel should be equipped with properly fitting footwear that provides adequate ankle support and traction for traversing rough terrain. Clothing should offer some protection in areas with dense vegetation or slash.

(3) Field personnel should be equipped with a compass, topographic map, flashlight, and first aid kit. In remote areas, extra food and a sleeping bag or blanket should be carried. If the route of travel crosses treacherous terrain or if location of study areas requires that individual observers work alone in remote locales, two-way radios should be used. As often as possible, study plots or transects should be close enough together so that another individual is within the general vicinity. In avian studies in Hawaiian rain forests, Scott et al. (1981a) have adopted a field plan in which, for safety reasons, independent but simultaneous counts are made by two observers. They feel that more accurate estimates of numbers can be achieved by dividing responsibilities for counting common species. However, in most situations, sampling by pairs or groups of observers is inefficient and distracting.

(4) A detailed map of the study area with the route of travel and specific hazards marked should be filed with a local individual or fellow employee not involved in the fieldwork. Radio frequencies and a schedule of censuses should be attached, with instructions for action should field personnel not return to their duty station or residence by an established time.

(5) Hazardous areas should be flagged or otherwise marked. Provided that habitat disturbance is minimal, pruning or clearing of vegetation to expose hazards may be warranted. Deviation from the defined straight line of travel is discouraged if the sampling scheme requires accurate measurements of distances and angles (e.g., the line transect method; Anderson et al. 1979); however, fewer observations of birds will be missed if physical obstacles or hazards are skirted.

(6) In planning fieldwork, care should be taken to avoid overestimating the area to be censused each morning; the pace must be slow enough so that observers are not forced to hurry over rough terrain. In addition, if the study area is not easily accessible from a road, field personnel should consider camping at or near the start of the census route on nights prior to censusing to reduce the amount of foot travel before daylight.

SELECTION OF A CENSUS TECHNIQUE

Difficulties are encountered in conducting avian field studies in rugged terrain regardless of the census technique selected. The objectives of the study to a large extent will dictate which technique will be the most efficient. For example, if the intent of a study is to determine species' densities in a montane habitat or to measure long-term effects of an environmental disturbance on bird populations, territory mapping would be the appropriate technique. A transect method would be more suitable in a study of seasonal changes in bird populations or of avian distribution along an elevational gradient.

Robbins (1978a) provides a summary and critique of the most widely used techniques for censusing forest birds. Other authors (e.g., J. T. Emlen 1971, 1977a; Shields 1979) have compared census methods. The researcher should weigh the efficiency and applicability of a particular technique for the objectives of his study in light of the additional difficulties posed by work in rugged terrain.

Territory Mapping Method

Territory mapping (International Bird Census Committee 1970) is the most widely used census technique. Its chief disadvantage, the amount of time required to set up a plot and conduct multiple census trips, is magnified in rugged terrain. To adequately conduct a census, it is essential that the observer traverse the plot; in some instances, this may require recrossing contours or areas with dense vegetation. Attempts have been made to overcome this drawback by establishing elongated census plots so that the centerline lies along a ridgetop (e.g., West Virginia censuses, Reeves 1980). However, the territory mapping plot then in effect becomes a strip transect, with year-to-year variation in the number of territories exaggerated by birds that move short distances, crossing plot boundaries from one year to another (Robbins 1978a).

Sampling error induced by terrain is minimal with this technique. The problem of double-recording of individuals is minimized by repeated coverage of an area. Distance estimates are not required, and the difficulties of detecting birds can be reduced by varying the route of travel on successive visits. Thus, territory mapping is probably the most accurate census technique in rugged terrain.

Transect Methods

In terms of time expended, the transect method is more efficient than the mapping method. This technique, which basically involves counting birds on one or both sides of a line, minimizes the amount and difficulty of walking. Transects can be established along or across contours. The number of observations of birds necessary to provide sufficient data for analysis is probably the best determinant of transect length (Anderson et al. 1979, Conner and Dickson 1980). If terrain or the size or shape of the study area precludes the establishment of long transects, several shorter transects can be used to obtain the desired base of data. A disadvantage of this method in rugged terrain is that, since observations are made while progressing along the transect, it is difficult for the observer to pay attention to his personal safety and to accurately detect and record distances to birds.

Of the various transect methods used for wildlife censuses (Eberhardt 1978), the strip transect method (Emlen 1977a, Conner and Dickson 1980) is the best suited for censusing avian populations in rugged terrain. Unless one is trying to correct for detectability (J. T. Emlen 1971), distance estimates are simplified because the observer need determine only whether a bird is within or outside the census strip boundary. Problems may arise, however, if physical obstacles or dense vegetation prevent adequate coverage of the strip at all points. It may be difficult to settle on a strip width that can be adequately covered and still allow sufficient sampling of rare species.

IPA or Point Count Method

Several researchers (Ferry 1974, Jorgensen 1974, Evans 1978, Whitcomb et al. 1979) have successfully used stationary counts to obtain indices of abundance for comparing bird populations. The basic technique, the IPA count (Indices Pontuels d'Abondance) developed by Ferry and Frochot (1970), involves a count of all birds heard or seen from a point. Subsequent modifications alter the effective area censused. Counts may be restricted to a defined area (fixed plot), or all birds may be counted, with estimates of the horizontal distance to each location used for more accurate calculation of the area censused (variable circular-plot, Reynolds et al. 1980).

The appeal of a technique involving stationary counts of birds is apparent in rugged terrain. When walking is separated from the actual counts, the observer is able to devote his full attention to detecting birds and still concentrate on his personal safety. A stationary observer is also less likely to influence bird activity. However, the effectiveness of this technique in rugged terrain depends on the type of count selected.

Undefined area counts.—Although this technique is the simplest to apply, the influence of topography on the detectability of birds and the area of coverage may make it impossible to combine or compare results from undefined area counts. Criticism has also been levied concerning the appropriateness of using these data for associations between presence or abundance of a bird species and habitat characteristics (Noon, in press). The area sampled varies with the species (e.g., the area of coverage is greater for a loud-voiced species such as the Red-eyed Vireo (*Vireo olivaceus*) than for a soft-voiced species such as the Black-and-white Warbler); thus coordination of vegetation samples with bird territory locations is difficult.

Fixed area counts.—The use of plots with a fixed or defined area enables the researcher to express count results as species' densities. However, the variability in physical factors and vegetation structure characteristic of rugged terrain may cause differences in the detectability of birds, making standardization of plot size difficult. Like fixed width transects, this technique is less efficient with regard to uncommon species, since only birds observed within the plot boundaries are counted.

Variable circular-plot counts.—In extensive surveys in Hawaiian rain forests where rugged terrain and dense vegetation make foot travel extremely difficult, Scott et al. (1981a) have found the variable circular (variable area)-plot technique to be the most effective for estimating bird numbers. This technique takes into account differences in detectability among species and among stations. The distance from the point to each bird observed is estimated and used to calculate the effective area covered for each species at each station. Thus, if topography or vegetation obscure detectability at a particular station, the limits of the area surveyed will be correspondingly smaller. Estimation of distances, which may be burdensome when made while moving along a transect, is more easily accomplished when the observer is stationary. However, because the area added to the circle of coverage increases with increasing distance from the observer, errors in estimates of distance may significantly affect the accuracy of the calculated densities. For this reason, the importance of training observers in estimating distances cannot be overemphasized.

CONCLUSIONS

It is clear that there are disadvantages associated with the use of any census technique in rugged terrain. The variable circular-plot method, currently being used in surveys of Hawaiian forest birds and in other studies, was developed for use in areas where terrain or vegetation hinder travel and detectability of birds, and can easily be adapted for use in other areas. It is likely that with more work in rugged terrain further refinement of present techniques or development of new techniques may effectively minimize potential sampling biases induced by terrain.

Studies in Avian Biology No. 6:316–321, 1981.

LIMITATIONS OF ESTIMATING BIRD POPULATIONS BECAUSE OF VEGETATION STRUCTURE AND COMPOSITION

HANS OELKE[1]

ABSTRACT.—The mapping method is difficult to use in many habitats. Moreover, there are no alternative census methods or means to calculate errors which can correct for difficulties in the census. Problems are most apparent in or near human population centers where an increasing number of vegetation types are unavailable for census work, for reasons of nature protection and possible economic damage.

Central European (German) bird censuses show (1) A standard mapping of wetlands with the use of the IBCC recommendations results in sharp, long lasting changes to fragile vegetation and significant disturbance of bird communities. (2) Changes in agriculture, as demonstrated by monocultures and increased sizes of farm fields, place large areas of the landscape out of reach. (3) Mosaic-like landscapes with distinct horizontal and vertical plant diversities can be censused with the mapping method providing individual error calculations are made for the "out-of-bond" subplots of the study area.

The IBCC bird census recommendations (Oelke 1974a) on the mapping method are the only internationally standardized census method. Among the recommendations suggested are the need for the position of the observer and that of the bird to be known as exactly as possible. When no topographic or physiognomic features are available, a grid should be established with points marked in 100 m intervals in open areas and in 50 m intervals where the vegetation is closed.

Unfortunately these recommendations cannot be followed in a number of vegetation types or in areas which are densely settled or used by people. This imposes limits to the estimation of bird numbers which must be taken into consideration when planning research.

PROBLEMS WITH VEGETATION

For many types of vegetation, difficulties are imposed by the nature of the habitat or by their economic value (Table 1). The large size and economic value of many critical vegetational types—in the sense of being sensitive to disturbance or of such limited extent as to be considered "endangered"—prevents calibration. A calibrating of the mapping method by nest search, line transects, point counts or best, color-banded populations is still possible. Many endangered habitats which show the vegetation criteria in Table 1 (1a–1c, partly 2c) are excluded for reasons of nature protection. Many nature reserves of the Federal Republic of Germany cannot be used for bird mapping. Examples are coastal boglands, natural inland lakes (such as Dümmer, Steinhuder Meer), seabird sanctuaries including peripheral dunes and marshes in Northwest Germany, and many unique plant communities (e.g., grasslands with orchids). Besides severe restrictions on access, the nature protection agencies have changed their attitude to research. They operate on the principle that protection of nature must be guaranteed, with research a secondary consideration (Erz et al. 1979). There are other administrative restrictions in the Federal Republic. All proposals to conduct research in nature reserves must be presented to authorized nature protection associations for consultation (§29 Act of Nature Conservation "Bundesnaturschutzgesetz" of 20 December 1976, or the adequate acts of the federal states of GFR). The associations include groups with diverse attitudes to nature. They include ornithological and bird watcher societies (e.g., Deutscher Bund für Vogelschutz), nature protection societies, historic-folkloristic groups (e.g., Heimatvereine), and the hunters' associations. This guarantees that approval for research will be delayed if not refused.

Thus the census of birds by mapping or other methods is not solely the decision of the research worker or the scientific institute.

SELECTED EXAMPLES

In a number of cases the vegetation structure is not compatible with hitherto applied bird census techniques. Form and range of these discrepancies will be evaluated by selected examples.

EFFECTS ON VEGETATION BY BIRD MAPPING

In the course of a breeding bird census (1961) and the monthly mapping (1960–1962) of a 13.1 ha bog (Wendesser Moor, county of Peine, Lower Saxony, Federal Republic of Germany), I censused an area 300 m in length by 30–60 m in width (Oelke 1963). The numerous, 28 main and 40 additional multi-hour visits, created 20 to 30 cm wide trails with a total length of approximately 1500 m throughout the inner part of the bog. Access to the outer parts of the bog were blocked by fences. The trails affected a zone of *Salix cinerea* with open areas of *Eriophorum*

[1] Zoologisches Institut, Universität Göttingen, D-34 Göttingen, Federal Republic of Germany.

TABLE 1
VEGETATION TYPES WHERE IT IS DIFFICULT OR IMPOSSIBLE TO USE THE MAPPING METHOD

Vegetation type/habitat	Common difficulties
1. Complex structures	
a. Climax woods (± natural woods) with uniform high plant cover in all strata, esp. deciduous woods, their successions (thicket types), man-made communities (e.g., mediterranean macchies)	Optical barriers; lacking or too wide-spaced roads/trails; unavoidable habitat manipulation when improving the census efficiency; too small study plots; disturbance or change of the bird community
b. Shore vegetation (e.g., reed beds, *Carices* communities, floating plant communities)	As 1a; additional danger of opening the study plots for human access; in some cases not to be entered (deep mud banks)
c. Habitats with a high horizontal plant species diversity (floristic island types), e.g., moors, dry grasslands with rare or endemic plant species	No immediate disturbance of avifauna, but sometimes irreparable damages to rare vegetation
d. Mosaic-like, cultural landscapes, esp. gardens	Visibility markedly reduced because of property lines (walls, hedges, fences), often disturbed by traffic or other noise
2. Uniform structures	
a. Monocultures and their successions in forestry (esp. thickets, 15–30 year-old coniferous forests)	As 1a; effects on birds not known because of low species and pair densities
b. High growing agricultural monocultures (e.g., banana, surgarcane, corn plantations)	Visibility reduced; little possibility of establishing smaller study plots; risk of economic damage
c. Low growing vegetation (e.g., *Carex* reeds, meadows, pastures, grain, sugar-beet, potato, oil seed, vegetable, flower fields)	Insufficient or—in the case of larger areas—lacking control routes; no additional census methods possible (e.g., nest searching); risk of economic damage
d. Water areas with uniform floating, or underwater vegetation (e.g., fish farms)	Visits only possible at the edges; study plots cannot be entered

angustifolium, Comarum palustre, Juncus conglomeratus and the dominant *Carex rostrata*. Even a stand of *Phragmites communis* was affected. The vegetation along the trails did not recover until 1968. The last traces of the trails within the rather uniform 50–75 cm high swamp disappeared in 1970. Mammals such as *Lepus europaeus, Ondatra zibethica, Vulpes vulpes, Capreolus capreolus,* as well as people used the trails for access and intensified the disturbance.

The impact on the vegetation had an adverse effect on the avifauna. The loss of plant cover reduced protection of breeding sites and split the uniform stand into patches (Table 2). The population decline of non-passerines shown in Table 2 is related to the disturbance but the decline of passerine species might be a normal fluctuation in population size.

The impact on the birds shown in Table 2 could be minimized by restricting observations to the periphery of the plot, by stopping all forms of nest search, and using blinds for studies of breeding or rare species. This means an increase in observation time.

PROBLEMS WITH MAPPING CAUSED BY AGRICULTURE

Most Central European sites are agricultural or urban. Forests and wetlands continue to decrease in area; the percentage cover of these habitats in the German Federal Republic (248,601 km^2) are: forests (28.7%), agricultural areas (53.0%), settlements (6.6%), traffic areas (4.7%), and water areas (1.8%). Although agricultural areas are the most extensive they have been neglected in bird censuses compared with woodlands. The proportion of woodland to agricultural areas studied by bird censuses (mapping) is 11:1 (numerically approximately 1000:100; Oelke 1974b, corrected for 1980 data).

Most bird watchers and ornithologists avoid agricultural areas because of the low species and pair densities. Compared with forest (450–500 breeding pairs, 40–60 species per km^2 on the average) the corresponding agricultural densities are lower (30–40 territorial males, 3–10 species per km^2) (Oelke 1963). Beside the small number of birds, it is difficult to inspect agricultural areas.

TABLE 2
EFFECTS OF INTENSIVE MAPPING (1960–1962) ON SPECIES NUMBER AND PAIR DENSITY IN THE SWAMP WENDESSER MOOR[a]

Species	Pairs/territorial birds 1961	Pairs/territorial birds 1962	Difference (%)
Little Grebe (*Podiceps ruficollis*)	1	1	—
Mallard (*Anas platyrhynchos*)	9	5	(−) 44.4
Gargany (*Anas querquedula*)	2	2	—
Teal (*Anas crecca*)	1	1?	?
Shoveler (*Anas clypeata*)	1	—	(−)100
Ferruginous Duck (*Aythya nyroca*)	1[b]	—	(−)100
Pheasant (*Phasianus colchicus*)	1	—	(−)100
Water Rail (*Rallus aquaticus*)	1	—	(−)100
Spotted Crake (*Porzana porzana*)	1	—	(−)100
Moorhen (*Gallinula chloropus*)	2	1–2	(−?)50
Coot (*Fulica atra*)	8	5	(−) 37.5
Lapwing (*Vanellus vanellus*)	2	—	(−)100
Snipe (*Gallinago gallinago*)	1	1	—
Tree Pipit (*Anthus trivialis*)	2	—	(−)100
Sedge Warbler (*Acrocephalus schoenobaenus*)	2	1?	(−) 50
Marsh Warbler (*A. palustris*)	1	—	(−)100
Reed Warbler (*A. scirpaeus*)	1	—	(−)100
Whitethroat (*Sylvia communis*)	1	—	(−)100
Willow Warbler (*Phylloscopus trochilus*)	1	—	(−)100
Blackbird (*Turdus merula*)	—	1	(+)100
Yellowhammer (*Emberiza citrinella*)	1	1	—
Reed Bunting (*E. schoeniclus*)	7	7–8	—
Magpie (*Pica pica*)	1	—	(−)100
Pairs/territorial birds	48	24–28	(−)42–50
Species	22	11	(−)50

[a] Weather conditions in 1960–62: relatively cold and rainy summer periods with more or less constant, 30–50 cm high water levels.
[b] Female illegally killed by hunters (Oelke 1962).

Agriculture in Europe is changing and the trend is towards larger and more uniform areas of production. In the Federal Republic, the average farm size rose from 8 ha (1960) to 18 ha (1975) and in the Democratic Republic of Germany, it increased from 280 ha to 1170 ha (Schultzke et al. 1979). The best German agricultural areas have field sizes of 100–300(–700) × 50–150 m in the loess belt of Hildesheim-Braunschweig-Hannover, Federal Republic, but blocks of 1000–1300(–1700) × 1500–2000(–3000) m occur in the loess belt of Halberstadt-Magdeburg, Democratic Republic.

The disadvantages of agricultural areas for bird census and especially for mapping procedures are many. It might be possible to observe from an average of 100–200 m distance those fields separated by field roads at intervals of 200–400 m. At times shorter distances are possible because of ditches, water lines, border rows, grassland strips, and along fields characterized by smaller strip sizes ("towel-like-fields"). Even this kind of observation is impossible on the state farm blocks. Regular traverse

TABLE 3
HABITAT TYPES AND AREA IN A MOSAIC-LIKE LANDSCAPE (FUHSE VALLEY, NW EDGE OF THE CITY OF PEINE, LOWER SAXONY, FEDERAL REPUBLIC OF GERMANY). CALCULATIONS (SMOOTHED) FOR 1980

Habitat	Visible complexes (number)	Total size (ha)
Alder swamp	7 (2 large plots)	38.4
Fuhse River	1	4.0
Old river beds (left after canalization)	10	3.3
Phragmititea reed	3	47.5
Meadows (unused by cattle)	1	80.0
Pastures (used by cattle)	4	14.5
Abandoned mining dump	1	4.4
Roads, trails[a]	37	6.1
	64	198.2

[a] Total length: approximately 20.2 km, including 0.5 km tar pavement, 2.7 km with compressed stone layer (1 km railway dam), 9.95 km grass roads, and 7.05 km small trails.

FIGURE 1. Aerial view of the mosaic-like landscape (Fuhse Valley, NW edge of the city of Peine, Lower Saxony, Federal Republic of Germany), April 6, April 15, 1980. By kind permission of Niedersächsisches Landesverwaltungsamt (Landesvermessung), no. 28/80/1708. For type and area of habitats see Table 3. Dark line = 500 m.

TABLE 4
LENGTH OF ROADS/TRAILS AND EDGE LINES IN THE DIFFERENT HABITATS OF THE STUDY AREA IN THE PEINE RIVER VALLEY (SEE TABLE 3)

Habitat	Length of roads, trails (m)	Edge lines (m)	Edge lines / Habitat size
Alder swamp	4000	10,500	273
River	2100	4200	1050
Old river beds	1550	6650	2015
Reeds	3550	14,750	310
Meadows	8450	15,700	196
Pastures	1550	5500	379

of a field is impossible because of crop damage. The time is near that no kind of mapping will be allowed on fields. Only transect methods along the rare rights-of-way through fields offer a solution. This will reduce detectability of birds and censuses will be less reliable. There will be zones within a field where vegetation will conceal some species. Quiet species are the main problem in these habitats.

RELIABILITY OF CENSUSES IN A MOSAIC-LIKE LANDSCAPE

While some habitats make censuses of any sort difficult, other habitats, including most mature woodlands of the boreal zone, tundra, steppe, and savannah habitats, are ideal for censusing because there are no "real" restrictions imposed by the vegetation. Greater difficulties arise in the mosaic-like landscapes which are typical around population centers of Europe. These landscapes are distinguished by diverse regional or local features. A variety of horizontal and vertical structures, vegetational elements, plots, human use, and ownership patterns are typical. This variety prevents the location of representative census plots. The best method might be to census a whole landscape and differentiate structural elements by summarizing similiar elements (participation method, after Puchstein 1966).

I should like to draw attention to the problems of a bird census that I started in 1960 on the river plain northwest of the city of Peine (50,000 inhabitants, situated between Hannover-Braunschweig, Lower Saxony, German Federal Republic). Size, habitats, length of access routes and edge lines are summarized in Tables 3 and 4. Although the landscape seems to be well developed, in the midst of suburbs encircling the river plain (Fig. 1), special protection of vegetation had to be observed: (a) securing the reed beds against a network of trails from the ever present number of walking people; (b) no entering of wetlands containing stands of rare plant species (*Carex* species); (c) keeping out of the meadows before mowing (i.e., between May–June); and (d) keeping out of alder swamps to protect particular plant associations (*Urtica urens* stands)—and for the safety of the observer against mud more than 6 m in depth.

These restrictions excluded access to the higher and lower parts of the different habitats (Table 5). When all visits carried out between 1960–1980 are considered, about 25% of the alder swamp and 30% of the reed beds had not been entered. This affected census results. In particular, errors in the estimation of density of rarer species are expected (Table 6).

The evaluation of bird densities in landscapes with many different types of vegetation has to allow for the many local or regional peculiarities. Botanical restrictions are only one feature with implications for bird censuses. Equally great restrictions may be imposed by the presence of certain species of animals. Endangered fauna of national or international significance may require a safety zone within neighbouring habitats which then cannot be censused. Ex-

TABLE 5
VISIBLE AND ENTERED PARTS OF MAPPED STUDY PLOTS (MOSAIC-LIKE LANDSCAPE IN THE PEINE RIVER VALLEY) IN RELATION TO ACOUSTICALLY CONTROLLED PARTS (AUDITORY BELT APPR. 50 M)

Habitat	Distance of visibility[a] (m)	Visible areas (% ha)[b]		Entered area (% ha)		Auditory area[c] (% ha)
		In theory	In reality	1960–1979	1980	
Alder swamp	15 (10–25)	50.8	31.2	2.6	0.8	52.1–100
River	50	100	100	—	—	100
Old river beds	2 (1–5)	100	9.4	18.2	3.0	100
Reeds	2 (1–5)	62.1	2.2	1.1	0.2	37.4–74.8
Meadows	50	98.1	52.8	0.6	0.1	52.8–100
Pastures	50	100	100	1.4	0.0	53.4–100

[a] Related to 50% visibility of a 1 × 1 m white cardboard during summer (May–August).
[b] Related to observations from all present edge lines (in theory) compared to present road and trail length (in reality).
[c] Related to auditory belts of 50 m (left numbers)–100 m (right numbers).

TABLE 6
NUMBER OF BIRD SPECIES WITH REDUCED VISIBILITY IN A 1975 MAPPING CENSUS (TABLE 5). MOSAIC-LIKE LANDSCAPE IN THE PEINE RIVER VALLEY, LOWER SAXONY. METHOD: INTERNATIONAL RECOMMENDATIONS OF THE IBCC (OELKE 1974A)

Habitat	Species	Pairs territorial birds	Critical[a] species	% of all pairs/terr. birds
Alder swamp (38.4 ha)	33	270	18	80.7
Reeds + old river beds (50.1 ha)	28	239	20	92.3
Meadows + pastures (94.5 ha)	8	9	—	—

[a] See text for explanation.

amples in the Federal Republic are: heronries of the Grey Heron (*Ardea cinerea*), breeding habitats of Bittern (*Botaurus stellaris*), Greylag Goose (*Anser anser*), White-tailed Eagle (*Haliaeetus albicilla*), Black Grouse (*Lyrurus tetrix*), Crane (*Grus grus*), and Golden Plover (*Pluvialis apricaria*); of specific beetles (*Dytiscus lapponicus, Carabus clathratus, Brephos parthenias*); other insects, especially butterflies; and such mammals as *Equus caballus* ssp., *Cervus elaphus, Bison bonasus,* and *Phoca vitulina.*

CONCLUSIONS

Vegetation limits the census of birds and especially the use of mapping techniques in many ways. The difficulties for an observer getting orientated in complex habitats and respecting the safety of vegetation and animals are only one side of the problem. There are problems not only between people, vegetation, and birds, but between birds and vegetation themselves. Bird species which have a wide area of habitat selection in Central Europe, e.g., Chaffinch (*Fringilla coelebs*), European Robin (*Erithacus rubecula*), Blackbird (*Turdus merula*), and Blackcap (*Sylvia atricapilla*), have different distribution patterns and densities in different parts of their range (Oelke 1980). The type of vegetation and the geographic site influence the species-specific detectability and will therefore affect density estimates.

Studies in Avian Biology No. 6:322–323, 1981.

SUMMARIZING REMARKS: ENVIRONMENTAL INFLUENCES

FRANK A. PITELKA[1]

My assignment is to comment on some biological highlights of papers in this session. But, without asking the organizers' permission, I'm going to do something quite different in the few minutes that are allowed me. You know, in university systems, we have the business of promoting faculty members, and when one is to be considered for tenure, we talk about a "midcareer evaluation." As a matter of fact, at this moment we are literally midstream or "midcareer" in this symposium, and it is coincident that I happen to be in this time slot. I'm going to take advantage of this coincidence to comment in a more general way on the drift of papers to now and on related things which I think it is important for us to bear in mind for the remainder of the symposium. The complex of motivations and methods in census work and the compartmentalization of the different elements and procedures that go into census work seem to have brought on an interval that I would like to call an orgy of cautionary noises about methods. We are wasting time, we are wasting a considerable amount of time, doing analyses which I will not deny are edifying to a certain degree, but which are really detours from the mainstream of effort that brings us together in the first place. This morning, for example, we had an excellent analysis of the degree to which one can depend on song signals in order to estimate the number of birds present. I hope that there is no censusor surviving this symposium who will go into the field and do work depending entirely on song signals. I never have. I was taught by Kendeigh not to do this in aught thirty-nine, and I don't know why we're worrying now about the consequences of depending simply on this source of evidence. That's one of several possible examples provided at this symposium of what I mean about analysis of a narrow methodology (in this case, counts of singing males) and the resulting cautionary noises. The data are subjected to statistical analysis that dignify them and command our attention when in fact one can never seriously depend on that class of data alone for conclusions about densities. And anyway, why aren't we talking more about objectives in the use of census data—what are the questions, what hypotheses are we testing? This is perspective obviously essential for the assessment of method, ultimately. And so we seem to be in something of a trap. First, there is a historical bit: ornithologists early were inspired by and took over techniques from plant ecology for density and abundance estimation. Plant ecology deals comfortably with stationary organisms, while we apply them to hilariously mobile organisms and then suffer the consequences of that mobility undercutting our efficiency and accuracy in data gathering. Second, there has been, of course, a tremendous increase in the diversity of analytical techniques that we can apply to field populations, and this has led to a preoccupation with the study of methods for their own sake. It appears now that we run the risk of exploring methods without adequately asking what the data do for us, or we run the risk of not presenting the full scale of data that one needs in reading a paper in order to evaluate the effectiveness of a prescribed method. There are examples both ways among papers coming into this symposium. So, because of difficulties due to mobility and the diversity of techniques used to cope with those difficulties, we are preoccupied with methodology without giving adequate time to *why* we are so preoccupied. It is a dilemma which arises out of the fact that we are gathered to discuss methods, and were we to discuss their utilities and merits adequately, we would of course have a symposium several times longer than this one. Nevertheless, it is quite clear now (as it was in several comments gently making the same point earlier), that we should repeatedly ask ourselves, why this or that critical study of method? By the end of yesterday I had the feeling we were sort of coming to a stage of self-immobilization, with criticisms of this method, that method, and with a growing inventory of methodologic shortcomings, all augmented to an alarming degree by the skepticism and pessimism of the statisticians. Are we to give up? Obviously not. Plot work and transect work will continue notwithstanding all the difficulties, the study of population phenomena in the field will continue. So, the point at issue is, why discuss methods without more attention to the biological utility of the resulting data? The published proceedings should include some sort of terminal assessment of methods and recommendations from this standpoint. There is another reason why this is important, beyond the basic study of avian populations. The committee involved in this symposium has recognized that the symposium volume when it appears is going to be an item of particular interest

[1] Museum of Vertebrate Zoology, University of California, Berkeley, California 94720.

to all environmental assessment agencies because of the degree to which birds figure in such work. They function very usefully as indicator organisms, and compared to other classes of animals, data on birds are relatively easy to gather. In this connection there is the fact that we are, these days, faced with formal court challenges to some of the data we gather and the ways we gather them. This may be the chief reason, ultimately, why we are here this week. For the reason, therefore, of the importance of this symposium to the environmental impact field as well as to basic avian ecology, the proceedings should include a strong terminal synthesis that will focus on both applied and basic aspects of our overall effort, and in particular on the fit of methods to objectives explicitly stated.

Studies in Avian Biology No. 6:324–325, 1981.

SUMMARIZING REMARKS: ENVIRONMENTAL INFLUENCES

KENNETH P. BURNHAM[1]

I will make a few comments on the eight individual papers of this session, followed with comments on the role of factors influencing bird counts and the use of counts to estimate bird abundance.

Anderson and Ohmart (1981) report on an extensive study which had a good study design. Data presentation was, however, inadequate and the estimation of density (J. T. Emlen's [1971] method was used) from counts and distances would be improved by using recently developed, comprehensive analysis methods. It would be very informative to present the data as graphs of estimated densities over time with 95% confidence intervals indicated. The data presentation in terms of kurtosis and skewness was uninformative. It was also an incorrect analysis to examine for a normal distribution because the data were first combined over factors such as season or habitat type and for these combined data to follow a normal distribution there would have to be no variation in bird density by season or habitat type. However, the authors' analysis showed that bird densities did vary by season and habitat type. The caveats in their discussion section should be memorized by ornithologists.

The paper by Best (1981) lacks a conceptual basis for relating the data (counts of birds) to the parameter of interest, bird density. The detectability profiles are based only on observed counts, with apparently no attempt made to estimate true density. These seasonal profiles, therefore, reflect a confounding of three factors: bird density, the rate of cue production, and the detectability of the cues. I believe Best is saying that these seasonal profiles are only useful as a qualitative basis for improved study design, in which case the confounding of these factors is not of concern. I agree with this idea, but wonder if such intensive studies are really needed to document what ornithologists probably already know about the optimal timing of bird studies.

Sampling in rugged terrain raises some theoretical problems about what to record for a distance in both line transect or circular plot sampling. Ms. Dawson's paper (1981) appropriately raises this question. I believe the guiding principle should be that we are sampling area (to the bird) in these methods; perhaps, therefore, the distance recorded should follow the contour of the landscape. This matter needs more thought. I find the author's comments regarding correcting distances for slope confusing, and I recommend using the actual line length and distances in the estimation of bird density. Conversion of the estimated density, $\hat{D}$, to total numbers, $\hat{N}$, in the sampled area is the problem. $\hat{N} = \hat{D}A$ should be used, where A is the actual habitat area available rather than the projected map area, A^*, of the study area because $A^* < A$ will hold in rugged terrain. Thus, taking A* from a map, which ignores the relief features of the study area, and using $\hat{N} = \hat{D}A^*$ will give a negatively biased estimate of N.

A typical, small scale, very limited ornithological study was summarized by Grue et al. (1981). Transect counts were done over a four week period in one impacted study plot and one control plot; there was no replication over years or plots. Thus, only very limited conclusions can be validly drawn from this study. Distance data were recorded in seven intervals; this would allow for a considerably more sophisticated data analysis than was done. Again, as is typical in ornithology, estimates of bird abundance are presented without any estimates of precision (i.e., standard errors). The state-of-the-knowledge allows a much better data analysis than is presented. Finally, comparing the number of detections (counts) as birds per 20 ha to the projected densities (estimates of D) as birds per 40 ha is ridiculous.

The paper by Oelke (1981) does not have, and basically does not need, statistical analysis of data. He reminds us of the practical difficulties of access to land (both legal and safety) and of the "big foot" effect, where an intensive study can change the biological community being studied and thereby render the results useless.

Richards (1981) presented the results of a worthwhile investigation and a refreshingly different one, because it is not just counting birds and finding that these counts depend on every conceivable influence. If estimation of the distances to detected birds depends entirely on hearing them, then it is important to understand the nature of this auditory cue. However, if the detection distance can be accurately obtained in some other manner, then the nature of the detection cue, especially attenuation of bird songs, is irrelevant.

There have been too many studies like the two papers of Robbins (1981b) and Skirvin (1981). The resources expended on these studies could be better used on other goals. Admittedly, all sorts

[1] U.S. Fish and Wildlife Service, Western Energy and Land Use Team 2625 Redwing Road, Fort Collins, Colorado 80521.

of factors effect the counts of birds. That is why it is necessary to "correct" these counts to an estimate of absolute density. This can be done using detection distance data from either line transect or circular plot sampling. The time to consider the factors effecting rate of cue production and the probability of detecting a cue is while designing the study. Stated simply, field work should only be done during "acceptable" conditions. The study of Robbins (1981b) is oriented to defining acceptable conditions. However, I maintain that it is pointless to try and precisely quantify such conditions for every species, habitat and season. Only general guidelines are needed, or feasible, and it should be possible for experienced ornithologists to provide such guidelines, in most cases, without further studies.

Judging from his paper, Skirvin (1981) has done a good job of data analysis in many respects, and the reporting of results is informative with the notable exception that no standard errors are given for means or density estimates. Surely these were available; they should be included in the paper. The paper provides a good argument against using counts as indices; the observed counts declined over a four hour morning period. By contrast, the density estimates (counts "adjusted" using detection distances) did not significantly decline during the same time period.

Some general points concerning the use of bird counts to estimate bird abundance that I want to emphasize are:

(1) Using just the count of birds detected (per unit effort) as an index abundance is neither scientifically sound nor reliable. Many papers in this symposium illustrate this fact, in effect, whether the authors so intended or not.
(2) It is necessary to adjust the study counts by the detection probability. Fortunately, this adjustment only requires appropriate detection distance data. The mathematical basis for this computation is now well understood and good estimation methods exist.
(3) Line transect and circular plot (distance) methods should only be used under conditions when the rate of cue production is high and these cues are very detectable. Then there are data analysis methods that eliminate the need for concern about the multitude of factors effecting detection probability. In effect, it becomes unnecessary to worry about all the reasons why birds are not always detected when they are away from the transect center line or the plot center.
(4) From a statistical viewpoint, there is no difference between bird density estimation based on counts and distances from line transect sampling and those based on circular plot sampling. Therefore, the basis for choosing between these two sampling methods is their appropriateness and feasibility in the field.
(5) Trustworthy, predetermined correction factors for counts of each species by habitat, year and observer are impossible to achieve.
(6) Data analysis and reporting of results from ornithological studies needs to be more rigorous. In particular, the precision of results needs to be reported, usually as the standard error of parameter estimates.

In his opening address to this Symposium, Dr. Callaham asked us to determine and compare the state-of-the-practice and the state-of-the-knowledge. There is a large gap between these two in ornithological studies. Specifically, there is approximately a ten year gap between data analysis and field procedures for line transect and circular plot studies. The state-of-the-practice is circa 1970 even though tremendous progress has been made in analysis methods in very recent years. A substantial gap also exists in application of other methodologies, such as capture-recapture and band recovery analysis and in the general level of sophistication of statistical analysis (and sometimes, design) of studies. The knowledge exists; ornithologists need to use it.

Studies in Avian Biology No. 6:326, 1981.

INTRODUCTORY REMARKS: OBSERVER VARIABILITY

RALPH J. RAITT,[1] CHAIRMAN

The topic of this session—observer variability, the contribution to variability of census results attributable to variability within and among the persons conducting the census—is clearly an important issue. In other sessions of the symposium the vexing problem of observer variability was mentioned a number of times, and the papers in this session, if they do not collectively deal with all of the parameters and modes of observer variability, certainly allude to and actually document a sufficient degree and number of types of such variability to prove that it should not be taken lightly in our efforts to improve methods of estimating numbers of birds.

Without taking a careful census of the sources of observer variability mentioned in the various contributions, I can think of at least the following: age, innate endowment, and past and present accident or illness as they affect observer vision and hearing; amount of experience with the techniques being employed and with the avifauna of the area and time of year of the census; and levels of physical condition and attentiveness during the census. Variation in these several sources may result in variation in at least the following abilities: detection of birds, species identification, and estimation of locations of birds, including their distance from the observer or line of transect. This multiplicity of both sources and modes of observer effects points up the complexity of the problem. One of the contributions of this session, it seems to me, is in illuminating that complexity; if we did not before, we now know at least the character of the problem.

That a large part of what several of the contributors to the session were able to say about observer effects was based on intuition, logic, and a minimum of quantitative observations, is an indication of the neglect that has been given to study of the subject. However, Scott et al. (1981) have led the way toward more systematic, statistical studies in their paper on observer variability in distance estimation. Obviously, more studies of this type are needed. But who will conduct them? They appear to require data that are difficult and expensive to obtain, namely comparable observations by a sizeable number of observers. A federal agency sponsoring a large censusing program provided the data for the studies of Scott et al. (1981) and of Kepler and Scott (1981), and it seems likely that any similar studies in the future will also perforce originate with such large organizations, or at least with data that they have collected and/or paid for.

In his summarizing remarks at the end of this session, McDonald points out that observer effects are all part of what statisticians have termed "measurement error" and that it is generally assumed that this type of error should be small relative to "sampling error" (error due to inherent variability of the system being measured). As indicated above, it unfortunately does not appear that measurement error is small in very many bird counts. McDonald gives some suggestions of ways to reduce it somewhat, and the papers of Emlen and DeJong (1981) and Kepler and Scott (1981) deal with other ways. None of these ways takes the form of a panacea. At least most of McDonald's suggestions would appear to result in no greater than minor improvements. The proposal of Emlen and DeJong (1981) is for a method not yet fully developed; it was received with what seemed a considerable amount of reservation, judging from the oral discussion following the presentation. The type of training program described by Kepler and Scott (1981) is probably practical only for a small number of well-funded large-scale census programs. These comments are not meant to derogate any of those suggestions and plans for potential partial solutions to the problem; all of them are beneficial and worthy of pursuit. The comments are merely an attempt to attain a realistic perspective on the problem; that this turns out to be a somewhat pessimistic perspective is only a reflection of the complexity of the problem.

The present situation, then, as regards observer effects seems to be that the problem has been identified and described qualitatively, but only a few of its aspects have been measured satisfactorily and only limited solutions have been proposed. My impression is that prospects for greatly reducing observer variability in the near future are poor. The sources of the variability are too many and too difficult to control. For the present and the immediate future, the actual design and practice of counts obviously should feature as many as feasible of the suggested ways of reducing observer variability, while research should be continued and expanded to explore quantitative aspects of the variability, so that rational comparison may be made of results of different observers. In this context, McDonald's suggestion of the possible applicability of the theory on "interviewer bias" is an intriguing possibility that will surely be explored.

[1] College of Arts and Sciences, Department of Biology, New Mexico State University, Box 3AF, Las Cruces, New Mexico 88003.

Studies in Avian Biology No. 6:327–333, 1981.

LIMITATION AND VARIABILITY IN HEARING ABILITY IN CENSUSING BIRDS

ANDRÉ CYR[1]

ABSTRACT.—Too few studies have dealt with the human observer's effect on census results. Factors limiting hearing include the physico-acoustical properties of the ear itself. The frequency levels heard, although they cover a large part of the range emitted by birds, are not perceived identically over their entire range. Age decreases the perception of high frequencies. Time interval, resolution of sound, binaurality, sound shadow effect, fatigue, and masking might all impair our perception to a higher degree than is usually believed and thus affect identification and the census results.

Heretofore, studies of variability in census taking have focused primarily on comparisons between observers without reference to a known bird population or known perceptible fraction of it. An experiment designed to compare the efficiency of observers to a known check sample tape recording shows that even audible sounds are easily overlooked, due either to lack of familiarity with a particular song by some observers or to the masking effect of simultaneous songs, or other factors. Some research topics are proposed to improve quality of hearing and efficiency in interpreting of bird songs to gain new insight into the observer's effect on census results.

Although hearing plays an important role in the life of birds (Hinde 1969, Thielcke 1976), it is also important to those who count them. Observers censusing terrestrial birds often spend 75% or more of their census time listening in order to localize or identify birds. The question here is: to what extent is it possible to use hearing ability and still be confident in our census results?

A census taker in the field faces many stimuli, emitted more or less simultaneously, and attempts to differentiate all these stimuli. We try to intercept messages sent primarily to other birds of the same or different species or to other animals, in addition to the information sent to us as potential predators. We try to intercept the information and correctly decode it for censusing purposes. Are our tools adequate to analyze and decode this information properly? To what extent do we succeed in doing so? What can we do to improve our success?

In the following, I will focus on the factors affecting the hearing variability of the observers. After reviewing the sparse literature involving comparisons between results obtained by different observers, I will suggest experiments that should be done to enhance our hearing ability and our knowledge of its drawbacks on potential results, and will report the results of a small number of such experiments.

LIMITATIONS OF HUMAN HEARING

Problems and limitations include the physical nature of the ear, the threshold of audibility, frequency discrimination, the sound shadow effect, fatigue, the masking effect, and environmental noises. Human ears (Burns 1973, Howard 1973) and those of birds (Pumphrey 1961, Schwartzkopff 1973) are anatomically different, but functionally about as efficient. Birds usually produce sound between 500 and 5000 Hz. Human ability to detect pure tones ranges from 16 to 20,000 Hz. We would conclude that we can perceive sounds produced by birds over almost all their frequency range, except for a few extreme species such as the Oilbird (*Steatornis caripensis*) and some other partly echo-locating species. But our ear remains more efficient between 2 and 6 kHz. Audiological measurements usually refer to pure tones, but we very seldom census these in the field. The limitations of our ear could be much greater than is usually believed, either from audiological measurements or census results. Thus, the first improvement depends on the inherent properties of the physical ear, the quality of which can be improved by practice: the more the tool is used, the more efficient it will be.

Threshold of audibility usually varies from person to person and even from day to day and hour to hour (Beranek 1954). After exposure to even a moderate noise level, slight temporary deafness occurs, which shifts the detection threshold upward, but age is the main factor affecting the threshold of audibility. As seen from Burns (1973:102), higher frequencies are lost faster with age than lower ones. Although sound localization can still be achieved by a single ear with fair accuracy, using intensity cues (Howard 1973), threshold of audibility could have a serious bearing on hearing efficiency for census taking. According to the goal and methodology of censusing, these problems might limit the participation in a particular program.

As compared to the human ear, that of a bird is capable of better resolution of sounds emitted at short time intervals, and birds can react to them accordingly (Pumphrey 1961, Knudson 1978). This is best exemplified by the duetting

[1] Département de Biologie, Université de Sherbrooke, Sherbrooke, Québec, Canada J1K 2R1.

in such birds as the Little Grebe (*Podiceps ruficollis*) (Thielcke and Blume 1973, Thielcke 1976). Hirsh (1959) reports that two brief sounds will be perceived as separate with only a few msec between them, but it will take intervals of up to 15–20 msec for the listener to report which of the two preceded the other. He says further that this result is independent of the nature of the sound, whether short or long or of high or low frequency. Henning (1966) found further that differences of 300 Hz are necessary for two sounds of high frequencies (10 kHz) to be discriminated correctly at a level of 75%. This time interval component is not of the utmost importance, for we do not need to react to single notes within a song, but it surely could reduce our faculty of song discrimination and might explain our reduced ability to discriminate between structurally comparable songs such as trills. Our ability to interpret census results correctly would be improved by new knowledge on the birds themselves: to what extent does a bird vary its singing within its range or between morphologically and structurally different habitats or within different bird communities?

Another problem lies in the sound shadow effect (Howard 1973). A sound reaching one ear laterally reaches the other slightly later, producing binaural cues that can be used to estimate the distance from which a sound is emitted. Casseday and Neff (1973) found that man uses different cues to localize pure tones of high and low frequency. Around 3–4 kHz, localization is more efficient than at lower or higher frequencies. At higher frequencies, intensity is used as the cue, whereas a time cue is used at lower frequencies, because of the relatively longer time lag of such a sound travelling from one ear to the other. Methodologies taking distances into account should perhaps avoid including species with high frequency utterances.

Fatigue may impair our hearing ability and is one source of systematic variability in that ability. Thus when designing our field experiments, the complexity of their application should be considered in view of this limiting factor (see Ramsey and Scott 1981a).

Masking is defined as the amount by which the threshold of detection of a sound is raised by the presence of another sound, the masker (Studebacker 1973). Fortunately, for census purposes, exposure to low frequency does not affect the threshold of detectability of high frequency sounds (Ward 1966). The contrary holds as well. Although not specifically studied in relation to census taking, some other factors that play a role in bird communication can surely affect our efficiency at locating and identifying some species. For example, Witkin (1977) showed that the directionality of the source as related to the receiver influences the receiver's ease of locating the source of bird communication. In censusing, the receiver is the observer but the problem remains, although little attention has been paid to that point (see also Wiley and Richards 1978).

The factors described above may variously limit our hearing ability, and systematic investigations are still needed with the census taker as the main study object. Studies on individuals as potential census takers should include objective examinations of the: (1) efficiency of bird identification at different levels of frequency and intensity, (2) pattern and speed of learning of bird songs in the ontogeny of a bird watcher, (3) number of song bouts needed for species identification, (4) parts of songs used as cues for identification, (5) effect of overlapping or masking on identification, (6) importance of the "out of range of birding" effect that occurs when an observer shifts from one locality to another, (7) effect of repetition on improvement of results, and (8) how these items vary in the application of different census methods. It will be important to formulate the problems carefully in order to compare the results with known check samples or parameters.

AN EXPERIMENT

Inasmuch as the effect of the human factor on census results is usually not correctly assessed, because of check sample bias, a test was designed to compare results obtained by different observers to a known sample. Observers (33) of varying quality, some of them currently involved in the Breeding Bird Survey, took part in the experiment. The aim was to examine the ability of these observers to discriminate sequences of species, species singing simultaneously and species from outside of the usual birding area of the observers. A total of 33 utterances from 12 species, arranged on a tape and delivered at intervals slightly longer than those heard at dawn hours, was played (Fig. 1). The observers had to identify the species and report them in sequence on a special checklist containing 36 species. The experiment was run twice.

The results of only 18 observers were kept for the analysis, because some observers did not complete the whole test during one or both runs or some have proved to be far from competent, identifying less than 20% of the birds. Table 1 shows how unrealistic were some estimates of bird numbers and comparisons between observers. Eighteen observers identified up to 27 species when only 12 were on tape, leading to discrepancies of up to 225% for number of species and 265% for number of individuals. The

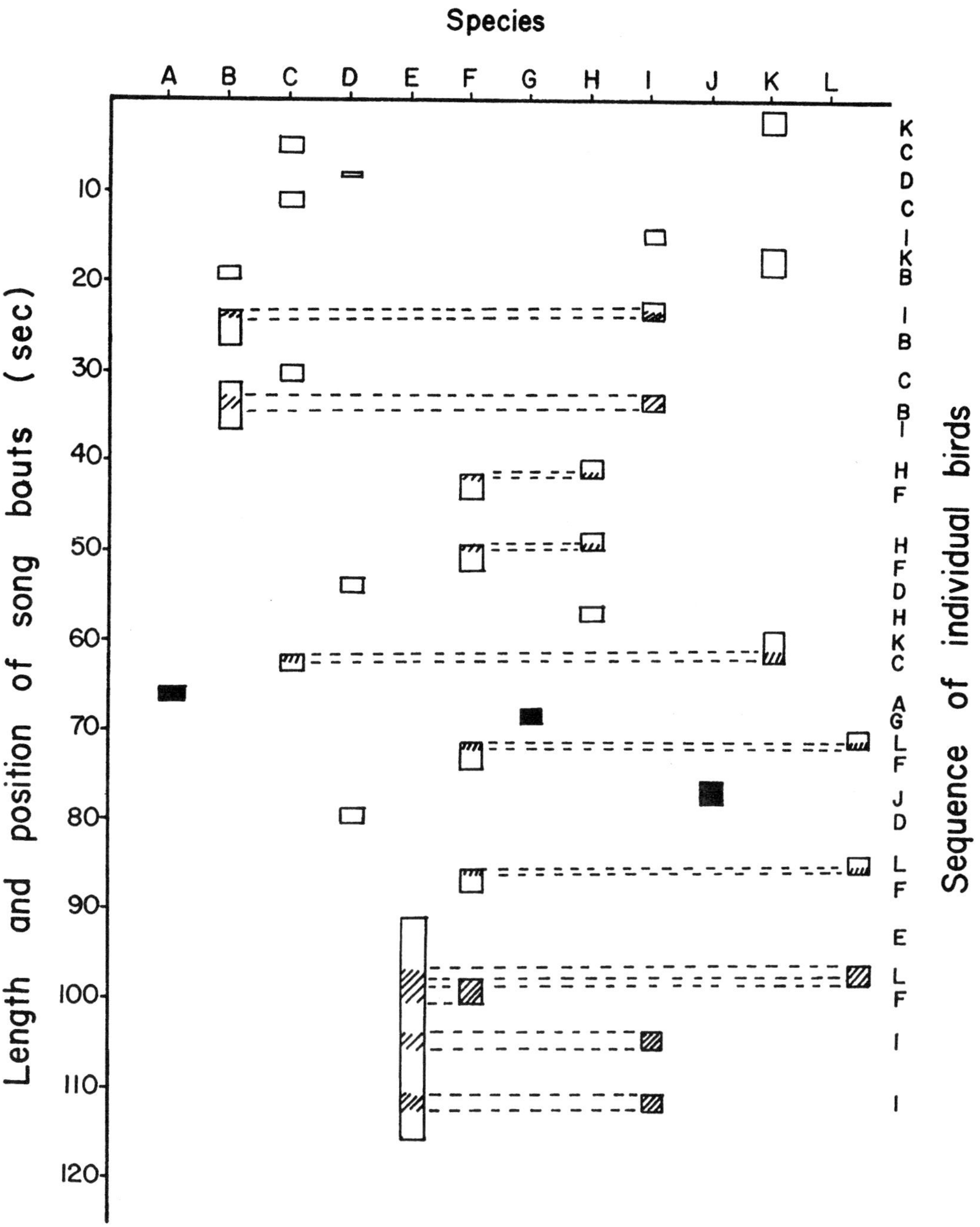

FIGURE 1. Arrangement of song bouts of 12 species on the tape recording used for the tests. Letters refer to the following species: A—*Nuttallornis borealis*; B—*Turdus migratorius*; C—*Catharus ustulatus*; D—*Catharus fuscescens*; E—*Vireo olivaceus*; F—*Vermivora peregrina*; G—*Dendroica virens*; H—*Dendroica pensylvanica*; I—*Wilsonia canadensis*; J—*Pinicola enucleator*; K—*Zonotrichia albicollis*; L—*Passerella iliaca*. Overlapping of songs is represented by broken lines and cross-hatching; black boxes are species singing only once on the recording.

TABLE 1
RESULTS OF ESTIMATES OF THE TAPE RECORDING CONTENT WHEN USING THE MAXIMUM, OR THE MEAN NUMBER OF BIRDS, OR THE RESULTS OBTAINED BY THE SUPPOSED BEST OBSERVER

	Number of birds						
Species	Max		Mean		Best		Tape
Nuttallornis borealis	2		0, 9		1		1
Turdus migratorius	5		1, 8		3		3
Catharus guttatus	4		1, 3		9		0
Catharus ustulatus	5		2, 8		4		4
Vermivora peregrina	4		0, 8		2		5
Dendroica petechia	3		0, 2		0		0
Dendroica magnolia	3		0, 2		3		0
Dendroica virens	1		0, 7		1		1
Dendroica pensylvanica	3		1, 5		3		3
Geothlypis trichas	2		0, 1		2		0
Wilsonia canadensis	6		2, 7		3		5
Pheucticus ludovicianus	3		0, 7		0		0
Pinicola enucleator	0		0		0		1
Pooecetes gramineus	3		0, 7		3		0
Zonotrichia albicollis	5		3, 2		3		3
Passerella iliaca	3		1, 2		0		3
Totals including errors							
Number of species[a]	27		27		14		12
Ratio over tape (%)		225		225		117	
Number of individuals (total)[a]	77		24, 3		27		29
Ratio over tape (%)		265		84		93	
Totals excluding errors							
Number of species[a]	11		11		10		12
Ratio over tape (%)		—		—		83	
Number of individuals (real)[a]	41		10, 1		24		29
Ratio over tape (%)		141		66		83	

[a] Species listing is incomplete, hence the discrepancy with the totals.

best overall estimate of the tape content was achieved with the results of the supposed best observer rather than the maximum or mean number of birds. But a 34% difference still occurs between the estimates by the best observer of species numbers with errors included and with errors excluded. This means that errors present partially cancel each other in estimates from census results. It is apparent that an unknown species may easily be unnoticed; for example, the summer song of the Pine Grosbeak (*Pinicola enucleator*) was unfamiliar to most of the observers, and it was unnoticed except by one observer in the first run. Some species, such as the Canada Warbler (*Wilsonia canadensis*) seem poorly known, being confused with 11 other species of birds. Confusion in counting the birds is also shown in the table.

Figure 2 shows that the number of individual birds correctly identified even on the second run tended to be directly correlated to the admitted use of hearing for bird identification by the observer (Spearman's $r = 0.52$, $P < 0.05$) and perhaps to the rating of the ability of the observer for the same purpose ($r = 0.40$, not significant). This means that using hearing more frequently increases the efficiency at identifying, as does practice. Repetition did improve the hearing and efficiency (Fig. 3). Although there seems to have been an improvement, the difference in the results between the runs was not significant ($\chi^2_{(11)} = 11.01$).

The masking phenomenon occurred in six overlapping singing situations. For example, the three Chestnut-sided Warblers (*Dendroica pensylvanica*) were correctly identified 3, 3, and 8 times in the first run and 5, 9, and 11 times out of 18 in the second. The first two songs of this warbler partly overlapped with one of a Tennessee Warbler (*Vermivora peregrina*), the last one did not overlap at all. The conclusion is self evident. Some of these conclusions do not pertain to hearing ability, but hearing ability is probably also correlated with species knowledge and training, due to the selectivity of response of the observers to the environmental stimuli (Lewis and Gower 1980). In fact, some observers mentioned not having heard the song of some of the species played on the tape!

In another study, we looked at the effect of different census methods on hearing ability. In this study, one observer censused the birds

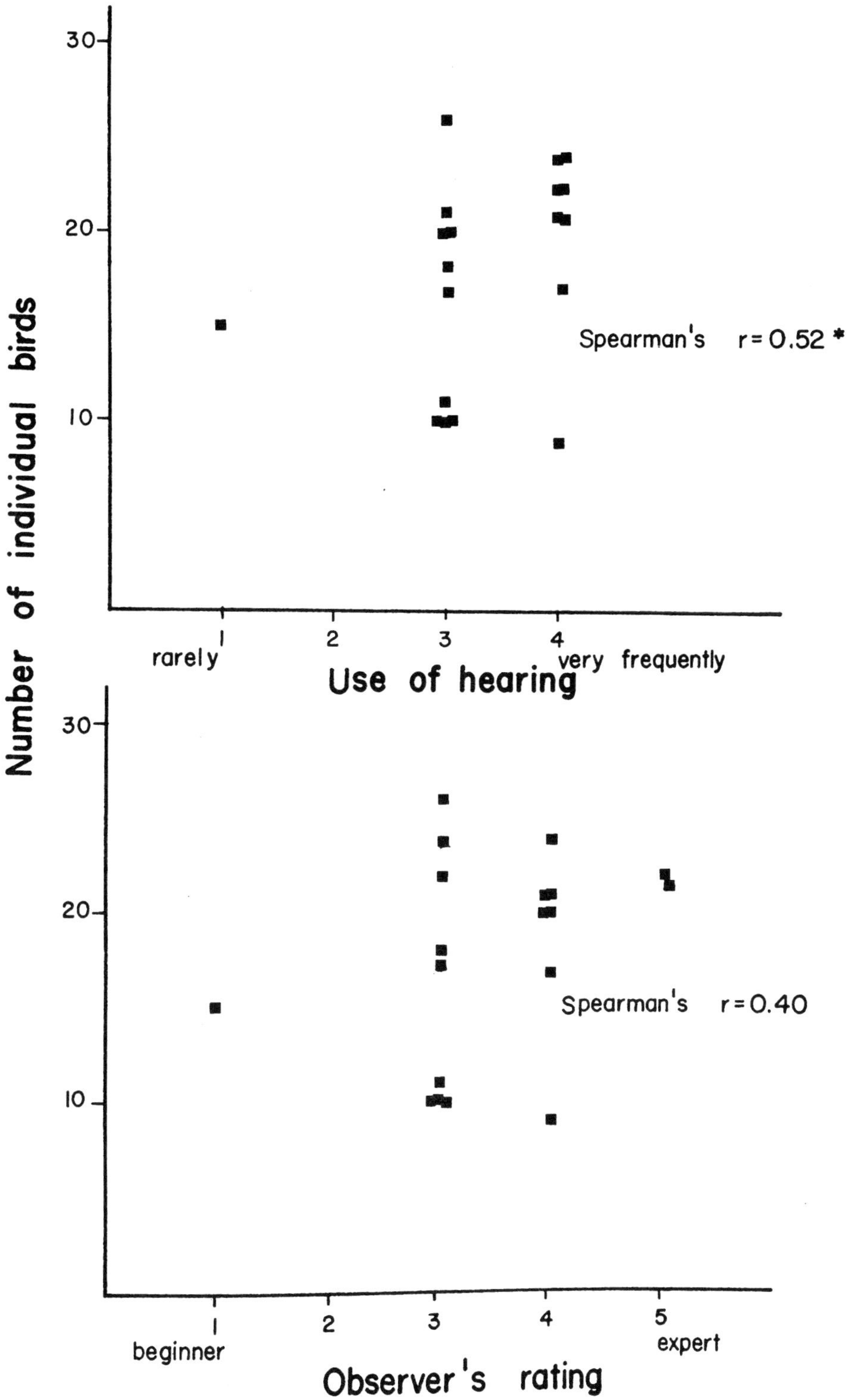

FIGURE 2. Relationship between the frequency of use of hearing by the observer or the observer's rating for field identification and the number of correctly identified individuals of birds over a possible maximum of 33 played on a tape.

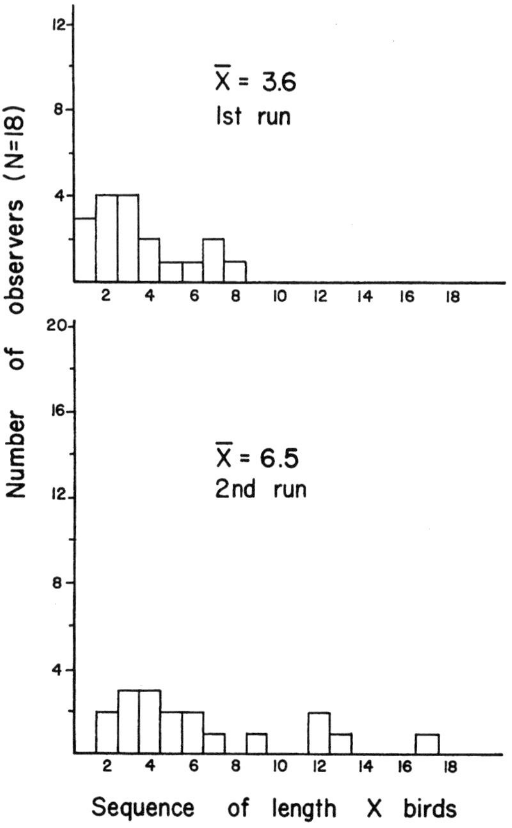

FIGURE 3. Effect of repetition on results of perception and identification of bird species. A sequence of, for example four, means that four individual adjacent birds were correctly identified along the played recording of 33 birds. The graphs show for each run of the test the number of observers that could identify correctly a maximum of χ birds in a sequence.

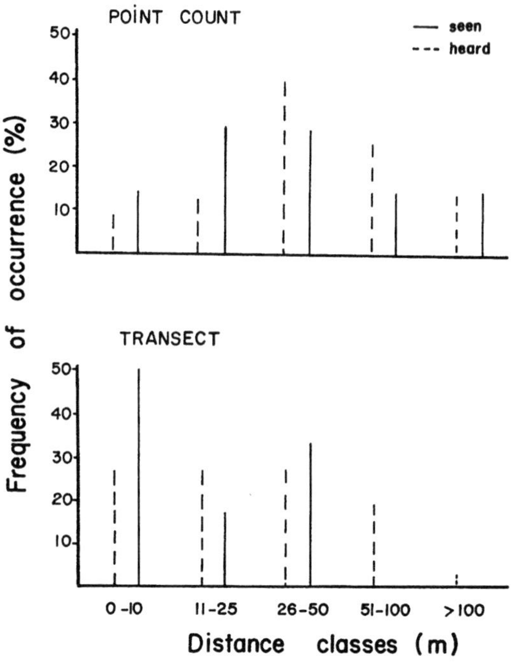

FIGURE 4. Importance of hearing in censusing and effect of a census method on hearing. The vertical bars represent the frequency of occurrence of birds (all species combined) heard or seen at different distances from the observer.

along a 3 km wooded path. The IPA method (point count method of Blondel et al. 1970) and the transect method were both used each week, but not simultaneously. The results of 16 weekly censuses from October to February were combined. Figure 4 shows that when the observer was walking he could not hear birds as far away as in point counts. On the other hand, standing for 10 minutes probably affected the activity of the birds near the observer. This figure also shows the importance of hearing in general, especially for detecting birds farther away from the observer. The relation between listening and looking would be much different in a breeding census situation. Further questions should be formulated to overcome hearing problems associated with different census methods.

DISCUSSION

Most researchers agree that observers do affect census results, whatever the census method used (Palmgren 1930, Enemar 1959). Examples of the qualities that could affect the census results are acuity of hearing, attentiveness, sensitivity in detecting individual birds, behavior of the observer on the terrain, emotional state, and others (Enemar 1962, Snow 1965, J. T. Emlen 1971, Best 1975, Berthold 1976, Enemar et al. 1978). For results obtained by the mapping technique, the error is often believed to be around 10% (see references in Berthold 1976), but preliminary evidence suggests that the error levels are probably higher than is usually and conveniently believed. Unfortunately, among the few studies involving comparisons of results obtained by different observers (Taylor 1965, Snow 1965, Enemar and Sjöstrand 1967, Enemar et al. 1978), only a few studies have made comparisons between results obtained simultaneously, or almost simultaneously (Carney and Petrides 1957; Enemar 1962, 1964; Hogstag 1967; Oelke et al. 1970; Jensen 1972). Unfortunately all of these studies except the one of Jensen compare the mean or maximum number of birds, or the results obtained by the supposed best observer for that plot. Studies are needed to evaluate the variability, the range of efficiency and the real impact of hearing ability on census results.

Audiology measurements do show differences from one individual to another, but most studies with humans are performed in relation to deafness (Martin 1975). To analyze this question more deeply, one would have to analyze observers from a psychological viewpoint to find out the reasons for the efficiency and performance of the different observers in species identification and in censusing. How, for example, can we explain the varying degrees of attentiveness of observers? How can we explain the different efficiencies in taking correct field notes? The lack of field or laboratory experiments on observers prevents further discussion.

We sought to answer the question: Are our perceptual tools able to analyze and decode properly the information sent by birds? We conclude that census takers need to improve the quality of their tools. It remains partly unknown how efficient we are at interpreting the emitted messages, in order to minimize the extent of the errors in census results. The extent to which we succeed at doing so needs to be reevaluated. Comparisons should be made with known check samples or parameters. Example of experiments would be to compare the results of observations with and without those obtained with a multi-microphone (multi-directionality) and a highly efficient recording device. It would be worthwhile to design tests in order to simulate the three dimensions for space locatability of bird songs played from a multiband recording device. Binaural hearing cues could be tested this way too. Other tests could simply play back especially arranged bird song sequences and ask the observer to identify species, or cues. Alterations of the songs could help to answer some of the above questions.

An increase in the number of cues used to perceive and identify the birds is definitely needed. One possibility is a wider use of sonograms, which have been overlooked in spite of their potential in the learning process, at least for those people who learn more visually than acoustically (Keith 1967, Beaver 1976). As in many other situations, progress depends on how one approaches the problem. Should we not also consider this strange creature, the census taker, as an object of scientific investigation!

ACKNOWLEDGMENTS

I wish to thank Ralph J. Raitt, J. Michael Scott, Patrick Bourgeron, C. John Ralph, and Jean-Marie Bergeron for generously criticizing different drafts of my manuscript. Thanks also to Nicole Vanasse who diligently retyped all the texts and Louis-Marie Thériault who made the drawings.

Studies in Avian Biology No. 6:334–340, 1981.

DISTANCE ESTIMATION AS A VARIABLE IN ESTIMATING BIRD NUMBERS FROM VOCALIZATIONS

J. MICHAEL SCOTT,[1] FRED L. RAMSEY,[2] AND CAMERON B. KEPLER[3]

ABSTRACT.—The accurate measurement of distances is basic to any accurate determination of bird densities. We used field studies to determine the accuracy of distance estimates to birds heard and not seen.

Under good field conditions observers were able to estimate the distance to birds heard and not seen to within ±10% (range of averages −9.1 to +6.3%). The range of all distance estimates was one-fourth to four times the measured value with 95% falling between 4/7 and 7/4. There were significant differences between observers ($P < .025$) and species ($P < .001$).

Simulation studies were used to determine the effect of measurement errors on the accuracy of density estimates.

Suggestions for reducing the bias in density estimates resulting from measurement errors include: (1) training observers; (2) flagging known distances; (3) using range finders; (4) explaining to observers the importance of their work; (5) minimizing the responsibilities of observers; and (6) using robust methods to analyze data.

The number of birds per unit area is being estimated with increasing frequency. The methods used include: 1) counting birds within a prescribed area and 2) recording all birds heard or seen in an undefined area. The first method requires that either the area of interest be marked (Anderson and Shugart 1974), with the presence or absence of a bird within that area recorded, or that an observer make repeated judgments as to whether an animal is within an area whose limits are estimated or measured (Fowler and McGinnes 1973). Distances employed have ranged from 10 to 400 m (Anderson and Shugart 1974, Robbins and Bystrak 1974) and assumptions of coverage range from all birds in the area to some unknown percentage. The second method requires that the initial detection distance to each animal be measured or estimated. Distances can be the flushing distance (Hayne 1949), right angle distance (J. T. Emlen 1971), Gates sighting angle and distance (Robinette et al. 1974) or the horizontal distance (Reynolds et al. 1980).

The accurate measurement of distances is essential to any accurate estimate of bird density. In fact, tape measure accuracy is the suggested standard (Burnham et al. 1980). Measurement of distances is frequently used in studies of nonmoving objects: plants, nests etc. In most studies of bird populations, however, observers use a range finder or simply estimate distances (J. T. Emlen 1971).

The accuracy of range finders varies with the model and the frequency of calibration, but they are indicated as being good to approximately 1% within 30 m and ±5% at distances between 100 and 300 m. Observers vary in their ability to estimate distances to objects that can be seen, but are felt to quickly become accurate within ±10–15% when estimating distances to birds that can be seen (Emlen 1977a).

Forest birds are more frequently heard than seen. In Hawaii, for example, the vast majority of all birds detected during surveys are never seen. In a random sample of 100 station counts made by 6 observers on Maui in 1980, we found that 894 of 1100 (81%) bird detections were made on the basis of sound alone. More surprising, on 37 of these 100 eight-minute count periods *all* detections were made on the basis of call notes or song: no birds were seen, even after the initial audio detections (Scott and Kepler, unpubl. data). Judgments as to the location of these birds could be made, and the distances then measured using tapes or a range finder. In practice, however, where more than two to three birds are detected during a count period (we recorded an average of 11 birds per count period on Maui), physical measurement of the distances, however desirable, becomes impractical. In addition to physical and time constraints, the concentration required to measure each detection distance makes it impossible to scan for additional birds. This results in a serious failure to detect other individuals present, and an underestimate of numbers. Thus detection distances must be estimated or severe restrictions placed on the number of birds to be recorded (Scott and Ramsey 1981b).

We have found that observers, after gaining confidence in their ability to estimate distances, make their estimates of distance without the aid of a range finder, or only use the range finder to measure the distance to one or two prominent objects. They then rely on known distances and

[1] U.S. Fish & Wildlife Service, Patuxent Wildlife Research Center, Mauna Loa Field Station, P.O. Box 44, Hawaii National Park, Hawaii 96718.

[2] Department of Statistics, Oregon State University, Corvallis, Oregon 97331.

[3] U.S. Fish & Wildlife Service, Patuxent Wildlife Research Center, Maui Field Station, 248 Kaweo Place, Kula, Hawaii 96790.

TABLE 1
OBSERVER DIFFERENCES IN MEASURED AND ESTIMATED DIFFERENCES

Observer	N	% Deviation
A	117	+6.3
B	32	+4.4
C	31	+3.2
D	74	+2.5
E	110	−1.8
F	103	−3.0
G	121	−3.3
H	124	−4.0
I	58	−4.1
J	143	−4.1
K	114	−4.7
L	48	−7.0
M	139	−7.6
N	111	−9.1

TABLE 2
PERCENT DEVIATION IN ESTIMATED FROM MEASURED DISTANCES FOR 15 SPECIES OF BIRDS

Species	N	% Deviation
Loxops coccineus coccineus	19	+9.0
Carpodacus mexicanus	50	+8.5
Vestiaria coccinea	172	+3.9
Phasianus sp.	3	+3.9
Meleagris gallopavo	4	+2.1
Himatione sanguinea sanguinea	102	−0.7
Loxops virens virens	206	−2.0
Cardinalis cardinalis	142	−2.6
Phaeornis obscurus obscurus	231	−4.5
Loxops maculatus mana	45	−5.7
Corvus tropicus	4	−6.6
Chasiempis sandwichensis sandwichensis	147	−8.7
Hemignathus wilsoni	46	−9.8
Leiothrix lutea	91	−10.1
Zosterops japonicus	63	−10.1

their own ability to estimate. With the importance placed on accurate distance measurements in the methods used to estimate bird densities (Burnham et al. 1980) and the fairly loose means by which these distances are actually estimated, we felt it was important to understand how accurately individuals can determine distances and how this accuracy might vary with distance, species, and between observers. We use this information to estimate the errors introduced in the calculation of bird densities and hence population sizes using estimated distances.

There are several possible sources of error. Distances can be mismeasured, under—or over-estimated, or rounded off to convenient figures (Gates et al. 1968, Anderson and Pospahala 1970, Robinette et al. 1974). In practice all three errors are made. The extent to which they are made and their effects on accurate measures of bird abundance are the topics of this paper. In addition to observer errors, the effects of screening by vegetation, wind, rain, temperature, and background noises vary from site to site and day to day, and there are also problems created by echoes, ventriloquism, or even the direction faced by the cue-emitting bird (Witkin 1977).

METHODS

In order to determine how accurately observers estimate distances, we asked small groups of four to six experienced observers to estimate the distance to birds heard but not seen. All participants had been carefully trained in distance estimation (see Kepler and Scott 1981 and Scott et al. 1981b). All observers were at a single station. Observers were preconditioned by estimating distances to reference points prior to the estimates to birds heard and not seen. One of the observers located a singing or calling bird, then identified it to the others. All independently estimated the distance to that bird, which was then located. The actual distance was measured using a range finder or tape measure. Estimations by observers who had seen the subject bird were eliminated.

Data for the accuracy of these distance measurements consist of 1325 ($=n$) distance pairs (x, y) where x = estimated detection distance and y = measured detection distance. Included were observations on 15 species made by 14 observers.

To measure the discrepancy between estimated (x) and measured (y) distances, we considered $Z = \log (x/y)$ which best met our criteria of a) variance is unrelated to average, b) observer and species effects are additive, and c) residuals from fitted model have a symmetric unimodal distribution. The model we used is a standard two-way ANOVA Model with interactions.

$$Z_{ijk} = \mu + y_i + \tau_j + \theta_{ij} + e_{ijk} \qquad (1)$$

where μ = reference level of Z
y_I = (fixed) effect of the ith observer; $I = 1, \ldots, m$ (=14)
τ_j = (fixed) effect of jth species; $j = 1, \ldots, s$ (=15)
θ_{ij} = fixed interaction terms

and

e_{ijk} = (random) departure of Z_{ijk} from its average.

The assumptions of the model are that the e_{ijk} are independently distributed with a mean of zero and a constant variance.

RESULTS

FIELD STUDIES

Estimated distances ranged from ¼ the measured distance to 4 times the measured distance. These represent extremes of the rather long-

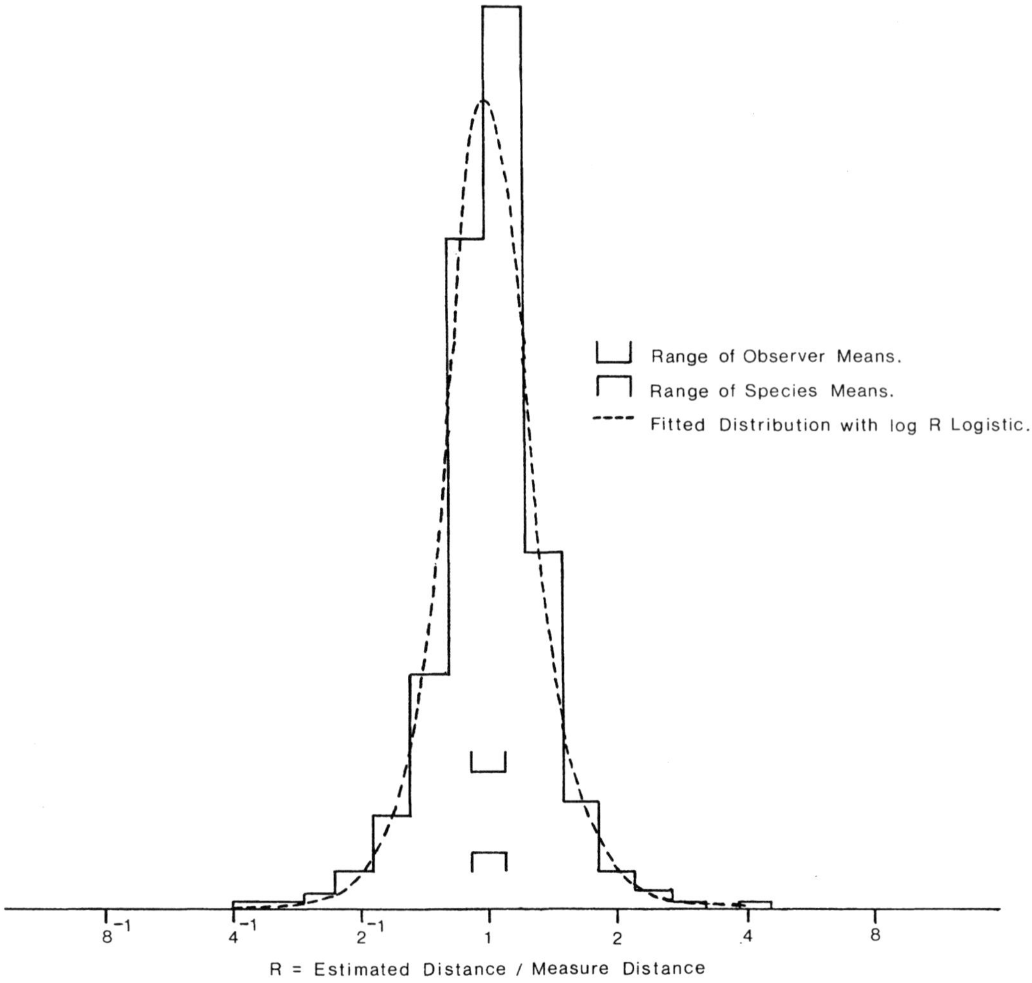

FIGURE 1. Distribution of ratios of estimated to measured distances.

tailed distribution of ratios of estimated to measured distances; 95% percent of these ratios were between 4/7 and 7/4.

There were statistically significant differences in the abilities of observers to estimate differences ($P < .025$), and the accuracy of distance estimates varied significantly with species ($P < .001$). The discrepancies are minor, however, in comparison with the overall variability of the results. Observer mean effects ranged from 9.1% below measured to 6.3% above measured, while species effects ranged from 10.1% below to 9.0% above measured distances (see Tables 1 and 2, respectively). Interactions were quite insignificant ($P > .40$). The reference level was not significantly different from zero ($P > .35$) indicating a lack of overall bias in the errors.

Figure 1 displays a histogram of the residuals from a least-squares fit to equation (1). The distribution appears reasonably symmetric and unimodal, but it possesses much more in the tail regions than should be expected of a normal curve. Thus the superimposed curve represents a logistic distribution with location parameter zero and scale parameter $\sigma = .152$ (estimated by the maximum likelihood method; see Johnson and Kotz, 1970).

Figure 2 shows a scatter plot of the same residual versus the measured distances. As the slightly negative slope to the regression line indicates, there is some small tendency for observers to overestimate short distances and underestimate larger distances.

SIMULATIONS

We conducted computer simulation studies to determine how distance estimation errors of the kind encountered in the above experiment might

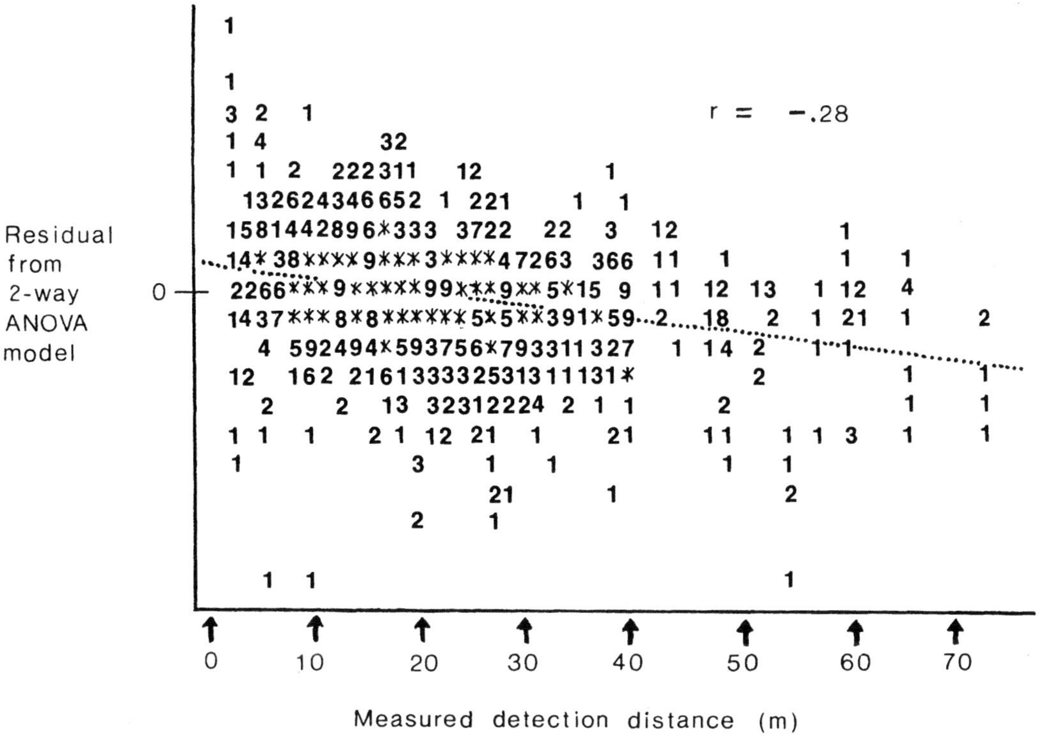

FIGURE 2. Distribution of the residuals from a least squares fit to equation (1) versus the measured distances. An asterisk indicates a cell count exceeding 9.

change estimates of density. In a typical run, "birds" are spread across a large circular region, according to a spatial Poisson process with known, constant density. The birds sing at random points in time, the "observer"—who occupies the circle's center—having a chance of detecting each call. The chance depends upon the distance separating bird and observer. (For a full explanation of the simulation model, see Ramsey, Scott and Clark, 1979). For each bird detected during a fixed observation period, we recorded its true distance, Y, from the station and then generated a random deviate, Z, from the logistic distribution with zero mean and scale σ. Then we took the estimated distance to be $X = Y \cdot \exp(Z)$.

Typically, a run produced 200–250 detections. We estimated density using both the true and estimated distances, grouped into distance classes. The method of estimating density (see Ramsey and Scott, 1979) is a variant of Emlen's (1971) technique for line transect data. Different runs were produced for different values of the scale factor, σ. Thus with σ small, there is very little estimation error in distances, while larger σ-values indicate larger magnitude errors. The model does not incorporate bias in the errors, nor does it feature a dependence of the relative error on the true detection distance.

Figure 3 contains the results of our simulations. Each run is represented by an arrow *from* the density estimate using true distances *to* the density estimate using estimated distances. These estimates are displayed relative to the actual density.

As should be anticipated, the density estimates become more severely corrupted as the magnitude of distance estimation errors increases. In most cases, because of a bias toward underestimation the result is to inflate the density estimate from its value using true distances (only 3 of the 32 arrows go down). The reason for this is traceable to the type of survey performed and to the density estimation procedure. The procedure, like most others, is quite sensitive to density of detections in strips "close to" the observer. With a variable circular plot design, the expected *numbers* of detections in concentric strips of equal radial increment increase dramatically with distance from the observer. Thus the very modest fraction of a large number of detections at intermediate distances which are

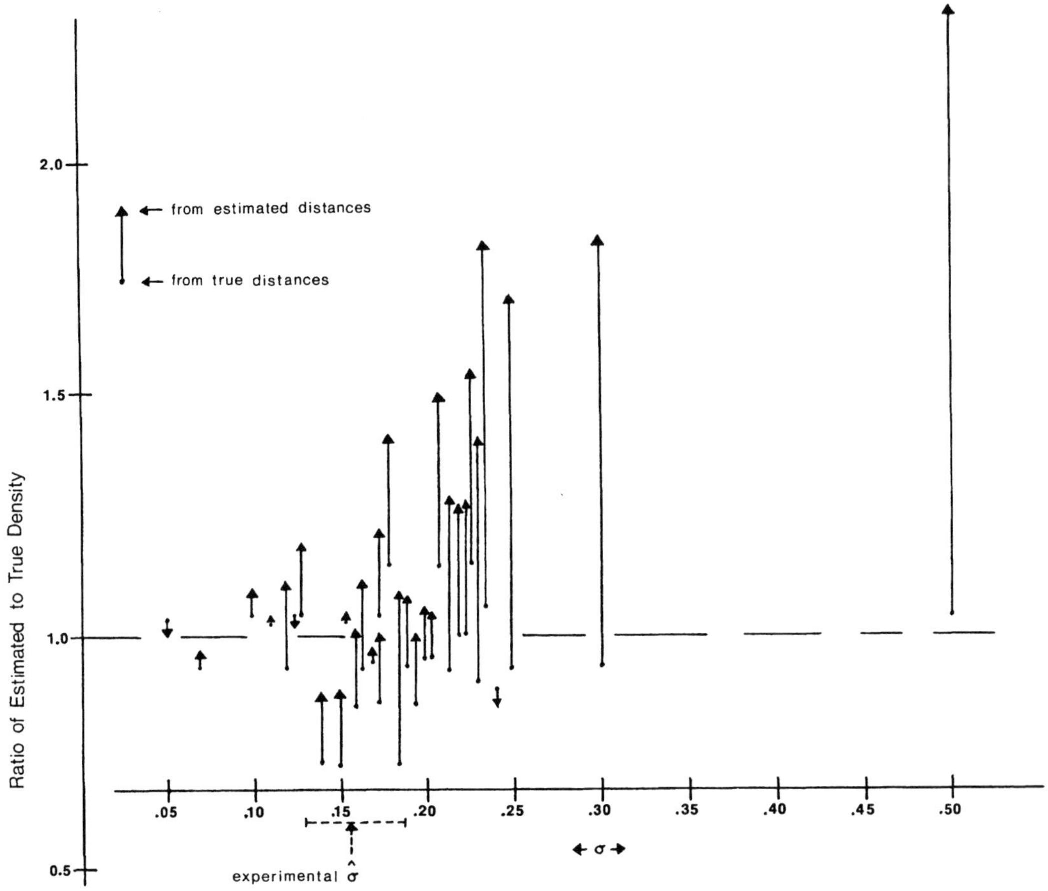

FIGURE 3. The effect on density estimation of errors in estimating distance. Data is from simulations with log-logistic error.

underestimated by the observer may comprise a substantial fraction of the total detections which the observer records as being close in. This effect is illustrated in Figure 4, where the $\sigma = .225$ run with 232 detections is displayed. The density versus radial distance class is plotted for the true and estimated distances. Nearly 50% of all detections were made of birds from 80–130 meters from the observer; less than 10% were from 0–50 meters. The few of the more distant birds whose detection distances have been underestimated substantially inflated the density in the near-observer region. Whether this is a severe practical problem is not clear nor is the extent that these biasis may be compensated for by birds that are completely missed (Mayfield 1981).

Returning to Figure 3, note that the experimentally determined scale factor of $\hat{\sigma} = .152$ is indicated along with a 95% confidence interval. Runs with σ in that interval typically had density estimates inflated by 20% as a result of the errors in distance estimation. However, in only one of these 14 runs did this create a density estimate more than 30% from the true density.

DISCUSSION

The ability of observers to estimate distances under the conditions of our experimental field trials was quite good. The range of observed accuracies is well within the 10 to 15% suggested by Emlen (1971; 1977). However, it is unknown to what degree observers may increase the error of their distance estimates under nonexperimental conditions. Variables that decrease this accuracy include the following: large number of birds and subsequent short time (0–20 seconds) to make distance estimates; often only one call or song (vs. several in experiment); and finally, birds heard when the observer is looking in a

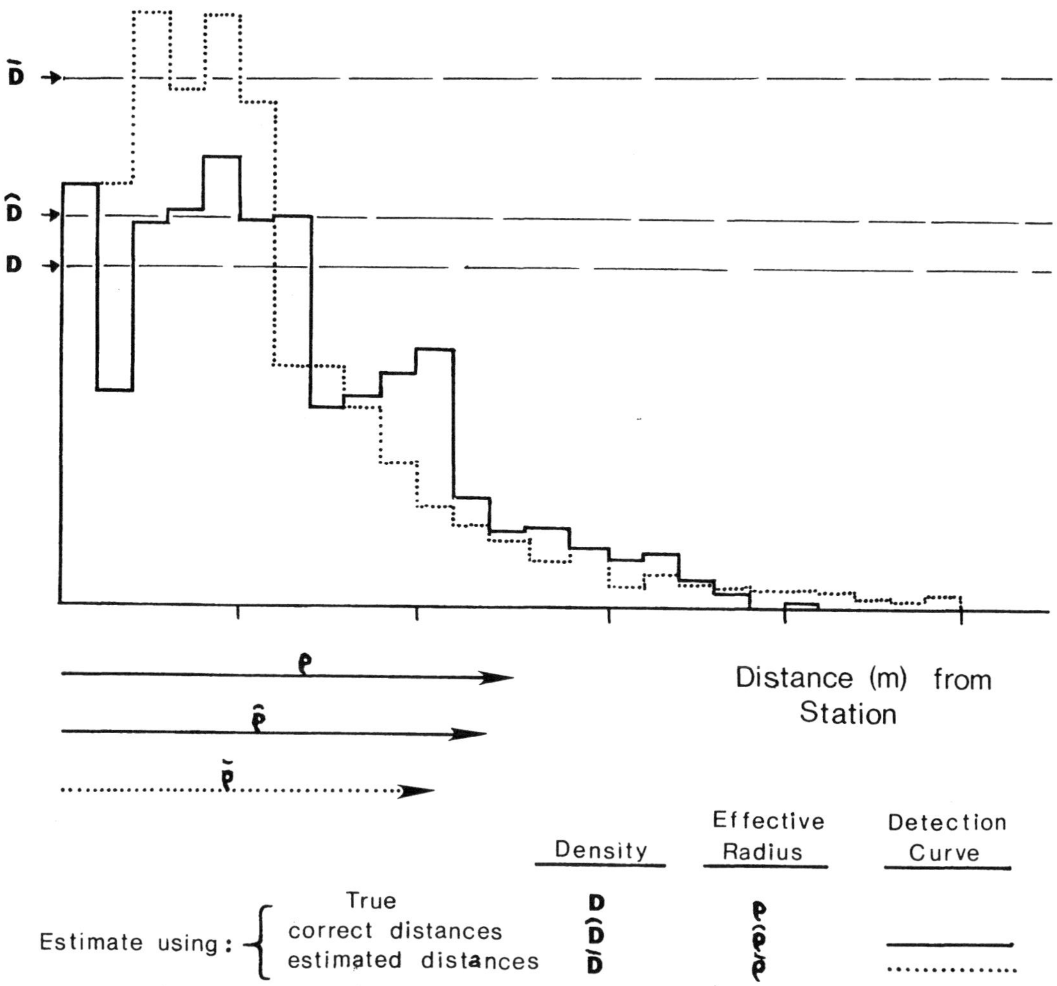

FIGURE 4. The effects of errors in distance estimation. Simulation with σ = .225 (n = 232 detections).

different direction. The degree to which this happens must be in part a function of the motivation of the observer to obtain reliable information. Use of distance references in the field and highly trained observers should help to reduce this source of error. The conditions under which we tested the reliability of distance estimates were very competitive.

We made no attempt to determine differences in ability to estimate distances to calls or songs of different types. Because of differences in the attenuation of sounds of varying quality (Richards 1981), the accuracy with which an observer estimates the distance to the bird giving a vocalization may vary with the type of vocalization given. (This aspect of the problem needs to be studied).

The increased percentage of underestimated distances within 18 m of the observers was interesting. Recent work has shown that for at least some calls of the Black-capped Chickadee (*Parus atricapillus*), the sound is not omnidirectional and is greatest to the front of the bird (Witkin 1977). The same was found to be true for the drumming of Ruffed Grouse (*Bonasa umbellus*), again with the highest intensity of sound being found in front of the bird (Archibald 1974). The assumption of our model is that birds are randomly oriented with regard to the observer. If in fact they are not, and those within 18 m tend to face the observer, then this could account for the tendency to underestimate these distances. The calls and songs would be louder and thus perceived as being closer. This hypothesis can be tested by comparing the accuracy of distance estimates for directional and omnidirectional vocalizations as a function of distance from the observer.

The error introduced in the area surveyed as a result of 10% errors in distance estimation is approximately 20% for circular plots and 10% for line transects.

RECOMMENDATIONS

In order to minimize the error introduced by distance estimation, we offer the following advice:

(1) Train all observers by having them estimate distances to objects, then verify the distances. Start with objects that can be seen, and work up to birds heard but not seen. Tape recording may also be used.

(2) Flag objects at known distances from sampling points and have observers use these as reference points.

(3) Use range finders to measure distances to additional reference points at each sampling point.

(4) Make measurements in feet (smaller rounding errors).

(5) Use robust methods to analyze data: (a) Lump distance estimates (categories may vary with species, season, and vegetation type); (b) Make certain that the limits of the lumping categories coincide with the natural rounding tendencies of observers (e.g., 50, 100, 150, 200 m etc.).

(6) Convince observers of the importance of obtaining accurate measurements, with the corollary of reinforcing their confidence in their abilities to do so.

ACKNOWLEDGMENTS

We thank the members of the 1978 and 1979 Hawaii Forest Bird Survey for their participation in the field experiments. We thank Robert Clark for assistance with the computer simulations. The simulations were supported by a grant from the Oregon State University Computer Center. Ralph Raitt, C. John Ralph and Sharon L. Scott made helpful comments on an early draft of this manuscript.

Studies in Avian Biology No. 6:341–345, 1981.

TESTS OF HEARING ABILITY

FRED L. RAMSEY[1] AND J. MICHAEL SCOTT[2]

ABSTRACT.—Hearing tests taken by 274 people at the symposium indicated large differences in hearing ability among active birders. Simulation of the detectability of birds for observers with hearing thresholds of 10, 20, 30, and 40 dB indicated differences in area effectively surveyed as large as an order of magnitude.

In order to increase the comparability of observers, we recommend testing all potential observers for hearing ability using pure tone tests from .5 to 8 kHz and eliminating all those with uncorrectable hearing thresholds of 20 dB or greater in the frequencies emitted by the species being surveyed.

The importance of hearing to birders has long been recognized (Saunders 1934; Mayfield 1966; Cyr 1981). Because of the great reliance placed on aural observations during bird counts (Kepler and Scott 1981), we felt it would be informative to determine the variation in hearing ability in active birders. Thus, we offered hearing tests to participants in this symposium. Two hundred seventy-four people took advantage of this opportunity.

METHODS

The test was a standard industrial type in which the hearing threshold—the lowest detectable volume in decibels (dB) of each ear—was determined for frequencies ranging from 0.5 to 6.0 kilohertz (kHz). Information was obtained on age, sex, number of years' birding experience, and the number of bird surveys conducted in the last year. Unfortunately, the question on numbers of bird surveys conducted was imprecisely phrased, so we were unable to fully use that information.

RESULTS

The hearing thresholds for six different age classes are shown in Figure 1. The decline in hearing ability with age, especially at higher frequencies, is clearly shown. The frequency distribution of hearing thresholds for these same individuals without regard to age class is shown in Figure 2. Mayfield (1966) considered a hearing loss of 0–15 dB insignificant; 15–30 dB, slight; 30–45 dB, mild; 45–60 dB, marked; 60–80 dB, severe; and 80 dB, extreme. Below 2 kHz, more than 90% of all individuals tested have a hearing threshold of 20 dB or less. Emlen and DeJong (1981) have suggested 20 dB as deficient hearing ability for counting birds.

As we mentioned, the question on number of bird surveys conducted was stated in such a way that several of the 274 participants reported conducting 300 or more bird surveys in the previous year. This indicates that the respondents misunderstood the question, but we still consider it an indicator of trouble that two-thirds of the surveys reported were conducted by observers with hearing thresholds of 20 dB or greater. The loss of information with reduced area surveyed is shown for one day's field effort using stations (Table 1) and transects (Table 2).

AN EXAMPLE

To appreciate the effect that hearing loss can have on an observer's ability to count birds, consider this simplified example. Assume that a bird is a directional sound source; i.e., the intensity of the sound pressure of its song concentrates in the direction the bird faces, as in Figure 3. The actual variation we use in this model assumes the intensity $I(\theta)$, at an angle of θ from the source direction is

$$(1) \qquad I(\theta) = E \times 10^{-.198\theta^2}/C, \quad -\pi < \theta < \pi,$$

$$\text{where } C = \int_{-\pi}^{\pi} 10^{-.198\theta^2}\, d\theta, \text{ and}$$

where E is the total energy in the sound wave. We assume further that this total energy ($E = 1.24 \times 10^{-10}$ watts) would provide an average intensity of 50 dB within a one meter cylinder surrounding the bird.

A general equation describing sound attenuation with distance is (Urick, Ch. 2 and 4)

$$(2) \quad N(x) = N(1) - 10 \log_{10}(x^s) - ax - bx.$$

Here N(x) is the number of dB at a distance of x meters from the bird. The number s is a spreading factor, and the term in which it occurs describes the spreading of the total energy over increasingly large areas as distance increases. With spherical spreading, $s = 2$; whereas $s = 1$ with cylindrical spreading, as might occur in a closed canopy situation. In practice, the spreading factor would be somewhere between 1 and 2, and we arbitrarily assumed it to be $s = 1.5$ for this example. With this choice, each doubling of distance results in a loss of 4.5 dB sound pressure. Martin and Marler (1977) arrived at a loss figure of 6 dB with each doubling of distance, which is a figure also used by Bowman (1979). The 6 dB figure corresponds to $s = 1.99$,

[1] Department of Statistics, Oregon State University, Corvallis, OR 97331.

[2] U.S. Fish and Wildlife Service, Mauna Loa Field Station, P.O. Box 44, Hawaii National Park, HI 96718.

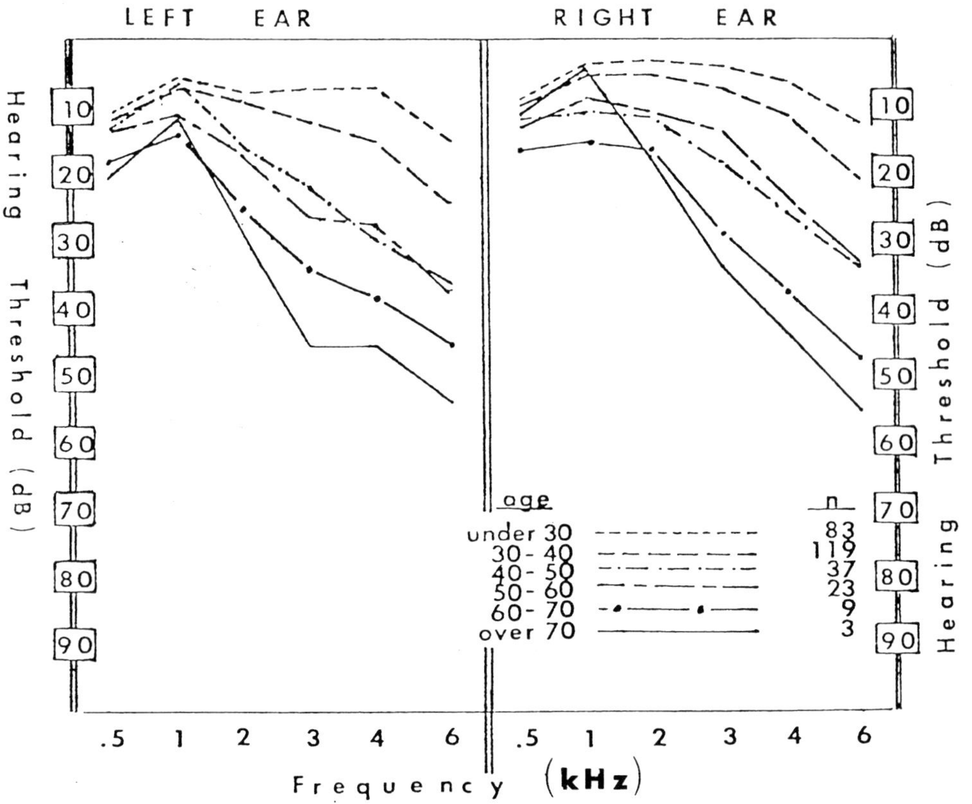

FIGURE 1. Age profiles of average hearing thresholds over the frequency range 0.5 to 6.0 kHz.

or virtually spherical spreading. Thus using s = 1.5 produces a model which is optimistic in that sound carries farther and hearing loss has less effect than it might have in practice. The final two terms in equation (2) describe absorption of energy by the medium and by vegetation, respectively. The constants a and b increase as the square of the frequency of the bird's song. However for the sake of generality we have made the model independent of frequency by taking a = b = 0.

With intensity spread as in equation (1), the actual sound intensity reaching the observer depends on the angle θ of the bird's orientation with respect to the bird-observer line. Taking this into account, we arrive at this simplified condition for song detection:

$$N(x) = 50 + 10 \log_{10}C - 1.98\theta^2 - 15 \log_{10}x \geq DT, \quad (3)$$

where DT is the observer's dB detection threshold, such as that measured on the standard hear-

TABLE 1
NUMBER OF BIRDS DETECTED WITH VARYING DENSITIES AND VARYING AREAS SURVEYED USING 15 CIRCULAR PLOTS AND ASSUMING PERFECT DETECTABILITY WITHIN EACH PLOT

Radius of area surveyed (m)	Area (km^2)	Density (birds/km^2)						
		25[a]	50	100	200	400	800	1600
5	.001	0.3	0.6	.1	.2	.5	1	2
10	.005	.1	.2	.5	1	2	3	6
20	.019	.5	1	2	4	8	15	30
40	.075	2	4	8	15	30	60	120
80	.302	8	15	30	60	120	241	483
160	1.210	30	60	120	241	483	965	1930

[a] Numbers are rounded to nearest tenth below 1 and to the nearest whole number above 1.

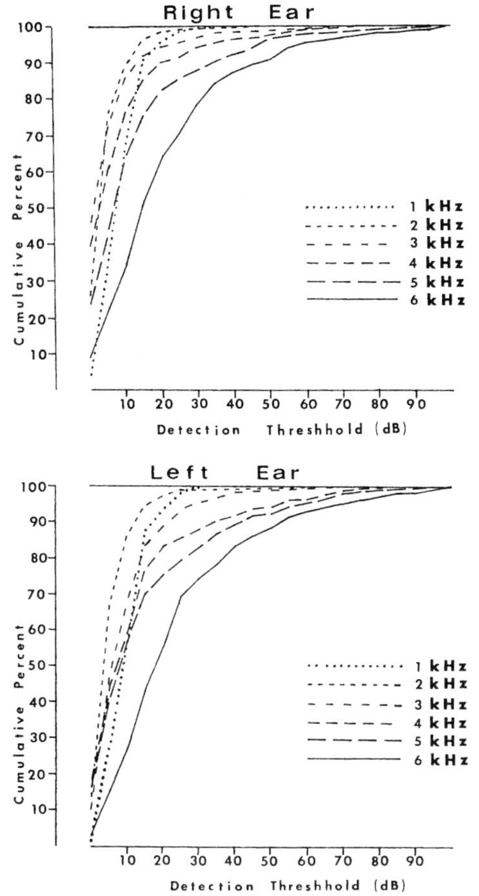

FIGURE 2. Distribution of detection thresholds for right (a) and left (b) ear for the 274 individuals who took the hearing test.

ing test that we offered to conference participants. Equation (3) ignores the fact that realistic signals arrive along with a certain amount of noise. In those situations, the signal-to-noise ratio must be greater than the observer's DT for that observer to make a detection. Thus, equation (3) represents an ideal situation with no noise.

Finally, we assume that the orientation angle θ has a uniform probability distribution on the angles $(-\pi, \pi)$. The resulting situation is this: letting

$$(4) \quad \psi(x, DT) = (1/\pi)\sqrt{28.69 - 7.58 \log_{10} x - DT/1.98},$$

then the probability of an observer with detection threshold DT being able to detect this bird at a distance of x meters is

$$(5) \quad \Pr\{\text{Detection} \mid x\} = \begin{cases} 0; & \text{if } \psi(x, DT) \leq 0 \\ \psi(x, DT); & \text{if } 0 < \psi(x, DT) \leq 1 \\ 1; & \text{if } \psi(x, DT) > 1 \end{cases}$$

Several of these song detection curves are plotted in Figure 4, using observer detection thresholds of 10, 20, 30, and 40 dB. An observer with DT $\geq$ 50 dB will be virtually deaf to this bird and must rely exclusively on visual detections during a survey.

DISCUSSION

The numbers of birds recorded by an observer in a survey are proportional to the effective area surveyed by the observer (Tables 1 and 2) (Ramsey et al. In press). We cannot use the results of the previous example to judge effective area surveyed without making further assumptions about the bird's song rate, the duration of the survey's count periods, the density of vegetation, background noise levels, and the form of the observer's visual detectability curves. However, it is reasonable to conclude from Figure 4 that hearing differences can result in differences as large as an order of magnitude in areas effectively surveyed. (In a hypothetical line transect survey where each bird sings once as the observer passes, the observer's effective area sur-

TABLE 2

NUMBER OF BIRDS DETECTED WITH VARYING DENSITIES AND VARYING WIDTHS ALONG A 1 KM TRANSECT

Area surveyed		Density (birds/km²)						
Area (km²)	Width (m)	25[a]	50	100	200	400	800	1600
.01	5	.3	.5	1	2	4	8	16
.02	10	.5	1	2	4	8	16	32
.04	20	1	2	4	8	16	32	64
.08	40	2	4	8	16	32	64	128
.16	80	4	8	16	32	64	128	256
.32	160	8	16	32	64	128	256	512

[a] Numbers are rounded to nearest tenth below 1 and to the nearest whole number above 1.

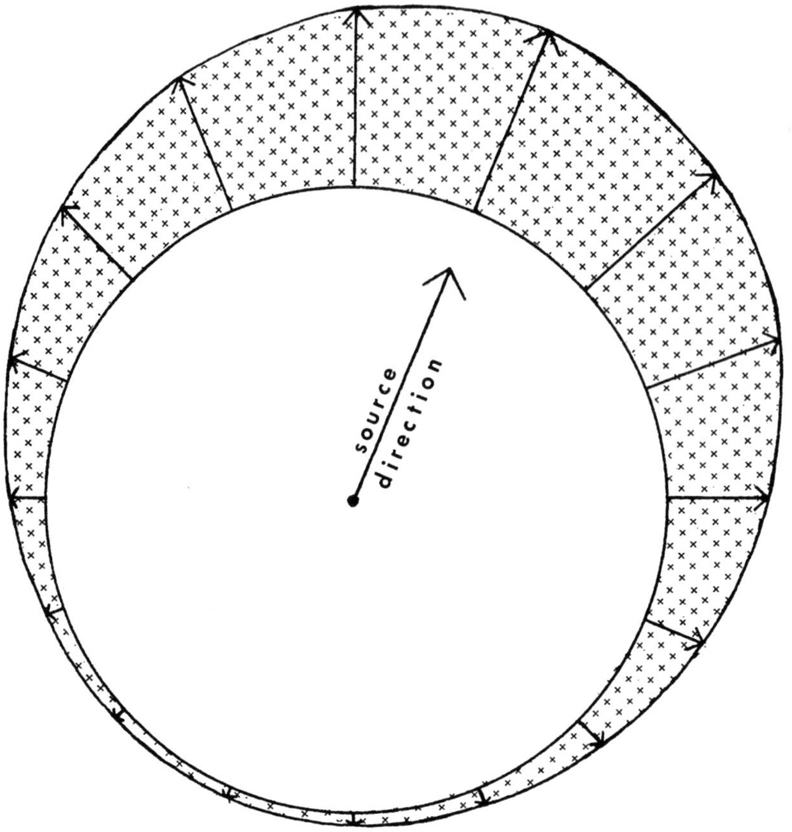

FIGURE 3. Idealized representation of sound transmission from a directional source.

veyed is the area under their detectability curve in the figure.) Surveys of terrestrial passerines typically record high percentages of audio only detections (Kepler and Scott 1981), thus differences in hearing ability will be reflected directly in the total number of detections.

The use of fixed area counts or simple counts may result in biased results when two observers of differing hearing ability are used. Hearing attenuation is not the same for all frequencies and is greater at higher frequencies. Thus with greater detection thresholds at higher frequencies, the high frequency emitters will be undersampled relative to low frequency emitters. This phenomenon could be very important and should be looked for. The use of variable area survey techniques such as the line transect or variable circular plot theoretically adjusts raw counts so that two observers with different hearing can still produce unbiased density estimates. However, the precision of the observer with the larger area is greater. Additionally, the numbers of species should increase as the hearing threshold decreases and the area surveyed increases.

SOLUTIONS

There is a tendency for observers as they grow older to become deficient in the higher frequencies first. Examination of the audiospectograms in Robbins et al. (1966) for the species listed by O'Meara et al. (1981) indicate that they have sounds ranging across at least 2 kHz and those with higher songs in the higher frequencies may range from 1 to above 6 kHz, e.g., white-eyed vireo (*Vireo griseus*). Thus, as observers lose their ability to perceive the higher frequencies, they may still be able to hear the lower pitched song portions. Their ability to identify these songs depends more on their field experience and the portion of the song which is discernable to them. Thus, field experience is an important variable to consider when evaluating (potential) observers. Also, differential attenuation of the high frequencies (Morton 1975) by the environment may make them relatively unimportant to all observers in detection and identification of distant songs. The noise level in some field situations may be sufficiently high to

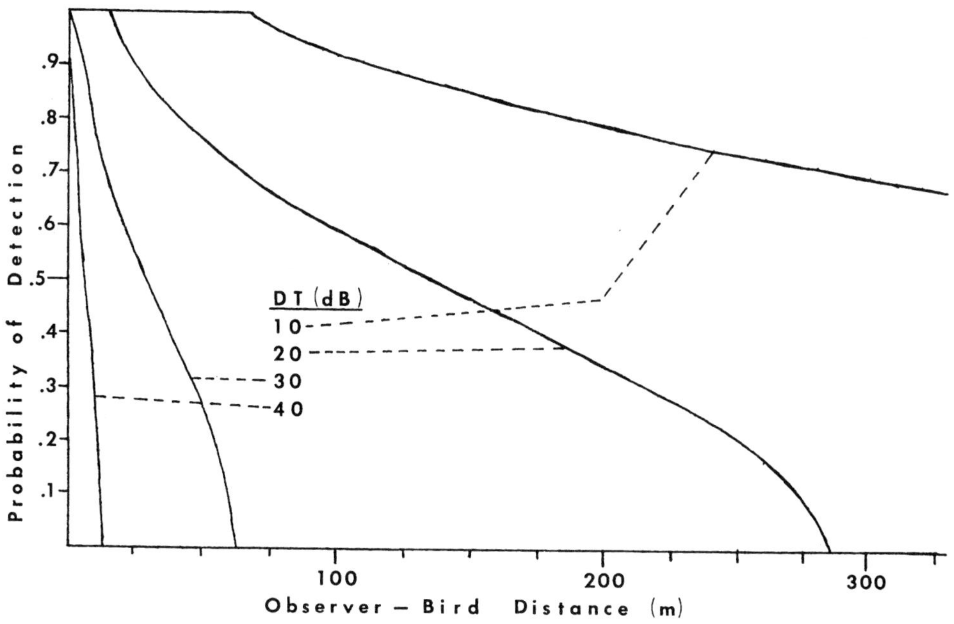

FIGURE 4. Detection curves for observer detection thresholds of 10, 20, 30, and 40 dB. Intensity of sound was 50 dB 1 m from the source.

mask songs until it is well above the thresholds of most observers. These effects combined with the long experience of older observers may act to reduce the disadvantage of hearing losses in the higher frequencies for at least some species.

One of the individuals we tested had his hearing threshold decrease by 20–30 dB when tested with a hearing aid. This suggests that one possible way for standardizing experienced observers with hearing problems would be to have them wear hearing aids which had been individually calibrated for a hearing threshold of say 10 dB within the frequencies emitted by the birds being counted.

In order to achieve the greatest possible coverage of an area at the lowest possible cost and increase the comparability and accuracy of observers, we recommend testing all potential observers for hearing ability using pure tone tests from .5 to 8 kHz (Kepler and Scott 1981). Observers who have hearing thresholds of 20 dB or greater in the frequencies emitted by the species of interest should be eliminated from the program (Emlen and DeJong 1981) or have their hearing corrected to 10 dB. All observers should then be tested for their ability to correctly identify species using simultaneous counts (Kepler and Scott 1981) and randomly presented sequences of song and calls at low sound levels (Cyr 1981; D. Richards, Pers. Comm.). Intensive field and laboratory training should then be used to correct any deficiencies (Kepler and Scott 1981).

ACKNOWLEDGMENTS

We thank Charles Greenlaw, Douglas Richards, and Cameron Kepler for helpful comments on the manuscript, and Eileen and Ron Hilton for conducting the hearing tests.

Studies in Avian Biology No. 6:346–352, 1981.

THE APPLICATION OF SONG DETECTION THRESHOLD DISTANCE TO CENSUS OPERATIONS

JOHN T. EMLEN AND MICHAEL J. DEJONG[1]

ABSTRACT.—Subjective estimates of the distance to unseen singing or calling birds are a serious source of error in all detection count operations where adjustments are required for variation in species conspicuousness or where areal denominators are needed for deriving absolute density values. The problem becomes acute where estimates are independently made by several observers differing in experience and estimating skills.

In this report we propose that measurements of the detection threshold distance (DTD) of a song, the maximum distance from which the song can be heard by an experienced observer with full hearing ability, can provide more reliable estimates than the elusive and highly subjective reference standards currently in use. We also hypothesize that DTD values will prove sufficiently objective and uniform when obtained in a standard manner to permit the preparation of reference tables applicable for census work by competent observers in a variety of situations.

To test this hypothesis we measured detection threshold distances for small samples of 12 bird species in one habitat type, the deciduous forests of southern Wisconsin. Mean distances ranged from 72 m for the Blue-gray Gnatcatcher to 186 m for the Wood Thrush. Coefficients of variation from the means for the various species ranged from 8.9 to 25.4%.

In a separate test we examined variation between observers. For this we played taped songs of 11 species at approximately natural volume levels and recorded the maximum distance at which they could be detected by 14 experienced observers with full normal hearing and four observers with variously deficient hearing. The field performance of these observers was then matched with their hearing performance in standard audiology tests in the laboratory and correlations sought for those with full and deficient hearing.

Environmental and observer variables affecting the distance to which songs can be heard are discussed. Some advantages and limitations of the proposed application to census operations are considered.

The songs of most birds are species-characteristic in features of pitch, song quality and pattern, and, when delivered on the home territory in the breeding season, usually also in intensity. Intensity varies greatly between species, however, and where songs are used in census work the distance at which a species can be heard will profoundly affect the area the observer covers and the number of birds he records. Recognizing this principle, and acutely aware of the crudeness of subjective estimates of distance to unseen birds, we attempted to test the hypothesis that the detection threshold distance of specific songs is uniform enough under specified field conditions to provide a reasonably reliable and objective base for calculating areas of coverage in census operations.

We felt that if the hypothesis were supported in a sample of species in one habitat type, and if the concept proved feasible under working field conditions, the sample could be extended to other species, other habitats, and other conditions to produce a body of data for constructing reference tables widely applicable in census operations. Detection threshold distances (DTDs) presented in these table and translated to detection areas (DAs) could then be applied directly as the denominators for the tallies of singing birds detected along a census route. Observers would be free of the onus of distance estimating and able to apply their full attention to detecting.

[1] Department of Zoology, University of Wisconsin, Madison, Wisconsin 53706.

The procedure as outlined in the following pages is formulated for singing males during the breeding (singing) season, and the density values obtained would be for songs and therefore still subject to adjustments for song frequency and sex ratio before translation into bird densities (Emlen 1977). The technique could conceivably be applied with appropriate modifications to other aural cues for census operations and in other seasons.

THE MODEL

Sound attenuates with distance according to definite physical laws, but the detection of a low-intensity or distant sound is an all-or-nothing phenomenon. Therefore, as in an audiologist's test for intensity thresholds of human hearing performance, all sounds in nature should, at least theoretically, be detected when they are above a critical threshold intensity or within a critical threshold distance, and go undetected when they are below that intensity or beyond that distance. Thus, in contrast to the gradual and continuous decline in detections postulated in earlier models, the curves of detection densities should, we believe, be level to the threshold distance and then drop precipitously (Curve a, Fig. 1). The gradual decline recorded in most field studies using strip-transect counts (Curve b, Fig. 1) are, we suggest, due to a combination of three factors: (1) the larger proportion of visual and call-note cues at short range in counts recording all detections, (2) the shorter time that the birds near the outer bounds of detectability

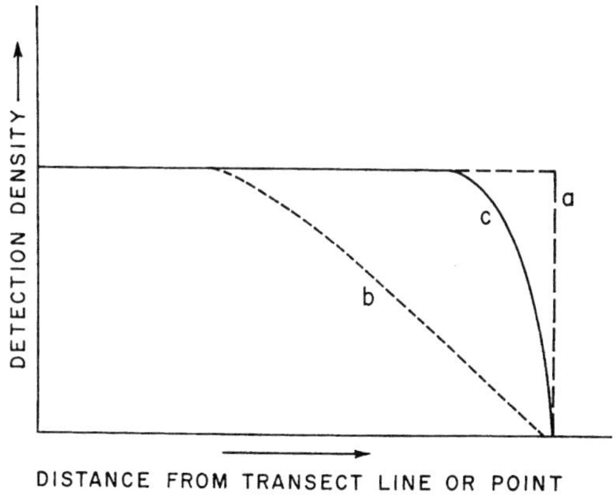

FIGURE 1. Curves of declining detection-density with increasing distance from the observer: **(a)** Theoretical pattern based on physical principles predicting an abrupt drop at the detection threshold point; **(b)** Characteristic sloping curve recorded in field studies based on subjective estimates of visual or aural detection distances; **(c)** Hypothetical curve based on the theoretical square pattern (a), rounded as a result of the environmental and acoustical variables characteristic of natural situations.

(Zone C in Figure 2) are within detection distance of the observer as he advances along the center line of the strip, and (3) the unreliability of subjective estimates of distance where the birds are no longer visible or readily locatable. We suspect that the actual pattern of decline for song cues in natural settings will depart from the square form and assume a rounded form (Curve c, Fig. 1) as a result of variations in habitat uniformity, song perch location and orientation, and factors associated with variable sound deterioration and background noise. These considerations are the subject of a further study by the authors.

MEASURING DETECTION THRESHOLDS

The distance at which a bird song becomes inaudible in a natural setting would be very difficult to calculate from physical parameters since it is subject to many complex and fluctuating variables in sound structure, transmission, and reception. It can, however, be measured empirically, and the variance in a set of measurements for a given species in a specified habitat under uniform conditions may be relatively small.

To test the prediction of reasonably small variances and the feasibility of reference tables for census operations, we measured the threshold distances for a series of common song birds in closed-canopy, deciduous forest habitats in southern Wisconsin. Two procedures were used: (1) measurements of the DTDs of "live" songs and calls by a single observer under standard conditions, and (2) measurements of the DTDs of standardized taped songs by a series of observers under a variety of conditions.

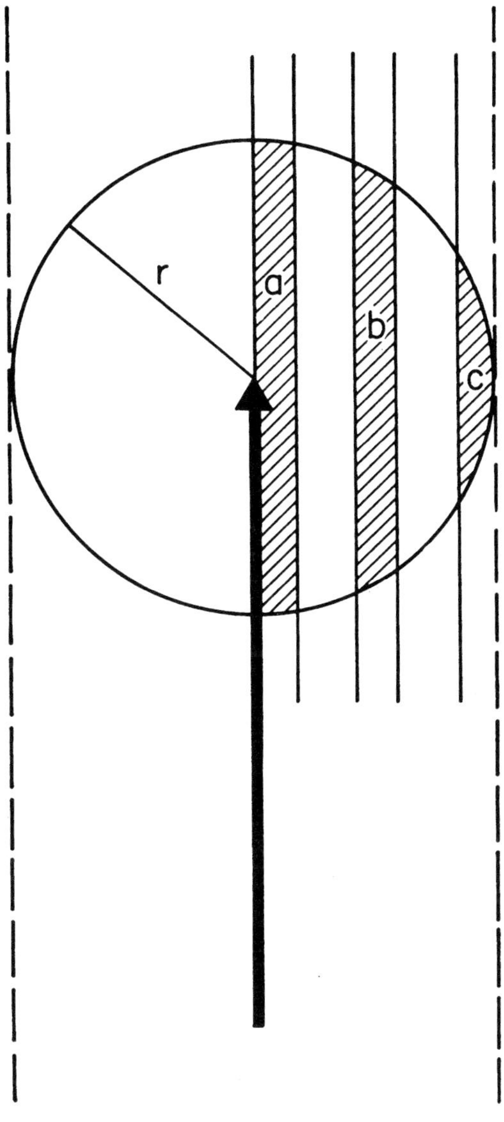

FIGURE 2. The length of time a bird is detectable within the area of detectability (circle) of an observer (arrow head) as he advances along a census trail declines rapidly in the peripheral zones (C) of a transect strip according to simple laws of geometry.

For the measurements of live songs the authors moved through the forest, separated ≥50 m, until one of us located a bird in full song. Communicating by portable two-way radios we then moved apart, one to the singing bird's location, the other (with demonstrated full hearing ability) away from the bird until the song could no longer be heard. While the first observer kept watch of the movements and activity of the bird and communicated these over the radio, the second moved back and forth to establish points where the

TABLE 1
DETECTION THRESHOLD DISTANCES[a] IN METERS FOR COMMON SONGBIRD SPECIES IN SOUTHERN WISCONSIN DECIDUOUS FORESTS[b]

Species	Observations (*n*)	Mean ($\bar{x}$)	SD	C.V.
Great Crested Flycatcher (*Myiarchus crinitus*)	6	98	15.4	15.7
Eastern Wood Pewee (*Contopus virens*)	14	124	23.1	18.6
White-breasted Nuthatch (*Sitta carolinensis*)	9	106	26.9	25.4
House Wren (*Troglodytes aedon*)	11	146	21.8	14.9
Gray Catbird (*Dumetella carolinensis*)	13	113	25.6	22.7
American Robin (*Turdus migratorius*)	7	150	21.3	14.2
Wood Thrush (*Hylocichla mustelina*)	7	186	16.5	8.9
Blue-gray Gnatcatcher (*Polioptila caerulea*)	6	72	6.4	8.9
Red-eyed Vireo (*Vireo olivaceus*)	15	135	24.8	18.4
American Redstart (*Setophaga ruticilla*)	7	93	10.7	11.5
Cardinal (*Cardinalis cardinalis*)	9	185	35.0	18.9

[a] Maximum distance at which vocalization was detected by an observer (M.D.) with full normal hearing as tested from 1000 to 8000 Hz by a licensed audiologist.
[b] All data were collected under favorable conditions (see text).

song (a) could still be heard consistently though very faintly, (b) could definitely be heard but only after prolonged and intensive listening, and (c) could not be heard. One or the other of us then paced the distance between the bird and the middle (b) point, also recording the height of the singer, a subjective appraisal of the foliage density between the bird and observer (from 0 to ++++) and a notation concerning any masking or interfering background noises of potential significance. All measurements were made during morning hours in July 1979 and 1980 under favorable conditions of light wind and low traffic noise.

In our tests with taped songs we placed an Electrovoice Sonocaster 1 speaker 5 m above the ground at the end of a 310 m long transect in a large tract of level and relatively uniform deciduous forest. A second speaker was placed 40 m in front of the first speaker (see Fig. 3). Each speaker used a separate but identical Kudelski DH amplifier and a Uher 4400 tape player with an operating speed of 38 cm/sec. The volume controls were set to give detection threshold distances for each test species roughly matching those already obtained from the measurements on living birds.

Fourteen experienced field ornithologists with full normal hearing to 8000 Hz and four with variously deficient hearing (tested by a licensed audiologist at the University of Wisconsin Hospital), were deployed successively at 26 listening stations spaced at 10 m intervals along the line. Each observer started at the farthest station and moved towards the speakers on signal, listening attentively for 10 minutes at each station and recording all the songs he could detect, live or taped. The taped songs, representing 11 familiar local species, were played in irregular sequences (so as to prevent anticipation of specific songs) and alternately from the two speakers (so as to counter any expectation the songs once detected at a distant station would be heard at all closer stations). The exact time of detection for each song was recorded by the observer and these times matched against the playing

TABLE 2
DETECTION THRESHOLD DISTANCES (MEAN, IN M, AND COEFFICIENT OF VARIATION AMONG OBSERVERS) OF TAPED SONG PLAYBACKS IN A CLOSED-CANOPY DECIDUOUS FOREST AT FOUR WIND SPEEDS (BACKGROUND WIND NOISE) AS RECORDED BY 14 OBSERVERS WITH FULL NORMAL HEARING

Date (1979) Wind speed No. observers	August 30 0.0–2.7 m/s 3		September 8 2.2–3.6 m/s 7		August 31 3.6–6.7 m/s 4		September 1 5.4–9.4 m/s 3		
	Mean	C.V.	Mean	C.V.	Mean	C.V.	Mean	C.V.	Mean C.V.
Crested Flyc.	180	11.1	149	20.7	110	22.3	77	19.9	18.5
Wood Pewee	190	13.9	174	3.1	138	3.6	123	20.4	10.3
W. B. Nuthatch	180	11.1	163	11.4	123	18.2	97	6.0	11.7
House Wren	190	10.5	190	13.2	143	3.5	153	13.6	10.2
Catbird	107	19.5	120	22.6	78	19.2	50	34.6	24.0
Robin	180	16.7	170	18.6	135	4.3	83	25.0	16.2
Wood Thrush	253	2.3	227	7.9	240	4.2	190	9.1	5.9
B. G. Gnatcatcher	63	48.2	77	22.1	80	17.7	50	34.6	30.7
R. E. Vireo	177	3.3	179	6.0	130	16.6	113	13.5	9.9
Redstart	120	22.0	128	15.2	88	5.7	90	0.0	10.7
Cardinal	267	10.8	244	11.0	225	10.6	213	2.7	8.8

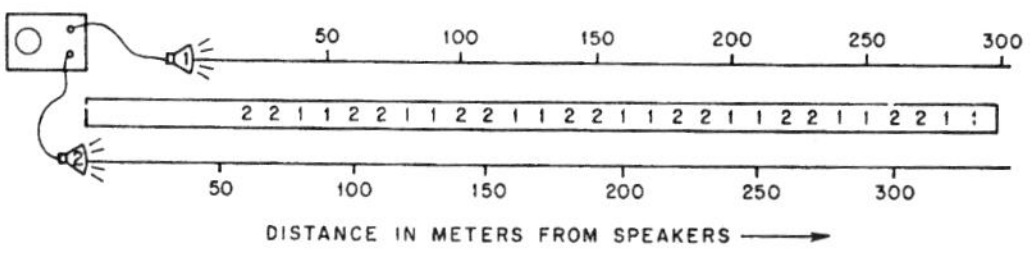

FIGURE 3. Arrangement of speakers and listening stations for the observer variability tests. Observers moved towards the speakers (right to left) listening for 10 minutes at each station while all 11 taped songs were played. The number at the station indicator shows which speaker was being used at the time. See text for more explanation.

time for confirmation that the recorded detections actually reflected a playing. All tests were run on clear mornings in August and September 1979, when the confusion created by natural song activity had subsided. Wind speeds and a subjective appraisal of distant traffic noises were recorded on each test day for correlation with observer hearing performance. The sound pressure level (SPL) of taped songs and background noise were measured at the test site under similar conditions in September 1980 using a General Radio Company Sound-Level Meter Type 1551-C.

RESULTS

Mean detection threshold distances for the 11 species tested directly ranged from 72 m for the high, thin lisping song of the Blue-gray Gnatcatcher to 186 m for the rich and varied song of the Wood Thrush (Table 1). The magnitude of the coefficients of variation (from 8.9 to 25.4%) in this series undoubtedly reflects in part the small sample sizes and the masking effects of variable background noise. Few clues were obtained concerning individual variation in song intensity within a species although some variation is suspected in several species, notably the Catbird and the Cardinal where one bird gave DTD readings consistently one-third less than the mean for the others.

The tests using playbacks of taped songs were conducted in order to analyze observer and environmental variables under standard sound intensity, and these detection-distance values consequently have no significance for further use as 'live' DTD values. Mean coefficients of variation for the 11 species in this series (based on 14 observers with full normal hearing acuity on 4 days with tree-top wind speeds <10 m/sec) ranged from 5.9 to 30.7% and averaged 15.8% (Table 2).

The hearing deficiencies revealed in the audiologist's test for four observers correlated with reduced field performance. The deficiency in one representative observer, classified by the audiologist as having "mild to moderate losses in high-frequency ranges characteristic of persons in their 60s and 70s but not appreciably affecting hearing in the human speech range" was reflected in DTD reductions of about 10% for the low-frequency song of the Cardinal and Wood Thrush, about 45% for the middle-frequency songs of the pewee and catbird, and about 75% for the high-frequency song of the gnatcatcher.

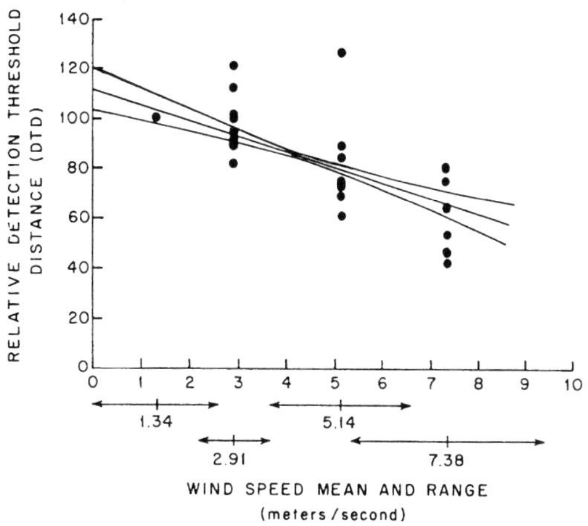

FIGURE 4. Masking effect of wind noise on detection threshold distance in deciduous forests. Data points represent mean DTD values for each of the 11 bird species on each of four days as recorded by the 14 observers with full normal hearing. Species DTD values for the windy days are relativized to the mean DTD for that species on the calmest day, thus all 11 species are located at the same point (100%) for the calmest day. The regression value (r) is -0.749 and the outer lines are 95% confidence limit boundaries.

Wind noise in the tree foliage had an appreciable effect on DTD values (Table 2). A regression analysis of average wind speeds against the mean DTDs for each species on the four test days revealed a significant negative value of -0.749 (Fig. 4).

VARIABLES AND THEIR CONTROL

The distance at which a bird song can be detected on a census count is subject to at least three major types of observer variables and three of environmental variables, all of which must be controlled as much as possible by restricting operations to specified situations or by applying appropriate adjustment factors.

Hearing acuity.—Observer variables associated with hearing deficiencies necessitate limiting participation in census work to observers proven to have full hearing acuity (detection at 20 dB or less in best ear in standard audiogram tests, Davis and Silverman 1960) at all frequency levels from 1000 to 8000 Hz. Exceptions to this rule may be made for skilled observers with

TABLE 3
PROCEDURE FOR CALCULATING ABSOLUTE DENSITIES FROM PREDETERMINED DTD VALUES IN A SAMPLE (HYPOTHETICAL) BIRD COMMUNITY CENSUSED BY THE TRANSECT COUNT METHOD[a]

				Density		
Species	DTD[b] (m)	DA[c] (ha)	Birds[d] heard	Heard[e] per km²	♂♂[f] per km²	Birds[g] per km²
Eastern Wood Pewee—song	124 ± 23	77±	5	6.5	13.0	26
Black-capped chickadee—call	87 ± 24	56±	10	17.8	35.6	36
House Wren—song	146 ± 22	102±	10	9.8	19.6	39
Gray Catbird—song	113 ± 28	63±	10	15.9	31.8	64
Wood Thrush—song	186 ± 17	122±	5	4.1	8.2	16
Blue-gray Gnatcatcher—song	72 ± 6	51±	5	9.8	19.6	39
Red-eyed Vireo—song	135 ± 25	87±	10	11.5	23.0	46
American Redstart—song	93 ± 11	65±	10	15.4	30.8	62
Cardinal—song	185 ± 38	146±	10	6.8	13.6	27

[a] For communities censused by the point method, the DA (detection area) is $\pi \times DTD^2 \times$ number of points on the census.
[b] DTD values taken from Table 1.
[c] Detection area (km^2) covered in 3 km of transect (2 DTD × 3,000/10,000).
[d] Number of songs heard on census (hypothetical).
[e] Density (songs heard per km^2).
[f] Density of males, adjusted for singing frequency (these values are for species in which only males sing and assume that 50% are singing or calling when the observer is within hearing range).
[g] Density of birds, adjusted for equal sex ratio.

slight deficiencies if appropriate adjustments are applied. The audiogram test should be taken two or more times in order to get a mean hearing level at each frequency since these can vary by 5 or more dB between successive tests (Peterson and Gross 1974). Censuses should not be undertaken when an observer is suffering temporary hearing impairment as with a bad cold.

Familiarity with songs and calls.—Familiarity with all of the songs anticipated in the census area is a prime requirement for participation in census work. A skilled observer can detect and identify small fragments of a threshold level song that a less experienced person might not even recognize as a bird sound. In the absence of recognized standards, the general and local experience of each participant in a census should be recorded.

Attentiveness.—Attentiveness is an elusive variable that can influence the performance of any census taker and can be controlled only by self-discipline. In our experience observers must always work alone in order to continuously apply full attention to the job at hand. To reduce distractions observers may wish to close their eyes at listening stations. Fatigue or physical discomfort should be minimized; three to five hours of full attentiveness on a census route is apparently close to the maximum for most observers.

Inability to fully control attentiveness probably accounts for the nonmatching performance commonly experienced when two or more competent observers work concurrently but independently along the same census route (cf. Lack 1976, Preston 1979). This psychological phenomenon of liminal and subliminal detection is presumably related to consciousness levels and will continue to elude clear definition or effective control.

Habitat structure.—The effects of habitat structure are reasonably predictable and therefore controllable. While every site is ultimately unique, a limited number of habitat types with similar sound-transmitting properties may be recognized, and separate reference tables of DTD values, such as the one presented in this paper for closed-canopy deciduous forests in summer (Table 3), must be constructed for each.

The position of a singing bird with respect to height above the ground and perch exposure can affect the attenuation of the signal and hence the distance to which it can be detected although its effect does not seem to be a severe problem for deciduous forests in the summer (Morton 1975, Martin and Marler 1977).

Meteorological conditions.—Temperature, relative humidity, fog, and particularly wind can influence the propagation characteristics of sound waves through a habitat and hence the distance at which songs can be detected. Theoretical considerations suggest that all of these should be standardized as much as possible by limiting operations to days when conditions are favorable (moderate) and similar. Standardizing operations with relations to time of day is one way to minimize these variables. Wind direction can profoundly affect DTD values in open habitat, but under a forest canopy its effects were negligible at tree top wind speeds of 5 or even

10 m/s. The effect of wind as a producer of background noise is discussed below.

Background noise.—Background or masking noise is a common and important variable in bird census work. Much of it is uncontrollable except by avoidance, and censuses should be attempted only when and where the background noise level is within acceptable limits of less than about 60 dB (20–20,000 Hz flat response). This requirement essentially eliminates census operations near highways, livestock concentrations or noisy machinery, or on days with appreciable falling rain or with tree-top winds in excess of about 10 m/s (22 mph).

The human ear and brain can, of course, discriminate and identify bird sounds at intensities far below background white noise levels, but discrimination, as with colors in vision, depends on contrast, and the nature (frequency structure) of background noises will have selective effects in masking different bird sound frequencies.

At tree-top winds between 3 and 10 m/s the masking effect of the noise in the foliage (essentially equal intensity between 50 and 6300 Hz) is predictable, and adjustments can be made for its effect on observer performance. Figure 4 shows that performance did not decline significantly on days with wind speeds less than 3 m/s but declined 20% on the days with winds about 5 m/s, and 35% on the day with winds ca. 7 m/s. Figure 5 compares the mean DTD values for all days after adjustments for the windy days were made (lower value) with the values obtained on the two calm days. It is clear that the two values are quite close for each species except that the adjusted values tend to be too large for birds normally detectable at long distances, and slightly too small for birds normally detectable at relatively short distances. This indicates that species detectable at relatively long distances are not masked by wind noise to so great an extent as species normally detectable only at shorter ranges. It may thus be necessary to apply species-specific adjustments for wind noise in DTD reference tables.

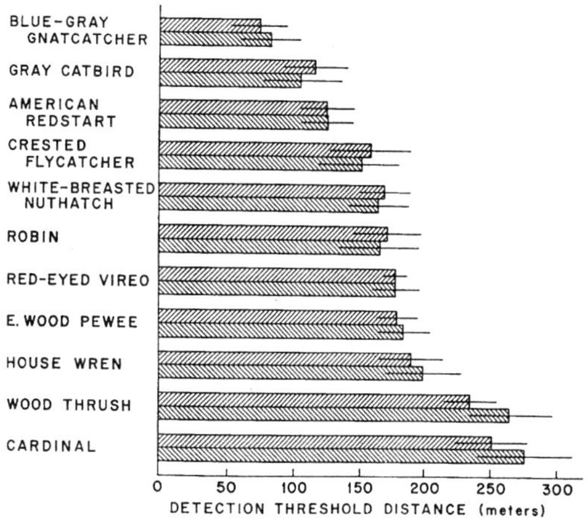

FIGURE 5. Effect of DTD value adjustments on taped songs for windy days. For each bird species, the lower value is the mean distance (meters) adjusted for windy days, while the upper value is the mean distance for only the two calm days. In each case the range of ± one standard deviation is indicated by the solid line.

CONCLUSIONS AND CAVEATS

The major advantages for using the distance-carrying properties of a bird's song as the indicator of area of coverage in transect census operations lie in the simplicity, directness, and potential for standardization of the procedure. With this method the count of each species along a transect route or at a series of points is simply divided by the area of coverage obtained from a reference table to give a density in singing birds per unit of survey area (Table 3). The values in the reference table will provide a common base for all census-takers who, following standardized instructions, would only have to count the birds they detect, and record the length of their transect or the number of their observation points. Problems of bird movements across inner belt or circle boundaries would be eliminated, and no detections would be excluded from the record by arbitrary boundary lines.

While the potential of a DTD method is considerable, a number of caveats must be voiced. First, the validity of the DTD method rests on three inadequately tested assumptions: (1) that the distance-carrying properties of a bird are species-characteristic, i.e., do not vary appreciably within and between individuals; (2) that habitats can be categorized into a reasonable number of types with respect to sound transmission properties; and (3) the observers with similar hearing acuity, experience and self-discipline are, indeed, similar in performance. Secondly, application would be restricted to reasonably flat and regular terrain and, as far as we now know, to song cues. If total birds per unit area is desired, adjustments must thus be made for the concurrent singing rate (incidence per unit of time) of each species (Emlen 1977a) and for sex ratios in species where only one sex sings. These are unsolved problems lying beyond the scope of this paper. Finally, the potential effectiveness of the method is restricted by two intrinsic features of the model; for strip counts by the reduced detection coverage near the lateral strip boundaries (see Fig. 2), and for circle counts by the proportionately greater area of the outer boundary zone with its large detectability variables. More research is needed to evaluate and resolve these complications.

ACKNOWLEDGMENTS

We thank the Cornell Laboratory of Ornithology for providing the high-fidelity tapes of bird songs used in our play-back experiments, and the 18 experienced observers who participated in our observer variability test. We are grateful to Dr. Richard Sauer for conducting the audiometry tests and providing comments on these tests. This is paper #101 in the journal series of the University of Wisconsin Arboretum.

Studies in Avian Biology No. 6:353–359, 1981.

THE ROLE OF OBSERVER BIAS IN THE NORTH AMERICAN BREEDING BIRD SURVEY

CRAIG A. FAANES[1] AND DANNY BYSTRAK[2]

ABSTRACT.—Ornithologists sampling breeding bird populations are subject to a number of biases in bird recognition and identification. Using Breeding Bird Survey data, these biases are examined qualitatively and quantitatively, and their effects on counts are evaluated. Differences in hearing ability and degree of expertise are the major observer biases considered. Other, more subtle influences are also discussed, including unfamiliar species, resolution, imagination, similar songs and attitude and condition of observers. In most cases, well-trained observers are comparable in ability and their differences contribute little beyond sampling error. However, just as hearing loss can affect results, so can an unprepared observer. These biases are important because they can reduce the credibility of any bird population sampling effort. Care is advised in choosing observers and in interpreting and using results when observers of variable competence are involved.

The ability of observers to discriminate among individuals of several breeding bird species aurally and visually is of paramount importance to the outcome of most bird population sampling efforts. A factor limiting the accuracy of most such efforts is observer ability to correctly identify songs of each species. Visual identification biases are certainly encountered, but do not appear to be as prevalent nor as important as aural biases. Because identification by song is predominant in most bird surveys, hearing acuity and training would be expected to affect results greatly. This factor should always be considered when selecting observers for bird population sampling.

The North American Breeding Bird Survey (BBS) was established in 1966 to provide an annual index of bird abundance and establish trends in continental and regional populations of most North American bird species (Robbins and Erskine 1975). The techniques of this survey have been discussed by Robbins and Van Velzen (1967). Normally only observers with a keen sense of hearing and a thorough knowledge of bird song are selected for BBS routes. In some instances, however, observers who are unfamiliar with many bird songs, or have a hearing loss are selected. Although data from these observers do not contribute to the intended purpose of the BBS, their results provide useful measures of observer bias.

METHODS

Data from 65 selected BBS routes conducted in the central and eastern U.S. and Canada were subjected to two tests for similarity using the Bray-Curtis (hereafter BC) similarity index (Clifford and Stephenson 1975:57). Many other such indices are available (Huhta 1979) but this one was used because it considers actual numbers of birds as well as presence and absence. One test (26 routes) compares qualified to unqualified observers and the other (39 routes) compares pairs of two comparably qualified observers. For both tests, four consecutive years of data were used from each route, with a change of observer occurring after the second, so that within-observer (internal), as well as between-observer similarities could be calculated. The quartets of years were chosen from the entire 14 years of the BBS, so any yearly biases are minimized.

In addition, results of eight BBS routes conducted in the central and eastern United States were used to examine some typical situations in more detail. Three routes were used to analyze biases resulting from hearing loss, and two each for observer training and song confusion in otherwise well-trained observers. The eighth route was run simultaneously by the authors to provide data on results from observers of equal ability.

RESULTS

UNEQUAL OBSERVERS

In editing BBS routes, it has become obvious that known "underqualified" (either from hearing loss or lack of training) observers record consistently lower species totals on the same routes than do qualified observers. However, qualified observers of equal expertise produce consistently similar species totals. Considering these facts, species total alone was used as the criterion for choosing unequal observers.

To evaluate the influence of underqualified observers on the BBS, we used results of 26 routes on which the change of observer was from qualified to underqualified or vice-versa. Table 1 shows the mean internal similarity, with standard deviations, of qualified and underqualified observers as well as the mean and standard deviation of the between-observer similarity.

In this test, the mean internal similarity of the qualified observers is only 0.80. This is also true in the equal observer test (Table 2). Considering that each index was calculated from data gathered on different runs conducted in adjacent years, the two major factors explaining the re-

[1] U.S. Fish and Wildlife Service, Northern Prairie Wildlife Research Center, Jamestown, North Dakota 58401.

[2] U.S. Fish and Wildlife Service, Migratory Bird and Habitat Research Laboratory, Laurel, Maryland 20811.

TABLE 1

MEAN YEAR-TO-YEAR SIMILARITY OF RESULTS OF 26 BBS ROUTES, EACH CONDUCTED FOR FOUR CONSECUTIVE YEARS, TWO YEARS EACH BY A QUALIFIED AND AN UNDERQUALIFIED OBSERVER

	Underqualified		Qualified
	Internal[a] similarity	Between[b] observer similarity	Internal[a] similarity
$\bar{x}$	.7415	.6360	.7958
σ	.0574	.0866	.0458

[a] Bray-Curtis Similarity Index of results of two runs of the same BBS route in adjacent years by the same observer.

[b] Bray-Curtis Similarity Index of results of two runs of the same BBS route conducted in adjacent years by different observers.

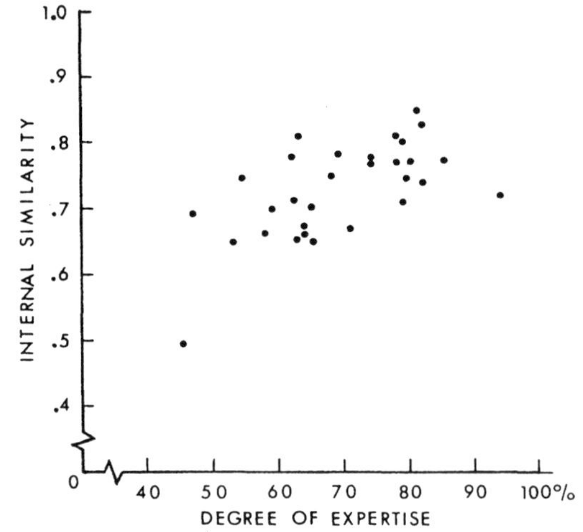

FIGURE 1. Relationship of degree of expertise to observer reliability. Degree of expertise is expressed as a ratio of species totals of underqualified to qualified observers on the same route. Reliability represents internal similarity of results of two separate runs of the same route by the underqualified observers.

maining 20% are probably sampling error and annual bird-population change. To test for sampling error alone, data from consecutive runs of the same route by the same observer are needed. A Maryland route was run by Bystrak on 3 consecutive days in 1969 and the mean of the three consequent BC indices was 0.84. This is a small sample, but it suggests that as little as 4% is the effect of annual change. Because the internal similarity indices of some of the observers in our sample were as high as 0.90, we feel that annual change contributing less than 4% is no doubt an insignificant consideration in this test. This is supported by the fact that few species ever show a significant population change from one year to the next (Bystrak 1981). Thus, sampling error appears to predominantly explain the 20%. Sampling error is influenced by several factors, such as weather, bird activity, noise, observer alertness and others. Further testing will be helpful in separating these factors and their magnitudes.

The mean internal similarity index of the qualified observers in this test (0.7958) was compared to the mean index of the qualified to underqualified observers (0.6360) and found to differ significantly ($P < 0.01$). This suggests that, in calculating population trends, it would be dangerous to compare an underqualified observer's results with those of a qualified observer. It is often suggested, however, that underqualified observers' results are reliable enough to be used in analyzing annual trends, presuming that they are at least producing a reliable index of those species they are recording. However, the mean internal similarity (cf. reliability) of the 26 underqualified observers (0.7415) was significantly different ($P < 0.01$) from that of the qualified observers (0.7958). The underqualified observers' own results from one year to the next are not even as comparable as those of two different qualified observers in adjacent years (0.7415—Table 1 vs. 0.7537—Table 2). We decided to investigate this further by testing the possibility that reliability increases as a function of expertise. We compared the internal similarity (BC index) of 30 underqualified observers with their apparent degree of expertise (Fig. 1). The correlation is significant ($\tau = .310$, $P = 0.016$) when tested with Kendall's coefficient of rank correlation. This strengthens our belief that it is safer to not include the results of underqualified observers whenever they can be identified as such.

TABLE 2

MEAN YEAR-TO-YEAR SIMILARITY OF RESULTS OF 39 BBS ROUTES, EACH CONDUCTED FOR FOUR CONSECUTIVE YEARS, THE FIRST TWO BY ONE QUALIFIED OBSERVER AND THE NEXT TWO BY A COMPARABLE OBSERVER

	Observer 1		Observer 2
	Internal[a] similarity	Between[b] observer similarity	Internal[a] similarity
$\bar{x}$	.7991	.7537	.7865
σ	.0490	.0707	.0624

[a] Bray-Curtis Similarity Index of results of two runs of the same BBS route in adjacent years by the same observer.

[b] Bray-Curtis Similarity Index of results of two runs of the same BBS route conducted in adjacent years by different observers.

Hearing loss

To explore the effects of hearing loss in more detail, we used results of three BBS routes, each

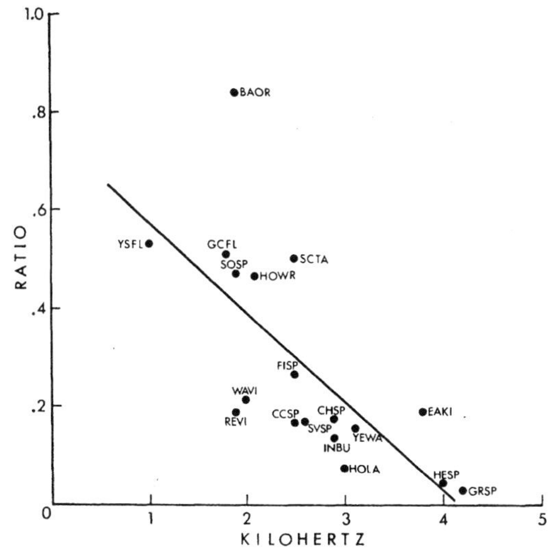

FIGURE 2. Relationship of bird-song frequency to hearing loss on two Wisconsin BBS routes. Ratio = i/u, where i = 3-year mean recorded by impaired observer and u = 3-year mean recorded by unimpaired observer ($y = 0.7575 - 0.1826$, $r^2 = 0.4931$). The 18 species are *Y*ellow-*s*hafted *Fl*icker (*Colaptes auratus*), *Ea*stern *Ki*ngbird (*Tyrannus tyrannus*), *G*reat *C*rested *Fl*ycatcher (*Myiarchus crinitus*), *Ho*rned *La*rk (*Eremophila alpestris*), *Ho*use *W*ren (*Troglodytes aedon*), *R*ed-*e*yed *Vi*reo (*Vireo olivaceus*), *Wa*rbling *Vi*reo (*Vireo gilvus*), *Ye*llow *Wa*rbler (*Dendroica petechia*), *Ba*ltimore *Or*iole (*Icterus galbula*), *Sc*arlet *Ta*nager (*Piranga olivacea*), *In*digo *Bu*nting (*Passerina cyanea*), *S*a*v*annah *Sp*arrow (*Passerculus sandwichensis*), *Gr*asshopper *Sp*arrow (*Ammodramus savannarum*), *He*nslow's *Sp*arrow (*Passerherbulus henslowii*), *Ch*ipping *Sp*arrow (*Spizella passerina*), *C*lay-*c*olored *Sp*arrow (*Spizella pallida*), *Fi*eld *Sp*arrow (*Spizella pusilla*) and *So*ng *Sp*arrow (*Melospiza melodia*). Italicized letters in common names indicate four-letter species codes.

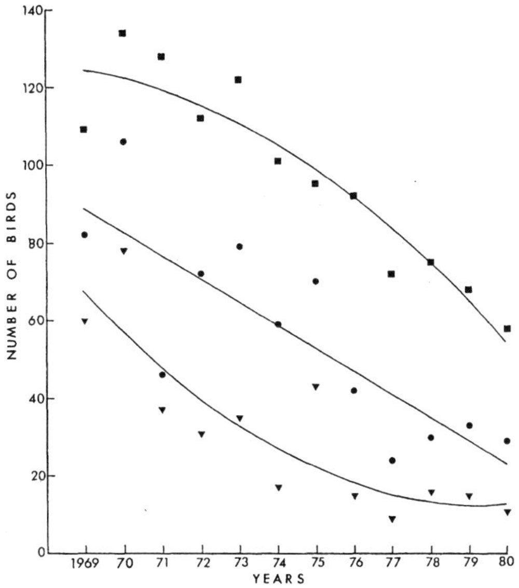

FIGURE 3. Twelve-year progression of hearing loss in a BBS observer, relative to frequency of bird song. Numbers of birds are annual totals in three frequency categories. Squares represent six low-pitched (1–2 kHz) species ($y = -0.44x^2 - 0.67x + 125.27$, $r^2 = 0.8861$). Circles represent six middle-pitched (2.1–2.9 kHz) species ($y = -5.92x + 94.45$, $r^2 = 0.6745$). Triangles represents six high-pitched (3.0–4.2 kHz) species ($y = 0.52x^2 - 11.66x + 78.36$, $r^2 = 0.7399$).

conducted in adjacent series of years by an observer with acute hearing and one without. We examined differences attributable to hearing loss and the progression of the loss as it relates to different song frequencies. The first two routes were run by an unimpaired observer for 3 years, followed by 12 years by an observer with an admitted progressive hearing loss. From these, we compared (Fig. 2) the interobserver ratio of 3-year mean counts of 18 species to the respective lowest frequencies of their typical songs (Robbins et al. 1966). The two observers' song perceptions were most similar at low frequencies, but the impaired observer had great difficulty in perceiving higher frequencies.

On these routes, the impaired observer's results were initially very similar to the unimpaired observer's, and thus offer a prime opportunity to study the effects of the progression of hearing loss during a 12-year period. The 18 species in Figure 2 were equally broken into three frequency categories, and the total annual counts for each plotted against time (Fig. 3). The three graphs imply that high-pitched songs were lost rapidly at first to a point where the birds were being recorded only visually or at close range, mid-frequency songs were lost at a steady rate, and low-pitched songs were stable longer, but also lost rapidly after time.

In the second example, we compared differences in mean species totals in several avian families on a BBS route covered by another pair of observers with differing hearing abilities (Table 3). We chose the five families portrayed because they represent a wide range of song frequencies and also vary considerably in conspicuousness. Totals were most similar for the Mimidae and Turdidae whose songs are generally repetitious, multi-syllabic, and of long duration and moderate frequency (Borror 1964), and least similar among the Tyrannidae and Parulidae whose songs are generally brief, simple and of higher frequency. As might be expected, the smallest differences among Tyrannids and

TABLE 3
TOTAL INDIVIDUALS OF FIVE BIRD FAMILIES REPORTED ON A WISCONSIN BBS ROUTE RUN IN SEVEN CONSECUTIVE YEARS BY TWO OBSERVERS WITH DIFFERENT HEARING ACUITY

Family	Good hearing (4-year means)	Poor hearing (3-year means)	% difference
Tyrannidae	55.75	21.33	−61.74
Mimidae	26.25	17.00	−35.24
Turdidae	54.25	55.67	+2.55
Vireonidae	43.00	19.67	−54.26
Parulidae	127.25	49.33	−61.23

Parulids were also in those species with the loudest, most repetitive, longest or lowest-frequency songs. For example, among Tyrannids the smallest difference was in Great Crested Flycatcher (*Myiarchus crinitus*), and among Parulids, Common Yellowthroat and Ovenbird (Table 4) were the least different.

Expertise

Degree of expertise is the second major contributor to observer inequality, hence the observer's ability to correctly identify and discriminate the species involved should always be considered in breeding bird population sampling. In this section, we offer a comparison of results obtained by observers of different levels of expertise. Also, some examples of confusion of similar songs by otherwise well-trained observers are given.

In the first example, a Maryland BBS route was covered for 4 years by qualified and comparable observers. The next 4 years, the route was covered by an observer who compared in age and hearing ability, but not in expertise. Four year mean counts of each species are used in this comparison (Table 5). In this example, three categories of species recognition are recognized: "sight," "song" and "both." The "song" species are further divided by song type: "easily learned," "intermediate" and "difficult to learn." The untrained observer recorded fewer birds overall but was consistent among "sight" species, "both" species and even "easily learned song" species. The "intermediate song" species showed a lower percentage and the "difficult song" species an even lower percentage. In examining BBS data, this low percentage for "difficult song" species appears to be typical for poorly trained observers. There is even some indication that, in many cases, poorly trained observers record the species they are familiar with in higher numbers than do well trained observers on the same route.

In another example, two observers of different levels of expertise conducted the same route in Wyoming three days apart in 1980. The results demonstrate 2 instances of misidentification of two pairs of birds with similar songs, Western and Willow Flycatchers (*Empidonax difficilis* and *E. traillii*) and Warbling Vireo (*Vireo gilvus*) and Cassin's Finch (*Carpodacus cassinii*). The experienced observer recorded 10 Western Flycatchers and 57 Warbling Vireos, both of which are common breeders in the area (Pettingill and Whitney 1965). The poorly trained observer recorded, instead, 12 Willow Flycatchers and 15 Cassin's Finches, neither of which is known to breed there (Pettingill and Whitney 1965). Because these songs are so similar, especially to an untrained observer, we have no doubt that the poorly trained observer misidentified them. The species totals differed on these two runs (52 vs. 34), in keeping with our observation that species total is the best single measure of degree of expertise.

Bird species with similar songs can present an

TABLE 4
TOTAL INDIVIDUAL WOOD WARBLERS REPORTED ON A WISCONSIN BBS ROUTE RUN IN SEVEN CONSECUTIVE YEARS BY TWO OBSERVERS WITH DIFFERENT HEARING ACUITY

Species	Good hearing (4-year means)	Poor hearing (3-year means)	% difference
Black-and-white Warbler (*Mniotilta varia*)	6.0	0.3	94.4
Golden-winged Warbler (*Vermivora chrysoptera*)	12.5	1.0	92.0
Nashville Warbler (*Vermivora ruficapilla*)	4.5	0.7	85.2
Yellow Warbler (*Dendroica petechia*)	20.2	6.3	68.6
Chestnut-sided Warbler (*Dendroica pensylvanica*)	15.7	3.7	76.6
Ovenbird (*Seiurus aurocapillus*)	13.5	10.0	25.9
Mourning Warbler (*Oporornis philadelphia*)	3.7	0.3	91.0
Common Yellowthroat (*Geothlypis trichas*)	43.5	26.3	39.5
Canada Warbler (*Wilsonia canadensis*)	1.2	0	100
American Redstart (*Setophaga ruticilla*)	6.2	0.7	89.2

TABLE 5
COMPARISON OF RESULTS OF QUALIFIED AND UNDERQUALIFIED OBSERVERS

	Species recognition method				
	Song				
	Difficult to learn	Intermediate	Easy to learn	Slight	Both
Qualified[a]	42.25	116.5	189.75	284.0	286.25
Underqualified[a]	6.75	63.25	156.25	210.0	243.0
Ratio	.16	.54	.82	.74	.95

[a] Numbers are 4-year means recorded on the same BBS route.

identification problem to many well-trained observers also, but usually on a small scale. In Maryland, the Ovenbird and the Kentucky Warbler (*Oporornis formosus*) are common nesting species with similar songs. Along a Maryland BBS route, a well trained observer familiar with both songs recorded means of 2.6 Kentucky Warblers and 1.4 Ovenbirds over five years. Another known well-trained observer covered this route for the following four years and recorded no Kentucky Warblers, but did record a mean of 3.5 Ovenbirds. The original observer resumed coverage and recorded 4-year means of 2.5 and 1.5 respectively. It thus appears that the second observer was combining the two. Counting birds by call note can produce similar differences between experienced observers because call notes of many species are confusing. For example, two comparable observers conducted another Maryland BBS route simultaneously, and obtained similar results for most species. However, one observer recorded 4 Common Flickers (*Colaptes auratus*) and 10 Red-bellied Woodpeckers (*Melanerpes carolinus*), while the other recorded 11 Common Flickers and 4 Red-bellied Woodpeckers. The call notes of these species are similar and distant birds can easily be confused.

EQUAL OBSERVERS

Comparable, qualified observers should, logically, produce similar results on an effort such as the BBS. The analysis programs used with the BBS have always used data from all qualified observers, regardless of changes of personnel on specific routes. To examine the results of comparable observers, we selected on the basis of similar species total, 39 routes on which changes of observers had occurred (Table 2). As with unequal observers (Table 1), 4-year periods were chosen, with the change occurring after the second year. Both sets of qualified observers produced mean BC indices of approximately 0.80. Unlike the unequal observers test, there is no significant difference between the internal similarities of the 2 groups (0.7991 vs. 0.7865). Surprisingly, however, the between-observer similarity (0.7537) was significantly different from both of the internal similarities ($P < 0.01$, $P < 0.05$ respectively). Although the differences are significant, they are slight, and because personnel changes are scattered throughout the 14 years of data, their effect is minimized. Robbins and Van Velzen (1969) analyzed BBS results for annual change using results of all qualified observers versus using only results of routes covered by the same observers and concluded that the increased sample size more than compensates for the small additional variability.

In late May 1980, we had the opportunity to conduct simultaneously a BBS route in the Turtle Mountains of North Dakota. We used this run to examine, in detail, the results of two equal observers when all other variables are the same. There were no significant differences between species totals (79 vs. 77, with 75 in common) or total individuals (1061 vs. 1141). Individual totals among species were quite similar, with no apparent difference in perception of most species. These similarities were readily apparent for both common and conspicuous species (e.g., Red-winged Blackbird, *Agelaius phoeniceus,* 97 vs. 96) and for uncommon species (e.g., Common Snipe, *Capella gallinago,* 4 vs. 3). The only major discrepancy was in the counts of Yellow Warbler (*Dendroica petechia*): 85 individuals at 41 stops, and 105 individuals at 48 stops.

Because observer bias did not appear to be a major factor affecting the outcome of this route, we investigated variability between ourselves by examining the raw data for Franklin's Gull (*Larus pipixcan*). There was no significant difference in our individual totals. Bystrak recorded 57 gulls at 16 stops; Faanes 55 gulls at 15 stops. Next, we examined the stops at which Franklin's Gulls were recorded, and found that we were not recording the same individuals at the same stops. The average error rate for individual totals was 3.7%, which suggests that our population figures were the same. The average error rate for each of the 50 stops was 18.9%, and at each stop where Franklin's Gulls were recorded, we usually obtained the same figure, or came within one bird of each other. The greatest amount of variability came from stops where eight or more individuals were observed. At several of these, our totals varied by nearly five birds, and there were several stops where one observer recorded the gulls and the other did not.

DISCUSSION

Hearing

As we and others have demonstrated, hearing is an important bias affecting counts of singing birds. We have found that, depending upon habitat, as high as 95% of the individual birds recorded on BBS routes are detected by hearing. Mayfield (1966) discussed hearing loss as it affects ornithologists, and showed that human males begin to lose perception of higher frequencies at age 32, and females at 37. The first frequencies to be affected are usually those above 4 kHz, the range of most bird songs. It is clear from our results that most bird population sampling is dependent on keen hearing ability for accurate results. An aspect of hearing ability not often considered is that of exceptionally good hearing. A few BBS recruits consistently report species and individual counts higher than those of well-trained observers with no known hearing impairments. In some instances, this may represent the influence of imagination, but in legitimate instances, such observers can innocently produce incomparable data.

Expertise and Training

Poorly trained observers present a problem that is potentially more difficult to deal with than hearing loss. Such observers are usually unreliable because they are inconsistent and given to incorrect identifications in addition to missing an unpredictable array of birds. The ability to learn bird songs is unfortunately highly variable and usually an individual process, yet few skilled observers did not benefit from earlier interaction with others. All observers respond to training and experience differently. Some are well trained in recognizing call notes as well as songs. In efforts such as the BBS, where more than singing males are sampled, this additional knowledge can bias results. This is also true for unusual song variations and song-trading, both of which are common, with examples too numerous to mention here. Even highly skilled observers find that there is much additional information to be learned about unusual bird vocalizations.

Subtle Biases

Although it is relatively easy to recognize observers with similar hearing or expertise, other subtle factors must be considered when striving for uniformity in sampling. Resolution, the ability to distinguish individual birds from a large number, such as in a dawn chorus, can produce higher counts, as can an active imagination that creates a second bird when one turns its head or sings from a different perch. Separating these two influences can often be difficult. Attitudes of observers can contribute to observer bias. Some are unaware of or unwilling to admit their shortcomings, and consider themselves well-trained. They are usually surprised to discover that they are unfamiliar with several species on a study plot or along a survey route. In these situations, data can be gathered incorrectly for years. A similar situation can occur when an observer moves to a new area and is unfamiliar with or unaware of new species or dialects. The condition of the observer can also be important. Most bird population sampling is conducted during early hours, so results are often a function of amount of sleep. It is likewise important to keep the effort reasonable because of individual differences in tolerance levels. A common complaint from BBS participants is that 50 stops is excessive. Proper rest notwithstanding, 4 hours appears too long for many to maintain the necessary high level of alertness.

SOLUTIONS

In order to work with the seemingly hopeless array of observer bias problems, the most important first step in any bird population sampling effort is to consider these problems and to what extent they will affect results. Next, an effort must be made to identify and overcome as many of these problems as possible.

Training is the logical solution in most cases, and should not only include a basic knowledge of identification cues but also as much familiarity with local variations and dialects as possible. If several observers are being used, it is fairly easy to equalize their abilities with concurrent field testing. Kepler and Scott (1981) describe an attempt to offset some observer bias in bird censusing with training sessions prior to initiation of fieldwork, and conclude that training is beneficial in arriving at more precise estimates. Such sessions help to acquaint observers with unknown, confusing, and similar songs, as well as to identify areas of weakness. Because of the large scale nature of the BBS, intensive training of observers would be excessively expensive and met with varying degrees of success. People differ widely in their abilities and speed in learning bird songs. Motivation plays an important role, and unmotivated individuals will probably never become particularly proficient. The best that can be hoped for on the BBS is quality control in choosing observers and in carefully screening results after they are submitted.

To overcome the problem of dawn chorus overloading or to increase the accuracy of most kinds of counts, a division of responsibility is useful. Scott and Ramsey (1981) found that by using two observers together and reducing the responsibility assigned to each, the accuracy of each observer was increased. This approach is, of course, not possible on the BBS because uniformity of coverage is crucial. It would be impossible, with the number of qualified observers available, to ensure two or more observers on every route.

Underqualified observers should not be a problem in small studies where control is easy or training is possible, but these observers can be numerous in large-scale projects such as the BBS. Enemar et al. (1978) postulated that observer variability in large scale census work involving many observers tends to produce an insignificant bias. In the BBS, where predominantly competent, comparable observers are involved, this appears to be true, especially when reports from the most obviously underqualified observers are eliminated. It is hoped that our examples, however, demonstrate the necessity for strict controls in small-scale studies.

ACKNOWLEDGMENTS

This paper benefitted from the constructive criticisms provided by D. H. Johnson, C. S. Robbins, S. D. Robbins and A. B. Sargeant.

Studies in Avian Biology No. 6:360–365, 1981.

PROBLEMS IN SEPARATING SPECIES WITH SIMILAR HABITS AND VOCALIZATIONS

CHANDLER S. ROBBINS[1] AND RICHARD W. STALLCUP[2]

ABSTRACT.—The possibilities for species misidentification based on vocalization or habitat association are high. However, the magnitude of the errors actually perpetrated is generally within an acceptable range in most types of bird survey work.

Examples of problems discussed are: congeners that are similar in appearance or in song (such as Chimney and Vaux's Swifts, *Chaetura pelagica, C. vauxi*; Hammond's, Dusky and Gray Flycatchers, *Empidonax hammondii, E. oberholseri, E. wrightii*; Willow and Alder Flycatchers, *E. traillii, E. alnorum*; Common and Fish Crows, *Corvus brachyrhynchos, C. ossifragus*); birds that are misidentified because they are not expected by the observer (House Finches, *Carpodacus mexicanus,* invading new areas of eastern U.S.); birds that imitate other species (especially Starling, *Sturnus vulgaris,* and Mockingbird, *Mimus polyglottos*); birds in mixed flocks; birds with geographic differences in vocalizations (Solitary Vireo, *Vireo solitarius*); woodpeckers that are only heard drumming; and nests or eggs that are misidentified.

Equally serious problems are the errors resulting from undetected species and from careless recording or failure to check manuscripts against original data. The quality of published count work can be improved considerably by (1) recognizing the problems that exist, (2) standardizing techniques for dealing with situations where not all birds can be identified, and (3) routinely applying all appropriate safeguards such as verification by mist netting and measuring, photography, tape recording or playback, additional observations, and careful verification of all entries in the final manuscript.

Errors of species identification are made not only by students, trainees, and field assistants, but by experienced field ornithologists as well. Even birds in the hand are often misidentified. Many misidentified birds have been stored in museum trays for decades before being "discovered." Banders are occasionally embarrassed to find they have misidentified a bird in the hand. Several years ago, after more than 30 years experience as a bird bander, the senior author netted and banded what he took to be a Common Yellowthroat (*Geothlypis trichas*); on release it gave the characteristic chip of a Mourning Warbler (*Oporornis philadelphia*). When recaptured later in the day the bird proved to be in typical immature female Mourning Warbler plumage; it simply had not been examined carefully.

Thus, we should realize it is not always someone else who is making incorrect identifications, and therefore we should be constantly alert to keep misidentifications to a minimum. We shall review several types of identification problems, then give some specific examples. We shall also briefly discuss how possibilities of misidentification should influence selection of a count technique. Finally, we shall add a few comments regarding errors that occur between the time a bird is observed and the time the report appears in print.

[1] Migratory Bird and Habitat Research Laboratory, U.S. Fish and Wildlife Service, Laurel, Maryland 20811.

[2] Box 533, Inverness, California 94937.

TYPES OF IDENTIFICATION PROBLEMS

Inexperienced observers.—The most obvious source of identification error is observers who are unfamiliar with the species, or with their songs, or with habitat requirements, or seasonal occurrence. A cautious inexperienced observer will miss species that are present; one not so cautious will include migrants with breeding or wintering species, or list distant birds of another habitat in the habitat being surveyed, or record species that are not present at all. Inexperienced observers are especially prone to mistake imitations by Starlings or Mockingbirds for the species being imitated, or to mistake vocal imitations of hawks uttered by jays.

Carelessness.—Carelessness can lead to misidentification, especially when an observer is under pressure. Examples include: 1) haste to complete a field trip, to catch up if behind schedule, or to record a large number of species in a measured time interval (such as a 3-minute BBS stop); 2) desire to record a bird that was seen too briefly or at too great a distance for positive identification; 3) failure to record a given species either because several other species were seen at the same time or because the observer was distracted before an observation was recorded; and 4) a snap judgment on a bird too briefly observed, a competitive urge, or an over-riding desire to excell.

Discomfort, fatigue.—Condition of the observer can certainly contribute to misidentifications. This topic is covered by Faanes and Bystrak (1981) in their discussion of observer variability and will not be belabored here.

Unfamiliar species or plumages, nests and eggs.—Some counters may perform very well for most of the species they encounter, yet may overlook or possibly misidentify a few species with which they may not be familiar or that they are not expecting (see Faanes and Bystrak 1981). Birds in juvenile plumage can be a problem during the breeding season because very few juvenile plumages are illustrated in field guides. Juveniles of some, but not all, species can be recognized by call notes that are similar to those of the adult. Many count takers are not familiar with nests and eggs of all the birds in their study plots, and if nests are located before incubation starts or after the nests have been abandoned there is danger of misidentification.

Racial variation.—Misidentification may result from the great spectrum in phenotypic appearance of various races of some species (e.g., Savannah Sparrow, *Passerculus sandwichensis*; some *Catharus* thrushes) that occur together during migration or on winter grounds. Field guides generally do not illustrate the full range of plumages.

Identification by habitat.—Dependence on printed material or popular belief about habitat requirements may lead to problems. Hammond's and Dusky Flycatchers, for instance, usually, but not always, nest in their "assigned" habitats. These species should always be double checked by plumage, behavioral, and vocal features. We must also be watchful for species expanding their "stated" ranges such as Anna's Hummingbird (*Calypte anna*), which is expanding its range to the north and east.

Dependence on field guides.—North American field guides, unfortunately, have an eastern bias. No field guide author or field guide artist has spent an appreciable time with live birds in the West, where geographic differences in appearance and vocalizations are much greater than in the East. Thus, technical points of identification of western species and races often receive secondary treatment. It would be impossible, for instance, for any inexperienced person to identify a Bell's Vireo (*Vireo bellii*) in California from the field guide pictures that are available. Field guides contain some inaccuracies and all contain some misleading pictures. Much of this may be a result of birds being painted from study skins, with resulting mistakes in shape and attitude. Field guides also lack new information that has become available since publication. Sophisticated articles, notes, and booklets on critical identification of many of the most difficult North American bird complexes and groups have been published over the last five years. Perhaps they will all be indexed or published together, but for now any person interested in accurate critical identification of birds should consult such journals as *Western Birds, British Birds, Continental Birdlife,* and *Western Tanager* (Los Angeles Audubon Society), or the British Trust for Ornithology Field Guide 17 (Prater et al. 1977).

Dependence on bird song tapes and records.—Because of tremendous geographic differences in songs and calls within a single species, no record or set of records has enough geographical treatment to solve all identification problems. Published recordings of many western species were no doubt taped in the East (e.g., Winter Wren, *Troglodytes troglodytes*). Some recordings have local dialects, recorded somewhere in the West, that are not representative of the vast majority of populations (e.g., Northern Pygmy Owl, *Glaucidium gnoma*). One should be especially careful with owls, wrens, blackbirds, and fringillids.

Birds with similar songs or calls.—There are a few species (to be discussed later) whose songs or calls are practically indistinguishable. There are many more that cause identification problems for relatively inexperienced census workers. This problem is compounded by occasional birds that give extremely atypical songs: for example, a Hooded Warbler (*Wilsonia citrina*) singing like a Yellow-bellied Flycatcher (*Empidonax flaviventris*).

Unfamiliar geographic area.—We know of instances where highly experienced census takers have made serious identification mistakes when working in different parts of the continent. Even familiar species may cause problems because of geographic or local dialects in their songs or call notes (e.g., Solitary Vireo; Red-winged Blackbird, *Agelaius phoeniceus*; Rufous-sided Towhee, *Pipilo erythrophthalmus*).

Hybrids.—Banders have found that hybrid birds are more frequent than had previously been supposed. Some hybrids sing the song of one parent or the other; some sing different or intermediate songs. Any aberrant song should, of course, be checked.

Mixed flocks.—A problem encountered frequently in winter (and even more so during migration) is a large flock in which not all species are individually identified. Some observers conservatively record only those individuals closely examined and accounted for at the same time. Other observers count or estimate the total number of birds in the flock and estimate the percentage composition of the flock on the basis of the sample that can be identified. Uniform procedures should be established.

Species overlooked in a flock.—In contrast to obvious mixed flocks, there is a problem with flocks that at first glance appear to contain only one species (e.g., Horned Larks, *Eremophila alpestris*; Red-winged Blackbirds, or Lapland Longspurs, *Calcarius lapponicus*). The second look can be very important.

SELECTED EXAMPLES OF IDENTIFICATION PROBLEMS

Hawks.—Accipiters pose a problem because of sexual size dimorphism and inconspicuousness (until the observer approaches close to the nest). The common woodland buteos (Red-tailed, *Buteo jamaicensis*; Red-shouldered, *B. lineatus*; and Broad-winged, *B. platypterus*, Hawks) also can be confused by an observer who is not familiar with their appearance, calls, and habitat use. Large falcons (Gyrfalcon, *Falco rusticolus*; Prairie Falcon, *F. mexicanus*; and Peregrine, *F. peregrinus*) are usually seen at a distance, making it important to learn their diagnostic shapes, wing beats, and flight characteristics.

Rails, cuckoos.—Experts have argued for years over calls of Black (*Laterallus jamaicensis*), Yellow (*Coturnicops noveboracensis*), and Virginia (*Rallus limicola*) Rails; and some calls of King (*R. elegans*) and Clapper (*R. longirostris*) Rails are hard to separate. We have seen Sora (*Porzana carolina*) and Virginia Rails at a nest utter what sounded to be identical alarm notes. Although the Black-billed (*Coccyzus erythropthalmus*) and Yellow-billed (*C. americanus*) Cuckoos have diagnostic calls, they also use calls consisting of a series of single notes that are extremely similar.

Owls.—Owls are more often missed than misidentified, simply through lack of nocturnal coverage. On the other hand, large owls flushed are difficult to identify. Screech Owls (*Otus asio*) may imitate Saw-whets (*Aegolius acadicus*), and nutria (*Myocastor coypu*) may sound like Long-ears (*Asio otus*). Winnowing Common Snipe (*Capella gallinago*) may be mistaken for Screech Owls.

Dark swifts, hummingbirds.—Chimney Swifts are rapidly colonizing the Southwest, and Vaux's Swifts are similarly expanding their breeding range in the Northwest, their winter range in the tropics, and in migration occur as far east as Louisiana. It is likely that Vaux's are overlooked as Chimneys in the East and Chimneys overlooked as Vaux's and even possibly Black Swifts (*Cypseloides niger*) in the West. Female and young male hummingbirds are notoriously hard to identify. Furthermore, east of the Rockies most observers incorrectly assume that all hummingbirds are Ruby-throated (*Archilochus colubris*).

Woodpeckers drumming.—A few species of woodpeckers (e.g., Pileated, *Dryocopus pileatus*; sapsuckers, *Sphyrapicus* spp.) have a characteristic pattern of drumming that can be used to identify them. Many others cannot be positively identified by their drumming because there is a greater difference in resonance between drumming substrates than among species. Additionally, some species of woodpeckers do not drum at all.

Flycatchers.—The recent range expansion of Wied's Crested Flycatchers (*Myiarchus tyrannulus*) may cause identification problems with Great Crested (*M. crinitus*) and Ash-throated (*M. cinerascens*) Flycatchers in the West. Some Olivaceous (*M. tuberculifer*) Flycatchers drift north after breeding and have overwintered successfully north of their breeding range (Luther et al. 1979). Few observers can confidently identify *Empidonax* flycatchers in the field by sight alone. The recent splitting of the Alder Flycatcher from the Willow poses a serious problem not only for banders, but also for the many observers who are unable to distinguish between the songs of these two species. Even more difficult are the songs of some of the western *Empidonax* species. During spring when on territory, each species of *Empidonax* has a totally diagnostic dawn and dusk (sometimes night) song or song series. Other, less intense songs and calls, which are given frequently during the day, can be most confusing. This is particularly true in the Hammond-Dusky-Gray Flycatcher group. It is imperative to hear true, complete songs as well as to compile clues from behavior, color, proportion, and habitat to identify these birds. A silent pewee (*Contopus* sp.) is more of a problem than a silent *Empidonax* flycatcher. Silent pewees are identified primarily on the basis of where they are found, rather than by appearance. Silent Coues' Flycatchers (*C. pertinax*) have more the appearance of pewees than they do of Olive-sides (*Nuttallornis borealis*).

Crows, chickadees.—Few observers are competent to separate the three North American crow species on sight, and many others cannot do so reliably by voice. To many observers, especially in the East, any chickadee is a Black-capped Chickadee (*Parus atricapillus*). Most misidentifications result from the observer not being aware of the existence of additional species. Some observers, who are aware that Black-capped Chickadees sometimes winter in the northern part of the breeding range of the Carolina Chickadee (*P. carolinensis*), habitually find Black-capped Chickadees every winter—even in those winters when there is no southward movement of Black-caps. The presence of small numbers of hybrid chickadees where the ranges of two species meet confuses the situation. In the Rocky Mountains, observers used to seeing only Black-capped Chickadees may assume that all chickadees they hear belong to this species. In autumn, molting Mountain Chickadees (*P. gambeli*) may completely lack a white eyebrow and are easily misidentified as Black-caps.

Thrushes.—The *Catharus* thrushes, especially when seen in poor light, can be a real problem. This problem is especially acute for color-blind persons.

Kinglets, vireos.—Golden-crowned (*Regulus satrapa*) and Ruby-crowned (*R. calendula*) Kinglets should be readily separable in the field, but the many reports of Ruby-crowned Kinglets in winter in the northern states suggest that inexperienced observers are reporting male Golden-crowned Kinglets as Ruby-crowns. Agitated Hutton's Vireos (*Vireo huttoni*) often move quickly and wing-flick persistently, thus appearing like Ruby-crowned Kinglets. The Solitary Vireo has a wide geographic range, including some strikingly different habitats in different parts of North America. There are not only conspicuous differences in plumage, but also striking differences in song. For example, some of the western Solitary Vireos sound more like Yellow-throated Vireos (*V. flavifrons*) than like eastern Solitaries. Another pair of vireos whose songs are practically indistinguishable are the Red-eyed (*V. olivaceus*) and Philadelphia (*V. philadelphicus*) Vireos. A fourth vireo problem relates to separation of immature White-eyed Vireos (*V. griseus*) with dark eyes from Bell's Vireos. Also, juvenile Warbling Vireos (*V. gilvus*) can be quite yellow below and green above, thus appearing like Philadelphia Vireos.

Warblers.—There are many instances of similarity in plumage or song in the wood warbler family, Parulidae. We mention just a few examples. Golden-winged (*Vermivora chrysoptera*) and Blue-winged (*V. pinus*) Warblers cannot be separated solely by song because hybrids and back-crosses are known to sing the typical songs of both parents. A general problem with most parulid warblers is that they typically have two quite different songs, one generally more diagnostic than the other. Cape May (*Dendroica tigrina*), Blackburnian (*D. fusca*), and Bay-breasted (*D. castanea*) Warblers and American Redstart (*Setophaga ruticilla*) have songs that are readily confused with each other. The Northern Parula (*Parula americana*) and Cerulean Warbler (*Dendroica cerulea*) songs frequently are confused, as are those of the Yellow (*D. petechia*) and Chestnut-sided (*D. pensylvanica*) Warblers. Songs of the Northern (*Seiurus noveboracensis*) and Louisiana (*S. motacilla*) Waterthrushes are confused by observers who are not thoroughly familiar with both, and another song similar to that of the Louisiana Waterthrush is given by the Swainson's Warbler (*Limnothlypis swainsonii*). We are unable to tell a MacGillivray's Warbler (*Oporornis tolmiei*) from a Mourning Warbler by its song.

Tanagers.—Tanagers can be a problem by either sight or sound if one is working in an area where more than one species may occur. The Summer Tanager (*Piranga rubra*) can even be passed up for a singing American Robin (*Turdus migratorius*), and a Scarlet Tanager (*P. olivacea*) with faint wing bars can be mistaken for a Western Tanager (*P. ludoviciana*).

Finches and sparrows.—Cassin's (*Carpodacus cassinii*) and Purple (*C. purpureus*) Finches are readily confused in parts of the mountainous West. An even more severe problem occurs in the Northeast where invading House Finches are being mistaken for the native Purple Finches. Possibilities of confusing different species of sparrows are almost unlimited. Problems in separating the so-called "grass" sparrows extend far beyond inability of some observers to hear their songs. Even when in plain sight, many sparrows are misidentified by people not thoroughly familiar with them. The Henslow's Sparrow (*Ammodramus henslowii*), because of its inconspicuous habits, is one of the birds most likely to be entirely overlooked. The tail patterns of wintering longspurs are often very difficult to see; however, these birds can be told by their diagnostic calls given during flight.

VULNERABILITY OF DIFFERENT CENSUS TECHNIQUES TO MISIDENTIFICATIONS

Count techniques vary considerably in their probability of error in species identification. Methods based on a single or brief visit or on input from inexperienced personnel are most subject to species identification errors. Most reliable are methods based on repeated visits over several days by different observers and especially by ornithologists experienced in the geographic area and in sampling methodology.

Any ranking of techniques according to vulnerability to species identification error is bound to reflect personal opinion. Having had personal experience in all of the following methods, we feel we are relatively unbiased in ranking them as follows, with methods least subject to error appearing first: (1) studies based on trapping and banding, and those in which banding is used as a supplementary tool; (2) censuses based on mapping of birds during a series of visits on different days (such as Breeding Bird Census and Common Birds Census); (3) Breeding Bird Atlas studies, in which the important records (confirmations) are based primarily on close or prolonged behavioral observations; (4) point counts and other variable circle counts, especially when they involve multiple visits; (5) transects that involve only a single visit to a particular area; and (6) Breeding Bird Survey, with its series of short single visits.

We consider that all of these breeding season methods are less subject to misidentifications than are census attempts at other seasons of the year. We rate the winter techniques as follows: (1) Winter Bird-Population Study with its numerous visits; (2) Winter Bird Survey, which is a single walking coverage; and (3) Audubon Christmas Bird Count.

We emphasize that the above ranking applies *only* to vulnerability to misidentifications, not to an overall appraisal of the techniques. Our rating of the Breeding Bird Survey at the bottom of the breeding season list does not reflect any lack of confidence in the Breeding Bird Survey. There undoubtedly are numerous misidentifications on the Breeding Bird Survey, but we believe these are very few compared to the total numbers of birds reported (which average close to 1000 individuals per year on each of the 1700 or more routes). The purpose of the Breeding Bird Survey is primarily to monitor changes in abundance over a period of years, and assuming that identification errors are small in relation to the total birds reported and of about the same annual magnitude, their effects on the Survey would be minimal. Much more serious are misidentifications on Christmas Bird Counts where unusual birds, rather than the common species, receive special emphasis.

For purposes of comparing avifaunas among different habitats or different geographic areas, a combination of mapping census and banding would presumably give the most accurate results because: (1) chance of misidentification is minimal, (2) presence of late migrants could be detected by examining for fat deposition on captured birds, and (3) few species would go undetected.

We believe the same general ranking given above for reliability of identification would also apply to completeness of an avifaunal survey. Again the Breeding Bird Survey would fall last among breeding season techniques because it is based on brief samples rather than an effort to observe a high percentage of the birds present. As an example, we cite a Maryland study in which two experienced observers covered a familiar 50-stop Breeding Bird Survey route and observed simultaneously without communicating with each other. At the end of the survey they compared their observations stop by stop, ignoring the number of individuals of each species recorded but comparing only the list of species each observer had noted during each 3-min stop. There was not a single stop out of the 50 at which the two observers had recorded the same list of species. On another day on a similar coverage of the same route, the results were the same: there was not a single stop at which both observers had noted the same list of species. It was not until the third trial that the first identical species lists were recorded (at two stops). On each of the three days the total number of species recorded by the two observers was almost identical, so this was not a matter of one observer being better trained or more alert than the other. It was simply a matter of concentration on birds heard, and chance as to which birds seen were noted by one or the other observer.

SUGGESTIONS FOR MINIMIZING SPECIES ERRORS

Procedural recommendations.—Because of observer variability, including differential rates of recognition in different species, it helps to follow a carefully planned procedure in the field, especially if comparisons are to be made among geographic regions, habitats, or years, and most especially if studies are to be made before and after habitat alteration. We suggest: (1) Rotate observers carefully among plots, transects, or points so that censuses will be as comparable as possible and so that as few species as possible will be overlooked or misidentified. (2) Train all observers in advance with appropriate books, skins, records, or tapes. (3) Follow this with field training to include familiarization with field conditions and with call notes that are not available for study on tapes or records. (4) Examine field records of all participants for comparability prior to and in early phases of the actual study. (5) Standardize handling of difficult problems such as woodpeckers that are only heard drumming, mixed flocks of birds, and species that cannot be identified by the observer.

Report preparation.—The following comments are based on Robbins' experience in editing Breeding Bird Census reports for *American Birds* for many years, and in subsequent checking of 43 years of these reports for a computerized data bank. We believe that one of the most serious sources of error is not in identification of birds in the field but in the various processes that take place at the desk after the close of the breeding season. Thus a special word of caution is in order, and this applies especially to the professionals and other experts who are experienced in census taking, but tend to be careless in report preparation.

Because of the high probability of error in interpreting results and preparing reports for publication, we urge all census field workers and compilers to follow these procedures: (1) start an initial draft of maps, tables, and text early, before the fieldwork is completed, so as to become aware of any potential identification problems or the need for additional fieldwork; (2) in mapping territories or computing densities, refer to your own or other prior fieldwork and be as consistent as you can with procedures used in prior years; and (3) check carefully for transcribing errors and especially for omissions and incorrect mathematical calculations. Frequently one or more species, sometimes common ones, are inadvertently omitted from manuscripts submitted for publication!

Studies in Avian Biology No. 6:366–371, 1981.

REDUCING BIRD COUNT VARIABILITY BY TRAINING OBSERVERS

CAMERON B. KEPLER[1] AND J. MICHAEL SCOTT[2]

ABSTRACT.—During avian surveys, an important variable that affects our ability to determine such basic bird population parameters as species diversity, distribution, density, and population size is the competency of the observers. Skills such as visual and audio acuity, familiarity with the birds being counted, motivation, and willingness to make identifications must all be seriously considered when designing any avian survey. The importance of training observers and ways to do so are discussed in detail.

During a 5-year survey of the forest birds of Hawaii, we developed a 3-week training program that attempts to minimize the variability inherent in a team that has included more than 20 observers.

Our training program included screening of applicants for the physical, psychological, and academic skills mentioned above. We then provide successful applicants with visual (study skins, field guides, slides, etc.) and auditory aids (tape recordings) of species anticipated during the survey. Time is then spent camping under field conditions to provide maximum exposure to the birds. Two to four new team members train with each experienced observer. Problem species are dealt with on an individual basis. Simultaneous counts are used to determine progress. As identification skills increase, the number of species is increased and distance estimates are added.

Among the many variables that stand between the avian biologist and an understanding of such basic population parameters as species diversity, distribution, density, and population size is the competence of the bird observer. It is clear that all birders are not equal in experience with nor ability to distinguish between or correctly identify all species that may occur in a study area (Enemar 1962, J. T. Emlen 1971, Berthold 1976, Carney and Petrides 1957). Visual acuity, color sensitivity, peripheral vision, and hearing acuity are all important physical characters that vary among observers. Such psychological factors as concentration, motivation, attention span, alertness, endurance, ability to remember nuances of color, shape, or sound, and willingness to make identifications can all critically affect census results. Even the same observer can vary from day to day, or hour to hour, in any or even all of these factors. Berthold (1976), citing numerous European studies, notes that variations in estimates of numbers among observers amounting to 50% or more are common for many, or in some cases even all, species in a study area. The complexity of this problem has led some biologists to forsake avian censuses altogether (Berthold 1976), and others to suggest that uniformity and comparability of results can be achieved only by using the same set of observers in different areas, seasons, or years.

The need for comparability among observers is thus great. That this can be achieved through actual training has only been alluded to in the past (Svensson 1977a, Kimball 1949), while the use of "experienced" observers has been more frequently suggested as the solution to observer problems (Carney and Petrides 1957). However, even experienced birders vary considerably in their abilities, and we suggest that their lack of comparability is still a serious problem. Without accurate identification of species all other efforts to improve the reliability of bird counts are fruitless.

During the course of the five year Hawaiian Forest Bird Survey we became increasingly aware of the importance of carefully selecting and training observers. We have attempted to reduce their inherent diversity by screening all applicants, then involving the selected team members in an intensive training program. We have found that training is a critically important component of our overall experimental design, and describe procedures in this paper that reduce observer variability. The methods we describe have evolved over a five year period.

THE TRAINING PROGRAM

SELECTING OBSERVERS

Observers vary consistently in such physical characteristics as peripheral vision, color blindness, and visual and aural acuity. All applicants to survey programs who are known to have been exposed to hearing stresses (scuba divers, hunters, ex-military men, etc.) should be carefully screened before acceptance. Because hearing loss is in part age-dependent, young observers are generally better than older ones. All potential team members should have hearing tests (audiograms), from 1 to 8 KHz, and those with

[1] U.S. Fish & Wildlife Service, Patuxent Wildlife Research Center, Maui Field Station, 248 Kaweo Place, Kula, Hawaii 96790.

[2] U.S. Fish & Wildlife Service, Patuxent Wildlife Research Center, Mauna Loa Field Station, P.O. Box 44, Hawaii National Park, Hawaii 96718.

serious hearing loss (20 db or greater?) within the range of bird vocalizations used for non-counting tasks or rejected. Years of experience with species may compensate for partial hearing loss and this should be considered when evaluating observers (Emlen and DeJong 1981). Faulty vision, if corrected, is not a problem. However, it may be appropriate to test for tunnel vision.

Psychological factors of known importance, such as motivation, attention span, and willingness to make identifications, should also be considered in the initial screening. For survey work in remote areas requiring camping, factors such as camping and hiking experience, and willingness to remain away from family or the comforts of civilization, should be weighed heavily in the selection process. While birding experience is a key selective factor, we have found that familiarity with the species occurring in the area is not, if the avifauna is simple. Inexperienced observers can be trained, and they learn rapidly.

Laboratory Training

All new observers should begin training indoors. This ordinarily involves an initial full day, followed by short review sessions during the field training period. Trainees should receive field guides and lists of anticipated species. Handouts that identify problem species or confusing groups of species can be provided. Other essential written materials include relevant publications, details of survey methodology, and background natural history information on the area's avifauna.

The heart of the laboratory session should involve an inspection of study skins of all anticipated species, including various morphs, age, and sex classes. Observers can then work with the skins in conjunction with field guides and other written materials. During this session informal spot quizzes and comparisons between similar species can be made. Slides detailing birds, forest types, terrain, and former survey activities are extremely useful, and provide a forum for informal tests that sharpen the participants' skills, especially after their initial exposure to the skins, and some field experience. This material needs to be available throughout the training period, and as observers learn more in the field, they can return to review it.

Tape recordings of songs and call notes should also be provided. Species can be arranged phylogenetically as well as by similarity of sound. When possible, training tapes should be of birds actually recorded in the study area to avoid confusing dialects. Observers need to listen to these tapes throughout the training session, and during the field season when questions arise.

Field Training

From the lab, trainees can be taken into the field and provided with maximum exposure to the birds by camping in the study area. Selection of a site with a small number of species will allow trainees to focus their attention. The first day divide trainees into small groups (two to five) led by an experienced observer. The leader can point out and identify as many birds as possible: "That's an Apapane, that's an . . . , notice the wing bars, etc." The initial objective is to maximize the trainee's exposure to a single set of species. This experience can be augmented at camp by listening to tapes and checking field guides and other identification aids. Trainees should be encouraged to ask questions. A frank statement that even the most experienced observer can't identify every bird is needed. This guided training should be continued as long as necessary, for it offers an excellent opportunity for the new observer to ask "what's that? Why? How does it differ from . . . ?"

Simultaneous counts

Trainees learn a reasonable number of calls and songs in the first day or two. After this the most important training exercise—simultaneous 5-minute counts—should be introduced. An experienced observer and 2 to 5 trainees simultaneously, and independently, record all species seen or heard from one spot (station). At the end of each count the different species lists are compared. These verbal exchanges strongly reinforce the data, and problem species are quickly identified whether they are missed or misidentified. We have consistently found that some common and conspicuous species are missed, even though trainees can readily identify them when they are pointed out. We call these birds "window species," for observers listen right through them without detecting them. Most observers have their own unique constellation of window species, and usually only realize this when they repeatedly fail to detect species that other observers consistently record. The window species are thereafter stressed on an individual basis, and trainees enthusiastically help each other.

Simultaneous counts are best conducted at the start of each day's work. Trainees can then identify species that are giving them difficulty, and work on them later that day. The whole process is simply one of focusing on an increasingly smaller number of problem species. The importance of mixing experienced and inexperienced observers can not be overstressed. Experienced observers provide instant feedback as to the identity of a bird, especially during camping, when everyone wakes up, eats, and drinks, to bird vocalizations. This provides maximum exposure at a time when there is no substitute for field experience.

Most bird survey teams are composed of a mix of new and old observers. Observers from previous seasons need some training, as skills erode with time, but normally do not need an entire training course. They can be brought into the training program at any time, or used as additional teachers. Trainees progress at different rates. Quick learners can be used as part-time trainers to increase their own accuracy and confidence. This is particularly valuable in afternoon sessions, where observers work "one-on-one" to point out to each other the window species discovered during the morning's simultaneous counts.

Simultaneous counts should be used continuously throughout the training period to identify problem species, either those with which trainees lack experience, are overlooking, or are ignoring, for whatever reasons. They are also an excellent measure of progress, provide important motivation, and prepare group members for the actual survey.

When trainees feel comfortable identifying and recording species, they should be asked to record the number of individuals detected during count periods. At this time tell trainees how to search an area around the station. Make sure that observers check directly overhead, rotate 360° if variable circular counts are used (Reynolds et al. 1980), and vary the distance being scanned. Searching at fixed distances is to be avoided as this biases density estimates.

It is important to shift training sites to add new species, terrain, and habitat types. This also improves training efficiency by reducing predictability and monotony. Trainees are usually ready to move to new areas every three or four days.

Distance estimation

Estimating or measuring the distance to detected birds is an integral part of most attempts to determine their numbers. It is critical that these measurements be as accurate as possible. Indeed, Burnham et al. (1980) suggest that "tape measure" precision is required. The use of range finders and flagging at known distances from the observer helps when birds can be seen. However, most observations are of birds heard and not seen (81% of all detections during the 1980 Maui Forest Bird Survey). This requires that distances be estimated. Under ideal conditions, practice can lead to ±10% accuracy (Scott et al. 1981b).

When trainees competently record species and individuals, it is time to introduce them to distance estimation. Working in groups of two to five, they should independently estimate distances to trees, rocks, and other clearly seen fixed objects, then measure the actual distance using tape measures or ranges finders. This exercise must be repeated until trainees are consistently within 10% to 15% of the measured distance. They can then begin estimating the distance to birds seen and heard, then to birds heard only. One observer identifies an unseen vocalizing bird, indicating its direction to the other trainees. They independently estimate its distance. One member of the group then locates it, moves to a point directly under it, and measures the linear distance back to the other observers. Care must be taken to make certain that the bird has not moved before its distance is measured, and that it is the individual originally identified. This exercise is combined with the simultaneous counts and identification work with selected species. Playbacks from hidden tape recorders might also be used to estimate distances (Emlen and DeJong 1981). Distances should be estimated with as many different calls, songs, and species as possible. After trainees achieve an average ±10% accuracy, distance estimation becomes a part of the simultaneous counts.

Beginning the survey

When trainees know the local birds and have mastered distance estimation, they should begin counting under actual survey conditions. Have two observers simultaneously census the same stations or strips to check their progress. When this exercise is concluded, check the field forms and ask the observers if they feel competent to collect real data—their own confidence is important. Quick learners may begin the actual survey while slower trainees continue parts of the training program. Any necessary additional training should be alternated with practice surveys until all trainees are ready to begin the actual counts. In Hawaii, the entire training program takes 12 to 15 days, with experienced observers from former years joining the training session on day 6.

During the actual survey, observers should refer to field guides, tapes, and other aids in order to remain sharp, and to sort out occasional unidentified birds. In the 1980 Hawaii Forest Bird Survey (on Maui) pairs simultaneously censused together (Scott and Ramsey 1981b). For the first four weeks, each pair consisted of an experienced and a new observer, and their interactions after each count provided instant feedback on uncertain species. Thus, in a very real sense, the initial training period extends throughout the field season.

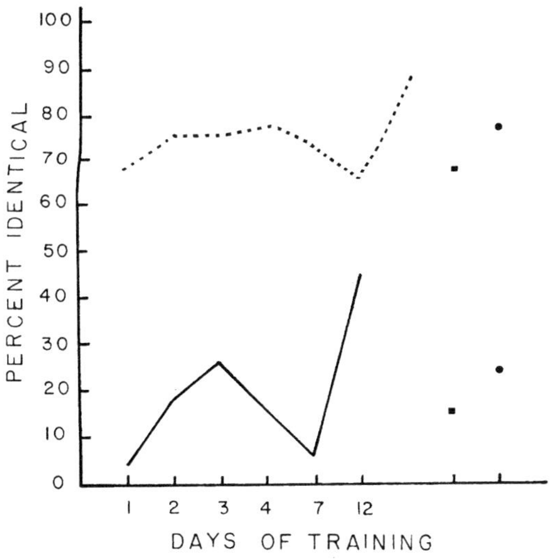

FIGURE 1. Measures of similarity between paired observers. The solid line indicates percent identical species lists, the dotted line the percentage of species simultaneously recorded on the same station counts, Maui Forest Bird Survey training session. The symbols represent the same measures for randomly selected station counts for all Maui (squares) and for ohia forest only (circles) during the actual survey.

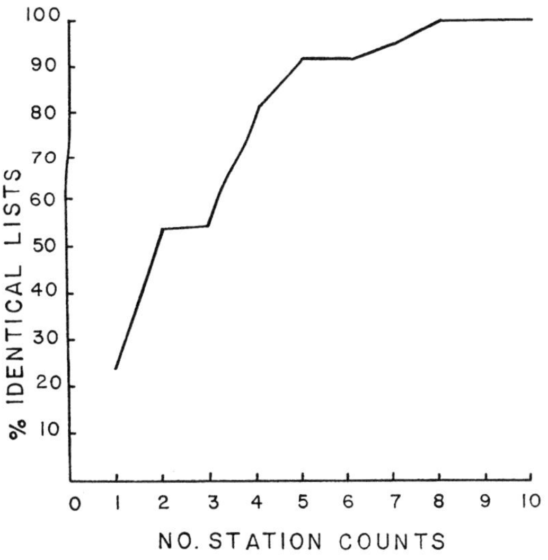

FIGURE 2. The frequency of identical species lists between paired observers as a function of the number of stations sampled, Maui Forest Bird Survey, 1980.

MEASURING OBSERVER VARIABILITY

The following data on bird count variability were taken from simultaneous 5-minute counts made during the 1980 Maui Forest Bird Survey training session, and from actual paired station counts during the survey. The training session counts were simultaneous in time and space, with all observers standing within about 3 m of a central point. They recorded all birds heard and seen, and, on day 7, added distance estimates to each observation. During the survey, observers stood 18.3 m apart, one 9.2 m upslope, the other 9.2 m downslope, from a central station, and this separation contributed importantly to the differences between them. Count periods were eight minutes, and distances to each detected bird were estimated.

Congruity of Species Lists

The simplest measure of similarity between observers is a comparison of species lists. The results can be unexpected. Robbins and Stallcup (1981) matched paired observers on 100 3-minute Breeding Bird Survey stations in Maryland and found no identical lists. Only when 150 stations were compared did they find any (2) that were the same.

The Maui Forest Bird Survey training program began with very low comparability (3%) on the first day. These observers were not naive on day 1. Two of them had participated in the Hawaii Forest Bird Survey for at least two years, on other islands and all others had studied skins, field guides, and tapes. They improved rapidly with training (Fig. 1), reaching a peak (46%) on the last day. The dip between days 3 and 7 corresponds to a period when observers were beginning to identify subtle call notes. Also, the training site was changed, distance estimates were added to the count on day 7, and training was interrupted by a weekend, all factors that tend to reduce comparability.

In a random selection of 50 paired station counts from the 1298 surveyed during the 1980 Maui Forest Survey, only 8 (16%) had identical lists (Fig. 1). This low figure resulted from a total number of species recorded per station that ranged from only three to nine ($\bar{x}$ = 5.6) in the simple Hawaiian ecosystems. The sample was drawn from a variety of dry and wet forest habitats. We wished to look at variability within a single prime forest ecosystem, and chose upper elevation (1372 to 1982 m) ohia (*Metrosideros collina*) forest inhabited by the Crested Honeycreeper—the same system in which our final training session was held. Thus we can directly compare survey results with training day 12. Four transects were randomly selected from the 13 that crossed this forest. Forty-five stations sampled the habitat; on 11 (24%) of them the paired observers had identical species lists (Fig. 1). The large difference between this and train-

TABLE 1

POOLED DATA FOR EIGHT 5-MINUTE COUNT PERIODS FROM DAY 12, MAUI SURVEY TRAINING SESSION, 20 MAY 1980. GROUPS A AND B WORKED INDEPENDENTLY IN SEPARATE AREAS, AND CAN NOT BE DIRECTLY COMPARED

	Group A Observer			Group B Observer		
Species	4	3	5	2	6	7
Amakihi[a]	10	8	12	17	16	18
Maui Creeper	25	24	31	31	27	28
Maui Parrotbill	2	0	1	0	1	0
Apapane	64	65	61	80	88	81
Crested Honeycreeper	23	26	17	22	21	24
Iiwi	16	11	9	22	18	15
Red-billed Leiothrix	5	4	3	9	12	8
Japanese White-eye	13	8	5	7	9	6

[a] See Table 2 for scientific names.

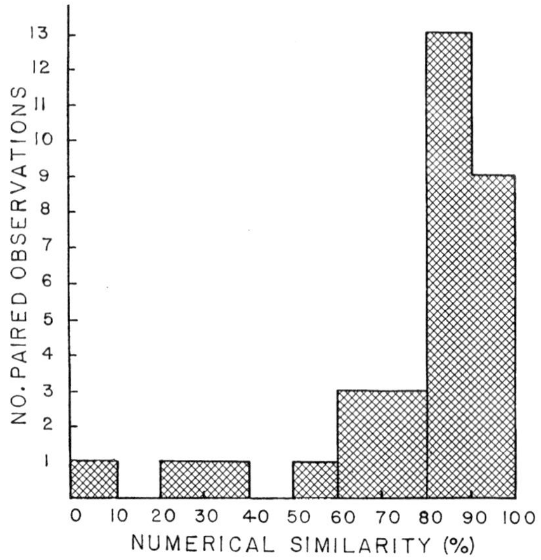

FIGURE 3. Numerical similarity between paired observers in the number of individuals recorded per species during the Maui Forest Bird Survey, 1980. The data set is presented in Table 2.

ing day 12 (46% identical lists) is largely a measure of the effect of the 18.4 m distance separating observers under actual survey conditions.

CONGRUITY OF LISTED SPECIES

A much clearer view of observer variability is provided by comparing the frequency at which species are recorded in common. During the training session this ranged from 67% (day 1) to 86% (day 12); daily sample sizes (number of species recorded, station pooled) ranged from 131 to 541, and totalled 1767. Of these, 1330 (75%) were simultaneously recorded by both observers. The improvement shown from day 1 to day 12 is significant ($\chi^2 = 34.56$, $P < 0.001$).

During the 50 randomly selected survey counts, species were recorded 280 times: observers found 189 (67%) of them during the same count. The remaining 91 were recorded by only one observer during a count period. Within the ohia forest, 10 species were recorded 289 times, 226 (78%) by both observers during each station count.

DECREASING VARIABILITY BY INCREASING COVERAGE

The above examples indicate that even in simple systems it is unrealistic to expect observers to obtain identical species lists. This is, however, far less important than the overall congruity between observers for listed species (67% in all habitats, 78% in ohia forest), and the relationship between species lists over repeated samples. By increasing sampling coverage the apparent differences, as reflected in a compari-

TABLE 2

SURVEY RESULTS FOR PAIRED OBSERVERS IN OHIA FOREST ON FOUR RANDOMLY SELECTED TRANSECTS, MAUI FOREST BIRD SURVEY, 1980. THE NUMBER OF STATIONS SAMPLED ON EACH TRANSECT WERE: TR. 3 (12), TR. 9 (9), TR. 10 (13), AND TR. 18 (11)

	Tr. 3 Observer		Tr. 9 Observer		Tr. 10 Observer		Tr. 18 Observer	
Species	1	2	3	4	3	1	4	5
Amakihi, *Loxops virens*	22	23	25	22	18	17	15	9
Maui Creeper, *Loxops maculata*	48	52	20	25	21	26	4	0
Maui Parrotbill, *Pseudonestor xanthophyrs*	0	0	1	1	3	2	0	0
Apapane, *Himatione sanguinea*	38	34	26	22	31	37	27	33
Crested Honeycreeper, *Palmeria dolei*	52	51	10	10	10	14	32	33
Iiwi, *Vestiaria coccinea*	10	10	25	31	5	22	18	13
Melodius Laughing-thrush, *Garrulax canorus*	0	0	0	0	0	0	1	3
Red-billed Leiothrix, *Leiothrix lutea*	65	40	23	20	28	33	17	14
Japanese White-eye, *Zosterops japonica*	23	19	11	13	5	9	15	12
Cardinal, *Cardinalis cardinalis*	0	0	6	6	0	0	0	0

son of single counts, decrease dramatically. In the 10-species ohia forest system, observers on each of the four transects recorded identical species when all counts within the habitat (from 9 to 13 per transect) were summed, with a single exception. In this instance one observer recorded a Maui Creeper (*Loxops maculatus*) missed by the other. Rare birds do, obviously, increase the likelihood of difference.

In Figure 2 we present data illustrating the number of samples needed to achieve identical species lists (excluding the creeper). The number of station counts are of adjacent pairs, triplets, etc. of stations along each transect. For example, three station comparisons include stations 1–3, 2–4, 3–5, etc. for each transect. Clearly, increasing coverage rapidly improves congruity between observers. It is a relatively simple matter to test for this effect under any set of habitat or species richness conditions, and such a test should be an integral part of one's experimental design. In the Maui example, a minimum of five counts per transect in prime habitat effectively overcomes much of the inherent variation between observers, even when they stand relatively far apart (18.3 m). We must stress, however, that merely increasing sample size is no substitute for a training program, and that coverage or sampling intensity may have to be increased considerably to detect rare or elusive species.

Numerical Differences among Observers

Bird observers detect differing numbers of individuals during their surveys, a fact that has provoked considerable concern (Berthold 1976). Fortunately, these numerical differences can be reduced by training. During the 5-minute simultaneous counts on training day 1, fully 41% of all species comparisons between observer pairs revealed differences of greater than 50%. By day 12 variation of this magnitude had decreased to 13%. During the same period the frequency of paired observations showing less than 20% difference increased from 26% to 52%. The numerical similarity between observers was impressive (Table 1), particularly for the common species. The highest percentage differences between observers generally occur with the rarer species.

We have 32 sets of paired observations for the Maui survey in ohia forest (Table 2), with from 2 to 105 individuals of each species recorded by the two observers. The majority (69%) of these sets agree within 80% (Fig. 3). At the low end, observers recorded a difference of greater than 60% for only three species. Important here is that two of these species, rare on their transects, were only found four times, and one of them (Maui Creeper) was not detected by one observer. There are times, however, when observer differences are extreme for relatively commonly encountered species (see Transect 10, Iiwi). Fortunately, such differences have been relatively rare (3% in this sample) in Hawaii, and their rarity is in part a function of the intensive training program and, perhaps, the simple fauna. Such differences also have another important cause, and this generally traces to an inequality in hearing acuity between observers. Thus an observer with "good ears" is sampling a larger area by hearing more distant birds. For the Iiwi on Transect 10, the median detection distance for Observer 1 was 58 m, while that for Observer 3 was only 22.9 m. Thus although Observer 3 only saw 23% of the birds recorded by Observer 1, he was sampling only 16% of the area. The effect of these differences in numbers is largely moderated when bird densities are calculated from distance estimates.

DISCUSSION

That differences exist among observers is indisputable, and they must be seriously addressed in any bird sampling program. Although many physical and psychological factors can not be eliminated, their impact can be greatly reduced by an extensive program that begins before observers are selected. Careful initial screening of applicants can eliminate the more obvious visual, aural, and psychological factors that increase observer variability. A rigorous observer training program further reduces inherent variation, but does not eliminate it. Such training must place heavy emphasis on distance estimation, for the ability to estimate distances accurately offsets inherent differences in aural and visual acuity when variable circle counts or certain linear transect counts are used. We must remain aware that observers are not perfect nor identical, and that we must seriously address and decrease this source of variation in order to minimize its impact upon our data, and hence our understanding of those population parameters we are attempting to evaluate. The reliability of "experienced" observers can be improved by pairing them and conducting simultaneous counts and thereby identifying problem species. Elimination of problem species for experienced observers might only take a day or two of work with audio and visual aids.

ACKNOWLEDGMENTS

The development of the training methods for the Hawaii Forest Bird Survey evolved over a five-year period. Its present format is the result of input from a large number of people. We particularly thank Philip Ashman, Tim Burr, Tonnie L. C. Casey, Jack Jeffery, and Peter Pyle for their valuable comments. Fred L. Ramsey has provided insightful comments and inspiration during the entire period.

Studies in Avian Biology No. 6:372–376, 1981.

THE INFLUENCE OF OBSERVER AND ANALYST EFFICIENCY IN MAPPING METHOD CENSUSES

RAYMOND J. O'CONNOR[1]

ABSTRACT.—Four observers of varied census experience conducted independent mapping method censuses of a 28.7 ha scrub habitat in the English Chiltern Hills during the 1977 and 1978 breeding seasons. Three trained analysts independently assessed each of the 1977 maps and showed a high degree of mutual consistency of interpretation, independent of field experience on the census plot. The four observers differed significantly as to the density of territorial clusters (all species pooled) they recorded but the absolute range of the four estimates was only 19%. A team of two very experienced field workers detected more birds than did a similarly experienced observer operating alone, and he in turn detected more birds than solo observers with no and two years previous census experience; these differences were partly explained by experienced workers spending longer on each field visit. Pairing of observer results across years eliminated the influence of these observer differences on the four estimates of the year on year change in bird density. The coefficient of concordance between the four observers' estimates of population changes was 0.64, based on data for 21 different species. Thus, population changes can be assessed accurately from mapping method data if the same observer is involved in both censuses and the analysts are properly trained but use of absolute densities requires consideration of observer field ability.

The mapping method (Enemar 1959) is widely regarded as the best available approximation to the true distribution/density of territorial birds in a census area. As such it has been used as a standard to calibrate other census methods, such as the French IPA system (Blondel et al. 1970) and to calibrate studies of census efficiency. There has, however, been little effort to assess the reproducibility of results acquired with the mapping method, despite the known existence of potentially serious sources of error in both fieldwork and interpretation components of the method (Svensson 1974b, Best 1975, Moss 1976). In the Common Birds Census (CBC) scheme of the British Trust for Ornithology—the major systematic users of the mapping method—the censuses are used primarily to compute an annual index of population change (Williamson and Homes 1964, Bailey 1967) and observer effort has been found to be sufficiently consistent from year to year to remove the effects of differences in census efficiency between observers (Taylor 1965). A number of field investigations have reported comparisons of observer census efficiency and broadly agree in reporting significant correlation but recognizable discrepancies between censuses (Snow 1965, Enemar 1962, Hogstad 1967, Enemar et al. 1978).

The present paper reports preliminary results of a systematic field trial of the effects of observer and analyst on the assessment of population densities and year-to-year population changes, using the mapping method of the BTO Common Birds Census scheme. The paper is a written version of a poster paper displayed at the present symposium, to make the principal findings of the study immediately available. The full findings will be described in a report on this and related matters in preparation for the UK Nature Conservancy Council (O'Connor and Marchant in prep.).

[1] British Trust for Ornithology, Beech Grove, Tring, Hertfordshire HP23 5NR, United Kingdom.

MATERIALS AND METHODS

The field experiment consisted of four observers independently censusing a common plot with their results subsequently interpreted independently by each of three analysts. The observers repeated their census of the plot a year later, to provide data on observer influence on estimation of population changes.

The observers and analysts were chosen to provide a cross-section of census experience (Table 1). All were competent ornithologists but their prior census work differed widely. One was previously familiar with the census plot chosen (Observer C), the other three being unfamiliar with that particular plot. Since Best (1975) suggested an observer familiar with the census plot produced more accurate interpretations of the registration maps later, the three analysts included one who also conducted field census on the plot (Observer A = Analyst W), one with a slight knowledge of the plot through conducting two sets of point count and belt transects there (Analyst X), and one totally unfamiliar with the plot except through sight of the habitat map for the plot. All three had, however, previous training in the interpretation of the IBCC and BTO guiding principles for map interpretation, it being established within the BTO experience that naive analysts do not follow these instructions adequately.

Fieldwork was conducted on a 28.7 ha census plot on Beacon Hill, within the National Nature Reserve at Aston Rowant (Oxfordshire, England). The Hill forms a prominence (244 m asl) extending northwest from the Chiltern escarpment which itself runs mainly southwest to northeast. The Reserve exists for its chalk grassland but the Hill carries much scrub and small woodland. Thus, the plot boundaries encompassed a wide range of habitats, from open well grazed chalk grassland to rough grass paddocks, de-

TABLE 1
INDIVIDUAL OBSERVER AND ANALYST EXPERIENCE OF MAPPING METHOD AND OF THE CENSUS PLOT AT ASTON ROWANT

OBSERVER EXPERIENCE

Observer	Previous field census experience	Previous experience of Aston Rowant
A	Variety of census plots over 10 years	Nil
B	Nil	Nil
C	Two years census work at Aston Rowant	2 years
D[a]	Variety of census plots over 10 years	Nil

ANALYST EXPERIENCE

Analyst	Previous CBC analysis experience, years	Knowledge of Aston Rowant census plot
W	9	Census work as Observer A above
X	4	IPA and belt transect work for this study
Y	0	Nil

[a] Two field workers operating as a team and in field together.

TABLE 2
TWO-WAY ANALYSIS OF VARIANCE FOR 1977 TERRITORY TOTALS WITH RESPECT TO OBSERVER AND ANALYST

DATA TABLE

Analyst	Observer A	B	C	D	Totals	Mean
W	292	259	239	305	1095	273.8
X	277	254	255	294	1080	270.0
Y	277	256	254	302	1089	272.2
Total	846	769	748	901	3264	
Mean	282.0	256.3	249.3	300.3		272.0

ANOVA TABLE

Source of variance	Degrees of freedom	Sum of squares	Mean square	F
Analyst	2	28.5	14.25	0.19
Observers	3	4986.0	1662.00	13.87*
Residuals	6	359.5	119.83	

* $P < 0.01$.

veloping scrub (particularly elder) and mature closed beech woodland. This diversity of habitat provided a dense and varied bird community further increased by the "leading line" effect of the Chilterns in bringing migrants to the census plot. The site thus provided a severe test of observer and analyst consistency, mitigated only by an abundance of numbered marker posts and a good network of paths across the plot. The general convexity of the hill precludes very distant sight and sound registrations except within an area of paddocks. There was also some loss of song registration on the southern edge of the plot, due to traffic noise from the adjacent M40 motorway.

Four observers conducted independent ten-visit mapping censuses on the plot in both 1977 and 1978, working to the Common Birds Census guidelines issued by the British Trust for Ornithology. As far as possible, clashes in visit dates and times were avoided by prior arrangement amongst the observers, but occasional spells of poor weather resulted in three cases of two observers on the site simultaneously and one case of three observers simultaneously present. Fieldwork was confined to fore-noon visits. Data from mapping visits were recorded on blank maps prepared for the study and incorporating sufficient detail to allow those observers new to the plot to position themselves accurately at all times. On completion of fieldwork the observers collated their visit map data to generate species maps. These were then photo-copied for systematic assessment by the analysts involved in the study.

The interpretation of the clustered data of these maps was performed in the standard manner defined by the published Common Birds Census Principles (Williamson et al. 1968). Three trained CBC analysts on the BTO staff independently assessed all species maps in 1977, each using his own copy of the maps. In 1978, when assessment of analyst variation was not desired, the map interpretation task was shared by the two analysts then available, each analyzing approximately 120 maps. The analysts differed substantially in experience and knowledge of the census plot (Table 1), this variation being part of the experimental test. However, all three analysts had previously been trained to adequate standards of compliance to the standards defined in Williamson et al. (1968) and were engaged in routine analysis of the annual Common Birds Census returns in parallel with the present study.

Other details of the study site, field procedure and analytical criteria will be documented in O'Connor and Marchant (in prep.).

RESULTS

INFLUENCES ON POPULATION DENSITY ESTIMATES

Table 2 presents the overall results of the 1977 fieldwork, without regard for the specific identity of the species mapped. With each analyst interpreting independently the mapped registrations of each of the four observers, the total range of cluster estimates amongst the three was only 3.8 or 1.4% of the average population of 272 clusters. By contrast, the range in estimates obtained from the four field workers was 51 clusters or 18.8% of the average estimate. Table 2 includes a formal analysis of variance of the data and shows that the differences between observers were statistically significant, whilst those between analysts were negligible.

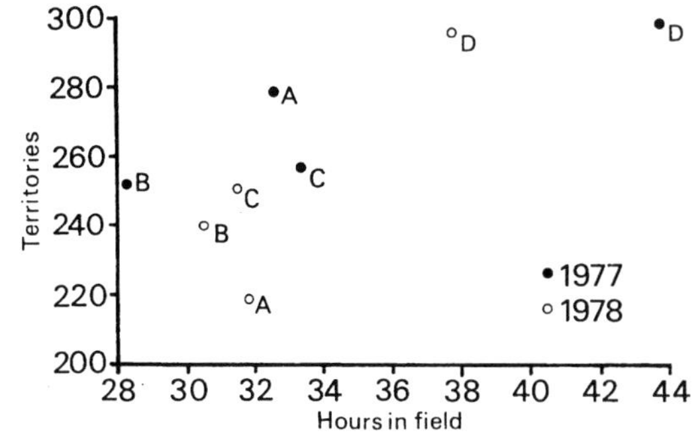

FIGURE 1. The relationship of territories assessed and time spent in field for a constant ten census visits. Letters indicate the observers described in Table 1.

Table 2 shows that observer D—a pair of very experienced census takers in the field together—provided registrations yielding the greatest number of clusters when analysed. The other three observers averaged 262.5 clusters against the maximum 300.3, the difference of 37.8 clusters having a confidence interval of 15.5 (Snedecor and Cochran 1967:301). Similarly, comparisons of the results obtained by the highly experienced solo observer A—mean of 282.0 clusters—against those of the less experienced observers B and C (cluster averages of 256 and 249 respectively) show a difference exceeding their LSD (ibid.) of 18.9 clusters. Thus, even amongst the solo workers extensive experience of the mapping method—in the form of CBC participation—can lead to an improved detection of breeding pairs on the plot.

Figure 1 suggests that the duration of fieldwork was a component in the better field performance of observer D. The CBC fieldwork guidelines do not set down specific targets for field time, though observer *consistency* of effort between years is requested. The figure shows that observer D spent substantially longer on the plot both in 1977 and in 1978 and that their link with greater cluster totals was reflected in an overall (across years and observers) correlation with field time. Since bird density on the site could (and did) vary between years there is no a priori requirement for overall correlation. Within each year the correlations were positive but not significant with only four data points (1977: $r = 0.802$, $P < 0.3$; 1978: $r = 0.870$, $P < 0.2$). Combining the within year correlations by z transformation (Snedecor and Cochran 1967) gave a pooled correlation corrected for differences in bird density of 0.840 ($P = 0.085$).

Influences on Population Monitoring

Although the analysis of Table 2 established the existence of significant differences of census efficiency between observers it left open the possibility that estimates of year on year population changes were independent of observer efficiency. That is, if observers differed between themselves as to absolute census efficiency but maintained those differences from year to year the resulting estimates of population change would be insensitive to observer ability. Figure 2 shows the population changes assessed by each of the four observers for 21 species with adequate sample sizes on the census plot in both 1977 and 1978. The four distributions are similar (Median test $\chi^2 = 2.46$, n.s.), indicating the observers showed no gross differential in census bias between years. A more powerful test for observer influence is to match observers across species since some species increased whilst others decreased on the census plot between the two years. Use of the non-parametric Friedman two-way analysis of variance (Siegel 1956) gave $\chi^2_r = 8.24$ ($0.05 < P < 0.10$). The data thus come close to demonstrating a slight statistical bias on the part of observer A (Fig. 2) but the effect is very slight and dependent on the collective analysis of all 21 species. More detailed analyses for individual species show that in no case was there evidence of observer bias in estimating species population changes.

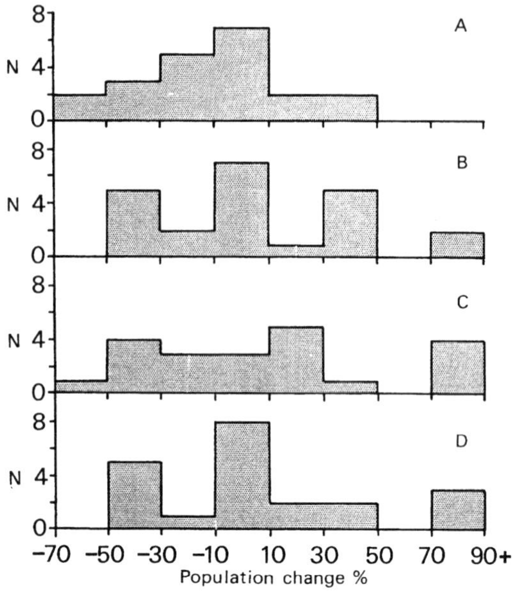

FIGURE 2. The distributions of estimates for 1977–78 population change in 21 species, as assessed by four independent observers A–D (described in Table 1). See text for statistical analyses.

An alternative analysis of the same problem asks not if the four observers differ in estimating population changes but whether these estimates show significant correlation across species. The agreement between the four field workers was therefore assessed by computing Kendall's coefficient of concordance (Siegel 1956) for the data of Figure 2. The coefficient obtained was W = 0.642 (χ^2 = 51.38, df = 20, $P < 0.001$), to be compared with the value of unity for perfect consistency. There was, therefore, very significant agreement between observers to the population changes undergone by this group of 21 common species.

DISCUSSION

The finding that analyst variation in the interpretation of the mapping method results was negligible is of some significance in the light of previous reports by Best (1975) and Svensson (1974b) who found major differences present. Svensson's study was the more substantial and reports a comparison of interpretation of a common set of species maps by 58 ornithologists of varied experience. Coefficients of variation in estimates for the six species tested ranged from 16 to 36% with some evidence of a slight increase (not statistically significant) with analyst experience. A major complaint of these workers was that they lacked habitat details for the test maps, a point relevant to Best's (1975) report of more accurate results from workers well acquainted with the plot. In the present study the analysts had access to the habitat maps, undoubtedly a factor in improving their performance relative to Svensson's workers. On the other hand, the present data provide no evidence that analyst W was systematically biased in interpretation procedure as a result of his field knowledge of the census plot. It must be remembered, though, that all three analysts had been trained to achieve consistent standards of interpretation of CBC, to allow them to undertake routine analysis of CBC returns, and this would appear to be the most important conclusion of this aspect of the present study.

Differences between observers were far more important than those between analysts but were nevertheless surprisingly small (about 19% in range Table 2) given the wide differences in observer census experience (Table 1). Enemar (1962) has previously compared the census efficiency of six ornithologists (one of whom had several years previous experience of the census plot) in the course of a single census visit and found considerable variation between observers, with a slight systematic bias in favour of the experienced observer. Within a ten-visit sequence, however, one would expect a reduction in variation because of the binomial cumulation of registrations against a fixed threshold for cluster acceptance (Svensson 1979a). On the average, any two of Enemar's (1962) observers had 75% of their birds in common, indicating a 50% visit efficiency. Such a value would fit Hogstad's (1967) study of four observers and Enemar et al.'s (1978) study of four and of three (in different years) observers. Variation amongst observers in this region of efficiency are greatly reduced by the process of visit cumulation (Svensson 1979a). Other multi-visit mapping censuses agree with the present findings as to relatively small overall variation in population estimates. Chessex and Ribant (1966) found a correlation of 0.990 between the results of two independent censuses of 21 species and Snow (1965) found that paired independent censuses of each of four farms in England were correlated at between 0.824 and 0.964. These reports thus agree with the findings here of significant but probably tolerable variations between different observers using the mapping method.

The link between observer efficiency and time in field indicated by Figure 1 is suggestive but, because of the confounding of variables present, not conclusive. Particularly interesting is the possibility that the confounding of time in field and previous census experience is genuine, with experienced observer's deliberately spending more time over each visit. Svensson (1979a) concluded that improving the effort of individual visits was the best option for improving the overall efficiency of a mapping census. In the same vein, Tomiałojć (1980) recommended concentrating on acquiring high quality registrations (simultaneous song, territorial boundary disputes, etc.) in enhancing census efficiency, a point met by greater time per visit.

The analysis of Figure 2 showed that observers were broadly consistent in their estimation of population changes between years, despite their differences in absolute efficiency. This is a particularly important validation of the large scale use of the mapping method for population monitoring, as in the BTO Common Birds Census, since it provides field data supporting the statistical evidence of between-year observer consistency provided by Taylor (1965). Nilsson (1977b) has previously reported that estimates of population changes for titmice (Paridae), Nuthatch (*Sitta europaea*) and Tree Creeper (*Certhia familiaris*) in two Swedish woodland plots were poorly correlated with changes assessed from intensive study of these populations (using intensive mapping, color-ringing, and nest searching). He identified the causes of these low correlations as phenological variation in breeding activity with respect to the census period. In Britain such variations are less pronounced (O'Connor 1980c) which presumably reduces a possible source of observer variation in assessing population change in the present study. The agreement between observers within the Aston Rowant data (Kendall concordance of 0.64) is very comparable with paired observer correlations of 0.65 and 0.88 for the changes assessed by three observers studying 13 species on a Swedish census plot (Enemar et al. 1978).

Overall, therefore, the present study indicates that with suitable training, map interpretation can be made highly consistent between individuals whilst observer pairing across years adequately eliminates the demonstrable observer bias. For population monitoring purposes, therefore, the mapping method is adequately accurate. For density assessment, on the other hand, observer differences must be taken into account, a point of particular importance when using mapping as a reference standard. Finally, it must be emphasized that the absolute efficiency of the mapping method—the proportions of territorial birds on the plot actually detected—has not been addressed at all by the work reported here.

ACKNOWLEDGMENTS

The fieldwork for this study was undertaken by L. A. Batten, A. J. Prater, P. Toynton and P. and E. Willson and their participation is gratefully acknowledged. Map interpretation was undertaken by L. A. Batten, P. A. Hyde and J. H. Marchant. R. K. Hicks and Caroline Hunt helped with the tasks of map preparation, sorting and collation of results. I am grateful to Drs. R. J. Hornby, P. M. North and to S. M. Taylor for advice in the design and analysis of the experiment. The study was commissioned by the Nature Conservancy Council under a contract with the British Trust for Ornithology for the supply of ornithological services; their support is gratefully acknowledged.

Studies in Avian Biology No. 6:377–381, 1981.

EFFECTS OF OBSERVERS USING DIFFERENT METHODS UPON THE TOTAL POPULATION ESTIMATES OF TWO RESIDENT ISLAND BIRDS

SHEILA CONANT,[1] MARK S. COLLINS,[2] AND C. JOHN RALPH[2]

ABSTRACT.—During a 5-week study of the Nihoa Millerbird and Nihoa Finch, we censused birds using these techniques: two line transect methods, a variable-distance circular plot method, and spot-mapping of territories (millerbirds only). Densities derived from these methods varied greatly. Due to differences in behavior, it appeared that the two species reacted differently to the observer. Millerbirds appeared to be attracted to a moving observer, perhaps to forage on insects; finches appeared to be attracted to a stationary observer in order to feed on seabird eggs temporarily abandoned during the count. Although these behaviors may be unusual, they dramatically demonstrate that no single census method will suffice for all species. The method that assures the least observer effect will provide the most accurate population estimate.

The ornithological literature contains few serious attempts to determine total species populations; the few cases are of very rare or insular populations, usually both. Because of the lack of immigration and emigration geographically inherent in island environments, the census of resident island bird populations can provide important insights into many questions of avian biology. We discuss here two such species' populations and demonstrate the potential of different census methods that result in quite different population estimates.

Nihoa Island, a volcanic high island remnant, is the easternmost of the Northwestern Hawaiian Islands, which largely make up the Hawaiian Islands National Wildlife Refuge. The island is 63.2 ha in area, rising to 273 m, with an average southward facing slope of 45°. The east, west, and north coasts are sheer cliffs, and the south coast consists of low (10–20 m) cliffs skirted by rock benches. There is one nearly inaccessible beach. The vegetation is very low, rarely exceeding 0.75 m high, and is largely made up of three shrub species (*Sida fallax, Solanum nelsoni,* and *Chenopodium oahuense*).

Nihoa is one of the only two sizable high islands (Necker is the other) of the Northwestern Hawaiian Islands. As such, it supports a biota that is unique in several ways, as compared to the refuge's atolls. Nihoa has endemic plants (Herbst *in* Clapp et al. 1977), several arthropod taxa (Beardsley 1966) and a very dense population of seabirds (including large numbers of shearwaters, petrels, and terns). Nihoa also has two endangered, endemic passerines, the Nihoa Millerbird (Sylviidae: *Acrocephalus familiaris kingi*) and the Nihoa Finch (Drepanididae: *Psittirostra ultima*). The Laysan Millerbird (*A. f. familiaris*), now extinct, was the only other native Hawaiian sylviid. The Nihoa Finch and its close relative the Laysan Finch (*P. cantans*), are two of the only four extant finch-billed drepanidids, all of which are endangered (USFWS 1980). Survival of these two finches on two tiny islands will be dependent on management programs that successfully prevent introduction of exotic biota and other forms of human disturbance.

The difficulty of successfully landing on Nihoa (approximately 50% of all attempts succeed), and the rigorous field conditions, explain why so little is known of the biology of the millerbird and the finch. Aside from a 12-day field expedition in 1969 by John Sincock of the U.S. Fish and Wildlife Service and Ernest Koska, then of the Hawaii Division of Fish and Game, only about 70 days have been spent in the study of the biota of Nihoa (Clapp et al. 1977). Periodic, usually annual, visits by USFWS biologists have rarely entailed more than a brief census of the two land birds. As a result, population estimates for the millerbird and finch have fluctuated greatly, in part no doubt a result of sampling intensity, as well as a reflection of population trends. Estimates of millerbirds have varied from as low as 41 to as high as 592. Estimates of finches have ranged from 1318 to 6686 (Sincock *in* Clapp et al. 1977). Recent (1977–1979) millerbird estimates by Sincock (pers. commun.) range from 127 ± 119% (95% confidence limits) to 490 ± 60%. Sincock's finch estimate for 1979 was 3612 ± 40%.

Obtaining total population estimates of these two resident island birds is desirable for at least two reasons: (1) knowledge of the total population size would enhance our understanding of population dynamics and limiting factors for these species, and (2) development and implementation of management plans would be aided by accurate information on total population numbers.

[1] Department of General Science, University of Hawaii, 2450 Campus Road, Honolulu, Hawaii 96822.

[2] USDA Forest Service, Institute of Pacific Islands Forestry, 1151 Punchbowl Street, Honolulu, Hawaii 96813.

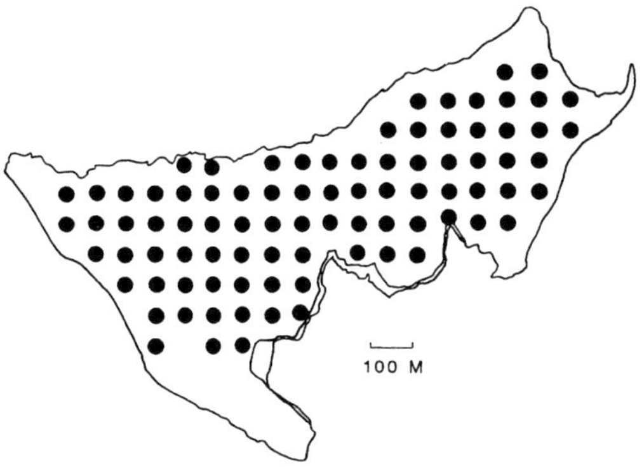

FIGURE 1. Map of Nihoa showing location of the 87 variable-distance circular plot stations.

The structural simplicity of the ecosystem and the extremely small and defined species' distributions contribute to the attractiveness of Nihoa as an experimental setting. It is possible here to test several methods of estimating avian population numbers, including the important variable, observer effect.

METHODS

Conant and Collins spent 31 May to 6 July 1980 on Nihoa. They censused using three methods: (1) 87 variable-distance circular plot stations (Reynolds et al. 1980), 82 of which were sampled twice for a total of 169 stations (Fig. 1); (2) 49 strip transects 76.2 m (250 ft) long (Fig. 2) were each sampled once in late June, using a variable-distance method (J. T. Emlen 1971); and (3) a fixed-distance strip method based upon that used previously by Sincock (pers. commun.) involving a total count of all birds within 3 m on either side of the observer along the 49 transects. The transects used in the last two methods were randomly established by Sincock in 1968 and have been censused annually since that time.

During studies of millerbirds' breeding behavior, Conant spot-mapped 20 breeding territories, 12 of

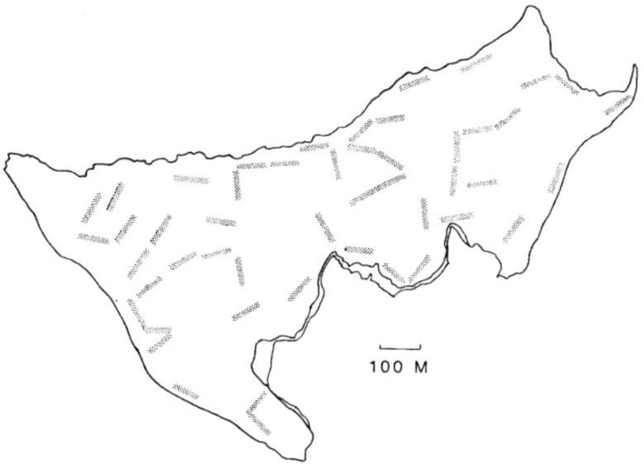

FIGURE 2. Map showing locations of the 49 strip transects.

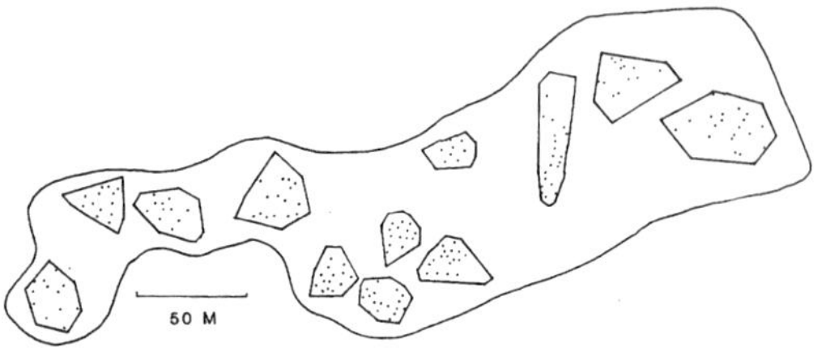

FIGURE 3. Map of 12 millerbird territories within an area thought to have no other resident birds.

which were within an area she thought to contain no resident birds other than the 12 breeding pairs (Fig. 3). Locations of either color-marked or behaviorally paired individuals were mapped over a 4-week period. A minimum of 14 observations of one of the pair members was required before we calculated the territory size. The range of registrations was 14 to 31. Two density estimates were derived from spot-map data using: (1) the average size of single territories, and (2) the total area occupied by the 12 territories and the immediately adjacent unoccupied areas (Fig. 3). Both values for territory size were extrapolated to the total island area and multiplied by two to arrive at a total population estimate. We judge that virtually the entire island has habitat suitable for breeding.

We calculated 95% confidence limits for all of the population estimates by the following formula:

$$\text{C.I.} = \hat{N} \pm \sqrt{\frac{d}{a}} \times 1.96$$

where $\hat{N}$ is the estimate of the total population, d is the density per ha, and a is the area surveyed in ha. This latter figure was derived directly in strip censuses, and calculated in variable-distance counts from the effective detection distance (Ramsey and Scott 1979). Significance levels between densities at different distances from the observer were calculated by the method of Ramsey and Scott (1979).

RESULTS

Total population estimates varied depending on which method was used (Fig. 4). Estimates

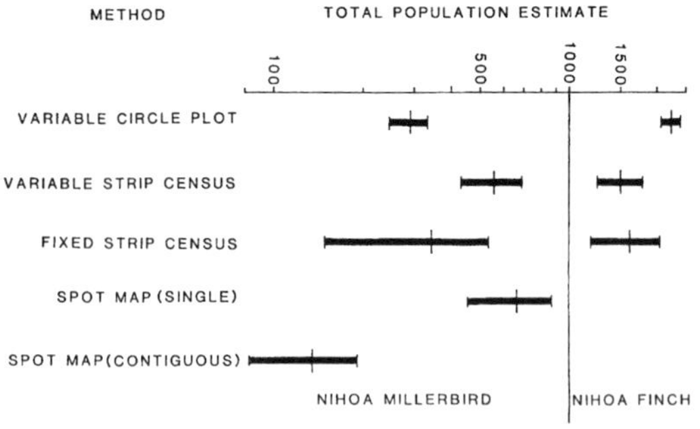

FIGURE 4. Total population estimates with 95% confidence limits for millerbirds and finches.

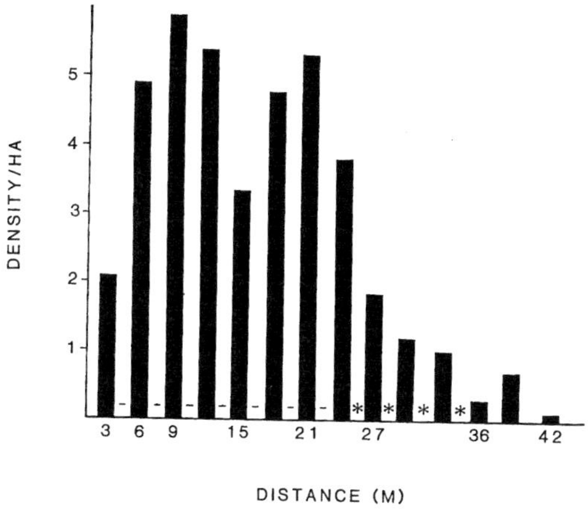

FIGURE 5. Millerbird densities resulting from variable-circular plot method. Distance measurements are the outer limit of each band in 3 m intervals. (* indicates a significant difference in densities between the band farther out and all those closer to the observer.)

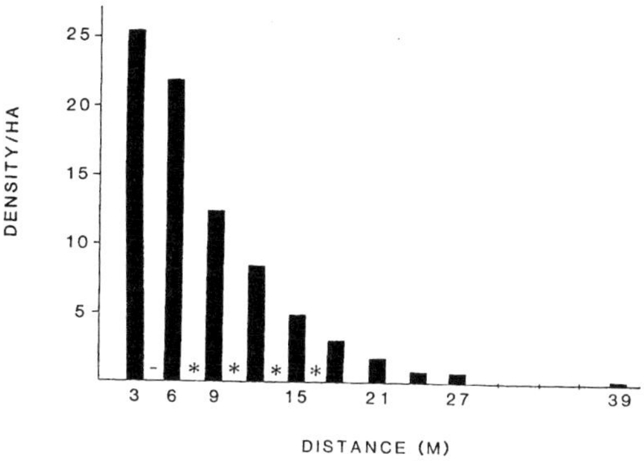

FIGURE 6. Finch densities resulting from variable-distance strip transect method. (*—see Fig. 5.)

of millerbirds varied from 133 to 659, and estimates of finches varied from 1499 to 2219 (Table 1). For the millerbird, highest total population estimates resulted from the variable strip and the variable-circle stations. For the finch, the highest density was from the variable-circle stations, and the lowest from the variable distance strips and the fixed-distance strips. The 95% confidence intervals calculated were smallest for the circular-plot stations and largest for the fixed-distance strips.

The variable-distance methods (circular-plot stations and variable-strip transects) provided data allowing us to identify different types of observer effects. The strip censuses were conducted by a moving observer, while in the station counts the observer was stationary. A species that is unaffected by the presence of an observer would show a reasonably flat distribution of individuals per ha out to the "basal radius" (see the "no attraction" curve in Figure 9). From this point outward fewer birds are detected, as some are overlooked. If birds are, for instance, repulsed by an observer, there would be significantly fewer close to the observer. With this in mind, we examined the patterns of abundance by the different methods.

With two census methods there appeared to be no observer effect: variable-circular plots of millerbirds (Fig. 5) and variable-strip censuses for the finch (Fig. 6). In both cases there was no significant increase in density out to the basal radius, which was 24 m in the millerbird and 9 m in the finch. In contrast, the variable-circle

TABLE 1
RESULTS OF VARIOUS CENSUS METHODS USED TO ESTIMATE DENSITIES AND TOTAL POPULATIONS OF THE NIHOA FINCH AND THE NIHOA MILLERBIRD

Method	No. of samples	No. birds in count	Effect. detect. dist. (m)	Size of plot (ha)	Area of plots (ha)	Dens. per ha	Total pop. est.	95% confid. limits (±)
Nihoa Finch								
Var.-dist. circle	169	597	18.6	0.11	18.3	32.6	2060	165
Var.-dist. strip	49	172	7.8	0.12	5.8	22.8	1443	247
Fixed-dist. strip	49	57	—	0.05	2.2	25.4	1608	418
Nihoa Millerbird								
Var.-dist. circle	169	187	27.8	0.24	41.1	4.5	287	41
Var.-dist. strip	49	91	11.4	0.17	8.5	8.5	537	124
Fixed-dist. strip	49	12	—	0.05	2.2	5.4	338	192
Spot map (single)	20	—	—	0.19[1]	3.8	10.4	659	205
Spot map (contig.)	12	—	—	0.95[1]	11.4	2.1	133	53

[1] Mean size of territory.

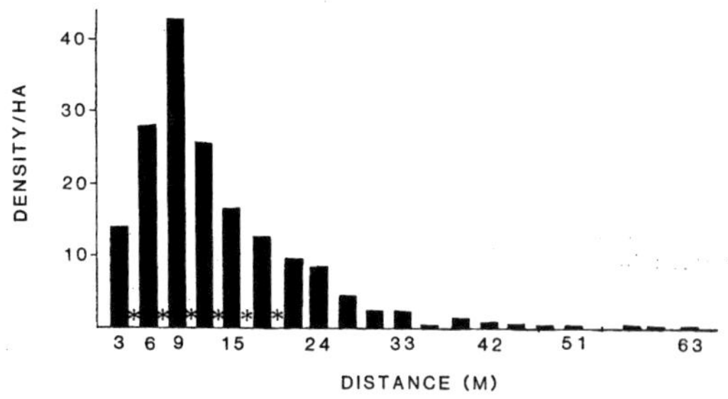

FIGURE 7. Finch densities resulting from variable-distance circle plot method. (*—see Fig. 5.)

plot censuses of finches with a stationary observer (Fig. 7) and the variable-strip censuses of millerbirds with a moving observer (Fig. 8) suggested an observer effect. In both cases, there are significantly ($P \leq 0.05$) fewer birds closer to the observer than at greater distances (peaking in the 6–9 m band). At first glance, it might appear that in both methods the birds were repulsed by the observer. However, we suggest that birds are actually attracted to the observer. The crucial point in arriving at this conclusion is that *the two censuses with probable observer effects both yielded the highest density estimates* of the species censused (Fig. 4). These results argue that the birds were attracted to the observer.

DISCUSSION

We hypothesize that the two patterns of density are derived as in Figure 9. When there is no

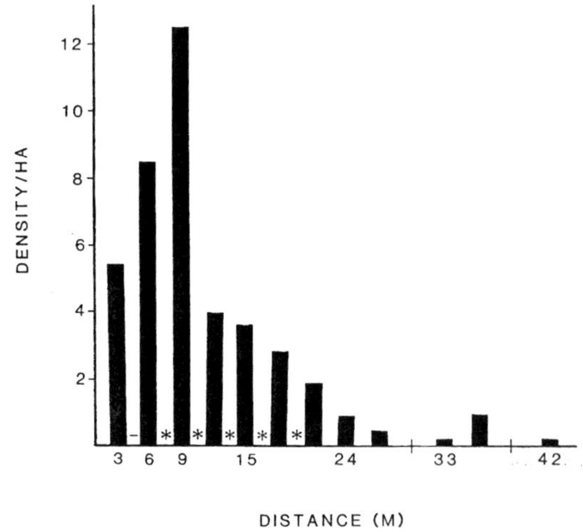

FIGURE 8. Millerbird densities resulting from variable-distance strip transect method. (*—see Fig. 5.)

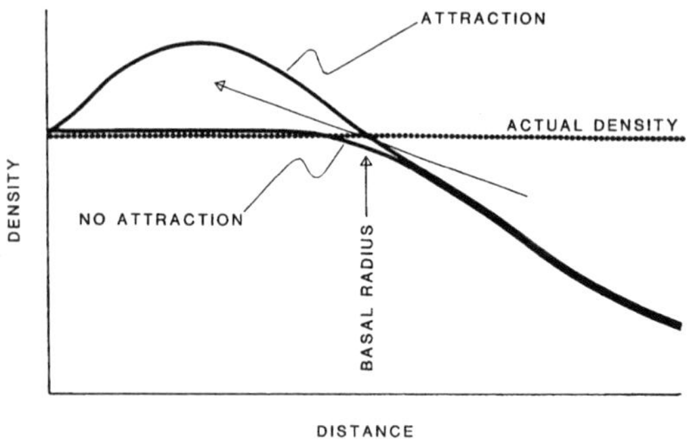

FIGURE 9. Hypothetical detection curves showing expected patterns with and without observer effect (attraction). Basal radius occurs at the effective detection distance.

observer effect, the detection curve should be without significant changes from the observer out to the basal radius, where it begins to drop off (see "no attraction" curve, Fig. 9). When birds are attracted to the observer, as we suggest, they approach the observer from an area near and beyond the basal radius (see "attraction" curve, Fig. 9). The question may be raised: What are the causes of the attraction? We suggest that it is a response to food resources.

The millerbird is an insectivorous bird, gleaning insects primarily from foliage, but also from stems, from litter, and on the soil surface. The finch is omnivorous, eating a considerable variety of vegetable material as well as the eggs of seabirds. The strong attraction of finches to exposed seabird eggs has been noted by several observers (e.g., Sincock, pers. commun., Clapp et al. 1977).

Based on observations of feeding behavior, and on examination of the lateral distribution of birds in relation to observers, we developed hypotheses to explain the apparent attraction of millerbirds to a moving observer and of finches to a stationary observer. We suggest that the millerbird is perhaps attracted to the insects flushed by a moving observer; hence the variable strip transects yielded the higher densities. The finch, on the other hand, should be attracted to a stationary observer because seabirds on nests near the observer will have left their eggs temporarily exposed during the count; thus the variable-circle plots yield the highest finch estimate. Finches may also be attracted to a moving observer, but are likely to be attracted to the area behind the observer, where seabirds are off the nest, so that birds drawn in behind the observer's path will not be counted.

Based on our interpretation of the data, we feel that the best population estimates for the two species result from the censuses without significant observer effect; that is, the variable-circle plot in the millerbird (287 ± 42 birds), and from the variable-strip census in the finch (1499 ± 250 birds). The fixed width strip census yielded densities with too large a variance to be useful because of the narrow area (6 m wide) surveyed. Spot-mapping is inadequate because: (1) it is extremely time-consuming relative to the sample size obtained; (2) it cannot document adequately "floaters" or those pairs that are relatively inconspicuous because of their stage of nesting; and (3) it is difficult to assign an accurate figure to the area surveyed (cf. "single" and "contiguous" spot-mapping in Fig. 4).

A major conclusion of this study is that the effects of an observer in bird censuses may be profound. Additionally, it is evident that the responses of birds to observers may vary depending on the species. An understanding of the birds' behavior permits the application of the least biased method to arrive at population estimates.

ACKNOWLEDGMENTS

The research was supported in part by grants and logistic assistance from the University of Hawaii Office of Research Administration, the U.S. Fish and Wildlife Service, the University of Hawaii Sea Grant Program, and the National Marine Fisheries Service. We are grateful to John Sincock for his comments on the manuscript as well as his help and guidance. James Baldwin provided statistical counseling.

Studies in Avian Biology No. 6:382–387, 1981.

LATERAL DETECTABILITY PROFILES FOR LINE TRANSECT BIRD CENSUSES: SOME PROBLEMS AND AN ALTERNATIVE

RICHARD L. HUTTO[1] AND SANDRA L. MOSCONI[2]

ABSTRACT.—We censused all land birds along a 1-km transect route through a riparian bottomland site in western Montana. Four censuses were conducted by one observer and four were conducted by two observers working together. The addition of a second observer: (1) increased the number of individuals detected for most of the bird species present; (2) increased the number of detections significantly more for rare than for common species; and (3) increased the number of detections significantly more at farther than at closer lateral distances. Considering the problems that exist with estimating true lateral detectability profiles, these results suggest that multiple observers might serve better as a method for dealing with detectability differences among species than a method involving the use of detectability profiles.

Reviews of methods for estimating the densities of land bird species (Kendeigh 1944, J. T. Emlen 1971, Dickson 1978, Shields 1979) all recognize the great utility of line transects. Despite the efficiency of such methods, the accuracy of the same methods is questionable. A major reason for inaccuracy stems from the fact that species differ in their conspicuousness. The proportion of individuals within a given transect width that is actually observed during a census differs among species, and is probably rarely 100% for any species (Järvinen 1978b).

In a seminal paper on census methods, J. T. Emlen (1971) addressed this problem of detectability differences among species and suggested that we record the lateral distance at which each individual is detected from the transect line. One can then plot the frequency of observations at various lateral distances and from this "detectability profile" determine the lateral distance at which detectability begins to decline for each species. The density of a given species is then based on the lateral width within which all individuals are assumed to be detected. We suspect that such detectability profiles are often inaccurate representations of the actual lateral detectabilities of many bird species because of the responses of birds to observers and because of the biased accumulation of detections that results from multiple use of a fixed transect route. In this paper we more clearly define these problems and investigate a possible alternative to the determination of lateral detectabilities.

Preston (1979) recently described how the number of individual birds detected increases with the number of observers, but he presented no data on whether the same proportionate increase occurs for all species and at all lateral distances from the transect line. If the bulk of additional observations comes from the greater lateral distances, or from the least conspicuous species, then additional observers would provide a simple method of minimizing the differences in detectability among species. In this paper "detectability" simply refers to the proportion of individuals of a given species that is likely to be detected within a given transect area. This usage is analogous to J. T. Emlen's (1971) "coefficient of detectability." Here we ask, "Do the additional observations that result from a second observer come disproportionately from inconspicuous species and/or from greater lateral distances?"

METHODS

We established a 1-km line transect in a heavily grazed riparian bottomland 8 km SW of Missoula, Montana (47°30′N, 114°6′W). The site was dominated by cottonwood (*Populus trichocarpa*) and ponderosa pine (*Pinus ponderosa*) in the overstory and by hawthorne (*Crataegus douglasii*), willow (*Salix alba*), dogwood (*Cornus canadensis*), snowberry (*Symphoricarpos albus*), and rose (*Rosa* sp.) in the understory.

We conducted eight censuses—four 2-observer censuses and four 1-observer censuses (each of us conducted two). In 2-observer censuses we generally walked within 5 m of one another and focused our attention in opposite directions. Since human voices seem to disturb birds little, if at all, we communicated vocally when a bird was detected to ensure that all detections were recorded and that no bird detected by both parties was recorded twice. All birds seen or heard were recorded by one of us and their lateral distances from the transect line were estimated with the aid of a "ranging-620" rangefinder. We conducted all censuses from 07:00–10:00 between 10 June and 1 July 1980, and alternated 1-observer censuses with 2-observer censuses to avoid biases that might be associated with time of season.

Singing males were recorded as a pair of birds and nonsinging individuals of the same species adjacent to such males were assumed to be mates and went un-

[1] Department of Zoology, University of Montana, Missoula, Montana 59812.

[2] Wildlife Biology Program, University of Montana, Missoula, Montana 59812 and USDI Fish and Wildlife Service, Denver, Colorado 80225.

TABLE 1
TOTAL DETECTIONS OF SPECIES BY LATERAL DISTANCE CATEGORIES AS RECORDED BY A SINGLE OBSERVER

Species	Lateral distance (m)									Total pairs
	0–5	5–10	10–15	15–20	20–25	25–30	30–40	40–60	>60	
Falco sparverius							1			1
Bonasa umbellus	1									1
Zenaida macroura	1	2			1		1	1		6
Megaceryle alcyon				1						1
Colaptes auratus	1	1	1	1			1			5
Melanerpes lewis		1								1
Sphyrapicus varius	1									1
Picoides pubescens		4	1							5
Tyrannus tyrannus	1	2		1	2	1	1	1		9
Empidonax traillii		2	2			2	1			7
Empidonax minimus			1							1
Contopus sordidulus	12	9	7	4	2	5	2			41
Pica pica		1								1
Parus atricapillus	1	1		1			1			4
Parus gambeli	1		1							2
Sitta carolinensis		2	2			1				5
Troglodytes aedon			1	2		1				4
Dumatella carolinensis	1		4	2	1		1			9
Turdus migratorius	9	8	5	5	2		3	1		33
Catharus fuscescens	1	2	2	3	3					11
Bombycilla cedrorum		1	1							2
Vireo solitarius			1	1			1			3
Vireo olivaceus	2	3	3	6	1	1	2			18
Vireo gilvus	1	4	5	2	2	1	2			17
Dendroica petechia	2	9	10	10	3	4	3			41
Dendroica coronata	1	2	1			1				5
Setophaga ruticilla	1	3	4	2	1	1	2			14
Icterus galbula		1	1					1		3
Molothrus ater	8	3	6	4	3	1	4			29
Piranga leudoviciana	1	1	1		1				1	5
Pheucticus melanocephalus		1		1	1	1		2		6
Melospiza melodia		1	1		2		1	1		6
Carduelis pinus	2									2
							Grand total			299
							Number of species			33

recorded. Nonsinging individuals observed away from the vicinity of singing males were recorded as single individuals unless there was evidence that they were paired (e.g., another nonsinging individual of the same species nearby, nest material in bill, and so forth).

Since all censuses were conducted within the same study plot, the generality of our results remains unknown.

RESULTS

The numbers of individuals of each species that were detected at various lateral distances for 1-observer and 2-observer censuses are given in Tables 1 and 2, respectively. Both the number of species and the number of individuals recorded were greater for the 2-observer censuses. Preston (1979) derived an empirical expectation from Lack's (1976) data that the number of birds observed ought to increase in proportion to the square root of the number of observers, but the increase recorded here (299 to 348) is less than expected (299 to 422) on that basis. We suspect that the less pronounced increase recorded here reflects differences in the habitat types involved. Lack (1976) worked in species-rich, tropical deciduous forests and it is not surprising that an additional observer might add proportionately more individuals there, where the species are generally more secretive and restricted in their vertical distributions (Lovejoy 1975).

To determine whether the difference in a species' abundance between 1- and 2-observer censuses was related to its commonness, we categorized a species as being uncommon if fewer than 3 pairs were recorded at any lateral distance after four 2-observer censuses. By this method, 13 species were categorized as "uncommon" and 25 species as "common." For

TABLE 2
Total Detections of Species by Lateral Distance Categories as Recorded by Two Observers

Species	Lateral distance (m)									Total pairs
	0–5	5–10	10–15	15–20	20–25	25–30	30–40	40–60	>60	
Falco sparverius							1		1	2
Bonasa umbellus	2									2
Zenaida macroura	1	1		1		1	1	2	2	9
Megaceryle alcyon									2	2
Colaptes auratus	1	2	1		1	1			2	8
Melanerpes lewis				1						1
Sphyrapicus varius			1							1
Picoides pubescens	2	1	1	2	1					7
Tyrannus tyrannus	2		1			1			2	6
Empidonax traillii	2	1				1	1	2		7
Empidonax minimus		1	1							2
Contopus sordidulus	9	7	7	7	5	3	8	2	1	49
Pica pica			1							1
Parus atricapillus		2	2			1		1		6
Parus gambeli	1		1							2
Sitta carolinensis	2		1				1	1		5
Troglodytes aedon			2			1	1	1		5
Dumatella carolinensis	2	1	1	1			1			6
Turdus migratorius	7	10	3	4		4	4	1		33
Catharus fuscescens				1	3	2	2		3	11
Bombycilla cedrorum	1	1		1			1			4
Vireo solitarius		1		1	1					3
Vireo olivaceus	1	5	3	2	3	1	2	1		18
Vireo gilvus	4	4	1	4	2	2	2	1		20
Dendroica petechia	3	5	8	10	3	4	4	2	4	43
Dendroica coronata		1			2					3
Seiurus noveboracensis								1		1
Oporornis tolmiei				1						1
Setophaga ruticilla	2	2	4	1	1	2	2			14
Icterus galba		2		1		1	1		1	6
Molothrus ater	9	4	3	4	3	4	6	7	2	42
Piranga leudoviciana	1	1	1		3		1			7
Pheucticus melanocephalus	1	1	1	1		2	1	2		9
Hesperiphona vespertina				1						1
Melospiza melodia			2		1		1			4
Carpodacus cassinii		1		1						2
Carduelis pinus	1			3						4
Carduelis tristis								1		1
							Grand Total			348
							Number of species			38

each species we then noted the percent change in abundance from 1- to 2-observer censuses. Any species that increased from zero observed had, of course, an infinite percent increase. Such increases from zero were conservatively labeled 100% if the increase was from 0 to 1, 200% if from 0 to 2, and so on.

Uncommon species showed an increase in numbers detected that was significantly greater than the increase recorded for common species (69.2 $\pm$ 48.0% vs. 22.5 $\pm$ 40.8%; approximation of t-test, $P < 0.05$, Sokal and Rohlf 1969). Thus, the increase in numbers of individuals detected with an additional observer is non-random; the additional detections come disproportionately from uncommon species.

We were also interested in whether the same proportionate increase in bird detections occurred at all lateral distances. The relationship between the number of detections and lateral distance category (<15 m, 15–30 m, >30 m) was significantly different between 1- and 2-observer censuses ($G = 10.51$, $P < 0.01$; Table 3); the increase in number of detections with two observers came disproportionately from the farthest lateral distance category. Continuing in this vein, we measured the direction and magnitude of change in detections from 1- to 2-ob-

TABLE 3
THE NUMBER OF DETECTIONS BY DISTANCE CATEGORIES FOR 1- AND 2-OBSERVER CENSUSES[a]

	Lateral distance		
	0–15 m	15–30 m	>30 m
1-observer censuses	173	91	35
2-observer censuses	154	108	86

[a] The data are significantly heterogeneous (G-test, $P < 0.01$).

TABLE 4
THE CHANGE IN NUMBER OF SPECIES WITH AN ADDITIONAL OBSERVER

	Lateral distance		
	<15 m	15–30 m	>30 m
Decrease	14	8	2
No change	16	13	17
Increase	8	17	19

server censuses for each species within each of the three lateral distance categories. The numbers of species that showed a decrease, no change, or an increase in number of detections upon the addition of a second observer are presented in Table 4. The direction of change is not statistically significantly related to lateral distance ($G = 7.9$, $P < 0.1$), but the trend for most species was for the number of detections to decrease or remain the same at close lateral distances, and remain the same or increase at farther lateral distances. The mean magnitudes of change in numbers of individuals detected with the addition of a second observer were +6.1%, +39.9%, and +79.8%, for the <15 m, 15–30 m, and >30 m categories, respectively.

DISCUSSION

Application of the variable-width strip transect method requires delineation of a "profile" of the detectability of each species—a plot of the number of observations of a given bird species against the lateral distance at which each individual is sighted from the transect line. At least two factors lead us to suspect that such profiles are often unrelated to the actual detectability of their respective species. The first involves poor sample sizes that accrue for the majority of species seen; the inflection point in their detectability profile may not even be recognizable. A look at any of the species in Table 1 with fewer than 30 observations (92% of the species) will illustrate the difficulty of pinpointing the lateral distance at which detectability begins to decline.

Secondly, when detectability profiles *can* be clearly delineated, they may be artifacts of the behavior of birds (Fig. 1). For example, some bird species may move toward or away from a moving observer or, as was often the case with vireos in our study, they may sing and be detected easily only when away from the observer. In other instances, the positions of individuals that are detected on each transect run may be fixed relative to a permanently positioned transect line. By way of example, consider the Willow Flycatcher (*Empidonax traillii*) (Table 1 or 2). We ordinarily observed two singing males, one always about 10 m from the transect and another always atop the same willow, about 30 m from the transect line. The lateral distances that we recorded for this species were unrelated to the species' lateral detectability but, instead, reflected where their song posts happened to be positioned relative to the transect line. The House Wren (*Troglodytes aedon*) observations are similarly biased. This problem is especially acute when sample sizes are low and the locations of singing individuals are not likely to change from day to day.

For these reasons we looked for an alternative to the use of detectability profiles. We expected that utilization of an additional observer during censuses would increase the number of birds observed (Lack 1976, Preston 1979), but wished to determine whether each species and each lateral distance category revealed the same proportionate increase in numbers observed. If the numbers of detections of the conspicuous species do not increase much by an additional observer but those of the inconspicuous species do, then additional observers would provide a simple method of minimizing the difference in detectability among species. Moreover, if the bulk of new observations are located at the greater lateral distances, then not only would the detectability differences among species be minimized, but the detectability of all species might be raised to an acceptable level within a belt transect that is wide enough to generate reasonable sample sizes.

In our study the addition of a second observer produced three results that are of interest:

(1) There was an increase in the number of individuals detected for most (61%) of the bird species present. Some species (29%) revealed no change in the number of detections, presumably because they are conspicuous and all the individuals that are present can be readily detected by a single observer. Four species (10%) actually revealed a decrease in the number of detections, which is not unexpected since we are dealing with mobile animals whose true abundances may vary from day to day.

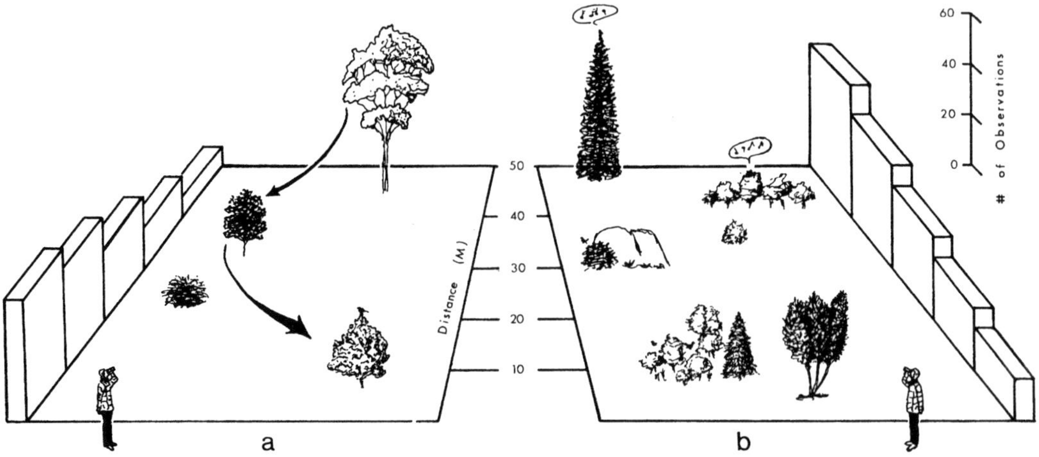

FIGURE 1. Two lateral detectability profiles that have been biased by the behavior of birds are represented by the block histograms. In (**a**), individuals of some species have been attracted to the observer, which would produce greatly inflated density estimates based on such a profile. In (**b**), individuals of other species are actually more conspicuous at farther than at closer lateral distances, which would produce an underestimation of true density.

(2) There was a significantly greater increase in the number of detections of rare than of common species. This result is of great interest since we are unable to categorize a species as conspicuous or inconspicuous on the basis of detectability profiles (because of the problems discussed earlier). Therefore, it becomes difficult to test whether the number of detections of inconspicuous and conspicuous species increase by the same proportionate amount. However, if we assume that, on average, inconspicuous species are rarer than conspicuous ones, we can conclude that the number of detections of inconspicuous species increased disproportionately more than for conspicuous species with the addition of a second observer.

(3) There was a significantly greater increase in the number of detections at farther than at closer lateral distances. This finding lends further support to the idea that the additional detections which resulted from use of a second observer came disproportionately from the least detectable species.

The mechanisms responsible for the increased detections upon the addition of a second observer are uncertain, but since the additional detections came disproportionately from rare species and from greater lateral distances, the simplest explanation is that two observers detect more birds by dividing their attention in different directions. The quick movement of an inconspicuous bird near an observer will be detected no matter what direction the observer is looking, while such movement at greater lateral distances will surely be missed by a single observer unless he or she is looking in the right place at the right time. This interpretation is consistent with the observation that the greatest proportion of additional detections came from the farther lateral distances. That proportionately few additional detections came from the abundant species probably means that a single person did well at detecting all individuals present and the second observer could add no additional detections (i.e., abundant species tend to be conspicuous). Alternatively, it is possible that observers become habituated or saturated at some point with the detections of abundant species and begin to ignore additional observations, thus adding relatively fewer detections of common species in larger samples. However, the sample area with two observers was no larger and, secondly, an additional observer should act to decrease such saturation effects and contribute additional detections if they existed. More definitive answers must await further work on this question.

It is interesting that the bulk of additional detections with a second observer were, in our case, recorded beyond 30 m (Table 3). Since fixed-width transects in forested areas are rarely wider than 30 m or so, the increase in detections that results from an additional observer may require a transect width wider than is practical for some habitat types.

The combined results suggest that use of an additional observer will minimize the differences in detectability among species and increase the overall detectability of each species within a fixed transect area. However, the accuracy of this and other methods needs to be established empirically from studies with banded birds where the true densities are known (see, for example, Järvinen et al. 1978a). The more usual comparisons of various transect methods with spot-map methods (Franzreb 1976, Dickson 1978, Järvinen et al. 1978b, Mikol et al. 1979) are interesting but still they do not reveal the accuracy of the transect methods because the spot-map method itself is subject to many possible inaccuracies (Svensson 1974b, Best 1975).

ACKNOWLEDGMENTS

We are most grateful to John Emlen, Olli Järvinen, and Fred Ramsey for their comments on the manuscript, to DeWayne Williams for preparing the figure, to Richard Rossignol for permission to conduct the study on his land, and to the USDI Fish and Wildlife Service, Denver, for their financial support.

Studies in Avian Biology No. 6:388–389, 1981.

SUMMARIZING REMARKS: OBSERVER VARIABILITY

LYMAN L. MCDONALD[1]

In general there are four sources of "error" or variation in scientific studies (Cochran 1977):

(1) *Sampling error* due to inherent variability between experimental units. Thus, if a study area is divided into quadrats and each member of a sample is censused perfectly, summary statistics will necessarily vary from sample to sample. Another sample will yield another estimate due to sampling error.
(2) *Measurement error* due to the lack of uniformity in the physical conduct of the study. The measurement procedure may be biased, imprecise or both biased and imprecise.
(3) *Missing data* due to the failure to measure some units in the sample.
(4) *Gross errors* introduced in coding, tabulating, typing and editing the results.

Usually the effects of these errors are completely confounded and the total variance cannot be separated into its components. An understanding of sampling error and its role in making inductive inferences is the basis of modern statistical inference procedures. Control of this source of error is at least partially the responsibility of the statistician. Control of the other three sources of error is primarily the responsibility of the researcher! All of the papers in this session on observer variability have as a first objective the control of measurement error, and I applaud their attempts to get a handle on this problem. Measurement errors may be modeled by statisticians but their control and reduction must come from careful experimental design. Consultation between the researcher and statistician before the study begins should be of value in controlling all potential sources of error.

In many fields of study the presence of measurement error is barely recognized and its influence is played down. For example, Box et al. (1978) state that "Usually only a small part of it (the total variance) is directly attributable to error in measurement." Many statisticians follow the rule of thumb that the measurement error should be "small" relative to the sampling error, especially in utilizing statistical procedures such as regression and correlation analysis. Considering the content of the papers in this session, measurement errors cannot be ignored in studies designed to measure terrestrial bird numbers. Furthermore, standard analysis procedures may not be applicable until this source of error is under control.

Robbins and Stallcup (1981) consider a particular type of measurement error, namely inaccurate lists of species present at a study site. For instance, they mention a study in Maryland in which there was not one stop out of 50 at which two observers had recorded the same list of species present. They also briefly address errors in the fourth class, that is, errors occurring between the time a bird is observed and the time the report appears in print. Cyr (1981) reports on an experiment to test the ability of observers to identify species from utterances recorded on a tape. Scott et al. (1981b) report on experiments to study the ability of observers to estimate distances and the effect of bias in this process on their estimates. Emlen and DeJong (1981) propose to attack this problem by determination of detection threshold distances for each species under standard conditions. Supposedly these detection threshold distances could then be used as the half-width of transect censuses and the radius of point-centered census plots. Unfortunately, their proposal is still subject to measurement errors. Two observers running the same census plots or lines at the same time will have different counts of birds heard even though they might be willing to use the same detection threshold distance. To quote Kepler and Scott (1981), "Thus, an observer with good ears is actually sampling a larger area by hearing more distant birds."

Errors or variance due to missing data are not directly addressed in these papers. However, Cyr's (1981) experiment suffers somewhat from this source. Thirty-three observers apparently started the experiment but in the end only the results of eighteen were analyzed. There may be good reasons to drop the data from those fifteen observers but it is obvious that their retention would produce different summary statistics. That is, missing data is a source of variation in scientific studies. For example, if Cyr had been able to retain two "inexperienced observers" instead of only one, or if the single inexperienced observer had been dropped, the results of his regression analyses in Figure 2 would likely have changed drastically.

The following are mentioned as general procedures which may help to increase the precision of bird studies by decreasing sampling error:

[1] Departments of Statistics and Zoology, University of Wyoming, Laramie, Wyoming 82071.

1. Use careful *stratification* of the study site. Ideally, we would like to stratify on bird density itself, but in practice one must stratify on variables which hopefully are highly correlated with bird density. For example, in the Red Desert of Wyoming the researcher might stratify on the proportion of area "covered" by shrubby vegetation. Regions with "high" cover form one stratum, regions with "medium" cover form another stratum, etc. and each stratum is sampled independently.
2. Measure covariates on the sample plots (transects) which may help explain variation in bird density. Again, in the Red Desert of Wyoming the researcher might divide the study area into rectangular quadrats and estimate bird density (the variate) and cover by shrubby vegetation (a covariate) on each plot. Bird density estimates may show a significant reduction in variation when "adjusted" for the cover values in a regression analysis.
3. Use systematic or cluster sampling plans which may speed up the data collection and hence enable one to increase sample sizes.

Continuing the list for the control of measurement errors I would suggest:

4. Use of double sampling with ratio or regression estimators. For example, the accuracy of the variable circular plot survey (Ramsey and Scott 1979) might be enhanced by double sampling where distances to birds are measured on a subset of the sample and distances are estimated for all birds. Perhaps the measured distances could be used to "calibrate" the estimated distances.
5. Refinement of the experimental design (i.e., the physical conduct of the study). All of the papers in this session fall into this category. For instance, Robbins and Stallcup (1981) recommend "methods based on repeated visits over several days by different observers . . . ," careful training of observers and careful examination of field records. Kepler and Scott (1981) also stress the value of training observers. Faanes and Bystrak (1981) stress the importance of choosing well-trained observers whose differences will contribute little beyond sampling error.

A few specific remarks should be made concerning some of the papers. Scott et al. (1981b) are very careful to point out that the 20% error of estimation in their computer simulation was due to errors in measurement of distance and that no other sources of error were simulated. Factors such as observer bias will in some cases tend to counteract the measurement error and in other cases tend to magnify the error. It is important to remember that the 20% figure is for only one component of the many possible sources of error. Undoubtedly other sources will be incorporated into future simulations. Also, sample sizes in Scott and Ramsey's simulation were fairly large (i.e., 200–250 birds). A reduction in the sample size will likely result in increased error. They report approximately 10% accuracy in distance estimates from their field work, but this figure is for the *mean* accuracy while individual estimates varied from −75% to +400%. Again, to obtain this accuracy in estimation of the mean distance, sample sizes will have to be approximately equal to those employed in their field study.

Faanes and Bystrak (1981) have stated that "In most cases, well-trained observers are comparable in ability and their differences contribute little beyond sampling error." I think that they would agree that their sample of well-trained observers is fairly small, namely a sample of two consisting of the two authors. The same criticism of small sample sizes is valid throughout most of their paper. Their inductive inferences may remain valid, but one would like to see a broader sample from the population of observers. It is also dangerous to compare observers when they conduct the survey in different years. The year effect and observer effect are completely confounded, and the strength of the inference is decreased.

In conclusion, many of the problems facing researchers in the estimation of bird density deal with observer variability. Similar problems exist in finite sampling theory under the heading of "interviewer bias," see for example Cochran (1977). There is an extensive literature on the control, reduction, and evaluation of interviewer bias. Perhaps review of that literature will provide new ideas for research on observer variability in the estimation of the numbers of birds.

Studies in Avian Biology No. 6:390, 1981.

SUMMARIZING REMARKS: OBSERVER VARIABILITY

DAVID E. DAVIS[1]

The authors of the papers merit commendation for tackling the problem of assumptions, a far more intricate task than had been expected. The central theme of this session is the observer, who makes mistakes and may be seriously handicapped in learning or seeing.

Robbins and Stallcup (1981) begin at the level of identification. They call attention to errors made by experienced observers. They present a valuable list of species likely to be confused. Lastly they present some criteria for deciding which census method is prone to errors of identification.

Scott et al. (1981b) examine the ability of people to estimate distances of the bird from the observer who may either see or hear the bird. They find, not surprisingly, that observers differ in ability to judge distances and also that species differ in characters that reveal distances. The combination of errors may result in considerable error in estimation of numbers. They provide a method of calculation of discrepancy between estimated and measured distances but it is sufficiently sophisticated to prevent use during a census. Perhaps a "field model" would be useful. The authors do not address the question "Are the deviations constant for observer and species?" But they do suggest some procedures to reduce deviations.

Kepler and Scott (1981) describe a training program which really adapts good teaching to the problem of errors by observer. Improvement during training was modest.

Cyr (1981) experimentally searches for deficiencies in ability to hear and identify birds. From audiograms he records differences in ability to hear different frequencies. He notes from the literature that older persons suffer loss of ability to hear high frequencies and finds that even young people have gaps in their ability. Such persons should not make counts. Cyr uses a tape for some experiments but one wonders about the fidelity of the tape. Hence, to what extent can conclusions drawn from such data be transferred to live birds?

Faanes and Bystrak (1981) examine abilities of trained (experienced) and untrained observers and find striking differences. Unfortunately they did not clearly separate differences due to training (e.g., learning a song) and physiological ability (e.g., hearing loss).

Emlen and DeJong (1981) determine the threshold distance at which a song or call can be heard by a young person with normal hearing. The distances can then be used in calculation of densities by a transect or point method.

As a group these papers indicate the problems of errors by observers but in only a few places suggest what to do. A drastic remedy for observer error is to eschew absolute densities and get relative counts. These make the encompassing assumption that the errors are the same throughout and cancel. But this remedy may be merely shifting from frying pan to fire.

Another remedy would be to eliminate rare species (which can't add much to a census) and to count difficult species by some special method devised for that species (e.g., a caprimulgid).

Still another remedy was hinted at: record the songs and then at leisure count and recount the birds, thereby reducing errors by observers. But the recording would have problems too.

The possibility that the observer causes error (e.g., cessation of song) was considered in other sessions but should be noted here because some observers will cause more than will others.

Lastly, I take this opportunity for a general injunction. Keep this method simple both in collection and calculation of data. I am editing a *Handbook of Census Methods for Terrestrial Vertebrates* and I find that the methods that are actually used (i.e., published) are simple, especially mathematically. My interpretation of this situation is that each particular author is primarily concerned with some particular topic (energy flow, management, habitat, etc.) and resists involvement in complicated census procedures. We can claim that the author is condoning inaccuracies and we may be correct, but he will nevertheless persist in the use of the simplest method.

[1] 777 Picacho Lane, Santa Barbara, California 93108.

Studies in Avian Biology No. 6:391, 1981.

INTRODUCTORY REMARKS: SAMPLING DESIGN

JARED VERNER,[1] CHAIRMAN

Probably no aspect of the study of avian ecology has shown so marked an increase in interest in recent years as the counting of birds. Accompanying this increase has been a growing awareness among field researchers that bird counting is a distressingly imprecise science. Variances in count data are extraordinarily high, partly because birds are so mobile and partly because so many different effects can bias the counts. The literature documents many of these sources of bias and addresses the question of how to design experiments to control bias and thereby reduce the variance in count data.

Sources of bias can be grouped conveniently into four categories: (1) *observer effects* include experience, acuity, and alertness, as well as number of observers; (2) *bird species effects* include species detectability, species density, timing of breeding, social or breeding system, and flocking habits; (3) *site selection effects* include the site selection procedure (random, stratified random, regular, or selective), site separation, number of sites, vegetation density, vegetation homogeneity, plot size or transect length, and terrain; and (4) *sampling schedule effects* include the season, time of day, duration of a single sampling period, duration of the overall sampling period, number of counts per site, the frequency of sampling, and weather constraints. Some of these effects are easily controlled by experimental design or modes of analysis, but others are not so easily controlled and offer a real challenge.

This session includes several papers and poster presentations. Most participants deal with transect or point counting methods for sampling bird communities, though some attention is given here to the mapping method, and one paper provides a useful overview of the capture-recapture method. Among these presentations, I have found data relating to experimental design for control of 15 of the 23 sources of bias I have categorized, but applications tend to be restricted to one sampling method. It is a fair assessment to say that we only scratch the surface of the many questions in experimental design.

Taken together in all their permutations, the many sources of bias listed, and others unlisted produce real world conditions that so far have defied our abilities to count accurately all species comprising avian communities thus far sampled. Our ingenuity in experimental design will undoubtedly go a long way toward improving this unhappy situation. We can certainly generate count data with low variances, and we may then draw conclusions within prescribed limits of confidence. However, low variances do not necessarily indicate accurate data. They can also result from a sampling design that controls only the magnitude of the effect of a given biasing factor, without eliminating the bias. Therefore, we will always err in the same direction, and within acceptable limits of variation, but still we will always err. The trick, if we insist on accurate counts of all species in a community, will be to determine for all species and habitats the directions and magnitudes of those errors. I see this as the essence of the bird counter's challenge.

[1] Pacific Southwest Forest and Range Experiment Station, Forest Service, U.S. Department of Agriculture, Fresno, California 93170.

Studies in Avian Biology No. 6:392–398, 1981.

EXPERIMENTAL DESIGN WHEN COUNTING BIRDS

DAVID G. DAWSON[1]

ABSTRACT.—The number of species discovered increases by a constant increment with each doubling of observer effort; this relationship can be used to standardize incomplete lists for studies of biogeography and species diversity. The use of frequency of occurrence, rather than the total number counted, prevents discrimination of differences in density at high population levels. The counts of many species approximate a Poisson distribution, for which chi-squared tests on the totals counted may be an adequate approximation; the difference between two samples that may be detected is inversely proportional to the square root of the total number of a species counted. Monitoring long-term changes in bird numbers is better done by point or transect counts than by territory mapping; a large number of sites should be visited under standardized conditions each year. Rare species or habitats are best sampled after a reconnaissance has established their distribution, not by random or systematic sampling of all habitats. Differences in bird populations between habitats can be studied with index techniques, provided that the habitats are not too diverse and that observers, times of day, season, and other sources of bias are taken into account. Territory mapping permits fine-scale definition of habitats. Estimates of absolute density, accurate enough for the study of energetic or trophic relationships, may be obtained from point, transect or mapping methods.

No single answer can be given to the question of how to count birds. Principles common to the design of all sampling schemes, such as the choice between random and systematic sampling, or the sampling units used, are adequately covered in standard texts (e.g., Cochran 1963), so this paper concentrates on the limitations of different counting methods in answering questions about bird numbers and diversity. My other two papers in this symposium give the background to the more commonly used point, transect and territory mapping techniques.

SPECIES RICHNESS AND SPECIES DIVERSITY

The number of species in an area is of interest in quantitative biogeography and conservation (MacArthur and Wilson 1967, Diamond 1975b). The same measure is also used in studies of species "diversity," where it is termed species "richness," one of the two components of diversity, the other being "equitability" (Tramer 1969). A problem for all uses is that the list of species increases with the time spent looking.

Several researchers have fitted empirical curves to this increase (Preston 1960, Caughley 1965, Järvinen and Väisänen 1977a, Ratowsky and Ratowsky 1979) and found a linear increase in the number of species with the logarithm of effort. This semilogarithmic relationship is supported by the results of my studies (Fig. 1). Robbins (1972) suggested that the relationship is closer for the square root or even the fourth root of the effort, but presented no data to support this.

The semilogarithmic equation may be written $S_n = S_1 + a \log n$, where S_n is the number of species discovered in n units of effort (e.g., points, km of transect, ha of mapping plot, or time), S_1 is the average number discovered in one unit, and a is a coefficient describing the rate of increase of S_n with increasing n. Coefficient a is estimated from $a = (S_t - S_1)/\log t$, where t is the number of units of effort put into a study area.

The average number of species discovered in a unit of effort, S_1, is also an estimate of the sum of the individual species' probabilities of discovery in a unit of effort. This will depend on the densities, d, of the individual species and on their susceptibility to discovery (conspicuousness or detectability), b. S_1 has been termed "mean richness" by Blondel (1977), but "richness" is conventionally used to denote the *total* number of species in an area by Tramer (1969), whom I follow. The semilogarithmic relationship between S_n and effort tells us that S_n increases quickly with increasing effort at first and then levels off to increase much more slowly at high n. Thus at large n the species list is nearly complete, much less dependent on variation in d and b, and a much better estimate of species "richness." The choice of a particular large value of n is arbitrary, but it is probably best to choose one towards the upper end of the values of n from the areas to be compared, to avoid excessive extrapolation. Frochot (1976) suggested a similar approach, but did not fit a curve to his results.

The application of this estimate of species richness can be illustrated with data from Blondel (1977). In *Quercus ilex* stands in Provençe, $t = 16$, $S_{16} = 23$, $S_1 = 10.7$, so a is 10.2 and S_{30} is estimated to be 25.8. Figure 2 compares Blondel's values of species diversity (which were derived from density estimates) with values of S_{30} calculated as above. Clearly S_{30} accounts for most of the variation in species diversity, a point made also by Tramer (1969),

[1] Ecology Division, Department of Scientific and Industrial Research, Private Bag, Lower Hutt, New Zealand.

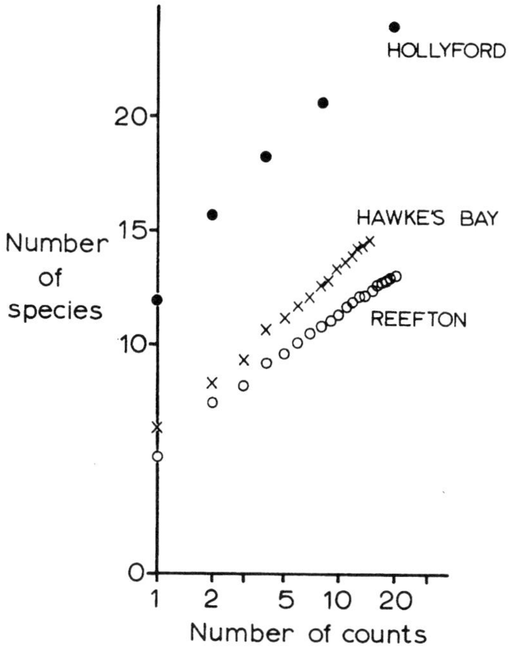

FIGURE 1. The increasing number of species detected with an increasing number of counts in three New Zealand studies using the five-minute point count (Dawson and Bull 1975): the total from 20 points in the Hollyford Valley and the average from 24 points in Hawke's Bay and 10 near Reefton (Dawson et al. 1978). All three relationships seem linear.

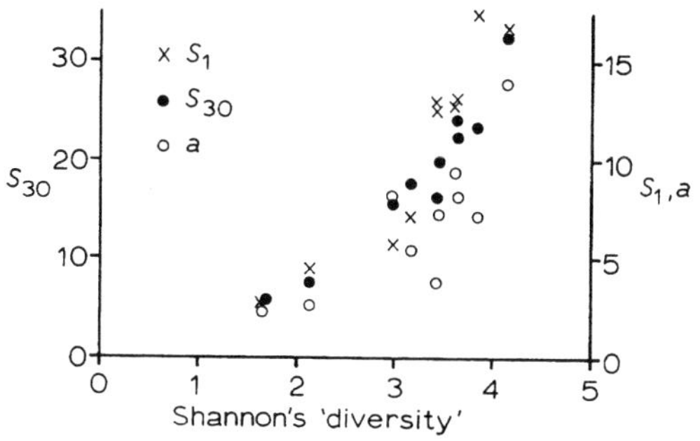

FIGURE 2. Relationship between Shannon's "diversity" (as calculated by Blondel [1977] from point counts and a conversion factor to density for each species) and the average number of species per point count (S_1), the slope of the semilogarithmic relationship (a), and the estimated species list for 30 E.F.P. counts (S_{30}). (The value of S_{30} for a given diversity varies little, so S_{30} is a good measure of diversity; a and S_1 are less good.)

Blondel (1975) and Cousins (1977). Thus point, transect or mapping samples readily yield an estimate of species richness, which is also well correlated with species diversity. This approach may be preferable to estimating b for each species and calculating diversity (as was done by Blondel 1975, Järvinen and Väisänen 1976a) and is certainly preferable to the assumption that b is constant (Lancaster and Rees 1979, Wilkinson and Guest 1977, and many mapping studies).

The coefficient of the semilogarithmic relationship, a, describes the rate at which species are added to the list as effort increases. This will, like S_1, depend on individual species' values of density and detectability, and each of these may vary with place or time. Thus it is difficult to read any meaning into the value of a, but like S_1 it tends to be large when the total number of species is large (Fig. 2).

If the equitability component of diversity is to be studied, estimates of density are needed (Tramer 1969, Taylor 1978), but the accuracy of density estimates from mapping, point or transect methods may be inadequate for this (Dawson 1981c).

Island biogeographic studies may compare the number of species in habitat patches of different size (MacArthur and Wilson 1967). Here two additional problems occur; as no area is sufficiently homogeneous for the species list at one point to be representative of the whole area, the list increases with effort in both space (Cousins 1977) and time. I believe that patches should be sampled in proportion to their area. This is impossible when the patches commonly range in size over several orders of magnitude, so the increase in the number of species with increasing area is likely to be underestimated. The problem should be studied to establish its magnitude and to show how it could be avoided.

MEASURES OF ABUNDANCE AND WHEN AND HOW TO USE THEM

Frequency of Occurrence of Each Species

The results for any given species may be reported either as the frequency with which it is recorded in a series of counts, or as the average number detected per count. Blondel (1975) noted the curvilinear relationship between these two measures (Fig. 3) but did not recognize that a curve could be fitted to it by assuming a Poisson or binomial distribution. In practice this relationship means that either frequency of occurrence or average number is an adequate measure for species which occur usually as one or none in each counting unit. On the other hand, frequency becomes an increasingly insensitive measure for species found in larger numbers. Robbins and Van Velzen (1974) claimed that frequency was more suitable for statistical testing with species found in colonies or flocks, but this

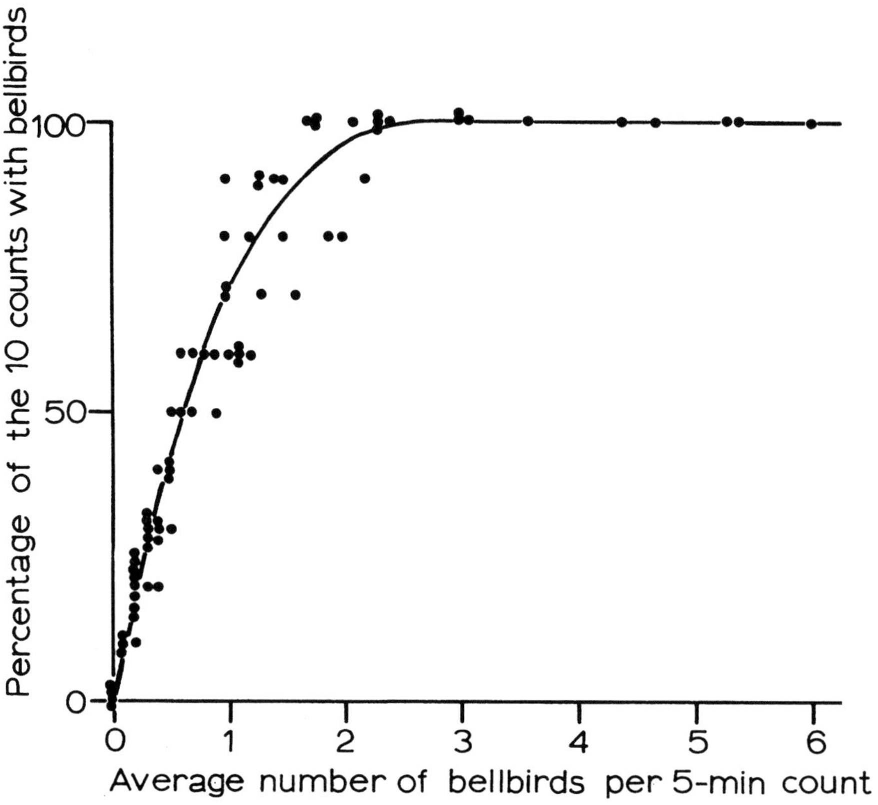

FIGURE 3. Relationship between the mean number of Bellbirds (*Anthornis melanura*) recorded in a series of counts and the percentage of the counts with bellbirds. The line is the expected percentage if Bellbird counts fit a positive binomial distribution with $K = 3$ (Elliott 1971). Other species are better described by Poisson or negative binomial models.

should not be so, as frequency uses less of the available information than do other measures (e.g., rank order).

Blondel et al. (1981) use frequency mainly to estimate the average number of species at a station, which they say is related to density (but see my discussion of this measure [my S_1] above). They claim that frequency is less affected by diurnal changes in detectability than is the average number, but this simply reflects the insensitivity of frequency and so is not an advantage.

Frequency better reflects density when the sampling time it is calculated from is shorter, but I can find no study of units less than five minutes and its deficiencies are still evident at this level (Fig. 3).

The Number Counted of Each Species

The numbers of individuals of a species counted in several sampling units (mapping plots, points, transects, or parts of a transect) may approximate a normal distribution, but typically the distribution is positively skewed, especially if the average number counted per unit is low (Elliott 1971), or if the species is confined to some parts of the sampling area (Dawson and Bull 1975, Ferry 1974, Gur'ev and Zubtwovkii 1974, Robbins and Van Velzen 1970). Given the large number of factors that may influence the numbers counted (Dawson 1981a), it is difficult to use the distribution of counts to obtain accurate details of the actual distribution of birds. For example, the use by Källander et al. (1977) of the average of positive records in five-minute stops as an index of flock size for a species is probably misleading.

If the counts are normally distributed and their variance is not correlated with their mean, parametric statistical techniques such as t-tests, regression and analysis of variance may be used (Conner and Dickson 1980). However, the variance of counts usually increases with increasing mean count (Fig. 4); the variance commonly approximates the mean, as is characteristic of a Poisson distribution (e.g., the Tomtit in Fig. 4).

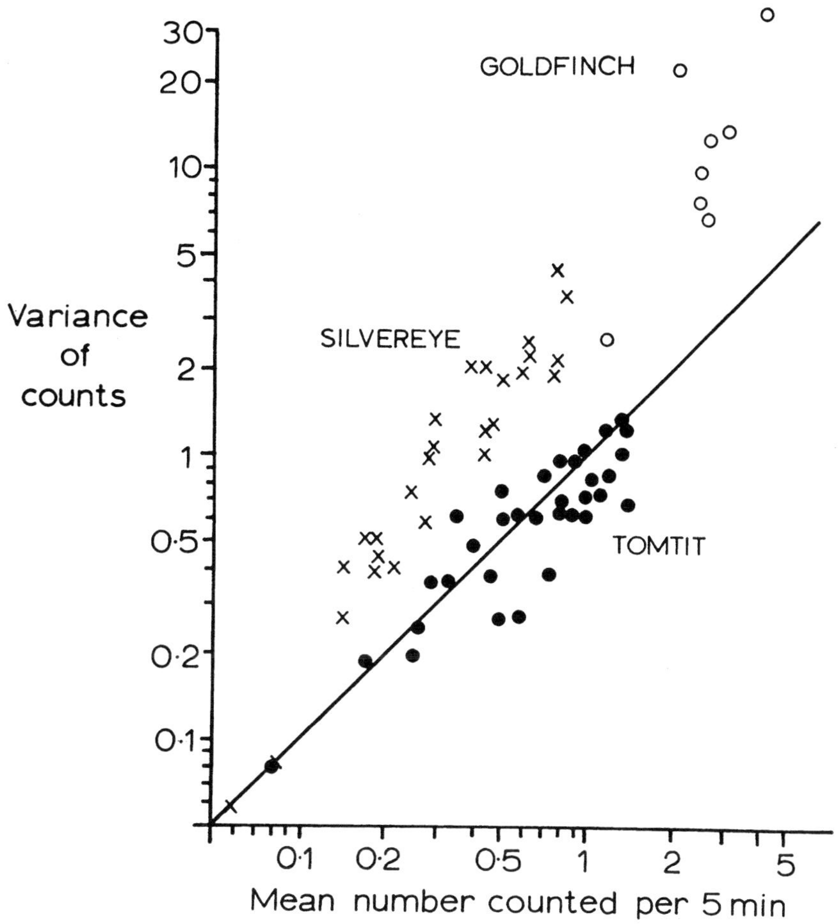

FIGURE 4. Relationship between the mean number counted per 5 minutes, and the variance of the counts, for three New Zealand species. Silvereyes (*Zosterops lateralis*) and Tomtits (*Petroica macrocephala*) were counted in forest on the Victoria Range, north Westland, with 12 to 42 counts contributing to each point (H. A. Best pers. comm.). Goldfinches (*Carduelis carduelis*) were counted in orchards and pastoral land in Hawke's Bay, with 96 counts contributing to each point. For the Tomtit, means approximately equal variances as in a Poisson distribution. The other two species' counts are more aggregated, so statistical tests assuming Poisson are not appropriate.

McCaughran and Jeffrey (1980) outline a binomial test for use with two small samples, and a large sample, normal approximation is available. Provided the total number of birds counted in each sample (nm) is greater than 30, a normal deviate is given by: $z = (m_1 - m_2)(m_1/n_1 + m_2/n_2)^{-1/2}$, where m is the mean number per unit and n is the number of sampling units counted (Elliott 1971). The special case where $n_1 = n_2$ is optimum for detecting differences, and the equation simplifies to $z = n^{1/2}(m_1 - m_2)(m_1 + m_2)^{-1/2}$. This is identical to a chi-squared test of equality of the total number counted in the two samples (nm):

$$\chi^2 = n(m_1 - m_2)^2/(m_1 + m_2) \ldots \quad (1)$$

observing that, for one degree of freedom, χ is a normal deviate.

Rearrangement of these equations gives a form which assists the design of counting studies. We define the overall mean number per count as:

$$m = (m_1 + m_2)/2 \ldots \quad (2)$$

and the percentage difference between the mean numbers in the two samples as:

$$d = 100|m_1 - m_2|/[(m_1 + m_2)/2]$$

using the overall average as the base for the percentage. This simplifies to

$$d = 200|m_1 - m_2|/(m_1 + m_2) \ldots \quad (3)$$

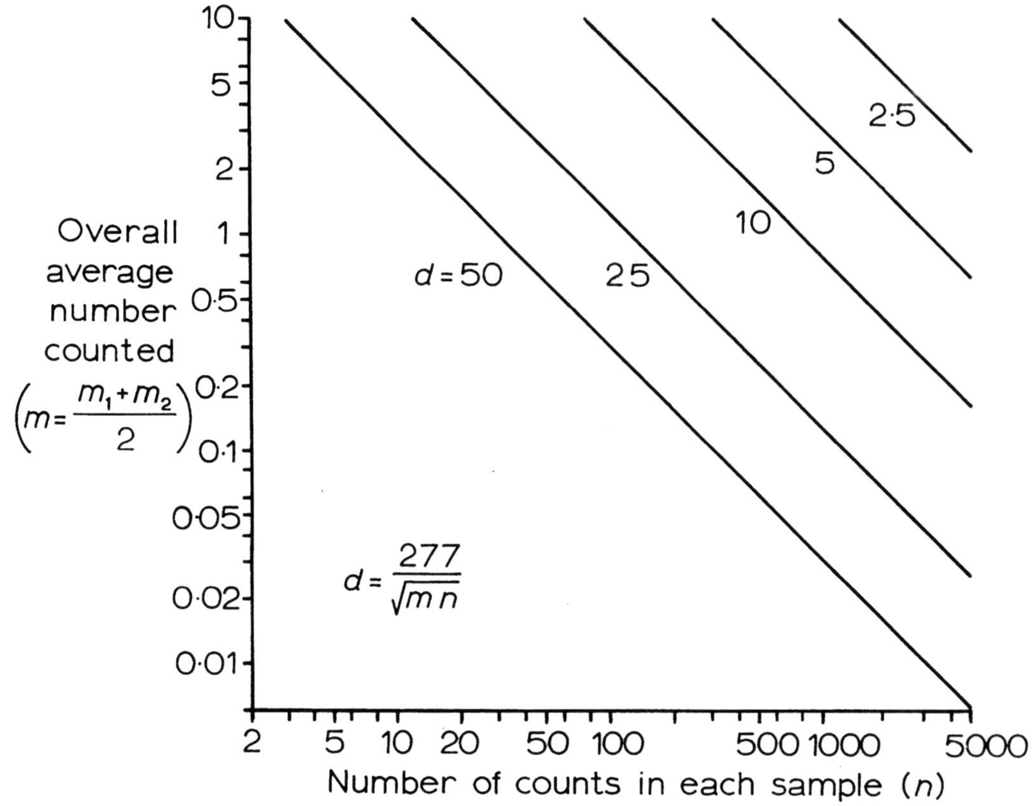

FIGURE 5. The percentage difference (d) between the counts obtained in two areas that can be detected with a chi-squared test for $P = 0.05$, given the number of counts in each area and the overall mean count. For example, if a species averages one per count, 770 counts are needed in each of two samples to permit a 10% difference to be detected in a chi-squared test; i.e., the average counts in the two areas could be as close as 0.95 and 1.05 before the test failed to reach the significance level.

Substitution in (1) for $m_1 - m_2$ from (3) and then for $m_1 + m_2$ from (2) gives:

$$\chi^2 = nd^2m/20{,}000.$$

For a result significant at the 5% level, χ^2 must be greater than 3.84: $3.84 < nd^2m/20{,}000$, or $n > 76{,}800/d^2m$; a form that permits estimation of the number of units needed to detect a given percentage difference between the samples when the average count of the species is known (Fig. 5). The requirement that $nm > 30$ corresponds to the area above and to the right of the line $d = 50$ in the graph. Dawson and Bull (1975) first proposed this method in their Table 3. The same relationship may be expressed as $d > 277(mn)^{-1/2}$, from which it can be seen that the percentage difference that may be detected decreases with the increasing total count of a species (nm), so that more units must be counted to detect the same difference with a rare species than are needed with a common one.

If a different number of counts is made in each sample, chi-squared tests approximate the normal deviate test (χ/z is less than 1.1) so long as the smaller sample is at least half the size of the larger one. Outside this range the normal deviate should be used.

Chi-squared tests have commonly been used outside of the restricted range discussed above—when the counts may not fit a Poisson distribution, for more than two samples, or for sample sizes differing more than twofold (Taylor 1965, Dawson et al. 1978)—with little or no justification that the assumptions of the test are met. Elliott (1971) advises the use of appropriate transformations in these situations to bring the distribution nearer normal and to stabilize the variance, or the use of nonparametric methods. The choice will depend on the complexity of analysis needed, as the number of factors readily taken into account by nonparametric methods (Siegel 1956) is limited. Robbins and Van Velzen (1969) used square-root transformations.

Empirical study shows that chi-squared tests remain a reasonable approximation provided the Lefkovich index of dispersion is in the range of −0.2 to +0.2 (Fig. 6); the test is conservative below −0.2 and optimistic above +0.2. Counts of many species lie within these limits, but for those that do not, routine use of transformations or nonparametric tests is not laborious in these days of electronic computers.

SAMPLING DESIGN

Comparison of Years

Studies that seek to monitor long-term changes in bird numbers have used a variety of techniques: mapping (Svensson 1974a, Williamson and Homes 1964), point counts (Robbins and Van Velzen 1967, Källander et al. 1977, Svensson 1977b) and line transects (Sammalisto 1974, Järvinen and Väisänen 1976c). All such studies seek an index of density rather than an estimate of the actual population, as the interest is simply in whether numbers increase or decrease each year. For this, point or transect counts are more cost-effective than mapping (Dawson 1981c).

An index of numerical change between years is better the more replicate sites are used to assess it, and the better control there is over other possible influences on the counts (e.g., time of year, observer, time of day, and weather; Cochran 1963, Taylor 1965, Robbins and Van Velzen 1969, Källander et al. 1977).

The Effect of a Treatment

Shields (1979) discusses the use of "control" areas and treated areas to distinguish the effect of a treatment from normal seasonal changes in density and conspicuousness. Here too replication is important, and Connor and Dickson (1980) discuss details of this for transect counts.

Comparison of Habitats

Point counts have been used to discover sites of conservation value and to document the distribution of rare species (Crook et al. 1977, Ramsey et al. in press, Svensson 1977a). A very large number of sites must be counted to achieve an accurate index of the abundance of a rare species; actual densities are even more difficult to assess. Figure 5 shows that the detection of a 50% difference in numbers between two areas would require 3000 units to be counted in each for a bird that is found in only one unit out of a hundred, but those same 3000 counts would permit a difference of 5% to be detected for a species that averages one per unit. The solution to this problem in sampling rare species is a two-stage process: (i) a reconnaissance survey to establish the distribution of the rare species, and

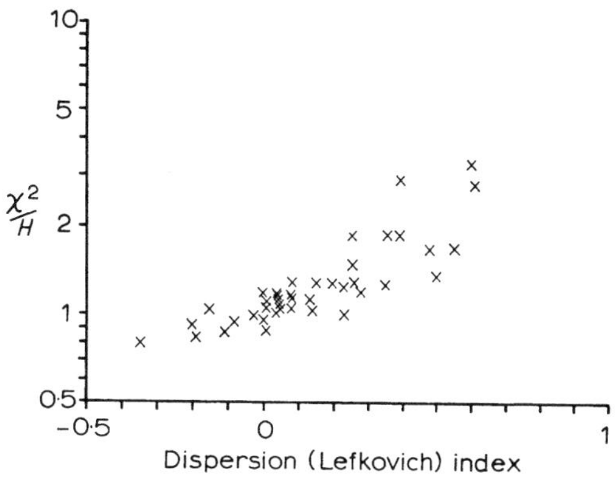

FIGURE 6. The effect of dispersion on the performance of chi-squared tests—a comparison of chi-squared tests with a more rigorous Kruskal-Wallis one-way analysis of variance. Each point is based on counts at nine counting stations for one species in the Victoria Range study of Figure 5. The Kruskal-Wallis "H" is distributed as chi-squared with 8 degrees of freedom (Siegel 1956), so the tests are equivalent when the fraction χ^2/H is unity. This is approximately so for Lefkovich (1966) indices of −0.2 to +0.2. The Lefkovich index is $1/45 \tan^{-1}$(variance/mean) -1.

(ii) counts of those areas alone. Similar reasoning can be applied to habitats as well as species: random or systematic sampling of large areas results in too many counts in the large habitats and too few in small ones. Again the solution is to establish the distribution of habitats first and then to sample each with a different intensity depending on its extent (Cochran 1963).

An index of differences in bird numbers between habitats can be less variable the more replicates it is based upon (Cochran 1963 gives methods for optimal allocation of effort within and between replicates). And other factors that influence the counts, such as the time of year, should be standardized or studied. For example, Dawson et al. (1978) used four observers to count in four study areas. On each visit each observer counted in all four areas (one per day) and all four areas were counted simultaneously on each of the four consecutive days. This design held observer, time of year, and weather more or less constant. Seasonal changes were studied by visiting the areas at two-month intervals through the year.

For some studies both time and place are factors in the sampling design. For example Williamson (1969) studied the change of habitat preferences with time and was able to demonstrate that Wren (*Troglodytes troglodytes*) numbers varied between years much more in suboptimal than in optimal habitat. Dawson et al.

(1978) suggested that some New Zealand forest birds show altitudinal migration.

Techniques giving an index of abundance are usually adequate for comparing habitats, but mapping techniques (Cousins 1977) permit greater precision of habitat definition than do point or transect counts. Another problem with using indices to study habitat selection is that detectability can vary between habitats (Dawson 1981a). An estimate of density may be needed to provide a correction factor.

Comparison of Species

To compare species with each other also requires estimates of density, as indices measure each species on a different scale. Densities may also be needed to examine species diversity (but see my discussion of this above), energetic or trophic relationships.

Estimates of density from mapping, or by means of the more cost-effective point or transect methods (Dawson 1981c) are imprecise, but may often be sufficient, given the magnitude of the errors in measuring other factors, such as food intake. More precise methods, such as capture-recapture (Frochot et al. 1977), or intensive observation of marked individuals, are also more time-consuming but may be needed if the imprecision of the estimate of density limits the accuracy of the overall calculation.

ACKNOWLEDGMENTS

A. E. Aldridge, P. C. Bull, M. Harrison, E. B. Spurr and R. H. Taylor made constructive comments on the manuscript.

Studies in Avian Biology No. 6:399–404, 1981.

OPTIMIZING SAMPLING FREQUENCY AND NUMBERS OF TRANSECTS AND STATIONS

CHARLES E. GATES[1]

ABSTRACT.—Five valid methods of calculating variances of average density are: (a) systematic sampling with multiple random starts; (b) systematic sampling with a single random start using either natural subunits or replication in time; (c) interpenetrating sampling; (d) direct estimation of $v(\hat{D})$; and (e) the jackknife method. Method (a) is "best," but highly impractical in many situations. Method (b) should prove very useful in those situations where the subunits are sufficiently long to provide reasonable density estimates from each subunit. Method (c) would appear useful in all situations with reasonable sample size. Methods (d) and (e) should prove useful where the subunits are so short that the individual densities are essentially meaningless. These methods are applicable for any method of determining density.

To ascertain the total length of transect needed to achieve a desired coefficient of variability, calculate $L_1 = (\mathrm{cv}_0(\hat{D}))^2 L_0/(\mathrm{cv}_1(\hat{D}))^2$ where $\mathrm{cv}_0(\cdot)$ and $\mathrm{cv}_1(\cdot)$ are the observed (in a preliminary survey) and desired c.v.'s, respectively, with LT lengths L_0 and L_1.

In optimizing the LTs with subunits (or stations) and multiple sampling dates, the larger the variance component associated with a particular source of variation the greater the number of levels of that factor required (for fixed sample size), ignoring costs. If costs are considered generalization is more difficult. Obviously, if it is much cheaper to take an additional station than to get to the transect, the effect is to tend to drive the solution to more stations per transect at the expense of transects.

The purpose of this paper is to discuss the design of sample surveys in line transect and related sampling methods. To set the stage I shall define briefly the line transect and related sampling methods, following the standardized terminology suggested by Eberhardt (1978). The *line transect* (LT) is a basic sampling method wherein an observer walks a randomly located straight line, observing the target species, whether song birds, ruffed grouse, deer, duck nests, plants or rocks. For convenience, I employ the terminology as if animals were the target species, even though the sampling method is more general. At a given sighting, the observer records one or more of the following statistics: right-angle (perpendicular) distance to the sighted individual(s), radial (sighting or flushing) distance to the sighted individual(s) and/or the sighting (flushing) angle. On the basis of these measurements and a number of assumptions (see Gates 1979), it is possible to estimate the total population in the sample area or, equivalently, the density of animals.

Closely related sampling methods include the *strip transect, line intercept* and *quadrat sampling*. A strip transect is similar to the LT except that all animals are counted within a predetermined width in which the observer is reasonably certain all animals have been seen; animals outside the strip are not counted. A quadrat is similar to the strip transect except that many small areas are censused rather than a small number of much larger strip transects. A line intercept is a line or a strip transect narrowed to the line itself. It is more commonly employed for plants and inanimate objects than for animals, although it could be used for dense populations of slow-moving animals, e.g., snails. Note that estimating densities by the line intercept and quadrat methods is considerably different from that by the line and strip transect methods. I will not discuss the former methods further and will not discuss estimation explicitly for any of the methods. I leave this discussion for others and note several recent LT reviews and announcements of general computer programs, e.g., Gates (1979) and Burnham et al. (1980).

The design of any experiment or survey is highly dependent on the variability exhibited by the variable under study. Thus computing a valid estimate of variance is a necessity. In the remainder of this paper I first discuss five ways of calculating the variance of the density estimates and consider approximations to reduce the coefficient of variation of estimated density, $s_{\hat{D}}/\hat{D}$, to a predetermined size. I then consider costs in conjunction with a more complex LT design consisting of a line with several stations or subunits, sampled over time. Data, possibly not densities, are available for each station-time period.

COMPUTING VARIANCE OF DENSITY

The principal difficulty with reducing variance of density estimates to manageable size is obtaining a LT of sufficient total length. A line or a strip transect must of necessity use a large amount of real estate, in order to minimize overlap and to assure sufficient length for estimation of the population density with precision. To achieve meaningful results for some species,

[1] Institute of Statistics, Texas A&M University, College Station, Texas 77843.

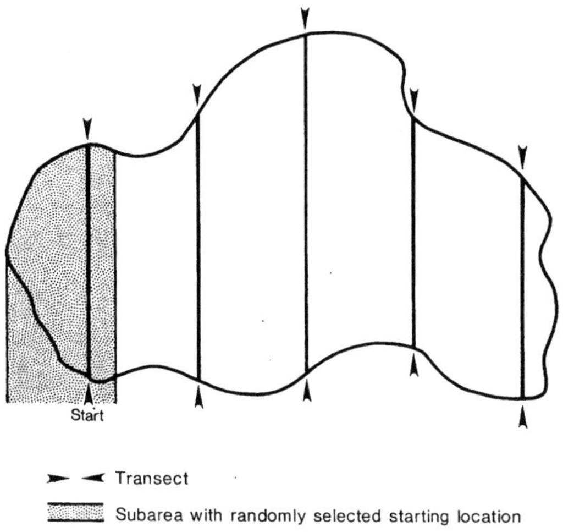

FIGURE 1. Systematic sampling with a single random start.

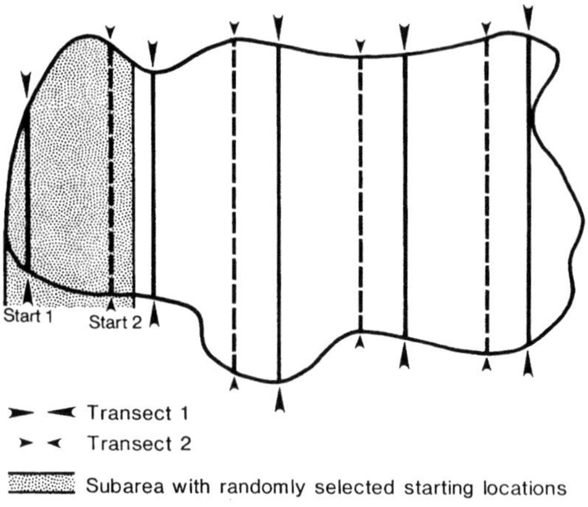

FIGURE 2. Systematic sampling with two random starts.

e.g., game birds, it may be necessary to have the line transect length 30, 40 or more km in length. If right-angle distances of only 100 meters to either side of the line are conservatively estimated (all of this is highly species-dependent, needless to say), then 6 km^2 would be utilized in a 30 km length.

Suppose the area being sampled is not sufficiently long (or wide) for 40 km of transect, e.g., an area 20 × 20 km. Then one could randomly locate in the sample area two transects of 20 km each (with restricted randomization such that there was no overlap). To ensure both that the entire area is representatively sampled and that there is no overlap, one could use *systematic sampling* (SS). For instance, one might select a random number between 1 and 10, say 5.2. This first selection determines directly the starting point; the second segment would start at 15.2 km and would be parallel to the first segment. If the SS were to be replicated in the true sense of the word, *two* random starting points would be required, say 5.2 and 7.3; thus the second portions of the transects would begin at 15.2 and 17.3 km from the base. The two techniques are called, respectively, SS with a single random start and SS with multiple random starts (Sukhatme 1954, Cochran 1977) (see Figs. 1 and 2).

Prior to discussing potential improvements in the design of a survey, a reasonably good estimate of variability of density estimate is required. Thus the estimation of variance must be discussed, which is related to the concept of replication. To the sampling purist, SS with at least two random starts would be required to have valid replication and thus valid variance estimation. My first reaction was in sympathy with this viewpoint, but on further reflection I moved away from that viewpoint. Some individuals would like to subsample without limit, dividing one large sample into more and more pieces, thus giving a large number of degrees of freedom for estimation of the variance. However, in using a line or strip transect, this "infinite" subdivision is not practical; if many subtransects were formed then most would have 0 animals sighted with a scattering of 1's, a very few 2's and so on. Such a situation would be totally impractical. To estimate density with any precision, large subtransects would have to be used. Natural subdivisions of the transects as shown in Figures 1 and 2 should be permissible. In fact, it may be necessary to clump adjacent subtransects to obtain a sufficient sample size for a reasonable estimate.

An objection of the sampling purists to using natural subdivisions or large fractions of single transects is that treatment of these subtransects as independent samples is incorrect. The theoretical difficulty is that, since these subtransects are physically close to one another, there may be large positive correlations among the dependent subtransects so that variance is underestimated. However, the situation does not concern me greatly because, unless the transects are very long, a high degree of variability will be associated with the estimation of density from each subtransect. In fact, the high variability ordinarily will swamp the positive correlation between adjacent subtransects. There is one important qualification in the use of SS that must be kept in mind. One should be certain the distance between parallel subunits does not coincide with some topographic feature, e.g., ridges.

This could prove disastrous in either estimation of density or variances.

On the other hand, if the transects are too long it is likely that heterogeneous habitat will be encountered. This introduces the topic of stratified sampling. One should stratify within each habitat type markedly influencing the density of animals. Using optimum allocation, habitats with either a greater density of the target species or increased variability will require a larger sample than otherwise. (Greater density leads to increased variance, everything else being held constant.) Similarly, habitats with reduced density or variability will require shorter transect lengths than otherwise. If we fail to stratify, then the lengths of transects in each habitat will be approximately proportional to the total area of each habitat (stratum), which will undoubtedly not be optimum. Further discussion will focus on optimizing surveys within strata or where stratification is not required.

If it is not feasible to replicate over space, it may be feasible to replicate over time. This is commonly done in LT sampling. Obviously, the time frame must be short enough so that significant mortality or recruitment could not have occurred, and ambient conditions should be similar. If density has changed, then an average density will be estimated with increased variability due to change in density.

However, transects need not be partitioned into either natural or artificial units to estimate variances of mean density. A legitimate sampling method for estimating sampling variances with one true replication is called *interpenetrating sampling* (Cochran 1977) and is closely related to the statistical jackknife method (to be described later). In interpenetrating sampling, the data are randomly sampled after collection. Suppose each sighting is randomly assigned to one of k subsamples. The density is then estimated from each subsample, where the number of observations will be a random variable. For b = 4 the LT length will be 25% of its former value. The variance is then determined from the densities of the individual groups, $\hat{D}_1, \hat{D}_2, \ldots, \hat{D}_k$, and is an unbiased estimate of $V(\hat{D})$ provided there is no correlation between the errors of measurement of any two sampling units in different groups. This condition would appear to be met in transect sampling. The disadvantage to the procedure is that if two individuals calculate the variance, even with the same number of subsamples, they will not obtain exactly the same answer. The method is not unique in that sense. Interpenetrating sampling is illustrated in Table 1. In the original population there were 40 sightings. Each sighting was randomly and independently assigned to one of four subsamples with the resulting subsample sizes, n, estimated intercepts, $\hat{f}(0)$, and densities shown in Table 1. The density, $\hat{D}$, was calculated assuming length of the line transect L = 100 km and distances recorded to the nearest meter.

TABLE 1
ILLUSTRATION OF INTERPENETRATING SAMPLING

Subsample	Number sightings, n	Intercept, $\hat{f}(0)$	$\hat{D}$
1	12	.301	72.2
2	12	.256	61.4
3	7	.177	24.8
4	9	.172	31.0
Total	40	.244	48.8

Another way (the "direct" method) of determining the variance of estimated density is to consider the general LT density estimator

$$\hat{D} = cn\hat{f}(0)$$

where c is the constant, $1/(2L)$. The variance of $\hat{D}$ may be written

$$V(\hat{D}) = c^2V[n\hat{f}(0)].$$

The expression in brackets is a product of variables. Using known information on the variance of a product of variables and that n and $\hat{f}(0)$ will be uncorrelated or very close to it yields

$$V(\hat{D}) = c^2[V(n)E^2\hat{f}(0) + V(\hat{f}(0))E^2(n)],$$

where $V(n)$ and $V(\hat{f}(0))$ are the variances of n and $\hat{f}(0)$, and $E(\hat{f}(0))$ and $E(n)$ the expected values of $\hat{f}(0)$ and n, respectively. If n is binomial, then $V(n) = NPQ$, $E(n) = NP$, where P is the probability of flushing an animal given that it is in the transect and $Q = 1 - P$. However, unless the animals truly flush independently of one another, it is unlikely that n will be binomially distributed. (It is more likely that n follows a negative binomial distribution.) Thus $V(n)$ and $V(\hat{f}(0))$ could be estimated empirically from natural subunits of a transect, although there seems to be no advantage in doing that over calculating the empirical variance of $\hat{D}$ from the $\hat{D}_i$ (as done in the interpenetrating sampling procedure).

Burnham et al. (1980) observe that for their recommended estimators, e.g., the Fourier series, the variance of $\hat{f}(0)$ is readily obtainable. Thus if one of those estimators is used, the only problem is in the calculation of $v(n)$. This quantity may always be calculated by empirical methods if natural subunits of a LT are available. If not, the binomial, the Poisson or negative binomial approximation to $v(n)$ would have to be used, depending on the user's best appraisal.

TABLE 2
ILLUSTRATION OF THE JACKKNIFE ESTIMATION OF D_J AND $v(D_J)$[a]

i	n_i	l_i	$n - i$	$L - l_i$	$\hat{D}_i$	$\hat{D}^{(i)}$
1	14	3	121	25	101.00	84.66
2	20	4	115	21	101.75	160.56
3	43	9	92	19	98.66	100.48
4	18	3	117	25	100.33	90.22
5	23	5	112	23	95.80	115.12
6	17	4	118	24	100.25	93.25

[a] Adapted from Burnham et al. (1980); $R = 6$, $L = 28$, $n = 135$, $D = 99.25$.

It is instructive to examine alternative methods of expressing the direct variance of $\hat{D}$ (replacing $E(n)$ and $E\{\hat{f}(0)\}$ by n and $\hat{f}(0)$, respectively):

$$v(\hat{D}) = \hat{D}^2\left[\frac{v(n)}{n^2} + \frac{v(\hat{f}(0))}{\hat{f}^2(0)}\right]$$

from which it follows that

$$cv(\hat{D}) = cv(\hat{n}) + cv(\hat{f}(0)),$$

where $cv(\cdot)$ and $v(\cdot)$ are the sample coefficient of variation and variance, respectively.

A special case of the direct method of calculating variance is to calculate the theoretical variance directly. For example, Gates et al. (1968) give

$$v(\hat{D}) = \frac{n}{(A\hat{P})^2}\left[\hat{Q} + \frac{n}{n - 2}\right]$$

where $\hat{P} = 2L/A\hat{\lambda}$, and A is the area of the study site. However, it is dangerous to use such variances, as they depend heavily on two assumptions—exponentiality of right angle sighting distances in this case—and on the strict independence of sightings. The failure of the assumptions will cause the estimated variance to underestimate the true variance by an unknown amount.

The fifth method for estimating variance of density is the jackknife method. The technique is illustrated by Burnham et al. (1980), whose Table 4 we modify and present here as Table 2. Basically, the method requires a series of natural subunits. The set of data from each subunit is omitted, one at a time, with the density estimated from the remaining data. These densities are called pseudovalues, $\hat{D}^{(i)}$, and are used to calculate the average density and ultimately $v(\hat{D})$:

$$\hat{D}^{(i)} = \frac{L\hat{D} - (L - l_i)\hat{D}_i}{l_i}$$

where l_i is the length of the ith subunit and $\hat{D}_i$ its density. Then

$$\hat{D}_J = \frac{1}{L}\sum_{i=1}^{R} l_i\hat{D}^{(i)}$$

and

$$v(\hat{D}_J) = \sum_{1}^{R} \frac{l_i(\hat{D}^{(i)} - \hat{D}_J)^2}{L(R - 1)},$$

where R is the number of subunits. For the data illustrated in Table 2, $\hat{D}_J = 107.85$ with $v(\hat{D}_J) = 130.60$. Thus 95% confidence intervals, using the t statistic with five degrees of freedom are 78.5 to 137.23. The chief disadvantage of this procedure is that computations are fairly heavy with a desk calculator. They are admirably adapted to the computer, however.

LENGTH OF LINE TRANSECT NEEDED

Given now that some legitimate estimate of sampling variance of density is computable, how can we improve our sampling in the next iteration? Gates et al. (1968) gave a procedure for estimating the length of line transect needed to reduce the ratio of $v(\hat{N})/\hat{N}$ to some predetermined value R for their parametric estimator. The difficulty with their expression is that it is highly dependent on the exponentiality of the right angle flushing distances.

A more general criterion would be to make the reasonable assumption that the product of LT length and the squares of the coefficients of

TABLE 3
MEAN SQUARE EXPECTATIONS FOR MULTIPLE STATIONS PER TRANSECT, SAMPLED AT VARIOUS TIME INTERVALS

Source of variation	Degrees of freedom	Mean square	Expected mean square
Transects	$t - 1$	M_t	$\sigma^2_e + s\sigma^2_{tw} + w\sigma^2_{s(t)} + ws\sigma^2_t$
Stations (T)	$t(s - 1)$	$M_{s(t)}$	$\sigma^2_e + w\sigma^2_{s(t)}$
Times	$w - 1$	M_w	$\sigma^2_e + s\sigma^2_{tw} + st\sigma^2_w$
Times × tran.	$(w - 1)(t - 1)$	M_{tw}	$\sigma^2_e + s\sigma^2_{tw}$
Residual	$t(s - 1)(w - 1)$	M_e	σ^2_e

variation (cv) of observed densities are proportional at different lengths:

$$L_0(cv_0(\hat{D}))^2 = L_1(cv_1(\hat{D}))^2$$

where $cv_0(\hat{D})$, L_0, $cv_1(\hat{D})$ and L_1 represent, respectively, the observed cv in a survey of a similar species in a similar habitat or small preliminary survey of length L_0 and the desired cv in the final survey with total length L_1. Solving for L_1, we have

$$L_1 = \frac{(cv_0(\hat{D}))^2 L_0}{(cv_1(\hat{D}))^2}.$$

This result is identical to that found by Burnham et al. (1980:35). Thus if a small survey is run with $cv = 0.3$ and $L = 3$ km, but a cv of 0.1 is desired, $L_1 = (0.3)^2 4/(0.1)^2 = 36$ km.

COST EFFECTIVE SAMPLING OF LTs WITH STATIONS

Next consider a more complex sampling plan wherein the observer has stations (stops or subunits) on the transect and may be interested in sampling on more than one occasion. How may he allocate his resources in some useful way? I shall make the assumption that the average of the variable being measured (not necessarily density) does not change markedly over time (if it does, then the problem degenerates to considering the optimal sampling within dates). Assume that the researcher has t transects, each with s stations (subunits) and samples on w occasions. The random model for the situation described is

$$y_{ijk} = \mu + t_i + s_{ij} + w_k + (tw)_{ik} + \epsilon_{ijk}$$

where y_{ijk} = observed value (e.g., density or calls per three minute time period), t_i = transect effect, s_{ij} = station (subunit) within transect effect, w_k = time effect, $(tw)_{ik}$ = transect by time interaction effect, ϵ_{ijk} = random residual.

The analysis of variance appropriate to this completely random model is shown in Table 3.

The mean square expectations do not provide a criterion per se. One possible criterion for improving the sampling procedure would be to minimize the variance of a transect mean. The variance of a transect mean, $V(\bar{T})$, is the expected mean square for the transect without the σ^2_t term, divided by the number of observations per transect, viz., sw. For fixed product sw, the minimization of $V(\bar{T})$ depends on the relative sizes of estimates of σ^2_{tw} and $\sigma^2_{s(t)}$ as the relative size of σ^2_e is immaterial. If σ^2_{tw} is much larger than $\sigma^2_{s(t)}$ then the transect should be sampled more often at the expense of sampling more stations. Conversely, if $\sigma^2_{s(t)}$ is much larger than σ^2_{tw} then more stations should be sampled at the expense of repeated sampling. If those two variance components are about the same size, then $s = w$ approximately. However, this is not a good criterion, as the number of transects is not considered and the cost of sampling is ignored. (One could optimize t and s by considering $(V(\bar{W}))$, variance of a time mean, but then no information is given on w.) It is undoubtedly more expensive to sample additional times than to sample additional stations.

Two common concepts involving costs in sampling invoke two different alternatives: (a) minimize cost subject to fixed variance or (b) minimize variance subject to fixed cost. Gates et al. (1975), with a model similar to the ANOVA model shown above, suggested specifically minimizing the variance of the overall mean subject to fixed cost. Consider a cost function such as

$$C = tc_t + wtc_w + wtsc_s$$

where t, w and s are defined as above and c_t = cost of establishing and maintaining a transect, c_w = average cost of traveling to a transect, and c_s = cost per station once the observer reaches the transect. The formal function for minimizing the overall variance subject to fixed cost, e.g., is

$$V(\bar{y}_{...}) + \lambda(C - tc_t - wtc_w - wstc_s)$$

where λ is a Lagrangian multiplier (Lindgren 1962:216–227) and

$$V(\bar{y}_{...}) = \frac{\sigma^2_e}{wls} + \frac{\sigma^2_{tw}}{tw} + \frac{\sigma^2_w}{w} + \frac{\sigma^2_{s(t)}}{st} + \frac{\sigma^2_t}{t}.$$

The minimization of this function requires the simultaneous solution of four non-linear equations in four unknowns (obtained by differentiating the previous expression with respect to s, t, w and λ, respectively). We need not show these but simply note that the equations cannot be solved directly, due to their non-linear nature, but must be solved by iteration. The procedure assumes that the variance components are known and treats s, t, w and λ as variables.

Gates et al. (1975) used $c_s = 0.3775$, $c_w = 5.442$ and $C_t = 1.00$ in a Mourning Dove (*Zenaida macroura*) survey in Texas, and concluded on the basis of analyzing several variables that the optimal design would be a very large number of transects, 8–13 stations per transect and one sampling time. When the number of sampling times was constrained to four, the optimum numbers of transects and stations/transect were about 170 and 5, respectively (vs. the original 91 transects and 20 stations/transect). Eventually 135 randomly-located transects with 15 stations each were established.

Modifications of the above technique would permit optimizing the number of transects and stations at a single sampling time or optimizing the number of transects and times for one station per transect. In the above development, w (or s) would be replaced by one and the number of non-linear equations would be reduced to three. The solutions would be a simplified version of the more general case.

A novel use of the procedure outlined would be to optimize the number of subunits for a lengthy transect for future similar work. Currently, it is not clear whether to have a small number of subunits with relatively small variance each or a large number of subunits to give more degrees of freedom for confidence limits but relatively large variances.

Studies in Avian Biology No. 6:405–408, 1981.

EFFECTS OF NUMBER OF CIRCULAR PLOTS ON ESTIMATES OF AVIAN DENSITY AND SPECIES RICHNESS

MICHAEL L. MORRISON,[1] R. WILLIAM MANNAN,[1] AND GEOFFREY L. DORSEY[2]

ABSTRACT.—We evaluated the effects of the number of census stations on estimates of avian density and species richness using the variable circular plot method. Data were collected in a variety of vegetation types, including coniferous and deciduous forests, shrublands, and uplands. Density estimates generally decreased with an increasing number of stations; however, density values in most habitats were fairly constant with as few as four stations censused four or five times each. The number of stations necessary to achieve a stable density estimate was related to the complexity of habitat. Inaccurate density estimates calculated for two stations on several sites were caused by gross underestimations of the effective detection distance for one or two species; these detection distances stabilized by four stations. Density estimates decreased slightly when sampling exceeded four stations, due to an increasing number of stations included in the calculations that reported zero values (i.e., species absent). The cumulative number of species observed increased with an increase in stations. We concluded that the optimum number of stations was a study specific problem, and should be based both on the complexity of the habitat and on the type of results needed. Guidelines for making such decisions are presented.

Of the methods available for censusing birds, the variable circular plot technique is one of the newest (Reynolds et al. 1980). A modification of J. T. Emlen's (1971) line transect method, this technique was developed to allow the observer to census from fixed points. Although applicable in all habitats, the method is especially suited to rough terrain.

Since the method is relatively new, little information is available for researchers to use in designing a study. Reynolds et al. (1980) noted that the number of stations necessary to calculate an accurate density for a species varied with the spatial distribution of individuals, its abundance, and its conspicuousness in various vegetation types. Thus, the researcher must be aware of the effects that the number of stations has on census results. This may be especially important when time is limiting and several different habitats must be assessed.

To establish guidelines for designing census procedures with this technique, we sought to determine the optimum number of stations for estimating density and species richness of birds. We approached this problem by analyzing data collected in several habitats in Oregon.

STUDY AREAS

Study areas ranged between 26 and 65 ha. We felt that 20 ha represented the smallest area that could be considered relatively continuous habitat; sites below that size are increasingly affected by the environment and associated avifauna of adjacent habitats. Therefore, our results should be interpreted as applying to areas of at least 20 ha.

[1] Oregon Cooperative Wildlife Research Unit, Oregon State University, Corvallis, Oregon 97331.

[2] Department of Fisheries and Wildlife, Oregon State University, Corvallis, Oregon 97331.

The following is a brief description of the six areas chosen for study. Each site will be referred to by its associated mnemonic (in italics) throughout the text. A more thorough description of the vegetative zones characteristic of our study areas was given by Franklin and Dyrness (1973).

Site 1: The Early-growth Clearcut site (4-yr post-planting) was located at 30 m elevation in the Oregon Coast Range about 30 km west of Corvallis, Benton County, Oregon. The 30 ha site had an average slope of 35% and a northern exposure. The area was clearcut logged (all commercial and noncommercial trees were cut) in 1972. After logging, the site was prepared for planting by broadcast burning (1972) followed by herbicide treatment (1973); Douglas fir (*Pseudotsuga menziesii*) seedlings were hand planted in 1975. During our study, the vegetation was characterized by a dense (48% cover) and ubiquitous shrub layer dominated by salmonberry (*Rubus spectabilis*), thimbleberry (*R. parviflorus*), vine maple (*Acer circinatum*), and salal (*Gaultheria shallon*). Dominants in the low shrub-herb layer included sword-fern (*Polystichum munitum*), tanzy ragwort (*Senecio vulgaris*), foxglove (*Digitalis purpurea*), pearly everlasting, (*Anaphalis margaritacea*), Oregon oxalis (*Oxalis oregana*), and various grasses. Douglas-fir had not yet assumed a position of dominance (4% cover) and with an average height of 1.2 m was severely suppressed by the shrub layer. Red alder (*Alnus rubra*) provided the only vertical diversification on the site. About five distinct patches of alder averaging 4.0 m in height were scattered about on areas of soil disturbance and collectively composed 8% of the total cover.

Site 2: The Plantation Clearcut (7-yr post-planting) was located near the previous site at about 300 m elevation. Comprising 26 ha, the site had a gentle slope and northeast exposure. Clearcut logging took place in 1970; site preparation and planting with Douglas-fir followed in 1972. Because of brush that was retarding conifer growth, the entire site was aerial sprayed with phenoxy herbicides (2,4-D and 2,4,5-T) in 1975. This treatment effectively eliminated red alder and greatly reduced the cover and vigor of shrubs. As a result,

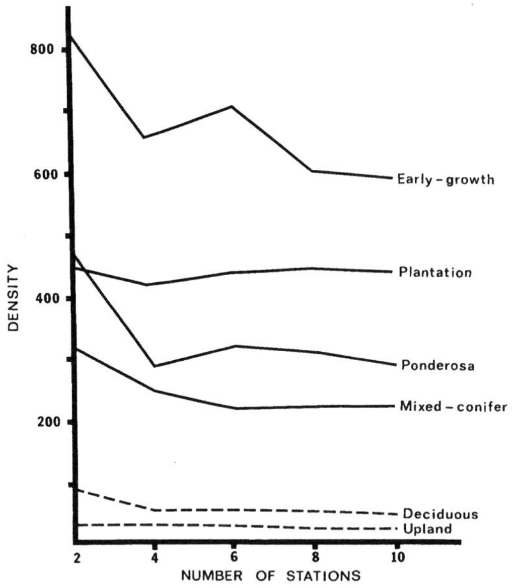

FIGURE 1. Density estimates obtained from census data collected at a varying number of stations for the six study sites (solid line = birds/40 ha; dashed line = birds/ha).

during our study, the site was dominated by a homogeneous cover (30%) of Douglas-fir. With an average height of 2.9 m, conifers supplied the only vertical structure on the site. Shrub cover (about 20%, primarily salmonberry, vine maple, and salal) was concentrated around the perimeter of the clearcut. The low shrub-herb layer was similar to that on the Early-growth Clearcut site.

Sites 3 and 4: The Deciduous Forest and Upland study areas were located on Miller Sands and West Sand Island, lower Columbia River, Oregon. Miller Sands, formed from deposition of dredged material, was located between river miles 22 to 25. West Sand Island, partially created from dredged material, was situated in Baker Bay, east of Cape Disappointment, Washington. A Deciduous Forest site and an Upland site were located on each island. Data were combined for our study (i.e., the two Deciduous Forest areas were lumped as were the two Upland areas) as no significant differences in avian communities were found between islands. The effective size of each study site was 35 ha.

In general, trees and shrubs occurred on low, mesic sites; higher elevations and drier soils typified Upland (forbs, grasses) plant communities. Red alder and willow (*Salix* spp.) composed 62% and 34%, respectively, of the Deciduous Forest overstory (over 5 m) on West Sand Island. Red alder (55%), willow (15%), and cottonwood (*Populus trichocarpa*) (16%) were the most abundant overstory trees on Miller Sands. Willow (46%), twinberry (*Lonicera involucrata*) (30%), and salmonberry (15%) were the dominant understory shrubs on West Sand Island. Salmonberry (36%) and elderberry (*Sambucus racemosa*) (25%) were the most abundant understory shrubs on Miller Sands. Principal habitat components for Miller Sands Upland were moss (37%) and forbs (24%). Grasses and sedges (26%), forbs (35%), and bare ground (20%) typified West Sand Island Upland.

Site 5: The Ponderosa Pine (*Pinus ponderosa*) study site was located in the southwestern Blue Mountains about 15 km north of John Day, Grant County, Oregon. It encompassed approximately 65 ha on a southwest-facing slope between 1450 and 1500 m elevation. Ponderosa pine seedlings and trees up to 10 m tall were abundant (55% canopy cover); large pine trees up to 1.2 m DBH and 30 to 40 m tall were scattered throughout the area and dominated the site. Western juniper (*Juniperus occidentalis*) was widespread (5% cover) but grew vigorously only on shallow, stony soils in forest openings. The edge of some of the openings also supported dense stands of mountain mahogany (*Cercocarpus ledifolius*) shrubs, 2 to 7 m tall (1% cover). Douglas-fir, grand fir (*Abies grandis*), and western larch (*Larix occidentalis*) were present but essentially confined to a cool, moist ravine on the southwest portion of the site (5% cover). Conspicuous species in the low shrub-herb layer included elk sedge (*Carex geyeri*), pinegrass (*Calamagrostis rubescens*), shinyleaf spiraea (*Spiraea betulifolia*), heart-leaf arnica (*Arnica cordifolia*), lupine (*Lupinus* spp.), and snowberry (*Symphoricarpos* spp.).

Site 6: The Mixed-conifer study site was located in the Wallowa-Whitman National Forest about 8 km west of Medical Springs, Union County, Oregon. The stand was approximately 45 ha in size and was situated on the southwest-facing slope of the Eagle Creek drainage. Elevation at the site ranged from 1500 to 1900 m. Although small openings were scattered throughout the stand, the site generally was densely forested (40–100% canopy closure) with trees of varied sizes providing a multi-layered canopy. The overstory was dominated by large (over 76 cm DBH, 25 to 35 m tall) Douglas-fir and ponderosa pine, while grand fir dominated the understory. Scouler willow (*Salix scouleriana*), 3 to 4 m tall, was sparsely distributed (less that 5% cover) throughout the area. Dominant species in the low shrub-herb layer were elk sedge, pinegrass, and heart-leaf arnica.

METHODS

We used the variable circular plot method (Reynolds et al. 1980) to census birds on each study area. Ten stations were established on each site. No station was closer than 100 m to the edge of adjacent habitat nor closer than 100 m to the next nearest station. Exceptions to these distances were made where a physical barrer (e.g., river) abruptly demarked the edge of the habitat. Censusing began at sunrise and ended 2 to 3 hrs later. On any given study site, count duration and number were the same at each station. Because of differences in vegetation structure and breeding phenologies of the birds between study sites, we varied counts per station (study site) from 4 to 5, and time per count from 8 to 10 minutes, respectively.

Bird densities were estimated from data taken at three combinations (replicates) each of 2, 4, 6, and 8 stations; only one estimate was possible for 10 stations. For example, for four stations, separate densities were derived from stations 1, 3, 5, 7; 2, 4, 6, 8;

TABLE 1
INTRA- AND INTERSTATION VARIATIONS IN DENSITY FOR A VARYING NUMBER OF STATIONS[a]

Site	Number of stations 2	4	6	8	10
Early-growth clearcut	ns	ns	ns	ns	ns
Plantation clearcut	*	ns	ns	ns	ns
Upland	ns	ns	ns	ns	ns
Deciduous forest	ns	ns	ns	ns	ns
Ponderosa pine	*	*	ns	ns	ns
Mixed-conifer	ns	ns	ns	ns	ns

[a] Nonsignificant interstation variations are indicated by horizontal lines; intrastation variation: ns = nonsignificant; * indicates $P < 0.05$.

TABLE 2
CUMULATIVE NUMBER OF SPECIES OBSERVED WITH A VARYING NUMBER OF STATIONS

Site	Number of stations 2	4	6	8	10
Early-growth clearcut	22	25	27	27	29
Plantation clearcut	17	21	24	25	26
Upland	15	21	23	23	27
Deciduous forest	15	18	20	20	22
Ponderosa pine	26	26	30	31	32
Mixed-conifer	19	25	25	26	27

and 1, 4, 7, 9. The effective radius of detection for each species was separately computed for each of the three groups of four stations. Densities and effective radii were computed for all possible combinations of 2, 4, 6, 8, and 10 stations for one study site, and results did not differ significantly from results based only on three combinations. A shortage of computer funds prevented similar treatment of results from other study sites.

The separate results from the three replicates were averaged to give an overall density estimate for each of the groups of stations. Analysis of variance (ANOVA) was used to determine if there were significant intra- or interstation variations in densities. The cumulative number of species observed also was determined for each group.

RESULTS

Densities generally decreased as the number of stations increased, until a stable point was reached with 4 or 6 stations, except for the Early-growth Clearcut site, where the density estimate did not stabilize until eight stations (Fig. 1). Significant interstation variations in densities were seen on only three sites—between 2 stations and 4 to 10 stations for the Early-growth Clearcut, Ponderosa Pine, and Deciduous Forest sites (Table 1). Although a qualitative judgment, these sites appeared to have a more patchy, or heterogeneous, vegetative structure than the other sites. Significant intrastation differences occurred when only 2 stations were used for density calculation, except for Ponderosa Pine, which exhibited such variation at 4 stations as well.

Although our data are too numerous to include here, perusal of results revealed that variation in the effective radius of detection of most species did not change appreciably with differing numbers of stations. Exceptions were obvious—on the Upland site, the effective radius for only one species, the Violet-green Swallow (*Tachycineta thalassina*), exhibited a marked fluctuation: for two stations, a radius of 10 m and a density of 8.0 birds per ha were calculated; whereas for 4 to 10 stations, the radius stabilized between 50 and 60 m, and the density between 1.5 and 2.0 birds per ha. The relatively high density estimate for this species, using only two stations, was responsible for the higher density value seen between two stations and 4 to 10 stations. A similar situation was responsible for the fluctuation noted on the Ponderosa Pine site: a radius of 15 m and a density of 349 birds per 40 ha were calculated for Chipping Sparrows (*Spizella passerina*); the radius was a constant 35 m, while the density ranged from 51 to 98 birds per 40 ha for 4 to 10 stations. Thus, for all sites, changes (errors) in the computed effective radius of birds were not the primary cause of interstation variation in density for four or more stations.

Not all species were observed at each station. If a species did not occur at a station, it was assigned a value of zero for that station. When densities were calculated for 2 or 4 stations, zero values had less of a dampening effect on overall density estimates as compared to their effect on 6 to 10 stations. Therefore, density estimates by species and overall site would be expected to decline with increasing numbers of stations until an equilibrium was reached between stations reporting a species and not reporting the species. This point should be attained sooner (fewer stations) on more homogeneous sites, where the dominant vegetation is more evenly distributed. This was evident for the Plantation and Early-growth Clearcut sites—no significant difference between stations was shown for the relatively homogeneous Plantation site, while on the more heterogeneous Early-growth Clearcut site, the density estimate did not stabilize until eight stations.

The cumulative number of species generally increased through 10 stations in all study areas (Table 2). It is not surprising that an increase in area censused would result in a more inclusive species list.

DISCUSSION

Our results show that two sources of error were responsible for unstable density estimates with a low number of stations. First, large errors with only two stations were caused by relatively small effective radii—an apparent result of small sample sizes. Second, after effective radii stabilized, densities continued to decline slightly as an increasing number of stations with zero values (i.e., species absent) were included in cumulative density calculations.

Following these results, we sought to establish broad guidelines which could be used to design a sampling scheme with the variable circular plot method. These guidelines fell into two general categories: (1) studies in which one needs only to estimate density based on the major components of the community and where estimates for minor (rare) species are relatively unimportant; and (2) studies in which the entire spectrum of the avifauna must be assessed regardless of relative densities. The former studies are often useful when simple baseline data must be collected from a variety of areas, and results need only to be within a fair degree of accuracy (e.g., inventories, study site selection). The latter involve projects where rare species are the object of concern, and/or differences within or between sites may be subtle (e.g., effects of various silvicultural treatments, guild analyses, community dynamics).

If a simple inventory is needed, one could place 2–4 stations in areas where the vegetation is relatively homogeneous; that is, a single physiognomic class of vegetation dominates the site with few pockets of obviously different vegetation. In patchy habitats, 4–6 stations would be indicated. However, if one needs to develop a detailed species list, 4–6 stations in homogeneous areas and 6–8 stations in more heterogeneous habitats would be required. The exact number of stations needed (e.g., 4 or 6) would depend upon the size of the study area and the extent of differing vegetation types. Caution should be used when placing only two stations; significant intrastation variations in densities were sometimes the result with such a low number of stations.

Placement (spacing) of stations is again dependent on the vegetative structure of the site. Even in homogeneous areas, few species were recorded at all 10 stations. Therefore, stations must be spaced throughout the area of interest and not concentrated along edges or points of easy access. We recommend a stratified random placement of stations (if the size of the study area permits). This is, if a site is 30% deciduous trees and 70% shrub cover, two stations should be placed in spots dominated by the tree component, and three or four stations scattered throughout the shrub cover.

The researcher should keep in mind that each study area should be censused four or five times during the period of interest (e.g., breeding season) to insure collection of an adequate sample size and to account for the phenologies of various species. Of course, data must always be scrutinized to determine whether the effective radii for rare species are reasonable.

We have thus shown that a study area need not be saturated with census stations before one can obtain a reasonably precise description of the avian community. It is possible to increase the scope of a study by careful placement of census stations. One might have adequate time to census two study areas in one morning or to collect other types of data while censusing.

ACKNOWLEDGMENTS

We thank E. C. Meslow and J. A. Crawford for reviewing the manuscript; Sherry Morrison, Jeanne Jensen, and Robin Lannan prepared several drafts. Financial support for the authors was provided by the following sources: MLM—U.S. Fish and Wildlife Service, Patuxent Wildlife Research Center, Cooperative Agreement 14-16-0009-78-977; RWM—USDA Forest Service, Pacific Northwest Forest and Range Experiment Station, La Grande, Oregon, Range and Wildlife Habitat Research Project USDA-FS-PNW-1701; GLD—lower Columbia River project, U.S. Army Contract No. DAC 57-78-C-0079. Special thanks go to Aaron Skirvin for allowing us to use his data for the Ponderosa Pine study area; support for his study was provided by USDA-FS-PNW Grant No. 40. This is Oregon State University Agricultural Experiment Station Technical Paper No. 5541.

Studies in Avian Biology No. 6:409–413, 1981.

LENGTH OF COUNT PERIOD AS A POSSIBLE SOURCE OF BIAS IN ESTIMATING BIRD DENSITIES

J. MICHAEL SCOTT[1] and FRED L. RAMSEY[2]

ABSTRACT.—Observers using point counts to estimate the numbers of birds in an area spend 2 to 20 minutes counting. The assumption inherent in this technique is that birds don't move during the count period. The degree to which this assumption is violated is determined by the length of the count period, and the speed at which birds move. The longer the count period and the greater the speed of the birds, the greater the potential for a biased estimate of density or other measures of abundance.

We use field observations and simulation studies to determine the effects of bird movement and length of count period on estimates of bird abundance.

The variable circular plot design (Reynolds et al. 1980) offers many advantages in bird surveys. With this design, transects are drawn through the region being surveyed. Observers proceed along the transects, conducting their surveys only at stations marked at regular intervals. The period of time during which observers survey on station is fixed in length. Observers record the distances from station to detected birds. These distances are used to estimate the area effectively surveyed for each bird species detected. Hence the "variable" circular plot, as opposed to circular plots of predetermined size outside of which observers ignore birds.

In practice, circular plot surveys (variable and fixed distance) have been conducted with count durations ranging from 2 min (Kimball 1949) to 20 min (Blondel et al. 1970). This paper examines some factors which influence how long the survey period at each station should be. We do not consider costs or total time available as constraints, but concentrate on those factors that affect the quality of the data obtained.

SOME THEORETICAL CONSIDERATIONS

The impossible (Preston 1979) ideal that an observer strives to obtain is an instantaneous picture of the birds and their locations surrounding a station. We recorded the time at which various species were first recorded in two Hawaiian forests, one with 14 and one with 5 species of birds. Figure 1 displays the cumulative number of species detected as a function of time on station, expressed as a percent of the total detected in 32 min. If a person stayed at a point long enough, all species in that particular habitat type would appear and be detected. This may be a convenient way to obtain a species list, but it would be difficult to convert waiting time to detection into densities or relative abundances. Examination of Figure 1 shows that, in both forest areas, about 80% of the species recorded in 32 min. of surveying were detected within the first 10–12 min. The rate of new detections declined steadily with time, indicating that more productive surveying can be accomplished by moving to a new location before all species are detected. Determining a good time to move is a difficult problem, compounded by the fact that cumulative numbers of detections vary from species to species.

In Figure 2 the 'Apapane (*Himatione sanguinea*) is the most mobile of the species shown while the 'Oma'o (*Phaeornis obscurus*) is the least mobile (C. J. Ralph pers. comm.). Fifty percent of all observations for the 'Oma'o were recorded with 1 min while it took 7 min to achieve this figure for the 'Apapane. We attribute these differences to 'Apapane that were beyond the area surveyed when the count was started moving to within the count area. This movement thus inflates the density estimate significantly.

Results such as these assist us in identifying advantages and disadvantages attending longer counting periods. Some of these are itemized below, and we consider in later sections ways to deal with them.

ADVANTAGES:

- A1. Birds inconspicuous because of their distance from the station have a higher chance of being detected.
- A2. Birds that vocalize infrequently will have a smaller chance of being missed during the count period.
- A3. Birds that react to the presence of the observer by becoming silent and immobile may resume more normal behavior.
- A4. The observer has more time to make careful identifications and to record distances accurately.
- A5. In an area of high bird density, the observer has more time to observe and record.

[1] U.S. Fish and Wildlife Service, Mauna Loa Field Station, P.O. Box 44, Hawaii National Park, Hawaii 96718.

[2] Department of Statistics, Oregon State University, Corvallis, Oregon 97331.

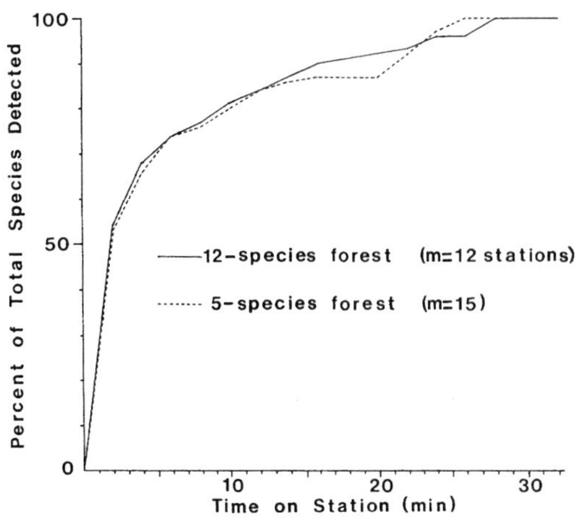

FIGURE 1. Cumulative percentages of bird species detected with increasing count duration in two Hawaiian forests. Fifteen 32-minute counts were conducted in the 5-species forest and 12 in the 12-species forest.

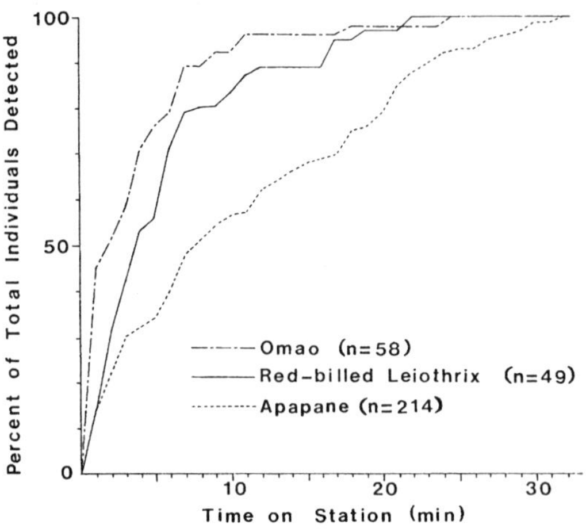

FIGURE 2. Cumulative percentages of total individual counts with increasing count duration for three Hawaiian species.

DISADVANTAGES:

D1. Birds that are initially beyond the range of detection have a greater chance of moving close enough to be detected *at distances too near to the observer to allow for accurate assessment of the area being covered.*

D2. The chance of recording a single bird more than once increases, because the bird may move or the observer may forget its location.

D3. The observer's ability to detect birds may decline because of boredom.

D4. The observer has greater freedom to allocate effort among species.

D5. There is more time for birds to be attracted by the observer's presence.

SOME PRACTICAL REMEDIES

Some of the disadvantages listed above can be neutralized by careful control over field technique. Consider D4, for example. The problem here concerns relatively common species. At a station where few species occur, the observer can spend nearly the entire count period locating members of these species. At a station with many species, however, the observer will tend to ignore a very common species after an initial count in order to concentrate on other species. Eight min at one station versus 2 min at another is not likely to yield density estimates reflecting the true situation. This tendency is a natural one among observers. It can be prevented by dividing among observers the responsibilities for counting common species, or by varying the species counted from station to station in a pattern that still gives ample coverage for the common species (Scott and Ramsey 1981a).

Factor D2 can be reduced by using a field form for each station that is essentially a map consisting of concentric circles drawn around a point (the station) on a line representing the transect. The observer turns around while surveying but keeps the form aligned with the transect at all times. As each bird is detected, its distance and direction are estimated and the observer enters on the form a four-letter species code at the resultant estimate of its position. If desired, the code may be underlined if the detection is by call, circled if by song, or unmarked if the detection is visual.

Factor D3 may be reduced by training the observers and impressing upon them the importance of their job (Kepler and Scott 1981). Another means to reduce boredom is to have two observers making simultaneous counts at the same or nearly same station. If an observer knows that his observations will be directly comparable with those of another observer he will tend to be more alert throughout the count period (Scott and Ramsey 1981b,. Kepler and Scott 1981). Factor D5 can be minimized by instructing observers to move quietly between stations and to make as little noise as possible while on station.

This leaves D1 as the principal constraint on the selection of long count periods. The effects of D1 can be minimized by using different length count periods for species that vary widely in rate of movement and conspicuousness.

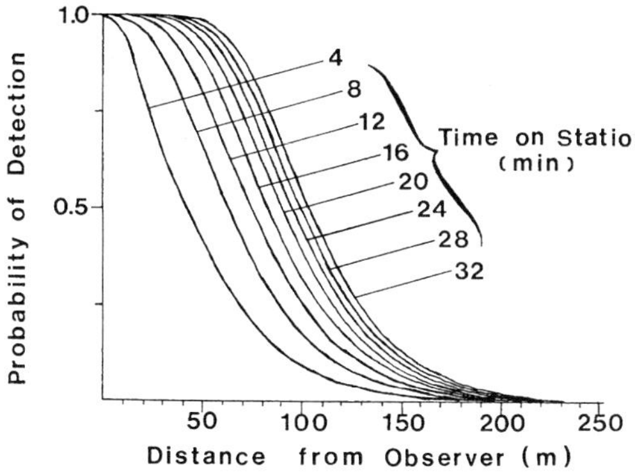

FIGURE 3. Detectability curves for surveys of different durations.

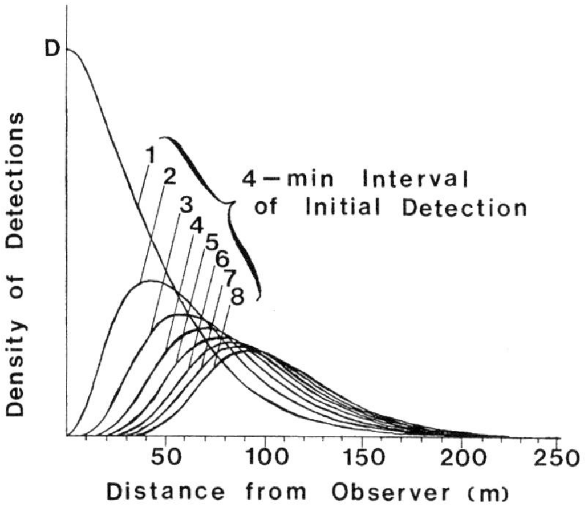

FIGURE 4. Theoretical densities based on initial detection distances, by 4-min periods in a 32-min survey.

How might we expect the movement of birds to affect survey results? As suggested in D1, an obvious consequence of longer count periods is that too many birds will be detected. Not only are those birds detected that were initially near station, but also new birds will approach near enough to the station to be detected. The 'Apapane in Figure 2 demonstrates this effect. What may not be obvious is where these recruits will first be detected and thus recorded for distance. At one extreme, if a species is highly detectable throughout a broad region around the station, we might expect that detection distances of recruits would all fall near the outer limits of the range of high detectability. This would create a "donut" pattern of higher observed density in a distance range removed from the station than exists near the station, where new recruits do not penetrate prior to detection. Such a pattern of detections could be discerned in data and appropriate corrections could be applied.

At the other extreme, fast moving species of low detectability would likely get quite near station before being detected. In this case detections of recruits might have the same distance patterns as do birds initially present and detected within the observer's range. No recognizable pattern of detections exists to distinguish this effect of bird movement, so corrections must be based solely on biological information. In the next section we will examine some simulated examples to see how movement might affect survey results.

SIMULATED MOBILITY: THE MODEL

We use a model developed by Ramsey et al. (in press), where an observer is stationed at the origin of a plane. A bird is randomly positioned in the plane. During a count period of 32 min, the bird is allowed to move along a straight line with speed, S, the direction of the line being chosen at random. Speed of movement is expressed as a proportion of the speed needed to travel one effective detection distance in 4 min. Every 20 sec., the observer performs a visual scan as follows: the distance, y, from bird to observer is measured and the observer has a chance for visually detecting the bird equal to

$$g_V(y) = 1 - \{1 - \exp[-(y/15)^2]\}^{1/24}.$$

Furthermore, each bird emits calls according to a random Poisson time process with an average of θ calls per min. The observer has a chance of making an audio detection of any call equal to

$$g_A(y) = \exp\{-(y/30)\}.$$

Here y is the distance at the time of its call, measured in meters.

A bird may be detected either visually or aurally at some time during the count period. If it is detected, its species and distance are recorded. This procedure is then repeated many times to simulate species densities.

It is possible to discuss the model's features in the absence of movement ($S = 0$), and assuming that we ignore multiple detections. The probability of a bird's being detected increases with time, regardless of its distance from the observer. Figure 3 displays eight of the detectability curves, corresponding to 4 min increments. Thus the cumulative number of detections is expected to increase with time (cf. Fig. 2). This would increase the corresponding estimate of density, were it not for the fact that the approaching birds, detected in the latter stages of the census, have greater average detection distances and therefore increase the estimate of effective area surveyed. (See Ramsey and Scott

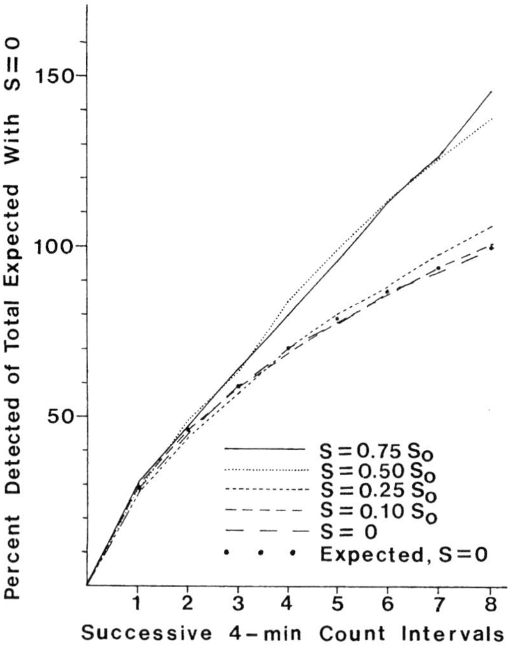

FIGURE 5. The effect of mobility on cumulative numbers of detections over successive 4-min count periods. The simulated speed of movement is expressed as a proportion of S_0, the average speed needed to travel one effective detection radius in one 4-min interval.

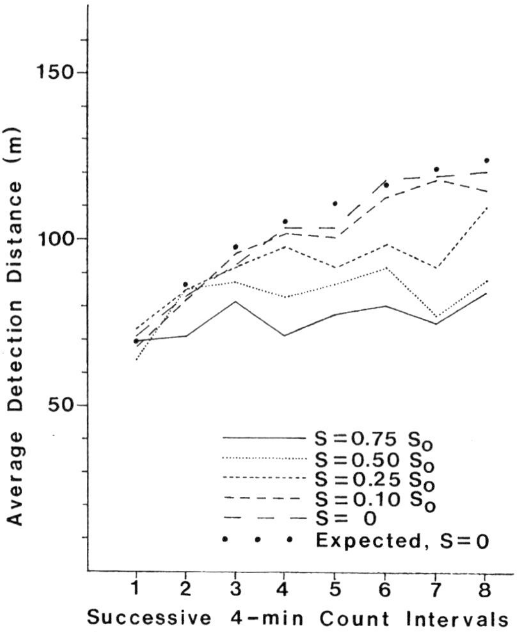

FIGURE 6. Simulated effect of bird mobility on average detection distance during successive 4-min counting periods. S is calculated as for Figure 5.

1981, for a description of the density estimation procedure used.) The theoretical situation is presented in Figure 4, where we show the density distributions of the birds detected in each of the 4 min. intervals. Note in particular how the average detection distance increases with longer census periods.

SIMULATED MOBILITY

The average speed with which a bird moves about its home range influences the probability of its coming within detection range of an observer. Figure 5 displays results of various simulated rates of movement on numbers of detections in successive 4 min counting intervals, in relation to expected numbers of detections with no movement. It is clear that increasing mobility increases the numbers of detections in later counting periods. With $S = 0.75\ S_0$, for example, the total simulated number of detections after 32 min of counting exceeded the number expected without bird movement by 45%. The effect of this, of course, is to overestimate bird densities.

The overestimate of density might be partially compensated for if birds that move into detection range are first detected farther away than expected for stationary birds. Figure 6 shows results of simulated effects of rates of movement on detection distances in successive 4 min. census intervals. Although average detection distances of moving birds increased in successive intervals, they did not keep up with expected values for stationary birds. And the effect decreased with increasing rates of movement. We tried simulations using several faster call rates with similar, but less marked, results. Apparently bird mobility at all rates simulated here allowed birds to get nearer to the observer than expected before detection. This effect would also tend to inflate density estimates of birds, and especially so for the higher mobility rates.

Next we combined the effects of mobility on numbers of detections on average detection distances to compute density estimates shown in Figure 7. For slower rates of movement (up to $S = 0.25S_0$), the combined effects of mobility do not appear to result in a marked overestimate of density compared with the expected for stationary birds. As the rate of movement increases, however, the net result is an increasing overestimation of density.

CONCLUSIONS

The results of simulation studies presented here show that bird mobility may seriously bias density estimates derived from variable circular

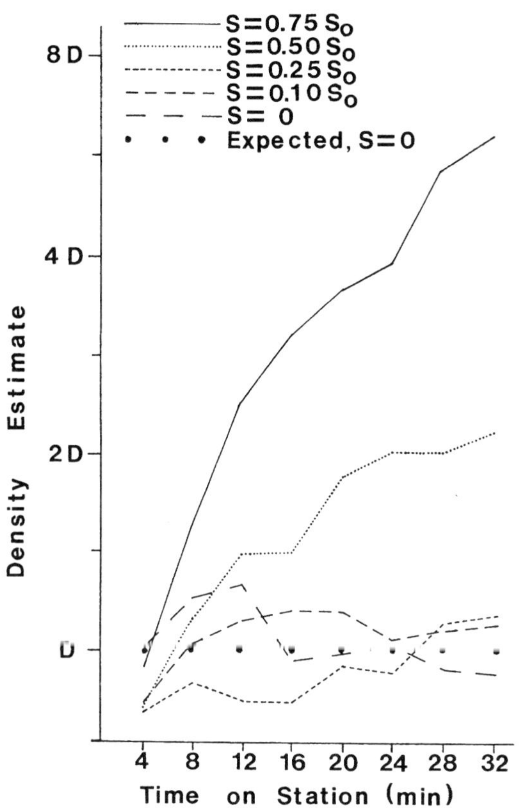

FIGURE 7. The simulated effect of bird mobility on estimates of density with increasing duration of counting period. *S* is calculated as for Figure 5.

plot surveys, especially for counts of longer duration. Because of bird mobility, an observer essentially surveys a much greater area than is indicated by detection distance information. The bias is even more serious for circular plots with a fixed radius, because density computations do not benefit from greater average detection distances that accompany longer count periods. Count periods of different lengths are required for species with dramatically different rates of movement. This may be handled by counting only birds of a similar mobility during a count period or not counting individuals of a species after a certain period, e.g., 1 min for swifts and swallows.

Counts of long duration are generally advantageous for sedentary species, particularly if they are rare or inconspicuous. The same may be true for territorial species, if the average defended area is small compared with the effective area surveyed from a given station. However, without considerable information on average rates of movement within a home range, and on call and song rates, it is not possible to use such counts for accurate estimates of the density of a very mobile species, or of a species with a territory that is large compared to the area effectively surveyed. The next logical step in our studies requires collection in the field of the information on mobility rates, song rates, and so on, needed to examine these relationships empirically.

ACKNOWLEDGMENTS

We thank Jared Verner for his insightful comments on an earlier draft of this manuscript. They improved it considerably. Programming assistance for the simulation studies was provided by Robert T. Clark and a grant for computer time was generously made available by the Oregon State University Computer Center. Cameron B. Kepler and Jack Jeffrey assisted in the field work.

Studies in Avian Biology No. 6:414–420, 1981.

POINT COUNTS WITH UNLIMITED DISTANCE

JACQUES BLONDEL,[1] CAMILLE FERRY[2] AND BERNARD FROCHOT[2]

ABSTRACT.—Point counts with unlimited distance give the number of species and quantitative informations either in the form of an index of abundance (IPA) or in frequencies (EFP). The number of points to count in any biotope should be increased according to the total number of species progressively recorded. It depends also on the number of individuals per species and of the sensitiveness of the tests we can apply to compare them. The mean richness (i.e., the average number of species per point) is an important parameter: coupled with the variance of the sample, it allows comparisons between biotopes; it is highly correlated with other parameters of the community: total richness, total density and H' diversity. An adequate dispersion of the points in both time and space in the biotope is a prerequisite for a reliable treatment of the data. The standardized recording of a definite set of environmental factors at each counting point allows one to analyze the correlations between the birds and the habitats.

Mapping method is the appropriate one to census breeding birds in one limited environment; the densities which it gives remain for us the standard of abundance. IPA is a good way of censusing and comparing bird communities in different homogeneous biotopes. EFP is the appropriate method to obtain at the least cost data on the structure of communities of extensive and patchy habitats.

A censusing technique should be chosen after (1) the aim of the study, (2) the scale of the habitats, (3) the manpower of the censusing team, and (4) the properties of the different methods have been determined.

Directly descended from an original transect censusing technique (Ferry and Frochot 1958), the point count methods presented here, IPA (Blondel et al. 1970, Ferry 1974, IBCC 1977) and EFP (Blondel 1975, 1977), are derived from a trivial observation: when a birdwatcher stands quietly for a moment in the field, on a spring morning, he notices a certain number of birds, mainly singing males. This gives him initial information on the bird community of the habitat. Standardization of the collection and the treatment of this information has led us to the point count methods for censusing birds. We stress that our technique is to record all available information, i.e., all the birds detectable, whatever their location. These are point counts with unlimited distance.

These methods were developed in consideration of three complementary censusing requirements: (1) To obtain quantitative results in a short time to permit counting birds simultaneously in several different habitats. (2) To obtain these data from samples, with a measure of the dispersion around the mean, so results can be objectively compared by statistical tests. (3) To be able to census birds in patchy habitats, where line transects are impossible to perform.

In this respect, it is important to stress that bird censusing is only a tool, which must be adapted to the aim of the research. We have found these point count methods to be suitable in a wide variety of situations in our research.

[1] CEPE/CNRS, B.P. 5051, F34033 Montpellier Cedex, France.
[2] C.E.O.B. Faculté de Sciences, Mirande 21 000, Dijon, France.

METHODS

IPA ("INDICE PONCTUEL D'ABONDANCE") METHOD

Data collection

The IPA method was first described by Blondel et al. (1970); the standardized procedure was published in English by the IBCC (1977). Data are collected at a fixed censusing spot, or station, which is well marked for relocation. Each station is counted twice in the breeding season, once within six weeks before and once within six weeks after the main time of settlement of the migrant species. This time varies with latitude and altitude. Counts are done early in the morning, only with little or no wind, only if not too cold, and only if not raining too hard. Each count lasts exactly 20 minutes, with the data separable into consecutive 5-minute periods. All birds seen and heard are recorded. Experience suggests that an observer's attentiveness and the conspicuousness of the birds limit the number of stations that can be counted in a morning to four or five.

For each species, counts are translated into a number of pairs, according to the following conventions: A singing male, a pair of birds, an occupied nest, or a family party are all counted as a pair of birds; a single bird seen or heard calling is one-half a pair. The higher of the two values, either from the first or the second count, is used as the IPA of that species for that spot and that breeding season. Taken together, the values for all species detected at a station comprise the "list of IPA" for that station and year.

Managing the data

The "list of IPA" of a point has little meaning by itself. It is just one sample, which must be combined with lists from other stations; together they comprise an IPA sample of a particular biotope. Such a sample of IPA will yield information about species abundances and species richness for the studied biotope.

For each species a mean IPA, with standard deviation, can be calculated as an index of abundance. These two parameters permit statistical comparison with the abundance of the same species in other habitats, censused in the same way, and assuming that the detectability of the species is the same in the different habitats (an assumption we have not tested). It must be stressed that the IPA is a strictly specific index; it is not possible to compare, or to add, IPAs of different species.

The sample of IPA lists allows us to assess the number of species (species richness) in the habitat. Two parameters of richness are measured. The average number of species per point is the *mean richness* ($\bar{s}$) for which we can compute the standard deviation; $\bar{s}$ can be compared statistically between different habitats. *Total richness* (S) is a cumulative parameter; the greater the number of censused spots, the nearer it will be to the actual richness of the bird community.

From IPA to density

As such, point counts in IPA do not yield densities; this is a drawback when we wish to know such parameters of the community as biomass or diversity. It is possible, however, to determine for each species a coefficient of conversion by which its IPA can be converted to a density estimate. We must simultaneously census the biotope by mapping and by IPA. (We use the word "biotope" as defined by Kendeigh (1961:6) "a topographic unit characterized by both uniform physical conditions, and uniform plant and animal life.") The coefficients of conversion thus obtained are specific to the species and the observer. Their use introduces nonmeasurable uncertainty, which precludes statistical comparisons between the densities.

EFP ("Echantillonnage Frequentiel Progressif") Method

The EFP, a frequency sampling method, was proposed and used later than the IPA (Blondel 1975, 1977). It differs from IPA on the following points: (1) The Point Count lasts 20 minutes, but instead of recording the number of birds of each species, one just records the species as present; (2) Each station is counted only once in the breeding season, instead of twice; and (3) This apparent simplification is an adaptation to both extensive and patchy environments, allowing the observer to census many points in a season (several hundred), but the dispersion of the points in the habitat and in the season must have been previously prepared to ensure a representative sample in space and time.

Since one records only the presence of each species, not all the singing males of each given species, it is possible to continue sampling points later in the day. Seven to nine points a day are feasible, instead of four or five as in IPA.

As for IPA, at the end of the field work season, one has for each biotope in the studied area a list of species, the EFP list, yielding parameters of richness, $\bar{s}$. One also has an index of abundance for each species, in this case frequency computed as the percent of the sampled points at which the species was recorded. The frequencies in different biotopes may be compared by appropriate tests.

It is not possible to infer densities from the frequencies. Although it has been shown that at lower densities the frequency of a given species is correlated with its density (Blondel 1975), this correlation breaks down with increasing density. At some high level of density of a species in the biotope, its frequency reaches 100%. Of course it can never increase thereafter, even with great increases in density. This drawback is not too important, however, because in our extensive studies we find that the mean richness ($\bar{s}$) derived from EFP data is, in all cases, highly correlated with the total density of the birds.

Describing the Habitat

Bird counts by the IPA or EFP method are useless unless correlated with environmental conditions. Up to now our experience is mainly limited to the study of "biotopes," i.e., areas homogeneous at least with respect to the main features of the habitat. The definition and localization of such biotopes are achieved according to preexisting maps (e.g., forestry, phytosociology). We must check in the field the exact structure of the biotopes, however, because the basic criteria are not always the same for an ornithologist as they are for a forester or a botanist. In practice the exact localization of biotopes to census remains to be done by the bird observer. In order to describe as accurately as possible the selected census area, we devised a simple optical apparatus, the "stratiscope" (Blondel and Cuvillier 1977), which permits a quantification of some relevant parameters of the biotope: number of layers, percent cover, and both horizontal and vertical structural diversities. These parameters often correlate well with such bird community variables as richness, overall density and the Shannon index of diversity (see also Blondel et al. 1973).

This method was devised for IPA censuses because it is very precise. It is so time-consuming that it must be done independently of the census work, at random locations in the biotope. This is a reasonable approach since it is assumed that the censused area is homogeneous. Time constraints preclude similar sampling of EFP sites. Instead, at EFP sites we record a set of habitat variables, immediately after completing the bird count. This is easily done by filling a precoded sheet, on which several sets of data are to be recorded: geography, topography, exposure, vegetation structure, vegetation form (with the dominant plant species) and particular ecological features. Details of this habitat description may be found in Blondel (1975, 1978).

SAMPLING DESIGN

Spacing and Timing of Point Counts

The representative sampling depends on an appropriate timing and spacing of counts in the study area.

In time, the frequency of singing by territorial males differs from one species to another. A poster in this symposium shows how reassessing our IPA data in two forests, month-by-month, has confirmed that birds with protracted breeding seasons have the same detectability over the season. On the other hand, residents with one

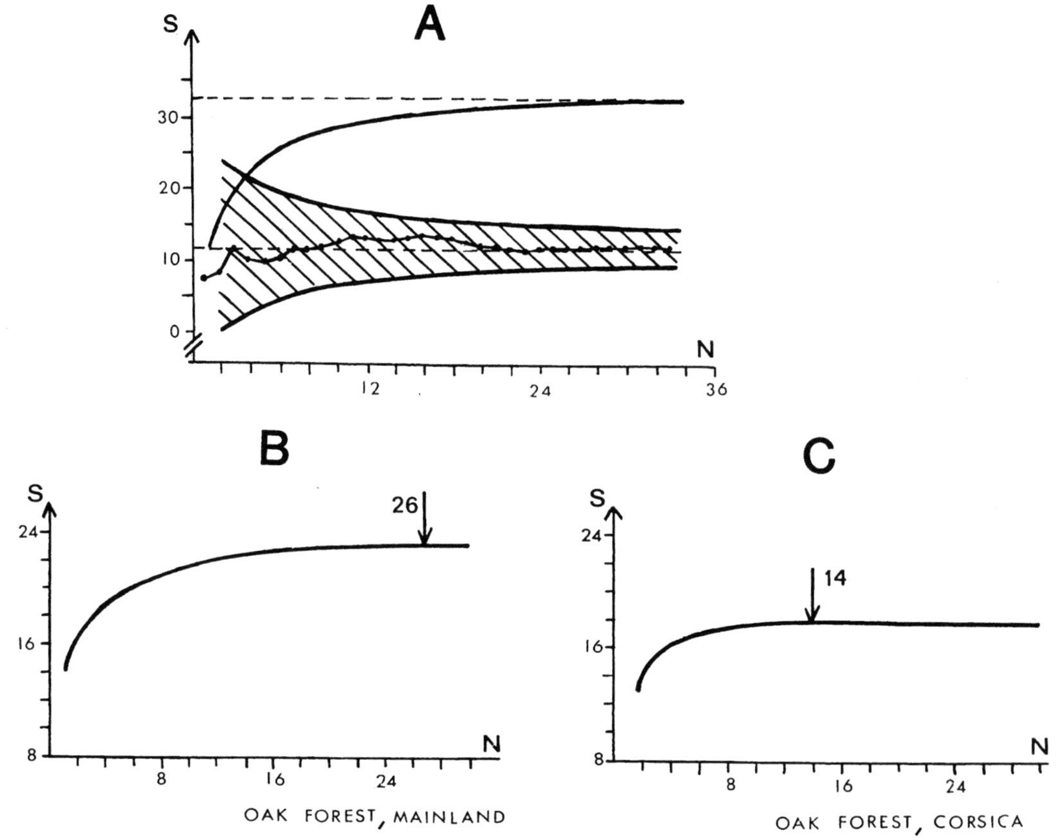

FIGURE 1. The upper figure (A) shows the two richnesses as a function of the number of IPA points, in a cedar forest in southern France. The total richness (*S*) is a cumulative parameter (upper curve); the mean richness ($\bar{s}$) is shown within its 95% confidence limits (lower curve). The two lower figures (B and C) show how the total richness is acquired in two similar old *Quercus ilex* forests. The arrows show the number of IPA points where the curve slope becomes zero.

early brood are better recorded at the beginning, and late migrants can be heard only during the second half of the counts. Considering only the higher value of the two counts in IPA sampling yields for the two last categories of birds a higher mean and a narrower variance.

In EFP sampling it is imperative that each area compared be counted by points equally spaced week after week, throughout the season. This helps to assure that measured differences in frequency for a given species reflect differences in abundance and not in detectability.

The counting stations can conveniently be regularly spaced by superimposing a metric grid upon the chosen biotope. It seems probable that the points thus located are not far from random spacing, because the grid will probably be independent of the inescapable, small differences within the biotope (e.g., old decaying trees and small glades). Points should be at least 200 m apart in IPA. This distance is evidently less than the detectability radius of some birds, but experience has shown that the differences between indices of abundance are equivalently estimated with or without overlapping of count station limits. In EFP sampling, our points are generally further apart, up to 1 km.

The order in which stations are counted may be drawn by lot, since station locations are determined before the field season.

The Number of Points Needed in a Given Biotope

The number of points will be chosen to yield sufficient information on the number of species and individuals present in a given biotope.

On the number of species.

Thirty-four IPAs were collected in a cedar (*Cedrus atlanticus*) forest in southern France.

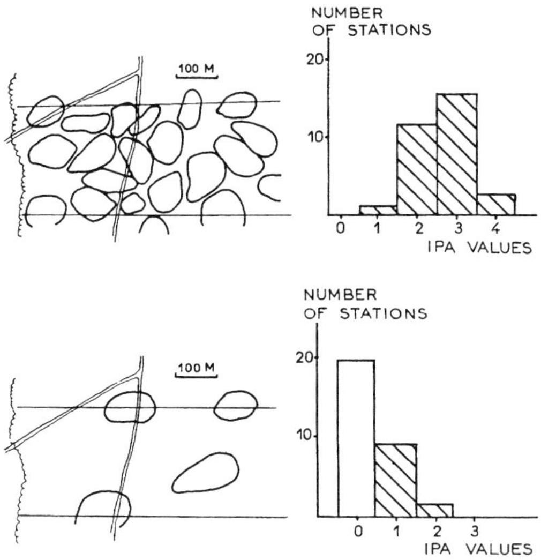

FIGURE 2. Territories and histograms of IPAs of Blue Tit (above), and Crested Tit (below), simultaneously censused in beech forest habitat by IPA and by a mapping plot of 15 ha.

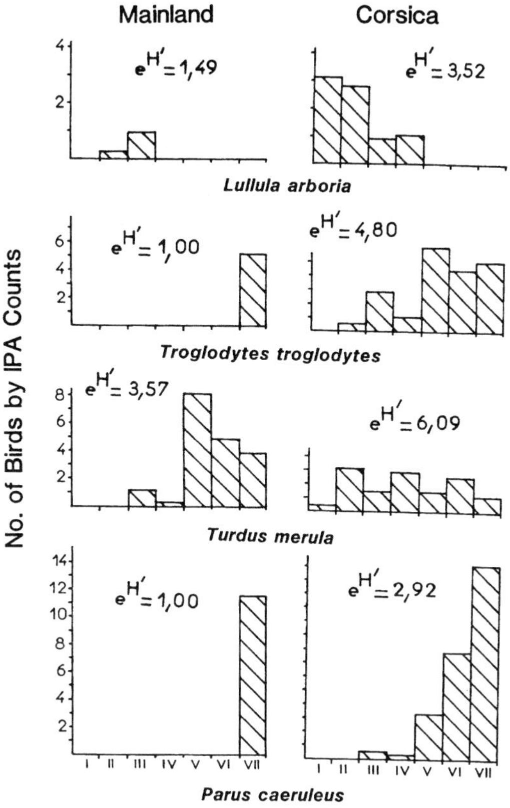

FIGURE 3. IPAs of four species in the seven stages of two gradients, from grassland to old forest. One of the mainland (left histogram) and Corsica (right histogram). The index of habitat breadth ($e^{H'}$) is the natural exponential of the Shannon Index calculated in nits from the IPA values (Ferry et al. 1976). The other species of the two gradients demonstrate the same trend.

Figure 1A shows the two components of richness assessed as a function of the number of points. Total richness (S) is 33 species, as shown by the upper curve. This curve was computed from a great number of permutations of the individual lists of IPAs; its slope is level from the 30th point on, suggesting that 33 species is near the actual total richness of the community. The forms of such curves differ markedly from one biotope to another, as shown by Figure 1B and 1C, so that the values of S are not comparable so long as the cumulative curves have not reached the same slope (Ferry 1976). Fifteen to 30 points in IPA sampling are necessary to obtain this result.

The mean richness ($\bar{s}$) of the cedar forest is 12.0 species (SD = 2.2). The lower curve of Figure 1A shows how the knowledge of this parameter is better assessed between narrowing limits of confidence as the sample size increases. This parameter is statistically comparable from one biotope to another, and this comparison is justified because it has been experimentally shown that $\bar{s}$ is correlated with S: for 23 bird communities (Blondel 1975) we find $\bar{s} = 0.43S + 0.56$ ($r = 0.94$, $P < 0.001$). In practice a dozen points in each biotope allows easy comparisons between $\bar{s}$ values.

We wish to stress here that mean richness ($\bar{s}$), besides its usefulness as a statistical parameter of the bird community, might have some biological significance, if we refer to the censusing procedure and assume that the hearing ability of the observer is of the same order of magnitude as that of the birds. Mean richness, being the number of species (mainly represented by singing males) found at any spot of the biotope, may give a rough estimate of the potential interspecific competition which would face a bird trying to settle in that habitat. This is because point counts with unlimited distance are based mainly on the recording of singing males and thus reflect the main natural mechanism of dispersal of territorial species in the environment.

To test this hypothesis we used the data from a beech forest censused by IPA method (Ferry 1974) at 30 stations. For each station, we calculated the number of species (point richness) and a point index of diversity ($H'\alpha$) from the IPAs converted to densities. The two parameters are highly correlated ($r = 0.60$ $P < 0.001$) by the equation: $\bar{s} = 9.57\ H'\alpha - 17.70$. This

TABLE 1
GUIDELINES SUGGESTED FOR SELECTING SAMPLING METHODS ACCORDING TO STUDY SCALE AND GOALS

Scale	Type of required information	Appropriate method
One habitat (homogeneous or not)	Species' densities Partitioning of territories in the habitat	Mapping plot
Two or more biotopes (or one biotope over several years)	Comparable parameters of the species (abundance) and the communities (richness, abundances, diversities)	IPA or EFP (anywhere) Line transects (in extensive, homogeneous biotopes)
Many habitats in a patchy area	Comparable parameters as above	EFP

confirms that $\bar{s}$ may be a simple first approximation of local diversity, of which Lloyd et al. (1968) write: "the average local diversity is the expected uncertainty of encounter that would confront an immigrant individual landing in a random plot in the community."

The preceding paragraphs concern the mean richness assessed by IPA methods. In EFP sampling, $\bar{s}$ is lower but highly correlated with the value yielded by IPA in the same place. Thus for 23 biotopes (Blondel 1975) the correlation between the two values is $r = 0.99$, with $\bar{s}$ EFP = 0.76 $\bar{s}$ IPA. Moreover, in EFP the mean richness ($\bar{s}$) is a reliable index of the total abundance in the community. For the same 23 biotopes, $\bar{s}$ was highly correlated with the total densities yielded by converted IPAs ($r = 0.91$, $P < 0.001$). And in fact in EFP, $\bar{s}$ takes the same value as the sum of the specific frequencies.

Finally the ratio $\bar{s}/S$ is probably of interest. For a sample of n EFP points it could theoretically vary from $1/n$ to 1. In practice, for 23 biotopes (Blondel 1975) it varied from 0.29 to 0.75. Moreover it is not correlated with total richness, nor with the total density of the bird community. If assessed from a random sample of points, the ratio $\bar{s}/S$ is low and might give an idea of the heterogeneity of the censused area. On the other hand, when calculated for biotopes assumed to be homogeneous, as the 23 cited communities, it might give an idea of the balance between inter- and intraspecific competition within the community, being lower when interspecific competition predominates, and higher when intraspecific competition is more important, as in isolated or insular communities. This hypothesis is enforced by the fact that in EFP the ratio $\bar{s}/S$ takes the same value as the mean specific frequency of all species ($\bar{F}$).

On the number of individuals

One of the main uses of IPA data is to compare, species-by-species, the abundance of birds between two or more biotopes (Ferry 1974). The feasibility and the sensibility of the comparison will depend on the number of points censused in each biotope.

A beech (*Fagus sylvatica*) forest was sampled simultaneously by mapping and by IPA methods. Figure 2 shows the dispersion of the territories of two species (based on the mapping plot) and the distribution of their samples of 30 IPAs. The Blue Tit (*Parus caeruleus*), a common species, had adjoining territories throughout the plot, and the values of its IPAs were distributed roughly normally. On the other hand, the Crested Tit (*Parus cristatus*), a rare species, had only three territories on the 15-ha plot. It was recorded in only one-third of the 30 counting spots; and the distribution graph of its IPAs was skewed. Such a relationship between the field dispersion pattern of territorial birds and the statistical distribution of their IPAs seems to hold well for the other species of the same biotope and also in other cases.

These experimental findings can help us to decide upon the number of points to count in a given biotope in order to compare species' abundances. Common birds, as the Blue Tit, are correctly tested by Student's t test, even for small samples, because their IPAs are normally distributed around the mean. Rare birds, with the distribution of their IPAs far from normal, must be tested either by nonparametric tests when the sample is small, or rather censused by a large sample (at least 30 stations) for one to be allowed to use tests on the mean and standard deviation.

With the EFP technique, we rely on frequencies to compare the numbers of individuals in different biotopes. Comparison of frequencies by χ^2 test necessitates that the absolute number of detections be high enough for the calculated value to be at least five. Thus for rare species large samples will be mandatory. Moreover, even if the test is applicable, its sensitivity will improve when the sample increases (graphical illustration in these proceedings by Dawson 1981b, figure 5). In practice, several dozen points per sample will be necessary to detect with confidence small differences between samples.

In summary, we suggest that a sample of a dozen stations in IPA, or twice as many points in EFP, will give a first idea of the bird community in a biotope, allowing abundance comparisons of the common species, and yielding such collective parameters as mean species richness and (in IPA) overall density (after conversion), and an index of species diversity. On the other hand, larger samples (30 stations in IPA, 40 or 50 points in EFP) are necessary for a good assessment of total richness, and fine abundance comparisons of most species.

How to Correlate the Data on Birds and Habitats

Simple correlations may be calculated by hand in most situations, from one biotope to another, if the sampling conditions are fulfilled. Figure 3 exemplifies the use of IPA values to assess the habitat breadth of individual species in two comparable gradients of habitats, one on the mainland and the other on the island of Corsica. Blondel and Frochot censused these gradients with 12 to 26 IPAs in each stage. This permitted us to calculate and compare indices of habitat breadth, because the stages had been chosen to ensure a good match between environmental variables in the two situations, as confirmed by "stratiscoping" the habitats. The broadening of habitat selection is a general characteristic of populations in insular situations (Ferry et al. 1976, Blondel 1979).

Complex relationships between the breeding birds and the environment may be studied by multivariate analyses. An example of such data processing will be found in Blondel (1976). A set of 340 EFP counts, collected in Mont Ventoux (southern France), gave data on 80 species of birds in 10 biotopes; the environmental parameters had been recorded at each point, as previously described. This permitted computation of a correspondence analysis between the presence of the breeding birds and the class value of twelve environmental variables. The location of the censusing stations had been stratified to ensure an equivalent sampling of the recognized biotopes, but during computer analysis the data for each point were interpreted independently to eliminate (or lessen) the bias of preadmitted partitioning of the ecological situations.

As yet we have not carried out counts with strictly random dispersion of points over an extensive area. Such a study is in progress in Provence, but results are not yet available. A short preliminary trial with two sets of 40 and 38 EFP points has been made in Burgundy (Grimoldi 1976). It confirmed that the various habitats had been sampled in proportion to their importance in the studied area, but the consequence was that the more restricted biotopes had been censused by only a few stations, with no interpretable results.

SELECTING A SAMPLING METHOD

No single method is most appropriate to all bird censusing studies. When we plan a study we must design the censusing work in accordance with the aim of the research, the characteristics of the area to be studied, and the manpower of the team. In most cases our aim will be to compare two or more systems or situations; in these cases great precision may not be necessary, and satisfactory results will be attained if data permit objective and reliable comparisons between or among systems and situations.

Table 1 proposes how to choose a censusing method after the scale of the study. For one given habitat, whether homogeneous or not, the mapping plot is the standard technique. It yields directly the number of breeding species, and for each of them a density estimate. Moreover, it is possible to correlate the location of the breeding territories with the ecological peculiarities which appear on the map. Certainly, in spite of the effort at international standardization made by the IBCC, an uncertainty remains in the number of "territories" identified on the maps (Svensson 1974b). But we may keep in mind that the notion of density does not refer to a fixed reality; the actual number of breeding pairs and other birds fluctuates during the course of a reproductive season. Mapping remains the reference technique, but we are aware of its main drawback, which is the fact that it yields values that are not objectively comparable from one plot to another.

When we need to census two or more biotopes, or the same biotope during several years, the advantages of the IPA method are obvious. It allows objective comparisons based on statistical parameters. Of course one could sample many mapping plots in an extensive biotope and compute the means and standard deviations of results. Besides the cost in field work, the many observers required for such a study would introduce the bias of their different field abilities; whereas several samples of IPAs may be covered in one season by one person or a small, homogeneous team.

At this scale (several biotopes), Point Counts should be compared with line transects. We have much experience with one kind of line transect (Ferry and Frochot 1958). The technique is suitable in extensive, homogeneous habitats, but it has a theoretical disadvantage compared to Point Counts. The relative index of abundance it yields is a function both of the time spent and the length of the route, whereas the figures derived from IPAs depend only on the time spent, which simplifies the interpretation.

Finally, EFP is an ideal method of censusing breeding birds in extensive and patchy areas, so long as sampling requirements are fulfilled. One cannot deal with densities, but at the species level the comparisons of the frequencies are objectively possible with large samples. EFP yields both richness values ($\bar{s}$ and S); and for the total abundance recall that mean richness is highly correlated with the total number of individuals. H' may be computed from the species' frequencies, as easily as from densities, even if its meaning is not so obvious as when it is derived from densities, but we note that Shannon's index is a robust one.

Another point of importance is the cost in time and manpower of the various techniques. One sample of a dozen IPAs requires less time than one mapping plot of 14 ha; about 12 "good" hours of spring mornings vs. 40. However, if we wish to compute densities, a mapping plot must be coupled with the IPA counting; thus for a single biotope Point Counts are not cheaper. However, one field worker can easily complete three or four samples of IPAs with one reference mapping plot in the time (one season) required for two plots that permit no comparison.

EFP is not much cheaper than IPA, even though more points are counted in a day, because a very large sample is necessary to apply tests with confidence to the frequency values obtained.

Finally, we have determined by a trial at censusing that the method of capture–recapture is much more time consuming than the other technique (400 h for 59 ha; Frochot et al. 1977).

CONCLUSION

Point Counts with unlimited distance have their own "niche" in the realm of bird censusing techniques. They give reliable and comparable parameters of abundance at the species level. They also yield collective parameters of the bird community, among them the mean richness, which has biological meaning as a measure of species packing in the habitat. Together with data on the biotope, they allow analysis of bird-habitat relationships. They must be considered as an appropriate tool for censusing breeding birds in many situations.

Studies in Avian Biology No. 6:421–425, 1981.

THE SPECIES-AREA RELATIONSHIP IN SPOT-MAP CENSUSING

TODD ENGSTROM[1]

ABSTRACT.—To approach an understanding of how plot size affects the results of bird censuses using the spot-mapping method, the species-area effect was studied in two Breeding Bird Censuses (BBCs) and a Winter Bird-Population Study (WBPS). For one BBC (BBC-79) and the WBPS (WBPS-79), a 58.3 ha plot of apparently uniform pine habitat was subdivided into nine subplots. To simulate progressively larger sample areas, all possible combinations of the subplot censuses were made. A study area of 20 to 25 ha contains an estimated 80% of the species observed on the 58.3 ha plot in both winter and the breeding season in this habitat. A BBC (BBC-80) conducted on a 20 ha plot in the same habitat one year later, resulted in 71% of the species observed in BBC-79 on 58.3 ha. Censuses of the subplots were highly variable especially in the winter. A statistical technique, rarefaction, was used to compare the estimated species accumulation curves of the three censuses. The total number of species found in the 20 ha BBC-80 was predicted well by the rarefaction curve of the 58.3 ha BBC-79.

Evaluation of bird populations using the spot-mapping method (Williams 1936) is appropriate when detailed knowledge of the distribution of birds within a habitat is required (Robbins 1978a). This is the method used in the Breeding Bird Census (BBC) and the Winter Bird Population Study (WBPS) sponsored by the National Audubon Society (Anon. 1937, Anon. 1947). In 1979, sixty-four WBPSs were conducted in 22 states and 1 Canadian province and 219 BBCs were conducted in 33 states and 4 Canadian provinces. One of the original goals of the BBC was to permit comparisons between the bird communities of "stable" habitats and those of recently disturbed habitats. Long-term studies were encouraged to provide information about changes in bird communities in relation to plant succession (Anon. 1937).

The BBC and WBPS are based on the premise that careful standardization of methods will produce comparable data. Edge effects, habitat uniformity, and the method of data collection are important variables to be considered before comparing census results (Berthold 1976). Other aspects of data collection that should be considered are time of day (Shields 1977), season (Järvinen et al. 1977b, Slagsvold 1977), between-observer variability (Enemar et al. 1978), map interpretation (Svensson 1974b), grid distance (International Bird Census Committee 1970) and census speed (Robbins 1972). The influence of plot size or sampling area on the results is also an important consideration.

Suggestions have been made for the "minimum" plot size to obtain an adequate representation of a bird community (Table 1). However, the quantitative relationship between plot size and the results of spot-mapping censuses has been approached only recently (Verner 1980a, Engstrom and James 1981).

The species-area effect is simply that species number increases with sampling area (Kilburn 1966). In this study, the species-area effect is discussed in relation to the spot-map censusing method as illustrated by three bird censuses: two BBCs and one WBPS. Two of them, WBPS-79 and BBC-79, were conducted on a 58.3 ha plot which was divided into subplots; the third, BBC-80, was conducted one year later on a 20-ha subset of the 58.3 ha plot. The relationship between plot size and WBPS-79 is investigated in a more detailed paper (Engstrom and James 1981). All three censuses will be used to discuss (1) optimal plot size in this habitat in winter and spring, (2) how plot size affects the comparability of census results, (3) differences between the WBPS and BBC, and (4) how the bird population dynamics of this forest habitat affect census results.

A statistical technique, rarefaction, is used to generate estimated species accumulation curves for the three studies. The rarefaction curves of BBC-79 and BBC-80 are compared as independent descriptions of the same community using different plot sizes. Then rarefaction is used as a means of comparing the results of censuses conducted on plots of different sizes.

METHODS

The WBPS-79 and BBC-79 were conducted on a 58.3 ha plot of apparently homogeneous mature longleaf pine (*Pinus palustris*) forest south of Thomasville, Georgia. The habitat is annually burned and has an open appearance. Some of the trees are 200 to 300 years old. The main plot was divided into nine subplots of 6.5 ha each. The central subplot is surrounded on four sides and the other subplots share either two or three sides each. The accuracy of evaluating territories along a subplot boundary is probably improved if the boundary is shared by another subplot. This nested subplot design may be a source of bias in the

[1] Department of Biological Science, Florida State University, Tallahassee, Florida 32306.

TABLE 1
RECOMMENDATIONS FOR CENSUS PLOT SIZE

Source	Plot size	
	Open habitat	Closed habitat
IBCC (1970)	40 to 60 ha	10 to 30 ha
Webster (1966)		8 to 12 ha
Anon. (1947)		no smaller than 8 ha (20 ha ideal)
Hall (1964)		not less than 6 ha, 8 ha or more best
Kendeigh (1944)	30 ha	20 ha
Kolb (1965)		plots <8 ha produce biased results
Verner (1980a)		plots <20 ha produce biased results

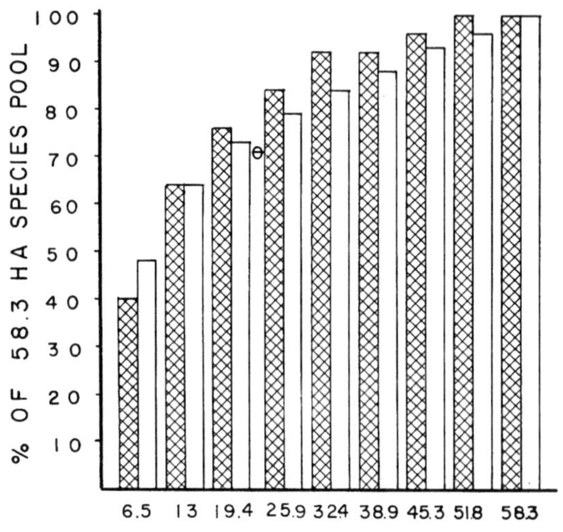

FIGURE 1. Proportional increase in species number with area. The average number of species for progressively larger sample areas can be determined by making all possible combinations of the 6.5-ha subplots. The average number of species for each simulated area as a proportion of the 58.3-ha plot species pool is represented for WBPS-79 (hatched bars) and BBC-79 (open bars). The bar-circle represents the percentage of the BBC-79 species pool observed on 20 ha in BBC-80.

subplot density estimates. However, I don't think that this strongly affected my results.

Independent censuses were made of each subplot according to the guidelines provided for the BBC (Anon. 1937, IBCC 1970) and the WBPS (Anon. 1947, IBCC 1970). Nine complete censuses were made in both the winter and spring. Each census took two days to complete. The effect of time of day on censusing was minimized by rotating the order in which the subplots were censused. Engstrom and James (1981) provide a more detailed description of the general procedure. BBC-80 was conducted on a 20 ha plot with no subplots during eight morning census trips.

All possible combinations of the nine subplot censuses were made to simulate the census results of progressively larger sample areas. For example, the nine 6.5 ha subplots permit 36 pairwise comparisons to make plots of 13 ha, 84 combinations to make 19.4 ha plots and so on. Each combination of subplots has an estimated number of species and individuals (or territories). It is important to recognize that the number of possible combinations varies between the different sized areas.

RESULTS

CENSUS RESULTS

The census results for WBPS-79 (Engstrom 1980b), BBC-79 (Engstrom 1980a) and BBC-80 (Engstrom 1981) are listed in Table 2. WBPS results are expressed in terms of the average number of individuals observed per trip, whereas BBC results are expressed in terms of territories as determined by evaluation of detection clusters and simultaneous records. This difference disallows direct comparison between the WBPS and the BBC.

Fourteen more species were seen during BBC-79 (39 species) than in WBPS-79 (25 species). A breakdown of the residency status of all species in both seasons (Table 2) reveals that 8 species (17%) occurred on the plot in the winter only, and 10 species (21%) were seen in the breeding season only. The remaining 29 species (62%) occur all year in the general vicinity. However, of the permanent resident species, 6 (13%) moved out of the plot during the winter and back again for the breeding season. Some species, such as the Red-headed Woodpecker, shifted habitats. Others occurred in flocks and were only observed flying over the plot in winter (e.g., Common Grackle). Some species might not have been detected because of decreased song and display.

ALL POSSIBLE COMBINATIONS OF SUBPLOTS

The average number of species for progressively larger sample areas can be derived from all possible combinations of the 6.5-ha subplots. The average number of species determined for each simulated sample area can be expressed as a percentage of the 58.3-ha species pool. This can be represented as a proportional increase in the number of species with increasing area for WBPS-79 and BBC-79 (Fig. 1). Within a plot size of 20 to 25 ha, approximately 80% of the species observed on the 58.3 ha plot would have

TABLE 2
THE NUMBER OF INDIVIDUALS IN WBPS-79 (58.3 HA), TERRITORIES IN BBC-79 (58.3 HA) AND BBC-80 (20 HA), AND RESIDENCY STATUS OF EACH SPECIES

Species	WBPS-79	BBC-79	BBC-80	Status[a]
Wood Duck (*Aix sponsa*)	+	2	2	WB
Bobwhite (*Colinus virginianus*)	–	2.5	2.5	BO
Mourning Dove (*Zenaida macroura*)	2	10.5	3	WB
Great Horned Owl (*Bubo virginianus*)	1	1	–	WB
Common Flicker (*Colaptes auratus*)	4	5	1.5	WB
Pileated Woodpecker (*Dryocopus pileatus*)	+	1	+	WB
Red-bellied Woodpecker (*Melanerpes carolinus*)	8	8.5	3.5	WB
Red-headed Woodpecker (*M. erythrocephalus*)	–	13.5	3.5	BO
Yellow-bellied Sapsucker (*Sphyrapicus varius*)	3	–	–	W
Red-cockaded Woodpecker (*Picoides borealis*)	17	5	1.5	WB
Hairy Woodpecker (*Picoides villosus*)	+	1	1	WB
Downy Woodpecker (*Picoides pubescens*)	+	1	+	WB
Eastern Kingbird (*Tyrannus tyrannus*)	–	3	+	B
Great Crested Flycatcher (*Myiarchus crinitus*)	–	13	4	B
Eastern Wood Pewee (*Contopus virens*)	–	8.5	4.5	B
Blue Jay (*Cyanocitta cristata*)	2	8	2	WB
Common Crow (*Corvus brachyrhynchos*)	–	2	–	WB
Tufted Titmouse (*Parus bicolor*)	–	1	–	WB
White-breasted Nuthatch (*Sitta carolinensis*)	7	5	2.5	WB
Brown-headed Nuthatch (*Sitta pusilla*)	7	7	4.5	WB
House Wren (*Troglodytes aedon*)	9	–	–	W
Carolina Wren (*Thryothorus ludovicianus*)	4	4	2.5	WB
Northern Mockingbird (*Mimus polyglottos*)	–	1	–	BO
Brown Thrasher (*Toxostoma rufum*)	–	3	1	BO
American Robin (*Turdus americanus*)	8	–	–	W
Eastern Bluebird (*Sialia sialis*)	3	3	2	WB
Loggerhead Shrike (*Lanius ludovicianus*)	1	1	+	WB
Solitary Vireo (*Vireo solitarius*)	2	–	–	W
Yellow-throated Vireo (*Vireo flavifrons*)	–	1.5	+	B
Yellow-rumped Warbler (*Dendroica coronota*)	2	–	–	W
Pine Warbler (*Dendroica pinus*)	11	10	6.5	WB
Palm Warbler (*Dendroica palmarum*)	2	–	–	W
Common Yellowthroat (*Geothlypis trichas*)	12	14	4.5	WB
Yellow-breasted Chat (*Icteria virens*)	–	11.5	2.5	B
Eastern Meadowlark (*Sturnella magna*)	5	7.5	3	WB
Red-winged Blackbird (*Agelaius phoeniceus*)	60	2	–	WB
Common Grackle (*Quiscalus quicala*)	–	1	–	BO
Brown-headed Cowbird (*Molothrus ater*)	–	5	4	BO
Orchard Oriole (*Icterus spurius*)	–	2	1	B
Summer Tanager (*Piranga rubra*)	–	4	1.5	B
Cardinal (*Cardinalis cardinalis*)	1	4	–	WB
Blue Grosbeak (*Guiraca caerulea*)	–	11	3.5	B
American Goldfinch (*Carduelis tristis*)	1	–	–	W
Indigo Bunting (*Passerina cyanea*)	–	14.5	6.5	B
Rufous-sided Towhee (*Pipilo erythrophthalmus*)	16	30	11	WB
Bachman's Sparrow (*Aimophila aestivalis*)	–	16.5	8	B
Swamp Sparrow (*Melospiza georgiana*)	1	–	–	W
Total species	25	39	27	
Total estimated density	189	245	94.5	

[a] WB = permanent resident, winter and breeding season; BO = permanent resident, breeding season only; W = winter resident only; B = breeding resident only.

been encountered in both winter and spring. Also note that the number of species observed in BBC-80 on 20 ha (27 species) is very close to the average number observed in BBC-79 (28 species) for the simulated 19.4 ha sample area.

RAREFACTION

Rarefaction is a statistical technique that can be used to generate a curve of the expected number of species in smaller random samples

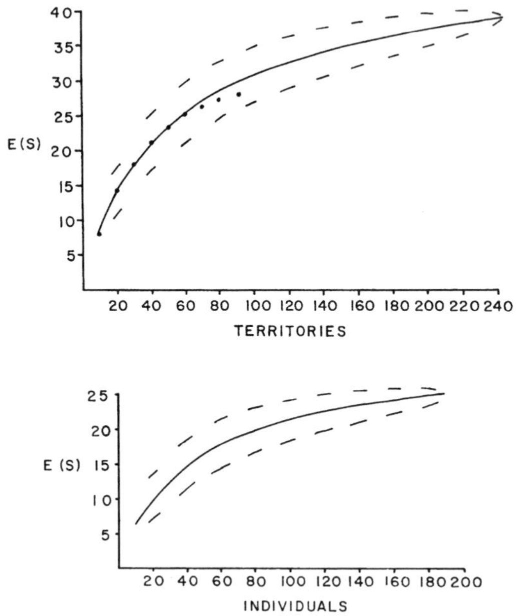

FIGURE 2. Rarefaction curves for BBC-79, BBC-80, and WBPS-79. The estimated number of species for any randomly drawn number of territories or individuals is represented by E(S). FIGURE 2a. These are the rearfaction curves for BBC-79 and BBC-80 combined. The heavy dots represent BBC-80. The solid line is BBC-79. The dashed line represents two standard deviations around BBC-79. The BBC-79 and BBC-80 curves were derived independently of each other. FIGURE 2b. The solid line represents the rarefaction curve for WBPS-79. The dashed line is two standard deviations about the curve.

than the original sample (Sanders 1968, Hurlbert 1971, Fager 1972, Simberloff 1978a). Given N individuals in S species, the expected number of species, $E(S)$, and its standard deviation can be calculated, A Fortran program is available upon request. Rarefaction makes it possible to compare the species richness of different-sized samples using the curves depicting the accumulation of S as a function of N.

Rarefaction curves were prepared for BBC-79 and BBC-80 (Fig. 2a) and WBPS-79 (Fig.2b). Again note that for the BBCs, the abscissa represents territories, but the abscissa of the WBPS curve is the average number of individuals per trip. The data for all three rarefaction curves are given in Table 2. The half-territories of the BBCs were rounded either up or down on a random basis to generate integers; pluses were omitted. The curve representing BBC-80 is well within two standard deviations of BBC-79, indicating that the rarefaction curve did not change much between the study years.

DISCUSSION

The species-area effect is a well-established ecological concept. An increase in species number can be attributable to an increase in habitat diversity with increasing area, by an increase in area per se, or by the "sampling effect," i.e., that larger areas may support more species simply because they contain larger samples of the source fauna (Connor and McCoy 1979). The longleaf pine forest was selected for this study because it appears to be very uniform. The subjective selection of a "representative" portion of a truly uniform habitat should not be difficult, yet all habitats have some degree of patchiness. In contrast to the uniform appearance, quantitative habitat descriptions and bird censuses of the subplots of WBPS-79 (Engstrom and James In press) and BBC-79 revealed substantial variation in habitat structure. This is simular to the results of Kilburn (1966) in a study of the species-area relationship of the plants in a jack pine (*Pinus banksiana*) forest. He found that even though the pine community was selected for its homogeneous appearance, "this apparent similarity was somewhat deceptive."

A large sampling area will reduce the effect of subjectively selecting a census plot in a patchy environment. In this open pine habitat during both WBPS-79 and BBC-79, a plot size of 20 to 25 ha was found to have roughly 80% of the species observed over an area 2 to 3 times larger. Censuses conducted on plots of less than 10 ha can be misleading because they tend to overestimate avian density (Verner 1981) or have more variable results (Engstrom and James In press). The BBC-80 on a 20 ha plot represented 71% of the species seen the year before on the 58.3 ha plot. Determination of sample area should be made on the basis of the amount of habitat available, the grid size, sampling speed and the area needed to obtain an adequate representation of the species pool. A census of 20 to 25 ha could be done effectively in early morning hours, and is the optimal plot size in this habitat.

Rarefaction was originally developed as a means of obtaining an estimate of diversity independent of sample size (Sanders 1968). The estimated species accumulation curves were thought to be habitat specific. James and Rathbun (MS) have generated rarefaction curves from many BBCs made in a wide variety of habitats. The curves are distinctive for each habitat and provide a good means of comparing the communities. The rarefaction curve for BBC-80 falls very close to the BBC-79 curve (Fig. 2a). We can predict the number of species expected for BBC-80 by arithmetically estimating the number of territories at 84 (245 territories from BBC-79 × 20 ha/58.3 ha) and then estimating the number of species (28 ± 2) by rarefaction. In fact, 27 species with 94 territories were detected on BBC-80.

Long-term studies of bird communities in different habitats over a large area are of great value for both theoretical ecology and resource management, if conducted in a systematic and comparable way. For example, Järvinen and Väisänen (1979a) used long-term censuses to explore the influence of climatic change, habitat alteration, and possible competition in the range dynamics of two pairs of congeners. Lynch and Whitcomb (1977) have used BBCs to look at species turnover rates in habitat islands. Järvinen (1979) made a quantitative test of European bird community stability along a north-south gradient using long-term spot-map censuses.

In conclusion, a number of points related to comparability of spot-map censuses can be made. These points include: (1) censuses conducted on plots of less than 10 ha cannot be reliably compared; (2) in this study, a plot size of 20 to 25 ha represents an optimal balance between minimizing censusing effort and providing an adequate sample of the bird community; (3) more information is needed on grid distances and censusing speeds in different habitats to improve spot-map method guidelines; and (4) rarefaction is a valuable method for comparing species richness of censuses conducted on plots of different size.

ACKNOWLEDGMENTS

I thank Tall Timbers Research Station, the U.S. Fish and Wildlife Service (Contract 14-16-0009-80-020), and Florida State University for financial support. I'm grateful to Tall Timbers Research Station for access to the tract of forest where the censuses were conducted and to Frances C. James, Jared Verner, and Olli Järvinen for many helpful comments on the manuscript.

Studies in Avian Biology No. 6:426–435, 1981.

CAPTURE-RECAPTURE MODELS: A REVIEW OF CURRENT METHODS, ASSUMPTIONS, AND EXPERIMENTAL DESIGN

K. H. POLLOCK[1]

ABSTRACT.—The current "state-of-the-art" in capture-recapture sampling methodology is reviewed. An emphasis is placed on model assumptions, model robustness to assumption failure, and in particular on the often neglected topic of study design.

For short term studies, there has been much recent work on closed population models that allow for unequal catchability of individual animals (heterogeneity and/or trap response). The closure assumption causes problems for many terrestrial bird population studies. A model which allows a closed population of "residents" plus some short term "wanderers" through the study area is considered.

For long term studies, open population models that assume equal catchability are discussed. They allow estimation of survival and birth rates as well as population sizes. Recent developments which allow some age-dependence of survival and capture probabilities are reported and related to some recent band recovery models. In terrestrial bird studies resighting of color-marked birds without capture could be potentially useful for estimation of survival rates.

Capture-recapture sampling has been widely used and abused for many years in the study of natural animal populations. Excellent reviews are given by Cormack (1968) and Seber (1973). Recently there has been a renewed interest in the assumptions behind capture-recapture sampling and especially in the assumption of "equal catchability" of all animals in the population. Some new models relaxing this assumption have been proposed, complete with estimators and tests of model fit (Pollock 1974, 1975a, 1975b; Burnham 1972; Burnham and Overton 1978, 1979). Of particular importance is the monograph for biologists by Otis et al. (1978).

Here I concentrate on reviewing the current theory from an applied statistician's viewpoint but for an audience mainly of biologists with some knowledge of statistical thinking. Assumptions, robustness to assumption failure, and design concepts will be emphasized with statistical complexities kept to a minimum. Extensive reference will be made to source materials that may be consulted for more technical details on a particular model or concept.

The review falls naturally into sections on closed and open population models. Here by open we mean *additions* (birth and/or immigration) into the population are allowed, as are *permanent deletions* (death and/or emigration) from the population. Closed then means that neither additions nor permanent deletions are permitted. These sections are followed by a general discussion of the important issues raised.

SOME DEFINITIONS

Typically a capture-recapture study is carried out in the following way. The population under study is sampled two or more times. Each time, every unmarked animal caught is uniquely marked (usually with a numbered leg band in bird studies); previously marked animals have their capture recorded and then most or all of the animals are released back into the population. Thus at the end of the study the experimenter has the complete capture history of each animal handled. Batch marks where all animals captured in a particular sample cannot be distinguished are sometimes used but provide much less information and should be avoided if practically feasible.

The typical capture-recapture study described then provides two distinct types of information: (1) information from the recovery of marked animals; and (2) information from comparing numbers of marked and unmarked animals captured at each sampling time. Data from (1) can be used to estimate survival rates, whereas data from (1) and (2) are necessary to estimate population size. Sometimes survival rate estimation is of primary concern and the type (2) information will not be collected. This is typical of the usual band recovery studies where banded birds are recovered dead by hunters or other persons. It is also true for studies where live banded birds are resighted without actual physical capture.

SHORT-TERM STUDIES, CLOSED-POPULATION MODELS

THE PETERSEN MODEL

This simplest form of capture-recapture experiment, which is also often called the Lincoln Index, has a long history (see Seber 1973:59). (Lincoln Index is really a misuse of the word "index," which usually refers to a measure of relative abundance (Caughley 1977:12). A sample of n_1 animals is caught, marked and released. Later a sample of n_2 animals is captured, of which m_2 have been marked.

[1] Department of Statistics, North Carolina State University, Box 5457, Raleigh, North Carolina 27650.

Intuitively one can derive an estimator of the population size (N) based on the notion that the ratio of marked to total animals in the sample should reflect the same ratio in the population so that

$$\frac{m_2}{n_2} \approx \frac{n_1}{N}$$

which gives the estimator ($\hat{N}$)

$$\hat{N} = \frac{n_1 n_2}{m_2}. \tag{1}$$

A modified version with less bias was originally given by Chapman (1951) as

$$\hat{N}_c = \frac{(n_1 + 1)(n_2 + 1)}{(m_2 + 1)} - 1, \tag{2}$$

with an unbiased estimate of its variance given by

$$\mathrm{Var}(\hat{N}_c) = \frac{(n_1 + 1)(n_2 + 1)(n_1 - m_2)(n_2 - m_2)}{(m_2 + 1)^2(m_2 + 2)}. \tag{3}$$

from Seber (1973:60).

These estimators are based on a model in which the following assumptions are made: (1) all animals are equally likely to be caught in each sample—the "Equal Catchability" Assumption; (2) the population is closed to additions or deletions—the "Closure" Assumption; and (3) marks (bands) are visible and are not lost—the "Zero Band Loss" Assumption.

Clearly these assumptions are not always satisfied in practice and in the following sections we consider in detail each assumption and models allowing some relaxation of them. It should be emphasized that usually we consider a general K-sample capture-recapture experiment with more than two samples. This is often referred to as the Schnabel Census. (Census has been an extremely misused term in the literature. A census is defined here to be a complete enumeration of a population, whereas capture–recapture methods involve sampling only part of a population.)

The Equal Catchability Assumption

This assumption is unlikely in most wildlife populations. Two general types of alternatives exist:

(1) Heterogeneity—the probability of capture in any sample is a property of the animal and may vary over the population. That is, animals may vary in capture probability according to age, sex, social status and many other factors.

(2) Trap Response—the probability of capture in any sample depends on the animal's prior history of capture. That is, animals may become "trap shy" or "trap happy" depending on the type of trapping method used.

Either one or both of these two types of alternatives may be acting in a particular animal population.

Here we discuss the following series of mutually exclusive models first considered by me (Pollock 1974) and later by Otis et al. (1978) in an excellent monograph for biologists interested in detailed study.

M_0: no trap response, no heterogeneity
M_b: trap response, no heterogeneity
M_h: no trap response, heterogeneity
M_{bh}: trap response, heterogeneity

M_0: The equal catchability model

This model does not allow heterogeneity or trap response of the individual animal capture probabilities and makes the following assumptions: (1) Every animal in the population has the same probability of capture at each sampling time, p_i; $i = 1, \ldots, K$; and (2) The probability of capture does not change over sampling times; $p_i = p$; $i = 1, \ldots, K$.

Otis et al. (1978) call the model using only assumption (1) above M_t, and it is the classical capture-recapture model for a closed population with a history going back to Schnabel (1938). A detailed statistical treatment is given by Darroch (1958) (who refers to it as Model A) and Seber (1973:164). An application involving Red-winged Blackbirds (*Agelaius phoeniceus*) is given by Hewitt (1967). M_0 requiring assumptions (1) and (2) is a special case of M_t with capture probabilities constant over time.

An approximate maximum likelihood (M. L.) estimator of N for mo Model M_0 is the solution of

$$\left(1 - \frac{M_{k+1}}{\hat{N}}\right) = (1 - \hat{p})^K, \tag{4}$$

which must be found iteratively. Note that $\hat{p} = n./K\hat{N}$ with M_{K+1} the number of distinct animals captured in the whole experiment and $n.$ the sum of the numbers of animals captured in each sample. The approximate estimated large sample variance is

$$\mathrm{Var}(\hat{N}) = \hat{N}\left[\frac{1}{(1 - \hat{p})^K} + (K - 1) - \frac{K}{(1 - \hat{p})}\right]^{-1}. \tag{5}$$

Otis et al. (1978) compute the M. L. estimator directly using a numerical method.

It should be emphasized that this estimator can be highly biased if there is unequal catchability. Heterogeneity causes a negative bias, while trap response can cause positive or negative bias depending on whether the animals are "trap shy" or "trap happy" respectively.

M_b: The trap response model

This model, allowing trap response but no heterogeneity, makes the following assumptions: (1) Every unmarked animal in the population has the same probability of capture (p) for all samples; and (2) Every marked animal in the population has the same probability of recapture (c) for all samples after it has been captured once.

An approximate M. L. estimator of N is the solution of

$$\left(1 - \frac{M_{K+1}}{\hat{N}}\right) = (1 - \hat{p})^K, \quad (6)$$

where $\hat{p}$ is now given by $M_{K+1}/\left(K\hat{N} - \sum_{i=1}^{K} M_i\right)$ and M_i is the number of marked animals available for capture in the ith sample. The M.L. estimator has an estimated approximate variance of

$$\widehat{\text{Var}}(\hat{N}) = \frac{\hat{N}(1 - (1 - \hat{p})^K)(1 - \hat{p})^K}{[1 - (1 - \hat{p})^K]^2 - (\hat{p}K)^2(1 - \hat{p})^{K-1}}. \quad (7)$$

Notice that the recapture probability, $\hat{c}$, does not appear in (6). In fact animals do not contribute any information for population size estimation after first capture. Thus this model is equivalent to the "removal" method (Zippin 1956, Seber 1973:309) in which an animal is considered removed by marking rather than physically removed.

Typically in the biological literature a linear regression method has been used to estimate N in removal studies. It has intuitive appeal and also is easy to compute. However, if computer programs are available it is probably better to use the M. L. estimator. In practice I have found that there is usually little difference between the two estimators.

The regression method is presented briefly here because of its intuitive appeal and because it will be helpful when we come to discuss M_{bh} below. It is based on the following expression

$$\begin{aligned} E(u_i \mid M_i) &= p(N - M_i) \\ &= pN - pM_i, \end{aligned} \quad (8)$$

where (8) can be described mathematically as follows. Given M_i, the expected or "average" catch of unmarked animals on day i (u_i) is a linear function of the number of marked animals in the population. In the language of fisheries where this technique has been most applied, we have a linear regression of catch (u_i) versus cumulative catch (M_i).

The regression estimators of N and p are

$$\tilde{N} = \bar{M} + \bar{u}/\tilde{p} \quad (9)$$

$$\tilde{p} = -\sum_{i=1}^{K} u_i(M_i - \bar{M}) / \sum_{i=1}^{K} (M_i - \bar{M})^2, \quad (10)$$

which are simple functions of the slope and intercept estimators in the linear regression.

M_h: The heterogeneity model

This model allows heterogeneity but no trap response and assumes that each animal has its own unique capture probability (p_j, $j = 1, \ldots, N$) *which remains constant over all the sampling times*. The p_j's are further assumed to be a random sample of size N from some probability distribution $F(p)$. This model was first considered by Burnham (1972) and later by Burnham and Overton (1978, 1979). Under this model the vector of capture frequencies ($f_1, f_2, \ldots, f_K$), composed of the numbers of animals captured 1, 2, . . . , K times, contains all the information for estimating N. In statistical parlance this is called a minimal sufficient statistic.

Difficulties exist in finding a satisfactory estimator for this model. Burnham originally tried modelling the capture probabilities as coming from a two parameter beta distribution (Johnson and Kotz 1970:37) and using M. L. estimation, but found it unsatisfactory.

A naive estimator of N is

$$\hat{N}_I = \sum_{i=1}^{K} f_i = M_{K+1}, \quad (11)$$

which is simply the number of distinct animals seen in the whole experiment. This estimator is highly biased unless the capture probabilities are very high or the number of samples very large. Burnham applied the "jackknife" technique for bias reduction originally proposed by Quenouille (1956) to this estimator and came up with a series of estimators which are given in Otis et al. (1978:109). He also proposed an objective technique for choosing which one to use on a particular data set. The estimators all have the form

$$\hat{N}_{hl} = \sum_{i=1}^{K} a_{il} f_i, \quad (12)$$

where the a_{il} are constants which take on different values depending on the order (l) of jackknife used. The naive initial estimator (11) is also of this form with $a_i = 1$ for all $i = 1, \ldots, K$. An approximate variance estimator for $\hat{N}_{hl}$ in (12) is

$$\widehat{\text{Var}}(\hat{N}_{hl}) = \sum_{i=1}^{K} a_{il}^2 f_i - \hat{N}_{jl} \quad (13)$$

Use of (12) and (13) for point and interval estimation is not without difficulties (Otis et al. 1978:37). However, this jackknife estimation procedure is the most robust method so far proposed for Model M_h.

M_{bh}: The trap response and heterogeneity model

This model allows for heterogeneity and trap response and assumes that each animal has its own unique pair of potential capture probabilities (p_j, c_j; $j = 1, \ldots, N$) with p_j and c_j referring to whether the animal is unmarked or marked. These probabilities are assumed *to remain constant over all sampling times.*

This model was first considered by me (Pollock 1974) and later it was developed further by Otis et al. (1978:40). They give an estimation procedure called the "generalized removal method." Here we describe their method intuitively by generalizing the linear regression method given at the end of the section above on M_b (the trap response model).

If heterogeneity is operating (as well as trap response), there is no longer a linear relationship between catch (u_i) and cumulative catch (M_i) and thus (8) is no longer valid. In fact we now have

$$E(u_i | M_i) = \bar{p}_i(N - M_i), \quad (14)$$

where $\bar{p}_i$ is the average conditional probability of capture in the ith sample for those animals not previously captured. We would expect the $\bar{p}_i$'s to decrease gradually because the animals with higher first capture probabilities would tend to be caught earlier than those with lower first capture probabilities.

Otis et al. (1978) also suggest that the biggest differences between the p_i's will be at the beginning of the experiment. This suggests that the regression should be more nearly linear if points corresponding to the earlier sampling times are excluded. An objective method of doing this is to first test if all the $\bar{p}_i$'s are equal. If so then we use the removal method (see discussion above) which is a regression using all the sample points. If not, then we test if $\bar{p}_2 = \bar{p}_3 = \ldots = \bar{p}_K$, and if this hypothesis is not rejected we use a regression based on all points except the first. We continue sequentially removing points until the remaining $\bar{p}_i$'s cannot be shown to be different.

Notice that while using fewer points in the regression gives us a more general model allowing heterogeneity it also means we are estimating N based on less information, so that standard errors for N under this model tend to be larger than under the removal model. Also a substantial negative bias on estimators of N can still exist if the heterogeneity is severe and the number of sampling times small. The problem with heterogeneity is that some animals may be essentially "invisible" because of their very small capture probabilities, and no model can deal with this extreme situation. This has implications for study design which will be considered below.

TABLE 1
CLOSED POPULATION MODELS

Model	Unequal catchability due to: Trap response	Heterogeneity	Time	Estimator available	References[a]
M_o				Yes	Darroch (1958)
M_b[b]	x			Yes	Zippin (1956)
M_h		x		Yes	Burnham and Overton (1978, 1979)
M_{bh}	x	x		Yes	
M_t			x	Yes	Schnabel (1938) Darroch (1958)
M_{tb}	x		x	No	
M_{th}		x	x	No	
M_{tbh}	x	x	x	No	

[a] Otis et al. (1978) is the definitive reference for biologists on all eight models.
[b] Manly (1977b) has developed an extension of M_b to allow for short-term "wanderers" through the study area.

Time variation in capture probabilities

In addition to the four models (M_o, M_b, M_h, M_{bh}) just considered, four others are possible by generalizing each to allow capture probabilities to vary over sampling times (M_t, M_{tb}, M_{th}, M_{tbh}). It should be emphasized that although these models are conceptually very important, because they may be appropriate for a practical study, they do not usually permit estimation of population size. The exception is M_t, the classic Schnabel Model, which was discussed above in the section on M_o (the equal catchability model). A summary of the 8 possible models and their assumptions is given in Table 1 to help the reader.

Model selection

Otis et al. (1978) present an important method of choosing a model from the 8 possible (M_o, M_t, M_b, M_{bt}, . . . , M_{tbh}) based on a series of goodness of fit tests. (They also provide a very detailed computer program, CAPTURE.) This complex procedure may often need to be used but with caution (see also Chapman 1980) because the tests are not independent and often have low power. If possible biological information should be used to reduce the number of models considered in a study. For example evidence may exist, based on the behavior of the animal concerned, that trap response is unlikely with the trapping method used. In this case any models which allow trap response could be eliminated from consideration (M_b, M_{bh}, M_{tbh}).

Some alternative approaches

Regression methods.—Tanaka (1951) and Marten (1970) have suggested different generalizations of the basic Schnabel Model (M_t) to allow "unequal catchability." Unfortunately these methods (based on regression techniques) are rather ad hoc and do not distinguish between heterogeneity and trap response. I see no compelling reason for using either method.

Frequency of capture methods.—Several authors (Craig 1953, Tanton 1965 and Eberhardt 1969) have suggested methods of allowing for unequal catchability (again with heterogeneity and trap response not being distinguished) based on frequency of capture methods. When heterogeneity alone is operating (M_h; see section above on heterogeneity model) we have seen that the frequencies of capture contain all of the information for estimation of N so that these methods then have some theoretical justification. If trap response or time variation in capture probabilities is present, I suspect the methods will not be very useful.

The Closure Assumption

This assumption that the population must be static over the sampling period is very important to the models proposed thus far. It allows a substantial weakening of the other major assumption of equal catchability of animals. As we shall see below, it is very difficult to allow for unequal catchability when we have the additional complexity of open population models.

Testing for closure

Pollock et al. (1974), assuming M_t as the basic model, derive a sequence of tests for closure versus the alternatives of deletions only, additions only, and both deletions and additions. Burnham (1972) derives a general closure test with Model M_h as the null hypothesis (see also Otis et al. 1978:66). The problems of these tests are: (1) they have low power for detecting departures from closure; and (2) there is difficulty in distinguishing closure from unequal catchability.

We shall consider the closure assumption further when we discuss study design below. The biologist must consider critically the reality of the closure assumption for his specific study.

A model allowing "wanderers"

In many short term studies, particularly on birds and small mammals, the closure assumption is a problem. Although the study may be short enough for no births or deaths to occur it is often difficult to prevent migration. Manly (1977b) develops a model based on some ideas of MacArthur and MacArthur (1974), which is a generalization of the trap response model (M_b). This model allows a proportion of the population to be permanent "residents" of the study area. These "residents" may be subject to trap response but not heterogeneity or time variation of their first capture probabilities. The remainder of the population are "wanderers" who are only in the population a short time. The arrival times of the "wanderers" are assumed to be uniformly distributed over the whole sampling period.

Manly (1977b) illustrates his model with some data on birds caught in mist nets by Terborgh and Faaborg (1973) in Puerto Rico. Estimators and their standard errors are given for the proportion of "residents" in the population, the number of "residents" in the population, and the rate of capture of "residents."

The model is appealing and should be investigated further as should the more general problem of trying to separate migration from births and deaths as violations of closure. Some potential problems exist with the model, however. These are possible heterogeneity or time variation of the first capture probabilities of the "residents," and possible trends in the arrival times of the "wanderers."

Manly (1977b:407) mentions that this second problem could be serious. Further the whole conceptual framework of the model depends on there being two extreme groups of animals, the "residents" and the short-term "wanderers." How long must a wandering animal have to be in the population to become a resident, which then emigrates?

The Zero Band Loss Assumption

If animals lose their tags, the number of recaptures will be too small resulting in a potentially serious overestimation of the population size. Seber (1973:93) gives a good review on methods of marking animals, failure of the assumption, and a method to estimate and adjust for mark loss using a double banding scheme. See also Caughley (1977:139). It is important to realize that tag loss will cause a decrease in precision of estimators even when it can be estimated and adjusted for (Pollock 1981b).

Study Design

Here we consider the design of short term studies where the objective is estimation of population size with perhaps a conversion to a density estimator. It is often difficult to obtain the exact area a population covers, although there is a large literature on the problem with most of it oriented towards small mammal studies (Otis et al. 1978:67).

In this type of study many potentially useful models exist. I feel that study design should be oriented around satisfaction of as many model assumptions as practically possible so that a simple and reasonably efficient model can be used for estimation. Traditional sample size calculations for a given precision are only partially useful because often the biologists must do a substantial amount of model selection *after* the study is completed.

Closure

In many short-term studies, especially on birds, this assumption is difficult to satisfy because of movement. In some cases careful choice of study areas with natural boundaries may reduce movement. Another method is to use a short sampling period for the whole study, but this must be balanced against the need for as many samples as possible.

The model of Manly (1977b) discussed above that allows for "wanderers," deserves further study because it allows some movement through the area. Other methods which allow movement (but not births or mortality) would be valuable but little has been developed at this time.

Time variation

If closure can be approximately guaranteed, the next most important aspect of the design process is to try to achieve constant capture probabilities over time. This makes available a range of potentially useful models which allow heterogeneity and/or trap response of the capture probabilities (M_o, M_b, M_h, M_{bh}).

The sampling times chosen should be as similar as is practically feasible. Effort, time of day, weather conditions and any other biologically important factors should be kept as constant as possible. There is once again a trade off in number of sampling periods. The capture probabilities will be easier to keep constant for a small number of periods, but the models obviously give more precise estimators with more periods, especially if heterogeneity is present (M_h, M_{bh}).

Trap response and heterogeneity

If closure and no time variation in capture probabilities can be guaranteed, then it is not strictly necessary to try to avoid trap response or heterogeneity. This is fortunate because in many practical studies some degree of trap response or heterogeneity is inevitable.

Often trap response is severe, especially in mist net studies (Manly 1979). "Baiting" of traps will often induce trap response but unfortunately the alternative may often be no captures! Sometimes use of resighting data (without capture) may avoid trap response.

Heterogeneity is likely to occur to some degree in almost all studies. Attempts to minimize it can be made by trying to ensure that all sections of the study area are sampled with equal intensity. Another approach is to try to stratify the data based on known sources of heterogeneity, such as age or sex. Unfortunately sample sizes often are not adequate to do this.

Band loss

Clearly one should attempt to avoid this problem by using a proven marking method. If a new method is being used, perhaps a pilot study should be considered or at the very least some attempt should be made to estimate mark loss through techniques like double banding (see also the section above dealing with the "Zero Band Loss Assumption").

Sample size considerations

Often sample sizes will be limited by practical problems but it is important for biologists to realize that small studies may be little better than none at all! Ideally one would like a study with approximately ten sampling periods and constant capture probabilities averaging at least 0.1 for the whole study. This enables reasonable identification of the correct model and good precision of the population size estimator under that model.

LONG-TERM STUDIES, OPEN-POPULATION MODELS

Often the biologist is interested in a long-term study in which it is not feasible to assume a closed population. Open population models are now of interest and they allow estimation of "survival" rates and "birth" rates as well as the population size at the different sampling times. Table 2 summarizes the models and their assumptions considered in this section. (It should be emphasized that in capture-recapture studies it is not possible to separate births from immigration or deaths from emigration so that "survival" and "birth" rate estimators may include migrants.)

THE JOLLY-SEBER MODEL

Assumptions

The most important open population model is the Jolly-Seber (J-S) Model, independently derived by Jolly (1965) and Seber (1965). A good intuitive introduction suitable for biologists is given by Cormack (1973). The best reference for detailed study of this model is Seber (1973:196). This model makes the following assumptions: (1) Every animal in the population (marked or unmarked) has the same probability (p_i) of being caught in the ith sample ($i = 1, \ldots, K$), given that it is alive and in the population when the sample is taken; (2) Every animal has the same probability (ϕ_i) of surviving from the ith to the

TABLE 2
OPEN POPULATION MODELS

Model	Assumptions	References
Jolly-Seber[a]	Equal capture and survival probabilities for all animals at each sampling time	Jolly (1965) Seber (1965) Seber (1973)
Manly-Parr	Equal capture probabilities but survival probabilities may vary with age of animal.	Manly and Parr (1968) Seber (1973)
Age-dependent	Identifiable age classes exist which may have different survival and capture probabilities	Pollock (1981b)

[a] Jolly (1979, 1981) and Crosbie (1979) have considered some restricted versions of the Jolly-Seber model, which could be very important to biologists.

$(i + 1)$th sample, given that it is alive and in the population immediately after the ith release ($i = 1, \ldots, K - 1$); (3) Marked animals do not lose their marks and all marks are reported on recovery; and (4) The actual time spent sampling occupies a short period.

Assumption (1) is the equal catchability assumption discussed in the short-term studies except that now it applies only to live animals. Also we now assume equal survival rates for all animals (2). These two assumptions and some alternatives to them will be discussed further in later sections. Clearly the Zero Band Loss Assumption (3) is still important. Assumption (4) is necessary because this is an open population model; otherwise it does not make logical sense to estimate parameters at a particular sampling time.

Parameter estimation

Here an intuitive discussion of parameter estimation will be given. Imagine to begin with that M_i, the number of marked animals in the population just before the ith sample, is known for all values $i = 2, \ldots, K$ (there are no marked animals at the time of the first sample so that $M_1 = 0$).

Obviously an intuitive estimator of N_i, the population size at time i, is the Petersen estimator discussed above under "The Petersen Model." If the model assumptions are valid, then

$$\frac{m_i}{n_i} \simeq \frac{M_i}{N_i},$$

which gives

$$\hat{N}_i = \frac{n_i M_i}{m_i}, \qquad (14)$$

where m_i and n_i are the marked and total numbers of animals captured in the ith sample respectively.

An estimator of the survival rate from sample i to sample $(i + 1)$ is M_{i+1}, which is the total number of marked animals in the population just before the $(i + 1)$th sample, divided by the total number of marked animals released after sample i, which is $M_i - m_i + R_i$. Note that R_i is the number of the n_i animals captured that are released.

$$\phi_i = \frac{M_{i+1}}{(M_i - m_i + R_i)} \qquad (15)$$

An intuitive estimator of the recruitment in time interval i to $(i + 1)$ is

$$\hat{B}_i = \hat{N}_{i+1} - \phi_i(\hat{N}_i - n_i + R_i). \qquad (16)$$

This is simply the estimated difference between the population size at time $(i + 1)$ (which is N_{i+1}) and the expected number of survivors from time i to time $(i + 1)$ (which is $\phi_i(N_i - n_i + R_i)$).

To complete this intuitive outline we need an estimator of the M_i because they are obviously unknown in an open population. This can be obtained by equating the two ratios

$$\frac{Z_i}{M_i - m_i} \simeq \frac{r_i}{R_i},$$

which are the future recovery rates of the two distinct groups of marked animals when $(M_i - m_i)$ are the marked animals *not seen* at i and R_i are the marked animals seen at i and then released for possible recapture. Note that Z_i and r_i are the members of $(M_i - m_i)$ and R_i, which are captured again at least once. The estimator of M_i is thus given by

$$\hat{M}_i = m_i + \frac{R_i Z_i}{r_i} \qquad (17)$$

and is defined only for $i = 2, \ldots, K - 1$. It follows that $\hat{N}_i$ in (14) is defined for $i = 2, \ldots, K - 1$; ϕ_i in (15) for $i = 1, \ldots, K - 2$; and $\hat{B}_i$ in (16) for $i = 2, \ldots, K - 2$.

A summary of these parameter estimators and their approximate large sample variances is given in Seber (1973:205), which should also be consulted for some detailed examples. Cormack (1964) presents a model for recaptures of marked animals by resighting without capture, which is actually a special case of the J-S Model. For that model it is possible to estimate survival rates but not population size.

Constant survival and/or constant capture probabilities

A very important new development is the work of Jolly (1979, 1981). He restricts the large number of parameters by assuming a constant survival rate and/or a constant capture rate *over the whole study*. If these assumptions are realistic, as they often are, then he shows that there can be large gains in precision of estimators. Crosbie (1979) has also considered these models and has developed a computer package to facilitate their use.

Unequal survival and catchability

With the added complexity of an open population model, it is difficult to build alternative models to the same degree as in the closed population case. We also have the added complexity of possible heterogeneity and trap response of the survival probabilities of different animals.

Trap response.—Robson (1969) and Pollock (1975b) have shown that it is possible to generalize the J-S model to allow for a trap response in survival and capture probabilities that lasts for a short time (typically only one period after initial capture). The estimators still have a similar intuitive form to the original J-S estimators. Tests for this type of temporary trap response are also given and involve contingency table chi-square tests.

Permanent trap response can have a very large influence on the J-S estimators. If animals are "trap shy" too few recaptures will be made, resulting in overestimation of population size whereas underestimation will result from "trap happy" animals. Survival and birth rate estimators can also be severely affected.

Heterogeneity.—Carothers (1973) and Gilbert (1973) have used simulation to study the influence of heterogeneity of capture probabilities on the J-S estimators. Serious negative bias of population size estimators can result, but survival estimators, although negatively biased, are much less affected. This has implications for the design of long-term studies which will be discussed below.

Heterogeneity of survival probabilities among individual animals has not been considered in any detail in the literature, although it obviously occurs in practice. Cormack (1972) states that the J-S estimators will be little affected by this type of heterogeneity. Based on some recent work, Pollock and Raveling (1981) agree with this assessment if an animal's survival probability is independent of its capture probability. This is probably a reasonable assumption in many practical studies with live recapture. However, in band recovery studies, in which animals are recovered dead, an animal's survival probabilitiy is clearly negatively related to its recovery probability. In this situation a negative bias on survival estimators results. Band recovery models will be considered below.

Age-dependent survival rates

A special type of heterogeneity of survival probabilities occurs when survival is a function of the age of the animal. Manly and Parr (1968) give a method of allowing for this when capture probabilities are assumed to be independent of age. The method is described in simple terms suitable for a biological audience and includes a worked example (see also Seber 1973:233).

Computer programs

Unless the number of samples is small, computation of the J-S estimators and their variances is time consuming. Some computer programs are available. Two simple programs are given by Davies (1971) and White (1971). For those who do many capture-recapture studies and desire to try a range of options on their data, a detailed program package called POPAN-2 is recommended. It was developed by Arnason and Baniuk (1978) and the manual provides much information on data management. It contains a particularly valuable section on methods of combining samples when capture probabilities are low (which is a common practical problem). The authors state that POPAN-2 is not really suitable for "one-shot" users, but they also validly point out that a biologist should be cautious about embarking on such a "one-shot" study anyway.

Generalization to Multiple Distinct Age Classes

For some species (especially of birds) several clearly identifiable age classes occur that are likely to have different survival rates and perhaps also different capture rates. I have shown that it is possible to generalize the J-S Model to allow these different age classes to have different survival and capture probabilities (Pollock 1981). The estimators take a form similar to those under the J-S Model and it is possible to test if the survival and capture probabilities are age dependent, using a series of contingency table chi-square tests. The model was found to be useful for some resighting data on neck-collared Giant Canada Geese (*Branta canadensis maxima*) where both young and adults (age one year or more) were neck-collared.

In this model (Pollock 1981b) we assume one capture period each "year" for K "years." (We use "year" to represent the period of time an animal remains in an age class. This will not necessarily represent a calendar year.) There are $(l + 1)$ distinguishable age classes of animals 0, 1, . . . , l which therefore move forward one class each "year." We further assume that each age class has a different capture rate in the ith sample and a different survival rate from the ith to the $(i + 1)$th sample ($i = 1, \ldots, K$). Immigration or emigration may occur for each age class of the population, but births may occur only into the zero age group. Thus when referring to survival, we really mean those animals that have not died or emigrated. Similarly, when referring to additions, we really mean births and immigration for the young animals ($V = 0$) but only immigration for the older animals ($V = 1, \ldots, l$). (It is not necessary to restrict oneself to one capture period per year. Stokes (pers. commun.) has investigated a model with several periods per year for use on American Woodcock (*Philohela minor*) data.)

I believe that this extension of the J-S Model will be useful to biologists. In most applications, the number of age classes will be only two or three because that is all that can be identified in the field.

The number of age classes should be kept as small as biologically reasonable so that sufficient numbers of each class can be marked to give reasonable precision to the estimators of population parameters. Restrictions on this general model, such as assuming that each age class has a constant survival rate over the whole study, could also be considered to increase precision of estimators. Unfortunately these restricted models do not give rise to simple intuitive estimators.

Band Recovery Models

The band recovery models in current use (Brownie et al. 1978) are closely related to the J-S Model and its generalization to allow for age-dependence (as discussed above), but now there are 100% "losses on capture." However the biological issues involved are beyond this review, which is intended to concentrate on live recapture studies.

Important references are Seber (1973:239) and Brownie et al. (1978) (written for biologists). The assumptions behind band recovery models are reviewed by Pollock and Raveling (1981). They are particularly interested in the effect of potential heterogeneity of survival probabilities.

Design of Long-Term Studies

Very little has been done on the design of capture-recapture studies that use open population models. Seber (1973:204) gives an ad hoc rule of thumb that m_i the number marked in each sample, and r_i, the number of animals released from the ith sample that are recovered, should be at least 10 for all values of i.

Manly (1971a) found that studies with small capture probabilities suffer from a problem in estimation of variances. Underestimates of parameters appear to be more accurate than they really are. This is because the parameter estimate is substituted in the variance formula to obtain a variance estimate.

Jolly (1965) pointed out that recaptures enter into the estimates in two distinct ways. First as the proportion of marked animals in the ith sample, and second as the ratio of future recaptures. This has design implications as Jolly (1965) realizes: "It might well, therefore, be advantageous to have a separate organization for recording future recaptures, Z_i and r_i, from that for releasing animals. Since it is necessary to distinguish only two classes of marked animals in the future recaptures with respect to each time i, a very simple code of marks might be used in specific situations, thus enabling untrained persons over a wide area to recapture, or possibly merely to observe, marked animals. Such a recapture system could proceed continuously, since the time at which an animal is recaptured is of no importance. Releases, on the other hand, would only be made at the particular times for which estimates $\hat{M}_i$ were required, the marking and releasing being done by more experienced staff."

Another method of increasing precision might be to use equally spaced sampling periods with the sampling periods being as similar as possible. The reason is that it may then be feasible to assume constant survival and/or capture rates over time and to use the restricted models of Jolly (1979, 1981), which were discussed above in the section on "*Constant survival and/or constant capture probabilities.*"

Any method of reducing heterogeneity and trap response of capture probabilities here is extremely important, because it is not possible to allow for them to the same degree as in closed population models. One design which has great potential, especially for some bird species, is to mark individuals so that they can be identified without capture. As usually only marked birds are considered for resighting, this design allows only estimation of survival rates. It does mean, however, that there is no problem with trap response and also it is known that heterogeneity does not have much influence on survival estimators (Carothers 1973). Cormack (1964) discusses an example of this type on the Fulmar (*Fulmarus glacialis*) and Pollock (1981b) gives an example on Giant Canada Geese where two age classes are neck-collared. The problems with this design are: (1) there is no estimator of population size unless the experimenter can in some way estimate the proportion of marked to total animals; (2) markings must be clearly visible and permanent, which can be a big problem with neck collars; and (3) some markings that are very visible may alter the survival rate of the animal.

GENERAL DISCUSSION

The sharp distinction drawn in this review between short-term and long-term studies is somewhat artificial. In practice a series of short-term studies may be carried out. One approach to analysis would be to analyze each short-term study using the closed population models which allow unequal catchability. Then all the sampling periods in each short-term study could be pooled and survival estimators between these short-term studies could be estimated using the Jolly-Seber Model. This approach allows population size estimation under models allowing unequal catchability while survival estimation, which is not so affected by unequal catchability, is under the Jolly-Seber Model.

A recurring problem is the shortage of models and techniques allowing for local movement. Cormack (1979) asserts that this may be the most important area for future research.

An approach to capture-recapture problems with some potential is the use of log linear models originally proposed by Fienberg (1972). Cormack (1981) has shown how this approach can be used practically for open and closed populations. Unfortunately this approach does not allow for heterogeneity or permanent trap response.

ACKNOWLEDGMENTS

I wish to thank Kenneth P. Burnham and David L. Otis for their thoughtful comments on an earlier version of this article.

Studies in Avian Biology No. 6:436–440, 1981.

IMPROVED POPULATION ESTIMATES THROUGH THE USE OF AUXILIARY INFORMATION

DOUGLAS H. JOHNSON[1]

ABSTRACT.—When estimating the size of a population of birds, the investigator may have, in addition to an estimator based on a statistical sample, information on one of several auxiliary variables, such as: (1) estimates of the population made on previous occasions, (2) measures of habitat variables associated with the size of the population, and (3) estimates of the population sizes of other species that correlate with the species of interest. Although many studies have described the relationships between each of these kinds of data and the population size to be estimated, very little work has been done to improve the estimator by incorporating such auxiliary information. A statistical methodology termed "empirical Bayes" seems to be appropriate to these situations. The potential that empirical Bayes methodology has for improved estimation of the population size of the Mallard (*Anas platyrhynchos*) is explored. In the example considered, three empirical Bayes estimators were found to reduce the error by one-fourth to one-half of that of the usual estimator.

The United States Fish and Wildlife Service (FWS) is charged by law with the authority and responsibility for migratory birds within the nation. Many species are protected by joint treaties with other nations: Great Britain (for Canada), Mexico, the Soviet Union, and Japan. One particular concern of the Fish and Wildlife Service is the regulation of hunting on game species. By late summer each year, regulations governing the hunting season during the subsequent fall and winter must be promulgated and published in the Federal Register.

In order to develop regulations that are consistent with the welfare of the game species, the FWS collects certain kinds of information about the status of those species (Martin et al. 1979). For waterfowl, which are of high interest to millions of hunters, the FWS each May conducts a survey of the population throughout the major breeding areas of North America. These surveys are done in cooperation with the Canadian Wildlife Service and various states and provinces. The survey is a complicated sample survey design (Martin et al. 1979), one sample unit being the transect, a linear route along which an aircraft is flown. Waterfowl are counted, according to species, within 0.2 km (⅛ mile) on either side of the aircraft. These counts are adjusted by the area covered, and by independently derived visibility rates, to estimate the density of waterfowl, by species, along each transect.

The sample counts are subject to fairly large variances, as well as possible biases. Although accurate population estimates are desired, improved precision through increased sample size is difficult to attain, because of the cost, time, and personnel requirements of the May surveys.

The purpose of this preliminary report is to examine the efficacy of a statistical methodology known as empirical Bayes for improving estimators of waterfowl density through the use of auxiliary information. The empirical Bayes methodology will be briefly surveyed. The kinds of auxiliary information considered are: (1) estimated population densities of the species of interest in previous years; (2) information on habitat variables that correlate with the density of the species; and (3) estimated densities of other species in the particular year.

METHODS

EMPIRICAL BAYES ESTIMATION

Assume we have a recurring problem of estimating a location parameter θ, for example, the average density of Mallards in eastern North Dakota. We have a statistic X, perhaps the average Mallard density of a sample of k transects, whose distribution depends on θ via the probability density function $f(x \mid \theta)$. Suppose that the situation recurs with various unknown values of θ. Let the distribution of θ be described by the probability density function $g(\theta)$. Suppose we have a sequence of n such situations, with observed statistics $x_1, x_2, \ldots, x_n$ and corresponding parameter values $\theta_1, \theta_2, \ldots, \theta_n$. We want to estimate the current value, θ_n; the current as well as previous statistics $x_1, \ldots, x_n$ are known to us.

The problem can be addressed from three points of view (Krutchkoff 1969). The *classical* approach incorporates the fact that X_n is sufficient for θ_n; therefore only the data for the current situation are used to estimate θ_n. For example,

$$\hat{\theta}_n = x_n.$$

This estimator is unbiased and has variance σ^2/k, where σ^2 is the variance of a single transect and k the number of transects that comprise the mean.

A strictly *Bayesian* approach would require that $g(\theta)$ be known *a priori*. The posterior distribution is then

$$p(\theta_n \mid x_n) = \frac{f(x_n \mid \theta)g(\theta)}{\int f(x_n \mid \theta)g(\theta)\, d\theta}.$$

A point estimator of θ_n can be taken as the mean of the posterior distribution:

$$\tilde{\theta}_n = E(\theta_n \mid x_n) = \frac{\int \theta f(x_n \mid \theta)g(\theta)\, d\theta}{\int f(x_n \mid \theta)g(\theta)\, d\theta}.$$

[1] U.S. Fish and Wildlife Service, Northern Prairie Wildlife Research Center, Jamestown, North Dakota 58401.

TABLE 1
DATA USED TO DEVELOP EMPIRICAL BAYES ESTIMATOR BASED ON PREVIOUS MALLARD DENSITIES

(1) Year (n)	(2) θ_n	(3) Sample 1	(4) Sample 2	(5) Sample 3	(6) x_n	(7) $s_n^2/3$	(8) $s^2/3$	(9) $\bar{x}$	(10) t^2	(11) $\hat{\theta}_n$
1958	11.0619	9.5162	15.4012	2.0451	8.9875	14.9353	14.9353	—	—	—
1959	3.9925	2.4834	5.2589	1.7530	3.1651	1.1405	8.0379	8.9875	—	—
1960	6.9193	8.3266	8.7649	8.8901	8.6605	0.0292	5.3683	6.0763	16.9502	8.0389
1961	6.4743	3.7140	8.1355	4.6994	5.5163	1.7960	4.4752	6.9377	10.7011	5.9354
1962	7.7613	7.7818	3.7898	10.0810	7.2175	3.3779	4.2558	6.5824	7.6392	6.9903
1963	12.5712	6.8975	13.2644	20.1620	13.4413	14.6701	5.9915	6.7094	5.8101	10.0236
1964	7.8600	5.2316	22.9385	2.4146	10.1949	41.2611	11.0300	7.8314	12.2012	9.0727
1965	8.9054	4.8844	10.4667	7.6523	7.6678	2.5969	9.9759	8.1690	10.9657	7.9066
1966	13.0822	7.3489	23.4924	7.9513	12.9309	27.9167	11.9693	8.1064	9.4306	10.2325
1967	9.0331	10.0919	10.3212	10.8092	10.4074	0.0447	10.7768	8.6424	10.8380	9.5274
1968	9.0470	6.4044	11.7682	13.7295	10.6340	4.7930	10.2329	8.8189	9.9453	9.7135
1969	8.2463	7.9337	8.0597	10.2578	8.7504	0.5694	9.4276	8.9839	9.2502	8.8683
1970	13.0868	7.5113	8.3892	20.8268	12.2424	18.4871	10.1245	8.9645	8.4139	10.4522
1971	13.0103	11.3584	9.2452	8.1886	9.5974	0.8683	9.4633	9.2166	8.5392	9.3972
1972	9.9792	25.1220	8.3740	7.2109	13.5690	33.4809	11.0645	9.2438	7.8927	11.0446
1973	8.8084	3.4951	6.6724	4.4483	4.8719	0.8861	10.4283	9.5322	8.5761	7.4292
1974	7.9099	4.6781	14.2320	3.6898	7.5333	11.2995	10.4796	9.2409	9.3617	8.4352
1975	7.9809	10.9896	12.3149	5.5662	9.6236	4.2619	10.1341	9.1404	8.9482	9.3670
1976	7.6175	9.3455	9.8446	11.4324	10.2075	0.3959	9.6216	9.1673	8.4348	9.6532
1977	4.0781	3.6468	6.2517	2.1881	4.0289	1.4126	9.2112	9.2220	8.0231	6.8045
1978	9.0618	7.7069	5.0626	6.9611	6.5769	0.6196	8.8020	8.9624	8.9493	7.7598
1979	12.0640	9.3060	8.7386	5.0049	7.6898	1.8272	8.4850	8.8488	8.7728	8.2596

The strictly Bayesian approach also ignores the earlier observations $x_1, \ldots, x_{n-1}$; instead it is necessary to assume that the prior distribution is completely specified.

In the *empirical Bayes* approach, we begin with the Bayes estimator $E(\theta_n | x_n)$, which is given in terms of the unknown prior distribution $g(\theta)$, and estimate it instead in terms of the data $x_1, x_2, \ldots, x_n$. There are several ways of doing this, each resting on different assumptions.

The superiority of empirical Bayes estimators was first suggested by Stein (1955) and James and Stein (1961), who considered the problem of estimating $n \geqslant 3$ independent normal means, each with variance one, and a quadratic loss function. For the set of n means, the maximum likelihood estimator X_n was inferior to $X_n[1 - b/(a + \Sigma X_i^2)]$, where b and a are selected constants and the summation is over all X's. Stein's procedure essentially shrinks the estimator away from the observed mean toward zero. Lindley (1962) recommended instead that they be shrunk toward the mean of all X's, and proposed the estimator

$$X_n[1 - (n - 3)/\Sigma(X_i - \bar{X})^2] + \bar{X}(n - 3)/\Sigma(X_i - \bar{X})^2.$$

Stein's estimator is a weighted average of the observed mean and zero; Lindley's estimator is a weighted average of the observed mean and the overall mean.

A wealth of estimators appropriate to more general situations have also been developed, and comparison with standard estimators has demonstrated their worth (e.g., Efron and Morris 1975). Despite the theoretical justification of empirical Bayes methods, their use has not been widespread.

THE EXAMPLE—MALLARDS IN EASTERN NORTH DAKOTA

Although numerous waterfowl species are counted during the May waterfowl surveys, the Mallard duck receives especial attention because of its abundance and prized status by hunters. For the immediate purpose of exploring alternative estimation procedures, this report treats only Mallard densities, and only in eastern North Dakota (FWS Strata 45 and 46 [Martin et al. 1979]).

Estimates are available annually 1958 to 1979. In each year a number (varying between 7 and 15) of transects were run in eastern North Dakota. To illustrate the empirical Bayes estimators developed here, it is desirable to know the "true" population parameter, against which the performance of various estimators can be judged. For our example, the average density of Mallards in all transects during a given year will be considered the true parameter. These values (θ) are given in the second column of Table 1. We randomly selected three of the transects to use as sample data; independent samples were drawn each year. The sample Mallard densities are given in columns 3–5 of Table 1.

The Mean Square Error (MSE) criterion will be used for comparing estimators. The MSE measures the average "closeness" of the estimator to the parameter being estimated. If we have n situations in which we develop two estimators E_i and F_i of an unknown parameter P_i ($i = 1, 2, \ldots, n$), then

$$\text{MSE}\ (E) = \sum_{i=1}^{n} (E_i - P_i)^2/n,$$

TABLE 2
DATA USED TO DEVELOP EMPIRICAL BAYES ESTIMATOR BASED ON WETLAND CONDITIONS

(1) Year (n)	(2) x_n	(3) $s_n^2/3$	(4) $s^2/3$	(5) W_n	(6) a	(7) b	(8) $\hat{\theta}_n(W)$	(9) Z_n^2	(10) $\hat{\hat{\theta}}_n(W)$	(11) θ_n
1958	8.9875	14.9353	14.9353	214.9	—	—	—	—	—	11.0619
1959	3.1651	1.1405	8.0379	88.5	—	—	—	—	—	3.9925
1960	8.6605	0.0292	5.3683	340.4	−.9115	.0461	14.7809	—	—	6.9193
1961	5.5163	1.7960	4.4752	64.5	2.2498	.0218	3.6559	12.7967	5.0343	6.4743
1962	7.2175	3.3779	4.2558	229.8	3.4757	.0175	7.4972	5.1959	7.3434	7.7613
1963	13.4413	14.6701	5.9915	357.1	3.4628	.0173	9.6406	4.7160	11.3146	12.5712
1964	10.1949	41.2611	11.0300	148.7	2.5472	.0245	6.1904	4.9667	7.4337	7.8600
1965	7.6678	2.5969	9.9759	303.3	3.7248	.0215	10.2458	7.3944	9.1484	8.9054
1966	12.9309	27.9167	11.9693	448.5	3.9539	.0190	12.4754	10.0027	12.6828	13.0822
1967	10.4074	0.0447	10.7768	480.5	3.8364	.0197	13.3022	7.5833	12.1066	9.0331
1968	10.6340	4.7930	10.2329	250.9	4.4417	.0164	8.5565	5.5348	9.2857	9.0470
1969	8.7504	0.5694	9.4276	495.8	4.6773	.0162	12.7093	6.7555	11.0567	8.2463
1970	12.2424	18.4871	10.1245	625.1	5.3669	.0126	13.2432	8.7660	12.7788	13.0868
1971	9.5974	0.8683	9.4633	452.6	5.5794	.0117	10.8748	5.7963	10.3896	13.0103
1972	13.5690	33.4809	11.0645	485.9	5.6382	.0112	11.0803	5.5015	11.9068	9.9792
1973	4.8719	0.8861	10.4283	221.3	5.4733	.0112	7.9519	5.3191	6.9115	8.8084
1974	7.5333	11.2995	10.4796	575.5	4.9832	.0131	12.5222	6.3527	10.6393	7.9099
1975	9.6236	4.2619	10.1341	539.0	5.5687	.0105	11.2282	7.1172	10.5662	7.9809
1976	10.2075	0.3959	9.6216	526.8	5.6950	.0099	10.9103	6.6681	10.6226	7.6175
1977	4.0289	1.4126	9.2112	220.7	5.7385	.0097	7.8793	6.1458	6.3384	4.0781
1978	6.5769	0.6196	8.8020	317.4	5.2116	.0106	8.5760	6.3954	7.7347	9.0618
1979	7.6898	1.8272	8.4850	487.9	5.0714	.0107	10.2919	6.4452	9.1686	12.0648

and E is a better estimator of P than F is if MSE $(E) <$ MSE (F). Mathematically, the MSE equals the variance of an estimator plus the square of its bias.

RESULTS

Suppose in a given year n, the true density of Mallards in eastern North Dakota is θ_n, that value having resulted as a random outcome of a process with probability density function $g(\theta)$. We have an estimator of θ_n, given by X_n, which we assume is normally distributed with mean θ_n and variance σ^2/k. That is,

$$X_n \sim N(\theta_n, \sigma^2/k)$$

In the present example X_n is the estimated density of Mallard pairs based upon a sample of k transects in eastern North Dakota. X_n is unbiased and its variance σ^2/k is estimated by the sample variance S_n^2/k, where k is the sample size in that year. The mean of the three samples, x_n, is given in column 6 of Table 1. The sample variance of this mean is presented in column 7. The accuracy of the classical estimator can be evaluated by comparing columns 2 and 6. The Mean Square Error of the classical estimator for all 22 years of data is

$$\Sigma\,(x_n - \theta_n)^2/22\ .$$

Thus MSE $(X_n) = 4.42$ for all years. We will use as a test period the years 1968–79, permitting 10 years of baseline data to be used to develop the procedure. The MSE during the test period is 6.53.

AN ESTIMATOR BASED ON PREVIOUS COUNTS

A simple empirical Bayes estimator may be obtained by assuming that the process that generated θ_n was itself normal, with unknown mean θ and unknown variance τ^2:

$$\theta_n \sim N(\theta, \tau^2).$$

Then the empirical Bayes estimator is a weighted average of the current X_n and the mean of the previous X's, $\bar{X} = (X_1 + X_2 + \ldots + X_{n-1})/(n - 1)$. The weights are simply the reciprocals of the respective variances.

$$\hat{\theta}_n = \frac{X_n k/S^2 + \bar{X}/t^2}{k/S^2 + 1/t^2}. \quad (1)$$

where $t^2 = \hat{\tau}^2 = \Sigma(X_i - \bar{X})^2/(n - 2)$ and S^2 is the pooled within-year variance estimator.

This empirical Bayes estimator involves the current year's estimate, X_n, and the average of the Mallard densities from previous years, $\bar{X}$. These cumulative averages are given in column 9 of Table 1. The variance among the previous years' Mallard densities, which is used in the weighting of the cumulative averages, is shown in column 10. The simple empirical Bayes estimate, from Equation 1, is shown in column 11.

Comparing columns 2, 6 and 11, it is seen that

the empirical Bayes estimator shrinks the sample mean, X_n, toward the cumulative mean, $\bar{X}$. This shrinkage on the average tends to produce an estimate closer to the true value, θ_n. The Mean Square Error for this estimator is 3.71 for all years and 4.48 for 1968–79 test period. The MSE for the test period is 31.4 percent lower than 6.53, the value for the classical estimator.

An Estimator Based on Wetland Conditions

A waterfowl biologist might balk at the procedure described above, despite the clear gain in accuracy it affords, because it includes averages of Mallard densities from all previous years. Biologists recognize that in some years the prairies are wet and the ponds are full, but in other years the prairies and the ponds are dry. Mallards are far more common in North Dakota during wet years than dry years; the correlation between Mallard density and pond index, also measured each May, for 1958–79 is 0.555. Accordingly, biologists would be reluctant to base an estimator of Mallard density during a wet year upon a cumulative mean involving dry years. The estimator proposed in this section overcomes this objection by incorporating information about wetland habitat conditions.

Suppose that the Mallard density in eastern North Dakota is related to the pond index W, in a particular year j, according to

$$\theta_j = \alpha + \beta W_j + \epsilon_j,$$

where $E(\epsilon_j) = 0$, $V(\epsilon_j) = \mu^2$. From the X's and W's of previous years, we can estimate α, β, and μ^2, by a, b, and m^2, respectively. The regression estimator of θ_n is thus given by

$$\hat{\theta}_n(W) = a + bW_n.$$

This estimator can be used in combination with the sample estimate in the current year according to:

$$\hat{\hat{\theta}}_n = \frac{X_n k/S^2 + (a + bW_n)/Z^2}{k/S^2 + 1/Z^2}.$$

In this formula, Z^2 is the variance of an individual value of θ predicted from W:

$$Z^2 = m^2[1 + (n - 1)^{-1} + (W_n - \bar{W})^2/\Sigma(W_i - \bar{W})^2] \quad (2)$$

where m^2 is the residual variance and is equal to

$$\begin{aligned} m^2 &= \Sigma[X_i - \hat{\theta}_i(W)]^2/(n - 3) \\ &= [\text{Var } X_i - b^2 \text{Var } \hat{\theta}_i(W)](n - 2)/(n - 3) \\ &= \text{Var } X_i(1 - r^2)(n - 2)/(n - 3) \\ &= t^2(1 - r^2)(n - 2)/(n - 3). \end{aligned}$$

Note that r^2 is the squared simple correlation coefficient between pond index (W) and Mallard density (X).

Returning to the 1958–79 data for Mallards in eastern North Dakota, we now consider the improvement possible by including information about wetland conditions. Table 2 displays the pertinent information. Columns 2 and 3 contain the sample mean and its variance for a particular year. The pond index is given in column 5. Columns 6 and 7 provide the intercept and slope for estimating Mallard density from pond index, based on the data from years prior to the current one. The estimate of θ, based on a, b, and W_n, is given in column 8, with associated variance in column 9. The empirical Bayes estimator is shown in column 10, to be compared to the true value in column 11.

The Mean Square Error of this estimator is 3.91 for all years and 5.05 for the 1968–79 test period. This estimator thus offers a 23% improvement in MSE over the classical one, but does not perform quite as well as the empirical Bayes estimator based on the overall mean of mallard densities.

An Estimator Based on Other Species

In addition to the Mallard, five other dabbling ducks are common in the prairies of eastern North Dakota. These are Gadwall (*Anas strepera*), American Wigeon (*A. americana*), Blue-winged Teal (*A. discors*), Northern Shoveler (*A. clypeata*) and Pintail (*A. acuta*). These six species tend to fluctuate together; the multiple correlation coefficient between Mallard density and the densities of other species is $R^2 = 0.62$. This value is appreciably higher than the R^2 between pond index and Mallard density, $R^2 = 0.31$.

The reasoning above suggests that the sample densities of other species in a particular year might be used to develop an estimator of the Mallard density that year. This estimator could be combined in an empirical Bayes manner with the sample Mallard density. The following result, incorporating only one other species, indicates the potential power of the method.

The single species most closely correlated with Mallard densities in Strata 45 and 46 during 1958–79 was the Pintail, with $r = 0.61$. A regression equation relating Mallard density (θ_n) to Pintail density from all transects (P_n) is given by

$$\hat{\theta}_n(P) = 5.7922 + 0.3421P_n.$$

Unlike previous analyses, this predictive equation was developed from the entire 22-year data set, rather than sequentially year by year. Table 3 displays the Mallard densities estimated from Pintail densities (column 5), the weighting factors obtained analogously to equation 2 (column 4), and the resulting empirical Bayes estimator (column 6).

TABLE 3
DATA USED TO DEVELOP EMPIRICAL BAYES ESTIMATOR BASED ON PINTAIL DENSITIES

(1) Year	(2) P_n	(3) θ_n	(4) Z_n^2	(5) $\hat{\theta}_n(P)$	(6) $\hat{\hat{\theta}}_n(P)$
1958	8.3321	11.0619	6.1140	8.6426	8.7428
1959	1.1431	3.9925	7.0121	6.1833	4.7771
1960	12.4357	6.9193	6.2156	10.0465	9.3028
1961	6.3318	6.4743	6.2263	7.9583	6.5375
1962	12.3693	7.7613	6.2104	10.0237	8.3586
1963	7.4014	12.5712	6.1531	8.3242	10.9168
1964	9.8798	7.8600	6.0999	9.1721	9.5363
1965	12.7789	8.9054	6.2443	10.1639	9.2030
1966	19.8332	13.0822	7.5264	12.5771	12.7137
1967	10.2286	9.0331	6.1055	9.2914	9.6950
1968	3.8462	9.0470	6.5137	7.1080	8.4795
1969	14.8464	8.2463	6.4834	10.8712	10.0070
1970	14.6725	13.0868	6.4590	10.8117	11.3689
1971	11.9692	13.0103	6.1816	9.8869	9.7725
1972	12.3579	9.9792	6.2095	10.0198	11.2956
1973	4.2776	8.8084	6.4521	7.2556	6.3445
1974	8.8351	7.9099	6.1025	8.8147	8.3431
1975	8.3193	7.9809	6.1144	8.6382	9.0090
1976	4.0097	7.6175	6.4897	7.1639	8.3899
1977	1.1215	4.0781	7.0168	6.1759	5.2476
1978	12.0574	9.0618	6.1876	9.9170	8.5382
1979	10.8530	12.0648	6.1235	9.5050	8.7438

The Mean Square Error of this estimator is 2.61 for all years and 3.26 for the test period. This latter value represents a 50% decrease in MSE compared to that of the ordinary mean. Although the estimator based on Pintail densities is not directly comparable to the others, because data from all years were used to develop each year's predictor, the potential worth of the estimator is nonetheless evident. Other species in addition to the Pintail could be used in an empirical Bayes manner, but I suspect a direct multivariate approach might prove more productive.

In a multivariate empirical Bayes approach the six individual species could be considered together as a 6-variate vector. Interest lies in estimating the entire vector, and the methods outlined in Efron and Morris (1972) can be used to develop empirical Bayes estimators that are better than the classical ones. Efron and Morris (1972:341) suggested that the multivariate approach will be preferable to a component-by-component univariate procedure if the variables are relatively highly correlated. This condition seems to be readily satisfied with the waterfowl density values.

DISCUSSION

This report has addressed the problem of improving the accuracy of waterfowl population estimates without additional sampling effort and the associated costs. The technique has been to invoke auxiliary information to develop a prior estimate of Mallard density. This prior value is combined with the estimate obtained by sampling to form an empirical Bayes estimate.

For the example considered here, an ordinary EB estimator, which uses the mean of earlier years as a prior estimate, was found to reduce the MSE by 31 percent for the 1968–79 test period. The implication is that the accuracy of the estimator of Mallard density in eastern North Dakota could be substantially improved simply through the use of EB estimation. Alternatively, the current precision could be maintained, but costs reduced, by sampling fewer transects and employing EB procedures.

We also considered an EB estimator based upon the relationship of Mallard density to an index of wetland conditions. This estimator proved, in the example, to be better (23% in MSE) than the classical one, but, perhaps surprisingly, it was not quite as accurate as the previous EB estimator.

The third estimator examined was based on the density of Pintails in each year. The predictive equation was derived from the entire 22-year sample, unlike the other estimators which used formulas incorporating only data from prior years. Thus the 50% reduction in MSE is not exactly comparable to the improvements obtained by the other estimators, but it illustrates the potential of the method.

The theory of empirical Bayes methods has existed for a quarter of a century. Despite a fairly well developed theory, relatively few practical applications have been made thus far, but this situation seems to be changing. I anticipate that EB procedures will have widespread uses in many fields before long.

Empirical Bayes procedures seem particularly promising for surveys of bird populations. Many surveys are conducted regularly, usually annually, accuracy is highly desired, and the sample data are often expensive or difficult to obtain. More research must be done to determine those problems the procedures can most profitably address. I suggest that EB estimators will be of greatest value in regular surveys of less common species, those that are the most difficult to measure, or those whose density can be best predicted from other available information.

ACKNOWLEDGMENTS

I appreciate the efforts of Richard S. Pospahala and Edgar L. Ferguson, who provided the data used to exemplify the methods, and of Stella G. Machado and Fred L. Ramsey, who commented on earlier drafts of the paper.

Studies in Avian Biology No. 6:441–444, 1981.

DENSITY ESTIMATION USING LINE TRANSECT SAMPLING

P. V. RAO,[1] K. M. PORTIER,[1] AND J. A. ONDRASIK[2]

ABSTRACT.—Line transect methods of estimating population density usually assume a fixed transect length. As a means of reducing the distance travelled by the observer, Rao and Ondrasik (1980) proposed a method based on a line transect of random length. In this method a length, L_0, and number, N_0, are fixed in advance and sampling is terminated as soon as either a distance L_0 is traversed or a number, N_0, of objects is sighted. A brief summary of the method including the density estimate and its variance is presented in the first part of the paper.

In the second part, a method of estimation of density for clustered populations is discussed. This method assumes (1) that the probability of sighting a cluster is a function of its size as well as its perpendicular distance from the transect line and (2) that not all objects in a cluster may be sighted by the observer. The estimate of the population density as well as estimates of other model parameters are obtained using maximum likelihood. The method is illustrated using artificially constructed data for a clustered population.

The use of line transect methods in estimating animal and plant population densities has received considerable attention in recent literature. Excellent reviews of the general subject area are found in Seber (1973), Eberhardt (1968) and Burnham et al. (1980).

With the few exceptions noted by Burnham et al. (1980 Appendix D), currently available density estimates from line transect methods use transects of fixed length and assume that sightings of objects are independent events. An obvious drawback of a sampling method based on a predetermined transect length is the possibility that it may be using an unnecessarily long transect to estimate density. Because cost of sampling is likely to increase with the length of the transect, it is desirable to consider estimation procedures based on random transect lengths, i.e., procedures which terminate as soon as a predetermined number of objects is sighted. Another drawback of most of the available line transect methods is due to the fact that many biological populations (e.g., coveys of quail, schools of porpoise and so on) aggregate into tight clusters. The assumption of independence is not reasonable for such populations, so density estimation procedures must account for the facts (1) that objects are sighted in groups and (2) that all objects in a group may not be seen by the observer.

The purpose of this paper is (1) to describe a recently developed method (Rao and Ondrasik 1980) that allows for the termination of sampling after a prespecified number of observations is made, and (2) to propose a model suitable for line transect sampling of clustered populations.

[1] Department of Statistics, University of Florida, Gainesville, Florida 32611.

[2] Boehringer Ingelheim Ltd., P. O. Box 368, Ridgefield, Connecticut 06877.

SAMPLING WITH RANDOM TRANSECT LENGTHS

A sampling plan which utilizes a predetermined length for the line transect will be referred to as a direct sampling plan. Many direct sampling density estimates use only the right angle distances of the objects and are based on the set of assumptions listed below (Seber 1973).

- A1. Objects are randomly and independently distributed over the area of interest at a rate (density) D objects per unit area.
- A2. Sightings of objects are independent events.
- A3. Objects are fixed, i.e., objects are immotile and no object is counted twice.
- A4. There exists a function $g(y)$ which is the probability of observing an object conditional on the existence of an object at right angle distance y from the transect.
- A5. $g(0) = 1$; i.e., objects on the transect line are observed with probability one.

INVERSE SAMPLING

In contrast to the direct sampling plan, one can consider an inverse sampling plan. In an inverse sampling plan, observation is terminated as soon as a prespecified number, N_0, of objects is sighted by the observer.

Rao and Ondrasik (1980) developed an estimation procedure suitable for the inverse sampling plan. Following Burnham and Anderson (1976), they assume that the conditional probability density, $f(y)$, of the perpendicular distance y is unknown. Utilizing assumptions A1 to A5 they estimate the population density to be

$$\hat{D}_1 = \frac{(N_0 - 1)\hat{f}(0)}{2l} \tag{1}$$

where l is the actual distance traversed and $\hat{f}(0)$ is an estimate of $f(0)$. Assuming the bias in $\hat{f}(0)$

to be relatively small, an approximation to the variance of $\hat{D}_1$ is given as

$$V(\hat{D}_1) = \frac{D^2}{N_0 - 1}[1 + C\cdot V(\hat{f}(0))], \quad (2)$$

where $C\cdot V(\hat{f}(0))$ is the coefficient of variation of $\hat{f}(0)$. While any reasonable estimate of $f(0)$ can be used in (1), the Fourier series estimator suggested by Crain et al. (1978) appears to be the most desirable. The monograph by Burnham et al. (1980) contains many examples of calculation of the Fourier series estimate for $f(0)$.

Combined Sampling

A disadvantage of the inverse sampling plan is the possibility that sampling may not terminate in a reasonable period of time. To overcome this drawback, Rao and Ondrasik (1980) developed the combined sampling plan in which sampling stops as soon as either a prescribed number, N_0, of objects is sighted or a prespecified length, L_0, of the transects is traversed. If n and l denote, respectively, the actual number of objects sighted and the actual distance traversed, then the combined sampling estimate of D has the form

$$\hat{D}_2 = \begin{cases} 0 & 0 \leq n \leq 1 \\ \dfrac{n\hat{f}(0)}{2L_0} & 1 < n < N_0 \\ \dfrac{(N_0 - 1)\hat{f}(0)}{2l} & n = N_0 \end{cases} \quad (3)$$

An approximation to the variance of $\hat{D}_2$ is

$$V(\hat{D}_2) = \frac{D^2}{N_0 - 2}\Bigg\{1 + C\cdot V(\hat{f}(0)) - e^{-n}C\cdot V(\hat{f}(0)) \left[\sum_{j=0}^{N_0-3}\left(\frac{j(N_0 - 2) - (j + 1)^2}{(j + 1)^2 j!}\right)n^j - (N_0 - 2)(2 - e^{-n})\right]\Bigg\} \quad (4)$$

SAMPLING CLUSTERED POPULATIONS

Anderson et al. (1976) note that density estimates for clustered populations can easily be obtained if assumptions A1 to A5 hold for clusters of objects rather than for individuals. It is clear in this case that existing methods of density estimation are directly applicable to the estimation of cluster density. If the number of objects in every sighted cluster can be determined without error, then an estimate for the density of objects is

$$\hat{\Delta} = \hat{D}\bar{s} \quad (5)$$

where $\hat{D}$ is an estimate of the cluster density and $\bar{s}$ is the average size of the observed clusters.

There are two reasons why the assumptions implied by the procedure suggested in the preceeding paragraph may not be reasonable when developing sampling models for clustered populations. First, the simple modification obtained by replacing the word "object" by the word "cluster" in A1 to A5 would imply that the probability of sighting a cluster depends only on its right angle distance. This may not be reasonable because the probability of sighting a larger cluster is likely to be greater than the probability of sighting a smaller cluster located at the same distance. Second, the sighting of a cluster may not necessarily mean that all of the objects comprising the cluster are seen and counted by the observer. A more reasonable assumption would be to let the probability of sighting an object belonging to a cluster depend on the distance to the cluster as well as the true cluster size.

Burnham et al. (1980, Appendix D) suggest a set of assumptions which imply that the *probability of sighting* a cluster depends on its size and distance. The set of assumptions listed below implicitly contains the assumption of Burnham et al. (1980) but also implies that the *number of objects seen* in a cluster depends on its (cluster) size and distance.

Assumptions

- B1. The clusters are randomly and independently distributed over the area of interest at a rate (density) of D clusters per unit area.
- B2. Sightings of clusters are independent events.
- B3. Clusters are fixed, *i.e.*, clusters are immotile and no cluster is counted twice.
- B4. The probability that a randomly chosen cluster is of size r is $p(r)$, $r = 1, 2, \ldots$
- B5. There exists a non increasing function $h(y)$, with $h(0) = 1$ and $0 \leq h(y) \leq 1$, such that the probability of sighting s objects in a cluster, conditional on a cluster of size $r \geq s$ being located at right angle distance y is

$$p(s \mid r, y) = \binom{r}{s}[h(y)]^s[1 - h(y)]^{r-s}$$
$$s = 0, 1, \ldots, r \quad (6)$$

An inspection of the assumptions B1 to B5 shows that B1, B2, and B3 are directly obtained from A1, A2, and A3. From assumption B5, the probability of sighting a cluster conditional on a cluster of size r being located at distance y is seen to be

$$1 - p(0 \mid r,y) = 1 - [1 - h(y)]^r, \quad (7)$$

which clearly depends on y and r. In particular,

TABLE 1
CONSTRUCTED DATA FOR INVERSE SAMPLING OF CLUSTERED POPULATION ($N_0 = 25$, $l = 25$ KM)

Perpendicular distance (y_i) (meters)	Observed cluster size (s_i)
1	1
3	2
7	1
10	1
2	3
5	5
4	1
7	2
15	1
22	1
6	1
3	6
2	1
12	1
28	3
9	2
18	1
36	7
17	6
5	1
4	1
3	1
8	2
3	4
13	1

the probability of sighting a cluster of size r at $y = 0$ is $1 - [1 - h(0)]^r = 1$.

The form of the cluster detection function is easily derived. Let $g(y)$ denote the probability of sighting a cluster conditional on the right angle distance y. Then

$$g(y) = \sum_{r=1}^{\infty} (1 - [1 - h(y)]^r)p(r)$$
$$= 1 - \sum_{r=1}^{\infty} [1 - h(y)]^r p(r) \quad (8)$$

If every cluster in the population has size 1, then the probability distribution of cluster size satisfies

$$p(1) = 1,$$

and

$$g(y) = 1 - [1 - h(y)]\, p(1) = h(y)$$

Thus $h(y)$ may be regarded as the probability of detecting a single object at distance y.

Under assumption B1 (see Seber 1973), the expected number of clusters seen in a transect of length l is θl, where,

$$\theta = 2cD \quad (9)$$

TABLE 2
MAXIMUM LIKELIHOOD ESTIMATES OF MODEL PARAMETERS

Parameter	Estimate	Standard error
Δ[a]	58.6 per km^2	21.0 per km^2
θ	1.00 per km	.04 per km
α	.709	.071
λ	59.3 per km	11.5 per km
D[a]	17.04 per km^2	4.32 per km^2
c[b]	.029	.005

[a] Estimate calculated using $\hat{\alpha}$, $\hat{\gamma}$ and $\hat{\theta}$.
[b] Estimate calculated using $\hat{\alpha}$ and $\hat{\gamma}$.

is the expected number of cluster sightings per unit length of the transect and

$$c = \int_0^{\infty} g(y)\, dy. \quad (10)$$

ESTIMATION OF DENSITY

Maximum likelihood estimation of the density of objects, Δ, is possible when $p(r)$ and $h(y)$ are completely specified. Clearly, the appropriate form of the likelihood function will depend upon the sampling plan. For example, suppose the sampling plan calls for sampling until N_0 clusters are sighted. If $(s_1, y_1), (s_2, y_2), \ldots, (s_{N_0}, y_{N_0})$ denote the sizes and right angle distances and l denotes the actual length of the transect traversed, then the likelihood function of the sample can be shown to have the form

$$L = \left(\frac{\theta^{N_0} l^{N_0 - 1} \exp(-\theta l)}{c^{N_0}(N_0 - 1)!}\right) \prod_{i=1}^{N_0} p(s_i \mid y_i) g(y_i), \quad (11)$$

where $p(s \mid y)$ is the conditional probability of sighting s objects at distance y:

$$p(s \mid y) = \sum_{r=1}^{\infty} p(s \mid r, y)p(r)$$
$$= \sum_{r=s}^{\infty} \binom{r}{s} [h(y)]^s [1 - h(y)]^{r-s} p(r) \quad (12)$$

Note that, in addition to θ and c, the likelihood function will contain parameters appearing in the specification of $p(r)$ and $h(y)$. The joint likelihood will have to be maximized using an appropriate iterative procedure.

EXAMPLE

Since real data to which the likelihood given by Eq. (11) is appropriate are not readily available in the literature, an artificially constructed data set will be used to illustrate the maximum likelihood estimation procedure. Suppose that

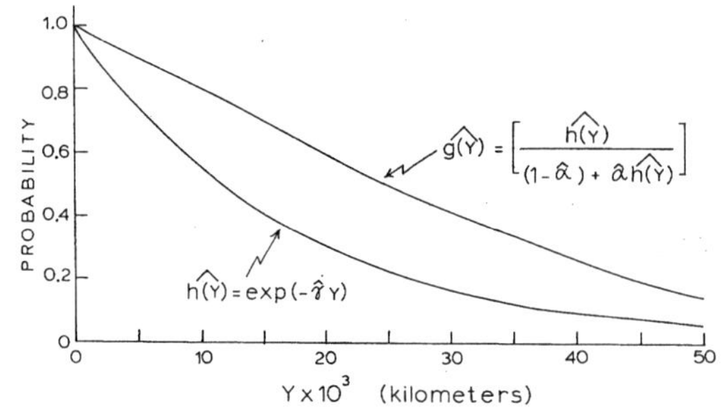

FIGURE 1. Estimated detection function $g(\hat{y})$ and function $h(\hat{y})$.

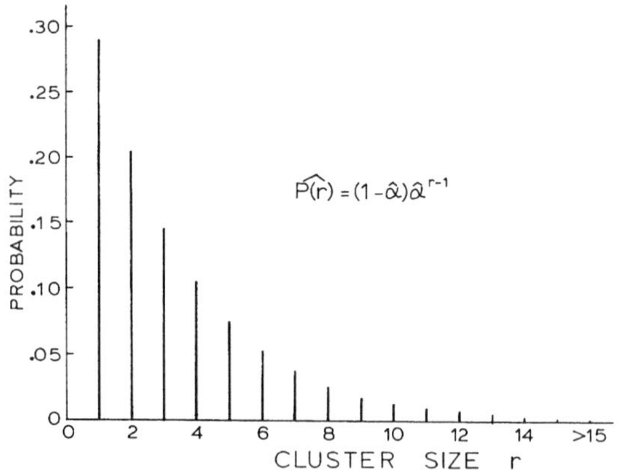

FIGURE 2. Estimated cluster size distribution.

a hypothetical inverse sampling plan designed to observe $N_0 = 25$ clusters resulted in the sighting distances y (in meters) and observed cluster sizes s presented in Table 1. Assume that a distance of $l = 25$ km was required to sight 25 clusters.

Assuming the geometric distribution

$$p(r) = (1 - \alpha)\alpha^{r-1} \quad r = 1, 2, \ldots \tag{13}$$

for cluster size and the exponential form

$$h(y) = \exp(-\gamma y) \quad y \geq 0, \gamma > 0 \tag{14}$$

for $h(y)$ in Eq. (8), it is easily seen that the cluster detection function has the form

$$g(y) = \exp(-\gamma y)/[(1 - \alpha) + (\alpha\exp(-\gamma y](15)$$

Similar calculations using (13) and (14) in (12) shows that

$$p(s|y) = (1 - \alpha)\alpha^{s-1}[g(y)]^s/ [(1 - \alpha) + \alpha\exp(-\gamma y)]. \tag{16}$$

Substituting Eq. (15) and Eq. (16) into Eq. (11) yields the likelihood function in terms of the parameters c, θ, α and γ. However, these parameters are not independent because substitution for $g(y)$ in Eq. (10) from Eq. (15) gives

$$c = -(\alpha\gamma)^{-1} \ln(1 - \alpha) \tag{17}$$

Therefore, the likelihood function, Eq. (8), must be maximized with respect to θ, α and γ. The estimate of the cluster density is (see Eq. (9)).

$$\hat{D} = \frac{\hat{\theta}}{2\hat{c}} \tag{18}$$

Finally, the estimate of Δ is obtained by noting the relationship

$$\Delta = DE(S),$$

where $E(S)$ is the expected cluster size. For the geometric distribution specified in Eq. (13) the expected cluster size is $(1 - \alpha)^{-1}$. Therefore the maximum likelihood estimate of Δ is

$$\hat{\Delta} = (1 - \hat{\alpha})^{-1}\hat{D}$$

where $\hat{D}$ is as in Eq. (18).

The maximization of the likelihood may be carried out using the FORTRAN based MAXLIK program (Kaplan and Elston 1978) designed to numerically find maximum likelihood estimates and their standard errors. Table 2 gives the estimates and their standard errors, based on data in Table 1.

The forms of $h(y)$ and the detection function $g(y)$, inserting the maximum likelihood estimates, $\hat{\gamma}$ and $\hat{\alpha}$, are given in Figure 1. As expected, the probability of sighting a cluster ($g(y)$) is greater than the probability of sighting an individual ($h(y)$) for all distances y.

Given, $\hat{\alpha}$, the estimated distribution of true cluster sizes is given in Figure 2. From this it is clearly seen that more than half of the clusters should have less than four individuals in them.

DISCUSSION

In conclusion it must be noted that the combined sampling method and the cluster sampling model presented in this paper are in a preliminary stage of their development. Many problems of practical importance have yet to be solved. For example, guidelines for the specification of L_0 and N_0 in a combined sampling plan need to be carefully formulated. Sensitivity of the cluster sampling model to errors in the specification of $p(r)$ and $h(y)$ must be investigated, and the possibility of developing a robust density estimator should be looked into. We are currently exploring solutions to some of these problems.

Studies in Avian Biology No. 6:445–449, 1981.

AVIAN CENSUSING WITH THE STRIP METHOD: A COMPUTER SIMULATION

RONALD W. ENGEL-WILSON, A. KURT WEBB, KENNETH V. ROSENBERG, ROBERT D. OHMART, AND BERTIN W. ANDERSON[1]

ABSTRACT.—Variable-width line transect censuses were simulated by computer to identify variation due to random fluctuation of density estimates inherent in the technique. Effects of transect length (457–1828 m), number of censuses (up to 15), and density value (4/40 ha to 201/40 ha) were tested. For moderately abundant species (35 to 37/40 ha), between six and nine censuses of transects, at least 914 m in length were sufficient to obtain accurate and consistent estimates. At least 15 censuses were needed for shorter transects. Estimates of rare and abundant species were equally close to known densities after 15 censuses, but estimates of rare species were too variable to be statistically reliable. Variation in estimates obtained by simulation are probably greater than those from real censuses.

Since J. T. Emlen (1971) popularized the variable-width line transect method for estimating bird densities in large tracts of habitat, thousands of transects have been walked, and millions of bird detections have been recorded. The reliability of this technique has been continually questioned, although some practical aspects of reliability have been addressed (Anderson and Ohmart 1977). Unfortunately, many questions cannot be answered using real data. In this paper, we attempt to answer some of these questions by simulating a simple habitat within which the number of species, their densities, and their degrees of detectability are known.

Specific questions that we address are: (1) How many censuses are needed to accurately estimate the true density in an area? (2) How long should a transect be? (3) Does the density of a species affect its density estimates?

The purpose of this paper is to explain that portion of the variance inherent in the line transect technique, assuming a random distribution of species in a habitat. Our feeling is that if one could find a uniform habitat with a known density of birds and a known number of species, find a perfect censuser, and keep weather conditions constant, one could calculate the minimum number of censuses of a known transect length needed to obtain a required degree of statistical confidence in the data. We have created such a situation using the computer simulation approach.

METHODS

Habitats of different sizes and different species compositions were generated (Fig. 1) using computer program RANGRID (available from the authors upon request). Each simulated habitat was 251 m wide with the center line as a transect. This gave census areas of 11.4 ha with a 457 m transect, 22.7 ha with a 914 m transect, and 45.4 ha with a 1829 m transect. Fifteen species, each with an equal density and represented by a letter, A to O, were randomly distributed. For this purpose, each habitat was treated as a grid 100 units wide by 300 units long, such that each individual bird occupied a single square equal to 0.33×10^{-5} times the area of each grid. No two birds occupied the same square. For each species, there was an equal probability of any individual occurring in any strip lateral to the transect line.

Each transect was censused fifteen times. For each census individuals were uniquely randomly distributed, and all species were equally detectable. Transects were censused by tallying all individuals of each species occurring within 15.2 m of the transect (A1), one-half of the detections 15.2 to 30.5 m (A2), one-fourth of the detections 30.5 to 61 m (A3), and one-eighth of the detections 61 to 126 m (A4) (Fig. 1). We next calculated the density of detections (birds per m^2) in each of the lateral strips out to 15.2, 30.5, 61, or 126 m. The area containing the highest density (transect length × lateral distance) was then extrapolated to the number per 40 ha.

Since each species had the same known density, we performed one-way analysis of variance on density estimates to find the minimum number of censuses needed to obtain a reliable estimate of that density. A nonsignificant F-statistic would indicate that the censuses were indeed estimating the same density. The only variable causing a difference in estimates was random variation in distributions. Using a χ^2 test, the mean density estimates of the 15 species were compared with their expected (known) densities for each of the three transect lengths and census replicates.

In a second set of simulations, sensitivity of the line transect method to different densities was tested using known densities ranging from 4/40 ha to 201/40 ha. Estimates using transect lengths of 457.2 m, 914.4 m, and 1828 m were also compared. Coefficients of variation were calculated for each estimate based on 15 censuses.

RESULTS

Means of density estimates based on sets of 3, 6, 9, and 15 censuses were calculated for the three transect lengths, as well as ranges of es-

[1] Department of Zoology and Center for Environmental Studies, Arizona State University, Tempe, Arizona 85281.

TABLE 1

DENSITY ESTIMATES (NO. PER 40 HA) FOR SPECIES OF KNOWN DENSITY AFTER 3, 6, 9, AND 15 CENSUSES OF A SIMULATED 457.2-M TRANSECT

Species	Known dens.	3 censuses[a]	6 censuses[a]	9 censuses[a]	15 censuses[a]
A	35	33 (59) 11–58	29 (25) 11–58	32 (12) 11–58	40 (13) 11–87
B	35	18 (25) 9–29	25 (19) 9–58	22 (15) 7–58	35 (21) 7–145
C	35	39 (41) 29–58	40 (29) 7–87	46 (22) 7–87	42 (14) 5–87
D	35	10 (13) 6–16	32 (44) 6–116	34 (31) 5–116	44 (22) 5–116
E	35	39 (41) 29–58	30 (17) 5–58	27 (11) 5–58	29 (10) 5–58
F	35	32 (61) 9–58	29 (25) 9–58	39 (28) 5–116	37 (26) 5–116
G	35	24 (20) 15–29	21 (9) 9–29	28 (14) 9–58	32 (13) 7–87
H	35	48 (42) 29–58	36 (19) 11–58	39 (26) 9–116	34 (16) 7–116
I	35	39 (41) 29–58	27 (19) 7–58	40 (25) 7–116	46 (19) 7–116
J	35	21 (33) 6–29	19 (12) 5–29	32 (20) 5–87	31 (12) 5–87
K	35	24 (38) 7–36	27 (10) 7–36	31 (18) 5–87	33 (13) 5–87
L	35	16 (28) 9–29	11 (9) 6–29	27 (19) 6–58	32 (13) 6–58
M	35	46 (52) 22–58	42 (18) 22–58	50 (20) 18–87	49 (16) 7–87
N	35	29 (0) 29–29	20 (11) 7–29	36 (20) 7–87	38 (13) 7–87
O	35	34 (54) 15–58	56 (41) 15–116	50 (28) 15–116	47 (18) 15–116

[a] For each number of censuses the columns are as follows, left to right: mean estimate, 95% confidence interval in parentheses, and range.

timates from individual censuses (Tables 1, 2, 3). In each case, lengthening the transect and/or increasing the number of censuses reduced the variation among estimates. For example, for a 914.4 m transect (Table 2) the estimate based on sets of three censuses ranged from 18 to 68/40 ha; after six censuses the range was 18 to 56/40 ha; after nine censuses the range was 28 to 50/40 ha; and after 15 censuses it was 33 to 44/40 ha. Transects 457.2 m long were more variable (Table 1) and 1828.8 m transects were less variable (Table 3).

Although mean estimates from multiple censuses came close to the known density (35 or 37/40 ha), estimates from single censuses varied greatly. Even for the 1828.8 m transect, single estimates ranged from 9 to 94/40 ha.

Table 4 lists the results of the χ^2 test and 1-way analysis of variance (ANOVA) for 15 equally abundant species after 3, 6, 9, 12, and 15 censuses. For transects of 914.4 m or longer, between six and nine censuses were sufficient to give estimates of the densities of the species that did not differ significantly from their true

TABLE 2

DENSITY ESTIMATES (NO. PER 40 HA) FOR SPECIES OF KNOWN DENSITY AFTER 3, 6, 9, AND 15 CENSUSES OF A SIMULATED 914.4-M TRANSECT

Species	Known dens.	3 censuses[a]	6 censuses[a]	9 censuses[a]	15 censuses[a]
A	37	21 (18) 15–29	24 (12) 15–44	28 (8) 15–44	38 (11) 15–87
B	37	20 (20) 15–29	34 (27) 15–73	36 (19) 15–73	40 (14) 15–87
C	37	34 (22) 29–44	26 (13) 8–44	32 (14) 8–58	35 (10) 8–73
D	37	53 (20) 44–58	38 (20) 8–58	39 (16) 8–73	36 (10) 8–73
E	37	18 (27) 7–29	18 (10) 7–29	31 (20) 7–87	36 (14) 7–87
F	37	20 (28) 12–33	19 (7) 12–33	30 (14) 12–58	36 (11) 12–73
G	37	42 (41) 25–58	46 (12) 25–58	48 (9) 25–58	43 (9) 22–73
H	37	50 (51) 33–73	42 (17) 29–73	36 (13) 15–73	40 (10) 15–73
I	37	54 (86) 18–87	48 (28) 18–87	46 (19) 18–87	41 (13) 6–87
J	37	68 (75) 44–102	56 (26) 29–102	50 (20) 15–102	42 (13) 15–102
K	37	57 (69) 25–73	53 (20) 25–73	48 (17) 15–73	44 (12) 15–73
L	37	34 (22) 29–44	29 (10) 15–44	39 (14) 15–73	37 (10) 15–73
M	37	20 (20) 15–29	25 (12) 15–44	28 (11) 15–58	33 (13) 15–102
N	37	39 (22) 29–44	41 (12) 25–58	36 (9) 18–58	36 (9) 18–73
O	37	29 (58) 29–58	41 (15) 29–58	39 (12) 15–58	39 (8) 15–58

[a] For each number of censuses, the columns are as follows, left to right: mean estimate, 95% confidence interval in parentheses, and range.

TABLE 3

DENSITY ESTIMATES (NO. PER 40 HA) FOR SPECIES OF KNOWN DENSITY AFTER 3, 6, 9, AND 15 CENSUSES OF A SIMULATED 1828.8-M TRANSECT

Species	Known dens.	3 censuses[a]	6 censuses[a]	9 censuses[a]	15 censuses[a]
A	37	20 (9) 15–22	27 (9) 15–36	30 (8) 15–51	38 (2) 15–73
B	37	44 (18) 36–51	31 (17) 10–51	35 (14) 10–65	35 (8) 10–65
C	37	18 (24) 10–29	17 (8) 9–29	30 (17) 9–73	35 (11) 9–73
D	37	44 (36) 29–58	43 (10) 29–58	41 (8) 29–58	41 (6) 22–65
E	37	61 (25) 15–73	50 (16) 31–73	46 (12) 20–73	40 (8) 20–73
F	37	44 (47) 22–58	40 (14) 22–58	43 (12) 22–73	40 (8) 22–73
G	37	27 (20) 22–36	31 (9) 22–44	30 (6) 22–44	34 (9) 18–87
H	37	35 (25) 24–44	34 (12) 22–51	35 (9) 22–51	33 (7) 13–51
I	37	29 (32) 22–44	26 (13) 11–44	37 (16) 11–73	34 (10) 11–73
J	37	27 (20) 22–36	28 (11) 13–44	31 (8) 13–44	33 (6) 13–58
K	37	27 (10) 22–29	32 (11) 22–51	38 (11) 22–65	37 (9) 18–65
L	37	34 (37) 25–51	34 (14) 22–51	34 (12) 15–58	36 (7) 15–58
M	37	68 (57) 51–94	52 (24) 18–94	50 (17) 18–94	42 (11) 11–94
N	37	29 (17) 22–36	34 (14) 22–58	32 (9) 16–58	35 (8) 16–73
O	37	32 (24) 22–44	36 (11) 22–51	35 (8) 22–51	35 (6) 20–51

[a] For each number of censuses, the columns are as follows, left to right: mean estimate, 95% confidence interval in parentheses, range.

densities ($P \geq .25$). For the 457.2 m transect, 15 censuses were barely sufficient ($P \geq .1$).

Results of the ANOVA were more complex. For transects of 914.4 m or longer, between six and nine censuses were also sufficient to reliably estimate a known density. With six or fewer censuses, a significant difference existed between censuses (F prob. $< .05$). For 457.2 m transects, lack of differences between estimates was a result of large variation within census replicates and not closeness of the estimates. Short transects, therefore, cannot be considered reliable or accurate.

Results of censuses using 14 known densities are shown for the three transect lengths in Table 5. After 15 censuses, mean estimates of both

TABLE 4

ACCURACY AND RELIABILITY OF COMPUTER-SIMULATED LINE TRANSECT ESTIMATES[a]

	Number of censuses	Known density	Mean estimate[b] χ^2	Mean estimate[b] P^c	Mean estimate[b] Min.–max.	Single reading estimate Max.–min.	ANOVA[d] F ratio	ANOVA[d] P of F
Transect length = 457.2 m;	3	35	60.860	.001	10–48	6–58	1.434	0.197
area sampled per transect = 11.3 ha	6	35	59.772	.001	11–56	5–116	1.499	0.132
	9	35	29.215	.005	22–50	5–116	0.835	0.630
	12	35	28.152	.01	24–50	5–116	1.021	0.435
	15	35	19.992	.1	29–49	5–145	0.765	0.707
Transect length = 914.4 m;	3	37	100.263	.001	18–68	7–102	2.621	0.013
area sampled per transect = 22.7 ha	6	37	56.739	.001	18–56	7–102	3.298	0.000
	9	37	17.879	.25	28–50	7–102	1.326	0.202
	12	37	8.949	.75	31–47	6–102	0.640	0.829
	15	37	5.066	.975	33–44	6–102	0.401	0.973
Transect length = 1828.8 m;	3	37	74.889	.001	18–68	10–94	3.993	0.001
area sampled per transect = 45.5 ha	6	37	34.390	.001	17–52	9–94	2.871	0.002
	9	37	14.781	.25	30–50	9–94	1.553	0.102
	12	37	4.695	.99	33–45	9–94	0.613	0.852
	15	37	3.477	.995	33–42	9–94	0.565	0.891

[a] All density estimates in number per 40 ha.

[b] Means are for the number of censuses listed; the minimum-maximum are from 15 species with equal densities.

[c] Probability of mean estimate; 14 degrees of freedom.

[d] Treatments are the density estimates for each of 15 species.

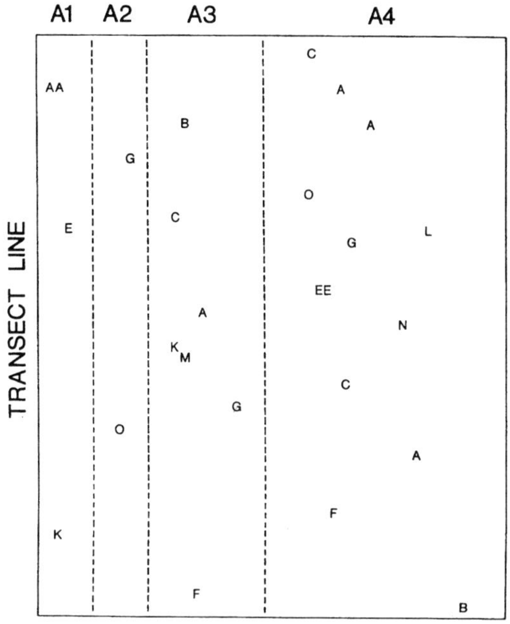

FIGURE 1. Portion of a simulated habitat 251 m wide. Each letter represents an individual bird of species A to O, which occupies an area equal to 0.33×10^{-5} of the total habitat. Letters are randomly distributed among the lateral strips, A1 to A4.

common and rare species were reasonably close to the true values. However, as either transect length or known density decreased, the coefficient of variation of mean estimates increased greatly.

DISCUSSION

When outside factors are controlled, the random variation inherent in the variable-width line transect technique ultimately limits the strength of such census data. To compare avian densities in different areas or over distinct time intervals, it is essential that differences due solely to such variation are eliminated. We have shown that line transect censusing can both accurately and consistently estimate a known density given the proper sample design. However, differences between, or changes in, the densities of very rare species become increasingly difficult to detect because of high coefficients of variation associated with those estimates.

We feel that the required sampling effort suggested by this model is conservative for several reasons. First, the computer simulation represented the maximum variation due to redistribution of birds between censuses. It assumed that the distribution of birds in a habitat during one census was independent of their distribution during any other census. In practice, this is probably not often the case. For moderately sedentary or territorial species, and especially during the breeding season, individual birds would not move great distances relative to an established transect line during a short time period. Therefore, consecutive density estimates during that period will often be less variable than those presented here.

A second assumption in our simulation was that all detections were counted within 15.2 m of a transect and only half were counted in the next 15.2 m strip. If, as real census data suggest, birds often are equally detectable in both the first and second 15.2 m strips, then the variability among consecutive estimates will again be reduced.

These hypotheses can be tested with further simulations and by examining large census data

TABLE 5
KNOWN DENSITY AND ESTIMATES (15 CENSUSES) AND COEFFICIENTS OF VARIATION OF MEANS

Known density/40 ha.			Est. density/40 ha.			C. V. of mean		
1828.8 m	914.4 m	457.2 m	1828.8 m (±95 CI)	914.4 m (±95 CI)	457.2 m (±95 CI)	1828.8 m	914.4 m	457.2 m
192	201	201	193 (22)	217 (26)	187 (53)	20.6	21.8	51.3
97	100	99	93 (16)	100 (21)	92 (34)	31.6	39.2	66.2
70	70	70	65 (8)	60 (15)	75 (22)	22.7	46.1	52.8
62	60	60	59 (14)	55 (20)	52 (13)	42.0	65.4	47.7
47	49	49	49 (9)	47 (15)	43 (19)	34.0	58.2	80.3
38	41	39	45 (8)	46 (13)	42 (12)	31.7	53.5	54.0
32	30	28	35 (7)	34 (10)	23 (9)	37.2	54.8	70.6
23	25	25	21 (5)	26 (10)	29 (15)	47.5	68.0	93.2
19	21	21	17 (4)	24 (7)	15 (9)	45.3	50.3	105.0
16	18	18	18 (5)	24 (8)	28 (13)	46.4	61.5	84.4
14	14	14	15 (6)	14 (6)	20 (11)	73.8	69.4	95.6
11	11	11	12 (6)	10 (4)	9 (6)	92.3	78.0	116.0
7	7	7	7 (3)	6 (4)	7 (6)	88.2	101.4	161.0
4	4	4	4 (2)	2 (2)	5 (5)	119.1	152.4	179.7

sets. Both will be the subjects of forthcoming papers, and the results presented here should be regarded as preliminary.

CONCLUSIONS

(1) Transects of 914 m or longer estimated known densities with greater accuracy and consistency than did 457 m transects.

(2) Between six and nine censuses of 914 m transects were sufficient to minimize random fluctuations in density estimates.

(3) Estimates of rare and abundant species were equally close to known values. However, high coefficients of variation for rare species' estimates reduced the statistical confidence in those estimates.

Studies in Avian Biology No. 6:450–451, 1981.

SUMMARIZING REMARKS: SAMPLING DESIGN

JAKE RICE[1]

Several major points addressed by the papers in this session on experimental design have already been discussed by previous summarizers. The fact that previous speakers make redundant parts of my summary of this session only serves to emphasize that these design considerations are essential points, demanding the attention of the biometrical and the ecological communities. Rather than repeat many of those points, I will reiterate just a few general comments here and then expand on some of the specific points raised by speakers in this session.

The frequent references to the basic points of experimental design emphasize both how fundamental and how straightforward these basics are. At their simplest, the purposes of experimental design are to maximize accuracy and precision through the minimization of bias and within-group variation. As numerous papers at this conference have amply documented, much work remains for biometricians interested in methods of counting bird populations, for there are many sources of bias and variance. Rather than list them yet again, I simply refer readers to Verner's introductory comments to this session, and the summary comments of the sessions on "Estimating Birds per Unit Area," "Comparison of Methods," and "Observer Variability."

Next I wish to focus on an additional caveat in design considerations: The variance and bias minimization must be achieved under constraints of fixed, and usually limiting funds and resources. Gates presented a brief introduction to optimization of resource use in allocating available time and personnel to counting efforts. Surprisingly little information is required to use those optimality calculations: (1) we must know the cost of laying out transects, mapping grids or stations; (2) we must know the cost of sampling the transect, grid or point (in practice, both of the above pieces of information, or at least rough estimates of them, are readily available); and (3) finally, we need to specify the desired power of our study. Although "power" is a statistical term, the basic point here is a biological one: how small a difference does one want to be able to detect, either for the management implications of the differences, or for the importance of the difference in evaluating theoretical predictions. It may be difficult for a researcher to specify precisely the magnitude of such differences, but usually all that is needed is a general estimate, and such estimates are usually possible.

The true difficulty in the optimal allocation problem is that one also needs an estimate of the inherent variance of the system under study. The importance of pilot studies in good experimental design was brought out by the papers of Gates (1981), Dawson (1981b), and Pollock (1981). Blondel et al. 1981 used a somewhat different approach, but their stress of confidence intervals for their population estimates stems from the importance of these same within-group variance sources. For some reason, field ecologists are loathe to conduct pilot studies, and even more reluctant to use in optimization models the knowledge of system variance provided by their own prior work or literature sources. It strikes me as strange that ecologists complain frequently about inadequate resources, and yet are reluctant to determine the best use of whatever resources they do have.

This reluctance cannot be due either to a lack of access to statistical consultants or to hesitancy about the use of computers and sophisticated mathematical algorithms. The same ecologists readily find statisticians to consult once they have data in hand, and they are then willing to conduct all sorts of complex statistical routines, such as principal components analysis, multiple regressions and the like. I feel that the reluctance of ecologists to involve design consultants in the initial planning of a study is due to their fear of getting advice akin to that one would derive from Dawson's Figure 5. This figure shows the intimidatingly large number of counts that are necessary to detect even a modest difference between numbers of uncommon species. We are generally afraid to know the magnitude of the Type II errors in our statistics. Other papers in this session also illustrate aspects of this propensity toward high Type II error rates in count studies, for example, the size of the confidence interval around a measure as rough as the mean number of species per I.P.A. count shown in the paper by Blondel et al. 1981. In this case, after more than 30 counts, the confidence interval still seems to be around four units (species) wide, whereas the mean is only slightly greater than 12 units (species). The paper by Morrison et al. (1981) illustrates the other side of the problem, in that their Table 2 shows that with few counts even "replicate" plots can dif-

[1] Center for Environmental Studies, Arizona State University, Tempe, Arizona 85281.

fer significantly (high Type I errors due to inadequate sampling effort, as they conclude) and their Table 1 shows that even after 10 stations, the coefficients of variation in their density estimates are greater than 10% in 4 of the 6 study areas and greater than 30% in two of them. Such variation in total avian density again indicates that probably many plots would be necessary to have much power for between plot comparisons of uncommon species.

In this light, two points in the paper by Pollock (1981) deserve special note. The first is that studies should be designed specifically to meet the assumptions of the models to be tested. If one begins a field project unaware of what uses will be made of the data, the study is premature; clearer research objectives need to be established before field work commences. Second, Pollock points out that often a small study is no better than no study at all. In many cases I would go even further and say that a small study, with an inherently high Type II error rate, is *worse* than no study at all. An inadequately supported study, doomed from the onset to produce low accuracy and a high likelihood of not detecting differences of a biologically significant magnitude, can be misused readily. Researchers rarely have any control over the uses made of their data. Regardless of how many qualifications may be put on the reports of studies known to be inadequate, we are all aware that such findings can easily be taken out of context. Until we are sure of the professional training (and unfortunately, in at least a few cases, the ethical standards) of the possible users of our studies, we must face seriously the ethics of conducting inadequately supported studies. The unfortunate part of the problem, however, is that withholding ecological findings known to be inadequate merely leads public and private agencies to make and implement environmental decisions in a vacuum. The wheels do not stop turning without our input.

An unrelated point worthy of additional comment is the issue of variance estimation. All papers in this session address this issue in one way or another. Everyone agrees that replication is essential for a good study, but the biologists and statisticians at this conference do not agree on the best means for conducting these replicates. The question of the relationship of statistical independence to repeated censuses of the same area especially needs further consideration, although both Dawson (1981b) and Gates (1981) argue that such replication is legitimate and Morrison et al. (1981) and Blondel et al. (1981) demonstrate its usefulness.

We also need to clarify exactly what the variance is. A tendency exists among the current generation of ecologists to assume that each data point and each difference we observe is the result of deterministic processes, including both biological factors and the sources of bias that Verner itemized in his introduction to this session. Some long-term studies, especially recent works of Wiens and Rotenberry in shrub-steppe habitats (Wiens 1977, Wiens and Rotenberry 1979), and our work on the lower Colorado River riparian systems (Rice et al. In press, Ohmart et al. MS), demonstrate some substantial stochastic components to many of the population parameters we are trying to estimate with our censuses. It is essential for design considerations that biologists quantify the true magnitudes of these stochastic processes, and help to separate them from the tangle of sources of bias in our studies.

Studies in Avian Biology No. 6:452–453, 1981.

SUMMARIZING REMARKS: SAMPLING DESIGN

FRANCES C. JAMES[1]

Strictly speaking, experimental design is concerned with treatments and controls, so this session should probably have been called "Sampling Design." The five papers are an interesting mix of approaches to designing bird census work. The overall message is that we need more validation of the accuracy of the methods, further standardization so that results can be compared, and more attention paid to variance in both the planning and analysis stages. In fact, variance is a biologically interesting statistic in itself. Progress could be made by concentrating on variance as the parameter of interest, thereby getting away from a typological emphasis on total or average species richness and substituting within- and between-habitat patterns in variance. Of course the biologically interesting variance would have to be separated from that attributable to error or bias inherent in the method. Another message in this set of papers is that we should pay more attention to the distribution of the data. Consider whether transformations are in order before making statistical comparisons that have underlying normality assumptions, or else use nonparametric methods of analysis. The field methods discussed in the first four papers are appropriate for broad-scale or long-term studies, and in that sense they are similar to the atlas projects or the Finnish line transects. The last paper, by Kenneth Pollock (1981), focuses on methods of studying a single population of one species over a long period. I think each paper is an excellent contribution towards the goal of characterizing avian populations in terms of species-individual, individuals-area, and species-area patterns.

The first paper, by David Dawson (1981b), discusses some limitations of point counts and transect methods. Nevertheless, he concludes that both are more suitable than territory mapping for broad surveys. I was surprised that after discussing the importance of understanding the underlying distribution of the data, he was willing to extrapolate the number of species observed, with different amounts of field effort, to obtain estimates of the total species richness of the habitat. The empirical observation of a generally linear increase in species with the logarithm of the field effort expended by the observer is interesting, but extrapolations of basically curvilinear phenomena make mathematicians uneasy. I think it is unwise to predict the species richness of an area beyond the data at hand. A preferable alternative exists for comparing the species richnesses of samples having different numbers of individuals. The procedure, rarefaction, is a distribution-free method of estimating the number of species that would have been present if fewer individuals had been observed (Heck et al. 1975). It is suitable for comparisons based on any field method. If density estimates are available, one can compare the species richnesses of areas of different size by first estimating the number of individuals that would have been present on equal-sized areas, and then, by rarefaction, finding how many species would have been present in samples of that size (Engstrom and James 1981).

One does not need density estimates to calculate the equitability or evenness of a community of birds. That requires only a list of the species and their relative abundances. Personally, I prefer graphs of the relative abundance patterns to calculations of indices such as J'. This index is usually close to 1 for bird communities and its value depends on the number of species in the list. This in turn depends on the sampling effort. So J' is not very sensitive to the evenness of the numbers of each species. Nevertheless, Dawson (1981b) is correct when he reminds us that the accuracy of density estimates made from point and transect samples is very difficult to determine.

Charles Gates (1981) discusses ways to plan the length of a transect survey, how many stations there should be along it, and how long the observer should stop at each station. The decisions are based on estimates of the variance in the results of a sample survey in the habitat in question. These estimates can be made in several ways. For instance, they could be based on the variation among transects replicated in either space or time. Or one long transect could be examined by "interpenetrating sampling," that is, estimating the variance of subsamples formed by selecting groups of individuals randomly from the larger sample. The optimal length of a transect will be the one for which the ratio of the variance in density to the density estimate itself is minimal. The author expands this problem to include estimates of the optimal number of stations, and the time to be spent at each. These procedures could save wasted effort spent either undersampling or oversampling an avifauna. I think that the variance estimates are

[1] Department of Biological Science, Florida State University, Tallahassee, Florida 34306.

probably of biological significance in themselves. If partitioned by habitat and season, they could be used to study the patchiness or heterogeneity of bird distribution. If analyzed along with quantitative data on the structure of the vegetation, they could be used to account for within-habitat patterns of distribution as well.

The paper by Jacques Blondel et al. (1981) describes two sampling procedures. They are being used to characterize broad-scale patterns of bird distribution in France, including analyses of correlates with quantitative vegetation data and implications for biogeographic theory. Both are based on point counts with unlimited distance to the birds. With the "Indice Ponctuel d'Abondance" (IPA), one visits several randomly selected points twice in a 30-day period in the habitat (biotope) in question, recording the birds heard or seen in 20 min at each point. Means and standard deviations are calculated for the higher of the two counts by species, and these values give an index to the abundance and variation in abundance of the species in that habitat. Also the average number of species per point and its standard deviation permit calculations of mean species richness ($\bar{s}$) for the habitat. On the basis of separate data obtained by spot-mapping, detection coefficients can be calculated for each species and each observer. These coefficients can then be used as weights to calculate densities from the IPA counts. With the second method, the "Echantillonage Frequentiel Progressif" (EFP) the observer visits a point only once and records only the presence or absence of the species. The frequency of each species (percentage of points at which the species was recorded) is considered to be an adequate index to its general abundance, except at very high densities. Clearly, comparisons of data from point counts should be based on nonparametric methods. I think the authors should consider whether median species richness might be a more appropriate statistic than mean species richness.

Michael Morrison et al. (1981) discuss a new method called the variable circular plot. They find that the number of stations required to obtain stable estimates of the density of birds varies with habitat, but that reasonably stable estimates can be obtained with only four stops. Of course the variation in the effective detection distance is bound to be a serious source of bias in the results, so the density estimates may be stable by being consistently inaccurate and this varies by species. The authors are aware of these problems, but they feel that the method is useful for inventories, especially in areas of rough terrain. The variable circular plot technique is reminiscent of the IPA technique described above. If it were validated by comparisons with spot-mapping, and standardized with the IPA method, comparisons between Old World and New World bird populations could be made.

If it were possible to make exact counts of the birds in an area, biologists would not need to wrestle with all these sampling problems. But given the complex multivariate nature of the sources of error and bias, plus the fact that the populations are open systems with no fixed boundaries, the problems will probably be with us into the indefinite future. It is good to know that we now have the attention of statisticians interested in applications of their theoretical work to the area of sampling. For a thorough treatment of this subject see Comak et al. (1979). This important book contains major articles on line transect and mark-recapture methods. If we can develop standardized, probabilistic methods that are of practical use considering the special nature of bird populations, then all that remains is to ask insightful questions and develop experimental tests.

Studies in Avian Biology No. 6:454, 1981.

INTRODUCTORY REMARKS: DATA ANALYSIS

FRED L. RAMSEY,[1] CHAIRMAN

This, the final regular session of the symposium, belongs almost exclusively to the statisticians. You will hear why some time-honored methods should be junked, why some should be retained and improved. You will be introduced to modern approaches to density estimation. And you will be led beyond the problem of finding one estimate of density to discussions of what to do with these estimates and how to do it—all under the heading, "Data Analysis."

To display the value of a statistician, we conducted an actual experiment in habitat modification. During the last coffee break, while you were outside, we went through the rows of the left side of this hall, placing candy canes on the floor next to every other chair. The right side of the hall was left as control, and we wanted to see if you folk are attracted by the candy. After break, I counted 147 of you sitting on the control side and 134 on the candy side. This result was so disappointing that I would not have mentioned it, had it not been that a gentleman in the balcony told me that there are actually 216 people seated on the candy side. Unfortunately, I couldn't see them all because many were bent over . . . picking up the candy!

The moral of this story, as David Dawson (1981) has told you before, is: simple counts of detections measure both abundance and conspicuousness, and it is not possible to separate the two without some additional measurement. Thomas Kuhn (1962) argues that a science is *not* determined by what it studies—ornithologists, ecologists, biologists, aeronautic engineers, architects, and painters all study birds; rather a science is determined by what *measurements* it takes. If our aim in this conference is to find ways to estimate abundance—absolute abundance, relative abundance, density, or derivatives such as diversity—then statisticians will be of little use unless we begin by taking the right measurements. Get hearing tests. Study bird songs. Determine how birds respond to observers. Measure detection distances! But don't rely on counts alone for scientific work.

[1] Department of Statistics, Oregon State University, Corvallis, Oregon 97331.

Studies in Avian Biology No. 6:455–460, 1981.

BREEDING BIRD CENSUSES USING SPOT-MAPPING TECHNIQUES UPON SAMPLES OF HOMOGENEOUS HABITATS

PAUL F. J. EAGLES[1]

ABSTRACT.—Breeding bird censuses using spot-mapping techniques upon samples of homogeneous habitats are widely used. Large scale measurement programs have been undertaken in Sweden, Great Britain, Canada and the United States. The method is rapidly becoming adopted as standard practice in a broad range of environmental planning and environmental impact assessment projects.

Several major underlying problems occur with the technique. First, the final result is not an arithmetic summation of the individual censuses. Therefore, after much field and analytical research time, only one "sample" results. Therefore, standard statistical tests can not be applied. Second, we do not have definitive experimental data on the effectiveness of this technique in measuring the actual avian population. Most researchers assume that a high percentage of the population is measured. Also, most assume that an underestimate of the actual population occurs. But, we are unsure of ourselves on this critical point. Up to now, no well defined definition of the term homogeneous has been given. A number of more minor sources or error are discussed.

Resolution of the major problems will require the development of innovative experiments that have not as of yet been undertaken.

Breeding bird censuses using spot-mapping techniques upon samples of homogeneous habitats have been conducted widely in both North America and Europe. The methodology has become relatively standardized over time due to its extensive use by avian population biologists (Williams 1936, Kendeigh 1944, Pough 1947, Pough 1950, Udvardy 1957, Enemar 1959, Williamson and Homes 1964, Hall 1964, Robbins 1970 and Van Velzen 1972). The method appears to be used with a considerable amount of confidence by many researchers.

A census is defined as a complete count of animals over a specified area at a specified point in time (Overton 1971). This technique might be more appropriately called a survey because at no time is the entire population measured.

The traditional applications of the technique for the study of various aspects of the population biology of birds have been recently augmented with the advent of environmental impact assessments. In many governmental jurisdictions the need for the ecological assessment of land use change has spurred field biologists in the search for census methods that offer the most advantageous combination of high accuracy and low cost. This method appears to offer such a combination to many environmental impact practitioners.

The method has been used in nationwide programs of population measurement in Britain (Batten and Marchant 1976), Sweden (Svensson 1978) and well as the United States and Canada (Van Velzen 1980). Examples of the use of this methodology for environmental assessment include: the measurement of the effects of residential development on avian populations (Aldrich and Coffin 1980), the species that reinvade reclaimed surface-mined land (Whitmore 1980), and the disturbances that affect breeding populations during the development of a new provincial park (Eagles 1976).

Because of the importance of the breeding period in the life cycle of most avian populations a detailed knowledge of the community at that time is often desirable.

The objectives of this paper are: to briefly summarize the methodology, to comment briefly on some census results, to critique its effectiveness, and to encourage the development of field experiments that will help clarify the various issues raised.

METHODS

The spot-mapping technique involves the repeated censusing of a sample of homogeneous habitat through the breeding season. The research plot, with a recommended minimum area of 10 ha, is traversed by an observer walking along transect lines on a 100 m grid in open habitats, such as fields, or a 50 m grid in denser habitats, such as thick forest. A minimum of eight censuses are done.

Each contact with a bird is marked as a registration on a map of the plot. Registrations that are indicative of territorial behavior, such as male song or boundary aggression, are particularly important. Each map registration contains coded information on the bird's identity, sex (if this can be determined), song (presence and type) and behavior. Standard behavioral observations include the giving of alarm reactions, the feeding of young, any aggressive reactions, the type of vocalizations and the type of activity (perching, flying, hopping, feeding, etc.). In the vast majority of cases it is male bird detection that occurs either by visual or aural means. Registrations of females help to confirm that a breeding pair is present. Constant effort is made to not include an individual as multiple reg-

[1] Department of Recreation, University of Waterloo, Waterloo, Ontario, Canada.

TABLE 1

SUMMARY OF H', S, J' FOR EASTERN NORTH AMERICA WITH MEANS ±2 STANDARD ERRORS FOR BREEDING-BIRD POPULATIONS IN NINE COMMUNITY TYPES (FROM TRAMER 1966)

Community	n[a]	H'[b]	S[c]	J'[d]
Marshes	15	1.79 ± 0.34	6.33 ± 1.32	0.718 ± 0.080
Grasslands	38	1.93 ± 0.24	5.74 ± 1.00	0.842 ± 0.034
Shrublands	26	3.14 ± 0.16	14.08 ± 2.31	0.848 ± 0.024
Deserts	6	3.25 ± 0.60	14.17 ± 5.68	0.884 ± 0.048
Coniferous forests	51	3.53 ± 0.14	17.43 ± 1.92	0.880 ± 0.014
Upland deciduous forests	62	3.82 ± 0.08	20.94 ± 1.34	0.879 ± 0.012
Mixed forests	30	3.92 ± 0.14	21.87 ± 2.76	0.893 ± 0.016
Floodplain deciduous forests	18	4.07 ± 0.16	24.22 ± 2.84	0.898 ± 0.020
Tropical woodlands	21	5.23 ± 0.24	55.14 ± 11.24	0.921 ± 0.012

[a] n = number of censuses in sample.
[b] $H' = -\Sigma p_i \log_2 p_i$.
[c] S = number of species.
[d] $J' = H'/\log_2 S$.

istrations without indicating that factor. During the analysis, a territory is assigned if there is a minimum of three valid registrations, that is, registrations on 37.5% of the site visits (3 out of 8).

The numerous details of the standardized methodology can be found in Hall (1946), Robbins (1970), Van Velzen (1972) and Eagles and Tobias (1978).

A large number of individual censuses, using this methodology, have been undertaken in North America. Many have been published through the years in *American Birds*. In Canada specifically, the Canadian Wildlife Service has published nearly comprehensive compilations of the Canadian studies (Erskine 1971, 1972, 1976a). Therefore, the original survey results are usually readily available for secondary analysis.

In order to look for similarities or patterns amongst the measured avifaunal populations in similar community types over broad geographical areas, the published results of these censuses were collected. An analysis of the number of species, diversity index, relative abundance and density in a variety of community types was done. The diversity index was calculated using the Shannon-Weiner formula (Tramer 1969).

This analysis, consultation of the literature and the author's personal experience with the methodology in the field, have been used as a basis for the critical analysis.

RESULTS

Tramer (1969) analyzed the results from 267 breeding bird censuses from eastern North America in a variety of vegetation community types (Table 1). A similar compilation of 70 censuses from southern Ontario (Table 2) revealed a pattern quite similar to that found by Tramer.

These analyses show that the number of breeding species (S), the diversity index (H') and the relative abundance (J') increase progressively along a sequence of community types. Figure 1 shows the diversity indices at a 95% confidence interval for each of the community types given in the tables. It appears that the avian populations in each of the vegetation community types can be structurally differentiated in this way. The confidence interval describes the situation for the final census results of a number of studies. That is, the single study is the sampled area and variation is the site-to-

TABLE 2

SUMMARY OF H', S, J' AND DENSITY FOR SOUTHERN ONTARIO WITH MEANS ±2 STANDARD ERRORS FOR BREEDING-BIRD POPULATIONS IN EIGHT COMMUNITY TYPES

Community	n[a]	H'[b]	S[c]	J'[d]	Average density in males/100 ha
Fields, pasture	11	1.85 ± 0.52	6.7 ± 1.8	.70 ± 0.15	184 ± 87
Sand dunes	6	2.42 ± 1.30	8.5 ± 5.3	.82 ± 0.25	79 ± 58
Urban	5	2.66 ± 0.43	8.4 ± 2.6	.88 ± 0.05	152 ± 269
Fields with trees	13	3.07 ± 0.34	14.5 ± 3.2	.81 ± 0.05	361 ± 96
Deciduous forest in urban ravines	7	3.57 ± 0.43	14.6 ± 4.1	.94 ± 0.01	230 ± 46
Upland coniferous forest	1	3.75	20	.87	356
Upland forest	15	3.82 ± 0.20	23.3 ± 2.8	.85 ± 0.04	601 ± 146
Lowland mixed forest	12	4.24 ± 0.34	26.8 ± 6.2	.91 ± 0.01	590 ± 86

[a] n = number of censuses in sample.
[b] $H' = -\Sigma p_i \log_2 p_i$.
[c] S = number of species.
[d] $J' = H'/\log_2 S$.

site and study-to-study variation. Presumably, the interval derives from the standard Student's t statistic.

DISCUSSION

The Census Summaries

Tables 1 and 2 and Figure 1 show quite clearly that the avian populations, as measured by this methodology, can be discriminated from one another according to the vegetative community type in which they occur. It is of course intuitively obvious to any student of avian populations that different species and populations are found in different vegetative complexes. What is surprising is the degree of similarity between some parameters of avian populations in similar vegetation communities in different areas.

These tables show the confidence interval around the means for a number of parameters. But these factors say nothing about the degree of statistical error found in each individual survey.

A few examples may serve to highlight the information found in the tables. The degree of similarity between the diversity index of upland deciduous forests in eastern North America (3.82 ± 0.08) and southern Ontario (3.82 ± 0.20) is striking. It appears that this forest type holds a certain avian species diversity, across a wide geographical area. The diversity found in tropical woodlands (5.23 ± 0.24) is significantly different from any other community type.

Major Problems

The final compilation of the avian population involves data from at least eight censuses that were done during the breeding season. During each census visit the avian activity evidence is marked as registrations on a field map. Later this information is transferred to a master map, one for each species. The evidence found on each successive visit is added to the master maps. The result of all the visits is one map for each species that represents the situation, which is assumed to be stable, prevailing during the time period of the study. The final result is not an arithmetic summation of the individual censuses. It is a temporally oriented, cumulative collection of the registrations for each species. The overall result is the number of territories, or the number of pairs, that occur on the research plot. Since, after considerable field and analytical research time, the result is essentially one "cumulative" sample, then standard statistical tests of variance and significance cannot be applied.

It is usual to consider that each master map is complete and accurate. But it must be recognized that each map has a statistical error attached to it, which represents the difference between the map and the actual situation. Presently we have no way of estimating the size of this statistical error.

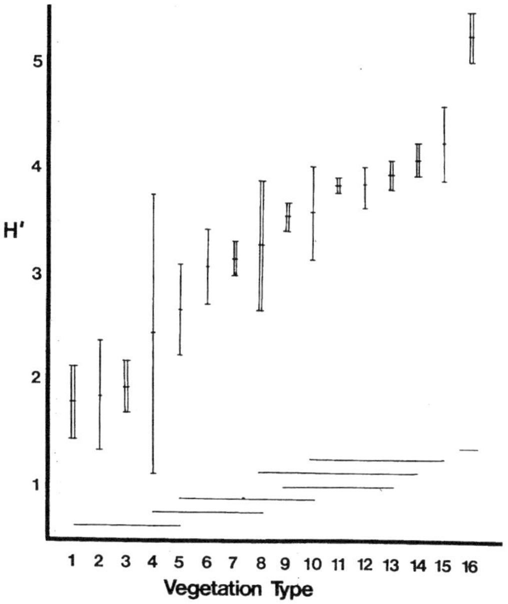

FIGURE 1. Diversity index means plotted against vegetation type. This figure graphically represents the data found in Tables 1 and 2. $H' = -\Sigma p_i \log_2 p$. Vegetation types represented by numbers: 1—marshes; 2—fields and pastures; 3—grasslands; 4—sand dunes; 5—urban; 6—fields with trees; 7—shrublands; 8—deserts; 9—coniferous forests; 10—deciduous forest in urban ravines; 11—upland deciduous forests; 12—upland deciduous forests; 13—mixed forests; 14—floodplain deciduous forests; 15—lowland mixed forests; 16—tropical woodlands. Means given with ±2SE. Means not signficant at 95% confidence level, according to Scheffe's Test, underlined together. Means with double lines from Table 1. Means with single line from Table 2.

This problem could be remedied if a larger number of censuses were conducted so that a number of compilations could be done independently. If 32 censuses were done, then four "samples," each composed of eight censuses, would be present. Statistical tests could then be conducted on the four "samples." This approach would involve the significant problem that the preconceived notions of territorial integrity that the researcher developed during the many hours spent on the research plot could spill over from one sample to the next. Also, the 32 trip level negates the use of the method if time and money are restricting factors.

This lack of statistical testing does not appear to have lowered the effectiveness of this method in the opinion of many researchers. This is obvious from its widespread use. But, it does leave an undesirable element of doubt.

Unfortunately, we do not have definitive experimental data on the effectiveness of this technique in the field. We do not know the actual percentage of the population that is being measured. Nevertheless, some researchers have used this method as a control for evaluating the accuracy of other census techniques (Stewart et al. 1952).

DesGranges (1980) plotted the cumulative number of species found in successive trips for six different community types. In all cases a similar curve resulted. The number of new species encountered was large in the first few visits and decreased in each consecutive visit. The curves levelled out from between 4 to 7 visits with an average of 5.8. Performance of a survey method for a bird can be defined according to the percentage of visits in which this individual was encountered. The overall performance is the average performance of six research plots. These values varied from 38.3% to 57.8%, with an average of 47.6 ± 6.8%. Therefore, it can be calculated that on the first visit to these research plots 47.6% of the observed population would be recorded. On the second visit more of the observed population would be recorded so that 72.5% could be considered to have been found. Therefore by the fourth visit 92.5% would be recorded and 99.8% by the tenth. These percentages deal with the observed individuals only. The total population, of which the observed is a part, is unknown.

Best (1975) did a comprehensive inventory of the individuals in a population of Field Sparrows (*Spizella pusilla*). Each individual was captured and marked and all nests were followed carefully. A standard spot-mapping census was undertaken and was submitted to five other ornithologists for analysis. Their compilations from the data estimated the population from 53 to 87% of the actual population. These analysts did not do any censusing in the field and therefore lacked any backup field knowledge. They relied solely on the field maps provided by the field observer. This study may indicate more about the significance of observer knowledge and memory of the plot than it does about census or analysis efficiencies.

Francis (1973) found that area counts underestimated the true population of Red-winged Blackbirds (*Agelaius phoeniceus*) by 27%. Since this species is relatively conspicuous on its breeding territories, it is suggestive that the underestimation may be even larger with inconspicuous and secretive species.

Stewart et al. (1952) estimated the accuracy of the method to be above 90% for most species, with an average of over 95%. Svensson (1979a) has calculated that the daily census results must be at least 40 to 60% efficient if the final compilation, using the 3 out of 8 rule, is to be considered acceptable.

Davis (1965) found a 30% difference between two aural censuses of male song in Rufous-sided Towhees (*Pipilo erythrophthalmus*) that were done in the same area but were approximately 15 minutes apart. This note reinforces the point that the starting points and the transect directions should be varied so as to "capture" portions of the plot at different times during the census period. Speirs and Orenstein (1975) and Best (1975) mention the importance of recording data on all the activities and behaviors of the birds in the research plot, not just the singing male registrations.

DesGranges (1980) maintains that this methodology is very accurate. Blondel (1969) states that if the methods are properly applied the margin of error will be 10% at the maximum.

Odum and Kuenzler (1955) studied four species in the field, Eastern Kingbird (*Tyrannus tyrannus*), Eastern Wood Pewee (*Contopus virens*), Eastern Meadowlark (*Sturnella magna*), and Orchard Oriole (*Icterus spurius*). They found that between 2 and 8 hours of field observation were required to reach the 1% level on the smoothed effort/yield curve. Beyond this point, each additional observation produced less than a 1% increase in the measured territory size. The average spot-mapping census which consists of 3 hours in the field, on eight separate occasions, will entail a period of 24 hours spent in the field. This is well above the time found to be necessary by Odum and Kuenzler. But it must be recognized that many species are much less conspicuous than the four studied by Odum and Kuenzler and the average census taker must deal with at least 15 species or more singing simultaneously on the research plot.

Preston (1979) discusses the theoretical basis of bird observation in the field. He suggests that bird-spotting can be considered to be a matter of chance. That chance is mediated by elements such as lighting, distance from the bird, foliage density, chance bird movement, and many other factors. He points out that the number of birds seen per hour increase proportionally with the square root of the number of observers. Preparatory work for the atlas of breeding birds in Great Britain and Ireland (Sharrock 1976) pointed out that in an area of 100 sq. km that 50% of the birds could be found after two hours, 75% after 10 hours, 87% in 16 hours and 100% could not be found even after 200 hours of field work. Both these works point out that even after 8 census visits (32 field hours) one observer will not reach the 100% detection level.

In practice, the same route is run in opposite directions by the same observer. This person develops an expectation of where the birds will be found and therefore distributes subsequent effort unevenly. This kind of variation is impossible to quantify. The results presented by DesGranges (1980), for example, are based on such a situation and therefore are potentially suspect because of the lack of independence between measurements.

Confidence intervals could be calculated for a single study if the field visits represented true replications based upon a random selection procedure (Ramsey, pers. comm.). This might necessitate the use of randomly selected observers.

Vegetation Analysis

There is no standard definition used for the concept of homogeneous habitat. A general trend seems to be developing, that groups vegetation communities into a number of quite generalized classes (Tables 1 and 2) (Van Velzen 1980). But it is obvious that any of these classes can be seen as being composed of a large number of different community types. In southern Ontario, Hills (1952) has shown that there are nine predominant forest types that vary according to the microclimate and soil moisture regimes. Recent work by Maycock and Beechey (pers. comm.) has expanded Hill's system into 150 vegetation types that occur in all of Ontario. But the breeding censuses tend to lump the vegetation types into only a few basic categories (Table 2).

This clumping of vegetation types need not be of concern as long as the detailed vegetation community composition data is included with the avian census data. This information can then be used to reclassify the vegetative community if it proves to be necessary at some future date.

A standardized vegetation analysis technique has been recommended for forested communities (James and Shugart 1970) but no such standard has yet come to the fore for non-forested communities.

Homogeneity can be considered to be a problem of mapping scale. Basically, the existing general community categories are those that are mapable at a 1:5000 scale. At a larger scale, the various sub-communities become visible. But it must be recognized that the bird population measurement is done on the ground at a 1:1 scale. If the vegetation communities are approached at this scale, definition becomes much more difficult because of the obvious lack of homogeneity.

This aspect of vegetation community mapping has not been systematically treated by avian population biologists up to now. In the future it would prove valuable to have general community categories defined for each of the North American biomes. This would result in the standardization of the reporting of the vegetation component of avian censuses and hopefully, in the development and acceptance of standard vegetation analysis methodologies.

Sources of Error

A number of sources of error are known with the spot-mapping technique. The most important ones are discussed below.

The territories of individuals may move through the time of the breeding season (Wiens 1969). Individual birds may die or otherwise abandon territories (Best 1975). The possibility of territorial infractions where individuals trespass on another territory raises the possibility of considerable confusion at the time of analysis if such an infraction was observed. There is always the problem of the presence of transient and non-breeding males on the research plot.

A variety of territory types occur in different species of birds (Schoener 1968). Each must be dealt with separately. Some species are non-territorial, such as Brown-headed Cowbird (*Molothrus ater*). Colonial nesters, such as the Great Blue Heron (*Ardea herodius*), pose unique problems. Species with very large territories, that is territories that are many multiples of the total plot size, can cause overestimates of population density if the entire plot is counted as one territory and not as just part of a territory. Small research plots, that is ones below the recommended minimum size, can cause this inflated density effect for even the intermediate-sized species such as the Eastern Meadowlark. These problems are discussed in Eagles and Tobias (1978).

The amount and intensity of singing and the overall conspicuousness varies considerably between species. After the incubation of the clutch begins, or the young hatch, the song level of the adults of many species decreases and therefore the males become less conspicuous. The conspicuousness may increase again briefly after fledging of the young occurs.

Any breeding census methodology of this type must be capable of dealing with those species that nest early or late in the season. When these species are well known in the community under study, then specific census can be done in the appropriate time of year. But the incubation time varies considerably between species and those that breed quickly. For example, these species may be under-sampled if censuses are widely spaced in time. Therefore considerable care must be exercised by the researcher in selecting dates.

Polygamous individuals obviously negate the assumption that the method measures the number of pairs on the plot.

As with any population sampling technique, it must be recognized that the entire population is not being measured. It is to be hoped that in the future field researchers will take on the job of finding the answers or suggesting solutions to the deficiencies pointed out in this paper. Most of them should be amenable to experimental investigation.

ACKNOWLEDGMENTS

Constructive criticism of an earlier draft of this paper was provided by Fred L. Ramsey and Tim Vetter. Their work is appreciated.

Studies in Avian Biology No. 6:461–465, 1981.

AN EVALUATION AND COMPARISON OF TECHNIQUES FOR ESTIMATING HOME RANGE AND TERRITORY SIZE

R. GLENN FORD[1] AND J. P. MYERS[2]

ABSTRACT.—Estimates of territory and home range size can yield widely varying results depending upon methods of data collection and analysis. To evaluate the merits of different methods we examine space-use patterns of Pectoral Sandpipers (*Calidris melanotos*) and Red Phalaropes (*Phalaropus fulicarius*) on their breeding ground. We use empirical data from these species to generate a series of computer-simulated home ranges. We then examine the efficiency of a non-probabilistic estimator of territory size (minimum convex polygon method) vs two probabilistic techniques, one parametric (Jennrich and Turner 1969) and one nonparametric (Ford and Krumme 1979), testing for their sensitivities to sample size and to temporal dependence between successive observations.

All methods are sensitive to temporal dependence and sample size, but the probabilistic techniques provide better estimates from small samples. Both the minimum convex polygon method and the parametric Jennrich-Turner technique overestimate area utilized by the species studied here, both of which deviated from a bivariate normal distribution. The Ford-Krumme approach provided the most accurate estimate of utilized area.

The size of areas utilized or defended by individual birds frequently is an important datum sought in avian ecology. In this paper we explore problems associated with territory and home range size estimation, with an emphasis on statistical estimates of utilized rather than defended areas. We will use data from two species of shorebirds, Pectoral Sandpipers and Red Phalaropes. These species exemplify opposite extremes in territoriality: Pectoral Sandpipers achieve virtual exclusive use of their defended areas (Pitelka 1959), whereas, Red Phalaropes breed non-territorially (Kistchinski 1975). We will consider the efficiency of different estimators of utilized area as affected by sample size, temporal dependence between observations, and by differences in spacing behavior.

METHODS

ESTIMATORS OF UTILIZED AREA

Three general approaches are used in estimating utilized area. The oldest and most widely used is to draw a polygon connecting those observations that appear to lie on the periphery and define the area bounded by connecting these points as the size of the home range, territory, or utilized area. Observations may be defined as peripheral because they lie adjacent to regions never utilized (Stefanski 1967), because they lie on a boundary which the bird will not transgress when flushed (Wiens 1969), or because they form part of the convex hull of all observation points, i.e., the smallest set of points that when connected contains all other points (Weeden 1967). For this analysis we employ the last definition because it is mathematically definable. Its most important aspect relative to other definitions (below) is that it lacks a probabilistic description of their use of space ("space-use") within the bounded area.

Jennrich and Turner (1969) fit a bivariate normal distribution to the array of location points obtained by passive observation and then calculate the area of a 95% probability ellipse. An alternative, nonparametric, probabilistic estimator is described by Ford and Krumme (1979). This method utilizes the distribution of frequency of distances generated by taking all observation points pairwise. Ford and Krumme use a computer optimization algorithm to generate a simulated space-use distribution with a discrete distribution of frequency of distances as similar as possible to the observed data. The simulated space-use pattern is then integrated to estimate the minimum area that will contain a specified proportion, usually 95%, of the animal's space-use. This index, called MAP (0.95), may be calculated from simulated or observed space-use distributions. The method provides both a nonparametric probabilistic area estimate and also gives a visualization of the shape of the distribution.

DATA COLLECTION

Breeding male Pectoral Sandpipers and breeding female Red Phalaropes were observed at Barrow, Alaska, during June 1976 and 1978, respectively. Opposite sexes were used because the Pectoral Sandpiper is polygynous (Pitelka et al. 1974) while the Red Phalarope is polyandrous (Schamel and Tracey 1977). Study areas were marked off in 50 m grids and the locations of individuals were recorded to a 10 m resolution at 1 min intervals for study sessions running 50 to 200 min per session. Each individual was tracked for multiple sessions. For this paper we use data from three Red Phalaropes and three Pectoral Sandpipers. For analysis, observations were lumped into 50 × 50-m units. Cell areas referred to in the text are multiples of these 50 × 50-m (0.25 ha) grid units.

MOVEMENT SIMULATOR MODEL

Although the analyses we present are based on sequential sighting data for real birds, we have not used the raw data themselves directly, instead using them to construct a stochastic model designed to mimic the movement and space-use patterns of individual birds. After verifying that the simulator accurately mimicked the data, we then generated a series of simulated data sets.

[1] Department of Biology, University of New Mexico, Albuquerque, NM 87131.

[2] Museum of Vertebrate Zoology, 2593 LSB, University of California, Berkeley, CA 94720.

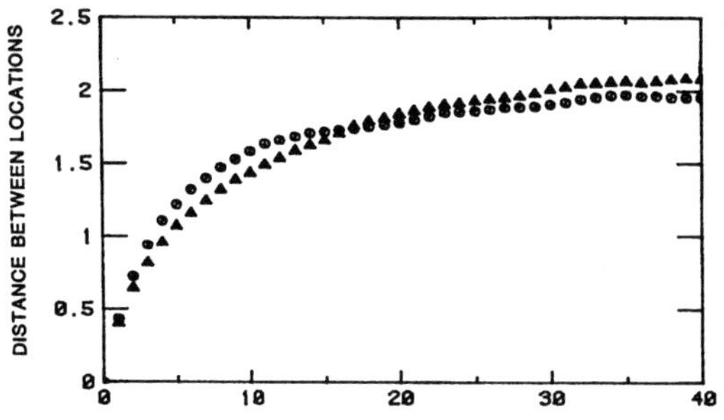

FIGURE 1. Average distances between observed locations as a function of the time separating observations. Time steps are 1 min, distance is in units of 10 m. Triangles are based on actual input data; circles are based on simulated movement sequences.

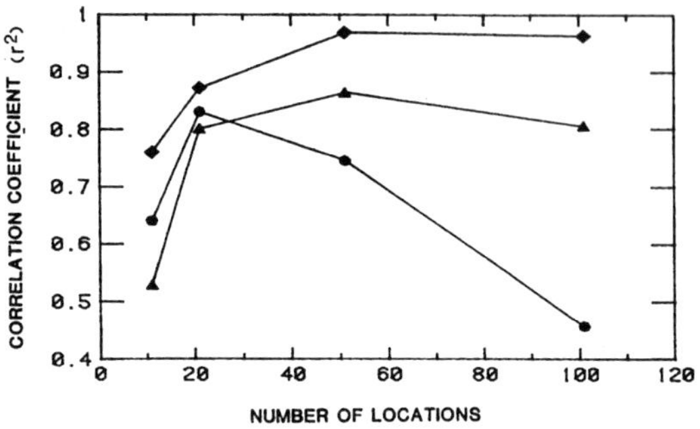

FIGURE 2. The correlation coefficient (r^2) of the regression of estimated individual utilized area size on actual 95% individual utilized area size as a function of sample size. Circles refer to polygon method, triangles to Jennrich-Turner index, diamonds to Ford-Krumme technique.

Two aspects of the movement patterns were considered to be important: the spatial distribution of sightings, and the relative frequency of distances moved between sightings. These distributions provide descriptions of two basic features of behavior that strongly influence the performance of space–use area estimators: (1) complicated nonuniform distributions; and (2) temporal dependence in movement pattern, usually with a series of relatively short movements interspersed with occasional longer movements. The result of (2) is that the expectation of distance moved between observations increases as the length of time between sightings increases.

The movement pattern is modelled as a Markov chain based on a $n \times n$ transition matrix where n is the number of grid cells contained in the territory. The transition probability from cell i to j is proportional to the joint probability of moving the distance D from the midpoint of i to the midpoint of j, $P(D)$, and the probability of moving to j, $P(j)$. We fit the model to a given individual by finding the sets of $P(D)$ and $P(j)$ which minimized the sum of squared differences between the observed and simulated frequencies of distances moved, and the observed and simulated frequencies of time spent in each cell. The minimization is carried out using a successive approximation computer algorithm.

The algorithm was successful at finding transition matrices that generated close fits of both movement rates and space-use for both species. Its success is shown by comparing the observed and simulated values for the average distance moved between locations as a function of the number of time steps between locations (Fig. 1).

TABLE 1
COMPARING THE ACCURACY OF UTILIZED AREA ESTIMATORS[a]

Estimate	Percent of true value ± 1 SD
Minimum convex polygon	
Pectoral Sandpiper	91 ± 12
Red Phalarope	200 ± 115
Jennrich-Turner	
Pectoral Sandpiper	129 ± 14
Red Phalarope	188 ± 26
Ford-Krumme	
Pectoral Sandpiper	101 ± 13
Red Phalarope	116 ± 23

[a] Values calculated from nine simulations per species, 251 locations per simulation, no temporal contingency between observations. Percent of true value was calculated for each simulation and averaged to obtain mean values ± 1 SD. Area is in grid units (0.25 ha).

TESTING THE ACCURACY OF AREA ESTIMATORS

Three sources of error can bias estimates of utilized area: sample size; temporal contingency between successive observations; and deviations from an assumed underlying distribution, such as bivariate normality. We examined the effects of these biases using output from the movement-mimic model described above. For each of the six individual real data sets, three from each species, we generated simulated observations involving 1000 locations per record. Each set was replicated three times, for a total of 18 simulated bird movement patterns. Three data sets were generated separately for examining the effect of temporal contingency: successive observations separated by 1 time step, 10 time steps, or completely independent in time. Curves of estimated utilized area versus number of observations were then generated for each data set using the different area estimators. We compared these estimates with the true MAP (0.95) values calculated directly from the simulated records.

RESULTS

Mean utilized area (MAP 0.95) for simulated Pectoral Sandpipers and Red Phalaropes were 18.4 and 14.6 grid units, respectively. These true values, calculated directly from the model output, are compared with estimates of the different indices in Table 1. For both species the Ford-

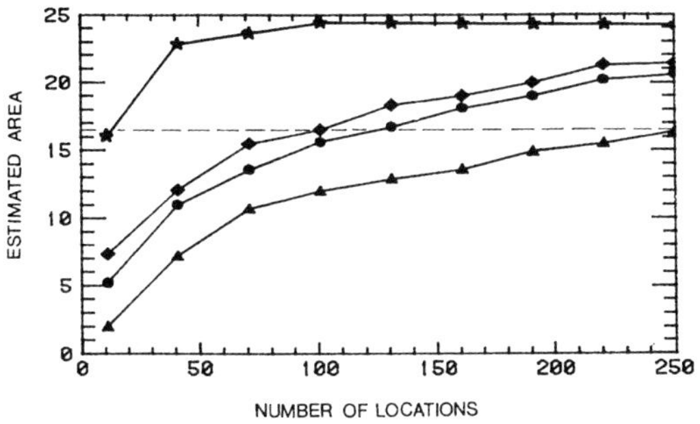

FIGURE 3. Estimated average utilized area in grid units (0.25 ha) as a function of sample size using the minimum convex polygon method. Triangles represent observation sequences with 1-min intervals, circles with 10-min intervals, and diamonds sequences where successive locations are independent. Stars represent the territorial flushing model described in text. Dashed line represent true values calculated directly from simulated data sets (MAP 0.95, see text).

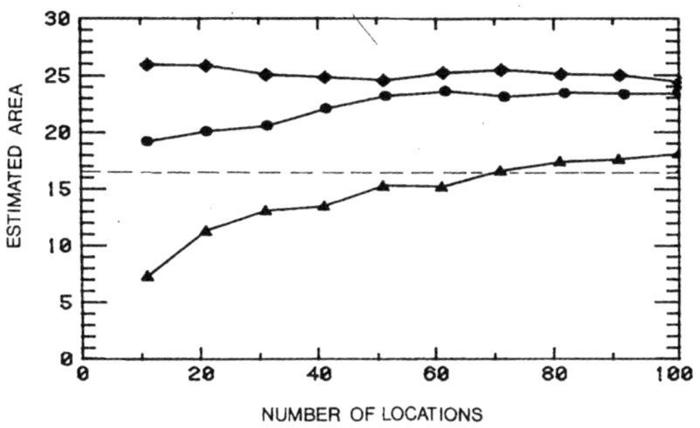

FIGURE 4. Same as Figure 3 but for the Jennrich-Turner index.

Krumme method gave the closest estimates to true values.

As an alternative to providing absolute estimates of utilized area, these indices may also be used to compare populations or individuals. Such a relative test requires only that the index be highly correlated with true size. To compare the estimators in this regard we correlated average index values with their corresponding true MAP (0.95) values for data sets with complete independence between observations.

Both probabilistic estimators become more accurate with increasing sample size (Fig. 2), and the nonparametric index is consistently more accurate than the parametric technique. The polygon method, by comparison, becomes less effective with increasing sample size. The downward trend in r^2 occurs as the index is progressively dominated by low probability regions in the periphery of the utilized area.

For samples without temporal contingency, the polygon method generates estimates that increase monotonically with increasing numbers of location records, but only slowly does it approach a limit (Fig. 3). The greater the contingency between successive observations, the slower the approach.

The very large number of observations required by the polygon method may be reduced significantly if the observer actively flushes the bird to its territorial boundary (Wiens 1969). This situation was modelled by generating distributions based on the same data set, but modified so that only grid cells bounded on at least one side by non-utilized area were included in the distribution. Assuming a "best case" of no contingency between observations and all perimeter cells equally likely, this method approaches an asymptote in fewer than 100 observations (Fig. 3).

The Jennrich-Turner method requires a much smaller sample size to achieve stability (Fig. 4). Stable estimates may be obtained with 10 or fewer observations per individual if successive observations are independent. With moderate temporal contingency, however, the index may require as many as 50 to 100 observations, and even more with high contingency.

The nonparametric method (Ford and Krumme 1979) performs similarly to the Jennrich-Turner method in its sensitivity to sample size, but it is somewhat more affected by temporal contingency (Fig. 5). At maximum contingency the index is strongly affected and the area estimates remain low even after 500 observations. For independent observations, however, the index is very stable, and as noted above it provides a better estimate of true value (Table 1; compare Figs. 4 and 5).

DISCUSSION

The polygon method is a poor choice for estimating utilized area despite its widespread use (Table 2). This index is strongly affected by sample size and requires prohibitively large sample sizes to reach an asymptote. The rate of approach to the asymptote is so slow even with little temporal dependence among observations that it is unlikely that many fieldworkers would be able to obtain the required data set. Further, there is virtually no way to determine what proportion of the asymptotic limit has been reached at a given sample size. Finally, the actual area estimated is much larger than true value. Thus this method presents serious interpretive problems, difficulties not offset by its seductive simplicity.

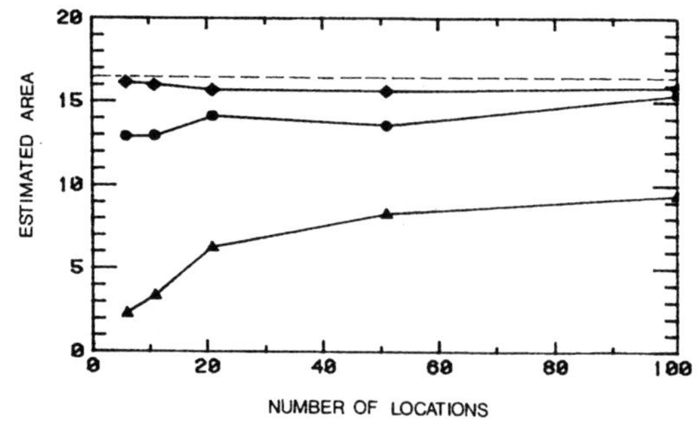

FIGURE 5. Same as Figure 3 but for the Ford-Krumme index.

The Jennrich-Turner method provides a stable area estimate based on a relatively small number of independent observations, and it appears to be the least sensitive method to temporal contingency. Its assumption of bivariate normality, however, creates problems. For the data analyzed here, the 95% use area estimated by the Jennrich-Turner technique is almost half again as great as the true value (Fig. 4). Such a bias may not occur if the distributions are more nearly normal. Although Zack and Falls (1978) found that song territories of male Ovenbirds (*Seirus aurocapillus*) are approximately bivariately normal, this assumption probably is not met widely. It is certainly violated by the data sets used in this study, as well as by space-use distributions of Buff-breasted Sandpipers (*Tryngites subruficollis*) (Ford and Krumme 1979) and Sanderlings (*Calidris alba*) (J. P. Myers, unpubl. data) on their wintering grounds. As a relative size index, the Jennrich–Turner method performs well even for non-normal distributions.

The Ford-Krumme method is superior to the Jennrich-Turner approach when normality is violated. It provides accurate estimates of absolute area even with small sample sizes, provided that successive observations are independent. It is also an effective tool for examining relative differences, showing consistently higher correlations with actual individual area values than do either alternative. Its principal disadvantage is its relative expense in terms of computer time.

CONCLUSIONS

For absolute estimates and relative comparisons of utilized areas, the probabilistic techniques of Jennrich and Turner (1969) and Ford and Krumme (1979) clearly surpass the minimum convex polygon technique. If the underlying distribution is already known to be bivariately normal, then the Jennrich-Turner parametric method offers an efficient estimator of utilized area. If, however, the underlying distribution is either not normal or is unknown, then the Ford-Krumme approach should be taken.

Our comparisons illustrate a clear basis for selecting the probabilistic nonparametric estimator (Table 2). But what if the parameter of interest is defended area, instead of utilized? For species where these are synonymous, results should be the same. But when defended area deviates from utilized area (Stefanski 1967, Zach and Falls 1979, Tryon and MacLean 1980), then the method of choice may change as a function of statistical factors we have yet to examine. Ultimately, then, as with so many other measurements made in avian ecology, the choice of technique depends not only on statistical considerations but also on a clear perception by the investigator of the questions under study, and

TABLE 2
RELATIVE ADVANTAGES OF DIFFERENT ESTIMATORS OF UTILIZED AREA

Estimator	Advantages	Disadvantages
Minimum convex polygon		
Passive observation	????	—requires enormous samples —highly sensitive to contingency —uninterpretable estimate
Territorial flushing	—simple, cheap	—overestimates utilized area
Jennrich-Turner	—calculation relatively simple, inexpensive —good predictor if bivariate normal —efficient with modest sample	—assumes bivariate normality —sensitive to contingency
Ford-Krumme	—assumes no distribution —excellent predictor even with small sample	—sensitive to contingency —expensive, requires computer

a sound appreciation of the system's biological details.

ACKNOWLEDGMENTS

We thank John Wiens and Dennis Heineman for their thoughtful comments on our analyses. Funding for computer work and field observations, respectively, was provided by NOAA-OCSEAP (Alaska) in grants to J. A. Wiens, Univ. New Mexico, and by DOE through grants to F. A. Pitelka, Univ. Calif., Berkeley.

Studies in Avian Biology No. 6:466–482, 1981.

LINE TRANSECT ESTIMATION OF BIRD POPULATION DENSITY USING A FOURIER SERIES

KENNETH P. BURNHAM,[1] DAVID R. ANDERSON[2] AND JEFFREY L. LAAKE[3]

ABSTRACT.—A general approach to the estimation of bird population density from line transect data is discussed. This method is based on a nonparametric statistical analysis technique: the Fourier series method. The Fourier series estimator is both robust and efficient; i.e., it is not dependent on specific distributional assumptions about the detection probability of birds at various perpendicular distances from the transect line to provide relatively precise density estimates. The method is especially easy to compute for ungrouped, perpendicular distances, but can also be applied to grouped data commonly taken when sampling birds. A comprehensive computer program, TRANSECT, implements the Fourier series method, under a variety of options, by conducting hypothesis testing and point and interval estimation of population density. Examples of the Fourier series method based on nongame breeding bird transect data are provided. Because results will only be as reliable as the data collected, brief guidelines on field procedures and sample size are given. Finally, comments on other methods of analysis of line transect data are presented.

Line transect sampling to estimate the abundance of biological populations has been in use for over 40 years. However, only within about the last 10 years have there been substantial efforts to apply line transect sampling to the problem of estimating abundance of nongame birds. Similarly, it has only been within recent years that the statistical properties of this method have been intensively studied. Line transect sampling is now (almost) an established method for estimating densities of some species of nongame birds, especially breeding birds. Reliable results appear possible if good field practices are used to collect the data and robust, efficient analysis methods are used to analyze these data. The objective of this paper is to bring to the attention of ornithologists a general, robust, reliable data analysis method for use with line transect sampling data.

Line transect sampling embodies the explicit recognition of the fact that the probability of detecting birds decreases with increasing distance from the transect line. Because of this, distance data to birds detected are recorded. Estimation of bird abundance involves using these distance data to "correct" the sample size for the detectability of birds. This can be viewed as a refinement on strip transect sampling which requires the assumption that all birds are detected within a fixed perpendicular distance (i.e., within the strip) of the transect line.

Strip transect sampling predates line transects; it was used as early as 1906 (as reported in Forbes and Gross 1921). The use of distance data to "correct" for missed birds seems to have first been suggested in the 1930s (see Gates 1979). During the 1930s and 1940s, faltering attempts were made to put line transect sampling, i.e., estimation based on distance data, on a mathematical basis: see e.g., Leopold's (1933) reference to King's work in the late 1920s and early 1930s, Breckenridge (1935), Colquhoun (1940a, 1940b), Colquhoun and Morley (1941), Webb (1942), Kendeigh (1944), Southern (1944) and Kelker (1945). None of these papers presented any real theory of line transect sampling or estimation methods. A pioneering paper by Hayne (1949) was the first significant attempt to formulate an estimator of animal density based on line transect sampling data (Hayne's estimator has not actually been used much with nongame bird data).

Rigorous, general development of line transect theory did not really start until the late 1960s. Key papers by Gates et al. (1968) and Eberhardt (1968) laid the initial foundations of line transect theory. During the 1970s, work progressed and culminated in a good general understanding of, and theory for, line transect sampling and estimation of population abundance. The most comprehensive single reference is Burnham, Anderson and Laake (1980); however, other relevant literature during that decade is Anderson and Pospahala (1970), Seber (1973, 1979), Kovner and Patil (1974), Burnham and Anderson (1976), Hayes (1977), Schweder (1977), Anderson et al. (1978), Eberhardt (1978, 1979), Pollock (1978), Anderson et al. (1979), Burnham (1979), Crain et al. (1979), Gates (1979), Ramsey (1979), Ramsey and Scott (1979), Quinn (1977, 1979, 1980) and Patil et al. (1979). This represents a significant output of fundamental theory on line transect (and closely related) sampling; unfortunately, it has not yet been incorporated into ornithological practice.

[1] U.S. Fish and Wildlife Service, 2625 Redwing Road, Ft. Collins, Colorado 80526.
[2] Utah Cooperative Wildlife Research Unit, Utah State University, Logan, Utah 84322.
[3] Inter-American Tropical Tuna Commission, Scripps Institution of Oceanography, La Jolla, California 92037.

During this same period (approximately the past 20 years), ornithologists have been increasingly concerned with transect sampling to estimate bird abundance. However, there has been almost no basic transect sampling theory developed or presented in the ornithological literature (Yapp 1956, is an exception, but Royama 1960, concluded that Yapp's theory is not applicable in practice). Ornithologists have concentrated on conducting field studies to understand the numerous factors influencing the detection of birds (such as time of day, weather, habitat, bird species, and observer differences): see for example, Amman and Baldwin (1960), Bergerud and Mercer (1966), Brewer (1972), Fowler and McGinnes (1973), Järvinen and Väisänen (1975, 1976b), Franzreb (1976, 1977), Myrberget (1976), Tilghman (1977), Hickey and Mikol (1979). However, in a properly designed and conducted line transect study such factors can be safely ignored during data analysis if a suitably "robust" estimation method is used.

The best known line transect method used in ornithological studies is that of J. T. Emlen (1971, 1977a). The data collection aspects of Emlen's method can be improved, in principle, by more precise recording of distances. However, the estimation aspects of Emlen's method were developed with no theoretical basis and can be greatly improved. They should be replaced by rigorously developed, well founded estimation methods. We discuss one such method in this paper.

LINE TRANSECT SAMPLING

BACKGROUND

A defined study area of known size, A, should be established before starting a transect study, especially if estimation of bird abundance at specific points in time is important (the alternative is to only compare changes in bird density over time). First, a set of transect lines must be established in the study area, along with a plan for sampling those lines. This constitutes an essential part of the study design and is of critical importance. Some comments on study design are presented in a different section of this paper. In general, one or more transect lines of fixed length are established in the area. Finally, the line is walked, at least once, and data on birds observed are recorded (replicates may, in fact, be different days of sampling the same line(s)).

In bird studies, it is common to establish a fixed distance, w, on either side of the line and only record birds observed within this distance. In strip transect sampling, all birds within the strip of length L and width $2w$ are assumed to have been observed (hence recorded). Thus, in strip transects, only the birds detected in the strip are counted. However, it is known that if w is large, detectability will decrease at increasing distances from the transect center line. Therefore, distance data on each bird, i.e., how far is it from the transect center line, must be recorded.

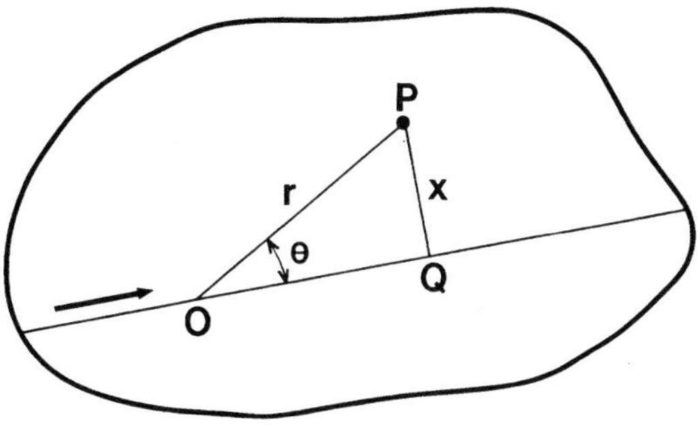

FIGURE 1. Diagram of the possible measurements that can be made for a detected object. The observer is at position O when an object is detected at position P, and Q is that point on the line perpendicular to the object. The sighting distance is r, the sighting angle is θ, and the perpendicular distance from the object to the center line is x. Note that the direction of the observer's travel, as shown by the arrow, is from O to Q.

Let there be R "replicate" lines, with lengths $l_1, l_2, \ldots, l_R$ and total length $L = l_1 + l_2 + \ldots + l_R$. We will sometimes treat the situation as if there were one overall line of length L. To facilitate further discussion, the following notation is defined: n_j = number of birds detected on line j, $j = 1, \ldots, R$; $n = n_1 + n_2 + \ldots + n_R$ = total count of birds for line length L; x = the recorded perpendicular distance from the transect (center) line to a detected bird—the total sample of such distances is x_i, $i = 1, \ldots, n$. Note that $x = r \cdot \sin(\theta)$; r = the sighting distance from the transect line to the bird; and θ = the sighting angle (see Fig. 1).

The basis for modeling line transect sampling is the concept that there is a decreasing probability of detection for birds at increasing distances from the transect line, and that this phenomenon can be represented by a "detection" function $g(x)$, where $g(x)$ = the probability of detecting a bird that is at perpendicular distance x from the transect line.

For strip transect sampling, it is assumed that $g(x) = 1$ for all distances less than w. However, for line transect sampling, the detection probability $g(x)$ decreases as perpendicular distance, x, increases.

In most ornithological field work, estimation of bird abundance has been based on the perpendicular distance data and we support that approach. However, these perpendicular distance data are often recorded by distance categories (i.e., as grouped data) rather than being recorded as exact measurements. The only justification for this is that the exact distances cannot always be determined; rather, the observer only knows that the bird, often heard rather than seen, is between some distance limits, such as 0 to 20 m or 20 to 50 m.

Assumptions

The goal of line transect sampling is to estimate the average density, D, of specified species in the study area. If N is the total number of birds in the area A, then $D = N/A$. A model is needed to relate the data to bird density, D, in order to derive a valid estimate of bird density. A model is just a set of assumptions; in their most concise form these are mathematical assumptions which, of course, have practical implications.

We recognize four basic assumptions in line transect sampling (in decreasing order of importance): (1) Birds directly on, or very near to, the line will always be detected; (2) There is no movement of birds in response to the observer and none are counted more than once during a given walking of the line; (3) All distance and angle data are recorded without measurement error; and (4) Sightings of different birds are statistically independent events.

In the abstract, line transect theory relates to sampling objects that do not move. This is reflected in assumption (2). However, movement that is random with respect to the location and movements of the observer causes no difficulty, provided the bird is counted only once during any one sampling of the transect and provided the distance to the transect line is accurately recorded when detection occurs. Assumption (2) will be violated by evasive movement, wherein birds move away from the transect line as the observer approaches, or by attraction of the birds toward the observer. Some degree of evasive movement is to be expected and can cause severe underestimation of bird density if it is extreme (e.g., even a moderate proportion of birds moving beyond the truncation distance, w).

Assumption (1) means that, if a bird is on, or very near, the line, the probability of detecting it is 1 (i.e., $g(0) = 1$). In practice, some birds will be missed during sampling. If they are well off the transect line, this causes no problems in estimating bird density. However, failure to detect birds that are on the line is a serious problem as regards density estimation. Birds that were on the line, but moved in response to the observer and then are missed, are also a problem. However, it is important to distinguish between these two situations as violating either assumption (1) or (2), respectively. Some degree of movement can be dealt with in the data analysis; i.e., moderate violation of assumption (2) may occur and transect sampling will still be useful. There is no way to deal with failure of assumption (1) from line transect data alone; failure to meet assumption (1) (all birds on the line are seen) directly and significantly biases *any* estimate of population density.

Assumption (3) is related to field techniques of distance measurement and the diligence of observers. If distances are to be recorded "exactly" (say to the nearest meter when $w = 100$ m), it is critical, for example, to avoid recording a distance between 1 and 10 m as zero (a not uncommon practice). It is necessary to have an objective way of measuring the distance (e.g., steel tape or pacing), otherwise the tendency is to record distances at convenient values like 5, 10, or 25 m. If the data are recorded by distance groups, then less rigor is needed because assumption (3) will be met if all birds detected are recorded in the correct distance category.

The primary way that assumption (4) is not met is if birds occur in distinct, small groups (large flocks of birds are not suitable for line transect sampling, anyway). We call this the case of birds (objects) occurring in "clusters." The proper treatment of clusters of birds is to regard the cluster itself as the object of interest and record only one distance per sighting, the distance to the cluster, and the cluster size. Standard line transect theory is then used to estimate the density of bird clusters and multiplying that estimate by the average cluster size in the total population gives the density of birds (see e.g., Burnham et al. 1980:192–194). Note, however, that the average cluster size observed from the actual line transect sampling may be a biased estimator of the true population average cluster size. In this case, estimation of average cluster size is not straightforward (see e.g., Burnham et al. 1980:192–194).

Approaches to Density Estimation

The total area sampled is $A = 2wL$. Let N be the total number of birds in this area. Given a properly designed study, an unbiased estimate of bird density is $\hat{D} = N/A$. For strip transect sampling, the total birds seen is $n = N$ (by assumption). But, for line transect sampling, N must be estimated as n/P, where P is the (average) probability of detecting a bird in the area sampled by the transect. This probability is related to the detection function; in fact,

$$P = \frac{1}{w} \int_0^w g(x)\, dx, \qquad (1)$$

which is the average value of $g(x)$ for $0 < x < w$ (J. T. Emlen 1971, 1977a, has called $1/P$ the "coefficient of detectability"). Define

$$a = \int_0^w g(x)\,dx \tag{2}$$

so that $P = a/w$. Then the estimate of D requires only an estimate of a (equivalent to estimating P because w is known):

$$\hat{D} = \hat{N}/A = \frac{n}{A\hat{P}} = \frac{n}{2wL\left(\frac{1}{w}\right)\hat{a}} = \frac{n}{2L\hat{a}},$$

$$\hat{D} = \frac{n}{2L\hat{a}}. \tag{3}$$

The estimate of this "correction factor" to account for birds missed that were off the transect center line depends on the recorded perpendicular distance data. The next step in this logical process is to derive the probability density function of the (random) variable x (=perpendicular distance). Seber (1973) has shown that

$$f(x) = \frac{g(x)}{a}, \tag{4}$$

where $f(x)$ represents the sampling distribution of x. Finally, by assumption (1), the probability of detecting a bird if it is on the transect center line is 1. Thus $g(0) = 1$, and from Eq. (4) we have

$$f(0) = \frac{1}{a} \tag{5}$$

Equation (5) provides a clear-cut relationship between the parameter a and the observed perpendicular distance data. Given any model for the detection function, or given a sampling model for perpendicular distances, there are many ways to estimate $f(0)$. Substituting Eq. (5) into Eq. (3) gives

$$\hat{D} = \frac{nf(0)}{2L}, \tag{6}$$

which is a general formula for estimating density, D.

The statistical estimation problem now is to specify a model, $f(x)$, of the sampling distribution of x (this is equivalent to modeling the detection function) and then derive an estimate of $f(0)$. A large variety of models for $f(x)$ have been used, for example, the negative exponential distribution (Gates et al. 1968), the half normal distribution (see, e.g., Gates 1979) and an exponential power series model which includes both these as special cases (Pollock 1978). Given the large variety of models for $f(x)$ and estimators for $f(0)$, reliable criteria are needed on which to base a choice of an estimator. It is not adequate or scientific to choose an estimator because one happens to like it or thinks it does well, or because one is comfortable with it. Finally, it is necessary to have an estimate of the sampling variance of the estimate of bird density. It is a major failing that most of the estimators in the biological literature have no associated estimates of precision.

CRITERIA FOR ROBUST ESTIMATION

The true detection function $g(x)$ is not known; moreover the work on line transect sampling in ornithology (and elsewhere) shows that the detection probability can vary due to numerous factors. Consequently, one cannot use a restrictive model for the detection probability and expect to get reliable estimates of density. A "robust" approach is needed; i.e., the estimator of bird density needs to be free of restrictive assumptions about the detection probability. The properties of $\hat{D}$ depend almost entirely on the estimator of $f(0)$, which depends, in turn, on the model chosen for the distribution of distances, and on the estimator used for $f(0)$. We have proposed several criteria that an estimator should satisfy in order to ensure reliable estimates of bird density from line transect sampling (Burnham et al. 1979).

Four criteria relate primarily to the properties of the assumed model for the sampling distribution of perpendicular distances. In order of importance these are: (1) *model robustness*; (2) *pooling robustness*; (3) *shape criterion*; and (4) *estimator efficiency*. Two additional criteria relate to promoting robustness of data analyses to common problems with transect distance data: (5) *data truncation*; and (6) *data grouping*. These last two criteria mean that the estimator of $f(0)$ should allow truncation of the data and should allow, or be developed to apply to, grouped data. Many line transect estimators in the literature are valid *only* for untruncated, ungrouped data.

Model robustness means that $f(x)$, the distribution of perpendicular distance data, is modeled with a general, flexible function, one that can take on a wide variety of shapes. Methods based on specific functional forms such as the negative exponential model are not *model robust* (see Burnham et al. 1980:162).

If an estimator is *pooling robust,* the fact that some birds off the line go undetected becomes totally irrelevant provided the basic assumptions are closely met. Data could be stratified by all possible factors likely to affect the detectability of birds and an estimate of density made for each strata. These separate estimates could then be combined into an estimate of total bird density;

this is a stratified estimator, $\hat{D}_s$. The alternative is to take the total set of data (n, $x_1, \ldots, x_n$ and L) and compute from this "pooled" data (pooled over replicate lines, observers and any other potential strata) an estimator, $\hat{D}_p$. An estimator of $f(0)$ is *pooling robust* if these two approaches produce the same estimate of density, i.e., $\hat{D}_s \equiv \hat{D}_p$. Thus, such an estimator is not affected by pooling the data over all the known and unknown factors that can effect the probability of detecting birds.

For line transect sampling of birds, it is very reasonable to assume that the detection function is 1 near the transect center line and hence has a "shoulder" near the line. This shoulder aspect of the shape of the detection function should be imposed on the model used to estimate $f(0)$. Mathematically, this is easy to do by specifying that the derivative of $f(x)$ at $x = 0$ is zero; hence $f'(0) = 0$ is the *shape criterion*. It means that the assumed detection function falls off very slowly near the transect line.

Criterion four, *estimator efficiency*, means that the estimator should have made the most use of the information in the distance data to estimate $f(0)$. An efficient estimator has a relatively small sampling variance. It is often easy to suggest ad hoc estimators; such ad hoc estimators are rarely efficient and are often badly biased.

The only general class of models that satisfy these four criteria are ones linear in their parameters, such as the polynomial:

$$f(x) = a_0 + a_1x + a_2x^2 + a_3x^3 + a_4x^4 \quad (7)$$

(see, for example, Gates and Smith 1980). However, $f(x)$ of Eq. (7) does not satisfy the *shape criterion* unless the parameter $a_1 = 0$. We considered the polynomial method, but found a better method for estimation of bird density. That method, the Fourier (pronounced *Foureay*) series estimator, is described in the next section.

Transect data in ornithology are generally taken with a finite truncation point, w. In other applications, w is often effectively infinite. Typically, it then will be necessary to delete a few "outliers" at extreme distances (which is why we presented criterion five). This is done by establishing a truncation point, w^*, and ignoring all data beyond distance w^*. This sort of data truncation may also be necessary with some species of birds. We distinguish between w^* and w because w^* is established after data collection while the transect width, w, is established before sampling. Such data truncation leads to more robust estimates of density (see e.g., Burnham et al. 1980:108–111). It is necessary to establish a truncation value w^* to apply the Fourier series estimator; however, it is entirely possible to take $w^* \equiv w$.

THE FOURIER SERIES ESTIMATOR

Ungrouped Data

The general estimator of density is

$$\hat{D} = \frac{n\hat{f}(0)}{2L}. \quad (8)$$

The estimator of $f(0)$ based on the Fourier series expansion is

$$\hat{f}(0) = \frac{1}{w^*} + \sum_{k=1}^{m} \hat{a}_k. \quad (9)$$

(The line length, L, and perpendicular distances must all be expressed in the same units.) The estimated Fourier coefficients $\hat{a}_1, \hat{a}_2, \ldots, \hat{a}_m$ are computed from the ungrouped distance data $x_1, x_2, \ldots, x_n$ using the formula

$$\hat{a}_k = \frac{2}{nw^*}\left[\sum_{i=1}^{n} \cos\left(\frac{k\pi x_i}{w^*}\right)\right]. \quad (10)$$

Consider the second coefficient, $\hat{a}_2$ (i.e., $k = 2$), for a survey where 45 ($=n$) birds were detected within a 100 m wide (on each side) line transect ($w = w^* = 100$), then

$$\hat{a}_2 = \frac{2}{45 \times 100}\left[\sum_{i=1}^{45} \cos\left(\frac{2 \times 3.1416x_i}{45}\right)\right].$$

(If the computations are to be done on a small calculator, be certain that the cosine function allows the argument to be in radians, not degrees). After simplification,

$$\hat{a}_2 = 0.000444\left[\sum_{i=1}^{45} \cos(0.1396x_i)\right].$$

The number of Fourier coefficients computed to estimate $f(0)$ is determined by choosing the first value of m such that

$$\frac{1}{w^*}\left[\frac{2}{n+1}\right]^{\frac{1}{2}} \geqslant |\hat{a}_{m+1}|, \quad (11)$$

where $|\hat{a}_{m+1}|$ is the absolute value of $\hat{a}_{m+1}$. Equation (11) is called a "stopping rule." This rule for selecting the number of terms in the Fourier series represents a tradeoff between achieving small bias and always having a large number of terms (m), thereby getting a large sampling variance, or between always having a small m and having a possibly biased estimator. Typically, m is only 1, 2, or 3 and rarely needs to be as large as 5 or 6. In fact, if the above rule indicates $m \geqslant 6$ in nongame bird applications, something is probably wrong with the perpendicular distance data x_i (such as large rounding errors or mistakes in recording the field measurement). If the data, through Eq. (11), indicate, say, $m = 2$, then $f(0)$ is estimated as

$$\hat{f}(0) = \frac{1}{w^*} + \hat{a}_1 + \hat{a}_2.$$

An estimate of the sampling variance of $\hat{D}$ is given by

$$\hat{\text{var}}(\hat{D}) = \hat{D}^2\left[\frac{\hat{\text{var}}(n)}{n^2} + \frac{\hat{\text{var}}(\hat{f}(0))}{(\hat{f}(0))^2}\right]. \quad (12)$$

Estimates of the sampling variances var(n) and var($\hat{f}$(0)) are discussed below.

The estimate of var($\hat{f}$(0)) is based on the estimated Fourier coefficients, $\hat{a}_1, \hat{a}_2, \ldots, \hat{a}_m$. The estimated variance-covariance matrix for these coefficients is

$$\hat{\text{cov}}(\hat{a}_k, \hat{a}_j) = \frac{1}{(n-1)}\left[\frac{1}{w^*}(\hat{a}_{k+j} + \hat{a}_{k-j}) - (\hat{a}_k\hat{a}_j)\right], \quad (13)$$

For $k > j$, use $\hat{\text{cov}}(\hat{a}_j, \hat{a}_k) \equiv \hat{\text{cov}}(\hat{a}_k, \hat{a}_j)$ and, for $k = j$, use $\hat{a}_0 \equiv 2/w^*$. Of course, for $k = j$, $\hat{\text{cov}}(\hat{a}_k,\hat{a}_j) = \hat{\text{var}}(\hat{a}_k)$.

Because the estimator of $f(0)$ (Eq. 9) is the sum of m Fourier coefficients (plus a constant term $1/w^*$), the sampling variance of $\hat{f}(0)$ is the sum of all the sampling variances and covariances of these m coefficients $\hat{a}_k$:

$$\hat{\text{var}}(f(0)) = \sum_{j=1}^{m}\sum_{k=1}^{m} \hat{\text{cov}}(\hat{a}_j, \hat{a}_k). \quad (14)$$

Equation (14) looks complex. However, consider the case $m = 2$; then Eq. (14) is the sum of the 4 elements in the 2×2 matrix

$$\begin{bmatrix} \hat{\text{cov}}(\hat{a}_1, \hat{a}_1) & \hat{\text{cov}}(\hat{a}_1, \hat{a}_2) \\ \hat{\text{cov}}(\hat{a}_2, \hat{a}_1) & \hat{\text{cov}}(\hat{a}_2, \hat{a}_2) \end{bmatrix},$$

or

$$\hat{\text{var}}(\hat{f}(0)) = \sum_{j=1}^{2}\sum_{k=1}^{2} \hat{\text{cov}}(\hat{a}_j, \hat{a}_k). \quad (15)$$

The sampling variance of n is harder to estimate, but it can be approached in several ways. We will illustrate the special case where the survey is conducted on R replicate lines of equal length, then

$$\hat{\text{var}}(n) = \frac{R}{R-1}\sum_{j=1}^{R}(n_j - \bar{n})^2 \quad (16)$$

where $\bar{n} = \frac{1}{R}\sum_{j=1}^{R} n_j$. Other approaches are found in Burnham et al. (1980:51–55).

Finally, the estimated standard error of $\hat{D}$ (written $\hat{\text{se}}(\hat{D})$) is the square root of the sampling variance of $\hat{D}$,

$$\hat{\text{se}}(\hat{D}) = \sqrt{\hat{\text{var}}(\hat{D})}.$$

While we do not recommend doing these computations by hand, it is certainly possible and could be accomplished in one to two hours for a typical data set. To avoid rounding errors during calculations and to allow many additional features of the data to be explored, we recommend the use of a computer program to perform the arithmetic (see the section: PROGRAM TRANSECT).

Grouped Data

In many surveys, it is convenient to take the perpendicular distance data only by intervals (say, 0–20 m, 20–50 m, 50–100 m and 100–200 m), instead of measuring and recording the exact distance for each individual. The intervals need not be equal in size and, while there can be as few as two groups, it is preferable to have at least four, and 5–8 groups is more reasonable.

Taking the field data by intervals as grouped data should not be an excuse for inexact field procedures. Each observation should be properly recorded in the correct interval. Also, distance data should be taken ungrouped (i.e., distances precisely known), if possible.

Data can be analyzed as grouped, even when the original data were recorded in the field as ungrouped measurements. There are a variety of advantages in grouping the data for analysis, especially when the ungrouped data contain rounding errors, bias, or other anomalies. Estimates of density from the grouped data will then be more reliable than those based on the original ungrouped data.

Whereas, for ungrouped data, the perpendicular distance data are $x_1, x_2, \ldots, x_n$. For the grouped case, the data are the counts of birds seen in each interval, say $n_1, n_2 \ldots, n_k$, corresponding to the 1st, 2nd, . . . , kth interval. These counts can be used to estimate $f(0)$ with the Fourier series procedure. The proper computations are quite difficult and cannot be done without a computer.

Program "Transect"

We developed a comprehensive computer program, TRANSECT, to facilitate the analysis of line transect data. Program documentation is given by Laake et al. (1979). TRANSECT provides a convenient and thorough analysis tool which eliminates tedious calculations. It provides a variety of options for describing the basic data and includes several estimators of density and graphical and statistical goodness of fit tests.

The program consists of a main routine and 57 subroutines; there are approximately 7200 statements. Numerous comment statements document major features of the program. The program is written in ANSI Fortran IV. It is very portable and has been successfully run on CDC, Burroughs, IBM, and DEC computer systems. The program, example data, and output are available from the SHARE Program Library Agency, P.O. Box 12076, Research Triangle Park, N.C. 27709, at a cost of approximately $40.00. Specifications for the tape (e.g., 7 or 9 track, 800 or 1600 bpi, etc.) and the program No. 3600-05-003-007 should be given at the time of ordering.

EXAMPLE APPLICATIONS OF THE FOURIER SERIES ESTIMATOR TO NONGAME BIRD TRANSECT DATA

This section provides some examples of the Fourier series estimator. Hopefully, these examples will help make the method more fully understood and will help illustrate the points previously discussed. The data used in these examples is from a study done under a contract with the U.S. Fish and Wildlife Service to estimate breeding bird densities on coal lands (Hickey and Mikol 1979). We selected a subset of the data from two species: the Western Meadowlark (*Sturnella neglecta*) and the Lark Bunting (*Calamospiza melanocorys*). The Fourier series estimator is illustrated in both ungrouped and grouped formats. (Not all of the capabilities of program TRANSECT are illustrated here; for further examples see Burnham et al. 1980:90–120).

UNGROUPED DATA

The analysis of line transect data with the Fourier Series estimator can be thought of in terms of eight steps. These eight steps are the same regardless if the data are grouped or ungrouped (although different mathematical methods are employed at several steps). They are as follows: (1) estimate the Fourier coefficients a_i from the data; (2) determine the number of terms (m) to be used; (3) estimate $f(0)$; (4) estimate variances and covariances of the $\hat{a}_i$; (5) estimate the variance of $\hat{f}(0)$; (6) estimate D; (7) estimate the variance of $\hat{D}$; and (8) examine the goodness of fit.

These eight steps will be illustrated with the output of TRANSECT using ungrouped, replicated data on the Lark Bunting. For these data, there were 209 total observations recorded on five separate occasions. The length of the transect was 1000 meters (for each sampling occasion). For each observation, the perpendicular distance was recorded in meters and only birds within 100 meters ($w = 100$) were noted. A sample histogram of the perpendicular distances is illustrated in Fig. 2. This histogram shows that the detection of Lark Buntings decreased considerably at distances of 80–100 m.

For a first analysis, the data from the five replications were pooled to make one estimate of bird density. The estimates of the Fourier coefficients, $\hat{a}_i$, (step 1) were calculated by program TRANSECT using Eq. (10); more specifically for these data

$$\hat{a}_k = \frac{2}{100 \times 209}\left[\sum_{i=1}^{209} \cos\left(\frac{k(3.1416)x_i}{200}\right)\right]$$

where x_i are the individual measurements of perpendicular distance. The stopping rule value used to determine the number of terms (step 2) is calculated by the lefthand side of Eq. (11) and has the value of 0.000975 in this example. This results in (m =) five terms being selected; the estimates, $\hat{a}_1, \ldots, \hat{a}_5$ are shown in Table 1. The estimate of $f(0)$ (step 3), as calculated by Eq. (9), is

$$\hat{f}(0) = \frac{1}{w} + \sum_{k=1}^{5} \hat{a}_k = 0.006894.$$

The estimates of the variances and covariances of the $\hat{a}_i$ (step 4) require that $2m$ terms be computed. The estimates are computed from Eq. (13). The estimates of the standard errors are given with the point estimates in Table 1. The covariances, such as between $\hat{a}_1$, and $\hat{a}_2$, are not shown on the output; however, the related quantities, the correlation coefficients between $\hat{a}_i$ and $\hat{a}_j$, are printed by TRANSECT (see Table 1). The estimate of the variance of $\hat{f}(0)$ (step 5) is computed using Eq. (14); the result for this example is given in Table 1.

The estimate of density (step 6) only requires basic arithmetic and in this case is

$$\hat{D} = \frac{209 \times 0.006894}{2 \times 5000} = 0.0001441.$$

This estimate is in terms of numbers per square meter because both line length, L, and distances, x_i, were in meters. In order to get the result in Table 1 ($\hat{D}$ in numbers per hectare), the estimate must be multiplied by the number of square meters in a hectare (10,000), which gives 1.44 birds per ha.

The estimate of var($\hat{D}$) (step 7) can be accomplished in a variety of ways, for a discussion of this see Burnham et al. (1980:51–55). In this case, the estimated variance of $\hat{D}$ (the density estimate) was calculated using Eq. (12), and the variance of n was calculated empirically from the five replicates using Eq. (16). The estimated standard error $\hat{D}$ is given in Table 1.

The chi square goodness of fit test of the estimator of $f(x)$ (step 8) is shown in Table 2. From

```
FREQUENCY      23.  30.  35.  19.  25.  21.  22.  23.   6.   5.
------------------------------------------------------------------
   35                    ****
   34                    ****
   33                    ****
   32                    ****
   31                    ****
   30               **** ****
   29               **** ****
   28               **** ****
   27               **** ****
   26               **** ****
   25               **** ****      ****
   24               **** ****      ****
   23          **** **** ****      ****           ****
   22          **** **** ****      ****      **** ****
   21          **** **** ****      **** **** **** ****
   20          **** **** ****      **** **** **** ****
   19          **** **** **** **** **** **** **** ****
   18          **** **** **** **** **** **** **** ****
   17          **** **** **** **** **** **** **** ****
   16          **** **** **** **** **** **** **** ****
   15          **** **** **** **** **** **** **** ****
   14          **** **** **** **** **** **** **** ****
   13          **** **** **** **** **** **** **** ****
   12          **** **** **** **** **** **** **** ****
   11          **** **** **** **** **** **** **** ****
   10          **** **** **** **** **** **** **** ****
    9          **** **** **** **** **** **** **** ****
    8          **** **** **** **** **** **** **** ****
    7          **** **** **** **** **** **** **** ****
    6          **** **** **** **** **** **** **** **** ****
    5          **** **** **** **** **** **** **** **** **** ****
    4          **** **** **** **** **** **** **** **** **** ****
    3          **** **** **** **** **** **** **** **** **** ****
    2          **** **** **** **** **** **** **** **** **** ****
    1          **** **** **** **** **** **** **** **** **** ****
-------------+----+----+----+----+----+----+----+----+----+----+--
INTERVAL           10.       30.       50.       70.       90.
END-POINTS  0.00        20.       40.       60.       80.       0.10E+03
```

FIGURE 2. Histogram of perpendicular distances (in meters) for the Lark Bunting data when each bird is treated as a separate, independent sighting.

Table 2, the Fourier series model does not provide a very good fit. In fact, whenever the Fourier series requires more than three terms it reflects, from our experience, some anomalies in the data.

In this example, it appears there is "heaping" of distances at convenient values. This heaping resulted from two problems: (1) the distances were recorded at convenient values (they were not really measured); and (2) the birds (territorial males) were sometimes observed in temporary "groups" of two or three, during territorial interactions, and yet each bird sighted was treated as a single independent observation. The first problem can be corrected by more accurate measurements or can be made less severe by grouping the data. The second problem occurred because the above analysis is improper and it was done here only to show the effect of ignoring clustering (i.e., violating assumption 4) and heaping.

To properly analyze these data, any cluster of birds must be treated as a single entity and the density of birds estimated in two stages. First, a density of clusters is estimated by the Fourier series method, then that cluster density is multiplied by the average size of Lark Bunting clusters. This reduces the sample size from 209 birds observed to 166 clusters of birds observed; in most cases the "cluster" is only one bird. A

```
FREQUENCY      22.  22.  26.  16.  22.  15.  14.  18.   6.   5.
------------------------------------------------------------------
          26             ****
          25             ****
          24             ****
          23             ****
          22   **** **** ****      ****
          21   **** **** ****      ****
          20   **** **** ****      ****
          19   **** **** ****      ****
          18   **** **** ****      ****           ****
          17   **** **** ****      ****           ****
          16   **** **** **** **** ****           ****
          15   **** **** **** **** **** ****      ****
          14   **** **** **** **** **** **** **** ****
          13   **** **** **** **** **** **** **** ****
          12   **** **** **** **** **** **** **** ****
          11   **** **** **** **** **** **** **** ****
          10   **** **** **** **** **** **** **** ****
           9   **** **** **** **** **** **** **** ****
           8   **** **** **** **** **** **** **** ****
           7   **** **** **** **** **** **** **** ****
           6   **** **** **** **** **** **** **** **** ****
           5   **** **** **** **** **** **** **** **** **** ****
           4   **** **** **** **** **** **** **** **** **** ****
           3   **** **** **** **** **** **** **** **** **** ****
           2   **** **** **** **** **** **** **** **** **** ****
           1   **** **** **** **** **** **** **** **** **** ****
--------------+----+----+----+----+----+----+----+----+----+----+--
INTERVAL          10.       30.       50.       70.       90.
END-POINTS 0.00       20.       40.       60.       80.       0.10E+03
```

FIGURE 3. Histogram of perpendicular distances (in meters) for the Lark Bunting data. In this analysis, the observations are of clusters of birds (each bird is not necessarily treated as a separate, independent sighting).

TABLE 1
SUMMARY OF THE DENSITY ESTIMATION FOR THE LARK BUNTING DATA WITH THE FOURIER SERIES ESTIMATOR (EACH BIRD IS TREATED AS A SEPARATE, INDEPENDENT SIGHTING)[a]

Parameter	Point estimate	SE	% C.V	95% C.I.	
A(1)	0.3447E-02	0.8819E-03	25.6	0.1718E-02	0.5175E-02
A(2)	−0.2635E-02	0.9009E-03	34.2	−0.4401E-02	−0.8692E-03
A(3)	0.1183E-02	0.9840E-03	83.2	−0.7456E-03	0.3111E-02
A(4)	−0.2425E-02	0.9930E-03	40.9	−0.4372E-02	−0.4792E-03
A(5)	−0.2676E-02	0.9679E-03	36.2	−0.4573E-02	−0.7785E-03
F(0)	0.6894E-02	0.2162E-02	31.4	0.2656E-02	0.1113E-01
D	1.441	0.4531	31.4	0.1830	2.699

Sampling correlation of estimated parameters

	1	2	3	4	5
1	1.000	0.335	−0.303	−0.036	−0.069
2	0.335	1.000	0.059	−0.161	0.043
3	−0.303	0.059	1.000	0.199	−0.062
4	−0.036	−0.161	0.199	1.000	0.107
5	−0.069	0.043	−0.062	0.107	1.000

[a] Density (D) units are numbers/ha.

TABLE 2

CHI-SQUARE GOODNESS OF FIT TEST FOR THE FOURIER SERIES ESTIMATOR (FROM FIG. 2) FIT TO THE LARK BUNTING DATA (EACH BIRD TREATED AS A SEPARATE, INDEPENDENT SIGHTING)[a]

Cell I	Cut points		Observed values	Expected values	Chi-square values
1	0.0	10.0	23.	17.56	1.69
2	10.0	20.0	30.	29.51	0.825E-02
3	20.0	30.0	35.	32.59	0.178
4	30.0	40.0	19.	22.89	0.662
5	40.0	50.0	25.	18.70	2.13
6	50.0	60.0	21.	25.73	0.871
7	60.0	70.0	22.	28.20	1.36
8	70.0	80.0	23.	18.69	0.992
9	80.0	90.0	6.	8.86	0.921
10	90.0	100.0	5.	6.26	0.255

[a] Total chi-square value = 9.063; degrees of freedom = 4; probability of a greater chi-square value = 0.05953485.

TABLE 4

CHI-SQUARE GOODNESS OF FIT TEST FOR THE FOURIER SERIES ESTIMATOR OF THE LARK BUNTING DENSITY WHEN THE OBSERVATIONS ARE CLUSTERS OF BIRDS[a]

Cell I	Cut points		Observed values	Expected values	Chi-square values
1	0.0	10.0	22.	19.49	0.324
2	10.0	20.0	22.	20.07	0.186
3	20.0	30.0	26.	20.84	1.28
4	30.0	40.0	16.	21.20	1.28
5	40.0	50.0	22.	20.58	0.987E-01
6	50.0	60.0	15.	18.70	0.732
7	60.0	70.0	14.	15.75	0.195
8	70.0	80.0	18.	12.36	2.57
9	80.0	90.0	6.	9.38	1.22
10	90.0	100.0	5.	7.64	0.912

[a] Total chi-square value = 8.793; degrees of freedom = 7; probability of a greater chi-square value = 0.26788825.

sample histogram of perpendicular distances is illustrated in Fig. 3. The summary of the density estimation from TRANSECT is provided in Table 3. In this analysis, the estimator only required two terms and the goodness of fit is acceptable (Table 4). The point estimate of cluster density ($\hat{D}$) is 1.938 clusters per hectare. In this example, the average cluster size ($\bar{c}$) can be calculated as the arithmetic mean; it is 1.259 with a standard error of 0.0498. This is an unbiased estimate of the true cluster size if the detection probability is independent of cluster size (Burnham et al. 1980:192–194). The estimated density of Lark Buntings is then

$$\hat{D} = \hat{D}_c{*}\bar{c} = 1.938 \times 1.259 = 2.440 \text{ birds/hectare.}$$

The standard error of $\hat{D}$ is calculated from

$$\hat{\mathrm{se}}(\hat{D}) = \hat{D}(\mathrm{cv}^2(\bar{c}) + \mathrm{cv}^2(\hat{D}_c))^{\frac{1}{2}} = 2.440\left[\left(\frac{0.0498}{1.259}\right)^2 + \left(\frac{0.2963}{1.938}\right)^2\right]^{\frac{1}{2}} = 0.3853.$$

The point estimate of density increased dramatically compared to the results in the previous, incorrect, analyses. More importantly, the coefficient of variation is reduced considerably. This results from the reduction in the number of terms in the model and the reduction in the "heaping."

An alternative method of estimating the variance of density is to make separate density estimates for the replicate transects and calculate the variance empirically. This approach will be illustrated using the same Lark Bunting data to estimate density of clusters. Data were collected on five separate dates by the same observer,

TABLE 3

SUMMARY OF DENSITY ESTIMATION OF CLUSTERS OF LARK BUNTINGS WITH THE FOURIER SERIES ESTIMATOR[a]

Parameter	Point estimate	SE	% C.V.	95% C.I.	
A(1)	0.3628E-02	0.1007E-02	27.8	0.1654E-02	0.5601E-02
A(2)	−0.1956E-02	0.1055E-02	54.0	−0.4024E-02	0.1128E-03
F(0)	0.1167E-01	0.1662E-02	14.2	0.8415E-02	0.1493E-01
D	1.938	0.2963	15.3	1.115	2.760

Sampling correlation of estimated parameters

	1	2
1	1.000	0.298
2	0.298	1.000

[a] Density(D) units are numbers/hectares.

```
FREQUENCY    6.   8.   3.   3.   3.   4.   2.   1.   2.   2.
-----------------------------------------------------------------
    8            ****
    7            ****
    6       **** ****
    5       **** ****
    4       **** ****                ****
    3       **** **** **** **** **** ****
    2       **** **** **** **** **** **** ****      **** ****
    1       **** **** **** **** **** **** **** **** **** ****
-----------+----+----+----+----+----+----+----+----+----+----+--

FREQUENCY    4.   4.   7.   6.   5.   5.   3.   3.   1.   0.
-----------------------------------------------------------------
    7                 ****
    6                 **** ****
    5                 **** **** **** ****
    4       **** **** **** **** **** ****
    3       **** **** **** **** **** **** **** ****
    2       **** **** **** **** **** **** **** ****
    1       **** **** **** **** **** **** **** **** ****
-----------+----+----+----+----+----+----+----+----+----+----+--

FREQUENCY    6.   5.   2.   4.   5.   3.   2.   6.   1.   2.
-----------------------------------------------------------------
    6       ****                               ****
    5       **** ****           ****           ****
    4       **** ****      **** ****           ****
    3       **** ****      **** **** ****      ****
    2       **** **** **** **** **** **** **** ****      ****
    1       **** **** **** **** **** **** **** **** **** ****
-----------+----+----+----+----+----+----+----+----+----+----+--

FREQUENCY    1.   3.   7.   2.   6.   1.   3.   4.   0.   1.
-----------------------------------------------------------------
    7                 ****
    6                 ****      ****
    5                 ****      ****
    4                 ****      ****           ****
    3            **** ****      ****      **** ****
    2            **** **** **** ****      **** ****
    1       **** **** **** **** **** **** **** ****      ****
-----------+----+----+----+----+----+----+----+----+----+----+--

FREQUENCY    5.   2.   7.   1.   3.   2.   4.   4.   2.   0.
-----------------------------------------------------------------
    7                 ****
    6                 ****
    5       ****      ****
    4       ****      ****                **** ****
    3       ****      ****      ****      **** ****
    2       **** **** ****      **** **** **** **** ****
    1       **** **** **** **** **** **** **** **** ****
-----------+----+----+----+----+----+----+----+----+----+----+--
INTERVAL        10.       30.       50.       70.       90.
END-POINTS 0.00      20.       40.       60.       80.       0.10E+03
```

FIGURE 4. Histograms of the perpendicular distances (in meters) for the five replicate transects of the Lark Bunting data; observations are clusters of birds.

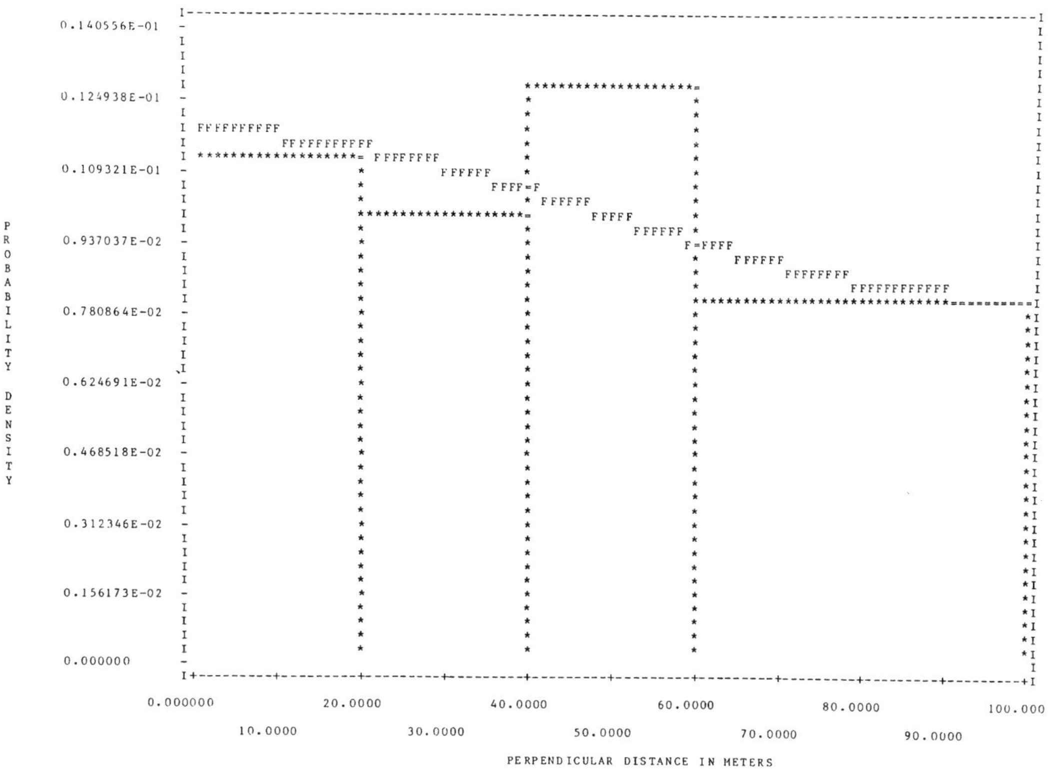

FIGURE 5. Graphical representation of the goodness of fit of the Fourier series estimator (from Table 6) for the Western Meadowlark data.

over a three week period. The histograms of the perpendicular distances for these five replicates are illustrated in Fig. 4. These histograms reflect a considerable amount of variability. This further illustrates the need for a *model* and *pooling robust estimator*.

The Fourier series estimator was applied to each of the five replicates and the results summarized in Table 5. The average density estimate and its standard error are compared to the density estimate for the pooled data. If all of the estimates required the same number of terms, then the average density estimate would be exactly the same as the pooled density estimate. This is not the case and there is a considerable amount of variability in the shape of the histograms and the number of terms used. However, because the Fourier series is also very *model robust,* the two point estimates are very close. This method of calculating a variance for density is quite desirable but it requires a substantial sample size like this example.

Grouped Data

Often there are times in which the exact perpendicular distance cannot be measured, such as in aerial surveys, and the data are collected by distance intervals. The analysis of such grouped data with the Fourier series is not simple, rather it requires the use of a computer program. However, the same eight steps described previously can still be used but the estimates of the model coefficients a_i and their variances and covariances are calculated through numerical methods.

The analysis of grouped data will be illustrated with meadowlark data from the same transects used to collect the Lark Bunting data. There were 90 total observations of perpendicular distance on the five separate occasions. These perpendicular distance data have been grouped into four intervals.

The estimates of the parameters, a_i, are calculated by TRANSECT using maximum likelihood estimation. First, a model with one parameter is estimated, then two and successively more parameters are used in the model. The significance of adding each additional term is tested. For this example, the one-term Fourier series model is selected as the appropriate model for these meadowlark data. The estimates of $f(0)$, bird density, and their standard errors, are

TABLE 5
SUMMARY OF DENSITY ESTIMATES OF THE CLUSTERS OF LARK BUNTING FOR THE FIVE REPLICATE TRANSECTS[a]

Replicate	Sample size (n)	Number of terms in Fourier series (m)	Estimate of $f(0)$	Density estimate
1	30	1	0.0129	1.934
2	28	0	0.0100	1.400
3	36	0	0.0100	1.800
4	38	2	0.0100	1.959
5	34	1	0.0155	2.630
Averaged	166	—	0.0117	1.944 (0.198)
Pooled	166	2	0.0117	1.938 (0.296)

[a] Also shown are the weighted average of the five density estimates and the pooled density estimate (standard errors are in parentheses).

computed from the a_i just as in the ungrouped case; these results are shown in Table 6. The chi square goodness of fit test for this one-term Fourier series model indicated a good fit (χ^2 = 1.757, 2 degrees of freedom, P = 0.415). Figure 5 shows the relative frequency histogram of the grouped data and the fitted one-term Fourier series model.

COMMENTS ON OTHER ANALYSIS METHODS

GENERAL COMMENTS

A variety of methods are available for the estimation of population size or density of biological populations. Here, we focus on variations of the line transect method, including the strip transect, in order to make some comparisons and suggestions.

Strip transects are merely very long, narrow quadrats and standard sampling theory applies. Strip transects do not require that distances be measured to estimate density. Line transect sampling offers two advantages over strip transects: (1) only animals on and near the centerline must be detected with certainty, and (2) the additional data taken at distances where the probability of detection is less than 1 can be used. This latter feature allows much more data to be used in the estimation of density. The ability to take the data as grouped greatly extends the applicability of the line transect procedure. In general, we recommend the use of line transect sampling over strip transect sampling in cases where both are appropriate.

The various approaches to density estimation using the sighting distance (r_i) and sighting angle (θ_i) are inferior to those based on perpendicular distances (x_i). Methods based on r_i and θ_i are quite sensitive to even small departures from the critical assumptions and these methods require additional assumptions as well as those required for estimation based on perpendicular distance data. The underlying models for the analysis of r_i and θ_i are very idealized and represent only crude approximations to the real situation. Finally, the estimators (e.g., Hayne's method, Hayne, 1949) are very sensitive to small sighting distances (i.e., the term $1/r_i$ will dominate the estimate if the ith sighting distance is quite small). We do not recommend the estimation of density based on sighting distances and angles. If these data are all that are available, convert them to perpendicular distances and proceed on that basis.

Good methods for the estimation of bird density must be based on the following conditions: (1) sound theoretical development, (2) *model robustness,* (3) *pooling robustness,* (4) the *shape criterion,* and (5) high *estimator efficiency.* Consideration of these criteria leads us to recommend the Fourier series estimator as a useful, omnibus procedure. Further information on other analysis methods is given by Burnham et al. (1980: Part 4).

Finally, we caution against the use of the numerous ad hoc procedures that have been sug-

TABLE 6
SUMMARY OF DENSITY ESTIMATION FOR THE GROUPED WESTERN MEADOWLARK DATA USING THE FOURIER SERIES ESTIMATOR[a]

Parameter	Point estimate	SE	% C.V.[b]	95% C.I.[b]	
A(1)	0.1807E-02	0.1528E-02	84.6	−0.1188E-02	0.4803E-02
F(0)	0.1181E-01	0.1528E-02	12.9	0.8812E-02	0.1480E-01
D	1.063	0.1969	18.5	0.5159	1.609

[a] Density (D) units are numbers/ha.

[b] Notes on variance calculations and confidence intervals: the confidence intervals for the coefficients $A(I)$ and $F(0)$ were constructed by assuming asymptotic normality and using the Z-value 1.96; the variance of n was estimated using replicate line lengths (var(N) = 1.43); the confidence interval for density was constructed with a t distribution with the degrees of freedom equal to the number of line lengths − 1; and the t-value with 4 degrees of freedom is 2.776. Squared coefficient of variation for n = 0.1759E-01. Squared coefficient of variation for $F(0)$ = 0.1675E-01. Percent of the variation of density attributable to the sampling variance of n = 51.22.

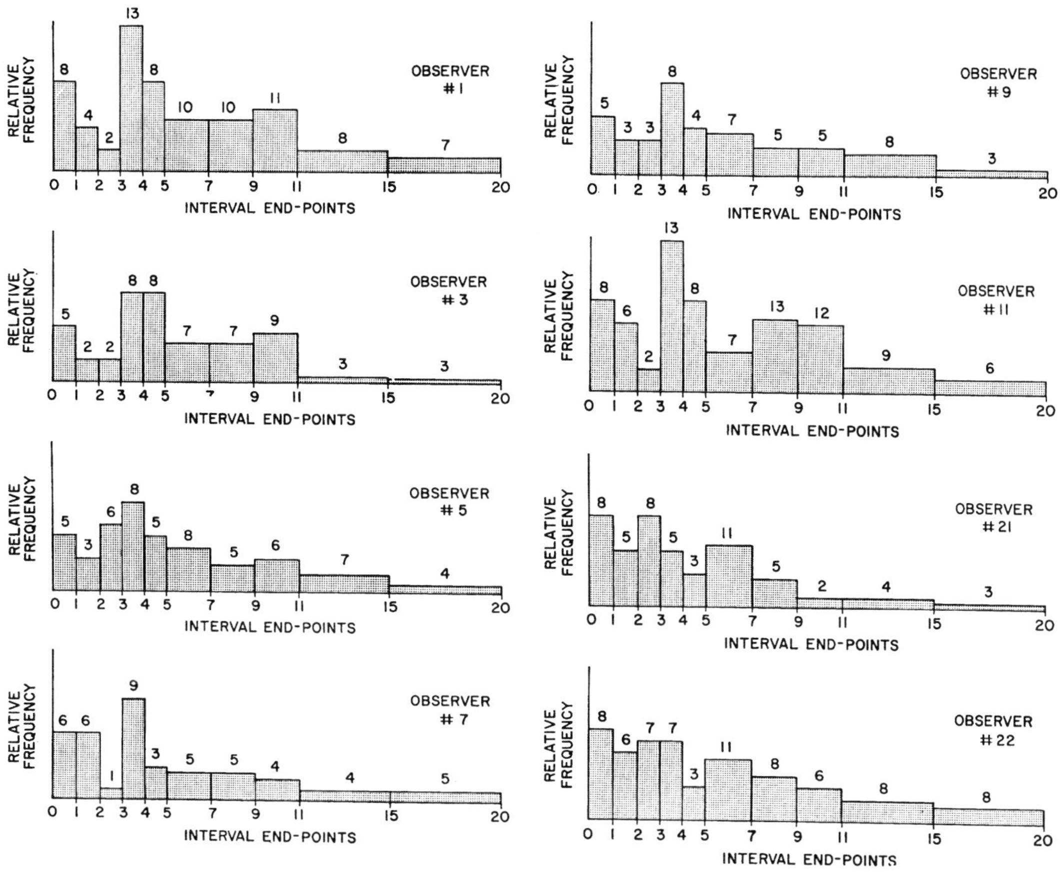

FIGURE 6. Eight histograms from the "stake" data. This illustrates the variability in results among observers.

gested in the ornithological literature. These methods do not have a firm foundation or basis. They often rest on unknown assumptions. Nothing is known of their properties or performance (although we feel confident that most are very inefficient), and estimates of bias and precision are not available. Without estimates of sampling variance, useful inferences concerning, and test hypotheses regarding bird density cannot be made.

The Number Seen (n) as an Index to Density

The number of birds seen (n) on a line transect survey is typically a poor index to density. Unless it is assumed that the detection function $g(x)$ is the same among observers, habitat types, bird species, time of day, or season of year, n is not a relative index to abundance. However, we illustrate below that it is not reasonable to make such an assumption.

Laake (1978) placed 150 wooden stakes (2.5 cm × 5 cm × 46 cm) in a 4 ha sagebrush-grass study area in northern Utah; density of stakes was 37.5 per ha. During the fall of 1977 and spring of 1978, undergraduate students walked a well marked 1000 m transect following the field methods detailed in Anderson et al. (1979). Eleven line transect surveys were conducted each year. Different observers conducted each survey except in one case, where two surveys were run by the same person. Stakes (driven in the ground) were placed randomly to avoid dependent sightings. The observers were carefully instructed and supervised and fatigue was probably only a minor factor because each survey could be completed in approximately two hours.

The number of stakes observed varied substantially among observers (Table 7): n ranged from 41 to 100 stakes. In each case, the true population size (N) of stakes was 150. Had n been used as an index to N or D, it would have been poor, suggesting a highly variable population size. The coefficient of variation for n (cv(n)) among observers was 23%. Examination of eight individual histograms (Fig. 6) indicates

TABLE 7
RESULTS OF 22 SURVEYS TO ESTIMATE THE DENSITY OF STAKES IN A 4 HA SAGEBRUSH-GRASS STUDY AREA IN UTAH, 1977–78

Observer	Number of stakes detected	Density estimate ($\hat{D}$)	$\hat{s}e(\hat{D})$
1	81	35.4	4.72
2	72	33.6	4.86
3	54	25.4	4.66
4	56	29.1	4.25
5	57	26.2	4.11
6	68	38.8	6.12
7	48	29.2	6.65
8	49	28.0	5.32
9	51	23.0	3.77
10	68	36.9	4.82
11	84	36.9	4.49
12	48	34.6	8.59
13	75	30.2	4.44
14	61	35.9	6.53
15	60	31.0	8.51
16	100	27.2	4.27
17	55	33.2	6.41
18	61	34.4	7.60
19	46	24.7	5.21
20	41	33.6	8.51
21	54	34.1	5.74
22	72	34.5	6.20
Average	61.9	31.6	5.72

that the underlying detection function $g(x)$ differs greatly among observers. In field studies, the detection function would surely also vary among habitat types and species of birds. These factors, of course, affect *n*, making it an unreliable index.

In contrast to an index, estimates of density were computed for the 22 surveys using the Fourier series method. This is a better procedure as it allows both $g(x)$ and n to vary and still provide a valid *estimate* of density (not just a crude index). In these 22 surveys, the range in estimates of density is only 23.0 to 38.8 stakes per ha and the coefficient of variation for the density estimate ($cv(\hat{D})$) is about half that for the index (14.1%).

The use of a good estimation procedure allows estimates of density for various observers, surveying various habitat types for various species of birds. Estimates of precision are available as are tests of model fit.

An important assumption in line transect sampling is that all birds on, or very near, the centerline of the transect are detected. Mathematically, this is $g(0) = 1$; that is, the probability of detecting an animal at zero distance is one (or 100%). Note from Table 7 that the average density estimate is 31.6 stakes per ha while we know the true density to be 37.5. This bias is at least partially due to the failure of the assumption that $g(0) = 1$. Field procedures must focus on this assumption, or bias will be likely. The use of, for example, dogs, two observers, and all available cues will aid in meeting this important assumption. We can expect the failure of this assumption to be most severe with inanimate objects (e.g., stakes) rather than birds which often respond to the observer.

REASONS WHY BIRDS ARE UNDETECTED

The literature has many examples showing specific reasons why birds are not detected during a line transect survey. Limitations of the observer are often cited as a primary cause such as, inexperience, poor eyesight or hearing, lack of interest or training, or fatigue. The physical setting represents another broad class of reasons why birds that are present go undetected, including habitat type, sun angle, time of day and wind or other inclement weather. The species of bird being surveyed may preclude detection at the greater distances (e.g., small, drab colored birds that do not flush or vocalize readily). Many studies have looked at factors that are associated with incomplete detection and methods proposed to help lessen the proportion of birds that are not detected.

The two reasons for considering why birds are detected (or missed) are: (1) to design and conduct better studies; and (2) to improve distance estimation when the recorded distance depends on the detection cue(s) (rather than being a directly measured distance). The latter case is illustrated when detection depends entirely on hearing a bird and the distance *estimation* is also entirely based on this detection cue. In essence, studies on reasons why birds are detected (or missed) should be aimed at collecting improved *data*.

When, however, these efforts are directed at improved data analysis, they are largely misguided. A properly conducted line transect survey will provide valid estimates of density *even if a very substantial fraction of the birds go undetected*. In fact, the theory for line transect sampling *deliberately* allows birds to go undetected away from the centerline. Only in strip transect surveys is it necessary for all birds in the strip to be detected. As an example, in the 22 surveys summarized in Table 7, only 27–67% of the stakes present were detected. Still, relatively good estimates of density were obtained. The specific *reasons* why birds are not detected is of no consequence in data analysis to estimate density.

PRACTICALITIES

There are 10 key points in line transect sampling that need special emphasis:

(1) The center line of the transect must be straight and well marked. The observer must be able to determine the position of the line at all times.

(2) Care must be taken to ensure that objects on the center line of the transect are always seen. In practice, that requirement often can be met if the observer travels carefully along the center of the line transect at all times.

(3) The width of the transect should be taken as quite large, or effectively unbounded.

(4) All measurements of distances and angles must be accurate. It would be best if a steel tape or other appropriate device were used to ensure a high degree of accuracy. Careless measurements and rounding errors lead to poor estimates of density and sampling variances.

(5) The three basic measurements should all be taken: perpendicular distance, sighting (flushing) distance, and sighting (flushing) angle.

(6) The measurements must be recorded separately for each segment or replicate line of the total transect length.

(7) A target goal should be established for the precision of D in terms of the coefficient of variation; then, the required line length should be determined. As a practical minimum, studies should be designed to ensure that at least 40 objects are seen ($n \geqslant 40$); it might be preferable if the length (L) were sufficient to allow the location of at least 60 to 80 objects ($n \geqslant 60$ to 80).

(8) A pilot survey should be made as an aid in planning the survey design. Often, a simple visit to the area to be surveyed, and basic biological information about the animal and its habits and habitat, will be sufficient to design a survey adequate to estimate density.

(9) The survey should be designed to ensure that the population to be surveyed is not correlated with the sample line transects (e.g., avoid transects running along roads, ridgetops, or stream bottoms).

(10) The survey should be conducted by competent, interested, and trained personnel. This is particularly relevant to points (1), (2), and (4) (above).

It was once thought that an observer could rove through an area and record only the sighting distance to each object detected. This is not true and will lead to very poor estimates of density. Furthermore, we do not recommend the use of sighting distances and sighting angles in estimation of density. These data are useful only as a check on the accuracy of the perpendicular distance data.

Birds on the centerline must be detected with certainty. The primary focus of the field survey must be on and near the centerline. The data near the centerline are far more important than those near w. For this reason, the accuracy of distance measurements near the center line is of critical importance. Measurement errors near w are less critical in influencing $f(0)$. We continue to see data where a large number of animals are recorded as if they had been detected exactly on the center line. These distances were inadvertently rounded to zero distance when, in fact, they may have been several meters from the actual centerline of the transect (see e.g., Burnham et al. 1980:105–108).

Unless the distances are properly measured, the "estimated" distances frequently are rounded to convenient numbers (i.e., 5, 10, 20, 50, or 100). Such data may benefit from grouping during the analysis, however, the group sizes and number of intervals become arbitrary. We believe a reliable survey must either: (1) be based on measured (i.e., steel tape) ungrouped distances; or (2) ensure that each bird is placed in the proper interval if the data are taken as grouped in the field.

It should be obvious that the detection of 5 or 10 birds is insufficient to compute a reasonable estimate of density or even a crude index to abundance. If samples of 40 birds or more cannot be obtained, the money and personnel resources intended for the study should probably be directed elsewhere.

Much of the literature on strip and line transects has focused on the many sources of variability in the detection process (such as observer-to-observer differences, observer fatigue, varying light intensity and relative direction, terrain, vegetation type and density, different auditory cues, season of the year and wind speed). These concerns have relatively little bearing on the data analysis if all animals on the line are detected and an estimation scheme that is *model robust* and *pooling robust* is employed.

If a large proportion of the birds move a substantial distance further from the line than their original location before being detected, then line transect sampling is simply not appropriate. If undetected movement is a minor problem, then there are other estimators somewhat robust to this (see Burnham et al. 1980:120–131).

Estimates of density without a measure of precision or sampling variation are of little value in either research or management programs. Such estimates are untrustworthy, not useful, and reflect poor procedure. Valid inferences or conclusions cannot be made without a good measure of the precision of the estimator. A variety of good procedures now exist for the careful and rigorous analysis of properly collected data. We see no excuse for using the many ad hoc approaches available in the literature.

ACKNOWLEDGMENTS

We thank S. A. Mikol and J. J. Hickey for letting us use some of their Lark Bunting and meadowlark data (from a much larger data set) to illustrate the Fourier series estimator. Gratitude is expressed to C. Snelling and M. Sieverin for assistance with manuscript preparation, and to C. Short for editorial assistance.

Studies in Avian Biology No. 6:483–487, 1981.

ANALYSIS OF BIRD SURVEY DATA USING A MODIFICATION OF EMLEN'S METHOD

FRED L. RAMSEY[1] AND J. MICHAEL SCOTT[2]

ABSTRACT.—This paper describes in general terms the data analysis procedures followed for the Hawaiian Forest Bird Survey. The method consists of first examining detection distances to estimate the Effective Areas Surveyed—a modification of Emlen's Coefficient of Detectability, then estimating density. The notion of Effective Area Surveyed is formulated to allow use of all detections in estimating density.

The Emlen method arises when a particular view of the detectability curve is held. Other views lead to other methods. The Emlen method has the kind of flexibility best able to deal with the particular problems of surveying birds.

To improve its efficiency, we present a modernized version of Emlen's method based on analysis of a Cumulative Detection Curve.

Previous papers in this symposium dealing with the methods of estimating population density have given little, none, or disparaging mention of a method first proposed by J. T. Emlen (1971). It has been termed inefficient, lacking in theoretical foundation, highly subjective and sensitive to arbitrary data groupings. Yet virtually every ornithological paper we have seen that actually attempts to estimate density—either from line transect or variable circular plot surveys—uses Emlen's method directly or in a modified form. Why? It has been suggested that the reason for this is that more efficient methods have not previously entered the ornithological literature. True as that may be, and welcome as the newer methods should be, the purpose of this paper is to demonstrate that Emlen's method need not, as a result, be discarded. It is quite possible to modify Emlen's method to counter criticisms while maintaining its conceptual framework. This, for the most part, we attempt here.

In the following sections we present the conceptual framework of Emlen's method; examine the coefficient of detectability yields to the effective area surveyed as a measure of sampling effort; describe a design for the data analysis of a large survey; and describe a general graphical method for interpreting results of line transect and variable circular plot surveys alike.

EMLEN'S METHOD AND THE CD

Perhaps the least understood feature of Emlen's method is the coefficient of detectability, or CD. Yet the CD, and its cousin—the Effective Area Surveyed (EAS), play an indispensible role in the data analysis.

Let us begin with an abstract view, as in Figure 1 below, of a target region $\mathcal{S}$ of habitat over which a species has uniform density D.

What this means is that the average number of birds, $\mathscr{E}(m)$, to be expected in any specific subregion $\mathscr{R}$ with area $A(\mathscr{R})$ is $\mathscr{E}(m) = D \cdot A(\mathscr{R})$. In particular, if N is the number of birds in the entire region, then we have

$$D = \mathscr{E}(m)/A(\mathscr{R}) = \mathscr{E}(N)/A(\mathcal{S}) \qquad (2.1)$$

Now place an observer at the point "O," say, counting birds. Suppose there are n birds detected, m of which are in the particular region $\mathscr{R}$. If we suppose there is perfect detectability in $\mathscr{R}$, then m is all the birds present in $\mathscr{R}$, so that $m/A(\mathscr{R})$ unbiasedly estimates D. Furthermore, $\hat{N} = [m/A(\mathscr{R})] \cdot A(\mathcal{S})$ unbiasedly predicts the number N in the entire region, in the sense that N and $\hat{N}$ have the same average.

What is commonly understood to be Emlen's method consists of the following steps: (1) determine, from detection distance data, a region $\mathscr{R}$ of perfect detectability (we refer this to a *basal* region.); (2) estimate density in the target region by the observed density in the basal region; and (3) calculate the coefficient of detectability as CD $= n/\hat{N}$.

Stated in this way, the method resembles closely that of Kelker (1945), with the exception that the basal region is predetermined by Kelker and determined from the data by Emlen. The criticisms leveled at the method are these: (1) the n-m birds detected outside the basal region are not used to estimate density, except insofar as their locations help determine $\mathscr{R}$; (2) the CD is influenced strongly by the limits of the target region, yet the limits are essential; (3) the method uses grouped data and is therefore sensitive to the grouping procedure; (4) the density estimate is not statistically efficient (Burnham et al. 1980); and (5) what does one do with a CD?

Defense of Emlen's method is based on clarifying several points. First, we replace the CD by an effective area measurement, the EAS.

[1] Department of Statistics, Oregon State University, Corvallis, Oregon 97331.

[2] U.S. Fish and Wildlife Service, Patuxent Wildlife Research Center, P.O. Box 44, Hawaii National Park, Hawaii 96718.

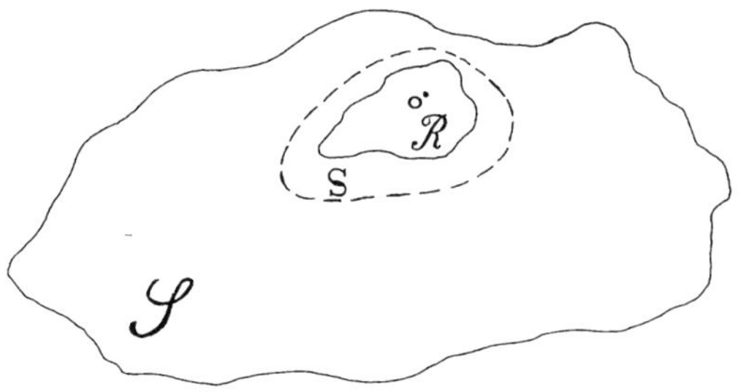

FIGURE 1. Conceptual Regions of an Emlen (Point) Survey.

Second we argue that this is not the proper time to estimate density. Then we present a simple, graphical technique for determining an estimate of EAS which does not require data grouping.

THE CD AND THE EAS

To begin, $n/\hat{N}$ estimates the probability of the observer's detecting a single bird positioned randomly in $\mathcal{S}$. Make $\mathcal{S}$ bigger and this probability must decrease. Make $\mathcal{S}$ unbounded, as several authors have done, and the probability is theoretically zero. To see what should be estimated, irrespective of the limits of the target region, consider the total number, n, of detections, which must lie between m and N. Hence its expectation lies between that of m and N, producing this result.

$$D \cdot A(\mathcal{R}) = \mathcal{E}(m) \leq \mathcal{E}(n) \leq \mathcal{E}(N) = D \cdot A(\mathcal{S})$$

Writing $\mathcal{E}(n) = D \cdot \mathcal{A}$, it is clear that $\mathcal{A}$ must be an area measurement intermediate between the basal area and the target area. It is an area representative of the observer's total survey effort, and is therefore defined to be the *Effective Area Surveyed* (EAS), (see Ramsey and Scott 1979, or Ramsey 1979). Thus

$$\mathcal{A} = \text{EAS} = E(n)/D, \tag{3.1}$$

which, if known, allows us to estimate density unbiasedly from all detections with $\tilde{D} = n/\text{EAS}$. It is sometimes theoretically convenient to view the EAS as the area of an effective *region* surveyed and to treat the whole procedure as one where the effective region is fully covered by the observer, while nothing is recorded outside of it. Such a region, S in Figure 1, is only, however, a hypothetical construct and perhaps should be deemphasized because of possible confusion with $\mathcal{R}$. At all costs, *avoid viewing EAS as the area of the basal region.*

Returning to (3.1), and incorporating (2.1), we get

$$\widehat{\text{EAS}} = \frac{\mathcal{E}(n)}{\mathcal{E}(m)} \cdot A(\mathcal{R}) \tag{3.2}$$

Theoretically, the EAS has this relation to the CD:

$$\text{CD} = \text{EAS}/A(\mathcal{S}).$$

The point here is that the EAS remains meaningful as $A(\mathcal{S})$ increases, whereas the CD does not.

ORGANIZING THE DATA ANALYSIS

Emlen argued that the CD should have some universality, being the same in regions of differing bird densities. By combining information from various sources, better estimates of CD's can be obtained. Here is how this works in an analysis of the results of a survey, except that we use the EAS.

The analysis proceeds in phases. In Phase 1, divide the target region into subregions according to a scale of detectability. At one end of the scale lies open grasslands. At the other lies dense forest with a high, closed canopy and thick understory. Between the extremes are classes reflecting how well one expects to detect birds visually and vocally. Lump together as a set all detections of a particular species in a particular detection class made by a particular observer. Further subdivisions should be made on factors such as time of day, weather, etc., which affect detectability, if these are not uniform during the survey.

In Phase 2 of the analysis, consider each set separately, producing with each a *detection curve,* such as in Figure 2. Here the density of observed detections is plotted against distance from the observer. Then comes this version of Emlen's method.

(I) Determine, from examination of the detection curve, a basal region $\mathcal{R}$ of near perfect detectability.

(II) Estimate the effective area surveyed by—see (3.2)—

$$\widehat{\text{EAS}} = \left(\frac{n}{m}\right) \cdot A(\mathcal{R}). \tag{4.1}$$

We emphasize here that *the purpose of looking at a detection curve is to estimate EAS, not density.* It should also be noted that one may use whatever auxiliary information one has available to judge what should be a suitable basal region. For example, a species which is attracted to the observer should not be allowed a basal region including only the area near the observer.

Phase 2 will produce estimates of EAS in many, but not all sets. A procedure for smoothing and filling in the missing EAS values is outlined in Ramsey and Scott (1979). It involves a weighted least-squares regression of the available values of $\log(\widehat{EAS})$ on variables indicating detectability class and observer. The fitted model is used to produce a full array of $\widehat{EAS}$ values for each observer in each habitat class, the whole procedure being done for each species. The full value of such a procedure is apparent when one realizes that this often gives EAS estimates in sets which began with very few or even no detections. Similarly, with rarer species and few detections, we are still able to use similarity with other, more abundant species to estimate EAS values. We have found, as Emlen suggested, that observer effects and detectability class effects are quite consistent from species to species.

Phase 3 consists of estimating population density. Suppose we wish to estimate the average density of some species in a given subregion of the target region. Divide the subregion as before into detectability strata, according to observer and detectability class. Determine, in each stratum, the total area (A_j) it occupies in the subregion, the total area (a_j) effectively surveyed, and the total number (n_j) of detections. The latter two are found by summing over pieces of transects or over stations, depending on how the survey was conducted. Then estimate the average density in the subregion to be

$$\hat{D} = \left(\sum_j n_j A_j / a_j\right) \Big/ \left(\sum_j A_j\right).$$

One expression which estimates the variability in $\hat{D}$ is

$$\widetilde{\mathrm{Var}}(\hat{D}) = \left[\sum_j n_j (A_j/a_j)^2\right] \Big/ \left[\sum_j A_j\right]^2.$$

This treats the effective areas as having been estimated without error and treats the numbers present as variables. Modifications may be made to recognize errors in EAS estimates. And in certain (management) situations, it may be preferable to hold fixed the numbers present.

SELECTING THE BASAL REGION

When J. T. Emlen (1971) proposed his method, he suggested that the basal region be found by inspecting the detection curve for a point of inflection, where density begins to decline rapidly with distance. Ramsey and Scott (1979) discussed several ways to formulate a rule that

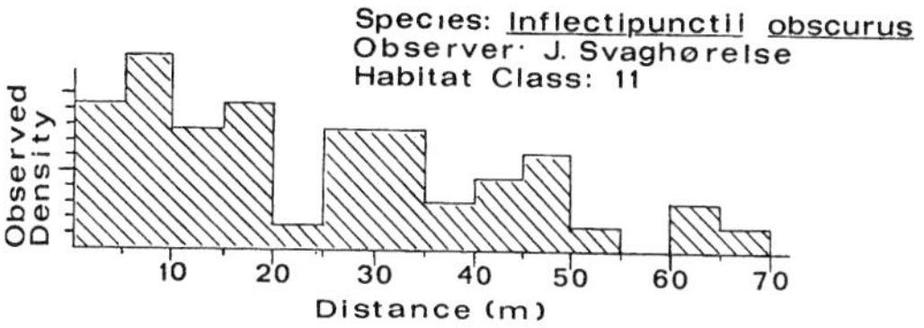

FIGURE 2. A plot of the density of detections versus the distance of detection for a hypothetical species and observer.

would replace "inspection" and settled on a scheme which uses likelihood ratios to judge if density is declining.

We emphasize that the purpose for devising such a rule was NOT simply to facilitate automatic data processing in a high speed computer. Detection curves should always be plotted and visually inspected. Only in this way can one understand the factors influencing detectability. The reason for the rule was to provide a method less subject to influence of random variations.

The likelihood ratio rule says that a basal region $\mathcal{R}$ should be expanded to include $\mathcal{R}^*$ if a statistical test finds no difference in density in the two regions. It incorporates a flexible critical ratio which may be chosen to provide balance between bias and variability in the resulting estimators. We choose a "conservative" cutoff value which usually underestimates density by 10–15% (see DeSante 1981), because this greatly reduces the possibility of seriously over-estimating density. (Our primary concern is with rare and endangered species, whose population sizes we do not want to over-estimate).

RELATED METHODS

There are a number of ways to estimate Effective Area Surveyed from detection distance data, Emlen's method being just one. Burnham et al. (1981) argue that the EAS bears a known relationship with the probability density function of detection distances (in line transects, but *squared* distances in circular plots), evaluated at zero distance. Ramsey (1979) suggests incorporating the EAS as a scale parameter in a flexible family of possible detectability curves. The choice of procedure here depends largely upon how one feels about the detectability curve. If one feels confident that detectability curves belong to a certain parametric family, then Ramsey's (1979) methods provide highly efficient estimators. If one is confident that all birds on transect (station) are detected but that detectability declines rapidly off transect (station), then the Burnham et al. (1981), non-parametric procedures might be best. However, if one feels

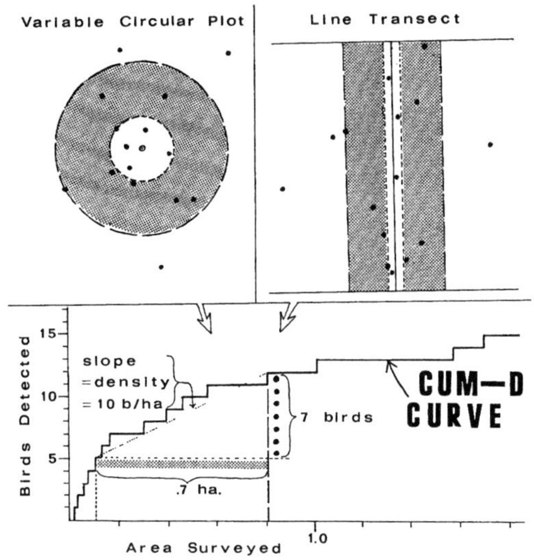

FIGURE 3. In this figure, a VCP and a LT survey (upper left and right, respectively) give the same CUM-D curve. Against Area on the abscissa, plot the number of detections made in that area around the observer. The slope through a part of the curve then gives the density of detections over the corresponding (shaded) region.

that there is some substantial region of uniform, near-perfect detectability, the modified Emlen technique is recommended.

EMLEN'S METHOD WITHOUT GROUPING

In this section we introduce a function which can be used to apply Emlen's method graphically to estimate EAS. The function is the CUMulative Detection Curve (CUM-D), which displays total numbers of detections as a function of area searched. It is applicable to both Line Transect surveys and to Variable Circular Plot surveys, as illustrated in Figure 3, where the two survey results at the top (dots represent detections) produce identical CUM-D curves. Specifically, we break up the region surveyed into zones of increasing area surrounding the observer's position(s). The CUM-D curve plots the total number of detections in a zone against the area of that zone. From the CUM-D curve, one may calculate the density of detected birds in any subzone. For example, the shaded regions have the same area, 0.7 ha, in Figure 3 and have the same number, 7, of detected birds. Therefore, the density of detections in the shaded region is—as illustrated—the slope of the CUM-D curve between the inside and outside areas.

Statisticians will tell you that division of the CUM-D curve by n will produce the "empirical distribution function" of the areas, $A_1, \ldots, A_n$, enclosed by the detections. Here

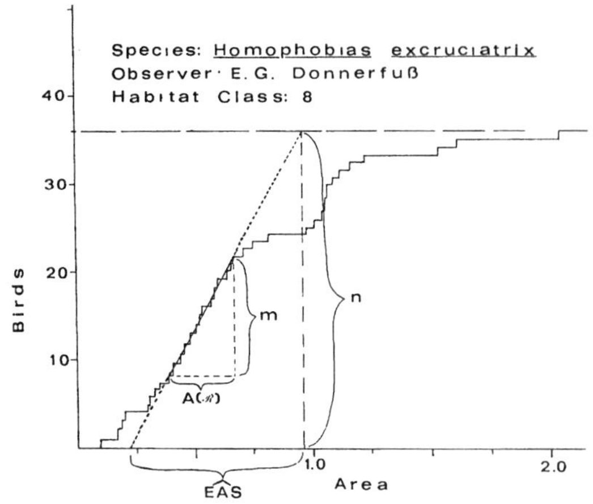

FIGURE 4. To estimate EAS, extend the maximum CUM-D curve slope up to the horizontal line of total detections, and extend it down to the horizontal axis. The $\widehat{EAS}$ is the area difference between these points.

$$A_j = \begin{cases} \pi \cdot R_j^2, \text{ in a VCP, where } R_j = j^{\text{th}} \text{ detection distance} \\ 2 \cdot L \cdot Z_j, \text{ in a } LT \text{ of length } L, \text{ where } Z_j = j^{\text{th}} \text{ right-angle distance.} \end{cases}$$

Note that the slope of the CUM-D curve at $A = 0$ is the critical parameter estimated by Burnham et al. (1981).

Because density of detections is highest in zones of highest detectability, we now offer a final version of Emlen's method, to wit:

(1) Determine a basal region $\mathscr{R}$ by seeking a zone of highest slope in the CUM-D curve.
(2) Estimate EAS by projecting the slope to "0" and "n" detections, as in Figure 4. (Equivalent to Equation (4.1).)
(3) Use EAS values (smoothed, if appropriate) to estimate densities.

Methods for selecting a largest slope and resultant properties of estimators are discussed in Wildman (pers. comm.).

DISCUSSION

There is no requirement that a basal region include the zone immediately surrounding the observer. In bird surveys, investigators have encountered observer avoidance problems in variable circular plot surveys where it might not be anticipated and in line transect surveys where it is to be expected. Indeed, it is almost inconceivable that birds would not react to the presence of an observer. This invalidates the assumption that detectability at zero distance is perfect ($g(0) = 1$). It need *not* preclude the determination of accurate estimates of population density.

The modified Emlen technique produces a basal region wherein observed densities are highest. Once obtained, the investigator still must relate that observed density to population density. Are birds moving away from the observer and then resuming normal behavior? Are birds near the observer simply making themselves undetectable? If so, how does the zone of avoidance compare with the zone of high detection? Is some fraction ($p = 1/2$, say, for females) of the population completely non-detectable? Are the birds moving over considerable distances in the time they are observable? Are distances misjudged because of ventriloquism? These are some of the questions one must confront in relating observed to true densities. Although the full shape of the CUM-D curve or the detection curve can be useful in the discussion, answers must, in the final analysis, be based largely on the biology of the target species.

There will be species for which these methods fail totally. But, there will also be those for which it works.

Studies in Avian Biology No. 6:488–491, 1981.

RESIDUAL EDGE EFFECTS WITH THE MAPPING BIRD CENSUS METHOD

J. H. MARCHANT[1]

ABSTRACT.—Edge effects on mapping census plots are an important consideration where density estimates are required. The IBCC recommended method for dealing with edge clusters is not fully efficient: in this study between 10% and 27% of edge clusters were found to have been wrongly included on census plots, leading to inflated estimates of density. The sources and implications of these results are discussed.

The mapping method (Enemar 1959, Williamson and Homes 1964) aims to index bird densities by means of selected census plots. One problem with this approach is that territories intersecting the chosen boundaries of the plot are only partially censused; edge effects may generate significant errors (Cousins unpubl. data and 1977).

Standard practice for the Common Birds Census is to include all "edge clusters" (clusters with some registrations inside, and some outside, the plot boundary). For the purpose of indexing population levels, edge effects are eliminated by the process of pairing plot cluster totals across years. Edge effects are similarly unimportant for ecological studies where the positions of registrations are being compared with the habitat structure of the plot. However, for comparison of cluster densities across habitats or between regions, standard measures of density are required and here edge effects are important. Clearly, a proportion of edge clusters should be included in the totals used for the calculation of cluster density, and the rest discarded. The recommendations of the International Bird Census Committee are that clusters should be included only if more than half of the registrations lie within the plot or on the boundary (IBCC 1969). (The application of this rule resulted in an average 3.7% of total territories being discarded from a sample of 20 farmland census plots in 1979.) The present study makes a preliminary assessment of the errors involved with this procedure, using data drawn from the 1979 Common Birds Census.

If a proportion of edge clusters is wrongly included or discarded, the relative error in the estimate of cluster density can be modelled approximately as shown below:

Let: A be the area of a plot with length of edge L; N be the total of clusters on the plot; n be the number of edge clusters; x be the number of wrongly assigned edge clusters; d be the characteristic linear dimension of a cluster, such that cluster area is proportional to d^2; p be the proportion of edge clusters wrongly assigned; and k, k', α be constants, such that $k = k'/\alpha$.

Assuming that—(1) d is much smaller than L; (2) the plot edge is not excessively convoluted; (3) territories are not clumped; (4) edge habitat is representative of the plot; and (5) all parts of the plot are included in one territory or another, then—

(1) the number of wrongly assigned clusters $x = pn$

(2) the number of edge clusters n is proportional to L/d, or

$$n = \frac{k'L}{d} \text{ and } x = \frac{k'pL}{d}, \text{ and}$$

(3) the total number of territories $N = \dfrac{\alpha A}{d^2}$

Thus the relative error in the cluster total

$$\frac{x}{N} = \frac{k'pL}{d} \times \frac{d^2}{A\alpha} = \frac{kpdL}{A}$$

MATERIALS AND METHODS

The relationships predicted by this model were examined using two large plots composed of independently-censused subplots.

Plot A was composed of two subplots (Fig. 1) to which census visits were made quite independently, although by the same team of two observers (one of whom was the author). The common boundary between the subplots ran along the center of a canal bank, wooded on both sides for much of its length; this was the best single feature of plot A in terms of number of territorial species and overall territory density. Plot B (Fig. 2) comprised five subplots, again censused independently, in this case by five separate observers. In contrast to plot A, the common subplot boundaries were ordinary hedgerows, roadways or wood-edges, differing little ornithologically from the external boundaries of the plot. For each plot species maps for each species were drawn up both for the whole plot as a single unit and for each subplot in isolation. After clustering had been completed, edge clusters were selectively discarded according to the IBCC recommendations.

By comparing the subplot species maps with those from the plot as a single unit, it was possible to identify

[1] Populations Section, British Trust for Ornithology, Beech Grove, Tring, Hertfordshire HP23 5NR, England.

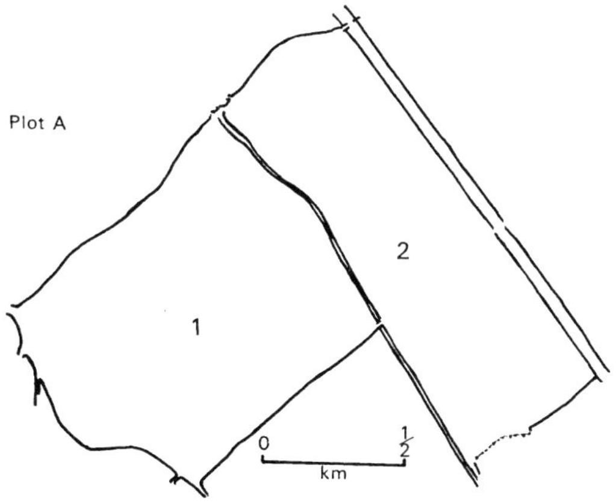

FIGURE 1. Map of plot A showing the distribution of subplots.

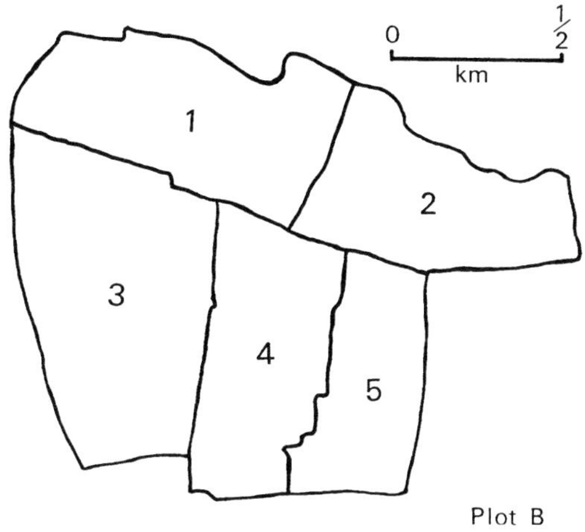

FIGURE 2. Map of plot B showing the distribution of subplots.

clusters included for both of two adjacent subplots. Such double-counted clusters would have given an inflated estimate of density on one of the subplots, were the subplots censused in isolation.

In order that the errors resulting from duplication could be considered a feature of the subplots rather than of their common boundaries, it was necessary to assign duplicated clusters to subplots on a 50:50 basis. Where the distribution of clusters changed markedly as a result of joining the subplots into a single unit, each decrease by one in the total number of clusters was assessed as a duplication, whether or not the subplot clusters leading to the duplication actually overlapped. In most cases, however, the duplicates were similarly positioned on the two subplots. Clusters present on the species maps for the whole plot, but on none of the subplots, were few in number and were not considered in the present study. Such clusters were an artifact of the double amount of visiting effort to the area of the subplot boundaries when considering the plot as a whole.

For each subplot, the number of clusters estimated to have been wrongly included was considered in relation to the length of common subplot boundary and to the total number of edge clusters included along those common boundaries. Further, data from plots A and B were combined so that species could be compared in the percentage of edge clusters double-counted.

RESULTS

On plot A thirty-six clusters were found to have been duplicated and these were assigned equally between the subplots as shown in Table 1. Since the common boundary of plot A bisected the best ornithological feature of the farm, the number of wrongly assigned clusters per kilometer was high. For each subplot an estimate was made for the proportion of edge clusters along the common boundary which were wrongly included. Note, however, that since the thirty-six duplicated clusters were divided equally between the subplots, these two estimates were not independent.

The results from plot B are shown in Table 2, the first part of which has equivalent headings to Table 1 for plot A. Ten duplicated clusters were assigned as shown.

Since there was a much lower frequency of clusters crossing the subplot boundaries on this plot than on plot A, owing to the poorer quality of the habitat for the birds, the number of wrongly assigned clusters per kilometer was much lower on plot B. However, the five estimates of the proportion of the total edge clusters which were wrongly assigned were reasonably consistent and encompassed the estimates from plot A in range. Again for plot B it must be noted that these estimates were not wholly independent.

Since for plot B the outside edges of the plot were similar in habitat to the internal, subplot boundaries, it was possible to estimate the total error in the number of clusters on each subplot due to wrongly included edge clusters. The results of this exercise are also shown in Table 2. These estimations were made on the basis of the number of wrongly assigned clusters per km, and not from the percentage of edge clusters which were wrongly assigned, because edge clusters around the outer boundary of plot B seemed to have been poorly recorded by the observers. It was not possible to make directly equivalent estimates for plot A, because the subplot boundary was so much richer in birds than the external boundaries of the plot.

According to the simple model, the relative error in the cluster totals should be proportional

TABLE 1
ERRORS FROM DUPLICATION OF CLUSTERS ON PLOT A, COMPOSED OF TWO SUBPLOTS

	Subplot 1	Subplot 2
Wrongly assigned clusters (all species)	18	18
Length of common subplot boundary (m)	952	952
Wrongly assigned clusters per km of boundary	18.9	18.9
Total edge clusters along subplot boundary	90	83
Wrongly assigned clusters/100 edge clusters	20%	21.7%

to cluster size, and inversely proportional to the edge:area ratio of each subplot. No cluster sizes were measured on plots A and B, but since the territory size of a species is known to be related to its body weight, the effect of increasing cluster size was examined using body weight as an approximate measure. Larger species tended to be double-counted more frequently, although the trend was not significant: a regression of percentage of total number of edge clusters which were duplicated for each species against log body weight gave $r = 0.053$, df = 54. An angular transformation was performed on the percentage data. As an alternative approach, the square root of the reciprocal of total cluster density was taken as a measure of d, since from the model

$$N \alpha \frac{A}{d^2} \quad \text{or} \quad d \alpha \sqrt{\frac{A}{N}}.$$

The relationship of this measure with the relative error in the estimation of total clusters is shown in Figure 3a for the five subplots of plot B. The correlation was short of significance, however ($r = 0.78$, df = 3, $P = 0.14$).

Figure 3b shows the relationship of the edge: area ratio of the five subplots to the relative error in cluster totals. No significant correlation was present ($r = -0.42$, df = 3) and, against the predictions of the model, the correlation coefficient was negative. However, a partial correlation allowing for the strong effect of d (Figure 3a) gave a positive but still not significant value ($r_p = 0.30$).

Finally, a further attempt was made to characterize the species most likely to lead to the mistaken inclusion of edge clusters. A subjective index of "ease of detection of CBC territories" was prepared from the combined results from four experienced fieldworkers who each scored ease of detection from 1 (difficult) to 5 (easy). There was, however, no correlation between this index and the percentage of edge clusters which were duplicated for each species, transformed to angles ($r = 0.04$, df = 29).

DISCUSSION

The results from both plots A and B suggest that a sizeable fraction (estimated at between 10% and 27%) of edge clusters included according to the IBCC recommendations for the mapping method will not strictly belong to the plot under consideration, and will thus lead to inflated estimates of cluster density. It is probable that observer biases towards censusing more

TABLE 2
ERRORS FROM DUPLICATION OF CLUSTERS ON PLOT B, COMPOSED OF FIVE SUBPLOTS

Subplot	1	2	3	4	5
Wrongly assigned clusters	1	0.5	3	3.5	2
Common subplot boundaries (m)	1280	798	1351	1898	1036
Wrongly assigned clusters/km	0.8	0.6	2.2	1.8	1.9
Edge clusters along subplot boundaries	4	5	11	14	16
Wrongly assigned clusters/100 edge clusters	25%	10%	27.3%	25%	12.5%
Total length of subplot edge (m)	2522	2213	2605	2316	1968
Total clusters on subplot	63	80	70	38	120
Estimated total of wrongly assigned clusters, based on no./km	2.0	1.4	5.8	4.3	3.8
Estimated total of wrongly assigned clusters, as % of total clusters	3.1%	1.7%	8.3%	11.2%	3.2%
Area of subplot (ha)	28.4	25.1	40.1	25.9	18.2

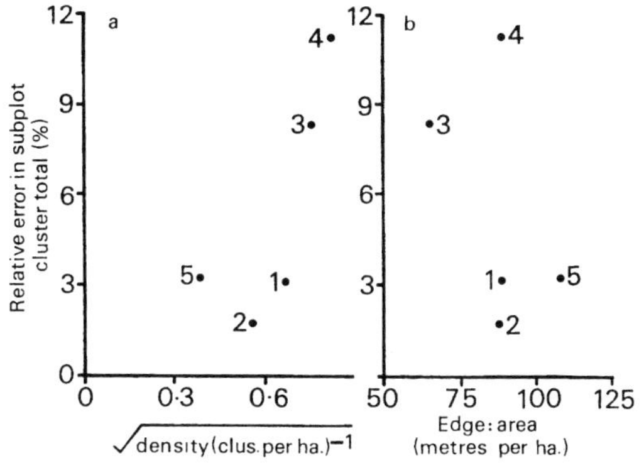

FIGURE 3. Data from plot B showing the relative error in subplot cluster total in relation to (a) a measure of cluster size (see text) and (b) the edge:area ratio of the subplot. Subplots are numbered 1–5.

thoroughly the areas within the boundary than just outside, and towards registering birds seen to cross the boundary as within the plot, are responsible for the inclusion of extra clusters.

Too few data were available to enable the thorough testing of the simple model but, assuming it to be a helpful one, it is apparent that the errors in cluster totals (and hence in density estimates) will be greatest where territory size is large and where the edge:area ratio of the plot is high. The implications of the present study for plot design are that the edge:area ratio should be minimized (as recommended by IBCC 1969), and that boundaries should as far as possible not be drawn through ornithologically rich areas, where as along the canal on plot A a significant number of clusters may be wrongly included.

The similarity between the two plots in the estimates of p, the proportion of the total of edge clusters which were wrongly included, suggests that it may be possible to estimate p in advance and thus produce from mapping census plots estimates of cluster density which are free from this source of edge error.

ACKNOWLEDGMENTS

I am grateful to P. A. Hyde for help with fieldwork on plot A, and to H. Lemon for coordinating coverage of plot B. Dr. R. J. O'Connor kindly commented on earlier drafts of this paper. Mrs. E. Murray, Miss C. Hunt and Miss C. Ferri typed drafts and provided artwork. This analysis was performed under contact with the Nature Conservancy Council.

Studies in Avian Biology No. 6:492–495, 1981.

THE EFFICIENCY OF THE MAPPING METHOD—A COMPUTER SIMULATION BASED ON FIELD DATA

MARTIN ERDELEN AND BEATE ERDELEN[1]

ABSTRACT.—A program CENSUS has been developed to simulate a population of stationary Willow Warbler males and its censusing by an observer using the mapping method. The song activity of the simulated birds was based on field data without any modification. Census efficiency increased with shorter distances between observer stops, longer listening time per stop, and greater hearing range of the observer. For comparable circumstances, good agreement was found between field and simulated census efficiencies: a census of 10 surveys under favourable conditions (many stops, long listening time, great hearing range) records 80 to 90% of the true population. It is concluded that the so far purely deterministic simulation can be regarded as a reliable base for further extensions including the introduction of stochastic factors such as bird movements, recording and mapping errors, etc.

Methodological research into bird census techniques is complicated by the fact that conditions in the field during a census can be controlled only to a very limited extent. The usual procedure of scientific experiments—to keep all factors constant except the one under investigation—can hardly be followed in the field: it is impossible to vary only one factor, e.g., the observer's speed, while keeping constant all other conditions such as time of day, time of year, weather, bird activity, etc. Thus, exact replications of a survey are not feasible.

To bypass these difficulties, we have developed a computer program for simulating the mapping method. (For a detailed description of this method see, e.g., the study of Enemar (1959), which also shows by a "provisional calculation" that song activity can influence census results.) In our simulation, we can vary systematically the factors under consideration without changing the other conditions. A further advantage is that the true size of the simulated population is known a priori, whereas in field studies the final result of the census itself often yields the only available 100% value used as a "yardstick" in efficiency calculations.

An essential feature of our approach has been to base the simulation on data as realistic as possible, and on a minimum number of generalizations and assumptions. In the construction of many ecological models, assumptions are introduced at an early stage, which often increases the range of results and applications. We felt, however, that the need for clarification in bird census methodology makes a completely empirical approach more desirable, and that—at least as a first step—"trivial" results on a safe basis are to be preferred over more "interesting" ones based on assumptions the validity of which is difficult or impossible to judge.

[1] Zoological Institute I, University of Cologne, Weyertal 119, D-5000 Cologne 41, Federal Republic of Germany.

METHODS

FIELD DATA

In 1973, 1974, and 1976 through 1978, breeding passerine bird censuses were conducted using the mapping method. The study area was the "Donatusfeld," a reforested former opencast brown coal mining area, 25 km southwest of the city of Cologne, Federal Republic of Germany. The Willow Warbler (*Phylloscopus trochilus*) was chosen for a more intensive study of singing behaviour and census efficiency: males of this species sing loud and regularly and can easily be observed on their singing posts. About 50% of the stationary males in the study area were banded with individual combinations of colour-rings. The population size ranged from 30 to 55 stationary males in different years. These numbers were obtained from many census surveys (16 to 39 per breeding season), from colorbanding, and from additional field observations; they can be regarded as very reliable estimates. In 1978, song activity of Willow Warbler males was measured during the four hours after sunrise by recording "song" or "no song" of individual birds for every half-minute interval. Most of the recorded males could be identified afterwards by their colour-rings. Song patterns, i.e., sequences of "+" and "−" signs, were obtained from 14 different males, and for a total of 52.3 hours.

SIMULATION

The program CENSUS was written in SIMULA, a programming language for simulation purposes which has been developed from ALGOL (for an introduction see Birtwistle et al. 1975). We simulated a population of 42 Willow Warbler males, which is in the range of real numbers for the study area. The spatial distribution of the birds was copied according to scale from the census maps; so was the observer's route. As we had to follow pre-established tracks during the surveys in the field to minimize disturbance, a somewhat irregular route resulted (Fig. 1). To keep the simulation as realistic as possible, we decided not to alter this route layout. However, other routes can also be simulated by simply changing the program input parameters. Each simulated bird was assigned an individual song pattern. We used the patterns exactly as recorded in the field, i.e., the sequences of "+" and "−" signs, without modifying or processing them by calculating average song probabilities or similar generalizations.

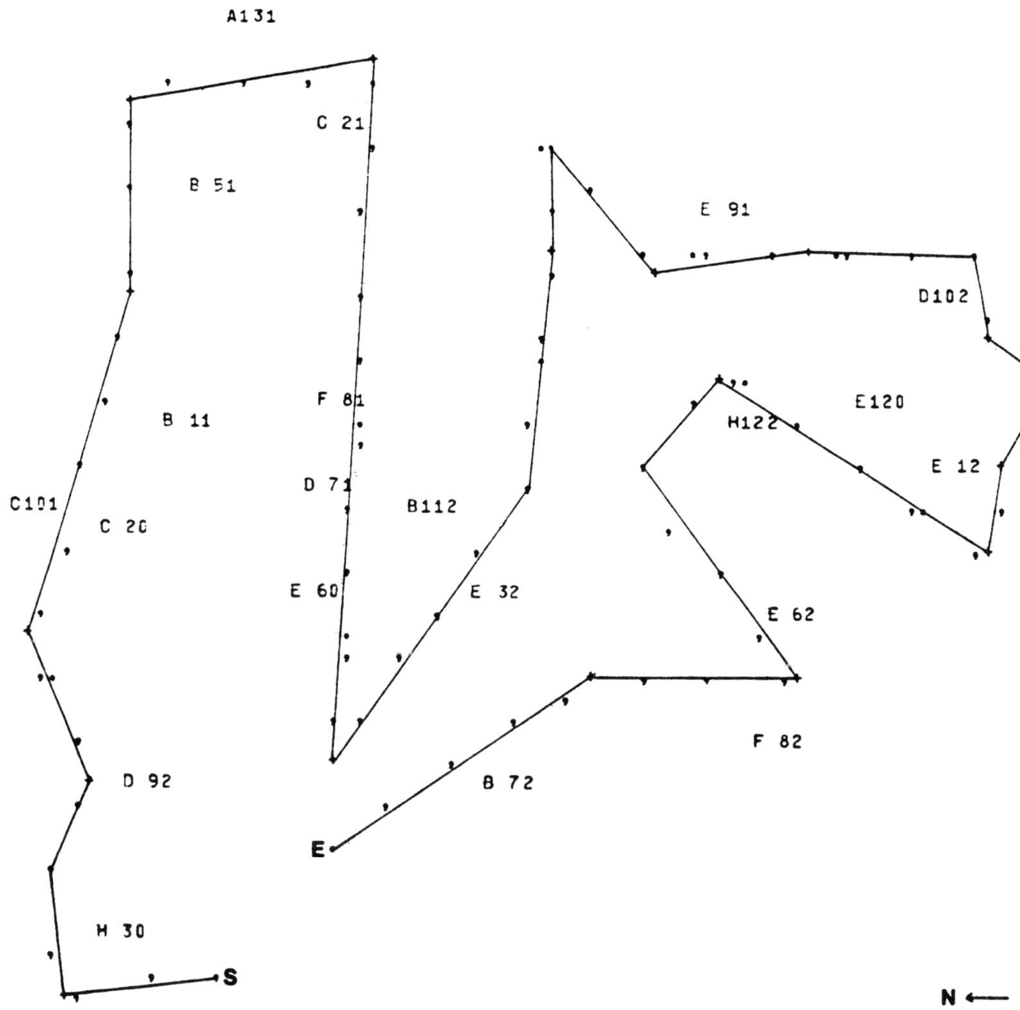

FIGURE 1. Computer output of simulated census. (Line: observer route; S, E: start and end points of route (walking direction alternated on successive surveys); full stops and commas: observer stops for both walking directions, respectively; figures: identification numbers of "males," e.g., 102 = second copy of bird no. 10; letters A, B, C, etc.: bird has been recorded on one, two, three etc. surveys; N: north direction of original study plot).

However, as data were available for 14 individuals only, we created two additional copies of each song pattern to obtain a total population of 42 birds. We made sure that birds with identical patterns were located far apart. Thus, the observer meets the original and the copies at different times and therefore has to deal with quite different sections of the full song pattern.

The procedure of a simulated survey is as follows: The observer starts at one end of the route, alternating start point and walking direction for successive surveys. He stops at fixed intervals and listens for a certain period. During this time, all males of the population are checked for occurrence within the observer's hearing range, and for their song activity. If they are within hearing range and are singing during the observer's stop, they are noted by the program as recorded for this survey. Each census consists of several surveys, the number of which can be varied. The computer output corresponds to the species map of the mapping method and shows the cumulative results of all census surveys (Fig. 1). Numbers identify individuals and letters indicate how often the bird has been recorded and where it is located. Further output shows the efficiency (number of recorded males as percentage of the true population) for each single survey as well as for the final result of the census after all surveys. As in the evaluation of real species maps, it is possible to set a lower threshold for acknowledgment of a stationary male; e.g., all recorded males with only

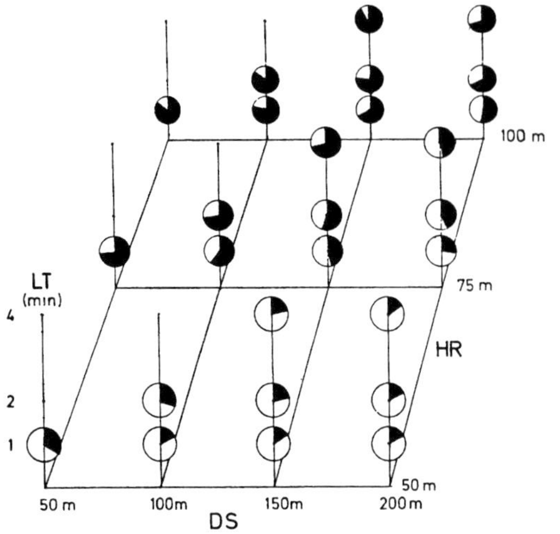

FIGURE 2. Efficiencies of simulated censuses with ten surveys, between 0 and 4 hours after sunrise. (DS: distance of observer stops; HR: hearing range of observer; LT: listening time per stop; black circle sectors: recorded proportion of total stationary male population).

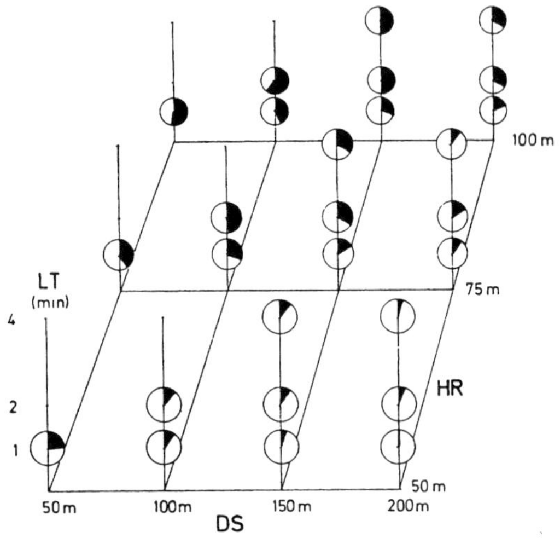

FIGURE 3. Efficiencies of simulated censuses with five early surveys, between 0 and 2 hours after sunrise. See Fig. 2 for abbreviations.

one or two registrations can be excluded from the final result as suspected migrants or floaters. Note that the simulation does not allow the birds to move around. The program has been designed to include this possibility; however, no field data were available on movement patterns of Willow Warblers, and for the reasons stated above, we did not want to introduce hypothetical assumptions at the present stage of the simulation. Neither did we include background noise by setting the recording probability below unity for singing birds within hearing range: every bird that can be recorded will be recorded. So far, the program is completely deterministic.

The following input parameters of the program can be varied: size of stationary male population; location of males; shape of observer's route; distance between observer's stops; listening time at each stop; hearing range of the observer; number of surveys per census; starting time of day for surveys; and minimum number of recordings for acknowledgment of a stationary male (as opposed to floaters).

As the sequence length of the available song pattern field data is limited, not all possible parameter combinations could be run: if, for instance, there are too many stops and too long listening times, the song data are used up before the observer has finished his survey.

The program was run on a Control Data Corporation Cyber 72 computer at the Regional Computing Center Cologne.

RESULTS

As the program CENSUS requires relatively large amounts of computer time and storage, only a limited number of runs could be conducted up to now. Additional runs with different parameter combinations will be necessary to allow a more precise investigation including regression analysis, etc. The results obtained so far are summarized in Figures 2 through 4. The recorded proportions of the true population range from 14 to 88% for censuses with 10 surveys (all on different days, between 0 and 4 hours after sunrise; Fig. 2); from 2 to 62% for censuses with 5 early surveys (0 to 2 hours after sunrise; Fig. 3); and from 0 to 52% for censuses

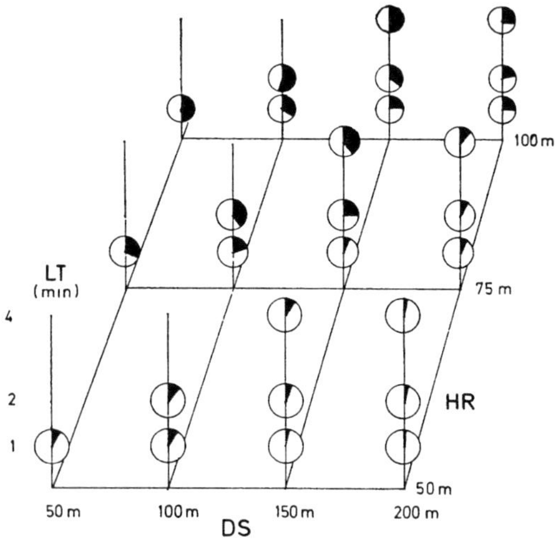

FIGURE 4. Efficiencies of simulated censuses with five late surveys, between 2 and 4 hours after sunrise. See Fig. 2 for abbreviations.

with 5 late surveys (2 to 4 hours after sunrise; Fig. 4). The mean efficiencies over all runs for the three groups of censuses are 49.8, 25.1, and 19.7%, respectively. These are not true averages, of course, since the parameters have been varied between runs. The figures can be used for comparison, however, because in all three sets of simulations the same parameter combinations have been used.

DISCUSSION

As our field surveys were conducted very thoroughly, their results are comparable to those parameter combinations with many stops and long listening time. The hearing range in our study area was about 100 m, due to a rather low vegetation height (less than 6 m), and due to the Willow Warbler's habit of singing from conspicuous tree tops. Ten surveys under these conditions yielded a census efficiency of 80 to 90% in the field. The census efficiency of the simulations with comparable parameter sets lay in the same range.

In agreement with intuitive expectations, census efficiency is increased with shorter distances between stops, longer listening time at each stop, and larger hearing range, which all increase the probability of spotting a bird while it sings.

The effect of the time of day on census results in general is also well known. The simulations confirm that even such minor shifts in survey time as from the first to the second two-hour period after sunrise can lead to a slight decrease in census efficiency. This result for the Willow Warbler, a marked morning singer, will also be true for other passerine species with similar daily song activity patterns.

A general argument against the usefulness of such a simulation might be that it does not lead to any new insights, as it only "can put out what has been put in." This is true inasmuch as the simulation is a deterministic one up to now. However, it renders possible a quantitative and replicable analysis of census efficiencies. Moreover, we considered as necessary the step of calibration against field results before proceeding towards a more realistic model. Only after having found—as we did—good agreement between results from the field and those of a simple, deterministic simulation, the introduction of stochastic factors such as movements of birds, background noise, mapping errors, etc., seems justified and worthwhile.

ACKNOWLEDGMENTS

Part of this study has been conducted within the Ph.D. thesis work of one of us (B.E.). We thank Prof. H. Engländer and Prof. A. Egelhaaf for advice and support.

Studies in Avian Biology No. 6:496–501, 1981.

METHODOLOGICAL STUDIES OF BREEDING BIRD SURVEYS IN NORTH AMERICA

A. R. SEN[1]

ABSTRACT.—Surveys are being annually conducted in U.S.A. and Canada to provide management with reliable data to detect and measure both short term (annual) and long term changes in abundance of non-game breeding birds. The data are collected by cooperators who make road side counts on predetermined stops on predetermined routes.

This paper deals with the methodological aspects of (1) estimating the density of birds of a given species and its error in a given region during a year (2) detecting short term or annual change in population density and (3) measuring relative abundance of a species between two years for a region based on the same sample points and the same routes covered during the period. The analysis was based on data for the Maritime Provinces of Canada.

It is shown that for some clumped species the estimated number of birds per route was highly skewed and the number of routes selected for sampling in a region was not adequate enough to ensure normality of mean number of birds per route. Logarithmic transformation of the data ensured approximate normality; it resulted in some gain in efficiency of the mean but considerable gain in efficiency of the variance.

For a few species, the estimate of relative change in number of birds per route was subject to high error and the statistical tests failed to detect real difference between the years. In such cases, tests based on transformed data proved more amenable to statistical treatment.

Surveys are being annually conducted in the U.S.A. and Canada for detecting and measuring changes in abundance of non-game breeding birds at the height of the breeding season. The data are collected by cooperators who make road-side counts on predetermined stops on predetermined routes according to a specified sampling scheme.

The methods which are being currently used are not sensitive and reliable enough to detect changes in a region for a number of species for which the distributions are highly skewed; also the sample size is not large enough to ensure normality of the mean. The present paper deals with the methodological aspects of: (1) estimating the abundance of birds of a given species and its error in a region during a year; (2) detecting differences in abundance among species during a year with a view to determining relative species abundance, annual changes in abundance for a particular species based on the same sample points in the same routes covered during the period; and (3) measuring relative change in abundance of a species in a region between two years to find how these vary for common and uncommon species. The average number of birds measured per route in a region will be defined as abundance for the region.

The analysis is based on data for the Maritime provinces for the two years 1971 and 1972.

DESIGN OF THE SURVEYS

The design of the breeding bird surveys is based on a stratified random sample of roadside counts (Robbins and Van Velzen 1967, 1969; Erskine 1970, 1973; Smith 1973). For operational convenience each degree block of latitude (8.5 km × 112.7 km each) was chosen as stratum. Within each stratum several transects or routes and the compass directions were chosen at random and without replacement. A point was first chosen at random within a degree block as the intersection of latitude and longitude determined with the help of a pair of random numbers. Next, the point which lay on a good road nearest the chosen point was used as the starting point of the route. Where there was no road in the immediate neighborhood, the chosen point was selected as the closest recognizable landmark on a passable route. The sampling is done by volunteer observers who travel along the selected routes making stops at regular intervals. All birds heard or seen at each of 50 three minute stops at 0.8 km (one-half mile) intervals along the routes are counted and recorded by observers. As far as practicable, the same observers were used on the same stops and on the same routes during the two years 1971–72.

RESULTS

ESTIMATES OF SKEWNESS—ITS EFFECT ON DENSITY ESTIMATES

The average number of birds counted per route for some of the common and uncommon species in the Maritime Provinces during 1971 and 1972 was used to estimate departures from normality by skewness ($g_1 = m_3/m_2^{3/2}$) and kurtosis ($g_2 = m_4/m_2^2 - 3$) where

$$m_2 = \Sigma(y_i - \bar{y})^2/n; \; m_3 = \Sigma(y_i - \bar{y})^3/n;$$
$$m_4 = \Sigma(y_i - \bar{y})^4/n,$$

y_i is the number of birds seen or heard along the ith route, $n(=41)$ is the number of comparable routes covered during the years.

The estimates of average number of birds counted per route, their standard error, skewness and kurtosis for the species are presented

[1] Migratory Birds Branch, Canadian Wildlife Service, Ottawa, Ontario, K1A 0E7 Canada.

TABLE 1
ESTIMATES OF AVERAGE NUMBER OF BIRDS PER ROUTE, STANDARD ERROR, SKEWNESS (G_1) AND KURTOSIS (G_2) IN THE MARITIMES DURING 1971 AND 1972[a]

Species	1971				1972			
	Average number of birds per route	Standard error	g_1	g_2	Average number of birds per route	Standard error	g_1	g_2
Common Snipe (*Capella gallinago*)	5.32	1.02	1.60*	2.11**	4.93	0.77	0.76*	−0.74*
Herring Gull (*Larus argentatus*)	37.12	13.26	3.73**	15.18**	54.41	25.74	5.70**	34.58**
Least Flycatcher (*Empidonax minimus*)	5.12	0.69	1.48*	2.99**	5.04	0.73	1.36*	2.71**
Tree Swallow (*Iridoprocne bicolor*)	21.21	2.71	1.68*	3.54**	18.73	2.75	2.48**	8.88**
Yellow-bellied Sapsucker (*Sphyrapicus varius*)	8.66	2.28	2.86**	10.18**	5.56	1.39	2.00**	3.37**
Bank Swallow (*Riparia riparia*)	24.15	6.50	2.17**	4.24**	18.05	5.25	2.95**	9.43**
Blue Jay (*Cyanocitta cristata*)	5.22	1.41	4.06**	20.07**	4.10	0.75	1.90**	4.79**
Common Crow (*Corvus brachyrhynchos*)	34.49	4.34	1.00**	3.95**	38.49	5.32	1.38*	1.25*
Robin (*Turdus migratorius*)	65.29	4.04	0.19	−0.63	60.73	4.30	0.37	−0.30
Starling (*Sturnus vulgaris*)	44.63	5.98	1.06*	1.00	44.27	6.10	1.00*	0.97
Common Grackle (*Quiscalus quiscula*)	24.66	6.93	3.81**	15.54**	26.05	7.22	3.86**	15.91**
White-throated Sparrow (*Zonotrichia albicollis*)	53.80	5.10	0.95*	0.41	57.80	5.39	1.18*	0.66

* $P < 0.05$.
** $P < 0.01$.
[a] Total number of comparable routes = 41.

in Table 1. For tests of significance of skewness and kurtosis for sample size $n = 41$, the one-tailed five percent and one percent significance levels of g_1 computed from a more accurate approximation given in Snedecor and Cochran (1977) may be employed. It is seen that the distribution of the numbers of birds was positively and highly skewed for almost all the species during both the years excepting Robin, Starling and White-throated Sparrow. This has the effect of increasing the variance of the annual index (mean count per route) and decreasing its precision. Owing to the marked positive skewness, the number of routes selected in the Maritimes is not large enough to ensure normality. Thus, tests based on normal theory will not be valid and tend to be less efficient and sensitive for detecting differences.

SAMPLE SIZE TO ENSURE NORMALITY

The question arises, "How large must n (the number of routes sampled within the Maritimes) be, so that the normal approximation is accurate for computing confidence limits of the annual indices (mean counts per route)?" For populations in which the principal deviation from normality consists of marked positive skewness and kurtosis, we have

$$n > 25G_1^2 - 1.64G_2 \qquad (1)$$

where G_1 and G_2 are Fisher's measure of skewness and kurtosis in the population and are estimated by g_1 and g_2, so that a 95% probability statement will be wrong not more than six percent of the time. In a personal correspondence Prof. Cochran reported that the above result (1) is due to Dr. Glen Bartsch and was derived later than his own (Cochran 1977:42) which assumes only marked skewness in the population.

For 9 of the 12 species considered in the study (excepting Robin, Starling and White-throated Sparrow) the sample size (number of routes) for estimating the abundance of birds during a year was less than that needed for the normal approximation. Hence for these species, the abun-

TABLE 2

ESTIMATES OF AVERAGE LOG (Y + 1), WHERE Y IS THE NUMBER OF BIRDS COUNTED ON A ROUTE, STANDARD ERROR, SKEWNESS (g'_1) and Kurtosis (g'_2) IN THE MARITIMES DURING 1971 AND 1972[a]

	1971				1972			
Species	Average log (y + 1)	Standard error	g'_1	g'_2	Average log (y + 1)	Standard error	g'_1	g'_2
Common Snipe	0.58	0.07	0.19	−1.14*	0.59	0.07	−0.12	−1.38**
Herring Gull	0.91	0.12	0.48	−0.77	1.02	0.13	0.23	−0.86
Least Flycatcher	0.67	0.05	−0.51	−0.15	0.65	0.06	−0.01	−0.96*
Tree Swallow	1.23	0.05	−0.12	−0.30	1.16	0.06	0.00	−0.43
Yellow-bellied Sapsucker	0.62	0.09	0.50	−0.93*	0.48	0.08	0.59	−1.09*
Bank Swallow	0.85	0.12	0.36	−1.12*	0.80	0.10	0.34	−0.75
Blue Jay	0.56	0.07	0.55	−0.05	0.54	0.06	0.18	−0.99*
Common Crow	1.40	0.08	−1.24*	2.03*	1.40	0.07	−1.58**	3.99**
Robin	1.78	0.03	−0.75	0.29	1.74	0.03	−0.73	0.29
Starling	1.44	0.08	−1.11**	0.99	1.37	0.09	−1.02**	0.19
Common Grackle	1.09	0.08	0.46	−0.01	1.13	0.08	0.28	0.56
White-throated Sparrow	1.65	0.05	−1.33**	3.80**	1.69	0.04	−0.29	0.72

* $P < 0.05$.
** $P < 0.01$.
[a] Total no. of comparable routes = 41.

dance of birds was estimated with lesser confidence unless sample sizes were increased considerably, which is not generally practicable nor feasible since it would be unwise to put pressure on the volunteer observers who are responsible for the collection of data.

LOG-TRANSFORMATION

When the data were transformed by taking $z = \log_e(y + 1)$ where y is the number of birds counted on a route, the distribution of z was approximately normal. Table 2 presents the picture using $\log_{10}(y + 1)$ instead of $\log_e(y + 1)$, which, as may be seen, will not affect our conclusions. Only 2 out of the 12 species (Table 2) showed significant skewness and kurtosis in both the seasons. In all cases, the number of routes selected was large enough (result (1)) to ensure normality of the mean on the logarithmic scale. Also, the estimates of the mean in the transformed scale were more precise having lower coefficient of variation (c.v.) to detect differences in the abundance of birds among species than on the original scale. Thus, with direct measurements (Table 3), there was no significant difference between the average number of Herring Gulls and Tree Swallows counted during 1971 (in fact the mean number of Herring Gulls appeared to be higher) but on the logarithmic scale, the average number of Tree Swallows counted was significantly higher than the number of Herring Gulls as shown by the 90% con-

TABLE 3

NINETY % CONFIDENCE INTERVAL OF AVERAGE NUMBER OF BIRDS COUNTED PER ROUTE ($\bar{y}$) AND AVERAGE LOG (y + 1) FOR THE DIFFERENT SPECIES IN THE MARITIMES DURING 1971 AND 1972[a]

	$\bar{y}$		Average log (y + 1)	
Species	1971	1972	1971	1972
Common Snipe	3.60– 7.04	3.63– 6.23	0.47–0.70	0.48–0.71
Herring Gull	14.79–59.45	11.06–97.76	0.70–1.11	0.81–1.24
Least Flycatcher	3.96– 6.28	3.81– 6.27	0.58–0.76	0.55–0.74
Tree swallow	16.65–25.77	14.10–23.36	1.14–1.32	1.07–1.25
Yellow-bellied Sapsucker	4.82–12.50	3.22– 7.90	0.47–0.77	0.34–0.62
Bank Swallow	13.20–35.10	9.21–26.89	0.65–1.04	0.63–0.98
Blue Jay	2.85– 7.59	2.84– 5.36	0.45–0.67	0.43–0.64
Common Crow	27.18–41.80	29.53–47.45	1.27–1.53	1.29–1.51
Robin	58.49–72.09	53.49–67.97	1.73–1.84	1.68–1.80
Starling	34.56–54.70	34.00–54.54	1.30–1.58	1.23–1.55
Common Grackle	12.99–36.33	13.89–38.21	0.96–1.22	1.00–1.26
White-throated Sparrow	45.21–62.39	48.72–66.88	1.57–1.73	1.63–1.77

[a] Total number of comparable routes = 41.

TABLE 4
EFFICIENCY OF THE ESTIMATES OF AVERAGE NUMBER OF BIRDS PER ROUTE AND OF VARIANCE ESTIMATES WITH RESPECT TO THE ESTIMATES MEAN (M) AND VARIANCE (V) BASED ON TRANSFORMED DATA BY SPECIES FOR THE MARITIMES FOR 1971 AND 1972[a]

	Percent efficiency			
	Average birds per route[a] ($\bar{y}$)		Variance	
Species	1971	1972	1971	1972
Common Snipe	87	89	14	17
Herring Gull	39	32	0.03	0.01
Least Flycatcher	95	94	32	28
Tree Swallow	95	94	34	30
Yellow-bellied Sapsucker	73	77	3	5
Bank Swallow	46	59	0.10	0.50
Blue Jay	89	92	16	23
Common Crow	82	89	8	16
Robin	99	111	66	60
Starling	76	66	4	1
Common Grackle	81	82	7	8
White-throated Sparrow	96	98	39	52

[a] Total no. of comparable routes = 41.

fidence limits. Similarly, there was no difference between the number of Herring Gulls and Common Crows during 1972 though there was evidence of a real difference on the transformed scale. Again, the number of starlings counted during 1971 was more than Common Grackles as would be evident from the results presented on the logarithmic scale though no such difference was noticeable on the original scale. Tree Swallow and Bank Swallow, Robin and Starling during 1972 may be cited as examples of other cases.

GAIN IN EFFICIENCY DUE TO TRANSFORMATION

We will now examine if there was any gain in efficiency when the means of the z's where $z = \log_e(y + 1)$ are transformed back into the original variates. This is important since number of birds counted per route can be readily interpreted and is, therefore, more useful to management than its logarithm. The efficiency of $\bar{y}$ with respect to mean m (efficient estimate of y) when the means of the logarithm are transformed back is approximately estimated by

$$\frac{\left(s^2 + \frac{s^4}{2}\right)}{e^{s^2} - 1} \tag{2}$$

where m is approximately equal to $e^{\bar{z}} + \frac{s^2}{2} - 1$ and s^2 is an unbiased estimate of σ^2(var z). This follows from a result due to Finney (1941). It would be seen from Table 4 that the efficiency of the direct sample mean was generally high (over 80%) except for Herring Gull, Yellow-bellied Sapsucker, Bank Swallow and Starling for which a logarithmic transformation is recommended. For other species, the mean was satisfactorily estimated by the direct sample mean. Since efficiency will be reduced with increasing values of σ^2, the above recommendation may not always be the right one and transformation will lead to increased efficiency of the mean for values of σ^2 exceeding two.

The efficiency of the direct estimates of population variance with respect to an efficient estimate v of the variance of the y population based on the transformed data has been obtained by Finney (1941). The efficiency is approximately estimated by

$$\frac{4s^2(t - 1)^2 + 2s^4(2t - 1)^2}{(t - 1)^2(t^4 + 2t^3 + 3t^2 - 4)} \tag{3}$$

where $t = e^{s^2}$.

The efficiency given in table 4 ranged from 0.01 to 32 percent except for Robin and White-throated Sparrow for which the efficiency was fairly high and ranged from 39 to 66 percent. The distribution of robins was approximately normal before transformation which explains the high efficiency of the direct estimates of the average number of birds counted per route. Thus, for all the species excepting Robin and White-throated Sparrow, the use of the direct estimate of variance of the y distribution proved very inefficient.

RELATIVE CHANGE IN ABUNDANCE

One of the main objectives of the study is the detection and measure of relative change in abundance of a species. In this paper we will consider the example of assessing relative change in abundance between two years. However, the results will apply equally to changes between two geographic locations though the latter comparison will be subject to more potential biases. Let x_i denote the number of birds counted by an observer on the ith route during a year and y_i, the corresponding number of birds counted in another year. Then the estimate of relative change in abundance $\hat{R}$ (expressed as percentage increase) between two years is given by

$$\hat{R} = \left(\frac{\bar{y} - \bar{x}}{\bar{x}}\right)100 = \left(\frac{\bar{y}}{\bar{x}} - 1\right)100 \tag{4}$$

where $\bar{x}$ = mean number of birds per route in the first year based on the sample of n routes,

TABLE 5

ESTIMATES OF MEAN DIFFERENCE IN BIRDS PER ROUTE, STANDARD ERRORS, t VALUES BOTH ON THE ORIGINAL AS WELL AS LOG SCALE FOR THE DIFFERENT SPECIES IN THE MARITIMES DURING 1971 AND 1972[a]

				$\bar{d}_1$ = (Ave log($\bar{y}$ + 1) − Ave log($\bar{x}$ + 1))		
Species	$\bar{d}(=\bar{y} - \bar{x})$	SE ($\bar{d}$)	t	$\bar{d}_1$	SE $\bar{d}_1$	t_1
Common Snipe	−0.30	0.94	−0.41	−0.01	0.06	−0.23
Herring Gull	17.29	20.32	0.85	0.12	0.08	1.50[b]
Least Flycatcher	−0.07	0.70	−0.11	−0.02	0.05	−0.43
Tree Swallow	−2.49	2.56	−0.97	−0.07	0.04	−1.75*
Yellow-bellied Sapsucker	3.10	1.51	2.06**	−0.14	0.06	−2.40**
Bank Swallow	−6.10	5.19	−1.18	−0.04	0.08	−0.55
Blue Jay	−1.12	1.30	−0.86	−0.02	−0.06	0.35
Common Crow	4.00	2.91	1.37	−0.00	0.04	0.03
Robin	−4.56	2.79	1.64[b]	−0.04	−0.02	2.36**
Starling	−0.37	3.46	−0.11	−0.05	0.04	−1.06
Common Grackle	1.39	1.33	1.05	0.04	0.04	1.02
White-throated Sparrow	4.00	2.16	1.85*	0.05	0.03	1.56[b]

* $P < 0.10$.
** $P < 0.05$.
[a] Total no. of comparable routes = 41.
[b] On verge of significance ($P < 0.10$).

and $\bar{y}$ = mean number of birds per route (using same routes as for $\bar{x}$) in another year.

TEST FOR DETECTION OF RELATIVE CHANGE IN ABUNDANCE

It is obvious from equation (4) that if the number of birds on the routes during one year is different from the number in another year on the same routes, $\hat{R}$ will be significantly different from zero. A test of the hypothesis $R = 0$ is equivalent to testing $\bar{Y} = \bar{X}$, where R, $\bar{X}$, $\bar{Y}$ are respectively the population versions of $\hat{R}$, $\bar{x}$ and $\bar{y}$ in the sample. We will, therefore use this test both on the original as well as on the transformed scale to detect change in abundance and hence in relative change in abundance between two years. The estimate of the change in abundance, $\bar{d}$, its standard error and the t value between two years is given by

$$\bar{d} = \bar{y} - \bar{x} \quad (5)$$

$$\text{s.e.}(\bar{d}) = [v(\bar{y}) + v(\bar{x}) - 2\,\text{cov}(\bar{x}, \bar{y})]^{\frac{1}{2}} \quad (6)$$

$$t = \bar{d}/\text{s.e.}(\bar{d}) \quad (7)$$

On the transformed scale, the corresponding estimates $\bar{d}_1$, s.e. ($\bar{d}_i$) and t_i—values are obtained by substituting $\log_{10}(y + 1)$ and $\log_{10}(x + 1)$ for y and x in the original scale.

It would be seen from Table 5 that the change in abundance measured by the mean difference in birds per route was significant for two of the species and was on the verge of significance ($P < 0.10$) for one; on the logarithmic scale, however, it was significant for three of the species and was on the verge of significance for two other species. Thus, log transformation was more sensitive to detect change in abundance between two years.

ESTIMATES OF RELATIVE CHANGE IN ABUNDANCE

Expression for the estimate of relative change in abundance $\hat{R}$ was given in equation (4). It can also be interpreted as the percentage increase in the number of birds observed between 1971 and 1972. It is easy to see that the estimate of the standard error (s.e.) of $\hat{R}$ would be given by

$$\text{s.e.}(\hat{R}) = 100\left(\frac{\bar{y}}{\bar{x}}\right)[v(\bar{y})/(\bar{y})^2 + v(\bar{x})/(\bar{x})^2 - 2\,\text{cov}(\bar{x}, \bar{y})/\bar{x}\bar{y}]^{\frac{1}{2}} \quad (8)$$

The estimates of relative change in abundance, standard error and coefficient of variation for the untransformed as well as transformed data are presented in Table 6. Note that for the transformed data equations (4) and (8) will be replaced by substituting $\bar{z}_1$, for $\bar{x}$ and $\bar{z}_2$ for $\bar{y}$ where $z_1 = \log_{10}(x + 1)$ and $z_2 = \log_{10}(y + 1)$. The conclusions are similar to those obtained in Table 5. Thus, for only three of the species i.e., Yellow-bellied Sapsucker, Robin and White-throated Sparrow, relative change in abundance from 1971 to 1972 was significant, whereas on the transformed scale decrease was, in addition, significant for two other species, i.e., Tree Swallow and Bank Swallow. Thus, the estimates of relative change in abundance on the transformed

TABLE 6

ESTIMATES OF RELATIVE CHANGE IN ABUNDANCE PER ROUTE (R × 100) DURING 1972 OVER 1971, STANDARD ERROR AND COEFFICIENT OF VARIATION (C.V.) WITH CORRESPONDING ESTIMATES BASED ON TRANSFORMED DATA FOR THE MARITIMES[a]

	Original data			Transformed data		
Species	Relative change in abundance (100R)	Standard error	C.V.	Relative change in abundance	Standard error	C.V.
Common Snipe	−8.11	16.77	2.07	1.71	9.81	5.74
Herring Gull	60.51	62.49	1.03	12.72	8.91	0.70
Least Flycatcher	−2.88	10.44	3.63	−3.53	8.03	2.28
Tree Swallow	−10.11	11.64	1.15	−5.98**	2.50	0.42
Yellow-bellied Sapsucker	−31.72**	12.60	0.40	−21.88**	8.32	0.38
Bank Swallow	−27.14	18.95	0.70	−5.12**	1.24	0.24
Blue Jay	−22.42	22.74	1.01	−4.00	14.58	3.75
Common Crow	12.43	9.47	0.76	0.09	2.64	30.25
Robin	−6.19*	3.37	0.54	−2.39[b]	1.44	0.60
Starling	−0.62	7.67	12.47	−3.28	3.10	0.95
Common Grackle	5.64	5.42	0.96	3.72	3.74	1.01
White-throated Sparrow	7.04[1]	4.67	0.66	2.94[b]	1.82	0.62

[a] Total no. of comparable routes = 41.
[b] Verges on significance ($P < 0.10$).
* $P < 0.10$.
** $P < 0.05$.

scale were more precise and able to detect changes between years.

The normal approximation for the ratio estimate $\bar{y}/\bar{x}$, and hence of relative change in abundance, was not realized for almost all the species (excepting Robin and White-throated Sparrow). Also, the sample size ($n = 41$) was not large enough to ensure that the c.v.'s of $\bar{y}$ and $\bar{x}$ are less than 0.1 (Table 1)—a necessary condition for the normal approximation to hold. However, when the data are transformed by taking logarithms, the c.v.'s of transformed means were less than 0.1 in almost all the cases (Table 2); also the sample size was large enough for the normal approximation to hold. The transformed data were, therefore, good enough to provide valid estimates of error for relative changes in the abundance of birds between 1971 and 1972.

DISCUSSION

The analysis of the data on breeding birds for the Maritime Provinces for 1971–72 has revealed that the abundance of birds for a number of clumped species was highly skewed. Also, in a majority of the cases, the number of routes selected in the region was not large enough to ensure normality of the number of birds per route so that tests based on normal theory tended to be less efficient and sensitive for detecting differences in abundance between any two species for a given season or between any two seasons for a given species.

Logarithmic transformation of the data ensured approximate normality; it resulted not only in some gain in efficiency of the mean but also considerable gain in efficiency of the variance. In all cases, the number of routes selected was large enough to ensure normality of the mean on the logarithmic scale. Also, the estimates of the mean on the transformed scale were more sensitive to detect real differences in the average number of birds counted per route among species than on the original scale.

One of the main objectives of the study was the detection and measure of relative change in abundance of a species between two years. It was shown that the test for detecting relative change in abundance was equivalent to detecting change in abundance.

For some of species, e.g., Tree Swallow and Bank Swallow, the estimates of relative change in number of birds per route were subject to high error and the statistical tests failed to detect real difference between the years. For these species tests based on logarithmic transformation proved valid and more precise to detect the change in abundance.

ACKNOWLEDGMENTS

Thanks are due to Professor F. L. Ramsey, Department of Statistics, Oregon State University and Dr. G. E. J. Smith, Canadian Wildlife Service for some helpful suggestions.

Studies in Avian Biology No. 6:502–508, 1981.

THE EFFECT OF GROUP SIZE ON LINE TRANSECT ESTIMATORS OF ABUNDANCE

TERRANCE J. QUINN II[1]

ABSTRACT.—Line transect methodology is appropriate for transect experiments where some measure of distance is made to the animal that is sighted or flushed. This methodology is extended to populations where animals are sighted in groups (schools, flocks, etc.). Thus, the probability of sighting increases as a function of the size of the group. The first method presented for such sighting data pools the data over group size and uses a line transect model that is appropriate to fit the data. The estimate of the number of groups is approximately unbiased provided a flexible estimator is chosen. The estimator of the number of individuals is the product of the estimated number of groups and the estimated average group-size. The estimated average group size must be weighted to account for the increased probability of sighting larger groups. The second method presented post-stratifies the sighting data by group size and then proceeds as in the first method. The two methods are evaluated theoretically and by computer simulation. The method of post-stratification produces estimates that are closer to the true value but have larger variances than the method of pooling.

Transect methods for the estimation of animal abundance have been carried out for many years on a variety of species. These surveys have usually been designed as strip transect surveys, defined by a fixed width from the transect line wherein all animals were thought to be seen, or index surveys, where all animals sighted are counted and the results are interpreted as relative indices between years or regions. Population estimates can be obtained only for the strip transect surveys and are calculated intuitively from extrapolating the number sighted in the strip to the entire population area. Although distances to sightings have been measured occasionally, they usually have been used for checking that all animals are sighted in the strip. Some heuristic estimators using distances have been developed (Amman and Baldwin 1960, J. T. Emlen 1971, see Gates 1979 for others) but lack of statistical formulation has prevented assessment of an estimator's properties.

Incorporation of measured distances into the experimental design of transect experiments forms a powerful technique for estimating abundance called a line transect experiment. The roots of its methodology are contained in statistical models for sampling theory and recent advances in non-parametric density estimation and robust estimation, as well-described in recent reviews (Eberhardt 1978, Gates 1979, Quinn and Gallucci 1980, Burnham et al. 1980). The focus of the methodology is to construct a sighting model from the measured distances to correct for animals that are overlooked. The sighting model $g(y)$ is the non-increasing probability of a sighting at perpendicular distance y from the transect line, and animals on the transect line are assumed to be sighted with probability 1 (i.e., $g(0) = 1$). The strip transect method is a special case of the more general line transect sampling methodology (Seber 1973).

When a population is made up of groups (i.e., schools, flocks, herds) of varying sizes, line transect methodology is still appropriate, but with some modification. The purpose of this paper is to describe and compare two methods to analyze data from populations where sightings are made in groups. The key concept in the methodology is that the probability of sighting is likely to be an increasing function of group size that need not be linear. Empirical experiments on porpoise populations support this assertion (R. Holt and J. Powers, in prep.).

The general estimation framework for line transect methodology is briefly reviewed below. Three sighting models are described which represent common classes of estimators for line transect sampling. The two methods of analyzing transect data from populations of groups are also discussed below. The first method is to pool the transect data over groups of all size classes in order to estimate the total number of groups. The estimator from this method is robust, because the pooled sighting model is self-weighted by the true relative abundance of each group-size class in the population (Quinn 1979, Burnham et al. 1980). In the second method, the total sample of n sightings of groups is partitioned into t group-size classes. This method is referred to as post-stratification, because the total sample is partitioned after the completion of the survey, rather than taking an independent sample of each group-size class. The number of groups in each class is estimated and the estimates are summed to get the total number of groups in the population. The salient estimation formulae for both methods of the total number of groups, the total number of individuals and their variability are presented from Quinn (1980). I will compare the two methods with computer simulation and theoretically. In the last section below, the results are discussed in terms of specific recom-

[1] International Pacific Halibut Commission, P.O. Box 5009, University Station, Seattle, Washington 98105.

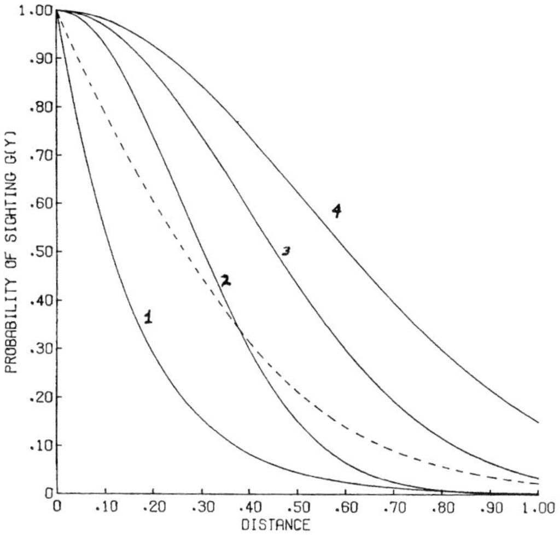

FIGURE 1. Plot of individual sighting models (solid lines) and the resultant pooled sighting model (dashed line) for the text example.

mendations for the planning and analysis of transect data from populations of groups.

ESTIMATION

For a line transect experiment on a population of individuals, the estimator of the number of individuals N in the population is given by

$$\hat{N} = \frac{An}{2L}\hat{c}^{-1} = \frac{An}{2L}\hat{f}(0), \qquad (1)$$

where A is the population area, L is the transect length, c is the effective half-width sampled [defined by the integral of $g(y)$], $f(0)$ is the probability density function of sightings evaluated at the origin, and a caret ($\hat{}$) indicates an estimate (Quinn and Gallucci 1980, Burnham et al. 1980). The number of sightings is extrapolated to the total number by the ratio of the population area and the effective area sampled $2L\hat{c}$ (Quinn and Gallucci 1980).

The estimated variance of (1) is

$$\hat{\mathrm{Var}}(\hat{N}) = \hat{N}^2[\hat{\mathrm{c.v.}}^2(n) + \hat{\mathrm{c.v.}}^2(\hat{c}^{-1}|\,n)]. \qquad (2)$$

The term $\hat{\mathrm{c.v.}}^2(n)$ is the estimated squared coefficient of variation [i.e., $\hat{\mathrm{Var}}(n)/n^2$] of the number of sightings obtained from subsampling or jackknifing. The term $\hat{\mathrm{c.v.}}^2(\hat{c}^{-1}|\,n)$ is the estimated squared coefficient of variation of the inverse of the estimated effective half-width, which is a derived formula from the sighting model (Quinn and Gallucci 1980, Burnham et al. 1980).

Three sighting models are used in this comparative study which are representative of available models. They are:

(1) The exponential model (EM)—a one-parameter model which postulates a sharp spiked decrease in sighting probability as distance from the transect increases; thought useful for flushing birds (Gates 1979, Eberhardt 1978).

TABLE 1
PARAMETERS AND MODELS USED IN THE SIMULATION STUDY[a]

Group-size class i	1	2	3	4
S_i	5	25	125	625
N_i/N	.4	.3	.2	.1
$c_i(\alpha \ln S_i)$	.161	.323	.484	.646
Sighting model $g_i(y)$	EM	HNM	HNM	HNM
$E(n_i/n)$	.2	.3	.3	.2

[a] $A = L = 1$; $N = 154.9$; $n = 50$.

(2) Fourier model (FOUR—a non-parametric approach from a Fourier series expansion of the probability density function $f(y)$; with the ability to assume a variety of non-spiked sighting curves (Burnham et al. 1980).

(3) Kelker model (KELK)—a version of the strip transect model as named after one of its earliest progenitors (see Gates 1979); a nonparametric approach, because no parameter of the model is estimated.

The mathematical representation of these sighting models and corresponding estimators is given in Quinn (1979). A fourth type of estimator, a generalized parametric approach, produces results similar to the Fourier series (Burnham et al. 1980).

METHODS FOR ANALYSIS OF GROUP SIGHTINGS

Let there be t classes of groups in the population where S_i is the number of individuals per group (group-size) for the ith size class. Let N_i be the true number of groups in the ith class and let $\Sigma N_i = N$. Suppose that a transect experiment is carried out and n total sightings occur with n_i sightings in class i. Each class has an associated sighting model $g_i(y)$ and effective half-width c_i. The group size S_i associated with each sighting is assumed to be determined without error.

The following example of such a population illustrates the important parameters of the experiment and forms the basis of the later computer simulation exercise. Four classes of groups in the population are constructed with true relative abundances N_i/N of 0.4, 0.3, 0.2, and 0.1 and represent group-sizes of 5, 25, 125, and 625 individuals. The exponential model (EM) is chosen for the underlying sighting model for the first class to represent the situation where sightings of small groups fall off rapidly at short distances from the transect line. The half-normal model (HNM) is chosen for the underlying sighting model for the three larger classes to represent the situation where sightings are fairly uniform at distances near the transect line and fall off smoothly at larger distances depending on the size of the group. The effective half-width c_i is chosen to be a logarithmic function of group-size in order to specify the scale of each individual curve (Fig. 1). The parameters which determine these relationships are shown in Table 1.

In general, the sighting model for groups of all sizes during the transect experiment is formed by weighting each individual sighting model by its relative abundance in the population (Quinn 1979). Applying this principle to the example produces the pooled sighting model for all sightings as shown in Figure 1. This pooled sighting model exhibits the heavier weighting of the first, more abundant class and has a shape that is functionally different from its component parts.

Method of Pooling

The first method of estimation for transect data pools the data over group-size classes. First, the estimated number of groups and its variance are calculated from (1) and (2), where N is redefined as the number of groups rather than individuals. A reliable estimator of N is obtained when the sighting model used for the pooled data approximates the unknown pooled sighting model.

Secondly, the estimated number of groups $\hat{N}_P$ is multiplied by an estimate of the average group-size $\bar{S} = \Sigma N_i S_i / N$ to estimate the total number of individuals in the population, i.e.,

$$\hat{T}_P = \hat{N}_P \hat{\bar{S}}. \tag{3}$$

The variance of $\hat{T}_P$ in (3) is the variance of a product (Seber 1973:7–9). The sample average group-size is not an unbiased estimate of $\bar{S}$ if there is a relationship between group-size and probability of sighting (or equivalently effective half-width), because larger groups are more likely to be in the sample than their presence in the population indicates. The estimate of $\bar{S}$ using (1) is

$$\hat{\bar{S}}_1 = \sum \hat{N}_i S_i \Big/ \sum \hat{N}_i = \sum n_i S_i \hat{c}_i^{-1} \Big/ \sum n_i \hat{c}_i^{-1}, \tag{4}$$

which is approximately unbiased when the n_i are near their expectations.

If the data are pooled over group-size, then (4) is not estimable, because each c_i is not estimated. One method of alleviating this problem is to: assume a functional relationship between c_i and S_i, i.e., $c_i = h(S_i)$(which may include a constant term); assume the mean sighting distance $\bar{y}_i$ is proportional to c_i for all classes; regress $\bar{y}_i$ against S_i to establish $\hat{h}$; and finally replace $\hat{c}_i^{-1}$ by $[\hat{h}(S_i)]^{-1}$ in (4). In particular, if c_i is proportional to the logarithm of group-size $\ln S_i$, then

$$\hat{\bar{S}}_2 = \sum \frac{n_i S_i}{\ln S_i} \Big/ \sum \frac{n_i}{\ln S_i} \tag{5}$$

results. An alternative estimator of $\bar{S}$, called $\hat{\bar{S}}_3$, is the average group size from sightings in a small interval about the transect line where groups of all sizes are likely to be seen. The estimators $\hat{\bar{S}}_2$ and two versions of $\hat{\bar{S}}_3$ using different intervals are evaluated by computer simulation below in the section dealing with comparison of the pooled and post-stratified methods.

Method of Post-Stratification

The second method of analysis is to partition the data by group-size. This method requires a sufficient number of sightings in each group-size class, say 25. The estimated number of groups in each class N_i is obtained from (1) using only the sightings from that class. Since the total sample of n sightings is stratified after the experiment is completed, this method is called post-stratification.

The intuitive post-stratified estimator of the total number of groups is

$$\hat{N}_s = \sum_{i=1}^{t} \hat{N}_i, \tag{6}$$

with estimated variance

$$\hat{\mathrm{Var}}(\hat{N}_s) = \sum_{i=1}^{t} \hat{\mathrm{Var}}(\hat{N}_i) + 2 \sum_{i<j} \hat{\mathrm{Cov}}(\hat{N}_i, \hat{N}_j). \tag{7}$$

The covariance terms are necessary because the n_i come from a multinomial distribution with parameters n and p_i^*, $i = 1, \ldots, t$, where p_i^* is the expected proportion of sightings $E(n_i)/E(n)$. Using results of conditional variance and covariance derived by Quinn (1980), the estimated variance of $\hat{N}_i$ is

$$\hat{\mathrm{Var}}(\hat{N}_i) = \hat{N}_i^2[\hat{\mathrm{c.v.}}^2(n_i \mid n) + \hat{\mathrm{c.v.}}^2(n) + \hat{\mathrm{c.v.}}^2(\hat{c}_i^{-1} \mid n_i)], \tag{8}$$

where

$$\hat{\mathrm{c.v.}}^2(n_i \mid n) = (n - n_i)/nn_i,$$

and the estimated covariance between $\hat{N}_i$ and $\hat{N}_j$ is

$$\hat{\mathrm{Cov}}(\hat{N}_i, \hat{N}_j) = \hat{N}_i \hat{N}_j [\hat{\mathrm{c.v.}}^2(n) - 1/n] \tag{9}$$

Methods of estimating c.v.$^2(n)$ are given by Quinn and Gallucci (1980). The estimates (8) and (9) are substituted into (7) for the estimated variance of $\hat{N}_s$.

Finally, the post-stratified estimate $\hat{T}_s$ of the total number of individuals is obtained by multiplying the estimated number of groups $\hat{N}_i$ for class i by its group-size S_i and adding up over groups, so that

$$\hat{T}_s = \sum \hat{N}_i S_i. \tag{10}$$

Its estimated variance is

$$\hat{\mathrm{Var}}(\hat{T}_s) = \sum S_i^2 \hat{\mathrm{Var}}(\hat{N}_i) + 2 \sum_{i<j} S_i S_j \hat{\mathrm{Cov}}(\hat{N}_i, \hat{N}_j). \tag{11}$$

Thus, the post-stratified estimator $\hat{T}_s$ does not require estimation of the average group-size $\bar{S}$ in contrast to $\hat{T}_p$. However, if S_i refers to a range of group-sizes, then this source of variability should be incorporated into (11) using (4), although its effect is likely to be minor compared to the variability of group-sizes over the entire population.

COMPARISON OF THE POOLED AND POST-STRATIFIED METHODS

In order to quantitatively compare the two methods, a computer simulation study was conducted using population parameters from the previous example, which are summarized in Table 1. The total number of sightings was fixed at 50, and the term c.v.$^2(n)$ was thus set to 0. This

TABLE 2
SIMULATION ESTIMATES OF THE NUMBER OF SCHOOLS

Estimator	Simulation average	Standard error (c.v.)		Root mean squared error $\sqrt{MSE}$[a]	
		Theoretical	Empirical	Theoretical	Empirical
Method of pooling					
EM	185.1	3.8 (.021)	3.6 (.019)	40.2	39.3
FOURIER	140.9	3.7 (.026)	4.3 (.031)	29.4	33.2
KELK	134.2	3.8 (.028)	4.4 (.033)	33.7	37.1
Method of post-stratification					
EM	200.7	5.1 (.025)	3.8 (.019)	57.6	52.5
FOURIER	154.3	5.6 (.036)	6.9 (.045)	39.2	48.3
KELK	146.1	4.8 (.033)	4.5 (.031)	34.7	32.7
True value	154.9				

[a] $\sqrt{MSE} = \sqrt{\hat{E}(\hat{N} - u)^2} = \sqrt{(n_r - 1)s^2 + (\bar{x} - u)^2}$
where n_r = number of replications; s = empirical or theoretical standard error; $\bar{x}$ = simulation average, and u = true parameter.

approach produces a smaller variance than a normal transect study where n is itself a random variable. However, the comparison of the two methods of analysis is still valid, because the term c.v.$^2(n)$ occurs equally in the variance expressions for both methods [Equations (2) and (8)]. The simulation was replicated 50 times to provide empirical means and standard errors for comparison with known or theoretical values. Further details concerning the mechanics of the simulation are found in Quinn (1979, 1980).

ESTIMATION OF THE NUMBER OF GROUPS

The simulation estimates of the number of groups N in the population are shown in Table 2 for the two methods. For the method of pooling, the EM estimator is positively biased, and the Fourier and Kelker estimators are negatively biased. The Fourier estimator is the least biased of the three. The spiked nature of the pooled sighting model (Fig. 1) causes underestimates to occur for estimators that assume a rounded shape near the origin (Crain et al. 1978, Quinn 1977). The EM estimate has the lowest coefficient of variation, followed by the Fourier and then the Kelker estimators. The root mean squared error, a convenient statistic incorporating the effects of variance and bias, favors the Fourier and then the Kelker estimator.

For the method of post-stratification, the Fourier estimator is the only estimator that produces an unbiased estimator. The Kelker and EM estimators produce under- and over-estimates, respectively. By examining the results of each group-size class (Table 3), the explanation for the bias is apparent. The EM overestimates the last three classes and correctly estimates the first class, producing an overall overestimate, in accord with the sighting models used for each group-size class (Table 1). The Fourier simulation average is unbiased for all classes, which produces an overall unbiased estimate. The Kelker estimate is negatively biased only for the first class, since the Kelker estimator performs poorly for spiked sighting models but reasonably well for rounded models (provided the truncation width is chosen small enough). The coefficients of variation show the same trends as for the method of pooling. The root mean squared error favors again the Fourier and then the Kelker estimator.

These results, which form a subset of a larger simulation study (Quinn 1980), suggest two general results. First, the simulation averages of a reasonable estimator such as the Fourier or the Kelker for the method of post-stratification generally are closer to the true parameter than for the method of pooling. This result is not unexpected, because the pooled sighting model generally has a more complicated shape and often a wider range of distances than the individual sighting models. Secondly, the theoretical and, generally, the empirical coefficients of variation for the method of post-stratification are larger than for the method of pooling (Table 2). This result is expected, because the number of sightings for each group-size class is substantially smaller than the total number of sightings, and the variance of an estimator of effective half-width is generally proportional to the inverse of the number of sightings (Quinn 1980, Burnham et al. 1980). The root expected mean squared error (Table 2) favors the method of pooling for the EM and Fourier, and either method for the Kelker.

These two generalizations from the simulation study have roots in theoretical relationships between sighting models and estimators. If an estimator has a functional form that is additive (Quinn 1980), then the method of pooling and the method of post-stratification produce identical estimators. This condition of additivity is satisfied by the Fourier and Kelker estimators, but only if their prespecified parameters are assumed constant for all classes. However, since the functional form of the sighting model for each class was different (e.g., Fig. 1), these prespecified parameters were not constant. In the simulation, these parameters were allowed to vary, so that each estimator could better estimate the number of groups in each class. Hence, simulation estimates for the post-stratified method were closer to the true parameter than for the pooling method.

The second generalization concerning the increased precision of the pooled estimator can also be verified theoretically by assuming that the precision of an estimator is proportional to the number of sightings, i.e.,

$$\text{c.v.}^2(\hat{c}^{-1} \mid n) = \sigma^2/n,$$

where σ_c^2 is an asymptotic constant dependent on the sighting model. This assumption appears to be reasonable by examination of the form of the variance estimator although non-parametric estimators are slightly less precise (Quinn 1980, Burnham et al. 1980, Eberhardt 1978). By substituting this relationship into (2) and (8) and using the Cauchy-Schwarz inequality, it can be shown that the theoretical coefficient of variation for the method of pooling is always less than or equal to that for the method of post-stratification (Quinn 1980, theorem 4). The only situation where the two are equal is when there is no relationship between the effective half-width and group-size.

The impact of the above results concerning transect estimation for grouped populations involves the trade-offs in accuracy (closeness to the true value) versus precision (as measured by the inverse of the coefficient of variation of the estimates). By post-stratifying the data, it is often possible to estimate each class accurately and, hence, the total number of groups N is estimated accurately. However, a single incorrect choice of a sighting model for a class leads to a biased estimate of N, and may not be detected by goodness-of-fit tests if there is a small number of sightings in the class. By pooling the data, the resultant sighting model may have a shape that is difficult to approximate by common sighting models, especially if the effective half-widths are substantially different. When a flexible model such as the Fourier is applied to both methods, the method of post-stratification is usually

TABLE 3
SIMULATION ESTIMATES OF N_i FOR EACH GROUP-SIZE CLASS[a]

Estimator	Group-size class (i) 1	2	3	4
EM	62.7	68.1	46.9	23.0
	4.1	3.5	2.3	1.5
	3.6	3.2	2.0	1.2
FOURIER	59.5	44.2	33.3	17.3
	4.8	3.0	2.1	1.5
	5.7	3.6	2.5	1.7
KELK	53.2	46.6	29.1	17.3
	3.9	3.0	1.9	1.3
	3.5	2.8	1.7	1.3
True value	62.0	46.5	31.0	15.5

[a] Reported for each group and estimator are the simulation average, theoretical standard error, and empirical standard error.

more accurate. On the other hand, the precision of the method of post-stratification, as compared to the method of pooling, becomes increasingly poor as greater differences occur in the effective half-widths.

The method of pooling is recommended for estimating the number of groups as long as an estimator derived from a flexible sighting model is chosen. In general, the bias of the estimator with the method of pooling is not large, and both the coefficient of variation and the mean-squared error are likely to be smaller than with the method of post-stratification.

ESTIMATES OF AVERAGE GROUP-SIZE AND THE TOTAL NUMBER OF INDIVIDUALS

In order to estimate the total number of individuals T for the method of pooling, the average group size $\bar{S}$ in the population must be estimated as an intermediate step as shown in the section above on methods for analysis of group sightings. The data from the simulation are used to compare estimators of $\bar{S}$ and to illustrate the magnitude of bias of the sample average group-size $\bar{s}$. The true mean group-size, $\bar{S}$, and expected sample average group-size, $E(\bar{s})$, are computed from the values in Table 1. Four estimates are computed from the simulation replications: $\bar{s}$, $\hat{\bar{S}}_2$ (the log-weighted estimator), and two estimates $\hat{\bar{S}}_3$ the first uses sightings in the interval $[0,\Delta]$, the second in the interval $[0,2\Delta]$, where 2Δ includes no more than 75% of the total sightings).

The results are straightforward (Table 4). The true and expected sample average group-sizes are radically different (97 as compared to 171). The simulation average group-size is close to its theoretical value. By correcting $\bar{s}$ for increasing

TABLE 4
ESTIMATES OF AVERAGE SCHOOL SIZE FOR THE SIMULATION STUDY

n	n_r	$\bar{S}$	$E(\bar{s})$	$\bar{s}$	$\hat{S}_2$	$\hat{S}_3$ Interval $[0,2\Delta]$[a]	$\hat{S}_3$ Interval $[0,\Delta]$[a]
50	50	97	171	170.3 ± 4.9	98.1 ± 3.1	127.0 ± 5.0	119.4 ± 6.6

[a] Δ is chosen so that the interval $[0,2\Delta]$ encompasses no more than 75% of the observations.

probability of sighting by deleting more and more sightings at large distances, the simulation average becomes closer to the true group-size. However, even $\hat{S}_3$ in the interval $[0,\Delta]$ is positively biased and has the highest coefficient of variation of the four estimates. The log-weighted estimator, $\hat{S}_2$, has the lowest coefficient of variation and no bias. Thus, when effective widths are proportional to the logarithm of group-size, $\hat{S}_2$ is the best estimator.

However, additional studies have shown that the estimator $\hat{S}_2$ is not robust to the relationship between effective width and group-size (Quinn 1980). The estimator $\hat{S}_3$ is fairly robust but usually biased upward. Thus, there appears to be a need for more efficient and robust approaches to the estimation of a weighted average group-size.

The final comparison between the two methods involves estimates of the total number of individuals from (3) and (10). As shown in Table 5, three estimation models are considered: the method of pooling using the efficient estimator $\hat{S}_2$, the method of pooling using the robust estimator $\hat{S}_3$, and the method of post-stratification. The estimates in Table 5 are calculated directly from the values in Tables 2, 3, and 4.

The same trends for estimating the total number of individuals are found as for estimating the number of groups. One interesting difference is

TABLE 5
POOLED AND POST-STRATIFIED ESTIMATORS OF THE TOTAL NUMBER OF INDIVIDUALS T[a]

Estimator	Pooled estimator $\hat{N}_p\hat{S}_3$	Pooled estimator $\hat{N}_p\hat{S}_2$	Post-stratified estimator $\hat{T}_s$
EM	22,092	18,160	22,254
	1312 (.059)	684 (.038)	932 (.042)
	1304 (.059)	674 (.037)	729 (.033)
FOURIER	16,816	13,824	16,378
	1036 (.062)	568 (.041)	949 (.058)
	1068 (.064)	607 (.044)	1087 (.066)
KELK	16,017	13,166	15,881
	1001 (.062)	558 (.042)	821 (.052)
	1035 (.065)	599 (.046)	814 (.051)

[a] Reported for each estimator are its estimate, theoretical and empirical standard errors, and coefficients of variation. True parameter is 15025.

that the differences in the coefficients of variation of $\hat{T}$ for the pooled estimators are not as large as for the coefficients of $\hat{N}$. The contribution to the coefficients of the average group-size predominates, because the range in group-size is two orders of magnitude.

The best estimator for T appears to be the precise estimator $\hat{N}_P\hat{S}_2$ using the Fourier series estimator. If a more robust estimator is desired, then $\hat{T}_s$ with the Fourier estimator is fairly robust and reasonably precise. However, it may not be possible to compute $\hat{T}_s$ when the number of sightings is small. Another robust estimator that may be used as a last resort is the pooled estimator $\hat{N}_P\bar{S}_3$ with the Kelker estimator, although this estimator is the least precise.

DISCUSSION

Based on simulation results and theoretical principles found here and in Burnham et al. (1980) and Quinn (1980), the following recommendations are given for a line transect sampling experiment of populations made up of groups.

1. The experiment should assure that a minimum of 50 groups are sighted for the method of pooling or 25 sightings per class for the method of post-stratification. Otherwise, criteria of accuracy and precision may not be met. If possible, a pilot study should be carried out to provide preliminary estimates of transect parameters. The preliminary parameters are necessary to calculate formulae for determining the amount of effort needed to be expended in the main experiment to achieve a certain level of precision (Quinn 1980, Burnham et al. 1980).

2. Critical assumptions of the line transect method are that all groups on the transect line are sighted and that there is no directional movement toward or away from the transect line. If possible, experimental design should include a test of these assumptions and ancillary experiments to develop correction factors if the assumptions are not met. Other assumptions of the line transect method are found in Burnham et al. (1980).

3. It is necessary to measure distances accurately to the geometric center of the group. Otherwise, the estimated number of groups is unreliable. The size of each group must also be

counted accurately. Otherwise, the estimated total number of individuals and average group-size are unreliable.

4. The method of pooling is recommended over the method of post-stratification for estimating the total number of groups, because of its increased precision, lower mean-squared error, larger number of sightings for hypothesis tests, and lack of arbitrary determinations about number of classes and sighting models for each class. Some stratification may be necessary, however, if the pooled sighting model has a complicated shape that is not well-represented by common sighting models in usage.

5. The recommended sighting model is the non-parametric Fourier series model based on these results and other studies (Burnham et al. 1980, Quinn 1980). The Kelker estimator should be used only as a last resort. One-parameter models such as the exponential model should not be used unless an extremely good fit to the data is produced. Generalized parametric estimators and other non-parametric estimators (Quinn and Gallucci 1980, Burnham et al. 1980) are often an acceptable alternative to the Fourier series.

6. The average group-size in the population must be estimated using a weighting procedure based on the relationship between the effective half-width and group-size. Current weighting procedures are largely empirical and have limitations of robustness. Better weighting procedures are needed.

Studies in Avian Biology No. 6:509–510, 1981.

SUMMARIZING REMARKS: DATA ANALYSIS

KENNETH H. POLLOCK[1]

First of all I should like to say that I have really enjoyed and benefited from this Symposium. I believe that it will stimulate important future research on techniques of sampling bird populations from both the biological and statistical perspectives. On a lighter note I mention that this is the first conference where I have seen participants use binoculars to read data slides!

With apologies to the authors of papers on other topics I would like to devote the major part of my summary to a discussion of the line transect and variable circular plot sampling techniques which have occupied central stage at this Symposium. It is unfortunate that we had to wait until so late in the program for three of the most important papers on these techniques.

The goal of these two techniques is to estimate the average density of birds in the study area. In each case the "effective area" sampled is estimated from the observer to bird distances. The same four basic assumptions (which obviously may not be realized in practice) apply: (1) Birds in the immediate vicinity of the observer (as he moves along the transect line or stands at the center point of the circular plot) will always be detected; (2) there is no movement of birds in response to the observer; (3) all measurements are made without error; and (4) sightings of different birds are independent of each other.

The first assumption is critical to density estimation and will cause a negative bias when it fails. It is perhaps more likely to be satisfied for variable circular plots because the observer is stationary and devoting his full attention to detecting birds. Movement will typically be away from the observer causing a negative bias on density estimates. On occasion, however, birds will move towards the observer causing a positive bias. The assumption of independent sightings will obviously be false for flocking birds but there we can consider the flock as a unit and extend the theory (as in Quinn's [1981] paper).

The detection function ($g(x)$) which relates the probability of a bird's detection to its distance from the observer (x) (usually defined in terms of perpendicular distance for line transects) is central to all methods of density estimation. In particular the assumption (1) which can be stated mathematically as $g(0) = 1$ is crucial. The nature of the detection function and how to use it is a point of rather fundamental disagreement.

Burnham et al. (1981) in their paper (and also Quinn 1981) take the approach of modelling the detection curve (using a Fourier Series) and assuming it is a decreasing function with distance. Ramsey and Scott (1981) (following J. T. Emlen 1971, 1977a) assume the detection function is constant and equal to one (all birds detected) for an appreciable distance from the observer and they concentrate on determining this distance. Ramsey and Scott (1981) state ". . . if one feels that there is some substantial region of uniform, near-perfect detectability, the modified Emlen technique is recommended." It seems to me, however, that the Fourier Series approach of Burnham et al. (1981) would work just as well on this type of detection function.

From a statistical modelling viewpoint I believe that line transects and variable circular plots are basically equivalent. From a biological viewpoint, however, they are very different. Line transects estimate the average density of birds in a long narrow strip so that Ramsey and Scott (1981) suggest they are less useful in variable habitats. They also suggest the disadvantage of line transects in rugged terrain where movement of the observer along a line is difficult, if not impossible! As mentioned earlier it may also be less likely that line transects will satisfy the assumption (1) of perfect detectability of birds in the vicinity of the observer. Line transects on the other hand have the important practical advantage of often covering much larger areas in the same period of time.

A problem not considered in any detail during the symposium is what to do about birds which occur in flocks (groups). Thus the paper by Quinn (1981) is particularly welcome. Although framed in terms of line transects, the methods and conclusions apply equally well to variable circular plots. Two approaches to the problem are considered and compared. The first approach uses a two stage procedure. All group sizes are pooled and the average detection function fitted using a flexible form like the Fourier series. This gives rise to an estimate of the density of groups. The estimate of the density of individual birds is the product of the estimated number of groups and the average group size. The estimated average group size must be weighted to account for the increased probability of sighting larger groups. The second approach "post stratifies" the data by group size and uses line transect theory to estimate the density of each group size separately. Quinn's (1981) conclusion is that overall the method of

[1] Department of Statistics, North Carolina State University, Box 5457, Raleigh, North Carolina 27650.

pooling is to be preferred. I believe the group size influence on the detection function should be the subject of more research.

Before concluding I should like to make a couple of points about some of the other papers. During this conference I have been a little worried by the wide use of the spot mapping method as a census with no sampling problems and was pleased to see that Eagles (1981) has given serious attention to its problems. The paper by Johnson (1981) is excellent. We all too often ignore auxiliary information (which costs nothing!) when we calculate our estimators.

In conclusion I would like to make some general comments. I feel that some biologists at this conference have been too optimistic about the use of indices of abundance. I know there are a lot of problems with methods of estimating absolute abundance but there are important assumptions behind indices as well and these have tended to be neglected. In particular I mention the large variability of indices and the assumption of direct proportionality to absolute density. There is a need for new statistical procedures to be made easily available to biologists through monographs and computer packages. The work of Burnham et al. (1980) on line transects will hopefully encourage similar efforts for other techniques. There is also the need for small workshops for biologists to study the new statistical procedures. Ecology graduate programs in the universities should also be encouraged to strengthen their courses on sampling methods. Finally, I would like to stress the need for biologists and statisticians to work closely together in the future.

Studies in Avian Biology No. 6:511–512, 1981.

SUMMARIZING REMARKS: DATA ANALYSIS

STUART L. PIMM[1]

The papers in this session cover a wide range of the techniques used by ecologists to census populations: spot-mapping (Eagles 1981), home range estimates (Ford and Myers 1981), line transect methods and their relatives (Ramsey and Scott 1981, Burnham et al. 1981, Quinn 1981) as well as techniques used to analyze these data once they are obtained (Sen 1981, Johnson 1981). The number of issues that these papers raise, however, is fewer. I shall consider three of them.

(1) The majority of papers in the symposium end with an estimate of density. Johnson's (1981) paper starts with such estimates. He points to an experience so common to the ecologist: one's experience is often ignored in the estimates of density one obtains and they often seem in conflict with it. Johnson shows that experience need not be ignored—experience of both the past densities and of the current situation can be incorporated into population estimates. Indeed they should be: there is a considerable improvement in the estimates obtained by incorporating prior knowledge. I consider Johnson's (1981) paper particularly innovative because it operates on data at a later stage than most of the techniques discussed elsewhere. It is, thus, an additional, rather than an alternative stage in obtaining populations estimates.

Sen's (1981) paper also addresses population estimates at a late stage—perhaps the last stage—when questions of the significance of population changes are to be answered. As he indicates, an appreciation of the unusual features of population estimates (they are often highly skewed) leads to a transformation of the estimates which greatly improves the power of the statistical tests used on the data. Both Sen (1981) and Johnson (1981) lead to a conclusion that population estimates are not ends in themselves, but are the inputs to subsequent analyses. And how these analyses are performed can be as crucial to the biological conclusions as how the estimates were obtained.

(2) Several of the papers in this session, as well as that of Pollock (1981), in the previous session, are "consumer's guides" to various techniques. They ask: is there a "best buy" among them? The answer is a combination of "yes" and "no." Certainly, some techniques are better than others. Ford and Myers (1981) show that probabilistic estimators of home range are better than the widely used minimum convex polygon (MCP) method. They used a computer-based movement generator that closely describes the space-use patterns of two species. Using this generator to produce simulated home ranges they examined the efficiency of two probabilistic (one parametric, the other not) and the non-probabilistic MCP method. They considered the non-parametric, probabilistic method to be the "best buy."

Similarly Burnham et al. (1981), Quinn (1981), and Ramsey and Scott (1981), evaluate the various forms of line transect techniques. The critical aspect of these studies is the sighting curve—how detection drops off with increasing distance from the observer. Various forms for this function have been suggested. Though each may be suitable in special cases, their inflexibility precludes their widespread use. Burnham et al. (1981) suggest fitting a Fourier series to the data. This approach has flexibility and a number of other desirable attributes required for accurate estimation of density and ease of use. The authors provide a fully documented program, called TRANSECT, which produces estimates in this way.

While line-transects have advantages for sampling small areas frequently, ecologists must often sample more extensively. Ramsey and Scott (1981) attack the problem of variable circular plot designs to this end. They evaluate the possible ways of performing these censusses. Finally, Quinn (1981) considers both the possible forms of the sighting curve, and the complications that arise when animals flock (or school or herd!). In such cases detectability will be a function of group size, and this function must be incorporated into the estimates.

The syntheses that these studies provide should not lead to too simple a view of "best buys." Clearly, some techniques are better than others. But there are still plenty of alternatives, each the best for a limited range of circumstances. The papers on line transect techniques and their modifications are complementary, each describing different field situations and observer needs. There is no global "best buy."

(3) My final comment stems directly from the previous one. If there is no single best line transect method, and, if line transects are but one of several possible census methods, and, moreover, if when I have obtained these estimates they still require additional processing, then my feelings are ones of despair. The necessary com-

[1] Department of Biological Sciences, Texas Tech University, Lubbock, Texas 79409, U.S.A.

puter programs are scattered over much of one continent, they differ widely in availability and documentation, and I suspect (from previous experience) that getting some of the programs to run on my computer will be anything but trivial. Dr. Rice, in his summary of the session on "Data Analysis," noted that biologists often use such sophisticated routines as principal component analyses and multiple regressions analyses once their data are in hand, but that they seem loath to approach the statistics/statistician at the sampling design stage. I suggest the inaccessibility of many of the techniques discussed in this symposium is the cause. While the techniques remain inaccesible, biologists, particularly those with an inherent fear of mathematics, computers and statistics, will certainly under-use these techniques. My plea is simple and will be unpopular with any programmer. If the statistical techniques are to be used they must be: (1) centrally located; (2) be implemented on a wide range of university systems; and (3) have input requirements and specifications of options that ecologists can easily comprehend. Many ecologists use SAS or SPSS; the package of sampling techniques that appears on these packages first, will whatever its statistical merits, be the one most widely used by biologists.

Studies in Avian Biology No. 6:513–521, 1981.

SCALE PROBLEMS IN AVIAN CENSUSING

JOHN A. WIENS[1]

ABSTRACT.—Bird populations and their habitats are not static, but vary in space and time on several scales of resolution. This variation is not smooth and continuous, but changes as a function of scale. The operation of biological processes is thus scale-dependent, and investigations conducted at one scale cannot evaluate the effects of processes that are most important at other scales; they may in fact produce misleading results. The patterns of habitat occupancy of breeding birds that are discerned by analyses spanning different spatial scales are different, for example, as are the factors that account for variations in bird community diversity. Temporal variations in population densities at local scales may also complicate the interpretation of bird-habitat associations, especially if populations do not completely pack the available habitat. Further, such temporal variations increase the likelihood that single samplings of densities will be incorrect and that patterns revealed by collections of such single censuses may be false. These problems are most severe in "quick and easy" studies that are conducted on ill-defined spatial scales and utilize single samplings of populations in time, but they beset carefully designed investigations as well. Only by recognizing that ecological processes operate with different intensities at different scales of space and time, and then attempting to match the scale of censusing or habitat evaluation with the scale(s) of operation of these processes, can we hope to derive a correct understanding of the patterns of nature.

Censuses of avian populations or evaluations of their habitat affinities can be conducted and analyzed on a variety of scales in space and time. Variations in population densities, for example, have been considered on spatial scales ranging from entire continents (e.g., Bock et al. 1977, 1978) to differences between areas within the same square kilometer (e.g., Wiens 1973, Wiens and Dyer 1975), and patterns of change in avian community diversity have been examined at local (Wiens and Rotenberry, In press a), regional (Rotenberry 1978), or continental scales (Schall and Pianka 1978, Short 1979). Similarly, patterns of avian habitat associations have been investigated at within-territory scales (Wiens 1969), over regions within the same general habitat type (Smith 1977), on continental, between-biome scales (e.g., Pitelka 1941), or by comparisons of the same habitat type on different continents (Cody 1975). In the temporal dimension, studies of habitat change may concentrate on seasonal patterns, successional transformations over decades or centuries, or large-scale changes in geological time. Populations may fluctuate in size over short time scales as consequences of reproduction, overwinter mortality, and emigration, follow multiyear cycles of abundance, or, over centuries, exhibit explosive expansions or decline to extinction.

It is thus possible to examine populations or habitat associations at several spatial or temporal scales of resolution. There is no "best" scale at which to operate; questions asked at one scale are inherently no better or worse than questions focused on some other scale (although some scales of resolution may yield answers more readily than others). It is just as legitimate, for example, to ask what factors contribute to yearly variations in the number of territories contained in a small plot in uniform habitat as it is to inquire what determines why the abundance of a species varies throughout its range, or how species are assembled into communities of different sizes in different biogeographic regions. The scale at which one asks questions, of course, dictates the scale at which one gathers information to answer the questions: broad surveys of continental abundance patterns, for example, are unlikely to provide information that will illuminate the dynamics of local populations.

Unfortunately, the spatial or temporal scale at which studies are conducted seems most often to be determined as a matter of convenience—whatever seems sufficient within the logistical constraints. A few local plots in various habitat types may be surveyed, for example, and the results then unhesitatingly generalized beyond the scale of space and time that has actually been sampled. The revealed patterns of population densities, habitat associations, and other local population phenomena are expected to hold over a much larger universe, so long as the basic conditions are relatively similar. We assume a scale–independent uniformitarianism in evolutionary and ecological processes. This approach, of drawing samples (censuses, vegetation surveys, etc.) in some unbiased fashion from a larger universe to determine what patterns exist, and then inferring that these patterns characterize the universe as a whole, is of course central to modern scientific methodology. So long as the samples are truly representative of the larger universe, this procedure is justified. However, if different processes operate to influence populations or communities at different scales, then the type of explanation

[1] Department of Biology, University of New Mexico, Albuquerque, New Mexico 87131.

that is appropriate at one scale may be quite inappropriate at another (Schopf 1979).

Consider, for example, patterns of variation in species diversity over a spectrum of spatial scales. Diversity generally increases as the geographic scale considered becomes larger, but the factors contributing to this increase are not the same at all scales (Cody 1975). At a local level, a certain *point diversity* level results from degree of overlap among the territories of different species occupying points within a plot. At a somewhat larger spatial scale, *α-diversity* measures the variety of species present in a given habitat. Variations in α-diversity thus are more a consequence of the suitability of conditions in different habitat types to different numbers of species than of the degree of territorial overlap among individuals of those species. As the geographic scale is expanded still further, however, habitats of different types will be included in the larger area. This habitat interspersion prompts a further increase in diversity (*β-diversity*) due to the addition of species restricted to only some of the habitat types. Finally, because similar habitats in different biogeographic regions may support different species for historical reasons, expansion of the geographic scale to include such regions results in a further change in diversity, this due to geographic species replacements (*γ-diversity*).

To see how the operation of different forces affecting diversity varies at different spatial scales, consider the two hypothetical patterns shown in Figure 1. In A, small-scale point diversity is relatively low, perhaps because some of the species are interspecifically territorial or because population densities are so low that individual territories of the different species are widely scattered. When one considers a somewhat larger area containing a single habitat type, however, diversity increases somewhat, as collectively several species are capable of occupying the habitat type (α-diversity). With further increases in scale, diversity rises dramatically, a consequence of the varied mosaic of habitat types in this landscape (as in a rural New England countryside) and the addition of new species characteristic of each of the habitat types (β-diversity). Finally, this region may be one that has undergone little biogeographic fragmentation, and thus as one encounters similar habitat types anywhere in the region, the species present are about the same—γ-diversity adds rather little to the overall diversity. In B, on the other hand, point diversity is initially relatively high, as nearly all of the species present in a local plot cover the entire plot and overlap completely with one another. Within-habitat (α) diversity is not much greater, because nearly all

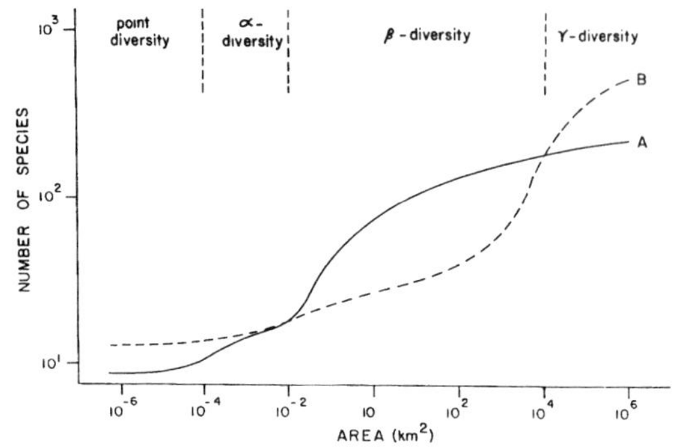

FIGURE 1. Changes in diversity as a function of increasing scale of geographical area surveyed. The two lines depict two different scenarios of diversity change, and demonstrate how different factors may affect diversity at different spatial scales. See text for explanation.

of the species that can occupy the habitat type occur at any point within that habitat. The α-diversity of B, however, is greater than that of A, perhaps reflecting a greater development of vertical vegetation profile. This habitat type may be rather widespread and unbroken over a fairly large geographic area; if that is the case, the accumulation of additional species with increasing area as a consequence of habitat interspersion (β-diversity) will be slow. At some larger geographic scale, however, areas of similar habitat that have a different biogeographic history (as a result, for example, of repeated fragmentation during glacial periods) differ in constituent species, and γ-diversity adds significantly to the overall species diversity.

The point of this exercise is to show that, because there are discontinuities in the operation or effectiveness of different processes affecting species diversity at different scales, extending interpretations of patterns revealed at any one scale uniformly to other scales is likely to produce incorrect conclusions and erroneous predictions. In the remainder of this paper, I wish to illustrate some additional problems associated with considering bird populations and communities at different scales in space and time, drawing from our studies in grassland and shrubsteppe systems. I can offer no definitive solutions to these problems, but by pointing them out I hope to foster greater care in the design and interpretation of avian surveys.

SCALE PROBLEMS IN SPACE

An example of the sort of contradictory results that can emerge from consideration of a question at different spatial scales comes from our attempts to define the patterns of habitat associations of grassland and shrubsteppe birds. In a "continental" scale analysis, we examined the patterns of correlation between the distribution and abundances of breeding bird species and a variety of habitat features over a series of sites ranging from tallgrass prairies in the eastern Great Plains to arid *Artemisia*-dominated shrubsteppes in the northwestern Great Basin (Rotenberry and Wiens, In press). At this scale of analysis, several bird species characteristic of tallgrass prairies (Dickcissels, *Spiza americana*; Grasshopper Sparrows, *Ammodramus savannarum*; Upland Sandpipers, *Bartramia longicauda*; Eastern Meadowlarks, *Sturnella magna*) exhibited strong correlations with single or multivariate measures of habitat structures. Another set of species (Sage Sparrows, *Amphispiza belli*; Sage Thrashers, *Oreoscoptes montanus*; Brewer's Sparrows, *Spizella breweri*), whose distributions are more or less restricted to the western shrubsteppe, was strongly associated with other features of habitat physiognomy. Several species that are widespread through the area we considered (most notably Western Meadowlarks, *Sturnella neglecta*), however, displayed *no* patterns of correlation with any of the habitat features that we measured. When we considered the habitat relationships of many of these same species at a more regional scale, restricting our study to shrubsteppe sites in the northwestern Great Basin (Wiens and Rotenberry 1981b), different patterns emerged. Here the species that are generally widespread through the shrubsteppe (Sage Sparrow, Sage Thrasher, Brewer's Sparrow) showed few significant correlations with variations in the structural configuration of the habitat, even though they had exhibited many clear correlations in the continental-scale analysis. On the other hand, species whose centers of abundance and distribution lie more in the steppe regions to the east, such as Western Meadowlarks, were strongly correlated with variations in several features of habitat physiognomy at this regional scale.

The patterns of habitat associations that emerged at these different scales of resolution were thus inconsistent and to some degree contradictory. Why? We have suggested (Wiens and Rotenberry 1981b) that these species are distributed among habitats over this grassland-shrubsteppe gradient more or less independently of one another, each following the dictates of its own habitat preferences and adaptations. The result is that species differ in the extent of their distribution over the habitat gradient: some species, such as Dickcissels or Sage Sparrows, occur over only a restricted portion of the gradient, while others, such as Western Meadowlarks, may encompass much of the spectrum of habitat conditions within their distribution (Fig. 2). When one surveys most of the gradient, as we did in our continental analysis, those species whose distributions cover only a restricted portion of the gradient will be likely to exhibit significant correlations, while the broadly distributed species will be more likely to vary in abundance independently of position on the gradient and thus produce few if any significant correlations. In a more restricted survey, such as our regional analysis in the shrubsteppe, only a portion of the habitat gradient is included. Here species distributed over most of the spectrum of conditions sampled (e.g., Sage Sparrows) reveal few clear associations with habitat features, while those occupying only a portion of the sampled gradient (e.g., Western Meadowlarks) do produce significant correlations and apparent patterns of habitat relationships. Obviously, neither of these scales of geographic resolution is "best" for studying the habitat relationships of these species, and in fact it is rather uncertain what the most appropriate scale of study might be for assessing the habitat affinities of the entire assemblage of bird species. It is clear, however, that our interpretation of habitat patterns among these birds is rather sensitive to the scale at which sampling and analysis are conducted.

Similar inconsistencies emerge if we consider an avian community attribute such as species diversity on several spatial scales. The diversity of an avifauna is a function of both the number of species present (its *richness*) and the equitability of the relative abundances of those species (*evenness*) (MacArthur 1965). When diversity is considered at a geographic scale including a wide range of habitat types in the northwestern U.S.A. (Washington, Oregon, and Idaho), the variations in diversity seem to be primarily a consequence of variations in evenness rather than richness (Rotenberry 1978). Within the shrubsteppe habitat type of this region, however, diversity variations may be associated with changes in either evenness or richness (Rotenberry and Wiens 1978). If the geographic scale is reduced still further, to include just a collection of locations in a restricted portion of this shrubsteppe habitat type, the variations in species diversity emerge as being a consequence primarily of differences in species richness (Wiens and Robenberry 1981b). Finally, at a local scale of resolution the variations in diversity of breeding birds among different study plots at a single shrubsteppe location in southeastern Washington are almost entirely related to differences in evenness, as nearly all of the plots contain the same three species (Wiens and Rotenberry, In press a). The factors that influence species richness and evenness, and thus the sorts of interpretations that one can place upon their variations, are complex (e.g., Rotenberry and Wiens 1980a), but it is nonetheless clear that unitary interpretations of variations in diversity without consideration of the spatial scale of the analysis are likely to be inaccurate.

SCALE PROBLEMS IN TIME

Population densities of breeding birds in local plots in many habitat types vary from year to year, often substantially (e.g., Wiens and Dyer 1975, Wiens and Rotenberry, In press a, Myers and Pitelka MS). Such variations complicate attempts to test ecological theories that assume equilibrium (Wiens 1977), although this complication is often circumvented by suggesting that the density variations represent close tracking of environmental conditions so as to maintain a shifting equilibrium between populations and their resources (e.g., Cody, In press). Whether population variations in local plots are largely independent of proximate local conditions or represent fine-tuned responses to variations in local conditions, the fact remains that such populations are not static through time, and this poses problems to analyses of avian distribution and abundance or habitat associations (see O'Connor 1981).

As an example of such temporal complications, we may again consider the patterns of habitat associations of breeding shrubsteppe birds (Wiens and Rotenberry 1981b). Bird populations and habitat features were measured at 14 locations during each of three years, so temporal as well as spatial patterns in bird-habitat associations could be analyzed. During the duration of this study, precipitation regimes varied from extremely dry to unusually wet. These variations were reflected in yearly changes in vegetational physiognomy, especially the coverage and stature of grasses and forbs, at the study locations. Despite these obvious habitat changes, the abundances of bird species over the region as a whole did not vary significantly between years, and variations in abundances were generally uncorrelated with annual variations in vegetation structure. The study sites were consistent from year to year in their relationships to one another as defined by overall habitat physiognomy, but varied independently of one another with respect to their bird populations. The strong patterns of yearly variation in habitat structure were thus not linked to variations in bird abundances and species occurrences and were independent as well of yearly dynamics of bird communities in nearby locations.

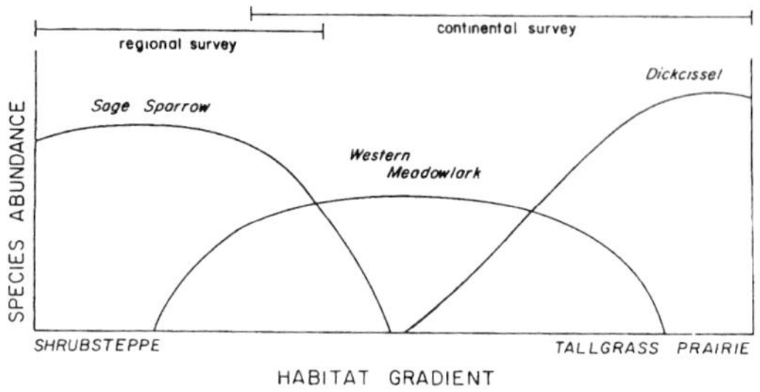

FIGURE 2. A schematic diagram of the distribution of three species along a habitat gradient from shrubsteppe through shortgrass and mixed-grass prairie to tallgrass prairie. The horizontal lines at the top indicate the span of the habitat gradient included in the regional and continental surveys described in the text.

These results argue in support of the view that populations may often vary independently of local conditions rather than tracking them closely, but how and why might this occur? We have sought a partial explanation of these variations by drawing an analogy with a checkerboard (Wiens and Dyer 1975, Rotenberry and Wiens 1980b). In this "checkerboard model" an area within a local habitat is represented by a checkerboard with walled edges. Checkers on this board represent the territories of individuals of a species. An area of, say, nine squares may represent a 10-ha study plot, within which populations are censused by counting the number of territories (checkers) or portions of territories included within the plot boundaries. The redistribution of individuals in the population resulting from migration and reestablishment of territories the following spring is simulated by shaking the checkerboard; this annual redistribution of territories is thus considered as a random process, subject to the constraint that territories may not overlap. We thus assume no site tenacity to previous breeding territories, and as well assume that territory size is fixed among all individuals within a population. Finally, we assume for simplicity that the number of checkers (individual territories) on the board as a whole remains unchanged from year to year. Altering these assumptions would in most cases only accentuate the conclusions we draw from the model. These conclusions may also apply in a general way to less mobile organisms, although the time scale of spatial rearrangement in the population will be much longer and the patterns will be more strongly influenced by reproduction and mortality than by individual movement.

If the available habitat in the region represented by the board is densely packed with individual territories (Fig. 3A), the process of yearly redistribution of individuals causes rela-

FIGURE 3. The "checkerboard model." Part A represents the distribution of territories of a species (= checkers) in a nearly saturated habitat (densely-packed board), while Part B depicts the dispersion of territories in a sparsely packed habitat. The habitat is assumed to be uniform, and thus the squares on the board do *not* represent habitat patches of different types. The solid line encompasses a sample plot established in the area depicted by the board as a whole and censused over three successive years. In A the number of territories included within the census plot (and thus our estimate of density) varies little between years, but in B there is substantial yearly variation in the census estimates, despite the fact that the total number of checkers on the board remains constant.

tively little change in the values recorded in the census of the smaller nine-square plot. If the board (habitat) is sparsely packed, on the other hand, the annual redistribution of individuals may produce substantial variations in the number of individuals occurring within the nine-square plot, and thus censused (Fig. 3B). In the latter situation, we will record substantial year-to-year variations in local density which, because the redistribution process is random, will be independent of any annual changes in the habitat, and which also will be likely to vary independently of yearly density changes in nearby plots in the same habitat. This, of course, is what we find in our shrubsteppe locations. It is also likely to be characteristic of endangered species, thus complicating efforts to define their true dynamics or habitat associations.

In a sense, this model demonstrates the close interplay between temporal and spatial scales in population censusing, for the considerable yearly variation recorded in plot censuses on the sparsely packed board, despite the actual constancy of the population occupying the board as a whole, indicates that the sample plot is too small. Had we used a plot the size of the entire board, our portrayal of population sizes would have been accurate. (This would be appropriate, of course, only if our initial questions were asked at a scale commensurate with that of the entire board.) On the densely packed board, on the other hand, a plot the size of nine squares is much more likely to provide a reasonably accurate estimate of population size and dynamics. Thus, the spatial scale at which such populations should be censused is at least partially a function of the degree to which individuals pack or saturate the available habitat. This observation may account for the general neglect of censusing methodology or census area size in much of the bird census work that has been aimed at testing ecological theory. Because this approach often presumes that populations are at equilibrium levels, saturating the available habitats, there seems little need for large survey plots, or indeed for much concern about sampling design at all. That habitats are in fact saturated (i.e., that populations are at "carrying capacity") is more often an unfounded assertion than a demonstrated fact, and indeed one might expect populations in many habitats to be below saturation levels frequently, especially in variable environments (Wiens 1977). If this is so, the spatial scale at which such populations should be censused is uncertain, and temporal changes in abundance recorded in censuses may be more apparent than real.

Real populations do change through time, however. Despite this, a good deal of the recent work in avian community ecology and population or habitat management follows the approach of conducting short surveys of a number of locations and then examining the collection of surveys for patterns, from which predictions or management policies are derived. Usually each location is surveyed only once (e.g., Cody 1968, 1974, 1978; Diamond 1972, 1975a; Terborgh and Faaborg 1980). It is thus implicitly assumed that a single census of a location provides an accurate statement of what is "normal" there, and that additional censuses would yield much the same results. The patterns that emerge from analyses of collections of such single surveys are, of course, considered to be accurate and real as well. Temporal variations in populations or habitats at any scale are thus ignored, perhaps because the underlying assumption that habitats are saturated and the ecological systems are in equilibrium generates confidence that the approach is robust.

Perhaps it is. Perhaps the patterns that emerge from analyses of such single-sample surveys are so reliable that the small amount of "noise" generated by temporal variation does not matter. To explore this possibility, I have conducted a rather simple exercise using censuses of breeding bird populations obtained in our work in North American grasslands; here I present just one example from the larger series of analyses that I conducted (Wiens 1981b). I had available 40 censuses taken at eight study locations; each location was censused for at least 2 years (usually 3 or 4), and in some cases two or three census plots were established in close proximity at a location. I used these values to test the hypothesis that the gradient of increasing annual precipitation among the eight locations should be accompanied by an increase in the total biomass of breeding birds of all species combined: more resources should support more avian biomass (Wiens 1974). This suggestion was tested in two different ways. In one, values for (a) the entire set of 40 censuses and (b) means for each location, averaged from all censuses taken there, were used to determine the "true" relation between total biomass and annual precipitation, using simple correlation procedures. The second test asked what sort of relationships might have been obtained had only a single census been taken in each location. To determine this, a single total biomass value was randomly selected for each location from the set of censuses actually taken there, and the correlation coefficient (r) between total biomass at the eight sites and annual precipitation calculated. A series of 200 iterations of this random census selection procedure yielded a frequency distribution of values of r. These portray the probability that a given value of r would be obtained by randomly drawing single censuses from the array actually available for each location.

The values for all 40 censuses considered together indicate that total biomass does indeed increase significantly with increasing annual precipitation ($r = 0.40$, $P = 0.01$), and a similar relationship is obtained when values for each site are averaged and then compared ($r = 0.75$, $P = 0.03$). The correlation coefficients for the randomly generated single-sample surveys cover a wide range of values (Fig. 4). Given the variation in total biomass at each of the locations, the probability that a set of single censuses would have detected a statistically significant positive relationship with annual precipitation is 0.185. Most sets of single-sample surveys would thus fail to demonstrate the "true" relationship, leading one to conclude (perhaps erroneously) that no such pattern exists in this system. Of course with single values for each location rather than an array of censuses, sample size is reduced from 40 to 8, and substantially larger values of r are required to demonstrate statistically significant relationships. Given a set of single samples from a relatively small number of locations, one might well be tempted to ignore statistical significance levels altogether and instead seek "biologically meaningful" trends. It is apparent from Figure 4 that single-sample surveys could easily provide a rather broad range of positive or negative trends, about which apparently logical biological explanations could then be developed. This suggests that the danger of drawing false conclusions from data obtained from a series of locations that are sampled only once is

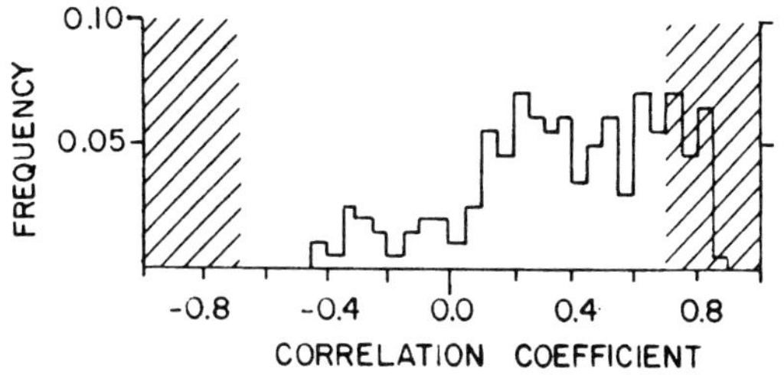

FIGURE 4. The frequency distribution of correlation coefficient values for 200 iterations of a procedure that randomly selects census values for each of eight locations from a larger series of censuses actually conducted at these sites. In this test, the correlations are between total avian biomass censused on the sites and the mean annual precipitation of the sites. The shaded area indicates correlation coefficients for which $P < 0.05$. The exercise simulates the effects of combining single surveys of a series of locations in an attempt to define broad patterns (see text).

very real, no matter how accurate the single censuses are.

CONCLUSIONS

To some extent these scale problems are simply consequences of inadequacies in sampling. The shifting patterns of habitat association as different portions of the grassland-shrubsteppe habitat spectrum are sampled (Fig. 2) may be a result of inappropriate sampling of the actual habitat ranges of the various species, and the density variations recorded on small plots in an unsaturated habitat (e.g., Fig. 3B) may represent sampling error rather than actual biological patterns. It is also possible that the patterns or dynamics seen in a "population" over a large area may simply be statistical artifacts resulting from the summation of a series of local populations whose dynamics vary independently of one another (Poole 1978). *If* processes were uniform in their actions and importance in time and space, such statistical problems could be resolved by adjusting the sampling intensity of scale until it became adequate. But it seems unlikely that such conditions of scale-independent uniformity often exist in nature. Rather, the processes that affect individuals, populations, and communities probably differ qualitatively as well as quantitatively at different scales in space or time. For some species, for example, the limits of geographic range at high latitudes may be set by physiological restrictions, while the low-latitude range boundaries may be determined by biotic interactions such as competition or predation (MacArthur 1972). Attempting to explain variations in local population densities or individual habitat selection by these factors, however, would be simplistic and probably incorrect. Local populations, in turn, may differ in their recruitment rates such that some produce a net excess of individuals while others are incapable of persisting without immigration (Wiens and Rotenberry 1981a). Combining these populations indiscriminately in analyses of patterns at larger geographic scales would obscure the real processes driving their dynamics. This concern with the scale-dependence of processes is by no means confined to censusing or ecological problems; one of the current controversies in evolutionary biology, for example, is over whether large-scale macroevolutionary patterns can be explained by simple extrapolation of small-scale microevolutionary (population) processes, or whether qualitatively different processes are at work (e.g., Bock 1979, Gould 1980, Eldredge and Cracraft 1980).

The problem that confronts us is really *not* that studies are conducted at a variety of different scales in space and time, but that such studies seem generally unaware of the ways in which the processes producing the patterns they seek to define may vary as a function of scale. The problems are most severe and readily observed in "quick and easy" studies, which pay little heed to the spatial scale on which sampling is conducted and usually follow the single-sample survey approach. But even in studies that are designed to produce census estimates of considerable accuracy, the problems posed by spatial and temporal scale are still pervasive. It is not simply a matter of paying attention to the statistical sufficiency of sample size in space or time. Because the processes that influence the population or community features of interest operate in different manners or with different intensities at different scales in space and time, it is quite possible that even well-designed sampling programs that appear to produce statistically sound results may nonetheless lead to erroneous conclusions about patterns and their causes if the scale of sampling does not coincide with the scale of the processes actually influencing the population.

The effects of scale differences in space are not independent of time-scale influences. As Elton and Miller (1954:478) observed, "it is because communities are never stable in time that the determination of their composition by sampling them in space is so perplexing." Something of the nature of these perplexing space-time effects can be seen from the generalized diagrams presented in Fig. 5. First, the sorts of processes that influence population or community patterns differ in space and time (Fig. 5A). Over small areas and short times, direct behavioral adjustments such as shifts in territory po-

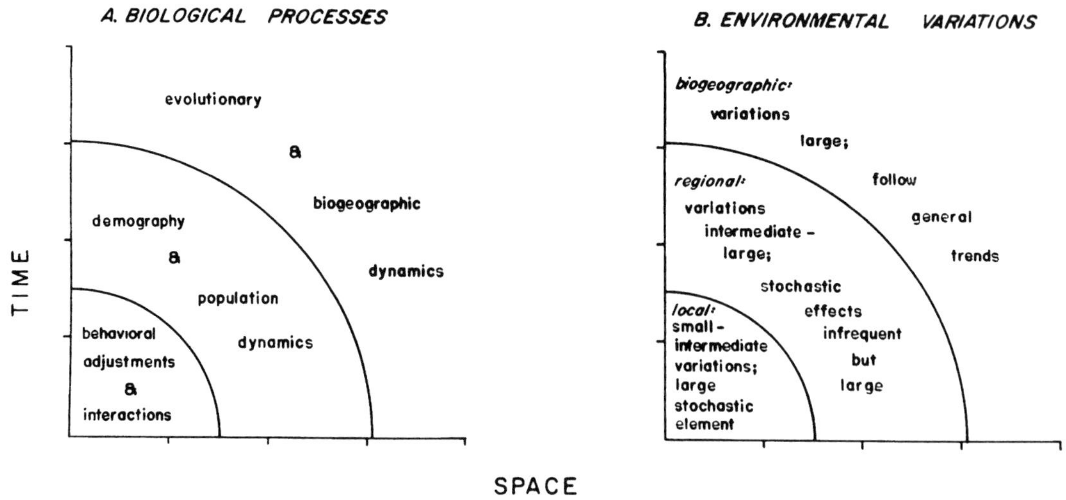

FIGURE 5. Space-time domains of features influencing populations and communities that change with changes in scale. Part A illustrates changes in the dominant biological processes that are operative in populations and communities at different scales in space and time, while Part B depicts the general features of environmental variability that may influence biota at different scales. The patterns shown here are not suggested to be unique or ubiquitous—one can easily think of alternative formulations or exceptions. However, they do indicate the difficulties in uncritically extrapolating processes or patterns from one scale to another. Time and space scales are intentionally dimensionless. The formulation was inspired by a similar treatment of marine plankton systems by Steele (1978).

sitions and sizes or modifications of habitat selection may influence population densities and habitat associations, and direct interactions between individuals of different species, such as competition or predation, may affect community composition. At a somewhat larger spatial scale and over longer time periods, features of the demography of the populations (e.g., reproduction, survivorship, age structure) may become important in dictating patterns, while over even longer periods of time and larger regions, processes of evolutionary change and speciation or the biogeography of range dynamics may be the major determinants of the patterns that emerge. Superimposed on this array of space-time processes, however, are the sorts of environmental variation that characterize natural systems (Fig. 5B). At a local level, the variations in, say, population densities that occur from place to place or over time may generally be small to intermediate (depending upon the sedentariness or reproduction/mortality schedules of the population), and may be quite sensitive to stochastic effects. These are the sorts of variations that may prevent local populations from attaining total habitat saturation or equilibrium. Variation among local populations may also be more tightly linked to localized environmental variation, as through ecotypic adaptation. Regionally, the major variations that influence populations may be somewhat more predictable (e.g., drought cycles) and of larger magnitude in their effects. The effects of stochastic environmental variations may be less important, but when they do occur their influences may often be spectacular and produce long-lasting effects (e.g., 100-year flood levels, record heat waves). Over large areas and long time periods, the environmental variations that affect populations and communities, such as glacial advances and retreats, may be quite large and produce fundamental reorganizations of the biota. Such variations usually follow well-defined trends over time and space, and thus are predictable in a general fashion.

Consideration of these scale relationships is complicated, however, by differences in the space-time scale to which different sorts of species are inherently adjusted. That bacteria and elephants operate in totally different domains in space and time is evident, but within a more coherent group, such as birds, there are still significant differences in the space-time domains of different species. Thus, the dynamics of a small, resident, habitat specialist species, such as an antbird or (in some areas) kinglets, may operate on the scale of small areas over short time periods. The dynamics of a large, migrant, habitat generalist (e.g., many raptors or seabirds), on the other hand, may be attuned to conditions expressed on completely different scales in space and time. Thus, an environmental perturbation that is quite localized in space and time might have profound effects upon a population of a small, resident, specialist (perhaps even leading to local extinction), but be of only minor consequence to the population dynamics of a large and widespread habitat generalist.

Does all of this mean that the situation is hopeless, that we should throw up our hands and become monks or molecular biologists? Hopefully not. The problems posed by the nonuniformity of processes and effects in space and time are formidable, but if we admit their existence and importance, perhaps we can begin to devise procedures of censusing populations, analyzing community patterns, and evaluating habitat relationships that are not so oblivious to the importance of scale. The first step is to attempt to select the size of a study area and the spatial scope and time duration of an investigation on the basis of something other than logistical constraints, or because someone else did it that way, or because some standardized "system" recommends it. These features of a study should be designed to mesh comfortably with the space and time scales of the organisms, the patterns of environmental variations, and the space-time zones of importance of particular biological processes, in the context of the objectives of the study or the questions being asked. Recent attempts to optimize the spatial scale of analyses, given some optimization criterion, such as maximizing diversity per unit area (Phipps 1975, Phipps and Cullen 1976), represent one possible approach, and spatial auto-correlation procedures (Sokal 1979) or time-domain analyses (Mulholland and Gowdy 1978) represent others. The ongoing discussion about optimizing the size of nature reserves (e.g., Diamond 1975b, Diamond and May 1976, Simberloff and Abele 1976, Whitcomb et al. 1976, Gilpin and Diamond 1980, Higgs and Usher 1980), while perhaps too often relying excessively on an incomplete model of island population and community dynamics, nonetheless conveys an encouraging sensitivity to the influences of spatial scale. Although a definitive solution to these scale problems is not in sight, it is apparent that we can no longer continue to ignore the problems produced by the nonuniformities of nature in space and time, or to think that they do not matter. Continued disregard of scale effects can only exacerbate the tendency to discern patterns in nature that may or may not be real, and to interpret them in ways that may or may not be true. Knowledge is likely to advance rather little from such an uncertain foundation.

ACKNOWLEDGMENTS

The research that led to these views has been supported by the National Science Foundation, most recently under Grant No. DEB 78-24364. John Rotenberry shared in the joys and frustrations of this research, and he and Rex Cates, Glenn Ford, Bea Van Horne, Bob Petruszka, Luke George, Dennis Heinemann, and Rich Bradley discussed these ideas and offered comments on the manuscript. Jean Ferner successfully confronted modern technology to prepare the final manuscript draft.

Studies in Avian Biology No. 6:522–532, 1981.

CENSUSING AND THE EVALUATION OF AVIAN HABITAT OCCUPANCY

JOHN A. WIENS AND JOHN T. ROTENBERRY[1]

ABSTRACT.—Determination of the habitat occupancy of bird populations is central to considerations of community structuring and niche relationships, as well as to intelligent management of those populations or habitats. The design of any population censusing program should thus include habitat measurement or evaluation whenever possible. We consider several methods of gathering habitat information along with censuses. Habitat measures may be obtained during station counts (e.g., roadside counts) by categorizing the habitat features within a defined area about each station, for example. Methods employing strip transect or plot procedures offer the potential for more detailed sampling and measurement of habitat features, which in turn permit more comprehensive analyses of bird-habitat associations. Applications of these approaches to breeding bird communities in grassland and shrubsteppe environments indicate that variations in features of habitat structure exhibit clear correlations with the distribution and abundance of several bird species, but that variation in habitat floristics (e.g., shrub species coverages) is also strongly associated with the density patterns of some species. These findings suggest that habitat evaluation schemes based upon only a few variables, or upon definition of a generally applicable "system" of habitat categorization, are not likely to produce sufficient detail to enable us to understand why the associations are important. Instead, consideration must be given to many habitat variables. Even if this is done, however, differences in the demographic structure of populations of a species in different habitats may complicate the interpretation of any bird-habitat relationships that seem apparent.

A major emphasis in avian ecology, as in ecology as a whole, is upon determining the distribution and abundance of species (Andrewartha and Birch 1954, Krebs 1978). It is this goal that drives us to be so concerned about properly estimating numerical abundance of populations and leads us to consider censusing methodology and analysis in such detail. But knowing the number of individuals of a species present in an area, or how abundance changes in time or space, is in a sense incomplete knowledge. In order to begin to understand *why* distribution and abundance vary in the ways they do, and in order to develop any means of making accurate predictions of future changes in population features, we must know how populations relate to the underlying habitat.

Habitat is thus the templet for ecological and evolutionary processes (Southwood 1977). In a basic or theoretical context, information about habitat is essential to any full understanding of the patterns of life history, adaptation, or behavior of a species (Rotenberry, In press), features that are expressed in modern ecology under the rubric "niche relationships." Similarly, habitat information is essential to interpreting community patterns. Alternative views, for example, suggest that bird species may be distributed along habitat gradients more or less independently of one another (Rotenberry and Wiens 1980a; Wiens and Rotenberry 1981b), or that interpopulational interactions such as competition produce distinctly nonrandom distributions of species assemblages along habitat gradients (Terborgh 1971, Cody 1974). In either case, habitat variation has a profound influence on the patterns that are observed, and to begin to distinguish such alternatives requires detailed knowledge of the habitat distributions of species and species assemblages.

In a more applied context, information about habitat relationships of populations is essential to their intelligent management, as it is almost invariably habitat conditions that are most directly and drastically influenced by human activities and resource demands. Habitat evaluation is therefore increasingly emphasized as an essential initial step (and at times the only step) in wildlife or environmental management. Several agencies are currently attempting to develop a unified habitat evaluation system that will permit a rapid and accurate determination of the relative value of land as wildlife habitat prior to development decisions (Flood et al. 1977, Whelan et al. 1979, Asherin et al. 1979, Ellis et al. 1979). Unfortunately, such habitat evaluation schemes are usually founded on the assumptions that habitat quality is a direct function of habitat diversity and that faunal diversity (especially bird species diversity, BSD) is a reliable index of the quality or "health" of the biota or is a good indicator of the relationship of wildlife to habitat conditions (Asherin et al. 1979, Thomas et al. 1979). This may lead those charged with resource management responsibilities to believe that areas with limited habitat diversity and low bird species diversity may be potentially suitable for resource development. Asherin et al. (1979), for example, found that BSD was closely related to both the complexity of vertical structuring of

[1] Department of Biology, University of New Mexico, Albuquerque, New Mexico 87131; present address of JTR: Department of Biological Sciences, Bowling Green State University, Bowling Green, Ohio 43403.

vegetation and the mixture of vegetation types: from this, they suggested that "resource development within a region will impact wildlife and wildlife habitat the least when that development is confined to large homogeneous areas with little vegetative stratification and relatively low cover type diversity" (1979:413). Such conclusions—indeed any management recommendations founded upon the premise that maintaining high bird species diversity will ensure proper wildlife management—are premature and ignore the many limitations of measures such as BSD (Balda 1975a, Wiens 1975, Thomas et al. 1979). These shortcomings, however, only point to the need for more thorough and careful consideration of the relationships of single species and species assemblages to the detailed features of their habitats in resource management (Willson 1974, Murton and Westwood 1974).

Knowledge of the habitat relationships of populations is thus important to both theoretical and applied pursuits. In view of this, it seems that the benefits to be gained from gathering information on habitat features along with censuses of bird populations are so great that they more than justify the additional effort required. The design stage for a project involving censusing of birds should thus include consideration of ways of obtaining appropriate quantitative habitat information. Our objectives here are to describe several ways in which habitat information may be gathered during census surveys, and to offer some brief comments on what sorts of habitat variables may be important to measure and how the resulting data may be analyzed. Our treatment is by no means intended as a review; instead, we draw heavily upon our own work on breeding bird communities in grassland and shrubsteppe systems.

COUPLING HABITAT MEASUREMENT WITH BIRD SURVEYS

There are a great many ways in which one may determine bird populations, as the contributions to this symposium demonstrate. Here we consider how habitat measurement may be combined with three different sorts of censusing procedures. These survey methods differ in the accuracy with which they enumerate bird populations, and the degree of resolution of habitat features generally varies concordantly.

Station Counts

One of the more widely employed count procedures is the station count or roadside count procedure employed in the North American Breeding Bird Survey (Robbins and Van Velzen 1967, 1974, 1979). The details of the counting method vary, but in general an observer follows a predetermined route on roads through an area, stopping at points 0.5 mi (0.8 km) apart and recording all individuals seen or heard within a 0.25-mi (0.4-km) radius circle during a 3-min observation period. The NABBS surveys have used 50 stations on each route; in our roadside surveys in relatively homogeneous grassland and shrubsteppe habitats (Wiens et al. 1972, Rotenberry and Wiens 1976) we used 25 stations. The roadside count method produces values that represent the frequency of occurrence of species among the stations and the overall number of individuals of each species recorded per count route; it does not permit an accurate determination of the density (individuals per unit area) of the species. Its primary value, therefore, is in charting broad continental or regional patterns of distribution or in assessing the relative change in the abundance or range of species over successive years.

Usually no information on habitat features is obtained during such breeding bird surveys. Peterson (1975) conducted a post facto analysis in which he assigned breeding bird census routes among 56 ecological regions covering North America, and then evaluated how species diversity varied among regions or with latitude. Such an analysis can reveal only the most general patterns of variation, however, and accordingly contributes rather little to our overall understanding of the habitat relationships of communities or individual bird species. There is considerable potential, however, for charting the general habitat affinities of bird species within a region and assessing temporal changes in habitat occupancy by recording even simple categorizations of habitats at the stations along a survey route. In a study in southern Wisconsin, for example, visual estimates of the relative coverage of various general habitat types were made within a 200 yd (183 m) radius of each station for 30–60 roadside surveys (Emlen and Wiens 1965, Wiens and Emlen 1966). These surveys were conducted primarily to assess the dynamics of Dickcissel (*Spiza americana*) distribution and abundance at the northern edge of the species' range during an "invasion" year (1964) and a "decline" year (1965), but the availability of even general habitat categorizations for the stations permitted a consideration of the patterns of habitat occupancy of the species and their changes as the distribution of the species in the state changed (Fig. 1).

In this study, observers simply estimated the occurrence of several major and easily categorized habitat types at each station as they conducted the bird census. More detailed habitat measurements could be obtained by sampling features at each station before or after the cen-

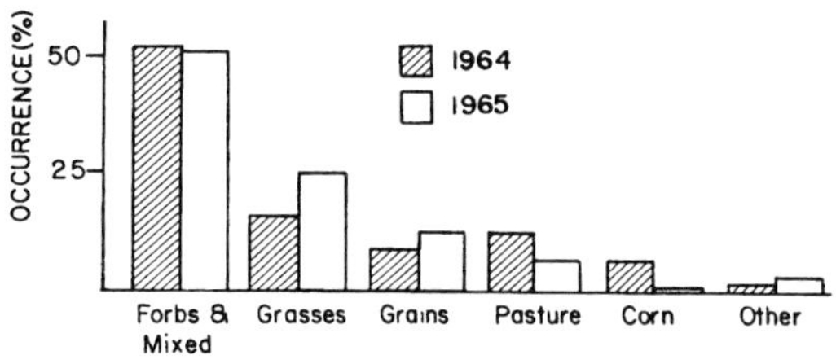

FIGURE 1. Distribution of Dickcissels in major habitat types in southern Wisconsin during 1964 and 1965, as measured by the percentage of all sightings occurring in the habitat types. "Forbs" includes alfalfa and other legumes. From Wiens and Emlen (1966).

sus is conducted, or by combining ground surveys with analyses of aerial imagery. Limitations on the quantitative accuracy of the census estimates derived using roadside survey techniques make it impractical to devote much time to obtaining very detailed and precise habitat measurements. Just as a series of roadside surveys can reveal trends in the distribution and abundance of species, however, they can also portray patterns of general habitat affinities if appropriate information on habitat features is gathered along with the census data.

Strip Transect Surveys

Strip transect surveys of various types provide more accurate and detailed census estimates of population densities in a more localized area than roadside counts, and therefore they can potentially provide the framework for more detailed habitat measurements and analyses. The sort of habitat sampling design that one follows will depend upon the overall goals of the investigation and the design of the strip transect survey; here we develop an example of one approach, drawn from our studies of bird assemblages in northwestern shrubsteppe habitats (Wiens and Rotenberry 1981b, Rotenberry and Wiens 1980b). We surveyed 14 plots at 9 locations, visiting each during the breeding seasons of 1977–1979. Bird densities were estimated on a linear transect placed in more or less uniform habitat at each plot, following the procedures of J. T. Emlen (1971, 1977a). Features of vegetation composition and structure were also recorded along each transect at the time the bird populations were censused. At 61-m intervals along the transect, 50-m tapes were laid out perpendicular to the transect on each side. Random numbers were then used to locate a sampling point in each 10-m interval of the tapes. Ten intervals along the transect were sampled in this manner, yielding 100 point samples of vegetation for each transect. Measures of coverage of different plant species and of physiognomic vegetation types, of several features of vertical and horizontal habitat structure, and of several indices of vertical and horizontal habitat patchiness or heterogeneity were then derived from the point samples taken at each plot. Two experienced observers could generally gather the vegetation information along a transect in 1–2 h.

The combination of bird censusing with habitat measurement permits us to evaluate not only the variations in abundances of species over the region sampled, but to begin to associate these variations with variations in habitat composition and structure, through various bivariate and multivariate correlational procedures. Table 1, for example, indicates the significant correlations between variations in the abundances of the two numerically dominant species in this system, Sage Sparrows (*Amphispiza belli*) and Brewer's Sparrows (*Spizella breweri*), and variations in single habitat features. These species exhibited relatively few significant correlations with the 20 measures of habitat structure or physiognomy, but apparently did vary in concert with variations in the coverages of several of the desert shrub species. Such observations hint at

TABLE 1
Correlations Between Bird Densities and Physiognomic Variables and Shrub Species Coverages over 14 Plots Sampled for Three Years in the Northern Great Basin[a]

Coverage	Sage sparrow	Brewer's sparrow
Rock	—	−0.47*
Shrub species diversity	−0.33*	−0.59***
Sagebrush (*Artemisia tridentata*)	0.61***	—
Hopsage (*Atriplex spinosa*)	—	−0.44**
Budsage (*Artemisia spinescens*)	—	−0.38*
Cottonthorn (*Tetradymia spinosa*)	−0.37*	—
Greasewood (*Sarcobatus vermiculatus*)	−0.53***	—

[a] Only significant correlations are shown: * = $P < 0.05$, ** = $P < 0.01$, *** = $P < 0.001$. From Wiens and Rotenberry (1981b).

possible causal relationships, and provide the starting point for more thorough investigations of the linkages between these birds and habitat features (Wiens, Cates, and Rotenberry, research in progress).

Plot Censuses

Some of the more reliable (and most time-consuming) avian census methods are based upon counting the number of individuals occupying a measured plot of ground by spot-mapping, mapping territorial locations, or some other procedure. Some of the most widely applied plot survey programs in North America have been the Breeding-Bird Census and the Winter Bird-Population Study, sponsored by the National Audubon Society. Each of these programs uses established plots from which estimates of species densities are obtained. In 1970, James and Shugart proposed a method of coupling quantitative habitat descriptions with these plot censuses that has been employed in a large number of subsequent censuses (James 1978). The method involves locating 5–10 0.1-acre (0.04-ha) circular plots at random within the study area. Within each sample plot, measures are then taken that enable one to calculate the density, basal area, and frequency of tree species, canopy height, shrub density, and percentage canopy cover for the study plot as a whole.

Although a fair number of breeding bird censuses have been taken incorporating James/Shugart habitat measurements, relatively few studies have attempted to analyze the accumulated data. Wamer and James (MS) conducted multivariate analyses of habitat associations using adjusted census results and habitat measures from such surveys, and Robbins (1978b) conducted both univariate and stepwise analyses of values from 80 deciduous and mixed woodland surveys to assess the habitat relations of selected bird species. Robbins added information on the latitude, precipitation, and extent of contiguous habitat for each stand to the James/Shugart measures in his analysis. He found that one of the strongest relationships that emerged was between habitat size and the overall abundance of breeding birds, leading him to recommend that the James/Shugart system be amended to include additional information on habitat size and precipitation.

The James/Shugart habitat description system works only in wooded habitats. In our work in more open grassland or steppe environments (Rotenberry and Wiens 1980a), we have followed a somewhat different approach to combining plot censuses of bird densities with habitat measurements. There we censused the populations of birds occupying 9.2–10.6-ha plots by mapping the territories of individuals using the "consecutive flush" procedure (Wiens 1969). Within these same plots, we sampled vegetation physiognomy at sample units that were located randomly within each 61 × 61-m block of the plot grid. At each sampling location, we recorded information on the coverage of various physiognomic categories of vegetation, the vertical and horizontal structuring of the habitat, and vertical and horizontal heterogeneity, using a combination of point samples and point-centered quarter samples (Cottam and Curtis 1956).

The resulting measures of habitat configuration can be analyzed at two levels of resolution. The most direct is simply to use bivariate and multivariate correlation procedures to examine the relations between variations in the densities of bird species or bird community attributes and variations in single habitat features, using both bird density values and habitat measurements for each entire plot. One of the analyses that we conducted was a Principal Components Analysis (PCA) of the habitat measures taken on the steppe plots. This analysis indicated that variation in habitat structure over the range of locations we considered (from tallgrass prairies in the eastern Great Plains to sagebrush shrubsteppe in the northern Great Basin) could be arrayed along three independent dimensions, representing variation in horizontal heterogeneity, variation in vertical heterogeneity, and variation in the abundance of forbs (chiefly wildflowers). The distributions of several bird species were significantly associated with these PCA vegetational axes, and these birds in fact were arrayed in "clusters" in the PCA-space (Fig. 2). Species that are normally considered "tallgrass prairie" birds, such as Dickcissels, Eastern Meadowlarks (*Sturnella magna*), Grasshopper Sparrows (*Ammodramus savannarum*), and Upland Sandpipers (*Bartramia longicauda*), reached their highest abundances on plots that exhibited the lowest horizontal heterogeneity and were high in vertical heterogeneity. Sage Sparrows and Sage Thrashers (*Oreoscoptes montanus*), birds more typical of arid shrubsteppe habitats, showed a similar response to increasing vertical patchiness but differed from the tallgrass species in their response to horizontal heterogeneity, joining with the remaining shrubsteppe species at the high-heterogeneity end of this gradient. "Shortgrass prairie" species such as Horned Larks (*Eremophila alpestris*), Lark Buntings (*Calamospiza melanocorys*), and McCown's Longspurs (*Calcarius mccowni*) did not differ in abundance with respect to changes in horizontal patchiness, but were negatively correlated with increasing vertical heterogeneity. The groupings of these species are not altogether unexpected,

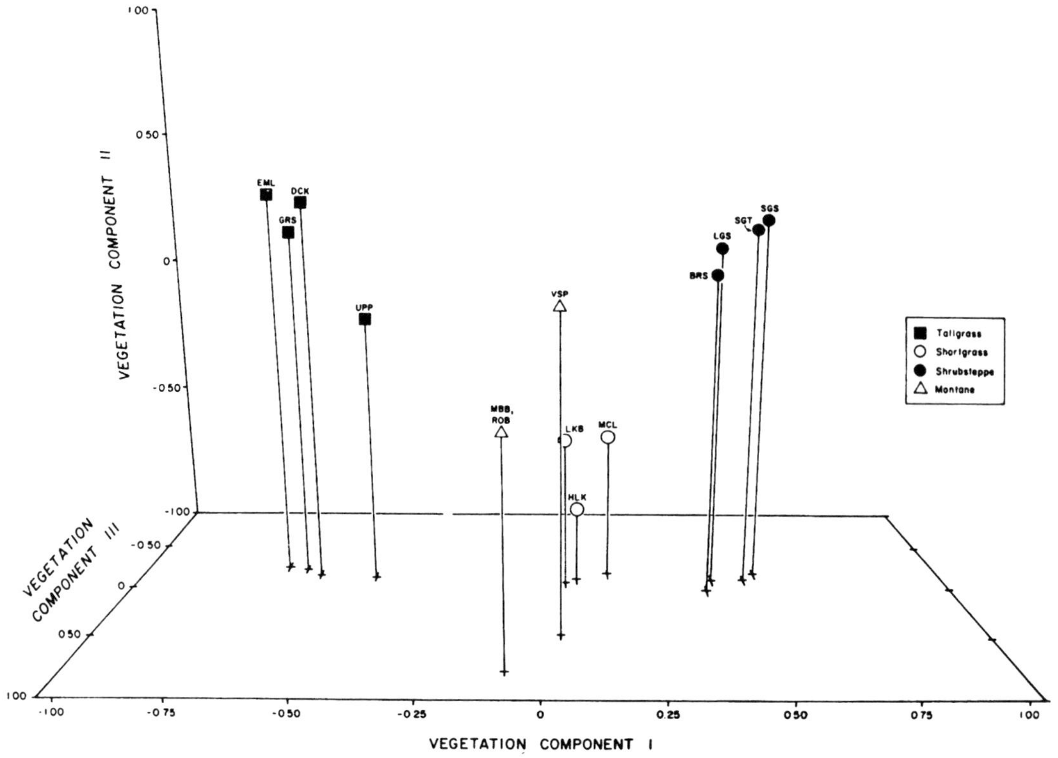

FIGURE 2. Correlations between bird species abundances and site factor scores on vegetational principal components for a series of locations in the North American Great Plains and western shrubsteppe. The axes represent the three major components derived from a Principal Components Analysis of features of habitat structure for the sites, scaled by the relative contribution of each component in accounting for variation in the total vegetation data set. Bird species codes are as follows: EML = Eastern Meadowlark, GRS = Grasshopper Sparrow, DCK = Dickcissel, UPP = Upland Sandpiper, MBB = Mountain Bluebird, ROB = American Robin, VSP = Vesper Sparrow, LKB = Lark Bunting, HLK = Horned Lark, MCL = McCown's Longspur, BRS = Brewer's Sparrow, LGS = Loggerhead Shrike, SGT = Sage Thrasher, SGS = Sage Sparrow. From Rotenberry and Wiens (1980a).

but the importance of variations in vertical and horizontal patchiness as components of habitat variation would not have been intuitively obvious without incorporation of the proper sorts of habitat measurements.

Such habitat analyses may be presented in other ways, some of which make the potential management applications (and the need for consideration of single-species responses in management) more apparent. As an example, one may use PCA to determine habitat gradients and then position the plots in the PCA-space according to their factor scores on the PCA axes. By then labelling each plot location with the density of a species, one may group together plots having similar densities to define isopleths or contours of abundance of a species in the PCA-space (Rotenberry and Wiens, In press). Because each plot is located in the PCA space according to its habitat features, it should be possible to predict how the position of a site might change were the habitat to be altered in some defined fashion. By relating this to the density contours of a species, one might then predict the patterns of response of the species to the habitat alteration. In the hypothetical case given in Figure 3, for example, an alteration that caused the plot to move in habitat space as indicated in Part A would likely result in an increase in the abundance of this species, while a different sort of change (part B) might be more likely to result in a decrease. Some other changes might foster the invasion of the species into a previously unoccupied area (part C), or lead to local extinction (part D).

Because our censusing procedure involves mapping the locations of individual territories within each plot, we can distinguish between vegetation sampling points falling within occupied portions of the plot and those occurring in

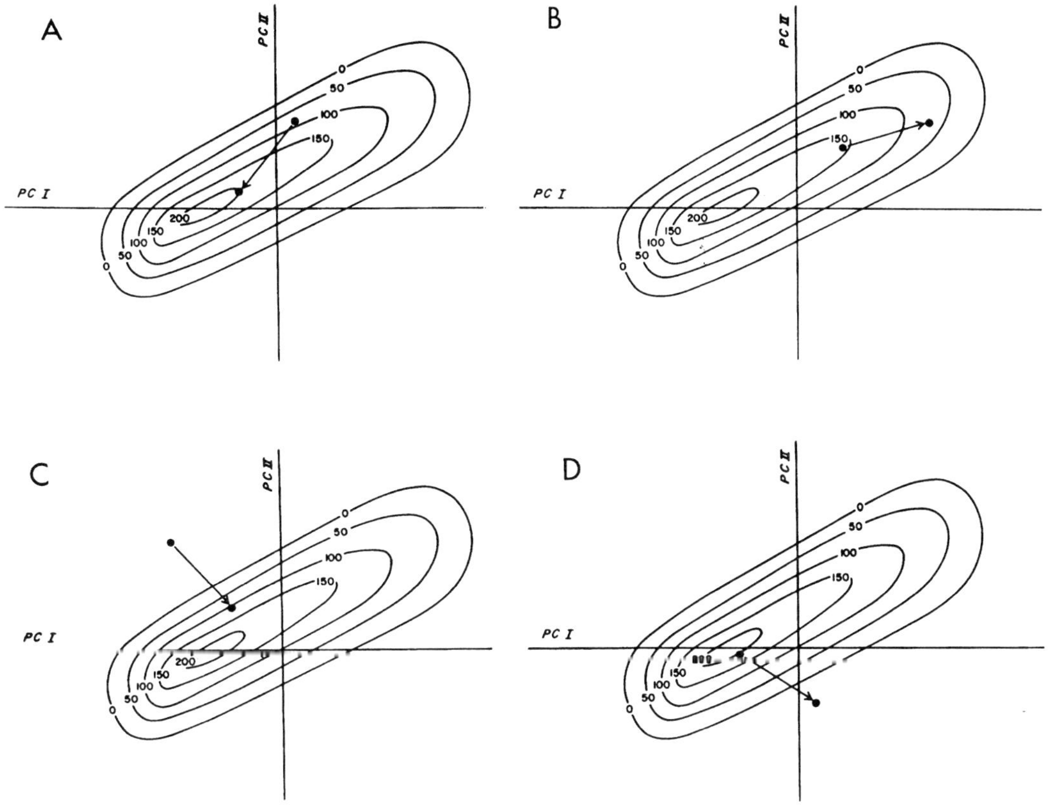

FIGURE 3. Hypothetical contours of species abundance patterns plotted in the environmental space defined by the first two principal components (PC I and PC II) of site-based environmental measures. The contours represent isopleths of density. The arrow denotes change in site characteristics as a result of habitat alterations. These changes may effect the following changes in the abundance of the species at the site: A = increase, B = decrease, C = local invasion, D = local extinction. From Rotenberry and Wiens (In press).

unoccupied portions. This permits a finer resolution of habitat associations of species, for if not all of a study plot is occupied by territories of a species, the average values for habitat features characterizing the territories of the species may deviate from those for the plot as a whole (Fig. 4). This level of resolution has been employed in analyses based upon the determination of mean habitat vectors of species in PCA (Anderson and Shugart 1974, Whitmore 1975, Rotenberry and Wiens 1980a), and Wiens (1973) used it to determine habitat differences associated with differences in the location of territories or the time of territorial establishment in two grassland bird species.

WHAT HABITAT FEATURES TO MEASURE

Given that one has decided that measuring or evaluating habitat is important and has defined a method of combining census surveys with habitat measurement, one still must determine which of the many measurable elements of the environment should be measured to characterize the habitat of a species or community. Those factors that are potentially important in influencing the distribution and abundance of species, or that might be coupled as direct or indirect selective forces to the adaptations of the species, are obvious candidates for inclusion in any measurement program, but it is far easier to speak of such general categories of habitat features than to define them precisely and determine how and at what scale they are to be measured. Since the suggestion of Lack (1933) and others nearly 50 yr ago that birds may select the habitats they occupy on the basis of the structural configuration or physiognomy of the habitat, most studies of bird-habitat relationships have emphasized such structural features (see Hildén 1965, Wiens 1969). Thus, for example, "each species requires a 'patch' of vegetation

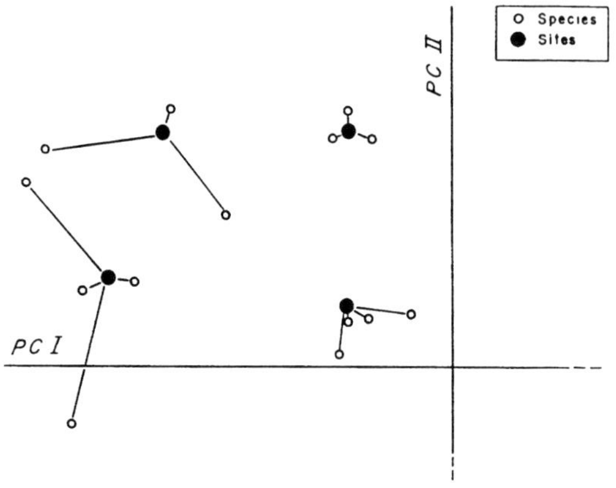

FIGURE 4. Hypothetical species and sites plotted in the environmental space defined by the first two principal components (PC I and PC II) of site-based environmental variables. The solid dot depicts a site's location, while each species is positioned in the PCA-space according to the environmental characteristics of only the areas actually occupied by the species at the site. Lines connect species to the sites on which they occur. From Rotenberry and Wiens (In press).

TABLE 2
PERCENTAGES OF TESTS CORRELATING THE DENSITIES OF BIRDS IN THREE MAJOR GROUPINGS WITH PHYSIOGNOMIC VARIABLES AND WITH SHRUB SPECIES COVERAGES THAT WERE SIGNIFICANT AT $P < 0.05$[a]

	Significant correlations with:	
Species group	Physiognomic features	Shrub species
Widespread shrubsteppe species	10	26
"Local" shrubsteppe species	13	20
"Grassland" species	24	16

[a] From Wiens and Rotenberry (1981b).

with a particular profile for its selected habitat, and . . . the variety of 'patches' within a habitat determines the variety of bird species breeding there" (MacArthur et al. 1962:167), or "habitat structure appears to be the major factor responsible for the complexity of associated bird communities" (Anderson 1979b:432). Such an emphasis has spawned a wide variety of habitat description schemes based upon physiognomy (e.g., Emlen 1956, 1977b; Dansereau et al. 1966; Wiens 1969). In general, however, most such systems do not consider the possible importance of plant species, independent of their physiognomy. Bevanger, for example, noted that "the structural complexity of the vegetation is a factor of prime importance for a bird community. There is therefore no point in devising a classification system of the same complexity as that used by phytosociologists for their plant communities" (1977:68), and DesGranges stated that "the influence of species composition of the vegetation on avifauna within a given habitat is only indirect. The species composition affects the physiognomy of the vegetation which, in turn, influences the composition of the avian community" (1980:5). Despite such assertions, several investigations have found that consideration of habitat structure alone provides only a partial explanation of the variations in distribution and abundance of species or the structuring of communities (Tomoff 1974, Ulfstrand 1975, Balda 1975a), and some of these studies have explicitly documented significant relationships with certain plant species. In the arid northwestern shrubsteppe systems that we studied (Wiens and Rotenberry 1981b), several of the bird species that are widespread in and characteristic of the shrubsteppe exhibited more significant relationships with the coverages of shrub species than with features of habitat physiognomy (Table 2), while bird species with localized distributions in the shrubsteppe also were correlated with a higher proportion of the floristic variables than of the physiognomic features. Species whose distributional patterns and habitat affinities lie more in the grassland regions to the east of the shrubsteppe, on the other hand, seemed more strongly associated with variations in habitat physiognomy than shrub species coverages. Despite their apparent association with several shrub species within the shrubsteppe region, the characteristic shrubsteppe species evidenced strong patterns of correlation with variations in several physiognomic habitat features when we considered them on a broader, "continental" level of analysis (Fig. 2; Rotenberry and Wiens 1980a). This suggests to us that at a large-scale, between-habitat level these birds may respond to elements of general habitat configuration, but within a habitat type their responses may be more strongly associated with the details of habitat floristics. To the extent that these findings apply to species in other systems, they complicate approaches to habitat analysis, for they suggest that in order really to understand the factors determining avian habitat occupancy patterns, we must evaluate both habitat structure and vegetational floristics. This has rarely been done.

The habitat measurement scheme that one follows also depends upon the overall objectives and scope of the study. We can distinguish three major approaches that seem to have dominated recent attempts to assess avian habitat patterns. One approach, exemplified by MacArthur and his followers (MacArthur and MacArthur 1961; MacArthur et al. 1962; Cody 1968, 1974) has alleged that avian habitat relationships can be understood from consideration of only a few easily measured habitat features. Indeed, MacArthur's early attempts to predict bird species diversity from variations in only the diversity of the vertical foliage profile met with sufficient success to lead to the adoption of this relationship as a tenet of at least some management schemes (e.g., Asherin et al. 1979), and Cody (1968) suggested that he could predict the niche patternings of species in grassland bird communities by examining just four measures of grassland habitat structure. Levins (1966) and Rosenzweig (1975) have clearly stated this view that by considering only a few key or "sufficient" parameters that incorporate the effects of a variety of lower-level parameters, one may gain a clear understanding of relationships and dispense with the need to measure a large number of parameters. This "few variables" approach has been somewhat reinforced by recent multiple regression analyses that have shown that a relatively small proportion of a larger set of habitat measures can provide good inductive models of variations in breeding bird populations (e.g., Robbins 1978b, Capen, In press).

A second approach features measurement of a large number of habitat variables. The initial attempts to quantify the association of bird species with many habitat features were those of Bond (1957) and Beals (1960), which provided at least some of the impetus for the subsequent analyses of Wiens (1969) and Emlen (1977b) (this approach should thus perhaps be termed the "Wisconsin approach"). The more recent development and application of multivariate techniques has facilitated the analysis of data on many habitat variables, and such investigations have generally been successful in distinguishing habitat features or more often suites of habitat features that are correlated with variations in the abundances of bird species or avian community attributes (e.g., James 1971; Anderson and Shugart 1974; Smith 1977; Rotenberry and Wiens 1980a; Wiens and Rotenberry 1981b; papers in Capen, In press).

The third basic approach is more strongly guided by management objectives. As pressures on natural resources have increased, the need for some form of evaluation of the suitability of habitats for wildlife has become increasingly apparent and urgent. In response, several habitat evaluation plans have been suggested (e.g., Whitaker et al. 1976, Thomas et al. 1976, Boyce 1978, Flood et al. 1978, Berry 1978, Whelan et al. 1979, Ellis et al. 1979, Asherin et al. 1979). While these systems vary in their details, they are similar in that: (1) each attempts to devise a habitat evaluation plan that will be broadly applicable to a wide variety of habitats and management objectives (i.e., a "unified" scheme); (2) each considers a moderate number of habitat measures, including features other than vegetation structure alone; (3) each aims to define an index or a small number of measures that will provide a good prediction of habitat suitability to wildlife as a whole; and (4) each relies heavily (some exclusively) on measures or rankings of features that are derived from literature sources, expert opinion, or aerial imagery—none places initial emphasis on direct field measurements, although some do intend that the habitat evaluations they produce serve as guides to designing the most efficient field studies in a follow-up phase.

Each of these approaches has a different frame of applicability and different limitations. The general habitat evaluation systems tend to be very general and to rely heavily upon indirect measures. They therefore lack sensitivity to conditions in local areas, and as population censuses are generally not taken at all, they are incapable of indexing the relations between variations in the distribution and abundance of species and habitat conditions with any real accuracy. Their emphasis upon development of a unitary approach to habitat evaluation is perhaps misdirected, as any single system is unlikely to be useful if the study objectives or the underlying organization of the biotic systems vary from study to study or place to place. The "few variables" approach is simple and easy, but provides little detailed information on the habitat relationships of species or local species associations (Anderson 1979b). Generally this approach is not combined with carefully conducted population censuses, so its chief applicability would seem to be in broad intercommunity comparisons involving variations in species lists rather than population densities. Historically, this approach has been followed especially by those who believe ecological communities to be saturated and species assemblages to be in equilibrium (Cody 1966, MacArthur 1972); if this is so, one might expect a small set of habitat features to determine the community patterns. The "many variables" approach, on the other hand, developed in the context of the philosophy that communities are composed of species that respond to ecological variations largely independently of one another (Curtis 1959), and more recent multivariate analyses of data gathered in this fashion provide support for this view (Rotenberry and Wiens 1980a; Wiens and Rotenberry 1981b; see also Wiens 1977). If in fact bird species do respond to habitat variations independently of one another, there is no reason to expect a few key variables to be equally important to all of the species present in an area, and consideration of a large number of habitat features, coupled with accurate censusing of the bird populations, becomes critical. This approach, however, is considerably more complex and time-consuming than either of the others. If one ultimately wishes to assess the significance of the patterns of habitat occupancy or of the distributional correlations with habitat features of a species, however, none of these approaches is sufficient; this requires more critical studies of habitat utilization, which may begin to reveal what the various habitat features actually mean to the birds (Verner 1975, Balda 1975).

THE ANALYSIS OF BIRD-HABITAT RELATIONSHIPS

The sorts of analyses that one performs in order to discern habitat relationships to bird species abundances are to a considerable extent conditioned by the approach to habitat measurement that has been followed. Obviously, if one has followed the "few variables" approach, data often may be analyzed using relatively uncomplicated methods. If many variables have been quantified, however, multivariate analytical techniques are likely to provide greater insights than less sophisticated procedures. Indeed, multivariate analyses may be of considerable value even if only a few variables have been quantified. Many of those techniques have been treated in detail in a recent symposium (Capen, In press); here we shall offer only a few brief comments on some of the more popular multivariate procedures.

Multiple regression or correlation analysis provides a relatively straightforward technique for coupling bird species abundance estimates with any number of measured habitat variables. Multiple regression models yield precise quantitative predictions of a species' density for given values of environmental measures, and as such can be a useful tool in a species management-oriented project (e.g., Robbins 1978b). Unfortunately, such precision is invariably gained at the cost of generality, as a model constructed for one species is seldom useful for another. Even for one species a model is useful only over the numerical range of habitat variables used in constructing the model; extrapolation beyond these ranges yields estimates of dubious reliability.

Although more often employed in the analysis of habitats in which a species is merely present or absent (e.g., James 1971, Whitmore 1975), discriminant function analysis (DFA) can be associated with estimates of relative abundance (i.e., species absent, rare, or common) to evaluate habitat variables with respect to their ability to distinguish such density classes (Anderson and Shugart 1974). DFA combines all measured variables into a linear function that is best able to separate the three abundance groups, taking into account all covariance relationships among the habitat variables. However, such an approach focuses only upon differences in habitat occupancy, and as a result may often overlook any other biologically important factors that do not otherwise contribute to these differences.

A variety of ordination technqiues may be used to extract patterns of covariance in habitat variables, and the resulting patterns may be interpreted as representing multidimensional environmental gradients. Species abundances may then be plotted along the gradients and significant correlations interpreted as representing species' responses to these derived habitat clines. The most commonly employed ordination is Principal Component Analysis (PCA) (e.g., Cody 1975, Rotenberry and Wiens 1980a); although many are available (e.g., Gauch and Whittaker 1972). In addition to ordination, PCA is also useful for reducing the number of habitat variables with which one need be concerned (by extracting covariance patterns) and summarizing the important points of species similarity or correlation matrices (i.e., identifying ecological groups; Nichols 1977). Unfortunately, PCA is not without hazards quite apart from its rather rigid statistical requirements and assumptions (Johnson, In press). Because by definition components are independent of one another there is a strong tendency to attribute each component to independent phenomena ("one component-one cause"), ignoring the fact that each original variable contributes at least in part to the construction of each component. In addition, there is no a priori expectation that the linear combinations of variables that PCA extracts (the components) are precisely the same combination that the birds deem important; hence, the absence of correlation does not necessarily imply an absence of habitat response.

Canonical correlation analysis extracts not only patterns of covariation in habitat variables but also patterns of covariation in species' abundances, with the purpose of maximizing the joint correlation between the two data sets. It is analogous to multiple regression or correlation analysis, only now more than one "dependent" variables are being considered simultaneously. While the notion of emphasizing correlations between habitat "components" and species "components" is intuitively appealing, the technique is beset with analytical difficulties. This, combined with apparently very stringent requirements for linearity of input values, limits its current usefulness in species-environment investigations (Gauch and Wentworth 1976).

We must emphasize that by the very nature of the kinds of data collected (field measurements of species' abundances and habitat variables), which are both subject to statistical sampling error, we are restricted to correlational analyses of one sort or another. While this is not to say that such interrelationships cannot be investigated experimentally, we must constantly bear in mind that we are defining habitat *correlations,* not habitat *selection,* and should hedge our biological inferences accordingly.

CONCLUDING COMMENTS

We hope that it is obvious from the above discussions that proper measurement and evaluation of avian habitat occupancy patterns is a tricky business, but that, despite this, conducting avian censuses without recording information on the associated habitat features leaves one with information that is of limited value. Even if one takes care to record both bird density and habitat characteristics following careful quantitative procedures, however, problems may still remain. First, proper analysis of such data may reveal intriguing and interpretable patterns of correlations, but such correlation does not necessarily imply that the relationships are directly causal and meaningful to the birds. It is a statistical triviality to point out that correlation does not imply causation, yet the ecological literature is replete with studies that, having demonstrated correlations, proceed to develop grandiose explanations of the adaptive significance of the patterns as if they were unquestionably true. Statistical correlation only shows that a particular pattern holds in the particular data set with a given degree of probability, and while these patterns may suggest many interesting and important potential linkages between birds and their habitats, to believe them proven, and to proceed to frame management policies upon them, would be premature.

A second problem is related to this. When we record density variations between habitats or fluctuations through time, we assume that these differences are directly related to underlying environmental (habitat) factors. This is implicit in correlational analyses of bird-habitat relationships. Without some knowledge of the demography of local populations, however, this assumption is not secure. Different habitat types may differ in their suitability or degree of optimality to a species. If the distribution of individuals among habitat types is some direct function of habitat suitability, as visualized in the

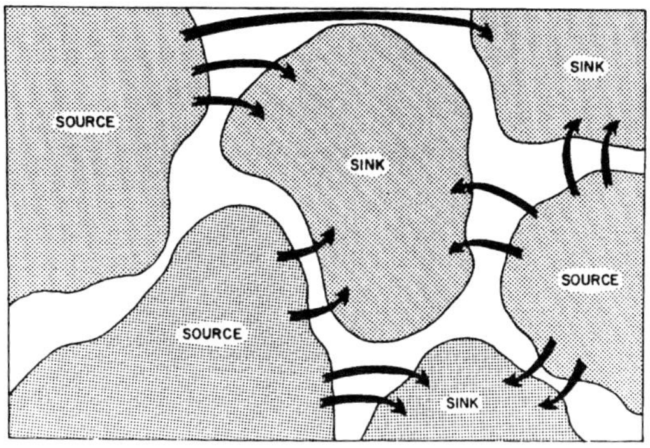

FIGURE 5. Hypothetical example of the "source-sink" structuring of populations. In "source" segments of the population, reproduction may exceed local carrying capacity, leading to net dispersal of individuals into other "sink" segments, in which reproduction alone cannot maintain population levels. The observed densities in either "source" or "sink" segments thus may not bear a close relationship to local habitat conditions.

Ideal Free Distribution model of Fretwell and Lucas (1969), then we may conclude that density variations between habitats do indeed index habitat suitability. But there are many reasons (e.g., territorial behavior, philopatry, time lags, and perhaps most important, the inability of individuals to exercise precisely optimal habitat choice, complicated by stochastic variations in environmental factors) not to expect an Ideal Free Distribution to be realized. This clouds interpretations of the observed density variations. Further, it is a mistake to believe that the size of any local population is at an equilibrium determined by local resource conditions. Instead, species' distributions may be broken into a mosaic of "source" and "sink" populations (Fig. 5). "Source" populations occupy habitat suitable for reproduction, and their output of offspring in fact exceeds the capacity of the local habitat, promoting dispersal. Here densities may be fairly stable through time, but the true suitability or productivity of the habitat is underestimated by considering only breeding density. "Sink" populations, on the other hand, may occupy habitat types that are generally unsuitable for reproduction or in which reproductive output is inadequate to maintain local population levels. These populations may be replenished by emigrants from the source populations, and individuals in sink population habitats may rapidly move into nearby source population habitats should vacancies arise. The densities and dynamics that characterize populations in these sink habitats thus vary not in response to local habitat conditions, but as consequences of events in the nearby source populations. Interpretation of population density-habitat correlations in populations structured in this manner would be difficult. Such a pattern seems evident in the Great Tit (*Parus major*) populations occupying woodlands and adjacent hedgerows in England (Krebs 1971; but see also Krebs and Perrins 1978, who suggest that these results may be equivocal), and may in fact be commonplace, especially where the interspersion of habitat types is close-knit. Unfortunately, detailed study of the demography of local populations is necessary to reveal the nature of such "source-sink" habitat occupancy patterns.

These problems call attention to the need for considerable care in the design and interpretation of ecological survey work. This symposium attests to the importance of developing and following rigorous methodology when censusing bird populations, and similar attention is demanded in the measurement and analysis of associated habitat features. As Elton and Miller (1954:474) observed some time ago, "because surveys must take up a great deal of time and labour and technical ingenuity, their aims should be clear, progressive and knit into ideas of dynamic ecology. They have to show a convincing reason for their existence, and not just accumulate a vague mass of field records."

ACKNOWLEDGMENTS

This synthesis and our research were supported by grants from the National Science Foundation, most recently Grant No. DEB-78-24364. Jean Ferner prepared the manuscript. John Emlen sparked the interests in avian habitat associations that have guided our research.

Studies in Avian Biology No. 6:533–537, 1981.

HABITAT CORRELATES OF BIRD DISTRIBUTION IN BRITISH CENSUS PLOTS

RAYMOND J. O'CONNOR[1]

ABSTRACT.—Recent studies based on the British Trust for Ornithology's Common Birds Census scheme are reviewed in relation to theories of habitat distribution. Studies of the Wren, Chiffchaff, and Kestrel provide evidence for a hierarchy of preferences between the available habitats, with less preferred ones coming into use at high population densities. For two additional species, the Great Tit and Yellowhammer, increased use of secondary habitats is associated with density-dependent reduction of reproductive success in the primary habitat. Bird species diversity in English woodlands is, as elsewhere, positively correlated with foliage height diversity, but populations of individual species are most stable in woodlands with species-specific vegetation profiles. These studies support theoretical models of population density as a determinant of habitat use and indicate a need to consider population levels in determining habitat preferences for individual species.

The dynamics of habitat use is currently one of the least studied areas of avian population ecology, presumably because few investigators have been able to afford to dedicate many years to the censusing of individual plots or habitats required for such investigations. Important theoretical work on the dynamics of bird populations between different habitats has consequently gone largely untested. Lidicker (1962) suggested that emigration from crowded habitats was an important component of population dynamics, as migrant individuals could expect to achieve greater fitness in other areas. Brown (1969) considered the implications of models in which exclusion from preferred habitats was the outcome of territorial behaviour by males successfully established in those habitats. These population models predicted that at certain densities in a preferred habitat, individuals should begin to move into an alternative habitat, a pattern found in Chaffinch (*Fringilla coelebs*) censuses by Glas (1960). The models also indicated that population densities should fluctuate more in less preferred habitats, a feature of titmice (Paridae) populations showing the buffer effect between secondary and primary habitats (Kluyver and Tinbergen 1953). However it is not necessary that birds in a secondary habitat be excluded from the preferred one through territorial behaviour as such: any form of scramble competition by which the breeding success of birds nesting in the preferred but crowded primary habitat could induce breeding in a secondary, less crowded habitat yielding a net gain in fitness (Fretwell and Lucas 1969) suffices.

The present paper uses the long-running Common Birds Census of the British Trust for Ornithology (BTO) to examine these ideas in the light of field data.

MATERIALS AND METHODS

The BTO commenced a programme of monitoring bird populations on farmland using a mapping method, in 1961 at the request of the then Nature Conservancy (now the Nature Conservancy Council). In 1963 fieldwork was extended to encompass woodland studies. Currently some 300 to 350 sites (approximately 100 farmland, 100 woodland, and the remainder mixed habitats of various sorts) are sampled annually by volunteer observers. The maps prepared by the fieldworkers are returned to Beech Grove, the Trust's headquarters, for central analysis, and the resulting counts of territories are summarised into a national index for (where possible) farmland and woodland separately or as a pooled national index where the individual samples for farmland and woodland are too scarce. Material from a particular observer is included in the index calculation each year only if the observer has contributed a census from the previous year; this procedure eliminates the effects of observer bias in the calculation of the index. Various tests have been conducted and confirm that differences between analysts have been largely eliminated by the training these professional workers receive before undertaking routine analysis of the Common Birds Census returns.

Each individual census worker submits a habitat map for his census plot when he first enters the Common Birds Census scheme. These maps are minimally marked to indicate location of general landscape features–copses and spinneys and other small woods, hedgerows, ponds, crops, etc.—and may run to great detail (pesticides and fertilizer applications, winter treatment, felling regime, etc.). He subsequently reports any habitat changes on a year-to-year basis, either by correspondence or by submitting a revised habitat map. In this way the results from any census plot for individual species or for all species can be related to the habitat of the census plot.

Several census workers are still surveying plots that they commenced in the early 1960s, but the majority of census workers contribute information from individual plots only for a matter of some years and then leave the scheme for a variety of reasons. With this information a very large number of plots have both habitat and bird data available for analysis. In using this information, however, it is necessary to allow for differences in censusing efficiency between observers. Results in preparation show that as much as 20% vari-

[1] British Trust for Ornithology, Beech Grove, Tring, Hertfordshire, United Kingdom.

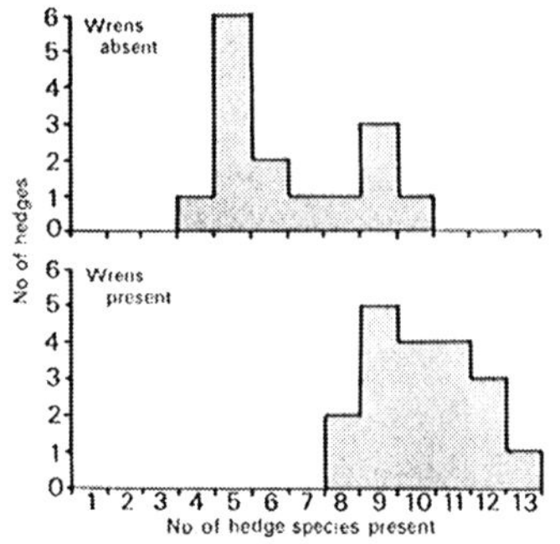

FIGURE 1. Wren habitat preferences in respect of arboreal species richness in field hedgerows. Data from P. Osborne.

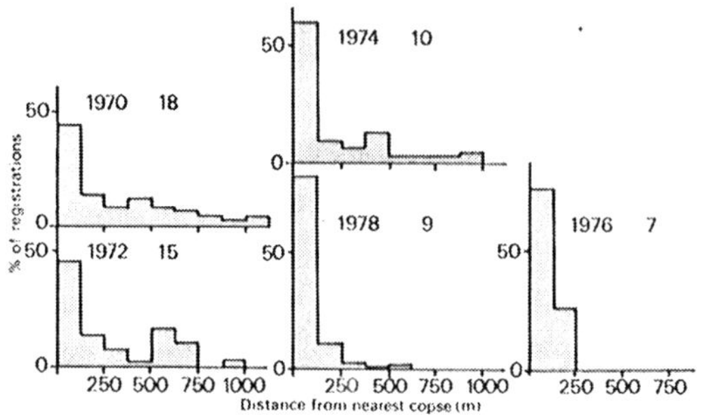

FIGURE 2. Distribution of Chiffchaff registrations with respect to copses in farmland, showing use of more remote song posts as population density increased. The figures to the right of each date give the species population level on the census plot in that year. Data from P. Osborne.

ation between individual census workers can occur on the same census plot (see Enemar 1962). This error does not affect the index calculations but is a source of error in the use of the census information in relation to habitat studies.

Against this background the present paper examines the habitat correlates of birds on the census plots studied in Britain. The extended time series information of bird populations available as a result of the running of the Common Birds Census scheme since the early 1960s allows rather more consideration to be given to temporal changes in habitat use than has been possible in previously reported short-term habitat studies.

RESULTS

Since 1961 bird populations in Britain have shown considerable variation, due largely to the occurrence of an exceptionally severe winter in 1962–63 and an only slightly less severe winter in 1978–79. Statistical analysis of the 1961–1978 population changes shows that for the majority of species winter conditions are limiting, though a few species are sensitive to summer weather (O'Connor unpublished data). For most species, population densities have varied two-to-three-fold over the two decades studied (Williamson and Batten 1977).

Williamson (1969) documented the changing habitat usage of the Wren (*Troglodytes troglodytes*) as it recovered in the immediate aftermath of the 1962–63 winter. In 1964 the commonest habitat element within Wren territories was woodland, but the use of this habitat rapidly saturated as the population grew. As the available woods were filled the second most common habitat used in 1964—streamside vegetation—began to fill up and saturate. Similarly, the third most popular habitat in 1964, gardens, began to fill up as the first two preferred habitats were less available. The least used habitat in 1964 was that of field hedgerows, and this clearly remained in a poor position throughout the 1964–67 population increase.

Even within this category of habitat it is possible to show the existence of more detailed microhabitat preference: Figure 1 shows that the Wren preferentially colonised those hedgerows that were rich in tree species, and were largely absent from those hedges containing rather few trees (P. Osborne, unpublished data). As these data were obtained in the late 1970s when Wren populations were rather high, they indicate that Wrens are showing considerable resistance to using species-poor hedgerows. There thus exists a distinct hierarchy of habitat preferences among the Wrens breeding in Britain, with the inferior habitats being colonised only under increasing population pressures.

Migrant species in Britain were, of course, less subject to population variation, escaping as they did the effects of the severe winters mentioned. Nevertheless, there is convincing evidence available that even migrant species show a similar hierarchy of habitat preferences. Figure 2 shows some data obtained for the Chiffchaff (*Phylloscopus collybita*) on a farm in Dorset: when the population of Chiffchaffs on the farm was low, as in 1976, practially all bird registrations were obtained within 250 m of a copse (a type of small wood) on the farm. At high population densities, on the other hand, much of the population was recorded on trees and hedgerows extending out from these copses at distances of up to 1 km from the nearest copse. The figure suggests that the Chiffchaff decreased on the farm during the study period and one might argue that this was due to habitat changes around the copses. In detail, however, the correlation of Chiffchaff positions with population density is greater than with time, both in the figure and within the full time series of data.

Why do these species show these population-dependent preferences in habitat usage? O'Connor (In press) has argued that resident species in Britain are largely *K*-selected in their population behaviour, in contrast to migrant species that are generally *r*-selected. On such a premise, the habitat use variations shown by resident species should be closely related to their reproductive fitness in the different habitats, and in some cases it has been possible to demonstrate that these exist. In the case of the Great Tit (*Parus major*) the long-running Wytham Woods study by the Oxford ecologists has shown that the Great Tit utilises farmland largely as a result of territorial exclusion from the preferred deciduous wood. Associated with this is a density-dependent reduction in clutch size within the woods themselves. The Common Birds Census data have been examined in conjunction with the BTO Nest Records scheme data on clutch size and breeding success, and the BTO Ringing Scheme data on dispersal and survival, to show that Great Tits nationally suffer similar clutch size and rearing success depressions in conditions of high population densities, and that the birds survive less well and disperse more widely under these conditions (O'Connor 1980d). These effects are also reflected for the Great Tit in apparent density-dependence within the woodland populations monitored by the Common Birds Census scheme, and are suggested for a number of other species (Williamson and Batten 1977).

A particularly interesting example of hierarchical use of habitats has been established in the case of the Yellowhammer (*Emberiza citrinella*) (O'Connor 1980a). In this study, Yellowhammer populations on farmland were shown to have levelled off after a period of recovery from the effects of the 1962–63 winter; on the other hand, the population of Yellowhammers recorded in woodland habitats continue to increase, suggesting the occurrence of an overspill effect such as described above for the Great Tit. By examining the clutch sizes recorded for Yellowhammer in farmland and woodland habitats, O'Connor (1980a) showed that clutch size decreases with population pressure, both in woodland and farmland, but more steeply for the latter. In this way, Yellowhammers attempting to breed in farmland suffer increasing clutch size depression, thereby eroding the advantage of breeding on farmland (the preferred habitat) in the first place and thus making it reproductively advantageous to move into the "inferior" but less crowded woodland habitat. In this way, a balance of advantage between habitat quality and population pressure generates a situation of "equality of fitness" for the two habitats, thus accounting for the observed pattern of population movements.

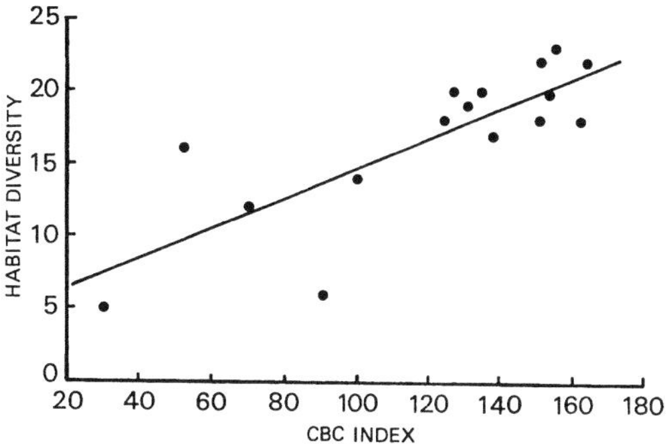

FIGURE 3. Diversity of breeding habitat use by Kestrels as a function of population density. Diversity assessed as number of standard nest record habitats cited in the annual sample. Data from British Trust for Ornithology nest record and Common Birds Census schemes.

A further example of changing habitat usage under increasing population pressures is provided by studies of the Kestrel (*Falco tinnunculus*) in Britain (O'Connor 1980b). This study showed that nonbreeding becomes increasingly frequent as the population rises, as might be expected for a carnivorous species subject to relatively little human persecution. This non-breeding is associated with an increasing use of a diversity of habitats, as recorded in the Nest Record cards for the species returned to the BTO since 1961 (Figure 3). Detailed examination of breeding performance by Kestrels in these different habitats showed that poorer habitats were increasingly used with the increase in population pressure. Thus, the hierarchy of habitat use by the Kestrel is directly linked to its breeding performance in the different habitats, providing a proximate explanation for the marked habitat preferences exhibited at low population densities. The increasing incidence of non-breeding demonstrated in this study at the highest Kestrel densities underlines this explanation.

Detailed study of the use made by Yellowhammers of particular "patches" on East Hammoon Farm in Dorset (see Williamson 1971) has shown an interesting trade-off between habitat quality, territory size, and population pressure. Under high population pressures, Yellowhammers can accept small territories provided they are of high quality, but can accept low quality territories only if they are extensive enough (O'Connor 1979 and unpublished data). Quite independent confirmation of this interpretation of the Dorset data comes from hedgerow cutback studies on Pendley Farm, Hertfordshire,

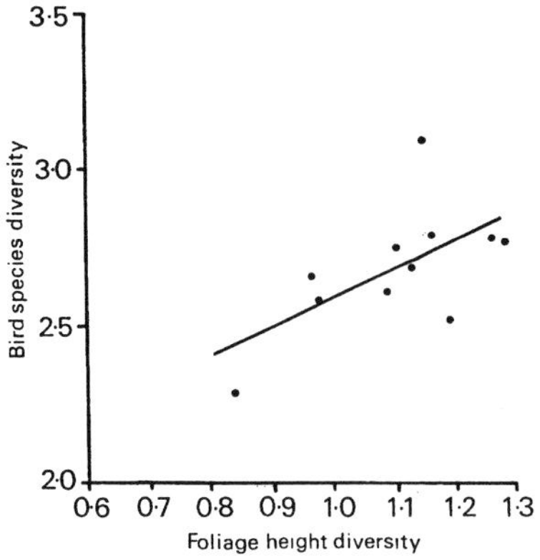

FIGURE 4. Bird species diversity in relation to foliage height diversity in 12 southern English woods censused by BTO members. Based on data in Pearson (1980). The regression equation is: $BSD = 0.92\ FHD + 1.67$; with $r = 0.726$ ($P < 0.02$).

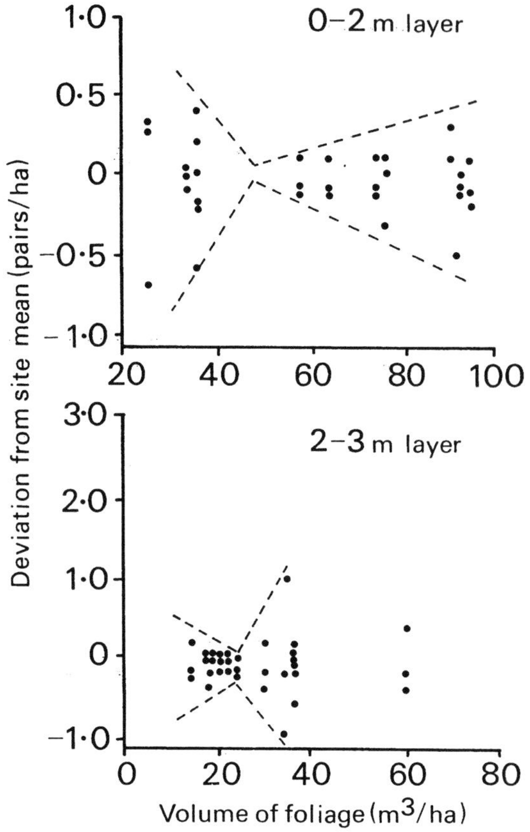

FIGURE 5. Population stability in relation to foliage volume in stated canopy layers for (top) European Robin and (bottom) European Blackbird. The vertical axis shows for each of the 12 woodland plots censused over three years the deviation of each annual population density from the three-year site mean. The dashed lines indicate the subjective limits of the data, to emphasize the reduction in the range of population densities apparent at particular foliage volumes (from Pearson 1980).

censused annually by BTO staff. Between 1975 and 1976 the internal hedgerows of the farm were severely pruned back to stump level. The Yellowhammers responded to the cut-back not by changing the location of the territories but by greatly enlarging their territory size on the farm (see Morgan and O'Connor 1980). Size and quality of territory are thus independent channels of adjustment to habitat quality and population pressure in Yellowhammers.

Examination of Common Birds Census data for a number of English woodlands has shown that their bird species diversities are correlated with foliage height diversity (Figure 4), in line with the results of previous studies (MacArthur and MacArthur 1961, Recher 1971, Moss 1978). However, the regression slope in Pearson's (1980) study is lower than those of the earlier studies. Whilst one may postulate several explanations, it is perhaps significant that the English data relate to years of generally high populations, when even poor habitats should be colonized if the models of Brown (1969) and Fretwell and Lucas (1969) are correct. To cope with such problems, Pearson (1980) has analyzed the variance of population densities in woods of different foliage structures and found that for individual species woods with structures either below or above some particular value contained more fluctuating populations than did woods at that value. Figure 5 illustrates the findings for the European Robin (*Turdus ericetorum*) and for the European Blackbird (*Turdus merula*) in relation to vegetation volumes within the 0–2 m and 2–3 m height zones, respectively. This approach identifies key characteristics of woods in which population densities are particularly stable, thus extending the analysis to take account of the population effects noted earlier.

DISCUSSION

The studies reviewed briefly here show that a variety of species display some form of hierarchical habitat preferences, as expected on the basis of Brown's (1969) theory. For some, such as Yellowhammer and Kestrel, the data suggest that utilization of the different habitats reflects the equality of fitness argument advanced by Fretwell and Lucas (1969). For others, such as the Great Tit, density-dependent reduction of clutch size and other components of reproductive success certainly occur within the preferred habitat (Krebs 1970, 1971; O'Connor 1980d), but fitness in the alternative habitat has not been measured.

These findings are particularly important for the analysis of habitat utilization by means of census studies. They indicate that different results may be obtained from censuses conducted in different years. This risk is particularly significant when one of the habitats being censused is a secondary or tertiary one for the species concerned. The distribution of the birds between habitats will reflect the total population pressure for the region, and may be changed suddenly following sharp changes in that pressure. Cawthorne and Marchant (1980) found that many English woodland species showed proportionally larger population decreases on farmland census plots than on woodland plots between 1978 and 1979, the intervening winter being unusually severe. They suggest that vacancies in woodland due to the death of resident birds were filled differentially through birds previously on farmland moving into the preferred woodland. Ringing data indicate that this hypothesis has a good basis in fact (unpublished data). Attempts to correlate bird densities with habitat features must therefore take into account the intraspecific competition for preferred habitats documented here.

ACKNOWLEDGMENTS

The Common Birds Census scheme is conducted by the British Trust for Ornithology under contract to the Nature Conservancy Council, and this generous funding is gratefully acknowledged. I am grateful to Patrick Osborne for the use of unpublished data.

Studies in Avian Biology No. 6:538–542, 1981.

CORRELATING HABITAT VARIABLES AND BIRDS

STANLEY H. ANDERSON[1]

ABSTRACT.—A brief overview of habitat correlation with birds by the use of multivariate statistics is presented. Examples taken from studies conducted in different forest habitats show that many species are correlated with macro features of the community such as habitat size and distance to the edge of the woods. Few data are available to distinguish clearly habitat where species are present or absent.

Analytic tests must be carefully selected. It is important that the assumptions of the tests are met. Care must be taken to determine that habitat variables found in one part of the range of the species are applicable to other parts. It is necessary to select habitat variables that discriminate between places where a species is found and is not found. To accomplish this goal, it is necessary to sample habitat or a species' territory. Finally, the results of correlation tests must be verified with field tests to assure their reliability.

Bird populations and species have often been associated with different plant communities (e.g., Adams 1908, Beecher 1942, Kendeigh 1948, Twomey 1945). Progression of bird habitat studies has led to more quantified studies in which actual features within the habitat have been associated with different birds. Stages in plant community succession have often been associated with changing bird species composition (Bond 1957, Anderson 1970a).

MacArthur and MacArthur (1961) indicated that they found bird species diversity associated with foliage height diversity. Although their studies have been disputed in different forest types, the work they conducted assisted ornithologists in defining habitat variables that characterize species habitat versus nonhabitat. Multivariate techniques came into vogue in the late 1960s. Using such techniques, James (1971) was able to show how a perceptual cue of the environment, called a "niche gestalt," could be defined for each bird species based on information from multivariate tests. For example, stepwise multiple regression indicates which variables, in association with others, appeared to be most commonly associated with the birds. Discriminant function analysis lists habitat variables that appear to be important in locations where bird species were found as compared with areas where the species were not found.

All multivariate analyses can only provide information based on input values. Thus, tests might well indicate variables that are important simply because they are measured in many areas in which the birds are found. In situations where biologists fail to measure important variables, results are distorted to reflect the inadequacies in input data.

Today, wildlife managers and land use planners seek to define habitat quantitatively for different species of birds. Such habitat classification schemes take specific characteristics that are identified as important to a particular population and classify habitat containing those features as optimal for that species. Some wildlife habitat classification schemes seek to prescribe means of trade-off whereby one area can be made suitable for a species as the original habitat is altered.

The purpose of this paper is to identify from field studies and published data, different features of the habitat that appear to be important to bird species. Most studies try to correlate individual species with habitat variables from data collected in one or several field seasons; very few completed projects, however, verify that the results obtained from the correlation studies are indeed important to the bird species. Such verification would often involve prescribed habitat alteration with observations following for several years. There are some before-after studies on habitat disturbance, such as strip mining or cutting of transmission line right-of-ways (Anderson 1979a). A study involving paired comparisons between burned and unburned areas in Michigan's Upper Peninsula (Anderson 1979b, 1979c) produced ambiguous results, primarily because it was not always known if the correct habitat features to prescribe management activities were being measured.

WHAT TYPE OF VARIABLES ARE MEASURED?

Data from different publications show that a number of features of the habitat, such as canopy volume, diameter at breast height (DBH), tree abundance, and ground cover, are commonly used. A comparison of data collected in several parts of the United States is made using stepwise multiple regression with bird abundance as the dependent variable and habitat factors as independent variables (Table 1). These data show which habitat variables are important. In these four study sites (the eastern deciduous forest in Tennessee and West Virginia, northern mixed forest in Michigan, and western deciduous forest in Oregon), forest size, distance to the edge, canopy volume, and trees per hectare

[1] Wyoming Cooperative Research Unit, U.S. Fish and Wildlife Service, Laramie, Wyoming 82071.

TABLE 1
NUMBER OF BIRD SPECIES SHOWING A CORRELATION WITH HABITAT FACTORS IN VARIOUS STATES BY MEANS OF STEPWISE MULTIPLE REGRESSION

Habitat factor	Mary-land	Minn.	Ore-gon	Tenn.
Trees per hectare	10	14	18	10
Canopy volume per hectare	20	10	17	15
Canopy cover per hectare	19	9	17	10
Snags per hectare	2	13	4	
Average trunk height	4	7	4	3
Logs	2	5		
Shrubs per hectare	11	16	6	8
Forest size	24	18	19	
Distance to edge	26	14	12	
Bird species considered	36	35	26	28

are associated with more species. Relatively few species are associated with features such as trunk height, logs, or snags. Michigan is an exception, as fire apparently left many snags that are associated with the birds coming into the area.

When variables used by biologists are examined, one can identify two classes. First, major features of the habitat such as habitat size, distance to the edge, canopy volume, DBH, and trees per hectare are correlated with many of the species. Such variables represent broad or "macro" habitat features and might be associated with the community as a whole. Second, features that can be associated with individuals, such as snags and logs, are not listed as correlators with large numbers of species. This "micro" level is associated with features that can potentially be correlated with individual species. We are often sampling macro features on a micro level, such as viewing the canopy through a camera on a tripod to measure openness or through a visual scope to determine the presence or absence of ground cover. This approach can provide useful data; however, when macro information are the only data required, they can often be sampled with aerial photos which are quicker and cheaper than tedious on-site sampling.

When biologists try to discover what habitat variables are associated with bird species, they often start by collecting habitat data in a circular grid around nest sites. Additional habitat data are collected in unoccupied areas. Correlation tests are then run between birds and variables in the two areas. Since results are based on the type of habitat data collected, if only macro data are collected little can be said about micro factors responsible for a species' presence. The

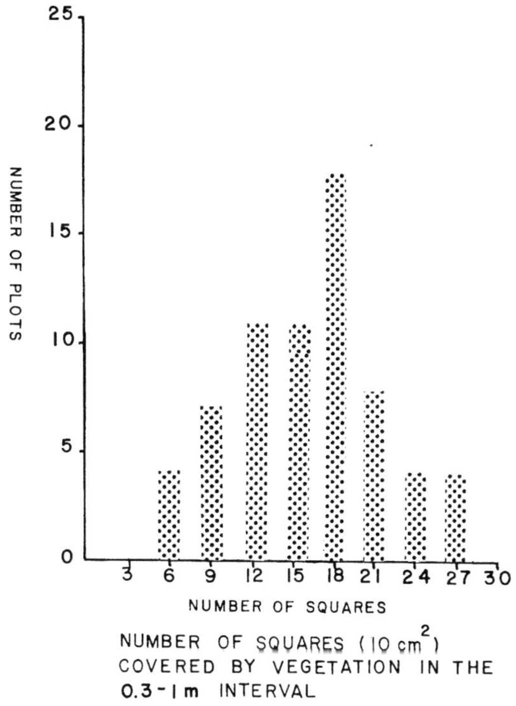

FIGURE 1. Number of 10 cm^2 squares covered by vegetation in the 0.3–1 m height interval on vertical drop cloth on plots in western Maryland.

type of habitat data collected also limits the investigator's ability to discuss optimal versus suboptimal habitat.

It is very important that habitat variables and sample sites be selected so that they discriminate between areas where a species is present or absent. For example, in an eastern deciduous forest of western Maryland, 56 of 68 sample sites are 120 ha or smaller. If habitat size is a discriminating factor and a species is present in forested areas up to 300 ha, this is not a good discriminator.

A better discriminator is one that shows some form of distribution between classes that can be used for discrimination. For example, to estimate foliage density, a drop cloth is used in the center of a 0.04 ha circle. Observation of the cloth indicates the number of 10 cm^2 circles covered by vegetation at different height intervals. At the 0.3–1 m interval, the study sites show a distribution of squares in each category (Fig. 2). As a result, this variable could be a potential discriminator for bird species.

Micro features of the habitat associated with individual species are difficult to distinguish. As yet, we probably do not have a good understanding of many of the micro features that can

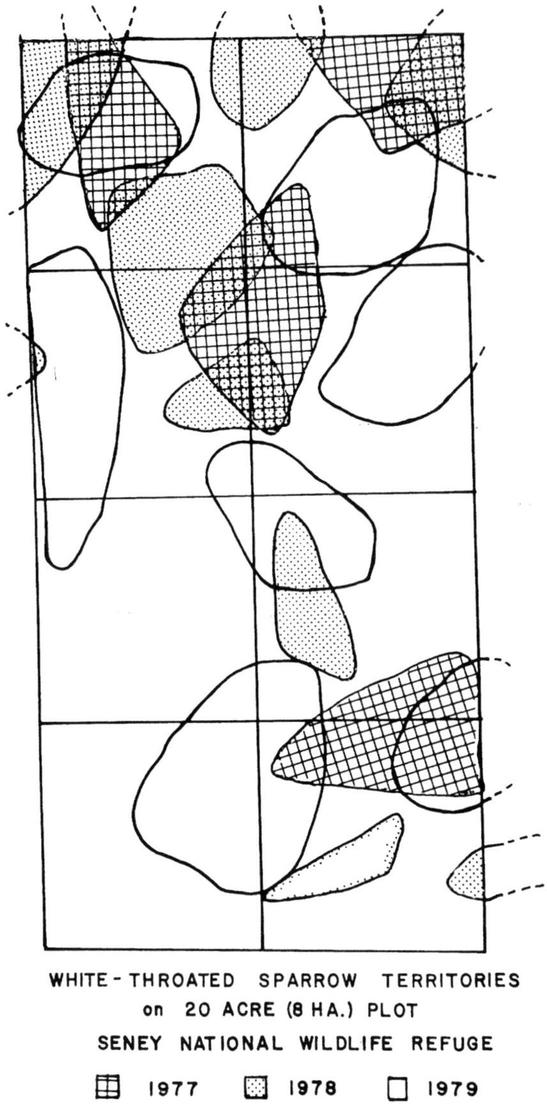

FIGURE 2. Shift in White-throated Sparrow territories on 8-ha (20-acre) plot in Michigan over three-year period.

be individually associated with species. It is true that studies such as those conducted by Anderson and Shugart (1974) indicate that the number of saplings is indicative of the number of Downy Woodpeckers (*Picoides pubescens*). The corollary may simply be that Downy Woodpeckers are found in early successional sequence or second-growth timber stands. We do know that Western Wood Pewees (*Contopus sordidulus*) are found in open sites in the forest where snags protrude. Clearly, studies on some endangered species, such as the Red-cockaded Woodpecker (*Picoides borealis*) (Jackson 1977) and Kirtland's Warbler (*Dendroica kirtlandii*) (Line 1964), identify habitat features that can be used to manage those populations. On the other hand, we do not have a good source of data on some of the more generalist or wide ranging forest species, such as the Red-eyed Vireo (*Vireo olivaceus*) or Scarlet Tanager (*Piranga olivacea*). General or community descriptions are often used to describe areas where these birds are found.

WHAT DOES HABITAT MEAN?

Habitat is an area where an organism's needs for survival are found. Many ecologists list food as an important resource that can be in limited supply. Some feel that differences in foraging behavior, habitat utilization, and morphological and temporal variation all occur because of competition for food. Sexual dimorphism is thought to occur in some species when males and females forage on different parts of trees or select different size prey items. If competition for food is responsible for differences in habitat utilization, methods of measuring habitat variables should indicate competition.

Studies of birds present along different habitat gradients, e.g., moisture (Smith 1977), altitude (Anderson 1970a), succession (Bond 1957), and competition (Cody and Walter 1976), show which species of birds inhabit different communities. Habitat structural characteristics are measured and correlated with species in those communities. Island biographic investigations report the absence of different species on smaller islands with presumably less habitat to allow all species to find food (Morse 1971). Such findings appear to be related to the habitat gradient studies where forest species are characterized as associated with dense understory, heavy canopy, or other variables that are part of the changing structure in forest succession.

Are habitat selection studies really measuring adaptability of forest birds? If so, then we should be able to show that each species is associated with unique features of the habitat. We should be able to identify habitats where species occur. Conner et al. (1975) states that selection of certain habitat types for nesting by woodpeckers probably reflects the prevalence of the habitat type more than the preference by any species. Still, requisite needs for a species must be satisfied. We must therefore identify those factors that allow a species to survive in an area.

With many bird and habitat sampling techniques, collection of habitat data for breeding birds does not always occur on the defended territory. For example, a study plot on a burned mixed forest site in the Upper Peninsula of

Michigan shows that White-throated Sparrow (*Zonotrichia albicollis*) territories do not generally abut one another. Because it is likely that a series of habitat features provide suitable areas for nesting sparrows, it is important that these features be sampled where the species actually uses the territory. Counting only singing males heard can be misleading. These sites must be compared with sites in which the species is not found. Random sampling therefore must be stratified to reflect a clear presence or absence before habitat sampling begins. It is also possible to partition data with presence or absence sets after sampling.

Territories do not remain constant from year to year (Figure 3). After three years of the study in the burned site of Michigan's Upper Peninsula, White-throated Sparrows showed territorial shifts that were presumably due to changes in the habitat structure and competition from other species following the fire. Thus, relocation of the habitat sampling units would be necessary each year to maintain sampling within the territory of the species.

Problems of study plot size also arise on sampling birds. Some census techniques do not adequately cover the larger bird territories. Habitat samples then do not yield accurate information about that species. In fact, some plots of 20 ha and less may not provide adequate microhabitat to sample.

Results are at times difficult to verify because objectives have not been clearly stated. Many studies looking at habitat selection in birds are very general in nature. A clear hypothesis is not distinguishable, although it may be there in vague terms. Biologists need to clearly define what they are trying to do. This is very important if they are to develop management guidelines. Managers cannot relate to such things as the total canopy volume of a deciduous forest to maintain a species. Often it is necessary for biologists to translate data results to useful information for managers. Managers would, however, be able to consider the total size of a habitat necessary to support a bird community. Furthermore, habitat classification schemes are being developed on the basis of inadequate data sources that do not clearly identify places where the bird is found with places where the bird is not found. Results do not identify optimum and suboptimum habitat and relate these results to field verification.

Another difficulty arises because of the dissimilarity of variables used in different areas. It is often very difficult to make comparisons among several study sites. Noon (in press) urges that similar features of the habitat be recorded in different studies, thus comparisons can be made. Although he suggests that the James-Shugart (1970) technique is a good start, he believes that further refinement is necessary to sample habitat adequately. Many of the results of the James-Shugart technique, however, do not provide good information on correlation with individual species because of the type of data collected. Microhabitat data necessary to discriminate individual species habitat are not available.

TABLE 2
HABITAT VARIABLES CONFORMING TO ASSUMPTIONS OF NORMALITY

	Ohio	Central Maryland	Western Maryland	Seney (unburned)
Foliage density 1	*		*	*
Foliage density 2		*		*
Foliage density 3	*	*	*	*
Foliage density 4		*		*
Number of trees	*		*	
Number of small trees		*	*	*
Number of medium trees	*	*	*	*
Number of large trees		*		
Ground cover	*	*	*	
Number of snags		*		
Tree height		*		
Average tree diameter		*	*	
Tree basal area	*		*	*

STATISTICAL EVALUATION

Multivariate analysis is used extensively to identify habitat variables important to bird species. Discriminant function analysis, a method of analyzing grouped multivariate data, is frequently used to identify important habitat variables. This technique is both predictive and explanatory. As a classifier, the technique aims at classification of individuals of unknown groups of membership. As a predictor, discriminant function analysis allows group separation by means of linear transformations (canonical analysis).

Recent evaluations of the canonical functions indicate that some assumptions of the tests are not met in utilizing the tests (Williams in press). For example, habitat variables do not always exhibit a normal distribution.

Evaluation of data from four study sites in the eastern deciduous forest (Table 2) indicates that only about half of the variables conform to the assumption of normality. This means that some results appearing in the literature that identify management criterion for birds may be in error. Correction of the deficiency in the discriminant function test is not always easy. Increasing sample size is apparently not the answer (Williams in press).

DISCUSSION

A number of macrohabitat features are felt to be important to birds. Robbins (1979) indicates the importance of habitat size to maintain populations of neotropical migrants in eastern deciduous forests. When the total area decreases below 810 ha, migrants begin to disappear. He shows how sections of forest habitat might be combined in different ways on tracts of land to attract the bird species.

Increased edge is a factor that does attract species, as shown by Lay (1938) and Johnston (1947). Such changes in forest habitat increase the diversity of the community by attracting different species. Most of those species attracted to the edge, however, are not migrant species (Anderson 1979a). Thus, edge may be good for resident species and poor for neotropical migrants.

Habitat structure is a very important component of the avian community. When comparisons between field and forest are made, distinct differences can be seen in the structure of the community. Likewise, structural differences within the successional seres in a forest can be recognized and correlations can be drawn between these structural features in the community. Actual comparisons of habitat structural components with the presence or absence of individual bird species, as may be done through a discriminant analysis, are difficult. It is often very hard to determine which exact features are responsible for a species being there. In fact, the total community structure could be the important component. Thus, it might not be possible to consider managing an individual species without looking at other populations within the community.

If the biologist's objectives are to prescribe a community management procedure, then macro level components may be the answer. Coupled with knowledge on the population biology of species, managers can maintain habitat to support communities of birds. When going beyond that to work with individual species, we may need to develop further sophistication in our measurement techniques to answer questions adequately.

If the objectives of the work are to classify habitat, then it is important to specify the level of involvement. Should we be discussing individual species? In most instances we cannot do so because we do not have the data. Quick surveys listing the habitat variables around an individual bird are inadequate to describe the general habitat of that species throughout its range. Some species have very similar requirements throughout their range whereas others show geographic variation (Noon et al. in press); however, such conclusions are reached only after considerable fieldwork.

After defining objectives, biologists need to select habitat variables to answer the questions asked. Then they must take a series of elimination steps to find habitat variables correlated with species. Breckenridge (1956) describes an interesting process of arriving at habitat characteristics associated with the Least Flycatcher (*Empidonax minimus*). He evaluates the abundance of trees, shrubs, and tree size. His attempt to correlate hazel stalks with the number of flycatchers is not effective. However, it leads to the suggestion that the degree of forest crown closure is related to flycatcher use. This approach discloses that the degree of openness just beneath the forest crown is the primary influencing factor. This factor is also related to the size of the forest in which the birds are observed.

Results from correlations drawn from data collected by biologists indicate that we have some information on communities of bird species. Thus by specifying general ecological features, such as forest size and degree of succession, we can discuss a form of community management. These data are useful when planning major community changes; we can specify which species might be eliminated when habitat size decreases.

At this time data are inadequate for managing all species in a community. Data repositories are not a substitute for fieldwork. Not only must we plan an adequate study at each site, but we must also continue to develop techniques to find parameters that define bird habitat relations.

Studies in Avian Biology No. 6:543–547, 1981.

MEASURING RESPONSES OF AVIAN COMMUNITIES TO HABITAT MANIPULATION

JARED VERNER[1]

ABSTRACT.—Increasing concern for the need to conserve our renewable natural resources, including birds, has resulted in the enactment of laws and the involvement of federal agencies to protect these resources. Past assessments of the effects of management activities on avian communities, and of the sampling procedures used, have been limited in approach and unsatisfactory in result. Recent research suggests that, in addition to sampling bird communities, relevant habitat features must be sampled. Multivariate statistical analyses of many sample plots is usually a preferred technique, trend estimates are usually preferable to density estimates, and the variable-diameter circular plot method is usually best suited for the inventory analyses needed by management.

Human activities of many sorts bring marked changes in the natural habitats of birds, resulting in changes in species composition and population densities. A growing concern that some of the ways we use land may result in irretrievable losses of some renewable resources, including birds, has led to the enactment of laws, at many levels of government, intended to assure wise stewardship of all renewable natural resources. At the federal level, for example, these laws include the Multiple Use-Sustained Yield Act of 1960, the Endangered Species Act of 1973, the Forest and Rangeland Renewable Resources Planning Act of 1974, and the National Forest Management Act of 1976. The laws recognize all wildlife, including birds, as valuable, renewable natural resources.

Because of these laws, many federal agencies are involved in a variety of bird studies (Hirsch et al. 1979). The studies have at least two common goals—to enable us to predict the effects of land or resource management projects on the composition of bird communities, and to monitor bird population trends in the community before and after project completion. It is impossible, of course, to monitor every management project adequately, but our ability to predict the effects of projects will improve in proportion to the monitoring accomplished. I believe it is imperative to ensure coordination among those involved in the effort, to minimize duplication, and to employ standardized methodologies.

This paper reviews and evaluates the state-of-the-art for predicting or assessing the effects of management activities on bird communities. The problems are assessed from the viewpoint of an applied ecologist constrained by the needs of management to find reliable and cost-effective methods for achieving goals.

[1] Pacific Southwest Forest and Range Experiment Station, Forest Service, U.S. Department of Agriculture, Fresno, California 93170.

PREDICTING AND ASSESSING PROJECT EFFECTS ON BIRD COMMUNITIES

Our ability to predict the effects of projects on bird communities is limited. This is particularly true in North America where, until recently, systematic, standardized, and continent-wide inventory programs encompassing all habitats and sampling both animal communities and their associated habitat elements have not been supported. Sampling has been disorganized, noncomprehensive, and has used different sampling procedures. The state-of-the-art is more advanced in Europe, especially in Scandinavia, where nationwide inventory and monitoring programs have received more attention (Lack 1937; Oelke 1966; Järvinen and Väisänen 1973, 1976c, 1977b; Sharrock 1976).

LIMITED SITE COMPARISONS

Until recently, the most common method for assessing and predicting project effects on bird communities has been to compare species' densities on treated and untreated sites. Sample sizes have typically been small, usually only one treated and one untreated site; or, for studies of successional changes, only one site each in several seral stages. Also, most of those earlier studies were done only after habitat treatment. Comparison was made between the treated site and a different, but usually nearby, untreated control site. This design assumes that prior to treatment the control site had an avian community not significantly different from that of the treated site in any sampling season. This may or may not have been true, but no test of this assumption can be made once a treatment is completed.

The optimal design, if a limited number of sites is to be sampled, involves both pre- and post-treatment sampling of the treated site and

> "an identical, untreated control area . . . The null hypothesis in this case would state that changes in bird species abundance in

> the treatment area over time would be identical to changes occurring in the control area. Contained within this design are controls for both space and time" (Conner and Dickson 1980).

The key assumption is that the control and treatment sites are identical. Each site should be sampled several years (my intuition suggests at least 5 years) prior to treatment to establish that patterns of annual variation do not differ between sites. Even then, the researcher has no measure of within-treatment variance, so valid generalizations to other similar habitats are limited.

The necessity for comparing communities before and after treatment has not been met because, until recently, no strong mandates to maintain all renewable resources were in force. Consequently, managers did not include in their decision-making process alternatives to incorporate a wide variety of resource needs. And researchers generally failed to provide managers the kinds of information required to accommodate the needs of birds, mammals, and other renewable resources considered in management prescriptions. This has resulted in a general lack of communication between researchers and managers. Researchers have not sought information on future projects, and managers have not brought projects to the attention of researchers. Fortunately, this situation is changing rapidly, and the result no doubt will be mutually beneficial.

Multiple Site Comparisons

The extensive literature on habitat selection by birds has been reviewed (Hildén 1965), and the value of an understanding of habitat selection in effective management of bird populations was summarized by Verner (1975). Factors important in habitat selection include food sources, nest sites, song posts, shelter, available water, nesting materials, watch-posts for insect hawking species, and general features of the terrain or vegetation. The data

> "generally demonstrate a great range of variability among bird species with respect to habitat selection. The data also show that for nearly all species our knowledge of specific factors eliciting positive habitat selection responses is woefully meagre. While I believe we must continue to research this problem on a species-by-species basis, I suggest that this is not the most fruitful approach to the immediate problem of managing wild lands in a manner that will minimize detrimental effects on populations of . . . birds. Studies most likely to yield information widely applicable in habitat management are integrative, multivariate analyses of bird species abundance and habitat variables" (Verner 1975:51).

This view is widely held (Anderson 1979, Niemi and Pfanmuller 1979, Rotenberry and Wiens in press and Shugart in press).

Assuming sufficient information on habitat selection by the bird species of a region, it should be possible to predict avian community composition for any site. This can be done by measuring habitat variables that provide proximal cues for habitat selection. Applications of multivariate statistical analyses to avian communities and their associated habitats show this to be a viable alternative to limited site-comparison methods. This is not to say, however, that it can take the place of intensive, individual species studies, especially rare species or those with very large home ranges. Detailed life history investigations must remain a vital part of our overall research effort to be fitted in where necessary. And we may not permit our confidence in the multivariate approach to dull our insistence on the need to test predictions.

The trend in North America toward applying multivariate statistics to avian community and habitat studies goes back to the work of Cody (1968), James (1971), and Anderson and Shugart (1974). Cody (1968) used discriminate function analyses to identify interspecific differences in habitat selection among species in grassland bird communities, and to identify those habitat variables contributing to the differences. James (1971) sampled 15 vegetational variables on 0.1 acre plots centered on 401 song perches of 46 species in a variety of habitats in Arkansas. Principal component and discriminate function analyses were used to establish habitat ordinations of the species along three dimensions representing gradients in vegetation structure. Anderson and Shugart (1974) used discriminate function analysis to order habitat variables "according to their strength in separating abundance categories for 13 of the more abundant bird species" in a primarily deciduous forest in Tennessee. They concluded that their results provided a basis for predicting changes in bird species composition as a result of habitat alterations.

Several multivariate techniques have been used recently by researchers studying the relationships of avian communities to habitats (Capen in press). Discriminate function analysis, principal components analysis, and cluster analysis have been used most often. But this field is in a dynamic state. The best methods are surfacing. Also, researchers need more experience in interpreting results from these sophisticated techniques—both the statistics and the biology.

If a time-efficient counting method is used to sample bird communities, a large number of sites can be included, and key habitat variables measured at each. Some standardization is desirable, as suggested by James and Shugart (1970). Multivariate approaches can yield information beyond that needed to evaluate effects of a management project on the bird community at a given site. As these data accumulate, our ability to predict impacts will improve. Significant insights into habitat selection by individual bird species will emerge, leading to understanding of the bases for regional differences in habitat selection within species. Clearly, such results can make profound contributions to basic knowledge, while still generating data needed by applied ecologists.

PRACTICAL CONSIDERATIONS

The task of maintaining all bird species, providing wise stewardship of other renewable resources, and also accommodating demands of resource users, is formidable. Managers need tools to do the job effectively within the constraints of tight budgets and limited personnel. For management considerations, trend estimates in bird populations may need to be the rule, because reliable trend data are more cheaply and quickly accumulated than reliable density data. And rather than intensive analyses of vegetation composition and structure on a study site, managers may be able to afford to measure only those parameters critical to predicting whether or not a site will be optimum, suitable, or only marginal for certain species. Parameter selection must, therefore, consider the speed and objectivity with which measurements can be taken. For example, diameter-at-breast-height (dbh) may be the preferred measurement as an index of tree height, foliage volume, or canopy diameter (Young 1977, Verner 1980b), even though bird species richness is more directly affected by tree height or volume.

However these challenges are met, it is obvious that much basic research remains to be done. This research needs to include:

- Quantitative sampling of bird communities and habitats;
- Identification of the habitat needs or preferences of birds on a species-by-species basis;
- Identification of effective habitat predictors of species richness and species occurrence; and
- The role of patch size, shape, and position in determining the make-up of avian communities.

Sampling Bird Communities

Many methods have been used to sample the composition of bird communities and to estimate densities of species in those communities. These topics have been dealt with in various papers presented at this symposium. We are far from consensus on which method is most appropriate in any given situation. The International Bird Census Committee recommended the mapping method as preferable for sampling communities of breeding birds (Svensson and Williamson 1970). The same procedure is standard for the National Audubon Society's annual Breeding Bird Census, with more than 100 censuses being reported annually in the pages of *American Birds*.

Various forms of transect counts are used widely including fixed-width and variable-width strips (review in J. T. Emlen 1971, 1977a). French scientists, for more than two decades, have made extensive use of timed counts from fixed points with unlimited boundaries (Ferry 1974; Blondel 1975, 1977). More recently, a method involving variable-diameter circular plot counts has been developed (Ramsey and Scott 1978, Reynolds et al. 1980) and tested extensively (J. M. Scott, pers. commun.). Nearly all bird species in most terrestrial habitats can be sampled by variations of these basic methods. However, the particular needs and constraints of land managers make some methods more appropriate than others.

The mapping method is applicable only to territorial populations. Moreover, it is poorly suited to large-scale inventories mandated by recent legislation. Resource agencies lack funds and manpower to sample enough plots to generate a sufficient data base for statistical analysis. Although many researchers believe the mapping method is the most accurate sampling procedure in wide use (but Berthold's [1976] thorough review casts much doubt on that belief), the small sample sizes possible mean that no confidence limits can be established. This could put the manager in an untenable position if a management decision based on such data comes under legal challenge.

The patchiness of vegetation makes it difficult to locate large, homogeneous areas of habitat. But large plots are needed to reduce the potential for a bias in the mapping method resulting from territories overlapping the edge of the study plot (Berthold 1976). The relationship be-

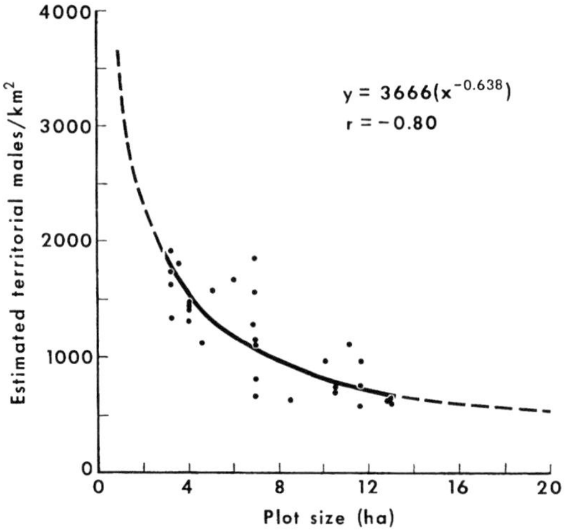

FIGURE 1. Relationship between plot size and estimated density of territorial males in California oak woodlands (Verner 1980a). Dashed portions of the curve are calculated extrapolations from the data.

tween mapping plot size and estimated density of birds in California oak woodlands (Fig. 1) is based on counts reported in *Audubon Field Notes* and *American Birds* from 1944 through 1976. Data were fitted to linear, exponential, and power curve regression models. Statistically significant ($P < 0.001$) negative correlations were found in each case ($r = -0.80$, -0.81, and -0.80, respectively). Figure 1 shows the power curve fit to the data, as this gave the best visual fit. These results suggest that bird density on 3.2 ha plots is about 2.7 times the density of birds on 13 ha plots, which may be true if the smaller plots are isolated patches that tend to attract a disproportionate share of birds. If, however, the data in Figure 1 reflect an overestimate of density on smaller plots, then plots of at least 20 ha appear to be required in California oak woodlands to escape this small plot effect. The character of most managed land does not provide homogeneous patches of habitat that large.

Transect methods lend themselves better to sampling more sites, because usually several transects can be counted in the same period of time required to complete one mapping survey. Conner and Dickson (1980) recommend fixed-width transects as the preferred technique to detect effects of habitat treatments on birds. I believe J. T. Emlen's (1971, 1977a) variable-width transect method is superior to a fixed-width method, however, because it compensates for species with variable detectability and for observer biases. Comparisons of density estimates obtained by mapping and by the variable-width strip transect method generally show lower estimates with the transects (J. T. Emlen 1971, Franzreb 1976, Dickson 1978). No method, however, yields completely accurate censuses of birds. Effective management *can* be based on reliable estimates of population trends, but those estimates should come from the same sampling method year-after-year (Conner and Dickson 1980).

The major limitation of the transect method is that, as for the mapping method, it is difficult to find large enough blocks of homogeneous habitat to contain the transects. Any given transect count commonly yields results reflecting considerable variation in habitat characteristics. This is incompatible with multivariate statistical analysis. Anderson et al. (1977) have attempted to minimize this problem by subdividing transects into 150-m segments and treating each segment as a separate sample. The obvious shortcoming of this solution is that bird counts in adjacent segments, and even some measured habitat variables, are not independent.

Counting of variable-diameter circular plots is probably the best compromise as a method for the enormous task facing land and resource managers. It is time-efficient, applicable to small patches of habitat, and can supply the trend data suited to management. It can also provide some information on species' densities. I agree with Shields (1979) that it is preferable to use a method having some potential for indexing absolute, rather than relative, abundance. Such information can be important to managers in the case of some species, and further work with the method should give us a better understanding of the relationship between real field densities and densities as computed by the method. Furthermore, we can gain the density information by this method with little or no more time than is required to obtain acceptable estimates of population trends by sampling only relative abundance.

The variable-diameter circular plot method is also well-suited to multivariate statistical analyses, because sampled plots are small enough that habitat structure can be kept reasonably uniform, and many sites can be counted in a short time. Statisticians recommend sample sizes five to ten times greater than the number of independent variables to be considered. Agencies constrained by limited time or personnel should be guided by the rule: results from multivariate analysis will be more valid if more time is given to accurate sampling of the independent variables, even if less accurate estimates of the dependent variables result (David Sharpnack, pers. commun.). In other words, sampling many sites a few times is better than sampling a few sites many times.

The variable-diameter circular plot method is not without its sources of error and bias in estimating bird densities. For example, one assumption of the method is that "The count period is short enough . . . that objects occupy fixed locations during the count" (Ramsey and Scott 1978). This, of course, is not true. If a bird is detected in the count area and leaves during the count, it is not deleted from the total of birds recorded. But if a bird that is not within detectability when the count begins later moves within detectable range during the count, it will be recorded. The effect of this according to S. L. Granholm (pers. commun.) is to inflate the count relative to ideal fixed locations assumed by the model. It also tends to create a "donut" effect in the distribution of detection distances relative to the observer, because such incoming birds generally are first detected some distance away, as they near the plot center (Ramsey and Scott 1978).

Finally, none of the methods considered here is well suited to sampling rare species, or species with large home ranges, because they yield insufficient numbers of observations to assess the effects of management on them. At least some of these species are among those recognized as especially sensitive to the sorts of changes humans have brought to natural ecosystems. The diurnal and nocturnal raptors are examples. An obvious solution is to apply a variety of census techniques.

CONCLUSIONS

In addition to sampling bird communities, it is essential to sample relevant habitat features. Research must focus attention on identifying habitat features that are good predictors of bird species presence. This process should consider the ease and accuracy with which selected features can be measured.

Analysis of the effects of habitat treatments on avian communities, based on pre- and post-treatment measurements on a limited number of treatment and control sites, has limited general application. With about the same time input, and well within the time constraints of a challenging Ph.D. dissertation, it is possible to carry out a study of many sites across a wide range of habitat conditions and to apply to the results a variety of multivariate statistical analyses. This can provide insights into habitat selection by birds and, at the same time, generate information upon which to base general predictions of a wide range of project impacts on bird communities. Given the probable course of secondary succession on a site disturbed by management, it should be possible to predict avian community structure well into the future. This is rapidly becoming an indispensable part of the planning process for public land management agencies.

Because no counting method equivalently samples the densities of all species, and because accurate density estimates of few, if any, species can be readily obtained, most assessments of management effects ought to rely on trend estimates only. Any well-conceived method is suitable for this, so long as sampling is standardized from plot-to-plot and from year-to-year.

The variable-diameter circular plot method of estimating bird densities is the best suited to most of the inventory work so badly needed by management agencies today. Even though the method may be used by management primarily for obtaining trend information, some density information may prove to be invaluable. Other methods may be superior in certain situations, and we must keep in mind the fact that some sensitive species are not adequately sampled by this or any other common census method. Intensive life history studies may be the only solution for some such species.

ACKNOWLEDGMENTS

The preparation of this paper has been significantly enhanced by draft reviews and discussions of major concepts with Richard T. Reynolds, Robert C. Szaro, and Jack W. Thomas. It is a pleasure to acknowledge their important contributions.

Studies in Avian Biology No. 6:548–553, 1981.

SURVEYING BIRDS IN THE TROPICS

JAMES R. KARR[1]

ABSTRACT.—Numerous difficulties plague the researcher as he/she sets out to determine the species composition and abundances of birds frequenting a specific area. Many of these difficulties have been minimized for temperate environments as censusing procedures have been improved in the past two decades. But procedures developed in temperate situations are often inadequate in tropical regions where avifaunas are composed of many rare species and, in addition, many "peculiarities" of species biology diverge from the "norm" of temperate avifaunas. Examples of these peculiarities include permanent occupation of territories, decreased levels of singing, secretive habits, extensive overlap in home ranges, and numerous species which wander over relatively large areas in search of mobile (e.g., army ants) or otherwise spatially patchy (e.g., fruits) food resources. The significance of these problems for censusing and procedures to improve the reliability of bird census data are described.

Many factors affect the degree to which census results reflect real densities of birds in a census area. These include both physical (weather, topography) and biotic (vegetation type, biology of birds) factors. For most terrestrial habitats in temperate regions acceptable census procedures have been developed and are in widespread use. However, even in temperate regions, selected species may be very difficult to census. The unusual pairing pattern and use of "territory" in Brown-headed Cowbirds (*Molothrus ater*) results in considerable difficulty in applying conventional census procedures. In grassland and marsh habitats, polygamous species may be abundant and make accurate censuses difficult. Because most birds are territorial and monogamous, however, the proportion of temperate birds for which peculiarities of natural history affect census results is low.

In sharp contrast, many tropical species exhibit "unusual" behavior and natural history attributes that significantly affect census accuracy. Consequently, researchers in tropical regions must use caution in selection of census methodology. It is too early to present a detailed and precise guide to censusing tropical birds. Rather, it is my intent here to discuss briefly many of the special circumstances obtaining in tropical habitats and their effects on accuracy of censuses. Since the greatest concentration of these peculiarities is in forest habitats, my discussion emphasizes forest birds.

THE ROLE OF PHYSICAL FACTORS

Weather and topography are the two most important physical factors affecting census accuracy. Primary weather factors that reduce census reliability are wind and rainfall. Wind is significant because it directly affects bird activity and because it reduces the ability of observers to hear vocalizations and detect movements of birds. Dry season winds are a problem in some areas. Occasionally these winds persist for extended periods, forcing one to avoid censusing or to census with results of less than optimum quality. Obviously, little else can be done to reduce the importance of this factor.

The other physical factor of major concern is rainfall. Lowland forest in the humid tropics often receives large quantities of rain. During late wet season several days of continuous rain may limit census opportunities. However, the effect of rain is usually less than expected from rainfall quantity because rains typically are concentrated in late afternoon due to their convectional origin. In contrast to the short, heavy rains of lowland areas, persistent light rain and fog in mountainous areas may limit census accuracy.

These problems are essentially the same as those in many temperate environments. To census or not must be determined by the field worker following guidelines formulated to minimize census activity during periods when census results are likely to be unreliable. However, strict adherence to such guidelines may result in periods without data.

THE ROLE OF VEGETATION

Type of vegetation on a study area is important in determining census accuracy. Vegetation density may inhibit one's ability to traverse the study plot and may make it difficult or impossible to observe birds. Both problems are common in early successional habitats and in grasslands. Grassland habitats in relatively wet areas present serious difficulties when grass heights exceed 5 m. Often this vegetation is impenetrable because of the sharp edges of grass blades. Late successional areas become impenetrable thickets in which movement results in noise levels sufficient to cause a reduction in bird activity. In habitats with very dense ground-level vegetation, trails can be cut to minimize distur-

[1] Department of Ecology, Ethology, and Evolution, 102A Vivarium, 606 E. Healey, University of Illinois, Champaign, IL 61820.

TABLE 1
TOTAL NUMBER OF BIRD SPECIES KNOWN FROM AREAS OF SEVERAL SIZES IN PANAMA AND ILLINOIS

Region	Area (km²)	Number of bird species
Panama	75,600	905
Illinois	144,700	390
Canal Zone	1424	560
East-central Illinois[a]	6050	292
Limbo Hunt Club study plot	0.02	205
Bottomland forest, Kickapoo State Park	0.02	70

[a] Three-county area of Champaign, Piatt, and Vermilion Counties.

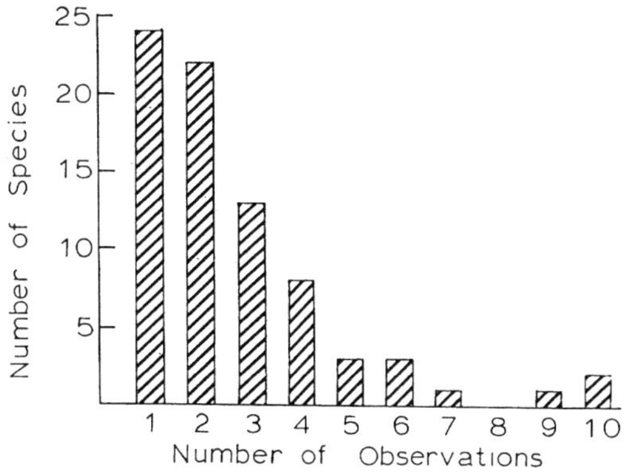

FIGURE 1. Number of observations for each of 77 rarely observed bird species at Limbo Hunt Club, Pipeline Road, Republic of Panama.

bance to vegetation while allowing the observer to pass throughout the area with relative ease and a minimum of disturbance. However, trails may require considerable work to keep them clear. I try to vary origin, direction of travel, and termination point for censusing along trails. This prevents secretive species from escaping the observer by the same behavior during each census. In my experience, density of undergrowth in forested areas is not sufficient to deter reliable censusing.

Major problems of censusing in forest are darkness in the undergrowth and extreme height of vegetation. When these factors are compounded with some topographic irregularity, censusing can be especially challenging. Tree heights above 30 m are not uncommon, and identification of small canopy species (e.g., hummingbirds and flycatchers) may thus be difficult or impossible. Consequently, the observer depends on vocalizations as a cue to bird presence to a greater extent than in many temperate habitats.

An abundance of "peculiar" plant life forms can also make censusing difficult. Dense epiphytes and lianas can limit unobstructed views of foraging birds. Flowering and fruiting plants (sites of major bird activity) in the forest canopy may be out of view from the ground.

The final difficulty that originates as a consequence of the nature of tropical vegetation is an extraordinary species richness. Barro Colorado Island, Panama, an area of 1450 ha, supports more than 1350 species of higher plants, including 652 woody species (Croat 1978). This wide array of species makes vegetation sampling and classification exceptionally difficult, especially in view of the many recent studies that show the importance of specific plant species in tropical (Howe 1977) and temperate environments (Holmes et al. 1979).

THE ROLE OF AVIAN NATURAL HISTORY

The most vexing problems in censusing tropical birds result from the birds themselves. Limitations of time and space prevent a detailed analysis of all relevant factors. At best, I can highlight a few of the more significant difficulties.

The most commonly cited characteristic of tropical forest avifaunas is their high species richness (Table 1). The number of species seen in a relatively small forest area in central Panama (Limbo Hunt Club) continues to increase after over a decade of intensive research. In two weeks of field work, I commonly record 100 to 110 species on that 2-ha study plot, and I have observed over 200 species on the study plot. South American forests are often richer.

This extraordinary species richness is combined with striking rarity for many species (Fig. 1). Rare species may be encountered only once or twice per year; rarest species are only encountered every few years. When mist nets are used to census forest undergrowth avifaunas, rare species (<2% of sample) constitute 75–85% of the species included in samples from Brazil (Novaes 1969) and Panama (Karr et al., In press).

Presence of species with very similar plumage compounds the problem of identification, especially when birds are seen as shadowy forms in dark undergrowth. In some cases males may be well-marked, but females are difficult to distinguish. Males of different species may be similar in nonbreeding plumages. Even voices may be similar, compounding the problem of distinguishing species.

One of the most popular census procedures used in temperate environments (spot-map or singing-male) is based on the assumption that most resident birds form monogamous pairs that defend clearly delineated territories (type A territories of Nice 1943). But in many tropical habitats many species do not defend type A territories (Karr 1971). My rather rough compilation suggested that most species (98%) in structurally simple habitats, such as grassland, defend exclusive territories. In more complex forest habitats, relatively fewer species (32%) defend exclusive territories. Even species that defend territories may do so in ways that are unusual. Plain-brown Woodcreepers (*Dendrocincla fuliginosa*) do not form permanent pair bonds; males and females establish territories that overlap extensively, but with little or no congruence in boundaries (Willis 1972).

Many unusual breeding systems exist that differ from conventional spacing and habitat-use patterns. The lekking behavior of the manakins (Foster 1977), hummingbirds (Stiles and Wolf 1979), and others are excellent examples. Intraspecific and interspecific flocking are common also. A tremendous diversity of spatial patterns is used by many flocking species. In some cases several species may co-defend territory boundaries, while other species joining the same mixed flock may not be territorial or may have entirely different territory boundaries (Munn and Terborgh 1979, Gradwohl and Greenberg 1980). The nature of flocks varies significantly and is often associated with type of food resource exploited and its spatial distribution (Karr 1971, Moriarty 1976).

Aggregations of individuals may involve single species (undergrowth tanager such as *Tachyphonus delatrii*) or mixed species (tanager-honeycreeper) flocks. Many types of interactions occur with varying spatial and temporal stability due to local variations in resource density and presumably other factors. Finally, even species that occupy type A territories may exhibit behavior different from that of ecologically similar species in North Temperate areas (e.g., Plain-brown Woodcreeper mentioned above).

Many species are more or less permanently mated and occupy territories all year. In this circumstance, it is not unusual for singing activity to be reduced throughout the year with obvious consequences for the probability of the observer recording the species.

To add further complicating factors, occupation of space may vary seasonally. This has been well documented for many migrant species that initially set up winter territories (Morton 1980) and later abandon territories to feed in mobile flocks. The latter behavior is apparently associated with periods when local food supplies may be unreliable and territory defense is not economical.

Temporal dynamics of tropical forest avifaunas also present significant problems. In addition to a variety of locally stable, but not classical territorial species, many tropical forest species travel over wide areas in search of their mobile and/or patchy food resources (Willis 1973). This makes reliable censuses on small study plots very difficult.

Often movements that are so common in tropical forest birds appear random when data bases are derived from short-term studies. However, many of these apparently random patterns are precisely timed movements. Local movements on diurnal, seasonal, and year-to-year time scales are clear (Karr, in press). Further, the patterns on these time scales are not always consistent among days, seasons, and years due to lean seasons that vary in severity or to bottlenecks in resource availability that limit species and their distributions. All in all, these patterns are exceedingly complex; they do not lend themselves to use of a single census procedure that is equally applicable for all species, seasons, years, and habitats.

Another temporal dynamic is associated with arrival and departure of migrants (Keast and Morton 1980). In some cases there are long distance migrants, while in other cases they may be local movements on altitudinal or other (e.g., rainfall) gradients. Transients also may be very common for short periods.

Another methodological problem is variability in census results from one period to another. I have censused a forest study plot on one morning and been hard pressed to detect more than 6 to 10 species of birds. On the very next day and under the same weather conditions, I may detect 50 or more species on the same area. This magnitude of variability defies easy classification and development of reliable census procedures.

High cicada densities also may create problems. During the dry season, the din of calling cicadas can effectively prevent any census activity that depends on hearing bird vocalizations. This may seem a trivial problem, but presence of the problem for weeks on end can result in major blanks in valuable and otherwise contiguous data records.

Intensity of predation pressure on tropical forest birds, especially during their breeding season, may have selected for cryptic behavior patterns. Many species are very effective at avoiding detection because of their secretive behavior. In addition, effectiveness of nest concealment also is a result of similar selective pressures. This limits success in searching for nests to confirm breeding on the study area. As an example, despite long studies by Willis on the Ocellated Antbird (*Phaenostictus mcleannani*) on Barro Colorado Island, Panama, he has still not discovered the nest of the species (Willis 1973).

SOLUTIONS

Regrettably, there are no simple solutions to the problem of censusing birds in tropical forest. The best advice is to use an array of procedures selected to provide the most comprehensive information for the objectives of the study. The problem is easiest to solve when the subject of the study is a small set of closely-related species. Under this circumstance, a procedure(s) can be selected to optimize quality of results.

However, when community level objectives are a top priority and/or when time is limited, I have grave doubts about the possibility of producing reliable census data across a wide spectrum of species. This pessimistic view is substantiated by comments and qualifications invariably included in papers on tropical avifaunas. Anyone anticipating attempts to census birds in tropical (especially forest) areas should carefully review their objectives in light of the comments and cautions of Orians (1969), Terborgh and Weske (1969), Howell (1971), Karr (1971, 1976c, In press), and Hespenheide (1980).

Several widely recognized census procedures have been used in tropical environments, including singing male, transect or trailside counts, mist nets and banding, and point counts. All have strengths and weaknesses.

Singing-Male Counts

This is, in my opinion, the least reliable procedure. Lack of breeding synchrony and limited singing activity of many species makes this procedure inappropriate. Several early studies (MacArthur et al. 1966; Howell 1971; Karr 1971, 1976c) depended heavily on this procedure. A large proportion of species is missed on any individual census, so there is a tendency to grossly underestimate species richness. In addition, use of this technique by persons inexperienced in identifying tropical birds also results in underestimates of species richness. Reliability of density estimates is no doubt low, although when supplemented with monitoring of banded birds, reliability may improve (Karr 1971, 1976c).

Due to the large number of rare and/or rarely encountered species in tropical areas, the suggestion, based on temperate work, that 5–6 censuses are sufficient to census an avifauna accurately must be viewed with caution. A more comprehensive census effort is essential.

Transect Counts

Transect counts also have been popular with researchers in tropical forest areas although they rarely include the type of corrections for sighting distance discussed by Emlen (1977a). Pearson (1977) used this procedure as did Orians (1969), Hespenheide (1980), and Fodgen (1972). All of these researchers recognize the inadequacy of a procedure which centers on the naive assumption that encounter probabilities are proportional to local density. Since species conspicuousness varies considerably (e.g., the tanager *Tachyphonus delatrii* vs. the wren *Microcerculus marginatus*), care must be used to standardize comparison of results from several areas. Variability among observers in knowledge of voices and sight identification may create real problems. Further, observer bias toward flocks (Hespenheide 1980) may significantly bias transect counts against solitary or quietly foraging species.

Mist Nets

Mist nets, in my opinion, are the best procedure available for "censusing" bird populations in tropical forest. They avoid the bias of inadequate knowledge of the resident avifauna and provide a random, unbiased sample of birds moving in the space sampled by nets. They do not, however, randomly sample the entire fauna. Species that walk on the ground, large and very small species, and species active at levels above net operation are undersampled. Very mobile species are captured out of proportion to their local density.

These disadvantages notwithstanding, I still feel more comfortable using nets to develop quantitative information on selected components of a tropical avifauna. Use of numbered or colored bands in combination with mist nets further enhances the value of mist-net counts. In addition, other data can be collected in concert with netting operations. Excellent examples of the use of bands to understand avian population dynamics include the detailed studies of antbirds by Willis (Willis and Oniki 1978), antwren flocks by Gradwohl and Greenberg (1980), and studies of moult by Fodgen (1972).

Point Counts

Point counts have rarely been used in tropical forest. Its only use to my knowledge was by MacArthur et al. (1966). MacArthur recognized the weakness of that study long ago. Recent census experience in temperate areas with this procedure suggests to me that its use should increase in the tropics. Without doubt, sample sizes and duration will have to be expanded considerably relative to the conventional use of point counts in temperate habitats.

All of these procedures have strengths and weaknesses, and thus must be used, and their results interpreted, with caution. The bottom line for tropical censuses is intimate knowledge of birds to be studied and design of a complex of census protocols selected to provide the greatest amount of information in the context of the purpose of the investigation. It is important to identify species with peculiarities in behavior or ecology. The situation is very different from that in the temperate zone, where most species have "normal" spacing and territorial systems. Census procedures, as well as interpretation of data, must reflect that reality. To attempt to establish a uniform protocol at this time would limit reliability of census data in years ahead.

DISCUSSION

Decisions about census procedures are perhaps the most important and complex decisions to be made by an ornithologist wanting to assess an avian population or community. During the past decade considerable effort has been made to examine techniques and biases in censusing birds in temperate regions. In contrast, systematic, comparative studies to evaluate census methodology in tropical areas are lacking; most tropical censuses have been conducted by researchers with limited time and a primary focus on research objectives unrelated to evaluation of census procedures. Short-term visits by temperate-based scientists are not likely to fill that gap in the near future. As a result, census results will often be less reliable than is desirable.

To minimize the problem created by inadequate information on census procedures in tropical areas, I suggest that four primary questions should be asked and carefully answered before censusing is initiated: Why? Who? What? How? (The same logic obtains for efforts to census birds at higher latitudes.)

Why?

Why is the research program being initiated? What are the study objectives and/or the specific hypothesis to be tested?

Who?

Emphasis in this question is determination of the species to be censused. Is it a single species or all of the birds in the assemblage? What are the important natural history attributes of the species in question? How will those attributes affect census results? In general, the who question will come second in studies of a disciplinary orientation, such as ecology or behavior. More applied efforts may have the "Who" question imposed by concern, for example, for specific rare or endangered species.

What?

The what question is concerned with the type of information needed to attain project objectives. Are absolute or relative densities required? How essential are data on sex and age structure? To what extent are data on foraging behavior and ecology or other natural history data required?

How?

Finally, the "How" question should be asked. Identification of suitable procedures must consider time and funds available for the study as well as information on study objectives and natural history of study organisms. At this point the researcher must evaluate the presumed reliability of his results from a variety of census procedures in the context of objectives, organisms, and environmental constraints.

Lack of knowledge of environmental constraints, species attributes, and census biases makes decisions about census procedures especially difficult in the tropics. At the very least, I urge caution in the uncritical acceptance and application in tropical areas of procedures designed for censusing in the temperate zone. Indeed, I have some doubts about the extent to which the common assumptions of the temperate-based procedures are satisfied by the biology of temperate-zone birds; those concerns have been reinforced by my tropical experience and are now being raised by others in this symposium. This is not to suggest that census efforts should be abandoned; rather, thoughtful evaluation of results must include assessment of census biases and reliability.

My own work in censusing tropical birds has led me to the following general approach:

(1) Use a composite of census procedures selected to provide the best possible data for a variety of species. Tropical habitats, especially tropical forest, are a microcosm of the most vexing problems for censusing terrestrial birds. As a corollary for this, recognize that the problems of accurate assessment of abundance for all species are overwhelming. I prefer to target my efforts to development of reliable information on a selected set of species (e.g., undergrowth avifauna with mist netting) even if it means little or no information on other species (e.g., canopy species) within the community.

(2) Select procedure(s) which do not depend on some seasonal phenomenon like breeding for their effectiveness.

(3) Identify exceptional species and use special procedures to improve knowledge of their

abundances if such knowledge is essential to project objectives.

(4) There is no substitute for knowledge of the organism under study.

(5) Keep in mind the constraints placed on these thoughts by variability in study objectives.

ACKNOWLEDGMENTS

The Smithsonian Tropical Research Institute, National Geographic Society, EARTHWATCH, and the University of Illinois Research Board have provided funds for my studies of birds in tropical regions. J. Blake and B. Loiselle made helpful comments on an early draft of the manuscript.

Studies in Avian Biology No. 6:554–558, 1981.

THE USEFULNESS OF ABSOLUTE ("CENSUS") AND RELATIVE ("SAMPLING" OR "INDEX") MEASURES OF ABUNDANCE

DAVID G. DAWSON[1]

ABSTRACT.—Territory mapping provides an estimate of density through repeated visits to an area, which lowers the chance of any resident birds being missed, but the usual estimate of the chance is obtained by circular reasoning, obscuring a severe bias for the less conspicuous species. Comparison with more detailed studies shows that territory mapping, point counts, and transect counts usually provide poor estimates of density. Given this, the Finnish technique of estimating density from transect counts of birds within and outside of an inner zone provides a useful compromise. For an index of density, point or transect counts are preferred because they can be applied to a wider range of species and seasons and seem more cost effective than territory mapping.

This paper compares the cost effectiveness of point, transect and territory mapping techniques for measuring absolute density and deriving indices of abundance. In another paper in this symposium I review various influences that affect the accuracy of point and transect counts. Here I first review territory mapping techniques in the same way, and then compare techniques in practical use.

INFLUENCES ON TERRITORY MAPPING

Territory or spot mapping techniques use data from several visits to a plot, on each of which the activity of each bird is recorded on a map. When the records of one species from all the visits are brought together, the information falls more or less easily into clusters corresponding to the territories of that species (Anonymous 1969, Enemar 1959, Williams 1936, Williamson 1964). The reasoning behind the method is that, if a bird has a probability q of not being counted on any one visit, this is reduced to q^n for n visits. If the value of n is high enough, very few territorial birds are missed.

In practice, one record is not enough to identify a territory; Svensson (1979a) recommends at least three 'registrations,' and shows that if q is less than about 0.5 the usually recommended number of 8–10 visits will reveal 90% or more of the clusters. When q is higher than 0.5, too many territories are missed. Accepting only two or one registrations as sufficient would allow more "territories" to be recognized, but would increase the risk of including chance clusters that do not correspond with territories. Increasing the number of visits is not an efficient way of identifying more territories of cryptic species, as the return per unit effort is small (for example, 16 visits are needed to identify 90% of territories if q is 0.7, and 52 visits are needed if q is 0.9). Territory mapping can therefore give an acceptable approximation to the true density with an acceptable amount of work only if the probability of detecting a bird each visit is high.

Estimates of q in the literature (DesGranges 1980, Enemar et al. 1978, Hogstad 1967, Seierstad et al. 1970, Slagsvold 1973c, Svensson 1978a, Williamson 1964) are bedevilled with circular reasoning: clusters are identified and are equated with territories, and q is then calculated from the number of visits without records in each "territory," divided by the total number of visits. If territories of birds with a high value of q are not recognized or are merged, q will be underestimated. Whatever the value of q, some territories will by chance have too few registrations to be recognized. This also underestimates q (Järvinen and Lokki 1978). The underestimation is by 0.05 at $q = 0.6$ and 10 visits and becomes rapidly worse as q increases. Thus, when the probability of missing a species on any one visit is high, good estimates can be obtained only from independent and thorough density estimates, such as from intensively studied color-ringed birds (Snow 1965).

The probability of missing a bird will also vary with season as well as with the age, reproductive, or physiological state of the bird (Slagsvold 1973c) and with other factors such as habitat, time of day, or weather (Dawson 1981a). Seierstad et al. (1970) and Slagsvold (1973c) tried with some success to overcome some of these problems by estimating q separately for each segment of the population, but the technique lacks rigor and, as Svensson (1979a) pointed out, there is rarely enough information from any one mapping "census" (8–10 visits) to give a good estimate of q. Without good estimates of q, the equation of map clusters with territories remains uncertain.

Other problems in estimating density from maps arise from the difficulty of identifying clusters and of allocating edge clusters. Best (1975) and Svensson (1974b) found considerable variation between different people's interpretation of species maps: coefficients of variation ranged from 15 to 36 for territorial passerines. Keeping

[1] Ecology Division, DSIR, Private Bag, Lower Hutt, New Zealand.

the same observers and standardizing the rules used for interpretation will remove some, but not all, of this variation.

That these problems may greatly influence the 'densities' estimated for most species is shown by studies comparing mapping results with more detailed studies (Bell et al. 1973, Diehl 1974, Enemar et al. 1978, Haukioja 1968, Jensen 1974, Mackowicz 1977, Mannes and Alpers 1975, Nilsson 1977b, Snow 1965). For a minority of species surveyed at the right time, the mapping method may yield a good density estimate, but in general it gives an index, not an estimate, of density.

Three studies where different observers were used in the same areas provide information on the precision of the mapping method as an index of numbers. I use the mean-to-variance (between observers) ratio as a measure of repeatability. Snow (1965) found large differences between farmland species, some (e.g., the Blackbird, *Turdus merula*) having means much less than the variance between observers and others (e.g., the Great Tit, *Parus major*) the converse. However, Frochot et al. (1977) and Enemar et al. (1978), confining their attention to small passerines in forest, found means almost always greater than variances. More work is needed to confirm this hint that mapping in forests may give an index of numbers that is quite insensitive to a change of observer.

THE STATISTICAL BEHAVIOUR OF THE COUNTS

Point Counts Compared with Transects

Järvinen (1978a) considered line transects to be superior to point counts in two respects. The average distance of detection in a line transect is linearly related to the effective area sampled, but in point counts the area is proportional to the square of the distance. This means that the same size of error in judging distance has a greater effect on density estimates from point counts than on those from transects; Emlen (1977a) suggested that distance estimation is facilitated in transect counts by the possibility of hearing a bird from a length of trail and thus getting a better 'fix' on it. Table 1 compares the repeatability (standard deviation) of estimates of distance for point counts and transects from the same areas; neither technique seems uniformly the more accurate, so one might prefer transects. Imprecision in distance estimates is a problem only if an estimate of area is needed to convert counts into densities.

Järvinen's second point was that the "saturation effect" (more cues being missed when there were more birds recorded in total) was not important in line transects. However, his evidence for this (Järvinen et al. 1978a, 1978b) is indirect, and the conclusion seems unlikely. Ramsey et al. (In press) preferred point counts to transects; they held that transects miss more of the birds close to the observers, but they did not state how they knew this.

TABLE 1
A Comparison of Distance Estimation in Point and Transect Counts (Transect Results in Brackets)

Species	Average percentage "Far" Records[a]	Standard deviation of the average[b]
Blackbird (*Turdus merula*)	50 (22)	12 (12)
Goldfinch (*Carduelis carduelis*)	45 (25)	3.3 (9)
Myna (*Acridotheres tristis*)	76 (58)	8 (6)
Skylark (*Alauda arvensis*)	56 (28)	16 (3)

[a] Data from 200-m transects and 5-minute point counts made in pastoral land in Hawke's Bay, New Zealand, March 1980.

[b] The standard deviation is of four estimates of the percentage, one for each day of the study and each based on 96 five-minute point counts or 80 transects. The variance of the percentages differs significantly only for the Skylark ($P < 0.05$ in F tests).

Dawson and Bull (1975) compared the results obtained in point counts and transects (0.7 km/h) in the same area. Their analysis was of the total counted as an index of density and showed that the two techniques were of similar value for demonstrating differences in mean values of the count. Subsequent more extensive work in pastoral and orchard land has supported this conclusion (Dawson and Robertson, unpubl.). Dawson and Bull (1975), Källandar et al. (1977), and Ramsey and Scott (1979) preferred point counts for work in forests because this method allowed undivided attention to be given to the birds and was probably much less affected by variation in terrain.

Ratowsky and Ratowsky (1979), working in Tasmanian forests, detected more species in 5 min walking at 3–6 km/h than in 5 min standing. Dawson and Robertson (unpubl.) have found the same in New Zealand farmland. Thus, if the main aim of the survey is to acquire a species list quickly, transect counts may be preferred.

Yapp (1956) examined the theoretical relationship between the speed of the observer, w, and of the bird, u, and the coefficient for converting counts into densities, k. His model is that $k \alpha (u^2 + w^2)^{-\frac{1}{2}}$. It is not valid for point counts ($w = 0$) because it deals with the entry of new birds into the observer's range, not with those initially there (Skellam 1958). For the same reason it will overestimate the density if the transect is short relative to the "effective radius" (Yapp 1956) of the birds. The effect of observer speed needs further study, and other effects might be expected—for example an observer will make more noise if he walks faster.

Estimation of Densities in Point and Transect Counts

Four approaches have been used to convert counts, c, into estimates of density, d. In the first, an independent "census" method is used to establish densities in the same places that counts are done, and k is estimated from $k = d/c$ (Ferry 1974, Gill 1980, Walankiewicz 1977). Given all the possible influences on k that I have discussed elsewhere in this symposium, such a calibration would be needed for a range of conditions, and can be no more than approximate without a prodigious amount of work. However, the small differences in k between habitats found by Gill (1980) give some hope. Of course, the usefulness of such estimates of k also depends on the accuracy of the independent estimate of density—a point discussed with regard to a mapping "census" in the first part of this paper.

The second approach is to collect information on the distance of detection of each bird and to establish for each species a distance up to which all individuals are detected (the "basal radius" of Ramsey and Scott 1979, see also the "specific strip" of Emlen 1977a and Balph et al. 1977). The records beyond this distance can then be discarded and the population estimate based upon those within the distance. Ramsey and Scott (1979) discuss criteria for estimating the basal radius, and it is clear that even their best method is subject to considerable bias and imprecision. Another problem with this technique is that it may necessitate discarding most of the records of some species, so that the density is based upon a small count.

The third method is to use the distribution of detection distances to give an estimate of a single effective distance, r: this is the "effective radius of detection" of Ramsey and Scott (1979), which is directly related to the effective area sampled and to my k by the equations $k = 1/\pi r^2$ for point counts, and $k = 1/r\,l$, where l is the length of a line transect. These equations permit density to be estimated from the total count. Ramsey and Scott (1979) estimated r via their basal radius r_b: $r = r_b(n_b/n)^{-\frac{1}{2}}$, where n_b is the basal number detected and n is the total number detected. This is subject to the same problems as their basal radius (see above) with an added error introduced in the conversion. J. T. Emlen (1971) estimated $1/k$ (his "coeffecient of detectability") by using a maximum count in bands close to the observer to estimate the number that there should have been in a wider band; the fraction of this estimated number that was actually counted gave his coefficient. Nilsson (1974a) attempted to get an independent estimate of the "effective radius of the birds" of Yapp (1956) using Brewer's (1972) method of estimating the "distance at which the number of near birds missed is equal to the number of far birds observed." However, this method too assumes that all the birds are observed in a band close to the observer and is therefore Emlen's technique in another form, so the close agreement between the two methods is no confirmation of either's accuracy. As both techniques depend on a basal density, they have similar problems to Ramsey and Scott's (1979) technique.

Järvinen and Väisänen (1975) described methods of estimating r in which the observations were simply classified into those within a main belt and the remainder. They used three models for the impairment of detectability with distance—exponential, linear, and half-normal—and found that their estimate of k differed little between the models, except when far records were less than 50% of the total. Most of their species had 60–90% far records (outside the 25 m inner belt they used) and for these their linear model estimator $r = w/(1 - f^{\frac{1}{2}})$, where w is the width of the near belt and f is the proportion of records that were far, is a reasonable approximation. There should be no great problem in applying similar reasoning to point counts.

The fourth method is to use the distribution of detection distances to give an estimate of density close to transect lines. Provided the lines are placed so as to sample the area randomly, this density will be representative. Burnham et al. (1980) give an excellent review of such methods.

The methods that use estimates of detection distances share several problems (see also Ramsey et al. 1979 and Burham et al. 1980).

1. Not all birds may be detected, even in the area close to the observer. My experience of counting birds in New Zealand native evergreen forests has provided plenty of anecdotal evidence that birds may easily be missed even when overhead. Emlen (1977a) tried to overcome this problem for breeding birds by estimating "cue frequency" of singing males in a similar way to the "efficiency" (q) of a mapping census, a technique that also has problems (see the first section of this paper). No one has suggested what to do about the near birds missed outside the breeding season.

2. Estimates of distance may be imprecise and biased, especially when they are based on a distant sound. I have found no published study of this fundamental problem, but Ramsey and Scott (1979), in estimating k for each observer, admitted its importance. Table 2 examines the

TABLE 2
THE DIFFERENCES BETWEEN FOUR OBSERVERS' ESTIMATION OF DISTANCE, AS SHOWN BY THE PERCENTAGE OF "FAR" RECORDS[a]

	Observer							
	A		B		C		D	
Species	Total	% far	Total	% far	Total	% far	Total	% far
Goldfinch	263	60	294	64	271	22	171	28
Blackbird	46	74	28	29	37	43	31	45
Myna	147	86	175	85	133	68	109	56
Skylark	82	72	60	77	74	35	63	40
All four	538	70	557	70	515	37	374	40

[a] From 96 five-minute counts by each observer in the study described in Table 1. There was a highly significant difference in the percentage of "far" records between the four observers for all four bird species ($P < 0.01$ in chi-squared tests). Small samples could account for some of the variation in the percentages (especially the low value for the blackbirds by observer B) but clearly the most important factor is a difference between the four observers' perception of distance.

difference between four New Zealand observers' estimation of the proportion of birds beyond 50 m, and suggests that the problem is significant.

3. The birds may react to the observer in such a way as to either increase or decrease the numbers detected close by, and so violate the assumption implicit in all techniques, that the observer has no influence on the probability of detection. Some published detection curves are suggestive of a movement away from the observer (e.g., Nuthatch, *Sitta europaea,* Nilsson 1974a, table I plot II; Wren, *Troglodytes troglodytes,* Hope Jones 1974), and a small amount of movement will not be obvious, but will still bias the estimate.

4. The birds may move into or out of the observer's range through their natural activity; none of the models allows for the movement of birds.

5. Estimates of r will be needed for all combinations of factors that I have shown elsewhere in this symposium to influence k, otherwise these factors must be held constant or a very robust estimator used (Burnham et al. 1980). Ramsey and Scott (1979) allow for k to vary with observer, habitat, and species, and standardized season, time of day, weather, and noise (J. M. Scott pers. comm.).

Attempts have been made to validate some estimates of k through independent estimation of densities. Emlen (1977a) compared densities obtained by his cue-frequency method with the mapping-method densities, but in fact his cue frequencies came from those same mapped territories and so were not independent. Järvinen et al. (1978a) in Lapland, and Järvinen et al. (1978b) in Poland compared the Finnish line transect with mapping, and Franzreb (1976) compared J. T. Emlen's (1971) technique with mapping. All three comparisons showed that the two techniques correlated well, with the transects usually giving slightly lower "densities" than mapping, but both approaches have their problems. Suffice it to say here that Järvinen and Väisänen's (1975) simple linear model may be as accurate as any, given the many possible sources of error. If one could be confident the errors were small, Burnham et al. (1980) offer a good range of transect techniques to choose from and a modification of Ramsey and Scott's (1979) point count method.

THE CHOICE OF TECHNIQUE

Territory mapping is normally suitable only for counting the stationary part of noncolonial passerine bird populations during the breeding season (Anonymous 1969), but transect or point counts are suitable for a wider range of species and seasons. If the problem being investigated requires estimates of density, none of the techniques reviewed in this paper can give an accurate answer for most species. More work is needed comparing these simple techniques with good estimates of density acquired from intensive study of marked populations, as most of the work to date has compared one imperfect measure with another. In the meantime, estimates of "density"—whether acquired from territory maps or from conversion of index counts—must be considered as subject to large and unknown errors. Many studies have assumed to the contrary, and without evidence, that territory mapping provides a good estimate of density (DesGranges 1980, Głowaciński and Weiner 1977, Walankiewicz 1977, Williamson 1964). Transects may be preferable to point counts when estimating density, as errors in calculating the effective area sampled are less, and a species list is acquired more quickly. Nevertheless, point counts are preferable in difficult terrain.

If the question being asked requires only an index of density, the choice depends on which technique has the lowest variability (Dawson and Bull 1975, Enemar et al. 1978). I have already concluded above that point and transect counts are about equally good on these grounds. Enemar et al. (1978) compared an index of abundance based on territory mapping with the total number of contacts achieved in the same study, and concluded that the two approaches were equally good. However, a territory-mapping survey is probably not the most efficient way of obtaining contacts. For example, in 8–10 two-hour visits to a study area using the point-counting technique of Dawson and Bull (1975), Dawson et al. (1978) acquired over 100 contacts for five species, whereas Enemar et al. (1978) averaged over 100 for only one species. If the point counts fit a Poisson distribution and are subject to a square root transformation to bring the variance to approximately 0.25, the mean count of four observers will have a standard error of about 0.025, or 1–3% of the mean for common species, but four mapping censuses would give standard errors at least 6–20% of the mean (Enemar et al. 1978, Snow 1965). Thus, point and transect counts seem to give a more cost-effective index of density than does territory mapping; it is a pity that there has been no direct comparison of the techniques to confirm this conclusion.

ACKNOWLEDGMENTS

A. E. Aldridge, P. C. Bull, M. Harrison, E. B. Spurr, and R. H. Taylor made constructive comments on the manuscript.

Studies in Avian Biology No. 6:559–565, 1981.

THE UNDEREXPLOITED POTENTIAL OF BIRD CENSUSES IN INSULAR ECOLOGY

YRJÖ HAILA[1] AND OLLI JÄRVINEN[1,2]

ABSTRACT.—Studies of insular ecology usually focus on the species composition of island communities, but using census data can lead to many novel insights and ways to look at island communities. One-visit censuses, with the possibility of covering a much larger number of islands per unit effort, may often be superior to more accurate but time consuming methods. Using empirical data from land bird communities in the Åland archipelago, we discuss the number of species on islands, species-abundnace distributions, and colonizing strategies, emphasizing methodological aspects.

Both empirical tests and several theoretical arguments support the conclusion that 80% or more of the species breeding on an island are observed in a one-visit census. The species missed are many of the rarities, but one-visit censuses often include non-territorial visitors. One-visit censuses can also be used for examining the species-abundnace distribution, which may give rise to interesting biological hypotheses. Quantitative data allow us to construct density-based *prevalence functions*, introduced here, instead of examining incidence functions based on presence-absence data. Prevalence functions and their annual variability lead to meaningful biological hypotheses about insular communities.

The paradigm of island biogeography (MacArthur and Wilson 1967) has attracted ecologists to study the species composition of insular communities; as a result, species-area curves, estimates of species turnover, and incidence functions have been derived (e.g., Diamond 1975a, Diamond and May 1976, Wilcox 1980), but substantial controversy has emerged as regards the biological interpretation and even reality of many patterns (e.g., Simberloff 1976, 1978b; Connor and McCoy 1979; Connor and Simberloff 1979; Gilbert 1980).

The basic problem in insular ecology is to understand the relative importance of different factors structuring island communities. There is no compelling reason why examining species lists should be the only, or even the dominant, method in studying island bird communities, for many relevant tests require quantitative data. Indeed, we argue here that censuses of island birds provide a remarkable potential that has not been exploited by more than a handful of ornithologists, such as: Blondel (1979) and Ferry et al. (1976) studying Corsica; Emlen (1977b) studying the Bahamas; Ricklefs and Cox (1978) studying taxon cycles in the West Indies; and Nilsson (1977a) studying bird communities on small islands in a Swedish lake.

Our emphasis here will be on methodological aspects rather than on final results, although it seems impossible to us to discuss methods without reference to the particular problems studied. As will be evident from our discussion, one-visit censuses are useful in many practical situations. This is surprising and may seem outrageous at first sight, but we argue that restricting the attention of insular ecologists to species lists makes it impossible to understand many real and interesting patterns in insular communities; it is therefore not a basic requirement that the census method used should be able to produce a complete species list. Our empirical data come primarily from a study in 1975–80 of land bird communities in the Åland archipelago, SW Finland; for additional data, see Haila et al. (1979, 1980a), Haila and Järvinen (1980) and Järvinen and Väisänen (1980).

THE NUMBER OF SPECIES ON ISLANDS

To generate a complete list of bird species breeding on an island certainly requires a long period of intensive surveying. A mapping census based on 8–12 visits may be a good substitute, even if the data do not give positive evidence of breeding. But censuses based on one or a few visits can also be useful, athough the results are less accurate than mapping or long-period surveys. Both direct tests and theoretical arguments support this claim.

DIRECT EMPIRICAL TESTS

We have studied land bird densities on more than 50 islands in the Åland archipelago in one or several breeding seasons using primarily censuses based on one visit. As one of the basic characteristics of insular communities is the number of species present, several tests have been conducted to find out the degree of completeness of our species lists.

The island Gåsholmen (about 14 ha) was censused in 1980, and the results were compared with a mapping census of the same island taken in the same year (Y. Haila and S. Kuusela, unpubl. data). The results (Table 1) can be summarized as follows. Of the 19 species judged to be territorial on the basis of mapping, 17 (89%)

[1] Department of Zoology, University of Helsinki, P. Rautatiekatu 13, SF-00100 Helsinki 10, Finland. Permanent address of both authors.

[2] Department of Biological Science, Florida State University, Tallahassee, Florida 32306.

TABLE 1
RESULTS OF MAPPING COMPARED WITH RESULTS OF A ONE-VISIT CENSUS ON GÅSHOLMEN (14 HA) IN 1980

Species	Mapping (territories)	One-visit (pairs)
Wryneck (*Jynx torquilla*)	1	1
White Wagtail (*Motacilla alba*)	3	2
Robin (*Erithacus rubecula*)	3	2
Thrush Nightingale (*Luscinia luscinia*)	1	1
Wheater (*Oenanthe oenanthe*)	2	1
Blackbird (*Turdus merula*)	3	1
Redwing (*T. iliacus*)	4	3
Icterine Warbler (*Hippolais icterina*)	—	1
Lesser Whitethroat (*Sylvia curruca*)	2	1
Whitethroat (*S. communis*)	1	1
Garden Warbler (*S. borin*)	1	1
Willow Warbler (*Phylloscopus trochilus*)	3	4
Goldcrest (*Regulus regulus*)	1	—
Spotted Flycatcher (*Muscicapa striata*)	2	2
Willow Tit (*Parus montanus*)	3	—
Blue Tit (*P. caruleus*)	1	1
Great Tit (*P. major*)	—	1
Red-backed Shrike (*Lanius collurio*)	—	1
Hooded Crow (*Corvus corone*)	—	1[a]
Starling (*Sturnus vulgaris*)	3	2
Chaffinch (*Fringilla coelebs*)	20	15
Greenfinch (*Carduelis chloris*)	1	1
Yellowhammer (*Emberiza citrinella*)	1	1
Total	56	39 + 5 non-territorial

[a] Observed to be a visitor from a neighboring island.

were observed in the one-visit census; 70% of the territorial males were observed in the single test census. In addition, the one-visit census revealed five non-territorial birds, including four species not accepted as territorial in the mapping. The non-territorial birds were presumably visitors from nearby islands lying no farther than about 100 m away (this was directly confirmed for one non-territorial species).

In another test (Y. Haila and S. Kuusela, unpubl. data.), Bockholmen, an island of 38 ha, was mapped and the results compared with a line transect of 800 m running from one end of the island to the other. Mapping revealed 33 territorial species, and the transect, covering not more than 40% of the island, included 26 of them. In addition, four of the seven missing species were observed during the census, but outside the transect belt. The one-visit census thus revealed 91% of the territorial species.

Finally, we report census results from two larger islands. Ulversö (almost 6 km^2) was surveyed during a period extending from late May to early July 1976, and we could thus establish the list of territorial land bird species with high accuracy; the species were breeders or sometimes probably single males. A transect census covering about 20% of the island revealed 53 species, or 84% of the total of 63 species. Slightly better results were obtained in 1979, but the total list may have been incomplete owing to a less efficient survey in that year. Transect censuses were also made on the "mainland" of Åland, which is an island of 970 km^2. A mere 214 km of transects, covering about 5% of the area, included 102 species, or 85% of the 120 species present (Haila et al. 1979).

Of course, the species missed in one-visit censuses are not a random sample from the actual community, but are usually species having very small populations. In the tests reported here (excluding those made on Ulversö and Åland, for which the relevant data are lacking), no species with four or more pairs was missed. Similarly, the results of one-visit censuses generally include visitors from surrounding areas, but they also are not a random subset of the species. In our tests, only one of the visitors had more than a single pair in the census.

THEORY

While empirical tests are necessary, they cannot be repeated on every island. We observed empirically that typical one-visit censuses capture 85–90% of the territorial species on islands of different sizes, but there are theoretical reasons for believing that this result is fairly general. Assume that: (1) Island bird communities have species-abundance distributions similar to those in mainland communities; (2) observation efficiency is similar on islands and on the mainland, usually 45–75% of the territorial pairs being observed on one visit (for data, see the compilation of Järvinen 1978b); and (3) a bird census can be approximated as a multinomial process (see Järvinen and Lokki 1978).

Simulations based on these assumptions and actual bird census data then show (Järvinen and Sammalisto 1973, Järvinen and Lokki 1978) that 80–100% of the species will be observed on a single visit. In general, the most convenient approach here would be rarefaction (e.g., Heck et al. 1975, Simberloff 1979), because that method gives directly the expected numbers of species in random samples from the actual community.

As density compensation is said to occur in many island bird communities, it is not clear how realistic our assumption (1) is, and it is also doubtful whether multinomiality can be regarded as a sufficiently realistic postulate (Järvinen and Lokki 1978, Kouki and Järvinen 1980).

The above assumptions can also be used to illustrate why the errors in species lists should indeed occur among the rare species. In consequence, what is gained by increasing the numbers of visits to an island is improved accuracy in recording the rare species in the island community.

An entirely different approach is to assume, following Preston (1962; see also May 1975), that assumption (1) is: the species-abundance distribution of insular communities is lognormal (for data, see below).

Without *any* assumptions on census efficiency, the effects of the incompleteness of the census on the species list can be analyzed, deriving maximum-likelihood estimates for the proportion of missing species on the basis of the theory of truncated lognormal distribution (Cohen 1961, Pielou 1975). In our censuses in 1976, 16 islands were surveyed completely. Data (numbers of pairs in each species observed in the one-visit censuses) for five larger ones among them gave the results in Table 2. For smaller islands, the estimates were usually smaller, but we doubt the validity of assumption (1) in these cases. The examination of our census data indicated that the assumption was valid in the above cases, although the sample sizes were not large enough to reveal other than gross deviations.

Finally, we may examine the theory of random sampling from communities having a specified species-abundance distribution (May 1975: 105–106, and related appendices) or the results of rarefaction as applied to bird census data (Engstrom 1981, James and Rathbun MS). Without going into details, all results indicate that a random sample of 50% from the community gives most of the species, and very rarely will more than 20% of the species be missing.

TABLE 2
NUMBER OF SPECIES OBSERVED ON FIVE ISLANDS WITH AN ESTIMATE OF PROPORTION OF PAIRS NOT OBSERVED

Island	Species	Maximum-likelihood estimate of proportion not observed
Gåsholmen (14 ha)	19	15%
Foderholmen (7 ha)	17	15%
Klobban (8 ha)	6	9%
Äskholm (3 ha)	8	4%
Börkholm (5 ha)	10	4%

We notice the following implications for island ecology:

(1) Omitting 10–20% of the species has little effect on species-area curves. In particular, if the best fit is a power function ($S = CA^z$, where S = number of species, A = area and C and z are fitted constants), omitting 10–20% of the species only depresses C by the same percentage and somewhat increases the error variance. Of course, if the percentage of species missed differs on different-sized islands, the problem is more serious.

(2) Species turnover is heavily affected by the immigrations and extinctions of small populations (Jones and Diamond 1976, Järvinen, In press). One-visit censuses cannot indicate anything but striking differences in species turnover because of numerous cases of "pseudo-turnover" (Lynch and Johnson 1974). Instead, quantifying numerical changes in all populations, called individual turnover by Järvinen (1978c), is possible. The turnover of species is just one aspect of the dynamic behavior of insular communities, and there is no reason why serious attempts should not be made to understand and examine patterns of quantitative population changes on islands.

(3) One-visit censuses do not give exact presence-absence lists of species, and therefore it is impossible to examine whether a certain species combination really does not occur in the archipelago or whether it is just missing owing to inadequate sampling. Again, we emphasize that studying quantitative patterns in the coexistence of possible competing species could often be a much more fruitful strategy than examining species lists. Besides, it is not clear that missing

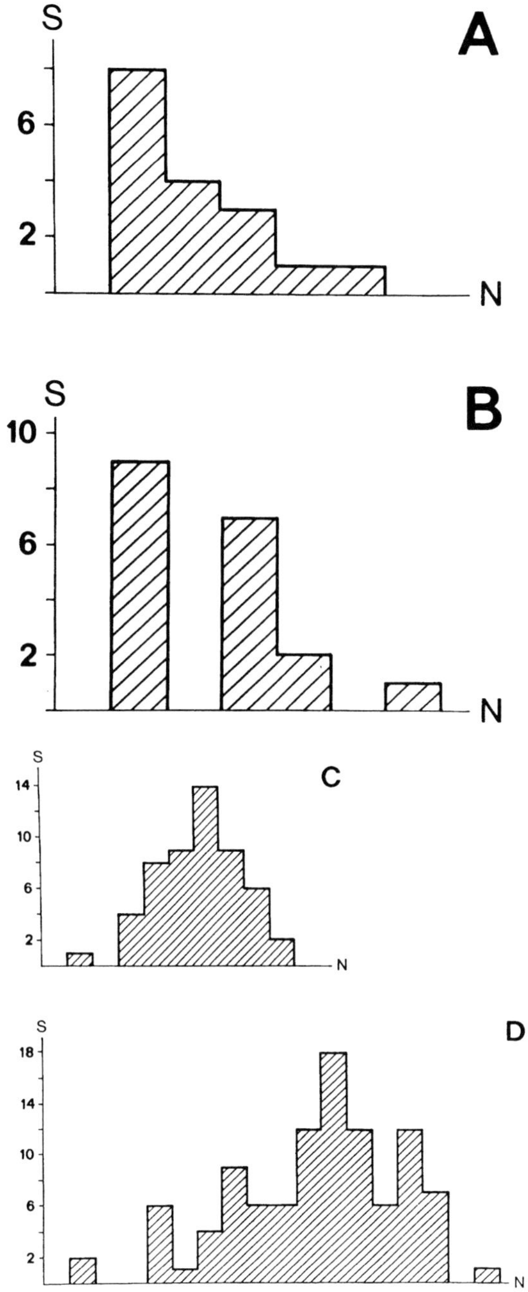

FIGURE 1. Species-abundance distributions on four islands in the Åland archipelago: A. Foderholmen (7 ha), B. Gåsholmen (14 ha), C. Ulversö (5.8 km²), and D. the mainland of Åland (970 km²). The number of pairs (N) is expressed on a logarithmic scale (base 2). D based on data from Haila et al. (1979); A–C our unpublished data.

combinations are particularly indicative of interspecific interactions.

(4) An important property of one-visit censuses is that they do not require an inordinate amount of time. While we readily acknowledge that absolutely accurate species lists are essential in studying certain problems of insular ecology and that more accurate census data are necessary for the solution of other problems, standardized one-visit censuses allow, for example, studying larger numbers of islands or repeating the census work more easily in several successive seasons in order to examine annual quantitative variability simultaneously in many insular communities. Of course, census work should be standardized as carefully as possible (for a review, see Berthold 1976). The important point is that the scope of questions can be broadened significantly if quantitative bird censuses are applied to insular ecology; the dynamic equilibrium model of MacArthur and Wilson (1967), whatever its merits and demerits, certainly does not exhaust all interesting island patterns.

Next we discuss the applicability of quantitative data to two central problems of insular ecology: species-abundance distributions and colonizing strategies.

SPECIES-ABUNDANCE DISTRIBUTIONS

It is well-known (Preston 1962, MacArthur and Wilson 1967, May 1975) that if the species-abundance distributions on different islands are lognormal, the relationship between area and the number of species can be well approximated by a power function, assuming that densities do not vary with island size. What has usually been done is that the available data on the numbers of species and area have been fitted with a power function, without even examining other alternatives (see Connor and McCoy 1979), and the resulting fit has then been interpreted in terms of the prevailing paradigm. This is, however, unjustified (Connor and McCoy 1979, Gilbert 1980).

One important role for censuses is obviously that they can provide solid data for testing the basic premise of lognormality (Preston 1962). As full data will be published elsewhere, only examples (Fig. 1) are given here. As visual inspection indicates, the small islands usually have species-abundance distributions resembling the log-series model, but the distribution approaches lognormality as island size increases. Notice, however, that sample sizes per island are too small to discriminate effectively between various alternatives. We do not discuss the smallest islands here, but make a brief comment on the fact that the species-abundance distributions of the larger islands seem to be lognormal except that they tend to be skewed to the left. This is interesting, because skewness constitutes a deviation from lognormality, which, according to May (1975), might be expected to be the rule. We give one possible explanation here, in order to show that quantitative data may broaden the scope of insular ecology. However, the following should not be interpreted as a final statement, because we have not given a rigorous statistical demonstration of our premise that skewness to the left is typical.

Suppose that skewness to the left is validly established as a typical pattern in the archipelago we have studied. In other words, there is a somewhat higher number of very small populations than May's (1975) statistical and probabilistic arguments would lead us to expect. An obvious hypothesis would be that the islands we have studied are actually not isolated from the standpoint of bird dispersal, but the birds use the whole archipelago more or less as a unit. Therefore, many of the islands may support very small populations that are not themselves self-sustaining but are a portion of a larger self-sustaining population in the archipelago and the nearby mainland. This idea can be tested on the basis of census data, if they are available. For example, if the whole archipelago is a unit, with no problems of dispersal between the islands, population increases on the islands occupied may be expected to lead to colonizations of new islands, and vice versa. This idea can also be applied to the Åland archipelago in its entirety, for it seems to be part of the mainland for many North European bird populations (Haila et al. 1979). An obvious alternative hypothesis, not supported by the data (Haila et al. 1979), would naturally be that colonizations and extinctions within the archipelago are not correlated with population dynamics elsewhere.

We wish to examine a methodological point here, contrasting mapping censuses with one-visit studies. As species-abundance distributions tend to be depicted by using the logarithms of population size, even considerable errors in estimating population size do not greatly distort the distribution. This conclusion can, of course, be tested; and several experiments comparing mapping with one-visit censuses have shown quite similar species-abundance distributions (e.g., Järvinen et al. 1978a, 1978b; see also Table 1).

Species-abundance distributions can lead to a number of relevant insights. For example, the positive relation between the number of species and island area may be due to three different causes: habitat diversity, area, and sampling (i.e., if there are more individuals, there should be more species). It is difficult to test these alternatives critically (e.g., Connor and McCoy 1979), but the species-abundance distributions observed in censuses immediately suggest one test. Distribution-free rarefaction (Simberloff 1979; James and Rathbun MS; Engstrom and James, In press; Engstrom 1981) can be used to study species richness in samples standardized to an equal size. This makes it possible to test the null hypothesis that all differences in the numbers of species on different islands are merely a result of differing numbers of individuals, i.e., the sampling effect.

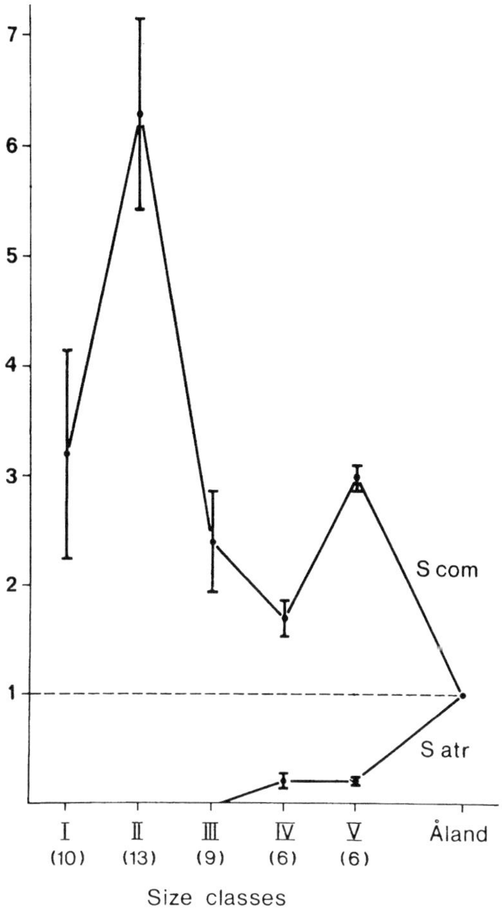

FIGURE 2. Prevalence functions for the Whitethroat, *Sylvia communis* (S com), and the Blackcap, *S. atricapilla* (S atr), in the Åland archipelago in 1976–79. Island classes were defined on the basis of the numbers of species (the group boundaries are 1–5, 6–10, 11–20, 21–30, and 31–60 spp.). Prevalence estimates, calculated for the pooled data of all islands in the same class, are shown with approximate estimates of SD (see Järvinen 1976). For details, see text.

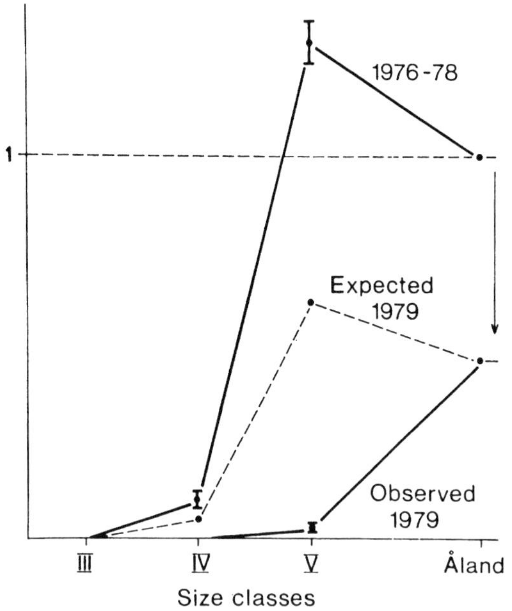

FIGURE 3. The prevalence function for the Fieldfare (*Turdus pilaris*) in the Åland archipelago in 1976–78 and 1979 (expected and observed). The arrow on the right shows the population decrease observed on the mainland of Åland. See text for explanation.

COLONIZING STRATEGIES: PREVALENCE FUNCTIONS

In his study of colonizing strategies of land birds, Diamond (1975a) constructed "incidence functions" on the basis of presence-absence data. An incidence function gives the frequency at which a certain species occurs on islands of different size classes in a certain archipelago. Incidence functions are necessarily fairly inaccurate because it is often impossible to study a sufficient number of islands in order to reduce the variance of the frequency estimates.

Census data can yield more information: while presence-absence data give only one record per island, quantitative censuses are able to indicate differences in population density among islands, and the number of records involved in such comparisons may often be great. The following new method, which we call the *prevalence function,* can be applied to analyzing census data. Prevalence functions are based on the average densities of a species on islands of different size classes. The average densities are compared with densities on the mainland (or an obvious species pool). If prevalence is 1, the species is equally abundant on the islands and on the mainland, and deviations from unity can be statistically tested.

In our study area the mainland of Åland is the obvious species pool. Comparing the average densities of a species on islands of different sizes with the density on Åland gives us an indication whether the species favors or avoids small islands—or, more accurately, whether it is comparatively numerous or scarce on small islands. Figure 2 gives an example comparing two European warblers, the Whitethroat (*Sylvia communis*) and the Blackcap (*S. atricapilla*). It is obvious that the former has comparably high densities on small and intermediate islands, while the latter is never found on the smallest islands and only rarely on the larger ones.

It is apparent that prevalence functions allow us to define different colonizing strategies; there may be species favouring small islands comprising patchy habitats, and there may also be species requiring large islands with continuous extensive tracts of habitat. The main point, however, is a deeper one.

An obvious hypothesis generally neglected in insular ecology is that the distribution pattern of a species in an archipelago is mainly determined by the availability of suitable habitat (Abbott 1980). It is feasible to refine prevalence functions if quantitative census data are available from different habitats (Y. Haila, O. Järvinen and S. Kuusela, in prep.). Our approach has been as follows. Nine broadly defined habitats were censused on the mainland of Åland (Haila et al. 1980a). As similar habitats characterize the archipelago studied by us in 1976–80, we can calculate the expected densities on the basis of the coverage of different habitats. In other words, prevalence functions can be so adjusted that gross habitat differences between the islands and the source area are eliminated.

For example, the two warblers in Figure 2 have an average density of 6.3 pairs/km^2 on the mainland of Åland (Haila et al. 1979). The high prevalence of Whitethroat on the islands is naturally interpreted as a consequence of the habitat composition of the islands (mosaic-like scrub is typical). Using the census data from different habitats on Åland, we would expect that the largest island studied separately by us, Ulversö, would support 25 pairs of Blackcap on the basis of the habitat composition of Ulversö. This is not the case, for we have observed one (1976, 1980) or two pairs (1979) there. Of course, the possibility of subtle habitat effects remains, because the match between insular and mainland habitats cannot be perfect and because our quantitative understanding of the autecological details of habitat selection is meager. In fact, the main reason eliminating Blackcap from most of the island forests seems to be a subtle habitat effect: the species favors tall luxuriant forests on the mainland of Åland, but the average height of the trees is less (10–12 m) on the islands than on the mainland (16–18 m) in parallel habitats. Our preliminary results indicate that in many cases an apparent specialization to small islands is merely a result of the wide coverage of a favored habitat on small islands, and vice versa.

Another example based on prevalence functions shows patterns in annual variability. We have censused certain transects in our source area in several years, and we have thus been able to follow annual fluctuations in populations. A marked decrease occurred in the numbers of the Fieldfare (*Turdus pilaris*) after the severe winter of 1978/79. Assuming that the decrease was proportional on the islands studied, we calculated an expected prevalence function for 1979 on the basis of our data from 1976–78. However, the losses suffered on the islands were (Fig. 3) disproportionately heavy, and the "colonizing strategy" of the species looks quite different before and after the population crash. We raise one possible hypothesis here, but do not pursue the point further: the observations are compatible with the hypothesis that in our study area islands are suboptimal for the Fieldfare compared with the source area, so that densities in our islands are a function of abundance on the mainland.

CONCLUSION

We conclude by stressing our take-home message. The theory of island biogeography has certainly led to an outburst of insular studies. In order to understand insular ecology, however, it is not necessary to restrict attention to qualitative presence-absence data. On the contrary, using quantitative census data can lead to many novel insights and ways to look at insular communities. The quantitative census methods should be chosen according to the actual needs of the biological problem studied, and here one-visit censuses, with a possibility to cover a much larger number of islands per unit effort, may often be superior to more accurate but time-consuming methods.

ACKNOWLEDGMENTS

This joint project was supported by grants from Alfred Kordelin Foundation, Oskal Öflund Foundation and the U.S. Educational Foundation. We are grateful to Larry Abele, Ilkka Hanski, Fran James, Jerry Niemi, Dan Simberloff, Kari Vepsäläinen, and John Wiens for many useful comments for the improvement of the manuscript. We also thank Seppo Kuusela for his help in the censuses.

Studies in Avian Biology No. 6:566–569, 1981.

STATISTICS FOR THE BIRDS

F. N. DAVID[1]

ABSTRACT.—There are a great many assumptions which go into bird counting. Most ornithologists feel that once they have acknowledged these assumptions, they are free to proceed with impunity in their analyses. The fallacy of this procedure is described, and comments made on idealized solutions to the problems.

In medieval times a favorite topic with theologians was estimating how many angels could stand on the head of a pin. It is with a certain relief that one notes the objective of this symposium is concerned with terrestrial matters, but in the writer's opinion, there is room for much more down-to-earthiness. For mathematical models and computer simulations are excellent in their proper place, but it should be pointed out that all models have to be based on data, and that no model is of any value if the data are not reliable. I propose therefore to look closely at data gathering as seen through the eyes of a statistician.

PURPOSE

Recent literature often refers to a census, so we should perhaps begin by asking: What is a census? Statistically, we mean the complete enumeration of all the objects under study. This is difficult enough with people, probably not possible with plants, almost certainly not possible with mammals, and especially not possible with birds—because of the three-dimensional effect introduced by the latter. So instead of thinking of a census, we think of estimating the number that we would get were we able to carry out the impossible and get a complete count. And as soon as estimation is the topic, it is necessary to decide: (1) which method of estimation to use; (2) what mathematical assumptions are involved in applying the method chosen (with the further thought that the mathematical assumptions must bear some relation to reality).

It is a commonplace among statisticians that before starting any investigation it is necessary to define, with as much precision as possible, the final objective. And so far as is possible, to state the causes, any one or all of which may cause a variation in the final result. A great deal is known about bird behavior, and any count has to consider the kind of bird, the time of day, the time of year, and the terrain, to mention only some of the factors known to have an influence. If the investigation is to determine only the number of kinds of bird, without requiring the numbers of each kind, then to a certain extent the emphasis of the enumeration procedure will be different. But for all investigations, the type of terrain is of importance.

MICRO-ENVIRONMENT

To reiterate, it is necessary to postulate the exact purpose of the investigation. A loosely worded statement such as, for example, "to find out the effect of spraying an area with insecticide on the bird population," has to be whittled down so that the kinds of bird are specified as well as the area. The area has then to be split up into roughly homogeneous sub-areas in much the same way that the counters of deer, for example, split up their basic areas. If one of these smaller areas should happen to be a field of grain or cotton, then the problem of specification of terrain is considerably easier than if it is a natural forest. This latter presents difficulties, so I will consider it chiefly in my further remarks.

The natural forest is not uniform but is built up of a number of micro-environments, in some of which it will be easy to see specified birds, and in others, it will be very difficult. If it is desired to make a bird count before and after some treatment, such as spraying, the makeup of the forest in terms of micro-environments may be of crucial importance. For the spraying may affect leaf cover, etc. and hence alter the basic conditions. Or again, if it is desired to compare two forested areas as far as bird counts are concerned, no reliable comparison can be made unless they are approximately the same in their micro composition. For, since the best we can do is some sort of sampling procedure, the basic conditions must be equivalent. This is true for all sampling procedures, of which there are many. For illustration let us consider a simple lattice.

Suppose a map of a forest. A random line is drawn across it and a random point is chosen in it. A line is drawn through the random point at right angles to the random line. Choose a distance l and draw a square lattice to cover the area studied with the lines a distance l apart. The crossover points of the lattice are commonly referred to as nodes, and these are the sampling points. At each node a description of the micro-environment in a circle of 30 yards (say) surrounding the point can be made. (The area

[1] Department of Statistics, University of California, Berkeley, California 94720, and USDA Forest Service, Pacific Southwest Forest & Range Experiment Station, Berkeley, California 94720.

need not necessarily be circular, provided the same size area is maintained for all nodes.) Data such as amount of ground cover, height and plane area of shrubs, dimensions of trees and their position, height of canopy and so on, may be recorded. This may all be done by eye, as in similar fashion is also the chance that any particular type of bird may be seen in such an environment. Clearly, some facets of the micro-environment will vary in importance, depending on the type of bird studied. Thus, for a bird of pedestrian habits, probably the ratio of shrub cover to the total area will be descriptive enough; while for another kind, the canopy will be of importance. Similar logic can be applied to *l*, the distance apart of the lines composing the lattice.

DETECTION

Statisticians may be pardoned if they treat with scepticism the counts of birds that ornithologists produce and that are the basis of their models. Obviously there are two main types of error—all the birds will not be counted, and some birds may be counted several times. Preston (1979) faces up to the first kind of error in his paper "The Invisible Birds," but the general tendency is to ignore the obvious flaws in the data and introduce a variety of mathematical assumptions that, it is hoped, will give validity to the conclusions drawn. The investigator remains forgetful of the truth that conclusions based on assumptions are dependent on those assumptions. (The next step in bird censuses should undoubtedly be to introduce a credibility function.) The probability of seeing a bird is dependent on the type of micro-environment, discussed in the previous paragraphs. In open grassland, possibly all the birds present will be seen. In some of the micro-environments in a forest, it will be easier to see some kinds of birds than others, so the chance of observing a bird in a given area will depend on its kind and on the composition of the environmental area. What is certain is that the count will not be a comprehensive one.

The problem of counting the same bird several times is usually dismissed in research papers (if it is mentioned at all) with the remark that care must be taken to avoid duplication of observations, etc.; yet it can be a significant source of error. For instance, the use of bird song or calls to identify the presence of a particular kind of bird is useful in that it indicates that at least one individual of that kind is in earshot. But no number of songs can definitely indicate more than that, unless they are accompanied by visual sightings.

INTRUSION OF THE OBSERVER

There have been enough recent references to Yapp's (1956) classic paper to send one back for a rereading. He wished to count Rooks (*Corvus frugilegus*), so he rode the side of a railway engine from Leamington Spa to a point south of Bicester—say about 20 miles—counting the birds within 50 feet of the line. The railway line passes through a variety of micro-environments even if we exclude, as he does, the tunnels and the cuttings. In the end he multiplied his count by two, because he could only see one side of the track, to arrive at the number of Rooks per square mile. But leaving all this aside, it is the effect of human intrusion which is unknown, but undoubtedly present. Do the birds all remain when they hear the train approaching? Are the birds attracted to the line instead of being dispersed over a larger area? These are only two of the many questions which have to be answered before giving a figure of number of Rooks per square mile.

Again, recent research papers have given bird counts obtained by driving a car and stopping at intervals for observations. Some birds like roads, as is instanced by the number of kestrels (*Falco* sp.) along many highways. Other birds will avoid roads because of the human intrusion. The counts are indicative that at least one bird of the kinds seen was present. But how representative the actual numbers counted are of the true numbers of birds in the vicinity, and whether all the different kinds of birds were seen, is a moot point.

The same remarks can be made regarding transect sampling. Apart from the fact that an observer moving through a forest will pass through a number of different micro-environments, his movement will make a noise. The count will therefore be indicative of the reaction of the birds to a human intrusion, rather than a partial unbiased estimate of the number of birds. Using the sampling lattice described earlier may offer the minimal human intrusion, although it will not eliminate it entirely. If there are twelve nodes then one would ideally require twelve observers. If the count is to be made at 10:00 the observers must of necessity be in place at a time sufficiently far beforehand for them to be accepted as part of the landscape at count time.

The tricky problem of counting the same bird twice is caused partly by the bird's natural three-dimensional activity, but partly also by the intrusion of the observer. Accordingly, with the lattice scheme it is important that the time of counting is restricted—say to five minutes—and ideally that all observers start at the same time. There seems to be no objection to there being

3 to 4 five-minute counts within 30 or 40 minutes, provided the counts are kept separate, since they will not be independent. The dimensions of the lattice will need to be such that it is unlikely that a bird startled by one observer can fly to within sight of another during the same five minute period.

MECHANICAL COUNTING

There is an agreement among ornithologists that some of them see, and therefore count, birds better than others, in spite of a frequent assumption that all the birds are counted. Moreover, they present a united front against any suggestion that possibly in this mechanical day and age, a mechanical means of counting would lead to greater consistency in enumeration; for cutting out the observer error would mean cutting out a source of variation. Thus, if we revert to the idea of a sampling lattice, when the region is surveyed to mark out the nodes and to obtain an idea of the micro-environment in an area around each node, the area itself could be marked out, and some form of mechanical counters operated by remote control installed. The remote control would rule out another source of variation, the human intrusion. Accordingly, it is suggested that experiments directed towards the mechanization of the counting process may be fruitful in producing, what should be the basis of the investigation, an unbiased count. If the chief deterrent to mechanization is not prejudice, but expense, one may ask whether it is preferable to obtain a few accurate counts or a plethora of inaccurate ones.

VARIATION

If the statement is accepted that time of year is a source of variation for the count, then the observer has to decide when this time shall be, bearing in mind that any conclusions drawn will be valid for the chosen period only. If it is accepted that the time of day is important, then this day-time interval should be narrowly postulated with reference to what is known of the bird's activity habits, for it will possibly not be so easy to see a bird in its inactive period as opposed to it active. There are also other possible sources of variation such as temperature and precipitation. A list of these has been given by W. M. Shields (1979). Once it has been determined over what period in the year and over what time of day the background conditions relating to activity can be considered to be more or less consistent, it is desirable that observations be taken on a number of successive days. The variability of temperature and the intervention of precipitation may be used appropriately to divide the series of days, thus providing further information.

MODEL BUILDING

It is not my purpose to introduce yet another mathematical model. Yet it is perhaps interesting to note gaps in estimation and difficulties still to be overcome, even with the simple lattice sampling design, so I will illustrate this for just one node. The extension to more than one node and to different micro-environments is straightforward.

Suppose there to be N birds in an area A. It will be assumed that throughout the period of observation there will be N birds. These will not necessarily be the same—some will fly out, others will fly in—but the total number is assumed constant. This number N is not known and we wish to estimate it.

Various assumptions can be made concerning how these N birds are dispersed over the area A. They may be distributed randomly, they may tend to go about in clusters, or they may tend to avoid each other. Whatever we assume will make very little difference to the estimate, N, of N, although it may alter the variance.

Consider an area, a, surrounding a given node. Suppose that an observation count is made in a for each of s successive days, and that these counts are $r_1, r_2, \ldots, r_s$. Some of these may be zero but not necessarily so. Write for the average count ($\bar{r}$) and the proportion of the total area sampled (P):

$$\bar{r} = \frac{1}{s}\sum_{i=1}^{s} r_i \qquad P = a/A$$

then

$$E(\bar{r}) = NPp$$

Where E stands for the mean value in repeated sampling, and p is the chance that a bird is observed in a.

It is worthwhile to point out that this p is not the chance of detection, which we are told increases with the number of birds in a. Instead, p depends on the micro-environment of a. Thus, for example, if a is a square, half covered with undergrowth and shrubs, then the chance of observing a pedestrian bird is one half, since the possible visibility is one half.

There are many ways of estimating N. The simplest is, possibly, to say:

$$N = \bar{r}/Pp$$

Because a is small compared with A, and therefore P is very small, the effective result will be that r_i, ($i = 1, 2, \ldots, s$) has a Poisson distribution whatever we assume for the dispersion of the N birds over A.

It will be noticed that p is important in the estimation of N. If it is routinely put equal to unity, then there will be underestimation of N in a number of cases. It would seem worthwhile to consider whether accurate estimation of p is possible from field observation, or whether we should try to include such estimation in the sampling scheme.

CONCLUSIONS

It is not the writer's desire to state flatly that all bird census data are invalid. However, merely acknowledging assumptions does nothing to pardon the researcher from the responsibility of eliminating the biases introduced by those assumptions.

Studies in Avian Biology No. 6:570–574, 1981.

ADAPTING GENERALIZED INSTRUCTIONS TO SPECIFIC SITUATIONS IN PLANNING COUNT WORK

A. J. ERSKINE[1]

ABSTRACT.—The decision to carry out a bird count leads to many other decisions, many of which lie well outside the interest or expertise of the would-be counter. The intensity of effort required scares off many potential candidates, but the amount required is debated even by experts. The need for precisely defined and measured areas and for detailed habitat descriptions are further stumbling blocks, upon which specialists disagree. Even the methods to be used are open to debate, depending on the purposes to which they are to be put. Many criticisms levelled at bird count work stem from misconceptions of objective, scale, or perspective, and no single set of guidelines will satisfy all needs.

This conference is about counting birds. Counting obviously means different things to different people, but the common denominator is "a systematic effort to count birds," for any of a wide variety of reasons. Within that general definition, bird-counting exercises may be as wide ranging as the cooperative Breeding Bird Survey (Bystrak 1981) or as confined as a study of a single plot, as rigidly standardized as the "Mettnau-Reit-Illmitz" migration monitoring program (Berthold and Schlenker 1975), or as unstructured as the Christmas Bird Counts (Arbib 1981). These diverse understandings of counting reflect an equally wide array of reasons for counting birds, among which may be mentioned the study of ecological relationships, the monitoring of trends relative to ongoing environmental changes, measuring impacts of land-use or pollution, management of directly exploited birds or pest species, or recreation. The last is by no means the least, since most of us who work in ornithology do so because we enjoy it, whether or not we make a living from it.

What sets counters apart from many bird watchers is their attempt to count birds systematically. All systems impose constraints, which in their turn call for decisions. Many of the decisions required of counters seem to have little to do with birds or our interest in them, since they revolve around the objectives, the methods, the results, or the interpretation of them. One of the most basic decisions is whether to pursue absolute numbers or densities or merely to obtain relative indices to populations. No satisfactory decision can be reached on this point until one has considered the resources of time, personnel, and equipment available to the counter, as well as the objectives in undertaking the census. Bird counting has come a long way in the 35 years since Kendeigh's (1944) review, but most counters still start out virtually in isolation, making their own decisions—and mistakes—the first time around.

[1] Canadian Wildlife Service, P.O. Box 1590, Sackville, New Brunswick E0A 3C0, Canada.

The purpose of this presentation is to explore some of the basic questions and constraints facing counters, and to outline what I believe to be the best types of decisions in dealing with them. Obviously this will be a biased viewpoint; I have never used transect counts systematically, and it will be obvious that I see mapping censuses of measured plots (Williams 1936, Enemar 1959) as fundamental to many censusing efforts. Furthermore, my remarks are focussed on censusing more or less sedentary populations, and not all my generalizations will fit counts of migrating birds.

BASIC QUESTIONS IN COUNTING

I referred already to the basic division of counting into absolute vs. relative counts. Except with a few very scarce and/or localized species, such as Kirtland's Warbler (*Dendroica kirtlandi*) and North Atlantic Gannet (*Morus bassanus*), whose entire breeding populations have been counted more or less directly, all censusing involves the counting of samples from a population. In absolute census counts, the samples are drawn from a defined area of (supposedly) known extent; relative counts are assumed to have sampled the same area each time, without the size of the area necessarily being known. If one method gives truly absolute results, these *should* be comparable with results from other absolute methods; unfortunately, the truism that "all things are relative" applies also to "absolute" census methods, to a greater or lesser extent. Results obtained by relative methods, however, can only be compared with others obtained by precisely the same methods and usually by the same observer. Migration counts in particular describe only the situation at one moment in time, so are always relative. Thus, one of the basic constraints here is comparability. One must decide if one's results need readily to be compared with those of other people, or if comparisons are to be made only among one's own samples.

Another basic conflict in counting is between standardization and practicality. The fact that so many methods have been discussed in this conference suggests that no one method readily meets all possible needs. One may ask: is it really possible to restrict counting to a few standardized methods? And, can any method or methods be flexible enough to accommodate the various objectives and situations without being so general as to defy meaningful comparison of results? One easy answer, previously arrived at by some people (e.g., Berthold 1976), is "yes!" to both questions. To such people, all that is needed is to settle on a few rigorously standardized and very intensive methods, and to insist on them to the exclusion of all others. The data so obtained will be comparable, and of irreproachable accuracy; but, unfortunately, most areas will remain unsampled, since most counters, whether professional or amateur, will not be able to put in the required intensity of effort, even if they are willing to accept the prescribed methods as the only satisfactory ones. There must be some compromise between scientific rigor and practical usefulness, but the range of acceptable compromise is not very wide.

Specialized knowledge is another stumbling block to many would-be bird counters. Ability to identify birds is obviously a prerequisite to any serious count effort, and some methods demand the identification of every "tweet," "cheep," "chip," or half-heard song as well as of those birds that happen to stray into view. Skills of bird identification by sound have spread widely since the Breeding Bird Survey started in 1965–68 (Robbins and Van Velzen 1967), and few bird-watchers now are unwilling to accept such challenges, within the constraints imposed by hearing and sight. Descriptions of habitats pose a much greater burden to most amateurs, and some professional biologists are unfamiliar with many of the more common plant species in areas where they have studied birds for years. Particularly in cooperative projects involving amateurs, the most successful are those that demand of the cooperators only to count birds. Rules governing when and where to count them are not nearly as much of a problem as are directions to perform other activities that do not interest most volunteer observers.

Statistics pose additional problems in censusing. Recent preoccupation with computers and complex statistical treatments has tended to focus on methods giving data that can be mechanically converted into coefficients that only a specialist can interpret. Most of this is not essential, and much of it may be counter-productive if it overemphasizes use of a large number of small and inherently highly variable samples merely because other methods give results less suitable for computers. The idea that count results without attached confidence limits are intrinsically inferior to those with them, regardless of the methods used, is sufficiently absurd to a dispassionate view that it should not be allowed to sway decisions on censusing.

EXAMPLES OF CONSTRAINTS REQUIRING DECISIONS

The examples that follow are hypothetical but all based in part on real situations. Amateurs usually start out by taking part in cooperative efforts like the Christmas Bird Count, and as they gain expertise some move on to individual projects. Because these are spare-time undertakings, their scope is limited to early mornings, evenings, weekends, or holidays. Repeated transect counts to monitor migration within a year or summer or winter numbers between years may satisfy some lone observers for a while (e.g., Erskine 1968). These relatively low-intensity methods are usually inherently variable, so the results tend to be difficult to compare with those of others. Few such projects are continued for long enough that the observer can usefully compare his (or her) own results accumulated over time, and a common result is a mass of summarized but unanalysed data. (The same thing has also resulted from some transect surveys by professionals . . .). A probably more useful exercise for a lone amateur is a mapping census of a plot, but here too the neophyte encounters problems. Without an assistant, the plot will probably be paced rather than measured, which in many habitats will be less accurate. The censuser may know nothing of vinyl flagging tape, and end up with inadequately marked grid lines, to the detriment both of the results and of his enjoyment. Except in southern and far western parts of North America, the usable census period is only 6 weeks or even less (cf. Erskine 1976b), so fitting the required eight counts into that period calls for more than one count each week, whatever the weather. Usually, this forces the use of a study area close to home, and small enough that it can be censused on weekdays outside of working hours, unless the censuser is eager enough to devote nearly every possible weekend morning through that period to the census. And as already noted, many amateurs give up when faced with the habitat description for the plot; if they cannot recruit a botanically minded friend at this stage, the census may never be written up and results lost. For some amateurs, the first mapping census attempt is also the last, at least partly for lack of quite elementary instructions to assist in decisions on matters other than censusing birds.

A second example is drawn from the field of graduate studies. A student wishing to explore, say, relationships of habitat and bird communities sets up census plots and conducts mapping censuses throughout the year. However, territories can only be mapped in breeding season, as birds tend not to be stationary nor to advertise their locations at other seasons. Accordingly, so as to have a common numerical basis throughout the year, the results are worked up only as mean numbers of each species per count in each season, which will be comparable among themselves, between seasons, or between years. Such results are amenable to statistical comparisons, but they have only limited comparability with those from other studies, as the density indices so obtained are substantially lower than the absolute values to be anticipated. A decision not to estimate territories of breeding birds may have been made merely because territories were thought too subjective to permit statistical analysis, as well as because they applied only to the breeding season. However, unless the field data are placed in a permanent repository, no one else will be able to reconstruct the (more or less) absolute density figures that could have been derived from such mapping censuses; and no one will be able to do so from the same familiarity with the areas as would the censuser. Often even the thesis omits the absolute density figures, and the subsequent publication—if and when it emerges—almost invariably lacks this basic information, because the censusing was looked on only as a means to the student's own ends. Students directly concerned with methodology are perhaps more likely to publish the actual census results, especially if these involve innovations; but they and their supervisors need to remember that the comparative data they obtained by established methods may turn out to be more useful to others than the innovations that justified their study.

Next we may turn to a consultant with, say, a contract to monitor the effects on birds of a spray program against forest insect pests. Spraying is timed to a particular stage of the emergence of insects or their larvae, usually sometime in the middle of the birds' breeding season. Count methods chosen have to give data that can be compared from one part of the season to another, and have to involve samples taken over sufficiently wide areas as to average out the inevitable unevenness in application of sprays. Use of a few large plots risks some being missed altogether or else overdosed, while having many small plots involves so much edge effect that the results may be nearly impossible to interpret; transects often seem the only solution. The results are frequently highly variable even when all counts are conducted by the same observer, and some changes that are obvious to the observer in the field may not show up in the results because of the swamping effect (nearby songs drown out distant ones at high but not lower densities). The end results are seldom satisfying either to the censuser or to the agency employing him, as only acute effects are documented to any conclusive extent. Most decisions on methodology for short-term monitoring have been made on an ad hoc basis, or empirically, in comparison to what worked or didn't work last time. And many people who have tried to monitor forest spray programs have turned to other work in frustration. Consultants, of course, are in that game to make a living; any extra expenditure on a project reduces their potential profits. When they set out to conduct, say, an inventory or monitoring of a bird population, they will do so with as few surveys and as poorly-paid staff as they can get away with, expecially if they can pull political strings to ensure that rival firms with higher competence and/or standards cannot compete for the contract. In one such case, a contract was awarded to a consulting firm, who sublet it to a graduate student, who passed the bulk of the actual field work to a "birding bum," who ended up abandoning the job—and departing without telling anyone—at the height of the breeding season; there was a gap of 2 weeks before another sub-sub-contractor could be found and put to work.

Lest I be accused of pointing a finger only at outside groups of counters, I would add that I have seen examples of most of the problems described in this section in government count projects as well—including some in my own. Likewise, I hope that the generalizations on decision-making in the next section will be of value to government counters as well as to the others.

DISCUSSION AND CONCLUSIONS

In a free world, and particularly in the private sector, counters typically select methods that fit their particular objectives, and we all hope this freedom continues. Virtually all questions regarding methodology involve some qualification regarding availability of resources, as a universal method could only be applied given unlimited resources, if at all. Some counting exercises, e.g., aerial surveys of pelagic seabirds, may occupy 10 or more people, with annual budgets of hundreds of thousands of dollars; obviously this is not something for the spare-time amateur, and it was not possible at all until the environmental impact stakes became big enough to cover it. Any method that requires more than one person at a time, or calls for more specialized equipment than the now-ubiquitous binoculars, or for knowledge over and above bird identification, will be less generally usable than a simpler one:

> #1: the simpler methods suffer fewer constraints than more complex ones, always provided they are adequately standardized.

Rigorous standardization is possible but generally impracticable, and some compromise is essential. The "bottom line" has to be set in terms of comparability, and particularly of comparability between different observers. Results of any count conducted by amateurs for recreation are likely to be used mainly by persons other than the counters; and any long-term comparisons of data from permanent plots (or transects) are likely to involve counts by different observers. Most methods not involving measured areas (i.e., relative methods) are especially sensitive to observer variability, owing to individual differences in acuity of hearing or sight. More intensive methods, and particularly the mapping census method, tend to be more easily replicated by other people. As resources usually are insufficient to allow exclusive use of reproducible methods on extensive areas, it is decidedly preferable to use them to calibrate other less intensive methods, which can be used more widely, than to rely solely on the latter in order to sample larger areas. Thus,

> #2: selection of methods for standardization must include some that can be replicated by other people, even if these must be coupled with quicker but more wide-ranging methods in most cases.

The reliability of the calibration methods usually arises out of intimate knowledge and understanding of the area and its birds, rather than from confidence limits or statistical tests. But if resources allow coverage of replicated baseline plots, even the statisticians may be kept happy (cf. Owens and Myres 1973).

No perfect method exists, so new methods must be tried, and innovation goes on apace, often at the expense of comparability. The graduate student has to show that he (or she) has thought up something new, and the consultant trying to stay afloat in the economic maelstrom won't weep if his rivals cannot use his data. But even these realists need to compare their new results with those of others, so it is to their advantage not to exclude existing methodology. No amount of standardization will or should be allowed to hamper those hardy spirits who wish to experiment with new or modified methods, but they need to be discouraged from "throwing out the baby with the bath water." A number of workers (e.g., Bell et al. 1973, Best 1975) have suggested discarding one or another method because it did not deal effectively with some species or group in which they were interested. All existing methods suffer from some problems, but most have merit for some or most groups of birds. Where innovation is needed, even more than in developing wholly new methods (e.g., Emlen 1971, 1977a), is in *supplementing* existing methods for those "difficult" species or groups poorly sampled at present (cf. Erskine 1974). This conference restricted its discussion largely to "terrestrial" birds, presumably because many major groups of aquatic birds—seabirds, herons and other colonial water birds, waterfowl—are already subjects of voluminous census literature, with special conferences to discuss them. Birds that defend only the nest site, or rely on flight songs for advertising, or nest in colonies, give rise to many of the difficulties encountered with mapping censuses, as well as other intensive methods covering only small areas. Because each method was developed to take advantage of some aspect of bird behaviour that lends itself to systematic counting, we cannot be surprised if not all species share that particular aspect; birds have adapted to their environments by a wide diversity of behavioural patterns. Some methods are flexible enough to be used in many habitats and situations, alone or in combination with other methods, while others fit only a few, specialized species. Birds have shown themselves adaptable in exploiting diverse environments, and counters have to be adaptable in supplementing established methods where necessary to deal with a species poorly sampled by the method of one's choice. Therefore,

> #3: innovation in census methods should be encouraged, but especially to supplement existing methods rather than to replace them.

It should be preferable to retain the practicality and comparability of an established method side-by-side with an innovation to cover a particular case. For example, most people doing mapping censuses also count the nests of Starlings (*Sturnus vulgaris*) and swallows (Hirundinidae), since those species do not defend all-purpose territories.

Most of what I have said so far is just common sense, but it needs to be said once again lest it be forgotten amid the complications of modern science. We need clear, simple procedures, covering even quite elementary points, especially if we are to encourage participation by interested amateurs. The scientific method implies that if you describe precisely what was done it can be duplicated, but whoever drafted that creed had never heard of inter-observer variability. So we must also use methods that are not sensitive to changes in observers, in case someone else, sometime, somewhere, might want or need to use our data in comparisons. This of course is very charitable, good for the soul as well as for science, but how does it help achieve the objectives for which the census project was started? It reduces the options open to, and thus the decisions required of, the amateur who wants to feel that his (or her) hobby has some spin-off value. For other workers, its chief value lies in ensuring that comparative baseline data exist when they are needed. If everyone "does their own thing" in their own way, comparisons become nearly impossible, and no data base is accumulated. This is pretty well what did happen with many North American waterfowl surveys in the 1950s and early 1960s, since procedures were poorly standardized and often not even written down (cf. Diem and Lu 1960, Dzubin 1969). The accumulation of data banks depends on comparable results (e.g., Erskine 1980), which depend on standardized methods.

Finally, despite all my emphasis on comparability, I am not so naïve as to believe that standardizing methodology will always lead to comparable results. People working with census data also have to make decisions as to what can and cannot be compared. The mapping census gives density values with quite low inter-observer variability, but I showed earlier (Erskine 1974) that the numbers of breeding species claimed varied in a much more subjective fashion. Obviously, decisions to compare diversity indices (which are based on numbers of species as well as their density) for counts conducted by different observers are often on shaky ground. Compilers and analysts also need to ensure that the data they compare are comparable.

Studies in Avian Biology No. 6:575–576, 1981.

SUMMARY OF THE SYMPOSIUM

JOHN T. EMLEN[1]

In bringing together a balanced mix of field biologists, statisticians, practitioners and theorists, the organizers of this symposium have been remarkably successful in achieving a unity of interest and purpose that has led to an outstandingly productive conference. The symposium was orchestrated around a central theme: how can we most effectively and efficiently obtain those elusive data we need for monitoring wild bird populations and for analysing the underlying mechanisms of population regulation? No definitive solutions were reached, but notable progress was made in defining problems and approaches. Conferees were repeatedly and forcefully reminded of the tremendous difficulties involved in obtaining accurate data in the field, and sternly admonished to adhere to appropriate statistical procedures in both the planning and analysis stages of their studies. Their response reveals a broad determination to meet these challenges vigorously and intelligently, and promises a surge of interest to develop effective solutions.

While the central emphasis of the conference was on methods, two full sessions and many poster displays were devoted to illustrating and evaluating recent progress in monitoring and mapping population trends and patterns. Studies of species and community densities on small plots using the now well-established mapping method were presented and critically examined for a variety of avian habitats. Population trends and range-boundary fluctuations were graphically displayed from data obtained in long-term monitoring programs using a variety of station count procedures, while patterns of density distribution were revealed in studies employing recently developed methods of counting detections within fixed or variable-width strips and point-centered circles along transects through large tracts of avian habitat. Other censusing techniques were discussed and illustrated including mist-netting, mark-recapture, and calling secretive birds to the observer with playbacks of recorded vocalizations.

Three sessions and a number of poster displays then focused on the evaluation of methods in common use, criticisms of traditional procedures, and suggetions for modifications and innovations. Results obtained by mapping, station counts and several types of transect surveys were compared and evaluated by field investigators and statisticians in a wide variety of situations including areas where the local populations had been marked and intensively studied.

Three more sessions and a large number of poster exhibits were devoted to the numerous and difficult problems encountered in obtaining reliable census data in the field, and principles to be followed in organizing field projects and analysing data sets. The field problems considered included aspects of species and individual variability, unpredictability of bird responses, interactions between birds and observers, seasonal and hourly changes in bird abundance, activity, and detectability, vegetation structure and terrain, sound attenuation with distance in various vegetation types, and variations in the experience, skill, hearing acuity, and attentiveness of observers. Programs for training field workers in identification and distance estimation were described and the possibility of replacing distance estimations with predetermined detection threshold distance values was considered. Several papers considered problems of designing projects and analysing census data.

In the final session attention was turned to applications of bird censusing in studies of habitat distribution, habitat responses, and community dynamics in situations ranging from scrub deserts to intensely cultivated farm lands, and from tropical forests to island archipelagos. Adaptation of procedure to the specifics of research objectives was discussed, and the potential of emerging quantitative techniques for analysing community structure and dynamics was explored.

An accomplishment of the symposium that will persist beyond the conference itself in the published proceedings is the guidance it will provide for investigators undertaking projects or organizing programs involving bird censuses. Stressing the importance of selecting the procedure most appropriate to the specific objectives of a project I have prepared a table summarizing my views on census problems and census methods as developed and discussed during the conference (Table 1).

Most projects will, I believe, fall into one of two problem areas identified in the first column of the table as population responses and population dynamics. These can then be assigned to one of a limited number of categories listed as project objectives in the second column. The third column, headed units of measurement, suggests that indices of relative abundance are

[1] Department of Zoology, Birge Hall, University of Wisconsin, Madison, Wisconsin 53706.

TABLE 1
BIRD CENSUS PROBLEMS AND METHODS

Problem areas	Project objectives	Units of measurement	Methods available and examples
Population behavior and responses — Comparisons	Temporal changes Year to year Season to season Before-after Site differences Different habitats Different sites Experimental-control	Detections per unit of effort — Indices of relative abundance	Mist net sampling Line counts No boundary (Aud. Xmas B.C.) (Fixed strip) (Variable strip) Point or station counts No boundary (IPA, NABBS) (Fixed radius) (Variable radius)
Population regulation and dynamics — Analyses	Density structure Species rank Equitability and diversity Subcommunity structure Density dynamics Biomass structure Trophic balance Energy flow	Birds per unit of area — Measures (Indices) of density — Instantaneous or cumulative	Mark-recapture (single species) Mapping (CBC, Aud. BBC) Line counts Fixed strip (Finn) Variable strip (USBLM-Ariz.) Point counts Fixed radius Variable radius (USFWS-Haw.)

adequate and preferable to density estimates for most if not all projects concerned with population responses (first problem area). In station counts, point counts or line counts (column 4) the units of observer effort could be units of time, distance, or area. Where properly standardized for speed of walking or duration of observation stops, time and distance will be equivalent and equally acceptable; area calculated from estimated detection distances could provide additional information on the relative abundance of species in the community, but because of the unsubstantial nature of most distance estimations should be avoided unless the additional information is clearly needed.

Absolute density values (birds per unit of area) may be necessary where species with differing detectabilities are to be compared as in studies dealing with numerical diversity, or combined as in studies involving consumer-resource, or predator-prey ratios. The methods currently available for density measurements are laborious, costly and limited in applicability (mark-recapture and territory mapping on small plots), or dependent on subjectively determined areal denominators (detection counts within estimated strip or circle boundaries). They are producing much valuable information, however, and must be promoted for the present as the best we have been able to devise.

Studies in Avian Biology No. 6:577–578, 1981.

APPENDIX I
TERMINOLOGY USED IN ESTIMATING NUMBERS OF BIRDS

C. JOHN RALPH[1]

Hundreds of papers, such as those presented in this volume, are produced annually by workers using various techniques. The purpose of this section is twofold: (1) to provide for readers unfamiliar with counting techniques a handy guide to the most common terms and methods used in the field; and (2) to attempt to set a consensus on the meanings of some terms that are too commonly used interchangeably in the literature. I have not attempted to set arbitrary standards here; rather, if two or more terms are commonly used to apply to the same method or unit, *and* if each is unambiguous, then I have indicated this synonymy.

I thank Dr. Frank Pitelka who, as chairman of the working group on terminology, synthesized the comments made on the original draft distributed at the symposium. He incorporated the comments of many people, applying his own erudition to the problems presented. Mike Scott contributed many suggestions at all stages. I thank David Anderson, James Baldwin, Kenneth Burnham, David E. Davis, John T. Emlen, David DeSante, Martin Erdelen, Charles Gates, Joseph J. Hickey, Frances James, Olli Jarvinen, Douglas H. Johnson, G. M. Jolly, Duncan Mac Lulich, Raymond O'Connor, Hans Oelke, Carol Pearson Ralph, Chandler S. Robbins, Larry D. Voorhees, Jared Verner, and many others for their comments.

Clear and precise statement of the method and units of measurement used in an estimation will facilitate communication between workers by preventing misunderstanding. It is hoped that the usage given here will gain currency among workers in the field.

accuracy (n): a measure of the closeness of an estimate to the actual value of the population parameter being estimated.

atlas (n): result of a comprehensive survey of a large geographical area that maps the occurrence (or occurrence and relative abundance) of species in subdivisions of that area; usually based on a grid of fixed intervals in kilometers or degrees latitude and longitude; restricted to a particular season of the year, usually the breeding season.

bias (n): the difference between the expected value of an estimator and the actual value of the parameter being estimated.

Breeding Bird Census: a census program of the National Audubon Society in North America that uses the Spot-mapping Method during the breeding season.

Breeding Bird Survey: a cooperative program of the U.S. Fish and Wildlife Service and the Canadian Wildlife Service for monitoring population changes in North American breeding birds using Index Method station counts.

Capture-recapture Method: a procedure involving the distinctive marking of individual birds and their subsequent recapture (or sighting) to estimate population size (and other population parameters).

census (n): a count of all individuals in a specified area over a specified time interval.

census (vb): the act or process of counting all birds within a specified area and estimating density or a total population for that area.

census efficiency (n): proportion of density as assessed by a census to actual population density (cf. detectability).

Christmas Bird Count (formerly "Census"): an annual project in the Americas of the National Audubon Society involving a one-day count of the individuals of all species observed within a 15 mile (24 km) diameter circle.

Common Birds Census (U.K.): a program of the British Trust for Ornithology using the Spot-mapping Method.

community (n): the aggregate of all populations within a defined area.

contact (n): a single field record of a bird by sight or sound (Syn. detection, cue, registration, observation).

count (n): a) the act or process of enumerating; b) number or sum total obtained by counting.

count (vb): to record the number of individuals or groups present in a population or population sample (cf. census and index).

density (n): the number of units (e.g., individual birds, pairs, groups, nests) per unit area (Syn. abundance; cf. frequency).

detectability (n): a measure of the conspicuousness of a species equal to the proportion of observed to actual units (individuals, territorial males, etc.) on a given area.

detection distance (n): the distance from the observer at which the individual bird or cluster of birds is seen or heard (radii in station

[1] USDA Forest Service, Institute of Pacific Islands Forestry, 1151 Punchbowl Street, Honolulu, Hawaii 96813.

counts; lateral or perpendicular distances in transect counts).

estimator (n): a function of sample data that describes or approximates a parameter.

Fixed-distance Methods: see Strip Transect Methods and Point Count Methods.

frequency (n): number of plots, stations, counts (visits), or intervals in which a species is detected; when expressed as a percent it becomes relative frequency (cf. density).

index (n): a) the proportional relation of counts of objects or signs associated with a given species to counts of that species on a given area; b) counts of individuals (e.g., at a feeding station) reflecting changes in relative abundance on a specified area, or just locally.

Index Methods: bird counting methods involving sampling that yield measures of relative abundance rather than density values.

line transect (n): a sampling route that bisects a surveyed area followed by an observer counting contacts over a measured distance.

Mapping Method: see Spot-mapping Method.

parameter (n): an unknown quantity (e.g., the number of robins in a fixed area).

Point Count Methods: counts of contacts recorded by an observer from a fixed observation point and over a specified time interval: *Fixed Distance (Radius) Point Counts* limited to birds within a single arbitrarily selected fixed distance; *Variable Distance (Radius) Point Counts* limited to birds within distances varying according to species-characteristic detection distances (Syn: Variable circular plot); and *Unlimited Distance Point Counts* all birds without limits, i.e., all detections recorded regardless of distance (e.g., the "Indices Ponctuels d'Abondance" developed in France). (Syn: Station Count Methods).

point transect (n): a transect along which the Point Count Method is used; no recordings are made between stations (as opposed to strip transects with continuous recordings).

population (n): the birds of one and only one species within a defined area.

precision (n): a measure of the sampling variability of an estimator (proportional to reciprocal of sampling variance).

quadrat (n): a small sample plot, usually square or rectangular.

relative abundance (n): a percent measure or index of abundances of individuals of all species in a community (Syn. [in Europe] dominance; cf. index, frequency, density).

relative frequency (n): See frequency.

species diversity (n): a measure of the variety of species in a community that takes into account the relative abundance of each species; also used for just species richness.

species richness (n): the number of species in a given area (cf. species diversity).

Spot-mapping Method: a census procedure that plots individuals seen or heard on a gridded map of the area surveyed, usually over a period of days or weeks in a season; and, relying on simultaneous observations, the subsequent demarcation of individual territories or home ranges by examination of the clustering of those observations. Used in Breeding Bird Census. (Syn. Territory-mapping Method) For further information on terms and procedures see Anon. (1969), Robbins (1970), or Svensson (1979).

station (n): the area within which observations made from a point are recorded by the observer (or often synonymous with "point," see Point Count Methods).

Strip Transect Methods: a procedure using a strip of land of fixed direction that is sampled visually and/or aurally by an observer; counts may be: *Fixed Distance (Width) Counts* limited to a strip of set width for all or specially chosen species; *Variable Distance (Width) Counts,* with different, species-specific widths that are determined to reflect detection attenuation; or *Unlimited Distance Counts,* all detections recorded regardless of distance. (Syn. belt-transect).

survey (n): an enumeration or index of the number of individuals in an area from which inferences about the population can be made (cf. census, count).

Territory-mapping Method: See Spot-mapping Method.

transect (n): a cross-section of an area along which the observer moves in a given direction; see line transect, point transect, and strip transect.

Variable Circular Plot: see Point Count Methods.

Variable-distance Methods: See Strip Transect Methods and Point Count Methods.

Winter Bird Population Study (U.S.): A Program of the National Audubon Society involving census of wintering birds by counting and mapping, but not depending on persisting territory or home range occupation (cf. Breeding Bird Census).

Studies in Avian Biology No. 6:579, 1981.

APPENDIX II
REPORT OF WORKING GROUP FOR THE REVISION OF CHRISTMAS BIRD COUNT PROCEDURES

ROBERT S. ARBIB, JR.[1] AND CARL E. BOCK[2]

As a sequel to the paper "The Christmas Bird Count: constructing an 'Ideal Model,'" a workshop was held during the symposium. Robert S. Arbib, Jr., chaired the workshop. Approximately 15 participants were present, and several others submitted comments and proposals subsequent to the meeting. Present or otherwise participating were C. E. Bock, D. Bystrak, B. Carlson, H. Cogswell, S. R. Drennan, D. James, D. Johnson, H. Kale, Jr., V. Kleen, R. LeValley, C. S. Robbins, K. Smith, W. Tucker, R. Tweit, W. Weber and S. Wilbur.

Discussions centered around the recommendations proposed in the paper. Among the topics of discussion, in an unstructured format, were: Would the proposed changes in procedure and reporting practice invalidate or antiquate data published in prior years? Were there alternate and more accurate ways by which to evaluate observer reliability? Could a standard be devised for the separate reporting of after-dark "owling" hours, and for bird feeder-watcher hours? What disposition is made of the original documentation of unusual observations? Would it not be useful to offer it for storage within the states in which they were recorded? Count costs, publishing costs, and participant fee structures were considered. It was agreed that various improved forms of observer training would be useful, such as club programs with slide quizzes on the estimation of flock numbers, difficult identifications, and even quizzes on difficult call notes.

The Chairman expressed hope that those present and other interested persons form a Christmas Bird Count Advisory Committee to the National Audubon Society to consider all such ways of refining and improving CBC data without forsaking the popular attractions that encourage participation, and to make further proposals concerning all aspects of the CBC. It was agreed that such a committee be formed, and that the persons present and those subsequently expressing interest would be invited to join. The Chairman expressed the hope that the form of an experimental "Ideal Model" be ready for the 1981–82 season, and that this be tested in a selected sample of counts.

It was understood by the participants that the committee would be, for the present, advisory only.

[1] National Audubon Society, 950 Third Avenue, New York, New York 10022.
[2] Department of Environmental Population and Organismic Biology, University of Colorado, Boulder, Colorado 80309.

Studies in Avian Biology No. 6:580–581, 1981.

APPENDIX III

REPORT OF WORKING GROUP[1] ON THE NEED FOR STANDARDIZED CENSUS METHODS

HARRY F. RECHER[2]

In an era of rapidly expanding human populations and the global impact of technology, the need to monitor bird populations worldwide must be accepted as critical. It is therefore essential that regional surveys use similar procedures, and that the storage of data be coordinated so that each set of surveys can be compared. Without standardization, trends in the abundance of birds, reflecting continental or global changes in environment, cannot be identified.

AMATEUR INVOLVEMENT

Two kinds of people count birds; professionals and amateurs. Distinguishing them is often difficult, but it is probably fair to say that professionals are paid for their effort and have specific research goals. The amateur is unpaid and views counting birds as a recreation. Nonetheless amateurs will find their activities more meaningful if they are part of a wide research effort with well-defined objectives. The amateur represents a pool of labor which enables work to be done on a scale that would otherwise not be possible; it is important to encourage their participation in research.

In America and Europe, amateurs already take part in regional surveys of birds and there are plans for a national bird survey in Australia. Although broadly similar in their intent, there are significant differences between the various regional surveys. For example, the Australians propose only to list the species of birds observed (Cullen 1980) while the American Breeding Bird Survey tallies individuals. With differences as great as this, it is impossible to compare results.

COMPARISON OF METHODS

Procedures do not need to be identical and professionals will continue to use a wide variety of techniques. The requirements of individual research projects effectively prevent adherence to a single set of rules. Some standardization could probably be achieved by the presentation of established procedures in a manual, but the critical need is to coordinate regional censuses that rely on amateurs to do individual counts.

Several methods are already used in regional surveys and have proven effective. Territory or spot mapping (e.g., Kendeigh 1944, Anon. 1970) is used in the British Common Bird Census (Williamson and Homes 1964). Fixed-distance transect counts are used in Finland (Järvinen and Väisänen 1977c), while the American Breeding Bird Survey is based on unlimited-distance station or point counts (Bystrak 1981).

Standard procedures for territory mapping and point counts are particularly desirable. In many ways, these two methods are complementary. The mapping method provides reasonably precise measures of breeding bird numbers, but is time consuming, labor intensive, and most useful for detailed and local studies. Adequate guidelines exist (Anon. 1970, Robbins 1970).

Point counts are suitable for the estimation of relative abundances and over broad areas. As illustrated by the Breeding Bird Survey, it is an effective way to use amateurs whose time is otherwise too limited to participate in censuses using the mapping method. Fixed-distance line transects are perhaps preferable if density estimates are required, but are not as efficient in the use of amateurs as point counts. In any event, using trained observers, unlimited-distance point counts can be converted to variable-distance censuses to give density figures. There are no recognized standard procedures for either the line transect or point count methods. We decided that unlimited-distance point counts were simplest to organize and would be easier to adapt to international requirements, than the line transect method. This does not preclude the possibility of another set of standards being developed for the line transect method.

GUIDELINES FOR POINT COUNTS

As with all survey methods, the unlimited-distance point count method is biased towards particular birds or groups of birds. It is also influenced by the age, sex or reproductive state of each bird, the season, habitat, time of day, weather, environmental noise, the observer and the number of birds being recorded. This emphasizes the need for guidelines. However, guidelines need to be flexible and allow for local differences in the avifauna, climate, and people.

---

[1] Working Group members: D. G. Dawson, J. T. Emlen, W. F Oelklaus, S. E. Svensson and S. A. Temple.

[2] Department of Vertebrate Ecology, The Australian Museum, P.O. Box A285, Sydney South, NSW 2000, Australia.

Allowing for such differences, the Committee suggests the following procedures for the fixed-distance point count method:

• A survey by the point count method should use a number of different areas or routes, done by foot or vehicle, along which the observer stops periodically and records birds seen or heard.

• All birds seen or heard are recorded regardless of their distance from the observer (with well-trained observers the distance can be recorded, allowing the calculation of density).

• Each count is done by a single observer.

• The area or route should be surveyed over a number of years by the same observer using approximately the same points.

• If different observers are used, the survey should be presented as if it was a different area or route.

• The time of day each area is surveyed should be the same for each count; different areas or routes can be surveyed at different times.

• The season when a count is conducted can be determined by climatic or environmental conditions.

• The recommended minimum number of points or stations on a survey area or route is twenty.

• The distance between stops can vary, but excessive overlap between stations should be avoided. One hundred meters is a suggested minimum distance between points.

• Five minutes should be spent at each point.

• Care must be taken to avoid counting the same bird(s) twice. A simple map on which the position of birds are noted as they are recorded helps to avoid repeat counts.

• Time of day, weather and environmental noise (e.g., aircraft, cicadas, running water) are recorded at each stop. Counts should be done under good weather conditions.

• A simple description of the habitat should be made for each point and abundant resources (e.g., nectar, grain) noted.

• Data should be recorded in a format suitable for computer coding.

The procedures suggested are a compromise. The members of the Committee had differing opinions on the need for a minimum number of stops (from no minimum to 50), the time spent at each stop (from 3 to 10 minutes), and on the need to describe the habitat at each station (from no need and a waste of time, to being an absolute necessity). Twenty was chosen as the minimum number of stations, as it was considered to be an adequate sample size, made efficient use of the time of a person wishing to spend a morning in the field, but was not so numerous as to be tiring. Five minutes was considered long enough to record the birds at a station without being hurried (as if only three minutes were available), but not so long as to cause undue problems with repeat counts. The suggested minimum distance (100 m) between points is based on experience in New Zealand (Dawson and Bull 1975), but needs to be tested in each region. The Committee unanimously agreed on the need to restrict the number of observers to one and to hold environmental conditions (e.g., time of day) as constant as possible for each count.

CONCLUSIONS

Clearly the guidelines will need refinement and should be presented for discussion at scientific meetings. Ultimately the procedures agreed upon will need to be endorsed by a recognized international body. It is considered that the International Ornithological Congress (Moscow 1982) is an appropriate venue for the presentation of recommendations with the view of their adoption as an international standard.

Studies in Avian Biology No. 6:582, 1981.

APPENDIX IV
NORTH AMERICAN ORNITHOLOGICAL ATLAS COMMITTEE FORMED

MIKLOS D. F. UDVARDY[1]

At the symposium a working group was appointed and a Committee (NAOAC) was formed by three Canadian and 17 USA delegates with Dr. Miklos D. F. Udvardy as convenor. The purposes of the Committee are to encourage regional atlas projects in as many states or provinces of Canada, USA and Mexico as possible, and to coordinate the regional projects to obtain uniformity of methods and to provide a communications center. Further, to promote, at an appropriate time, a North American Breeding Bird Atlas Project. The Committee resolved that the UTM system should be adopted, and a 10 km square grid established uniformly for all of North America, so that each new atlas project will mesh with the overall design. The issuing of a Newsletter has also been decided.

This event is an important step also from the point of view of biogeography and conservation of the biosphere of our continent. The UTM grid will provide uniformity and it is hoped that in time all biogeographical data would be recorded in this decimal system. Such a grid has been established in the British Isles already in the 1950s resulting in the Atlas of the British Flora (1962), of the breeding birds of Britain and Ireland (1976) and several ongoing projects in biogeography. Hence it spread to continental Europe where, besides several published or ongoing national atlas projects the European Invertebrate, Floral and Breeding Bird Atlas surveys are exemplarily coordinated international biogeographical efforts. Australia, New Zealand, and several African countries also have their atlas projects though owing to the size of some of the areas some countries use a latitude—longitude based grid.

A regional atlas project is either the combination of field work and compiling and mapping of previously existing (herbaria, museum specimens, published papers) data or, to discount distributional and density fluctuations characteristic of plants and animals, it is based on 5 to 10 years of field work of volunteers guided by specialists. The need of central coordination and animation arose as the number of atlas projects already started or going in the USA and Canada approaches twenty. The Committee asks interested geographers, ornithologists and potential other volunteers to contact the editor of the Newsletter, Dr. M. D. F. Udvardy, Department of Biological Sciences, California State University, Sacramento CA 95819 or Dr. Chandler S. Robbins, U.S. Fish & Wildlife Service, Laurel, MD 20811, co-convenors of the Committee.

[1] Dept. of Biological Sciences, California State University, Sacramento, California 95819.

Studies in Avian Biology No. 6:583, 1981.

APPENDIX V
REPORT OF WORKING GROUP ON THE NEED FOR A MANUAL OF COUNTING METHODS

LARRY F. PANK

The time lag between publication and utilization of new techniques to estimate densities of avian populations has been excessive. A potential solution to the problem was posed to the working group in the form of a question, "Is a manual on censusing bird populations needed?"

It was the consensus of the members that:

1. Delayed implementation of new techniques was attributable to inadequate searches for the dispersed literature, insecurity with the level of statistics, and a lack of appreciation for the advantages in accuracy, precision, effort and assumptions associated with the new techniques.
2. Current, comprehensive and practical references to both the techniques and associated computer programs already exist, indicating that the major barrier is related to the volume of the references and the diversity of the literature sources.
3. A comprehensive manual would consolidate the literature; however, the publication of ongoing research would probably outdate the manual before printing.
4. A more practical and efficient approach would be to publish an easily updated guide that would enable users to direct themselves to the pertinent publications.

The committee unanimously recommended the guide (above alternative #4) over the manual. Formats ranging from abstracted references to arrays containing species, methods of observation, and habitats that lead to cells containing the pertinent references (Gates 1979) were considered for the manual. Although the use of arrays has potential, effort was directed at developing the logic for a key based on the differences (i.e., assumptions) between the estimators and the user's input on the particular population and census environment.

The following brief example was included to stimulate the future development of a comprehensive dichotomous key (* indicates references for generalized computer programs):

1a. Species is easily captured or individually identified and subsequently reidentified 2

1b. Species is difficult or impossible to capture 5

2a. Assumptions of population (demographic) closure are not violated between samples (no immigration, birth and differential mortality or emigration between marked [identifiable] and unmarked [unidentifiable] individuals) .... *White et al. (1978*), Otis et al. (1978)*

2b. Assumptions of population closure may be violated 3

3a. Death and/or emigration may occur. Birth and/or immigration does not occur .... *Jolly (1965), Darroch (1959)*

3b. Both death and/or emigration and birth and/or immigration occurs 4

4a. Demographic changes occur uniformly across all population classes (sex, age, locality) .... *Arnason and Baniuk (1980*), Brownie et al. (1978*), Begon (1979), Seber (1973), Jolly (1965), Pollock (1975)*

4b. Demographic changes differ between population classes .... *Brownie et al. (1978*), Pollock (1981), Stokes (1980)*

5a. Species is hunted .... *Paulik and Robson (1969), Dupont (1976)*

5b. Species is not hunted 6

6a. Species is conspicuous in its habitat 7

6b. Species is inconspicuous .... *No reliable estimator available.*

7a. Survey restricted to the breeding season (labor intensive) .... *Williams (1936), International Bird Census Committee (1970, 1977)*

7b. Survey is not restricted to the breeding season (labor efficient) 8

8a. Habitat or terrain is difficult to traverse .... *Reynolds et al. (1980)*

8b. Habitat and terrain are easily traversed .... *Burnham et al. (1980), Gates (1979, 1980*), Laake et al. (1979*)*

Studies in Avian Biology No. 6:584, 1981.

APPENDIX VI

REPORT OF THE WORKING GROUP TO IDENTIFY FUTURE RESEARCH NEEDS[1]

JARED VERNER[2]

The conference brought into focus many challenging research needs related to the counting of birds. Our purpose here is to itemize some of the more important ones and to place them in some sort of priority. We agree that before any research related to assessment of bird numbers is begun, investigators have an obligation to decide where they are headed and why—that is, goals must be precisely defined. This involves a determination of what is meant by "bird population" and what properties of that population we must estimate. There can be no excuse for using any but the most parsimonious methods and sampling designs.

The following list is not exhaustive but should serve to convey the magnitude and scope of the tasks remaining. We have identified four broad areas for consideration, and for the first we have itemized a number of more specific research needs. Questions pertaining to this research area must be answered before we can have confidence that we are estimating with reasonable accuracy the actual numbers of birds of various species in a community. We take this opportunity to thank those who voluntarily participated in our committee's discussions. Their thoughtful contributions were invaluable to us as we prepared the following list.

LIST OF RESEARCH NEEDS

(1) Quantify the magnitude of the bias and variance in estimating real numbers of birds for the various counting methods, species-by-species, by habitat condition, by time of day, by season, by sex and age class, and so on.

- A. Expand our research to include the whole year.
- B. Quantify seasonal activity budgets of birds, particularly as they affect production of cues used to detect them.
- C. Determine the accuracy of distance and direction estimates.
- D. Determine species-specific and individually-specific song intensities in relation to effective detection distance.
- E. Evaluate the effects of floaters on count results and their interpretation.
- F. Devise census methods that are sensitive to movements of locally settled birds within and between territories.
- G. Develop research methods for counting special groups, such as raptors, flocking species, and roost aggregations.
- H. Apply a variety of standard methods to populations (preferrably banded) for which the total numbers are known so that errors in estimation can be evaluated and biases can be understood.

(2) Study the statistical distribution of the counts, so that suitable methods of analysis can be performed.

(3) Determine effective and realistic standardized methods for quantifying habitats and other environmental factors in relation to avian communities. This could include examination of the possibilities of aerial photo interpretation and remote sensing data, in addition to on-the-ground measurements.

(4) Assess the kinds of questions that can be addressed with data from extensive indexing surveys, such as the Christmas Bird Counts, Breeding Bird Surveys, Atlases, and Migration Counts.

---

[1] Working Group members: David Hussell, Frances James; Frank Pitelka; and Robert Szaro.

[2] USDA Forest Service, Pacific Southwest Forest and Range Experiment Station, 2801 East Sierra Ave., Fresno, California 93710.

Studies in Avian Biology No. 6:585–586, 1981.

APPENDIX VII
POSTER PAPERS PRESENTED AT SYMPOSIUM

BIASES IN ESTIMATING A DIFFERENCE IN POPULATION DENSITIES USING SINGING BIRD SURVEYS

Jonathan Bart, Ohio Cooperative Wildlife Research Unit, The Ohio State University, 1735 Neil Avenue, Columbus, OH 43210.

PLOT SIZE AND THE ACCURACY OF BIRD CENSUS RESULTS IN CONIFER FORESTS OF YOSEMITE NATIONAL PARK

Edward C. Beedy, Department of Zoology, University of California, Davis, CA 95616.

SEASONAL CHANGES IN THE DETECTABILITY OF BREEDING BIRDS

J. Blondel, C. Ferry, and B. Frochot, Centre d'Etude Ornithologiques de Bourgage, Faculté de Science Mirande University, 21000 Dijon, France.

HOW TO DETERMINE THE DURATION OF OBSERVATION IN STATION COUNTS (UNLIMITED DISTANCE)

J. Blondel and B. Frochot, Centre d'Etude Ornithologiques de Bourgagne, Faculté de Science Mirande University, 21000 Dijon, France.

THE CALIFORNIA RAPTOR SURVEY PROGRAM: AN EVALUATION OF THE TECHNIQUE

Charles T. Collins and Paula Peters, Department of Biology, California State University, Long Beach, CA 90840.

A COMPUTER PROGRAM TO SIMULATE AURAL CENSUSING TECHNIQUES FOR BOBWHITE QUAIL

Robert J. Cooper, Envirosphere Company, 145 Technology Park, Atlanta, GA 30092.

HABITAT PREFERENCES OF GEORGIA CAPRIMULGIDS BASED ON AURAL INFORMATION

Robert J. Cooper, Envirosphere Company, 145 Technology Park, Atlanta, GA 30092.

USEFULNESS AND BIASES IN DATA COLLECTED ON BIRDS OF QUEBEC, CANADA, BY NON-STANDARDIZED METHODS

André Cyr and Jacques Larivée, Université de Sherbrooke, Sherbrooke, Quebec J1K 2R1 Canada.

THE APPLICATION OF AVIAN ACTIVITY ISOPLETHS TO TERRITORY MAPPING CENSUS TECHNIQUES

Kenneth R. Dixon and J. Edward Gates, University of Maryland, Appalachian Environmental Laboratory, Frostburg State College Campus, Frostburg, MD 21532.

DIVERSITY MEASUREMENTS AS APPLIED TO AVIAN POPULATIONS ALONG THE SACRAMENTO RIVER SYSTEM

R. M. Engeman, C. P. Stone, W. E. Dusenberry, and M. Hehnke, U.S. Fish & Wildlife Service, Denver Wildlife Research Center, Denver, CO 80225.

METHODS FOR RELATING THE DIVERSITY OF BREEDING BIRD COMMUNITIES TO HABITAT HETEROGENEITY IN ISOLATED WOODLOTS

K. E. Freemark, Carleton University, Ottawa, Canada K1S 5B6.

STANDARD FORMS AND INSTRUCTIONS FOR COUNTS OF MIGRATING RAPTORS

Mark R. Fuller, Migratory Bird and Habitat Research Laboratory, Patuxent Wildlife Research Center, Laurel, MD 20811.

LINETRAN, A GENERAL COMPUTER PROGRAM FOR ANALYZING LINE TRANSECT DATA

Charles E. Gates, Institute of Statistics, Texas A&M University, College Station, TX 77843.

WHICH SPECIES ARE BREEDING SPECIES AND HOW MANY CIRCULAR PLOT CENSUSES ARE NECESSARY TO DETECT THEM?

Stephen L. Granholm, Department of Zoology, University of California, Davis, CA 95616.

PROGRAM TRANSECT: A COMPREHENSIVE ANALYSIS PACKAGE FOR LINE TRANSECT DATA

J. L. Laake, K. P. Burnham, and D. R. Anderson, Utah Cooperative Wildlife Research Unit, Utah State University, Logan, UT 84322.

BREEDING BIRD CENSUSES IN FREQUENTLY BURNED PITCH PINE-SCRUB OAK FOREST

Trevor L. Lloyd-Evans, Manomet Bird Observatory, Manomet, MA 02345.

MEDIAN STRIP METHOD OF CENSUS

D. A. MacLulich, 26 Stewart Street, Strathroy, Ontario N7G 3K9, Canada.

THE USE OF VARIABLE CIRCULAR PLOT SURVEYS IN COMPARING BIRD DENSITIES IN MINED AND UNMINED HABITATS IN NORTH FLORIDA

David S. Maehr, School of Forest Resources and Conservation, 118 Newins-Ziegler Hall, University of Florida, Gainesville, FL 32611.

ANALYSIS OF "INSTANTANEOUS" COUNTS FOR ESTIMATION OF ABSOLUTE ABUNDANCE

Lyman McDonald, Departments of Zoology and Statistics, The University of Wyoming, Laramie, WY 82071.

DIFFERENCES IN OBSERVER EFFICIENCY WITH THE MAPPING METHOD

R. J. O'Connor and J. H. Marchant, British Trust for Ornithology, Beech Grove, Tring, Hertfordshire, England HP23 5NR.

BIRD SPECIES DIVERSITY AND GUILD STRUCTURE OF A TROPICAL FALLOW RICEFIELD

David R. Osborne, Department of Zoology, Miami University, Oxford, OH 45056.

USING LINE TRANSECTS TO ESTIMATE DENSITY OF TERRITORIAL MALES IN NON-ROADSIDE HABITAT

D. L. Otis, Section of Bird Damage Control, U.S. Fish & Wildlife Service, Building 16, Denver Federal Center, Denver, CO 80225.

ANNUAL VARIATION IN BREEDING BIRD CENSUSES

John R. Probst, Connie P. Cooper, and Donald Rakstad, North Central Forest Experiment Station, USDA Forest Service, St. Paul, MN 55108.

A ONE-SPECIES TEST OF THE SPOT-MAP CENSUS METHOD

R. R. Roth and J. T. Paul, Jr., Department of Entomology and Applied Ecology, University of Delaware, Newark, DE 19711.

A THREE-DIMENSIONAL ANALYSIS OF AVIAN HABITAT FROM A BIRD'S EYE VIEW

James E. Skaley, Resource Information Laboratory, Cornell University, Ithaca, NY 14853.

RIGHTS OF BIRDS AND RIGHTS OF WAY

Jim Stapleton, John Burroughs Nature Sanctuary, P.O. #220, West Park, NY 12493, and Erik Kiviat, Smithtown Road, Fishkill, NY 12524.

INTENSIVE PLOT CENSUS IN A NORTHERN HARDWOODS FOREST

F. W. Sturges and R. T. Holmes, Division of Science and Math, Shepherd College, Shepherdstown, WV 25443.

COMPARISON OF VARIABLE CIRCLE AND SPOT-MAP METHODS IN DESERT RIPARIAN AND SCRUB HABITATS

Robert C. Szaro and Martin D. Jakle, Rocky Mountain Forest and Range Experiment Station, Forestry Sciences Lab., Arizona State University, Tempe, AZ 85281.

APPLICATIONS OF THE BREEDING BIRD SURVEY (BBVS) IN BASELINE AND MONITORING STUDIES

Larry S. Thompson, Department of Natural Resources and Conservation, 32 So. Ewing, Helena, MT 59601.

EXCESSES IN AVIAN SURVEY METHODS USED FOR THE PREDICTION OF IMPACTS

William T. Tucker and Steven J. Brodhead, United Engineers & Constructors, Inc., 100 Summer Street, Boston, MA 02110.

VARIATION IN EFFECTIVENESS OF COUNTS OF TWO SPECIES USING TRAPPING-BANDING AND VISUAL OBSERVATION IN A WINTER BIRD-POPULATION STUDY

R. C. Tweit, R and J Associates, 3810 N. Romero Road, #23, Tucson, AZ 85705.

USE OF THE BREEDING BIRD SURVEY METHOD TO STUDY ALTITUDINAL DISTRIBUTION OF BIRDS

Wayne C. Weber, 303-9153 Saturna Drive, Burnaby, B.C., Canada V3J 7K1.

ESTIMATING BIRD DENSITIES USING FIXED DISTANCE POINT COUNTS

Daniel A. Welsh and R. Kenyon Ross, Canadian Wildlife Service, Ontario Region, 1725 Woodward Drive, Ottawa, Ontario K1G 3Z7.

LIMITATIONS OF THE ROADSIDE BREEDING BIRD SURVEY IN DETERMINING SPECIES OCCURRENCE AND POPULATION CHANGE IN LOCAL AREAS (WITH EMPHASIS ON USE IN IMPACT ASSESSMENT)

K. J. Wilson, D. W. Treasure, and T. A. Gatz, U.S. Water and Power Resources Service, Biology Branch, Bismarck, ND 58502.

LEARNING LARGE AREA CENSUS METHODS

Barbara L. Winternitz, The Colorado College, Colorado Springs, CO 80903.

A SIMPLE METHOD FOR ANALYZING TRENDS IN BREEDING-BIRD SURVEY DATA

John L. Zimmerman, Division of Biology, Kansas State University, Manhattan, KS 66506.

Studies in Avian Biology No. 6:587–590, 1981.

APPENDIX VIII
LIST OF SYMPOSIUM PARTICIPANTS

Lawrence R. Abbott
Ehrenburg, Arizona

A. S. Adegoke
Kaduna, Nigeria

Fred J. Alsop, III
Kingsport, Tennessee

Debby Amos
Anchorage, Alaska

Bertin W. Anderson
Blythe, California

Daniel W. Anderson
Davis, California

David R. Anderson
Logan, Utah

Kathleen S. Anderson
Manomet, Massachusetts

Stanley H. Anderson
Laurel, Maryland

Walter Anderson
Colusa, Georgia

John Annear
Corvallis, Oregon

Robert Arbib
New York, New York

Wayne J. Arendt
Palmer, Puerto Rico

Patricia C. Arrowood
Davis, California

Joe Artmann
Arvada, Colorado

Larry Axelrod
Northridge, California

Pat Baird
Anchorage, Alaska

Kenneth N. Baker
Philadelphia, Pennsylvania

James A. Baldwin
Honolulu, Hawaii

Reginald Barrett
Berkeley, California

Thomas W. Barry
Alberta, Canada

Jonathan Bart
Columbus, Ohio

Michael Basham
San Jose, California

Oron L. Bass, Jr.
Homestead, Florida

Fillibus Bature
Kaduna, Nigeria

Elizabeth Bauer
Colorado Springs, Colorado

Dean Bazzi
Goleta, California

Carl Becker
Springfield, Illinois

Edward C. Beedy
Davis, California

Louis Best
Ames, Iowa

Hugh Black
Washington, District of Columbia

Robert J. Blohm
Crofton, Maryland

Lawrence J. Blus
Corvallis, Oregon

Sandra Bobroft
no address given

Carl F. Bock
Boulder, Colorado

Gerard C. Boere
Utrecht, Netherlands

Susan Bon
New Ipswich, New Hampshire

John W. Brack
McKinleyville, California

Dawn Breese
Volcano, Hawaii

Evelyn Bull
La Grande, Oregon

Kenneth Burnham
Ft. Collins, Colorado

Timothy A. Burr
Honolulu, Hawaii

Daniel C. Butler
Wheat Ridge, Colorado

Danny Bystrak
Laurel, Maryland

Mayo Call
Littleton, Colorado

Robert Z. Callaham
Berkeley, California

Mary Candee
Bolinas, California

Richard Cannings
Vancouver, B.C., Canada

David E. Capen
Burlington, Vermont

Barbara A. Carlson
Upland, California

Daniel Cary
Coral Gables, Florida

John Cary
Madison, Wisconson

Charles Chase, III
Denver, Colorado

Douglas T. Cheeseman, Jr.
Cupertino, California

Chris Chouteau
Oakland, California

David S. Christie
St. John, Canada

Steven P Christman
Gainesville, Florida

Dennis M. Christopherson
Billings, Montana

Richard L. Clawson
Columbia, Missouri

Howard L. Cogswell
Hayward, California

Gerald Collier
San Diego, California

Charles T. Collins
Long Beach, California

Mildred C. Comar
Atherton, California

Sheila Conant
Honolulu, Hawaii

Peter F Connell
New York, New York

Peter G. Connor
Bodega Bay, California

James Cook
Ontario, California

Robert J. Cooper
Norcross, Georgia

Blair A. Csuti
Sacramento, California

Eric Cummins
Olympia, Washington

André Cyr
Quebec, Canada

Bernard Daemen
Voorburg, The Netherlands

David E. Davis
Santa Barbara, California

Mark Davis
Battle Mountain, Nevada

William E. Davis
Foxboro, Massachusetts

D. G. Dawson
Lower Hutt, New Zealand

Deanna K. Dawson
Laurel, Maryland

Anthony R. DeGange
Arcata, California

Michael De Jong
Madison, Wisconsin

Penelope Delevoryas
San Jose, California

David F DeSante
Stinson Beach, California

Jean-Luc DesGranges
Quebec, Canada

Barbara Diehl
Wysockiego, Poland

Kenneth R. Dixon
Frostburg, Maryland

Richard Dolbeer
Sandusky, Ohio

Randall Downer
no address given

Susan Roney Drennan
New York, New York

William E. Dusenberry
Denver, Colorado

Paul F J. Eagles
Ontario, Canada

David M. Ealey
Alberta, Canada

Daniel K. Edwards
Seasite, Oregon

Jan Ekman
Goteburg, Sweden

Raymond Ekstrom
no address given

William Eldridge
Anchorage, Alaska

Douglas Ellis
Eldridge, California

John T. Emlen
Madison, Wisconsin

Richard M. Engeman
Denver, Colorado

Todd Engstrom
Tallahassee, Florida

Martin Erdelen
West Germany

Curtis Erickson
Olympia, Washington

Richard A. Erickson
Antioch, California

A. J. Erskine
Sackville, Canada

Chuck Evans
Fresno, California

Keith Evans
Columbia, Missouri

Michael U. Evans
San Diego, California

Craig A. Faanes
Jamestown, North Dakota

J. Bruce Falls
Toronto, California

Gary A. Falxa
Ventura, California

Camille Ferry
Dijon, France

Elmer J. Finck
Manhattan, Kansas

George Finney
Ottawa, Canada

Jan M. Fischer
Anderson, California

Robert J. Fischer
Portland, Oregon

John Fleckenstein
Fargo, North Dakota

Mary Anne Flett-Lambert
Walnut Creek, California

Glenn R. Ford
Albuquerque, New Mexico

Susie Formenti
Morgan Hill, California

Kathleen E. Franzreb
Elk Grove, California

Ron Freeman
San Diego, California

Kathryn Freemark
Ottawa, Canada

Ron Friesz
no address given

Jeffrey B. Froke
San Juan Capistrano, California

Mark R. Fuller
Laurel, Maryland

Edward O. Garton
Moscow, Idaho

Charles E. Gates
College Station, Texas

Robert J. Gates
Idaho Falls, Idaho

Thomas Gatz
Bismarck, North Dakota

Paul H. Geissler
Laurel, Maryland

Robert Gill, Jr.
Anchorage, Alaska

Ernest Gluesing
Denver, Colorado

Sharon Goldwasser
Tucson, Arizona

Anthony Gomez
Riverside, California

J. Paul Goossen
Manitoba, Canada

Paul Gorenzel
Davis, California

Rod Goss
Porterville, California

Patrick J. Gould
Anchorage, Alaska

Steve Granholm
Davis, California

C. Val Grant
Logan, Utah

Randall L. Gray
Davis, California

Richard Greer
North Rim, Arizona

Laura Greffinius
Bolinas, California

James P. Griffing
Williams, Arizona

Christian E. Grue
Laurel, Maryland

Paul C. Guilkey
Berkeley, California

John Robert Gulley
Alberta, Canada

Pam Gunther
Skykomsh, Washington

Perry Haaland
Coral Gables, Florida

James L. Hainline
Tulelake, California

Daniel J. Hall
Fremont, California

Frank Hall
Stockton, California

Judith Latta Hand
Pacific Palisades, California

Colleen M. Handel
Anchorage, Alaska

Robert B. Hansen
Fresno, California

Roger D. Harris
Berkeley, California

Terry Hart
Sunnyvale, California

Grenville Hatch
La Jolla, California

Charles J. Henny
Corvallis, Oregon

Bob Henry
Tucson, Arizona

Gary Hensler
Laurel, Maryland

Patrick Herzog
Manitoba, Canada

Joseph J. Hickey
Madison, Wisconsin

Olavi Hildén
Helsinki, Finland

Herbert O. Hill
Las Vegas, Nevada

Susan Hills
Seattle, Washington

Jim C. Hitchcock
Nevada City, California

Mary E. Hogan
Anchorage, Alaska

David Hunt
Cornwall, England

David J. T. Hussell
Ontario, Canada

Richard L. Hutto
Missoula, Montana

Douglas James
Fayetteville, Arkansas

Frances C. James
Tallahassee, Florida

Ross D. James
Ontario, Canada

Olli Järvinen
Helsinki, Finland

Douglas H. Johnson
Jamestown, North Dakota

R. Roy Johnson
Tucson, Arizona

Stephen Johnson
Fairbanks, Alaska

George M. Jolly
Scotland

Jim Jones
Toronto, Canada

Paul D. Jorgensen
La Mesa, California

Nancy Elaine Joste
Salyer, California

Michael R. Joyce
Anchorage, Alaska

Gary Kaiser
no address given

Herbert W. Kale
Vero Beach, Florida

James R. Karr
Champaign, Illinois

Mark Kasprzyk
Ehrenbert, Arizona

Darrel Kaufman
no address given

Tom Keeney
Rosemeade, California

Barry Keller
Pocatello, Idaho

Cherry Keller
Slidell, Louisiana

Robert L. Kelley
Coral Gables, Florida

Sean T. Kelly
Reynoldsburg, Ohio

Charles A. Kemper
Chippewa Falls, Wisconsin

Charles Kennedy
Tucson, Arizona

Cameron B. Kepler
Kula, Hawaii

Winifred Kessler
Moscow, Idaho

Rod King
Anchorage, Alaska

Ronald E. Kirby
Washington, District of Columbia

Vernon M. Kleen
Springfield, Illinois

Ronald J. Klein
Portland, Oregon

Fritz Knopf
Ft. Collins, Colorado

Connie Kolb
Morro Bay, California

James R. Koplin
Douglas, Arizona

Ernest Kosaka
Kaneohe, Hawaii

Steve Kovack
Pt. Mugu, California

Lynne D. Krasnow
Anchorage, Alaska

Jun-Young Kwack
Seoul, Korea

Stephen A. Laymon
Yreka, California

Linda Leddy
Manomet, Massachusetts

Eugene Le Febvre
Carbondale, Illinois

William Lehnhausen
Anchorage, Alaska

Lora Leschner
Olympia, Washington

Ron LeValley
McKinleyville, California

Thomas S. Litwin
Ithaca, New York

Trevor L. Lloyd-Evans
Manomet, Massachusetts

Gordon Loery
Morris, Connecticut

K. J. Lyons
Manitoba, Canada

Anne F Maben
Guam

Duncan A. Mac Lulich
Ontario, Canada

David S. Maehr
Gainesville, Florida

David A. Manuwal
Seattle, Washington

Wayne R. Marion
Gainesville, Florida

W. H. Marshall
no address given

John Martin
Huntsville, Utah

R. P Martin
no address given

Barbara W. Massey
Long Beach, California

Harold F Mayfield
Waterville, Ohio

Donald McCrimmon
Ithaca, New York

Lyman McDonald
Laramie, Wyoming

Beverly McIntosh
Long Beach, California

Julie K. Meents
Blythe, California

John F Mehner
Staunton, Virginia

Charles Meslow
Corvallis, Oregon

James Metz
Vernal, Utah

L. Richard Mewaldt
San Jose, California

Pete Meyers
Berkeley, California

Sharon Mikol
Lander, Wyoming

Karen Miller
Sacramento, California

Pam Miller
Olympia, Washington

Robert L. Miller
Delmar, New York

Steve Miller
Kent, Washington

Robert Montgomery
Dundee, Illinois

Michael Morrison
Corvallis, Oregon

J. P Myers
Berkeley, California

Luis G. Naranjo
Colombia, South America

William A. Neitro
Aloha, Oregon

Jim Nichols
Laurel, Maryland

Lars Nilsson
Linloping, Sweden

Barry R. Noon
Laurel, Maryland

Bob Oakleaf
Lander, Wyoming

Raymond O'Connor
Hertfordshire, England, UK

Hans Oelke
Göttingen, Germany

Bill Oelklaus
Portland, Oregon

Robert D. Ohmart
Tempe, Arizona

John Oldemeyer
no address given

Timothy E. O'Meara
Gainesville, Florida

Edward J. O'Neil
Tulelake, California

David Osborne
Oxford, Ohio

David Otis
Lakewood, Colorado

Oscar T. Owre
Coral Gables, Florida

Tom Palmer
Sacramento, California

Larry F Pank
Hilo, Hawaii

Peter A. Pearce
New Brunswick, Canada

Stephen Penland
Lynnwood, Washington

Allan Pfister
no address given

Michael Phelan
Fort Collins, Colorado

Mark Phillips
Battle Mountain, Nevada

Charles W. Philpot
Riverside, California

Wayne R. Pilz
Albuquerque, New Mexico

Stuart L. Pimm
Lubbock, Texas

Frank A. Pitelka
Berkeley, California

Kenneth Pollock
Raleigh, North Carolina

John Polson
Saskatchewan, Canada

Mark A. Ports
Elko, Nevada

Dennis M. Power
Santa Barbara, California

John R. Probst
St. Paul, Minnesota

Susan E. Quinlan
Anchorage, Alaska

Terrance J. Quinn, II
Seattle, Washington

Ralph J. Raitt
Las Cruces, New Mexico

C. John Ralph
Honolulu, Hawaii

Fred L. Ramsey
Corvallis, Oregon

P V. Rao
Gainesville, Florida

Martin G. Raphael
Salyer, California

Harry F Recher
Sydney, Australia

Roland L. Redmond
Missoula, Montana

Ronald L. Refsnider
Coon Rapids, Minnesota

Richard T. Reynolds
Fort Collins, Colorado

Jake Rice
Tempe, Arizona

Kenneth W. Rice
Anchorage, Alaska

Douglas G. Richards
Honolulu, Hawaii

Merle Richmond
Vancouver, Washington

Robert E. Ricklefs
Philadelphia, Pennsylvania

M. G. Ridpath
Winnellie, Australia

John P. Rieger
San Diego, California

Lyman V Ritter
Fresno, California

Virginia Rivers
no address given

Chandler S. Robbins
Laurel, Maryland

Michael Robbins
Yreka, California

Bay Roberts
no address given

Mary N. Rodgers
Sacramento, California

Elizabeth Rodrick
Olympia, Washington

Jerry A. Roppe
Portland, Oregon

Kenneth V Rosenberg
Tempe, Arizona

John Rotenberry
Albuquerque, New Mexico

Roland R. Roth
Newark, Delaware

Paul Rothbart
Battle Mountain, Nevada

Terrell P. Salmon
Davis, California

Michael O. Samuel
Moscow, Idaho

Sharon Sarappo
Becker, Minnesota

Wayne Sarappo
Becker, Minnesota

Jean-Pierre Savard
British Columbia, Canada

Peter Sawyer
no address given

Mary Lou Schnoes
Minneapolis, Minnesota

Betty A. Schreiber
Los Angeles, California

Ralph Schreiber
Los Angeles, California

Len Schumann
East Lansing, Michigan

Richard Schwendinger
Denver, Colorado

J. Michael Scott
Volcano, Hawaii

Thomas Scott
San Diego, California

Henri Seibert
Athens, Ohio

A. R. Sen
Ottawa, Ontario, Canada

Mark L. Shaffer
Washington, District of Columbia

David A. Sharpnack
Berkeley, California

George P. Sheppard
St. George, Utah

Steven L. Sheriff
Columbia, Maryland

James T. Shiflett
Lyman, Washington

Tom Shoemaker
Ft. Collins, Colorado

Jeffrey Short
Panama City, Florida

Kath Shurcliff
South Australia

James E. Skaley
Ithaca, New York

John R. Skalski
Richland, Washington

Aaron A. Skirvin
Corvallis, Oregon

Carey S. Smith
no address given

Charles R. Smith
Ithaca, New York

Graham W. Smith
Boise, Idaho

Kimberly G. Smith
Bodega Bay, California

Neal G. Smith
Panama, Central America

Norman S. Smith
Tucson, Arizona

Peter C. Sorenson
Meeker, Colorado

Arthur L. Sowls
Arcata, California

Paul F Springer
Arcata, California

John A. Sproul, Jr.
El Paso, Texas

Alexander Sprunt, IV
Tavernier, Florida

Jim Stapleton
West Park, New York

Allen R. Stickley, Jr.
Bowling Green, Kentucky

Peter A. Stine
Portland, Oregon

Robert Stoecker
Boulder, Colorado

Lynne Stokes
Seabrook, Maryland

Charles P. Stone
Lakewood, Colorado

Mark A. Strong
Portland, Oregon

F W. Sturges
Shepherdstown, West Virginia

Joseph Suchecki
Northbrook, Illinois

Soren E. Svensson
Lund, Sweden

Chris Swarth
Oakland, California

Jeff Swinebroad
Washington, District of Columbia

Robert C. Szaro
Tempe, Arizona

Lasio S. Sziji
Pomona, California

John Taapken
Baraboo, Wisconsin

John Y. Takekawa
Moscow, Idaho

Warwick R. Tarboton
Nylstrom, South Africa

William E. Taylor
Escanaba, Michigan

Stanley A. Temple
Madison, Wisconsin

Carl Thelander
San Francisco, California

Glenn D. Therres
Baltimore, Maryland

Larry S. Thompson
Helena, Montana

Steve Thompson
Hines, Oregon

Nancy G. Tilghman
Amherst, Maryland

Steven G. Torres
Tucson, Arizona

David L. Trauger
Washington, District of Columbia

Donald W. Treasure
Bismarck, North Dakota

Declan Troy
Fairbanks, Alaska

William T. Tucker
Kingston, New Hampshire

Robert C. Tweit
Tucson, Arizona

Miklos D. F Udvardy
Sacramento, California

Risto A. Väisänen
Finland

Brad Valentine
Fresno, California

Charles van Riper, III
Davis, California

Jared Verner
Fresno, California

Larry D. Voorhees
Oak Ridge, Tennessee

F. Vuilleumier
New York, New York

Judith L. Wagner
Los Altos, California

Robert B. Waide
San Juan, Puerto Rico

Jeffrey Walters
Clifton Springs, New York

Brian Walton
Santa Cruz, California

Richard E. Warner
Davis, California

Wayne C. Weber
Burjaby B.C., Canada

Doug Wechsler
Seattle, Washington

Charles Weise
Milwaukee, Wisconsin

Daniel Welsh
Ontario, Canada

Ted Wenzel
Butte, Montana

Harold Werner
Three Rivers, California

Frank Wernette
Stockton, California

Karen West
Madison, Wisconsin

Robin White
Madison, Wisconson

E. G. White-Swift
Coosbay, Oregon

Robert C. Whitmore
Morgantown, West Virginia

John A. Wiens
Albuquerque, New Mexico

Harold A. Wier
San Diego, California

Sanford R. Wilbur
Ojai, California

Byron K. Williams
Laurel, Maryland

Norman A. Williams
Newfoundland, Canada

Anne I. Wilson
Los Altos Hills, California

Karen Wilson
Deer Lodge, Montana

Jon Winter
Santa Rosa, California

Barbara L. Winternitz
Colorado Springs, Colorado

Richard A. Wolinski
Ann Arbor, Michigan

Pyong-Oh Won
Seoul, Korea

Christopher Wood
Weston, Connecticut

Helen S. Wood
Colorado Springs, Colorado

Patty Worthing
San Bruno, California

Ellen Yeoman
Livermore, California

Jan Young
Logan, Utah

Richard Zembal
Laguna Niguel, California

Michael C. Zicus
Bemidji, Minnesota

John L. Zimmerman
Manhattan, Kansas

Jim Zook
Phoenix, Arizona

LITERATURE CITED

COMPILED BY STEPHEN R. SABO[1,2]

ABBOTT, I. 1980. Theories dealing with the ecology of landbirds on islands. Adv. Ecol. Res. 11:329–371.

ABLE, K. P., AND B. R. NOON. 1976. Avian community structure along elevational gradients in the northeastern United States. Oecologia (Berl.) 26:275–294.

ADAMS, C. C. 1908. The ecological succession of birds. Auk 25:109–153.

ADAMS, P. M. 1977. Effects of anticholinergic and cholinesterase blocking drugs on appetitive behavior under different deprivation conditions. Life Sci. 21:129–136.

ALBERS, P. H. 1976. Determining population size of territorial Red-winged Blackbirds. J. Wildl. Manage. 40:761–768.

ALDRICH, J. W. 1963. Geographic orientation of American Tetraonidae. J. Wildl. Manage. 27:529–545.

ALDRICH, J. W., AND R. W. COFFIN. 1980. Breeding-bird populations from forest to suburbia after thirty-seven years. Am. Birds 34:3–7.

ALDRICH, J. W., AND J. S. WESKE. 1978. Origin and evolution of the eastern House Finch population. Auk 95:528–536.

ALERSTAM, T. 1978. Analysis and a theory of visible bird migration. Oikos 30:273–349.

ALERSTAM, T., AND S. ULFSTRAND. 1972. Radar and field observations of diurnal bird migration in south Sweden, autumn, 1971. Ornis Scand. 3:99–139.

ALEXANDER, H. G. 1931. The effect of severe weather on bird song. Br. Birds 25:97–101.

ALEXANDER, W. C. 1973. Abundance and habitat preference of birds on the slopes of Grandfather Mountain, North Carolina, April–November. J. Mitchell Soc. 89:92–100.

ALLEN, P. F., AND P. R. SIME. 1943. Hawk census on a Texas panhandle highway. Wilson Bull. 55:29–39.

ALLARD, H. A. 1930. The first morning song of some birds of Washington, D.C.; its relation to light. Am. Nat. 64:436–469.

AMERICAN ORNITHOLOGISTS' UNION. 1957. Check-list of North American birds. Fifth ed. American Ornithologists' Union, Baltimore, Md.

AMERICAN ORNITHOLOGISTS' UNION. 1973. Thirty-second supplement to the American Ornithologists' Union check-list of North American birds. Auk 90:411–419.

AMERICAN ORNITHOLOGISTS' UNION. 1976. Thirty-third supplement to the American Ornithologists' Union check-list of North American birds. Auk 93: 875–879,

AMMAN, G. D., AND P. H. BALDWIN. 1960. A comparison of methods for censusing woodpeckers in spruce-fir forests of Colorado. Ecology 41:699–706.

ANDERSON, A. 1963. Patagial tags for waterfowl. J. Wildl. Manage. 27:284–288.

ANDERSON, B. W., R. W. ENGEL-WILSON, D. WELLS, AND R. D. OHMART. 1977. Ecological study of southwestern riparian habitats: techniques and data applicability. Pp. 146–155 *in* R. R. Johnson and D. A. Jones (tech. coords.). Importance, preservation and management of riparian habitat: a symposium (Tucson, Ariz., July 9, 1977). U.S. For. Serv., Gen. Tech. Rep. RM-43, Fort Collins, Col.

ANDERSON, B. W., AND R. D. OHMART. 1977. Climatological and physical characteristics affecting avian population estimates in Southwestern riparian communities using transect counts. Pp. 193–200 *in* R. R. Johnson and D. A. Jones (tech. coords.). Importance, preservation and management of riparian habitat: a symposium (Tucson, Ariz., July 9, 1977). U.S. For. Serv., Gen. Tech. Rep. RM-43, Fort Collins, Col.

ANDERSON, B. W., AND R. D. OHMART. 1981a. Comparisons of avian census results using variable distance transect and variable circular plot techniques. Pp. 186–192 *in* C. J. Ralph and J. M. Scott (eds.). Estimating the numbers of terrestrial birds. Stud. Avian Biol. 6.

ANDERSON, B. W., R. D. OHMART, AND JAKE RICE. 1981b. Seasonal changes in avian densities and diversities. Pp. 262–264 *in* C. J. Ralph and J. M. Scott (eds.). Estimating the numbers of terrestrial birds. Stud. Avian Biol. 6.

ANDERSON, D. R., K. P. BURNHAM, AND B. R. CRAIN. 1978. A log-linear model approach to estimation of population size using the line transect sampling method. Ecology 59:190–193.

ANDERSON, D. R., J. L. LAAKE, B. R. CRAIN, AND K. P. BURNHAM. 1979. Guidelines for line transect sampling of biological populations. J. Wildl. Manage. 43:70–78.

ANDERSON, D. R., AND R. S. POSPAHALA. 1970. Correction of bias in belt transect studies of immotile objects. J. Wildl. Manage. 34:141–146.

ANDERSON, D. R., AND R. T. STERLING. 1974. Population dynamics of molting Pintail drakes banded in south-central Saskatchewan. J. Wildl. Manage. 38:266–274.

ANDERSON, D. W., AND I. T. ANDERSON. 1976. Distribution and status of Brown Pelicans in the California Current. Am. Birds 30:3–12.

ANDERSON, S. H. 1970a. Forest birds of western Oregon. Ph.D. thesis, Oregon State Univ., Corvallis.

ANDERSON, S. H. 1970b. The avifaunal composition of Oregon white oak stands. Condor 72:417–423.

ANDERSON, S. H. 1972. Season variations in forest birds of western Oregon. Northwest Sci. 46:194–206.

ANDERSON, S. H.. 1979a. Changes in forest bird species composition caused by transmission-line corridor cuts. Am. Birds 33:3–6.

ANDERSON, S. H. 1979b. Habitat assessment for breeding bird populations. Trans. N. Am. Wildl. Nat. Res. Conf. 44:431–445.

ANDERSON, S. H. 1979c. Habitat structure, succession, and bird communities. Pp. 9–21 *in* M. DeGraff (tech. coord). Proceedings of workshop, management of northcentral and northeastern forests for nongame birds, U.S. For. Serv., Gen. Tech. Rep. NC-51, St. Paul, Minn.

[1] U.S. Fish and Wildlife Service, Mauna Loa Field Station, P.O. Box 44, Hawaii National Park, Hawaii 96718.

[2] The assistance of Sharon Scott, Ann Tuscany, and J. Michael Scott is gratefully acknowledged.

ANDERSON, S. H. 1981. Correlating habitat variables and birds. Pp. 538–542 *in* C. J. Ralph and J. M. Scott (eds.). Estimating the numbers of terrestrial birds. Stud. Avian Biol. 6.

ANDERSON, S. H., AND H. H. SHUGART. 1974. Habitat selection of breeding birds in an east Tennessee deciduous forest. Ecology 55:828–837.

ANDREWARTHA, H. G. 1971. Introduction to the study of animal populations. Univ. Chicago Press, Chicago.

ANDREWARTHA, H. G., AND L. C. BIRCH. 1954. The distribution and abundance of animals. Univ. of Chicago Press, Chicago, Ill.

ANDRILENAS, P. A., AND T. R. EICHERS. 1977. Evaluation of pesticide supplies and demand for 1977. U.S. Dept. Agric., Econ. Res. Ser., Agric. Econ. Rep. 336.

ANKNEY, C. D. 1975. Neckbands contribute to starvation in female Lesser Snow Geese. J. Wildl. Manage. 39:825–826.

ANONYMOUS. 1937. Announcement of the Breeding Bird Census. Bird Lore 39:147–150.

ANONYMOUS. 1947. Announcement of the Winter Bird-Population Study. Aud. Field Notes 1:165–166.

ANONYMOUS. 1950. Instructions for making bird population studies. Aud. Field Notes 4:183–187.

ANONYMOUS. 1969. Recommendations for an international standard for a mapping method in bird census work. Bird Study 16:249–254.

ANONYMOUS. 1970. Recommendations for an international standard for a mapping method in bird census work. Pp. 49–52 *in* S. Svensson (ed.). Bird census work and environmental monitoring. Bull. Ecol. Res. Comm. 9, Swedish Nat. Sci. Res. Counc., Stockholm.

ANT, H., AND H. ENGLKE. 1973. Die Naturschutzgebiete der Bundesrepublik Deutschland. Bundesanstalt für Vegetationskunde, Naturschutz and Landschaftspflege, Bonn-Bad Godesberg.

ARBIB, R. S. 1967. Considering the Christmas census. Aud. Field Notes 21:39–42.

ARBIB, R. S. 1972. On the art of estimating numbers. Am. Birds 26:706–712.

ARBIB, R. S. 1981. The Christmas Bird Count: constructing an "ideal model." Pp. 30–33 *in* C. J. Ralph and J. M. Scott (eds.). Estimating the numbers of terrestrial birds. Stud. Avian Biol. 6.

ARMBRUSTER, M. J., T. S. BASKETT, W. R. GOFORTH, AND K. C. SADLER. 1978. Evaluating call-count procedures for measuring local Mourning Dove populations. Trans. Missouri Acad. Sci. 12:75–90.

ARMSTRONG, E. A. 1963. A study of bird song. Oxford Univ. Press, London.

ARMSTRONG, E. A. 1973. A study of bird song. Dover, New York.

ARNASON, A. N. 1973. The estimation of population size, migration rates and survival in a stratified population. Res. Popul. Ecol. 15:1–8.

ARNASON, A. N., AND L. BANIUK. 1978. POPAN-2, a data maintenance and analysis system for recapture data. Charles Babbage Res. Cen., St. Pierre, Man.

ARNASON, A. N., AND L. BANIUK. 1980. A computer system for mark-recapture analysis of open populations. J. Wildl. Manage. 44:325–332.

ARNASON, A. N., AND N. S. KREGER. 1973. User's manual, POPAN-I: A program to calculate Jolly-Seber estimates from mark-recapture data with pooling methods to increase the precision. Univ. Manitoba, Dept. Computer Science Rep. 76, Winnipeg.

ARTMANN, J. W. 1971. Capturing Sharp-tailed Grouse hens by using taped chick distress calls. J. Wildl. Manage. 35:557–559.

ARTMANN, J. W. 1977. Woodcock status report, 1976. U.S. Fish and Wildl. Serv., Spec. Sci. Rep. Wildl. 209.

ASHERIN, D. A., H. L. SHORT, AND J. E. ROELLE. 1979. Regional evaluation of wildlife habitat quality using rapid assessment methodologies. Trans N. Am. Wildl. Nat. Res. Conf. 44:404–424.

ASTROM, G. 1976. Environmental influences on the daily song activity of the Reed Bunting, *Emberiza schoeniclus*. Zoon, Suppl. 2. Uppsala Univ., Sweden.

AUSTIN, G. T. 1976. Behavioral adaptations of the Verdin to the desert. Auk 93:245–262.

BAGG, A. M. 1950. Minimum temperatures and maximum flights. Connecticut Valley hawk flights—September 1949. Bull. Mass. Aud. Soc. 34:75–80.

BAILEY, N. T. J. 1951. On estimating the size of mobile populations from recapture data. Biometrika 38:293–306.

BAILEY, R. S. 1967. An index of bird population changes on farmland. Bird Study 14:195–209.

BAILEY, R. W., AND K. T. RINELL. 1968. West Virginia wild Turkey. West Virginia Dept. Nat. Resour., Div. Game and Fish., Bull. 6.

BAIRD, K. E. 1974. A field study of the King, Sora, and Virginia Rails at Cheyenne Bottoms in west central Kansas. M.S. thesis, Fort Hays Kansas State Coll., Hays.

BAKER, B. W. 1977. An analysis of the Breeding Bird Survey in northwestern California. M.S. thesis, Humboldt State Univ., Arcata, Calif.

BALDA, R. P. 1975a. Vegetation structure and breeding bird diversity. Pp. 59–80 *in* D. R. Smith (tech. coord.). Proceedings of the symposium on management of forest and range habitats for nongame birds. U.S. For. Serv., Gen. Tech. Rep. WO-1, Washington, D.C.

BALDA, R. P. 1975b. The relationship of secondary cavity nesters to snag densities in western coniferous forests. U.S. For. Serv., Wildl. Hab. Tech. Bull. 1, Albuquerque, N.M.

BALDWIN, S. P., S. C. KENDEIGH, AND R. W. FRANKS. 1932. The protection of hawks and owls in Ohio. Ohio J. Sci. 32:403–424.

BALPH, M. H., L. C. STODDART, AND D. F. BALPH. 1977. A simple technique for analyzing bird transect counts. Auk 94:606–607.

BANKS, A., AND R. W. RUSSELL. 1967. Effects of chronic reductions in acetylcholinesterase activity on serial problem-solving behavior. J. Comp. Physiol. Psychol. 64:262–267.

BARLOW, J. C., AND R. R. JOHNSON. 1969. The Gray Vireo, *Vireo vicinior* Coues (Passeriformes: Vireonidae), in the Sierra del Carmen, Coahuila, Mexico. Can. J. Zool. 47:151–152.

BARNES, C. P., AND F. J. MARSCHNER. 1933. Natural

land-use areas of the United States. U.S. Dept. Agric., Bur. Agric. Econ.

BART, J. 1977. Winter distribution of Red-tailed Hawks in central New York State. Wilson Bull. 89:623–625.

BART, J. 1979. Effects of acephate and sevin on forest birds. J. Wildl. Manage. 43:544–549.

BARTELT, P. E., AND C. J. ORDE. 1976. A raptor survey and trapping technique in Clay County, South Dakota. South Dakota Bird Notes 28:24–30.

BARTONEK, J. C., AND C. W. DANE. 1964. Numbered nasal discs for waterfowl. J. Wildl. Manage. 28:688–692.

BARWICK, L. H., D. H. AUSTIN, AND L. E. WILLIAMS. 1970. Roosting of young Turkey broods during summer in Florida. Proc. Annu. Conf. Southeast. Assoc. Game and Fish Comm. 24:231–243.

BASKETT, T. S., M. J. ARMBRUSTER, AND M. W. SAYRE. 1978. Biological perspectives for the Mourning Dove call-count survey. Trans. N. Am. Wildl. Nat. Res. Conf. 43:163–180.

BATTEN, L. A., AND J. H. MARCHANT. 1976. Bird population changes for the years 1973–74. Bird Study 23:11–20.

BAUMGARTNER, F. M. 1939. Territory and population in the Great Horned Owl. Auk 56:274–282.

BEALS, E. W. 1960. Forest bird communities in the Apostle Islands of Wisconsin. Wilson Bull. 72:156–181.

BEAMAN, M., AND C. GALEA. 1974. The visible migration of raptors over the Maltese Islands. Ibis 116:419–431.

BEARDSLEY, J. W. 1966. Insects and other terrestrial arthropods from the Leeward Hawaiian Islands. Proc. Hawn. Entomol. Soc. 19:157–185.

BEATTY, W. H. 1977. Attracting Screech Owls. Redstart 44:102–104.

BEAVER, J. C. 1976. Sonograms as aids in bird identification. Am. Birds 30:899–903.

BEDNAREK, R., AND C. S. DAVIDSON. 1967. Influence of spraying with carbaryl on nest success in a sample of bird boxes on Cape Cod. Bird-Banding 38:66–72.

BEDNARZ, J. C. 1979. I. Productivity, nest sites, and habitat of Red-shouldered and Red-tailed Hawks in Iowa. II. Status, habitat utilization, and management of Red-shouldered Hawks in Iowa. M.S. thesis, Iowa State Univ., Ames.

BEEBE, S. B. 1974. Relationships between insectivorous hole-nesting birds and forest management. Multilith, Yale Univ. Sch. For. and Environ. Stud., New Haven, Conn.

BEECHER, W. J. 1942. Nesting birds and the vegetation substrate. Chicago Ornithol. Soc., Chicago.

BEGON, M. 1979. Investigating animal abundance: capture-recapture for biologists. Univ. Park Press, Baltimore.

BELL, B. D., C. K. CATCHPOLE, AND K. J. CORBETT. 1968. Problems of censusing Reed Buntings, Sedge Warblers and Reed Warblers. Bird Study 15:16–21.

BELL, B. D., C. K. CATCHPOLE, K. J. CORBETT, AND R. J. HORNBY. 1973. The relationship between census results and breeding populations of some marshland passerines. Bird Study 20:127–140.

BELL, R. E. 1964. A second triangulation method for counting Barred Owls. Wilson Bull. 76:292–294.

BELLROSE, F. C. 1955. A comparison of recoveries from reward and standard bands. J. Wildl. Manage. 19:71–75.

BELLROSE, F. C. 1976. Ducks, geese and swans of North America. Stackpole Books, Harrisburg, Pa.

BELLROSE, F. C., T. G. SCOTT, A. S. HAWKINS, AND J. B. LOW. 1961. Sex ratios and age ratios in North American ducks. Ill. Nat. Hist. Surv. Bull. 27:391–474.

BENNITT, R. 1951. Some aspects of Missouri quail and quail hunting, 1938–1948. Missouri Cons. Comm. Tech. Bull. 2.

BENT, A. C. 1938. Life histories of North American birds of prey (part 2): Falconiformes and Strigiformes. U.S. Natl. Mus. Bull. 170.

BERANEK, L. L. 1954. Acoustics. McGraw-Hill, New York.

BERGERUD, A. T., AND W. E. MERCER. 1966. Census of the Willow Ptarmigan in Newfoundland. J. Wildl. Manage. 30:101–113.

BERGGREN, V., AND J. WAHLSTEDT. 1977. The sound repertoire of the Great Gray Owl, *Strix nebulosa*. Vår Fågelvärld 36:243–249.

BERNES, C. 1980. The Environmental Monitoring Programme in Sweden. Bull. SNV PM 1327, Swedish Protection Board, Solna.

BERRY, K. H. 1978. The east Mojave desert: an example of inventory design and habitat analysis. Pp. 495–519 *in* A. Marmelstein (gen. chmn.). Classification, inventory and analysis of fish and wildlife habitat. U.S. Fish and Wildl. Serv., Biol. Serv. Prog. OBS-78/76, Washington, D.C.

BERRY, R. B. 1971. Peregrine Falcon population survey, Assateague Island, Maryland, fall, 1969. Raptor Res. 5:31–43.

BERRY, R. B., AND F. P. WARD. 1975. Autumn migration of Peregrine Falcons on Assateague Island, 1970–1972. Pp. 57–60 *in* J. R. Murphy, C. M. White, and B. E. Harrell (eds.). Population status of raptors. Raptor Res. Rep. 3.

BERTHOLD, P. 1976. Methoden der Bestandserfassung in der Ornithologie: Übersicht und kritische Betrachtung. J. Ornithol. 117:1–69.

BERTHOLD, P., AND U. QUERNER. 1979. Über Bestandsentwicklung und Fluktuationsrate von Kleinvogelpopulationen: Fünfjährige Untersuchungen in Mitteleuropa. Ornis Fenn. 56:110–123.

BERTHOLD, P., AND R. SCHLENKER. 1975. Das "Mettnau-Reit-Illmitz-Programm"—ein langfristiges Vogelfangprogramm der Vogelwarte Radolfzell mit vielfältiger Fragestellung. Vogelwarte 28:97–123.

BEST, L. B. 1975. Interpretational errors in the "mapping method" as a census technique. Auk 92:452–460.

BEST, L. B. 1977. Territory quality and mating success in the Field Sparrow (*Spizella pusilla*). Condor 79:192–204.

BEST L. B. 1978. Field Sparrow reproductive success and nesting ecology. Auk 95:9–22.

BEST, L. B. 1981. Seasonal changes in detection of individual bird species. Pp. 252–261 *in* C. J. Ralph and J. M. Scott (eds.). Estimating the numbers of terrestrial birds. Stud. Avian Biol. 6.

BEVANGER, K. 1977. Proposal for a new classification of Norwegian bird communities. Biol. Conserv. 11:67–78.

BEZZEL, E., F. LECHNER, AND H. RANFTL. 1980. Arbeitsatlas der Brutvögel Bayerns. Themen der Zeit. Vol. 4. Kilda, Greven, W. Ger.

BEZZEL, E., AND H. UTSCHICK. 1979. Die Rasterkartierung von Sommervogelbeständen—Bedeutung und Grenzen. J. Ornithol. 120:431–440.

BIBBY, C. J. 1978. A heathland bird census. Bird Study 25:87–96.

BILDSTEIN, K. L. 1978. Behavioral ecology of Red-tailed Hawks (*Buteo jamaicensis*), Rough-legged Hawks (*B. lagopus*), Northern Harriers (*Circus cyaneus*), American Kestrels (*Falco sparverius*) and other raptorial birds wintering in south central Ohio. Ph.D. thesis, Ohio State Univ., Columbus.

BILDSTEIN, K. L. 1979a. Fluctuations in the numbers of Northern Harriers (*Circus cyaneus hudsonius*) at communal roosts in south central Ohio. Raptor Res. 13:40–46.

BILDSTEIN, K. L. 1979b. Trends in the number of diurnal raptors seen on recent North and South Carolina Christmas Bird Counts. Chat 43:82–85.

BIRKHEAD, T. R. 1974. A census of rookeries in the Leeds area. Naturalist 1974 (2):65–69.

BIRTWISTLE, G. M., O.-J. DAHL, B. MYHRHAUG, AND K. NYGAARD. 1975. SIMULA begin. Petrocelli/ Charter, New York.

BISHOP, J. A., AND P. M. SHEPPARD. 1973. An evaluation of two capture-recapture models using the technique of computer simulation. Pp. 235–252 *in* M. S. Bartlett and R. W. Hiorns (eds.). The mathematical theory of the dynamics of biological populations. Academic, New York.

BLACK, C. T., AND G. L. ZORB. 1965. Cereal leaf beetle control and wildlife in Michigan. Mich. Dept. Cons., Res. Dev. Rep. 33.

BLANA, H. 1978. Die Bedeutung der Landschaftsstruktur für die Vogelwelt. Beitr. Avifauna Rheinland, vol. 12. Gesellschaft Rheinischer Ornithologen, Düsseldorf.

BLOKPOEL, H., AND W. J. RICHARDSON. 1978. Weather and spring migration of Snow Geese across southern Manitoba. Oikos 30:350–363.

BLONDEL, J. 1969. Méthodes de dénombrement des populations d'oiseaux. Pp. 97–151 *in* M. Lamotte and F. Bourlière (eds.). Problèmes d'écologie: l'échantillonnage des peuplements animaux des milieux terrestres. Masson, Paris.

BLONDEL, J. 1975. L'analyse des peuplements d'oiseaux, élément d'un diagnostic écologique. 1. La méthode des échantillonnages fréquentiels progressifs (E.F.P.). Terre et Vie 29:533–589.

BLONDEL, J. 1976. L'influence des reboisements sur les communautés d'oiseaux, l'exemple du Mont-Ventoux. Ann. Sci. Forest 33:221–245.

BLONDEL, J. 1977. The diagnosis of bird communities by means of frequential sampling (E.F.P.). Pol. Ecol. Stud. 3(4):19–26.

BLONDEL, J. 1978. L'avifaune du Mont-Ventoux, essai de synthèse biogéographique et écologique. Terre et Vie 32, supp. 1:111–145.

BLONDEL, J. 1979. Biogéographie et écologie. Masson, Paris.

BLONDEL, J. 1980. Ecologie et gestion de l'espace naturel, l'apport du "modèle-oiseaux." *In* Ecologie et développement. C.N.R.S., Paris.

BLONDEL, J., AND R. CUVILLIER. 1977. Une méthode simple et rapide pour décrire les habitats d'oiseaux : le stratiscope. Oikos 29:326–331.

BLONDEL, J., C. FERRY, AND B. FROCHOT. 1970. La méthode des indices ponctuels d'abondance (IPA) ou des relevés d'avifaune par "stations d'écoute." Alauda 38:55–71.

BLONDEL, J., C. FERRY, AND B. FROCHOT. 1973. Avifauna et végétation, essai d'analyse de la diversité. Alauda 41:63–84.

BLONDEL, J., C. FERRY, AND B. FROCHOT. 1981. Point counts with unlimited distance. Pp. 414–420 *in* C. J. Ralph and J. M. Scott (eds.). Estimating the numbers of terrestrial birds. Stud. Avian Biol. 6.

BLONDEL, J., AND B. FROCHOT. 1981. How to determine the duration of observation in station counts (unlimited distance). P. 585 (title only) *in* C. J. Ralph and J. M. Scott (eds.). Estimating the numbers of terrestrial birds. Stud. Avian Biol. 6.

BOCK, C. E. 1979. Joseph Grinnell and the Christmas Bird Count. Am. Birds 33:693–694.

BOCK, C. E., AND J. H. BOCK. 1974. Geographical ecology of the Acorn Woodpecker: abundance versus diversity of resources. Am. Nat. 108:694–698.

BOCK, C. E., AND J. H. BOCK. 1978. Response of birds, small mammals, and vegetation to burning sacaton grasslands in southeastern Arizona. J. Range Manage. 31:296–300.

BOCK, C. E., J. H. BOCK, AND L. W. LEPTHIEN. 1977. Abundance patterns of some bird species wintering on the Great Plains of the U.S.A. J. Biogeogr. 4:101–110.

BOCK, C. E., AND L. W. LEPTHIEN. 1974. Winter patterns of bird species diversity and abundance in the United States and southern Canada. Am. Birds 28:556–562.

BOCK, C. E., AND L. W. LEPTHIEN. 1975a. A Christmas Count analysis of woodpecker abundance in the United States. Wilson Bull. 87:355–366.

BOCK, C. E., AND L. W. LEPTHIEN. 1975b. Patterns of bird species diversity revealed by Christmas Counts versus Breeding Bird Surveys. West. Birds 6:95–100.

BOCK, C. E., AND L. W. LEPTHIEN. 1967a. A Christmas Count analysis of the Fringillidae. Bird-Banding 47:264–272.

BOCK, C. E., AND L. W. LEPTHIEN. 1976b. Growth in the eastern House Finch population, 1962–1971. Am. Birds 30:791–792.

BOCK, C. E., AND L. W. LEPTHIEN. 1976c. Population growth in the Cattle Egret. Auk 93:164–166.

BOCK, C. E., AND L. W. LEPTHIEN. 1976d. Synchronous eruptions of boreal seed-eating birds. Am. Nat. 110:559–571.

BOCK, C. E., J. B. MITTON, AND L. W. LEPTHIEN. 1978. Winter biogeography of North American Frigilladae (Aves): a numerical analysis. Syst. Zool. 27:411–420.

BOCK, C. E., AND T. L. ROOT. 1981. The Christmas Bird Counts and avian ecology. Pp. 17–23 *in* C.

J. Ralph and J. M. Scott (eds.). Estimating the numbers of terrestrial birds. Stud. Avian Biol. 6.

BOCK, C. E., AND R. S. SMITH. 1971. An analysis of Colorado Christmas Counts. Am. Birds 25:945–947.

BOCK, W. J. 1979. The synthetic explanation of macroevolutionary change—a reductionist approach. Pp. 20–69 *in* J. H. Schwartz and H. B. Rollins (eds.). Models and methods in evolutionary theory. Bull. Carnegie Mus. Nat. Hist. 13.

BOEKER, E. L. 1971. Golden Eagle population studies in the Southwest. Condor 73:463–467.

BOEKER, E. L. 1974. Status of Golden Eagle surveys in the western states. Wildl. Soc. Bull. 2:46–49.

BOEKER, E. L., AND E. B. BOLEN. 1972. Winter Golden Eagle populations in the Southwest. J. Wildl. Manage. 36:477–484.

BOEKER, E. L., AND T. D. RAY. 1971. Golden Eagle populations in the Southwest. Condor 73:463–467.

BOEKER, E. L., AND V. E. SCOTT. 1969. Roost tree characteristics for Merriam's Turkey. J. Wildl. Manage. 33:121–124.

BOGUCKI, Z. 1977. The ornithological atlas project in Poland. Pol. Ecol. Stud. 3(4):299–303.

BOHL, W. H. 1956. Experiments in locating wild Chukar partridges by use of recorded calls. J. Wildl. Manage. 20:83–85.

BOND, R. R. 1957. Ecological distribution of breeding birds in upland forests of southern Wisconsin. Ecol. Monogr. 27:351–384.

BONNEY, R. E. 1979. Wintering Peregrine Falcon populations in the eastern United States, 1940–1975: a Christmas Bird Count analysis. Am. Birds 33:695–697.

BORCHELT, P. L. 1975. The organization of dustbathing components in Bobwhite quail (*Colinus virginianus*). Behaviour 53:217–237.

BORCHELT, P. L., AND L. DUNCAN. 1974. Dustbathing and feather lipid in Bobwhite (*Colinus virginianus*). Condor 76:471–472.

BORROR, D. J. 1948. Analysis of repeat records of banded White-throated Sparrows. Ecol. Monogr. 18:412–430.

BORROR, D. J. 1964. Songs of the thrushes (Turdidae), wrens (Troglodytidae), and mockingbirds (Mimidae) of eastern North America. Ohio J. Sci. 64:195–207.

BOSWELL, R. S. 1974. A comparative study of the population biology of two sub-species of Great Horned Owl (*Bubo virginianus occidentalis, Bubo virginianus virginianus*). M.S. thesis, Bowling Green State Univ., Bowling Green, Ohio.

BOWMAN, R. I. 1979. Adaptive morphology of song dialects in Darwin's finches. J. Ornithol. 120:353–380.

BOX, G. E. P., W. G. HUNTER, AND J. S. HUNTER. 1978. Statistics for experimenters. Wiley, New York.

BOYCE, S. G. 1978. Management of forests for timber and related benefits (DYNAST-TM). U.S. For. Serv., Res. Pap. SE-184.

BOYD, H. 1956. Statistics of the British population of the Pink-footed Goose. J. Anim. Ecol. 25:253–273.

BRADLEY, R. 197. Making animal sound recordings. Am. Birds 31:279–285.

BRAUN, C. E., R. W. HOFFMAN, AND G. E. ROGERS. 1976. Wintering areas and winter ecology of White-tailed Ptarmigan in Colorado. Col. Div. Wildl., Spec. Rep. 38.

BRAUN, C. E., R. K. SCHMIDT, AND G. E. ROGERS. 1973. Census of Colorado White-tailed Ptarmigan with tape-recorded calls. J. Wildl, Manage. 37:90–93.

BRECKENRIDGE, W. J. 1935. A bird census method. Wilson Bull. 47:195–197.

BRECKENRIDGE, W. J. 1956. Measurements of the habitat niche of the Least Flycatcher. Wilson Bull. 68:47–51.

BREWER, R. 1967. Bird populations of bogs. Wilson Bull. 79:371–395.

BREWER, R. 1972. An evaluation of Winter Bird-Population Studies. Wilson Bull. 84:261–277.

BREWER, R. 1978. A comparison of three methods of estimating winter bird populations. Bird-Banding 49:252–261.

BREWER, R., AND L. SWANDER. 1977. Life history factors affecting the intrinsic rate of natural increase of birds of the deciduous forest biome. Wilson Bull. 89:211–231.

BROWN, D. E., C. L. COCHRAN, AND T. E. WADDELL. 1978. Using call-counts to predict hunting success for Scaled Quail. J. Wildl. Manage. 42:281–297.

BROWN, D. E., AND D. H. ELLIS. 1977. Status summary and recovery plan for the Masked Bobwhite. U.S. Fish and Wildl. Serv., Off. Endang. Spec., Albuquerque, N.M.

BROWN, J. L. 1969. The buffer effect and productivity in tit populations. Am. Nat. 103:347–354.

BROWN, L. 1971. African birds of prey. Houghton Mifflin, Boston.

BROWN, L. 1974. Data required for effective study of raptor populations. Pp. 9–20 *in* F. N. Hamerstrom, B. E. Harrell, and R. R. Olendorff (eds.). Management of raptors. Raptor Res. Rep. 2.

BROWN, L. H. 1977. The status, population structure, and breeding dates of the African Lammergeier. Raptor Res. 11:49–80.

BROWN, L. H., AND D. AMADON. 1968. Eagles, hawks and falcons of the world. Country Life, Hamlyn Publ. Group, London.

BROWN, R. L. 1976a. Mearn's Quail census technique. Final Report. Arizona Game and Fish Dept., Res. Div.

BROWN, W. H. 1964. Population changes in the Red-shouldered and Red-tailed Hawks. Iowa Bird Life 34:82–88.

BROWN, W. H. 1971. Winter population trends in the Red-shouldered Hawk. Am. Birds 25:813–817.

BROWN, W. H. 1973. Winter population trends in the Marsh, Cooper's and Sharp-shinned Hawks. Am. Birds 27:6–7.

BROWN, W. H. 1975. Winter population trends in the Bald Eagle. Am. Birds 29:12–14.

BROWN, W. H. 1976b. Winter population trends in the Black and Turkey Vultures. Am. Birds 30:909–912.

BROWNIE, C., D. R. ANDERSON, K. P. BURNHAM, AND D. S. ROBSON. 1978. Statistical inference from band recovery data: a handbook. U.S. Fish and Wildl. Serv., Resour. Publ. 131.

BROWNIE, C., AND D. S. ROBSON. 1976. Models allowing age-dependent survival rates for band-return data. Biometrics 32:305–323.

BULL, E. L. 1978. Roosting activities of a male Pileated Woodpecker. Murrelet 59:35–36.

BULL, E. L. 1981. Indirect estimates of abundance of birds. Pp. 76–80 *in* C. J. Ralph and J. M. Scott (eds.). Estimating the numbers of terrestrial birds. Stud. Avian Biol. 6.

BULL, P. C., P. D. GAZE, AND C. J. R. ROBERTSON. 1978. Bird distribution in New Zealand—a provisional atlas 1969–1976. Ornithol. Soc. N.Z., Wellington.

BULLER, R. J. 1967. Sandhill Crane study in the central flyway. U.S. Fish and Wildl. Serv., Spec. Sci. Rep. Wildl. 113.

BUNDESMINISTERIUM FÜR ERNÄHRUNG, LANDWIRTSCHAFT UND FORSTEN. 1976. Bundesnaturschutzgesetz. Bonn.

BUNDESMINISTERIUM FÜR ERNÄHRUNG, LANDWIRTSCHAFT UND FORSTEN. 1977. Blickpunkt Landwirtschaft. Bonn.

BURNHAM, K. P. 1972. Estimation of population size in multiple capture-recapture studies when capture probabilities vary among animals. Ph.D. thesis, Oregon State Univ., Corvallis.

BURNHAM, K. P. 1979. A parametric generalization of the Hayne estimator for line transect sampling. Biometrics 35:325–336.

BURNHAM, K. P., AND D. R. ANDERSON. 1976. Mathematical models for nonparametric inferences from line transect data. Biometrics 32:325–336.

BURNHAM, K. P., D. R. ANDERSON, AND J. L. LAAKE. 1979. On robust estimation from line transect data. J. Wildl. Manage. 43:992–996.

BURNHAM, K. P., D. R. ANDERSON, AND J. L. LAAKE. 1980. Estimation of density from line transect sampling of biological populations. Wildl. Monogr. 72.

BURNHAM, K. P., D. R. ANDERSON, AND J. L. LAAKE. 1981. Line transect estimation of bird population density using a Fourier series. Pp. 466–482 *in* C. J. Ralph and J. M. Scott (eds.). Estimating the number of terrestrial birds. Stud. Avian Biol. 6.

BURNHAM, K. P., AND W. S. OVERTON. 1978. Estimation of the size of a closed population when capture probabilities vary among animals. Biometrika 65:625–633.

BURNHAM, K. P., AND W. S. OVERTON. 1979. Robust estimation of population size when capture probabilities vary among animals. Ecology 60:927–937.

BURNS, W. 1973. Noise and man. Lippincott, Philadelphia.

BURTON, J. A. (ed.) 1973. Owls of the world. Dutton, New York.

BUSSE, P. 1973. Dynamics of numbers of some migrants caught at Polish Baltic coast 1961–70. (In Polish with English summary.) Notatki Ornithologicze 14:1–38.

BYSTRAK, D. 1971. How to prepare a winter range map from Christmas Count data. Am. Birds 25:952–956.

BYSTRAK, D. 1979. The Breeding Bird Survey. Sialia 1:74–79, 87.

BYSTRAK, D. 1981. The North American Breeding Bird Survey. Pp. 34–41 *in* C. J. Ralph and J. M. Scott (eds.). Estimating the numbers of terrestrial birds. Stud. Avian Biol. 6.

BYSTRAK, D., AND S. R. DRENNAN. (eds.) 1975. Ten new early winter distribution maps from Christmas Bird Count data. Am. Birds 29:603–611.

BYSTRAK, D., C. S. ROBBINS, S. R. DRENNAN, AND R. ARBIB. (eds.) 1974. Wintering areas of bird species potentially hazardous to aircraft. Nat. Aud. Soc., New York.

CADE, T. J. 1969. The status of the Peregrine and other falconiforms in Africa. Pp. 289–321 *in* J. J. Hickey (ed.). Peregrine Falcon populations: their biology and decline. Univ. Wisconsin Press, Madison.

CALL, D. J. 1975. Seasonal hawk population densities in southeastern South Dakota. Proc. S.D. Acad. Sci. 54:172–177.

CALL, M. W. 1978. Nesting habitats and surveying techniques for common western raptors. U.S. Bur. Land Manage., Tech. Note TN-316, Denver, Col.

CANTLE, P. C. 1978. Avian population densities and species diversity on reclaimed strip-mined land in east-central Texas. M.S. thesis, Texas A&M Univ., College Station.

CAPEN, D. E. (ed.) In press. Proceedings of workshop: the use of multivariate statistics in the study of wildlife habitat (Burlington, Vermont, April 23–25, 1980). U.S. For. Serv., Gen. Tech. Rep. RM-, Fort Collins, Col.

CARDIFF, S. W. 1978. Status of the Elf Owl in California. Calif. Dept. Fish and Game, Sacramento.

CARNEY, S. M., AND G. A. PETRIDES. 1957. Analysis of variation among participants in pheasant cock-crowing censuses. J. Wildl. Manage. 21:392–397.

CAROTHERS, A. D. 1971. An examination and extension of Leslie's test of equal catchability. Biometrics 27:615–630.

CAROTHERS, A. D. 1973. The effects of unequal catchability on Jolly-Seber estimates. Biometrics 29:79–100.

CAROTHERS, A. D. 1979. Quantifying unequal catchability and its effect on survival estimates in an actual population. J. Anim. Ecol. 48:863–869.

CAROTHERS, S. W., AND R. R. JOHNSON. 1971. A summary of the Verde Valley breeding bird survey, 1970. Ariz. Game and Fish Dept., Rep. FW-16-10, Pp. 46–64. Phoenix, Ariz.

CARRIER, W. D., AND W. E. MELQUIST. 1976. The use of a rotor-winged aircraft in conducting nesting surveys of Ospreys in northern Idaho. Raptor Res. 10:77–83.

CASSEDAY, J. H., AND W. D. NEFF. 1973. Localization of pure tones. J. Acoust. Soc. Am. 54:365–372.

CATCHPOLE, C. K. 1973. The functions of advertising song in the Sedge Warbler (*Acrocephalus schoenobaenus*) and the Reed Warbler (*A. scirpaceus*). Behaviour 46:300–320.

CATLING, P. M. 1971. Spring migration of Saw-whet Owls at Toronto, Ontario. Bird-Banding 42:110–114.

CAUGHLEY, G. 1965. A method of comparing the numbers of species in areas covered by different periods of observation. Emu 65:115–118.

CAUGHLEY, G. 1977. Analysis of vertebrate populations. Wiley, New York.

CAWTHORNE, R. A., AND J. H. MARCHANT. 1980. The effects of the 1978/79 winter on British bird populations. Bird Study 27:163–172.

CHANCELLOR, R. D. (ed.) 1977. World conference on birds of prey, Vienna 1975. Int. Council Bird Pres., Hampshire, U.K.

CHAPMAN, D. G. 1951. Some properties of the hypergeometric distribution with applications to zoological censuses. Univ. Calif. Publ. Stat. 1:131–160.

CHAPMAN D. G. 1980. Review of "Statistical inference from capture data on closed animal populations" by Otis et al. (1978). Biometrics 36:362.

CHESSEX, C., AND J. P. RIBAUT. 1966. Evolution d'une avifaune suburbaine et test d'une méthode de recensement. Nos Oiseaux 28:193–211.

CINK, C. L. 1975. Population densities of Screech Owls in northeastern Kansas. Kansas Ornithol. Soc. 26:13–16.

CLAPP, R. B., E. KRIDLER, AND R. R. FLEET. 1977. The natural history of Nihoa Island, northwestern Hawaiian Islands. Atoll Res. Bull. 207, Smithsonian Inst., Washington, D.C.

CLARK, J. M. 1978. Population dynamics of breeding Woodcock in Canaan Valley, West Virginia. M.S. thesis, West Virginia Univ., Morgantown.

CLARK, R. B. 1947. Seasonal fluctuations in the song of the Sky-lark. Br. Birds 40:34–43.

CLARK, W. S. 1973. Cape May Point raptor banding station—1972 results. EBBA News 36:150–165.

CLARK, W. S. 1976. Cape May Point raptor banding station—1974 results. N. Am. Bird Band. 1:5–13.

CLIFFORD, H. T., AND W. STEPHENSON. 1975. An introduction to numerical classification. Academic, New York.

COCHRAN, W. G. 1963. Sampling techniques. Wiley, New York.

COCHRAN, W. G. 1977. Sampling techniques. Third ed. Wiley, New York.

CODY, M. L. 1966. The consistency of intra- and inter-continental bird species counts. Am. Nat. 100:371–376.

CODY, M. L. 1968. On the methods of resource division in grassland bird communities. Am. Nat. 102:107–147.

CODY, M. L. 1974. Competition and the structure of bird communities. Monogr. Pop. Biol. 7, Princeton Univ. Press, Princeton, N.J.

CODY, M. L. 1975. Towards a theory of continental species diversities: bird distributions over Mediterranean habitat gradients. Pp. 214–257 *in* M. L. Cody and J. M. Diamond (eds.). Ecology and evolution of communities. Belknap, Cambridge, Mass.

CODY, M. L. 1978. Habitat selection and interspecific territoriality among the sylviid warblers of England and Sweden. Ecol. Monogr. 48:351–396.

CODY, M. L. In press. Species packing in insectivorous bird communities: density, diversity, and productivity. Proc. Int. Ornithol. Congr. 17.

CODY, M. L., AND J. M. DIAMOND. (eds.) 1975. Ecology and evolution of communities. Belknap, Cambridge, Mass.

CODY, M. L., AND H. WALTER. 1976. Habitat selection and interspecific interactions among mediterranean sylviid warblers. Oikos 27:210–238.

COHEN, A., H. S. PETERS, AND L. E. FOOTE. 1960. Calling behavior of Mourning Doves in two Midwest life zones. J. Wildl. Manage. 24:203–212.

COHEN, A. C. 1961. Tables for maximum likelihood estimates: singly truncated and singly censored samples. Technometrics 3:535–541.

COLE, L. C. 1954. Some features of random population cycles. J. Wildl. Manage. 18:2–24.

COLQUHOUN, M. K. 1940a. The density of woodland birds determined by the sample count method. J. Anim. Ecol. 9:53–67.

COLQUHOUN, M. K. 1940b. Visual and auditory conspicuousness in a woodland bird community: a quantitative analysis. Proc. Zool. Soc. Lond. 110:129–148.

COLQUHOUN, M. K., AND A. MORLEY. 1941. The density of downland birds. J. Anim. Ecol. 10:35–46.

CONANT, S., M. S. COLLINS, AND C. J. RALPH. 1981. Effects of observers using different methods upon the total population estimates of two resident island birds. Pp. 377–381 *in* C. J. Ralph and J. M. Scott (eds.). Estimating the numbers of terrestrial birds. Stud. Avian Biol. 6.

CONDER, P. 1977. Legal status of birds of prey and owls in Europe. Pp. 189–193 *in* R. D. Chancellor (ed.). World conference on birds of prey, Vienna 1975. Int. Council Bird Pres.

CONFER, J. L., T. KAARET, AND L. JONES. 1979. Effort, location, and the Christmas Bird Count tally. Am. Birds 33:690–692.

CONNER, R. N., AND J. G. DICKSON. 1980. Strip transect sampling and analysis for avian habitat studies. Wild. Soc. Bull. 8:4–10.

CONNER, R. N., R. G. HOOPER, H. S. CRAWFORD, AND H. S. MOSBY. 1975. Woodpecker nesting habitat in cut and uncut woodland in Virginia. J. Wildl. Manage. 39:144–150.

CONNOR, E. F., AND E. D. MCCOY. 1979. The statistics and biology of the species-area relationship. Am. Nat. 113:791–833.

CONNOR, E. F., AND D. SIMBERLOFF. 1979. The assembly of species communities: chance or competition? Ecology 60:1132–1140.

CORMACK, R. M. 1964. Estimates of survival from the sighting of marked animals. Biometrika 51:429–438.

CORMACK, R. M. 1966. A test for equal catchability. Biometrics 22:330–342.

CORMACK, R. M. 1968. The statistics of capture-recapture methods. Ocean. Mar. Biol. Annu. Rev. 6:455–506.

CORMACK, R. M. 1972. The logic of capture-recapture estimates. Biometrics 28:337–343.

CORMACK, R. M. 1973. Commonsense estimates from capture-recapture studies. Pp. 225–234 *in* M. S. Bartlett and R. W. Hiorns (eds.). The mathematical theory of the dynamics of biological populations. Academic, New York.

CORMACK, R. M. 1979. Models for capture-recapture. Pp. 217–255 *in* R. M. Cormack, G. P. Patil, and D. S. Robson (eds.). Sampling biological populations. Stat. Ecol. Ser., vol. 5. Int. Co-op. Publ. House, Fairland, Md.

CORMACK, R. M. 1981. Loglinear models for capture-recapture experiments on open populations. *In* R. W. Hiorns and D. Cooke (eds.). The mathematical theory of the dynamics of biological populations. Academic, London (in press).

CORMACK, R. M., G. P. PATIL, AND D. S. ROBSON. (eds.) 1979. Sampling biological populations. Stat. Ecol. Ser., vol. 5. Int. Co-op. Publ. House, Fairland, Md.

CORTELYOU, R. C. 1978. The first eleven years of Breeding Bird Surveys in Nebraska. Nebraska Bird Rev. 46:38–62.

COTTAM, C. 1956. Uses of marking animals in ecological studies: marking birds for scientific purposes. Ecology 37:674–681.

COTTAM, G., AND J. CURTIS. 1956. The use of distance measurements in phytosociological sampling. Ecology 37:451–460.

COUSINS, S. 1977. Sample size and edge effect on community measures of farm bird populations. Pol. Ecol. Stud. 3(4):27–35.

COWARDIN, L. M., AND K. F. HIGGINS. 1967. Visibility, movement, and behavior of waterfowl on a river habitat in Minnesota. Trans. N. Amer. Wildl. Nat. Resour. Conf. 32:301–315.

COX, P. R. 1944. A statistical investigation into birdsong. Br. Birds 38:3–9.

CRAIG, C. C. 1953. On the utilization of marked specimens in estimating populations of flying insects. Biometrika 40:170–176.

CRAIG, T. H. 1978. A car survey of raptors in southeastern Idaho, 1974–76. Raptor Res. 12:40–45.

CRAIGHEAD, J. J., AND F. C. CRAIGHEAD. 1956. Hawks, owls, and wildlife. Wildl. Manage. Inst., Washington, D.C.

CRAIN, B. R., K. P. BURNHAM, D. R. ANDERSON, AND J. L. LAAKE. 1978. A Fourier series estimator of population density for line transect sampling. Utah State Univ. Press, Logan.

CRAIN, B. R., K. P. BURNHAM, D. R. ANDERSON, AND J. L. LAAKE. 1979. Nonparametric estimation of population density for line transect sampling using Fourier series. Biometr. J. 21:731–748.

CRAVEN, S. R. 1979. Some problems with Canada Goose neckbands. Wildl. Soc. Bull. 7:268–273.

CRISSEY, W. F. 1963. Exploitation of migratory waterfowl populations in North America. Pp. 105–122 *in* J. J. Swift (ed.). Proceedings of first European meeting on wildfowl conservation. Nature Conservancy, London.

CROAT, T. B. 1978. Flora of Barro Colorado Island. Stanford Univ. Press, Stanford, Calif.

CROMER, J. R. 1978. Analysis of eleven years of banding data from a free-flying, resident flock of Canada geese (*Branta canadensis*) on Old Hickory Lake, Tennessee. Tennessee Wildl. Resour. Agency, Tech. Rep. 77-20.

CROOK, I. G., H. A. BEST, AND M. HARRISON. 1977. A survey of the native bird fauna of forest in the proposed beech project area of North Westland. Proc. N.Z. Ecol. Soc. 24:113–126.

CROSBIE, S. F. 1979. The mathematical modeling of capture-mark-recapture experiments on open animal populations. Ph.D. thesis, Univ. of Otago, Dunedin, N.Z.

CULLEN, M. 1980. I. Rolling bird survey. Atlas Newsletter 16, Royal Australian Ornithologists Union.

CUNNINGHAM-VAN SOMEREN, B. R. 1979. Kakamega (Kakamegoes) Forest area, nature reserve and Yala Nature reserve and environs. Checklist of birds, Sect. Ornithol., Nat. Mus. Kenya, Nairobi. MS.

CURIO, E. 1959. Verhaltensstudien am Trauerschnäpper. Z. Tierpsychol., Beih. 3.

CURTIS, J. T. 1956. The modification of mid-latitude grasslands and forests by man. Pp. 721–736 *in* W. L. Thomas (ed.). Man's role in changing the face of the earth. Vol 2. Univ. Chicago Press, Chicago, Ill.

CURTIS, J. T. 1959. The vegetation of Wisconsin. Univ. Wisconsin Press, Madison, Wis.

CYR, A. 1977. Beziehungen zwischen Strukturdiversität und Vogelpopulationen in der Umgebung des Verdichtungsraumes von Saarbrücken. Dissertation, Saarbrücken, W. Ger.

CYR, A. 1981. Limitation and variability in hearing ability in censusing birds. Pp. 327–333 *in* C. J. Ralph and J. M. Scott (eds.). Estimating the numbers of terrestrial birds. Stud. Avian Biol. 6.

CZEKALA, N. M., AND B. L. LASLEY. 1977. A technical note on sex determination in monomorphic birds using faecal steroid analysis. Int. Zool. Yearb. 17:209–211.

DAHLSTEN, D. L., AND W. A. COPPER. 1979. The use of nesting boxes to study the biology of the Mountain Chickadee (*Parus gambeli*) and its impact on selected forest insects. Pp. 217–260 *in* J. G. Dickson, R. N. Conner, R. R. Fleet, J. A. Jackson, and J. C. Kroll (eds.). The role of insectivorous birds in forest ecosystems. Academic, New York.

DANIEL, C., AND R. LAMAIRE. 1974. Evaluating effects of water resource developments on wildlife habitat. Wildl. Soc. Bull. 2:114–118.

DANSEREAU, P., P. F. BUELL, AND R. DAGON. 1966. A universal system for recording vegetation. II. A methodological critique and an experiment. Sarracenia 10:1–64.

DARLEY, J. A. 1971. Sex ratio and mortality of the Brown-headed Cowbird. Auk 88:560–566.

DARROCH, J. N. 1958. The multiple-recapture census. I. Estimation of a closed population. Biometrika 45:343–359.

DARROCH, J. H. 1959. The multiple-recapture census. II. Estimation when there is immigration or death. Biometrika 46:336–351.

DASSMANN, R. F., AND A. S. MOSSMANN. 1962. Road strip counts for estimating numbers of African ungulates. J. Wildl. Manage. 26:101–104.

DAVID, F. N. 1981. Statistics for the birds. Pp. 566–569 *in* C. J. Ralph and J. M. Scott (eds.). Estimating the numbers of terrestrial birds. Stud. Avian Biol. 6.

DAVIES, R. G. 1971. Computer programming in quantitative biology. Academic, London.

DAVIS, C. V. 1969. A 1968 Breeding Bird Survey for Montana. Montana Acad. Sci. Proc. 29:1–12.

DAVIS, D. E. 1937. A cycle in Northern Shrike emigrations. Auk 54:43–49.

DAVIS, D. E. 1970. Evaluation of techniques for measuring mortality. J. Wildl. Dis. 6:365–375.

DAVIS, D. E. 1974. Emigrations of Northern Shrikes 1959–1970. Auk 54:43–49.

DAVIS, H., AND S. R. SILVERMAN. 1960. Hearing and deafness. Holt, Rinehart and Winston, New York.

DAVIS, J. 1958. Singing behavior and the gonad cycle of the Rufous-sided Towhee. Condor 60:308–336.

DAVIS, J. 1965. The "singing male" method of censusing birds: a warning. Condor 67:86–87.

DAWSON, D. G. 1981a. Counting birds for a relative measure (index) of density. Pp. 12–16 *in* C. J. Ralph and J. M. Scott (eds.). Estimating the numbers of terrestrial birds. Stud. Avian Biol. 6.

DAWSON, D. G. 1981b. Experimental design when counting birds. Pp. 392–398 *in* C. J. Ralph and J. M. Scott (eds.). Estimating the numbers of terrestrial birds. Stud. Avian Biol. 6.

DAWSON, D. G. 1981c. The usefulness of absolute ("census") and relative ("sampling" or "index") measures of abundance. Pp. 554–558 *in* C. J. Ralph and J. M. Scott (eds.). Estimating the numbers of terrestrial birds. Stud. Avian Biol. 6.

DAWSON, D. G., AND P. C. BULL. 1975. Counting birds in New Zealand forests. Notornis 22:101–09.

DAWSON, D. G., P. J. DILKS, P. D. GAZE, J. G. R. MCBURNEY, AND P. R. WILSON. 1978. Seasonal differences in bird counts in forests near Reefton, South Island, New Zealand. Notornis 25:257–278.

DAWSON, D. K. 1981. Sampling in rugged terrain. Pp. 311–315 *in* C. J. Ralph and J. M. Scott (eds.). Estimating the numbers of terrestrial birds. Stud. Avian Biol. 6.

DEHAVEN, R. W. 1973. Winter population trends of the Starling in California. Am. Birds 27:836–838.

DEKKER, D. 1979. Characteristics of Peregrine Falcons migrating through central Alberta, 1969–1978. Can. Field-Nat. 93:296–302.

DELAMORE, M. 1979. Further investigations of the Spotted Owl (*Strix occidentalis*) in Six Rivers National Forest. Tech. Rep., Six Rivers Natl. Forest, Calif.

DEMENTIEW, G. P., AND N. A. GLADKOW. (eds.) 1951–1954. Ptitsy Sovietskogo Soyusa (Birds of the Soviet Union). 6 vol. Nauka, Moscow.

DESANTE, D. F. 1981. A field test of the variable circular-plot censusing technique in a California coastal scrub breeding bird community. Pp. 177–185 *in* C. J. Ralph and J. M. Scott (eds.). Estimating the numbers of terrestrial birds. Stud. Avian Biol. 6.

DESGRANGES, J.-L. 1980. Avian community structure of six forest stands in La Mauricie National Park, Quebec. Can. Wildl. Serv., Occ. Pap. 41.

DEWEESE, L. R., AND C. J. HENNY, R. L. FLOYD, K. A. BOBAL, AND A. W. SCHULTZ. 1979. Response of breeding birds to aerial sprays of trichlorfon (dylox) and carbaryl (sevin 4-oil) in Montana forests. U.S. Fish and Wild. Serv., Spec. Sci. Rep. Wildl. 224.

DHONT, A. A. 1966. A method to establish the boundaries of bird territories. Gerfault 56:404–408.

DIAMOND, J. M. 1972. Avifauna of the eastern highlands of New Guinea. Publ. Nuttall Ornithol. Club 12.

DIAMOND, J. M. 1975a. Assembly of species communities. Pp. 342–444 *in* M. L. Cody and J. M. Diamond (eds.). Ecology and evolution of communities. Belknap, Cambridge, Mass.

DIAMOND, J. 1975b. The island dilemma: lessons of modern biogeographic studies for the design of natural reserves. Biol. Conser. 7:129–146.

DIAMOND, J. M., AND R. M. MAY. 1976. Island biogeography and the design of natural reserves. Pp. 163–186 *in* R. M. May (ed.). Theoretical ecology. Principles and applications. Saunders, Philadelphia, Pa.

DICKSON, J. G. 1978. Comparison of breeding bird census techniques. Am. Birds 32:10–13.

DIEHL, B. 1974. Results of a breeding bird community census by the mapping method in a grassland ecosystem. Acta Ornithol. 14:362–376.

DIEHL, B. 1976. The effect of ambient temperature on density-dependent processes in a population of the Red-backed Shrike, (*Lanius collurio* L.) Bull. Pol. Sci. 24:711–718.

DIEHL, B. 1981. Bird populations consist of individuals differing in many respects. Pp. 225–229 *in* C. J. Ralph and J. M. Scott (eds.). Estimating the numbers of terrestrial birds. Stud. Avian Biol. 6.

DIEM, K. L., AND K. H. LU. 1960. Factors influencing waterfowl censuses in the parklands, Alberta, Canada. J. Wildl. Manage. 24:113–133.

DIETER, M. P., AND J. L. LUDKE. 1975. Studies on combined effects of organophosphates and heavy metals in birds. I. Plasma and brain cholinesterase in coturnix quail fed methyl mercury and orally dosed with parathion. Bull. Environ. Contam. Toxicol. 13:257–262.

DIXON, W. J., AND M. B. BROWN. (eds.) 1979. BMDP-79. Biomedical computer programs. P-series. Univ. Calif. Press, Berkeley.

DIXON, W. J., AND F. J. MASSEY. 1957. Introduction to statistical analysis. Second ed. McGraw-Hill, New York.

DOANE, C. C., AND P. W. SCHAEFER. 1971. Aerial applications of insecticides for control of the gypsy moth with studies on the effects on non-target insects and birds. Connecticut Agric. Exp. Stn., Bull. 724.

DOERR, P. D. 1968. Nesting activities and migratory status of some Goshawks in northeastern Colorado. M.S. thesis, Colorado State Univ., Fort Collins.

DOERR, P. D., AND J. H. ENDERSON. 1965. Index of abundance of the Goshawk in Colorado in winter. Auk 82:284–285.

DOLBEER, R. A., AND R. A. STEHN. 1979. Population trends of blackbirds and starlings in North America, 1966–76. U.S. Fish and Wildl. Serv., Spec. Sci. Rep. Wildl. 214.

DOLTON, D. D. 1977. Mourning Dove status report, 1976. U.S. Fish and Wildl. Serv., Spec. Sci. Rep. Wildl. 208.

DORNEY, R. S., D. R. THOMPSON, J. B. HALE, AND R. F. WENDT. 1958. An evaluation of Ruffed Grouse drumming counts. J. Wildl. Manage. 22:35–40.

DORST, J. 1962. The migration of birds. Houghton Mifflin, Cambridge, Mass.

DORST, J. 1974. The life of birds. Vol. 1. Columbia Univ. Press, New York.

DOW, D. D. 1970. Indexing population densities of the Cardinal with tape-recorded song. Wilson Bull. 82:83–91.

DOWSETT, R. J. 1979. Zambian bird atlas: 1978 report. Bull. Zambian Ornithol. Soc. 11:44–48.

DOWSETT, R. J., AND F. DOWSETT-LEMAIRE. 1979. An atlas of the birds of Malawi. Nyala 5:93–96.

DRENNAN, S. R. 1981. The Christmas Bird Count: an overlooked and underused sample. Pp. 24–29 *in* C. J. Ralph and J. M. Scott (eds.). Estimating the numbers of terrestrial birds. Stud. Avian Biol. 6.

DRÖSCHER, V. B. 1974. Sie töten und sie lieben sich. Naturgeschichte socialen Verhaltens, Hamburg.

DUNNE, P. J., AND W. S. CLARK. 1977. Fall hawk movement at Cape May Point, New Jersey—1976. New Jersey Audubon 3:114–124.

DUNNET, G. M., AND J. C. OLLASON. 1978. The estimation of survival rate in the Fulmar (*Fulmarus glacialis*). J. Anim. Ecol. 47:507–520.

DUPONT, W. A. 1976. A stochastic method for estimating animal abundance from catch-effort data. Ph.D. thesis. Johns Hopkins Univ. School of Hygiene and Public Health, Baltimore, Md.

DURANGO, S. 1956. Territory in the Red-backed Shrike (*Lanius collurio*). Ibis 98:476–484.

DYBBRO, T. 1976. De danske ynglefugles utbredelse. Dansk Ornithol. Forening, Copenhagen.

DZUBIN, A. 1969. Assessing breeding populations of ducks by ground counts. Pp. 178–230 *in* Saskatoon Wetlands Seminar. Can. Wildl. Serv., Rep. Ser. 6.

EAGLES, P. F. J. 1976. A study of breeding-bird populations in Bronte Creek Provincial Park, M.Sc. thesis, Univ. of Guelph, Guelph, Ont.

EAGLES, P. F. J. 1981. Breeding bird censuses using spot-mapping techniques upon samples of homogeneous habitats. Pp. 455–460 *in* C. J. Ralph and J. M. Scott (eds.). Estimating the numbers of terrestrial birds. Stud. Avian Biol. 6.

EAGLES, P. F. J., AND T. N. TOBIAS. 1978. A replication of a breeding bird census. Am. Birds 32:14–17.

EATON, S. W., T. L. MOORE, AND E. N. SAYLOR. 1970. A ten year study of the food habits of a northern population of wild Turkeys. Sci. Stud. 26:43–64.

EBERHARDT, L. L. 1968. A preliminary appraisal of line transects. J. Wildl. Manage. 32:82–88.

EBERHARDT, L. L. 1969. Population estimates from recapture frequencies. J. Wildl. Manage. 33:28–39.

EBERHARDT, L. L. 1978. Transect methods for population studies. J. Wildl. Manage. 42:1–31.

EBERHARDT, L. L. 1979. Line-transects based on right-angle distances. J. Wildl. Manage. 43:768–774.

EDELSTAM, C. 1972. The visible migration of birds at Ottenby, Sweden. Vår Fågelvärld, Suppl. 7:1–360.

EDWARDS, D. K., G. L. DORSEY, AND J. A. CRAWFORD. 1981. A comparison of three avian census methods. Pp. 170–176 *in* C. J. Ralph and J. M. Scott (eds.). Estimating the numbers of terrestrial birds. Stud. Avian Biol. 6.

EDWARDS, W. R., AND R. R. GRABER. 1968. Responses of avians to methyl parathion in a hayfield. Pp. 53–59 *in* Summaries of presentations, 20th Illinois Custom Spray Operators Training School. Ill. Nat. Hist. Surv., Urbana.

EFRON, B., AND C. MORRIS. 1972. Empirical Bayes on vector observations: an extension of Stein's method. Biometrika 59:335–347.

EFRON, B., AND C. MORRIS. 1975. Data analysis using Stein's estimator and its generalizations. J. Am. Stat. Assoc. 70:311–319.

EISENMANN, E. 1971. Range expansion and population increase in North and Middle America of the White-tailed Kite (*Elanus leucurus*). Am. Birds 25:529–536.

EKMAN, J. 1979. Coherence, composition and territories of winter social groups of the Willow Tit (*Parus montanus*) and the Crested Tit (*P. cristatus*). Ornis Scand. 10:56–68.

EKMAN, J. 1981. Problems of unequal observability. Pp. 230–234 *in* C. J. Ralph and J. M. Scott (eds.). Estimating the numbers of terrestrial birds. Stud. Avian Biol. 6.

EKMAN, J., G. CEDERHOLM, AND C. ASKENMO. In press. Spacing and survival in winter groups of Willow Tit (*Parus montanus*) Conrad and Crested Tit (*P. cristatus* L.)—a removal study. J. Anim. Ecol.

ELDER, J. B. 1956. Analysis of whistling patterns in Eastern Bobwhite (*Colinus v. virginianus* L.). Iowa Acad. Sci. Proc. 63:639–651.

ELDER, J. B., AND C. HENDERSON. 1969. Field appraisal of ULV batex mosquito control applications on fish and wildlife. U.S. Fish and Wildl. Serv. and Ohio Dept. Nat. Res., Spec. Rep.

ELDREDGE, N., AND J. CRACRAFT. 1980. Phylogenetic patterns and the evolutionary process. Columbia Univ. Press, New York.

ELGOOD, G. H., C. H. FRY, AND R. J. DOWSETT. 1973. African migrants in Nigeria. Ibis 115:1–45, 375–411.

ELLIOTT, J. M. 1971. Some methods for the statistical analysis of samples of benthic invertebrates. Freshw. Biol. Assoc. Sci. Publ. 25.

ELLIS, J. A., J. N. BURROUGHS, M. J. ARMBRUSTER, D. L. HALLETT, P. A. KORTE, AND T. S. BASKETT. 1979. Appraising four field methods of terrestrial habitat evaluation. Trans. N. Am. Wildl. Nat. Res. Conf. 44:369–379.

ELLMAN, G. L., K. D. COURTNEY, V. ANDREAS, AND R. M. FEATHERSTONE. 1961. A new rapid colorimetric determination of acetylcholinesterase activity. Biochem. Pharmacol. 7:88–95.

ELTON, C. S., AND R. S. MILLER. 1954. The ecological survey of animal communities: with a practical system of classifying habitats by structural characters. J. Ecol. 42:460–496.

EMLEN, J. T. 1938. Midwinter distribution of the American Crow in New York State. Ecology 19:264–275.

EMLEN, J. T. 1940. The midwinter distribution of the Crow in California. Condor 42:287–294.

EMLEN, J. T. 1941. Cliff Swallow colonies of the central Sacramento Valley in 1941. Condor 43:248.

EMLEN, J. T. 1956. A method for describing and comparing avian habitats. Ibis 98:565–576.

EMLEN, J. T. 1971. Population densities of birds derived from transect counts. Auk 88:323–342.

EMLEN, J. T. 1973. Vocal stimulation in the Great Horned Owl. Condor 75:126–127.

EMLEN, J. T. 1974. An urban bird community in Tucson, Arizona: derivation, structure, regulation. Condor 76:184–197.

EMLEN, J. T. 1977a. Estimating breeding season bird densities from transect counts. Auk 94:455–468.

EMLEN, J. T. 1977b. Land bird communities of Grand

Bahama Island: the structure and dynamics of an avifauna. Ornithol. Monogr. 24.

EMLEN, J. T., AND M. J. DEJONG. 1981. The application of song detection threhold distance to census operations. Pp. 346–352 *in* C. J. Ralph and J. M. Scott (eds.). Estimating the numbers of terrestrial birds. Stud. Avian Biol. 6.

EMLEN, J. T., AND J. A. WIENS. 1965. The Dickcissel invasion of 1964 in southern Wisconsin. Passenger Pigeon 27:51–59.

EMLEN, S. T. 1971. The role of song in individual recognition in the Indigo Bunting. Z. Tierpsychol. 28:241–246.

ENDERSON, J. H. 1964. A study of the Prairie Falcon in the central Rocky Mountain region. Auk 81:332–352.

ENDERSON, J. H. 1965. Roadside raptor counts in Colorado. Wilson Bull. 77:82–83.

ENDERSON, J. H. 1969. Migratory populations of the Peregrine Falcon: coastal migration data as population indices for the Peregrine Falcon. Pp. 275–280 *in* J. J. Hickey (ed.). Peregrine Falcon populations: their biology and decline. Univ. Wisconsin Press, Madison.

ENDERSON, J. H., F. A. COLLEY, AND J. STEVENSON. 1970. Aerial eagle count in Colorado. Condor 72:112.

ENEMAR, A. 1959. On the determination of the size and composition of a passerine bird population during the breeding season. Vår Fågelvärld, Suppl. 2:1–114.

ENEMAR, A. 1962. A comparison between the bird census results of different ornithologists. Vår Fågelvärld 21:109–120.

ENEMAR, A. 1964. Ett försök att mäta fyra ornitologers frmåga att uppfatta och registrera flyttfågelsträcket i Falsterbo. (A preliminary estimation of the reliability of the registration work of four ornithologists watching autumn migration.) Vår Fågelwärld 23:1–25

ENEMAR, A. 1966. A ten-year study of the size and composition of a passerine bird community. Vår Fågelvärld, Suppl. 4:47–94.

ENEMAR, A., S.-G. HÖJMAN, P. KLAESSON, AND L. NILSSON. 1976. The relationship between census results and the breeding population of birds in subalpine birch forests. Ornis Fenn. 53:1–8.

ENEMAR, A., P. KLAESSON, AND B. SJÖSTRAND. 1979. Accuracy and efficiency of mapping territorial Willow Warblers *Phylloscopus trochilus*: a case study. Oikos 33:176–181.

ENEMAR, A., AND B. SJÖSTRAND. 1967. The strip survey as a complement to study area investigations in bird census work. Vår Fågelvärld 26:111–130.

ENEMAR, A., AND B. SJÖSTRAND. 1970. Bird species densities derived from study area investigations and line transects. Pp. 33–37 *in* S. Svensson (ed.). Symposium: Bird census work and environmental monitoring. Bull. Ecol. Res. Committee 9. Swedish Nat. Sci. Res. Coun., Stockholm.

ENEMAR, A., B. SJÖSTRAND, AND S. SVENSSON. 1978. The effect of observer variability on bird census results obtained by a territory mapping technique. Ornis Scand. 9:31–39.

ENGEL-WILSON, R. W., A. K. WEBB, K. V. ROSENBERG, R. D. OHMART, AND B. W. ANDERSON. 1981. Avian censusing with the strip method: a computer simulation. Pp. 445–449 *in* C. J. Ralph and J. M. Scott (eds.). Estimating the numbers of terrestrial birds. Stud. Avian Biol. 6.

ENGSTROM, R. T. 1980a. Breeding Bird Census number 64. Am. Birds 34:61.

ENGSTROM, R. T. 1980b. Winter Bird-Population Study number 15. Am. Birds 34:29–30.

ENGSTROM, R. T. 1981. The species-area relationship in spot-map censusing. Pp. 421–425 *in* C. J. Ralph and J. M. Scott (eds.). Estimating the numbers of terrestrial birds. Stud. Avian Biol. 6.

ENGSTROM, R. T. In press. Breeding Bird Census. Am. Birds.

ENGSTROM, R. T., AND F. C. JAMES. 1981. Plot size as a factor in Winter Bird-Population Studies. Condor 83:34–41.

ERDELEN, B. 1979. Untersuchungen zum Kartierungsverfahren bei Brutvogel-Bestandsaufnahmen. Dissertation, Köln, W. Ger.

ERDELEN, M. 1978. Quantitative Beziehungen zwischen Avifauna und Vegetationsstruktur. Dissertation, Köln, W. Ger.

ERDELEN, M., AND B. ERDELEN. 1981. The efficiency of the mapping method—a computer simulation based on field data. Pp. 492–495 *in* C. J. Ralph and J. M. Scott (eds.). Estimating the numbers of terrestrial birds. Stud. Avian Biol. 6.

ERSKINE, A. J. 1970. The co-operative breeding bird survey in Canada, 1966–1969. Can. Wildl. Serv., Prog. Notes 15.

ERSKINE, A. J. 1971. A preliminary catalogue of bird census studies in Canada. Can. Wildl. Serv., Prog. Notes 20.

ERSKINE, A. J. 1972. A preliminary catalogue of bird census plot studies in Canada, part 2. Can. Wildl. Serv., Prog. Notes 30.

ERSKINE, A. J. 1973. The co-operative breeding bird survey in Canada, 1972. Can. Wildl. Serv., Prog. Notes 32.

ERSKINE, A. J. 1974. Problems associated with bird populations not adequately sampled by the mapping census method. Acta Ornithol. 14:197–203.

ERSKINE, A. J. 1976a. A preliminary catalogue of bird census plot studies in Canada, part 3. Can. Wildl. Serv., Prog. Notes 59.

ERSKINE, A. J. 1976b. Chronology of nesting in urban birds as a guide to timing of censuses. Am. Birds 30:667–672.

ERSKINE, A. J. 1978. The first ten years of the co-operative Breeding Bird Survey in Canada. Can. Wildl. Serv., Rep. Ser. 42.

ERSKINE, A. J. 1980. A preliminary catalogue of bird census plot studies in Canada, part 4. Can. Wildl. Serv., Prog. Notes 112.

ERSKINE, A. J. 1981. Adapting generalized instructions to specific situations in planning count work. Pp. 570–574 *in* C. J. Ralph and J. M. Scott (eds.). Estimating the numbers of terrestrial birds. Stud. Avian Biol. 6.

ERSKINE, J. S. 1968. Winter birds of Wolfville, Nova Scotia: 1948–1968. Nova Scotia Mus., Occ. Paper 7, Sci. Ser. 5.

ERWIN, R. M., AND J. C. OGDEN. 1979. Multiple-factor influences upon feeding flight rates at wading bird colonies. Proc. Colonial Waterbird Group 3:225–234.

ERZ, W., K. HAARMANN, AND P. PRETSCHER. 1979. Katalog der Naturschutzgebiete in der Bundesre-Publik Deutschland. Kilda, Greven, W. Ger.

EVANS, D. L. 1975. Fall owl migration at Duluth, Minnesota. Loon 47:57.

EVANS, D. L. 1980. Multivariate analyses of weather and fall migration of Saw-whet Owls at Duluth, Minnesota. M.S. thesis, North Dakota State Univ., Fargo.

EVANS, K. E. 1978. Forest management opportunities for songbirds. Trans. N. Am. Wildl. Nat. Res. Conf. 43:69–77.

EVANS, K. E., AND D. K. DAWSON. 1976. The Breeding Bird Survey in Missouri—1967 to 1975. Bluebird 43:12–22.

EVANS, P. R., AND G. W. LATHBURY. 1973. Raptor migration across the Straits of Gibraltar. Ibis 115:572–585.

EVENDEN, F. G. 1980. Live birds at the White House. Am. Birds 34:700–702.

EYRING, E. F. 1946. Jungle acoustics. J. Acoust. Soc. Amer. 18:257–270.

FAABORG, J., T. DEVRIES, C. B. PATTERSON, AND C. R. GRIFFIN. 1980. Preliminary observations on the occurrence and evolution of polyandry in the Galapagos Hawk (*Buteo galapagoensis*). Auk 97:581–590.

FAANES, C. A., AND D. BYSTRAK. 1981. The role of observer bias in the North American Breeding Bird Survey. Pp. 353–359 *in* C. J. Ralph and J. M. Scott (eds.). Estimating the numbers of terrestrial birds. Stud. Avian Biol. 6.

FAGER, E. W. 1972. Diversity: a sampling study. Am. Nat. 106:293–310.

FALK, L. L. 1979. An examination of observers' weather sensitivity in Christmas Bird Count data. Am. Birds 33:688–689.

FALLA, R. A., R. B. SIBSON, AND E. G. TURBOTT. 1970. A field guide to the birds of New Zealand. Collins, London.

FALLS, J. B. 1969. Functions of territorial song in the White-throated Sparrow. Pp. 207–232 *in* R. A. Hinde (ed.). Bird vocalizations, their relations to current problems in biology and psychology. Cambridge Univ. Press, Oxford.

FALLS, J. B. 1978. Bird song and territorial behavior. Pp. 61–89 *in* L. Krames, P. Pliner, and T. Alloway (eds.). Aggression, dominance, and individual spacing. Plenum, New York.

FALLS, J. B. 1981. Mapping territories with playback: an accurate census method for songbirds. Pp. 86–91 *in* C. J. Ralph and J. M. Scott (eds.). Estimating the numbers of terrestrial birds. Stud. Avian Biol. 6.

FALLS, J. B., AND R. J. BROOKS. 1975. Individual recognition by song in White-throated Sparrows. II. Effects of location. Can. J. Zool. 53:1412–1420.

FARNER, D. S. 1949. Age groups and longevity in the American Robin: Comments, further discussion, and certain revisions. Wilson Bull. 61:68–81.

FARNER, D. S. 1955. Birdbanding in the study of population dynamics. Pp. 397–499 *in* A. Wolfson (ed.). Recent studies in avian biology. Univ. Illinois Press, Urbana.

FELLER, W. 1970. An introduction to probability theory and its applications. Third ed. Vol. 1. Wiley, New York.

FENNEMAN, N. M. 1931. Physiography of western United States. McGraw-Hill, New York.

FENNEMAN, N. M. 1938. Physiography of eastern United States. McGraw-Hill, New York.

FERRY, C. 1974. Comparison between breeding bird communities in an oak forest and a beech forest, censused by the IPA method. Acta Ornithol. 14:302–309.

FERRY, C. 1976. Un test facile pour savoir si la richesse mesurée d'un peuplement se rapproche de sa richesse réelle. Jean-Le-Blanc 15:21–28.

FERRY, C., J. BLONDEL, AND B. FROCHOT. 1976. Plant successional stage and avifaunal structure on an island. Proc. Int. Ornithol. Congr. 16:643–653.

FERRY, C., AND B. FROCHOT. 1958. Une méthode pour dénombrer les oiseaux nicheurs. Terre et Vie 12:85–102.

FERRY, C., AND B. FROCHOT. 1970. L'avifaune nidificatrice d'une forêt de Chênes pédonculés en Bourgogne: étude de deux successions écologiques. Terre et Vie 24:153–250.

FERRY, C., B. FROCHOT, AND Y. LERUTH. 1981. Territory and home range of the Blackcap (*Sylvia atricapilla*) and some other passerines, assessed and compared by mapping and capture-recapture. Pp. 119–120 *in* C. J. Ralph and J. M. Scott (eds.). Estimating the numbers of terrestrial birds. Stud. Avian Biol. 6.

FIELD, M. 1971. Hawk Cliff raptor banding station—first annual report, 1971. Ont. Bird Banding 7:56–75.

FIENBERG, S. E. 1972. The multiple recapture census for closed populations and incomplete 2^k contingency tables. Biometrika 59:591–603.

FINLEY, R. B. 1965. Adverse effects on birds of phosphamidon applied to a Montana forest. J. Wildl. Manage. 29:580–591.

FINNEY, D. J. 1941. On the distribution of a variate whose logarithm is normally distributed. J. Roy. Stat. Soc. 7B:155–161.

FINNEY, G. H., P. CADIEUX, AND E. SILIEFF. 1980. The co-operative breeding bird survey in Canada, 1979. Can. Wildl. Serv., Prog. Notes 111.

FITCH, H. S., AND R. O. BARE. 1978. A field study of the Red-tailed Hawk in eastern Kansas. Trans. Kansas Acad. Sci. 81:1–13.

FITZPATRICK, J. W. 1973. Response by a Long-eared Owl to Barred Owl calls. Wilson Bull. 85:334–335.

FLOOD, B. S., M. E. SANGSTER, R. D. SPARROWE, AND T. S. BASKETT. 1977. A handbook for habitat evaluation procedures. U.S. Fish and Wildl. Serv., Res. Pub. 132.

FOGDEN, M. P. L. 1972. The seasonality and population dynamics of equatorial forest birds in Sarawak. Ibis 114:307–343.

FOOTE, L. E., H. S. PETERS, AND A. L. FINKER. 1958. Design tests for Mourning Dove call-count sampling in seven southeastern states. J. Wildl. Manage. 22:402–408.

FORBES, S. A., AND A. O. GROSS. 1921. The orchard

birds of an Illinois summer. Ill. Nat. Hist. Surv. Bull. 14:1–8.
FORBUSH, E. H. 1927. Birds of Massachusetts and other New England States, Part II. Land birds from bob-whites to grackles. Mass. Dept. Agric., Norwood.
FORBUSH, E. H. 1929. Birds of Massachusetts and other New England states, Part III. Land birds from sparrows to thrushes. Mass. Dept. Agric., Norwood.
FORD, R. G., AND D. W. KRUMME. 1979. The analysis of space-use patterns. J. Theor. Biol. 76:125–155.
FORD, R. G., AND J. P. MYERS. 1981. An evaluation and comparison of techniques for estimating home range and territory size. Pp. 461–465 *in* C. J. Ralph and J. M. Scott (eds.). Estimating the numbers of terrestrial birds. Stud. Avian Biol. 6.
FORSMAN, E. 1976. A preliminary investigation of the Spotted Owl in Oregon. M.S. thesis, Oregon State Univ., Corvallis.
FORSMAN, E. D., E. C. MESLOW, AND M. J. STRUB. 1977. Spotted Owl abundance in young versus old-growth forests, Oregon. Wildl. Soc. Bull. 5:43–47.
FOSTER, M. 1965. An early reference to the technique of owl calling. Auk 82:651–653.
FOSTER, M. L. 1977a. Status of the Salt Marsh Yellowthroat (*Geothlypis trichas sinuosa*) in the San Francisco Bay area, California, 1975–1976. Calif. Dept. Fish and Game, Nongame Wildl. Invest. Final Rep., Sacramento.
FOSTER, M. S. 1977b. Odd couples in manakins: A study of social organization and cooperative breeding in *Chiroxiphia linearis*. Am. Nat. 111:845–853.
FOWLE, C. D. 1972. Effects of phosphamidon on forest birds in New Brunswick. Can. Wildl. Serv., Rep. Ser. 16.
FOWLER, D. K., AND B. S. MCGINNES. 1973. A circular plot method of censusing post-breeding bird populations. Proc. Annu. Conf. Southeast. Assoc. Game and Fish Comm. 27:237–243.
FOWLER, D. L., AND J. N. MAHAN. 1978. The pesticide review, 1977. U.S. Dept. Agric., Agric. Stab. Conserv. Serv., Washington, D.C.
FRANCIS, W. J. 1973. Accuracy of census method of territorial Red-winged Blackbirds. J. Wildl. Manage. 37:98–102.
FRANKEL, A. I., AND T. S. BASKETT. 1961. The effect of pairing on cooing of penned Mourning Doves. J. Wildl. Manage. 25:372–384.
FRANKLIN, J. F., AND C. T. DYRNESS. 1973. Natural vegetation of Oregon and Washington. U.S. For. Serv., Gen. Tech. Rep. PNW-8, Portland, Ore.
FRANZREB, K. E. 1976. Comparison of variable strip transect and spot-map methods for censusing avian populations in a mixed-coniferous forest. Condor 78:260–262.
FRANZREB, K. E. 1977. Inventory techniques for sampling avian populations. U.S. Bur. Land Manage., Tech. Note 307, Denver, Col.
FRANZREB, K. E. 1981a. The determination of avian densities using the variable-strip and fixed-width transect surveying methods. Pp. 139–145 *in* C. J. Ralph and J. M. Scott (eds.). Estimating the numbers of terrestrial birds. Stud. Avian Biol. 6.
FRANZREB, K. E. 1981b. A comparative analysis of territorial mapping and variable-strip transect censusing methods. Pp. 164–169 *in* C. J. Ralph and J. M. Scott (eds.). Estimating the numbers of terrestrial birds. Stud. Avian Biol. 6.
FRANZREB, K. E., AND R. D. OHMART. 1978. The effects of timber harvesting on breeding birds in a mixed-coniferous forest. Condor 80:431–441.
FRASER, J. D. 1978. Bald Eagle reproductive surveys: accuracy, precision, and timing. M.S. thesis, Univ. Minnesota, Minneapolis.
FREEMARK, K. E., C. R. COOPER, G. H. FINNEY, AND A. J. ERSKINE. 1979. The co-operative breeding bird survey in Canada. Can. Wildl. Serv., Prog. Notes 99.
FRENCH, J. M., AND J. R. KOPLIN. 1977. Distribution, abundance, and breeding status of Ospreys in northwestern California. Pp. 223–240 *in* J. C. Ogden (ed.). Transactions of the North American Osprey research conference. U.S. Natl. Park Serv., Trans. and Proc. Ser. 2.
FRETWELL, S. D., AND H. L. LUCAS. 1969. On territorial behavior and other factors influencing habitat distribution in birds. I. Theoretical development. Acta Biotheor. 19:16–36.
FREUND, R. J., AND P. D. MINTON. 1979. Regression methods, a tool for data analysis. Dekker, New York.
FRINGS, H., M. FRINGS, J. JUMBER, R. BUSNEL, J. GIBAN, AND P. GRAMET. 1958. Reactions of American and French species of *Corvus* and *Larus* to recorded communication signals tested reciprocally. Ecology 39:126–131.
FROCHOT, B. 1976. Les méthodes utilisées pour dénombrer les oiseaux. Pp. 47–69 *in* J. Ruwet (ed.). Problèmes liés à l'étude et à la gestion de la faune des Hautes-Fagues et de la Haute-Ardenne. Univ. Liège, Belgium.
FROCHOT, B., D. REUDET, AND Y. LERUTH. 1977. A comparison of preliminary results of three census methods applied to the same population of forest birds. Pol. Ecol. Stud. 3(4):71–75.
FULLER, M. R. 1979. Hawk count research. Newsl. Hawk Migr. Assoc. N. Am. 4:1–3.
FULLER M. R. In press. Raptor research priorities for the 80's. Raptor Res.
FULLER, M. R., AND G. S. CHRISTENSON. 1975. An evaluation of techniques for capturing raptors in east-central Minnesota. Raptor Res. 10:9–19.
FULLER, M. R., AND J. A. MOSHER. 1981. Methods of detecting and counting raptors: a review. Pp. 235–246 *in* C. J. Ralph and J. M. Scott (eds.). Estimating the numbers of terrestrial birds. Stud. Avian Biol. 6.
FULLER, M. R., AND C. S. ROBBINS. 1979. Report forms. Newsl. Hawk Migr. Assoc. N. Am. 4:1–2.
FYFE, R. W., AND R. R. OLENDORFF. 1976. Minimizing the dangers of nesting studies to raptors and other sensitive species. Can. Wildl. Serv., Occ. Pap. 23.
FYFE, R. W., S. A. TEMPLE, AND T. J. CADE. 1976. The 1975 North American Peregrine Falcon survey. Can. Field-Nat. 90:228–273.
GAINES, D. 1974a. Distribution, density and habitat requirements of the California Yellow-billed Cuckoo

in the Sacramento Valley: 1972–73. Calif. Dept. Game and Fish, Sacramento.

GAINES, D. 1974b. Review of the status of the Yellow-billed Cuckoo in California: Sacramento Valley populations. Condor 76:204–209.

GAINES, D. A. 1977. The valley riparian forests of California: their importance to bird populations. Pp. 57–85 *in* A. Sands (ed.). Riparian forests in California: their ecology and conservation. Inst. Ecol. Publ. 15, Davis, Cal.

GALLIZIOLI, S. 1964. Results of a brief investigation of the Masked Bobwhite in Sonora, Mexico. Arizona Game and Fish Dept., Spec. Invest., Phoenix.

GALUSHIN, V. M. 1974. Synchronous fluctuations in populations of some raptors and their prey. Ibis 116:127–134.

GALUSHIN, V. M. In press. Changes in population status and nest range distribution of falconiforms in the USSR since 1950. Raptor Res.

GARCIA, E. R. 1979. A survey of the Spotted Owl in southwestern Washington. Pp. 18–28 *in* P. P. Schaeffer and S. M. Ehlers (eds.). Proceedings of symposium on owls of the West: their ecology and conservation. Nat. Aud. Soc. West. Ed. Cen., Tiburon, Cal.

GARSON, P. J., AND M. L. HUNTER. 1979. Effects of temperature and time of year on the singing behavior of Wrens *Troglodytes troglodytes* and Great Tits *Parus major*. Ibis 121:481–487.

GATES, C. E. 1969. Simulation study of estimators for the line transect sampling method. Biometrics 25:317–328.

GATES, C. E. 1979. Line transect and related issues. Pp. 71–154 *in* R. M. Cormack, G. P. Patil, and D. S. Robson (eds.). Sampling biological populations. Stat. Ecol. Ser. vol. 5. Int. Co-op. Publ. House, Fairland, Md.

GATES, C. E. 1980. LINETRAN, a general computer program for analyzing line-transect data. J. Wildl. Manage. 44:658–661.

GATES, C. E. 1981. Optimizing sampling frequency and numbers of transects and stations. Pp. 399–404 *in* C. J. Ralph and J. M. Scott (eds.). Estimating the numbers of terrestrial birds. Stud. Avian Biol. 6.

GATES, C. E., T. L. CLARK, AND K. E. GAMBLE. 1975. Optimizing Mourning Dove breeding population surveys in Texas. J. Wildl. Manage. 39:237–242.

GATES, C. E., W. H. MARSHALL, AND D. P. OLSON. 1968. Line transect method of estimating grouse population densities. Biometrics 24:135–145.

GATES, C. E., AND P. W. SMITH. 1980. An implementation of the Burnham-Anderson distribution-free method of estimating wildlife densities from the line transect data. Biometrics 36:155–160.

GATES, C. E., AND W. B. SMITH. 1972. Estimation of density of Mourning Doves from aural information. Biometrics 28:345–359.

GATES, J. M. 1966. Crowing counts as indices to cock pheasant populations in Wisconsin. J. Wildl. Manage. 30:735–744.

GAUCH, H. G., AND T. R. WENTWORTH. 1976. Canonical correlation analysis as an ordination technique. Vegetatio 33:17–22.

GAUCH, H. G., AND R. H. WHITTAKER. 1972. Comparison of ordination techniques. Ecology 53:868–875.

GEIS, A. D. 1972. Use of banding data in migratory game bird research and management. U.S. Fish and Wildl. Serv., Spec. Sci. Rep. Wildl. 154.

GEIS, A. D., AND E. L. ATWOOD. 1961. Proportion of recovered waterfowl bands reported. J. Wildl. Manage. 25:154–159.

GEIS, A. D., R. I. SMITH, AND J. P. ROGERS. 1971. Black Duck distribution, harvest characteristics, and survival. U.S. Fish and Wildl. Serv., Spec. Sci. Rep. Wildl. 139.

GEISSLER, P. H., AND B. R. NOON. 1981. Estimates of avian population trends from the North American Breeding Bird Survey. Pp. 42–51 *in* C. J. Ralph and J. M. Scott (eds.). Estimating the numbers of terrestrial birds. Stud. Avian Biol. 6.

GIBBONS, J. D. 1971. Nonparametric statistical inference. McGraw-Hill, New York.

GILBERT, F. S. 1980. The equilibrium theory of island biogeography: fact or fiction? J. Biogeogr. 7:209–235.

GILBERT, R. O. 1973. Approximations of the bias in the Jolly-Seber capture-recapture model. Biometrics 29:501–526.

GILES, R. H. 1970. The ecology of a small forested watershed treated with the insecticide malathion-S^{35}. Wildl. Monogr. 24.

GILL, B. J. 1980. Abundance, feeding, and morphology of passerine birds at Kowhai Bush, Kaikoura, New Zealand. N.Z. J. Zool. 7:235–246.

GILL, R. 1979. Status and distribution of the California Clapper Rail (*Rallus longirostris obsoletus*). Calif. Fish and Game 65:36–49.

GILPIN, M. E., AND J. M. DIAMOND. 1980. Subdivision of nature reserves and the maintenance of species diversity. Nature 285:567–568.

GLAHN, J. F. 1974. Study of breeding rails with recorded calls in northcentral Colorado. Wilson Bull. 86:206–214.

GLAS, P. 1960. Factors governing density in the Chaffinch (*Fringilla coelebs*) in different types of wood. Arch. neerl. Zool. 13:466–472.

GLINSKI, R. L. 1976. Birdwatching etiquette: the need for a developing philosophy. Am. Birds 30:655–657.

GLIWICZ, J. 1970. Relation between trappability and age of individuals in a population of the Bank Vole. Acta Theriol. 15:15–23.

GLOVER, F. A. 1948. Winter activities of wild Turkey in West Virginia. J. Wildl. Manage. 12:416–427.

GŁOWACIŃSKI, Z., AND J. WEINER. 1977. Energetics of bird communities in successional series of a deciduous forest. Pol. Ecol. Stud. 3(4):147–175.

GOLDMAN, P. 1973. Song recognition by Field Sparrows. Auk 90:106–113.

GOUDY, W. H., R. C, KLETZLY, AND J. C. RIEFFENBERGER. 1970. Characteristics of a heavily hunted Woodcock population in West Virginia. Trans. N. Am. Wildl. Nat. Resour. Conf. 35:183–194.

GOULD, G. I. 1974. The status of the Spotted Owl in California. Calif. Dept. Fish and Game, Sacramento, and U.S. For. Serv.

GOULD, G. I. 1975. Yuma Clapper Rail study—censuses and habitat distribution 1973–1974. Calif.

Dept. Fish and Game, Wildl. Manage. Branch Admin. Rep. 75-2, Sacramento.
GOULD, G. I. 1977. Distribution of the Spotted Owl in California. West. Birds 8:131–146.
GOULD, G. I. 1979. Status and management of Elf and Spotted Owls in California. Pp. 86–97 *in* P. P. Schaeffer and S. M. Ehlers (eds.) Proceedings of symposium on owls of the West: their ecology and conservation. Natl. Aud. Soc. West. Ed. Cen., Tiburon, Calif.
GOULD, S. J. 1980. The promise of paleobiology as a nomothetic, evolutionary discipline. Paleobiology 6:96–118.
GRABER, R. R., AND J. S. GOLDEN. 1960. Hawks and owls: population trends from Illinois Christmas counts. Ill. Nat. Hist. Surv., Biol. Note 41.
GRABER, R. R., AND J. W. GRABER. 1963. A comparative study of bird populations in Illinois, 1906–1909 and 1956–1958. Bull. Ill. Nat. Hist. Surv. 28:383–528.
GRADWOLD, J., AND R. GREENBURG. 1980. The formation of antwren flocks on Barro Colorado Island, Panama. Auk 97:385–395.
GRANT, J. 1966. The Barred Owl in British Columbia. Murrelet 47:39–45.
GRAY, L. 1961. Banding Sharpshins at Point Pelee. EBBA News 24:25–26.
GREEN, D. M., AND J. A. SWETS. 1974. Signal detection theory and psychophysics. Krieger, New York.
GREENLAW, J. S., AND J. SWINEBOARD. 1967. A method for constructing and erecting aerial-nets in a forest. Bird-Banding 38:114–119.
GREENWOOD, R. J., AND W. C. BAIR. 1974. Ice on waterfowl markers. Wildl. Soc. Bull. 2:130–134.
GREMPE, G. 1973. Brutvogelbestandaufnahmen auf dem Alten Friedhof in Rostock (1962–1972). Ornithol. Rundbr. Mecklenb. N.F. 14:37–41.
GRIER, J. W. 1974. Reproduction, organochlorines, and mercury in northwestern Ontario Bald Eagles. Can. Field-Nat. 88:469–475.
GRIER, J. W. 1977. Quadrat sampling of a nesting population of Bald Eagles. J. Wildl. Manage. 41:438–443.
GRIER, J. W., AND G. D. HAMILTON. 1978. Aerial census of Bald Eagles in the West Patricia Planning Area. West Patricia Land Use Plan. Ontario Min. Nat. Res., Wildl. Tech. Rep. 7.
GRIER, J. W., J. M. GERRARD, G. D. HAMILTON, AND P. A. GRAY. 1981. Aerial visibility bias and survey techniques for nesting Bald Eagles in northwestern Ontario. J. Wildl. Manage. 45:83–92.
GRIESE, H. J., R. A. RYDER, AND C. E. BRAUN. 1980. Spatial and temporal distribution of rails in Colorado. Wilson Bull. 92:96–102.
GRIMOLDI, D. 1976. Essais de dénombrements en milieux hétérogènes par la méthode des EFP. Jean-Le-Blanc. 15:38–54.
GRINNELL, J. 1922. The role of the "accidental." Auk 39:373–380.
GROSS, A. O. 1948. *Troglodytes aedon aedon* Vieillot: Eastern House Wren. Pp. 113–114 *in* A. C. Bent (ed.). Life histories of North American nuthatches, wrens, thrashers, and their allies. U.S. Natl. Mus. Bull. 195.
GRUBB, T. C. 1975. Weather-dependent foraging behavior of some birds wintering in a deciduous woodland. Condor 77:175–182.
GRUBB, T. C. 1977. Weather-dependent foraging behavior of some birds wintering in a deciduous woodland: horizontal adjustments. Condor 79:271–274.
GRUBB, T. C. 1978. Weather-dependent foraging rates of wintering woodland birds. Auk 95:370–376.
GRUBB, T. G., D. A. MANUWAL, AND C. M. ANDERSON. 1975. Nest distributions and productivity of Bald Eagles in western Washington. Murrelet 56: 2–6.
GRUE, C. E. 1977. The impact of powerline construction on birds in central Arizona. M.S. thesis, Northern Arizona Univ., Flagstaff.
GRUE, C. E., R. P. BALDA, AND C. D. JOHNSON. 1981. Diurnal activity patterns and population estimates of breeding birds within a disturbed and undisturbed desert-scrub community. Pp. 287–291 *in* C. J. Ralph and J. M. Scott (eds.). Estimating the numbers of terrestrial birds. Stud. Avian Biol. 6.
GRUE, C. E., AND B. K. SHIPLEY. 1981. Interpreting population estimates of birds following pesticide applications—behavior of male Starlings exposed to an organophosphate pesticide. Pp. 292–296 *in* C. J. Ralph and J. M. Scott (eds.). Estimating the numbers of terrestrial birds. Stud. Avian Biol. 6.
GULLION, G. W. 1966. The use of drumming behavior in Ruffed Grouse population studies. J. Wildl. Manage. 30:717–729.
GUR'EV, V. N., AND N. E. ZUBTSOVSKII. 1974. On the accuracy of winter bird counts on linear transects. Ekologiya 1:100–102.
GUTHERY, F. S. 1975. Footprint measurements of Canadian Sandhill Cranes. J. Wildl. Manage. 39:447–448.
HACKMAN, C. D., AND C. J. HENNY. 1971. Hawk migration over White Marsh, Maryland. Chesapeake Sci. 12:137–141.
HAILA, Y., AND O. JÄRVINEN. 1980. Bird communities in a Finnish archipelago 50 years ago and now: a general survey. Pp. 151–157 *in* H. Oelke (ed.). Bird census work and nature conservation. Dacherverband Deutscher Avifaunisten, Göttingen, W. Ger.
HAILA, Y., AND O. JÄRVINEN. 1981. The underexploited potential of bird censuses in insular ecology. Pp. 559–565 *in* C. J. Ralph and J. M. Scott (eds.). Estimating the numbers of terrestrial birds. Stud. Avian Biol. 6.
HAILA, Y., O. JÄRVINEN, AND R. A. VÄISÄNEN. 1979. Effect of mainland population changes on the terrestrial bird fauna of a northern island. Ornis Scand. 10:48–55.
HAILA, Y., O. JÄRVINEN, AND R. A. VÄISÄNEN. 1980a. Habitat distribution and species associations of land bird populations on the Åland Islands, SW Finland. Ann. Zool. Fenn. 17:86–105.
HAILA, Y., O. JÄRVINEN, AND R. A. VÄISÄNEN. 1980b. Effects of changing forest structure on long-term trends in bird populations in SW Finland. Ornis Scand. 11:12–22.
HALL, B. P., AND R. E. MOREAU. 1970. An atlas of speciation in African passerine birds. Brit. Mus. (Nat. Hist.), London.
HALL, G. A. 1964. Breeding-Bird Censuses—why and how. Aud. Field Notes 18:413–416.

HALL, M. A. 1979. Annotated computer output for regression analysis. BMDP Tech. Rep. 48, Health Sci. Comp. Fac., Univ. Calif., Los Angeles.

HAMERSTROM, F. 1969. A Harrier population study. Pp. 367–383 *in* J. J. Hickey (ed.). Peregrine Falcon populations: their biology and decline. Univ. Wisconsin Press, Madison.

HAMMERSLEY, J. M. 1953. Capture-recapture analysis. Biometrika 40:265–270.

HANCOCK, D. 1964. Bald Eagles wintering in the southern Gulf Islands, British Columbia. Wilson Bull. 76:111–120.

HANSEN, K. 1977. The Greenland White-tailed Eagle. Pp. 73–74 *in* R. D. Chancellor (ed.). World conference on birds of prey, Vienna 1975. Int. Council Bird Pres.

HARWOOD, M. (ed.) 1975. Proceedings of the North American hawk migration conference, 1974. Hawk Migr. Assoc. of N. Am., Washington Depot, Ct.

HARWOOD, M., AND A. C. NAGY. 1977. Assessing the hawk counts at Hawk Mountain. Pp. 69–75 *in* J. C. Ogden (ed.). Transactions of the North American Osprey research conference. U.S. Natl. Park Serv., Trans. and Proc. Ser. 2.

HAUGH, J. R. 1975. Local ephemeral weather conditions and their effects on hawk migration routes. Pp. 72–84 *in* M. Harwood (ed.). Proceedings of the North American hawk migration conference, 1974. Hawk Migr. Assoc. of N. Am., Washington Depot, Ct.

HAUGH, J. R., AND T. J. CADE. 1966. The spring hawk migration around the southeastern shore of Lake Ontario. Wilson Bull. 78:88–110.

HAUKIOJA, E. 1968. Reliability of the line survey method in bird census, with reference to Reed Bunting and Sedge Warbler. Ornis Fenn. 45:105–113.

HAVERSCHMIDT, F. 1946. Observations on the breeding habits of the Little Owl. Ardea 34:214–246.

HAYASHI, C., M. ISHIDA, Y. OISHI, K. TAKATA, J. TOYOSHIMA, S. HADA, AND R. HORIGUCHI. 1966. Estimation of size of mobile populations (I). Sample survey for the estimation of the number of hares in a district. Ann. Inst. Stat. Math. 14:63–86.

HAYES, R. 1977. A critical review of line transect sampling. M.S. thesis, Univ. Edinburgh, Edinburgh, Scotland.

HAYNE, D. W. 1949. An examination of the strip census method for estimating animal populations. J. Wildl. Manage. 13:145–157.

HEATH, J. (ed.) 1970. Provisional atlas of the insects of the British Isles. Part 1. Lepidoptera Rhopalocera. Biol. Records Ctr., Huntingdon, Eng.

HECK, K. L., G. VAN BELLE, AND D. SIMBERLOFF. 1975. Explicit calculation of the rarefaction diversity measurement and the determination of sufficient sample size. Ecology 56:1459–1461.

HEIMERDINGER, M. A., AND R. C. LEBERMAN. 1966. The comparative efficiency of 30 and 36 mm mesh mist nets. Bird-Banding 37:280–285.

HEINTZELMAN, D. S. 1975. Autumn hawk flights: the migrations in eastern North America. Rutgers Univ. Press, New Brunswick, N.J.

HEINTZELMAN, D. S. 1979. Hawks and owls of North America. Universe, New York.

HEINZ, G. H., E. F. HILL, W. H. STICKEL, AND L. F. STICKEL. 1979. Environmental contaminant studies by the Patuxent Wildlife Research Center. Pp. 9–35 *in* E. E. Kenaga (ed.). Avian and mammalian wildlife toxicology. Am. Soc. Testing and Materials, STP 693.

HENNESSY, S. P. 1978. Ecological relationships of accipiters in northern Utah—with special emphasis on the effects of human disturbance. M.S. thesis, Utah State Univ., Logan.

HENNING, G. B. 1966. Frequency discrimination of random-amplitude tones. J. Acoust. Soc. Am. 39:336–339.

HENNY, C. J., AND D. W. ANDERSON. 1979. Osprey distribution, abundance, and status in western North America: III. The Baja California and Gulf of California population. Bull. So. Calif. Acad. Sci. 18:89–106.

HENNY, C. J., AND K. P. BURNHAM. 1976. A reward band study of Mallards to estimate band reporting rates. J. Wildl. Manage. 40:1–14.

HENNY, C. J., M. A. BYRD, J. A. JACOBS, P. D. MCCLAIN, M. R. TODD, AND B. F. HALLA. 1977. Mid-Atlantic coast Osprey population: present numbers, productivity, pollutant contamination, and status. J. Wildl. Manage. 41:254–265.

HENNY, C. J., J. A. COLLINS, AND W. J. DEIBERT. 1978a. Osprey distribution, abundance, and status in western North America: II. The Oregon population. Murrelet 59:14–25.

HENNY, C. J., D. J. DUNAWAY, R. D. MALLETTE, AND J. R. KOPLIN. 1978b. Osprey distribution, abundance, and status in western North America: I. The northern California population. Northwest Sci. 52:261–272.

HENNY, C. J., AND A. P. NOLTEMEIER. 1975. Osprey nesting populations in the coastal Carolinas. Am. Birds 29:1073–1079.

HENNY, C. J., M. M. SMITH, AND V. D. STOTTS. 1974. The 1973 distribution and abundance of breeding Ospreys in the Chesapeake Bay. Chesapeake Sci. 15:125–133.

HERMAN, S. G. 1971. Peregrine Falcon decline in California. Am. Birds 25:818–820.

HESPENHEIDE, H. S. 1980. Bird community structure in two Panama forests: Residents, migrants, and seasonality during the nonbreeding season. Pp. 227–237 *in* A. Keast and E. S. Morton (eds.). Migrant birds in the Neotropics: ecology, behavior, distribution and conservation. Smithsonian Inst. Press, Washington, D.C.

HEWITT, O. H. 1963. Development of a roadside count method of census of breeding Redwinged Blackbirds (*Agelaius phoeniceus*). Angew. Ornithol. 1:123–125.

HEWITT, O. H. 1967. A road-count index to breeding populations of Red-winged Blackbirds. J. Wildl. Manage. 31:39–47.

HIATT, R. W. 1944. A raptor census in Montana. Am. Midl. Nat. 31:384–388.

HICKEY, J. J. 1942. Eastern population of the Duck Hawk. Auk 59:176–204.

HICKEY, J. J. 1955. Letter to the editor. Wilson Bull. 67:144–145.

HICKEY, J. J. (ed.) 1969. Peregrine Falcon populations: their biology and decline. Univ. Wisconsin Press, Madison.

HICKEY, J. J., AND D. W. ANDERSON. 1969. The Peregrine Falcon: life history and population literature. Pp. 3–42 *in* J. J. Hickey (ed.). Peregrine Falcon populations: their biology and decline. Univ. Wisconsin Press, Madison.

HICKEY, J. J., AND S. A. MIKOL. 1979. Estimating breeding-bird densities on coal lands in Montana and Wyoming. U.S. Fish and Wildl. Serv., Western Energy and Land Use Team, Rep. 79/03, Fort Collins, Col.

HICKMAN, G. L. 1972. Aerial determination of Golden Eagle nesting status. J. Wildl. Manage. 36:1289–1292.

HICKS, D. L., D. T. ROGERS, AND G. I. CHILD. 1966. Autumnal hawk migration through Panama. Bird-Banding 37:121–123.

HICKS, L. E. 1933. A statistical study of Ohio raptorial birds. Wilson Bull. 45:139.

HIGBY, L. W. 1975. The eagle survey in Wyoming. Pp. 97–102 *in* J. R. Murphy, C. M. White, and B. E. Harrell (eds.). Population status of raptors. Raptor Res. Rep. 3.

HIGGINS, R. M. 1979. Temperature-related variation in the duration of morning song of the Song Thrush (*Turdus ericetorum*). Ibis 121:333–335.

HIGGS, A. J., AND M. B. USHER. 1980. Should nature reserves be large or small? Nature 285:568–569.

HILDÉN, O. 1965. Habitat selection in birds: a review. Ann. Zool. Fenn. 2:53–75.

HILDÉN, O. 1979. Lintujen maastotuntemuksen kehittyminen ja linjalaskennat. (Effects of the advances in field ornithology on bird censuses.) Lintumies 14:9–14.

HILDÉN, O. 1981. Sources of error involved in the Finnish line transect method. Pp. 152–159 *in* C. J. Ralph and J. M. Scott (eds.). Estimating the numbers of terrestrial birds. Stud. Avian Biol. 6.

HILL, E. F. 1971. Toxicity of selected mosquito larvicides to some common avian species. J. Wildl. Manage. 35:757–762.

HILL, E. F. 1979. Cholinesterase activity in Japanese quail dusted with carbaryl. Lab. Anim. Sci. 29:349–352.

HILLS, G. A. 1952. The classification and evaluation of site for forestry. Ont. Dept. Lands and Forests, Res. Rep. 24.

HILTON, J. R. 1977. The legal status of birds of prey in other parts of the world. Pp. 193–197 *in* R. D. Chancellor (ed.). World conference on birds of prey, Vienna 1975. Int. Council Bird Pres.

HINDE, R. A. 1969. Bird vocalizations. Univ. Press, Cambridge.

HINZ, J. E. 1969. A life history of the Great Horned Owl in southeastern Stearns County, Minnesota. M.A. thesis, St. Cloud State Univ., St. Cloud, Minn.

HIRSCH, A., W. B. KROHN, D. L. SCHWEITZER, AND C. H. THOMAS. 1979. Trends and needs in federal inventories of wildlife habitat. Trans. N. Am. Wildl. Nat. Res. Conf. 44:340–358.

HIRSH, I. J. 1959. Auditory perception of temporal order. J. Acoust. Soc. Am. 31:759–767.

HIRST, S. M. 1969. Road-strip census techniques for wild ungulates in African woodland. J. Wildl. Manage. 33:40–48.

HJORT, C., AND C.-G. LINDHOLM. 1978. Annual bird ringing totals and population fluctuations. Oikos 30:387–392.

HODGES, J. I., J. G. KING, AND F. C. ROBARDS. 1979. Resurvey of the Bald Eagle breeding population in southeast Alaska. J. Wildl. Manage. 43:219–221.

HOFFMAN, D. M. 1968. Roosting sites and habits of Merriam's Turkeys in Colorado. J. Wildl. Manage. 32:859–866.

HOGSTAD, O. 1967. Factors influencing the efficiency of the mapping method in determining breeding bird populations in conifer forests. Nytt. Mag. Zool. (Oslo)14:125–141.

HOLLANDER, M., AND D. A. WOLFE. 1973. Nonparametric statistical methods. Wiley, New York.

HOLMES, R. T., R. E. BONNEY, AND S. W. PACALA. 1979. Guild structure of the Hubbard Brook bird community: a multivariate approach. Ecology 60:512–520.

HOLMES, R. T., AND F. W. STURGES. 1975. Bird community dynamics and energetics in a northern hardwoods ecosystem. J. Anim. Ecol. 44:175–200.

HOPE JONES, P. 1974. Transect sampling of upland bird populations in north Wales. Acta Ornithol. 14:354–360.

HOPKINS, D. A. 1975. The New England hawk watch. Pp. 137–146 *in* M. Harwood (ed.). Proceedings of the North American Hawk Migration Conference, 1974. Hawk Migr. Assoc. of N. Am., Washington Depot, Ct.

HOWARD, I. P. 1973. The spatial senses. Pp. 273–290 *in* E. C. Carterette and M. P. Friedman (eds.). Handbook of perception. Vol. 3. Academic, New York.

HOWARD, R. P., L. O. WILSON, AND F. B. RENN. 1976. Relative abundance of nesting raptors in southern Idaho. Raptor Res. 10:120–128.

HOWE, H. F. 1977. Bird activity and seed dispersal of a tropical wet forest tree. Ecology 58:539–550.

HOWELL, J. C. 1951. The roadside census as a method of measuring bird populations. Auk 68:334–357.

HOWELL, J. C. 1973. The 1971 status of 24 Bald Eagle nest sites in east-central Florida. Auk 90:678–680.

HOWELL, J. C., AND G. M. HEINZMAN. 1967. Comparisons of nesting sites of Bald Eagles in central Florida from 1930 to 1965. Auk 84:602–603.

HOWELL, T. R. 1971. An ecological study of the birds of the lowland pine savanna and adjacent rain forest in northeastern Nicaragua. Living Bird 10:185–242.

HUHTA, V. 1979. Evaluation of different similarity indices as measures of succession in arthropod communities of the forest floor after clear-cutting. Oecologia (Berl.) 41:11–23.

HULTÉN, E. 1950. Atlas of the distribution of vascular plants in northwest Europe. Generalstabens Litografiska Förlag, Stockholm.

HUMPHREY, P. S., D. BRIDGE, AND T. E. LOVEJOY. 1968. A technique for mist netting in the forest canopy. Bird-Banding 39:43–50.

HUNGERFORD, K. E. 1953. A Ruffed Grouse drumming count technique for northern Idaho conditions. Univ. Idaho For., Wildl. and Range Expt. Sta. Res. Note 10.

HURLBERT, S. H. 1971. The nonconcept of species diversity: a critique and alternate parameters. Ecology 52:577–586.

HUSSELL, D. J. T. 1981a. The use of migration counts for monitoring bird population levels. Pp. 92–102 *in* C. J. Ralph and J. M. Scott (eds.). Estimating the numbers of terrestrial birds. Stud. Avian Biol. 6.

HUSSELL, D. J. T. 1981b. Migrations of the Least Flycatcher in southern Ontario. J. Field Ornithol. 52, in press.

HUSSELL, D. J. T., AND C. J. RISLEY. 1978. Monitoring bird populations by migration sampling. Long Point Bird Obs., Port Rowan, Ont.

HUSSELL, D. J. T., AND R. W. STAMP. 1965. Movements of Black-capped Chickadees at Long Point, Ontario, during the spring of 1962. Bird-Banding 36:71–80.

HUSSELL, D. J. T., AND J. WOODFORD. 1961. The use of a Heligoland trap and mist-net at Long Point, Ontario. Bird-Banding 32:115–125.

HUTCHESON, K. 1970. A test for comparing diversities based on the Shannon formula. J. Theor. Biol. 29:151–154.

HUTTO, R. L., AND S. L. MOSCONI. 1981. Lateral detectability profiles for line transect bird censuses: some problems and an alternative. Pp. 382–387 *in* C. J. Ralph and J. M. Scott (eds.). Estimating the numbers of terrestrial birds. Stud. Avian Biol. 6.

IMHOF, T. I. 1976. Alabama birds. Univ. Alabama Press, University, Ala.

INTERNATIONAL BIRD CENSUS COMMITTEE (I.B.C.C.). 1969. Recommendations for an international standard for a mapping method in bird census work. Bird Study 16:248–255.

INTERNATIONAL BIRD CENSUS COMMITTEE (I.B.C.C.). 1970. An international standard for a mapping method in bird census work recommended by the International Bird Census Committee. Aud. Field Notes 24:722–726.

INTERNATIONAL BIRD CENSUS COMMITTEE (I.B.C.C.). 1977. Censusing breeding birds by the IPA method. Pol. Ecol. Stud. 3(4):15–17.

IRVING, L. 1960. Birds of Anaktuvuk Pass, Kobuk, and Old Crow: a study in arctic adaptation. U.S. Natl. Mus. Bull. 217.

JACKMAN, S. M. 1974. Woodpeckers of the Pacific Northwest: their characteristics and their role in the forests. M.S. thesis, Oregon State Univ., Corvallis.

JACKSON, J. A. 1977. Determination of the status of Red-cockaded Woodpecker colonies. J. Wildl. Manage. 41:448–452.

JAMES, F. E. 1971. Ordinations of habitat relationships among breeding birds. Wilson Bull. 83:215–236.

JAMES, F. C. 1978. On understanding quantitative surveys of vegetation. Am. Birds 32:18–21.

JAMES, F. C., AND H. H. SHUGART. 1970. A quantitative method of habitat description. Aud. Field Notes 24:727–736.

JAMES, F. C., AND H. H. SHUGART. 1974. The phenology of the nesting season of the American Robin (*Turdus migratorius*) in the United States. Condor 76:159–168.

JAMES, W., AND C. STEIN. 1961. Estimation with quadratic loss. Proc. 4th Berkeley Symposium Math. Stat. and Prob. 1:361–379.

JÄRVINEN, O. 1976. Estimating relative densities of breeding birds by the line transect method. II. Comparison between two methods. Ornis Scand. 7:43–48.

JÄRVINEN, O. 1978a. Estimating relative densities of land birds by point counts. Ann. Zool. Fenn. 15:290–293.

JÄRVINEN, O. 1978b. Species-specific census efficiency in line transects. Ornis Scand. 9:164–167.

JÄRVINEN, O. 1978c. Turnover in north European bird communities. Pp. 72–73 *in* XVII Int. Ornithol. Congr., Symposia Abstracts.

JÄRVINEN, O. 1979. Geographical gradients of stability in European land bird communities. Oecologia 38:51–69.

JÄRVINEN, O. In press. Dynamics of north European bird communities. Proc. Int. Ornithol. Congr. 17.

JÄRVINEN, O., K. KUUSELA, AND R. A. VÄISÄNEN. 1977a. Effects of modern forestry on the numbers of breeding birds in Finland in 1945–1975. (In Finnish with English summary.) Silva Fennica 11:284–294.

JÄRVINEN, O., AND J. LOKKI. 1978. Indices of community structure in bird censuses based on a single visit: effect of variation in species efficiency. Ornis Scand. 9:87–93.

JÄRVINEN, O., AND L. SAMMALISTO. 1973. Indices of community structure in incomplete bird censuses when all species are equally detectable. Ornis Scand. 4:127–143.

JÄRVINEN, O., AND R. A. VÄISÄNEN. 1973. Species diversity of Finnish birds, I: Zoogeographical zonation based on land birds. Ornis Fenn. 50:93–125.

JÄRVINEN, O., AND R. A. VÄISÄNEN. 1975. Estimating relative densities of breeding birds by the line transect method. Oikos 26:316–322.

JÄRVINEN, O., AND R. A. VÄISÄNEN. 1976a. Between-year component of diversity in the communities of breeding land birds. Oikos 27:34–39.

JÄRVINEN, O., AND R. A. VÄISÄNEN. 1976b. Estimating relative densities of breeding birds by the line transect method. IV. Geographical constancy of the proportion of main belt observations. Ornis Fenn. 53:87–91.

JÄRVINEN, O., AND R. A. VÄISÄNEN. 1976c. Finnish line transect censuses. Ornis Fenn. 53:115–118.

JÄRVINEN, O., AND R. A. VÄISÄNEN. 1977a. How complete are the species lists of breeding birds in censuses of large areas? Ornis Fenn. 54:160–165.

JÄRVINEN, O., AND R. A. VÄISÄNEN. 1977b. Long-term changes of the north European land bird fauna. Oikos 29:225–228.

JÄRVINEN, O., AND R. A. VÄISÄNEN. 1977c. Line transect method: a standard for field-work. Pol. Ecol. Stud. 3(4):11–15.

JÄRVINEN, O., AND R. A. VÄISÄNEN. 1977d. Recent quantitative changes in the populations of Finnish land birds. Pol. Ecol. Stud. 3(4):177–188.

JÄRVINEN, O., AND R. A. VÄISÄNEN. 1977e. Constants and formulae for analyzing line transect data. Mimeograph, Helsinki.

JÄRVINEN, O., AND R. A. VÄISÄNEN. 1978a. Ecological zoogeography of north European waders, or why do so many waders breed in the North? Oikos 30:496–507.

JÄRVINEN, O., AND R. A. VÄISÄNEN. 1978b. Long-term changes of the most abundant south Finnish forest birds during the past 50 years. J. Ornithol. 119:441–449.

JÄRVINEN, O., AND R. A. VÄISÄNEN. 1978c. Miksi aikaisemmin tehtyjä pesimälinnuston laskentoja kannattaa toistaa? (Trends in the numbers of breeding forest birds in south Finland in the period 1936–77.) Lintumies 13:1–6.

JÄRVINEN, O., AND VÄISÄNEN. 1978d. Population trends of Finnish Starlings in the 1970s. Ornis Fenn. 55:80–82.

JÄRVINEN, O., AND R. A. VÄISÄNEN. 1978e. Recent changes in forest bird populations in northern Finland. Ann. Zool. Fenn. 15:279–289.

JÄRVINEN, O., AND R. A. VÄISÄNEN. 1979a. Climatic changes, habitat changes, and competition: dynamics of geographical overlap in two pairs of congeneric bird species in Finland. Oikos 33:261–271.

JÄRVINEN, O., AND R. A. VÄISÄNEN. 1979b. Changes in bird populations as criteria of environmental changes. Holarctic Ecol. 2:75–80.

JÄRVINEN, O., AND R. A. VÄISÄNEN. 1980. Quantitative biogeography of Finnish land birds as compared with regionality in other taxa. Ann. Zool. Fenn. 17:67–85.

JÄRVINEN, O., AND R. A. VÄISÄNEN. 1981. Methodology for censusing land bird faunas in large regions. Pp. 146–151 *in* C. J. Ralph and J. M. Scott (eds.). Estimating the numbers of terrestrial birds. Stud. Avian Biol. 6.

JÄRVINEN, O., R. A. VÄISÄNEN, AND A. ENEMAR. 1978a. Efficiency of the line transect method in mountain birch forest. Ornis Fenn. 55:16–23.

JÄRVINEN, O., R. A. VÄISÄNEN, AND Y. HAILA. 1976. Estimating relative densities of breeding birds by the line transect method. III. Temporal constancy of the proportion of main belt observations. Ornis Fenn. 53:40–45.

JÄRVINEN, O., R. A. VÄISÄNEN, AND Y. HAILA. 1977b. Bird census results in different years, stages of the breeding season and times of the day. Ornis Fenn. 54:108–118.

JÄRVINEN, O., R. A. VÄISÄNEN, AND W. WALANKIEWICZ. 1978b. Efficiency of the line transect method in central European forests. Ardea 66:103–111.

JENNRICH, R. I., AND F. B. TURNER. 1979. Measurement of non-circular home range. J. Theor. Biol. 22:227–237.

JENSEN, H. 1974. The reliability of the mapping method in marshes with special reference to the internationally accepted rules. Acta Ornithol. 14:378–385.

JENSEN, J. 1972. Ornithologiske undersøgelser i Kagsmosen, 7. Danske Fugle 7:239–247.

JOHNSGARD, P. A., AND R. DISILVESTRO. 1976. Seventy-five years of changes in Mallard-Black Duck ratios in eastern North America. Am. Birds 30:905–908.

JOHNSON, D. H. 1981. Improved population estimates through the use of auxiliary information. Pp. 436–440 *in* C. J. Ralph and J. M. Scott (eds.). Estimating the numbers of terrestrial birds. Stud. Avian Biol. 6.

JOHNSON, D. H. In press. The use and misuse of statistics in wildlife habitat studies. Pp. 436–440 *in* D. E. Capen (ed.). Proceedings of workshop on the use of multivariate statistics in studies of wildlife habitat. U.S. For. Serv., Gen. Tech. Rep.

JOHNSON, D., AND J. H. ENDERSON. 1972. Roadside raptor census in Colorado—winter 1971–72. Wilson Bull. 84:489–490.

JOHNSON, D. H., AND A. B. SARGEANT. 1977. Impact of Red Fox predation on the sex ratio of prairie Mallards. U.S. Fish and Wildl. Serv., Wildl. Res. Rep. 6.

JOHNSON, N. L., AND S. KOTZ. 1970. Distributions in statistics: continuous univariate distributions—2. Houghton Mifflin, Boston.

JOHNSON, R. R., B. T. BROWN, L. T. HAIGHT, AND J. M. SIMPSON. 1981. Playback recordings as a special avian censusing technique. Pp. 68–75 *in* C. J. Ralph and J. M. Scott (eds.). Estimating the numbers of terrestrial birds. Stud. Avian Biol. 6.

JOHNSON, R. R., C. T. HAIGHT, AND J. M. SIMPSON. 1979. Owl populations and species status in the southwestern United States. Pp. 40–59 *in* P. P. Schaeffer and S. M. Ehlers (eds.). Proceedings of symposium on owls of the West: their ecology and conservation. Natl. Aud. Soc. West. Ed. Cen., Tiburon, Calif.

JOHNSON, V. P. 1947. Breeding birds of the forest edge in Illinois. Condor. 49:45–53.

JOLLY, G. M. 1965. Explicit estimates from capture-recapture data with both death and immigration-stochastic model. Biometrika 52:225–247.

JOLLY, G. M. 1979. A unified approach to mark-recapture stochastic models, exemplified by a constant survival rate model. Pp. 277–282 *in* R. M. Cormack, G. P. Patil, and D. S. Robson (eds.). Sampling biological populations. Stat. Ecol. Ser., vol. 5. Int. Co-op Publ. House, Fairland, Md.

JOLLY, G. M. 1981a. Mark-recapture models with parameters constant in time. Biometrics 37, in press.

JOLLY, G. M. 1981b. Mark-recapture—what next? Pp. 137–138 *in* C. J. Ralph and J. M. Scott (eds.). Estimating the numbers of terrestrial birds. Stud. Avian Biol. 6.

JONES, H. L., AND J. M. DIAMOND. 1976. Short-time-base studies of turnover in breeding bird populations on the California Channel Islands. Condor 78:526–549.

JONES, S. 1979. Habitat management series for unique or endangered species: The accipiters—Goshawk, Cooper's Hawk, Sharp-shinned Hawk. U.S. Bur. Land Manage., Rep. 17, Tech. Note 335.

JORGENSEN, O. H. 1974. Results of IPA-censuses on Danish farmland. Acta Ornithol. 14:310–321.

JUREK, R. M. 1975. Survey of Yuma Clapper Rails and California Black Rails along Coachella Canal, Imperial County, May 1975. Calif. Dept. Fish and Game, Sacramento.

KÄLLANDER, H., S. G. NILSSON, AND S. SVENSSON. 1977. The Swedish winter bird census programme. Pol. Ecol. Stud. 3(4):77–88.

KÄLLANDER, H., AND O. RYDÉN. 1974. Inter-observer differences in studies of visible migration at Falsterbo. Ornis Scand. 5:53–62.

KÄLLANDER, H., O. RYDÉN, AND C. WEIKERT. 1972. Unterschiede in der Beobachtungs-Effektivität bei der Registrierung vom Küsten-Seevogelzug. Vogelwarte 26:303–310.

KAPLAN, E. B., AND R. C. ELSTON. 1978. A subroutine package for maximum likelihood estimation (MAXLIK). Inst. Statistics Ser. 823, Dept. Biostatistics, Univ. North Carolina, Chapel Hill.

KARR, J. R. 1971. Structure of avian communities in selected Panama and Illinois habitats. Ecol. Monogr. 41:207–233.

KARR, J. R. 1976a. On the relative abundances of migrants from the north temperate zone in tropical habitats. Ecol. Monogr. 46:457–481.

KARR, J. R. 1976b. Seasonality, resource availability, and community diversity in tropical bird communities. Am. Natur. 110:973–994.

KARR, J. R. 1976c. Within- and between-habitat avian diversity in African and Neotropical lowland habitats. Ecol. Monogr. 46:457–481.

KARR, J. R. 1979. On the use of mist nets in the study of bird communities. Inland Bird Banding 51:1–10.

KARR, J. R. 1980. Geographical variation in the avifaunas of tropical forest undergrowth. Auk 97:283–298.

KARR, J. R. 1981a. Surveying birds in the tropics. Pp. 548–553 *in* C. J. Ralph and J. M. Scott (eds.). Estimating the numbers of terrestrial birds. Stud. Avian Biol. 6.

KARR, J. R. 1981b. Surveying birds with mist nets. Pp. 62–67 *in* C. J. Ralph and J. M. Scott (eds.). Estimating the numbers of terrestrial birds. Stud. Avian Biol. 6.

KARR, J. R. In press. Turnover dynamics in a tropical continental avifauna. Proc. Int. Ornith. Cong. 17.

KARR, J. R., D. W. SCHEMSKE, AND N. BROKAW. In press. Temporal variation in the undergrowth bird community of a tropical forest. Pp. 000–000 *in* E. G. Leigh and A. S. Rand (eds.). Seasonal rhythms in a tropical forest. Smithsonian Inst., Washington, D.C.

KAUTZ, J. E. 1977. Effects of Band-tailed Pigeon behavior on estimates of population parameters. M.S. thesis, Univ. Br. Columbia, Vancouver.

KEARNEY, T. H., AND R. H. PEEBLES. 1960. Arizona flora. Univ. Calif. Press, Berkeley.

KEAST, A., AND E. S. MORTON. (eds.). 1980. Migrant birds in the neotropics: ecology, behavior, distribution and conservation. Smithsonian Inst. Press, Washington, D.C.

KEITH, J. O., AND M. S. MULLA. 1966. Relative toxicity of five organophosphorus mosquito larvicides to Mallard ducks. J. Wildl. Manage. 30:553–563.

KEITH, L. B., AND E. C. MESLOW. 1968. Trap response by Snowshoe Hares. J. Wildl. Mange. 32:795–801.

KEITH, S. 1967. Review of "Birds of North America: a guide to field identification" by Chandler S. Robbins, Bertel Brunn and Herbert S. Zim. Wilson Bull. 79:251–254.

KELKER, G. H. 1945. Measurement and interpretation of forces that determine populations of managed deer herds. Ph.D. thesis, Univ. Michigan, Ann Arbor.

KELSALL, J. P., AND J. R. CALAPRICE. 1972. Chemical content of waterfowl plumage as a potential diagnostic tool. J. Wildl. Manage. 36:1088–1097.

KENAGA, E. E. 1965. Are birds increasing in numbers? Bull. Entomol. Soc. Amer. 11:81–83.

KENDEIGH, S. C. 1941. Territorial and mating behavior of the House Wren. Ill. Biol. Monogr. 18:1–120.

KENDEIGH, S. C. 1944. Measurement of bird populations. Ecol. Monogr. 14:67–106.

KENDEIGH, S. C. 1948. Bird populations and biotic communities in northern lower Michigan. Ecology 29:101–114.

KENDEIGH, S. C. 1961. Animal ecology. Prentice Hall, Englewood Cliffs, N.J.

KENNARD, J. H. 1976. A biennial rhythm in the winter distribution of the Common Redpoll. Bird-Banding 47:231–237.

KENNARD, J. H. 1977. Biennial rhythm in Purple Finch migration. Bird-Banding 48:155–157.

KENNEDY, R. S. 1977a. Notes on the biology and population status of the Monkey-eating Eagles of the Philippines. Wilson Bull. 89:1–20.

KENNEDY, R. S. 1977b. The status of the Osprey in Tidewater, Virginia, 1970–71. Pp. 121–133 *in* J. C. Ogden (ed.). Transactions of the North American Osprey research conference. U.S. Natl. Park Serv., Trans. and Proc. Ser. 2.

KEPLER, C. B., AND J. M. SCOTT. 1981. Reducing count variability by training observers. Pp. 366–371 *in* C. J. Ralph and J. M. Scott (eds.). Estimating the numbers of terrestrial birds. Stud. Avian Biol. 6.

KESSEL, B. 1957. A study of the breeding biology of the European Starling (*Sturnus vulgaris*) in North America. Am. Midl. Nat. 58:257–331.

KILBURN, P. D. 1966. Analysis of the species-area relation. Ecology 47:831–843.

KIMBALL, J. W. 1949. The crowing count pheasant census. J. Wildl. Manage. 13:101–120.

KING, J., F. C. ROBARD, AND C. J. LENSINK. 1972. Census of the Bald Eagle breeding population in south-eastern Alaska. J. Wildl. Manage. 36:1292–1295.

KIRKLEY, J. S., AND M. A. SPRINGER. 1980. Nesting populations of Red-tailed Hawks and Great-horned Owls in central Ohio. Raptor Res. 14:22–28.

KISTCHINSKI, A. A. 1975. Breeding biology and behavior of the Grey Phalarope (*Phalaropus fulicarius*) in eastern Siberia. Ibis 117:285–301.

KLIMKIEWICZ, M. K., AND C. S. ROBBINS. 1978. Standard abbreviations for common names of birds. N. Am. Bird Bander 3:16–25.

KLIMKIEWICZ, M. K., AND J. K. SOLEM. 1978. The breeding bird atlas of Montgomery and Howard Counties, Maryland. Maryland Birdlife 34:3–39.

KLUYVER, H. M., AND L. TINBERGEN. 1953. Territory and regulation of density in titmice. Arch. neerl. Zool. 10:265–286.

KNUDSEN, E. I. 1978. Strategies for sound localization in birds. J. Acoust. Soc. Am. 64:S4.

KOLB, H. 1965. The Audubon Winter Bird-Population Study. Aud. Field Notes 19:432–434.

KORPIMÄKI, E. 1978. Breeding biology of the Starling *Sturnus vulgaris* in western Finland. Ornis Fenn. 55:93–104.

KOUKI, J., AND O. JÄRVINEN. 1980. On single-visit censuses of peatland birds. (In Finnish with English summary.) Ornis Fenn. 57:134–136.

Kovner, J. L., and S. A. Patil. 1974. Properties of estimators of wildlife population density for the line transect method. Biometrics 30:225–230.

Kozicky, E. L. 1952. Variations in two spring indices of male Ring-necked Pheasant populations. J. Wildl. Manage. 16:429–437.

Kozicky, E. L., T. A. Bancroft, and D. G. Homeyer. 1954. An analysis of Woodcock singing ground counts, 1948–1952. J. Wildl. Manage. 18:259–266.

Krebs, C. J. 1978. Ecology. The experimental analysis of distribution and abundance. Second ed. Harper and Row, New York.

Krebs, J. R. 1970. Regulation of numbers in the Great Tit (Aves: Passeriformes). J. Zool. (Lond.) 162:317–333.

Krebs, J. R. 1971. Territory and breeding density in the Great Tit, (*Parus major* L.). Ecology 52:2–22.

Krebs, J., and C. Perrins. 1978. Behavior and population regulation in the Great Tit (*Parus major* L.). Pp. 23–47 *in* F. J. Ebling and D. M. Stoddart (eds.). Population control by social behaviour. Inst. Biology, London.

Kreger, N. S. 1973. A simulation study of Jolly's estimates for animal populations when sampling intensity is low. M.S. thesis, Univ. Manitoba, Winnipeg.

Kroodsma, D. E. 1976a. The effect of large song repertoires on neighbor "recognition" in male Song Sparrows. Condor 78:97–99.

Kroodsma, D. E. 1976b. Reproductive development in a female songbird: differential stimulation by quality of male song. Science 192:574–575.

Kroodsma, D. E., and L. D. Parker. 1977. Vocal virtuosity in the Brown Thrasher. Auk 94:783–785.

Krutchkoff, R. G. 1969. An introduction to empirical Bayes. Pp. 1–15 *in* T. A. Atchison and H. F. Martz (eds.). Proceedings of the symposium on empirical Bayes estimation and computing in statistics. Texas Tech. Univ. Math. Ser. 6.

Kruyfhooft, C. 1964. Serious decline of the Peregrine Falcon in Belgium during the autumn migration. Pp. 70–73 *in* R. D. Chancellor (ed.). Working conference on birds of prey and owls, London. Int. Council Bird Pres.

Küchler, A. W. 1964. Manual to accompany the map: potential natural vegetation of the conterminous United States. Amer. Geogr. Soc., Spec. Publ. 36.

Kuhn, T. S. 1962. The structure of scientific revolutions. Univ. Chicago Press, Chicago, Ill.

Kuroda, N. 1977. Observations of territorial life in breeding season of a pair of Jungle Crow (*Corvus macrorhynchos*) in city Tokyo. Misc. Rep. Yamashina Inst. Ornithol. 9(2), Pt. 49.

Kurtz, P. J. 1977. Dissociated behavioral and cholinesterase decrements following malathion exposure. Toxicol. Appl. Pharmacol. 42:589–594.

Laake, J. L. 1978. Line transect estimators robust to animal movement. M.S. thesis, Utah State Univ., Logan.

Laake, J. L., K. P. Burnham, and D. R. Anderson. 1979. User's manual for program TRANSECT. Utah State Univ. Press, Logan.

Lack, D. 1933. Habitat selection in birds, with special reference to the effects of afforestation on the Breckland avifauna. J. Anim. Ecol. 2:239–262.

Lack, D. 1935. The breeding bird population of British heaths and moorland. J. Anim. Ecol. 4:43–51.

Lack, D. 1937. A review of bird census work and bird population problems. Ibis 79:369–395.

Lack, D. 1943. The life of the Robin. Witherby, London.

Lack, D. 1954. The natural regulation of animal numbers. Oxford Univ. Press, London.

Lack, D. 1976. Island biology. Univ. Calif. Press, Berkeley.

Lahti, E. 1972. Nest sites and nesting habitats of the Ural Owl *Strix uralensis* in Finland during the period 1870–1969. Ornis Fenn. 49:91–97.

Lamoreaux, R. J., and L. W. Newland. 1977. The fate of organophosphorus pesticides in the environment. Biol. Conserv. 11:59–66.

Lancaster, R. K., and W. E. Rees. 1979. Bird communities and the structure of urban habitats. Can. J. Zool. 57:2358–2368.

Langslow, D. R. 1977. Movements of Black Redstarts between Britain and Europe as related to occurrences at observatories. Bird Study 24:169–178.

Langslow, D. R. 1978. Recent increases of Blackcaps at bird observatories. Br. Birds 71:345–354.

Laperriere, A. J., and A. O. Haugen. 1972. Some factors influencing calling activity of wild Mourning Doves. J. Wildl. Manage. 36:1193–1199.

Larson, D. 1980. Increase in the White-tailed Kite populations of California and Texas—1944–1978. Am. Birds 34:689–690.

Laskey, A. R. 1944. A study of the Cardinal in Tennessee. Wilson Bull. 56:27–44.

Lawrence, L. de K. 1967. A comparative life-history study of four species of woodpeckers. Ornithol. Monogr. 5.

Lay, D. W. 1938. How valuable are woodland clearings and birdlife? Wilson Bull. 45:254–256.

Leach, H. R., J. M. Brode, and S. J. Nicola. 1974. At the crossroads. Calif. Dept. Fish and Game.

Leclerq, J. 1967. Pour des atlas de répartition des insectes de l'Europe occidentale, oeuvre coopérative. Cmpt. Rend. Séan. Soc. Biogéogr. 386:69–81.

Leclerq, J. (ed.) 1970. Atlas provisoire des insectes de Belgique. Cartes 1 à 100. Faculté des sciences agronomiques de l'état, Zoologie Generale et Faunistique. Gembloux, France.

Lefkovitch, L. P. 1966. An index of spatial distribution. Res. Popul. Ecol. 8:89–92.

Legrand, G. 1978. Recherche sur l'écologie des peuplements d'oiseaux nicheurs du Massif des Maures (Var). Thesis, Univ. Saint Jerôme, Marseille.

Lehninger, A. L. 1975. Biochemistry. Second ed. Worth, New York.

Lehtonen, L. 1979. Kvantitatiivisen linnustotutkimuksen metodeista ja niiden luotettavuudesta. (Über die Methoden der quantitativen Vogelforschung und deren Zuverlässigkeit.) Lintumies 14:1–8.

Leighton, F. A., J. M. Gerrard, P. Gerrard, D. W. A. Whitfield, and W. J. Maher. 1979. An aerial census of Bald Eagles in Saskatchewan. J. Wildl. Manage. 43:61–69.

LEOPOLD, A. 1933. Game management. Scribner, New York.

LEOPOLD, A. 1942. A raptor tally in the Northwest. Condor 44:37–38.

LEOPOLD, A., AND A. E. EYNON. 1961. Avian daybreak and evening song in relation to time and light intensity. Condor 63:269–293.

LEOPOLD, A. S., T. RINEY, R. MCCAIN, AND L. TEVIS. 1951. The Jawbone deer herd. Calif. Div. Fish and Game, Game Bull. 4.

LEOPOLD, L. B., F. E. CLARKE, B. B. HANSHAW, AND J. R. BALSLEY. 1971. A procedure for evaluating environmental impact. U.S. Geol. Surv., Circ. 645.

LEPTHIEN, L. W., AND C. E. BOCK. 1976. Winter abundance patterns of North American kinglets. Wilson Bull. 88:483–485.

LESLIE, P. H. 1952. The estimation of population parameters by means of the capture-recapture method. II. The estimation of total numbers. Biometrika 39:363–388.

LESLIE, P. H. 1958. Statistical appendix. J. Anim. Ecol. 27:84–86.

LESLIE, P. H., AND D. CHITTY. 1951. The estimation of population parameters from data obtained by means of the capture-recapture method. I. Maximum likelihood equations for estimating the death rate. Biometrika 38:269–292.

LESLIE, P. H., D. CHITTY, AND H. CHITTY. 1953. The estimation of population parameters from data obtained by means of the capture-recapture method. III. An example of the practical applications of the method. Biometrika 40:137–169.

LEVIN, H. S., AND R. L. RODNITZKY. 1976. Behavioral effects of organophosphate pesticides in man. Clin. Toxicol. 9:391–405.

LEVINS, R. 1966. The strategy of model building in population biology. Am. Sci. 54:421–431.

LEVY, S. H., J. J. LEVY, AND R. A. BISHOP. 1966. Use of tape recorded female quail calls during the breeding season. J. Wildl. Manage. 30:426–428.

LEWIS, D. B., AND D. M. GOWER. 1980. Biology of communication. Wiley, New York.

LIDICKER, W. Z. 1962. Emigration as possible mechanism permitting the regulation of population density below carry capacity. Am. Nat. 96:29–33.

LIGON, D. J. 1968. The biology of the Elf Owl *Micrathene whitneyi*. Univ. Michigan Mus. Zool., Misc. Publ. 36.

LINCOLN, F. C. 1930. Calculating waterfowl abundance on the basis of banding returns. U.S. Dept. Agric., Circ. 118.

LINCOLN, G. A. 1975. Bird counts either side of Wallace's line. J. Zool. (Lond.) 177:349–361.

LINDBERG, P. 1977. The Peregrine Falcon in Sweden. Pp. 329–338 *in* R. D. Chancellor (ed.). World conference on birds of prey, Vienna 1975. Int. Council Bird Pres.

LINDGREN, B. W. 1962. Statistical theory. Macmillan, New York.

LINDLEY, D. V. 1962. Discussion. J. Royal Stat. Soc. B24:285–287.

LINE, L. 1964. The bird worth a forest fire. Aud. Mag. 66:370–375.

LINSKENS, H. F., M. J. M. MARTENS, H. J. G. M. HENDRICKSEN, A. M. ROESTENBERG-SINNIGE, W. A. J. M. BROUWERS, A. L. H. C. VAN DER STAKK, AND A. M. J. STRIK-JANSEN. 1976. The acoustic climate of plant communities. Oecologia (Berl.) 23:165–177.

LISH, J. W., AND J. C. LEWIS. 1975. Status and ecology of Bald Eagles wintering in Oklahoma. Proc. Annu. Conf. Southeast. Assoc. Game and Fish Comm. 29:415–423.

LLOYD, M., J. H. ZAR, AND J. KARR. 1968. On the calculation of information-theoretical measures of diversity. Am. Midl. Nat. 79:257–272.

LOVEJOY, T. E. 1975. Bird diversity and abundance in Amazon forest communities. Living Bird 13:127–191.

LOWE, J. I. 1956. Breeding density and productivity of Mourning Doves on a county-wide basis in Georgia. J. Wildl. Manage. 20:428–433.

LUDKE, J. L., E. F. HILL, AND M. P. DIETER. 1975. Cholinesterase (ChE) response and related mortality among birds fed ChE inhibitors. Arch. Environ. Contam. Toxicol. 3:1–21.

LUDWIG, J. P. 1967. Band loss—its effect on banding data and apparent survivorship in the Ring-billed Gull populations of the Great Lakes. Bird-Banding 38:309–323.

LUTHER, J. S., G. MCCASKIE, AND J. DUNN. 1979. Third report of the California Bird Records Committee. West. Birds 10:169–187.

LUTTICH, S. N., L. B. KEITH, AND J. D. STEPHENSON. 1971. Population dynamics of the Red-tailed Hawk (*Buteo jamaicensis*) at Rochester, Alberta. Auk 88:75–88.

LUU-MAU THANH. 1962. Les déplacements humains; human displacements. Monaco.

LYNCH, J. F., AND N. K. JOHNSON. 1974. Turnover and equilibria in insular avifaunas, with special reference to the California Channel Islands. Condor 76:370–384.

LYNCH, J. F., AND R. F. WHITCOMB. 1977. Effects of the insularization of the eastern deciduous forest on avifaunal diversity and turnover. Pp. 461–490 *in* A. Marmelstein (proj. lead.). Proceedings of a national symposium on classification, inventory and analysis of fish and wildlife habitat. U.S. Fish and Wildl. Serv., Phoenix, Ariz.

MACARTHUR, R. H. 1965. Patterns of species diversity. Biol. Rev. (Cambridge) 40:510–533.

MACARTHUR, R. H. 1972. Geographical ecology. Harper & Row, New York.

MACARTHUR, R. H., M. L. CODY, AND H. F. RECHER. 1966. On the relation between habitat selection and species diversity. Am. Nat. 100:319–322.

MACARTHUR, R. H., AND A. T. MACARTHUR. 1974. On the use of mist nets for population studies of birds. Proc. Nat. Acad. Sci., U.S.A. 71:3230–3233.

MACARTHUR, R. H., AND J. W. MACARTHUR. 1961. On bird species diversity. Ecology 42:594–598.

MACARTHUR, R. H., J. W. MACARTHUR, AND J. PREER. 1962. On bird species diversity. II. Prediction of bird census from habitat measurements. Am. Nat. 96:167–174.

MACARTHUR, R. H., AND E. O. WILSON. 1967. The theory of island biogeography. Princeton Univ. Press, Princeton, N.J.

MACDONALD, S. D. 1968. The courtship and territorial behavior of Franklin's race of the Spruce Grouse. Living Bird 7:5–25.

MACKOWICZ, R. 1977. The influence of the biology of the River Warbler (*Locustella fluviatilis* Wolf.) on the effectiveness of the mapping method. Pol. Ecol. Stud. 3(4):89–93.

MADER, W. J. 1979. Breeding behavior of a polyandrous trio of Harris' Hawks in southern Arizona. Auk 96:776–788.

MAEHR, D. S. 1980. Avian abundance and habitat preference on new habitats created by phosphate mining. M.S. thesis, Univ. Florida, Gainesville.

MANLY, B. F. J. 1970. A simulation study of animal population estimation using the capture-recapture method. J. Appl. Ecol. 7:13–29.

MANLY, B. F. J. 1971a. A simulation study of Jolly's method for analysing capture-recapture data. Biometrics 27:415–424.

MANLY, B. F. J. 1971b. Estimates of a marking effect with capture-recapture sampling. J. Appl. Ecol. 8:181–189.

MANLY, B. F. J. 1977a. A simulation experiment on the application of the jackknife with Jolly's method for the analysis of capture-recapture data. Acta Theriol. 22:215–223.

MANLY, B. F. J. 1977b. The analysis of trapping records for birds trapped in mist nets. Biometrics 33:404–410.

MANLY, B. F. J., AND M. J. PARR. 1968. A new method of estimating population size, survivorship, and birth rate from capture-recapture data. Trans. Soc. Br. Entomol. 18:81–89.

MANNES, P., AND R. ALPERS. 1975. Errors of census methods concerning birds breeding in nest boxes. J. Ornithol. 116:308–314.

MANOLIS, T. D. 1977. California Black Rail breeding season survey in central California, 1977. Calif. Dept. Fish and Game, Endang. Wildl. Prog., Sacramento.

MANOLIS, T. D. 1978. Status of the Black Rail in central California. West. Birds 9:151–158.

MARCHANT, J. H. 1981. Residual edge effects with the mapping bird census method. Pp. 488–491 *in* C. J. Ralph and J. M. Scott (eds.). Estimating the numbers of terrestrial birds. Stud. Avian Biol. 6.

MARION, W. R. 1974a. Ecology of the Plain Chachalaca in the lower Rio Grande Valley of Texas. Ph.D. thesis, Texas A&M Univ., College Station.

MARION, W. R. 1974b. Status of the Plain Chachalaca in south Texas. Wilson Bull. 86:200–205.

MARION, W. R., T. E. O'MEARA, AND D. S. MAEHR. 1981. Use of playback recordings in sampling elusive or secretive birds. Pp. 81–85 *in* C. J. Ralph and J. M. Scott (eds.). Estimating the numbers of terrestrial birds. Stud. Avian Biol. 6.

MARION, W. R., AND R. A. RYDER. 1975. Perch-site preferences of four diurnal raptors in northeastern Colorado. Condor 77:350–352.

MARION, W. R., AND J. D. SHAMIS. 1977. An annotated bibliography of bird marking techniques. Bird-Banding 48:42–61.

MARSHALL, J. T. 1939. Territorial behavior of the Flammulated Screech Owl. Condor 41:71–78.

MARTEN, G. G. 1970. A regression method for mark-recapture estimates with unequal catchability. Ecology 51:291–295.

MARTEN, G. G. 1972. The remote sensing approach to censusing. Res. Popul. Ecol. 14:36–57.

MARTEN, K., AND P. MARLER. 1977. Sound transmission and its significance for animal vocalization. I. Temperate habitats. Behav. Ecol. Sociobiol. 2:271–290.

MARTEN, K., D. QUINE, AND P. MARLER. 1977. Sound transmission and its significance for animal vocalization. II. Tropical forest habitats. Behav. Ecol. Sociobiol. 2:291–302.

MARTIN, D. J. 1974. Copulatory and vocal behavior of a pair of Whiskered Owls. Auk 91:619–624.

MARTIN, F. N. 1975. Introduction to audiology. Prentice-Hall, Englewood Cliffs, N.J.

MARTIN, F. W., R. S. POSPAHALA, AND J. D. NICHOLS. 1979. Assessment and population management of North American migratory birds. Pp. 187–239 *in* J. Cairns, G. P. Patil, and W. E. Waters (eds.). Environmental biomonitoring, assessment, prediction, and management—certain case studies and related quantitative issues. Stat. Ecol. Ser., vol. 11. Int. Co-op. Publ. House, Fairland, Md.

MARTINSON, R. K. 1966. Proportion of recovered duck bands that are reported. J. Wildl. Manage. 30:264–268.

MARTINSON, R. K., AND J. A. MCCANN. 1966. Proportion of recovered goose and brant bands that are reported. J. Wildl. Manage. 30:856–858.

MATHISEN, J. E. 1968. Effects of human disturbance on nesting of Bald Eagles. J. Wildl. Manage. 32: 1–6.

MATHISEN, J. E. 1977. The status of Ospreys in the Chippewa National Forest. Pp. 175–179 *in* J. C. Ogden (ed.). Transactions of the North American Osprey research conference. U.S. Natl. Park Serv., Trans. and Proc. Ser. 2.

MATHISEN, J. E., AND A. MATHISEN. 1968. Species and abundance of diurnal raptors in the panhandle of Nebraska. Wilson Bull. 80:479–486.

MATHISEN, J. E., D. J. SORENSON, L. D. FRENZEL, AND T. C. DUNSTAN. 1977. A management strategy for Bald Eagles. Trans. N. Am. Wildl. Nat. Res. Conf. 42:88–92.

MATTOX, W. R., W. R. HEINRICH, J. OAR, S. J. BELARDO, K. E. RIDDLE, AND T. M. SMYLIE. 1980. West Greenland Peregrine Falcon survey, 1978. Arctic 33:199–202.

MAY, R. M. 1975. Patterns of species abundance and diversity. Pp. 81–120 *in* M. L. Cody and J. M. Diamond (eds.). Ecology and evolution of communities. Belknap, Cambridge, Mass.

MAYFIELD, H. F. 1953. A census of Kirtland's Warbler. Auk 70:17–20.

MAYFIELD, H. F. 1960. The Kirtland's Warbler. Cranbrook Inst. Sci., Bloomfield Hills, Mich.

MAYFIELD, H. 1966. Hearing loss and bird song. Living Bird 5:167–175.

MAYFIELD, H. F. 1981. Problems in estimating population size through counts of singing males. Pp. 220–224 *in* C. J. Ralph and J. M. Scott (eds.). Estimating the numbers of terrestrial birds. Stud. Avian Biol. 6.

MCCAUGRAN, D. A., AND R. JEFFREY. 1980. Estimation of the audio index of relative abundance of Band-tailed Pigeons. J. Wildl. Manage. 44:204–209.

MCCLELLAND, R. R. 1973. Autumn concentrations

of Bald Eagles in Glacier National Park. Condor 75:121–123.

McClure, H. E. 1939. Cooing activity and censusing of the Mourning Dove. J. Wildl. Manage. 3:323–328.

McClure, H. E. 1945. Comparison of census methods for pheasants in Nebraska. J. Wildl. Manage. 9:38–45.

McCluskey, D. C., J. W. Thomas, and E. C. Meslow. 1977. Effects of aerial application of DDT on reproduction in House Wrens and Mountain and Western Bluebirds. U.S. For. Serv., Pac. Northwest For. and Range Exp. Stn., Res. Pap. PNW-228, Portland, Ore.

McEwen, L. C., C. W. Hall, P. Johnson, and Nebraska Game, Forestation and Parks Commission. 1965. Effects on wildlife of rangeland spraying of malathion for grasshopper control. P: 37 *in* Effects of pesticides on fish and wildlife. U.S. Fish and Wildl. Ser., Cir. 226.

McGahan, J. 1968. Ecology of the Golden Eagle. Auk 85:1–12.

McGowan, J. D. 1975. Nesting habits and reproductive success of Goshawks in interior Alaska. Pp. 147–152 *in* J. R. Murphy, C. M. White, and B. E. Harrell (eds.). Population status of raptors. Raptor Res. Rep. 3.

McKay, W. D. 1976. Nesting density and distribution of Red-tailed Hawks and Great Horned Owls on a quadrangle in unglaciated Ohio. M.S. thesis, Ohio State Univ., Columbus.

McLeod, J. M. 1967. The effect of phosphamidon on bird populations in jack pine stands in Quebec. Can. Field-Nat. 81:102–106.

Mehrotra, K. N., Y. P. Beri, S. S. Mira, and A. Phokela. 1967. Physiological effects of malathion on the House Sparrow, *Passer domesticus* L. Indian J. Exp. Biol. 5:219–221.

Mendelssohn, H., and U. Paz. 1977. Mass mortality of birds of prey caused by azodrin, an organophosphorus insecticide. Biol. Conserv. 11:163–170.

Merikallio, E. 1946. Über regionale Verbreitung und Anzahl der Landvögel in Süd- und Mittelfinnland, besonders in deren östlichen Teilen, im Lichte von quantitativen Untersuchungen. I–II. Ann. Zool. Soc. 'Vanamo' 12(1):1–143; 12(2):1–119.

Merikallio, E. 1951. On the numbers of land-birds in Finland. Acta. Zool. Fenn. 65:1–16.

Merikallio, E. 1958. Finnish birds, their distribution and numbers. Fauna Fenn. 5:1–181.

Merriam, M. 1911. A text-book on the method of least squares. Eleventh ed. Wiley, New York.

Michener, H., and J. R. Michener. 1935. Mockingbirds, their territories and individualities. Condor 37:97–140.

Mikol. S. A., J. J. Hickey, J. R. Cary, A. K. Stratman, and M. E. Lardy. 1979. Comparison of bird-transect estimators in sagebrush and grassland habitats. Pp. 1–76 *in* J. J. Hickey and S. A. Mikol (eds.). Estimating breeding-bird densities on coal lands in Montana and Wyoming. U.S. Fish and Wildl. Serv., Biol. Serv. Prog., Fort Collins, Col.

Miller, A. H., and L. Miller. 1951. Geographic variation of the Screech Owls of the deserts of western North America. Condor 53:161–177.

Miller, F. L., R. H. Russell, and A. Gunn. 1975. Distribution and numbers of Snowy Owls in Melville, Eglinton, and Byam Martin Islands, Northwest Territories, Canada. Raptor Res. 9:60–64.

Miller, L. 1930. The territorial concept in the Horned Owl. Condor 32:290–291.

Misztal, A. 1974. The population and nesting of Red-tailed Hawks (*Buteo jamaicensis*), and Great Horned Owls (*Bubo virginianus*), on the Wisconsin Till Plains of western Ohio. M.S. thesis, Ohio State Univ., Columbus.

Mitchell, R. T., H. P. Blagbrough, and R. C. VanEtten. 1953. The effects of DDT upon the survival and growth of nestling songbirds. J. Wildl. Manage. 17:45–54.

Moisan, G., R. I. Smith, and R. K. Martinson. 1967. The Green-Winged Teal: its distribution, migration, and population dynamics. U.S. Fish and Wildl. Serv., Spec. Sci. Rep. Wildl. 100.

Monroe, B. 1970. 1969 Kentucky summer Breeding Bird Survey. Kentucky Warbler 46:43–45.

Monroe, B. L. 1978. Analysis of Kentucky's breeding birds: declining species. Kentucky Warbler 54:19–26.

Moreau, R. 1972. The Palaearctic-African bird migration systems. Academic, London.

Morgan, R. A., and R. J. O'Connor. 1980. Farmland habitat and Yellowhammer distribution in Britain. Bird Study 27:155–162.

Moriarty, D. 1976. The adaptive nature of bird flocks: a review. Biologist 58:67–69.

Morrison, M. L., R. W. Mannan, and G. L. Dorsey. 1981. Effects of number of circular plots on estimates of avian density and species richness. Pp. 405–408 *in* C. J. Ralph and J. M. Scott (eds.). Estimating the numbers of terrestrial birds. Stud. Avian Biol. 6.

Morrison, M. L., and R. D. Slack. 1977. Population trends and status of the Olivaceous Cormorant. Am. Birds 31:954–959.

Morse, D. H. 1971. The foraging of warblers isolated on small islands. Ecology 52:216–278.

Morton, E. S. 1975. Ecological sources of selection on avian sounds. Am. Nat. 109:17–34.

Morton, E. S. 1980. Adaptations to seasonal changes by migrant land birds in the Panama Canal Zone. Pp. 437–453 *in* A. Keast and E. S. Morton (eds.). Migrant birds in the Neotropics: ecology, behavior, distribution, and conservation. Smithsonian Inst. Press, Washington, D.C.

Moss, D. 1976. Woodland song-bird populations and growth of nesting Sparrowhawks. Ph.D. thesis, Univ. Edinburgh, Edinburgh, Scot.

Moss, D. 1978. Diversity of woodland song-bird populations. J. Anim. Ecol. 47:521–527.

Moulding, J. D. 1976. Effects of a low-persistence insecticide on forest bird populations. Auk 93:692–708.

Mudd, D., L. Boe, and R. Bugert. 1980. Evaluation of wildlife habitat developed on government project lands along the Snake River in Washington. Wash. Game Dept., Olympia.

Mudd, D., T. Clausing, L. Boe, and R. Bugert. 1979. Evaluation of wildlife habitat developed on government project lands along the Snake River in Washington. Wash. Game Dept., Olympia.

Mueller, H. C., and D. D. Berger. 1961. Weather and fall migration of hawks at Cedar Grove, Wisconsin. Wilson Bull. 73:171–192.

Mueller, H. C., and D. D. Berger. 1967a. Fall migration of Sharp-skinned Hawks. Wilson Bull. 79:397–415.

Mueller, H. C., and D. D. Berger. 1967b. Some observations and comments on the periodic invasions of Goshawks. Auk 84:183–191.

Mueller, H. C., and D. D. Berger. 1967c. Wind drift, leading lines, and diurnal migration. Wilson Bull. 79:50–63.

Mueller, H. C., and D. D. Berger. 1968. Sex ratios and measurements of migrant Goshawks. Auk 85:431–436.

Mueller, H. C., and D. D. Berger. 1969. Navigation by hawks migrating in spring. Auk 86:35–40.

Mueller, H. C., and D. D. Berger. 1970. Prey preferences in the Sharp-skinned Hawk. The roles of sex, experience, and motivation. Auk 84:452–457.

Mueller, H. C., and D. D. Berger. 1973. The daily rhythm of hawk migration at Cedar Grove, Wisconsin. Auk 90:591–596.

Mueller, H. C., D. D. Berger, and G. Allez. 1977. The periodic invasions of Goshawks. Auk 94:652–663.

Mueller, H. C., D. D. Berger, and G. Allez. 1979. Age and sex differences in size of Sharp-skinned Hawks. Bird-Banding 50:34–44.

Mulholland, R. J., and C. M. Gowdy. 1978. Determination of sampling intervals for ecological process. Pp. 71–86 *in* H. H. Shugart (ed.). Time series and ecological processes. Soc. Industr. Appl. Math., Philadelphia.

Munn, C. A., and J. W. Terborgh. 1979. Multi-species territoriality in neotropical foraging flocks. Condor 81:338–347.

Munro, J. A., and I. M. Cowan. 1947. A review of the bird fauna of British Columbia. Br. Col. Prov. Mus., Spec. Pub. 2.

Murphy, J. R., F. J. Camenzind, D. G. Smith, and J. B. Weston. 1969. Nesting ecology of raptorial birds in central Utah. Brigham Young Univ. Sci. Bull., Biol. Ser. 10:1–36.

Murphy, J. R., C. M. White, and B. E. Harrell. (eds.) 1975. Population status of raptors. Proceedings of the Conference on Raptor Conservation Techniques. Ft. Collins, Col. 1973. Raptor Res. Rep. 3.

Murray, J. J. 1946. The altitudinal distribution of birds in the Virginia mountains. Raven 17:70–75.

Murton, R. K., R. J. P. Thearle, and J. Thompson. 1972. Ecological studies of the feral pigeon (*Columba livia* var.). I. Population, breeding biology, and methods of control. J. Appl. Ecol. 9:835–874.

Murton, R. K., and N. J. Westwood. 1974. Some effects of agricultural change on the English avifauna. Br. Birds 67:41–69.

Myers, J. P. 1980. Sanderlings at Bodega Bay: facts, inferences, and shameless speculations. Wader Study Group Bull., in press.

Myrberget, S. 1976. Field tests of line transect methods for grouse. Norw. J. Zool. 24:307–317.

Myrberget, S., and S. A. Strömme. 1974. The effect of terrain structures on population estimates found by King's method. Norw. J. Zool. 22:273–277.

Nagy, A. C. 1977. Population trend indices based on 40 years of autumn counts at Hawk Mountain Sanctuary in north-eastern Pennsylvania. Pp. 243–253 *in* R. D. Chancellor (ed.). World conference on birds of prey, Vienna 1975. Int. Council Bird Pres.

Nankinov, D. 1977. Attempt at censusing Corvidae and diurnal birds of prey in winter. Pol. Ecol. Stud. 3(4):19–192.

Neff, D. J. 1968. The pellet-group count technique for big game trend, census, and distribution: a review. J. Wildl. Manage. 32:597–614.

Neidermeyer, W. J., and J. J. Hickey. 1977. The Monk Parakeet in the United States, 1970–1975. Am. Birds 31:273–278.

Nettleship, O. N. 1976. Census techniques for seabirds of arctic and eastern Canada. Can. Wildl. Serv., Occ. Pap. 25.

Newman, J. R., W. H. Brennan, and L. M. Smith. 1977. Twelve-year changes in nesting patterns of Bald Eagles (*Haliaeetus leucocephalus*) on San Juan Island, Washington. Murrelet 58:37–39.

Newton, I. 1979. Population ecology of raptors. Buteo Books, Vermillion, S.D.

Newton, I., M. Marquiss, D. N. Weir, and D. Moss. 1977. Spacing of Sparrowhawk nesting territories. J. Anim. Ecol. 46:425–441.

Newton, I., E. Meek, and B. Little. 1978. Breeding ecology of the Merlin in Northumberland. Br. Birds 71:376–398.

Nice, M. M. 1934. A hawk census from Arizona to Massachusetts. Wilson Bull. 46:93–95.

Nice, M. M. 1937. Studies in the life history of the Song Sparrow. I. Trans. Linn. Soc. N.Y. 4:1–246.

Nice, M. M. 1943. Studies in the life history of the Song Sparrow. II. Trans. Linn. Soc. N.Y. 6:1–328.

Nichols, J. D., B. R. Noon, S. L. Stokes, and J. E. Hines. 1981. Remarks on the use of mark-recapture methodology in estimating avian population size. Pp. 121–136 *in* C. J. Ralph and J. M. Scott (eds.). Estimating the numbers of terrestrial birds. Stud. Avian Biol. 6.

Nichols, S. 1977. On the interpretation of principal component analysis in ecological contexts. Vegetatio 34:191–197.

Nie, N. H., C. H. Hull, J. G. Jenkins, K. Steinbrenner, and D. H. Brent. 1975. SPSS: Statistical package for the social sciences. McGraw-Hill, New York.

Nielsen, B. P., and S. Christensen. 1970. Observations on the autumn migration of raptors in the Lebanon. Ornis Scand. 1:65–73.

Niemi, G. J., and L. Pfanmuller. 1979. Avian communities: Approaches to describing their habitat associations. Pp. 154–178 *in* R. M. DeGraaf (tech. coord.). Proceedings of workshop, management of northcentral and northeastern forests for nongame birds. U.S. For. Ser., Gen. Tech. Rep. NC-51, St. Paul. Minn.

Nilsson, S. G. 1974a. Methods of estimating bird population densities during the winter. Ornis Scand. 5:37–46.

NILSSON, S. G. 1974b. Seasonal changes in song-activity of birds in two woodland areas in south Sweden. Vår Fågelvärld 33:218–221.

NILSSON, S. G. 1977a. Density compensation and competition among birds breeding on small islands in a south Swedish lake. Oikos 28:170–176.

NILSSON, S. G. 1977b. Estimates of population density and changes for titmice, Nuthatch, and Treecreeper in southern Sweden—an evaluation of the territory mapping method. Ornis Scand. 8:9–16.

NILSSON, S. G. 1977c. Några mål och medel vid fågelinventeringar. (Some aims and methods in bird census work.) Anser 16:117–122.

NISBET, I. C. T., AND T. C. SMOUT. 1957. Autumn observations on the Bosphorus and Dardanelles. Ibis 99:483–499.

NOLAN, V. 1978. The ecology and behavior of the Prairie Warbler, *Dendroica discolor*. Ornithol. Monogr. 26.

NOON, B. R. In press. Techniques for sampling avian habitat. Pp. 000–000 *in* D. E. Capen (ed.). Proceedings of workshop: the use of multivariate techniques in the study of wildlife habitat. U.S. For. Serv.

NOON, B. R., AND K. P. ABLE. 1978. A comparison of avian community structure in the northern and southern Appalachian Mountains. Pp. 98–117 *in* R. M. DeGraff (tech. coord.). Proceedings of workshop: management of southern forests for nongame birds. U.S. For. Serv., Gen. Tech. Rep. SE-14.

NOON, B. R., D. K. DAWSON, D. B. INKLEY, C. S. ROBBINS, AND S. H. ANDERSON. 1980. Consistency in habitat preference of forest bird species. Trans. N. Am. Wildl. Nat. Res. Conf. 45:226–244.

NORBERG, A. 1977. An ecological theory on foraging time and energetics and choice of optimal food searching methods. J. Anim. Ecol. 46:511–529.

NORTH, P. M. 1978. Statistical methods in ornithology. Ph.D. thesis, Univ. Kent, Canterbury.

NOVAES, F. C. 1969. Análise ecológica de uma avifauna de região do Rio Acará, Estado do Pará. Boletin do Museu Paraense Emílio Goeldi (Nová Serie) 69:1–52.

NOWICKI, T. 1974. A census of Screech Owls using tape-recorded calls. Jack-Pine Warbler 52:98–101.

NUNNELEY, S. A. 1964. Analysis of banding records of local populations of Blue Jays and of Redpolls at Granby, Massachusetts. Bird-Banding 35:8–22.

OBA, N., AND K. OTO. 1976. Consideration of the mechanism of impaired vision seen in patients with chronic ocular diseases due to organophosphorus pesticides. Jap. Rev. Clin. Ophthalmol. 70:660–661.

OBERHEU, J. C. 1977. The status of the American Osprey on National Wildlife Refuges. Pp. 51–67 *in* J. C. Ogden (ed.). Transactions of the North American Osprey research conference. U.S. Natl. Park Serv., Trans. and Proc. Ser. 2.

OBERHOLSER, H. C. 1905. Notes on the mammals and summer birds of western North Carolina. Biltmore Forest School, Biltmore, N.C.

O'BRIEN, R. D. 1967. Insecticides: action and metabolism. Academic, New York.

O'CONNOR, R. J. 1979. Population dynamics of some farmland species. Br. Trust Ornithol. News 98:9–10.

O'CONNOR, R. J. 1980a. Population regulation in the Yellowhammer (*Emberiza citrinella*) in Britain. Pp. 190–200 *in* H. Oelke (ed.). Bird census work and nature conservation. Proceedings of the VI International Conference on Bird Census Work. Univ. Göttingen.

O'CONNOR, R. J. 1980b. The breeding dynamics of British Kestrels (*Falco tinnunculus*). Ibis 122:420–421.

O'CONNOR, R. J. 1980c. The effects of census date on the results of intensive Common Birds Census surveys. Bird Study 27:126–136.

O'CONNOR, R. J. 1980d. Pattern and process in Great Tit (*Parus major*) populations in Britain. Ardea 68:165–183.

O'CONNOR, R. J. 1981a. Habitat correlates of bird distribution in British census plots. Pp. 533–537 *in* C. J. Ralph and J. M. Scott (eds.). Estimating the numbers of terrestrial birds. Stud. Avian Biol. 6.

O'CONNOR, R. J. 1981b. The influence of observer and analyst efficiency in mapping method censuses. Pp. 372–376 *in* C. J. Ralph and J. M. Scott (eds.). Estimating the numbers of terrestrial birds. Stud. Avian Biol. 6.

O'CONNOR, R. J. In press. Comparisons between migrant and non-migrant birds in Britain. *In* D. J. Aidley (ed.). Animal migration. Soc. Exper. Biol. Symposium Series.

O'CONNOR, R. J., AND R. K. HICKS. 1980. The influence of weather conditions on the detection of birds during Common Birds Census fieldwork. Bird Study 27:137–151.

ODUM, E. P. 1950. Review of 50th Christmas Bird Count. Wilson Bull. 62:227.

ODUM, E. P., AND E. J. KUENZLER. 1955. Measurement of territoriality and home range size in birds. Auk 72:128–136.

OECH, V., AND L. OECH. 1960. The use of recorded bird songs as bird-censusing technique. Flicker 32:46–47.

OELKE, H. 1962. Moorente (*Aythyanyroca*) 1961 Brutvogel in Mittelniedersachsen. J. Ornithol. 103:215–218.

OELKE, H. 1963. Die Vogelwelt des Peiner Moränen- und Lößgebietes. Dissertation, Göttingen.

OELKE, H. 1966. Thirty-five years of breeding-bird census work in Europe. Aud. Field Notes 20:635–642.

OELKE, H. 1967. Breeding-bird censuses 9–13 and 29–33 of the southern Canadian Shield. Aud. Field Notes 21:617–621, 629–634.

OELKE, H. 1970. Empfehlungen für eine international standardisierte Kartierungsmethode bei siedlungsbiologischen Vogelbestandsaufnahmen. Orn. Mitt. 22:124–128.

OELKE, H. 1974a. Siedlungsdichte. Pp. 33–34 *in* P. Berthold, E. Bezzel, and G. Thielcke (eds.). Praktische Vogelkunde. Dachverband Deutscher Avifaunisten, Greven.

OELKE, H. 1974b. Bird census work in the Federal Republic of Germany—present status. Acta Ornithol. 14:192–196.

OELKE, H. 1977. Methoden der Bestandserfassung von Vögeln: Nestersuche—Rivierkartierung. Orn. Mitt. 29:151–166.

OELKE, H. 1980. The bird structure of the central

European spruce forest biome—as regarded from breeding bird censuses. Pp. 201–209 *in* H. Oelke (ed.). Bird census work and nature conservation. Dachverband Deutscher Avifaunisten, Göttingen.

OELKE, H. 1981a. Limitations of estimating bird populations because of vegetation structure and composition. Pp. 316–321 *in* C. J. Ralph and J. M. Scott (eds.). Estimating the numbers of terrestrial birds. Stud. Avian Biol. 6.

OELKE, H. 1981b. Limitations of the mapping methods. Pp. 114–118 *in* C. J. Ralph and J. M. Scott (eds.). Estimating the numbers of terrestrial birds. Stud. Avian Biol. 6.

OELKE, H., J. SCHIERER, AND W. WINKELMANN. 1970. Winter bird censuses of an oak-pine forest in north-west Germany, a methodological study. Pp. 40–46 *in* S. Svennson (ed.). Bird census work and environmental monitoring. Ecol. Res. Comm., Bull. 9.

OELKLAUS, W. F. 1976. Chukar partridge dispersion along the middle and lower Snake and Columbia Rivers. M.S. thesis, Univ. Idaho, Moscow.

OETTING, R. B., AND C. C. DIXON. 1975. Winter nest densities and success at Oak Hammock Marsh, Manitoba. Wildl. Soc. Bull. 3:166–171.

OJANEN, M., M. ORELL, AND E. MERILÄ. 1978. Population decrease of Starlings in northern Finland. Ornis Fenn. 55:38–39.

OLENDORFF, R. R., AND M. N. KOCHERT. 1977. Land management for the conservation of birds of prey. Pp. 294–313 *in* R. D. Chancellor (ed.). World conference on birds of prey, Vienna 1975. Int. Council Bird Pres.

OLIPHANT, L. W., AND W. J. P. THOMPSON. 1978. Recent breeding success of Richardson's Merlin in Saskatchewan. Raptor Res. 12:34–39.

OLSEN, D. 1980. Whooping Crane recovery plan. U.S. Fish and Wildl. Serv., Washington, D.C.

O'MEARA, T. E. 1981. A field test of two density estimators for transect data. Pp. 193–196 *in* C. J. Ralph and J. M. Scott (eds.). Estimating numbers of terrestrial birds. Stud. Avian Biol. 6.

OOSTING, H. 1956. The study of plant communities. Second ed. Freeman, San Francisco.

ORELL, M., AND M. OJANEN. 1980. Numbers of Starlings *Sturnus vulgaris* breeding in northern Finland still low in 1978–79. Ornis Fenn., 57:133–134.

ORIANS, G. H. 1958. A capture-recapture analysis of a shearwater population. J. Anim. Ecol. 27:71–86.

ORIANS, G. H. 1961. Social stimulation within blackbird colonies. Condor 63:330–337.

ORIANS, G. H. 1969. The number of bird species in some tropical forests. Ecology 50:783–801.

ORTEGO, B. 1979. An owl survey technique. Louisiana Ornith. Soc. News 86:3–5.

OTIS, D. L., K. P. BURNHAM, G. C. WHITE, AND D. R. ANDERSON. 1978. Statistical inference from capture data on closed animal populations. Wildl. Monogr. 62.

OVERTON, W. S. 1971. Estimating the numbers of animals in wildlife populations. Pp. 403–455 *in* R. H. Giles (ed.). Wildlife management techniques. Third ed. Wildl. Soc., Washington, D.C.

OVERTON, W. S., AND D. E. DAVIS. 1969. Estimating the numbers of animals in wildlife populations. Pp. 403–455 *in* R. H. Giles (ed.). Wildlife management techniques. Wildl. Soc., Washington, D.C.

OWEN, R. B., AND J. W. MORGAN. 1975. Influence of night-lighting and banding on Woodcock movements. Wildl. Soc. Bull. 3:77–79.

OWENS, R. A., AND M. T. MYRES. 1973. Effects of agriculture upon populations of native passerine birds of an Alberta fescue grassland. Can. J. Zool. 51:697–713.

PALMGREN, P. 1930. Quantitative Untersuchungen über die Vogelfauna in den Wäldern Südfinnlands, mit besonderer Berücksichtigung Ålands. Acta Zool. Fenn. 7:1–218.

PARKER, J. W. 1975. Populations of the Mississippi Kite in the Great Plains. Pp. 159–172 *in* J. R. Murphy, C. M. White, and B. E. Harrell (eds.). Population status of raptors. Raptor Res. Rep. 3.

PATIL, G. P., C. TAILLIE, AND R. L. WIGLEY. 1979. Transect sampling methods and their application to the deep-sea red crab. Pp. 51–75 *in* J. Cairns, G. P. Patil, and W. E. Waters (eds.). Environmental biomonitoring, assessment, prediction, and management—certain case studies and related quantitative issues. Stat. Ecol. Ser., vol. 11. Int. Co-op Publ. House, Fairland, Md.

PATTERSON, T. L., AND L. PETRINOVICH. 1978. Territory size in the White-crowned Sparrow (*Zonotrichia leucophrys*): measurement and stability. Condor 80:97–98.

PAULIK, G. J., AND D. S. ROBSON. 1969. Statistical calculations for change-in-ratio estimators of population parameters. J. Wildl. Manage. 33:1–27.

PAYNE, R. B., AND K. PAYNE. 1977. Social organization and mating success in local song populations of Village Indigobirds, *Vidua chalybeata*. Z. Tierpsychol. 45:113–173.

PEARCE, P. A., D. B. PEAKALL, AND A. J. ERSKINE. 1976. Impact on forest birds of the 1975 spruce budworm spray operation in New Brunswick. Can. Wildl. Serv., Prog. Note 62.

PEARSON, D. L. 1977. A pantropical comparison of bird community structure on six lowland forest sites. Condor 79:232–244.

PEARSON, I. 1980. The prediction of bird species diversity and territory density from foliage profile measurements in southern English woodlands. B. Sc. thesis, Univ. East Anglia, E. Anglia, England.

PENNYCUICK, C. J. 1975. The use of motor-gliders for studying raptor migration. Pp. 65–71 *in* M. Harwood (ed.). Proceedings of the North American Hawk Migration Conference, 1974. Hawk Migr. Assoc. of N. Am., Washington Depot, Ct.

PENNYCUICK, C. J. 1976. Breeding of the Lappet-faced and White-headed Vultures on the Serengeti Plains, Tanzania. E. African Wildl. J. 14:67–84.

PERRING, F. H., AND S. M. WALTERS. (eds.) 1962. Atlas of the British flora. Nelson, London.

PETERSEN, L. 1979. Ecology of Great Horned Owls and Red-tailed Hawks in southeastern Wisconsin. Dept. Natural Resources, Tech. Bull. 111, Madison, Wis.

PETERSON, A. P. G., AND E. E. GROSS. 1974. Handbook of noise measurement. Seventh ed. General Radio, Concord, Mass.

PETERSON, S. R. 1975. Ecological distribution of

breeding birds. Pp. 22–38 *in* D. R. Smith (tech. coord.). Proceedings of the symposium on management of forest and range habitats for nongame birds. U.S. For. Serv., Gen. Tech. Rep. WO-1, Washington, D.C.

PETRABORG, W. H., E. G. WELLEIN, AND V. E. GUNVALSON. 1953. Roadside drumming counts, a spring census method for Ruffed Grouse. J. Wildl. Manage. 17:292–295.

PETRIDES, G. A. 1944. Sex ratios in ducks. Auk 61:564–571.

PETTINGILL, O. S., AND N. R. WHITNEY. 1965. Birds of the Black Hills. Cornell Lab. Ornithol., Spec. Publ. 1.

PHELAN, F. J. S., AND R. J. ROBERTSON. 1978. Predatory responses of a raptor guild to changes in prey density. Can. J. Zool. 56:2565–2572.

PHILLIPS, A. R., J. T. MARSHALL, AND G. MONSON. 1964. Birds of Arizona. Univ. Ariz. Press, Tucson.

PHIPPS, J. B. 1975. Bestblock: optimizing grid size in biogeographic studies. Can. J. Bot. 53:1447–1452.

PHIPPS, J. B., AND J. CULLEN. 1976. Centres of diversity quantified—a maximum variance approach to a biogeographic problem. Vegetatio 31:147–159.

PICOZZI, N. 1978. Dispersion, breeding, and prey of the Hen Harrier (*Circus cyaneus*) in Glen Dye, Kincardineshire. Ibis 120:498–509.

PIELOU, E. C. 1975. Ecological diversity. Wiley, New York.

PILLMORE, R. E., E. L. FLICKINGER, AND M. L. RICHMOND. 1971. Forest spraying of zectran and its safety to wildlife. J. For. 69:721–727.

PINOWSKI, J., Z. BOGUCKI, AND L. TOMIAŁOJĆ. (eds.) 1977. Bird census and atlas studies. Pol. Ecol. Stud. 3(4).

PINOWSKI, J., AND K. WILLIAMSON. (eds.) 1974. Proceedings of the second meeting of the European Ornithological Atlas Committee. Acta Ornithol. 14(32–37).

PITELKA, F. A. 1941. Distribution of birds in relation to major biotic communities. Am. Midl. Nat. 25:113–137.

PITELKA, F. A. 1959. Numbers, breeding schedule, and territoriality in Pectoral Sandpipers of northern Alaska. Condor 61:233–264.

PITELKA, F. A., R. T. HOLMES, AND S. F. MACLEAN. 1974. Evolution and ecology of social organization in arctic sandpipers. Am. Zool. 14:185–204.

PLATT, J. B. 1971. A survey of nesting hawks, eagles, falcons, and owls in Curlew Valley, Utah. Great Basin Nat. 31:51–65.

PLAZA, P. D. 1978. Distribution of selected North American picids determined by computer mapping. Am. Birds 32:912–922.

POLLOCK, K. H. 1974. The assumption of equal catchability of animals in tag-recapture experiments. Ph.D. thesis, Cornell Univ., Ithaca, New York.

POLLOCK, K. H. 1975a. A k-sample tag-recapture model allowing for unequal survival and catchability. Biometrika 62:577–583.

POLLOCK, K. H. 1975b. Building models of capture-recapture experiments. Statistician 25:253–260.

POLLOCK, K. H. 1978. A family of density estimators for line-transect sampling. Biometrics 34:475–478.

POLLOCK, K. H. 1981a. Capture-recapture models: a review of current methods, assumptions, and experimental design. Pp. 426–435 *in* C. J. Ralph and J. M. Scott (eds.). Estimating the numbers of terrestrial birds. Stud. Avian Biol. 6.

POLLOCK, K. H. 1981b. Capture-recapture models allowing for age-dependent survival and capture rates. Biometrics 37, in press.

POLLOCK, K. H., AND D. G. RAVELING. 1981. The assumptions of modern bird banding models with emphasis on heterogeneous survival rates of population segments. J. Wildl. Manage., in press.

POLLOCK, K. H., D. L. SOLOMON, AND D. S. ROBSON. 1974. Tests for mortality and recruitment in a K-sample tag-recapture experiment. Biometrics 30:77–87.

POOLE, R. W. 1978. Some relative characteristics of population fluctuations. Pp. 18–23 *in* H. H. Shugart (ed.). Time series and ecological processes. Soc. Industr. Appl. Math., Philadelphia, Pa.

POPE, G. G., AND P. WARD. 1972. The effects of small applications of an organophosphorus poison, fenthion, on the weaver-bird *Quelea quelea*. Pestic. Sci. 3:197–205.

POSTOVIT, H. R. 1979. Population estimates of breeding raptors in North Dakota badlands. M.S. thesis, North Dakota State Univ., Fargo.

POSTUPALSKY, S. 1974. Raptor reproductive success: some problems with methods, criteria, and terminology. Pp. 21–32 *in* F. N. Hamerstrom, B. E. Harrell, and R. R. Olendorff (eds.). Management of raptors. Raptor Res. Rep. 2.

POUGH, R. H. 1947. How to take a breeding bird census. Audubon 49:290–297.

POUGH, R. H. 1950. Comment faire un recensement d'oiseaux nicheurs? Terre et Vie 4:203–217.

POWELL, G. V. N., AND D. C. GRAY. 1980. Dosing free-living nestling Starlings with an organophosphate pesticide, famphur. J. Wildl. Manage. 44: 918–921.

PRATER, A. J. 1979. Trends in accuracy of counting birds. Bird Study 26:198–200.

PRATER, A. J., J. H. MARCHANT, AND J. VUORINEN. 1977. Guide to the identification and ageing of holarctic waders. Br. Trust for Ornithol., Tring, U.K.

PRESTON, F. W. 1960. Time and space and the variation of species. Ecology 41:611–627.

PRESTON, F. W. 1962. The canonical distribution of commonness and rarity. Ecology 43:185–215, 410–432.

PRESTON, F. W. 1979. The invisible birds. Ecology 60:451–454.

PRESTON, F. W. 1980. Noncanonical distributions of commonness and rarity. Ecology 61:88–97.

PRESTT, I. 1969. Recent population trends among British raptors. Pp. 283–287 *in* J. J. Hickey (ed.). Peregrine Falcon populations: their biology and decline. Univ. Wisconsin Press, Madison.

PREVOST, Y. A., R. P. BANCROFT, AND N. R. SEYMOUR. 1978. Status of the Osprey in Antigonish County, Nova Scotia. Can. Field-Nat. 92:294–297.

PRODON, R., AND J. D. LEBRETON. 1981. Breeding avifauna of a mediterranean succession: the Holm Oak and Cork Oak series in the eastern Pyrénées. I.

Analysis and modelling of the structure gradient. Oikos, in press.

Pruett-Jones, S. G., M. Pruett-Jones, and R. L. Knight. 1980. The White-tailed Kite in North and Middle America: current status and recent population changes. Am. Birds 34:682–688.

Puchstein, K. 1966. Zur Vogelökologie gemischter Flächen. Vogelwelt 87:161–176.

Pumphrey, R. J. 1961. Sensory organs: hearing. Pp. 69–86 *in* A. J. Marshall (ed.). Biology and comparative physiology of birds. Academic, New York.

Purdue, J. R., C. C. Carpenter, D. L. Marcellini, and R. F. Clarke. 1972. Spring migration of Swainson's Hawk and Turkey Vulture through Veracruz, Mexico. Wilson Bull. 84:92–93.

Purroy, F. J. 1974. Breeding communities of birds in the beech and fir forests of the Pyrenees. Acta Ornithol. 14:294–300.

Quaintance, C. W. 1938. Content, meaning, and possible origin of male song in the Brown Towhee. Condor 40:97–101.

Quenouille, M. H. 1956. Notes on bias reduction. Biometrika 43:353–360.

Quinn, T. J. 1977. The effects of aggregation on line transect estimation of population abundance with application to marine mammal population. M.S. thesis, Univ. Washington, Seattle.

Quinn, T. J. 1979. The effects of school structure on line transect estimators of abundance. Pp. 473–491 *in* G. P. Patil and M. L. Rosenzweig (eds.). Contemporary quantitative ecology and related ecometrics. Stat. Ecol. Ser., vol. 12. Int. Co-op. Publ. House, Fairland, Md.

Quinn, T. J. 1980. Sampling for the abundance of schooling populations with line-transect, mark-recapture, and catch-effort methods. Ph.D. thesis, Univ. Washington, Seattle.

Quinn, T. J. 1981. The effect of group size on line transect estimators of abundance. Pp. 502–508 *in* C. J. Ralph and J. M. Scott (eds.). Estimating the numbers of terrestrial birds. Stud. Avian Biol. 6.

Quinn, T. J., and V. F. Gallucci. 1980. Parametric models for line transect estimators of abundance. Ecology 61:293–302.

Ralph, C. J. 1981. An investigation of the effect of seasonal activity levels on avian censusing. Pp. 265–270 *in* C. J. Ralph and J. M. Scott (eds.). Estimating the numbers of terrestrial birds. Stud. Avian Biol. 6.

Ramsey, F. L. 1979. Parametric models for line transect surveys. Biometrika 66:505–512.

Ramsey, F. L., and J. M. Scott. 1978. Use of circular plot surveys in estimating the density of a population with Poisson scattering. Tech. Rep. 60, Dept. Statistics, Oregon State Univ., Corvallis.

Ramsey, F. L., and J. M. Scott. 1979. Estimating population densities from variable circular plot surveys. Pp. 155–181 *in* R. M. Cormack, G. P. Patil, and D. S. Robson (eds.). Sampling biological populations. Stat. Ecol. Ser., vol. 5. Int. Co-op. Publ. House, Fairland, Md.

Ramsey, F. L., and J. M. Scott. 1981. Analysis of bird survey data using a modification of Emlen's method. Pp. 483–487 *in* C. J. Ralph and J. M. Scott (eds.). Estimating the numbers of terrestrial birds. Stud. Avian Biol. 6.

Ramsey, F. L., and J. M. Scott. 1981. Tests of hearing ability. Pp. 341–345 *in* C. J. Ralph and J. M. Scott (eds.). Estimating the numbers of terrestrial birds. Stud. Avian Biol. 6.

Ramsey, F. L., J. M. Scott, and R. T. Clark. In press. Statistical problems arising from surveys of rare and endangered forest birds. Proc. Int. Stat. Inst. (Manila, P.I.) 42.

Rao, P. V., K. M. Portier, and J. A. Ondrasik. 1981. Density estimation using line transect sampling. Pp. 441–444 *in* C. J. Ralph and J. M. Scott (eds.). Estimating the numbers of terrestrial birds. Stud. Avian Biol. 6.

Rao, P. V., and J. A. Ondrasik. 1980. Density estimation using inverse sampling on line transects. Tech. Rep. 154, Dept. Statistics, Univ. Florida, Gainesville.

Ratcliffe, D. A. 1962. Breeding density in the Peregrine, *Falco peregrinus,* and Raven, *Corvus corax.* Ibis 104:13–29.

Ratcliffe, D. A. 1972. The Peregrine population of Great Britain in 1971. Bird Study 19:117–156.

Ratowsky, A. V., and D. A. Ratowsky. 1977. The birds of the Mt. Wellington range, Tasmania. Emu 77:19–22.

Ratowsky, A. V., and D. A. Ratowsky. 1978. A survey of the birds of the Mt. Wellington range, Tasmania, during the non-breeding months. Emu 78:223–226.

Ratowsky, A. V., and D. A. Ratowsky. 1979. A comparison of counting methods to obtain bird species numbers. Notornis 26:53–61.

Raynor, G. S. 1975. Techniques for evaluating and analyzing Christmas Bird Count data. Am. Birds 29:626–633.

Raynor, G. S. 1980. Bobwhite populations changes on Long Island, New York, 1959–1978. Am. Birds 34:691–694.

Rea, A. M. 1973. Turkey Vultures casting pellets. Auk 90:209–210.

Rea, A. M. 1977. Historic changes in the avifauna of the Gila River Indian Reservation, central Arizona. Ph.D. thesis, Univ. Arizona, Tucson.

Readshaw, J. L. 1968. Estimates of the size of winter flocks of the Pied Currawong, *Streperu graculina* (Shaw), from mark-recapture data—a new approach. Austral. J. Zool. 16:27–35.

Recher, H. F. 1971. Bird species diversity: a review of the relationship between species number and environment. Proc. Ecol. Soc. Austral. 6:135–152.

Recher, H. F. 1977. Ecology of co-existing White-cheeked and New Holland Honeyeaters. Emu 77:136–142.

Redmond, R. L., T. K. Bicak, and D. A. Jenni. 1981. An evaluation of breeding season census techniques for Long-billed Curlews (*Numenius americanus*). Pp. 197–201 *in* C. J. Ralph and J. M. Scott (eds.). Estimating the numbers of terrestrial birds. Stud. Avian Biol. 6.

Reese, J. 1975. Osprey nesting success in Eastern Bay, Maryland. Chesapeake Sci. 16:56–61.

Reeves, K. 1980. Breeding Bird Census of maple-cherry-birch forest. Am. Birds 34:47–48.

Reischl, P., G. A. Van Gelder, and G. Karas. 1975. Auditory detection behavior in parathion-treated squirrel monkeys (*Saimiri sciureus*). Toxicol. Appl. Pharmacol. 34:88–101.

RENAUD, W. E., AND G. J. WAPPLE. 1977. A review of Saskatchewan Christmas Bird Counts: 1942–1976 (Part 1). Blue Jay 35:224–239.

REPKING, C. F., AND R. D. OHMART. 1977. Distribution and density of Black Rail populations along the lower Colorado River. Condor 79:486–489.

REYNOLDS, R. T. 1975. Distribution, density, and productivity of three species of accipiter hawks in Oregon. M.S. thesis, Oregon State Univ., Corvallis.

REYNOLDS. R. T. 1978. Food and habitat partitioning in two groups of coexisting *Accipiter*. Ph.D. thesis, Oregon State Univ., Corvallis.

REYNOLDS, R. T., J. M. SCOTT, AND R. A. NUSSBAUM. 1980. A variable circular-plot method for estimating bird numbers. Condor 82:309–313.

RHEINWALD, G. 1977. Atlas der Brutverbreitung westdeutscher Vogelarten. Dachverband Deutscher Avifaunisten, Bonn.

RICE, D. W., AND K. W. KENYON. 1962. Breeding distribution, history, and populations of North Pacific albatrosses. Auk 79:365–386.

RICE, J., B. W. ANDERSON, AND R. D. OHMART. 1980. Seasonal habitat selection by birds in the lower Colorado River valley. Ecology 61:1402–1411.

RICHARDS, D. G. 1981a. Alerting and message components in songs of Rufous-sided Towhees. Behaviour, in press.

RICHARDS, D. G. 1981b. Environmental acoustics and censuses of singing birds. Pp. 297–300 *in* C. J. Ralph and J. M. Scott (eds.). Estimating the numbers of terrestrial birds. Stud. Avian Biol. 6.

RICHARDS, D. G. 1981c. Estimation of distance of singing conspecifics by the Carolina Wren. Auk, 98:127–133.

RICHARDS, D. G., AND R. H. WILEY. 1980. Reverberations and amplitude fluctuations in the propagation of sound in a forest: implications for animal communication. Am. Nat. 115:381–399.

RICHARDS, J. H., AND K. I. FUNG. 1969. Atlas of Saskatchewan. Univ. Saskatchewan, Saskatoon.

RICHARDSON, W. J. 1974. Multivariate approaches to forecasting day-to-day variations in the amount of bird migration. Pp. 309–329 *in* S. A. Gauthreaux (ed.). Proceedings of Conference on Biological Aspects of the Bird/Aircraft Collision Problem. Clemson Univ., Clemson, S. Car.

RICHARDSON, W. J. 1975. Autumn hawk migration in Ontario studied with radar. Pp. 47–58 *in* M. Harwood (ed.). Proceedings of the North American Hawk Migration Conference, 1974. Hawk Migr. Assoc. of N. Am., Washington Depot, Ct.

RICHARDSON, W. J. 1978. Timing and amount of bird migration in relation to weather: a review. Oikos 30:224–272.

RICHMOND, M. L., C. J. HENRY, R. L. FLOYD, R. W. MANNAN, D. M. FINCH, AND L. R. DEWEESE. 1979. Effects of sevin-4-oil, dimilin, and orthene on forest birds in northeastern Oregon. U.S. For. Serv., Pac. Southwest For. and Range Exp. Stn., Res. Pap. PSW-148, Berkeley, Calif.

RICKLEFS, R. E. 1973. Fecundity, mortality, and avian demography. Pp. 366–435 *in* D. S. Farner (ed.). Breeding biology of birds: Proceedings of a symposium on breeding behavior and reproductive physiology in birds. Nat. Acad. Sci., Washington, D.C.

RICKLEFS, R. E., AND G. W. COX. 1978. Stage of taxon cycle, habitat distribution, and population density in the avifauna of the West Indies. Am. Nat. 112:875–895.

RICKLEFS, R. E., AND F. R. HAINSWORTH. 1968. Temperature dependent behavior of the Cactus Wren. Ecology 49:227–233.

ROBBINS, C. S. 1964. Measurement of a known mortality rate in a small bird population. Pp. 89–90 *in* Pesticide-wildlife Studies, 1963. U.S. Fish and Wildl. Serv., Cir. 199.

ROBBINS, C. S. 1970. Recommendations for an international standard for a mapping method in bird census work. Aud. Field Notes 24:723–726.

ROBBINS, C. S. 1972. An appraisal of the Winter Bird-Population Study technique. Am. Birds 26:688–692.

ROBBINS, C. S. 1975. A history of North American hawk watching. Pp. 29–40 *in* M. Harwood (ed.). Proceedings of the North American Hawk Migration Conf., 1974. Hawk Migr. Assoc. of N. Am., Washington Depot, Ct.

ROBBINS, C. S. 1977. Bird atlasing in the United States. Pol. Ecol. Stud. 3(4):325–328.

ROBBINS, C. S. 1978a. Census techniques for forest birds. Pp. 142–163 *in* R. M. DeGraff (tech. coord.). Proceeding of the workshop on management of southern forests for nongame birds. U.S. For. Serv., Gen. Tech. Rep. SE-14, Asheville, N.C.

ROBBINS, C. S. 1978b. Determining habitat requirements of nongame species. Trans. N. Am. Wildl. Nat. Res. Conf. 43:57–68.

ROBBINS, C. S. 1979a. Effects of forest fragmentation on bird populations. Pp. 198–212 *in* R. M. DeGraaf and K. E. Evans (comps.). Management of north central and northeastern forests for nongame birds. U.S. For. Serv., Gen. Tech. Rep. NC-51, St. Paul, Mn.

ROBBINS, C. S. 1979b. Instructions for conducting breeding bird survey routes. Offset, Migratory Bird and Habitat Research Laboratory, U.S. Fish and Wildl. Serv., Laurel, Md.

ROBBINS, C. S. 1981a. Bird activity levels related to weather. Pp. 301–310 *in* C. J. Ralph and J. M. Scott (eds.). Estimating the numbers of terrestrial birds. Stud. Avian Biol. 6.

ROBBINS, C. S. 1981b. Effect of time of day on bird activity. Pp. 275–286 *in* C. J. Ralph and J. M. Scott (eds.). Estimating the numbers of terrestrial birds. Stud. Avian Biol. 6.

ROBBINS, C. S. 1981c. Reappraisal of the Winter-Bird-Population Study technique. Pp. 52–57 *in* C. J. Ralph and J. M. Scott (eds.). Estimating the numbers of terrestrial birds. Stud. Avian Biol. 6.

ROBBINS, C. S., B. BRUUN, AND H. S. ZIM. 1966. Birds of North America. Western, New York.

ROBBINS, C. S., AND D. BYSTRAK. 1974. The Winter Bird Survey of central Maryland, U.S.A. Acta Ornithol. 14:254–271.

ROBBINS, C. S., AND A. J. ERSKINE. 1975. Population trends in nongame birds in North America. Trans. N. Am. Wildl. Nat. Resour. Conf. 40:288–293.

ROBBINS, C. S., AND R. W. STALLCUP. 1981. Prob-

lems in separating species with similar habits and vocalizations. Pp. 360–365 *in* C. J. Ralph and J. M. Scott (eds.). Estimating the numbers of terrestrial birds. Stud. Avian Biol. 6.
ROBBINS, C. S., AND W. T. VAN VELZEN. 1967. The breeding bird survey, 1966. U.S. Bur. Sport Fish. and Wildl., Spec. Sci. Rep. Wildl. 102.
ROBBINS, C. S., AND W. T. VAN VELZEN. 1969. The breeding bird survey, 1967 and 1968. U.S. Fish and Wildl. Serv., Spec. Sci. Rep. Wildl. 124.
ROBBINS, C. S., AND W. T. VAN VELZEN. 1970. Progress report on the North American breeding bird survey. Pp. 22–30 *in* S. Svensson (ed.). Bird census work and environmental monitoring. Bull. Ecol. Res. Comm. 9. Swedish Nat. Sci. Res. Counc., Stockholm.
ROBBINS, C. S., AND W. T. VAN VELZEN. 1974. Progress report on the North American breeding bird survey. Acta Ornithol. 14:170–191.
ROBBINS, S. 1971. Wisconsin Breeding Bird Survey: 1966–70. Passenger Pigeon 33:115–136.
ROBBINS, S. In press. Birds of Wisconsin.
ROBEL, R. J. 1969. Movements and flock stratification within a population of Blackcocks in Scotland. J. Anim. Ecol. 38:755–763.
ROBEL, R. J., D. J. DICK, AND G. F. KRAUSE. 1969. Regression coefficients used to adjust Bobwhite quail whistle count data. J. Wildl. Manage. 33:662–668.
ROBEL, R. J., F. R. HENDERSON, AND W. JACKSON. 1972. Some Sharp-tailed Grouse population statistics from South Dakota. J. Wildl. Manage. 36:87–98.
ROBERTS, H. B., AND G. S. LIND. 1977. Status of the American Osprey in Oregon. Pp. 215–222 *in* J. C. Ogden (ed.). Transactions of the North American Osprey research conference. U.S. Natl. Park. Serv., Trans. and Proc. Ser. 2.
ROBERTS, J. L. 1979. Observations of the migration of raptors and other large soaring birds in Bulgaria, 1975–1978. Ibis 121:301–312.
ROBERTS, J. O. L. 1971. Survival among some North American wood warblers. Bird-Banding 42:165–185.
ROBINETTE, W. L., C. M. LOVELESS, AND D. A. JONES. 1974. Field tests of strip census methods. J. Wildl. Manage. 38:81–96.
ROBINS, J. D. 1971. Differential niche utilization in a grassland sparrow. Ecology 52:1065–1070.
ROBSON, D. S. 1969. Mark-recapture methods of population estimation. Pp. 120–140 *in* N. L. Johnson and H. Smith (eds.). New developments in survey sampling. Wiley, New York.
ROBSON, D. S. 1979. Approximations to some mark-recapture sampling distributions. Pp. 257–275 *in* R. M. Cormack, G. P. Patil, and D. S. Robson (eds.). Sampling biological populations. Stat. Ecol. Ser., vol. 5. Int. Co-op. Publ. House, Fairland, Md.
ROBSON, D. S., AND H. A. REGIER. 1964. Sample size in Petersen mark-recapture experiments. Trans. Amer. Fisheries Soc. 93:215–226.
ROBSON, D. S., AND W. D. YOUNGS. 1971. Statistical analysis of reported tag-recaptures in the harvest from an exploited population. Biometrics Unit, Bull. 396-M, Cornell Univ., Ithaca, N.Y.
ROFF, D. A. 1973. An examination of some statistical tests used in the analysis of mark-recapture data. Oecologia (Berl.) 12:35–54.
ROGERS, D. T., AND M. A. DAUBER. 1976. A study of overwintering raptors in Alabama. Alabama Bird Life 24:10–12.
ROGERS, D. T., AND M. A. DAUBER. 1977. Status of the Red-shouldered Hawk in Alabama. Alabama Bird Life 25:19.
ROGERS, R. R., AND G. HUNT. 1975. Peregrines on South Padre Island: Recent years. Pp. 61–65 *in* J. R. Murphy, C. M. White, and B. E. Harrell (eds.). Population status of raptors. Raptor Res. Rep. 3.
ROOT, R. B. 1967. The niche exploitation pattern of the Blue-gray Gnatcatcher. Ecol. Monogr. 37:317–350.
ROOT, T. L., M. A. HOLMGREN, AND R. W. ANDREWS. 1980. The winter abundance patterns of some songbirds near the 100th meridian in the southern United States. Southwest Nat., in press.
ROSAHN, J. W. 1980. Bobwhite populations—a 10-year study. Am. Birds 34:695–697.
ROSENE, W., AND D. W. LAY. 1963. Disappearance and visibility of quail remains. J. Wildl. Manage. 27:139–142.
ROSENFIELD, R. N., AND D. L. EVANS. 1980. Migration incidence and sequence of age and sex classes of the Sharp-shinned Hawk. Loon 52:66–69.
ROSENZWEIG, M. L. 1975. On continental steady states of species diversity. Pp. 121–140 *in* M. L. Cody and J. M. Diamond (eds.). Ecology and evolution of communities. Belknap, Cambridge, Mass.
ROTENBERRY, J. T. 1978. Components of avian diversity along a multifactorial climatic gradient. Ecology 59:693–699.
ROTENBERRY, J. T. In press. Why measure bird habitat? *In* D. E. Capen (ed.). Proceedings of workshop: the use of multivariate statistics in studies of wildlife habitat. U.S. For. Serv., Gen. Tech. Rep. RM-, Fort Collins, Col.
ROTENBERRY, J. T., AND J. A. WIENS. 1976. A method for estimating species dispersion from transect data. Am. Midl. Nat. 95:64–78.
ROTENBERRY, J. T., AND J. A. WIENS. 1978. Nongame bird communities in northwestern rangelands. Pp. 32–46 *in* R. M. DeGraaf (ed.). Proceedings of workshop on nongame bird management in the coniferous forests of the western United States. U.S. For. Serv., Gen. Tech. Rep. PNW-64, Portland, Ore.
ROTENBERRY, J. T., AND J. A. WIENS. 1980a. Habitat structure, patchiness, and avian communities in North American steppe vegetation: a multivariate analysis. Ecology 61:1228–1250.
ROTENBERRY, J. T., AND J. A. WIENS. 1980b. Temporal variations in habitat structure and shrub-steppe bird dynamics. Oecologia 47:1–9.
ROTENBERRY, J. T., AND J. A. WIENS. In press. A new approach to principal component analysis of bird/habitat relationships: new factor scores from old data. *In* D. E. Capen (ed.). Proceedings of workshop, the use of multivariate statistics in studies in wildlife habitat. U.S. For. Serv., Gen. Tech. Rep. RM-, Fort Collins, Col.
ROTRAMEL, G. L. 1973. The development and appli-

cation of the area concept of biogeography. Syst. Zool. 22:227–232.

ROWAN, M. K. 1964. Relative abundance of raptorial birds in Cape Province. Ostrich 35:224–227.

ROYAMA, T. 1959. Test of an automatic nest recorder. Br. Birds 52:295–302.

ROYAMA T. 1960. The theory and practice of line transects in animal ecology by means of visual and auditory recognition. Rep. Yamashina Inst. Ornithol. and Zool. 2(14):1–17.

RUDEBECK, G. 1963. Studies on some palaearctic and arctic birds in their winter quarters in South Africa. 4. Birds of prey. (Falconiformes). Pp. 418–453 *in* B. Hanströme, P. Brinck, and G. Rudebeck (eds.). South Africa animal life: results of the Lund University expedition in 1950–51. Vol. 9. Almquist and Wiksell, Stockholm.

RUSCH, D. H., E. C. MESLOW, P. D. DOERR, AND L. B. KEITH. 1972. Response of Great Horned Owl populations to changing prey densities. J. Wildl. Manage. 36:282–296.

RUSHMORE, F. M. 1973. Techniques for calling sapsuckers and finding their nesting territories. U.S. For. Serv., Res. Pap. NE-281, Upper Darby, Pa.

RUSSELL, S. M., AND R. R. JOHNSON. 1973. Appendix A: remarks for inventory of birds in the study area. Pp. 76–81 *in* E. F. Haase (comp.). Draft environmental study—Gila River from the confluence of the Salt River downstream to Gillespie Dam. U.S. Army Corps Eng., Los Angeles, Calif.

SABO, S. R. 1980. Niche and habitat relations in subalpine bird communities of the White Mountains of New Hampshire. Ecol. Monogr. 50:241–259.

SAFRIEL, U. 1968. Bird migration at Elat, Israel. Ibis 110:283–320.

SALT, G. W. 1957. An analysis of avifaunas in the Teton Mountains and Jackson Hole, Wyoming. Condor 59:373–393.

SAMMALISTO, L. 1974. On the organization of the Finnish winter bird census, and results of the winter 1971/1972. Acta Ornithol. 14:246–248.

SAMSON, F. B. 1978. Vocalizations of Cassin's Finch in northern Utah. Condor 80:203–210.

SAMSON, F. B. 1980. Use of montane meadows by birds. Pp. 113–129 *in* R. M. DeGraff and N. G. Tilghman (eds.). Workshop proceedings, management of western forests and grasslands for nongame birds. U.S. For. Serv., Gen. Tech. Rep. INT-86, Ogden, Ut.

SANDERS, H. L. 1968. Marine benthic diversity: a comparative study. Am. Nat. 102:243–282.

SAPPINGTON, J. N., AND J. A. JACKSON. 1973. A technique for mist netting birds at communal roosts. Inland Bird Banding News 45:102–106.

SAUNDERS, A. A. 1929. Bird Song. N.Y. State Museum Handbook 7, Albany.

SAUNDERS, A. A. 1934. Losing the bird songs. Auk 51:503–506.

SAUROLA, P. 1976. Finnish raptor censuses. Ornis Fenn. 53:135–139.

SAUROLA, P. 1978. Mitä rengastusaineisto kertoo kottaraisesta? (Finnish ringing and recovery data of Starling, *Sturnus vulgaris*.) Lintumies 13:90–98.

SAYRE, M. W., R. D. ATKINSON, T. S. BASKETT, AND G. H. HESS. 1978. Reappraising factors affecting Mourning Dove perch cooing. J. Wildl. Manage. 42:884–889.

SCATTERGOOD, L. W. 1954. Estimating fish and wildlife populations: a survey of methods. Pp. 273–285 *in* O. Kempthorne, T. A. Bancroft, G. W. Gowen, and J. L. Lush (eds.). Statistics and mathematics in biology. Iowa State Coll. Pres, Ames.

SCHALL, J. J., AND E. R. PIANKA. 1978. Geographical trends in numbers of species. Science 201:679–686.

SCHAMEL, D., AND D. TRACEY. 1977. Polyandry, replacement clutches, and site-tenacity in the Red Phalarope (*Phalaropus fulicarius*) at Barrow, Alaska. Bird Banding 48:314–325.

SCHEMSKE, D. W., AND N. BROKAW. In press. Treefalls and the distribution of understory birds in a tropical forest. Ecology.

SCHIERMANN, G. 1930. Studien über Siedlungsdichte im Brutgebiet. J. Ornithol. 78:137–180.

SCHIERMANN, G. 1934. Studien über Siedlungsdichte im Brutgebiet II. Der brandenburgische Kiefernwald. J. Ornithol. 82:455–486.

SCHIFFERLI, A., P. GÉROUDET, AND R. WINKLER. 1980. Verbreitungsatlas der Brutvögel der Schweiz. Schweizerische Vogelwarte, Sempach, Switz.

SCHLEIDT, W. M. 1973. Tonic communication: continual effects of discrete signals in animal communication systems. J. Theor. Biology 42:359–386.

SCHNABEL, Z. E. 1938. Estimation of the total fish population in a lake. Amer. Math. Monthly 45:348–352.

SCHNELL, G. D. 1967. Population fluctuations, spatial distribution, and food habits of Rough-legged Hawks in Illinois. Kans. Ornithol. Soc. Bull. 18:21–18.

SCHNELL, G. D. 1969. Communal roosts of wintering Rough-legged Hawks (*Buteo lagopus*). Auk 86:682–690.

SCHOENER, T. W. 1968. Size of feeding territories among birds. Ecology 49:123–141.

SCHOPF, T. J. M. 1979. Evolving paleontological views on deterministic and stochastic approaches. Paleobiology 5:337–352.

SCHREIBER, R. W., AND E. A. SCHREIBER. 1973. Florida's Brown Pelican population: Christmas Bird Count analyses. Am. Birds 27:711–715.

SCHULTZ, V., AND R. J. MUNCY. 1957. An analysis of variance applicable to transect population data. J. Wildl. Manage. 21:274–278.

SCHULTZE, A., E. BLANK, J. BÜNSTORF, M. GEUTING, K.-G. KRAUTER, E. KROSS, R. SCHÖNBACH, AND U. SCHRÖDER. 1979. Terra Geographie 9/10. Klett, Stuttgart, W. Ger.

SCHWARTZKOPFF, J. 1973. Mechanoreception. Pp. 417–474 *in* D. S. Farner and J. R. King (eds.). Avian biology. Vol. 3. Academic, New York.

SCHWEDER, T. 1977. Point process models for line transect experiments. Pp. 221–242 *in* J. R. Barra, F. Brodeau, G. Romier, and B. Van Cutsem (eds.). Recent developments in statistics. North Holland Publ., New York.

SCOTT, J. M., J. D. JACOBI, AND F. L. RAMSEY. 1981a. Avian surveys of large geographical areas: a systematic approach. Wildl. Soc. Bull. 9, in press.

SCOTT, J. M., AND F. L. RAMSEY. 1981a. Length of

count period as a possible source of bias in estimating bird numbers. Pp. 409–413 *in* C. J. Ralph and J. M. Scott (eds.). Estimating the numbers of terrestrial birds. Stud. Avian Biol. 6.

Scott, J. M., and F. L. Ramsey. 1981b. The effect of abundant species on the ability of observers to make accurate counts of birds. Auk 99, in press.

Scott, J. M., F. L. Ramsey, and C. B. Kepler. 1981b. Distance estimation as a variable in estimating bird numbers. Pp. 334–340 *in* C. J. Ralph and J. M. Scott (eds.). Estimating the numbers of terrestrial birds. Stud. Avian Biol. 6.

Seabloom, R. W., G. L. Pearson, L. W. Oring, and J. R. Reilly. 1973. An incident of fenthion mosquito control and subsequent avian mortality. J. Wildl. Dis. 9:18–20.

Seber, G. A. F. 1965. A note on the multiple recapture census. Biometrika 52:249–259.

Seber, G. A. F. 1970a. Estimating time-specific survival and reporting rates for adult birds from band returns. Biometrika 57:313–318.

Seber, G. A. F. 1970b. The effects of trap response on tag-recapture estimates. Biometrics 26:13–22.

Seber, G. A. F. 1973. The estimation of animal abundance. Hafner, New York and Griffin, London.

Seber, G. A. F. 1979. Transects of random length. Pp. 183–192 *in* R. M. Cormack, G. P. Patil, and D. S. Robson (eds.). Sampling biological populations. Stat. Ecol. Ser., vol. 5. Int. Co-op. Publ. House, Fairland, Md.

Seber, G. A. F. 1980. Some recent advances in the estimation of animal abundance. Washington Sea Grant, Tech. Rep. WSG 80-1. Div. Marine Res., Univ. Wash., Seattle.

Seierstad, S., Seierstad, A., and I. Mysterud. 1970. Generalization of the standard check method for survey efficiency estimation. Ornis Scand. 1:1–9.

Sen, A. R. 1981. Methodological studies of Breeding Bird Surveys in North America. Pp. 496–501 *in* C. J. Ralph and J. M. Scott (eds.). Estimating the numbers of terrestrial birds. Stud. Avian Biol. 6.

Shake, W. F., and J. P. Mattsson. 1975. Three years of cowbird control: an effort to save the Kirtland's Warbler. Jack-Pine Warbler 53:48–53.

Shannon, C. E. 1948. A mathematical theory of communication. Bell Syst. Tech. J. 27:379–423, 623–656.

Sharrock, J. T. R. 1976. The atlas of breeding birds in Britain and Ireland. Poyser, Berkhamsted, England.

Sharrock, J. T. R., Z. Bogucki, and L. Yeatman. 1977. Introduction and minutes of third meeting of the European Ornithological Atlas Committee. Pol. Ecol. Stud. 3(4):257–259.

Shelford, V. E. 1945. The relation of Snowy Owl migration to the abundance of the Collared Lemming. Auk 62:592–596.

Shields, W. M. 1977. The effect of time of day on avian census results. Auk 94:380–383.

Shields, W. M. 1979. Avian census techniques: an analytical review. Pp. 23–51 *in* J. G. Dickson, R. N. Conner, R. R. Fleet, J. C. Kroll, and J. A. Jackson (eds.). The role of insectivorous birds in forest ecosystems. Academic, New York.

Short, J. J. 1979. Patterns of alpha-diversity and abundance in breeding bird communities across North America. Condor 81:21–27.

Shugart, H. H. In press. An overview of multivariate methods and their application to studies of wildlife habitat. *In* D. E. Capen (ed.). Proceedings of workshop, the use of multivariate statistics in studies of wildlife habitat (Burlington, Vermont, April 23–25, 1980). U.S. For. Serv., Gen. Tech. Rep. RM-, Fort Collins, Col.

Sidney, F. C., R. D. Mallette, J. C. Borneman, and R. S. Dalen. 1968. Third cooperative survey of the California Condor. Calif. Fish and Game 54:297–303.

Siegel, S. 1956. Non-parametric statistics for the behavioral sciences. McGraw-Hill, New York.

Siegfried, W. R. 1966. Relative abundance of raptorial birds in the south-western Cape. Ostrich 37:42–44.

Silvy, N. J., and R. J. Robel. 1967. Recordings used to help trap booming Greater Prairie Chickens. J. Wildl. Manage. 31:370–373.

Simberloff, D. S. 1976. Species turnover and equilibrium island biogeography. Science 194:572–578.

Simberloff, D. S. 1978a. Use of rarefaction and related methods in ecology. Pp. 150–165 *in* K. L. Dickson, J. Cairns, and R. J. Livingston (ds.). Biological data in water pollution assessment: quantitative and statistical analyses. Am. Soc. Testing and Materials, STP 652.

Simberloff, D. S. 1978b. Using island biogeographic distributions to determine if colonization is stochastic. Am. Nat. 112:713–726.

Simberloff, D. S. 1979. Rarefaction as a distribution-free method of expressing and estimating diversity. Pp. 159–176 *in* J. F. Grassle, G. P. Patil, W. Smith, and C. Taillie (eds.). Ecological diversity in theory and practice. Stat. Ecol. Ser., vol. 6. Int. Co-op Publ. House, Fairland, Md.

Simberloff, D. S., and L. G. Abele. 1976. Island biogeography theory and conservation practice. Science 191:285–286.

Siminski, D. P. 1976. A study of Great Horned Owl (*Bubo virginianus*) population density with recorded calls in northwestern Ohio. M.S. thesis, Bowling Green State Univ., Bowling Green, Ohio.

Simmons, K. E. L. 1951. Raptor migration in the Suez area: autumn 1949—spring 1950. Ibis 93:402–406.

Simpson, M. B. 1972. Saw-whet Owl population of North Carolina's southern Great Balsam Mountains. Chat 36:39–47.

Sindelar, C. 1977. Reproduction in Wisconsin Ospreys. Pp. 167–173 *in* J. C. Ogden (ed.). Transactions of the North American Osprey research conference. U.S. Natl. Park Serv., Trans. and Proc. Ser. 2.

Skellam, J. G. 1958. The mathematical foundations underlying the use of line transects in animal ecology. Biometrics 14:385–400.

Skirvin, A. A. 1981. Effect of time of day and time of season on the number of observations and density estimates of breeding birds. Pp. 271–274 *in* C. J. Ralph and J. M. Scott (eds.). Estimating the numbers of terrestrial birds. Stud. Avian Biol. 6.

SKUTCH, A. F. 1945. The migration of Swainson's and Broad-winged Hawks through Costa Rica. Northwest Sci. 19:80–89.

SLAGSVOLD, T. 1973a. Critical remarks on bird census work performed by means of the mapping method. Norw. J. Zool. 21:29–31.

SLAGSVOLD, T. 1973b. Variation in the song activity of passerine forest bird communities throughout the breeding season, special regard to the Song Thrush *Turdus philomelos* Brehm. Norw. J. Zool. 21:139–158.

SLAGSVOLD, T. 1973c. Estimation of density of the Song Thrush *Turdus philomelos* Brehm by different methods based upon singing males. Norw. J. Zool. 21:159–172.

SLAGSVOLD, T. 1975. Breeding time of birds in relation to latitude. Norw. J. Zool. 23:213–218.

SLAGSVOLD, T. 1976a. Annual and geographical variation in the time of breeding of the Great Tit *Parus major* and the Pied Flycatcher *Ficedule hypoleuca* in relation to environmental phenology and spring temperature. Ornis Scand. 7:127–145.

SLAGSVOLD, T. 1976b. Arrival of birds from spring migration in relation to vegetational development. Norw. J. Zool. 24:161–173.

SLAGSVOLD, T. 1977. Bird song activity in relation to breeding cycle, spring weather, and environmental phenology. Ornis Scand. 8:197–222.

SMEENK, C. 1974. Comparative-ecological studies of some east African birds of prey. Ardea 62:1–97.

SMITH, C. F. 1978. Distributional ecology of Barred and Great Horned Owls in relation to human disturbance. M.S. thesis, Univ. Connecticut, Storrs.

SMITH, D. G., AND C. D. MARTI. 1976. Distributional status and ecology of Barn Owls in Utah. Raptor Res. 10:33–44.

SMITH, D. G., AND J. R. MURPHY. 1978. Biology of the Ferruginous Hawk in central Utah. Sociobiology 3:79–98.

SMITH, D. G., AND J. R. MURPHY. 1979. Breeding responses of raptors to jackrabbit density in the eastern Great Basin Desert of Utah. Raptor Res. 13:1–14.

SMITH, G. E. J. 1973. The measurement of relative changes in breeding populations for non-game birds. Biometrics Section, Can. Wildl. Serv., Ms. Rep. 1.

SMITH, K. G. 1977. Distribution of summer birds along a forest moisture gradient in an Ozark watershed. Ecology 5:810–819.

SMITH, K. G. 1979. The effect of weather, people, and time on 12 Christmas Bird Counts, 1954–1973. Am. Birds 33:698–702.

SMITH, M. H., R. H. GARDNER, J. B. GENTRY, D. W. KAUFMAN, AND M. H. O'FARRELL. 1975. Density estimation of small mammal populations. Pp. 25–53 *in* F. B. Golley, K. Petrusewicz, and L. Ryszkowski (eds.). Small mammals: their productivity and population dynamics. Cambridge Univ. Press, Cambridge.

SMITH, N. G. 1980. Hawk and vulture migrations in the neotropics. Pp. 56–65 *in* A. Keast and E. S. Morton (eds.). Migrant birds in the neotropics: ecology, behavior, distribution and conservation. Smithsonian Inst. Press, Washington, D.C.

SMITH, P. M. 1974. Yuma Clapper Rail study, Mohave County, Arizona, 1973. Calif. Dept. Fish and Game, Spec. Wildl. Invest. Prog. Rep., Sacramento.

SMITH, R. H., AND S. GALLIZIOLI. 1965. Predicting hunter success by means of a spring call count of Gambel Quail. J. Wildl. Manage. 29:806–813.

SMITH, R. L. 1959. The songs of the Grasshopper Sparrow. Wilson Bull. 71:141–152.

SMITH, S. M. 1978. The "underworld" in a territorial sparrow: adaptive strategy for floaters. Am. Nat. 122:571–582.

SNEATH, P. H. A., AND R. R. SOKAL. 1973. Numerical taxonomy. Freeman, San Francisco.

SNEDECOR, G. W., AND W. G. COCHRAN. 1967. Statistical methods. Sixth ed. Iowa State Univ. Press, Ames.

SNOW, D. W. 1965. The relationship between census results and the breeding population of birds on farmland. Bird Study 12:287–304.

SNYDER, D. P. 1950. Bird communities in the coniferous forest biome. Condor 52:17–27.

SNYDER, N. F. R., H. A. SNYDER, J. L. LINCER, AND R. T. REYNOLDS. 1973. Organochlorines, heavy metals, and the biology of North American accipiters. BioScience 23:300–305.

SOKAL, R. R. 1979. Ecological parameters inferred from spatial correlograms. Pp. 167–196 *in* G. P. Patil and M. L. Rosenzweig (eds.). Contemporary quantitative ecology and related ecometrics. Stat. Ecol. Ser., vol. 12. Int. Co-op. Publ. House, Fairland, Md.

SOKAL, R. R., AND F. J. ROHLF. 1969. Biometry: the principles and practice of statistics in biological research. Freeman, San Francisco.

SOUTHERN, H. N. 1944. A transect census of pigeons. J. Anim. Ecol. 13:134–139.

SOUTHERN, W. E. 1963. Winter populations, behavior, and seasonal dispersal of Bald Eagles in northwestern Illinois. Wilson Bull. 75:42–55.

SOUTHERN, W. E. 1964. Additional observations on winter Bald Eagle populations: including remarks on biotelemetry techniques and immature plumages. Wilson Bull. 76:121–137.

SOUTHWOOD, T. R. E. 1977. Habitat, the templet for ecological strategies? J. Anim. Ecol. 46:337–365.

SPEIRS, J. M., AND R. ORENSTEIN. 1975. Bird populations in forests of Ontario County. Ont. Field Biol. 29:1–24.

SPENCER, J. S., AND H. W. THORNE. 1972. Wisconsin's 1968 timber resource—a prospective. U.S. For. Serv., Res. Bull. NC-15, St. Paul, Minn.

SPOFFORD, W. R. 1969. Hawk Mountain counts as population indices in northeastern America. Pp. 323–332 *in* J. J. Hickey (ed.). Peregrine Falcon populations: their biology and decline. Univ. Wisconsin Press, Madison.

SPRINGER, M. A. 1978. Foot surveys versus owl calling surveys: a comparative study of two Great Horned Owl censusing techniques. Inland Bird Banding News 50:83–92.

STAHLDECKER, D. W. 1975. Trends in wintering diurnal raptor populations from central Colorado Christmas bird counts. Am. Birds 29:936–940.

STAMM, D. D., D. E. DAVIS, AND C. D. ROBBINS. 1960. A method of studying wild bird populations by mist netting and banding. Bird-Banding 31:115–130.

STATISTISCHES BUNDESAMT. 1979. Statistisches Jahrbuch 1979 für die Bundesrepublik Deutschland. Stuttgart, Mainz.

STAUFFER, D. F., AND L. B. BEST. 1980. Habitat selection by birds of riparian communities: evaluating effects of habitat alteration. J. Wildl. Manage. 44:1–15.

STEARNS, E. I. 1949. The study of hawks from a blimp. Wilson Bull. 61:110.

STEARNS, F. W. 1970. Animal census and other applications of remote sensing in wildlife management and habitat research in forested areas. Pp. 194–198 *in* Range and wildlife habitat evaluation—a research symposium. U.S.D.A. Misc. Publ. 1147, Washington, D.C.

STEEL, R. G. D., AND J. H. TORRIE. 1960. Principles and procedures of statistics. McGraw-Hill, New York.

STEELE, J. H. 1978. Some comments on plankton patches. Pp. 1–20 *in* J. H. Steele (ed.). Spatial pattern in plankton communities. Plenum, New York.

STEFANSKI, R. A. 1967. Utilization of the breeding territory in the Black-capped Chickadee. Condor 69:259–267.

STEIN, C. 1955. Inadmissibility of the usual estimator for the mean of a multivariate normal distribution. Proc. 3rd Berkeley Symp. Math. Stat. Prob. 1:197–206.

STENGER, J., AND J. B. FALLS. 1959. The utilized territory of the Ovenbird. Wilson Bull. 71:125–140.

STEVENSON, H. M., AND A. STUPKA. 1948. The altitudinal limits of certain birds of the mountains of the southeastern United States. Migrant 19:33–60.

STEWART, P. A. 1954. The value of the Christmas Bird Counts. Wilson Bull. 66:184–195.

STEWART, P. A. 1973. Estimating numbers in a roosting congregation of blackbirds and starlings. Auk 90:353–358.

STEWART, P. A. 1980. Population trends of Barn Owls in North America. Am. Birds 34:698–700.

STEWART, R. E., AND J. W. ALDRICH. 1951. Removal and repopulation of breeding birds in a spruce-fir forest community. Auk 68:471–482.

STEWART, R. E., J. B. COPE, C. S. ROBBINS, AND J. W. BRAINERD. 1952. Seasonal distribution of bird populations at the Patuxent Research Refuge. Amer. Midl. Nat. 47:257–363.

STILES, F. G., AND L. L. WOLF. 1979. Ecology and evolution of lek mating behavior in the Long-tailed Hermit Hummingbird. Ornithol. Monogr. 27.

STIRLING, I., AND J. F. BENDELL. 1966. Census of Blue Grouse with recorded calls of a female. J. Wildl. Manage. 30:184–187.

STJERNBERG, T. 1979. Breeding biology and population dynamics of the Scarlet Rosefinch *Carpodacus erythrinus*. Acta Zool. Fenn. 157:1–88.

STOCEK, R. F., AND P. A. PEARCE. 1978. The Bald Eagle and Osprey in the Maritime Provinces. Can. Wildl. Serv., Wildl. Toxicol. Div. Misc. Rep. 37.

STODDARD, H. L. 1931. The Bobwhite Quail: its habits, preservation, and increase. Scribner's, N.Y.

STOKES, S. L. 1980. The Jolly-Seber method applied to age stratified populations. 1980 Abstracts, p. 227. Joint Statistical Meetings of the American Statistical Association and Biometric Society. Am. Stat. Assoc., Washington, D.C.

STRESEMANN, E., AND L. A. PORTENKO. 1960. Atlas der Verbreitung palaearktischer Vögel. Akademie-Verlag, Berlin.

STROLL, R. J. 1980. Indices to Ruffed Grouse hunting success in Ohio. Wildl. Soc. Bull. 8:24–28.

STUDEBACKER, G. A. 1973. Auditory masking. Pp. 117–154 *in* J. Jerger (ed.). Modern developments in audiology. Academic, New York.

SUKHATME, P. V. 1954. Sampling theory of surveys with applications. Iowa State Univ. Press, Ames.

SULZBACH, D., AND F. COOKE. 1978. Elements of nonrandomness in mass-captured samples of Snow Geese. J. Wildl. Manage. 42:437–440.

SULZBACH, D. S., AND F. COOKE. 1979. Demographic parameters of a nesting colony of Snow Geese. Condor 81:232–235.

SVENSSON, S. 1974a. The Swedish breeding bird census. Acta Ornithol. 14:220–224.

SVENSSON, S. 1974b. Interpersonal variation in species map evaluation in bird census work with the mapping method. Acta Ornithol. 14:322–338.

SVENSSON, S. 1975. Handledning för Svenska häckfågeltaxeringen. Zool. Inst., Lund, Sweden.

SVENSSON, S. 1977a. Land use planning and bird census work with particular reference to the application of the point sampling method. Pol. Ecol. Stud. 3(4):99–117.

SVENSSON, S. 1977b. Population trends of common birds in Sweden. Pol. Ecol. Stud. 3(4):207–213.

SVENSSON, S. 1977c. The Swedish ornithological atlas, progress report 1974–1975. Pol. Ecol. Stud. 3(4):309–311.

SVENSSON, S. E. 1978a. A simplified territory mapping technique for estimating bird numbers on bogs. Vår Fågelfärld 37:9–19.

SVENSSON, S. E. 1978b. Census efficiency and number of visits to a study plot when estimating bird densities by the territory mapping method. J. Appl. Ecol. 16:61–68.

SVENSSON, S. E. 1978c. Efficiency of two methods for monitoring bird population levels: breeding bird censuses contra counts of migrating birds. Oikos 30:373–386.

SVENSSON, S. E. 1979a. Census efficiency and number of visits to a study plot when estimating bird densities by the territory mapping method. J. Appl. Ecol. 16:61–68.

SVENSSON, S. E. 1979b. Svensk fågelatlas 1974–1978: halvtidsrapport. Vår Fågelvärld 38:114–123.

SVENSSON, S. E. 1980. Comparison of bird census methods. *In* H. Oelke (ed.). Bird census work and nature conservation. Dachverband Deutscher Avifaunisten, Göttingen.

SVENSSON, S. E. 1981. Do transect counts monitor abundance trends in the same way as territory mapping in study plots? Pp. 209–214 *in* C. J. Ralph and J. M. Scott (eds.). Estimating the numbers of terrestrial birds. Stud. Avian Biol. 6.

SVENSSON, S., AND K. WILLIAMSON. 1970. Recommendations for an international standard for a mapping method in bird census work. Pp 49–52 *in* S. Svensson (ed.). Bird census work and environmental monitoring. Bull. Ecol. Res. Comm. 9, Swedish Nat. Sci. Res. Counc., Stockholm.

SWARTZ, L. G., W. WALKER, D. G. ROSENEAU, AND A. M. SPRINGER. 1975. Populations of Gyrfalcons on the Seward Peninsula, Alaska 1968–1972. Pp. 71–75 *in* J. R. Murphy, C. M. White, and B. E. Harrell (eds.). Population status of raptors. Raptor Res. Rep. 3.

SWENSON, J. E. 1979. Factors affecting status and reproduction of Ospreys in Yellowstone National Park. J. Wildl. Manage. 43:595–601.

SWINEBROAD, J. 1964. Net-shyness and Wood Thrush populations. Bird-banding 35:196–202.

SYKES, P. W. 1979. Status of the Everglade Kite in Florida 1968–1978. Wilson Bull. 91:495–511.

TABER, R. D., AND I. COWAN. 1969. Capturing and marking wild animals. Pp. 277–317 *in* R. H. Giles (ed.). Wildlife management techniques. Third ed. Wildl. Soc., Washington, D.C.

TALBOT, L. M. 1970. The counts of game animals in large territories in North America and Africa. Trans. Int. Congr. Game Biol. 9:47–53.

TANAKA, R. 1951. Estimation of vole and mouse populations on Mount Ishizuchi and on the uplands of southern Shikoku. J. Mammal. 32:450–458.

TANNER, J. T. 1955. The altitudinal distribution of birds in a part of the Great Smoky Mountains. Migrant 26:37–40.

TANTON, M. T. 1965. Problems of live trapping and population estimation for the Wood Mouse, *Apodemus sylvaticus* (L.). J. Anim. Ecol. 34:1–22.

TAYLOR, L. R. 1978. Bates, Williams, Hutchinson—a variety of diversities. Symp. Roy. Entomol. Soc. Lond. 9:1–18.

TAYLOR, R. C. 1980. Migration of the Ringed Plover (*Charadrius hiaticula*). Ornis Scand. 11:30–42.

TAYLOR, S. M. 1965. The Common Birds Census—some statistical aspects. Bird Study 12:268–286.

TAYLOR, S. M. 1966. Recent quantitative work on British bird populations: a review. Statistician 16:119–170.

TEMPLE, S. A., AND B. L. TEMPLE. 1976. Avian population trends in central New York State, 1935–1972. Bird-Banding 47:238–257.

TERBORGH, J. 1971. Distribution on environmental gradients: theory and preliminary interpretation of distributional patterns in the avifauna of the Cordillera Vilcabamba, Peru. Ecology 52:23–40.

TERBORGH, J. 1977. Bird species diversity on an Andean elevational gradient. Ecology 5:1007–1019.

TERBORGH, J. W., AND J. FAABORG. 1973. Turnover and ecological release in the avifauna of Mona Island, Puerto Rico. Auk 90:759–779.

TERBORGH, J. W., AND J. FAABORG. 1980a. Factors affecting the distribution and abundance of North American migrants in the eastern Caribbean region. Pp. 145–155 *in* A. Keast and E. S. Morton (eds.). Migrant birds in the neotropics. Smithsonian Inst. Press, Washington, D.C.

TERBORGH, J. W., AND J. FAABORG. 1980b. Saturation of bird communities in the West Indies. Am. Nat. 116:178–195.

TERBORGH, J., AND J. S. WESKE. 1969. Colonization of secondary habitats by Peruvian birds. Ecology 50:765–782.

TEXEIRA, R. M. (ed.) 1980. Atlas van de Nederlandse Broedfogels. Veren. Natuurmonum, Deventer, Netherland.

THIELCKE, G. A. 1976. Bird Sounds. Univ. Michigan Press, Ann Arbor.

THIELCKE, G. A., AND D. BLUME. 1973. Biologie der Vogelstimmen I. Klett-Verlag, München. Record 17 cm, No. 79311.

THIOLLAY, J. M. 1977. Importance des populations de rapaces migrateurs en Mediterranée occidentale. Alauda 45:115–121.

THIOLLAY, J. M. 1978. Les migrations de rapaces en Afrique occidentale: adaptations écologiques aux fluctuations saisonnières de production des écosystems. Terre et Vie 32:89–133.

THIOLLAY, J. M. 1980. Spring hawk migration in eastern Mexico. Raptor Res. 14:13–20.

THOMAS, J. W., R. G. ANDERSON, C. MASER, AND E. L. BULL. 1979a. Snags. Pp. 60–77 *in* J. W. Thomas (ed.). Wildlife habitats in managed forests—the Blue Mountains of Oregon and Washington. U.S. For. Serv., Agric. Handb. 553, Washington, D.C.

THOMAS, J. W., C. MASER, AND J. E. RODIEK. 1979b. Wildlife habitats in managed rangelands—the Great Basin of southeastern Oregon. Edges. U.S. For. Serv., Gen. Tech. Rep. PNW-85, Portland, Ore.

THOMAS, J. W., R. J. MILLER, H. BLACK, J. E. RODIEK, AND C. MASER. 1976. Guidelines for maintaining and enhancing wildlife habitat in forest management in the Blue Mountains of Oregon and Washington. Trans. N. Am. Wildl. Nat. Res. Conf. 41:452–476.

THOMPSON, L. 1980. Circle West wildlife monitoring study. Second annual report. Montana Dept. Nat. Res., Circle West Tech. Rep. 6.

THORPE, W. H. 1961. Bird song. Cambridge Univ. Press, Cambridge.

TIAINEN, J., J.-L. MARTIN, T. PAKKALA, J. PIIROINEN, T. SOLONEN, M. VICKHOLM, AND E. VIROLAINEN. 1980. Efficiency of the line transect and point count methods in a south Finnish forest area. Pp. 107–113 *in* H. Oelke (ed.). Bird census work and nature conservation. Dachverband Deutscher Avifaunisten, Göttingen, W. Ger.

TIAINEN, J., AND T. SOLONEN. 1979. Erään estelä-suomalaisen alueen lintuyhteisö. (The breeding bird community in a south Finnish study area.) Lintumies 14:97–103.

TILGHMAN, N. G. 1977. Problems in sampling songbird populations in southeastern Wisconsin woodlots. M.S. thesis, Univ. Wisconsin, Madison.

TILGHMAN, N. G., AND D. H. RUSCH. 1981. Comparison of line-transect methods for estimating breeding bird densities in deciduous woodlots. Pp. 202–208 *in* C. J. Ralph and J. M. Scott (eds.). Estimating the numbers of terrestrial birds. Stud. Avian Biol. 6.

TINBERGEN, N. 1939. The behavior of the Snow Bunting in spring. Trans. Linn. Soc. 5:1–94.

TINBERGEN, N. 1960. The Herring Gull's world: a study of the social behavior of birds. Basic Books, New York.

TITUS, K., AND J. A. MOSHER. In press. Nest site habitat selected by woodland hawks in the central Appalachians. Auk.

TODD, R. L. 1976. Wildlife surveys and investigations: nongame investigations. Ariz. Game and Fish Dept., Performance Rep., Phoenix.

TOMIAŁOJĆ, L. 1980. The combined version of the mapping method. Pp. 92–106 *in* H. Oelke (ed.). Bird census work and nature conservation. Proceedings VI Int. Conf. Bird Census Work/IV Meeting Eur. Orn. Atlas Committ. (Göttingen 1979). Dachverband Deutscher Avifaunisten, Göttingen, W. Ger.

TOMLINSON, R. E. 1968. Reward banding to determine reporting rate of recovered Mourning Dove bands. J. Wildl. Manage. 32:6–16.

TOMLINSON, R. E. 1972. Current status of the endangered Masked Bobwhite Quail. Trans. N. Am. Wildl. Nat. Res. Conf. 37:294–311.

TOMLINSON, R. E., AND R. L. TODD. 1973. Distribution of two western Clapper Rail races as determined by responses to taped calls. Condor 75:177–183.

TOMOFF, C. S. 1974. Avian species diversity in desert scrub. Ecology 55:396–403.

TRAMER, E. J. 1969. Bird species diversity: components of Shannon's formula. Ecology 50:927–929.

TRAMER, E. J. 1974. An analysis of the species density of U.S. landbirds during winter using the 1971 Christmas Bird Count. Am. Birds 28:563–567.

TRYON, P. R., AND S. F. MACLEAN. 1980. Use of space by Lapland Longspurs breeding in arctic Alaska. Auk 97:509–520.

TUBBS, C. R. 1974. The buzzards. David and Charles, North Pomfret, Vt.

TWOMEY, A. C. 1945. The bird population of an elm-maple forest, with special reference to aspection, territorialism, and coaction. Ecol. Monogr. 15:175–205.

TYLER, W. M. 1946. *Cyanocitta cristata bromia* Oberholser: Northern Blue Jay. Pp. 32–52 *in* A. C. Bent (ed.). Life histories of North American jays, crows, and titmice. U.S. Natl. Mus. Bull. 191.

UDVARDY, M. D. F. 1957. An evaluation of quantitative studies in birds. Cold Spring Harb. Symp. Quant. Biol. 22:301–311.

UDVARDY, M. D. F. 1969. Dynamic zoogeography. Van Nostrand Reinhold, New York.

UDVARDY, M. D. F. 1970. Danske ynglefulge i fortid og nutid (a review). Auk 87:593–596.

UDVARDY, M. D. F. 1981. An overview of grid-based atlas works in ornithology. Pp. 103–109 *in* C. J. Ralph and J. M. Scott (eds.). Estimating the numbers of terrestrial birds. Stud. Avian Biol. 6.

ULFSTRAND, S. 1958. The annual fluctuations in the migration of the Honey-Buzzard (*Pernis apivorus*) over Falsterbo. (In Swedish, with English summary). Vår Fågelvärld 17:118–144.

ULFSTRAND, S. 1975. Bird flocks in relation to vegetation diversification in a south Swedish coniferous plantation during winter. Oikos 26:65–73.

ULFSTRAND, S., G. ROOS, T. ALERSTAM, AND L. OSTERDAHL. 1974. Visible bird migration at Falsterbo, Sweden. Vår Fågelvärld, Suppl. 8:1–244.

UNITED STATES DEPARTMENT OF THE INTERIOR. 1971. Trends in populations of raptors in North America. Hearings of the House of Representatives, Comm. on Merchant Marine and Fisheries, Serial 92–14: 97–109.

UNITED STATES FISH AND WILDLIFE SERVICE. 1980. Republication of lists of endangered and threatened species and correction of technical errors in final rules. Fed. Reg. 45(99):33767–33781.

UNITED STATES FISH AND WILDLIFE SERVICE AND CANADIAN WILDLIFE SERVICE. 1977. North American bird banding manual. Vol. 2. Bird banding techniques. U.S. Fish and Wildl. Serv., Washington, D.C.

URICK, R. S. 1967. Principles of underwater sound for engineers. McGraw-Hill, New York.

VAN DER WEYDEN, W. J. 1975. Vocal affinities of the African and European Scops Owls *Otus scops* (Strigidae). Bull. de l'I.F.A.N.T. 35 ser A no. 3.

VAN RIPER, C. 1978. The breeding ecology of the Amakihí (*Loxops virens*) and Palila (*Psittirostra bailleui*) on Mauna Kea, Hawaii. Ph.D. thesis. Univ. Hawaii, Honolulu.

VAN RIPER, C. 1980. Observations on the breeding of the Palila *Psittirostra bailleui* of Hawaii. Ibis 122:462–475.

VAN TYNE, J., AND A. J. BERGER. 1959. Fundamentals of ornithology. Wiley, New York.

VAN VELZEN, W. T. 1966. Maryland's part in the 1966 Breeding Bird Survey. Maryland Birdlife 22:88–93.

VAN VELZEN, W. T. 1972. Breeding-bird census instructions. Am. Birds 26:927–931.

VAN VELZEN, W. T. 1980. Forty-third breeding-bird census. Am. Birds 34:41–44.

VERNER, J. 1965. Time budget of the male Long-billed Marsh Wren during the breeding season. Condor 67:125–139.

VERNER, J. 1975. Avian behavior and habitat management. Pp. 39–58 *in* D. R. Smith (tech. coord.). Proceedings, symposium on management of forest and range habitats for nongame birds. U.S. For. Serv., Gen. Tech. Rep. WO-1, Washington, D.C.

VERNER, J. 1980a. Birds of California oak habitats—management implications. Pp. 246–264 *in* T. R. Plumb (tech. coord.). Proceedings symposium on the ecology, management, and utilization of California oaks. U.S. For. Serv., Gen. Tech. Rep. PSW-44, Berkeley, Calif.

VERNER, J. 1980b. Bird communities of mixed-conifer forests of the Sierra Nevada. Pp. 198–223 *in* R. M. DeGraff (tech. coord.). Proceedings of workshop, management of western forests and grasslands for nongame birds. U.S. For. Serv., Gen. Tech. Rep. Int-86, Ogden, Utah.

VERNER, J. 1981. Measuring responses of avian communities to habitat manipulation. Pp. 543–547 *in* C. J. Ralph and J. M. Scott (eds.). Estimating the numbers of terrestrial birds. Stud. Avian Biol. 6.

VERNER, J., AND M. M. MILLIGAN. 1971. Responses of White-crowned Sparrows to playback of recorded songs. Condor 73:56–64.

VIAN, W. E., AND C. W. BLIESE. 1974. Observations on population changes and on the behavior of the Bald Eagle. Nebraska Bird Rev. 42:46–55.

VISHER, S. S. 1954. Climatic atlas of the United States. Belknap, Cambridge, Mass.

VON HAARTMAN, L. 1951. Successive polygamy. Behaviour 3:256–274.

VON HAARTMAN, L. 1956. Territory in the Pied Flycatcher *Muscicapa hypoleuca*. Ibis 98:460–475.

von Haartman, L. 1957. Adaptation in hole-nesting birds. Evolution 11:339–347.

von Haartman, L. 1963. The nesting times of Finnish birds. Proc. Int. Ornithol. Congr. 13:611–619.

von Haartman, L. 1971. Population dynamics. Pp. 391–459 *in* D. S. Farner and J. R. King (eds.). Avian Biology. Vol 1. Academic, New York.

von Haartman, L. 1978a. Severe decrease in a population of Starlings. Ornis Fenn. 55:40–41.

von Haartman, L. 1978b. Ny nedgång i starpopulationen i holkar på Lemsjöholm. (Further decrease of the Starling population at Lemsjöholm.) Ornis Fenn. 55:83.

von Knorring, J. 1978. Nedgang i en starpopulation i Salo. (Decrease in a Starling population). Ornis Fenn. 55:82.

Voous, K. H. 1960. Atlas of European birds. Nelson, London.

Walankiewicz, W. 1977. A comparison of the mapping method and I.P.A. results in Białowieźa National Park. Pol. Ecol. Stud. 3(4):119–125.

Wallace, G. O. 1970. An evaluation of the roadside count technique utilized in the Breeding Bird Survey. Ph.D. thesis, Univ. Tennessee, Knoxville.

Walter, H. 1979. Eleonora's Falcon. Adaptations to prey and habitat in a social raptor. Univ. Chicago Press, Chicago.

Ward, F. P., and R. B. Berry. 1972. Autumn migrations of Peregrine Falcons on Assateague Island, 1970–71. J. Wildl. Manage. 36:484–492.

Ward, W. D. 1966. Temporary threshold shift in males and females. J. Acoust. Soc. Am. 40:478–485.

Watson, D. 1977. The Hen Harrier. Poyser, London.

Webb, W. L. 1942. Notes on a method for censusing Snowshoe Hare populations. J. Wildl. Manage. 6:67–69.

Weber, W. C., and J. B. Theberge. 1977. Breeding bird survey counts as related to habitat and date. Wilson Bull. 89:543–561.

Webster, J. D. 1966. An analysis of Winter Bird-Population Studies. Wilson Bull. 78:458–461.

Weeden, J. S. 1965. Territorial behavior of the Tree Sparrow. Condor 67:193–209.

Weeden, J. S., and J. B. Falls. 1959. Differential responses of male Ovenbirds to recorded songs of neighboring and more distant individuals. Auk 76:343–351.

Weller, M. W., I. C. Adams, and B. J. Rose. 1955. Winter roosts of Marsh Hawks and Short-eared Owls in central Missouri. Wilson Bull. 67:189–193.

Welty, J. C. 1975. The life of birds. Second ed. Saunders, Philadelphia.

Wetmore, A. 1939. Notes on the birds of Tennessee. Proc. U.S. Nat. Mus. 86:175–243.

Wetmore, S. P., and D. I. Gillespie. 1977. Osprey populations in Labrador and northeastern Quebec. Pp. 87–88 *in* J. C. Ogden (ed.). Transactions of the North American Osprey research conference. U.S. Natl. Park Serv., Trans. and Proc. Ser. 2.

Whelan, J. B., A. R. Tipton, J. F. Williamson, P. R. Johansen, J. P. McClure, and N. D. Cost. 1979. A comparison of three systems for evaluating forest wildlife habitat. Trans. N. Am. Wildl. Nat. Res. Conf. 44:392–403.

Whisler, S., and D. Horn. 1977. An investigation into the occurrence and habits of Northern Spotted Owls (*Strix occidentalis caurina*) in the Ukiah District, BLM, April–June 1977. Bur. Land Manage., Ukiah, Calif.

Whitaker, A. B. 1972. An improved mist net rig for use in forests. Bird-Banding 43:108.

Whitaker, G. A., E. R. Roach, and R. H. McCuen. 1976. Inventorying habitats and rating their value for wildlife species. Proc. Annu. Conf. Southeast., Assoc. Game and Fish Comm. 30:590–601.

Whitcomb, D. A. 1974. Characteristics of an insular Woodcock population. Michigan Dept. Nat. Resour., Wildl. Div. Rep. 2720.

Whitcomb, R. F., J. F. Lynch, P. A. Opler, and C. S. Robbins. 1976. Island biogeography and conservation: strategy and limitations. Science 193:1030–1032.

Whitcomb, R. F., C. S. Robbins, J. F. Lynch, B. L. Whitcomb, M. K. Klimkiewicz, and D. Bystrak. 1979. Effect of forest fragmentation on avifauna of the eastern deciduous forest. *In* Forest island dynamics in man-dominated landscapes. Springer-Verlag, New York, in press.

White, C. M. 1965. Roadside raptor count through Utah, Colorado, and Kansas. Kansas Ornithol. Soc. 16:17–19.

White, C. M., T. D. Ray, and L. W. Sowl. 1977. The 1970–1972–1974 raptor surveys along the trans-Alaska oil pipeline. Pp. 222–228 *in* R. D. Chancellor (ed.). World conference on birds of prey, Vienna 1975. Int. Coun. Bird Pres.

White, C., and S. Sherrod. 1973. Advantages and disadvantages of the use of rotor-winged aircraft in raptor surveys. Raptor Res. 7:97–104.

White, E. G. 1971. A versatile FORTRAN computer program for the capture-recapture stochastic model of G. M. Jolly. J. Fish. Res. Bd. Canada 28:443–446.

White, G. C., D. R. Anderson, K. P. Burnham, and D. L. Otis. 1981. A primer for application and use of capture-recapture methods: A workbook. Los Alamos Sci. Lab., Los Alamos, N.M.

White, G. C., K. P. Burnham, D. L. Otis, and D. R. Anderson. 1978. User's manual for program CAPTURE. Utah State Univ. Press, Logan.

Whitfield, D. W. A., J. M. Gerrard, W. J. Maher, and D. W. Davis. 1974. Bald Eagle nesting habitat, density and reproduction in central Saskatchewan and Manitoba. Can. Field-Nat. 88:399–408.

Whitmore, R. C. 1975. Habitat ordination of passerine birds in the Virgin River Valley, southwestern Utah. Wilson Bull. 87:65–74.

Whitmore, R. C. 1980. Reclaimed surface mines as avian habitat islands in the eastern forest. Am. Birds 34:13–14.

Whitney, N. R. 1967. 1967 Breeding Bird Survey. So. Dakota Bird Notes 19:81–83.

Wiemeyer, S. N. 1977. Reproductive success of Potomac River Ospreys, 1971. Pp. 115–119 *in* J. C. Ogden (ed.). Transactions of the North American Osprey research conference. U.S. Natl. Park Serv., Trans. and Proc. Ser. 2.

Wiens, J. A. 1969. An approach to the study of ecological relationships among grassland birds. Ornithol. Monogr. 8.

Wiens, J. A. 1973. Interterritorial habitat variation

in Grasshopper and Savannah Sparrows. Ecology 54:877–884.

WIENS, J. A. 1974. Habitat heterogeneity and avian community structure in North American grasslands. Am. Midl. Nat. 91:195–213.

WIENS, J. A. 1975. Avian communities, energetics, and functions in coniferous forest habitats. Pp. 226–265 *in* D. R. Smith (tech. coord.). Proceedings of the symposium on management of forest and range habitats for nongame birds. U.S. For. Serv., Gen. Tech. Rep. WO-1, Washington, D.C.

WIENS, J. A. 1977. On competition and variable environments. Am. Sci. 65:590–597.

WIENS, J. A. 1981a. Scale problems in avian censusing. Pp. 513–521 *in* C. J. Ralph and J. M. Scott (eds.). Estimating the numbers of terrestrial birds. Stud. Avian Biol. 6.

WIENS, J. A. 1981b. Single-sample surveys of communities: are the revealed patterns real? Am. Nat. 117:90–98.

WIENS, J. A., AND M. I. DYER. 1975. Rangeland avifaunas: their composition, energetics, and role in the ecosystem. Pp. 146–182 *in* D. R. Smith (tech. coord.). Proceedings of the symposium on management of forest and range habitats for nongame birds. U.S. For. Serv., Gen. Tech. Rep. WO-1, Washington, D.C.

WIENS, J. A., AND J. T. EMLEN. 1966. Post-invasion status of the Dickcissel in southern Wisconsin. Passenger Pigeon 28:63–69.

WIENS, J. A., AND J. T. ROTENBERRY, 1979. Diet niche relationships among North American grassland and shrubsteppe birds. Oecologia 36:253–292.

WIENS, J. A., AND J. T. ROTENBERRY. 1981a. Censusing and the evaluation of avian habitat occupancy. Pp. 522–532 *in* C. J. Ralph and J. M. Scott (eds.). Estimating the numbers of terrestrial birds. Stud. Avian Biol. 6.

WIENS, J. A., AND J. T. ROTENBERRY. 1981b. Habitat associations and community structure of birds in shrubsteppe environments. Ecol. Monogr. 51:21–41.

WIENS, J. A., AND J. T. ROTENBERRY. In press. Bird community structure in cold shrub deserts: competition or chaos? Proc. Int. Ornithol. Congr. 17.

WIENS, J. A. J. T. ROTENBERRY, AND J. F. WARD. 1972. Avian populations at IBP grassland biome sites: 1971. U.S. IBP Grassland Biome Tech. Rep. 205.

WILBUR, S. R. 1978a. The California Condor, 1966–76: a look at its past and future. U.S. Fish and Wildl. Serv., N. Am. Fauna 72.

WILBUR, S. R. 1978b. Volunteer participation in California Condor surveys. Wildl. Soc. Bull. 6:157–159.

WILCOX, B. A. 1980. Insular ecology and conservation. Pp. 95–117 *in* M. E. Soulé and B. A. Wilcox (eds.). Conservation biology: an evolutionary-ecological perspective. Sinauer, Sunderland, Mass.

WILEY, R. H., AND D. G. RICHARDS. 1978. Physical constraints on acoustical communication in the atmosphere: implications for the evolution of animal vocalizations. Behav. Ecol. Sociobiol. 3:69–94.

WILEY, R. H., AND D. G. RICHARDS. 1981. Adaptations for acoustic communication in birds: sound transmission and signal detection. *In* D. E. Kroodsma and E. H. Miller (eds.). Ecology and evolution of acoustic communication in birds. Academic, New York, in press.

WILKINSON, G. B., AND R. GUEST. 1977. Forest birds of the Hope catchment, Lake Sumner State Forest Park, North Canterbury. Notornis 24:161–166.

WILLIAMS, A. B. 1936. The composition and dynamics of a beech-maple climax community. Ecol. Monogr. 6:317–408.

WILLIAMS, G. G. 1950. Weather and spring migration. Auk 67:52–65.

WILLIAMS, H. W. 1961. The influence of physical and biological factors on the rally call of the Chukar Partridge (*Alectoria graeca*) with regard to the use of the call as a census method. West. Assoc. State Game and Fish Comm. 41:117–129.

WILLIAMS, K. In press. Discriminant analysis in wildlife research: theory and application. *In* D. E. Capen (ed.). Proceedings of workshop, the use of multivariate statistics in the study of wildlife habitat. U.S. For. Serv., Gen. Tech. Rep. RM-, Fort Collins, Col.

WILLIAMS, R. J. 1957. The Great Blue Heron colonies of Wisconsin. Passenger Pigeon 19:51–66.

WILLIAMSON, K. 1964. Bird census work in woodland. Bird Study 11:1–22.

WILLIAMSON, K. 1969. Habitat preferences of the Wren on English farmland. Bird Study 16:53–59.

WILLIAMSON, K. 1971. A bird census study of a Dorset dairy farm. Bird Study 18:80–96.

WILLIAMSON, K. 1972. The relevance of the mapping census technique to the conservation of migratory bird populations. Wildlife Res. Rep. 2. (Population Ecology of Migratory Birds, Symposium.) Washington, D.C.

WILLIAMSON, K. 1975. Birds and climatic change. Bird Study 22:143–164.

WILLIAMSON, K., R. S. BAILEY, AND L. A. BATTEN. 1976. Guiding principles for the analysis of Common Birds Census returns. Br. Trust Ornithol., Tring, U.K.

WILLIAMSON, K., AND L. A. BATTEN. 1977. Ecological implications of the Common Birds Census. Pol. Ecol. Stud. 3(4):237–244.

WILLIAMSON, K., AND R. C. HOMES. 1964. Methods and preliminary results of the Common Birds Census, 1962–63. Bird Study 11:240–256.

WILLIAMSON, R., AND K. WILLIAMSON. 1973. The bird community of yew woodland at Kingley Vale, Sussex. Br. Birds 66:12–23.

WILLIS, E. O. 1972. The behavior of Plain-brown Woodcreepers, (*Dendrocincla fuliginosa*). Wilson Bull. 84:377–420.

WILLIS, E. O. 1973. The behavior of Ocellated Antbirds. Smithsonian Contrib. Zool. 144.

WILLIS, E. O., AND Y. ONIKI. 1978. Birds and army ants. Annu. Rev. Ecol. Syst. 9:243–263.

WILLSON, M. F. 1974. Avian community organization and habitat structure. Ecology 55:1017–1029.

WILLSON, M. F., AND D. J. MORIARITY. 1976. Bird species diversity in forest understory: analysis of mist net samples. Oecologia (Berl.) 25:373–379.

WIMMER, R. B. 1954. An ecological study of the Mourning Dove in Emporia, Kansas, 1953. M.S. thesis, Kansas State Teachers Coll., Emporia.